Classics in Mathematics

Herbert Federer Geometric Measure Theory

Springer
Berlin
Heidelberg
New York
Barcelona
Budapest
Hong Kong
London
Milan
Paris
Santa Clara
Singapore
Tokyo

Herbert Federer was born on July 23, 1920, in Vienna. After emigrating to the US in 1938, he studied mathematics and physics at the University of California, Berkeley. Affiliated to Brown University, Providence since 1945, he is now Professor Emeritus there.

The major part of Professor Federer's scientific effort has been directed to the development of the subject of Geometric Measure Theory, with its roots and applications in classical geometry and analysis, yet in the functorial spirit of modern topology and algebra. His work includes more than thirty research papers published between 1943 and 1986, as well as this book.

Herbert Federer

Geometric Measure Theory

Reprint of the 1969 Edition

Springer

Herbert Federer
(Professor Emeritus)
Department of Mathematics
Brown University
Providence, RI 02912
USA

Originally published as Vol. 153 of the
Grundlehren der mathematischen Wissenschaften

Cataloging-in-Publication Data applied for

Die Deutsche Bibliothek - CIP-Einheitsaufnahme

Federer, Herbert:
Geometric measure theory / Herbert Federer. - Reprint of the 1969 ed. - Berlin ; Heidelberg ; New York ; Barcelona ; Budapest ; Hong Kong ; London ; Milan ; Paris ; Santa Clara ; Singapore ; Tokyo : Springer, 1996
(Grundlehren der mathematischen Wissenschaften ; Vol. 153) (Classics in mathematics)
ISBN 3-540-60656-4
NE: 1. GT

Mathematics Subject Classification (1991): 53C65, 46AXX

ISBN 3-540-60656-4 Springer-Verlag Berlin Heidelberg New York

SPIN 10485236 41/3144- 5 4 3 2 1 0 – Printed on acid-free paper

Herbert Federer

Geometric Measure Theory

Springer-Verlag Berlin · Heidelberg · New York 1969

Herbert Federer

Florence Pirce Grant University Professor
Brown University, Providence, Rhode Island

Geschäftsführende Herausgeber:

Prof. Dr. B. Eckmann

Eidgenössische Technische Hochschule Zürich

Prof. Dr. B. L. van der Waerden

Mathematisches Institut der Universität Zürich

 Library of Congress Catalog Card Number 69-16846
Printed in Germany
Title No. 5136

To my friends
Charles Donald Shane
Mary Lea Shane

Preface

During the last three decades the subject of geometric measure theory has developed from a collection of isolated special results into a cohesive body of basic knowledge with an ample natural structure of its own, and with strong ties to many other parts of mathematics. These advances have given us deeper perception of the analytic and topological foundations of geometry, and have provided new direction to the calculus of variations. Recently the methods of geometric measure theory have led to very substantial progress in the study of quite general elliptic variational problems, including the multidimensional problem of least area.

This book aims to fill the need for a comprehensive treatise on geometric measure theory. It contains a detailed exposition leading from the foundations of the theory to the most recent discoveries, including many results not previously published. It is intended both as a reference book for mature mathematicians and as a textbook for able students. The material of Chapter 2 can be covered in a first year graduate course on real analysis. Study of the later chapters is suitable preparation for research. Some knowledge of elementary set theory, topology, linear algebra and commutative ring theory is prerequisite for reading this book, but the treatment is selfcontained with regard to all those topics in multilinear algebra, analysis, differential geometry and algebraic topology which occur.

The formal presentation of the theory in Chapters 1 to 5 is preceded by a brief sketch of the main theme in the Introduction, which contains also some broad historical comments.

A systematic attempt has been made to identify, at the beginning of each chapter, the original sources of all relatively new and important material presented in the text. References to literature on certain additional topics, which this book does not treat in detail, appear in the body of the text. Some further related publications are listed only in the bibliography. All references to the bibliography are abbreviated in square brackets; for example [C 1] means the first listed work by C. Carathéodory.

The index is supplemented by a list of basic notations defined in the text, and a glossary of some standard notations which are used but not defined in the text.

I wish to thank Brown University and the National Science Foundation for supporting my work on this book, and to express appreciation of the efforts of my fellow mathematicians who helped with the project. Frederick J. Almgren Jr. and I had many stimulating discussions, in particular about his ideas presented in Section 5.3. Casper Goffman posed some interesting questions which inspired part of 4.5.9. Katsumi Nomizu showed me an elegant treatment of 5.4.13. William K. Allard read the whole manuscript with great care and contributed significantly, by many valuable queries and comments, to the accuracy of the final version. John E. Brothers, Lawrence R. Ernst, Josef Král, Arthur Sard and William P. Ziemer read parts of the manuscript and supplied very useful lists of errata.

The editors and personnel of Springer-Verlag have been unfailingly cooperative in all phases of publication of this book. I am grateful, in particular, to David Mumford for inviting me to contribute my work to the Grundlehren series, and to Klaus Peters for planning all the necessary arrangements with utmost consideration.

Herbert Federer

Providence, Rhode Island
January 1969

Contents

CHAPTER THREE

Rectifiability

CHAPTER FOUR

Homological integration theory

CHAPTER FIVE

Applications to the calculus of variations

Introduction

A systematic description of the material in this book is provided by the table of contents, the introductions to individual chapters, and the index. Here we wish to comment more informally on our general plan, to emphasize and motivate the central concepts, and to mention briefly some additional topics not discussed in the text. This introduction is not a logical prerequisite for the chapters which follow – quite the opposite – but it may give the reader some helpful first impressions of our subject.

Among the basic tools of geometric measure theory are the methods of multilinear algebra presented in Chapter 1, and the techniques of the general theory of integration developed in Chapter 2. We use exterior and alternating algebras to discuss oriented m dimensional vectorsubspaces of n dimensional Euclidean space $\mathbf{R}^n$, in particular the tangent spaces of m dimensional rectifiable sets and currents, which are the principal objects of study in Chapters 3 and 4. We employ symmetric algebras to treat higher differentials of mappings, for example in Whitney's extension theorem and the theory of analytic sets in Chapter 3, and in the theory of strongly elliptic systems of second order partial differential equations in Chapter 5. Our exposition of general measure theory features equally the set theoretic approach of Carathéodory and the function lattice approach of F. Riesz and P.J. Daniell. It includes not only the fundamental facts of Lebesgue integration, but also numerous additional topics like the theory of Suslin sets, the theory of covariant measures on homogeneous spaces, results about derivates based on generalizations of the covering theorems of Vitali and Besicovitch, and primary properties of Hausdorff type measures.

The n dimensional Lebesgue measure $\mathscr{L}^n$ over $\mathbf{R}^n$ is uniquely characterized by Borel regularity, invariance under Euclidean isometries, and the condition

$$\mathscr{L}^n\{x\colon 0 \leq x_i \leq 1 \text{ for } i=1, \ldots, n\}=1.$$

However for any two positive integers $m<n$ there exist many distinct so-called m dimensional measures over $\mathbf{R}^n$. All of them are Borel regular, invariant under the Euclidean group, and assign to every subset of an m dimensional submanifold of class 1 of $\mathbf{R}^n$ the correct number hallowed by the universal accord of geometers. But these measures disagree

violently in their evaluation of more general subsets of $\mathbf{R}^n$, even though each measure is defined by an eminently reasonable geometric construction. The two most important m dimensional measures are the Hausdorff measure $\mathscr{H}^m$ and the integralgeometric measure $\mathscr{I}_1^m$. If $A \subset \mathbf{R}^n$, then $\mathscr{H}^m(A)$ is the greatest lower bound of the set of all t such that $0 \le t \le \infty$ and for every $\varepsilon > 0$ there exists a countable covering G of A with

$$\sum_{S \in G} \alpha(m)\,[\operatorname{diam}(S)/2]^m \le t$$

and $\operatorname{diam}(S) < \varepsilon$ for $S \in G$; here $\alpha(m) = \mathscr{L}^m(\mathbf{R}^m \cap \{x: |x| < 1\})$. We define $\mathscr{I}_1^m(A)$ similarly, restricting the members of G to be Borel sets and replacing the above summand by

$$\int \mathscr{L}^m[p(S)]\, d\theta^*_{n,m}\, p / \beta_1(n,m)$$

where $\theta^*_{n,m}$ is an invariant measure over the space of all orthogonal projections mapping $\mathbf{R}^n$ onto $\mathbf{R}^m$, and $\beta_1(n,m)$ is a suitable constant. In case A is a Borel set we infer the generalized Crofton formula

$$\mathscr{I}_1^m(A) = \int \nu(A \cap W)\, d\mu\, W / \beta_1(n,m)$$

where μ is an invariant measure over the space of all $n-m$ dimensional affine subspaces of $\mathbf{R}^n$, and $\nu(E)$ equals the number of elements (possibly ∞) of any set E.

Much of geometric measure theory during the first half of this century consisted of detailed studies of certain peculiar sets of Cantor type for which various m dimensional measures disagree. From this analysis of pathology there gradually evolved, thanks largely to the pioneering genius of A. S. Besicovitch, a pattern of structure. The following central theorem was proved for the special case where $m=1$ and $n=2$ by Besicovitch, then for general m and n by the author: *Every Borel set A of $\mathbf{R}^n$ with $\mathscr{H}^m(A) < \infty$ contains a Borel set B such that*

$$\mathscr{I}_1^m(B) = \mathscr{H}^m(B), \qquad \mathscr{I}_1^m(A \sim B) = 0,$$

B can be covered by a countable family of m dimensional submanifolds of class 1 of $\mathbf{R}^n$, and

$$\mathscr{H}^m[(A \sim B) \cap C] = 0$$

for every m dimensional submanifold C of class 1 of $\mathbf{R}^n$. It follows that $\mathscr{H}^m(A) \ge \mathscr{I}_1^m(A)$, and that

$$\mathscr{H}^m(A) = \mathscr{I}_1^m(A) \text{ if and only if } \mathscr{H}^m(A \sim B) = 0,$$

in which case A is called $(\mathscr{H}^m, m)$ rectifiable. In Section 3.3 we establish several results of this type, which provide usable criteria for rectifiability.

Our work often involves the behavior of rectifiable sets under Lipschitzian maps. The main results on this topic, which we study in Section 3.2, may be summarized as follows. *Suppose* $n \geq m \geq k \leq l$ *are positive integers,* X *is an* $(\mathscr{H}^m, m)$ *rectifiable Borel set of* $\mathbf{R}^n$, $Y \subset \mathbf{R}^l$, *and the map* $f: X \to Y$ *satisfies a Lipschitz condition. Then:*

(1) *At* $\mathscr{H}^m$ *almost all points* x *in* X *the set* X *has an approximate tangent vectorspace of dimension* m *and the function* f *has an approximate differential* $\operatorname{ap} D f(x)$.

(2) *If* Y *is* $(\mathscr{H}^k, k)$ *rectifiable, then* $f^{-1}\{y\}$ *is* $(\mathscr{H}^{m-k}, m-k)$ *rectifiable for* $\mathscr{H}^k$ *almost all* y, *and*

$$\int_Y \mathscr{H}^{m-k}(f^{-1}\{y\})\, d\mathscr{H}^k y = \int_X j(x)\, d\mathscr{H}^m x$$

where $j(x) = \|\wedge_k \operatorname{ap} D f(x)\|$ *is the* k *dimensional approximate Jacobian of* f *at* x.

(3) *If* $\varepsilon > 0$ *and* $\mathscr{H}^{m-k}(f^{-1}\{y\}) \geq \varepsilon$ *for* $\mathscr{H}^k$ *almost all* y *in* Y, *then* Y *is* $(\mathscr{H}^k, k)$ *rectifiable.*

The intuitive idea of an m dimensional domain of integration in $\mathbf{R}^n$ can be given several distinct precise and useful formulations. One may consider a bounded Borel set A of $\mathbf{R}^n$ with $\mathscr{H}^m(A) < \infty$. One may study the corresponding Radon measure $\mathscr{H}^m \llcorner A$ defined by the formula

$$(\mathscr{H}^m \llcorner A)(S) = \mathscr{H}^m(S \cap A) \text{ for } S \subset \mathbf{R}^n.$$

Assuming that A is $(\mathscr{H}^m, m)$ rectifiable one may additionally postulate an $\mathscr{H}^m \llcorner A$ summable function ξ which relates to $\mathscr{H}^m$ almost every point x in A a simple m vector $\xi(x)$ such that $|\xi(x)|$ is a positive integer (the multiplicity at x) and the m dimensional vectorsubspace of $\mathbf{R}^n$ associated with $\xi(x)$ equals the set of all approximate tangent vectors of A at x. Then the integral

$$\int_A \Phi[x, \xi(x)]\, d\mathscr{H}^m x$$

exists for every continuous $\Phi: \mathbf{R}^n \times \wedge_m \mathbf{R}^n \to \mathbf{R}$ such that

$$\Phi(x, t\alpha) = t\,\Phi(x, \alpha) \text{ whenever } x \in \mathbf{R}^n,\ \alpha \in \wedge_m \mathbf{R}^n,\ 0 < t \in \mathbf{R},$$

and this integral depends linearly on Φ. Differential forms of degree m and class ∞ on $\mathbf{R}^n$ correspond to those parametric integrands Φ for which $\Phi(x, \alpha)$ is linear with respect to α and infinitely differentiable with respect to x. The linear operator mapping each differential form onto the above integral is the m dimensional rectifiable current

$$(\mathscr{H}^m \llcorner A) \wedge \xi$$

with the mass

$$\mathbf{M}[(\mathscr{H}^m \llcorner A) \wedge \xi] = \int_A |\xi(x)|\, d\mathscr{H}^m x.$$

The general concept of current, defined as a real valued linear function on the space of differential forms of class ∞, continuous with respect to convergence of forms in a topology involving derivatives of all orders, was introduced by G. De Rham for use in the theory of harmonic forms. His work was closely related to the development of distribution theory by L. Schwartz. Independently the notion of generalized surface, defined as a real valued continuous linear function on the space of all continuous parametric integrands, was introduced by L. C. Young for use in the calculus of variations. The theory of rectifiable currents was developed by W. H. Fleming and the author.

Many geometric problems involve boundary relations between m and $m-1$ dimensional domains of integration. In the theory of currents the boundary operator ∂ is defined as the dual of exterior differentiation. If T is any m dimensional current in $\mathbf{R}^n$, then ∂T is the $m-1$ dimensional current such that

$$(\partial T)(\psi) = T(d\psi)$$

for all differential forms ψ of degree $m-1$ and class ∞ with compact support in $\mathbf{R}^n$, where $d\psi$ is the exterior derivative of ψ. Of course, rectifiability of T does not imply rectifiability of ∂T. However the main results of Chapter 4 concern integral currents, whereby we mean those rectifiable currents T for which ∂T is also rectifiable. Modern versions of the classical theorems of Gauss, Green and Stokes are propositions which describe the structure of ∂T in terms of the structure of T. A simple example is the formula expressing the boundary of an oriented m dimensional simplex in terms of its oriented $m-1$ dimensional faces. A striking application of our theory is the Gauss-Green type theorem discovered by E. De Giorgi and the author, which makes use of a measure theoretic concept of exterior normal. Analytic and holomorphic chains, supported by subvarieties of $\mathbf{R}^n$ and $\mathbf{C}^n$, are locally integral currents.

It should be understood that the m dimensional Hausdorff measure of the support of an m dimensional integral current need not be finite, as shown by the example in 4.2.25.

Assuming that K is a compact Lipschitz neighborhood retract in $\mathbf{R}^n$, we adapt Whitney's idea of flat norm to metrize the space of all m dimensional integral currents with support in K as follows: The distance between T_1 and T_2 equals the greatest lower bound of the set of numbers

$$\mathbf{M}(T_1 - T_2 - \partial S) + \mathbf{M}(S)$$

corresponding to all $m+1$ dimensional integral currents S with support in K. One of the most important facts about integral currents proved in Section 4.2 is the compactness property: *If* $0 < c < \infty$, *then the set of all those integral currents T, for which*

$$\operatorname{spt} T \subset K \quad \text{and} \quad M(T)+M(\partial T) \leq c,$$

is compact with respect to the above metric. This result was discovered in the special case when $m=n$ and T has multiplicity 1 by E. De Giorgi, in the general case by W. H. Fleming and the author. Another fundamental result is the deformation theorem, which yields basic isoperimetric estimates and links the theory of integral currents with the classical integral homology theory of K.

In order to apply the theory of integral currents to the calculus of variations we consider a positive parametric integrand Φ satisfying the ellipticity condition in 5.1.2, which is implied by the classical parametric regularity condition of Legendre. We prove existence theorems of the following types:

(1) *For every $m-1$ dimensional integral cycle R which bounds in K there exists an m dimensional integral current $T=(\mathscr{H}^m \llcorner A) \wedge \xi$ such that* spt $T \subset K$, $\partial T = R$ *and T minimizes the integral*

$$\langle \Phi, T \rangle = \int_A \Phi[x, \xi(x)] \, d\mathscr{H}^m x$$

in competition with all integral currents with support in K and boundary R.

(2) *In every m dimensional integral homology class of K there exists a cycle T which minimizes $\langle \Phi, T \rangle$ in competition with all members of that homology class.*

For example we obtain solutions of the m dimensional problem of Plateau by letting Φ be the area integrand of degree m, so that $\Phi(x, \alpha) = |\alpha|$, hence $\langle \Phi, T \rangle = \mathbf{M}(T)$.

Since the existence of Φ minimizing currents is now assured under very general hypotheses, present research directs itself to the *problem of smoothness of a Φ minimizing current T: Does smoothness of K and Φ imply smoothness of most of* spt $T \sim$ spt ∂T? In Sections 5.3 and 5.4 we give a detailed exposition of recent work on this problem. Throughout the classical period in the development of the calculus of variations, until about ten years ago, it was generally thought that solutions of regular variational problems could have no interior singularities, in other words, that spt $T \sim$ spt ∂T must be a smooth m dimensional submanifold of $\mathbf{R}^n$. However, R. Thom constructed a 14 dimensional compact analytic manifold K with a 7 dimensional integral homology class which cannot be represented as the image of the fundamental class of any 7 dimensional compact oriented manifold L under a continuous map of L into K; in this case the support of a minimizing cycle T furnished by the existence theorem (2) fails to be a manifold. Also, the author discovered that the current corresponding to any complex r dimensional holomorphic subvariety of an open subset of $\mathbf{C}^s$ minimizes $2r$ dimensional area; here $m=2r$, $n=2s$ and the set of singular points can have real dimension $2(r-1)=m-2$.

The problem of smoothness of minimizing currents has not yet been completely solved, but some very significant partial results have been obtained, first for the area integrand by E. De Giorgi and E. R. Reifenberg, then for general elliptic integrands by F. J. Almgren. Localization reduces the problem to the case when $\operatorname{spt} T \subset \operatorname{Int} K$. The higher differentiability theorem of C. B. Morrey, proved in Section 5.2, implies that a relatively open subset B of $\operatorname{spt} T \sim \operatorname{spt} \partial T$ is a highly smooth submanifold of $\mathbf{R}^n$ if and only if the tangent spaces of spt T vary Hölder continuously on B. In Section 5.3 we verify such Hölder continuity under certain conditions by the powerful new method of Almgren and prove:

(1) $\mathscr{H}^m$ *almost all those points of* $\operatorname{spt} T \sim \operatorname{spt} \partial T$, *at which the multiplicity of* T (= the density of $\|T\|$) *is lowersemicontinuous, have neighborhoods* W *such that* $W \cap \operatorname{spt} T$ *is a smooth* m *dimensional manifold.*

(2) *In case* $m=1$ *there are no singular points.*

(3) *In case* $m=n-1$ *the set of all singular points has* m *dimensional Hausdorff measure* 0.

It follows from (1) that the regular points form a dense relatively open subset of $\operatorname{spt} T \sim \operatorname{spt} \partial T$.

In Section 5.4 we show that *area minimizing currents have no interior singularities in case* $m=n-1 \leq 6$.

The study of singularities of minimizing currents is likely to remain a challenging and fruitful field of research in the years to come.

Several modified versions of the variational problem discussed above have been studied. We refer to Reifenberg's work on the problem of Plateau in [R 2, 3], which was also treated in [MCB 4, Chapter 10], and Almgren's work on general elliptic variational problems in [AF 6]. The most important modifications are replacement of our integral $\langle \Phi, T \rangle$ by the integral

$$\int_A \Phi[x, \xi(x)/|\xi(x)|] \, d\mathscr{H}^m x$$

and, frequently, use of boundary conditions involving homologies with more general coeffizient groups or deformations relative to a given subset. For the minimizing solutions of the modified problems it has been shown, without restriction on m, that the interior singularities form a set with m dimensional Hausdorff measure 0.

All these settings of the variational problem differ fundamentally from the classical approach in which one studies, for some given m dimensional manifold U oriented by an m vectorfield χ, the integrals

$$\int_U \Phi[f(u), [\textstyle\bigwedge_m D f(u)] \, \chi(u)] \, d\mathscr{H}^m u$$

corresponding to all differentiable maps $f: U \to \mathbf{R}^n$ satisfying suitable boundary conditions. Unfortunately no general theorems asserting the existence of minimizing maps have been proved for $m>2$. In particular, the classical version of the m dimensional problem of Plateau has not been solved for $m>2$. The mapping approach to the variational problem yielded a complete solution for $m=1$ and led to many important results for $m=2$, including the famous work of J. Douglas on the problem of Plateau. Expositions of the classical theory may be found in [RA 2], [MCS], [CR], [MCB 4, Chapter 9] and [NJ]. We note that even for $m=2$ and $n=3$ there are some geometrically significant differences between the results of the classical and new theories. For example there exists an area minimizing 2 dimensional integral current T in $\mathbf{R}^3$ such that ∂T is the oriented simple closed curve constructed in [FL 1], spt $T \sim$ spt ∂T is an analytic 2 dimensional manifold with infinite genus pictured in [AF 4, p. 4], and T does not correspond to a surface of any of the types considered by Douglas. One may hope for the future development of a general variational theory for mappings, related to homotopy theory, but this will require the addition of radically new tools to the standard machinery.

Much of the theory of Lebesgue area, a measure theoretic study of continuous maps motivated by the classical calculus of variations, has already been extended to arbitrary dimension m in [F 7, 8, 9, 12, 16, 17] and [DF].

CHAPTER ONE

Grassmann algebra

This chapter presents a systematic account of Grassmann (exterior) algebra, with emphasis on aspects useful for geometric measure theory, and with strict adherence to the principles of naturality. The reader is assumed to be familiar with the category of vector spaces and linear maps, but no knowledge of multilinear algebra (or determinants) is presupposed. The field of scalars will be the field **R** of real numbers, except where another field is explicitly specified. Of course much of the theory is applicable more generally, even to modules.

The development of exterior algebra was begun over a century ago by H. Grassmann, and received its most significant impetus from the works of É. Cartan. A sketch of the classical history may be found in [BO, Livre II, Chapitre III]. Among the more recent improvements adopted here is the treatment of the exterior algebra as a Hopf algebra; the diagonal map was introduced in [CC] under the name "analyzing mapping". The concepts of mass and comass originated in [WH 4], the proof of Wirtinger's inequality is taken from [F 20], and the content of 1.4.5 from [F 15].

The two concluding sections of this chapter treat symmetric algebra by methods analogous to those used for exterior algebra in the earlier part. The definition of polynomial function was taken from [GG].

1.1. Tensor products

1.1.1. A function f which maps a cartesian product

$$V_1 \times V_2 \times \cdots \times V_n$$

of n vectorspaces $V_1, V_2, \ldots, V_n$ into some other vectorspace W is called n **linear** if and only if, for any i and any $v_j \in V_j$ corresponding to all $j \neq i$, the function on V_i carrying x into

$$f(v_1, \ldots, v_{i-1}, x, v_{i+1}, \ldots, v_n)$$

is a linear map of V_i into W.

By the **tensor product** of $V_1, \ldots, V_n$ one means a vectorspace

$$V_1 \otimes V_2 \otimes \cdots \otimes V_n$$

together with a particular n linear map

$$\mu\colon V_1 \times V_2 \times \cdots \times V_n \to V_1 \otimes V_2 \otimes \cdots \otimes V_n$$

which are jointly characterized as follows:

For each n linear map f of $V_1 \times \cdots \times V_n$ into any vectorspace W there exists a unique linear map g of $V_1 \otimes \cdots \otimes V_n$ into W such that $f = g \circ \mu$.

It is customary to write

$$v_1 \otimes v_2 \otimes \cdots \otimes v_n$$

in place of $\mu(v_1, v_2, \ldots, v_n)$, whenever $v_j \in V_j$ for $j = 1, \ldots, n$.

The uniqueness (up to a linear isomorphism) of the tensor product follows immediately from the above characterization. To prove its existence one considers the one to one map

$$\phi\colon V_1 \times V_2 \times \cdots \times V_n \to F,$$

where F is the vectorspace consisting of those real valued functions on $V_1 \times \cdots \times V_n$ which vanish outside some (varying) finite set, and $\phi(v_1, \ldots, v_n)$ is the function with value 1 at $(v_1, \ldots, v_n)$ and value 0 elsewhere in $V_1 \times \cdots \times V_n$. Letting G be the vectorsubspace of F generated by all elements of the two types

$$\phi(v_1, \ldots, v_{i-1}, x, v_{i+1}, \ldots, v_n) + \phi(v_1, \ldots, v_{i-1}, y, v_{i+1}, \ldots, v_n)$$
$$- \phi(v_1, \ldots, v_{i-1}, x + y, v_{i+1}, \ldots, v_n)$$

and

$$\phi(v_1, \ldots, v_{i-1}, c\,v_i, v_{i+1}, \ldots, v_n) - c\,\phi(v_1, \ldots, v_{i-1}, v_i, v_{i+1}, \ldots, v_n)$$

with $c \in \mathbf{R}$, one defines $V_1 \otimes \cdots \otimes V_n = F/G$ and takes μ to be the composition of ϕ and the canonical map of F onto F/G.

The construction of tensor products is natural in the sense that *for any linear maps*

$$f_1\colon V_1 \to V_1', \ldots, f_n\colon V_n \to V_n'$$

there exists a unique linear map

$$f_1 \otimes \cdots \otimes f_n\colon V_1 \otimes \cdots \otimes V_n \to V_1' \otimes \cdots \otimes V_n'$$

such that

$$(f_1 \otimes \cdots \otimes f_n)(v_1 \otimes \cdots \otimes v_n) = f_1(v_1) \otimes \cdots \otimes f_n(v_n)$$

whenever $v_j \in V_j$ for $j = 1, \ldots, n$.

1.1.2. One frequently uses the following linear isomorphisms:

For each permutation λ of $\{1, \ldots, n\}$,

$$V_1 \otimes \cdots \otimes V_n \simeq V_{\lambda(1)} \otimes \cdots \otimes V_{\lambda(n)},$$

where $v_1 \otimes \cdots \otimes v_n$ is mapped onto $v_{\lambda(1)} \otimes \cdots \otimes v_{\lambda(n)}$.

For $m<n$,

$$(V_1 \otimes \cdots \otimes V_m) \otimes (V_{m+1} \otimes \cdots \otimes V_n) \simeq V_1 \otimes \cdots \otimes V_n,$$

where $(v_1 \otimes \cdots \otimes v_m) \otimes (v_{m+1} \otimes \cdots \otimes v_n)$ is mapped onto $v_1 \otimes \cdots \otimes v_n$.

For each vectorspace V the scalar multiplication is a bilinear map of $\mathbf{R} \times V$ into V, inducing the isomorphism $\mathbf{R} \otimes V \simeq V$, where $c \otimes x$ is mapped onto $c\,x$.

If $V \simeq P \oplus Q$ (direct sum), then

$$V \otimes W \simeq (P \otimes W) \oplus (Q \otimes W).$$

In fact, if $f: V \to P$, $\phi: P \to V$, $g: V \to Q$, $\psi: Q \to V$ are linear maps for which

$$f \circ \phi = \mathbf{1}_P, \quad g \circ \psi = \mathbf{1}_Q, \quad \phi \circ f + \psi \circ g = \mathbf{1}_V,$$

then $f \otimes \mathbf{1}_W$, $\phi \otimes \mathbf{1}_W$, $g \otimes \mathbf{1}_W$, $\psi \otimes \mathbf{1}_W$ are linear maps for which

$$(f \otimes \mathbf{1}_W) \circ (\phi \otimes \mathbf{1}_W) = \mathbf{1}_{P \otimes W}, \quad (g \otimes \mathbf{1}_W) \circ (\psi \otimes \mathbf{1}_W) = \mathbf{1}_{Q \otimes W},$$
$$(\phi \otimes \mathbf{1}_W) \circ (f \otimes \mathbf{1}_W) + (\psi \otimes \mathbf{1}_W) \circ (g \otimes \mathbf{1}_W) = \mathbf{1}_{V \otimes W}.$$

It follows that *if B_j is a basis of V_j for each j, then the elements $b_1 \otimes \cdots \otimes b_n$, with $b_j \in B_j$, form a basis of $V_1 \otimes \cdots \otimes V_n$*. Consequently

$$\dim(V_1 \otimes \cdots \otimes V_n) = \prod_{j=1}^{n} \dim V_j.$$

1.1.3. As an *illustrative example* we consider the special case where each V_j is the set of all real valued functions on some set S_j, W is the set of all real valued functions on the cartesian product $S_1 \times \cdots \times S_n$, and

$$f(v_1, \ldots, v_n)(s_1, \ldots, s_n) = v_1(s_1) \cdot v_2(s_2) \ldots v_n(s_n)$$

whenever $v_j \in V_j$ and $s_j \in S_j$ for $j=1, \ldots, n$. Here the corresponding linear map g is a monomorphism, as seen by induction with respect to n; in case $n=2$ one readily verifies that

$$\sum_{k=1}^{m} v_{1,k} \otimes v_{2,k} \notin \ker g$$

whenever $v_{j,1}, v_{j,2}, \ldots, v_{j,m}$ are linearly independent elements of V_j, for $j=1,2$. We also observe that g is an epimorphism if and only if at least $n-1$ of the n sets S_j are finite.

1.1.4. We will use the natural linear transformation

$$\phi\colon \operatorname{Hom}(V,\mathbf{R})\otimes W \to \operatorname{Hom}(V,W);$$

for each linear function $\alpha\colon V\to \mathbf{R}$ and each $y\in W$, the linear function $\phi(\alpha\otimes y)\colon V\to W$ maps $x\in V$ onto $\alpha(x)\cdot y\in W$. Making use of a basis of W, one readily sees that ϕ is always a monomorphism, and that ϕ is an epimorphism unless $\dim V=\infty=\dim W$.

We also recall that, if $n=\dim V<\infty$ and $e_1,\dots,e_n$ form a basis of V, then the **dual basis** of $\operatorname{Hom}(V,R)$ consists of the real valued linear functions $\omega_1,\dots,\omega_n$ characterized by the conditions

$$\omega_i(e_i)=1,\qquad \omega_i(e_j)=0 \text{ for } i\neq j.$$

1.2. Graded algebras

1.2.1. For the purpose of this book a **graded algebra** will be a vectorspace A with a specified direct sum decomposition

$$A=\bigoplus_{m=0}^{\infty} A_m$$

and a bilinear function (multiplication) $\mu\colon A\times A\to A$ such that

$$\mu(A_m\times A_n)\subset A_{m+n}$$

whenever m and n are nonnegative integers; we ordinarily use a product notation, like $x\cdot y$, in place of $\mu(x,y)$.

Usually the multiplication will be **associative**, and there will be a linear isomorphism $\mathbf{R}\simeq A_0$ mapping 1 onto a unit element of the ring A.

Several, but not all, of the algebras considered will satisfy the **anticommutative law:**

$$\eta\cdot\xi=(-1)^{mn}\xi\cdot\eta \text{ for } \xi\in A_m,\ \eta\in A_n.$$

1.2.2. If A and B are graded algebras, then the **graded tensor product**

$$A\otimes B=\bigoplus_{m=0}^{\infty}\ \bigoplus_{p+q=m} A_p\otimes B_q$$

can be made a graded algebra with either of the following two standard definitions of multiplication:

(1) To obtain the **commutative product** $A\otimes B$, let

$$(a\otimes b)\cdot(c\otimes d)=(a\cdot c)\otimes(b\cdot d)$$

whenever $a\in A$, $b\in B$, $c\in A$, $d\in B$.

(2) To obtain the **anticommutative product** $A \otimes B$, let

$$(a \otimes b) \cdot (c \otimes d) = (-1)^{qr} (a \cdot c) \otimes (b \cdot d)$$

whenever $a \in A_p$, $b \in B_q$, $c \in A_r$, $d \in B_s$.

These two definitions are motivated by the simple fact that the commutative product of two commutative algebras is a commutative algebra, while the anticommutative product of two anticommutative algebras is an anticommutative algebra. (Nevertheless it is sometimes also useful to consider the commutative product of two anticommutative algebras!)

The anticommutative products $A \otimes B$ and $B \otimes A$ are isomorphic; the standard isomorphism maps $a \otimes b$ onto $(-1)^{pq} b \otimes a$, whenever $a \in A_p$ and $b \in B_q$. (The analogous isomorphism of commutative products omits the factor $(-1)^{pq}$.)

For every graded algebra A there is a unique linear map

$$\Phi\colon A \otimes A \to A$$

such that $\Phi(x \otimes y) = x \cdot y$ whenever $x, y \in A$. In case A is an associative commutative (anticommutative) algebra, then Φ is a graded algebra homomorphism of the commutative (anticommutative) product $A \otimes A$ into A.

1.2.3. We shall now construct, for each vectorspace V, a particular graded algebra

$$\otimes_* V = \bigoplus_{n=0}^{\infty} \otimes_n V,$$

called the **tensor algebra** of V. We let

$$\otimes_0 V = \mathbf{R}, \qquad \otimes_1 V = V, \qquad \otimes_2 V = V \otimes V, \qquad \ldots;$$

in general $\otimes_m V$ is the m fold tensor product with all m factors equal to V. We define multiplication in $\otimes_* V$ so that its restriction to $\otimes_m V \times \otimes_n V$ is simply the (bilinear) composition

$$\otimes_m V \times \otimes_n V \to \otimes_m V \otimes \otimes_n V \simeq \otimes_{m+n} V.$$

One readily verifies the associative law and the fact that the element 1 of $\otimes_0 V$ is a unit element of the ring $\otimes_* V$.

Among all graded associative algebras with a unit, whose direct summand of index 1 is linearly isomorphic to V, the tensor algebra $\otimes_* V$ is characterized (up to isomorphism) by the following universal mapping property:

For every graded associative algebra A with a unit element, each linear map of V into A_1 can be uniquely extended to a unit preserving algebra homomorphism of $\otimes_* V$ into A, carrying $\otimes_m V$ into A_m for each m.

Finally we take note of the naturality of the construction $\otimes_*$: *Each linear map $f: V \to V'$ can be uniquely extended to a unit preserving algebra homomorphism*

$$\otimes_* f: \otimes_* V \to \otimes_* V'.$$

Moreover f is the direct sum of the linear maps

$$\otimes_m f: \otimes_m V \to \otimes_m V'.$$

1.3. The exterior algebra of a vectorspace

1.3.1. In the associative tensor algebra $\otimes_* V$ we consider the **two sided ideal** $\mathfrak{A}V$ generated by all the elements $x \otimes x$ in $\otimes_2 V$ corresponding to $x \in V$. The quotient algebra

$$\wedge_* V = \otimes_* V / \mathfrak{A}V$$

is called the **exterior algebra** of the vectorspace V. Clearly $\mathfrak{A}V$ is a homogeneous ideal, in fact

$$\mathfrak{A}V = \bigoplus_{m=2}^{\infty} (\otimes_m V \cap \mathfrak{A}V)$$

and therefore

$$\wedge_* V = \bigoplus_{m=0}^{\infty} \wedge_m V$$

where

$$\wedge_m V = \otimes_m V / (\otimes_m V \cap \mathfrak{A}V);$$

in particular $\wedge_0 V = \mathbf{R}$ and $\wedge_1 V = V$. The elements of $\wedge_m V$ are called ***m*-vectors of** V. The multiplication in $\wedge_* V$ is called **exterior multiplication** and denoted by the wedge symbol $\wedge$. It follows that if $v_1, \ldots, v_m \in V$, then the canonical homomorphism maps the product $v_1 \otimes \cdots \otimes v_m \in \otimes_m V$ onto the product

$$v_1 \wedge \cdots \wedge v_m \in \wedge_m V.$$

Clearly $\wedge_m V$ is the vectorspace generated by all such products.

If x and y belong to V, then

$$x \otimes y + y \otimes x = (x+y) \otimes (x+y) - x \otimes x - y \otimes y \in \mathfrak{A}V,$$

hence $y \wedge x = -x \wedge y$. Therefore

$$\begin{aligned}(v_{p+1} \wedge \cdots \wedge v_{p+q}) \wedge (v_1 \wedge \cdots \wedge v_p) \\ = (-1)^{pq} (v_1 \wedge \cdots \wedge v_p) \wedge (v_{p+1} \wedge \cdots \wedge v_{p+q})\end{aligned}$$

whenever $v_1, \ldots, v_{p+q} \in V$, which implies that *the anticommutative law holds for exterior multiplication.*

Among all anticommutative associative algebras with a unit, whose direct summand of index 1 is linearly isomorphic to V, the exterior algebra $\wedge_* V$ is characterized (up to isomorphism) by the following property:

For every anticommutative associative algebra A with a unit element, each linear map of V into A_1 can be uniquely extended to a unit preserving algebra homomorphism of $\wedge_* V$ into A, carrying $\wedge_m V$ into A_m for each m.

Such an extension is unique because the algebra $\wedge_* V$ is generated by $V \cup \{1\}$. To prove its existence, we recall that each linear map of V into A_1 can be extended to an algebra homomorphism h of $\otimes_* V$ into A; since A is anticommutative and $\mathbf{R}$ has characteristic different from 2, $a^2 = 0$ whenever $a \in A_1$, hence $\mathfrak{A} V \subset \ker h$, and h is divisible by the canonical homomorphism of $\otimes_* V$ onto $\wedge_* V$.

The construction $\wedge_*$ is natural: *Each linear map $f\colon V \to V'$ can be uniquely extended to a unit preserving algebra homomorphism*

$$\wedge_* f\colon \wedge_* V \to \wedge_* V'.$$

Moreover $\wedge_* f$ is the direct sum of the linear maps

$$\wedge_m f\colon \wedge_m V \to \wedge_m V'.$$

1.3.2. *The function $\wedge_*$ converts direct sums of vectorspaces into anticommutative products of algebras: If $V \simeq P \oplus Q$, then $\wedge_* V \simeq \wedge_* P \otimes \wedge_* Q$.*

In fact, if $f\colon V \to P$, $\phi\colon P \to V$, $g\colon V \to Q$, $\psi\colon Q \to V$ are linear maps for which

$$f \circ \phi = \mathbf{1}_P, \quad g \circ \psi = \mathbf{1}_Q, \quad \phi \circ f + \psi \circ g = \mathbf{1}_V,$$

then there is a unique unit preserving algebra homomorphism

$$\alpha\colon \wedge_* V \to \wedge_* P \otimes \wedge_* Q \text{ (anticommutative product)}$$

such that $\alpha(v) = f(v) \otimes 1 + 1 \otimes g(v)$ whenever $v \in V$; moreover the composition β of algebra homomorphisms

$$\wedge_* P \otimes \wedge_* Q \xrightarrow{\wedge_* \phi \otimes \wedge_* \psi} \wedge_* V \otimes \wedge_* V \xrightarrow{\Phi} \wedge_* V$$

(where Φ is induced by the multiplication of $\wedge_* V$) is inverse to α, because $\beta \circ \alpha$ and $\alpha \circ \beta$ induce identity maps on the direct summands of degree 1.

Since $\wedge_* \mathbf{R} = \mathbf{R} \oplus \mathbf{R}$, it follows that, *if $e_1, e_2, e_3, \ldots$ form a basis of V, then the products*

$$e_\lambda = e_{\lambda(1)} \wedge e_{\lambda(2)} \wedge \cdots \wedge e_{\lambda(m)}$$

corresponding to all increasing m termed sequences λ form a basis of $\wedge_m V$.
In case V has finite dimension n, this implies that

$$\dim \wedge_m V = \binom{n}{m} \text{ for } m \leq n, \quad \wedge_m V = \{0\} \text{ for } m > n.$$

In fact $\wedge_m V$ has a basis equipotent with the set

$$\Lambda(n, m)$$

of all increasing maps of $\{1, \ldots, m\}$ into $\{1, \ldots, n\}$.

1.3.3. The **diagonal map** of $\wedge_* V$ is the unit preserving algebra homomorphism

$$\Psi: \wedge_* V \to \wedge_* V \otimes \wedge_* V \text{ (anticommutative product)}$$

such that $\Psi(v) = v \otimes 1 + 1 \otimes v$ whenever $v \in V$.

For $v_1, \ldots, v_m \in V$ we compute the product

$$\Psi(v_1 \wedge \cdots \wedge v_m) = \prod_{i=1}^{m} (v_i \otimes 1 + 1 \otimes v_i)$$

using the rules

$$(v_i \otimes 1) \cdot (1 \otimes v_j) = (v_i \otimes v_j) = -(1 \otimes v_j) \cdot (v_i \otimes 1).$$

The result can be conveniently expressed in terms of the notion of **shuffle of type** (p, q), meaning a permutation of $\{1, 2, \ldots, p+q\}$ which is increasing on each of the two sets $\{1, \ldots, p\}$ and $\{p+1, \ldots, p+q\}$. Letting $\mathrm{Sh}(p, q)$ be the set of all shuffles of type (p, q), we find that the product equals

$$\sum_{p=0}^{m} \sum_{\sigma \in \mathrm{Sh}(p, m-p)} \operatorname{index}(\sigma) \cdot (v_{\sigma(1)} \wedge \cdots \wedge v_{\sigma(p)}) \otimes (v_{\sigma(p+1)} \wedge \cdots \wedge v_{\sigma(m)}).$$

The **index** of any permutation σ equals $(-1)^N$, where N is the number of pairs (i, j) such that $i < j$ and $\sigma(i) > \sigma(j)$.

The diagonal map Ψ of $\wedge_* V$ is associative, which means that the following diagram is commutative:

$$\begin{array}{ccc} & \wedge_* V & \\ {}^{\Psi}\swarrow & & \searrow{}^{\Psi} \\ \wedge_* V \otimes \wedge_* V & & \wedge_* V \otimes \wedge_* V \\ \downarrow{\scriptstyle \Psi \otimes 1} & & \downarrow{\scriptstyle 1 \otimes \Psi} \\ (\wedge_* V \otimes \wedge_* V) \otimes \wedge_* V & \simeq & \wedge_* V \otimes (\wedge_* V \otimes \wedge_* V) \end{array}$$

In fact the two vertical compositions of algebra homomorphisms agree on V, mapping v onto $v \otimes 1 \otimes 1 + 1 \otimes v \otimes 1 + 1 \otimes 1 \otimes v$.

The *diagonal map* Ψ *of* $\wedge_* V$ *is anticommutative*, which means that $\alpha \circ \Psi = \Psi$, where α is the automorphism of the algebra $\wedge_* V \otimes \wedge_* V$ which maps $x \otimes y$ onto $(-1)^{pq} y \otimes x$ whenever $x \in \wedge_p V$ and $y \in \wedge_q V$. This is true because $\alpha \circ \Psi$ and Ψ agree on V.

The diagonal map is a natural transformation: If f is a linear map of V into a vectorspace V', with diagonal map Ψ', then

$$\Psi' \circ \wedge_* f = (\wedge_* f \otimes \wedge_* f) \circ \Psi .$$

1.3.4. We conclude this section by defining and computing the **determinant** of a linear map $f\colon V \to V$, where $\infty > \dim V = n$. Since $\dim \wedge_n V = 1$, there exists a unique real number $\det(f)$ such that

$$(\wedge_n f)\,\xi = \det(f) \cdot \xi \quad \text{whenever} \quad \xi \in \wedge_n V.$$

Relative to any choice of base vectors $e_1, \ldots, e_n$ of V, the endomorphism f can be described by the matrix a consisting of real coefficients $a_{i,j}$ such that

$$f(e_i) = \sum_{j=1}^{n} a_{i,j}\, e_j \quad \text{for} \quad j = 1, \ldots, n.$$

Then we find that

$$(\wedge_n f)(e_1 \wedge \cdots \wedge e_n) = f(e_1) \wedge \cdots \wedge f(e_n) = \sum_{\lambda} \left(\prod_{i=1}^{n} a_{i,\lambda(i)} \right) e_\lambda ,$$

where the summation is over the set of all permutations λ of $\{1, \ldots, n\}$, and since $e_\lambda = \operatorname{index}(\lambda) \cdot e_1 \wedge \cdots \wedge e_n$ we obtain

$$\det(f) = \sum_{\lambda} \operatorname{index}(\lambda) \prod_{i=1}^{n} a_{i,\lambda(i)} .$$

If g is another endomorphism of V, then

$$\wedge_n (g \circ f) = (\wedge_n g) \circ (\wedge_n f), \quad \text{hence} \quad \det(g \circ f) = \det(g) \cdot \det(f).$$

Again using base vectors $e_1, \ldots, e_n$ of V we associate with each permutation λ of $\{1, \ldots, n\}$ the endomorphism $\phi(\lambda)$ of V which maps e_i onto $e_{\lambda(i)}$. Since ϕ and det are multiplicative homomorphisms, so is $\operatorname{index} = \det \circ \phi$.

1.4. Alternating forms and duality

1.4.1. An m linear function f which maps the m fold cartesian product V^m of a vector space V into some other vectorspace W, is called **alternating** if and only if $f(v_1, \ldots, v_m) = 0$ whenever $v_1, \ldots, v_m \in V$ and $v_i = v_j$ for some $i \neq j$. We let

$$\wedge^m (V, W)$$

be the vectorspace of all m linear alternating functions (forms) mapping V^m into W. If $f \in \wedge^m(V, W)$ and $g: \bigotimes_m V \to W$ is the corresponding linear function, then $\mathfrak{A}\, V \cap \bigotimes_m V \subset \ker g$, hence there exists a unique linear function $h: \wedge_m V \to W$ such that

$$f(v_1, \ldots, v_m) = h(v_1 \wedge \cdots \wedge v_m) \text{ whenever } v_1, \ldots, v_m \in V.$$

Thus associating h with f, we obtain the linear isomorphism

$$\wedge^m(V, W) \simeq \operatorname{Hom}(\wedge_m V, W).$$

Moreover there is an obvious linear isomorphism

$$\operatorname{Hom}(\wedge_m V, W) \simeq \operatorname{Hom}^m(\wedge_* V, W),$$

where the right side means the set of those linear maps of $\wedge_* V$ into W which vanish on $\wedge_n V$ whenever $n \neq m$. The above isomorphisms remain true for $m = 0$ with the convention $\wedge^0(V, W) = W$. We define

$$\wedge^*(V, W) = \bigoplus_{m=0}^{\infty} \wedge^m(V, W).$$

Most frequently we shall deal with the case when $W = \mathbf{R}$; we therefore abbreviate

$$\wedge^m(V, \mathbf{R}) = \wedge^m V, \; \wedge^*(V, \mathbf{R}) = \wedge^* V.$$

The elements of $\wedge^m V$ are called **m-covectors** of V.

In an extension of the usual notation

$$\langle \xi, h \rangle = h(\xi) \text{ for } \xi \in \wedge_m V, \; h \in \operatorname{Hom}(\wedge_m V, W),$$

we shall also write $\langle \xi, f \rangle = \langle \xi, h \rangle = \langle \xi, k \rangle$ whenever $f \in \wedge^m(V, W)$ and $k \in \operatorname{Hom}^m(\wedge_* V, W)$ correspond to h under the above isomorphisms.

Each linear map $f: V \to V'$ induces a dual linear map

$$\wedge^*(f, W): \wedge^*(V', W) \to \wedge^*(V, W)$$

which is the direct sum of the linear maps

$$\wedge^m(f, W): \wedge^m(V', W) \to \wedge^m(V, W)$$

characterized by the equations

$$\langle \xi, \wedge^m(f, W)\, \phi \rangle = \langle (\wedge_m f)\, \xi, \phi \rangle$$

for $\xi \in \wedge_m V$ and $\phi \in \wedge^m(V', W)$.

We abbreviate $\Lambda^*(f, \mathbf{R}) = \Lambda^* f$.

For example, in case $V' = V$ with $\infty > \dim V = n$, then

$$(\Lambda^n f)\,\phi = \det(f) \cdot \phi \quad \text{whenever} \quad \phi \in \Lambda^n(V, W),$$

because, for each $\xi \in \Lambda_n V$,

$$\langle \xi, \Lambda^n(f, W)\,\phi \rangle = \langle (\Lambda_n f)\,\xi, \phi \rangle = \langle \det(f) \cdot \xi, \phi \rangle = \langle \xi, \det(f) \cdot \phi \rangle.$$

1.4.2. Whenever W is an (ungraded) algebra over $\mathbf{R}$, we shall use the diagonal map Ψ of $\Lambda_* V$ to turn the graded vectorspace $\Lambda^*(V, W)$ into a graded algebra, called the **alternating algebra of** V **with coefficients in** W. Recalling that

$$\Lambda^*(V, W) \simeq \bigoplus_{m=0}^{\infty} \operatorname{Hom}^m(\Lambda_* V, W)$$

we define, for $\alpha \in \operatorname{Hom}^p(\Lambda_* V, W)$ and $\beta \in \operatorname{Hom}^q(\Lambda_* V, W)$, the product $\alpha \wedge \beta \in \operatorname{Hom}^{p+q}(\Lambda_* V, W)$ to be the composition

$$\Lambda_* V \xrightarrow{\Psi} \Lambda_* V \otimes \Lambda_* V \xrightarrow{\alpha \otimes \beta} W \otimes W \xrightarrow{\nu} W,$$

where $\nu(s \otimes t) = s \cdot t$ for $s, t \in W$. Taking account of the associativity, anticommutativity and naturality of Ψ, one easily verifies:

If the multiplication of W is associative, then $\wedge$ is associative.

If the multiplication of W is commutative, then $\wedge$ is anticommutative.

A multiplicative unit element of W acts also as a unit element for $\wedge$.

For each linear map $f: V \to V'$, the linear map $\Lambda^*(f, W)$ is a multiplicative homomorphism.

The alternating product $\alpha \wedge \beta \in \Lambda^{p+q}(V, W)$ of $\alpha \in \Lambda^p(V, W)$ and $\beta \in \Lambda^q(V, W)$ is defined through the isomorphisms $\Lambda^m(V, W) \simeq \operatorname{Hom}^m(\Lambda_* V, W)$. However, the shuffle formula for $\Psi(v_1 \wedge \cdots \wedge v_m)$ leads immediately to the following explicit formula for $\alpha \wedge \beta$:

$$\begin{aligned}(\alpha \wedge \beta)&(v_1, \ldots, v_{p+q}) \\ &= \sum_{\sigma \in \operatorname{Sh}(p,q)} \operatorname{index}(\sigma) \cdot \alpha(v_{\sigma(1)}, \ldots, v_{\sigma(p)}) \cdot \beta(v_{\sigma(p+1)}, \ldots, v_{\sigma(p+q)})\end{aligned}$$

whenever $v_1, \ldots, v_{p+q} \in V$. Assuming the multiplication of W to be associative, one readily obtains by induction a similar formula for the product of m alternating forms

$$\alpha_i \in \Lambda^{p(i)}(V, W)$$

corresponding to $i = 1, \ldots, m$. Abbreviating

$$s(i) = \sum_{j \leq i} p(j),$$

one defines a **shuffle of type** $[p(1), \ldots, p(m)]$ as a permutation of the set $\{1, \ldots, s(m)\}$ which is increasing on each of the m sets $\{s(i-1)+1, \ldots, s(i)\}$, and one finds that

$$(\alpha_1 \wedge \cdots \wedge \alpha_m)(v_1, \ldots, v_{s(m)}) = \sum_{\sigma \in \mathrm{Sh}[p(1), \ldots, p(m)]} \operatorname{index}(\sigma) \prod_{i=1}^{m} \alpha_i(v_{\sigma[s(i-1)+1]}, \ldots, v_{\sigma[s(i)]})$$

whenever $v_1, \ldots, v_{s(m)} \in V$.

Therefore, in case $p(i)=1$ for $i=1, \ldots, m$, we have

$$(\alpha_1 \wedge \cdots \wedge \alpha_m)(v_1, \ldots, v_m) = \sum_{\sigma} \operatorname{index}(\sigma) \prod_{i=1}^{m} \alpha_i(v_{\sigma(i)}),$$

where the summation extends over all permutations σ of $\{1, \ldots, m\}$. In particular, if $\alpha_i(v_j)=0$ whenever $j<i$, we obtain $\prod_{i=1}^{m} \alpha_i(v_i)$.

1.4.3. Next we take $W=\mathbf{R}$ and observe that, *if $\omega_1, \ldots, \omega_n$ are linearly independent in $\wedge^1 V$, and $m \leq n$, then the products*

$$\omega_\lambda = \omega_{\lambda(1)} \wedge \cdots \wedge \omega_{\lambda(m)}$$

corresponding to all $\lambda \in \Lambda(n, m)$ are linearly independent in $\wedge^m V$. In fact, choosing $e_j \in V$ so that $\langle e_j, \omega_j \rangle = 1$ and $\langle e_j, \omega_i \rangle = 0$ whenever $i \neq j$, we find that $\langle e_\lambda, \omega_\lambda \rangle = 1$ and $\langle e_\mu, \omega_\lambda \rangle = 0$ whenever $\lambda \neq \mu \in \Lambda(n, m)$. *In case $\infty > \dim V = n$, then the products ω_λ form a basis of $\wedge^m V$,* because in this case

$$\dim \wedge^m V = \dim \wedge_m V = \binom{n}{m};$$

we also note the equations

$$\phi = \sum_{\lambda \in \Lambda(n,m)} \langle e_\lambda, \phi \rangle \, \omega_\lambda \quad \text{for } \phi \in \wedge^m V,$$

$$\xi = \sum_{\lambda \in \Lambda(n,m)} \langle \xi, \omega_\lambda \rangle \, e_\lambda \quad \text{for } \xi \in \wedge_m V.$$

The coefficients $\langle e_\lambda, \phi \rangle$ and $\langle \omega_\lambda, \xi \rangle$ are called **Grassmann coordinates** of ϕ and ξ, and will usually be denoted ϕ_λ and ξ_λ.

1.4.4. The identity $\operatorname{Hom}(V, \mathbf{R}) = \wedge^1 V$ leads to a unique unit preserving algebra homomorphism

$$\Omega \colon \wedge_* \operatorname{Hom}(V, \mathbf{R}) \to \wedge^* V$$

such that $\Omega(\alpha) = \alpha$ whenever $\alpha \in \operatorname{Hom}(V, \mathbf{R})$. We see that *$\Omega$ is a monomorphism*, because for each choice of base elements of $\operatorname{Hom}(V, \mathbf{R})$, Ω maps their m-fold exterior products, which form a base of $\wedge_m \operatorname{Hom}(V, \mathbf{R})$, onto their alternating products in $\wedge^m V$, which are linearly independent.

Moreover Ω *is an epimorphism in case* $\infty > \dim V = n$, because then $\wedge_m \operatorname{Hom}(V, \mathbf{R})$ and $\wedge^m V$ both have dimension $\binom{n}{m}$. If $\infty = \dim V$, then $\operatorname{im} \Omega$ does not contain $\wedge^m V$ for any $m \geq 2$.

(While the identity $\operatorname{Hom}(V, W) = \wedge^1(V, W)$ allows one to define a similar algebra homomorphism of $\wedge_* \operatorname{Hom}(V, W)$ into $\wedge^*(V, W)$, for any associative algebra W with a unit element, this homomorphism is not injective unless $\dim W = 1$; it is surjective whenever $\dim V < \infty$. On the other hand, in case W is also commutative, one obtains a W linear homomorphism

$$\wedge_*^W \operatorname{Hom}(V, W) \to \wedge^*(V, W),$$

which is an isomorphism is case $\dim V < \infty$. Here $\wedge_*^W$ means the exterior algebra constructed with W replacing $\mathbf{R}$ as coefficient ring.)

1.4.5. For any two vectorspaces V and W the *commutative product* $\wedge^* V \otimes \wedge_* W$ is an associative algebra, which is neither commutative nor anticommutative (unless $V \simeq \mathbf{R}$ or $W \simeq \mathbf{R}$). However *the subalgebra*

$$A = \bigoplus_{m=0}^{\infty} \wedge^m V \otimes \wedge_m W$$

is commutative.

Assuming $\dim V = n < \infty$ we recall (1.1.4) the natural isomorphism

$$\wedge^1 V \otimes W \simeq \operatorname{Hom}(V, W),$$

whose inverse Γ can be computed as follows: If $e_1, \ldots, e_n$ and $\omega_1, \ldots, \omega_n$ are dual basic sequences of V and $\wedge^1 V$, then to each $f \in \operatorname{Hom}(V, W)$ corresponds

$$\Gamma(f) = \sum_{i=1}^{n} \omega_i \otimes f(e_i) \in \wedge^1 V \otimes W.$$

Using the multiplication in A we compute the divided m-th power

$$\Gamma(f)^m / m! = \sum_{\lambda \in \Lambda(n, m)} \omega_\lambda \otimes (\wedge_m f)\, e_\lambda = \Gamma(\wedge_m f),$$

where the symbol Γ on the right designates the inverse of the natural isomorphisms

$$\wedge^m V \otimes \wedge_m W \simeq \wedge^1(\wedge_m V) \otimes \wedge_m W \simeq \operatorname{Hom}(\wedge_m V, \wedge_m W).$$

In case $V = W$ we define $\operatorname{trace} \in \operatorname{Hom}(A, \mathbf{R})$ so that

$$\operatorname{trace}(\phi \otimes \xi) = \langle \xi, \phi \rangle \text{ for } \phi \in \wedge^m V,\ \xi \in \wedge_m V,$$

and observe that

$$\operatorname{trace}[\zeta \cdot \Gamma(\mathbf{1}_{\wedge_j V})] = \binom{n-m}{j} \operatorname{trace}(\zeta)$$

for $\zeta \in \bigwedge^m V \otimes \bigwedge_m V$ and $j=0,1,\ldots,n-m$. Moreover

$$\operatorname{trace}[\Gamma(f)^n/n!] = \operatorname{trace}[\Gamma(\bigwedge_n f)] = \det(f)$$

for $f \in \operatorname{Hom}(V,V)$. Using the binomial theorem we find that, whenever $t \in \mathbf{R}$,

$$\Gamma(t\,\mathbf{1}_V - f)^n/n! = [t\,\Gamma(\mathbf{1}_V) - \Gamma(f)]^n/n! = \sum_{m=0}^{n} t^{n-m}\,\Gamma(\mathbf{1}_{\bigwedge_{n-m}V})(-1)^m\,\Gamma(\bigwedge_m f),$$

hence the value of the **characteristic polynomial** of f at t equals

$$\det(t\,\mathbf{1}_V - f) = \sum_{m=0}^{n} t^{n-m}(-1)^m \operatorname{trace}[\Gamma(\bigwedge_m f)].$$

Similarly one obtains the formula

$$\det(\mathbf{1}_V + t\,f) = \sum_{m=0}^{n} t^m \operatorname{trace}[\Gamma(\bigwedge_m f)].$$

Hereafter we abbreviate $\operatorname{trace}[\Gamma(f)] = \operatorname{trace}(f)$.

If $f, g \in \operatorname{Hom}(V,V)$, then

$$\operatorname{trace}(f \circ g) = \operatorname{trace}(g \circ f);$$

in fact this equation is bilinear, and in the special case when $\Gamma(f) = \alpha \otimes v$, $\Gamma(g) = \beta \otimes w$ it holds because

$$\Gamma(f \circ g) - \Gamma(g \circ f) = \alpha(w)\,\beta \otimes v - \beta(v)\,\alpha \otimes w.$$

If $f \in \operatorname{Hom}(V,V)$, then $\operatorname{trace}(\bigwedge^1 f) = \operatorname{trace}(f)$.

1.5. Interior multiplications

1.5.1. These operations are bilinear maps

$$\lrcorner : \bigwedge_p V \times \bigwedge^q(V,W) \to \bigwedge^{q-p}(V,W)$$
$$\llcorner : \bigwedge_q V \times \bigwedge^p V \to \bigwedge_{q-p} V$$

defined for $p \leq q$, and characterized by the conditions:

$$\langle \xi, \eta \lrcorner \phi \rangle = \langle \xi \wedge \eta, \phi \rangle \text{ whenever } \xi \in \bigwedge_{q-p} V,\ \eta \in \bigwedge_p V,\ \phi \in \bigwedge^q(V,W);$$
$$\langle \zeta \llcorner \alpha, \beta \rangle = \langle \zeta, \alpha \wedge \beta \rangle \text{ whenever } \zeta \in \bigwedge_q V,\ \alpha \in \bigwedge^p V,\ \beta \in \bigwedge^{q-p} V.$$

The interior multiplications $\lrcorner$ and $\llcorner$ may be constructed by essentially dual procedures as follows:

Right exterior multiplication by η maps $\bigwedge_{q-p} V$ into $\bigwedge_q V$; the induced map

$$\operatorname{Hom}(\bigwedge_q V, W) \simeq \bigwedge^q(V,W) \to \operatorname{Hom}(\bigwedge_{q-p} V, W) \simeq \bigwedge^{q-p}(V,W)$$

carries ϕ onto $\eta \lrcorner \phi$.

The diagonal map Ψ of $\wedge_* V$ and the map $h \in \mathrm{Hom}^p(\wedge_* V, \mathbf{R})$ corresponding to α lead to the composition

$$\wedge_* V \xrightarrow{\Psi} \wedge_* V \otimes \wedge_* V \xrightarrow{h \otimes 1} \mathbf{R} \otimes \wedge_* V \simeq \wedge_* V$$

which carries ζ onto $\zeta \llcorner \alpha$. Letting $k \in \mathrm{Hom}^{q-p}(\wedge_* V, \mathbf{R})$ correspond to β, we derive the characteristic condition from the commutativity of the diagram:

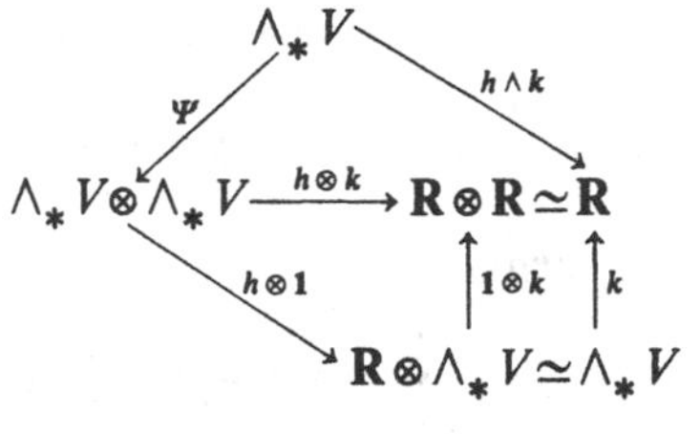

We note that, whenever $r+s \leq t$,

$$(\xi \wedge \eta) \lrcorner \phi = \xi \lrcorner (\eta \lrcorner \phi) \text{ for } \xi \in \wedge_r V,\ \eta \in \wedge_s V,\ \phi \in \wedge^t V;$$

$$\zeta \llcorner (\alpha \wedge \beta) = (\zeta \llcorner \alpha) \llcorner \beta \text{ for } \xi \in \wedge_t V,\ \alpha \in \wedge^r V,\ \beta \in \wedge^s V.$$

1.5.2. If $n = \dim V < \infty$, $e_1, \dots, e_n$ and $\omega_1, \dots, \omega_n$ are dual bases of V and $\Lambda^1 V$, $\lambda \in \Lambda(n,p)$ and $\mu \in \Lambda(n,q)$, the characteristic conditions immediately yield the following values of the interior products $e_\lambda \lrcorner \omega_\mu$ and $e_\mu \llcorner \omega_\lambda$:

In case $\operatorname{im} \lambda \not\subset \operatorname{im} \mu$, then

$$e_\lambda \lrcorner \omega_\mu = 0 \quad \text{and} \quad e_\mu \llcorner \omega_\lambda = 0.$$

In case $\operatorname{im} \lambda \subset \operatorname{im} \mu$, then

$$e_\lambda \lrcorner \omega_\mu = (-1)^M \omega_\nu \quad \text{and} \quad e_\mu \llcorner \omega_\lambda = (-1)^N e_\nu,$$

where $\nu \in \Lambda(n, q-p)$, $\operatorname{im} \lambda \cup \operatorname{im} \nu = \operatorname{im} \mu$, and

$$M = \text{the number of pairs } (i,j) \in \operatorname{im} \lambda \times \operatorname{im} \nu \text{ with } i<j,$$

$$N = \text{the number of pairs } (i,j) \in \operatorname{im} \lambda \times \operatorname{im} \nu \text{ with } i>j.$$

Note that $M+N = p(q-p)$.

In particular, in case μ is the identity map of $\{1, \dots, n\}$, one finds that right interior multiplication by ω_μ, and left interior multiplication by e_μ, give linear isomorphisms

$$\mathbf{D}_p\colon \wedge_p V \simeq \wedge^{n-p} V \quad \text{and} \quad \mathbf{D}^p\colon \wedge^p V \simeq \wedge_{n-p} V.$$

Moreover $\mathbf{D}_p$ and $\mathbf{D}^{n-p}$ are inverse to each other. Note that these isomorphisms depend only on the dual base vectors e_μ and ω_μ of $\wedge_n V$ and $\wedge^n V$.

The final equations of 1.5.1 imply, whenever $r+s\leq n$, that

$$\mathbf{D}_{r+s}(\xi\wedge\eta)=\xi\,\lrcorner\,\mathbf{D}_s\eta \text{ for } \xi\in\bigwedge_r V,\ \eta\in\bigwedge_s V;$$

$$\mathbf{D}^{r+s}(\alpha\wedge\beta)=(\mathbf{D}^r\alpha)\llcorner\beta \text{ for } \alpha\in\bigwedge^r V,\ \beta\in\bigwedge^s V.$$

1.5.3. *If $v\in V$ and $\alpha\in\bigwedge^1 V$ with $\langle v,\alpha\rangle=1$, then*

$$\phi=v\,\lrcorner\,(\phi\wedge\alpha)+(v\,\lrcorner\,\phi)\wedge\alpha$$

whenever $\phi\in\bigwedge^j V$ with $j\geq 1$. One readily verifies this equation after expressing $\phi=\beta\wedge\alpha+\psi$ with $\beta\in\bigwedge^{j-1}V$, $\psi\in\bigwedge^j V$, $v\,\lrcorner\,\beta=0$, $v\,\lrcorner\,\psi=0$. The equation implies that

$$\phi\wedge\alpha=0 \textit{ if and only if } \phi=\beta\wedge\alpha \textit{ for some } \beta\in\bigwedge^{j-1}V,$$

$$v\,\lrcorner\,\phi=0 \textit{ if and only if } \phi=v\,\lrcorner\,\gamma \textit{ for some } \gamma\in\bigwedge^{j+1}V,$$

and yields the direct sum decomposition

$$\bigwedge^j V=\{\beta\wedge\alpha\colon \beta\in\bigwedge^{j-1}V\}\oplus\{v\,\lrcorner\,\gamma\colon \gamma\in\bigwedge^{j+1}V\}.$$

We also note the dual equation:

$$\xi=(v\wedge\xi)\llcorner\alpha+v\wedge(\xi\llcorner\alpha) \text{ for } \xi\in\bigwedge_j V.$$

1.6. Simple *m*-vectors

1.6.1. An element of $\bigwedge_m V$ is called **simple** (or **decomposable**) if and only if it equals the exterior product of m elements of V. We shall see that there is a close connection between simple m-vectors and m dimensional vectorsubspaces of V.

With each $\xi\in\bigwedge_m V$ we associate the vectorsubspace

$$T=V\cap\{v\colon \xi\wedge v=0\}.$$

Assuming $\xi\neq 0$, we claim that $k=\dim T\leq m$ *and for any base vectors* $e_1,\ldots,e_k$ *of T there exists a $\xi'\in\bigwedge_{m-k}V$ such that*

$$\xi=e_1\wedge\cdots\wedge e_k\wedge\xi'.$$

We observe that it suffices to verify this assertion in case $n=\dim V<\infty$, and choose $e_{k+1},\ldots,e_n\in V$ so that $e_1,\ldots,e_n$ form a basis of V. Expanding

$$\xi=\sum_{\lambda\in\Lambda(n,m)}\xi_\lambda e_\lambda$$

and multiplying by e_i, where $i\leq k$, we find that $\xi_\lambda=0$ unless $i\in\operatorname{im}\lambda$, because the products $e_\lambda\wedge e_i$ with $i\notin\operatorname{im}\lambda$ are linearly independent.

We deduce the following four corollaries:

A nonzero m-vector ξ *is simple if and only if its associated subspace T has dimension m; in this case* ξ *equals the exterior product of m suitable base vectors of T.*

The associated subspaces of two nonzero simple m-vectors ξ *and* η *are equal if and only if* $\xi = c\,\eta$ *with* $0 \neq c \in \mathbf{R}$.

If ξ *is a nonzero simple m-vector and* η *is a nonzero simple n-vector, then* $\xi \wedge \eta \neq 0$ *if and only if the subspace associated with* $\xi \wedge \eta$ *is the direct sum of the two subspaces associated with* ξ *and* η.

The subspace associated with a nonzero simple m-vector ξ *is contained in the subspace associated with a nonzero simple n-vector* η *if and only if* $\eta = \xi \wedge \zeta$ *for some* $\zeta \in \wedge_{n-m} V$.

1.6.2. The above association maps the set of all nonzero simple m-vectors of V onto the **Grassmann manifold**

$$\mathbf{G}(V, m)$$

of all m **dimensional subspaces of** V. With respect to this map, ξ and η are equivalent if and only if $\xi = c\,\eta$ for some nonzero real number c.

One also considers the following somewhat finer equivalence relation on the set of all nonzero simple m-vectors of V: ξ and η are equivalent if and only if $\xi = c\,\eta$ for some *positive* number c. The equivalence classes now obtained are called **oriented** m **dimensional subspaces of** V, and the identification space will be denoted

$$\mathbf{G}_0(V, m).$$

We shall abbreviate

$$\mathbf{G}(\mathbf{R}^n, m) = \mathbf{G}(n, m), \qquad \mathbf{G}_0(\mathbf{R}^n, m) = \mathbf{G}_0(n, m).$$

1.6.3. An element of $\wedge^m V$ is called **simple** (or **decomposable**) if and only if it equals the alternating product of m elements of $\wedge^1 V$. In case $n = \dim V < \infty$, the isomorphism (1.4.4)

$$\Omega\colon \wedge_* \wedge^1 V \simeq \wedge^* V$$

shows that simple m-covectors behave just like simple m-vectors; in particular, for $0 \neq \phi \in \wedge^m V$, the associated subspace

$$\wedge^1 V \cap \{\alpha\colon \phi \wedge \alpha = 0\}$$

has dimension $\leq m$, with equality holding if and only if ϕ is simple. Moreover the isomorphisms $\mathbf{D}_p$ and $\mathbf{D}^p$ defined in 1.5.2 preserve simplicity. *If T is the subspace associated with a nonzero simple p-vector* ξ, *then the subspace associated with the simple $n-p$ covector $\mathbf{D}_p(\xi)$ equals*

$$\wedge^1 V \cap \{\alpha\colon \langle v, \alpha\rangle = 0 \text{ for all } v \in T\},$$

the annihilator of T. This is obvious because

$$\mathbf{D}_p(e_1 \wedge \cdots \wedge e_p) = (-1)^{p(n-p)}\, \omega_{p+1} \wedge \cdots \wedge \omega_n.$$

Since the annihilator of the intersection of two subspaces equals the vector sum of their annihilators, we obtain the following corollary:

If T and U are the subspaces associated with simple nonzero p- and r-vectors ξ and η, then

$$\dim(T \cap U) = p + r - n \text{ if and only if } \mathbf{D}_p(\xi) \wedge \mathbf{D}_r(\eta) \neq 0;$$

when these conditions hold, then

$$T \cap U \text{ is associated with } \mathbf{D}^{2n-p-r}[\mathbf{D}_p(\xi) \wedge \mathbf{D}_r(\eta)].$$

1.6.4. Regarding the dual pairing of simple m-vectors and simple m-covectors we shall prove:

If $e_1, \ldots, e_m \in V$, $\xi = e_1 \wedge \cdots \wedge e_m \neq 0$ and $\alpha_1, \ldots, \alpha_m \in \wedge^1 V$, then

$$\langle \xi, \alpha_1 \wedge \cdots \wedge \alpha_m \rangle = \det(f),$$

where f is the endomorphism of the subspace associated with ξ such that

$$f(e_i) = \sum_{j=1}^{m} \langle e_i, \alpha_j \rangle\, e_j \text{ for } i = 1, \ldots, m.$$

By the naturality of $\langle\ ,\ \rangle$ we may assume that $e_1, \ldots, e_m$ form a base of V, and choose $\omega_j \in \wedge^1 V$ so that $\langle e_j, \omega_j \rangle = 1$ and $\langle e_i, \omega_j \rangle = 0$ whenever $j \neq i$. Then $(\wedge^* f)\,\omega_j = \alpha_j$, hence

$$\alpha_1 \wedge \cdots \wedge \alpha_m = (\wedge^* f)(\omega_1 \wedge \cdots \wedge \omega_m) = \det(f) \cdot (\omega_1 \wedge \cdots \wedge \omega_m),$$
$$\langle \xi, \alpha_1 \wedge \cdots \wedge \alpha_m \rangle = \det(f) \cdot \langle \xi, \omega_1 \wedge \cdots \wedge \omega_m \rangle = \det(f).$$

1.6.5. Here we recall 1.5.2 and use the linear maps

$$P\colon V \to V \times V,\ Q\colon V \to V \times V,\ g\colon V \to V \times V,\ f\colon V \times V \to V,$$
$$P(x) = (x, 0),\ Q(x) = (0, x),\ g(x) = (x, x),\ f(x, y) = x - y \text{ for } x, y \in V.$$

If $\xi \in \wedge_k V$ and $\eta \in \wedge_l V$ with $k + l \geq n$, then

$$(\langle \xi, \wedge_k P \rangle \wedge \langle \eta, \wedge_l Q \rangle) \llcorner \langle \omega_1 \wedge \cdots \wedge \omega_n, \wedge^n f \rangle$$
$$= (-1)^{(n-k)l} \langle \mathbf{D}^{2n-k-l}(\mathbf{D}_k \xi \wedge \mathbf{D}_l \eta), \wedge_{k+l-n} g \rangle.$$

We observe that a change of dual bases multiplies both members of this equation by the determinant of the corresponding automorphism of V.

To verify the equation in case ξ and η are simple with $\mathbf{D}_k\xi\wedge\mathbf{D}_l\eta\neq 0$ we choose $e_1,\ldots,e_n$ and $\omega_1,\ldots,\omega_n$ so that

$$\xi=e_1\wedge\cdots\wedge e_k,\qquad \eta=e_{n-l+1}\wedge\cdots\wedge e_n,$$

and we compute

$$\begin{aligned}&\mathbf{D}^{n-k+n-l}(\mathbf{D}_k\xi\wedge\mathbf{D}_l\eta)=\xi\llcorner\mathbf{D}_l\eta\\ &=(e_1\wedge\cdots\wedge e_k)\llcorner\omega_1\wedge\cdots\wedge\omega_{n-l}=e_{n-l+1}\wedge\cdots\wedge e_k.\end{aligned}$$

We also let $a_j=2^{-1}(e_j,-e_j)$ and $b_j=(e_j,e_j)$, note that

$$P(e_j)\wedge Q(e_j)=a_j\wedge b_j \text{ for } j\in\{1,\ldots,n\},$$

and infer that, for all $\phi\in\bigwedge^{k+l-n}(V\times V)$,

$$\begin{aligned}&\langle P(e_1)\wedge\cdots\wedge P(e_k)\wedge Q(e_{n-l+1})\wedge\cdots\wedge Q(e_n),(\omega_1\circ f)\wedge\cdots\wedge(\omega_n\circ f)\wedge\phi\rangle\\ &=\langle P(e_1)\wedge\cdots\wedge P(e_{n-l})\wedge a_{n-l+1}\wedge\cdots\wedge a_k\wedge b_{n-l+1}\wedge\cdots\wedge b_k\\ &\qquad\qquad\wedge Q(e_{k+1})\wedge\cdots\wedge Q(e_n),(\omega_1\circ f)\wedge\cdots\wedge(\omega_n\circ f)\wedge\phi\rangle\\ &=(-1)^{(n-k)(k+l-n)+n-k}\langle b_{n-l+1}\wedge\cdots\wedge b_k,\phi\rangle.\end{aligned}$$

1.6.6. Assuming that V and W are vectorspaces over the field $\mathbf{C}$ of complex numbers we define

$$\textstyle\bigwedge^m_{\mathbf{C}}(V,W)$$

as the subset of $\bigwedge^m(V,W)$ consisting of those forms ϕ for which

$$\phi(v_1,\ldots,v_{j-1},c\,v_j,v_{j+1},\ldots,v_m)=c\,\phi(v_1,\ldots,v_m)$$

whenever $j\in\{1,\ldots,m\}$, $c\in\mathbf{C}$ and $v_1,\ldots,v_m\in V$. We also let

$$\textstyle\bigwedge^*_{\mathbf{C}}(V,W)=\bigoplus_{m=0}^{\infty}\bigwedge^m_{\mathbf{C}}(V,W).$$

Clearly $\bigwedge^*_{\mathbf{C}}(V,\mathbf{C})$ is a $\mathbf{C}$ subalgebra of $\bigwedge^*(V,\mathbf{C})$.

Complex conjugation is an automorphism of $\bigwedge^*(V,\mathbf{C})$. To each $\alpha\in\bigwedge^m(V,\mathbf{C})$ correspond $\sigma,\tau\in\bigwedge^m(V,\mathbf{R})$ such that $\alpha=\sigma+\mathbf{i}\,\tau$, $\bar\alpha=\sigma-\mathbf{i}\,\tau$, hence

$$\alpha\wedge\bar\alpha=\sigma\wedge\sigma+\tau\wedge\tau \text{ for even } m,\quad \alpha\wedge\bar\alpha=-2i\,\sigma\wedge\tau \text{ for odd } m.$$

If $\varepsilon_1,\ldots,\varepsilon_n$ and $\alpha_1,\ldots,\alpha_m$ are dual $\mathbf{C}$ bases of V and $\bigwedge^1_{\mathbf{C}}(V,\mathbf{C})$, and if $\alpha_j=\sigma_j+\mathbf{i}\,\tau_j$ with $\sigma_j,\tau_j\in\bigwedge^1(V,\mathbf{R})$, then

$$\varepsilon_1,\mathbf{i}\,\varepsilon_1,\ldots,\varepsilon_n,\mathbf{i}\,\varepsilon_n \quad\text{and}\quad \sigma_1,\tau_1,\ldots,\sigma_n,\tau_n$$

are dual $\mathbf{R}$ bases for V and $\bigwedge^1(V,\mathbf{R})$. Moreover the products

$$\alpha_{\lambda(1)}\wedge\cdots\wedge\alpha_{\lambda(p)}\wedge\bar\alpha_{\mu(1)}\wedge\cdots\wedge\bar\alpha_{\mu(q)}$$

corresponding to all $\lambda \in \Lambda(n,p)$, $\mu \in \Lambda(n,q)$ with $p+q=m$ form a **C** base of $\wedge^m(V,\mathbf{C})$, and those products which correspond to $p=m$, $q=0$ form a **C** base of $\wedge^m_{\mathbf{C}}(V,\mathbf{C})$. We also note that

$$\sigma_1 \wedge \tau_1 \wedge \cdots \wedge \sigma_n \wedge \tau_n = (\mathrm{i}/2)^n \alpha_1 \wedge \bar{\alpha}_1 \wedge \cdots \wedge \alpha_n \wedge \bar{\alpha}_n$$
$$= (\mathrm{i}/2)^n (-1)^{n(n-1)/2} \alpha_1 \wedge \cdots \wedge \alpha_n \wedge \bar{\alpha}_1 \wedge \cdots \wedge \bar{\alpha}_n.$$

If f is any **C** *linear endomorphism of V, then* $\det(f) \geq 0$. To prove this we choose bases as above and observe that

$$(\alpha_1 \circ f) \wedge \cdots \wedge (\alpha_n \circ f) = d\alpha_1 \wedge \cdots \wedge \alpha_n$$

for some complex number d (the **C** determinant of f), hence

$$(\bar{\alpha}_1 \circ f) \wedge \cdots \wedge (\bar{\alpha}_n \circ f) = \bar{d}\bar{\alpha}_1 \wedge \cdots \wedge \bar{\alpha}_n,$$
$$(\wedge^{2n} f)(\sigma_1 \wedge \tau_1 \wedge \cdots \wedge \sigma_n \wedge \tau_n) = d\,\bar{d}\sigma_1 \wedge \tau_1 \wedge \cdots \wedge \sigma_n \wedge \tau_n$$

and $\det(f) = d\,\bar{d} = |d|^2 \geq 0$.

A nonzero simple m vector $\xi \in \wedge_m V$ is called **complex** if and only if the **R** vectorsubspace of V associated with ξ is a **C** vectorsubspace of V. It follows that ξ is complex if and only if m is even, say $m=2p$, and

$$\xi = r\, v_1 \wedge \mathbf{i}\, v_1 \wedge \cdots \wedge v_p \wedge \mathbf{i}\, v_p$$

for some $r \in \mathbf{R}$ and $v_1, \ldots, v_p \in V$. Moreover sign(r) is uniquely determined by ξ, because for any two **C** bases $v_1, \ldots, v_p$ and $w_1, \ldots, w_p$ of the vectorspace associated with ξ there exists a **C** linear automorphism g of this vectorspace which maps v_j onto w_j, hence

$$w_1 \wedge \mathbf{i}\, w_1 \wedge \cdots \wedge w_p \wedge \mathbf{i}\, w_p = \det(g)\, v_1 \wedge \mathbf{i}\, v_1 \wedge \cdots \wedge v_p \wedge \mathbf{i}\, v_p,$$

with $\det(g) > 0$ according to the preceding paragraph. We term ξ **positive** in case $r > 0$.

1.7. Inner products

1.7.1. We recall that the bilinear functions

$$B\colon V \times V \to \mathbf{R}$$

are in one to one correspondence with the linear maps

$$\beta\colon V \to \wedge^1 V$$

through the connecting formula

$$B(x,y) = \langle x, \beta(y) \rangle \quad \text{for } x, y \in V.$$

One calls B **symmetric** if and only if $B(x,y) = B(y,x)$ whenever $x, y \in V$; in this case we call the corresponding linear map β a **polarity**.

An **inner product** is a symmetric bilinear function B satisfying the condition

$$B(x, x)>0 \text{ if and only if } 0 \neq x \in V.$$

When discussing a vectorspace V with a particular fixed inner product B, we generally use the dot product notation $x \bullet y$ in place of $B(x, y)$, and also define the **norm**

$$|x|=(x \bullet x)^{\frac{1}{2}}.$$

For example we usually write

$$x \bullet y=\sum_{i=1}^{n} x_i y_i \text{ for } x=(x_1, \ldots, x_n) \text{ and } y=(y_1, \ldots, y_n) \in \mathbf{R}^n.$$

Assuming henceforth that $\bullet$ is an inner product for V, we use the fact that

$$t^2(x \bullet x)+2t(x \bullet y)+(y \bullet y)=(t x+y) \bullet (t x+y) \geq 0$$

whenever $t \in \mathbf{R}$ to obtain the inequalities

$$x \bullet y \leq |x| \cdot |y|, \quad \text{hence} \quad |x+y| \leq |x|+|y|$$

for $x, y \in V$; both inequalities are strict in case x and y are linearly independent.

A sequence $v_1, \ldots, v_p$ satisfying the conditions $v_i \bullet v_i=1$, and $v_i \bullet v_j=0$ for $i \neq j$, is called **orthonormal**. For every linearly independent sequence $u_1, \ldots, u_p \in V$ there exists an orthonormal sequence $v_1, \ldots, v_p$ such that, for $k=1, \ldots, p$ the sets $\{u_1, \ldots, u_k\}$ and $\{v_1, \ldots, v_k\}$ generate the same vector subspace of V. Therefore, in case $\dim V<\infty$, V has an orthonormal base.

We metrize V so that the distance between x and y equals $|x-y|$. It follows that V is boundedly compact if and only if $\dim V<\infty$.

1.7.2. A linear map $f: V \rightarrow V'$, where V' is another vectorspace with an inner product (also denoted by $\bullet$) is called an **orthogonal injection** if and only if $f(x) \bullet f(y)=x \bullet y$ whenever $x, y \in V$. The set of all orthogonal injections of $\mathbf{R}^m$ into $\mathbf{R}^n$ will be denoted

$$\mathbf{O}(n, m).$$

Moreover $\mathbf{O}(n)=\mathbf{O}(n, n)$ is the orthogonal group of $\mathbf{R}^n$.

In case $\dim V<\infty$, the polarity corresponding to the inner product of V is a linear isomorphism of V onto $\wedge^1 V$, and one endows $\wedge^1 V$ with the inner product which makes this polarity orthogonal. The resulting norms on V and $\wedge^1 V$ are dual, in the sense that

$$|\alpha|=\sup\{\langle x, \alpha\rangle: x \in V, |x| \leq 1\}$$

whenever $\alpha \in \wedge^1 V$. Moreover the polarity maps each orthonormal basis of V onto the dual basis of $\wedge^1 V$.

1.7.3. *For each symmetric bilinear function* $S: V \times V \to \mathbf{R}$, *with* $\dim V < \infty$, *there exists an orthonormal base* $e_1, \ldots, e_n$ *of* V *such that*

$$S(e_i, e_i) \geq S(e_j, e_j) \quad \text{and} \quad S(e_i, e_j) = 0 \quad \text{for } i < j.$$

Proceeding inductively, one may choose e_i in the compact set

$$C_i = \{x\colon\ |x| = 1,\ x \bullet e_k = 0 \text{ whenever } k < i\}$$

so that $S(e_i, e_i) \geq S(x, x)$ for $x \in C_i$; for $i < j$ the fact that

$$|e_i + t\,e_j|^{-1}(e_i + t\,e_j) \in C_i$$

whenever $t \in \mathbf{R}$ implies $S(e_i, e_j) = 0$.

An endomorphism f of V is called

symmetric in case $f(x) \bullet y = x \bullet f(y)$ for $x, y \in V$,

skewsymmetric in case $f(x) \bullet y = -x \bullet f(y)$ for $x, y \in V$.

Every symmetric or skewsymmetric endomorphism f is *orthogonally reducible* in the following sense: *If* W *is a subspace of* V, *with the orthogonal complement*

$$W' = V \cap \{x\colon\ x \bullet y = 0 \text{ for all } y \in W\},$$

then $f(W) \subset W$ *implies* $f(W') \subset W'$. In case $\dim V < \infty$ it follows that V is the direct sum of finitely many mutually orthogonal subspaces W which are *minimal* with respect to the property that $f(W) \subset W$.

In case f *is symmetric, these minimal subspaces have dimension* 1. One associates with f the symmetric bilinear function S such that $S(x, y) = f(x) \bullet y$ for $x, y \in V$, and obtains an orthonormal base $e_1, \ldots, e_n$ of V such that $f(e_i) \bullet e_j = 0$ for $i \neq j$, hence $f(e_i) = \lambda_i e_i$ with $\lambda_i \in R$.

In case f *is skewsymmetric, the minimal subspaces have dimension* ≤ 2. This is true because $f^2 = f \circ f$ is symmetric, and $f^2(e_i) = \lambda_i e_i$ implies that f maps the subspace generated by e_i and $f(e_i)$ into itself. As a corollary one obtains the following representation of alternating 2-forms:

If $\phi \in \wedge^2 V$, *with* $\dim V < \infty$, *then there exists an orthonormal sequence* $\omega_1, \omega_2, \ldots, \omega_{2m-1}, \omega_{2m} \in \wedge^1 V$ *and a sequence of nonnegative numbers* $\lambda_1, \ldots, \lambda_m$ *such that*

$$\phi = \sum_{j=1}^{m} \lambda_j\, \omega_{2j-1} \wedge \omega_{2j}.$$

This is trivial if $\dim V \leq 2$. To obtain the general case we consider the skewsymmetric endomorphism f of V such that $f(x) \bullet y = \phi(x, y)$ for

$x, y \in V$, we decompose V into the direct sum of mutually orthogonal subspaces $W_1, \ldots, W_s$ with $f(W_j) \subset W_j$ and $\dim W_j \leq 2$, and observe that $\phi(x,y)=0$ whenever $x \in W_j$, $y \in W_k$, $j \neq k$.

If f: $V \to V'$ is a linear map of inner product spaces, with $\dim V < \infty$, *then V has an orthonormal base $e_1, \ldots, e_n$ such that $f(e_i) \cdot f(e_j) = 0$ for $i \neq j$. In case* $\dim V \leq \dim V'$, *there exists a symmetric endomorphism g of V and an orthogonal injection h: $V \to V'$ such that $h \circ g = f$.* Choosing e_i adapted to the symmetric bilinear function S such that $S(x,y) = f(x) \cdot f(y)$ for $x, y \in V$, one may define

$$g(e_i) = |f(e_i)| \cdot e_i \text{ for all } i,$$

and choose h so that

$$h(e_i) = |f(e_i)|^{-1} \cdot f(e_i) \text{ whenever } f(e_i) \neq 0,$$

while h maps $\ker f$ orthogonally into the orthogonal complement of $\operatorname{im} f$ in V'. Similarly, *in case* $\dim V \geq \dim V'$ *there exists a symmetric endomorphism k of V' and* (*see* 1.7.4) *an orthogonal projection p: $V \to V'$ such that $k \circ p = f$*; this assertion may be proved by applying the preceding proposition to f^*.

1.7.4. Now suppose V and V' are finite dimensional vector spaces with inner products, and with the corresponding polarities β and β'. With each linear map f: $V \to V'$ one associates the **adjoint** linear map f^*: $V' \to V$ by means of the commutative diagram

$$\begin{array}{ccc} V & \xrightarrow{\beta} & \Lambda^1 V \\ {\scriptstyle f^*}\uparrow & & \uparrow{\scriptstyle \Lambda^1 f} \\ V' & \xrightarrow{\beta'} & \Lambda^1 V' \end{array}$$

or equivalently by the condition

$$x \cdot f^*(y) = f(x) \cdot y \text{ for } x \in V, \ y \in V'.$$

If g: $V' \to V''$ is also linear, then $(g \circ f)^* = f^* \circ g^*$.

In case $V = V'$, f is symmetric if and only if $f^* = f$; f is skewsymmetric if and only if $f^* = -f$.

Always $f^{**} = f$. The endomorphisms $f^* \circ f$, $f \circ f^*$ of V, V' are symmetric.

We observe that f is an orthogonal injection if and only if $f^* \circ f = \mathbf{1}_V$. In case f is an orthogonal injection, we call f^* an **orthogonal projection.** Hence a linear map g: $V' \to V$ is an orthogonal projection if and only if $g \circ g^* = \mathbf{1}_V$. We shall frequently consider the set

$$\mathbf{O}^*(n,m) = \{f^* \colon f \in \mathbf{O}(n,m)\}$$

of all orthogonal projections of $\mathbf{R}^n$ onto $\mathbf{R}^m$. In particular, to each $\lambda \in \Lambda(n, m)$ corresponds the map $\mathbf{p}_\lambda \in \mathbf{O}^*(n, m)$ such that

$$\mathbf{p}_\lambda(x) = (x_{\lambda(1)}, \ldots, x_{\lambda(m)}) \text{ for } x = (x_1, \ldots, x_n) \in \mathbf{R}^n.$$

If W is any m dimensional vectorsubspace of $\mathbf{R}^n$, with the inclusion map $h: W \to \mathbf{R}^n$, then $\wedge^m h$ is an epimorphism, hence $(\wedge^m h)\,\omega_\lambda \neq 0$ for some $\lambda \in \Lambda(n, m)$, where $\omega_1, \ldots, \omega_n$ form the standard base of $\wedge^1 \mathbf{R}^n$; since ω_λ generates $\operatorname{im} \wedge^m \mathbf{p}_\lambda$, it follows that $\wedge^m(\mathbf{p}_\lambda \circ h) \neq 0$, $\mathbf{p}_\lambda \circ h$ is an isomorphism, hence

$$W \cap \ker \mathbf{p}_\lambda = \{0\}.$$

If $f: V \to W_1$, $g: V \to W_2$ are orthogonal projections and W_3 is an inner product space such that

$$\dim W_1 + \dim W_2 - \dim V \geq \dim W_3,$$

then there exist orthogonal projections $p: W_1 \to W_3$, $q: W_2 \to W_3$ with $p \circ f = q \circ g$. In fact, since $\dim(\operatorname{im} f^* \cap \operatorname{im} g^*) \geq \dim W_3$, there exist orthogonal injections $u: W_3 \to W_1$, $v: W_3 \to W_2$ with $f^* \circ u = g^* \circ v$, and we can take $p = u^*$, $q = v^*$.

1.7.5. Next we discuss the manner in which *inner products for the spaces* $\wedge_m V$ *are induced by the given inner product for* V: The polarity $\beta: V \to \wedge^1 V$ can be uniquely extended to a unit preserving algebra homomorphism $\gamma: \wedge_* V \to \wedge^* V$, which is the direct sum of linear maps

$$\gamma_m: \wedge_m V \to \wedge^m V.$$

Composing γ_m with $\wedge^m V \simeq \wedge^1(\wedge_m V)$, we obtain linear maps

$$\beta_m: \wedge_m V \to \wedge^1(\wedge_m V)$$

which satisfy the condition

$$\langle \xi, \beta_m(\eta) \rangle = \langle \eta, \beta_m(\xi) \rangle \text{ for } \xi, \eta \in \wedge_m V.$$

It suffices to verify that this holds true if ξ and η are simple, say

$$\xi = v_1 \wedge \cdots \wedge v_m \text{ and } \eta = w_1 \wedge \cdots \wedge w_m,$$

in which case the permutation formula for the alternating product of 1-forms (1.4.2) gives

$$\begin{aligned}\langle \xi, \beta_m(\eta) \rangle &= \langle v_1 \wedge \cdots \wedge v_m, \beta(w_1) \wedge \cdots \wedge \beta(w_m) \rangle \\ &= \sum_\sigma \operatorname{index}(\sigma) \prod_{i=1}^m v_{\sigma(i)} \bullet w_i = \sum_\sigma \operatorname{index}(\sigma^{-1}) \prod_{j=1}^m v_j \bullet w_{\sigma^{-1}(j)} \\ &= \langle w_1 \wedge \cdots \wedge w_m, \beta(v_1) \wedge \cdots \wedge \beta(v_m) \rangle = \langle \eta, \beta_m(\xi) \rangle.\end{aligned}$$

Thus β_m is a polarity, and we define a symmetric bilinear function $\bullet$ on $\wedge_m V \times \wedge_m V$ by the formula

$$\xi \bullet \eta = \langle \xi, \beta_m(\eta) \rangle \quad \text{for } \xi, \eta \in \wedge_m V.$$

We shall soon see that $\xi \bullet \xi > 0$ in case $\xi \neq 0$, so that $\bullet$ is in fact an inner product for $\wedge_m V$.

The above permutation formula shows that if some w_i is orthogonal to all v_j, then $(v_1 \wedge \cdots \wedge v_m) \bullet (w_1 \wedge \cdots \wedge w_m) = 0$.

For any $v_1, \ldots, v_m \in V$ we can express $v_i = u_i + w_i$, where u_i belongs to the subspace generated by $\{v_k \colon k < i\}$ and w_i is orthogonal to this subspace; then $v_1 \wedge \cdots \wedge v_m = w_1 \wedge \cdots \wedge w_m$ and the permutation formula implies

$$(v_1 \wedge \cdots \wedge v_m) \bullet (v_1 \wedge \cdots \wedge v_m) = \prod_{i=1}^{m} v_i \bullet w_i \leq \prod_{i=1}^{m} v_i \bullet v_i,$$

equality holding if and only if $v_1, \ldots, v_m$ are mutually orthogonal.

Therefore, if $e_1, \ldots, e_n$ form an orthonormal base for V, then the base vectors e_λ of $\wedge_m V$, corresponding to $\lambda \in \Lambda(n, m)$, are likewise orthonormal. For any m-vectors ξ and η the representations

$$\xi = \sum_{\lambda \in \Lambda(n, m)} \xi_\lambda e_\lambda, \qquad \eta = \sum_{\lambda \in \Lambda(n, m)} \eta_\lambda e_\lambda$$

and the bilinearity of $\bullet$ lead to the formula

$$\xi \bullet \eta = \sum_{\lambda \in \Lambda(n, m)} \xi_\lambda \eta_\lambda .$$

In case $\xi = \eta \neq 0$, we obtain

$$\xi \bullet \xi = \sum_{\lambda} (\xi_\lambda)^2 > 0.$$

One now readily estimates the norm of the exterior product of a p-vector ξ and a q-vector η:

In case ξ or η is simple, then $|\xi \wedge \eta| \leq |\xi| \cdot |\eta|$.

In case both ξ and η are simple and nonzero, equality holds if and only if the subspaces associated with ξ and η are orthogonal.

Always

$$|\xi \wedge \eta| \leq \binom{p+q}{p}^{\frac{1}{2}} |\xi| \cdot |\eta| .$$

To prove the last inequality we represent

$$\xi = \sum_{\lambda \in \Lambda(n, p)} \xi_\lambda e_\lambda, \qquad \eta = \sum_{\mu \in \Lambda(n, q)} \eta_\mu e_\mu$$

with $\xi_\lambda, \eta_\mu \in \mathbf{R}$, we define

$$S(v) = \{(\lambda, \mu) \colon \lambda \in \Lambda(n, p),\ \mu \in \Lambda(n, q),\ e_\lambda \wedge e_\mu = \pm e_v\}$$

whenever $\nu \in \Lambda(n, p+q)$, and observe that

$$\begin{aligned}
|\xi \wedge \eta|^2 &\le \sum_{\nu \in \Lambda(n,\,p+q)} \Big(\sum_{(\lambda,\,\mu) \in S(\nu)} |\xi_\lambda \eta_\mu| \Big)^2 \\
&\le \sum_{\nu \in \Lambda(n,\,p+q)} \operatorname{card} S(\nu) \sum_{(\lambda,\,\mu) \in S(\nu)} (\xi_\lambda \eta_\mu)^2 \\
&\le \binom{p+q}{p} \sum_{\lambda \in \Lambda(n,\,p)} (\xi_\lambda)^2 \sum_{\mu \in \Lambda(n,\,q)} (\eta_\mu)^2 .
\end{aligned}$$

The maps β, γ_m, β_m occurring in the preceding construction are related by the commutative diagram:

$$\begin{array}{ccccc}
 & & \bigwedge_m V & & \\
 & \swarrow{\scriptstyle \bigwedge_m \beta} & \downarrow{\scriptstyle \gamma_m} & \searrow{\scriptstyle \beta_m} & \\
\bigwedge_m \bigwedge^1 V & \xrightarrow{\;\Omega\;} & \bigwedge^m V & \simeq & \bigwedge^1 \bigwedge_m V
\end{array}$$

In case $\dim V < \infty$, the maps β_m and γ_m are linear isomorphisms, and one endows $\bigwedge^1 \bigwedge_m V$ and $\bigwedge^m V$ with inner product so that β_m and γ_m become orthogonal. If $e_1, \ldots, e_n$ and $\omega_1, \ldots, \omega_n$ are dual orthogonal bases of V and $\bigwedge^1 V$, then $\gamma_m(e_\lambda) = \omega_\lambda$ for $\lambda \in \Lambda(n, m)$. Also

$$|\langle \xi, \phi \rangle| \le |\xi| \cdot |\phi| \text{ whenever } \xi \in \textstyle\bigwedge_m V,\ \phi \in \bigwedge^m V;$$

equality holds if and only if $\gamma_m(\xi)$ and ϕ are linearly dependent.

1.7.6. Suppose $f: V \to V'$ is a linear map, where V and V' are finite dimensional inner product spaces. From the commutative diagram

$$\begin{array}{ccccccc}
\bigwedge_m V & \xrightarrow{\bigwedge_m \beta} & \bigwedge_m \bigwedge^1 V & \simeq & \bigwedge^m V & \simeq & \bigwedge^1 \bigwedge_m V \\
\uparrow{\scriptstyle \bigwedge_m f^*} & & \uparrow{\scriptstyle \bigwedge_m \bigwedge^1 f} & & \uparrow{\scriptstyle \bigwedge^m f} & & \uparrow{\scriptstyle \bigwedge^1 \bigwedge_m f} \\
\bigwedge_m V' & \xrightarrow{\bigwedge_m \beta'} & \bigwedge_m \bigwedge^1 V' & \simeq & \bigwedge^m V' & \simeq & \bigwedge^1 \bigwedge_m V'
\end{array}$$

we see that $\bigwedge_m f^* = (\bigwedge_m f)^*$.

In case $V = V'$, $\det(f^*) = \det(f)$ and $\operatorname{trace}(f^*) = \operatorname{trace}(f)$. If f is symmetric, so is $\bigwedge_m f$. If f is skewsymmetric, then $(\bigwedge_m f)^* = (-1)^m \bigwedge_m f$. If f is orthogonal, so is $\bigwedge_m f$, hence $\det(f)^2 = 1$.

In general, if f is an orthogonal injection, so is $\bigwedge_m f$. If f is an orthogonal projection so is $\bigwedge_m f$.

The **norm of** f is defined by the formula

$$\|f\| = \sup \{ |f(x)| : x \in V \text{ and } |x| \le 1 \}.$$

It follows that $\|f\| = \|f^*\| = \|\bigwedge^1 f\|$ and

$$\|\textstyle\bigwedge_m f\| = \|\bigwedge_m f^*\| = \|\bigwedge^m f\| \le \|f\|^m \text{ for all } m.$$

To prove this we choose an orthonormal base $e_1, \ldots, e_n$ of V such that $f(e_i) \bullet f(e_j) = 0$ for $i \neq j$; then the m-vectors $(\wedge_m f) e_\lambda$ corresponding to $\lambda \in \Lambda(n, m)$ are mutually orthogonal, hence $\xi \in \wedge_m V$ implies

$$|(\wedge_m f)\xi|^2 = \Big|\sum_\lambda \xi_\lambda (\wedge_m f) e_\lambda\Big|^2 = \sum_\lambda (\xi_\lambda)^2 \, |(\wedge_m f) e_\lambda|^2 \leq |\xi|^2 \, \|f\|^{2m}.$$

In case $m = \dim V$, then $\|\wedge_m f\| = |(\wedge_m f)\xi|$ for every $\xi \in \wedge_m V$ with $|\xi| = 1$; hence $\|\wedge_m f\| > 0$ if and only if f is a monomorphism.

In case $m = \dim V'$, then $\|\wedge^m f\| = |(\wedge^m f)\phi|$ for every $\phi \in \wedge^m V'$, with $|\phi| = 1$; hence $\|\wedge_m f\| > 0$ if and only if f is an epimorphism.

In case $V = V'$ and $m = \dim V$, then $\|\wedge_m f\| = |\det(f)|$.

1.7.7. If $\xi \in \wedge_m \mathbf{R}^n$ and ξ is simple, then

$$|\xi| = \sup\{|(\wedge_m g)\,\xi| : g \in \mathbf{O}^*(n, m)\}.$$

In fact $g \in \mathbf{O}^*(n, m)$ implies $\|\wedge_m g\| = 1$, hence $|(\wedge_m g)\,\xi| \leq |\xi|$. In case $\xi \neq 0$ we can choose $f \in \mathbf{O}(n, m)$ so that $\operatorname{im} f$ is the subspace T associated with ξ, hence $f \circ f^* | T = \mathbf{1}_T$ and $(\wedge_m f \circ \wedge_m f^*)\,\xi = \xi$; since $\|\wedge_m f\| = 1$, $|(\wedge_m f^*)\,\xi| \geq |\xi|$.

Similarly, if $\phi \in \wedge^m \mathbf{R}^n$ and ϕ is simple, then

$$|\phi| = \sup\{|(\wedge^m f)\,\phi| : f \in \mathbf{O}(n, m)\}.$$

1.7.8. Given $n = \dim V < \infty$ and $E \in \wedge_n V$ with $|E| = 1$ we define linear maps

$$*: \wedge_p V \to \wedge_{n-p} V, \qquad *\xi = E \llcorner \gamma_p(\xi) \text{ for } \xi \in \wedge_p V,$$

$$*: \wedge^p V \to \wedge^{n-p} V, \qquad *\phi = \gamma_{n-p}(E \llcorner \phi) \text{ for } \phi \in \wedge^p V.$$

These maps equal $\mathbf{D}^p \circ \gamma_p$ and $\gamma_{n-p} \circ \mathbf{D}^p$, where $\mathbf{D}^p$ is defined as in 1.5.2 with $e_\mu = E$, $\mu = (1, \ldots, n)$; using the notation introduced there we find that

$$*e_\lambda = (-1)^N e_\nu \quad \text{and} \quad *\omega_\lambda = (-1)^N \omega_\nu,$$

$$*e_\nu = (-1)^M e_\lambda \quad \text{and} \quad *\omega_\nu = (-1)^M \omega_\lambda.$$

It follows that

$$**\xi = (-1)^{p(n-p)} \xi, \qquad \xi \wedge *\eta = (\xi \bullet \eta)\, e_{(1, \ldots, n)},$$

$$**\phi = (-1)^{p(n-p)} \phi, \qquad \phi \wedge *\psi = (\phi \bullet \psi)\, \omega_{(1, \ldots, n)},$$

$$\langle *\xi, *\phi \rangle = \langle \xi, \phi \rangle, \qquad \langle \xi, *\phi \rangle = (-1)^{p(n-p)} \langle *\xi, \phi \rangle$$

whenever $\xi, \eta \in \wedge_p V$ and $\phi, \psi \in \wedge^p V$.

1.7.9. Whenever V and W are inner product spaces, the space $V\otimes W$ can be given an inner product such that

$$(v\otimes w)\bullet(v'\otimes w)=(v\bullet v')\cdot(w\bullet w') \text{ for } v,v'\in V \text{ and } w,w'\in W.$$

Clearly this equation characterizes a unique symmetric bilinear function. Moreover, if $v_1,\dots,v_n\in V$ and $w_1,\dots,w_s\in W$ are orthonormal sequences, then the vectors $v_i\otimes w_j$ are orthonormal. Therefore we have indeed defined an inner product.

Assuming $\dim V=n<\infty$ and $f,g\in\operatorname{Hom}(V,W)$, hence $\Gamma(f)$, $\Gamma(g)\in\bigwedge^1 V\otimes W$ according to 1.4.5, we will apply the above definition with V replaced by $\bigwedge^1 V$ to compute $\Gamma(f)\bullet\Gamma(g)$. Choosing dual orthonormal bases $e_1,\dots,e_n$ and $\omega_1,\dots,\omega_n$ of V and $\bigwedge^1 V$ we obtain

$$\begin{aligned}\Gamma(f)\bullet\Gamma(g)&=\left[\sum_{i=1}^n\omega_i\otimes f(e_i)\right]\bullet\left[\sum_{j=1}^n\omega_j\otimes g(e_j)\right]\\&=\sum_{i=1}^n f(e_i)\bullet g(e_i)=\sum_{i=1}^n e_i\bullet(f^*\circ g)e_i\\&=\operatorname{trace}\left[\sum_{i=1}^n\omega_i\otimes(f^*\circ g)e_i\right]=\operatorname{trace}(f^*\circ g).\end{aligned}$$

Replacing f,g by $\bigwedge_m f$, $\bigwedge_m g$ we find that

$$\Gamma(\textstyle\bigwedge_m f)\bullet\Gamma(\bigwedge_m g)=\operatorname{trace}[\bigwedge_m(f^*\circ g)]$$

for every positive integer m; in particular

$$\Gamma(\textstyle\bigwedge_n f)\bullet\Gamma(\bigwedge_n g)=\det(f^*\circ g),$$

$$|\Gamma(\textstyle\bigwedge_n f)-\Gamma(\bigwedge_n g)|^2=\det(f^*\circ f)+\det(g^*\circ g)-2\det(f^*\circ g).$$

We note the inequalities

$$\|f\|\le|\Gamma(f)|\le n^{\frac{1}{2}}\,\|f\|.$$

Also, if s and t are orthogonal automorphisms of V and W, then $(\bigwedge^1 s)\otimes t$ is an orthogonal automorphism of $\bigwedge^1 V\otimes W$ mapping $\Gamma(f)$ onto $\Gamma(t\circ f\circ s)$, hence $|\Gamma(f)|=|\Gamma(t\circ f\circ s)|$.

Hereafter we abbreviate $\Gamma(f)\bullet\Gamma(g)=f\bullet g$, $|\Gamma(f)|=|f|$.

1.7.10. In case V is a finite dimensional inner product space we define the **discriminant** and the **trace** of a bilinear function $B\colon V\times V\to\mathbf{R}$ by letting

$$\operatorname{discr}(B)=\det(f),\qquad \operatorname{trace}(B)=\operatorname{trace}(f)$$

where f is the linear endomorphism of V such that

$$f(x)\bullet y=B(x,y) \text{ for } x,y\in V.$$

1.7.11. Assuming $n=\dim V<\infty$, $\sigma\in\wedge_k V$, $\tau\in\wedge_m V$, $\lambda=k+m-n\geq 0$, σ and τ are simple, $|\sigma|=1=|\tau|$, S and T are the vectorsubspaces of V associated with σ and τ, we consider here the linear map

$$f\colon S\times T\to V,\quad f(x,y)=x-y \text{ for } (x,y)\in S\times T.$$

We will prove that (see 1.7.8)

$$\|\wedge_n f\|=2^{\lambda/2}\,|(*\sigma)\wedge(*\tau)|.$$

For this purpose we select:

dual orthonormal bases $e_1,\dots,e_n$ and $\omega_1,\dots,\omega_n$ of V and $\wedge^1 V$ such that $e_i\in S$ for $i\leq k$;

dual orthonormal bases $e'_1,\dots,e'_n$ and $\omega'_1,\dots,\omega'_n$ of V and $\wedge^1 V$ such that $e'_i\in T$ for $i>n-m$.

Since $\dim(S\cap T)\geq\lambda$ we may also require that

$$e_i=e'_i,\text{ hence } \omega_i=\omega'_i,\text{ for } n-m<i\leq k.$$

Using the orthonormal basis of $S\times T$ consisting of the vectors

$$2^{-\frac{1}{2}}(e_i,e_i)\text{ and } 2^{-\frac{1}{2}}(e_i,-e_i)\text{ with } n-m<i\leq k,$$

$$(e_i,0)\text{ with } i\leq n-m,\qquad (0,e'_i)\text{ with } i>k,$$

and observing that $(e_i,e_i)\in\ker f$ for $n-m<i\leq k$, we compute

$$\begin{aligned}\|\wedge_n f\|&=\left|(\wedge_n f)\left[\bigwedge_{i=1}^{n-m}(e_i,0)\wedge\bigwedge_{i=n-m+1}^{k}2^{-\frac{1}{2}}(e_i,-e_i)\wedge\bigwedge_{i=k+1}^{n}(0,e'_i)\right]\right|\\
&=2^{\lambda/2}\,|e_1\wedge\cdots\wedge e_k\wedge e'_{k+1}\wedge\cdots\wedge e'_n|\\
&=2^{\lambda/2}\,|\langle e_1\wedge\cdots\wedge e_k\wedge e'_{k+1}\wedge\cdots\wedge e'_n,\omega_1\wedge\cdots\wedge\omega_n\rangle|\\
&=2^{\lambda/2}\,|\langle e'_{k+1}\wedge\cdots\wedge e'_n,\omega_{k+1}\wedge\cdots\wedge\omega_n\rangle|\\
&=2^{\lambda/2}\,|\langle e'_1\wedge\cdots\wedge e'_n,\omega'_1\wedge\cdots\wedge\omega'_k\wedge\omega_{k+1}\wedge\cdots\wedge\omega_n\rangle|\\
&=2^{\lambda/2}\,|\omega'_1\wedge\cdots\wedge\omega'_k\wedge\omega_{k+1}\wedge\cdots\wedge\omega_n|\\
&=2^{\lambda/2}\,|\omega'_1\wedge\cdots\wedge\omega'_{n-m}\wedge\omega_{k+1}\wedge\cdots\wedge\omega_n|=2^{\lambda/2}\,|(*\tau)\wedge(*\sigma)|.\end{aligned}$$

1.7.12. If f is an endomorphism of a finite dimensional inner product space, then

$$2\operatorname{trace}(\wedge_2 f)=(\operatorname{trace} f)^2-\operatorname{trace}(f\circ f),$$

$$\operatorname{trace}[\wedge_2(f+f^*)]=2(\operatorname{trace} f)^2-\operatorname{trace}(f\circ f)-\operatorname{trace}(f^*\circ f).$$

To verify the first formula we expand both sides of the equation

$$\det(1+tf)\det(1-tf)=\det(1-t^2 f\circ f)$$

in powers of t, using 1.4.5, and compare the coefficients of t^2. We obtain the second formula by applying the first to $f+f^*$.

1.7.13. Here we assume that V is a **Hilbert space,** which means that V has an inner product $\bullet$ and V is complete relative to the metric with value

$$|x-y|=[(x-y)\bullet(x-y)]^{\frac{1}{2}} \text{ for } (x,y)\in V\times V.$$

If $a\in V$, C is a nonempty closed convex subset of V and

$$d=\operatorname{dist}(a,C)=\inf\{|a-x|:\ x\in C\},$$

then there exists a unique $c\in C$ with $d=|a-c|$. To prove this we consider for $0<\varepsilon\in\mathbf{R}$ the nonempty closed set

$$C_\varepsilon=C\cap\{x:\ |x-a|^2\leq d^2+\varepsilon^2\}$$

and observe that if $x,y\in C_\varepsilon$, then $(x+y)/2\in C$ and

$$\begin{aligned} d^2+\varepsilon^2 &\geq (|x-a|^2+|y-a|^2)/2 \\ &=(|x+y-2a|^2+|x-y|^2)/4\geq d^2+(|x-y|/2)^2, \end{aligned}$$

hence $|x-y|\leq 2\varepsilon$; thus $\operatorname{diam} C_\varepsilon\leq 2\varepsilon$. Since $C_\delta\subset C_\varepsilon$ for $0<\delta<\varepsilon$ it follows that $\bigcap\{C_\varepsilon:\ \varepsilon>0\}$ consists of a single point $c\in C$. Moreover, *in case C is a vectorsubspace of V, then*

$$(a-c)\bullet x=0 \text{ \textit{whenever} } x\in C,$$

because $c+t\,x\in C$ for all $t\in\mathbf{R}$, and

$$|a-(c+t\,x)|^2=|a-c|^2-2t(a-c)\bullet x+t^2|x|^2$$

is smallest for $t=0$.

For every continuous linear map $f\colon V\to\mathbf{R}$ there exists a unique $u\in V$ such that $f(v)=u\bullet v$ for all $v\in V$. To prove this in case $f\neq 0$ we choose $a\in V$ with $f(a)=1$, apply the preceding proposition with $C=\ker f$, let

$$u=|a-c|^{-2}(a-c)$$

and infer that if $v\in V$, then $v-f(v)(a-c)\in C$,

$$(a-c)\bullet[v-f(v)(a-c)]=0,\qquad u\bullet v=f(v).$$

1.8. Mass and comass

1.8.1. Consider a finite dimensional inner product space V, with the induced dual inner products and norms (denoted $\bullet$ and $|\ |$) on the spaces of m-vectors and m-covectors (see 1.7.5). In addition to these Euclidean norms $|\ |$ we shall use *another pair of dual norms* (denoted $\|\ \|$) *on* $\wedge_m V$ *and* $\wedge^m V$. These other norms are defined as follows:

For each $\phi \in \wedge^m V$, the **comass of** ϕ is

$$\|\phi\| = \sup\{\langle \xi, \phi \rangle : \ \xi \in \wedge_m V, \ \xi \text{ is simple, } |\xi| \le 1\}.$$

Always

$$|\phi| \ge \|\phi\| \ge \binom{\dim V}{m}^{-\frac{1}{2}} |\phi|.$$

Moreover $|\phi| = \|\phi\|$ if and only if ϕ is simple.

For each $\xi \in \wedge_m V$, the **mass of** ξ is

$$\|\xi\| = \sup\{\langle \xi, \phi \rangle : \ \phi \in \wedge^m V, \ \|\phi\| \le 1\}.$$

Always

$$|\xi| \le \|\xi\| \le \binom{\dim V}{m}^{\frac{1}{2}} |\xi|.$$

Moreover $|\xi| = \|\xi\|$ if and only if ξ is simple.

An alternate, more direct, characterization of the mass norm may be derived, using some elementary properties of convex sets (for which see [B F, pages 5, 9] or [B O, Livre V, Chapitre II, § 1] or [E G 2, pages 23, 35]). Since $\wedge^m V \simeq \wedge^1 \wedge_m V$, the set

$$C = \wedge_m V \cap \{\xi : \ \|\xi\| \le 1\}$$

is the convex hull of the compact connected set

$$S = \wedge_m V \cap \{\xi : \ \xi \text{ is simple and } |\xi| \le 1\},$$

hence C consists of all finite sums

$$\sum_{i=1}^{N} c_i \xi_i \text{ with } \xi_i \in S, \ c_i > 0, \quad \sum_{i=1}^{N} c_i = 1$$

and

$$N \le \dim \wedge_m V = \binom{\dim V}{m}.$$

It follows that *for each* $\xi \in \wedge_m V$ *there exist simple m-vectors* $\xi_1, \ldots, \xi_N$ *with*

$$\xi = \sum_{i=1}^{N} \xi_i, \quad \|\xi\| = \sum_{i=1}^{N} |\xi_i|, \quad N \le \binom{\dim V}{m}.$$

Consequently

$$\|\xi\| = \inf\left\{\sum_{i=1}^{N} |\xi_i|:\ \xi_i \text{ are simple and } \xi = \sum_{i=1}^{N} \xi_i\right\}.$$

If $\xi \in \wedge_p V$ and $\eta \in \wedge_q V$, then $\|\xi \wedge \eta\| \le \|\xi\| \cdot \|\eta\|$.

If $\phi \in \wedge^p V$ and $\psi \in \wedge^q V$, then

$$\|\phi \wedge \psi\| \le \binom{p+q}{p} \|\phi\| \cdot \|\psi\|;$$

in case ϕ or ψ is simple, then $\|\phi \wedge \psi\| \le \|\phi\| \cdot \|\psi\|$.

If $f: V \to V'$ is a linear map of finite dimensional inner product spaces, then

$$\|(\wedge^m f)\phi\| \le \|f\|^m \cdot \|\phi\| \quad \text{for } \phi \in \wedge^m V',$$
$$\|(\wedge_m f)\xi\| \le \|f\|^m \cdot \|\xi\| \quad \text{for } \xi \in \wedge_m V.$$

1.8.2. We assume here that V is a vectorspace over the field $\mathbf{C}$ of complex numbers, with a **Hermitian product** H. Thus

$$H:\ V \times V \to \mathbf{C}$$

is bilinear with respect to $\mathbf{R}$ and satisfies the conditions

$$H(v, \mathbf{i}\, w) = \mathbf{i}\, H(v, w), \qquad H(w, v) = \overline{H(v, w)}, \qquad H(v, v) > 0 \text{ in case } v \neq 0,$$

where $v, w \in V$, $\mathbf{i}^2 = -1$ and $\overline{}$ is complex conjugation. Expressing

$$H = B + \mathbf{i}\, A,$$

where B and A are real valued functions, we find that B *is an inner product* (hereafter denoted $\bullet$) and A *is an alternating* 2-*form*. Moreover

$$|H(v, w)| \le |v| \cdot |w| \text{ for } v, w \in V;$$

equality holds if and only if v and w are linearly dependent over $\mathbf{C}$. This follows from the inequality $|v \bullet (c\, w)| \le |v| \cdot |c\, w|$, when $c \in \mathbf{C}$ is chosen so that $|c| = 1$ and $H(v, c\, w) \in \mathbf{R}$.

For example the vectorspace $\mathbf{C}^\nu$ has the Hermitian product

$$H(v, w) = \sum_{j=1}^{\nu} \bar{v}_j\, w_j \text{ for } v, w \in \mathbf{C}^\nu.$$

The inner product B corresponds to the standard inner product of $\mathbf{R}^{2\nu}$ under the canonical isomorphism $\mathbf{C}^\nu \simeq \mathbf{R}^{2\nu}$. If $Z_1, \ldots, Z_\nu \in \wedge^1(\mathbf{C}^\nu, \mathbf{C})$ are the usual coordinate functions on $\mathbf{C}^\nu$, then

$$A = (\mathbf{i}/2) \sum_{j=1}^{\nu} Z_j \wedge \bar{Z}_j \in \wedge^2(\mathbf{C}^\nu, \mathbf{C}).$$

We shall now compute the comass of the μ'th exterior power $A^\mu \in \wedge^{2\mu} V$, where $\mu \le \dim_{\mathbf{C}} V$, by proving **Wirtinger's inequality:**

If $\xi \in \wedge_{2\mu} V$ *and* ξ *is simple, then*

$$\langle \xi, A^\mu \rangle \le \mu! \, |\xi|;$$

equality holds if and only if there exist $v_1, \ldots, v_\mu \in V$ *such that*

$$\xi = v_1 \wedge (\mathbf{i}\, v_1) \wedge \cdots \wedge v_\mu \wedge (\mathbf{i}\, v_\mu).$$

Consequently $\|A^\mu\| = \mu!$

We assume that $|\xi| = 1$.

In case $\mu = 1$, we let $\xi = v \wedge w$, where v and w are orthonormal. Then $H(v, w) = \mathbf{i}\, A(v, w)$, hence

$$\langle \xi, A \rangle = A(v, w) = H(\mathbf{i}\, v, w) = (\mathbf{i}\, v) \bullet w \le 1;$$

equality holds if and only if $\mathbf{i}\, v = w$.

In case $\mu > 1$, we consider the 2μ dimensional subspace T associated with ξ, the inclusion map $f\colon T \to V$, and the 2-form $(\wedge^2 f) A \in \wedge^2 T$. Then we choose dual orthonormal bases $e_1, \ldots, e_{2\mu}$ and $\omega_1, \ldots, \omega_{2\mu}$ of T and $\wedge^1 T$, and nonnegative numbers $\lambda_1, \ldots, \lambda_\mu$ such that

$$(\wedge^2 f) A = \sum_{j=1}^{\mu} \lambda_j (\omega_{2j-1} \wedge \omega_{2j}).$$

Noting that $\lambda_j = A(e_{2j-1}, e_{2j}) \le 1$ for each j, and that $\xi = \varepsilon\, e_1 \wedge \cdots \wedge e_{2\mu}$ with $\varepsilon = \pm 1$, we compute

$$(\wedge^{2\mu} f) A^\mu = \mu! \, \lambda_1 \ldots \lambda_\mu \, \omega_1 \wedge \cdots \wedge \omega_{2\mu}, \qquad \langle \xi, A^\mu \rangle = \varepsilon\, \mu! \, \lambda_1 \ldots \lambda_\mu \le \mu!;$$

equality holds if and only if $\varepsilon = 1$ and $\lambda_j = 1$, hence $e_{2j} = \mathbf{i}\, e_{2j-1}$, for each j.

1.8.3. Very little appears to be known about the structure of the convex sets

$$\wedge^m(R^n) \cap \{\phi\colon \|\phi\| \le 1\}.$$

What are their extreme points?

1.8.4. *Suppose* S *and* T *are mutually orthogonal subspaces of an inner product space* V, $s\colon S \to V$ *and* $t\colon T \to V$ *are the inclusion maps,* $\xi \in \operatorname{im} \wedge_p s$ *and* $\eta \in \operatorname{im} \wedge_q t$. *The equation*

$$\|\xi \wedge \eta\| = \|\xi\| \cdot \|\eta\|$$

holds if either ξ *or* η *is simple.* In case ξ is simple one can choose an orthonormal sequence $\omega_1, \ldots, \omega_p \in \wedge^1 V$ with

$$\langle \xi, \omega_1 \wedge \cdots \wedge \omega_p \rangle = \|\xi\| \quad \text{and} \quad T \subset \ker \omega_i \text{ for } i = 1, \ldots, p,$$

as well as select $\psi \in \Lambda^q V$ with $\|\psi\|=1$ and $\langle\eta,\psi\rangle=\|\eta\|$; it follows that

$$\langle \xi\wedge\eta, \omega_1\wedge\cdots\wedge\omega_p\wedge\psi\rangle = \|\xi\|\cdot\|\eta\|$$

with $\|\omega_1\wedge\cdots\wedge\omega_p\wedge\psi\|\leq 1$, hence $\|\xi\wedge\eta\|\geq\|\xi\|\cdot\|\eta\|$. *I do not know whether the above equation holds always* (in case neither ξ nor η is simple).

1.9. The symmetric algebra of a vectorspace

1.9.1. Proceding similarly as in 1.3, we now consider in the tensor algebra $\otimes_* V$ of any vectorspace V the *two sided ideal* $\mathfrak{B}\, V$ generated by the elements

$$x\otimes y - y\otimes x \in \otimes_2 V$$

corresponding to all $x, y\in V$. The quotient algebra

$$\odot_* V = \otimes_* V/\mathfrak{B}\, V$$

is called the **symmetric algebra** of V. Clearly

$$\mathfrak{B}\, V = \bigoplus_{m=2}^{\infty} (\otimes_m V \cap \mathfrak{B}\, V)$$

is a homogeneous ideal, hence

$$\odot_* V = \bigoplus_{m=2}^{\infty} \odot_m V,$$

where

$$\odot_m V = \otimes_m V/(\otimes_m V \cap \mathfrak{B}\, V);$$

in particular $\odot_0 V = \mathbf{R}$ and $\odot_1 V = V$. The multiplication in $\odot_* V$ will be denoted by the symbol $\odot$. Therefore $\odot_m V$ is the vectorspace generated by all the products $v_1\odot\cdots\odot v_m$ corresponding to $v_1, \ldots, v_m \in V$, and we see from the definition of $\mathfrak{B}\, V$ that *the symmetric multiplication* $\odot$ *is commutative.*

Among all commutative associative graded algebras with a unit, whose direct summand of index 1 is isomorphic with V, the symmetric algebra $\odot_* V$ is characterized (up to isomorphism) by the following property:

For every commutative associative graded algebra A with a unit element, every linear map of V into A_1 can be uniquely extended to a unit preserving algebra homomorphism of $\odot_* V$ into A, carrying $\odot_m V$ into A_m for each m.

It follows that *every linear map $f\colon V \to V'$ can be uniquely extended to a unit preserving algebra homomorphism*

$$\odot_* f\colon \odot_* V \to \odot_* V',$$

which is the direct sum of the linear maps

$$\odot_m f\colon \odot_m V \to \odot_m V'.$$

1.9.2. The functor $\odot_*$ *converts direct sums of vector spaces into commutative products of algebras:*

$$\odot_*(P \oplus Q) \simeq \odot_* P \otimes \odot_* Q.$$

If V has a basis consisting of a single element x, then $\odot_m V$ has a basis consisting of the m'th symmetric power

$$x^m = x \odot \cdots \odot x \ (m \text{ factors}).$$

Therefore, *if V has a basis consisting of $e_1, \ldots, e_n$, then a basis of $\odot_m V$ is formed by the products*

$$e^\alpha = (e_1)^{\alpha_1} \odot \cdots \odot (e_n)^{\alpha_n}$$

corresponding to all n termed sequences α of nonnegative integers with

$$\Sigma\, \alpha = \sum_{i=1}^{n} \alpha_i = m;$$

designating the set of all such sequences by

$$\Xi(n, m)$$

we conclude that

$$\dim \odot_m V = \operatorname{card} \Xi(n, m) = \binom{m+n-1}{n-1}$$

1.9.3. Suppose k and m are positive integers.

For $t = (t_1, \ldots, t_k) \in \mathbf{R}^k$ and $v = (v_1, \ldots, v_k) \in V^k$ the k-nomial theorem (which holds in every commutative ring) implies

$$(t_1 v_1 + \cdots + t_k v_k)^m / m! = \sum_{\alpha \in \Xi(k,m)} t^\alpha v^\alpha / \alpha!$$

where $v^\alpha = (v_1)^{\alpha_1} \odot (v_2)^{\alpha_2} \odot \cdots \odot (v_k)^{\alpha_k}$ and

$$t^\alpha = \prod_{i=1}^{k} (t_i)^{\alpha_i}, \qquad \alpha! = \prod_{i=1}^{k} (\alpha_i!).$$

Taking $k=m$ and letting T be the set of all functions mapping $\{1, \dots, m\}$ into $\{1, -1\}$ we obtain the **polarization formula**

$$\sum_{t\in T} \prod_{i=1}^{m} t_i \Big(\sum_{j=1}^{m} t_j v_j\Big)^m \Big/ m! = 2^m v_1 \odot \cdots \odot v_m;$$

in fact the above sum equals

$$\sum_{\alpha\in\Xi(m,m)} \Big[\sum_{t\in T} \prod_{i=1}^{m} (t_i)^{1+\alpha_i}\Big] v^\alpha/\alpha!,$$

and the bracketed term equals 0 whenever $\alpha_i=0$ for some i, because summation over T is invariant under the permutation of T mapping t onto $(t_1, \dots, t_{i-1}, -t_i, t_{i+1}, \dots, t_m)$.

1.9.4. The **diagonal map** of $\odot_* V$ is the unit preserving algebra homomorphism

$$\Upsilon\colon \odot_* V \to \odot_* V \otimes \odot_* V \text{ (commutative product)}$$

such that $\Upsilon(v) = v\otimes 1 + 1\otimes v$ whenever $v\in V$.

For $v_1, \dots, v_m \in V$ we compute the product

$$\Upsilon(v_1\odot\cdots\odot v_m) = (v_1\otimes 1 + 1\otimes v_1)\odot\cdots\odot(v_m\otimes 1 + 1\otimes v_m)$$
$$= \sum_{p=0}^{m} \sum_{\sigma\in \mathrm{Sh}(p,m-p)} (v_{\sigma(1)}\odot\cdots\odot v_{\sigma(p)})\otimes(v_{\sigma(p+1)}\odot\cdots\odot v_{\sigma(m)}).$$

Therefore, if $e_1, \dots, e_n$ form a basis of V and $\alpha\in\Xi(n,m)$, then

$$\Upsilon(e^\alpha) = \sum_{p=0}^{m} \sum_{\alpha\ge\beta\in\Xi(n,p)} \binom{\alpha_1}{\beta_1}\cdot\ldots\cdot\binom{\alpha_m}{\beta_m} e^\beta\otimes e^{\alpha-\beta},$$

hence

$$\Upsilon(e^\alpha/\alpha!) = \sum_{p=0}^{m} \sum_{\alpha\ge\beta\in\Xi(n,p)} [e^\beta/\beta!]\otimes[e^{\alpha-\beta}/(\alpha-\beta)!].$$

We observe (in analogy with 1.3.3) that *the diagonal map Υ of $\odot_* V$ is associative and commutative*, and that it *is a natural transformation.*

1.10. Symmetric forms and polynomial functions

1.10.1. An m linear function f, which maps the m fold cartesian product V^m of a vectorspace V into some other vectorspace W, is called **symmetric** if and only if

$$f(v_{\sigma(1)}, \dots, v_{\sigma(m)}) = f(v_1, \dots, v_m)$$

whenever $v_1, \dots, v_m \in V$ and σ is a permutation of $\{1, \dots, m\}$. We let

$$\odot^m(V, W)$$

be the vectorspace of all m linear symmetric functions (forms) mapping V^m into W. We shall use the linear isomorphism

$$\odot^m(V, W) \simeq \mathrm{Hom}(\odot_m V, W),$$

where the corresponding $f \in \odot^m(V, W)$ and $h \in \mathrm{Hom}(\odot_m V, W)$ are related by the condition

$$f(v_1, \dots, v_m) = h(v_1 \odot \cdots \odot v_m) \quad \text{whenever} \quad v_1, \dots, v_m \in V;$$

when this is the case we write

$$\langle \xi, f \rangle = \langle \xi, h \rangle = h(\xi) \quad \text{for} \quad \xi \in \odot_m V.$$

Moreover there is an obvious linear isomorphism

$$\mathrm{Hom}(\odot_m V, W) \simeq \mathrm{Hom}^m(\odot_* V, W),$$

where the right member means the set of all those linear maps of $\odot_* V$ into W which vanish on $\odot_n V$ whenever $n \neq m$.

We define $\odot^0(V, W) = W$,

$$\odot^*(V, W) = \bigoplus_{m=0}^{\infty} \odot^m(V, W),$$

and abbreviate $\odot^m(V, \mathbf{R}) = \odot^m V$, $\odot^*(V, \mathbf{R}) = \odot^* V$.

Each linear map $f: V \to V'$ induces a dual linear map

$$\odot^*(f, W): \odot^*(V', W) \to \odot^*(V, W)$$

which is the direct sum of the linear maps

$$\odot^m(f, W): \odot^m(V', W) \to \odot^m(V, W)$$

characterized by the equations

$$\langle \xi, \odot^m(f, W)\phi \rangle = \langle (\odot_m f)\xi, \phi \rangle$$

for $\xi \in \odot_m V$ and $\phi \in \odot^m(V', W)$.

We abbreviate $\odot^*(f, \mathbf{R}) = \odot^* f$.

1.10.2. Whenever W is an (ungraded) algebra over $\mathbf{R}$, we shall use the diagonal map Υ of $\odot_* V$ (in analogy with 1.4.2) to turn the graded vectorspace $\odot^*(V, W)$ into the graded **algebra of symmetric forms of V with coefficients in W.** For

$$\phi \in \mathrm{Hom}^p(\odot_* V, W) \quad \text{and} \quad \psi \in \mathrm{Hom}^q(\odot_* V, W)$$

we define the symmetric product

$$\phi \odot \psi \in \mathrm{Hom}^{p+q}(\odot_* V, W)$$

to be the composition

$$\odot_* V \xrightarrow{\Upsilon} \odot_* V \otimes \odot_* V \xrightarrow{\phi \otimes \psi} W \otimes W \xrightarrow{\nu} W,$$

where ν corresponds to the multiplication of W.

If W is associative, or commutative, or has a unit element, then $\odot^*(V, W)$ has the same property. Each induced map $\odot^*(f, W)$ is a homomorphism of algebras.

From the shuffle formula for Υ we see that

$$(\phi \odot \psi)(v_1, \ldots, v_{p+q}) = \sum_{\sigma \in \mathrm{Sh}(p,q)} \phi(v_{\sigma(1)}, \ldots, v_{\sigma(p)}) \cdot \psi(v_{\sigma(p+1)}, \ldots, v_{\sigma(p+q)})$$

whenever $\phi \in \odot^p(V, W)$, $\psi \in \odot^q(V, W)$ and $v_1, \ldots, v_{p+q} \in V$.

If $e_1, \ldots, e_n$ and $\omega_1, \ldots, \omega_n$ form dual bases of V and $\mathrm{Hom}(V, \mathbf{R}) = \odot^1 V$, *then the products*

$$\omega^\alpha = (\omega_1)^{\alpha_1} \odot \cdots \odot (\omega_n)^{\alpha_n} \in \odot^m V$$

corresponding to all $\alpha \in \Xi(n, m)$ form a basis of $\odot^m V$, which is dual to the base $\{e^\alpha/\alpha!: \alpha \in \Xi(n, m)\}$ of $\odot_m V$.

In the definition of the symmetric product of forms the multiplication of an algebra W may be replaced by any bilinear map $\mu: W_1 \times W_2 \to W_3$; thus one obtains

$$\phi \odot \psi \in \odot^{p+q}(V, W_3) \text{ for } \phi \in \odot^p(V, W_1),\ \psi \in \odot^q(V, W_2).$$

1.10.3. Proceeding just as in 1.5.1 we define for $p \leq q$ the **interior multiplications**

$$\lrcorner : \odot_p V \times \odot^q(V, W) \to \odot^{q-p}(V, W), \qquad \llcorner : \odot_q V \times \odot^p V \to \odot_{q-p} V.$$

If $e_1, \ldots, e_n$ and $\omega_1, \ldots, \omega_n$ are dual bases of V and $\odot^1 V$, and if $\alpha \in \Xi(n, p)$ and $\beta \in \Xi(n, q)$, then

$$e^\alpha \lrcorner (\omega^\beta/\beta!) = \omega^{\beta-\alpha}/(\beta-\alpha)!, \qquad (e^\beta/\beta!) \llcorner \omega^\alpha = e^{\beta-\alpha}/(\beta-\alpha)!$$

in case $\alpha \leq \beta$; otherwise these interior products equal 0.

1.10.4. A map $P: V \to W$ is called a **homogeneous polynomial function of degree** m if and only if there exists a form $\phi \in \odot^m(V, W)$ such that

$$P(x) = \langle x^m/m!, \phi \rangle \text{ for } x \in V.$$

The polarization formula (1.9.3) shows that $P=0$ if and only if $\phi=0$. Thus $\odot^m(V,W)$ *is linearly isomorphic with the vectorspace of all homogeneous polynomial functions of degree m mapping V into W.*

More generally, a map $P: V \to W$ is called a **polynomial function** if and only if there exists an integer $M \geq 0$ and forms $\phi_m \in \odot^m(V,W)$ corresponding to $m=0, \dots, M$ such that

$$P(x)=\sum_{m=0}^{M} \langle x^m/m!, \phi_m \rangle \text{ for } x \in V.$$

Since this formula implies

$$P(t\,x)=\sum_{m=0}^{M} t^m \langle x^m/m!, \phi_m \rangle \text{ for } x \in V \text{ and } t \in \mathbf{R},$$

we see that $P=0$ if and only if $\phi_m=0$ for $m=0, \dots, M$. Thus $\odot^*(V,W)$ *is linearly isomorphic with the vectorspace of all polynomial maps of V into W,* and we can define

$$\text{degree } P=\sup(\{m\colon \phi_m \neq 0\} \cup \{0\})$$

whenever P and $\phi_0, \dots, \phi_M$ are related as above. Moreover, *in case W is an algebra, the symmetric product* $\odot$ *corresponds under the preceding isomorphism to pointwise multiplication in the function space* W^V, hence *the algebra* $\odot^*(V,W)$ *is isomorphic with the algebra of polynomial maps of V into W; in fact whenever* $\phi \in \odot^p(V,W)$ and $\psi \in \odot^q(V,W)$ the shuffle formula shows that, for $x \in V$,

$$\begin{aligned}&\langle x^{p+q}/(p+q)!, \phi \odot \psi \rangle \\ &= \text{card Sh}(p,q) \cdot \langle x^p, \phi \rangle \cdot \langle x^q, \psi \rangle/(p+q)! = \langle x^p/p!, \phi \rangle \cdot \langle x^q/q!, \psi \rangle.\end{aligned}$$

Whenever P and $\phi_0, \dots, \phi_M$ are related as above we use the binomial theorem to obtain for $x, v \in V$ the **Taylor formula**

$$P(x+v)=\sum_{m=0}^{M}\sum_{i=0}^{m} \langle v^i/i! \odot x^{m-i}/(m-i)!, \phi_m \rangle = \sum_{i=0}^{M} \langle v^i/i!, S_i(x) \rangle,$$

where

$$S_i(x)=\sum_{m=i}^{M} x^{m-i}/(m-i)! \,\lrcorner\, \phi_m.$$

We observe that S_i is a polynomial function mapping V into $\odot^i(V,W)$; in 3.1.11 it will be identified with the i-th differential of P. Here we also note that

$$S_i(x+v)=\sum_{m=i}^{M} v^{m-i}/(m-i)! \,\lrcorner\, S_m(x),$$

because the right member of this equation equals

$$\sum_{m=i}^{M} v^{m-i}/(m-i)! \lrcorner \sum_{j=m}^{M} x^{j-m}/(j-m)! \lrcorner \phi_j$$

$$= \sum_{j=i}^{M} \sum_{m=i}^{j} (v^{m-i}/(m-i)! \odot x^{j-m}/(j-m)!) \lrcorner \phi_j$$

$$= \sum_{j=i}^{M} (v+x)^{j-i}/(j-i)! \lrcorner \phi_j = S_i(v+x).$$

To represent the polynomial function P in terms of a basis $e_1, \ldots, e_n$ of V, we recall 1.9.3 and find that

$$P\left(\sum_{i=1}^{n} t_i e_i\right) = \sum_{m=0}^{M} \sum_{\alpha \in \Xi(n,m)} \langle e^\alpha/\alpha!, \phi_m \rangle t^\alpha$$

whenever $t \in \mathbf{R}^n$.

1.10.5. In analogy with 1.7.5, any given inner product of V can be used to construct inner products on the spaces $\odot_m V$, by extending the given polarity to an algebra homomorphism of $\odot_* V$ into $\odot^* V$. If $e_1, \ldots, e_n$ are orthonormal in V, then the products $(\alpha!)^{-\frac{1}{2}} e^\alpha$ corresponding to all $\alpha \in \Xi(n,m)$ are orthonormal in $\odot_m V$. The norms corresponding to such inner products satisfy the inequality

$$|v_1 \odot \cdots \odot v_m| \le (m!)^{\frac{1}{2}} |v_1| \cdot \cdots \cdot |v_m| \text{ for } v_1, \ldots, v_m \in V.$$

Other useful norms can be constructed by adapting the method of 1.8.1 as follows: Whenever V and W are normed vectorspaces and $\phi \in \odot^m(V, W)$ we define

$$\|\phi\| = \sup\{|\phi(v_1, \ldots, v_m)| : v_i \in V \text{ and } |v_i| \le 1 \text{ for } i = 1, \ldots, m\}.$$

Clearly $\|(a_1 \odot \cdots \odot a_j) \lrcorner \phi\| \le |a_1| \cdot \cdots \cdot |a_j| \cdot \|\phi\|$ in case $j \le m$ and $a_1, \ldots, a_j \in V$. The shuffle formula shows that

$$\|\phi \odot \psi\| \le \binom{p+q}{p} \|\phi\| \cdot \|\psi\|$$

for $\phi \in \odot^p(V, W)$ and $\psi \in \odot^q(V, W)$, provided W is a normed algebra with $|w \cdot z| \le |w| \cdot |z|$ for $w, z \in W$; moreover

$$\|\phi^k\| = k! \, \|\phi\|^k \text{ for } \phi \in \odot^1(V, W), \; k = 1, 2, 3, \ldots$$

in case W is associative and $|w^k| = |w|^k$ for $w \in W$. We also define

$$\|\xi\| = \sup\{\langle \xi, \phi \rangle : \phi \in \odot^m V \text{ and } \|\phi\| \le 1\}$$

whenever $\xi \in \bigodot_m V$, and readily verify that

$$\|\xi \odot \eta\| \leq \|\xi\| \cdot \|\eta\| \text{ for } \xi \in \bigodot_p V,\ \eta \in \bigodot_q V;$$

$$\|x^k\| = |x|^k \text{ for } x \in V,\ k = 1, 2, 3, \ldots.$$

If V and $\bigodot^1 V$ have dual basic sequences $e_1, \ldots, e_n$ and $\omega_1, \ldots, \omega_n$ with $|e_i| = \|\omega_i\| = 1$ for $i = 1, \ldots, n$, then

$$\alpha! \leq \|\omega^\alpha\| \leq (\Sigma\,\alpha)! \text{ and } \alpha!/(\Sigma\,\alpha)! \leq \|e^\alpha\| \leq 1 \text{ for } \alpha \in \Xi(n, m).$$

It may also be shown that $\|\xi\| > 0$ for $\xi \in \bigodot_m V \sim \{0\}$, and (compare 1.8.1) that $\|\xi\|$ equals the infimum of all finite sums

$$\sum_{j=1}^{N} |v_{1,j}| \cdot \cdots \cdot |v_{m,j}|$$

corresponding to $v_{i,j} \in V$ with

$$\xi = \sum_{j=1}^{N} (v_{1,j} \odot \cdots \odot v_{m,j});$$

if $\dim V < \infty$ one can take $N \leq \dim \bigodot_m V$.

For every homogeneous polynomial function $P: V \to W$ of degree m we define

$$\|P\| = \sup\{|P(x)|: x \in V,\ |x| \leq 1\}.$$

Choosing $\phi \in \bigodot^m(V, W)$ so that $P(x) = \langle x^m/m!, \phi\rangle$ whenever $x \in V$, we observe that

$$m!\ \|P\| \leq \|\phi\| \leq m^m\ \|P\|$$

as a consequence of the polarization formula; for the case when V is an inner product space it was shown in [H 1] that $m!\ \|P\| = \|\phi\|$. Recalling 1.9.3 we also obtain

$$\left|P\left(\sum_{j=1}^{k} t_j v_j\right)\right| \leq \sum_{\alpha \in \Xi(k, m)} |t^\alpha/\alpha!| \cdot |\langle v^\alpha, \phi\rangle|$$

$$\leq \|\phi\| \sum_{\alpha \in \Xi(k, m)} |t|^\alpha/\alpha! = \|\phi\| \left(\sum_{j=1}^{k} |t_j|\right)^m \Big/ m!$$

provided $|v_j| \leq 1$ for $j = 1, \ldots, k$.

1.10.6. Assuming that V and W are inner product spaces with $\dim V = n < \infty$, we first endow $\bigodot^m V$ with the inner product such that the polarity described in the first paragraph of 1.10.5 is an orthogonal isomorphism mapping $\bigodot_m V$ onto $\bigodot^m V$, then use the method of 1.7.9 to define an inner product on

$$\bigodot^m(V, W) \simeq [\bigodot^m V] \otimes W.$$

Now suppose $e_1, \ldots, e_n$ and $\omega_1, \ldots, \omega_n$ are dual orthonormal bases of V and $\odot^1 V$, hence $\alpha!^{-\frac{1}{2}} e^\alpha$ and $\alpha!^{-\frac{1}{2}} \omega^\alpha$ corresponding to $\alpha \in \Xi(n,m)$ form dual orthonormal bases of $\odot_m V$ and $\odot^m V$. With each ϕ in $\odot^m(V,W)$ we associate

$$\sum_{\alpha \in \Xi(n,m)} \alpha!^{-\frac{1}{2}} \omega^\alpha \otimes \langle \alpha!^{-\frac{1}{2}} e^\alpha, \phi \rangle$$

in $[\odot^m V] \otimes W$ and compute

$$|\phi|^2 = \sum_{\alpha \in \Xi(n,m)} |\langle \alpha!^{-\frac{1}{2}} e^\alpha, \phi \rangle|^2.$$

Letting $\mathscr{S}(n,m)$ be the set of all functions mapping $\{1, \ldots, m\}$ into $\{1, \ldots, n\}$ and observing that for each $\alpha \in \Xi(n,m)$ there exist precisely $m!/\alpha!$ sequences $s \in \mathscr{S}(n,m)$ such that

$$\alpha(j) = \operatorname{card}\{i \colon s(i) = j\} \text{ for } j \in \{1, \ldots, m\},$$

we obtain the formula

$$m! \, |\phi|^2 = \sum_{s \in \mathscr{S}(n,m)} |\langle e_{s(1)} \odot \cdots \odot e_{s(m)}, \phi \rangle|^2.$$

It follows that

$$|\langle x^m/m!, \phi \rangle| \leq m!^{-\frac{1}{2}} |x|^m |\phi| \text{ for } x \in V.$$

Recalling 1.10.5 we see that

$$\|\phi\| \leq m^m m!^{-\frac{1}{2}} |\phi| \quad \text{and} \quad |\phi| \leq m!^{-\frac{1}{2}} n^{m/2} \|\phi\|.$$

CHAPTER TWO

General measure theory

In this chapter we treat mainly that part of measure theory which is valid on spaces with relatively little geometric structure. The definition of "measure" adopted in 2.1 was derived from Carathéodory's concept of "outer measure" through omission of his axiom requiring additivity on metrically separated sets. Relations between measure and topology are discussed in 2.2 and 2.3. In 2.4 we take up the process of Lebesgue integration, by which we mean integration with respect to an arbitrary measure. The particular measures also bearing Lebesgue's name are characterized in 2.5.17, 2.6.5 as well as 2.7.16(1), 2.10.7, 2.10.35. Connections between integration and linear operations (wherein we include the Radon-Nikodym theorem) are discussed in 2.5, first for an arbitrary lattice of real valued functions, then for continuous functions on a locally compact space. The Fubini theorem follows in 2.6. In 2.7 we develop the theory of invariant measures over homogeneous spaces of locally compact groups. Metric structure begins to play a more important role in the theory of covering and derivation, 2.8 and 2.9, which includes several modern versions of the differentiation theory originated by Vitali and Lebesgue. Finally in 2.10 we take up Carathéodory's construction of measures, by which he achieved the significant extension of measure theory to lower dimensional subsets of the space, obtaining for example a reasonable notion of area for a two-dimensional surface in Euclidean three-space. In the present chapter we discuss only those consequences of Carathéodory's construction which are closely related to the basic definitions, leaving the more involved problems for Chapter 3.

Much of this material is classical and has been discussed in numerous excellent books like [S], [L 3, Chapters III, VI], [C 2], [W 1, Chapter 2], [B 0, Livre VI], which contain ample references covering the early history of the subject. Here it should therefore suffice to list authorship credits for more recent or less known contributions.

The material for 2.1.6 was taken from [U], the proof of 2.2.2 from [M R 2]. Parts of the treatment of Suslin sets in 2.2.6 to 2.2.14 were borrowed from [K U, Vol. 1, Chapter 3, 3rd edition], and some parts are new. The proof of 2.2.16 uses techniques from [M D] and [M S].

The ideas for 2.5.11 and the proof of 2.7.14 were taken from [HA]. Incorporated in 2.8.2 to 2.8.7 are important refinements of Banach's approach to the Vitali covering theorem which were achieved in [M 2]. In 2.8.9 to 2.8.15 we simplify and extend to more general spaces the theorem of [B4] about spherical ball coverings in $\mathbf{R}^n$; the generalization of this theorem obtained in [M3] is quite different from ours, though some lemmas are similar. Theorem 2.9.17 resulted from simplification of [F4, § 6] in [MIR]. The material of 2.9.22, 23 originated in [G], [P2], [A]. The measure $\mathscr{C}^m$ was introduced in [C1], $\mathscr{H}^m$ and $\mathscr{S}^m$ in [HF], $\mathscr{G}^m$ in [GR1, 2], $\mathscr{Q}_1^m$ in [MR1] and [F4]. The measure $\mathscr{I}_1^m$ was defined in [FA] and has sometimes been called Favard measure, but the first theorems about it were proved in [N1, 2] and [F3, 4, 5]. $\mathscr{I}_\infty^m$ originated in [MI] and [NE]. $\mathscr{T}^m$ and the measures $\mathscr{Q}_t^m$, $\mathscr{I}_t^m$ with $1<t<\infty$ are new. Densities of the type considered in 2.10.19 were studied first in [B1], and also [SP1]. The proofs of 2.10.22, 23 were taken from [D2, 1]. A special case of Theorem 2.10.25 appeared in [E], a general version in [F10]. The constructions of 2.10.28, 29 originated in [BEM], [BM]. The proof of 2.10.31 was borrowed from [HH1]. Ideas from [BL, p. 50], [HH2], [ME2] led to Theorems 2.10.32, 36, 38. Theorem 2.10.47 was discovered in [B8].

This exposition is self-contained, requiring no previous knowledge of measure or integration theory, only some familiarity with elementary point set topology.

2.1. Measures and measurable sets

2.1.1. We shall use the **extended real number system**

$$\overline{\mathbf{R}} = \mathbf{R} \cup \{\infty\} \cup \{-\infty\}$$

with the obvious ordering, and with the algebraic operations defined wherever possible so as to make them continuous with respect to the order topology of $\overline{\mathbf{R}}$; for instance $\infty + \infty = \infty$ and $x \cdot \infty = \infty$ for $x > 0$, but $\infty - \infty$ and $0 \cdot \infty$ are undefined.

However, for certain purposes it is very convenient to use a special convention which assigns the value 0 to the product of 0 and ∞; such usage will be specifically indicated when it occurs.

Every set $S \subset \overline{\mathbf{R}}$ has a greatest lower bound $\inf S$ and a least upper bound $\sup S$. If S is nonempty, then $\inf S \leq \sup S$; however $\inf \varnothing = \infty$ and $\sup \varnothing = -\infty$.

We write $r_n \uparrow s$ as $n \uparrow \infty$ ($r_n \downarrow s$ as $n \uparrow \infty$) if and only if r is a nondecreasing (nonincreasing) sequence with limit s in $\overline{\mathbf{R}}$.

We let sign $t = 1$ for $t > 0$, sign $t = -1$ for $t < 0$, and sign $0 = 0$.

With each function f, such that $\operatorname{im} f \subset \overline{\mathbf{R}}$, we associate the functions f^+ and f^- by the formulae

$$f^+(x)=\sup\{f(x),0\} \quad \text{and} \quad f^-(x)=-\inf\{f(x),0\}$$

for all x; then f^+ and f^- are nonnegative,

$$f=f^+-f^-, \qquad |f|=f^++f^-.$$

Similarly we define $(\operatorname{sign} f)(x)=\operatorname{sign} f(x)$ for all x.

We form **numerical sums**

$$\sum_A f, \quad \text{also denoted} \quad \sum_{x\in A} f(x),$$

where f is an $\overline{\mathbf{R}}$ valued function whose domain contains A, in three successive steps. In case A is finite, one readily verifies, by induction with respect to the cardinal number of A, the existence of one and only one summation operator $\sum_A$, applied to nonnegative functions, such that $\sum_\varnothing f=0$ and the following two conditions hold:

(1) *If* $a\in A$, $f(a)\geq 0$ *and* $f(x)=0$ *whenever* $a\neq x\in A$, *then* $\sum_A f=f(a)$.

(2) *If* $f(x)\geq 0$ *and* $g(x)\geq 0$ *whenever* $x\in A$, *then*

$$\sum_A (f+g)=\sum_A f+\sum_A g.$$

The passage from finite to arbitrary sets A is effected by the definition:

(3) *If* $f(x)\geq 0$ *whenever* $x\in A$, *then*

$$\sum_A f=\sup\{\sum_B f\colon\ B \text{ is a finite subset of } A\}.$$

Finally the restriction to nonnegative functions is removed by the convention:

(4) $\sum_A f=\sum_A f^+-\sum_A f^-$

We observe that the conditions (1) and (2) remain true in case A is infinite, and that the summation operator has the following further properties:

(5) $\sum_A f\in\overline{\mathbf{R}}$ *if and only if* $\sum_A f^+<\infty$ *or* $\sum_A f^-<\infty$;

$\sum_A f\in\mathbf{R}$ *if and only if* $\sum_A |f|\in\mathbf{R}$.

(6) *If* $0\neq c\in\mathbf{R}$, *then* $\sum_A c f=c\sum_A f$.

(7) *If* $\sum_A f+\sum_A g\in\overline{\mathbf{R}}$, *then* $\sum_A (f+g)=\sum_A f+\sum_A g$.

(8) *If* $f(x)\leq g(x)$ *for* $x\in A$, *and either* $-\infty<\sum_A f$ *or* $\sum_A g<\infty$, *then*

$$\sum_A f\leq\sum_A g.$$

(9) *If* $\sum_A f\in\overline{\mathbf{R}}$ *and* $h\colon A\to Y$, *then* $\sum_A f=\sum_{y\in Y}\sum_{h^{-1}\{y\}} f$.

(10) *If* $\sum_A f\in\overline{\mathbf{R}}$ *and* $A=U\times V$, *then*

$$\sum_A f=\sum_{u\in U}\sum_{v\in V} f(u,v)=\sum_{v\in V}\sum_{u\in U} f(u,v).$$

(11) *If* $f\colon U\times V\to\{t\colon 0\leq t\leq\infty\}$ *and* $g\colon U\to\mathbf{R}$, *then*

$$\sum_{(u,v)\in U\times V} g(u)f(u,v)=\sum_{u\in U} g(u)\sum_{v\in V} f(u,v);$$

here we use the special convention $0\cdot\infty=0$.

(12) *If* $\sum_A f\in\mathbf{R}$, *then* $A\cap\{x\colon f(x)\neq 0\}$ *is countable.*

(13) *If* $\sum_A f\in\overline{\mathbf{R}}$, *then, for every increasing sequence of sets* B_n *with union* A, $\sum_{B_n} f\to\sum_A f$ *as* $n\to\infty$.

We note that (7) may be derived from (2) and the equation

$$f^+ +g^+ +(f+g)^- =f^- +g^- +(f+g)^+,$$

which holds unless $f(x)$ and $g(x)$ are opposite infinities for some $x\in A$. (8) follows from (7), because either $g=f+(g-f)$ or $f=g+(f-g)$. (9) may be derived using (1)–(4) and also (6), (7). (10) is obtained by applying (9) to the projections. (11) follows from (10), with $f(u,v)$ replaced by the positive and negative parts of $g(u)f(u,v)$. (12) and (13) are consequences of (3) and (4).

2.1.2. For each class X we let $\mathbf{2}^X$ be the class of all subsets of X.

We say that ϕ **measures** X, *or that* ϕ *is a* **measure over** X, *if and only if* X *is a set,* $\phi\colon \mathbf{2}^X\to\overline{\mathbf{R}}\cap\{t\colon 0\leq t\leq\infty\}$ *and*

$$\phi(A)\leq\sum_{B\in F}\phi(B) \text{ whenever } F\subset\mathbf{2}^X,\ F \text{ is countable},\ A\subset\bigcup F.$$

It follows that $\phi(\varnothing)=0$, because $\bigcup\varnothing=\varnothing$, and that $A\subset B\subset X$ implies $\phi(A)\leq\phi(B)$.

All the uses of measures depend on additivity with respect to certain partitions of the space X. We begin by considering the simplest case, the partition $\{A, X \sim A\}$ corresponding to any set $A \subset X$. Following Carathéodory, we say that

A is a ϕ **measurable set** *if and only if $A \subset X$ and*

$$\phi(T) = \phi(T \cap A) + \phi(T \sim A) \textit{ whenever } T \subset X.$$

Since $\phi(T)$ never exceeds the sum on the right, ϕ measurability of a set A may be proved by showing that

$$\phi(T) \geq \phi(T \cap A) + \phi(T \sim A) \text{ whenever } \phi(T) < \infty.$$

A simple example is the **counting measure** over X, which associates with each subset of X its number of elements (∞ in case the subset is infinite). All subsets of X are measurable with respect to the counting measure. However, with respect to most other interesting measures, nonmeasurable sets do exist (see 2.1.6, 2.2.4, 2.5.15, 2.7.17).

For every measure ϕ over X, the sets $\varnothing$ and X are ϕ measurable. These may be the only ϕ measurable sets, as seen from the example where $\phi(\varnothing) = 0$, $\phi(A) = 1$ whenever $\varnothing \neq A \subset X$. Of course this example is not representative of the situation normally occurring in analysis and geometry [see 2.3.2 (9), 2.10.1, 2.5.2].

With any measure ϕ over X, and any set $Y \subset X$, one associates another measure

$$\phi \llcorner Y$$

over X by the formula

$$(\phi \llcorner Y)(A) = \phi(Y \cap A) \text{ for } A \subset X.$$

All ϕ measurable sets are also $\phi \llcorner Y$ measurable. Use of $\phi \llcorner Y$ often simplifies the discussion of the behavior of ϕ on subsets of Y.

Every function $f\colon X \to Y$ induces a map $f_\#$ which associates with each measure ϕ over X the measure

$$f_\#(\phi)$$

over Y by the formula

$$(f_\# \phi) B = \phi(f^{-1} B) \text{ for } B \subset Y.$$

One readily verifies that *$f^{-1}(B)$ is ϕ measurable if and only if B is $f_\#(\phi \llcorner A)$ measurable for every $A \subset X$.*

2.1.3. Theorem. *Suppose ϕ measures X.*

(1) *If A is a ϕ measurable set, so is $X \sim A$.*

(2) *If F is a countable, nonempty family of ϕ measurable sets, then $\bigcup F$ and $\bigcap F$ are ϕ measurable.*

(3) *If $A_1, A_2, A_3, \dots$ are disjoint ϕ measurable sets, then*

$$\phi\left(\bigcup_{i=1}^{\infty} A_i\right)=\sum_{i=1}^{\infty}\phi(A_i).$$

(4) *If $B_1 \subset B_2 \subset B_3 \subset \cdots$ form an increasing sequence of ϕ measurable sets, then*

$$\phi\left(\bigcup_{i=1}^{\infty} B_i\right)=\lim_{i\to\infty}\phi(B_i).$$

(5) *If $C_1 \supset C_2 \supset C_3 \supset \cdots$ form a decreasing sequence of ϕ measurable sets, with $\phi(C_1)<\infty$, then*

$$\phi\left(\bigcap_{i=1}^{\infty} C_i\right)=\lim_{i\to\infty}\phi(C_i).$$

(6) *If $\phi(A)=0$, then A is ϕ measurable.*

(7) *If A is a ϕ measurable set and $B\subset X$, then*

$$\phi(A)+\phi(B)=\phi(A\cap B)+\phi(A\cup B).$$

Proof. (1) and (6) are trivial.

If A and B are ϕ measurable sets, then

$$\begin{aligned}\phi(T)&=\phi(T\cap A)+\phi(T\sim A)\\&=\phi(T\cap A)+\phi[(T\sim A)\cap B]+\phi[(T\sim A)\sim B]\\&\geq\phi[T\cap(A\cup B)]+\phi[T\sim(A\cup B)]\end{aligned}$$

whenever $T\subset X$, hence $A\cup B$ is ϕ measurable. Using induction and (1), we conclude that (2) holds in case F is finite.

To prove (3) we note that, for each positive integer j,

$$\phi\left(\bigcup_{i=1}^{\infty} A_i\right)\geq\phi(B_j),\ \text{where}\ B_j=\bigcup_{i=1}^{j} A_i,$$

and verify by induction that

$$\phi(B_j)=\sum_{i=1}^{j}\phi(A_i);$$

in fact $B_{j+1}\cap A_{j+1}=A_{j+1}$, $B_{j+1}\sim A_{j+1}=B_j$, hence

$$\phi(B_{j+1})=\phi(A_{j+1})+\phi(B_j).$$

Now we obtain (4) by applying (3) to the disjoint ϕ measurable sets

$$A_1=B_1,\quad A_i=B_i\sim B_{i-1}\ \text{for}\ i>1.$$

Then (5) follows from (4) with $B_i=C_1\sim C_i$, because

$$C_1=\bigcap_{i=1}^{\infty} C_i\cup\bigcup_{i=1}^{\infty} B_i,\quad \phi(C_1)\leq\phi\left(\bigcap_{i=1}^{\infty} C_i\right)+\lim_{i\to\infty}[\phi(C_1)-\phi(C_i)].$$

Next we show that (2) holds in case F is countably infinite, say F consists of $S_1, S_2, S_3, \ldots$. Then

$$\bigcup F = \bigcup_{j=1}^{\infty} B_j, \text{ where } B_j = \bigcup_{i=1}^{j} S_i$$

form an increasing sequence of ϕ measurable sets. Whenever $\phi(T) < \infty$ we can apply (4) and (5) with ϕ replaced by $\phi \llcorner T$ to obtain

$$\phi(T \cap \bigcup F) + \phi(T \sim \bigcup F) = (\phi \llcorner T) \bigcup_{j=1}^{\infty} B_j + (\phi \llcorner T) \bigcap_{j=1}^{\infty} (X \sim B_j)$$

$$= \lim_{j \to \infty} (\phi \llcorner T) B_j + \lim_{j \to \infty} (\phi \llcorner T)(X \sim B_j) = \phi(T).$$

Thus $\bigcup F$ is ϕ measurable. Use of (1) completes the proof of (2).

The statement (7) may be verified by applying the definition of measurability with $T = B$ and with $T = A \cup B$.

2.1.4. Assuming that ϕ measures X, we say that B *is a* ϕ **hull of** A *if and only if* $A \subset B \subset X$, B *is* ϕ *measurable and*

$$\phi(T \cap A) = \phi(T \cap B) \text{ for every } \phi \text{ measurable set } T.$$

We note that, *if* $A \subset B$, B *is* ϕ *measurable and* $\phi(A) = \phi(B) < \infty$, *then* B *is a* ϕ *hull of* A; in fact, for every ϕ measurable set T,

$$\phi(T \cap B) = \phi(B) - \phi(B \sim T) \le \phi(A) - \phi(A \sim T) = \phi(T \cap A).$$

2.1.5. A measure ϕ over X is called **regular** *if and only if for each set* $A \subset X$ *there exists a* ϕ *measurable set* B *such that*

$$A \subset B \quad \text{and} \quad \phi(A) = \phi(B).$$

The following five statements are true *in case* ϕ *is regular:*

(1) *For every increasing sequence* $A_1 \subset A_2 \subset A_3 \subset \cdots$ *of subsets of* X,

$$\phi\left(\bigcup_{i=1}^{\infty} A_i\right) = \lim_{i \to \infty} \phi(A_i).$$

(2) *If* $\phi(A) < \infty$, *then* A *has a* ϕ *hull.*

(3) *If* $A \cup B$ *is* ϕ *measurable and* $\phi(A) + \phi(B) = \phi(A \cup B) < \infty$, *then* A *and* B *are* ϕ *measurable.*

(4) *If* $\phi(X) < \infty$, $f: X \to Y$ *and* C *is an* $f_\# \phi$ *measurable set, then* $f^{-1}(C)$ *is* ϕ *measurable.*

(5) *If $\phi(S) < \infty$, then the class of all $\phi \llcorner S$ measurable sets equals*

$$\{(B \cap S) \cup C\colon B \text{ is } \phi \text{ measurable, } C \subset X \sim S\}.$$

To prove (1) we choose ϕ measurable sets C_i containing A_i with $\phi(C_i) = \phi(A_i)$, note that the ϕ measurable sets

$$B_i = \bigcap_{j=i}^{\infty} C_j$$

satisfy the conditions $A_i \subset B_i \subset C_i$ and $B_i \subset B_{i+1}$, and infer from 2.1.3 (4) that

$$\phi\left(\bigcup_{i=1}^{\infty} A_i\right) \leq \phi\left(\bigcup_{i=1}^{\infty} B_i\right) = \lim_{i\to\infty} \phi(B_i) = \lim_{i\to\infty} \phi(A_i).$$

The statement (2) follows trivially from 2.1.4.

To obtain (3) we choose a ϕ measurable set C such that

$$A \subset C \subset A \cup B \quad \text{and} \quad \phi(A) = \phi(C),$$

use 2.1.3 (7) to infer that

$$\phi(C \cap B) + \phi(C \cup B) = \phi(C) + \phi(B) = \phi(A) + \phi(B) = \phi(A \cup B) = \phi(C \cup B),$$

hence $\phi(C \cap B) = 0$, $\phi(C \sim A) = 0$, A is ϕ measurable.

We deduce (4) from (3) with $A = f^{-1}(C)$, $B = f^{-1}(Y \sim C)$.

To prove (5) we assume that A is a $\phi \llcorner S$ measurable subset of S, choose ϕ hulls S' and A' of S and A, with $A' \subset S'$, and compute

$$\begin{aligned} \phi[(A' \cap S) \sim A] &= (\phi \llcorner S)(A' \sim A) = (\phi \llcorner S)\, A' - (\phi \llcorner S)\, A \\ &= \phi(S \cap A') - \phi(A) = \phi(S' \cap A') - \phi(A') = 0; \end{aligned}$$

hence $B = A' \sim [(A' \cap S) \sim A]$ is ϕ measurable and $B \cap S = A$.

Simple examples show that none of the above five propositions need to hold in case ϕ is irregular. (1) and (2) fail when X is an infinite set, $\phi(\emptyset) = 0$, $\phi(A) = 1$ for each finite $A \subset X$, $\phi(A) = 2$ for each infinite $A \subset X$. (3) and (4) fail when $\operatorname{card}(X) = 3$, $\phi(\emptyset) = 0$, $\phi(X) = 2$, $\phi(A) = 1$ for every nonempty proper subset A of X. (5) fails when $\operatorname{card}(X) = 3$, $\operatorname{card}(S) = 2$, $\phi(A) = \operatorname{card}(S)$ for $A \subset S$, $\phi(A) = 2$ for $A \not\subset S$.

Most measures which we shall encounter will be regular. Moreover *one associates with an arbitrary measure ϕ over X a regular measure γ by the formula*

$$\gamma(A) = \inf\{\phi(B)\colon A \subset B \text{ and } B \text{ is } \phi \text{ measurable}\}$$

for $A \subset X$; if A is ϕ measurable, then A is γ measurable and $\phi(A) = \gamma(A)$; if A is γ measurable and $\gamma(A) < \infty$, then A is ϕ measurable.

2.1.6. We shall now discuss the connection between the problem whether all subsets of a given set can be measurable with respect to some nontrivial finite measure, and the question about the existence of inaccessible cardinal numbers.

By an **Ulam number** we mean a cardinal α with the following property:

If ϕ measures X, $\phi(X)<\infty$, every subset of X is ϕ measurable, $\phi\{x\}=0$ for each $x\in X$, and card $X\leq\alpha$, *then $\phi(X)=0$.*

This property is equivalent with the following:

If ψ is a measure, F is disjointed, $\psi(\bigcup F)<\infty$, the union of every subfamily of F is ψ measurable, $\psi(A)=0$ for each $A\in F$, and card $F\leq\alpha$, *then* $\psi(\bigcup F)=0$.

A family of sets is called **disjointed** if and only if any two distinct members of the family have an empty intersection. One deduces the first property from the second by taking

$$F=\{\{x\}: x\in X\} \quad \text{and} \quad \psi=\phi;$$

the second from the first by taking

$$X=F, \quad \phi(G)=\psi(\bigcup G) \text{ whenever } G\subset F.$$

Clearly $\aleph_0$ is an Ulam number, and the class of all Ulam numbers is an initial segment in the well ordered class of all cardinal numbers. This segment is very large, as shown by the two propositions:

(1) *If S is a set of Ulam numbers and* card S *is an Ulam number, then the least upper bound of S is an Ulam number.*

(2) *If a cardinal β is the successor of an Ulam number α, then β is an Ulam number.*

We shall formulate the proofs in terms of von Neumann's model of the ordinal and cardinal numbers [K, Appendix].

To verify (1) we take ψ and F as above with

$$\text{card } F\subset\bigcup S \ (=\text{the least upper bound of } S),$$

and express F as the union of mutually disjoint subfamilies F_s, corresponding to $s\in S$, whose cardinals are Ulam numbers, hence $\psi(\bigcup F_s)=0$. Moreover the cardinal of the disjointed family

$$F'=\{\bigcup F_s: s\in S\}$$

is an Ulam number, the union of every subfamily of F' equals the union of some subfamily of F, and we conclude that

$$\psi(\bigcup F)=\psi(\bigcup F')=0.$$

To verify (2) we assume α infinite, take ϕ and X as above with $X=\beta$, choose for each $x\in\beta$ a univalent map f_x of x into α, and define

$$U(b,a)=\{x\colon\ f_x(b)=a\}\ \text{ whenever }\ b\in\beta,\ \alpha\in\alpha.$$

For each $a\in\alpha$ the sets $U(b,a)$ corresponding to all $b\in\beta$ are mutually disjoint, hence

$$\{b\colon\ \phi\, U(b,a)>0\}\ \text{ is countable;}$$

consequently

$$\operatorname{card}\{(b,a)\colon\ \phi\, U(b,a)>0\}\le\aleph_0\cdot\alpha=\alpha<\beta.$$

Therefore we can now choose $b\in\beta$ so that

$$\phi\, U(b,a)=0\ \text{ for all }\ a\in\alpha,$$

and infer, since these sets $U(b,a)$ are mutually disjoint and α is an Ulam number, that

$$\phi\bigcup_{a\in\alpha}U(b,a)=0;$$

moreover $\phi(b)=0$ because $\operatorname{card} b\le\alpha$, and the equation

$$\beta=b\cup\{b\}\cup\{x\colon\ b\in x\in\beta\}=b\cup\{b\}\cup\bigcup_{a\in\alpha}U(b,a)$$

allows us to conclude $\phi(\beta)=0$.

We now recall that a cardinal number α is called **accessible** if and only if either the class of all cardinals less than α has a largest element, or this class has a subset S whose least upper bound equals α and for which $\operatorname{card} S<\alpha$. The preceding two propositions show that *if there exist any cardinal numbers which are not Ulam numbers, the smallest such number cannot be accessible.* Since the exclusion of uncountable inaccessible cardinals which are sets is a postulate consistent with the usual axioms of set theory (see [TA]), one might reasonably require all cardinals occurring in measure theory to be Ulam numbers.

2.2. Borel and Suslin sets

2.2.1. A set F is called a **Borel family** with respect to a set X if and only if $\varnothing\in F\subset\mathbf{2}^X$ and the following three conditions hold:

(1) *If* $A\in F$, *then* $X\sim A\in F$.

(2) *If* $G\subset F$ *and* G *is countable, then* $\bigcup G\in F$.

(3) *If* $\varnothing\neq G\subset F$ *and* G *is countable, then* $\bigcap G\in F$.

Clearly these conditions are redundant, because (1) implies that (2) and (3) are equivalent. $\mathbf{2}^X$ and $\{\varnothing, X\}$ are the largest and smallest Borel families (with respect to X). The intersection of any set of Borel families is a Borel family. For every $S \subset 2^X$ there exists a *smallest Borel family containing S*, called the *Borel family generated by S.*

Similarly, if $S \subset 2^X$, then *S is contained in a smallest family F satisfying* (2) *and* (3). *In case*

$$A \in S \quad \textit{implies} \quad X \sim A \in F,$$

then F equals the Borel family generated by S. To prove this we observe that the set

$$H = \{A: \ A \in F \text{ and } X \sim A \in F\}$$

contains S, and is closed to countable unionization and intersection; therefore $F \subset H$, and F satisfies the condition (1).

The argument of the preceding paragraph remains valid with (2) replaced by the weaker condition:

(2)′ *If $G \subset F$, G is countable and G is disjointed, then $\bigcup G \in F$.*

In fact if $A_1, A_2, A_3, \ldots$ belong to H, then

$$\bigcup_{i=1}^{\infty} A_i = \bigcup_{i=1}^{\infty} [A_i \cap \bigcap_{j<i} (X \sim A_j)] \in F.$$

In case X is the space of a topology T, then the members of the Borel family generated by T are, by definition, the **Borel sets** associated with T. If every open set is the union of a countable family of closed sets, for instance in case T is induced by a metric, then the Borel sets form the smallest class satisfying (2) and (3), and containing the class of all closed sets; they also form the smallest class satisfying (2)′ and (3), and containing T.

One calls G a **Borel partition** of A if and only if G is a countable disjointed family of Borel sets with $\bigcup G = A$.

If ϕ measures X, then the class of all ϕ measurable sets is a Borel family.

Whenever ϕ measures the space of a topology T one defines the **support** of ϕ as the closed set

$$\operatorname{spt} \phi = X \sim \bigcup \{V: \ V \in T \text{ and } \phi(V) = 0\}.$$

The equation $\phi(X \sim \operatorname{spt} \phi) = 0$ holds often (2.2.5, 2.2.16) but not always (2.5.15).

2.2.2. Theorem. *Suppose ϕ measures a metric space X, all open sets are ϕ measurable, and B is a Borel set.*

(1) *If $\phi(B) < \infty$ and $\varepsilon > 0$, then B contains a closed set C for which*

$$\phi(B \sim C) < \varepsilon.$$

(2) *If B is contained in the union of countably many open sets V_i with $\phi(V_i)<\infty$, and if $\varepsilon>0$, then B is contained in an open set W for which*

$$\phi(W\sim B)<\varepsilon.$$

Proof. We let $\psi=\phi\llcorner B$ and consider the class F of all subsets A of X with the property: For every $\varepsilon>0$, A contains a closed set C for which

$$\psi(A\sim C)<\varepsilon.$$

We verify that if $A_1, A_2, A_3, \ldots \in F$, then also

$$\bigcap_{i=1}^{\infty} A_i\in F \quad\text{and}\quad \bigcup_{i=1}^{\infty} A_i\in F;$$

given $\varepsilon>0$, we choose closed sets $C_i\subset A_i$ with $\psi(A_i\sim C_i)<\varepsilon\, 2^{-i}$, and estimate

$$\psi\Big(\bigcap_{i=1}^{\infty} A_i\sim\bigcap_{i=1}^{\infty} C_i\Big)\le\psi\bigcup_{i=1}^{\infty}(A_i\sim C_i)<\sum_{i=1}^{\infty}\varepsilon\, 2^{-i}=\varepsilon,$$

$$\lim_{n\to\infty}\psi\Big(\bigcup_{i=1}^{\infty} A_i\sim\bigcup_{i=1}^{n} C_i\Big)=\psi\Big(\bigcup_{i=1}^{\infty} A_i\sim\bigcup_{i=1}^{\infty} C_i\Big)\le\psi\bigcup_{i=1}^{\infty}(A_i\sim C_i)<\varepsilon.$$

Moreover $\bigcap_{i=1}^{\infty} C_i$ and $\bigcup_{i=1}^{n} C_i$ are closed.

Thus F satisfies the conditions (2) and (3) of 2.2.1. Since F contains the class of all closed sets, it contains the class of all Borel sets. In particular $B\in F$, as asserted in (1).

To prove (2) we choose closed sets $C_i\subset V_i\sim B$ with

$$\phi[(V_i\sim C_i)\sim B]=\phi[(V_i\sim B)\sim C_i]<\varepsilon\, 2^{-i},$$

note that $B\cap V_i\subset V_i\sim C_i$, and take $W=\bigcup_{i=1}^{\infty}(V_i\sim C_i)$.

2.2.3. A measure ϕ over a topological space X is called **Borel regular** *if and only if all open sets are ϕ measurable and each subset A of X is contained in a Borel set B for which $\phi(A)=\phi(B)$.*

If ϕ is a Borel regular measure and A is a ϕ measurable set for which $\phi(A)<\infty$, then there exist Borel sets B and D

$$D\subset A\subset B \quad\text{and}\quad \phi(B\sim D)=0;$$

in fact one may first choose $B\supset A$ with $\phi(A)=\phi(B)$, then choose a Borel set $E\supset B\sim A$ with $\phi(E)=\phi(B\sim A)=0$, and take $D=B\sim E$.

Therefore, *if ϕ is a Borel regular measure over a metric space, the statements* (1) *and* (2) *of* 2.2.2 *hold for every ϕ measurable set B.*

If ϕ is Borel regular and A is a Borel set, then $\phi \llcorner A$ is Borel regular.

We also observe that if ϕ is any measure over a topological space X such that all Borel subsets of X are ϕ measurable, and if

$$\psi(A) = \inf\{\phi(B)\colon A \subset B \text{ and } B \text{ is a Borel set}\}$$

whenever $A \subset X$, then ψ is a Borel regular measure, and $\psi(A) = \phi(A)$ in case A is a Borel set.

2.2.4. Theorem. *If ϕ is a Borel regular measure over a complete, separable metric space X, $0 < \phi(A) < \infty$, and $\phi(\{x\}) = 0$ whenever $x \in A$, then A has a ϕ nonmeasurable subset.*

Proof. We consider the class Γ of all closed subsets C of A for which $\phi(C) > 0$, hence $\operatorname{card}(C) = 2^{\aleph_0}$. Noting that $\operatorname{card}(\Gamma) \le 2^{\aleph_0}$, we wellorder Γ so that, for each $C \in \Gamma$, the set Γ_C of all predecessors of C has cardinal less than $2^{\aleph_0}$. By induction with respect to this wellordering we define functions f and g on Γ such that, for each $C \in \Gamma$, $f(C)$ and $g(C)$ are distinct elements of

$$C \sim [f(\Gamma_C) \cup g(\Gamma_C)];$$

this is possible because

$$\operatorname{card}[f(\Gamma_C) \cup g(\Gamma_C)] = 2 \operatorname{card}(\Gamma_C) < 2^{\aleph_0} = \operatorname{card}(C).$$

Since both $\operatorname{im} f$ and $A \sim \operatorname{im} f$ (which contains $\operatorname{im} g$) meet every member of Γ, neither set contains any member of Γ. If these two sets were ϕ measurable, both would have ϕ measure 0, hence $\phi(A) = 0$. Therefore either A or $\operatorname{im} f$ is ϕ nonmeasurable.

2.2.5. By a **Radon measure** we mean a measure ϕ, over a *locally compact Hausdorff space* X, with the following three properties:

If K is a compact subset of X, then $\phi(K) < \infty$.

If V is an open subset of X, then V is ϕ measurable and

$$\phi(V) = \sup\{\phi(K)\colon K \text{ is compact, } K \subset V\}.$$

If A is any subset of X, then

$$\phi(A) = \inf\{\phi(V)\colon V \text{ is open, } A \subset V\}.$$

We observe that, in case ϕ is a Radon measure, then

$$\phi(X \sim \operatorname{spt} \phi) = 0,$$

because the set $X \sim \operatorname{spt} \phi$ is open, and each of its' compact subsets is contained in the union of finitely open sets with ϕ measure 0, hence has ϕ measure 0.

Next we prove the approximation theorem: *If ϕ is a Radon measure, A is ϕ measurable, $\phi(A) < \infty$ and $\varepsilon > 0$, then A contains a compact set K with $\phi(A \sim K) < \varepsilon$.* We choose open sets V and W for which

$$A \subset V, \quad \phi(V \sim A) < \varepsilon/2, \quad V \sim A \subset W, \quad \phi(W) < \varepsilon/2,$$

and also choose a compact set $C \subset V$ with $\phi(V \sim C) < \varepsilon/2$; then $C \sim W$ is a compact subset of $V \sim W \subset A$, and

$$\phi[A \sim (C \sim W)] \le \phi(V \sim C) + \phi(W) < \varepsilon.$$

[Comparing this argument with the theory developed in 2.2.1, 2.2.2 and 2.2.3, we get our first illustration of how much simpler Radon measures are than arbitrary Borel regular measures!]

2.2.6. Let $\mathscr{P}$ be the set of all positive integers. The set

$$\mathscr{N} = \mathscr{P}^{\mathscr{P}}$$

of all infinite sequences of positive integers is a cartesian product with all factors equal to $\mathscr{P}$; using the discrete topology on each factor $\mathscr{P}$ we put the cartesian product topology on $\mathscr{N}$. This topology can also be induced by a metric; the distance between two sequences m and n equals

$$\sum_{i=1}^{\infty} 2^{-i} |m_i - n_i| / (1 + |m_i - n_i|).$$

The metric space $\mathscr{N}$ is complete and separable.

The space $\mathscr{N}$ is the union of the sets

$$\mathscr{N}_j = \mathscr{N} \cap \{n \colon n_1 = j\}$$

corresponding to $j \in \mathscr{P}$. Each set $\mathscr{N}_j$ is open and closed in $\mathscr{N}$, and homeomorphic with $\mathscr{N}$.

The cartesian product of countably many factors all equal to $\mathscr{N}$,

$$\mathscr{N}^{\mathscr{P}} = (\mathscr{P}^{\mathscr{P}})^{\mathscr{P}} \simeq \mathscr{P}^{\mathscr{P} \times \mathscr{P}}$$

is homeomorphic with $\mathscr{N}$.

2.2.7. A map $f \colon X \to Y$, where X and Y are metric spaces, is called **Lipschitzian** if and only if exists a finite positive number M such that

$$\operatorname{distance}[f(a), f(b)] \le M \cdot \operatorname{distance}(a, b)$$

whenever $a, b \in X$; one refers to M as a **Lipschitz constant** for f. Every Lipschitzian function has a least Lipschitz constant, denoted

$$\operatorname{Lip}(f).$$

We say that f is **locally Lipschitzian** if and only if each point of X has a neighborhood U such that $f|U$ is Lipschitzian.

In case X is a convex subset of a normed vectorspace, f is Lipschitzian with $\operatorname{Lip}(f) \leq M$ *if and only if*

$$\limsup_{z \to x} \operatorname{dist}[f(x), f(z)]/|x-z| \leq M \quad \textit{for } x \in X.$$

To prove the sufficiency of this condition we suppose $a, b \in X$, $\mu > M$, let

$$S = \{t\colon\ 0 \leq t \leq 1, \operatorname{dist}[f(a), f(a+t(b-a))] \leq \mu\, t\, |b-a|\}$$

and note that $\tau = \sup S \in S$ because f is continuous at $a+\tau(b-a)$. If $\tau < 1$ there would exist t with $\tau < t < 1$ and

$$\operatorname{dist}[f(a+\tau(b-a)), f(a+t(b-a))] < \mu(t-\tau)\, |b-a|,$$

hence $\tau < t \in S$. Therefore $1 \in S$.

2.2.8. Theorem. *For every complete, separable, nonempty metric space X there exists a locally Lipschitzian map g of $\mathcal{N}$ onto X.*

Proof. Proceeding by induction with respect to k we associate with every finite sequence $s_1, \ldots, s_k$ of positive integers a nonempty closed subset

$$E(s_1, \ldots, s_k)$$

of X, whose diameter does not exceed 2^{-k-2}, subject to the conditions

$$X = \bigcup_{j \in \mathcal{P}} E(j) \quad \text{and} \quad E(s_1, \ldots, s_k) = \bigcup_{j \in \mathcal{P}} E(s_1, \ldots, s_k, j).$$

Then there corresponds to each $n \in \mathcal{N}$ the sequence of closed sets

$$E(n_1) \supset E(n_1, n_2) \supset E(n_1, n_2, n_3) \supset \cdots$$

whose diameters approach 0, and whose intersection consists of exactly one point of X, which we call $g(n)$.

Clearly $\operatorname{im} g = X$. For any two distinct points $m, n \in \mathcal{N}$ with distance $(m, n) < \frac{1}{4}$ there exists $k \in \mathcal{P}$ such that

$$2^{-k-2} \leq \operatorname{distance}(m, n) < 2^{-k-1}, \text{ hence } m_i = n_i \text{ for } i \leq k,$$

$$\operatorname{distance}[g(m), g(n)] \leq \operatorname{diam} E(m_1, \ldots, m_k) \leq 2^{-k-2}$$

2.2.9. A variation of the preceding argument yields the proposition: *If X is a complete, nonempty metric space without isolated points, then X has a Borel subset Γ which is homeomorphic to $\mathcal{N}$.* For each $k \in \mathcal{P}$ we construct disjoint nonempty open sets $U(s)$ with closures $E(s)$ and diameters less than 2^{-k-2} corresponding to all sequences $s \in \mathcal{P}^k$; in passing from k to $k+1$ we require that

$$U(s_1, \ldots, s_k) \supset \bigcup_{j \in \mathcal{P}} E(s_1, \ldots, s_k, j).$$

Defining g as before, we readily see that g maps $\mathcal{N}$ homeomorphically onto the Borel set

$$\Gamma = \bigcap_{k \in \mathcal{P}} \bigcup_{s \in \mathcal{P}^k} U(s).$$

In particular when $X = \mathbf{R}$ one can choose intervals $U(s)$ in such a way that $\mathbf{R} \sim \Gamma$ is countable.

The functions g constructed in this manner are similar to the classical homeomorphism mapping $\mathcal{N}$ onto the set of all irrational numbers between 0 and 1, which associates with $n \in \mathcal{N}$ the limit of the continued fraction (see [PO, p. 103])

$$\frac{1|}{|n_1} + \frac{1|}{|n_2} + \frac{1|}{|n_3} + \cdots.$$

2.2.10. Consider a topological space X and the projection

$$p: X \times \mathcal{N} \to X, \quad p(x, n) = x \text{ whenever } x \in X, \ n \in \mathcal{N}.$$

By a **Suslin subset of** X we can mean the *p image of some closed subset of $X \times \mathcal{N}$.*

If X is a Hausdorff space and $f: \mathcal{N} \to X$ is continuous, then $\operatorname{im} f$ is a Suslin subset of X; in fact $\operatorname{im} f$ equals the p image of the closed set

$$(X \times \mathcal{N}) \cap \{(x, n): x = f(n)\}.$$

For each nonempty Suslin subset S of a complete, separable metric space X there exists a continuous map h of $\mathcal{N}$ onto S. Taking C closed in $X \times \mathcal{N}$ with $p(C) = S$, we note that $X \times \mathcal{N}$ and C are complete, separable metric spaces, apply 2.2.8 to obtain a locally Lipschitzian map g of $\mathcal{N}$ onto C, and let $h = p \circ g$.

If Y is a Hausdorff space, S is a Suslin subset of some complete, separable metric space, and $f: S \to Y$ is continuous, then $f(S)$ is a Suslin subset of Y. Choosing h as above we see that

$$f \circ h: \mathcal{N} \to Y \text{ is continuous}, \quad f(S) = \operatorname{im}(f \circ h).$$

If $f: X \to Y$ is continuous and S is a Suslin subset of Y, then $f^{-1}(S)$ is a Suslin subset of X. In fact, if C is closed in $Y \times \mathscr{N}$ and $p_Y(C) = S$, then

$$f^{-1}(S) = p_X\{(x, n): (f(x), n) \in C\}.$$

Next we study relations between Suslin sets and Borel sets. Let F be the family consisting of all sets $p(C)$ *such that C is closed in $X \times \mathscr{N}$ and $p|C$ is univalent.*

If X is a metric space and S is open in X, then $S \in F$; we observe that

$$A = (X \times \mathbf{R}) \cap \{(x, t): t \operatorname{dist}(x, X \sim S) = 1\}$$

is closed in $X \times \mathbf{R}$, use 2.2.9 and 2.2.6 to construct a closed subset B of $\mathscr{N}$ and a univalent continuous map ψ of B onto $\mathbf{R}$, hence infer

$$C = (X \times \mathscr{N}) \cap \{(x, n): n \in B, (x, \psi(n)) \in A\}$$

is closed in $X \times \mathscr{N}$, $p|C$ is univalent and $p(C) = S$.

If S_1, S_2, S_3 are disjoint members of F, then

$$V = \bigcup_{j=1}^{\infty} S_j \in F;$$

recalling 2.2.6 we select closed sets $C_j \subset X \times N_j$ so that $p|C_j$ is univalent and $p(C_j) = S_j$, hence infer

$$D = \bigcup_{j=1}^{\infty} C_j \text{ is closed in } X \times \mathscr{N}, \ p|D \text{ is univalent, } p(D) = V.$$

If $S_1, S_2, S_3, \ldots$ belong to F, then

$$W = \bigcap_{j=1}^{\infty} S_j \in F;$$

choosing closed sets $C_j \subset X \times \mathscr{N}$ so that $p|C_j$ is univalent and $p(C_j) = S_j$, we see that under the projection

$$X \times \mathscr{N}^{\mathscr{P}} \to X$$

the set W is the one to one image of the closed set

$$\{(x, n_1, n_2, n_3, \ldots): (x, n_j) \in C_j \text{ whenever } j \in \mathscr{P}\}.$$

Using 2.2.1 we conclude: *If X is a metric space, then every Borel set of X belongs to F; therefore every Borel set of X is a Suslin set.*

Similarly one sees that if G is a countable, nonempty family of Suslin subsets of X, then $\bigcup G$ and $\bigcap G$ are Suslin subsets of X.

Two subsets P and Q of a topological space Y are said to be **Borel separated** if and only if there exist disjoint Borel sets A and B of Y with $P \subset A$ and $Q \subset B$. Clearly:

If G and H are countable families of subsets of Y such that P and Q are Borel separated whenever $P\in G$ and $Q\in H$, then $\bigcup G$ and $\bigcap H$ are Borel separated.

Assuming that *X is a complete, separable metric space, Y is a Hausdorff space and $f\colon X\to Y$ is continuous*, we will prove:

(1) *If C and D are closed subsets of X such that $f(C)$ and $f(D)$ are disjoint, then $f(C)$ and $f(D)$ are Borel separated in Y.*

(2) *If C is a closed subset of X and $f|C$ is univalent, then $f(C)$ is a Borel subset of Y.*

For this purpose we define $E(s_1,\ldots,s_k)$ as in 2.2.8. The denial of (1) would allow the inductive construction of sequences $m,n\in\mathcal{N}$ such that for each $k\in\mathcal{P}$ the sets

$$P_k=f[C\cap E(m_1,\ldots,m_k)] \quad\text{and}\quad Q_k=f[D\cap E(n_1,\ldots,n_k)]$$

fail to be Borel separated; the intersections

$$C\cap\bigcap_{k\in\mathcal{P}}E(m_1,\ldots,m_k) \quad\text{and}\quad D\cap\bigcap_{k\in\mathcal{P}}E(n_1,\ldots,n_k)$$

would consist of single points $u\in C$ and $v\in D$, hence $f(u)$ and $f(v)$ would be distinct points with disjoint neighborhoods A and B in Y; then the continuity of f at u and v would imply $P_k\subset A$ and $Q_k\subset B$ for large k.

From (1) it follows that *if $S_1,S_2,S_3,\ldots$ are Suslin subsets of X such that $f(S_1),f(S_2),f(S_3),\ldots$ are disjoint, then there exist disjoint Borel sets $B_1,B_2,B_3,\ldots$ of Y with $f(S_j)\subset B_j$ for $j=1,2,3,\ldots$.*

To deduce (2) we partition X for each k into the Borel sets

$$A(s)=A(s_1,\ldots,s_{k-1})\cap E(s)\sim\bigcup_{i<s_k}E(s_1,\ldots,s_{k-1},i)$$

corresponding to all $s=(s_1,\ldots,s_k)\in\mathcal{P}^k$ [here $A(\emptyset)=X$], note that the sets $f[C\cap A(s)]$ are disjoint, and choose disjoint Borel sets $B(s)$ of Y with

$$f[C\cap A(s)]\subset B(s)\subset B(s_1,\ldots,s_{k-1})\cap\operatorname{Clos}f[C\cap E(s)];$$

we conclude that

$$T=\bigcup_{n\in\mathcal{N}}\bigcap_{k\in\mathcal{P}}B(n_1,\ldots,n_k)=\bigcap_{k\in\mathcal{P}}\bigcup_{s\in\mathcal{P}^k}B(s)$$

is a Borel set of Y, and that $T=f(C)$ because

$$f(C)\subset\bigcup_{n\in\mathcal{N}}\bigcap_{k\in\mathcal{P}}f[C\cap A(n_1,\ldots,n_k)]\subset T,$$

$$T\subset\bigcup_{n\in\mathcal{N}}\bigcap_{k\in\mathcal{P}}\operatorname{Clos}f[C\cap E(n_1,\ldots,n_k)]\subset f(C).$$

From (2) it follows that *if A is a Borel set of X and $f|A$ is univalent, then $f(A)$ is a Borel set of Y.*

We further infer from (1) and (2) that, *for any complete, separable metric space X, the class of Borel sets of X coincides with the class F of all one to one projection images of closed subsets of $X \times \mathcal{N}$, and with the class of those Suslin sets A of X for which $X \sim A$ is also a Suslin set.*

2.2.11. Here we *construct a Suslin set which is not a Borel set.* First we observe:

If Z is a topological space with a countable base $U(1), U(2), U(3), \ldots$, then the set

$$C = (Z \times \mathcal{N}) \cap \{(z,n) \colon z \notin \bigcup_{i \in \mathcal{P}} U(n_i)\}$$

is closed in $Z \times \mathcal{N}$, and every closed subset of Z occurs among the slices

$$C_n = \{z \colon (z,n) \in C\}.$$

Taking $Z = Y \times \mathcal{N}$ we see that

$$S = (Y \times \mathcal{N}) \cap \{(y,n) \colon (y,m,n) \in C \text{ for some } m \in \mathcal{N}\}$$

is a Suslin subset $Y \times \mathcal{N}$, and every Suslin subset of Y occurs among the slices

$$S_n = \{y \colon (y,n) \in S\} = \{y \colon (y,m) \in C_n \text{ for some } m \in \mathcal{N}\}.$$

Letting $Y = \mathcal{N}$ we use the diagonal map of $\mathcal{N}$ to infer that

$$T = \{n \colon (n,n) \in S\}$$

is a Suslin subset of $\mathcal{N}$. Moreover its complement

$$\mathcal{N} \sim T = \{n \colon n \notin S_n\}$$

is not a Suslin subset of $\mathcal{N}$, because the assumption $\mathcal{N} \sim T = S_m$ would imply the equivalence of $m \notin S_m$ and $m \in S_m$. Therefore T is not a Borel subset of $\mathcal{N}$.

Finally we conclude with the help of 2.2.9 that every complete, nonempty metric space without isolated points has a Suslin subset which is not a Borel subset.

2.2.12. Theorem. *Suppose ϕ measures a topological space X, all closed subsets of X are ϕ measurable, and S is a Suslin subset of X. If $\phi(T) < \infty$ and $\varepsilon > 0$, then S contains a closed set C for which $\phi(T \cap S \sim C) \le \varepsilon$. Therefore S is ϕ measurable.*

Proof. Replacing ϕ by $\phi \llcorner T$, we assume $\phi(X) < \infty$, and we define the regular measure γ by the formula

$$\gamma(A) = \inf\{\phi(B) \colon A \subset B,\ B \text{ is } \phi \text{ measurable}\} \text{ whenever } A \subset X.$$

Suppose Z_0 is closed in $X \times \mathcal{N}$, with $p(Z_0)=S$. Corresponding to $i\in\mathscr{P}$ we inductively select $m_i\in\mathscr{P}$ and closed sets $Z_i\subset Z_0$ so that

$$Z_i=Z_{i-1}\cap\{(x,n): n_i\leq m_i\}, \quad \gamma[p(Z_{i-1})]-\gamma[p(Z_i)]<\varepsilon\, 2^{-i};$$

this is possible by 2.1.5(1) because γ is regular and

$$\bigcup_{j=1}^{\infty} p(Z_{i-1}\cap\{(x,n): n_i\leq j\})=p(Z_{i-1}).$$

Next we define the closed set

$$C=\bigcap_{i=1}^{\infty}\text{Closure}\, p(Z_i).$$

Since $\gamma(X)<\infty$, and $\gamma(S)-\gamma[p(Z_i)]<\varepsilon$ whenever $i\in\mathscr{P}$, we see that

$$\gamma(C)=\lim_{i\to\infty}\gamma[\text{Closure}\, p(Z_i)]\geq\gamma(S)-\varepsilon.$$

Then we observe that

$$K=\mathcal{N}\cap\{n: n_i\leq m_i \text{ whenever } i\in\mathscr{P}\} \text{ is compact},$$

$$\bigcap_{i=1}^{\infty} Z_i=Z_0\cap(X\times K).$$

To complete the proof we will show that $C\subset S$, by verifying that

$$C=p[Z_0\cap(X\times K)].$$

Clearly $C\supset p[Z_0\cap(X\times K)]$. To prove the opposite inclusion we assume

$$a\in X\sim p[Z_0\cap(X\times K)], \quad \text{hence} \quad (\{a\}\times K)\cap Z_0=\varnothing,$$

and secure open sets $V\subset X$, $W\subset\mathcal{N}$ for which

$$a\in V, \quad K\subset W, \quad (V\times W)\cap Z_0=\varnothing.$$

Then we choose $i\in\mathscr{P}$ so that

$$\text{distance}(K,\mathcal{N}\sim W)>2^{-i}.$$

Since $(x,n)\in Z_i$ implies $(n_1,\ldots,n_i,1,1,1,\ldots)\in K$, hence

$$\text{distance}(K,n)\leq 2^{-i}, \quad n\in W,$$

we obtain

$$Z_i\subset(X\times W)\sim(V\times W)=(X\sim V)\times W,$$

and conclude that $p(Z_i)\subset X\sim V$, $a\notin\text{Closure}\, p(Z_i)$.

2.2.13. Combining the results of 2.2.10 and 2.2.12 we obtain the proposition: *If $f: X\to Y$ is continuous, X is a complete, separable metric space, Y is a Hausdorff space, μ measures Y, and every closed subset of Y*

is μ measurable, then the f image of every Borel subset of X is μ measurable. The problem of proving this proposition was the historical motivation for the creation of the theory of Suslin sets.

2.2.14. By a **Borel function** we mean a map $f: X \to Y$ such that X and Y are topological spaces and $f^{-1}(E)$ is a Borel subset of X whenever E is an open set of Y; it follows that $f^{-1}(E)$ is a Borel subset of X whenever E is a Borel subset of Y, because

$$2^Y \cap \{E: f^{-1}(E) \text{ is a Borel set}\}$$

is a Borel family.

The **characteristic function of a subset** A **of** X is the map $f: X \to \mathbf{R}$ such that

$$f(x)=1 \text{ for } x \in A, \quad f(x)=0 \text{ for } x \in X \sim A.$$

We observe that the Borel subsets of a topological space are those subsets whose characteristic functions are Borel functions.

We can represent any map $f: X \to Y$ as composition of the two maps

$$g: X \to X \times Y, \quad g(x)=(x, f(x)) \text{ for } x \in X,$$

$$h: X \times Y \to Y, \quad h(x,y)=y \text{ for } (x,y) \in X \times Y.$$

If f is a Borel function, X and Y are complete separable metric spaces, and A is any Suslin subset of X, then

$$g(A)=(A \times Y) \cap (f \times \mathbf{1}_Y)^{-1}\{(y,y): y \in Y\}$$

is a Suslin subset of $X \times Y$, hence $f(A)=h[g(A)]$ is a Suslin subset of Y.

2.2.15. A class Φ of functions, mapping a set X into a metric space Y is called a **Baire class** provided it satisfies the condition: *If $f_1, f_2, f_3, \ldots \in \Phi$ and*

$$g(x)=\lim_{i \to \infty} f_i(x) \in Y \textit{ for each } x \in X,$$

then $g \in \Phi$.

In case X is a topological space, the **Baire functions on** X **with values in** Y are the members of the smallest Baire class Φ such that every continuous map of X into Y belongs to Φ. Since the Borel functions form a Baire class [see the proof of 2.3.2(6)], *every Baire function is a Borel function.*

If X is a metric space, then every Borel function with values in $\mathbf{R}$ *is a Baire function.* To prove this one verifies that the class of all real valued Baire functions on X is an algebra, and one considers the family F of all those subsets of X whose characteristic function is a Baire function.

If C is the characteristic function of a nonempty closed subset A of X, then

$$C(x) = \lim_{i\to\infty} 2^{-i\,\text{distance}(x,A)} \quad \text{whenever } x \in X,$$

hence $A \in F$; since F is a Borel family, it consists of all Borel subsets of X. For any Borel function $f: X \to \mathbf{R}$ and every positive integer i, the characteristic functions $g_{i,j}$ of the Borel sets

$$\{x: j/i \le f(x) < (j+1)/i\}$$

corresponding to $j \in \mathbf{Z}$ are Baire functions, and so is the function

$$h_i = \sum_{j\in\mathbf{Z}} (j/i)\, g_{i,j};$$

since $h_i(x) \to f(x)$ as $i \to \infty$ whenever $x \in X$, f is a Baire function.

The preceding proposition is not valid for functions with values in an arbitrary separable metric space Y; in order that every Borel function mapping any metric space X into Y be a Baire function it is necessary and sufficient that for every finite subset Z of Y and every $\varepsilon > 0$, Z be contained in the ε neighborhood of the image of some continuous map of $\mathbf{R}$ into Y.

2.2.16. Theorem. *If ϕ measures the metric space X, $\phi(X) < \infty$ and all open subsets of X are ϕ measurable, then the support of ϕ is separable. Moreover, in case X has a dense subset whose cardinal is an Ulam number, then $\phi(X \sim \operatorname{spt} \phi) = 0$.*

Proof. For each positive integer n we choose a maximal subset A_n of $\operatorname{spt}\phi$ such that the distance between any two distinct points of A_n exceeds $2/n$. The open balls with radius $1/n$ and centered at the points of A_n are disjoint, and each has positive ϕ measure; therefore A_n is countable. Since each point of $\operatorname{spt}\phi$ is within $2/n$ of some point of A_n, we find that $\bigcup_{n=1}^{\infty} A_n$ is a countable dense subset of $\operatorname{spt}\phi$.

In case X has a dense subset whose cardinal is an Ulam number α, then the topology of X has a base with cardinal α, hence we can choose a family H, whose members are open sets with ϕ measure 0, such that

$$X \sim \operatorname{spt}\phi = \bigcup H \quad \text{and} \quad \operatorname{card} H \le \alpha.$$

We wellorder H and define, for each $V \in H$ and each positive integer n, the closed set $C_n(V)$ consisting of all those points which do not belong to any member of H preceding V, and whose distance from $X \sim V$ is greater than or equal to $1/n$. Clearly

$$\bigcup H = \bigcup_{n=1}^{\infty} \bigcup F_n, \quad \text{where} \quad F_n = \{C_n(V): V \in H\}.$$

Since any two points belonging to distinct members of F_n are at least $1/n$ apart, we see that the union of every subfamily of F_n is closed. From 2.1.6 we conclude that $\phi(\bigcup F_n)=0$ for each n.

2.2.17. *If $f: X \to Y$ is a continuous map of locally compact Hausdorff spaces, f is proper* $[f^{-1}(K)$ is compact for every compact $K \subset Y]$, *X is the union of some countable family of compact sets and ϕ is a Radon measure over X, then $f_\# \phi$ is a Radon measure over Y.* Leaving other parts of the proof as an exercise, we will show that whenever $B \subset Y$ and $\varepsilon > 0$ there exists an open subset W of Y such that

$$B \subset W \quad \text{and} \quad (f_\# \phi)(W) \leq \varepsilon + (f_\# \phi)(B).$$

Since $Y \sim \operatorname{im} f$ is open we consider only the case when $B \subset \operatorname{im} f$. We observe that $\operatorname{im} f$ is contained in the union of a sequence of open subsets $\varnothing = V_0 \subset V_1 \subset V_2 \subset \cdots$ of Y with compact closures, choose open subsets U_i of X for $i \geq 1$ so that

$$A_i = f^{-1}(B \cap V_i \sim V_{i-1}) \subset U_i, \phi(U_i) \leq 2^{-i}\varepsilon + \phi(A_i),$$

note that $W_i = V_i \sim f[f^{-1}(\operatorname{Clos} V_i) \sim U_i]$ is open in Y with

$$B \cap V_i \sim V_{i-1} \subset W_i \quad \text{and} \quad f^{-1}(W_i) \subset U_i,$$

hence infer $B \subset W = \bigcup \{W_i : i = 1, 2, 3, \ldots\}$ and

$$\phi[f^{-1}(W)] \leq \sum_{i=1}^{\infty} \phi(U_i) \leq \varepsilon + \sum_{i=1}^{\infty} \phi(A_i)$$

$$= \varepsilon + \sum_{i=1}^{\infty} [\phi \llcorner f^{-1}(B)][f^{-1}(V_i) \sim f^{-1}(V_{i-1})] = \varepsilon + \phi[f^{-1}(B)].$$

2.3. Measurable functions

2.3.1. When ϕ measures X, phrases containing the expression "ϕ **almost**" are often used to express the fact that certain exceptional sets have ϕ measure 0. For example a sentence of the form

"... for ϕ almost all x."

means that

$$\phi(X \sim \{x : \ldots\}) = 0.$$

In particular, if f and g are functions, then $f(x) = g(x)$ for ϕ almost all x if and only if $\phi(X \sim \{x : f(x) = g(x)\}) = 0$; in this case we may also use the equivalent phrase "f and g are ϕ almost equal". Similarly, we say that

... for ϕ almost all x in A if and only if $\phi(A \sim \{x : \ldots\}) = 0$;

B contains ϕ almost all of A if and only if $\phi(A \sim B) = 0$.

2.3.2. Assuming that ϕ measures X, and Y is a topological space, we say that f *is a* ϕ **measurable function** *if and only if* f *is a function whose domain contains* ϕ *almost all of* X, *whose image is contained in* Y, *and for which* $f^{-1}(E)$ *is* ϕ *measurable whenever* E *is an open subset of* Y.

For ease in applications we have allowed the domain of f to differ from X by a set of ϕ measure 0, but one could of course always extend f to a ϕ almost equal function with domain X, without changing anything of importance.

We observe that, for every function f which maps ϕ almost all of X into Y, the class

$$2^Y \cap \{E: f^{-1}(E) \text{ is } \phi \text{ measurable}\}$$

is a Borel family. Therefore:

(1) *If* f *is* ϕ *measurable, then* $f^{-1}(E)$ *is* ϕ *measurable whenever* E *is a Borel subset of* Y.

(2) *If* $S \subset 2^Y$ *and all open subsets of* Y *belong to the smallest Borel family containing* S, *and if* $f^{-1}(E)$ *is* ϕ *measurable whenever* $E \in S$, *then* f *is* ϕ *measurable.*

From (2) one obtains, for example, the criterion that an $\overline{\mathbf{R}}$ valued function f is ϕ measurable if and only if

$$\{x: f(x) < t\} \text{ is } \phi \text{ measurable whenever } t \in \mathbf{Q}.$$

Of course $\mathbf{Q}$ may be replaced by any dense subset of $\overline{\mathbf{R}}$, and $<$ by $>$, or $\leq$, or $\geq$.

By virtue of the last remark in 2.1.2, Theorem 2.2.12 implies:

(3) *If* f *is* ϕ *measurable, then* $f^{-1}(E)$ *is* ϕ *measurable whenever* E *is a Suslin subset of* Y.

Further general properties of ϕ measurable functions are:

(4) *If* f *is a* ϕ *measurable function with values in* Y, *and* $g: Y \to Z$ *is a Borel function, then* $g \circ f$ *is* ϕ *measurable.*

(5) *In case* Y *is the cartesian product of countably many spaces* Y_i, *whose topologies have countable bases, and* $p_i: Y \to Y_i$ *are the projections, then a map* f *into* Y *is* ϕ *measurable if and only if all the functions* $p_i \circ f$ *are* ϕ *measurable.*

(6) *If* $f_1, f_2, f_3, \ldots$ *are* ϕ *measurable functions with values in a metric space* Y, *and if*

$$\lim_{k \to \infty} f_k(x) = g(x) \in Y \text{ for } \phi \text{ almost all } x,$$

then g *is a* ϕ *measurable function.*

To prove (6), we suppose E is open in Y, let E_i be the open subset of E consisting of all points whose distance from $Y \sim E$ exceeds $1/i$, and verify that $g^{-1}(E)$ is ϕ almost equal to

$$\bigcup_{i=1}^{\infty} \bigcup_{j=1}^{\infty} \bigcap_{k=j}^{\infty} f_k^{-1}(E_i).$$

From (4) and (5) it follows that the class all $\overline{\mathbf{R}}$ valued ϕ measurable functions is closed to the basic numerical operations. For instance:

A function f with values in $\overline{\mathbf{R}}$ is ϕ measurable if and only if f^+ and f^- are ϕ measurable; if f is ϕ measurable, so is $|f|$.

If f and g are ϕ measurable functions with values in $\mathbf{R}$ (in $\overline{\mathbf{R}}$), so are $f+g$, $f-g$, $f\cdot g$ (provided their domains contain ϕ almost all of X).

If F is a countable set of $\overline{\mathbf{R}}$ valued ϕ measurable functions and

$$l(x)=\inf\{f(x)\colon f\in F\}, \qquad u(x)=\sup\{f(x)\colon f\in F\}$$

whenever $x\in\bigcap\{\operatorname{dmn} f\colon f\in F\}$, then l and u are ϕ measurable.

If $f_1, f_2, f_3, \ldots$ are $\overline{\mathbf{R}}$ valued ϕ measurable functions, so are the lower and upper limit functions mapping x onto

$$\liminf_{i\to\infty} f_i(x)=\sup_j \inf_{i>j} f_i(x), \qquad \limsup_{i\to\infty} f_i(x)=\inf_j \sup_{i>j} f_i(x).$$

Next we establish two extremely useful criteria for the measurability of functions with values in metric spaces.

(7) *An* $\overline{\mathbf{R}}$ *valued function* f *is* ϕ *measurable if and only if*

$$\phi(T)\ge\phi(T\cap\{x\colon f(x)\le a\})+\phi(T\cap\{x\colon f(x)\ge b\})$$

whenever $T\subset X$ *and* $-\infty<a<b<\infty$.

(8) *In case* Y *is metrized by* ρ, *a function* f *with values in* Y *is* ϕ *measurable if and only if*

$$\phi(T)\ge\phi[T\cap f^{-1}(A)]+\phi[T\cap f^{-1}(B)]$$

whenever $T\subset X$, $A\subset Y$, $B\subset Y$ *and* $\operatorname{dist}(A,B)=\inf\rho(A\times B)>0$.

First we show how (8) can be obtained from (7). Assuming the second half of (8) we let C be any nonempty closed subset of Y and define

$$g(y)=\operatorname{dist}(C,y)=\inf\rho(C\times\{y\}) \text{ for } y\in Y.$$

Whenever $-\infty<a<b<\infty$ our hypothesis is applicable with

$$A=\{y\colon g(y)\le a\} \quad\text{and}\quad B=\{y\colon g(y)\ge b\},$$

hence the criterion (7) with f replaced by $g\circ f$ shows that $g\circ f$ is a ϕ measurable function. Therefore $f^{-1}(C)=(g\circ f)^{-1}\{0\}$ is a ϕ measurable set.

To prove (7) we must verify that the stated condition implies that, for any $r \in \mathbf{R}$, the set

$$A = \{x : f(x) \le r\}$$

is ϕ measurable. Suppose $T \subset X$ and $\phi(T) < \infty$. Defining

$$B_i = T \cap \{x : r + (i+1)^{-1} \le f(x) \le r + i^{-1}\}$$

for every positive integer i, we will check the following proposition:

If K is a finite set of positive integers, whose elements are either all even or all odd, then

$$\phi(\bigcup\{B_k : k \in K\}) \ge \sum_{k \in K} \phi(B_k).$$

Proceeding by induction with respect to the cardinal number of K, we assume that this number exceeds 1, and let j be the largest element of K; then

$$\sup f(B_j) < \inf f(\bigcup\{B_k : j > k \in K\}),$$

hence

$$\phi(\bigcup\{B_k : k \in K\}) \ge \phi(B_j) + \phi(\bigcup\{B_k : j > k \in K\}),$$

and we may take the inductive step from $K - \{j\}$ to K.

From the above proposition we infer that

$$\infty > 2\phi(T) \ge \sum_{i=1}^{\infty} \phi(B_i).$$

Consequently, for each $\varepsilon > 0$ there exists a positive integer n such that

$$\varepsilon > \sum_{i=n}^{\infty} \phi(B_i) \ge \phi(T \cap \{x : r < f(x) \le r + n^{-1}\}),$$

hence

$$\begin{aligned} &\phi(T \cap A) + \phi(T \sim A) - \varepsilon \\ &\le \phi(T \cap \{x : f(x) \le r\}) + \phi(T \cap \{x : f(x) > r + n^{-1}\}) \le \phi(T). \end{aligned}$$

We conclude this discussion with the remark that, *in case ϕ measures a topological space X, the identity map $\mathbf{1}_X$ is a ϕ measurable function if and only if all open subsets of X are ϕ measurable sets*; in particular we may apply (8) to obtain **Carathéodory's criterion:**

(9) *In case ϕ measures a metric space X, all open subsets of X are ϕ measurable if and only if*

$$\phi(A \cup B) \ge \phi(A) + \phi(B)$$

whenever $A \subset X$, $B \subset X$ and $\operatorname{dist}(A, B) > 0$.

2.3.3. Theorem. *If ϕ measures X, $f: X \to \overline{\mathbf{R}} \cap \{y \geq 0\}$ is ϕ measurable, and $r_1, r_2, r_3, \ldots$ are positive numbers for which*

$$\lim_{n\to\infty} r_n = 0 \quad \text{and} \quad \sum_{n=1}^{\infty} r_n = \infty,$$

then there exist ϕ measurable sets $A_1, A_2, A_3, \ldots$ with characteristic functions $g_1, g_2, g_3, \ldots$ such that

$$f(x) = \sum_{n=1}^{\infty} r_n\, g_n(x) \quad \textit{whenever} \quad x \in X.$$

Proof. We inductively define

$$A_n = \{x: f(x) \geq r_n + \sum_{j<n} r_j\, g_j(x)\}.$$

2.3.4. A *set* is called **countably ϕ measurable** if and only if it is expressible as the union of some countable family of ϕ measurable sets, each with finite ϕ measure.

A *function* f with values in a topological vectorspace is termed **countably ϕ measurable** if and only if f is ϕ measurable and $\{x: f(x) \neq 0\}$ is countably ϕ measurable.

2.3.5. Lusin's theorem. *If ϕ is a Borel regular measure over a metric space X [or a Radon measure over a locally compact Hausdorff space X], f is a ϕ measurable function with values in a separable metric space Y, A is a ϕ measurable set for which $\phi(A) < \infty$, and $\varepsilon > 0$, then A contains a closed [compact] set C such that $\phi(A \sim C) < \varepsilon$ and $f|C$ is continuous.*

Proof. For each positive integer i we represent Y as the union of countably many disjoint Borel subsets $Y_{i,1}, Y_{i,2}, Y_{i,3}, \ldots$ with diameters less that i^{-1}, let

$$A_{i,j} = A \cap f^{-1}(Y_{i,j})$$

and apply 2.2.3 [or 2.2.5] to obtain disjoint closed [compact] sets

$$E_{i,j} \subset A_{i,j} \quad \text{with} \quad \phi(A_{i,j} \sim E_{i,j}) < \varepsilon\, 2^{-i-j}.$$

Inasmuch as

$$\phi\left(A \sim \bigcup_{j=1}^{\infty} E_{i,j}\right) = \sum_{j=1}^{\infty} \phi(A_{i,j} \sim E_{i,j}) < \varepsilon\, 2^{-i},$$

we can secure a positive integer $J(i)$ for which

$$\phi(A \sim D_i) < \varepsilon\, 2^{-i} \quad \text{with} \quad D_i = \bigcup_{j=1}^{J(i)} E_{i,j}.$$

Picking $y_{i,j} \in Y_{i,j}$ we define a continuous function g_i on the closed [compact] set D_i by letting

$$g_i(x) = y_{i,j} \text{ whenever } x \in E_{i,j},\ j = 1, \ldots, J(i).$$

Noting that distance $[g_i(x), f(x)] < i^{-1}$ for $x \in D_i$, we consider the closed [compact] set

$$C = \bigcap_{i=1}^{\infty} D_i \quad \text{with} \quad \phi(A \sim C) \leq \sum_{i=1}^{\infty} \phi(A \sim D_i) < \varepsilon,$$

on which the sequence of continuous functions $g_i | C$ converges uniformly to $f | C$, and conclude that $f | C$ is continuous.

2.3.6. The following two statements are corollaries of Lusin's theorem:

In case X is countably ϕ measurable, f is ϕ almost equal to some Borel function mapping X into Y. (However, note the example in 2.5.10!)

In case $Y = \mathbf{R}$ and X is normal, there exists (by Tietze's extension theorem) *a continuous map $h: X \to \mathbf{R}$ such that $h | C = f | C$, hence*

$$\phi(A \cap \{x: h(x) \neq f(x)\}) < \varepsilon.$$

We also observe that in Lusin's theorem *the hypothesis that Y be separable may be replaced by the hypothesis that Y have a dense subset whose cardinal is an Ulam number*, because then 2.2.16 implies that

$$Z = \operatorname{spt} f_{\#}(\phi \llcorner A) \text{ is closed and separable,}$$

$$\phi[A \sim f^{-1}(Z)] = f_{\#}(\phi \llcorner A)[Y \sim Z] = 0,$$

hence f is ϕ almost equal to a map into the separable space Z.

2.3.7. Egoroff's theorem. *Suppose $f_1, f_2, f_3, \ldots$ and g are ϕ measurable functions with values in a separable metric space Y. If $\phi(A) < \infty$,*

$$f_n(x) \to g(x) \text{ as } n \to \infty \text{ for } \phi \text{ almost all } x \text{ in } A,$$

and $\varepsilon > 0$, then there exists a ϕ measurable set B such that $\phi(A \sim B) < \varepsilon$ and

$$f_n(x) \to g(x), \text{ uniformly for } x \in B, \text{ as } n \to \infty.$$

Proof. Letting ρ be the metric of Y we define

$$C_{i,j} = \bigcup_{n=j}^{\infty} \{x: \rho[f_n(x), g(x)] \geq 2^{-i}\}$$

whenever i, j are positive integers, and infer from 2.1.3 (5), with ϕ replaced by $\phi \llcorner A$, that

$$\phi(A \cap C_{i,j}) \downarrow \phi\left(A \cap \bigcap_{k=1}^{\infty} C_{i,k}\right) = 0 \text{ as } j \uparrow \infty$$

for each i. Hence we may choose $J(i)$ so that $\phi(A \cap C_{i,J(i)}) < \varepsilon\, 2^{-i}$, and take

$$B = X \sim \bigcup_{i=1}^{\infty} C_{i,J(i)}.$$

2.3.8. Assuming that ϕ measures X, and Y is a *complete normed vectorspace* (**Banach space**) we now consider the vectorspace

$$\mathbf{A}(\phi, Y)$$

consisting of those ϕ measurable functions f with values in Y for which there exists a separable subspace Z of Y such that

$$\phi[f^{-1}(Y \sim Z)] = 0.$$

[In view of 2.2.16 and 2.3.6 this separability condition holds automatically in case X is countably ϕ measurable and Y has a dense subset whose cardinal is an Ulam number. I would not know how to prove without the separability condition that $\mathbf{A}(\phi, Y)$ is closed to addition.] We define

$$|f|_\phi = \inf\{r : \phi\{x : |f(x)| > r\} \le r\}$$

for $f \in \mathbf{A}(\phi, Y)$; using 2.1.3 (4) we find that

$$\phi\{x : |f(x)| > |f|_\phi\} \le |f|_\phi.$$

Since $|\;|_\phi$ is readily seen to satisfy the triangle inequality

$$|f+g|_\phi \le |f|_\phi + |g|_\phi \text{ for } f, g \in \mathbf{A}(\phi, Y),$$

the map sending (f, g) into $|f-g|_\phi$ is a translation invariant pseudometric on $\mathbf{A}(\phi, Y)$; the pseudodistance $|f-g|_\phi$ equals 0 if and only if f and g are ϕ almost equal; it may have the value ∞.

Sequences which converge with respect to this pseudometric are termed **convergent in ϕ measure**. From 2.3.7 we see that, in case $\phi(X) < \infty$, convergence ϕ almost everywhere implies convergence in ϕ measure. While the converse implication is false, we will obtain partial converses in 2.3.9 and 2.3.10, while proving the completeness of the pseudometric space $\mathbf{A}(\phi, Y)$.

The function $|\;|_\phi$ is not homogeneous; in case f is the characteristic function of A, then

$$|cf|_\phi = \inf\{c, \phi(A)\} \text{ for } 0 \le c < \infty.$$

Moreover the sets

$$\mathbf{B}(0, \varepsilon) = \{f : |f|_\phi \le \varepsilon\}$$

need not be convex; in case X is the union of finitely many disjoint ϕ measurable sets $S_1, \ldots, S_n$ with $\phi(S_i) \leq \varepsilon$ for $i = 1, \ldots, n$, then the convex hull of $\mathbf{B}(0, \varepsilon)$ equals $\mathbf{A}(\phi, Y)$; in fact, if $f \in \mathbf{A}(\phi, Y)$ and s_i is the characteristic function of S_i, then

$$|n\, s_i\, f|_\phi \leq \phi(S_i) \leq \varepsilon \text{ for } i = 1, \ldots, n, \quad \text{and} \quad f = \sum_{i=1}^{n} n^{-1}(n\, s_i\, f).$$

However it is true, for each $c \in \mathbf{R}$, that

$$|c\, f|_\phi \leq \sup\{|f|_\phi, |c| \cdot |f|_\phi\} \text{ whenever } f \in \mathbf{A}(\phi, Y),$$

hence $c\, f \to 0$ as $f \to 0$. Also, in case $f \in \mathbf{A}(\phi, Y)$ and

$$\phi\{x\colon f(x) \neq 0\} < \infty,$$

then $c\, f \to 0$ as $c \to 0$ in $\mathbf{R}$; in fact, given $\varepsilon > 0$, we can choose a positive integer n for which $\phi\{x\colon |f(x)| > n\} < \varepsilon$, and conclude that $|c\, f|_\phi \leq \varepsilon$ whenever $|c| < \varepsilon/n$.

2.3.9. Theorem. *If $g_n \in \mathbf{A}(\phi, Y)$ for $n = 1, 2, 3, \ldots$ and*

$$\sum_{n=1}^{\infty} |g_{n+1} - g_n|_\phi < \infty,$$

then for ϕ almost all x the sequence $g_1(x), g_2(x), g_3(x), \ldots$ converges to a point $h(x)$ in Y, and

$$|g_n - h|_\phi \to 0 \text{ as } n \to \infty.$$

Proof. We define

$$A_n = \{x\colon |g_{n+1}(x) - g_n(x)| \leq |g_{n+1} - g_n|_\phi\},$$

$$B_i = \bigcap_{n=i}^{\infty} A_n, \quad r_i = \sum_{n=i}^{\infty} |g_{n+1} - g_n|_\phi,$$

and observe that

$$\phi(X \sim A_n) \leq |g_{n+1} - g_n|_\phi, \quad \phi(X \sim B_i) \leq r_i,$$

$$\phi\left(X \sim \bigcup_{i=1}^{\infty} B_i\right) = \lim_{i \to \infty} \phi(X \sim B_i) = 0.$$

Now, if $x \in B_i$, then

$$\sum_{n=i}^{\infty} |g_{n+1}(x) - g_n(x)| \leq r_i < \infty,$$

hence the points $g_n(x)$ form a Cauchy sequence in Y, which converges to some point $h(x)$ because Y is complete; moreover

$$|h(x) - g_i(x)| \leq \sum_{n=i}^{\infty} |g_{n+1}(x) - g_n(x)| \leq r_i.$$

Thus $g_n(x) \to h(x)$ as $n \to \infty$ for ϕ almost all x, and

$$\phi\{x\colon |h(x)-g_i(x)|>r_i\} \le \phi(X \sim B_i) \le r_i,$$

whence $|h-g_i|_\phi \le r_i$.

2.3.10. Corollary. *Every Cauchy sequence in* $\mathbf{A}(\phi, Y)$ *has a subsequence which converges both* ϕ *almost everywhere and in* ϕ *measure; hence* $\mathbf{A}(\phi, Y)$ *is complete.*

Proof. Given a Cauchy sequence $f_1, f_2, f_3, \ldots$ in $\mathbf{A}(\phi, Y)$, we can pick positive integers $N(n)$ so that

$$|f_i - f_j|_\phi \le 2^{-n} \text{ whenever } i \ge j \ge N(n),$$

with $N(n+1) \ge N(n)$, and apply the preceding theorem with $g_n = f_{N(n)}$.

2.4. Lebesgue integration

2.4.1. Assuming that ϕ measures X, we will now study the integration of $\overline{\mathbf{R}}$ valued ϕ measurable functions. For this purpose we will first treat certain functions whose images are countable subsets of $\mathbf{R}$, next use these special functions to define the upper and lower integrals of arbitrary functions, and then develop the fundamental convergence and continuity properties of the ϕ integral.

We say that u *is a* ϕ **step function** *if and only if* u *is a* ϕ *measurable function whose image is a countable subset of* $\mathbf{R}$ *and for which*

$$\sum_{y \in \mathbf{R}} y \cdot \phi(u^{-1}\{y\}) \in \overline{\mathbf{R}};$$

here we use the convention $0 \cdot \infty = 0$.

The mapping associating the above sum with each ϕ step function u is obviously $\mathbf{R}$ homogeneous; we shall prove that it is also additive, in the following sense:

If u *and* v *are* ϕ *step functions, and if*

$$\sum_{y \in \mathbf{R}} y \cdot \phi(u^{-1}\{y\}) + \sum_{y \in \mathbf{R}} y \cdot \phi(v^{-1}\{y\}) = z \in \overline{\mathbf{R}},$$

then $u+v$ *is a* ϕ *step function and*

$$\sum_{y \in \mathbf{R}} y \cdot \phi[(u+v)^{-1}\{y\}] = z.$$

Clearly the image of $u+v$ is countable, and each of the two countable families

$$\{u^{-1}\{r\}\colon r \in \operatorname{im} u\}, \quad \{v^{-1}\{s\}\colon s \in \operatorname{im} v\}$$

consists of disjoint ϕ measurable sets whose union equals ϕ almost all of X. Abbreviating

$$f(r,s)=\phi[u^{-1}\{r\}\cap v^{-1}\{s\}] \text{ for } r,s\in\mathbf{R},$$

we find with the help of 2.1.1 (11), (7), (9) that

$$\begin{aligned} z &= \sum_{r\in\mathbf{R}} r\sum_{s\in\mathbf{R}} f(r,s)+\sum_{s\in\mathbf{R}} s\sum_{r\in\mathbf{R}} f(r,s)\\ &= \sum_{(r,s)\in\mathbf{R}\times\mathbf{R}} r f(r,s)+\sum_{(r,s)\in\mathbf{R}\times\mathbf{R}} s f(r,s)\\ &= \sum_{(r,s)\in\mathbf{R}\times\mathbf{R}} (r+s) f(r,s)\\ &= \sum_{t\in\mathbf{R}} t \sum_{\{(r,s):\, r+s=t\}} f = \sum_{t\in\mathbf{R}} t\,\phi[(u+v)^{-1}\{t\}]. \end{aligned}$$

2.4.2. Suppose f is any function mapping ϕ almost all of X into $\overline{\mathbf{R}}$. We call u *an* **upper [lower] function for** f *if and only if u is a ϕ step function and*

$$u(x)\geq f(x)\ \ [u(x)\leq f(x)] \ \ \textit{for } \phi \textit{ almost all } x.$$

Then we define the **upper [lower] ϕ integral of** f, denoted

$$\int^* f\,d\phi \quad [\int_* f\,d\phi],$$

as the *infimum [supremum] of the set of numbers*

$$\sum_{y\in\mathbf{R}} y\cdot\phi(u^{-1}\{y\})$$

associated with all the upper [lower] functions u for f.

We observe that, if there exists no upper [lower] functions for f, then $\int^* f\,d\phi=\infty$ $[\int_* f\,d\phi=-\infty]$.

We say that f *is a* ϕ **integrable function** *if and only if f is a ϕ measurable function whose upper and lower ϕ integrals are equal*; in this case we define the ϕ **integral of** f,

$$\int f\,d\phi=\int^* f\,d\phi=\int_* f\,d\phi\in\overline{\mathbf{R}}.$$

The ϕ integral of f will sometimes be designated by the alternate notations:

$$\int f(x)\,d\phi\,x, \quad \phi(f), \quad \phi_x[f(x)], \quad \langle f,\phi\rangle.$$

By a ϕ **summable function** we mean a ϕ *integrable function whose ϕ integral is finite* (belongs to $\mathbf{R}$).

2.4.3. Theorem. *Suppose f and g map ϕ almost all of X into $\overline{\mathbf{R}}$.*

(1) $\int_* f\,d\phi=-\int^*(-f)\,d\phi$.

(2) *If $f(x)\leq g(x)$ for ϕ almost all x, then $\int^* f\,d\phi\leq\int^* g\,d\phi$.*

(3) *If* $f(x) \geq 0$ *for* ϕ *almost all* x, *then* $\int^* f\,d\phi \geq 0$.

(4) *If* $\int^* f\,d\phi < \infty$, *then* $\int^* f^+\,d\phi < \infty$ *and* $f(x) < \infty$ *for* ϕ *almost all* x.

(5) *If* $0 < c < \infty$, *then* $\int^* (cf)\,d\phi = c\int^* f\,d\phi$.

(6) *If* $\int^* f\,d\phi + \int^* g\,d\phi < \infty$, *then* $\int^*(f+g)\,d\phi \leq \int^* f\,d\phi + \int^* g\,d\phi$.

(7) $\int_* f\,d\phi \leq \int^* f\,d\phi$.

Proof. (1)−(5) are trivial. In view of (4) the hypothesis of (6) implies that the domain of $f+g$ contains ϕ almost all of X; if the conclusion failed we could choose upper functions u and v for f and g so that

$$\int^*(f+g)\,d\phi > \sum_{y\in\mathbf{R}} y\cdot\phi(u^{-1}\{y\}) + \sum_{y\in\mathbf{R}} y\cdot\phi(v^{-1}\{y\}),$$

and the addition formula proved in 2.4.1 would show $u+v$ to be an upper function for $f+g$, whose associated sum would be less than the upper integral of $f+g$.

The assumption $\int_* f\,d\phi > \int^* f\,d\phi$ would imply by virtue of (1), (6), (3) that

$$0 > \int^* f\,d\phi - \int_* f\,d\phi = \int^* f\,d\phi + \int^*(-f)\,d\phi \geq \int^*(f-f)\,d\phi \geq 0.$$

2.4.4. Theorem.

(1) *If* f *is a* ϕ *step function, then* $\int f\,d\phi = \sum_{y\in\mathbf{R}} y\,\phi(f^{-1}\{y\})$.

(2) *If* f *is a* ϕ *measurable function such that* $f(x) \geq 0$ *for* ϕ *almost all* x, *then* $\int f\,d\phi \geq 0$.

(3) $\int (cf)\,d\phi = c\int f\,d\phi$ *whenever* $0 \neq c \in \mathbf{R}$.

(4) *If* $\int f\,d\phi + \int g\,d\phi \in \overline{\mathbf{R}}$, *then* $\int (f+g)\,d\phi = \int f\,d\phi + \int g\,d\phi$.

(5) *If* f *and* g *are* ϕ *measurable functions with* $f(x) \leq g(x)$ *for* ϕ *almost all* x, *and if either* $\int g\,d\phi < \infty$ *or* $\int f\,d\phi > -\infty$, *then*

$$\int f\,d\phi \leq \int g\,d\phi.$$

(6) $\int f\,d\phi = \int f^+\,d\phi - \int f^-\,d\phi$.

(7) *If* f *is* ϕ *integrable, then* $|\int f\,d\phi| \leq \int |f|\,d\phi$.

(8) *If* f *is* ϕ *summable, so is* $|f|$.

Proof. If f is a ϕ step function, then f is an upper and lower function for itself, hence

$$\int^* f\,d\phi \leq \sum_{y\in\mathbf{R}} y\cdot\phi(f^{-1}\{y\}) \leq \int_* f\,d\phi;$$

therefore (1) follows from 2.4.3(7).

Next we prove that each ϕ almost everywhere nonnegative ϕ measurable function f is ϕ integrable. If $\phi(f^{-1}\{\infty\})>0$ we use the fact that

$$\int_* f\,d\phi \geq n\,\phi(f^{-1}\{\infty\}) \text{ for every positive integer } n,$$

because n times the characteristic function of $f^{-1}\{\infty\}$ is a lower function for f, in conjunction with 2.4.3 (7) to infer that $\int f\,d\phi=\infty$. On the other hand, if $f(x)<\infty$ for ϕ almost all x, we can show that

$$\int^* f\,d\phi \leq t \int_* f\,d\phi \text{ whenever } 1<t<\infty,$$

by constructing a lower function u for f such that $t\,u$ is an upper function for f; since ϕ almost all of the set $P=\{x: f(x)>0\}$ is contained in the union of the disjoint ϕ measurable sets

$$A_n=\{x: t^n \leq f(x)<t^{n+1}\}$$

corresponding to all integers n, we can take $u(x)=t^n$ for $x\in A_n$, $u(x)=0$ for $x\in X\sim P$.

(3) follows from 2.4.3 (1), (5).

In case $-\infty<\int f\,d\phi+\int g\,d\phi<\infty$ we obtain from 2.4.3 (6), (7), (1) the three inequalities

$$\int f\,d\phi+\int g\,d\phi \geq \int^*(f+g)\,d\phi \geq \int_*(f+g)\,d\phi \geq \int f\,d\phi+\int g\,d\phi,$$

hence the conclusion of (4). In case $\int f\,d\phi+\int g\,d\phi=-\infty$, we still obtain the first two inequalities, which then suffice to give the conclusion. In case $\int f\,d\phi+\int g\,d\phi=\infty$ we similarly use the last two inequalities.

To treat (5) in case $\int g\,d\phi<\infty$ we first infer from 2.4.3 (4) that $f(x)\leq g(x)<\infty$ for ϕ almost all x, hence $g-f$ is a ϕ almost nonnegative ϕ measurable function; it then follows from (2), (3), (4) that

$$\int g\,d\phi \geq \int g\,d\phi+\int (f-g)\,d\phi=\int f\,d\phi.$$

If $\int f^+\,d\phi-\int f^-\,d\phi\in\overline{\mathbf{R}}$, the conclusion of (6) follows from (3), (4). On the other hand, if f is ϕ integrable, so are f^+ and f^- by (2), and 2.4.3 (4), (1) imply that either $\int f\,d\phi<\infty$ and $\int f^+\,d\phi<\infty$, or $\int(-f)\,d\phi<\infty$ and $\int f^-\,d\phi<\infty$; in both cases $\int f^+\,d\phi-\int f^-\,d\phi\in\overline{\mathbf{R}}$.

From (4) and (2) we see that, for every ϕ measurable function f,

$$\int |f|\,d\phi=\int f^+\,d\phi+\int f^-\,d\phi.$$

Comparing this equation with (6) we obtain (7) and (8).

2.4.5. Suppose $A\subset X$ and f is the characteristic function of A. It follows from 2.4.4 (1) that

$$\int f\,d\phi=\phi(A) \text{ in case } A \text{ is } \phi \text{ measurable},$$

and that always

$$\int^* f\,d\phi \geq \phi(A) \geq \int_* f\,d\phi.$$

Moreover, if ϕ is regular, then $\int^* f\,d\phi = \phi(A)$.

2.4.6. Fatou's lemma. *If $f_1, f_2, f_3, \ldots$ are nonnegative ϕ measurable functions, then*

$$\liminf_{n\to\infty} \int f_n\,d\phi \geq \int \liminf_{n\to\infty} f_n(x)\,d\phi\,x.$$

Proof. Since the ϕ integral of the inferior limit function can be computed by means of lower functions, we need only verify the following statement:

If $y_1, \ldots, y_m$ are finite positive numbers and $A_1, \ldots, A_m$ are disjoint ϕ measurable sets such that

$$\liminf_{n\to\infty} f_n(x) \geq y_i \text{ for } x \in A_i \text{ and } i = 1, \ldots, m,$$

then

$$\liminf_{n\to\infty} \int f_n\,d\phi \geq \sum_{i=1}^m y_i \cdot \phi(A_i).$$

Suppose $0 < t < 1$. Each set A_i is the union of the nondecreasing sequence of ϕ measurable sets

$$B_{i,n} = A_i \cap \{x : f_k(x) > t\,y_i \text{ for all } k \geq n\}$$

corresponding to $n = 1, 2, 3, \ldots$. Consequently

$$\int f_n\,d\phi \geq \sum_{i=1}^m t\,y_i\,\phi(B_{i,n}) \to t \sum_{i=1}^m y_i\,\phi(A_i) \text{ as } n \to \infty.$$

The asserted statement now follows because t can be chosen arbitrarily close to 1.

2.4.7. Lebesgue's increasing convergence theorem. *If $f_1, f_2, f_3, \ldots$ are ϕ measurable functions such that*

$$0 \leq f_n(x) \leq f_{n+1}(x)$$

for $x \in X$ and $n = 1, 2, 3, \ldots$, then

$$\lim_{n\to\infty} \int f_n\,d\phi = \int \lim_{n\to\infty} f_n(x)\,d\phi\,x.$$

Proof. 2.4.4 (2), (5) imply $\leq$, while Fatou's lemma implies $\geq$.

2.4.8. Corollary. *If K is a countable set, and a nonnegative ϕ measurable function f_k is associated with each $k \in K$, then*

$$\sum_{k\in K} \int f_k\,d\phi = \int \sum_{k\in K} f_k(x)\,d\phi\,x.$$

2.4.9. Lebesgue's bounded convergence theorem. *Suppose h is a ϕ summable function.* If $f_1, f_2, f_3, \ldots$ *and g are ϕ measurable functions such that*

$$|f_n(x)| \leq h(x) \text{ for } n=1,2,3,\ldots, \quad f_n(x) \to g(x) \text{ as } n \to \infty$$

whenever $x \in X$, then

$$\int |f_n - g|\, d\phi \to 0, \quad \text{hence} \quad \int f_n\, d\phi \to \int g\, d\phi, \text{ as } n \to \infty.$$

Proof. Applying Fatou's lemma to the sequence of nonnegative ϕ measurable functions $2h - |f_n - g|$, which converge pointwise to $2h$, we find that

$$\begin{aligned} 0 &\geq \int 2h\, d\phi - \liminf_{n\to\infty} \int (2h - |f_n - g|)\, d\phi \\ &= \limsup_{n\to\infty} \int |f_n - g|\, d\phi \geq \limsup_{n\to\infty} \left|\int f_n\, d\phi - \int g\, d\phi\right|, \end{aligned}$$

because f_n and g are ϕ summable, being dominated by h.

2.4.10. Whenever $A \subset X$ and f is an $\overline{\mathbf{R}}$ valued function whose domain contains ϕ almost all of A, we define

$$\int_A^* f\, d\phi = \int^* f_A\, d\phi, \quad \int_{*A} f\, d\phi = \int_* f_A\, d\phi, \quad \int_A f\, d\phi = \int f_A\, d\phi,$$

where $f_A(x) = f(x)$ for $x \in A$, $f_A(x) = 0$ for $x \in X \sim A$.

If $g(x) \geq 0$ for ϕ almost all x, and

$$\psi(A) = \int_A^* g\, d\phi \quad \text{whenever } A \subset X,$$

then ψ measures X, and all ϕ measurable sets are ψ measurable. Moreover, in case g is ϕ measurable we see from 2.3.3, 2.4.8, 2.4.4(6) that

$$\int f\, d\psi = \int f\, g\, d\phi$$

for every ϕ measurable $\overline{\mathbf{R}}$ valued function f.

In case X is $\mathbf{R}$ or $\overline{\mathbf{R}}$, and $-\infty \leq a \leq b \leq \infty$, one abbreviates

$$\int_{\{x:\, a < x \leq b\}} f\, d\phi = \int_a^b f\, d\phi.$$

Therefore $\int_a^b f\, d\phi + \int_b^c f\, d\phi = \int_a^c f\, d\phi$ whenever $a \leq b \leq c$ and $\{x:\, x \leq b\}$ is ϕ measurable.

2.4.11. Theorem. *If f is ϕ summable and $\varepsilon > 0$, then there exists a $\delta > 0$ such that*

$$\int_A |f|\, d\phi < \varepsilon \text{ whenever } A \text{ is } \phi \text{ measurable and } \phi(A) \leq \delta.$$

Proof. Since $g_n(x)=\inf\{|f(x)|,n\}\uparrow|f(x)|$ as $n\uparrow\infty$ for ϕ almost all x, Theorem 2.4.7 allows us to choose a positive integer n for which

$$\int(|f|-g_n)\,d\phi<\varepsilon/2.$$

Then we observe that

$$\int_A|f|\,d\phi=\int_A(|f|-g_n)\,d\phi+\int_A g_n\,d\phi<\varepsilon/2+n\,\phi(A)$$

for every ϕ measurable set A, and take $\delta=\varepsilon/(2n)$.

2.4.12. By a **seminorm** on a vectorspace V we mean a function σ on V such that

$$0\le\sigma(x)\le\infty,\qquad \sigma(c\,x)=|c|\cdot\sigma(x),\qquad \sigma(x+y)\le\sigma(x)+\sigma(y)$$

whenever $x,y\in V$ and $c\in\mathbf{R}$; here $0\cdot\infty=0$. It follows that

$$\{x\colon \sigma(x)<\infty\}\text{ is a vectorsubspace of }V.$$

(In case $0<\sigma(x)<\infty$ for $0\neq x\in V$, σ is a **norm**.)

A basic fact about seminorms is the **Hahn-Banach theorem**: *If σ is a seminorm on V and f is a linear function mapping a vectorsubspace W of V into $\mathbf{R}$ so that $f\le\sigma|W$, then f can be extended to a linear map g of V into $\mathbf{R}$ with $g\le\sigma$.* Using Hausdorff's maximal principle one reduces this problem to the case when V is spanned by W and a single vector ξ; since

$$f(x)+f(y)=f(x+y)\le\sigma(x+y)\le\sigma(x-\xi)+\sigma(y+\xi)$$

whenever $x,y\in W$, one finds that

$$\sup\{f(x)-\sigma(x-\xi)\colon x\in W\}\le\inf\{\sigma(y+\xi)-f(y)\colon y\in W\};$$

choosing $c\in\mathbf{R}$ so that c lies between the above sup and inf, one defines

$$g(w+t\,\xi)=f(w)+t\,c\text{ whenever }w\in W,\ t\in\mathbf{R}.$$

This theorem has the corollary: *For each seminorm σ on V and each $v\in V$ there exists a linear map g of V into $\mathbf{R}$ such that $g(v)=\sigma(v)$ and $g\le\sigma$.* One takes $f(t\,v)=t\,\sigma(v)$ whenever $t\in\mathbf{R}$.

The classical seminorms on the space $\mathbf{A}(\phi,Y)$ [see 2.3.8] are the functions $\phi_{(p)}$ defined for $1\le p\le\infty$ by the formulae

$$\phi_{(p)}(f)=\left(\int|f|^p\,d\phi\right)^{1/p}\text{ in case }1\le p<\infty,$$

$$\phi_{(\infty)}(f)=\inf\{r\colon \phi\{x\colon |f(x)|>r\}=0\}.$$

Associated with $\phi_{(p)}$ is the **Lebesgue space**

$$\mathbf{L}_p(\phi,Y)=\mathbf{A}(\phi,Y)\cap\{f\colon \phi_{(p)}(f)<\infty\text{ and }f\text{ is countably }\phi\text{ measurable}\}.$$

We abbreviate $\mathbf{L}_p(\phi,\mathbf{R})=\mathbf{L}_p(\phi)$. [In case $p<\infty$ the condition "f is countably ϕ measurable" is redundant.]

The fact that $\phi_{(p)}$ is a seminorm is obvious in case $p=1$ or $p=\infty$; for $1<p<\infty$ it will be proved in 2.4.15.

If $u_1, u_2, u_3, \ldots \in \mathbf{A}(\phi, Y)$ and $u_n(x) \to v(x)$ for ϕ almost all x, then

$$\liminf_{n\to\infty} \phi_{(p)}(u_n) \geq \phi_{(p)}(v);$$

this follows from 2.4.6 in case $1 \leq p < \infty$, and is trivial in case $p=\infty$.

If $f \in \mathbf{L}_p(\phi, Y)$ and $1 \leq p < \infty$, then

$$r^p \phi\{x\colon |f(x)|>r\} \leq \int |f|^p \, d\phi \text{ whenever } 0 \leq r < \infty;$$

choosing r so that $r^{p+1} = \int |f|^p \, d\phi$ we obtain

$$|f|_\phi \leq \phi_{(p)}(f)^{p/(p+1)}.$$

Moreover $|f|_\phi \leq \phi_{(\infty)}(f)$.

Every $\phi_{(p)}$ Cauchy sequence $f_1, f_2, f_3, \ldots$ is also a $|\ |_\phi$ Cauchy sequence, hence has by 2.3.10 a subsequence $g_1, g_2, g_3, \ldots$ which converges ϕ almost everywhere to a function $h \in \mathbf{A}(\phi, Y)$; it follows that

$$\phi_{(p)}(f_n - h) \leq \liminf_{m\to\infty} \phi_{(p)}(f_n - g_m) \text{ for each } n,$$

and hence that $\phi_{(p)}(f_n - h) \to 0$ as $n \to \infty$. Therefore:

$\mathbf{L}_p(\phi, Y)$ *is complete with respect to* $\phi_{(p)}$.

We also observe that *the functions with finite image form a* $\phi_{(p)}$ *dense subset of* $\mathbf{L}_p(\phi, Y)$.

According to the above conventions, $f \in \mathbf{L}_1(\phi, Y)$ if and only if $f \in \mathbf{A}(\phi, Y)$ and $\int |f| \, d\phi \in \mathbf{R}$; we shall now construct

$$\int f \, d\phi \in Y$$

in such a way that

$$\alpha\left(\int f \, d\phi\right) = \int (\alpha \circ f) \, d\phi$$

whenever $\alpha \in Y^*$, the space of all continuous real valued linear functions on Y, with the dual norm

$$|\alpha| = \sup\{\langle y, \alpha\rangle\colon y \in Y, |y| \leq 1\}.$$

For this purpose we consider the continuous linear maps

$$\mathbf{L}_1(\phi, Y) \xrightarrow{\sigma} (Y^*)^* \xleftarrow{\tau} Y$$

such that, for $f \in \mathbf{L}_1(\phi, Y)$, $\alpha \in Y^*$, $y \in Y$,

$$\langle \alpha, \sigma(f)\rangle = \int (\alpha \circ f) \, d\phi, \text{ hence } |\sigma(f)| \leq \phi_{(1)}(f),$$

$$\langle \alpha, \tau(y)\rangle = \langle y, \alpha\rangle, \text{ hence } |\tau(y)| = |y|.$$

We observe that $\sigma^{-1}(\operatorname{im} \tau)$ is a $\phi_{(1)}$ closed vectorsubspace of $\mathbf{L}_1(\phi, Y)$ and

$$\tau^{-1} \circ \sigma\colon \sigma^{-1}(\operatorname{im} \tau) \to Y$$

is a continuous linear map; moreover

$$\sigma^{-1}(\operatorname{im}\tau)=\mathbf{L}_1(\phi,Y)$$

because the vectorsubspace

$$\mathbf{L}_1(\phi,Y)\cap\{f\colon \operatorname{im} f \text{ is finite}\}$$

is obviously contained in $\sigma^{-1}(\operatorname{im}\tau)$, and is $\phi_{(1)}$ dense in $\mathbf{L}_1(\phi,Y)$. Therefore we may define

$$\int f\,d\phi=(\tau^{-1}\circ\sigma)\,f$$

whenever $f\in\mathbf{L}_1(\phi,Y)$, and infer that $|\int f\,d\phi|\le\int|f|\,d\phi$. [Of course the construction becomes simpler when $\operatorname{im}\tau=(Y^*)^*$, in which case Y is called **reflexive**; in particular, every finite dimensional normed vectorspace is reflexive.]

2.4.13. Suppose u, v, α, β are positive numbers, with $\alpha+\beta=1$. Since $-\log$ is a strictly convex function,

$$\log(\alpha\,u+\beta\,v)\ge\alpha\log(u)+\beta\log(v).$$

Applying exp we obtain the inequality

$$\alpha\,u+\beta\,v\ge u^\alpha\,v^\beta$$

relating the **arithmetic** and **geometric means** of u, v with weights α, β. Equality holds if and only if $u=v$.

2.4.14. Hölder's inequality. *If* $p\ge1$, $q\ge1$, $p^{-1}+q^{-1}=1$, *then*

$$|\int f\,g\,d\phi|\le\phi_{(p)}(f)\cdot\phi_{(q)}(g)$$

whenever $f\in\mathbf{L}_p(\phi)$ *and* $g\in\mathbf{L}_q(\phi)$; *equality holds if and only if* $|f|^p\operatorname{sign} f$ *and* $|g|^q\operatorname{sign} g$ *are* ϕ *almost linearly dependent.*

Proof. In case $\phi_{(p)}(f)>0$ and $\phi_{(q)}(g)>0$ we consider the functions

$$F=|f|^p/\int|f|^p\,d\phi,\qquad G=|g|^q/\int|g|^q\,d\phi$$

and apply 2.4.13 with $\alpha=p^{-1}$, $\beta=q^{-1}$ to obtain

$$\phi_{(p)}(f)^{-1}\,\phi_{(q)}(g)^{-1}\int|f\,g|\,d\phi=\int F^\alpha\,G^\beta\,d\phi$$

$$\le\int(\alpha\,F+\beta\,G)\,d\phi=\alpha\int F\,d\phi+\beta\int G\,d\phi=1.$$

2.4.15. Minkowski's inequality. *If* $1<p<\infty$, *then*

$$\phi_{(p)}(f+g)\le\phi_{(p)}(f)+\phi_{(p)}(g)$$

whenever $f,g\in\mathbf{L}_p(\phi,Y)$; *equality holds if and only if there exist nonnegative numbers* a, b *such that* $a+b=1$ *and*

$$|f(x)|=a\,|f(x)+g(x)|,\ |g(x)|=b\,|f(x)+g(x)| \text{ for } \phi \text{ almost all } x.$$

[In case Y is an inner product space, equality holds if and only if f and g are ϕ *almost positively proportional.]*

Proof. We see that $f+g\in\mathbf{L}_p(\phi,Y)$ because

$$|f+g|^p\le(|f|+|g|)^p\le 2^p(|f|^p+|g|^p).$$

Letting $q=p/(p-1)$ we also compute

$$\phi_{(q)}(|f+g|^{p-1})=\phi_{(p)}(f+g)^{p-1}$$

Then we observe that

$$|f+g|^p=|f+g|\cdot|f+g|^{p-1}\le|f|\cdot|f+g|^{p-1}+|g|\cdot|f+g|^{p-1}$$

and use 2.4.14 to obtain

$$\phi_{(p)}(f+g)^p\le[\phi_{(p)}(f)+\phi_{(p)}(g)]\cdot\phi_{(p)}(f+g)^{p-1}.$$

2.4.16. Theorem. *Suppose either* (i) $1<p<\infty$ *and* $q=p/(p-1)$; *or* (ii) $p=1$ *and* $q=\infty$; *or* (iii) $p=\infty$ *and* $q=1$.

If $f: X\to\mathbf{R}$ *is countably* ϕ *measurable, then*

$$\phi_{(p)}(f)=\sup\{\textstyle\int f\,u\,d\phi:\ u\in\mathbf{L}_{(q)}(\phi) \text{ and } \phi_{(q)}(u)=1\};$$

moreover the supremum is unchanged when u is also required to satisfy the condition $\operatorname{sign}u=\operatorname{sign}f$; *in case* $p<\infty$ *and* $0<\phi_{(p)}(f)<\infty$, *the supremum is attained by the function*

$$u=[|f|/\phi_{(p)}(f)]^{p-1}\operatorname{sign}f.$$

Proof. We first consider the case when $0<\phi_{(p)}(f)<\infty$.

The supremum does not exceed $\phi_{(p)}(f)$, by Hölder's inequality if $1<p<\infty$, trivially otherwise.

If $p<\infty$ and $u=[|f|/\phi_{(p)}(f)]^{p-1}\operatorname{sign}f$, then

$$\textstyle\int f\,u\,d\phi=\int|f|^p\,d\phi/\phi_{(p)}(f)^{p-1}=\phi_{(p)}(f);$$

in case $p=1$, $u=\operatorname{sign}f$, hence $\phi_{(\infty)}(u)=1$; in case $1<p<\infty$,

$$\textstyle\int|u|^q\,d\phi=\int[|f|/\phi_p(f)]^p\,d\phi=1.$$

If $p=\infty$, let $A_1, A_2, A_3, \ldots$ be disjoint ϕ measurable sets, each of finite and positive ϕ measure, with union $\{x: f(x)\neq 0\}$, let a_i be the characteristic function of A_i, and let $\varepsilon_1, \varepsilon_2, \varepsilon_3, \ldots$ be positive numbers with sum 1; then the function

$$u=\sum_{i=1}^{\infty} \varepsilon_i\, \phi(A_i)^{-1}\, a_i \operatorname{sign} f$$

satisfies the conditions $\int |u|\, d\phi=1$, $\operatorname{sign} u=\operatorname{sign} f$ and

$$\int u f\, d\phi=\sum_{i=1}^{\infty} \varepsilon_i\, \phi(A_i)^{-1} \int_{A_i} |f|\, d\phi .$$

If $0<t<\phi_{(\infty)}(f)$, we can choose $A_1 \subset \{x: |f(x)|>t\}$, and ε_1 close to 1, making $\int u f\, d\phi > \varepsilon_1 t$ close to t.

To treat the case when $\phi_{(p)}(f)=\infty$ we represent f as pointwise limit of a sequence of bounded ϕ measurable functions g_n for which

$$\phi\{x: g_n(x)\neq 0\}<\infty \text{ and } |g_n| \leq |g_{n+1}| \leq |f| .$$

Then $\phi_{(p)}(g_n) \uparrow \phi_{(p)}(f)=\infty$ as $n \uparrow \infty$, by 2.4.7. Moreover, for each n, $\phi_{(p)}(g_n)<\infty$, hence

$$\begin{aligned} \phi_{(p)}(g_n) &= \sup\{\textstyle\int g_n u\, d\phi: \phi_{(q)}(u)=1\} \\ &\leq \sup\{\textstyle\int f u\, d\phi: \phi_{(q)}(u)=1\}, \end{aligned}$$

because $g_n u \leq |f| \cdot |u| = f \cdot (u \cdot \operatorname{sign} u \cdot \operatorname{sign} f)$.

2.4.17. Assuming that *ϕ measures X and f is a real valued ϕ measurable function*, one readily verifies the following facts with the help of Hölder's inequality and Lebesgue's convergence theorems:

$P=\{p: \phi_{(p)}(f)<\infty\}$ is a subinterval of $\overline{\mathbf{R}}$.

The function mapping $p \in \operatorname{Closure} P$ onto $\phi_{(p)}(f)$ is continuous.

$U=\{u: 0<\phi_{(1/u)}(f)<\infty\}$ is an interval; here $1/0=\infty$.

The function mapping $u \in U$ onto $\log \phi_{(1/u)}(f)$ is convex.

If $\phi(X)=1$, then $\phi_{(s)}(f) \leq \phi_{(t)}(f)$ for $1 \leq s<t \leq \infty$.

2.4.18. Theorem. *Suppose u maps X into Y, and f maps $u_{\#}(\phi)$ almost all of Y into $\overline{\mathbf{R}}$.*

(1) *If $f \circ u$ is ϕ measurable, then f is $u_{\#}(\phi)$ measurable and*

$$\int (f \circ u)\, d\phi=\int f\, d(u_{\#} \phi).$$

(2) *If f is $u_{\#}(\phi)$ measurable, ϕ is regular and $\phi(X)<\infty$, then $f \circ u$ is ϕ measurable.*

Proof. The first conclusion of (1) is evident from the last assertion in 2.1.2. The second conclusion of (1) is obvious in case f is the characteristic function of a subset of Y; it then follows from 2.3.3 and 2.4.8 in case f is nonnegative, and from 2.4.4 (6) in the general case.

The statement (2) is a consequence of 2.1.5 (4).

2.4.19. Jensen's inequality. *If $0<\phi(X)<\infty$, Y is a Banach space, C is a closed convex subset of Y and F is a continuous nonnegative convex function on C, then*

$$F[\int g\, d\phi/\phi(X)] \leq \int (F \circ g)\, d\phi/\phi(X)$$

for every $g \in \mathbf{L}_{(1)}(\phi, Y)$ with $\phi[X \sim g^{-1}(C)]=0$.

In case im g is finite this assertion is obvious. In case C and F are bounded we use 2.4.9 to deduce the inequality through $\phi_{(1)}$ approximation of g by functions with finite images in C. In the general case we consider for each positive integer n the closed convex set

$$C_n = C \cap \{y: |y| \leq n \text{ and } F(y) \leq n\},$$

let $A_n = g^{-1}(C_n)$, infer from the preceding case with ϕ replaced by $\phi \llcorner A_n$ that

$$F[\int_{A_n} g\, d\phi/\phi(A_n)] \leq \int_{A_n} (F \circ g)\, d\phi/\phi(A_n),$$

and note that $\phi(X \sim A_n) \to 0$ as $n \to \infty$ by 2.1.3 (5).

2.5. Linear functionals

2.5.1. By a **lattice of functions on** X we mean a set L whose elements are *functions mapping X into* $\mathbf{R}$, and which satisfies the following condition: *If $0 \leq c < \infty$, and if f and g belong to L, so do $f+g$, cf,* $\inf\{f, g\}$ *and* $\inf\{f, c\}$; *in case $f \leq g$, also $g-f$ belongs to L.* Here we use the conventions

$$\inf\{f, g\}(x) = \inf\{f(x), g(x)\}, \quad \inf\{f, c\}(x) = \inf\{f(x), c\}$$

for all $x \in X$; similarly $f \leq g$ if and only if $f(x) \leq g(x)$ for all $x \in X$.

We observe that if f and g belong to L with $g \geq 0$, so does $\sup\{f, g\} = f + g - \inf\{f, g\}$; therefore f^+ and $f^- = f^+ - f$ belong to L.

If L is a lattice of functions on X, so is the subset

$$L^+ = L \cap \{f: f \geq 0\}.$$

Examples of such lattices are: the set of all functions which map X into $\mathbf{R}$; the subset of those functions which are continuous with respect to a topology on X; the subset of those functions which are summable with respect to a measure over X.

Whenever ϕ is a measure over a space X and L is a sublattice of the lattice of all ϕ summable functions, then the ϕ integral gives a map of L into $\mathbf{R}$ (associating $\int f\,d\phi$ with $f \in L$) which preserves linear structure [2.4.4 (3), (4)], ordering [2.4.4 (5)] and increasing convergence [2.4.7]. The next theorem shows that these properties completely characterize the process of Lebesgue integration.

2.5.2. Theorem. *Suppose L is a lattice of functions on X, λ maps L into $\mathbf{R}$, and satisfies the following conditions whenever $f, g, h_1, h_2, h_3, \ldots$ belong to L:*

$$\lambda(f+g)=\lambda(f)+\lambda(g);$$

$$0 \le c < \infty \quad \textit{implies} \quad \lambda(cf)=c\,\lambda(f);$$

$$f \ge g \quad \textit{implies} \quad \lambda(f) \ge \lambda(g);$$

$$h_n \uparrow g \textit{ as } n \uparrow \infty \quad \textit{implies} \quad \lambda(h_n) \uparrow \lambda(g) \textit{ as } n \uparrow \infty.$$

Then there exists a measure ϕ over X such that

$$\lambda(f)=\int f\,d\phi \textit{ for every } f \in L.$$

Proof. Note that $f \in L^+$ implies $\lambda(f) \ge \lambda(0 \cdot f)=0$.

For any $A \subset X$, let us say that *a sequence $f_1, f_2, f_3, \ldots$ suits A* if and only if $f_n \in L^+$, $f_n \le f_{n+1}$ for $n=1, 2, 3, \ldots$ and

$$\lim_{n\to\infty} f_n(x) \ge 1 \text{ whenever } x \in A;$$

then we define

$$\phi(A)=\inf\{\lim_{n\to\infty} \lambda(f_n)\colon f_1, f_2, f_3, \ldots \text{ suits } A\}.$$

The function ϕ *measures* X, because if

$$A \subset \bigcup_{i=1}^{\infty} B_i \subset X$$

and if, for each i, the sequence $f_{i,1}, f_{i,2}, f_{i,3}, \ldots$ suits B_i, then the sequence of functions

$$g_n = \sum_{i=1}^{n} f_{i,n}$$

suits A, and

$$\lambda(g_n)=\sum_{i=1}^{n} \lambda(f_{i,n}) \le \sum_{i=1}^{\infty} \lim_{m\to\infty} \lambda(f_{i,m}).$$

If $g \in L^+$, $A \subset X$, $g(x) \le 1$ for $x \in A$, $g(x)=0$ for $x \in X \sim A$, then

$$\lambda(g) \le \phi(A).$$

In fact, if $f_1, f_2, f_3, \ldots$ suits A, then $h_n=\inf\{f_n, g\} \uparrow g$ as $n \uparrow \infty$, hence

$$\lambda(g)=\lim_{n\to\infty} \lambda(h_n) \le \lim_{n\to\infty} \lambda(f_n).$$

Next we apply 2.3.2 (7) to show that *every member f of L^+ is a ϕ measurable function.* Assuming $T \subset X$ and $-\infty < a < b < \infty$, we must verify the inequality

$$\phi(T) \geq \phi(T \cap \{x\colon f(x) \leq a\}) + \phi(T \cap \{x\colon f(x) \geq b\}).$$

Since this is trivial if $a < 0$, we assume $a \geq 0$, suppose $g_1, g_2, g_3, \ldots$ suits T, and let

$$h = (b-a)^{-1}[\inf\{f, b\} - \inf\{f, a\}], \qquad k_n = \inf\{g_n, h\}.$$

Inasmuch as $0 \leq k_{n+1} - k_n \leq g_{n+1} - g_n$ and

$$h(x) = 1 \text{ whenever } f(x) \geq b, \qquad h(x) = 0 \text{ whenever } f(x) \leq a,$$

we find that the two sequences of functions k_n and $g_n - k_n$ suit the two sets

$$B = T \cap \{x\colon f(x) \geq b\} \quad \text{and} \quad A = T \cap \{x\colon f(x) \leq a\};$$

we conclude that

$$\lim_{n\to\infty} \lambda(g_n) = \lim_{n\to\infty} [\lambda(k_n) + \lambda(g_n - k_n)] \geq \phi(B) + \phi(A).$$

Now we shall prove that $\lambda(f) = \int f\, d\phi$ for $f \in L^+$. Letting

$$f_t = \inf\{f, t\} \text{ for } t \geq 0,$$

we observe that, if $\varepsilon > 0$ and k is a positive integer, then

$$0 \leq f_{k\varepsilon}(x) - f_{(k-1)\varepsilon}(x) \leq \varepsilon \text{ for } x \in X,$$

$$f_{k\varepsilon}(x) - f_{(k-1)\varepsilon}(x) = \varepsilon \text{ whenever } f(x) \geq k\,\varepsilon,$$

$$f_{k\varepsilon}(x) - f_{(k-1)\varepsilon}(x) = 0 \text{ whenever } f(x) \leq (k-1)\,\varepsilon.$$

Consequently

$$\begin{aligned}\lambda(f_{k\varepsilon} - f_{(k-1)\varepsilon}) &\geq \varepsilon\, \phi\{x\colon f(x) \geq k\,\varepsilon\}\\ &\geq \int (f_{(k+1)\varepsilon} - f_{k\varepsilon})\, d\phi\\ &\geq \varepsilon\, \phi\{x\colon f(x) \geq (k+1)\,\varepsilon\} \geq \lambda(f_{(k+2)\varepsilon} - f_{(k+1)\varepsilon})\end{aligned}$$

and, summing with respect to k from 1 to n, we find that

$$\lambda(f_{n\varepsilon}) \geq \int (f_{(n+1)\varepsilon} - f_\varepsilon)\, d\phi \geq \lambda(f_{(n+2)\varepsilon} - f_{2\varepsilon}).$$

Since $f_{n\varepsilon} \uparrow f$ as $n \uparrow \infty$, $\lambda(f) \geq \int (f - f_\varepsilon)\, d\phi \geq \lambda(f - f_{2\varepsilon})$, and therefore, since $f_\varepsilon \downarrow 0$ as $\varepsilon \downarrow 0$, $\lambda(f) = \int f\, d\phi$.

Finally, if $f \in L$, then $f^+ \in L^+$ and $f^- \in L^+$, hence

$$\lambda(f) = \lambda(f^+) - \lambda(f^-) = \int f^+\, d\phi - \int f^-\, d\phi = \int f\, d\phi.$$

2.5.3. There is usually more than one measure ϕ for which the conclusion of the preceding theorem holds (see 2.5.4, 2.5.14). However, *every such ϕ satisfies the equation*

$$\phi\{x: f(x)>t\} = \lim_{h\to 0+} h^{-1}\lambda[\inf\{f, t+h\} - \inf\{f, t\}]$$

whenever $f \in L^+$ and $t>0$. Thus λ uniquely determines the restriction of ϕ to the class F_0 of all these sets $\{x: f(x)>t\}$; using 2.1.3 (4) and (5), we infer the same for the classes

$$F_1 = \left\{\bigcup_{n=1}^{\infty} A_n\colon A_1 \subset A_2 \subset A_3 \subset \cdots \text{ belong to } F_0\right\},$$

$$F_2 = \left\{\bigcap_{n=1}^{\infty} A_n\colon A_1 \supset A_2 \supset A_3 \supset \cdots \text{ belong to } F_1,\ \phi(A_1)<\infty\right\}.$$

While these properties are shared by all measures ϕ satisfying the conclusion of 2.5.2, *the particular measure ϕ constructed in our proof may be characterized by the following additional property, which we call L* **regularity:**

$$\phi(A)<\infty \textit{ if and only if, for some } W,$$
$$A \subset W \in F_2 \quad \textit{and} \quad \phi(A) = \phi(W) < \infty.$$

We shall presently establish this property, and also show that our L regular measure ϕ satisfies the four further conditions:

If $\phi(A)<\infty$ and A is ϕ measurable, then $\phi(A)$ equals the supremum of the set of numbers $\phi(K)$ corresponding to all sets of the type

$$K = \{x: h(x) \geq 1\} \sim E \subset A, \textit{ with } h \in L^+ \textit{ and } E \in F_1.$$

For each nonnegative ϕ summable function u, and each $\varepsilon>0$, there exists a function $k \in L^+$ such that $\int |k-u|\, d\phi < 4\varepsilon$.

T is contained in the union of a countable family of sets with finite ϕ measure if and only if the characteristic function of T is less than or equal to the sum of a series of functions in L^+.

B is ϕ measurable if and only if $B \cap \{x: f(x) \geq 1\}$ is ϕ measurable for every $f \in L^+$.

Given that $\phi(A)<\infty$, we can choose sequences $f_{i,1}, f_{i,2}, f_{i,3}, \ldots$ which suit A and for which

$$\phi(A) = \lim_{i\to\infty} \lim_{j\to\infty} \lambda(f_{i,j}),$$

and define

$$g_{i,j} = \inf\{f_{1,j}, \ldots, f_{i,j}\} \in L^+,$$
$$B_{i,j} = \{x: g_{i,j}(x) > 1 - i^{-1}\} \in F_0;$$

then $g_{i+1,j} \le g_{i,j} \le g_{i,j+1}$, $B_{i+1,j} \subset B_{i,j} \subset B_{i,j+1}$,

$$(1-i^{-1})\,\phi(B_{i,j}) \le \lambda(g_{i,j}) \le \lambda(f_{i,j}),$$

$$A \subset V_i = \bigcup_{j=1}^{\infty} B_{i,j} \in F_1, \quad V_i \supset V_{i+1},$$

$$\phi(V_i) = \lim_{j\to\infty} \phi(B_{i,j}) \le (1-i^{-1})^{-1} \lim_{j\to\infty} \lambda(f_{i,j}),$$

$$A \subset W = \bigcap_{i=1}^{\infty} V_i \in F_2, \quad \phi(W) = \lim_{i\to\infty} \phi(V_i) \le \phi(A).$$

Now suppose also that A is ϕ measurable, a is the characteristic function of A, and $\varepsilon > 0$; choosing i and j so that

$$\phi(A \sim B_{i,j}) < \varepsilon \quad \text{and} \quad \lambda(h) \le \phi(A) + \varepsilon \quad \text{with} \quad h = (1-i^{-1})^{-1} g_{i,j},$$

we find that $B_{i,j} \subset C = \{x\colon h(x) \ge 1\}$,

$$\int (h-a)^- \, d\phi \le \phi(A \sim C) \le \phi(A \sim B_{i,j}) < \varepsilon,$$

$$\phi(C \sim A) \le \int |h-a| \, d\phi = \int (h-a) \, d\phi + 2 \int (h-a)^- \, d\phi < 3\varepsilon;$$

selecting $E \in F_1$ so that $C \sim A \subset E$ and $\phi(E) < 3\varepsilon$ we obtain

$$C \sim E \subset A \quad \text{with} \quad \phi[A \sim (C \sim E)] \le \phi(A \sim C) + \phi(E) < 4\varepsilon.$$

In case $u = a$ we take $k = h$; then we immediately pass to the case when u is a finite linear combination of characteristic functions, and use increasing convergence to treat any ϕ summable $u \ge 0$.

The third condition follows immediately from the definition of $\phi(T)$, and is helpful in verifying the fourth condition by means of 2.1.2 and 2.1.3 (2).

To see that the *fourth hypothesis of* 2.5.2 *does not follow from the first three*, we consider the **example** where X is the set of all positive integers, L is the set of all convergent sequences of real numbers, and

$$\lambda(f) = \lim_{n\to\infty} f(n) \quad \text{whenever} \quad f \in L.$$

2.5.4. We now illustrate the preceding theory by two simple examples (which depend on 2.5.17):

(1) Let $p\colon \mathbf{R}^2 \to \mathbf{R}$, $p(x) = x_1$ for $x = (x_1, x_2) \in \mathbf{R}^2$, let L_1 be the set of all functions $u \circ p$ corresponding to continuous $u\colon \mathbf{R} \to \mathbf{R}$, and let

$$\lambda_1(u \circ p) = \int_0^1 u \, d\mathscr{L}^1 \quad \text{whenever} \quad u \circ p \in L_1.$$

The L_1 regular measure ϕ_1 associated with λ_1 satisfies the equation

$$\phi_1(A) = \mathscr{L}^1[p(A) \cap \{t\colon 0 \le t \le 1\}] \quad \text{whenever} \quad A \subset \mathbf{R}^2.$$

(2) Let L_2 be the set of all continuous real valued functions on $\mathbf{R}^2$, and let

$$\lambda_2(f)=\int_0^1 f(t,0)\,d\mathscr{L}^1 t \text{ whenever } f\in L_2.$$

The L_2 regular measure ϕ_2 associated with λ_2 satisfies the equation

$$\phi_2(A)=\mathscr{L}^1[\xi^{-1}(A)\cap\{t:\ 0\leq t\leq 1\}] \text{ whenever } A\subset\mathbf{R}^2,$$

where $\xi: \mathbf{R}\to\mathbf{R}^2$, $\xi(t)=(t,0)$ for $t\in\mathbf{R}$.

We observe that all Borel sets of $\mathbf{R}^2$ are ϕ_2 measurable, but not all are ϕ_1 measurable; for instance $\mathbf{R}^2\cap\{x:\ x_2\geq 0\}$ is not ϕ_1 measurable. Also $L_1\subset L_2$, $\lambda_1=\lambda_2|L_1$ hence the conclusion of 2.5.2 holds with both L_1, λ_1, ϕ_1 and L_1, λ_1, ϕ_2; of course ϕ_2 is not L_1 regular.

2.5.5. Theorem. *Suppose L is a lattice of functions on X, μ maps L into $\mathbf{R}$, and satisfies the following conditions whenever $f, g, h_1, h_2, h_3, \ldots$ belong to L:*

$$\mu(f+g)=\mu(f)+\mu(g);$$

$$0\leq c<\infty \quad \textit{implies} \quad \mu(cf)=c\,\mu(f);$$

$$\sup\mu(L\cap\{k:\ 0\leq k\leq f\})<\infty;$$

$$h_n\uparrow g \textit{ as } n\uparrow\infty \quad \textit{implies} \quad \mu(h_n)\to\mu(g) \textit{ as } n\to\infty.$$

Let μ^+ and μ^- be the functions on L^+ such that

$$\mu^+(f)=\sup\mu(L\cap\{k:\ 0\leq k\leq f\}),$$

$$\mu^-(f)=-\inf\mu(L\cap\{k:\ 0\leq k\leq f\})$$

whenever $f\in L^+$. Then there exist L^+ regular measures ψ^+ and ψ^- over X such that

$$\mu^+(f)=\int f\,d\psi^+ \quad \textit{and} \quad \mu^-(f)=\int f\,d\psi^- \quad \textit{whenever } f\in L^+,$$

$$\mu(f)=\int f\,d\psi^+-\int f\,d\psi^- \quad \textit{whenever } f\in L.$$

Moreover, if $f\in L^+$, there exists a function h on X, which is the lower limit of a sequence of functions $k_n\in L^+$ with $k_n\leq f$, and for which

$$h(x)=f(x) \quad \textit{for } \psi^+ \textit{ almost all } x,$$

$$h(x)=0 \quad \textit{for } \psi^- \textit{ almost all } x.$$

Proof. If $f\in L^+$, then $f\geq g\in L^+$ implies $f\geq f-g\in L^+$, hence

$$\mu(g)-\mu^-(f)\leq\mu(g)+\mu(f-g)\leq\mu(g)+\mu^+(f);$$

therefore $\mu^+(f)-\mu^-(f)\leq\mu(f)\leq-\mu^-(f)+\mu^+(f)$, so that

$$\mu(f)=\mu^+(f)-\mu^-(f).$$

Next suppose f and g belong to L^+; if $f+g \ge h \in L^+$, then

$$f \ge k = \inf\{f, h\} \in L^+ \quad \text{and} \quad g \ge h-k \in L^+,$$

hence

$$\mu^+(f)+\mu^+(g) \ge \mu(k)+\mu(h-k)=\mu(h);$$

therefore $\mu^+(f)+\mu^+(g) \ge \mu^+(f+g)$; since the opposite inequality is obvious, we conclude that the function μ^+ is additive on L^+. Clearly μ^+ is positively homogeneous and monotone.

To show that μ^+ preserves increasing convergence, suppose $h_n \uparrow g$ as $n \uparrow \infty$, where g and all h_n belong to L^+; if $g \ge k \in L^+$, then $f_n = \inf\{h_n, k\} \uparrow k$ as $n \uparrow \infty$, hence

$$\mu(k) = \lim_{n\to\infty} \mu(f_n) \le \lim_{n\to\infty} \mu^+(h_n);$$

therefore $\mu^+(h_n) \uparrow \mu^+(g)$ as $n \uparrow \infty$.

Accordingly 2.5.2 and 2.5.3, applied with $\lambda = \mu^+$, furnish an L^+ regular measure ψ^+ over X such that $\mu^+(f) = \int f\, d\psi^+$ whenever $f \in L^+$. Similarly, since $\mu^- = (-\mu)^+$, we obtain an L^+ regular measure ψ^- over X such that $\mu^-(f) = \int f\, d\psi^-$ whenever $f \in L^+$.

Moreover, if $f \in L^+$, $f \ge g_n \in L^+$ and $\mu(g_n) \to \mu^+(f)$ as $n \to \infty$, then

$$\int |f-g_n|\, d\psi^+ = \mu^+(f)-\mu^+(g_n) \le \mu^+(f)-\mu(g_n)$$

and

$$\int |g_n|\, d\psi^- = \mu^-(g_n) = \mu^+(g_n)-\mu(g_n) \le \mu^+(f)-\mu(g_n)$$

approach 0 as n approaches ∞. According to 2.4.12 the sequence $g_1, g_2, g_3, \ldots$ converges in ψ^+ measure to f, and in ψ^- measure to 0. Applying 2.3.10 twice, we obtain a subsequence $k_1, k_2, k_3, \ldots$ such that

$$\lim_{n\to\infty} k_n(x) = f(x) \quad \text{for } \psi^+ \text{ almost all } x,$$

$$\lim_{n\to\infty} k_n(x) = 0 \qquad \text{for } \psi^- \text{ almost all } x.$$

Finally we define $h(x) = \liminf_{n\to\infty} k_n(x)$ for all $x \in X$.

2.5.6. A function μ satisfying the conditions of 2.5.5 will be called a **Daniell integral** on L. By a **monotone Daniell integral** we mean a function λ of the type considered in 2.5.2. In 2.5.7, 2.5.13 and 2.5.12 (which deals with vector valued functions) we shall see that Daniell integrals often appear in the guise of linear functions which are bounded with respect to the basic seminorms of classical analysis.

The initial part of the proof of 2.5.5 shows that the additivity of μ is sufficient for the equivalence

$$\mu^+(f) < \infty \text{ if and only if } \mu^-(f) < \infty$$

whenever $f \in L^+$; moreover one easily verifies that, *in case* L *is a vector space,*

$$\mu^+(f)+\mu^-(f)=\sup\{|\mu(k)|: k\in L \text{ and } |k|\le f\}.$$

One calls $\mu^+ + \mu^-$ the **variation integral,** and $\psi^+ + \psi^-$ the **variation measure,** corresponding to μ.

It is sometimes convenient to designate $\mu(f)$ by one of the alternate notations

$$\langle f,\mu\rangle, \quad \int f\,d\mu, \quad \int f(x)\,d\mu\,x, \quad \mu_x[f(x)].$$

2.5.7. Theorem. *Suppose* ϕ *measures* X, $1\le p\le\infty$ *and*

$$\mu: L=\mathbf{L}_p(\phi)\to\mathbf{R}$$

is a linear function for which $M=\sup\{|\mu(f)|: \phi_{(p)}(f)=1\}<\infty$. *Furthermore suppose*

either (i) $1<p<\infty$, $q=p/(p-1)$;

or (ii) $p=1$, $q=\infty$ *and* X *is countably* ϕ *measurable.*

or (iii) $p=\infty$, $q=1$ *and, whenever* $f_1, f_2, f_3, \ldots$ *and* g *belong to* L,

$$f_n\uparrow g \text{ as } n\uparrow\infty \quad \text{implies} \quad \mu(f_n)\to\mu(g) \text{ as } n\to\infty.$$

Then there exists a function $k\in\mathbf{L}_q(\phi)$ *such that*

$$\mu(f)=\int f k\,d\phi \text{ whenever } f\in L;$$

moreover $\phi_{(q)}(k)=M$, *and* k *is* ϕ *almost unique.*

Proof. Observing that μ is a Daniell integral on L, we define μ^+ and μ^- as in 2.5.5.

We let U be the subset of $\mathbf{L}_q(\phi)$ consisting of all those nonnegative functions u for which

$$\mu^+(f)\ge\int f u\,d\phi \text{ whenever } f\in L^+.$$

If $u\in U$, then $\int f u\,d\phi\le M\,\phi_{(p)}(f)$ for all $f\in L^+$; it follows from 2.4.16 that $\phi_{(q)}(u)\le M$.

If u and v belong to U, so does $w=\sup\{u,v\}$; in fact, if g is the characteristic function of $\{x: u(x)\ge v(x)\}$, then $f\in L^+$ implies

$$\mu^+(f)=\mu^+(fg)+\mu^+(f-fg)\ge\int f g u\,d\phi+\int(f-fg)v\,d\phi=\int f w\,d\phi.$$

If $u_1, u_2, u_3, \ldots$ belong to U and $u_n\uparrow v$ as $n\uparrow\infty$, then v also belongs to U.

We partially order U, agreeing that $u<v$ if and only if

$$\phi\{x: u(x)>v(x)\}=0 \quad\text{and}\quad \phi\{x: u(x)<v(x)\}>0,$$

and we define an increasing function $\tau: U \to \{t: 0 \leq t \leq M\}$ by the formulae

$$\tau(u) = \phi_{(q)}(u) \text{ in the cases (i) and (iii);}$$

$$\tau(u) = \sum_{i=1}^{\infty} 2^{-i} \phi(A_i)^{-1} \textstyle\int_{A_i} u \, d\phi \text{ in case (ii),}$$

where $A_1, A_2, A_3, \ldots$ are ϕ measurable sets, each of finite ϕ measure, with union X. From the properties of U established above, we deduce the existence of a function $k_+ \in U$ for which

$$\tau(k_+) = \sup \operatorname{im} \tau, \text{ hence never } k_+ < u \in U.$$

We claim that $\mu^+(f) = \int f k_+ \, d\phi$ whenever $f \in L^+$. If not, we could choose $g \in L^+$ with $\mu^+(g) > \int g k_+ \, d\phi$, and secure, by 2.4.16, a function r such that

$$0 \leq r \in \mathbf{L}_q(\phi), \quad \{x: r(x) > 0\} = \{x: g(x) > 0\},$$

$$\mu^+(g) - \textstyle\int g k_+ \, d\phi > \int g r \, d\phi > 0.$$

Defining the Daniell integral ν on L by the formula

$$\nu(f) = \mu^+(f) - \textstyle\int f(k_+ + r) \, d\phi \text{ for } f \in L,$$

we could use 2.5.5 to obtain L^+ regular measures σ^+ and σ^-, and a function $h \in L^+$, such that

$$\nu(f) = \textstyle\int f \, d\sigma^+ - \int f \, d\sigma^- \text{ whenever } f \in L,$$

$$0 \leq h \leq g, \quad \textstyle\int h \, d\sigma^+ = \nu^+(g), \quad \int h \, d\sigma^- = 0.$$

Letting s be the characteristic function of the ϕ measurable set

$$S = \{x: h(x) > 0\},$$

for which $\sigma^-(S) = 0$, we would find that

$$\mu^+(f) - \textstyle\int f(k_+ + s r) \, d\phi$$

$$= \mu^+(f - f s) - \textstyle\int (f - f s) k_+ \, d\phi + \nu(f s) \geq \nu^+(f s) \geq 0$$

whenever $f \in L^+$, hence $k_+ + s r \in U$. From the maximal nature of k_+ it would follow that

$$0 = \phi\{x: s(x) r(x) > 0\} = \phi\{x: h(x) > 0\},$$

hence $\phi_{(p)}(h) = 0$, $\mu^+(h) = 0$, $\nu(h) = 0$, which is incompatible with $\nu(h) = \nu^+(g) > 0$.

Similarly, since $\mu^- = (-\mu)^+$, we construct a function $k_- \in \mathbf{L}_q(\phi)$ for which

$$\mu^-(f) = \textstyle\int f k_- \, d\phi \text{ whenever } f \in L^+.$$

Clearly the principal conclusion of the theorem holds for $k = k_+ - k_-$, and the postscript follows from 2.4.16.

2.5.8. We observe that case (iii) of the preceding theorem implies the following form of the **Radon-Nikodym theorem:**

If ϕ measures X and μ is a Daniell integral on $\mathbf{L}_\infty(\phi)$, then there exists a function $k \in \mathbf{L}_1(\phi)$ such that

$$\mu(f) = \int f k \, d\phi \quad \textit{whenever} \quad f \in \mathbf{L}_\infty(\phi).$$

We must verify that $M < \infty$. Choosing a sequence of functions $f_n \in \mathbf{L}_\infty(\phi)$ with $\phi_{(\infty)}(f_n) \le 1$ and $|\mu(f_n)| \to M$ as $n \to \infty$, we let

$$g(x) = \sup\{|f_n(x)|: n = 1, 2, 3, \ldots\} \text{ for } x \in X,$$

infer that $g \in \mathbf{L}_\infty(\phi)$ with $\phi_{(\infty)}(g) \le 1$ and

$$|\mu(f_n)| \le \mu^+(|f_n|) + \mu^-(|f_n|) \le \mu^+(g) + \mu^-(g)$$

for $n = 1, 2, 3, \ldots$, hence conclude $M \le \mu^+(g) + \mu^-(g) < \infty$.

2.5.9. We shall use the following corollary of case (ii) of Theorem 2.5.7:

If λ is a monotone Daniell integral on L, ϕ is the L regular measure associated λ, X is countably ϕ measurable, and μ is a Daniell integral on L such that

$$|\mu(f)| \le \lambda(|f|) \quad \textit{whenever} \quad f \in L,$$

then there exists a ϕ measurable function k such that

$$\mu(f) = \int f k \, d\phi \quad \textit{whenever} \quad f \in L;$$

moreover k is ϕ almost unique, and $\phi_{(\infty)}(k) \le 1$.

The inference is trivial in case $L = \mathbf{L}_1(\phi)$. In general L may be a proper sublattice of $\mathbf{L}_1(\phi)$, but 2.5.3 shows that L is $\phi_{(1)}$ dense in $\mathbf{L}_1(\phi)$; therefore μ has a unique $\phi_{(1)}$ continuous extension μ_1 to $\mathbf{L}_1(\phi)$, for which

$$|\mu_1(f)| \le \phi_{(1)}(f) \text{ whenever } f \in \mathbf{L}_1(\phi).$$

2.5.10. Here we briefly discuss the necessity of the hypothesis "*X is countably ϕ measurable*" in case (ii) of 2.5.7. Omitting this hypothesis, we ask whether there always exists a ϕ measurable function k such that $\mu(f) = \int f k \, d\phi$ whenever f is ϕ summable; we do not require k to be countably ϕ measurable.

Example 2.5.11 shows that in general the answer is no.

However the answer is obviously yes when the omitted hypothesis is replaced by the following condition: *There exists a family G of disjoint ϕ measurable sets with finite ϕ measure such that every set with finite ϕ*

measure is ϕ almost contained in the union of some countable subfamily of G. Applying the theorem to each of the measures $\phi \llcorner A$ corresponding to $A \in G$, we obtain a function k_A vanishing outside A; then we let k be the sum of all the functions k_A.

In case ϕ is any Borel regular measure over a separable metric space X, the above condition is implied (see 2.2.3) *by the continuum hypothesis*

$$\aleph_1 = 2^{\aleph_0};$$

wellordering the class C of all closed sets with finite ϕ measure so that every proper initial segment is countable, we let G be the family of sets obtained by subtracting from each member of C the union of its predecessors.

A rather unexpected situation arises when the preceding result (assuming the continuum hypothesis) is applied to the following particular **example** (which depends on 2.10.2, 2.10.7, 2.10.11, 2.5.17, 2.6.5):

$$X = \mathbf{R}^2, \quad \phi = \mathscr{H}^1, \quad p: \mathbf{R}^2 \to \mathbf{R}, \quad p(x) = x_1 \text{ for } X \in \mathbf{R}^2,$$

$$\mu(f) = \int \sum_{p^{-1}\{t\}} f \, d\mathscr{L}^1 t \leq \int |f| \, d\mathscr{H}^1$$

whenever f is $\mathscr{H}^1$ summable. Reflecting briefly upon the geometric significance of a corresponding function k, the reader will agree that k cannot be $\mathscr{L}^2$ measurable. Of course such a function k is useless for any properly geometric investigation. I believe that this example clearly shows the superiority of the constructive process of derivation (2.9), which unfortunately applies only in case ϕ is locally finite.

The above condition also holds in case ϕ is a Radon measure. Here we let G be a maximal disjointed family of compact sets K such that

$$K \cap V = \varnothing \text{ or } \phi(K \cap V) > 0 \text{ for every open } V \subset X.$$

Since every set with finite ϕ measure is ϕ almost contained in the union of some countable family of compact sets, we need only verify that for each compact set C the family

$$G_C = G \cap \{K : K \cap C \neq \varnothing\}$$

is countable and covers ϕ almost all of C; now C is contained in some open set V with compact closure, hence $\phi(V) < \infty$, and G_C is contained in the countable set

$$G \cap \{K : \phi(K \cap V) > 0\};$$

if $\phi(C \sim \bigcup G_C) > 0$, then $C \sim \bigcup G_C$ would contain a compact set Y with $\phi(Y) > 0$, and consideration of the family $G \cup \{\operatorname{spt}(\phi \llcorner Y)\}$ would show that G is not maximal.

2.5.11. Example. For $j \in \{1, 2\}$ let Y_j be a set with cardinal number $\aleph_j$, let α_j be the counting measure over Y_j, and let β_j be the measure over Y_j such that

$$\beta_j(A)=0 \text{ whenever } A \subset Y_j \text{ and } \operatorname{card} A < \aleph_j,$$

$$\beta_j(A)=1 \text{ whenever } A \subset Y_j \text{ and } \operatorname{card} A = \aleph_j.$$

Over the cartesian product $X = Y_1 \times Y_2$ consider the product measures (see 2.6) $\alpha_1 \times \beta_2$ and $\beta_1 \times \alpha_2$, the function lattice

$$L = \mathbf{L}_1(\alpha_1 \times \beta_2) \cap \mathbf{L}_1(\beta_1 \times \alpha_2),$$

and the linear functionals λ and μ on L such that

$$\lambda(f) = \int f \, d(\alpha_1 \times \beta_2) + \int f \, d(\beta_1 \times \alpha_2), \qquad \mu(f) = \int f \, d(\alpha_1 \times \beta_2)$$

whenever $f \in L$; clearly $\mu(|f|) \le \lambda(|f|)$. Letting ϕ be the L regular measure associated with λ, we see from 2.5.3 that $\mathbf{L}_1(\phi) = L$, and we will show that *there exists no ϕ measurable function k such that*

$$\mu(f) = \int f k \, d\phi \textit{ whenever } f \in L.$$

Suppose k were such a function. For each $u \in Y_1$ the characteristic function f of $\{u\} \times Y_2$ would belong to L, and

$$1 = \mu(f) = \int_{\{u\} \times Y_2} k \, d\phi = \int k(u, v) \, d\beta_2 \, v,$$

the β_2 measurable set $\{v : k(u, v) > 0\}$ would have positive β_2 measure, its complement in Y_2 would have β_2 measure 0, hence

$$\operatorname{card}\{v : k(u, v) \le 0\} \le \aleph_1.$$

It would follow that

$$\operatorname{card}\{(u, v) : k(u, v) \le 0\} \le (\operatorname{card} Y_1) \cdot \aleph_1 = \aleph_1.$$

On the other hand, for each $v \in Y_2$ the characteristic function g of $Y_1 \times \{v\}$ would belong to L, and

$$0 = \mu(g) = \int_{Y_1 \times \{v\}} k \, d\phi = \int k(u, v) \, d\beta_1 \, u,$$

hence $\{u : k(u, v) \le 0\}$ would be nonempty. This would imply that

$$\operatorname{card}\{(u, v) : k(u, v) \le 0\} \ge \operatorname{card} Y_2 = \aleph_2.$$

2.5.12. Theorem. *Suppose:*

(1) *E is a vectorspace normed by ν, and separable.*

(2) *E^* is the space of all ν continuous linear maps of E into* $\mathbf{R}$, *with the norm ν^* (dual to ν) such that*

$$\nu^*(\alpha) = \sup\{\langle y, \alpha \rangle : \nu(y) = 1\} \textit{ whenever } \alpha \in E^*,$$

and with the weak topology generated by all sets

$$E^* \cap \{\alpha:\ a < \langle z, \alpha\rangle < b\}$$

corresponding to $z \in E$ and $a, b \in \mathbf{R}$.

(3) *L is a lattice of functions on X, and L^+ contains a countable set K for which*

$$\sum_{f\in K} f(x) \geq 1 \ \textit{whenever}\ x \in X.$$

(4) *Ω is a vectorspace of functions mapping X into E, such that*

$$f\in L,\ y\in E \quad \textit{implies}\quad f\cdot y\in\Omega;$$

$$\omega\in\Omega,\ \alpha\in E^* \quad \textit{implies}\quad \alpha\circ\omega\in L,\ \nu\circ\omega\in L;$$

$$\omega\in\Omega,\ \nu\circ\omega\geq f\in L^+ \ \textit{implies there exists}\ \xi\in\Omega\ \textit{with}$$

$$\nu\circ\xi = f,\ (\nu\circ\omega)\cdot\xi = f\cdot\omega.$$

(5) *T is a linear map of Ω into $\mathbf{R}$ such that, whenever $f\in L^+$ and $\xi_1, \xi_2, \xi_3, \ldots, \in \Omega$,*

$$\lambda(f) = \sup T(\Omega\cap\{\omega:\ \nu\circ\omega\leq f\}) < \infty,$$

$$\nu\circ\xi_n \downarrow 0 \ \textit{as}\ n\uparrow\infty \quad \textit{implies}\quad T(\xi_n)\to 0 \ \textit{as}\ n\to\infty.$$

Then λ is a monotone Daniell integral on L^+ and, if ϕ is the L^+ regular measure associated with λ, there exists a ϕ measurable function k with values in the weakly topologized space E^* such that $\nu^*\circ k$ is ϕ measurable,

$$\nu^*[k(x)] = 1 \ \textit{for}\ \phi\ \textit{almost all}\ x,$$

$$T(\omega) = \int \langle \omega(x), k(x)\rangle\, d\phi\, x \ \textit{whenever}\ \omega\in\Omega.$$

Such a function k is ϕ almost unique. Every member of Ω is ϕ measurable. For each ϕ measurable function η with values in E, the real valued function $\langle \eta, k\rangle$ is ϕ measurable; in case $\nu\circ\eta$ is ϕ summable, so is $\langle\eta, k\rangle$.

(One calls λ the **variation integral** corresponding to T, and ϕ the **variation measure.**)

Proof. To show that λ is additive, we suppose

$$f\in L^+,\quad g\in L^+,\quad \omega\in\Omega,\quad \nu\circ\omega\leq f+g;$$

letting $h = \inf\{\nu\circ\omega, f\}$, we choose $\xi\in\Omega$ with

$$\nu\circ\xi = h,\quad (\nu\circ\omega)\cdot\xi = h\cdot\omega,$$

and infer that $\nu\circ(\omega-\xi) = (\nu\circ\omega) - h \leq g$, hence

$$T(\omega) = T(\xi) + T(\omega-\xi) \leq \lambda(f) + \lambda(g);$$

therefore $\lambda(f+g)\leq\lambda(f)+\lambda(g)$; the opposite inequality is obvious.

Clearly λ is positively homogeneous and monotone. To prove that λ preserves increasing convergence, we suppose $h_n \uparrow g$ as $n \uparrow \infty$, where g and all h_n belong to L^+; for any $\omega \in \Omega$ such that $\nu \circ \omega \leq g$, we let $f_n = \inf\{h_n, \nu \circ \omega\}$, choose $\xi_n \in \Omega$ with

$$\nu \circ \xi_n = f_n, \quad (\nu \circ \omega) \cdot \xi_n = f_n \cdot \omega,$$

and infer that $\nu \circ (\omega - \xi_n) = (\nu \circ \omega) - f_n \downarrow 0$ as $n \uparrow \infty$, hence

$$T(\omega) = \lim_{n \to \infty} T(\xi_n) \leq \lim_{n \to \infty} \lambda(h_n);$$

therefore $\lambda(h_n) \uparrow \lambda(g)$ as $n \to \infty$.

Next we choose a countable dense subset D of E such that D is a vectorspace over the rational field $\mathbf{Q}$. For each $y \in D$ we define

$$\mu_y\colon L^+ \to \mathbf{R}, \quad \mu_y(f) = T(f \cdot y) \text{ whenever } f \in L^+,$$

note that $\nu \circ (f \cdot y) = \nu(y) \cdot f$, $\mu_y(f) \leq \nu(y) \cdot \lambda(f)$, hence that μ_y is a Daniell integral on L^+; since X is countably ϕ measurable we obtain from 2.5.9 a function $k_y \in \mathbf{L}_\infty(\phi)$ for which

$$\mu_y(f) = \int f\, k_y \, d\phi \text{ whenever } f \in L^+;$$

moreover k_y is ϕ almost unique, and $\phi_{(\infty)}(k_y) \leq \nu(y)$. We readily infer that ϕ almost all points x of X satisfy the conditions

$$k_{ry+sz}(x) = r\, k_y(x) + s\, k_z(x), \quad |k_y(x)| \leq \nu(y)$$

for all $y, z \in D$ and $r, s \in \mathbf{Q}$; whenever this is the case, the map sending y into $k_y(x)$ is $\mathbf{Q}$ linear and ν bounded, hence has a unique extension

$$k(x) \in U^* = E^* \cap \{\alpha\colon \nu^*(\alpha) \leq 1\}.$$

Associating $\beta_y \in U^*$ with $y \in D$ so that $\langle y, \beta_y \rangle = \nu(y)$, we observe that

$$\nu(z) = \sup\{\langle z, \alpha \rangle\colon \alpha \in U^*\} = \sup\{\langle z, \beta_y \rangle\colon y \in D\}$$

whenever $z \in E$, because $\langle z, \beta_y \rangle = \nu(y) - \langle y - z, \beta_y \rangle \geq \nu(y) - \nu(y - z)$; consequently

$$\{z\colon \nu(z-p) < \delta\} = \bigcup_{n=1}^{\infty} \bigcap_{y \in D} \{z\colon \langle z - p, \beta_y \rangle < \delta - 1/n\}$$

whenever $p \in D$ and $\delta > 0$. We infer that every member ω of Ω is ϕ measurable; in fact $\beta_y \circ \omega \in L$ for each $y \in D$, and E is separable.

We observe that the sets

$$U^* \cap \{\alpha\colon a < \langle y, \alpha \rangle < b\}$$

corresponding to $y \in D$ and $a, b \in \mathbf{Q}$ form a countable family generating the weak topology on U^*. Therefore, since each of the functions $\langle y, k \rangle = k_y$

is ϕ measurable, so is k. Moreover $v^* \circ k$ is ϕ measurable because

$$v^*[k(x)] = \sup\{\langle y, k(x)\rangle : y \in D,\ v(y) < 1\}$$

for $x \in \operatorname{dmn} k$.

Using 2.3.2 (4), (5) we derive the remaining conclusions about ϕ measurability and summability from the fact that the map

$$\langle\ ,\ \rangle : E \times U^* \to \mathbf{R}$$

is continuous with respect to the product of the v topology on E and the weak topology on U^*, because

$$|\langle z, \alpha\rangle - \langle p, q\rangle| \le v(z-p) + |\langle p, \alpha - q\rangle|$$

whenever $z, p \in E$ and $\alpha, q \in U^*$.

We now define

$$G = \Omega \cap \{\omega : T(\omega) = \textstyle\int \langle \omega, k\rangle \, d\phi\},$$

and will show that $G = \Omega$.

Clearly $f \cdot y \in G$ whenever $f \in L^+$, $y \in D$.

Next we shall prove that the conditions

$$\omega \in \Omega, \qquad f \in L^+, \qquad \{x : \omega(x) \neq 0\} \subset S = \{x : f(x) \ge 1\}$$

imply $\omega \in G$. Since ω is ϕ measurable and $v \circ \omega$ is ϕ summable we can, for any $\varepsilon > 0$, choose $y_1, \ldots, y_m \in D$ and disjoint ϕ measurable sets T_0, $T_1, \ldots, T_m$ with union S such that

$$v[\omega(x) - y_i] < \varepsilon \text{ whenever } i = 1, \ldots, m \text{ and } x \in T_i,$$

$$\textstyle\int_{T_0} (v \circ \omega)\, d\phi < \varepsilon.$$

Letting t_i be the characteristic function of T_i we use 2.5.3 to obtain $h_i \in L^+$ with

$$v(y_i) \cdot \textstyle\int |t_i - h_i| \, d\phi < \varepsilon/m \text{ for } i = 1, \ldots, m.$$

We infer that

$$\xi = \sum_{i=1}^m t_i \cdot y_i \text{ is } \phi \text{ measurable}, \qquad \eta = \sum_{i=1}^m h_i \cdot y_i \in G,$$

$$\int v \circ (\omega - \xi)\, d\phi \le \varepsilon + \sum_{i=1}^m \varepsilon\, \phi(T_i) \le \varepsilon[1 + \phi(S)],$$

$$\int v \circ (\xi - \eta)\, d\phi \le \sum_{i=1}^m v(y_i) \int |t_i - h_i|\, d\phi \le \varepsilon,$$

$$\begin{aligned} |T(\omega) - \textstyle\int \langle \omega, k\rangle\, d\phi| &= |T(\omega - \eta) - \textstyle\int \langle \omega - \eta, k\rangle\, d\phi| \\ &\le 2 \textstyle\int v \circ (\omega - \eta)\, d\phi \le 2\varepsilon[2 + \phi(S)]. \end{aligned}$$

Hence our free choice of ε allows is to conclude that $\omega \in G$.

Finally, for an arbitrary $\omega \in \Omega$, we consider the sequence of functions

$$g_n = \inf\{\nu \circ \omega, n^{-1}\} \in L^+$$

and choose $\xi_n \in \Omega$ with $\nu \circ \xi_n = g_n$, $(\nu \circ \omega) \cdot \xi_n = g_n \cdot \omega$. Then $\omega - \xi_n \in \Omega$, $n\, g_n \in L^+$ and

$$\{x\colon (\omega - \xi_n)(x) \neq 0\} \subset \{x\colon n\, g_n(x) = 1\},$$

hence $\omega - \xi_n \in G$ by the preceding paragraph; therefore

$$T(\omega) - \int \langle \omega, k \rangle \, d\phi = T(\xi_n) - \int \langle \xi_n, k \rangle \, d\phi .$$

Moreover the right member of the last equation approaches 0 as n approaches ∞, because $|\langle \xi_n, k \rangle| \leq \nu \circ \xi_n \downarrow 0$ as $n \uparrow \infty$.

Having shown that

$$T(\omega) = \int \langle \omega, k \rangle \, d\phi \leq \int (\nu \circ \omega) \cdot (\nu^* \circ k) \, d\phi$$

whenever $\omega \in \Omega$, we infer that $\phi\{x\colon \nu^*[k(x)] < 1\} = 0$, because otherwise we could choose $f \in L^+$ with

$$\int f \cdot (\nu^* \circ k) \, d\phi < \int f \, d\phi = \lambda(f),$$

and then pick $\omega \in \Omega$ so that $\nu \circ \omega \leq f$ and

$$T(\omega) > \int f \cdot (\nu^* \circ k) \, d\phi \geq \int (\nu \circ \omega) \cdot (\nu^* \circ k) \, d\phi .$$

2.5.13. Riesz's representation theorem. *Suppose X is a locally compact Hausdorff space and L is the set of all those continuous maps $f\colon X \to \mathbf{R}$ whose support*

$$\operatorname{spt} f = \operatorname{Closure}\{x\colon f(x) \neq 0\}$$

is compact. If $\mu\colon L \to \mathbf{R}$ is a linear function such that

$$\sup \mu(L \cap \{g\colon 0 \leq g \leq f\}) < \infty \quad \text{whenever } f \in L^+,$$

then μ is a Daniell integral on L, hence there exist L regular measures ψ^+ and ψ^- over X such that

$$\mu(f) = \int f \, d\psi^+ - \int f \, d\psi^- \quad \text{whenever } f \in L .$$

Proof. To verify the last hypothesis of 2.5.5, we suppose $g, h_1, h_2, h_3, \ldots$ belong to L^+ and $h_n \uparrow g$ as $n \uparrow \infty$. Since some neighborhood of $\operatorname{spt} g$ is completely regular, there exists a function $f \in L^+$ such that $f(x) = 1$ whenever $x \in \operatorname{spt} g$; then (see 2.5.6)

$$c = \sup\{|\mu(k)|\colon k \in L \text{ and } 0 \leq k \leq f\} < \infty .$$

For each $\varepsilon > 0$, the intersection of the compact sets

$$S_n = \{x: g(x) \geq h_n(x) + \varepsilon\}$$

is empty; since $S_n \supset S_{n+1}$ for all n, it follows that $S_n = \emptyset$ when n is sufficiently large; but $S_n = \emptyset$ implies

$$0 \leq g - h_n \leq \varepsilon f, \quad |\mu(g - h_n)| \leq \varepsilon c.$$

2.5.14. Suppose X, L are as in 2.5.13, λ is a monotone Daniell integral on L, and ϕ is the L regular measure associated with λ. From 2.5.3 we infer that

$$\phi(A) = \inf\{\phi(V): A \subset V, V \text{ is open and } \phi \text{ measurable}\}$$

for every $A \subset X$; in case A is ϕ measurable and $\phi(A) < \infty$, then also

$$\phi(A) = \sup\{\phi(K): K \subset A, K \text{ is compact and } \phi \text{ measurable}\}.$$

Moreover, if K is any compact subset of X, then $\phi(K) < \infty$; however we shall see in 2.5.15 that K need not be ϕ measurable. Thus ϕ may fail to be a Radon measure. Since the ϕ measurable open sets form a base for the topology of X, ϕ is a Radon measure in case the topology of X has a countable base. In 2.7.15 we will prove that ϕ is a Radon measure in case ϕ is covariant with respect to a group action.

A Radon measure $\tilde{\phi}$, such that

$$\int f \, d\tilde{\phi} = \lambda(f) = \int f \, d\phi \text{ whenever } f \in L,$$

may always be constructed as follows: For every open set V let

$$\alpha(V) = \sup \lambda(L^+ \cap \{f: f \leq 1 \text{ and } \operatorname{spt} f \subset V\});$$

for every $A \subset X$ let

$$\tilde{\phi}(A) = \inf\{\alpha(V): A \subset V, V \text{ is open}\}.$$

Clearly $\tilde{\phi}$ $V) = \alpha(V) \leq \phi(V)$ for every open set V, hence

$$\tilde{\phi}(A) \leq \phi(A) \text{ whenever } A \subset X.$$

If $f \in L^+$, $s > 0$, $U = \{x: f(x) > s\}$ and $t > s$, then

$$\alpha(U) \geq h^{-1} \lambda[\inf\{f, t+h\} - \inf\{f, t\}]$$

whenever $h > 0$; consequently $\tilde{\phi}(U) = \phi(U)$. Next we verify that, for each sequence of open sets V_n,

$$W = \bigcup_{n=1}^{\infty} V_n \text{ implies } \alpha(W) \leq \sum_{n=1}^{\infty} \alpha(V_n);$$

if $f \in L^+$, $f \leq 1$, $\operatorname{spt} f \subset W$, there exist finitely many functions $g_1, \ldots, g_N \in L^+$ with $g_n \leq 1$, $\operatorname{spt} g_n \subset V_n$ and

$$\sum_{n=1}^{N} g_n(x) = 1 \text{ whenever } x \in \operatorname{spt} f,$$

hence $\lambda(f) = \sum_{n=1}^{N} \lambda(g_n f) \leq \sum_{n=1}^{N} \alpha(V_n)$. It follows that $\tilde{\phi}$ measures X. To show that each open set V is $\tilde{\phi}$ measurable, we suppose $\tilde{\phi}(T) < \infty$ and $\varepsilon > 0$; choosing an open set W containing T with $\alpha(W) < \tilde{\phi}(T) + \varepsilon$, and a function $f \in L^+$ with $f \leq 1$, $\operatorname{spt} f \subset W \cap V$, $\lambda(f) > \alpha(W \cap V) - \varepsilon$, we find that

$$\tilde{\phi}(T) + \varepsilon > \alpha(W) \geq \lambda(f) + \alpha(W \sim \operatorname{spt} f) \geq \tilde{\phi}(T \cap V) - \varepsilon + \tilde{\phi}(T \sim V).$$

It is now obvious that $\tilde{\phi}$ has all the required properties.

The last paragraph of 2.5.10 is applicable with ϕ replaced by $\tilde{\phi}$. Hence we can modify 2.5.12 for the present lattice L, removing the hypothesis (3), and replacing ϕ measurability by $\tilde{\phi}$ measurability in the conclusions; however k need not be $\tilde{\phi}$ almost unique, only locally so. Moreover the hypothesis (5) is here redundant, because the argument used in 2.5.13 may be applied with $g - h_n$, c, μ replaced by $\nu \circ \xi_n$, $\lambda(f)$, T.

I do not know whether the replacement of ϕ by $\tilde{\phi}$ is really necessary for the removal of the hypothesis (3) in 2.5.12.

The function lattice described in 2.5.13 is often denoted

$$\mathscr{K}(X).$$

The **support of a Daniell integral** μ on $\mathscr{K}(X)$ is the closed set

$$\operatorname{spt} \mu = X \sim \bigcup \{W\colon W \text{ is open, } \mu(f) = 0 \text{ whenever } \operatorname{spt} f \subset W\}.$$

2.5.15. Example. Suppose $Y = \{y\colon -1 \leq y \leq 1\}$, J is uncountable, $X = Y^J$ with the cartesian product topology, $p_j\colon X \to Y$ is the projection corresponding to $j \in J$, ξ is the point of X such that $p_j(\xi) = 0$ for all $j \in J$, λ is the Daniell integral on $\mathscr{K}(X)$ such that

$$\lambda(f) = f(\xi) \text{ whenever } f \in \mathscr{K}(X),$$

and ϕ is the $\mathscr{K}(X)$ regular measure associated with λ.

The compact set $\{\xi\}$ is not ϕ measurable because

$$\phi(X) = \phi(\{\xi\}) = \phi(X \sim \{\xi\}) = 1.$$

To prove the last equation one may use the fact that for each $f \in \mathscr{K}(X)$ there exists (by the Weierstrass-Stone theorem) a countable set $S \subset J$ such that $f(x)$ depends only on the coordinates $p_j(x)$ corresponding to

$j \in S$; this property subsists when f is the limit of a sequence of functions in $\mathscr{K}(X)$; in particular, if

$$X \sim \{\xi\} \subset \{x: f(x) \geq 1\},$$

we can pick $i \in J \sim S$, then infer from

$$\varnothing \neq \{x: p_i(x) \neq 0\} \subset X \sim \{\xi\},$$

and the fact that $f(x)$ is independent of $p_i(x)$, that $f(\xi) \geq 1$.

2.5.16. Suppose g maps $\mathbf{R}$ into a complete space Y metrized by ρ. Whenever $a \leq b$ the **variation** (or **length**) **of** g **from** a **to** b, denoted

$$\mathbf{V}_a^b g, \quad \text{or} \quad \mathbf{V}_{x=a}^{\;\;b} g(x),$$

is defined as the supremum of the set of numbers

$$\sum_{i=1}^{n} \rho[g(t_i), g(t_{i+1})]$$

corresponding to all finite sequences $a = t_1 \leq t_2 \leq \cdots \leq t_n \leq t_{n+1} = b$.

In case $a > b$ one defines $\mathbf{V}_a^b g = -\mathbf{V}_b^a g$.

We note that $\mathbf{V}_a^b g + \mathbf{V}_b^c g = \mathbf{V}_a^c g$ whenever $a \leq b \leq c$.

In case $\mathbf{V}_a^b g < \infty$ whenever $-\infty < a < b < \infty$, one may consider the nondecreasing function

$$s: \mathbf{R} \to \mathbf{R}, \quad s(x) = \mathbf{V}_0^x g \text{ for } x \in \mathbf{R}.$$

For each $x \in \mathbf{R}$, $s(x-) \leq s(x) \leq s(x+)$, where

$$s(x-) = \sup\{s(t): t < x\} = \lim_{t \to x-} s(t),$$

$$s(x+) = \inf\{s(t): t > x\} = \lim_{t \to x+} s(t),$$

and the completeness of Y implies the existence of

$$g(x-) = \lim_{t \to x-} g(t) \quad \text{and} \quad g(x+) = \lim_{t \to x+} g(t),$$

(because $\rho[g(t), g(u)] \leq s(x-) - s(t)$ whenever $t < u < x$) with

$$s(x) - s(x-) = \rho[g(x-), g(x)], \qquad s(x+) - s(x) = \rho[g(x), g(x+)].$$

Therefore g and s have the same points of discontinuity, and to each point x of discontinuity corresponds the nonempty open interval

$$\{z: s(x-) < z < s(x+)\}.$$

Since the intervals corresponding to distinct points of discontinuity are disjoint, they form a countable family.

Similarly we find that

$$\text{either } \sup \operatorname{im} s = \infty \text{ or } g(\infty) = \lim_{x \to \infty} g(x) \in Y;$$

$$\text{either } \inf \operatorname{im} s = -\infty \text{ or } g(-\infty) = \lim_{x \to -\infty} g(x) \in Y.$$

Since $\rho[g(t), g(u)] \le |s(t) - s(u)|$ whenever $t, u \in \mathbf{R}$ there exists a unique Lipschitzian function

$$G: \operatorname{im} s \to Y \quad \text{with} \quad \operatorname{Lip}(G) \le 1 \quad \text{and} \quad G \circ s = g.$$

In case Y is a Banach space with a norm inducing ρ, we can extend G to a Lipschitzian map

$$H: \mathbf{R} \to Y \quad \text{with} \quad \operatorname{Lip}(H) = \operatorname{Lip}(G),$$

by defining

$$H[\lambda\, s(x-) + (1-\lambda)\, s(x)] = \lambda\, g(x-) + (1-\lambda)\, g(x),$$

$$H[\lambda\, s(x) + (1-\lambda)\, s(x+)] = \lambda\, g(x) + (1-\lambda)\, g(x+)$$

whenever $x \in \mathbf{R}$ and $0 \le \lambda \le 1$, and

$$H(z) = g(-\infty) \text{ whenever } -\infty < z \le \inf \operatorname{im} s,$$

$$H(z) = g(\infty) \text{ whenever } \sup \operatorname{im} s \le z < \infty.$$

We observe that $\operatorname{im} g = \operatorname{im} G$ is separable.

The assumption that Y be a Banach space is not as restrictive as it may seem, because for any metric space Y we can consider the Banach space E of all bounded real valued continuous functions on Y with the norm

$$\nu(f) = \sup \operatorname{im} |f| \text{ whenever } f \in E,$$

choose any base point $\eta \in Y$, and construct the isometric embedding

$$F: Y \to E, \quad F(y)(w) = \rho(y, w) - \rho(\eta, w) \text{ for } y, w \in Y.$$

2.5.17. Suppose $g: \mathbf{R} \to \mathbf{R}$ with

$$\mathbf{V}_a^b g < \infty \text{ whenever } -\infty < a < b < \infty.$$

Whenever $f \in \mathscr{K}(\mathbf{R})$ and $-\infty < a \le b < \infty$ we consider for each $\delta > 0$ the set of numbers

$$\sum_{i=1}^{n} f(u_i) \cdot [g(t_{i+1}) - g(t_i)]$$

corresponding to all finite sequences

$$a = t_1 \le u_1 \le t_2 \le u_2 \le t_3 \le \cdots t_n \le u_n \le t_{n+1} = b$$

such that $t_{i+1}-t_i<\delta$ for $i=1,\ldots,n$; the diameter of this set does not exceed

$$2\sup\{|f(u)-f(v)|\colon |u-v|\le\delta\}\cdot \mathbf{V}_a^b g,$$

hence approaches 0 as δ approaches 0; consequently the above sums converge, as $\delta\to 0+$, to a real number, denoted

$$\textstyle\int_a^b f\,dg, \quad\text{or}\quad \int_a^b f(x)\,d_x g(x).$$

The operation just described is called **Riemann-Stieltjes integration;** it is linear with respect to f, and satisfies the inequality

$$\textstyle|\int_a^b f\,dg|\le\sup\{|f(x)|\colon a\le x\le b\}\cdot\mathbf{V}_a^b g.$$

We note that $\int_a^b f\,dg+\int_b^c f\,dg=\int_a^c f\,dg$ whenever $a\le b\le c$.

Defining the linear function $\mu\colon \mathscr{K}(\mathbf{R})\to\mathbf{R}$ so that

$$\textstyle\mu(f)=\int_a^b f\,dg$$

whenever $f\in\mathscr{K}(\mathbf{R})$ and $\operatorname{spt} f\subset\{x\colon a\le x\le b\}$, we see from 2.5.13, 2.5.6, 2.5.3 that there exist Borel regular measures ψ^+ and ψ^- satisfying the conditions:

(1) $\mu(f)=\int f\,d\psi^+-\int f\,d\psi^-$ whenever $f\in\mathscr{K}(\mathbf{R})$,

(2) $\psi^+\{x\colon a<x<b\}+\psi^-\{x\colon a<x<b\}\le\mathbf{V}_a^b g$ for $a<b$.

From these conditions it may be inferred that

$$\textstyle\int_a^b f\,dg=\int_{\{x:\,a\le x\le b\}} f\,d\psi^+-\int_{\{x:\,a\le x\le b\}} f\,d\psi^-$$

provided g is continuous at a and b, $a<b$, $f\in\mathscr{K}(\mathbf{R})$; in particular

$$g(b)-g(a)=\psi^+\{x\colon a\le x\le b\}-\psi^-\{x\colon a\le x\le b\}.$$

It follows that, if g is a continuous function, then equality holds in (2).

We shall refer to μ, ψ^+, ψ^- as the **Daniell integral and measures induced by** g. Two such functions g_1 and g_2 induce the same Daniell integral if and only if there exists a real number c such that

$$g_1(x)-g_2(x)=c \text{ for all but countably many } x \text{ in } \mathbf{R}.$$

In case the function g is nondecreasing, the induced Daniell integral μ is monotone, hence $\psi^-=0$.

For example, if $g(x)=x$ for all $x\in\mathbf{R}$, then μ is the Riemann integral, and ψ^+ is the 1 dimensional **Lebesgue measure**

$$\mathscr{L}^1$$

over $\mathbf{R}$; this is the Borel regular measure characterized by the property that

$$b-a=\mathscr{L}^1\{x\colon a\le x\le b\} \text{ whenever } -\infty<a<b<\infty.$$

2.5.18. We note the following additional simple facts concerning Riemann-Stieltjes integration:

(1) *If* $f: \mathbf{R} \to \mathbf{R}$ *and* $g: \mathbf{R} \to \mathbf{R}$ *are continuous,* $-\infty < a < b < \infty$,

$$\mathbf{V}_a^b f < \infty \quad \text{and} \quad \mathbf{V}_a^b g < \infty,$$

then

$$\int_a^b f\,dg + \int_a^b g\,df = f(b)\,g(b) - f(a)\,g(a).$$

In fact, for $a = t_1 \leq t_2 \leq \cdots \leq t_{n+1} = b$,

$$\sum_{j=1}^{n} f(t_{j+1})[g(t_{j+1}) - g(t_j)] + \sum_{j=1}^{n} g(t_j)[f(t_{j+1}) - f(t_j)]$$
$$= f(b)\,g(b) - f(a)\,g(a).$$

(2) *If* f, g, h *map* $\mathbf{R}$ *into* $\mathbf{R}$, f *and* g *are continuous,* g *is increasing,* $-\infty < a < b < \infty$ *and*

$$\mathbf{V}_{g(a)}^{g(b)} h < \infty,$$

then

$$\int_a^b (f \circ g)\,d(h \circ g) = \int_{g(a)}^{g(b)} f\,dh.$$

In fact, both integrals are limits of the sums

$$\sum_{j=1}^{n} f[g(t_j)] \cdot (h[g(t_{j+1})] - h[g(t_j)]).$$

Therefore $g_\#$ *maps the measures induced by* $h \circ g$ *onto the measures induced by* h.

(3) *If* ϕ *measures* X, $\phi(X) < \infty$, u *is an* $\mathbf{R}$ *valued* ϕ *measurable function and*

$$g(t) = \phi\{x: u(x) < t\} \quad \text{whenever } t \in \mathbf{R},$$

then g *is nondecreasing and*

$$\int (f \circ u)\,d\phi = \int_a^b f\,dg$$

whenever $f \in \mathscr{K}(\mathbf{R})$, $\operatorname{spt} f \subset \{t: a \leq t \leq b\}$. (One calls g the ϕ **distribution function** of u.)

For the proof we define ψ^+ as in 2.5.17 and observe that

$$u_\#(\phi)\{t: a < t < b\} = g(b) - g(a) = \psi^+\{t: a < t < b\}$$

whenever $-\infty < a < b < \infty$ and g is continuous at a. Consequently $u_\#(\phi)$ and ψ^+ agree on all Borel sets, and the asserted equation follows from 2.4.18.

2.5.19. For any lattice L of functions on a set X, the set L^* of all Daniell integrals on L is a vectorspace. The inclusion of L^* in the cartesian product $\mathbf{R}^L$, the set of all real valued functions on L, endows L^* with its so-called **weak topology.** Also there is a map $\delta: X \to L^*$ which associates with each $x \in X$ the monotone Daniell integral δ_x such that

$$\delta_x(f) = f(x) \text{ whenever } f \in L.$$

The vectorspace generated by the image of δ is dense in L^*.

In case X is a locally compact Hausdorff space and $L = \mathscr{K}(X)$, the map δ embeds X homeomorphically in $\mathscr{K}(X)^*$. If X is compact, then the closure of the convex hull of the image of δ equals the set of those monotone Daniell integrals on $\mathscr{K}(X)$ whose associated measures have the value 1 at X. For each closed subset S of X,

$$\mathscr{K}(X)^* \cap \{\mu: \operatorname{spt} \mu \subset S\}$$

is relatively closed in $\mathscr{K}(X)^*$. If $M: \mathscr{K}(X) \to \{t: 0 \leq t < \infty\}$, then

$$\mathscr{K}(X)^* \cap \{\mu: \mu^+ + \mu^- \leq M\} \text{ is compact}.$$

If X is a compact Hausdorff space, then the above defined weak topology coincides with the weak topology [see 2.5.12 (2)] of $\mathscr{K}(X)^*$ as the conjugate space of the vectorspace $\mathscr{K}(X)$ relative to the norm ν defined by the formula

$$\nu(f) = \sup \operatorname{im} |f| \text{ whenever } f \in \mathscr{K}(X).$$

Suppose $u: X \to Y$ is a continuous map of locally compact Hausdorff spaces. If u is proper, there exists a unique continuous linear map $u_\#$ such that the following diagram is commutative:

$$\begin{array}{ccc} X & \xrightarrow{u} & Y \\ \downarrow{\scriptstyle \delta^X} & & \downarrow{\scriptstyle \delta^Y} \\ \mathscr{K}(X)^* & \xrightarrow[u_\#]{} & \mathscr{K}(Y)^*. \end{array}$$

In fact $\langle f, u_\#(\mu)\rangle = \langle f \circ u, \mu\rangle$ for $f \in \mathscr{K}(Y)$, $\mu \in \mathscr{K}(X)^*$. Applying 2.4.18 to the $\mathscr{K}(X)$ regular measure ϕ associated with a monotone Daniell integral λ on $\mathscr{K}(X)$, we obtain

$$\langle f, u_\#(\lambda)\rangle = \int f \, d(u_\# \phi) \text{ for } f \in \mathscr{K}(Y);$$

however $u_\#(\phi)$ need not be $\mathscr{K}(Y)$ regular.

2.5.20. Here we apply Theorem 2.5.12 to the **decomposition of Daniell integrals.** Suppose X and Y are compact metric spaces and

$$\mu \text{ is a Daniell integral on } \mathscr{K}(X \times Y).$$

We take $L = \mathscr{K}(X)$, $E = \mathscr{K}(Y)$ with the norm

$$\nu(\eta) = \sup\{|\eta(y)|: y \in Y\} \text{ for } \eta \in \mathscr{K}(Y),$$

and let Ω be the vectorspace of all ν continuous maps of X into $\mathscr{K}(Y)$. Using the linear isomorphism

$$S: \Omega \simeq \mathscr{K}(X \times Y)$$

such that $S(\omega)(x, y) = \omega(x)(y)$ whenever $\omega \in \Omega$, $x \in X$, $y \in Y$, we define

$$T = \mu \circ S: \Omega \to \mathbf{R}.$$

Then the theorem yields a Radon measure ϕ over X, and a ϕ measurable function k with values in the space of Daniell integrals on $\mathscr{K}(Y)$ such that

$$\mu(u) = \int k(x)_y[u(x, y)]\, d\phi\, x$$

whenever $u \in \mathscr{K}(X \times Y)$. It may be inferred that

$$\operatorname{spt} k(x) \subset \{y: (x, y) \in \operatorname{spt} \mu\} \text{ for } \phi \text{ almost all } x.$$

Now suppose γ is a Daniell integral on $\mathscr{K}(Y)$ and $g: Y \to X$ is continuous. We consider the continuous map

$$h: Y \to X \times Y, \quad h(y) = (g(y), y) \text{ for } y \in Y,$$

and the Daniell integral $\mu = h_\#(\gamma)$ on $\mathscr{K}(X \times Y)$. If $f \in \mathscr{K}(Y)$ the preceding result is applicable with $u(x, y) = f(y)$, hence $\mu(u) = \gamma(u \circ h) = \gamma(f)$, and yields the formula

$$\gamma(f) = \int \langle k(x), f \rangle\, d\phi\, x.$$

Also, since $\operatorname{spt} \mu \subset \operatorname{im} h$, $\operatorname{spt} k(x) \subset g^{-1}\{x\}$ for ϕ almost all x. Moreover it may be verified that ϕ is the $\mathscr{K}(X)$ regular measure associated with $g_\#(\gamma^+ + \gamma^-)$.

2.6. Product measures

2.6.1. When α measures X and β measures Y, we define the function

$$\alpha \times \beta: \mathbf{2}^{X \times Y} \to \overline{\mathbf{R}} \cap \{t: t \geq 0\}$$

so that, for any $S \subset X \times Y$, $(\alpha \times \beta) S$ is the infimum of the numbers

$$\sum_{j=1}^{\infty} \alpha(A_j) \cdot \beta(B_j)$$

corresponding to all sequences of α measurable sets A_j and β measurable sets B_j with

$$S \subset \bigcup_{j=1}^{\infty} A_j \times B_j.$$

(Here we use the convention $0 \cdot \infty = \infty \cdot 0 = 0$.)

Clearly $\alpha \times \beta$ measures $X \times Y$, and

$$(\alpha \times \beta)(A \times B) \leq \alpha(A) \cdot \beta(B)$$

whenever A is α measurable and B is β measurable; in fact $\alpha \times \beta$ is the largest measure satisfying this inequality. One calls $\alpha \times \beta$ the **Cartesian product** of α and β.

2.6.2. Fubini's theorem. *Suppose α measures X, and β measures Y.*

(1) *$\alpha \times \beta$ is a regular measure over $X \times Y$.*

(2) *If A is an α measurable set and B is a β measurable set, then $A \times B$ is an $\alpha \times \beta$ measurable set and*

$$(\alpha \times \beta)(A \times B) = \alpha(A) \cdot \beta(B).$$

(3) *If S is a countably $\alpha \times \beta$ measurable set, then*

$$\{x : (x, y) \in S\} \text{ is } \alpha \text{ measurable for } \beta \text{ almost all } y,$$

$$\{y : (x, y) \in S\} \text{ is } \beta \text{ measurable for } \alpha \text{ almost all } x,$$

$$(\alpha \times \beta) S = \int \alpha\{x : (x, y) \in S\}\, d\beta\, y = \int \beta\{y : (x, y) \in S\}\, d\alpha\, x.$$

(4) *If f is an $\alpha \times \beta$ integrable and countably $\alpha \times \beta$ measurable function (in particular, if f is $\alpha \times \beta$ summable), then*

$$\int f\, d(\alpha \times \beta) = \iint f(x, y)\, d\alpha\, x\, d\beta\, y = \iint f(x, y)\, d\beta\, y\, d\alpha\, x.$$

Proof. We let F be the family of those sets $S \subset X \times Y$ whose characteristic function c_S satisfies the repeated integrability condition

$$\rho(S) = \iint c_S(x, y)\, d\alpha\, x\, d\beta\, y \in \overline{\mathbf{R}}.$$

Twice applying 2.4.8 and 2.4.9 we see that:

If $S_1, S_2, S_3, \ldots$ are disjoint members of F, then

$$\sum_{j=1}^{\infty} \rho(S_j) = \rho\Big(\bigcup_{j=1}^{\infty} S_j\Big), \text{ hence } \bigcup_{j=1}^{\infty} S_j \in F.$$

If $S_1 \supset S_2 \supset S_3 \supset \cdots$ belong to F, and $\rho(S_1) < \infty$, then

$$\lim_{j \to \infty} \rho(S_j) = \rho\Big(\bigcap_{j=1}^{\infty} S_j\Big), \text{ hence } \bigcap_{j=1}^{\infty} S_j \in F.$$

We also consider the families

$$P_0=\{A\times B\colon A \text{ is } \alpha \text{ measurable, } B \text{ is } \beta \text{ measurable}\},$$

$$P_1=\{\textstyle\bigcup G\colon G\subset P_0,\ G \text{ is countable}\},$$

$$P_2=\{\textstyle\bigcap H\colon \varnothing\neq H\subset P_1,\ H \text{ is countable}\},$$

and observe:

If $A\times B\in P_0$, then $\rho(A\times B)=\alpha(A)\cdot\beta(B)$. Hence $P_0\subset F$.

If $A\times B\in P_0$ and $C\times D\in P_0$, then

$$(A\times B)\cap(C\times D)=(A\cap C)\times(B\cap D)\in P_0,$$

and

$$(A\times B)\sim(C\times D)=[(A\sim C)\times B]\cup[(A\cap C)\times(B\sim D)]$$

is the union of two disjoint members of P_0. It follows that every member of P_1 is the union of a *disjointed* countable subfamily of P_0. Hence $P_1\subset F$.

The intersection of any two members of P_1 belongs to P_1. Therefore every member of P_2 is the intersection of a decreasing sequence of members of P_1.

Next we shall prove: *For any* $S\subset X\times Y$,

$$(\alpha\times\beta)S=\inf\{\rho(V)\colon S\subset V\in P_1\}$$

and there exists a set W *such that*

$$S\subset W\in P_2 \quad \textit{and} \quad (\alpha\times\beta)S=(\alpha\times\beta)W=\rho(W).$$

First, if $A_1\times B_1,\ A_2\times B_2,\ A_3\times B_3,\ \ldots$ belong to P_0 and

$$S\subset V=\bigcup_{j=1}^{\infty} A_j\times B_j,$$

then $c_V\le\sum_{j=1}^{\infty} c_{A_j\times B_j}$, hence

$$\rho(V)\le\sum_{j=1}^{\infty}\rho(A_j\times B_j)=\sum_{j=1}^{\infty}\alpha(A_j)\cdot\beta(B_j);$$

moreover equality holds in case the sets $A_j\times B_j$ are disjoint. Second, in case $(\alpha\times\beta)S<\infty$, we choose $V_1\supset V_2\supset V_3\supset\cdots$ so that $S\subset V_j\in P_1$ with

$$(\alpha\times\beta)S=\lim_{j\to\infty}\rho(V_j),$$

and take $W=\bigcap_{j=1}^{\infty} V_j$. In case $(\alpha\times\beta)S=\infty$, we take $W=X\times Y$.

Now, to prove (2), suppose $A\times B\in P_0$. Since

$$\alpha(A)\cdot\beta(B)=\rho(A\times B)\le\rho(V) \text{ whenever } A\times B\subset V\in P_1,$$

we see that $(\alpha\times\beta)(A\times B)=\alpha(A)\cdot\beta(B)$. Moreover, if $T\subset X\times Y$ then $T\subset U\in P_1$ implies that $U\cap(A\times B)$ and $U\sim(A\times B)$ are disjoint members of P_1, hence

$$(\alpha\times\beta)[T\cap(A\times B)]+(\alpha\times\beta)[T\sim(A\times B)]$$
$$\le\rho[U\cap(A\times B)]+\rho[U\sim(A\times B)]=\rho(U).$$

Consequently $A\times B$ is $\alpha\times\beta$ measurable.

It follows from (2) that every member of P_2 is $\alpha\times\beta$ measurable. Thus (1) holds because every subset of $X\times Y$ is contained in some member of P_2 with equal $\alpha\times\beta$ measure.

If $(\alpha\times\beta)S=0$, then S is contained in some set W with $\rho(W)=0$, hence $\rho(S)=0$.

If $(\alpha\times\beta)S<\infty$ and S is $\alpha\times\beta$ measurable, then S is contained in some set $W\in P_2$ with $(\alpha\times\beta)S=(\alpha\times\beta)W=\rho(W)$, hence $(\alpha\times\beta)(W\sim S)=0$. Therefore $\rho(W\sim S)=0$,

$$\alpha\{x\colon (x,y)\in S\}=\alpha\{x\colon (x,y)\in W\} \text{ for } \beta \text{ almost all } y,$$

hence $\rho(S)=\rho(W)=(\alpha\times\beta)S$. Now (3) follows immediately.

Next we note that (4) reduces to (3) in case f is a characteristic function, apply 2.3.3 and 2.4.8 in case f is nonnegative, and use 2.4.4(6) to complete the proof.

2.6.3. The first two of the following three examples demonstrate the need for all the hypotheses in 2.6.2(3) and (4). The third example shows that the family F occurring in the proof need not be a Borel family.

(1) Suppose $X=Y=\mathbf{R}$, $\alpha=\mathscr{L}^1$, $\beta=$ counting measure,

$$S=\{(x,x)\colon 0\le x\le 1\},$$

and f is the characteristic function of S. Then $\alpha\times\beta$ is Borel regular, hence S is $\alpha\times\beta$ measurable and f is $\alpha\times\beta$ integrable, but

$$\textstyle\iint f(x,y)\,d\alpha\,x\,d\beta\,y=\int 0\,d\beta\,y=0, \qquad \iint f(x,y)\,d\beta\,y\,d\alpha\,x=\int_0^1 1\,d\alpha\,x=1.$$

Hence f and S fail to be countably $\alpha\times\beta$ measurable, and

$$\textstyle\int f\,d(\alpha\times\beta)=\infty.$$

(2) Suppose $X=Y=\mathbf{R}$, $\alpha=\beta=\mathscr{L}^1$ and

$$f(x,y)=(4xy-x^2-y^2)(x+y)^{-4} \text{ whenever } x>0 \text{ and } y>0,$$

$$f(x,y)=0 \text{ whenever } x\le 0 \text{ or } y\le 0.$$

For each $y>0$ we use the fundamental theorem of the elementary calculus (see 2.9.20) to compute

$$\textstyle\int_0^a f(x,y)\,d\mathscr{L}^1\,x=(a^2-a\,y)(a+y)^{-3} \text{ whenever } a>0;$$

since

$$|f(x,y)| \le (4y+1+y^2)\,y^{-4} \text{ whenever } 0 \le x \le 1,$$
$$|f(x,y)| \le (4y+1+y^2)\,x^{-2} \text{ whenever } x > 1,$$

and $\int_1^\infty x^{-2}\,d\mathscr{L}^1 x = 1$, we see that $\int_0^\infty |f(x,y)|\,d\mathscr{L}^1 x < \infty$, hence

$$\int_0^\infty f(x,y)\,d\mathscr{L}^1 x = \lim_{a\to\infty} \int_0^a f(x,y)\,d\mathscr{L}^1 x = 0.$$

This equation, and the symmetry of f, imply that

$$\iint f(x,y)\,d\mathscr{L}^1 x\,d\mathscr{L}^1 y = 0 = \iint f(x,y)\,d\mathscr{L}^1 y\,d\mathscr{L}^1 x.$$

However f is not $\mathscr{L}^2$ integrable, because if it were then

$$\begin{aligned}\int f\,d\mathscr{L}^2 &= \lim_{a\to\infty} \int_{\{(x,y):\,0<x<a,\,0<y<ma\}} f\,d\mathscr{L}^2 \\ &= \lim_{a\to\infty} \int_0^{ma}\int_0^a f(x,y)\,d\mathscr{L}^1 x\,d\mathscr{L}^1 y \\ &= \lim_{a\to\infty} \int_0^{ma} (a^2 - a\,y)(a+y)^{-3}\,d\mathscr{L}^1 y \\ &= \lim_{a\to\infty} a^2\,m(a+m\,a)^{-2} = m(1+m)^{-2}\end{aligned}$$

for every positive number m, which is absurd.

(3) Suppose A_1 and A_2 are disjoint α measurable sets with

$$\alpha(A_1) = \alpha(A_2) > 0,$$

B is a β nonmeasurable subset of Y, and

$$S = [A_1 \times B] \cup [A_2 \times (Y \sim B)], \qquad T = A_1 \times Y.$$

Then $S \in F$, $T \in P_0 \subset F$, but $S \cap T = A_1 \times B \notin F$.

2.6.4. The bilinear map

$$\mathbf{L}_1(\alpha) \times \mathbf{L}_1(\beta) \to \mathbf{L}_1(\alpha \times \beta),$$

which associates with $(f,g) \in \mathbf{L}_1(\alpha) \times \mathbf{L}_1(\beta)$ the function h such that

$$h(x,y) = f(x)\,g(y) \text{ whenever } (x,y) \in X \times Y,$$

induces according to 1.1.1 and 1.1.3 *a linear monomorphism*

$$\mathbf{L}_1(\alpha) \otimes \mathbf{L}_1(\beta) \to \mathbf{L}_1(\alpha \times \beta).$$

The image of this monomorphism is $(\alpha\times\beta)_{(1)}$ *dense in* $\mathbf{L}_1(\alpha\times\beta)$; in fact, recalling the proof of 2.6.2 we see by successive consideration of the classes P_0, P_1, P_2 that the characteristic function of every $\alpha \times \beta$ measurable set with finite $\alpha\times\beta$ measure belongs to the closure of the image; hence the density follows from 2.3.3, 2.4.8, 2.4.4 (6). We now define the monotone Daniell integrals λ, μ, ν by the formulae

$$\lambda(f)=\int f\,d\alpha \text{ for } f\in\mathbf{L}_1(\alpha), \quad \mu(g)=\int g\,d\beta \text{ for } g\in\mathbf{L}_1(\beta),$$

$$\nu(h)=\int h\,d(\alpha\times\beta) \text{ for } h\in\mathbf{L}_1(\alpha\times\beta),$$

and observe that *the following diagram is commutative:*

$$\begin{array}{ccc} \mathbf{L}_1(\alpha)\otimes\mathbf{L}_1(\beta) & \rightarrow & \mathbf{L}_1(\alpha\times\beta) \\ \downarrow{\scriptstyle \lambda\otimes\mu} & & \downarrow{\scriptstyle \nu} \\ \mathbf{R}\otimes\mathbf{R} & \simeq & \mathbf{R} \end{array}$$

This shows that the cartesian product of two measures is associated with a suitably completed tensor product of the corresponding Daniell integrals, and motivates the use of $\alpha\otimes\beta$ as an alternate notation for $\alpha\times\beta$. The same philosophy is also illustrated by the following proposition (whose proof is straightforward):

If X, Y are locally compact Hausdorff spaces and λ, μ are arbitrary Daniell integrals on $\mathscr{K}(X)$, $\mathscr{K}(Y)$, then there exists a unique Daniell integral ν on $\mathscr{K}(X\times Y)$ such that the following diagram is commutative:

$$\begin{array}{ccc} \mathscr{K}(X)\otimes\mathscr{K}(Y) & \rightarrow & \mathscr{K}(X\times Y) \\ \downarrow{\scriptstyle \lambda\otimes\mu} & & \downarrow{\scriptstyle \nu} \\ \mathbf{R}\otimes\mathbf{R} & \simeq & \mathbf{R} \end{array}$$

In case λ, μ are monotone, and α, β are the associated $\mathscr{K}(X)$, $\mathscr{K}(Y)$ regular measures, then $\alpha\times\beta$ is the $\mathscr{K}(X\times Y)$ regular measure associated with ν.

2.6.5. If α measures X, β measures Y and

$$r\colon X\times Y\simeq Y\times X, \quad r(x,y)=(y,x),$$

then $r_\#(\alpha\times\beta)=\beta\times\alpha$. If, in addition, γ measures Z and

$$s\colon (X\times Y)\times Z\simeq X\times(Y\times Z), \quad s((x,y),z)=(x,(y,z)),$$

then $s_\#((\alpha\times\beta)\times\gamma)=\alpha\times(\beta\times\gamma)$. The cartesian multiplication of measures is thus commutative and associative. Therefore, when α_j measures X_j for $j=1,\ldots,n$, we are able to define (by induction with respect to n) the measure

$$\alpha_1\times\cdots\times\alpha_n \quad \text{over} \quad X_1\times\cdots\times X_n,$$

using an arbitrary association of the factors.

For example, since $\mathbf{R}^n$ is the cartesian product of n equal factors $\mathbf{R}$, we define the n **dimensional Lebesgue measure**

$$\mathscr{L}^n$$

over $\mathbf{R}^n$ as the cartesian product of n factors $\mathscr{L}^1$ (see 2.5.17). It follows that the usual isomorphism

$$\mathbf{R}^{m+n} \simeq \mathbf{R}^m \times \mathbf{R}^n \text{ maps } \mathscr{L}^{m+n} \text{ onto } \mathscr{L}^m \times \mathscr{L}^n.$$

Since $\mathscr{L}^1$ is Borel regular, so is $\mathscr{L}^n$.

Applying induction with respect to n by means of Fubini's theorem, we see that $\mathscr{L}^n\{x: p(x)=0\}=0$ *for every nonzero polynomial map* $p: \mathbf{R}^n \to \mathbf{R}$.

2.6.6. Here we discuss **infinite cartesian products** of monotone Daniell integrals on lattices of continuous functions on compact Hausdorff spaces, with total measures 1.

Suppose J is an infinite set, and to each $j \in J$ corresponds a compact Hausdorff space X_j and a monotone Daniell integral μ_j on $\mathscr{K}(X_j)$ with $\mu_j(1)=1$. Let X be the cartesian product of all the spaces X_j.

Let A be the class of all finite subsets of J. With each $a \in A$ we associate the cartesian product X_a of the spaces X_j corresponding to $j \in a$, the projection p_a of X onto X_a, and the cartesian product μ_a of the Daniell integrals μ_j corresponding to $j \in a$ (constructed by the method of 2.6.4); moreover p_a induces a monomorphism of $\mathscr{K}(X_a)$ into $\mathscr{K}(X)$, with image

$$L_a = \{f \circ p_a : f \in \mathscr{K}(X_a)\}.$$

Whenever $a \subset b \in A$, the projection $p_a^b: X_b \to X_a$ shows that

$$L_a = \{(f \circ p_a^b) \circ p_b : f \in \mathscr{K}(X_a)\} \subset L_b;$$

moreover $p_{a\,\#}^b(\mu_b) = \mu_a$ because

$$\mu_b(f \circ p_a^b) = \mu_a(f) \cdot \mu_{b \sim a}(1) = \mu_a(f) \text{ for } f \in \mathscr{K}(X_a).$$

Accordingly there exists a unique linear map

$$\gamma: L = \bigcup_{a \in A} L_a \to \mathbf{R}$$

such that $\gamma(f \circ p_a) = \mu_a(f)$ whenever $a \in A$, $f \in L_a$. Since L is dense in $\mathscr{K}(X)$, with respect to the norm ν defined in 2.5.19, and since $|\gamma| \le \nu$, the function γ has a unique continuous linear extension

$$\mu: \mathscr{K}(X) \to \mathbf{R} \text{ with } |\mu| \le \nu,$$

which is a Daniell integral by 2.5.13. We note that μ is uniquely characterized by the condition:

$$p_{a\,\#}(\mu) = \mu_a \text{ whenever } a \in A.$$

We observe that in the preceding argument the passage from finite to infinite products depends only on the representation of X as inverse limit of the spaces X_a, and on the consistency conditions $p_{a\,\#}^b(\mu_b) = \mu_a$.

The same type of reasoning can be applied to arbitrary **inverse limits** of Daniell integrals on lattices of continuous functions on compact Hausdorff spaces.

2.6.7. If α, β are Radon measures over $\mathbf{R}$, $-\infty<a<b<\infty$,

$$f(r)=\alpha\{t: t<r\}<\infty \quad \text{and} \quad g(r)=\beta\{t: t\leq r\}<\infty$$

for $a\leq r\leq b$, then

$$\int_{\{r: a<r\leq b\}} f\, d\beta+\int_{\{r: a\leq r<b\}} g\, d\alpha=f(b)\, g(b)-f(a)\, g(a).$$

In fact, applying 2.6.2 (3) with $S=\{(x,y): a\leq x<y\leq b\}$ we obtain the equations

$$\begin{aligned}(\alpha\times\beta)\, S&=\int_{\{y: a<y\leq b\}}[f(y)-f(a)]\, d\beta\, y\\ &=\int_{\{y: a<y\leq b\}} f\, d\beta-f(a)[g(b)-g(a)],\\ (\alpha\times\beta)\, S&=\int_{\{x: a\leq x<b\}}[g(b)-g(x)]\, d\alpha\, x\\ &=g(b)[f(b)-f(a)]-\int_{\{x: a\leq x<b\}} g\, d\alpha.\end{aligned}$$

2.7. Invariant measures

2.7.1. Letting X be a locally compact Hausdorff space, and G a locally compact Hausdorff topological group, we assume that X is a **homogeneous space** of G through the transitive left action

$$A: G\times X\rightarrow X;$$

this means that A is continuous, that

$$A(g\, h, x)=A[g, A(h, x)] \text{ for } g\in G,\ h\in G,\ x\in X,$$

and that for each $x\in X$ the map

$$A^x: G\rightarrow X, \quad A^x(g)=A(g, x) \text{ whenever } g\in G,$$

is open with $\operatorname{im} A^x=X$.

It follows that for each $g\in G$ the map

$$A_g: X\rightarrow X, \quad A_g(x)=A(g, x) \text{ whenever } x\in X$$

is a homeomorphism, with $(A_g)^{-1}=A_{g^{-1}}$ and $A_e=\mathbf{1}_X$, where e is the identity of G. Moreover, for each $x\in X$ the **isotropy subgroup**

$$G^x=G\cap\{g: A(g, x)=x\}$$

is closed in G, and A^x induces a homeomorphism

$$G/G^x \simeq X,$$

which maps each coset $g\,G^x$ onto $A(g,x)$.

Every closed subgroup S of G is the isotropy subgroup for some such action A, because one may take $X = G/S$ and $A(g, h\,S) = g\,h\,S$ whenever $g, h \in G$; then $G^S = S$.

A Daniell integral μ on $\mathscr{K}(X)$ is called a χ **covariant integral** if and only if $\chi\colon G \to \mathbf{R}$ and

$$\mu(u \circ A_g) = \chi(g)\,\mu(u) \quad \text{whenever } u \in \mathscr{K}(X),\ g \in G.$$

In case $\mu \neq 0$, it follows that χ is a continuous homomorphism of G into the multiplicative group $\mathbf{R} \sim \{0\}$. According to 2.5.19, χ covariance of μ is equivalent to the condition

$$A_{g\,\#}(\mu) = \chi(g) \cdot \mu \quad \text{whenever } g \in G.$$

In particular, if $\chi(g) = 1$ for all $g \in G$, one calls μ an **invariant integral** (with respect to the action A of G on X).

The $\mathscr{K}(X)$ regular measure associated with a monotone χ covariant (invariant) integral is termed a χ **covariant (invariant) measure.**

From 2.5.6 one sees that, if μ is χ covariant, then $\mu^+ + \mu^-$ is $|\chi|$ covariant.

It is often convenient to use the abbreviations:

$$u \circ A_g = u_g, \qquad A(g,x) = g\,x.$$

2.7.2. Two important special cases occur when $X = G$ and the multiplication of G is used to define

$$\text{either (1) } A(g,x) = g\,x, \quad \text{or} \quad (2)\ A(g,x) = x\,g^{-1}$$

whenever $g \in G$ and $x \in G$.

In case (1), A_g is left multiplication by g. To this action correspond the so-called **left** χ **covariant (invariant) integrals** and **measures.** In particular, a nonzero monotone left invariant integral (measure) is called a **left Haar integral (measure)** of G.

In case (2), A_g is right multiplication by g^{-1}. To this action correspond the so-called **right** χ **covariant (invariant) integrals** and **measures.** In particular, a nonzero monotone right invariant integral (measure) is called a **right Haar integral (measure)** of G.

These two cases are related by the commutative diagram

$$\begin{array}{ccc} G & \xrightarrow{\text{left multiplication by } g} & G \\ \rho\downarrow & & \downarrow\rho \\ G & \xrightarrow{\text{right multiplication by } g^{-1}} & G \end{array}$$

where $\rho(x)=x^{-1}$ for $x\in G$. We see that μ is a left χ covariant integral if and only if $\rho_*(\mu)$ is a right χ covariant integral.

2.7.3. We shall say that G **acts uniformly on** X if and only if the topology of X is induced by some uniformity with a G invariant basis B; by saying that B is G invariant we mean that, whenever $\beta\in B$ and $g\in G$,

$$(x, y)\in\beta \text{ if and only if } (g\,x, g\,y)\in\beta.$$

One easily verifies that G acts uniformly on X if and only if for each neighborhood V of a point x in X there exists a neighborhood W of x such that

$$g\in G,\, W\cap A_g(W)\neq\varnothing \quad \text{implies} \quad A_g(W)\subset V;$$

when this condition holds one may take as B the family of the sets

$$\bigcup_{g\in G} A_g(V)\times A_g(V)$$

corresponding to all nonempty open subsets V of X.

2.7.4. Theorem. *If G acts uniformly on X, then there exists a nonzero monotone invariant integral on $\mathscr{K}(X)$.*

Proof. Whenever $u\in\mathscr{K}(X)^+$ and $0\neq v\in\mathscr{K}(X)^+$ we let

$$C(u, v)$$

be the set of all maps $\xi\colon G\to\{t\colon 0\le t<\infty\}$ for which

$$\{g\colon \xi(g)>0\} \quad \text{is finite and} \quad u\le\sum_{g\in G}\xi(g)\cdot(v\circ A_g),$$

and we define the so-called Haar ratio

$$(u\!:\!v)=\inf\{\textstyle\sum_G \xi\colon \xi\in C(u, v)\}.$$

Obviously $C(u, v)\neq\varnothing$ and $(u\!:\!v)\ge\sup\operatorname{im} u/\sup\operatorname{im} v$,

$$(u\circ A_h\!:\!v)=(u\!:\!v) \text{ for } h\in G,$$

$$(c\,u\!:\!v)=c(u\!:\!v) \text{ for } 0<c<\infty,$$

$$(u_1+u_2\!:\!v)\le(u_1\!:\!v)+(u_2\!:\!v),$$

$$u_1\le u_2 \quad \text{implies} \quad (u_1\!:\!v)\le(u_2\!:\!v).$$

Moreover, if $u, v, w \in \mathscr{K}(X)^+$ are nonzero, then

$$(u:w) \leq (u:v) \cdot (v:w)$$

because $\xi \in C(u, v)$ and $\eta \in C(v, w)$ implies

$$u \leq \sum_{g \in G} \xi(g) \sum_{h \in G} \eta(h)(w \circ A_h \circ A_g) = \sum_{k \in G} \zeta(k)(w \circ A_k)$$

$$\text{with } \zeta(k) = \sum_{hg=k} \xi(g)\,\eta(h), \quad \sum_G \zeta = \sum_G \xi \cdot \sum_G \eta.$$

Consequently

$$1/(w:u) \leq (u:v)/(w:v) \leq (u:w).$$

We now choose a particular nonzero $w \in \mathscr{K}(X)^+$, and consider the cartesian product P of the compact intervals

$$\{t\colon 0 \leq t \leq (u:w)\}$$

corresponding to all $u \in \mathscr{K}(X)^+$. Whenever $0 \neq v \in \mathscr{K}(X)^+$ we define

$$p_v \in P, \quad p_v(u) = (u:v)/(w:v) \text{ for } u \in \mathscr{K}(X)^+,$$

and with each $\beta \in B$ (see 2.7.3) we associate the closed set

$$S(\beta) = \text{Closure}\{p_v\colon (\operatorname{spt} v) \times (\operatorname{spt} v) \subset \beta\}.$$

If $\beta_1, \beta_2, \beta_3 \in B$ and $\beta_1 \cap \beta_2 \supset \beta_3$, then $S(\beta_1) \cap S(\beta_2) \supset S(\beta_3) \neq \varnothing$. Accordingly, since P is compact, there exists a point

$$\lambda \in \bigcap_{\beta \in B} S(\beta).$$

This function λ turns out to be a nonzero invariant integral on $\mathscr{K}(X)^+$. The required properties of λ (see 2.5.2, 2.5.13, 2.7.1) follow from like properties of the approximating functions p_v. In view of the facts concerning Haar ratios listed above, the only nontrivial part of the verification occurs in showing that

$$\lambda(u_1 + u_2) \geq \lambda(u_1) + \lambda(u_2) \text{ whenever } u_1, u_2 \in \mathscr{K}(X)^+.$$

To prove this inequality we choose $f \in \mathscr{K}(X)^+$ with

$$\operatorname{spt} u_1 \cup \operatorname{spt} u_2 \subset \{x\colon f(x) > 0\}.$$

Given any $\varepsilon > 0$, we define $s, r_1, r_2 \in \mathscr{K}(X)^+$ so that

$$s = u_1 + u_2 + \varepsilon f, \quad r_j s = u_j \text{ and } \operatorname{spt} r_j = \operatorname{spt} u_j \text{ for } j \in \{1, 2\},$$

and use the uniform continuity of r_1, r_2 to obtain $\beta \in B$ with

$$|r_j(x) - r_j(y)| \leq \varepsilon \text{ whenever } (x, y) \in \beta,\ j \in \{1, 2\}.$$

For any $v \in S(\beta)$, $a \in \operatorname{spt} v$, $\xi \in C(s, v)$ we define

$$\xi_j(g) = [r_j(g^{-1}a) + \varepsilon]\,\xi(g) \text{ whenever } g \in G,\ j \in \{1, 2\},$$

and infer that

$$u_j(x)=r_j(x)\,s(x)\leq \sum_{g\in G} r_j(x)\,\xi(g)\,v(g\,x)\leq \sum_{g\in G} \xi_j(g)\,v(g\,x)$$

because $v(g\,x)\neq 0$ implies $(g\,x, a)\in\beta$, $(x, g^{-1}\,a)\in\beta$; therefore $\xi_j\in C(u_j, v)$ and

$$(u_1:v)+(u_2:v)\leq \sum_G \xi_1+\sum_G \xi_2\leq (1+2\varepsilon)\sum_G \xi$$

because $r_1+r_2\leq 1$. It follows that

$$p_v(u_1)+p_v(u_2)\leq (1+2\varepsilon)\,p_v(s)\leq (1+2\varepsilon)[p_v(u_1+u_2)+\varepsilon\, p_v(f)]$$

whenever $v\in S(\beta)$. Since $\lambda\in$ Closure $S(\beta)$, we conclude that

$$\lambda(u_1)+\lambda(u_2)\leq (1+2\varepsilon)[\lambda(u_1+u_2)+\varepsilon\,\lambda(f)].$$

2.7.5. Corollary. *There exist left and right Haar integrals of G.*

Proof. Taking B as the family of sets

$$\{(x, y):\ x^{-1}\,y\in V\}$$

corresponding to all neighborhoods V of e, one sees that the action 2.7.2 (1) of G on G is uniform.

Similarly one treats the action 2.7.2 (2) by using the sets

$$\{(x, y):\ x\,y^{-1}\in V\}.$$

2.7.6. Lemma. *If λ is a left Haar integral and μ is a right Haar integral, then there exists a continuous positive real valued function k on G such that*

(1) $k(y)\,\mu(v)=\lambda_x\, v(y\,x^{-1})$ *whenever* $y\in G$, $v\in\mathscr{K}(G)$,

(2) $\mu(k\,u)=\lambda(u)$ *whenever* $u\in\mathscr{K}(G)$.

Proof. Using Fubini's theorem we compute

$$\lambda(u)\,\mu(v)=\mu_y\,\lambda_x[u(x)\,v(y)]=\mu_y\,\lambda_x[u(y\,x)\,v(y)]$$
$$=\lambda_x\,\mu_y[u(y\,x)\,v(y)]=\lambda_x\,\mu_y[u(y)\,v(y\,x^{-1})]=\mu_y[u(y)\cdot\lambda_x\,v(y\,x^{-1})].$$

If we choose any particular nonzero $v\in\mathscr{K}(G)^+$, we can use the equation in (1) to define a positive continuous function k on G, and our computation implies (2) for all $u\in\mathscr{K}(G)$. However k is uniquely characterized by (2), hence does not really depend on the choice of v. It follows that (1) holds for all $v\in\mathscr{K}(G)$.

2.7.7. Theorem. *Any two left (right) Haar integrals of G are constant positive multiples of each other.*

Proof. Suppose λ_1, λ_2 are left Haar integrals and μ is a right Haar integral. According to 2.7.6 there exist positive continuous functions k_1, k_2 on G such that

$$\mu(k_1\, u) = \lambda_1(u),\ \mu(k_2\, u) = \lambda_2(u) \text{ whenever } u \in \mathscr{K}(G).$$

Then $f = k_1/k_2$ is a positive continuous function such that

$$\lambda_1(u) = \mu(k_2\, f\, u) = \lambda_2(f\, u) \text{ whenever } u \in \mathscr{K}(G).$$

If $g \in G$, then

$$\lambda_2(f\, u) = \lambda_1(u) = \lambda_1(u \circ A_g) = \lambda_2[f \cdot (u \circ A_g)] = \lambda_2[(f \circ A_g^{-1}) \cdot u]$$

whenever $u \in \mathscr{K}(G)$, hence $f = f \circ A_g^{-1}$. Thus f is constant on G.

2.7.8. Theorem. *There exists a continuous positive real valued function Δ_G with the following properties:*

(1) *Every left Haar integral of G is right Δ_G covariant.*

(2) *Every right Haar integral of G is left Δ_G covariant.*

(3) *If λ is a left Haar integral of G, then*

$$\lambda_x[u(x^{-1})\, \Delta_G(x^{-1})] = \lambda(u) \textit{ whenever } u \in \mathscr{K}(G).$$

(4) *If μ is a right Haar integral of G, then*

$$\mu_x[u(x^{-1})\, \Delta_G(x)] = \mu(u) \textit{ whenever } u \in \mathscr{K}(G).$$

Moreover Δ_G is uniquely characterized by each of these properties; it is called the **modular function** *of G.*

Proof. In view of 2.7.7 we need only consider one particular left Haar integral λ, and the corresponding (see 2.7.2) right Haar integral $\mu = \rho_{\#}(\lambda)$. As Δ_G we may take the function k of 2.7.6, because

$$k(y)\, \mu(v) = \mu_x\, v(y\, x) \text{ whenever } y \in G,\ v \in \mathscr{K}(G),$$

hence μ is left k covariant, $\lambda = \rho_{\#}(\mu)$ is right k covariant by 2.7.2, while

$$\lambda[(k \circ \rho) \cdot (u \circ \rho)] = \mu(k\, u) = \lambda(u), \qquad \mu[k \cdot (u \circ \rho)] = \lambda(u \circ \rho) = \mu(u)$$

whenever $u \in \mathscr{K}(G)$.

2.7.9. Theorem. *Suppose ν is a Daniell integral on $\mathscr{K}(G)$ and χ is a continuous homomorphism of G into the multiplicative group* $\mathbf{R} \sim \{0\}$.

(1) *ν is left χ covariant if and only if there exists a left Haar integral λ and a real number r such that*

$$\nu(u) = r\, \lambda(u\, \chi^{-1}) \textit{ whenever } u \in \mathscr{K}(G).$$

(2) *ν is right χ covariant if and only if there exists a right Haar integral μ and a real number r such that*

$$\nu(u)=r\,\mu(u\,\chi) \quad \textit{whenever} \quad u\in\mathscr{K}(G).$$

(3) *ν is left χ covariant if and only if ν is right $\Delta_G\,\chi^{-1}$ covariant.*

Proof. To verify (1) we define

$$\sigma(u)=\nu(u\,\chi) \quad \text{whenever} \quad u\in\mathscr{K}(G),$$

compute

$$\sigma(u\circ A_g)=\nu_x[u(g\,x)\,\chi(x)]=\chi(g)^{-1}\,\nu[(u\,\chi)\circ A_g]$$

for $g\in G$, and infer that ν is left χ covariant if and only if σ is left invariant. If σ is left invariant, so are σ^+ and σ^-, hence σ^+ and σ^- are nonnegative multiples of a left Haar integral.

We obtain (2) from (1) by means of $\rho_\#$ (see 2.7.2).

Regarding (3) we observe that if ν is left χ covariant, then σ is right Δ_G covariant by 2.7.8 (1), hence

$$\begin{aligned}\nu_x[u(x\,g^{-1})]=\sigma_x[u(x\,g^{-1})\,\chi^{-1}(x)]&=\chi^{-1}(g)\,\sigma_x[u(x\,g^{-1})\,\chi^{-1}(x\,g^{-1})]\\ &=\chi^{-1}(g)\,\Delta_G(g)\,\sigma(u\,\chi^{-1})=(\chi^{-1}\,\Delta_G)(g)\,\nu(u)\end{aligned}$$

whenever $u\in\mathscr{K}(G)$ and $g\in G$, so that ν is right $\chi^{-1}\,\Delta_G$ covariant. The converse follows by use of $\rho_\#$.

2.7.10. Lemma. *To each $x\in X$ and each left Haar integral μ of the isotropy group G^x corresponds a linear epimorphism*

$$\mu^x\colon \mathscr{K}(G)\to\mathscr{K}(X)$$

with the following properties:

(1) $\mu^x(u)(g\,x)=\mu_s[u(g\,s)]$ *whenever* $u\in\mathscr{K}(G)$, $g\in G$.

(2) $\operatorname{spt}\mu^x(u)=A^x(\operatorname{spt}u)$ *whenever* $u\in\mathscr{K}(G)^+$.

(3) $\mu^x[(v\circ A^x)\cdot u]=v\cdot\mu^x(u)$ *whenever* $u\in\mathscr{K}(G)$, $v\in\mathscr{K}(X)$.

(4) *If* $u\in\mathscr{K}(G)$, $h\in G$, $u_h(y)=u(h\,y)$ *for* $y\in G$, *then*

$$\mu^x(u_h)=\mu^x(u)\circ A_h.$$

(5) *For each $v\in\mathscr{K}(X)$ there exist functions $f\in\mathscr{K}(G)^+$ with*

$$\operatorname{spt}v\subset A^x\{g\colon f(g)>0\},$$

and this condition implies that

$$\mu^x\left[\left(\frac{v}{\mu^x(f)}\circ A^x\right)\cdot f\right]=v.$$

(6) *If ν is a (monotone) Daniell integral on $\mathscr{K}(X)$, then $\nu \circ \mu^x$ is a (monotone) Daniell integral on $\mathscr{K}(G)$. Moreover, ν is χ covariant if and only if $\nu \circ \mu^x$ is left χ covariant.*

Proof. For each $u \in \mathscr{K}(G)$ the map

$$w: G \to \mathbf{R}, \quad w(g) = \mu_s[u(g\,s)] \text{ whenever } g \in G,$$

is continuous, because u is uniformly continuous; moreover

$$w(g\,t) = w(g) \text{ whenever } t \in G^x,$$

because μ is left invariant; hence there exists a unique continuous map

$$\mu^x(u): X \to \mathbf{R} \quad \text{with} \quad \mu^x(u) \circ A^x = w.$$

The assertions (1), (2), (3), (4) are now easily verified.

Regarding (5), we infer the existence of suitable functions f from the fact that the sets

$$A^x\{g: u(g) > 0\},$$

corresponding to $u \in \mathscr{K}(G)^+$, form a base for the topology of X. Moreover, for any such f, we see that

$$\operatorname{spt} v \subset \{y: \mu^x(f)(y) > 0\},$$

hence $v = w \cdot \mu^x(f)$ with $w \in \mathscr{K}(X)$, and (3) implies $\mu^x[(w \circ A^x) \cdot f] = v$. Finally (6) follows from (1), (4), (5).

2.7.11. Theorem. *Suppose χ is a continuous homomorphism of G into the multiplicative group $\mathbf{R} \sim \{0\}$, and $x \in X$.*

(1) *Nonzero χ covariant integrals on $\mathscr{K}(X)$ exist if and only if χ satisfies* **Weil's condition:**

$$\chi(t)\,\Delta_{G^x}(t) = \Delta_G(t) \text{ whenever } t \in G^x.$$

(2) *Any two nonzero χ covariant integrals are constant real multiples of each other.*

Proof. We observe that (2) follows from 2.7.10(6), 2.7.9(1) and the fact that $\operatorname{im} \mu^x = \mathscr{K}(X)$.

To prove (1) we first suppose ν is a nonzero χ covariant integral on $\mathscr{K}(X)$, infer from 2.7.10(6) that $\lambda = \nu \circ \mu^x$ is a nonzero left χ covariant integral on $\mathscr{K}(G)$, from 2.7.9(3) that λ is right $\Delta_G\,\chi^{-1}$ covariant, and choose $u \in \mathscr{K}(G)$ so that $\lambda(u) \neq 0$. Given $t \in G^x$, we define

$$w(y) = u(y\,t^{-1}) \text{ for } y \in G,$$

and compute

$$\mu^x(w)(g\,x) = \mu_s[u(g\,s\,t^{-1})] = \Delta_{G^x}(t)\,\mu^x(u)(g\,x)$$

for $g \in G$, hence $\mu^x(w) = \Delta_{G^x}(t)\,\mu^x(u)$ and

$$\Delta_{G^x}(t)\,\lambda(u) = \lambda(w) = \Delta_G(t)\,\chi(t)^{-1}\,\lambda(u).$$

Next, to prove the other half of (1), we assume Weil's condition and use 2.7.9 to choose a left χ covariant and right $\Delta_G\,\chi^{-1}$ covariant integral λ on $\mathscr{K}(G)$. For any $u \in \mathscr{K}(G)$ we obtain from 2.7.10 a $w \in \mathscr{K}(G)^+$ such that

$$\mu^x(w)(g\,x) = 1 \text{ whenever } g \in \operatorname{spt} u,$$

and apply 2.7.8 (3) in computing

$$\begin{aligned}
&\lambda_g[w(g) \cdot \mu^x(u)(g\,x)] = \mu_s\,\lambda_g[w(g)\,u(g\,s)] \\
&= \mu_s(\Delta_G(s)^{-1}\,\chi(s)\,\lambda_g[w(g\,s^{-1})\,u(g)]) \\
&= \lambda_g(u(g)\,\mu_s[\Delta_{G^x}(s^{-1})\,w(g\,s^{-1})]) = \lambda_g[u(g) \cdot \mu^x(w)(g\,x)] = \lambda(u).
\end{aligned}$$

Therefore $\ker \mu^x \subset \ker \lambda$, there exists a linear function

$$v\colon \mathscr{K}(X) \to \mathbf{R} \text{ with } v \circ \mu^x = \lambda,$$

and we see from 2.7.10 (5), (6) that v is a χ covariant integral.

2.7.12. *Weil's condition holds whenever χ is positive and the isotropy group G^x is compact*, because every continuous homomorphism of a compact group into the multiplicative group of positive numbers is constant with value 1. In this case the map A^x is proper. If λ is a left χ covariant integral of G, then $A^x_\#(\lambda)$ is a χ covariant integral on $\mathscr{K}(X)$, which equals the function v constructed in 2.7.11 provided μ is chosen so that $\mu(1) = 1$, because

$$\begin{aligned}
[A^x_\#(\lambda) \circ \mu^x]\,u &= \lambda[\mu^x(u) \circ A^x] = \lambda_g[\mu_s\,u(g\,s)] \\
&= \mu_s[\lambda_g\,u(g\,s)] = \mu_s[\lambda(u)] = \lambda(u)
\end{aligned}$$

whenever $u \in \mathscr{K}(G)$.

A group G is called **unimodular** if and only if the left and right Haar integrals of G coincide; this is equivalent to the condition that the modular function Δ_G is constant with value 1. In particular, *every compact group is unimodular.*

For each closed subgroup S of G, Weil's condition implies that the homogeneous space G/S *admits a nonzero invariant integral if and only if $\Delta_S = \Delta_G | S$*. In particular, *this equation holds whenever S is normal*, because in this case every Haar integral of the group G/S is G invariant (alternately, because G acts uniformly on G/S). Considering the special case when $S = \ker \Delta_G$, one finds that $\ker \Delta_G$ *is unimodular.*

2.7.13. Theorem. *Suppose γ is the $\mathscr{K}(X)$ regular measure associated with a monotone Daniell integral ν, $x \in X$, β is the $\mathscr{K}(G^x)$ regular measure associated with a left Haar integral μ of G^x, and α is the $\mathscr{K}(G)$ regular measure associated with the monotone Daniell integral $\nu \circ \mu^x$ (see 2.7.10). Then there exists a linear epimorphism*

$$\beta^x\colon \mathbf{L}_1(\alpha) \to \mathbf{L}_1(\gamma)$$

such that, for each α summable function u,

$$\beta^x(u)(g\,x) = \int u(g\,s)\,d\beta\,s \text{ whenever } g \in G, \quad \int \beta^x(u)\,d\gamma = \int u\,d\alpha.$$

Proof. *With each* $u\colon G \to \mathbf{R}$ we associate

$$T(u)\colon E(u) = \{g\,x\colon \int u(g\,s)\,d\beta\,s \in \mathbf{R}\} \to \mathbf{R},$$

$$T(u)(g\,x) = \int u(g\,s)\,d\beta\,s \text{ whenever } g\,x \in E(u).$$

We consider the class Φ of those functions u for which

$$\gamma[X \sim E(u)] = 0 \quad\text{and}\quad \int T(u)\,d\gamma = \int u\,d\alpha \in \mathbf{R}.$$

Referring to 2.7.10 we see that $\mathscr{K}(G) \subset \Phi$ and $\mu^x = T | \mathscr{K}(G)$.

Next we follow the method of 2.5.3 to prove that the characteristic function of every α measurable set with finite α measure belongs to Φ. If $w \in \mathscr{K}(G)^+$, $t > 0$ and u is the characteristic function of $\{g\colon w(g) > t\}$, then

$$u_h = h^{-1}[\inf\{w, t+h\} - \inf\{w, t\}] \uparrow u \text{ as } h \downarrow 0,$$

and Lebesgue's increasing convergence theorem implies

$$T(u_h) \uparrow T(u) \text{ as } h \downarrow 0, \quad E(u) = X, \; u \in \Phi.$$

Similarly we treat the characteristic function of any set with finite α measure in $F_j[\mathscr{K}(G)]$, for $j = 1, 2$. Therefore, if u is the characteristic function of any set with α measure 0, there exists a $w \in \Phi$ with $w \geq u$ and $\int w\,d\alpha = 0$, hence $T(w)$ and $T(u)$ are γ almost equal to 0, hence $u \in \Phi$. Moreover all α measurable sets with finite α measure are obtained by subtracting sets of α measure zero from sets in $F_2[\mathscr{K}(G)]$.

Now it becomes clear that Φ equals the class of all α summable functions with domain G, and that β^x may be defined by extension of $T | \Phi$.

Finally, to prove that β^x is an epimorphism, it will suffice to show that the characteristic function of every γ measurable set with compact closure belongs to $T(\Phi)$. Assuming that C is a compact subset of X, we use 2.7.10(5) to obtain

$$f \in \mathscr{K}(G)^+ \quad\text{with}\quad C \subset A^x\{g\colon f(g) > 0\}.$$

If v is the characteristic function of

$$\{y\colon z(y)>t\}\subset C \text{ with } z\in\mathscr{K}(X) \text{ and } t>0,$$

then $v_h=h^{-1}[\inf\{z,t+h\}-\inf\{z,t\}]\uparrow v$ as $h\downarrow 0$, with $\operatorname{spt} v_h\subset C$, hence

$$v_h=\mu^x(u_h) \quad\text{with}\quad u_h=([v_h/\mu^x(f)]\circ A^x)\cdot f\in\mathscr{K}(G)^+,$$

$$u_h\uparrow u=([v/\mu^x(f)]\circ A^x)\cdot f \text{ as } h\downarrow 0,\ T(u)=v,$$

$$\int u_h\,d\alpha=\int v_h\,d\gamma\uparrow\int v\,d\gamma \text{ as } h\downarrow 0,\ u\in\Phi,\ v\in\operatorname{im}\beta^x.$$

Hereafter one treats in succession the characteristic functions of those subsets of C which belong to $F_j[\mathscr{K}(X)]$, which have γ measure 0, which are γ measurable.

2.7.14. Theorem. *Every left (right) χ covariant measure over G is a Radon measure.*

Proof. Suppose α is the $\mathscr{K}(G)$ regular measure associated with a monotone left χ covariant integral λ. In view of 2.5.3, 2.5.14 we need to show that, if U is any nonempty open subset of G with compact closure, then U is α measurable. Letting u be the characteristic function of U we define

$$\Phi=\mathscr{K}(G)^+\cap\{f\colon f\le u\}$$

and choose a countable subset Ψ of Φ such that

$$\sup\lambda(\Phi)=\sup\lambda(\Psi).$$

Furthermore we consider the closed subgroup

$$S=G\cap\{s\colon f(t\,s)=f(t) \text{ whenever } f\in\Psi,\ t\in G\}$$

and the action of G on the quotient space $X=G/S$ by left multiplication, with the isotropy group $G^x=S$ at $x=S$.

For each $f\in\Psi$ we define the pseudometric ρ_f on G by the formula

$$\rho_f(g,h)=\sup\{|f(t\,g)-f(t\,h)|\colon t\in G\}$$

whenever $g,h\in G$; since $\rho_f(g\,s_1,h\,s_2)=\rho_f(g,h)$ whenever $s_1,s_2\in S$, there exists a pseudometric σ_f on X such that $\sigma_f\circ(A^x\times A^x)=\rho_f$. The uniform continuity of f on G implies continuity of ρ_f on $G\times G$, hence continuity of σ_f on $X\times X$ with respect to the quotient topology. If $0<r<\sup\operatorname{im} f$ and $g\in G$, then

$$\{h\colon \rho_f(g,h)\le r\} \text{ is compact,}$$

because there exists $t\in G$ with $f(t\,g)>r$, hence

$$\rho_f(g,h)\le r \text{ implies } f(t\,h)>0,\ h\in t^{-1}\cdot\operatorname{spt} f.$$

Choosing a positive function ε on Ψ with $\sum_{\Psi} \varepsilon = 1$, we shall now prove that

$$\sum_{f \in \Psi} \varepsilon(f) \cdot \sigma_f$$

is a metric inducing the quotient topology of X. It suffices to show that, if $g \in G$ and W is an open neighborhood of g in G with $W \cdot S = W$, then the family of compact sets

$$\{h: \rho_f(g,h) \le r \text{ and } h \notin W\}$$

which correspond to $f \in \Psi$ and $0 < r \le \sup \operatorname{im} f$, does not have the finite intersection property. The alternative is the existence of a point h belonging to every set of this family, which would imply both $h \notin W$ and

$$f(t\,g) = f(t\,h) \text{ for } f \in \Psi \text{ and } t \in G, \; g^{-1} h \in S, \; h \in g\,S \subset W.$$

Next we observe that

$$S = \bigcap_{f \in \Psi} \{h: \rho_f(e,h) = 0\} \text{ is compact.}$$

We also choose a left Haar integral μ of S, and let β be the associated $\mathscr{K}(S)$ regular measure.

Since X is metrizable and $\operatorname{Clos} A^{\times}(U)$ is compact, the topology of $A^{\times}(U)$ has a countable base. Consequently Φ has a countable subset Ω such that

$$A^{\times}(U) = \bigcup_{\omega \in \Omega} A^{\times}\{g: \omega(g) > 0\} = \bigcup_{\omega \in \Omega} \{y: \mu^{\times}(\omega)(y) > 0\}$$

(see 2.7.10), hence

$$U \subset \bigcup_{\omega \in \Omega} \{g: [\mu^{\times}(\omega) \circ A^{\times}](g) > 0\}.$$

Letting p be the characteristic function of the α measurable set

$$P = \bigcap_{f \in \Phi} \{g: f(g) = 0\},$$

we note that $P \cdot S = P$, hence

$$p(g\,s) = p(g) \text{ whenever } g \in G \text{ and } s \in S.$$

If $\omega \in \Omega$, then $f \in \Phi$ implies

$$p\,\omega + f \le \sup\{\omega, f\} \in \Phi, \qquad \int p\,\omega\,d\alpha + \lambda(f) \le \sup \lambda(\Phi);$$

therefore $\int p\,\omega\,d\alpha = 0$ and

$$\begin{aligned} \int p \cdot [\mu^{\times}(\omega) \circ A^{\times}]\,d\alpha &= \int p(g) \int \omega(g\,s)\,d\beta\,s\,d\alpha\,g \\ &= \iint p(g\,s)\,\omega(g\,s)\,d\alpha\,g\,d\beta\,s = \int \left(\int p\,\omega\,d\alpha\right) d\beta\,s = 0 \end{aligned}$$

because Δ_G and χ map the compact subgroup S onto $\{1\}$. We infer that $\alpha(P \cap U)=0$, and conclude

$$U=(X \sim P) \cup (P \cap U) \text{ is } \alpha \text{ measurable.}$$

2.7.15. Corollary. *Every χ covariant measure over X is a Radon measure.*

Proof. We readopt the hypothesis of 2.7.10, with the additional assumption that ν is χ covariant. It follows from 2.7.10 (6) that $\nu \circ \mu^x$ is left χ covariant, hence from 2.7.14 that α is a Radon measure.

For any open subset V of X with compact closure, we can choose an open subset U of G with compact closure such that $A^x(U)=V$. Then the characteristic function u of U is α measurable, hence $\beta^x(u)$ is a γ measurable function and

$$V=\{y\colon \beta^x(u)(y)>0\} \text{ is } \gamma \text{ measurable.}$$

2.7.16. Here we will illustrate the theory of invariant integration by several classical examples. For this purpose we recall the following elementary geometric concepts:

A subset S of a vectorspace V is an **affine subspace** if and only if $\alpha v+\beta w \in S$ whenever $v \in S$, $w \in S$, $\alpha \in \mathbf{R}$, $\beta \in \mathbf{R}$ and $\alpha+\beta=1$; in case $0 \in S$, this means that S is a linear subspace of V.

A function f on a vectorspace V into a vectorspace V' is an **affine map** if and only if

$$f(\alpha v+\beta w)=\alpha f(v)+\beta f(w)$$

whenever $v \in V$, $w \in V$, $\alpha \in \mathbf{R}$, $\beta \in \mathbf{R}$ and $\alpha+\beta=1$; this is equivalent to the condition that f is an affine subspace of $V \times V'$; in case $f(0)=0$ it means that f is linear.

The class $\mathscr{A}(V,V')$ of all affine maps of V into V' is a vectorspace. Using the linear epimorphism

$$L\colon \mathscr{A}(V,V') \to \operatorname{Hom}(V,V'), \quad L(f)(v)=f(v)-f(0),$$

we construct the linear isomorphism

$$M\colon \mathscr{A}(V,V') \simeq \operatorname{Hom}(V,V') \oplus V', \quad M(f)=(L(f), f(0)).$$

In case $V=V'$, L preserves functional composition, and the univalent map

$$\tau\colon V \to \mathscr{A}(V,V), \quad \tau_z(v)=z+v \text{ for } z,v \in V,$$

transforms vector addition into composition of maps. We observe that, whenever $g \in \operatorname{Hom}(V, V)$ and $z \in V$,

$$M^{-1}(g, z) = \tau_z \circ g, \qquad g \circ \tau_z = \tau_{g(z)} \circ g.$$

Furthermore $\operatorname{im} \tau = \{f \colon L(f) = \mathbf{1}_V\}$.

If $\dim V < \infty$, then $\dim \mathscr{A}(V, V) < \infty$ and all norms on $\mathscr{A}(V, V)$ induce the same locally compact topology. In particular we can choose an inner product on V and norm $\operatorname{Hom}(V, V)$ as in 1.7.6. Then we see that composition of linear maps is a continuous operation because

$$f \circ g - h \circ k = (f - h) \circ g + h \circ (g - k), \qquad \|f \circ g\| \leq \|f\| \cdot \|g\|$$

whenever $f, g, h, k \in \operatorname{Hom}(V, V)$. Moreover the automorphisms form a dense open subset of $\operatorname{Hom}(V, V)$, called the **linear group of** V, and inversion is a continuous operation because

$$f^{-1} - g^{-1} = f^{-1} \circ (g - f) \circ g^{-1},$$

hence

$$\|f^{-1}\| \cdot (1 - \|g - f\| \cdot \|g^{-1}\|) \leq \|g^{-1}\|,$$

whenever f, g are linear automorphisms of V. Using the maps L, M, τ we infer that functional composition is a continuous operation in $\mathscr{A}(V, V)$, and that inversion is continuous on the **affine group of** V, the open dense subset of univalent maps in $\mathscr{A}(V, V)$, which equals

$$L^{-1}(\text{linear group of } V).$$

It should be clear from the preceding remarks that the topological assumptions of our general theory are satisfied in each of the following particular **examples:**

(1) *$\mathscr{L}^1$ is the unique (left and right) Haar measure of the additive group of* $\mathbf{R}$ *such that* $\mathscr{L}^1\{t \colon 0 < t < 1\} = 1$. More generally, since the additive group of $\mathbf{R}^n$ is the direct sum of n factors equal to $\mathbf{R}$, *$\mathscr{L}$ is the unique (left and right) Haar measure over* $\mathbf{R}^n$ *such that*

$$\mathscr{L}^n\{x \colon 0 < x_i < 1 \text{ for } i = 1, \ldots, n\} = 1.$$

We verify next that, *for every linear endomorphism f of* $\mathbf{R}^n$,

$$\mathscr{L}^n[f(S)] = |\det f| \cdot \mathscr{L}^n(S) \text{ whenever } S \subset \mathbf{R}^n.$$

In case $\det f = 0$, $\operatorname{im} f$ is a proper subspace of $\mathbf{R}^n$, hence has $\mathscr{L}^n$ measure 0. In case $\det f \neq 0$, the assertion to be proved is equivalent to the equation

$$f_\#(\mathscr{L}^n) = |\det f|^{-1} \cdot \mathscr{L}^n.$$

If $z \in \mathbf{R}^n$ and $T \subset \mathbf{R}^n$, then

$$\begin{aligned} f_\#(\mathscr{L}^n)[z+T] &= \mathscr{L}^n[f^{-1}(z+T)] \\ &= \mathscr{L}^n[f^{-1}(z)+f^{-1}(T)] = \mathscr{L}^n[f^{-1}(T)] = f_\#(\mathscr{L}_n)[T]; \end{aligned}$$

thus $f_\#(\mathscr{L}^n)$ is a Haar measure of $\mathbf{R}^n$, hence

$$f_\#(\mathscr{L}^n) = \chi(f)\cdot\mathscr{L}^n \text{ with } \chi(f)>0.$$

The class $H=\{f\colon \chi(f)=|\det f|^{-1}\}$ is a subgroup of $\mathbf{GL}(n,\mathbf{R})$, **the (general) linear group of $\mathbf{R}^n$**, because χ and $|\det|^{-1}$ are homomorphisms. In case f is orthogonal, then f maps every open ball with center 0 onto itself, hence $\chi(f)=1$, $f\in H$. In case there exists positive numbers $c_1,\ldots,c_n$ such that

$$f(x)=(c_1 x_1,\ldots,c_n x_n) \text{ whenever } x\in\mathbf{R}^n,$$

then

$$\begin{aligned} &f\{x\colon 0<x_i<1 \text{ for } i=1,\ldots,n\} \\ &= \{x\colon 0<x_i<c_i \text{ for } i=1,\ldots,n\}, \end{aligned}$$

hence $\chi(f)=(c_1\cdot c_2\cdot\cdots\cdot c_n)^{-1}$, $f\in H$. Since all maps of these two special types generate $\mathbf{GL}(n,\mathbf{R})$, according to 1.7.3, we conclude that $H=\mathbf{GL}(n,\mathbf{R})$. *Thus $\mathscr{L}^n$ is $|\det|^{-1}$ covariant with respect to the action of $\mathbf{GL}(n,\mathbf{R})$ on $\mathbf{R}^n$.* (This action is not transitive, but a transitive action results through replacement of $\mathbf{R}^n$ by $\mathbf{R}^n \sim \{0\}$, which really does not matter because $\mathscr{L}^n\{0\}=0$.)

In case $n\geq 2$ the special linear group

$$\mathbf{SL}(n,\mathbf{R}) = \mathbf{GL}(n,\mathbf{R}) \cap \{f\colon \det(f)=1\}$$

acts transitively on $\mathbf{R}^n\sim\{0\}$, and $\mathscr{L}^n$ is invariant with respect to the action of $\mathbf{SL}(n,\mathbf{R})$. It follows that *no nonzero Radon measure ϕ over $\mathbf{R}^n\sim\{0\}$ can be invariant with respect to the action of* $\mathbf{GL}(n,\mathbf{R})$, because ϕ would be $\mathbf{SL}(n,\mathbf{R})$ invariant, hence a positive multiple of $\mathscr{L}^n$.

We define

$$\alpha(n)=\mathscr{L}^n\{x\colon |x|<1\}$$

and observe that, whenever $a\in\mathbf{R}^n$ and $r>0$,

$$\alpha(n)\,r^n=\mathscr{L}^n\{x\colon |x-a|<r\}$$

because $\mathscr{L}^n$ is invariant under translation by a, and multiplication by r has determinant r^n. Moreover $\alpha(1)=2$, and for $n\geq 2$ Fubini's Theorem yields the recursion formula

$$\begin{aligned} \alpha(n) &= \textstyle\int_{-1}^{1} \mathscr{L}^{n-1}\{y\colon |y|<(1-t^2)^{\frac{1}{2}}\}\, d\mathscr{L}^1 t \\ &= \alpha(n-1)\textstyle\int_{-1}^{1}(1-t^2)^{(n-2)/2}\, d\mathscr{L}^1 t. \end{aligned}$$

We also define $\alpha(0)=1$. In 3.2.13 we will compute $\alpha(n)$ in terms of Euler's function Γ.

The $\mathscr{L}^n$ measure of a **rectangular box**

$$Q = \{x\colon 0 \le (x-a) \bullet e_i \le c_i \text{ for } i=1,\dots,n\},$$

where $a \in \mathbf{R}^n$, $c_i \ge 0$, and $e_1, \dots, e_n$ form an orthonormal base of $\mathbf{R}^n$, equals the product $c_1 \cdot \dots \cdot c_n$.

If S is the convex hull of $n+1$ points $v_0, \dots, v_n$ of $\mathbf{R}^n$, then

$$\mathscr{L}^n(S) = |(v_1 - v_0) \wedge \dots \wedge (v_n - v_0)|/n!.$$

Since the effect of a linear endomorphism f or $\mathbf{R}^n$ is to multiply both sides of this equation by $|\det f|$, and both sides are invariant under translations, it suffices to verify the equation in case $v_0 = 0$ and $v_1, \dots, v_n$ form the standard orthonormal base of $\mathbf{R}^n$; but then

$$S = \mathbf{R}^n \cap \left\{x\colon x_j \ge 0 \text{ for } j \in \{1,\dots,n\},\ \sum_{j=1}^{n} x_j \le 1\right\}$$

and using 2.6.2, 2.9.20 one inductively computes

$$\mathscr{L}^n(S) = \int_0^1 \mathscr{L}^{n-1}\left\{y\colon y_j \ge 0 \text{ for } j \in \{1,\dots,n-1\},\ \sum_{j=1}^{n-1} y_j \le 1-t\right\} d\mathscr{L}^1 t$$

$$= \int_0^1 (1-t)^{n-1}/(n-1)!\, d\mathscr{L}^1 t = 1/n!$$

(2) *Suppose U and V are vectorspaces with*

$$\dim U = m \le \dim V = n < \infty.$$

The linear group G of V acts (by functional composition) *on the vectorspace $X = \operatorname{Hom}(U,V)$* of all linear maps of U into V. *If ϕ is a Haar measure of the additive group of X, then ϕ is $|\det|^{-m}$ covariant with respect to the action of G.* In fact, using a base $e_1, \dots, e_m$ of U we construct a G action preserving isomorphism $X \simeq V^m$ by mapping $f \in X$ onto $(f(e_1), \dots, f(e_m)) \in V^m$. We observe that the *monomorphisms form an open subset of X whose complement has ϕ measure* 0, because f fails to be univalent if and only if it satisfies the polynomial equation

$$|f(e_1) \wedge \dots \wedge f(e_m)|^2 = 0.$$

Moreover G acts transitively on the set of monomorphisms.

In case $U = V$, ϕ is a left $|\det|^{-n}$ covariant measure over G. According to 2.7.9 we can define a left Haar integral λ of G by the formula

$$\lambda(u) = \int u \cdot |\det|^n\, d\phi \quad \text{whenever } u \in \mathscr{K}(G).$$

We also know that ϕ is right $\Delta_G\, |\det|^n$ covariant over G. On the other hand the linear isomorphism

$$\textstyle\bigwedge^1\colon \operatorname{Hom}(V,V) \simeq \operatorname{Hom}(\bigwedge^1 V, \bigwedge^1 V)$$

maps G onto the group G' of all linear automorphisms of $\wedge^1 V$, and $(\wedge^1)_\# \phi$ is a Haar measure of the additive group of $\operatorname{Hom}(\wedge^1 V, \wedge^1 V)$, hence $(\wedge^1)_\# \phi$ is a left $|\det|^{-n}$ covariant measure over G'. For each $g \in G$ the diagram

$$\begin{array}{ccc} G & \xrightarrow{\text{right multiplication by } g^{-1}} & G \\ \downarrow{\scriptstyle \wedge^1} & & \downarrow{\scriptstyle \wedge^1} \\ G' & \xrightarrow{\text{left multiplication by } (\wedge^1 g)^{-1}} & G' \end{array}$$

is commutative, with $|\det(\wedge^1 g)^{-1}|^{-n} = |\det g|^n$; therefore ϕ is right $|\det|^n$ covariant. We conclude that $\Delta_G = 1$, *G is unimodular.*

(3) *Suppose* $W \subset V$ *are vectorspaces with*

$$0 < \dim W = m < \dim V = n < \infty,$$

and G is the group of those linear automorphisms of V which map W onto W. Letting X be the vectorsubspace of $\operatorname{Hom}(V, V)$ generated by G, we shall prove first that every Haar measure γ of the additive group of X is χ covariant, where

$$\chi(g) = |\det(g|W)|^{-m} |\det g|^{m-n} \text{ whenever } g \in G.$$

For this purpose we choose a subspace Z of V so that $V = W \oplus Z$ and construct a left G action preserving isomorphism

$$\xi \colon X \simeq \operatorname{Hom}(W, W) \times \operatorname{Hom}(Z, V), \quad \xi(f) = (f|W, f|Z) \text{ whenever } f \in X.$$

If ϕ, ψ are Haar measures of the additive groups of $\operatorname{Hom}(W, W)$, $\operatorname{Hom}(Z, V)$, then $\xi(G)$ is an open set whose complement has $\phi \times \psi$ measure 0. We know from (2) that ϕ is $|\det|^{-m}$ covariant with respect to left action of the linear group of W, and ψ is $|\det|^{m-n}$ covariant with respect to left action of the linear group of V. Therefore left multiplication by $g \in G$ maps $\phi \times \psi$ onto

$$|\det(g|W)|^{-m} \phi \times |\det g|^{m-n} \psi = \chi(g) \cdot (\phi \times \psi).$$

Therefore every Haar measure γ of X is right $\Delta_G \chi^{-1}$ covariant over G. On the other hand $\wedge^1$ maps X onto the vectorspace X' generated by the group G' of those linear automorphisms of $\wedge^1 V$ which map

$$W' = \wedge^1 V \cap \{u \colon u|W = 0\}$$

onto W', and $(\wedge^1)_\# \gamma$ is a Haar measure of the additive group of X', hence $(\wedge^1)_\# \gamma$ is left χ' covariant over G', where

$$\chi'(h) = |\det(h|W')|^{m-n} \cdot |\det h|^{-m} \text{ whenever } h \in G'.$$

For each $g \in G$ the commutative diagram in (2) now implies

$$\Delta_G(g)\,\chi^{-1}(g) = \chi'(\wedge^1 g^{-1}),$$

and since $\det(\wedge^1 g) = \det(g) = \det(g|W)\det(\wedge^1 g|W')$ it follows that

$$\Delta_G(g) = |\det(g|W)|^{-n} \cdot |\det g|^m.$$

In particular, G *is not unimodular.*

(4) *For* $0 < m < n$, *the action of* $\mathbf{GL}(n, \mathbf{R})$ *on* $\mathbf{G}(n, m)$ (see 1.6.2) *admits no* χ *covariant integral, for any homomorphism* χ *of* $\mathbf{GL}(n, \mathbf{R})$ *into the multiplicative group* $\mathbf{R} \sim \{0\}$. In fact, if $W \in \mathbf{G}(n, m)$ and $V = \mathbf{R}^n$, the isotropy group $\mathbf{GL}(n, \mathbf{R})^W$ is the group studied in (3), hence Weil's condition becomes:

$$\chi(t) \cdot |\det(t|W)|^{-n} \cdot |\det t|^m = 1 \quad \text{for } t \in \mathbf{GL}(n, \mathbf{R})^W.$$

Since $\{g: \det g = 1\}$ is the commutator subgroup of $\mathbf{GL}(n, \mathbf{R})$, $\chi(g) = 1$ whenever $\det g = 1$. Therefore Weil's condition is violated by any $t \in \mathbf{GL}(n, \mathbf{R})^W$ for which $\det t = 1$ and $\det(t|W) \neq 1$.

However, *the action of* $\mathbf{GL}(n, \mathbf{R})$ *on the space of all nonzero simple m vectors of* $\mathbf{R}^n$ *admits a* $|\det|^{-m}$ *covariant integral.* In fact, if W is the m dimensional subspace associated with a nonzero simple m vector x, the isotropy group at x consists of all those endomorphisms g of $\mathbf{R}^n$ for which $g(W) = W$ and $\det(g|W) = 1$, hence is a normal subgroup of the group studied in (3), and it follows from 2.7.12 that the modular function of the present isotropy group equals $|\det|^m$.

(5) *If* G *is the affine group of* $\mathbf{R}^n$, *then*

$$\Delta_G(g) = |\det L(g)|^{-1} \quad \textit{whenever } g \in G.$$

To prove this we note first that $\ker L|G = \operatorname{im} \tau$ is abelian, hence $\Delta_G|\ker L = 1$. Then we observe that $\mathscr{L}^n$ is $|\det \circ L|^{-1}$ covariant with respect to the usual action of G on $\mathbf{R}^n$, with $G^0 = \mathbf{GL}(n, \mathbf{R})$, hence $|\det \circ L|^{-1}$ and Δ_G agree on $\mathbf{GL}(n, \mathbf{R})$.

For $0 < m < n$, *there exists no* χ *covariant integral with respect to the action of* G *on the space* X *of all* m *dimensional affine subspaces of* $\mathbf{R}^n$. In fact, choosing an m dimensional vectorsubspace W of $\mathbf{R}^n$ we now find that the isotropy group G^W equals $M^{-1}(H \times W)$, where

$$H = \mathbf{GL}(n, \mathbf{R}) \cap \{h: h(W) = W\},$$

and argue as above to prove that the modular function of G^W has the value $\det[L(g)|W]^{-1} \Delta_H[L(g)]$ at g. Since H is the group studied in (3) [under the name "G"], we can show as in (4) that Weil's condition fails in the present case.

(6) Since the **orthogonal group** $\mathbf{O}(n)$ is a compact subset of $\operatorname{Hom}(\mathbf{R}^n, \mathbf{R}^n)$, there exists a unique *left and right Haar measure* θ_n *over* $\mathbf{O}(n)$ *such that* $\theta_n[\mathbf{O}(n)]=1$. For every transitive action A of $\mathbf{O}(n)$ on a homogeneous space X, and each $x \in X$, the measure $(A^x)_* \theta_n$ over X is invariant with

$$[(A^x)_* \theta_n] X = 1.$$

In particular we thus obtain the $\mathbf{O}(n)$ invariant measures

$$\theta_{n,m} \text{ over } \mathbf{O}(n,m), \quad \theta^*_{n,m} \text{ over } \mathbf{O}^*(n,m), \quad \gamma_{n,m} \text{ over } \mathbf{G}(n,m).$$

Whenever $n \geq m$ *are positive integers and* $1 \leq t < \infty$, *there exists a positive number* $\beta_t(n,m)$ *such that*

$$\left(\int |(\textstyle\bigwedge_m p)\,\xi|^t \, d\theta^*_{n,m}\, p\right)^{1/t} = \beta_t(n,m) \cdot |\xi|$$

for every simple m-vector ξ *of* $\mathbf{R}^n$. In fact the integral is positively homogeneous of degree t with respect to ξ, the group $\mathbf{O}(n)$ acts transitively on the set of all simple m vectors of norm 1, and the invariance of $\theta^*_{n,m}$ implies the integrals corresponding to ξ and $(\bigwedge_m g)\,\xi$ are equal whenever $g \in \mathbf{O}(n)$.

If $\xi \in \bigwedge_m \mathbf{R}^n$, $\eta \in \bigwedge_k \mathbf{R}^n$, $m+k \leq n$ *and* ξ, η *are simple, then*

$$\int_{\mathbf{O}(n)} |\langle \xi, \textstyle\bigwedge_m g \rangle \wedge \eta|^t \, d\theta_n\, g = \beta_t(n,m)^t \, \beta_t(n-k,m)^{-t} \, |\xi|^t \, |\eta|^t.$$

To prove this we assume $|\xi| = 1 = |\eta|$, choose $q \in \mathbf{O}^*(n, n-k)$ with

$$\mathbf{R}^n \cap \{v\colon v \wedge \eta = 0\} = \ker q = \operatorname{im}(\mathbf{1}_{\mathbf{R}^n} - q^* \circ q),$$

and observe that, for any $\zeta \in \bigwedge_m \mathbf{R}^n$,

$$|\zeta \wedge \eta| = |\langle \zeta, \textstyle\bigwedge_m (q^* \circ q) \rangle \wedge \eta| = |\langle \zeta, \textstyle\bigwedge_m q \rangle|,$$

$$\beta_t(n-k,m)^t \, |\zeta \wedge \eta|^t = \int |\langle \zeta, \textstyle\bigwedge_m (r \circ q) \rangle|^t \, d\theta^*_{n-k,m}\, r.$$

Using Fubini's theorem and the fact that $r \circ q \in \mathbf{O}^*(n,m)$ for each $r \in \mathbf{O}^*(n-k,m)$, we infer

$$\begin{aligned} &\beta_t(n-k,m)^t \int |\langle \xi, \textstyle\bigwedge_m g \rangle \wedge \eta|^t \, d\theta_n\, g \\ &= \iint |\langle \xi, \textstyle\bigwedge_m (r \circ q \circ g) \rangle|^t \, d\theta_n\, g \, d\theta^*_{n-k,m}\, r \\ &= \int \beta_t(n,m)^t \, d\theta^*_{n-k,m}\, r = \beta_t(n,m)^t. \end{aligned}$$

As a corollary we obtain the equation

$$\beta_t(n,m)/\beta_t(n-k,m) = \beta_t(n,k)/\beta_t(n-m,k),$$

because $\mathbf{O}(n)$ is unimodular and

$$|\langle \xi, \textstyle\bigwedge_m g \rangle \wedge \eta| = |\xi \wedge \langle \eta, \textstyle\bigwedge_k g^{-1} \rangle| \text{ for } g \in \mathbf{O}(n).$$

In 3.2.13 we will compute $\boldsymbol{\beta}_t(n,m)$ in terms of Euler's function $\boldsymbol{\Gamma}$.

We also define $\boldsymbol{\beta}_\infty(n,m)=1$. From 2.4.12 and 1.7.7 we see that, *if* $\xi\in\bigwedge_m \mathbf{R}^n$, ξ *is simple and*

$$f(p)=|\langle\xi,\bigwedge_m p\rangle| \text{ for } p\in\mathbf{O}^*(n,m),$$

then $[\boldsymbol{\theta}^*_{n,m}]_{(t)}(f)=\boldsymbol{\beta}_t(n,m)\cdot|\xi|$ *whenever* $1\le t\le\infty$.

(7) *Suppose G is the group of all isometries* (distance preserving transformations) *of* $\mathbf{R}^n$. Then

$$G^0=G\cap\{g\colon g(0)=0\}=\mathbf{O}(n),$$

because for $g\in G^0$, $x\in\mathbf{R}^n$, $y\in\mathbf{R}^n$, $t\in\mathbf{R}$ the equations

$$|g(x)-g(y)|^2=|x-y|^2,\qquad |g(x)|^2=|x|^2,\qquad |g(y)|^2=|y|^2$$

show, when expanded in terms of inner products, that

$$g(x)\bullet g(y)=x\bullet y,\qquad |g(x)+g(y)-g(x+y)|^2=0,$$
$$|g(t\,x)-t\,g(x)|^2=0, \text{ hence } g\in\mathbf{O}(n).$$

We infer that $G=A(V,V)\cap\{g\colon L(g)\in\mathbf{O}(n)\}=\operatorname{im}\boldsymbol{\tau}\circ\mathbf{O}(n)$. Since $\mathscr{L}^n$ is G invariant and $\mathbf{O}(n)$ is compact, *G is unimodular*; furthermore we may apply 2.7.10 to *define a Haar integral* λ *on* $\mathscr{K}(G)$ *by the formula*

$$\lambda(u)=\int\theta_n^0(u)\,d\mathscr{L}^n=\iint u(\boldsymbol{\tau}_z\circ s)\,d\boldsymbol{\theta}_n\,s\,d\mathscr{L}^n\,z$$

whenever $u\in\mathscr{K}(G)$.

If X is the space of all m dimensional affine subspaces of $\mathbf{R}^n$, *and* $p\in\mathbf{O}^*(n,n-m)$, *then a G invariant integral* μ *on* $\mathscr{K}(X)$ *is given by the formula*

$$\mu(v)=\iint v[s(p^{-1}\{y\})]\,d\boldsymbol{\theta}_n\,s\,d\mathscr{L}^{n-m}\,y$$

whenever $v\in\mathscr{K}(X)$. To prove this we consider the maps

$$\xi\colon\mathbf{O}(n)\times\mathbf{R}^{n-m}\to X,\qquad \xi(s,y)=s(p^{-1}\{y\}),$$
$$\delta\colon X\to\mathbf{R},\qquad \delta(W)=\operatorname{distance}(W,0).$$

Since $\operatorname{im}p^*=\operatorname{im}(p^*\circ p)$ is the orthogonal complement of $\ker p$,

$$p^{-1}\{y\}=\boldsymbol{\tau}_{p^*(y)}\ker p \text{ whenever } y\in\mathbf{R}^{n-m},$$
$$\boldsymbol{\tau}_z\ker p=\boldsymbol{\tau}_{(p^*\cdot p)z}\ker p \text{ whenever } z\in\mathbf{R}^n,$$
$$\delta[(s\circ\boldsymbol{\tau}_z)\ker p]=|p(z)| \text{ whenever } z\in\mathbf{R}^n,\ s\in\mathbf{O}(n),$$
$$(\delta\circ\xi)(s,y)=|y| \text{ whenever } y\in\mathbf{R}^{n-m},\ s\in\mathbf{O}(n),$$

we see that ξ is continuous and open, $\operatorname{im} \xi = X$, δ is continuous, ξ is proper, hence $\gamma = \xi_\# (\theta_n \times \mathscr{L}^{n-m})$ is a Radon measure over X. Then we use the fact that

$$t\,\xi(s, y) = \xi(t\,s, y), \qquad \tau_z\,\xi(s, y) = \xi(s, y + (p \circ s^{-1})\,z)$$

whenever $s \in \mathbf{O}(n)$, $y \in \mathbf{R}^{n-m}$, $t \in \mathbf{O}(n)$, $z \in \mathbf{R}^n$ to verify that γ is invariant under the action of G.

Since $s[p^{-1}\{y\}] = (p\,s^{-1})^{-1}\{y\}$ for $s \in \mathbf{O}(n)$, $y \in \mathbf{R}^n$, and the function mapping s onto $p\,s^{-1}$ carries θ_n onto $\theta^*_{n,n-m}$ [see (6)], we obtain the alternate formula

$$\mu(v) = \iint v(q^{-1}\{y\})\, d\theta^*_{n,n-m}\, q\, d\mathscr{L}^{n-m}\, y.$$

2.7.17. If G is a compact topological group, S is a countably infinite subgroup of G, and T is a subset of G which has exactly one point in common with each left coset of S, then T is nonmeasurable with respect to any Haar measure of G. In fact the countably many sets $T s$ corresponding to $s \in S$ are disjoint, have equal measures, and add up to G. If T were measurable, so would the sets $T s$, and G would have infinite measure.

This construction was used by Vitali (with $G = \mathbf{R}/\mathbf{Z}$, $S = \mathbf{Q}/\mathbf{Z}$) in the first discovery of a nonmeasurable set.

2.7.18. *If γ is a χ covariant measure over X, $v \in \mathbf{L}_1(\gamma)$ and $h \in G$, then*

$$\int |v(g\,x) - v(h\,x)|\, d\gamma\, x \to 0 \quad \text{as} \quad G \ni g \to h.$$

In fact, given $\varepsilon > 0$ we can choose $w \in \mathscr{K}(X)$ with $\int |v - w|\, d\gamma < \varepsilon$ and infer that, for every $g \in G$,

$$|v_g - v_h| \leq |w_g - w_h| + |v - w|_g + |v - w|_h,$$

$$\int |v_g - v_h|\, d\gamma \leq \int |w_g - w_h|\, d\gamma + \chi(g)\,\varepsilon + \chi(h)\,\varepsilon;$$

moreover $\chi(h) \to \chi(g)$ and $\int |w_g - w_h|\, d\gamma \to 0$ as $g \to h$, because $\operatorname{spt} w$ is compact and w is uniformly continuous.

2.8. Covering theorems

2.8.1. In this section we consider a fixed *space X with a metric ρ*, and let

$$\mathbf{B}(a, r) = \{x\colon\ \rho(a, x) \leq r\}, \qquad \mathbf{U}(a, r) = \{x\colon\ \rho(a, x) < r\}$$

be the closed and open **balls** with center $a \in X$ and radius $r > 0$. We also assume that *F is a family of closed subsets of X, ϕ measures X, every open subset of X is ϕ measurable*, and *every bounded subset of X has finite ϕ measure.*

We say that F **covers** A **finely** if and only if for any $a \in A$, $\varepsilon > 0$ there exists a set $B \in F$ with $a \in B \subset \mathbf{U}(a, \varepsilon)$.

We say that F is ϕ **adequate for** A if and only if for each open subset V of X there is a countable disjointed subfamily G of F with

$$\bigcup G \subset V \quad \text{and} \quad \phi[(V \cap A) \sim \bigcup G] = 0.$$

The work of this section has the following motivation: In order to derive global from local properties of ϕ, with respect to a subset A of X, one frequently needs to find a countable subfamily G of a given covering F of A such that

$$\sum_{B \in G} \phi(B)$$

can be used to estimate $\phi(A)$. This sum provides an *upper estimate* for $\phi(A)$ if G **covers** ϕ **almost all of** A, which means that $\phi(A \sim \bigcup G) = 0$, or more generally if A can be covered by sets obtained from the members of G through enlargement increasing ϕ measure by at most a known constant factor. On the other hand, assuming all members of G are contained in some neighborhood of A whose ϕ measure is not much greater than $\phi(A)$, the above sum provides a *lower estimate* for $\phi(A)$ if G is disjointed, or more generally if there is a known upper bound for the number of members of G to which any point of A may belong. Such applications of the covering theorems proved here will occur in 2.9, 2.10, 3.3 and 4.2.

2.8.2. Theorem. *Suppose K is a countable family of subsets of X, and to each $A \in K$ corresponds a number $\sigma(A)$ with $0 < \sigma(A) < 1$ and the following property: For every open subset W of X there exists a countable disjointed subfamily H of F such that*

$$\bigcup H \subset W \quad \textit{and} \quad \phi[(W \cap A) \sim \bigcup H] \leq \sigma(A) \cdot \phi(W \cap A).$$

Then F is ϕ adequate for $\bigcup K$.

Proof. Since X is the union of countably many bounded sets, we may assume that each member of K is bounded. We also construct a sequence $B_1, B_2, B_3, \ldots$ in which every member of K occurs infinitely often.

Given any open subset V of X, we define open sets W_j and finite disjointed subfamilies G_j of F by induction, starting with $W_0 = V$ and $G_0 = \phi$, in such a way that

$$W_j = W_{j-1} \sim \bigcup G_{j-1}, \qquad \bigcup G_j \subset W_j$$

and

$$\phi[(W_j \cap B_j) \sim \bigcup G_j] \leq \sigma(B_j)^{\frac{1}{2}} \, \phi(W_j \cap B_j)$$

whenever $j \geq 1$; this is possible because $\bigcup G_{j-1}$ is closed, F has a countable disjointed subfamily H for which $\bigcup H \subset W_j$ and

$$\phi[(W_j \cap B_j) \sim \bigcup H] \leq \sigma(B_j)\, \phi(W_j \cap B_j),$$

$\phi(B_j) < \infty$ and the members of H are ϕ measurable, hence G_j can be chosen as a sufficiently large finite subfamily of H.

It follows that

$$G = \bigcup_{j=1}^{\infty} G_j$$

is a countable disjointed subfamily of F with

$$\bigcup G \subset V \text{ and } V \sim \bigcup G = \bigcap_{j=1}^{\infty} W_j.$$

Moreover G covers ϕ almost all of $V \cap \bigcup K$; in fact

$$\lim_{j \to \infty} \phi(A \cap W_j) = 0 \text{ for each } A \in K,$$

because $\phi(A) < \infty$, $\sigma(A)^{\frac{1}{2}} < 1$ and there exist infinitely many positive integers j for which $B_j = A$, hence

$$\phi(A \cap W_{j+1}) = \phi[(A \cap W_j) \sim \bigcup G_j] \leq \sigma(A)^{\frac{1}{2}}\, \phi(A \cap W_j).$$

2.8.3. Corollary. *If F is ϕ adequate for each member of a countable family K, then F is ϕ adequate for $\bigcup K$.*

2.8.4. Theorem. *If δ is a nonnegative bounded function on F, and $1 < \tau < \infty$, then F has a disjointed subfamily G such that for each $T \in F$ there exists an $S \in G$ with*

$$T \cap S \neq \varnothing \text{ and } \delta(T) \leq \tau\, \delta(S).$$

Proof. We consider the class Ω of all disjointed subfamilies H of F with the following property: *Whenever $T \in F$,*

either $T \cap S = \varnothing$ for all $S \in H$,

or $T \cap S \neq \varnothing$ and $\delta(T) \leq \tau\, \delta(S)$ for some $S \in H$.

Applying Hausdorff's maximal principle [K, p. 33] we choose $G \in \Omega$ so that G is not a proper subset of any member of Ω.

Let $K = F \cap \{T: T \cap \bigcup G = \varnothing\}$. If $K \neq \varnothing$ we could select $W \in K$ so that

$$\tau\, \delta(W) \geq \sup\{\delta(T): T \in K\},$$

and verify that $G \cup \{W\} \in \Omega$, contrary to the maximal choice of G.

For use in the following corollaries and their applications we associate with each member S of F its δ, τ **enlargement**

$$\hat{S} = \bigcup \{T\colon T \in F,\ T \cap S \neq \emptyset,\ \delta(T) \leq \tau\,\delta(S)\}.$$

2.8.5. Corollary. $\bigcup F \subset \bigcup \{\hat{S}\colon S \in G\}$.

2.8.6. Corollary. *If F covers A finely and H is any finite subset of G, then*

$$A \sim \bigcup H \subset \bigcup \{\hat{S}\colon S \in G \sim H\}.$$

Proof. Since $\bigcup H$ is closed, each point of $A \sim \bigcup H$ belongs to some member T of F for which $T \cap \bigcup H = \emptyset$. Therefore T meets some $S \in G \sim H$ with $\delta(T) \leq \tau\,\delta(S)$, hence $T \subset \hat{S}$.

2.8.7. Theorem. *If F covers A finely, δ is a nonnegative bounded function on F, $1 < \tau < \infty$, $1 < \lambda < \infty$, and*

$$\phi(\hat{S}) < \lambda\,\phi(S)$$

whenever $S \in F$ and $\hat{S}$ is the δ, τ enlargement of S, then F is ϕ adequate for A.

Proof. In view of 2.8.3 we may assume that A is bounded.

Given any bounded open subset V of X, we observe that

$$F' = F \cap \{S\colon S \subset V\}$$

covers $V \cap A$ finely, and apply 2.8.4–2.8.6 to obtain a disjointed subfamily G of F' such that

$$(V \cap A) \sim \bigcup H \subset \bigcup \{\hat{S}\colon S \in G \sim H\}$$

for every finite subfamily H of G. Since $\phi(V) < \infty$ and

$$S \subset V \text{ with } \phi(S) > 0 \text{ whenever } S \in G,$$

the family G is countable and

$$\sum_{S \in G} \phi(\hat{S}) \leq \lambda \sum_{S \in G} \phi(S) \leq \lambda\,\phi(V) < \infty.$$

It follows that for every $\varepsilon > 0$ there exists a finite subfamily H of G with

$$\sum_{S \in G \sim H} \phi(\hat{S}) < \varepsilon,$$

hence $\phi[(V \cap A) \sim \bigcup G] \leq \phi[(V \cap A) \sim \bigcup H] < \varepsilon$.

2.8.8. The results of 2.8.4–2.8.7 are applied most frequently with

$$\delta(S) = \operatorname{diam}(S) = \sup \rho(S \times S) \text{ for } S \in F.$$

(Other choices of δ occur in [F 11, p. 296] and [M 2, p. 215].) In case F is a family of closed balls and $\delta = \operatorname{diam}$, then

$$S = \mathbf{B}(a, r) \in F \text{ implies } \hat{S} \subset \mathbf{B}[a, (1+2\tau) r],$$

hence the inequality hypothesized in 2.8.7 holds whenever ϕ satisfies the **diametric regularity** condition:

$$\phi \mathbf{B}[a, (1+2\tau) r] < \lambda \cdot \phi \mathbf{B}(a, r).$$

Sufficient for diametric regularity is the validity of an estimate

$$0 < u \le s^{-n} \phi \mathbf{B}(a, s) \le v < \infty$$

which is applicable both with $s = r$ and with $s = (2\tau + 1) r$ whenever $\mathbf{B}(a, r) \in F$; examples of such measures ϕ are $\mathscr{L}^n$ over $\mathbf{R}^n$ [see 2.7.16(1)] and, more generally, $\mathscr{H}^n$ over any n dimensional Riemannian manifold [see 3.2.49]. However diametric regularity may hold in the absence of such an estimate, for instance for the Borel regular measure ϕ over $\mathbf{R}$ such that

$$\phi(B) = \int_B \sup \{|x|^{-\frac{1}{2}}, 1\} \, d\mathscr{L}^1 x$$

whenever B is a Borel subset of $\mathbf{R}$.

The preceding covering theorems are very useful in the presence of some regularity hypothesis, and suffice for most applications. However the study of some important problems, where no such regularity hypothesis is available, demands the use of another type of covering theorems, to be discussed in 2.8.9–2.8.14. There we will impose no special hypothesis on the measure ϕ, but require that F be a family of closed balls, that every point of A be the center (not only an element!) of some member of F, and that the metric ρ satisfy the geometric condition described in 2.8.9.

2.8.9. By saying that ρ is **directionally ξ, η, ζ limited at** A we mean that $A \subset X$, $\xi > 0$, $0 < \eta \le \frac{1}{3}$, ζ is a positive integer and the following condition holds:

If $a \in A$, $B \subset A \cap \mathbf{U}(a, \xi) \sim \{a\}$, *and if*

$$\rho(x, c)/\rho(a, c) \ge \eta$$

whenever $b \in B$, $c \in B$, $b \neq c$, $\rho(a, b) \ge \rho(a, c)$, $x \in X$ *with*

$$\rho(a, x) = \rho(a, c), \ \rho(x, b) = \rho(a, b) - \rho(a, c),$$

then card $B \le \zeta$.

The utility of this concept will appear in 2.8.11. We now illustrate its geometric significance by two examples:

An important *special case* (motivating the term "directional") occurs when *X is a finite dimensional vectorspace and*

$$\rho(x, y) = \nu(x-y) \text{ for } x, y \in X,$$

where ν *is any norm on* X, *and* $A = X$. Taking

$$x = a + [\nu(c-a)/\nu(b-a)] \cdot (b-a)$$

one finds that

$$\rho(x, c)/\rho(a, c) = \rho[(b-a)/\nu(b-a), (c-a)/\nu(c-a)]$$

is the distance between the directions from a to b and c. Since these directions belong to the compact set of all ν unit vectors, the above condition holds for each $\eta > 0$ with a suitable ζ; here $\xi = \infty$. (In case ν is strictly convex, in particular if ν comes from an inner product, the point x is uniquely determined by a, b, c.)

Another interesting *special case* (whose study led me to formulate the general concept) arises when X is a *Riemannian manifold* (of class ≥ 2) with its usual metric (see [HE, p. 51] or [KN, p. 157]) and A is any compact subset of X. Assuming that $\mathbf{U}(a, \xi)$ is a normal neighborhood of $a \in A$, and that the restriction of Exp_a to the ball of radius ξ has the Lipschitz constant λ, we find that if $b = \mathrm{Exp}_a(\beta)$ and $c = \mathrm{Exp}_a(\gamma)$, then

$$\rho(a, b) = |\beta|, \quad \rho(a, c) = |\gamma|, \quad x = \mathrm{Exp}_a[(|\gamma|/|\beta|) \cdot \beta],$$
$$\rho(x, b)/\rho(a, c) \leq \lambda \, |\beta/|\beta| - \gamma/|\gamma||.$$

We thus associate with each $b \in B$ the direction $\beta/|\beta|$ of the shortest geodesic from a to b, and infer from the compactness of the set of all unit tangent vectors at the points of A that a suitable ζ exists for each $\eta > 0$.

Since discussion of families of closed balls involves constant reference to centers and radii, and since neither center nor radius of a ball need be unique, we shall first prove certain propositions about subsets P of

$$X \times \{r : 0 < r < \infty\},$$

and later deduce results about the corresponding families

$$\{\mathbf{B}(a, r) : (a, r) \in P\}.$$

Assuming $1 < \tau < \infty$, we say that such a set P is τ **controlled** if and only if, *for any two distinct elements* (a, r) *and* (b, s) *of* P,

$$\text{either } \rho(a, b) > r > s/\tau \text{ or } \rho(a, b) > s > r/\tau.$$

[Note that this implies $a \neq b$ and $\mathbf{B}(a, r) \neq \mathbf{B}(b, s)$.]

2.8.10. Lemma. *If* $1<\tau<\infty, 0<\mu<\infty$ *and*

$$Q\subset X\times\{r\colon 0<r<\mu\},$$

then Q *has a subset* R *satisfying the conditions:*

(1) *Whenever* (a, r) *and* (b, s) *are distinct elements of* R,

$$\rho(a, b)>r+s.$$

(2) *Whenever* $(a, r)\in Q$ *there exists* $(b, s)\in R$ *with*

$$\rho(a, b)\leq r+s \text{ and } s>r/\tau.$$

Proof. We consider the class Ω of all those subsets R of Q which satisfy (1) as well as the condition:

(3) *Whenever* $(a, r)\in Q$,
either $\rho(a, b)>r+s$ *for all* $(b, s)\in R$
or $\rho(a, b)\leq r+s$ *and* $s>r/\tau$ *for some* $(b, s)\in R$.

Observing that $\varnothing\in\Omega$, and the union of every nest contained in Ω is a member of Ω, we infer from Hausdorff's maximal principle [K, p. 33] that Ω has a maximal member (with respect to inclusion).

Such a maximal member R of Ω satisfies (2), because otherwise (3) would imply

$$K=Q\cap\{(a, r)\colon \rho(a, b)>r+s \text{ for all } (b, s)\in R\}\neq\varnothing,$$

hence we could choose $(c, t)\in K$ with

$$\tau t>\sup\{r\colon (a, r)\in K\},$$

and verify that $R\cup\{(c, t)\}\in\Omega$, contrary to the maximality of R.

2.8.11. Theorem. *Suppose* ρ *is directionally* ξ, η, ζ *limited at* A *and*

$$1<\tau<2-\eta, \quad \eta+\tau/(2-\eta)+\tau(\tau-1)<1;$$

such numbers τ *exist because.* $0<\eta\leq\frac{1}{3}$.

If P *is a* τ *controlled subset of* $A\times\{r\colon 0<r<\infty\}$, *if* $(a, r)\in P$, *and if*

$$\rho(a, b)<\xi,\ \rho(a, b)\leq r+s \text{ and } s>r/\tau$$

whenever $(b, s)\in P$, *then* card $P\leq 2\zeta+1$.

Proof. Since $\eta+1/(2-\eta)\leq\frac{1}{3}+\frac{3}{5}<1$, any number slightly larger than 1 is a permissible choice for τ.

With $k=(2-\eta)/\tau$ we define

$$P_1 = P \cap \{(b,s)\colon 0<\rho(a,b)\le kr\},$$
$$P_2 = P \cap \{(b,s)\colon \rho(a,b)>kr\}.$$

For $j=1$ or 2, projection induces a one to one correspondence between P_j and the set

$$B_j = \{b\colon (b,s)\in P_j \text{ for some } s\},$$

because P is τ controlled. We shall verify that B_j satisfies the hypothesis of the condition in 2.8.9, and then conclude that

$$\operatorname{card} P_j = \operatorname{card} B_j \le \zeta.$$

Whenever (b,s) and (c,t) are distinct members of P_j, with $\rho(a,b)\ge \rho(a,c)$, and for any x as in 2.8.9, we see that

$$\rho(b,c)\le \rho(b,x)+\rho(x,c)=\rho(a,b)-\rho(a,c)+\rho(x,c),$$

hence

$$\rho(x,c)\ge \rho(a,c)+\rho(b,c)-\rho(a,b).$$

In case $j=1$ we use the inequalities

$$\rho(b,c)>\inf\{s,t\}>r/\tau, \qquad kr\ge \rho(a,b),$$
$$\rho(a,c)>\inf\{r,t\}>r/\tau, \qquad k\tau-1=1-\eta>0$$

to obtain

$$\rho(x,c)>\rho(a,c)+r/\tau-kr=\rho(a,c)-(r/\tau)(k\tau-1)$$
$$>\rho(a,c)\cdot[1-(1-\eta)]=\rho(a,c)\cdot\eta.$$

In case $j=2$ we note that $\rho(a,b)\le r+s$, $\rho(a,c)>kr$,

$$\rho(b,c)>s \text{ or } \rho(b,c)>t>s/\tau, \text{ hence } \rho(b,c)-s>t-\tau t,$$
$$\rho(a,c)>t \text{ or } \rho(a,c)>r>t/\tau, \text{ hence } \rho(a,c)>t/\tau,$$

and infer that

$$\rho(x,c)\ge \rho(a,c)-r+\rho(b,c)-s>\rho(a,c)-r-t(\tau-1)$$
$$>\rho(a,c)\cdot[1-k^{-1}-\tau(\tau-1)]>\rho(a,c)\cdot\eta.$$

Thus $\rho(x,c)/\rho(a,c)\ge\eta$, whether $j=1$ or 2, as required.

2.8.12. Corollary. *If* $0<\mu<\xi/2$ *and* Q *is a* τ *controlled subset of* $A\times\{r\colon 0<r<\mu\}$, *then the family*

$$\{\mathbf{B}(a,r)\colon (a,r)\in Q\}$$

is the union of $2\,\zeta+1$ *disjointed subfamilies.*

Proof. We define subsets Q_j and R_j of Q by induction, starting with $Q_0 = Q$ and $R_0 = \varnothing$, and applying 2.8.10 for $j > 1$ with

$$Q_j = Q_{j-1} \sim R_{j-1}$$

to obtain $R_j \subset Q_j$ satisfying the conditions:

(1) *Whenever (a, r) and (b, s) are distinct elements of R_j,*

$$\rho(a, b) > r + s.$$

(2) *Whenever $(a, r) \in Q_j$ there exists $(b, s) \in R_j$ with*

$$\rho(a, b) \leq r + s \text{ and } s > r/\tau.$$

From (1) we see that, for each j,

$$\{\mathbf{B}(a, r) \colon (a, r) \in R_j\} \text{ is disjointed}.$$

We shall complete the proof by showing that

$$Q_{2\zeta+2} = \bigcap_{j=1}^{2\zeta+1} (Q \sim R_j) = \varnothing.$$

In fact, if $(a, r) \in Q_{2\zeta+2}$ we could use (2) to select

$$(b_j, s_j) \in R_j \text{ with } \rho(a, b_j) \leq r + s_j \text{ and } s_j > r/\tau$$

corresponding to $j = 1, 2, \ldots, 2\zeta+1$ and obtain the τ controlled set

$$P = \{(a, r)\} \cup \{(b_j, s_j) \colon j = 1, 2, \ldots, 2\zeta+1\}$$

with card $P = 2\zeta + 2$, because the pairs (b_j, s_j) are distinct from (a, r) and belong to the disjoint sets R_j. However, since

$$\rho(a, b_j) \leq r + s_j < 2\mu < \xi,$$

our theorem would imply that card $P \leq 2\zeta + 1$.

2.8.13. Lemma. *If $1 < \tau < \infty$, $0 < \mu < \infty$ and*

$$P \subset X \times \{r \colon 0 < r < \mu\},$$

then P has a τ controlled subset Q such that

$$\{a \colon (a, r) \in P \text{ for some } r\} \subset \bigcup \{\mathbf{B}(a, r) \colon (a, r) \in Q\}.$$

Proof. We consider the class Ω of all τ controlled subsets Q of P with the following property: *Whenever $(b, s) \in P$,*

$$\textit{either } \rho(a, b) \leq r \textit{ for some } (a, r) \in Q,$$
$$\textit{or } \rho(a, b) > r > s/\tau \textit{ for all } (a, r) \in Q.$$

Applying Hausdorff's maximal principle we find that Ω has a maximal member.

For such a maximal member Q of Ω the conclusion of the lemma holds, because otherwise

$$K = P \cap \{(b,s) \colon \rho(a,b) > r \text{ for all } (a,r) \in Q\} \neq \varnothing,$$

hence we could choose $(c,t) \in K$ with

$$\tau t > \sup\{s \colon (b,s) \in K\},$$

and verify that $Q \cup \{(c,t)\} \in \Omega$, contrary to the maximality of Q.

2.8.14. Theorem. *If ρ is directionally ξ, η, ζ limited at A, $0 < \mu < \xi/2$, F is a family of closed balls with radii less than μ, and each point of A is the center of some member of F, then A is contained in the union of $2\zeta + 1$ disjointed subfamilies of F.*

Proof. We choose τ as in 2.8.11, apply 2.8.13 with

$$P = \{(a,r) \colon a \in A, \mathbf{B}(a,r) \in F\}$$

to obtain a τ controlled subset Q of P such that A is covered by the family

$$G = \{\mathbf{B}(a,r) \colon (a,r) \in Q\},$$

and infer from 2.8.12 that G is the union of $2\zeta + 1$ disjointed subfamilies.

2.8.15. Corollary. *In case A is separable and*

$$\inf\{r \colon \mathbf{B}(a,r) \in F\} = 0 \textit{ whenever } a \in A,$$

then F is ϕ adequate for A.

Proof. We will show that the hypothesis of 2.8.2 holds with

$$K = \{A\}, \quad \sigma(A) = 2\zeta/(2\zeta + 1).$$

For any open subset W of X, each point of $A \cap W$ is the center of some member of the family

$$F' = F \cap \{\mathbf{B}(a,r) \colon a \in A,\ r > 0 \text{ and } \mathbf{B}(a,r) \subset W\},$$

hence 2.8.14 implies that $A \cap W$ is contained in the union of $2\zeta + 1$ disjointed subfamilies $H_1, H_2, \ldots, H_{2\zeta+1}$ of F'. Since A is separable, each H_j is countable, hence $\bigcup H_j$ is ϕ measurable. The inequality

$$\sum_{j=1}^{2\zeta+1} \phi(A \cap \bigcup H_j) \geq \phi(A \cap W)$$

allows us to choose an integer j for which

$$\phi(A \cap \bigcup H_j) \geq \phi(A \cap W)/(2\zeta + 1), \text{ hence}$$

$$\phi[(A \cap W) \sim \bigcup H_j] \leq [1 - 1/(2\zeta + 1)]\,\phi(A \cap W).$$

2.8.16. We now define some basic concepts underlying the axiomatic theory of derivation to be discussed in 2.9.

By a **covering relation** we mean a subset of

$$\{(x, S): x \in S \subset X\}.$$

Whenever C is a covering relation and $Z \subset X$, we let

$$C(Z) = \{S: (x, S) \in C \text{ for some } x \in Z\}.$$

We say that C is **fine at** x if and only if

$$\inf\{\operatorname{diam}(S): (x, S) \in C\} = 0.$$

By a ϕ **Vitali relation** we mean a covering relation V such that $V(X)$ *is a family of Borel sets, V is fine at each point of X*, and the following condition holds:

If $C \subset V$, $Z \subset X$ and C is fine at each point of Z, then $C(Z)$ has a countable disjointed subfamily covering ϕ almost all of Z.

Examples of ϕ Vitali relations will be provided by the Theorems 2.8.17 – 2.8.19. (Other examples occur in [M 3, §5] and [H M]).

Whenever a covering relation C is fine at some point x, and f is an $\overline{\mathbf{R}}$ valued function we shall use the alternate notations

$$(C)\limsup_{x} f, \qquad (C)\limsup_{S \to x} f(S)$$

to designate

$$\limsup_{\varepsilon \to 0+} \{f(S): (x, S) \in C, \operatorname{diam} S < \varepsilon, S \in \operatorname{dmn} f\}.$$

Similarly we define $(C)\liminf$ and $(C)\lim$.

2.8.17. Theorem. *If V is a covering relation, $V(X)$ is a family of bounded closed sets, V is fine at each point of X, δ is a nonnegative function on $V(X)$, $1 < \tau < \infty$ and* (see 2.8.4)

$$0 \le (V)\limsup_{S \to x} [\delta(S) + \phi(\hat{S})/\phi(S)] < \infty$$

for ϕ almost all x in X, then V is a ϕ Vitali relation.

Proof. Suppose $C \subset V$, $Z \subset X$ and C is fine at each point of Z. For every positive integer n the relation

$$C_n = C \cap \{x, S): x \in Z, \delta(S) + \phi(\hat{S})/\phi(S) < n\}$$

is fine at each point of the set

$$A_n = Z \cap \{x: 0 \le (V)\limsup_{S \to x} [\delta(S) + \phi(\hat{S})/\phi(S)] < n\},$$

and 2.8.7 implies that $C_n(A_n)$ is ϕ adequate for A_n. Since the union of all the sets A_n contains ϕ almost all of Z, it follows from 2.8.2 that $C(Z)$ is ϕ adequate for Z.

2.8.18. Theorem. *If X is the union of a countable family K such that ρ is directionally limited at each member of K, and if X is separable, then*

$$V=\{(x,\mathbf{B}(x,r)):\ x\in X,\ 0<r<\infty\}$$

is a ϕ Vitali relation.

Proof. Suppose $C\subset V$, $Z\subset X$ and C is fine at each point of Z. For each $A\in K$, 2.8.15 implies that $C(Z)$ is ϕ adequate for $A\cap Z$. It follows from 2.8.2 that $C(Z)$ is ϕ adequate for Z.

2.8.19. Theorem. *Suppose X is separable and $P_1, P_2, P_3, \ldots$ are Borel partitions of X such that each member of P_1 is bounded, each member of P_j is the union of some subfamily of P_{j+1}, and*

$$\lim_{j\to\infty} \sup\{\operatorname{diam}(S):\ S\in P_j\}=0.$$

Then $V=\{(x,S):\ x\in S\in P_j$ for some $j\}$ is a ϕ Vitali relation.

Proof. If $C\subset V$, $Z\subset X$ and $Z\subset\bigcup C(Z)$, then each point of Z belongs to some maximal member of $C(Z)$; moreover any two distinct maximal members of $C(Z)$ are disjoint.

2.8.20. In [C 2, pp. 689–692] it is shown that if $X=\mathbf{R}^2$ and F consists of all rectangles $J\times K$, where J and K are compact nondegenerate subintervals of $\mathbf{R}$, then $\{(x,S):\ x\in S\in F\}$ is not an $\mathscr{L}^2$ Vitali relation.

2.9. Derivates

2.9.1. Throughout this section we assume: *X is metrized by ρ;*

M is the class of all Borel regular measures ψ over X such that every bounded set has finite ψ measure;

$\phi\in M$ and V is a ϕ Vitali relation (see 2.8.16).

With each $\psi\in M$ we associate another measure ψ_ϕ by defining

$$\psi_\phi(A)=\inf\{\psi(B):\ B \text{ is a Borel set and } \phi(A\sim B)=0\}$$

whenever $A\subset X$. Some elementary facts about ψ_ϕ, and the relation between ψ_ϕ and ψ, will be proved in 2.9.2. Evidently $\psi_\phi\le\psi$.

To describe the local and global connections between any member ψ of M and the basic measure ϕ, we shall prove that (2.9.5) the V **derivate**

$$\mathbf{D}(\psi, \phi, V, x) = (V) \lim_{S \to x} \psi(S)/\phi(S)$$

exists for ϕ almost all x, and that (2.9.7) the indefinite ϕ integral of this derivate equals ψ_ϕ. Other topics in the general theory will be the "fundamental theorem of the calculus" (2.9.8) and approximate continuity (2.9.13).

Then, following the general theory, we show that under some special conditions certain results can be sharpened (2.9.15), and even partially extended to the case when $\psi \notin M$ and some Borel sets may be ψ nonmeasurable (2.9.17); this extension will be useful in 3.3.4.

Finally (2.9.19 – 2.9.23) the derivation of measures is applied to the differentiation of functions with locally finite variation.

2.9.2. Theorem. *For each $\psi \in M$ there exists a Borel set B such that*

$$\psi_\phi = \psi \llcorner B \quad \text{and} \quad \phi(X \sim B) = 0.$$

Therefore $\psi_\phi \in M$ and $\psi - \psi_\phi = \psi \llcorner (X \sim B) \in M$. Moreover, $\psi = \psi_\phi$ if and only if every set with ϕ measure 0 *has ψ measure* 0 (in which case ψ is said to be **absolutely continuous** with respect to ϕ).

Proof. Obviously ψ_ϕ measures X and, by 2.3.2(9), all Borel sets are ψ_ϕ measurable.

Each bounded Borel set A contains a Borel set B for which

$$\psi_\phi(A) = \psi(B) \quad \text{and} \quad \phi(A \sim B) = 0.$$

We will show that

$$\psi_\phi(S) = \psi(B \cap S) \quad \text{whenever } S \subset A.$$

First, if S is a Borel subset of A, then

$$\psi_\phi(S) \le \psi(B \cap S), \quad \psi_\phi(A \sim S) \le \psi(B \sim S),$$

$$\psi_\phi(S) + \psi_\phi(A \sim S) = \psi_\phi(A) = \psi(B) = \psi(B \cap S) + \psi(B \sim S),$$

hence $\psi_\phi(S) = \psi(B \cap S)$. Next, for an arbitrary subset S of A there exists a Borel set T such that

$$S \subset T \subset A, \quad \phi(S) = \phi(T), \quad \psi(S) = \psi(T),$$

hence 2.1.4 implies that T is both a ϕ hull and a ψ hull of S, and therefore

$$\psi_\phi(S) = \psi_\phi(T) = \psi(B \cap T) = \psi(B \cap S).$$

Since X is the union of a countable disjointed family of bounded Borel sets, the conclusion of the theorem is now evident.

2.9.3. Lemma. *If* $\alpha \in M, \beta \in M, 0 < c < \infty$ *and*

$$A \subset \{x\colon (V) \liminf_x \alpha/\beta < c\},$$

then $\alpha_\phi(A) \le c\, \beta_\phi(A)$.

Proof. For any $\varepsilon > 0$ we can apply the definition of β_ϕ in conjunction with Theorem 2.2.2 to obtain an open set W for which

$$\phi(A \sim W) = 0 \quad \text{and} \quad \beta(W) \le \beta_\phi(A) + \varepsilon.$$

Then the covering relation

$$C = V \cap \{(x, S)\colon S \subset W \text{ and } \alpha(S)/\beta(S) < c\}$$

is fine at each point of $A \cap W$, hence $C(A \cap W)$ has a countable disjointed subfamily G covering ϕ almost all of $A \cap W$, and we conclude that

$$\alpha_\phi(A) = \alpha_\phi(A \cap W) \le \alpha(\bigcup G) = \sum_{S \in G} \alpha(S)$$
$$\le c \sum_{S \in G} \beta(S) \le c\, \beta(W) \le c\,[\beta_\phi(A) + \varepsilon].$$

2.9.4. Corollary. *If* $\psi \in M$ *and* $0 < c < \infty$, *then*

$$A \subset \{x\colon (V) \liminf_x \psi/\phi < c\} \quad \textit{implies} \quad \psi_\phi(A) \le c\, \phi(A),$$
$$A \subset \{x\colon (V) \limsup_x \psi/\phi > c\} \quad \textit{implies} \quad \psi_\phi(A) \ge c\, \phi(A).$$

Proof. $\phi_\phi = \phi$, and

$$(V) \limsup_x \psi/\phi > c \text{ implies } (V) \liminf_x \phi/\psi < 1/c.$$

2.9.5. Theorem. *If* $\psi \in M$, *then*

$$0 \le \mathbf{D}(\psi, \phi, V, x) < \infty \quad \textit{for } \phi \textit{ almost all } x.$$

Proof. We define $C = V \cap \{(x, S)\colon \phi(S) = 0\}$,

$$P = \{x\colon C \text{ is fine at } x\}, \qquad Q = \{x\colon (V) \limsup_x \psi/\phi = \infty\},$$
$$R(a, b) = \{x\colon (V) \liminf_x \psi/\phi < a < b < (V) \limsup_x \psi/\phi\}$$

and observe that those points, where the derivate fails to exist and be finite, belong to the union of P, Q and the sets $R(a, b)$ corresponding to all pairs of rational numbers $a < b$.

Since $C(P)$ has a countable disjointed subfamily G covering ϕ almost all of P, we find that

$$\phi(P) \le \phi(\bigcup G) = \sum_{S \in G} \phi(S) = 0.$$

For any bounded subset A of Q the second implication of 2.9.4 shows that

$$\infty > \psi_\phi(A) \geq c\,\phi(A) \text{ whenever } 0 < c < \infty,$$

hence $\phi(A)=0$; we infer that $\phi(Q)=0$.

For any bounded subset A of $R(a,b)$ it follows from 2.9.4 that

$$b\,\phi(A) \leq \psi_\phi(A) \leq a\,\phi(A) < \infty$$

hence $\phi(A)=0$ because $a<b$; we conclude that $\phi[R(a,b)]=0$.

2.9.6. Lemma. *If $\psi \in M$, then $\mathbf{D}(\psi,\phi,V,\cdot)$ is a ϕ measurable function.*

Proof. Recalling 2.3.2(7) we suppose $0<a<b<\infty$ and A, B are bounded sets such that

$$A \subset \{x\colon \mathbf{D}(\psi,\phi,V,x) < a\}, \qquad B \subset \{x\colon \mathbf{D}(\psi,\phi,V,x) > b\}.$$

Choosing Borel sets A' and B' for which

$$A \subset A', \quad \phi(A)=\phi(A'), \quad \psi_\phi(A)=\psi_\phi(A'),$$
$$B \subset B', \quad \phi(B)=\phi(B'), \quad \psi_\phi(B)=\psi_\phi(B'),$$

we see from 2.1.4 and 2.9.4 that

$$\psi_\phi(A' \cap B') = \psi_\phi(A \cap B') \leq a\,\phi(A \cap B') = a\,\phi(A' \cap B'),$$
$$\psi_\phi(A' \cap B') = \psi_\phi(A' \cap B) \geq b\,\phi(A' \cap B) = b\,\phi(A' \cap B'),$$

hence $\phi(A' \cap B')=0$, and we conclude that

$$\phi(A \cup B) = \phi[(A \cup B) \cap A'] + \phi[(A \cup B) \cap B'] \geq \phi(A) + \phi(B).$$

2.9.7. Theorem. *If $\psi \in M$ and A is a ϕ measurable set, then A is ψ_ϕ measurable and*

$$\psi_\phi(A) = \int_A \mathbf{D}(\psi,\phi,V,x)\,d\phi\,x.$$

Proof. A is contained in a Borel set B with $\phi(B \sim A)=0$, hence $\psi_\phi(B \sim A)=0$.

According to 2.9.6 the sets

$$Z = \{x\colon \mathbf{D}(\psi,\phi,V,x)=0\}, \qquad W = \{x\colon \mathbf{D}(\psi,\phi,V,x)=\infty\}$$

are ϕ measurable; from 2.9.4 and 2.9.5 it follows that

$$\psi_\phi(Z) = 0 = \int_Z \mathbf{D}(\psi,\phi,V,x)\,d\phi\,x,$$
$$\phi(W)=0, \quad \psi_\phi(W) = 0 = \int_W \mathbf{D}(\psi,\phi,V,x)\,d\phi\,x.$$

Moreover, whenever $1<t<\infty$, $A\sim(Z\cup W)$ is the union of the disjoint ϕ measurable (by 2.9.6) sets

$$P_n=A\cap\{x:\ t^n\le \mathbf{D}(\psi,\phi,V,x)<t^{n+1}\}$$

corresponding to all integers n, hence 2.9.4 yields the inequalities

$$\begin{aligned}\psi_\phi(A)&=\sum_{n\in\mathbf{Z}}\psi_\phi(P_n)\le\sum_{n\in\mathbf{Z}}t^{n+1}\,\phi(P_n)\\&\le\sum_{n\in\mathbf{Z}}t\int_{P_n}\mathbf{D}(\psi,\phi,V,x)\,d\phi\,x=t\int_A\mathbf{D}(\psi,\phi,V,x)\,d\phi\,x\end{aligned}$$

and

$$\begin{aligned}\psi_\phi(A)&=\sum_{n\in\mathbf{Z}}\psi_\phi(P_n)\ge\sum_{n\in\mathbf{Z}}t^{n}\,\phi(P_n)\\&\ge\sum_{n\in\mathbf{Z}}t^{-1}\int_{P_n}\mathbf{D}(\psi,\phi,V,x)\,d\phi\,x=t^{-1}\int_A\mathbf{D}(\psi,\phi,V,x)\,d\phi\,x.\end{aligned}$$

2.9.8. Theorem. *If f is an $\overline{\mathbf{R}}$ valued ϕ measurable function such that*

$$\int_A|f|\,d\phi<\infty$$

for every bounded ϕ measurable set A, then

$$(V)\lim_{S\to x}\int_S f\,d\phi/\phi(S)=f(x)\ \textit{for}\ \phi\ \textit{almost all}\ x.$$

Proof. Assuming f nonnegative we define $\psi\in M$ by the formula

$$\psi(A)=\int_A^* f\,d\phi\quad\text{whenever } A\subset X,$$

and infer from 2.9.7 that

$$\int_A f(x)\,d\phi\,x=\psi(A)=\psi_\phi(A)=\int_A\mathbf{D}(\psi,\phi,V,x)\,d\phi\,x$$

for every ϕ measurable set A, hence $f(x)=\mathbf{D}(\psi,\phi,V,x)$ for ϕ almost all x.

2.9.9. Corollary. *If f is a ϕ measurable function with values in a separable normed vectorspace Y, and if*

$$\int_A|f|\,d\phi<\infty$$

for every bounded ϕ measurable set A, then

$$(V)\lim_{S\to x}\int_S|f(z)-f(x)|\,d\phi\,z/\phi(S)=0$$

for ϕ almost all x. [The set of all points x for which this equation holds is called the (ϕ,V) **Lebesgue set** of f.]

Proof. With each $x \in X$ we associate the set $T(x)$ of all points $y \in Y$ for which

$$(V)\lim_{S\to x} \textstyle\int_S |f(z)-y|\, d\phi\, z/\phi(S) = |f(x)-y|.$$

We will show that $T(x)=Y$ for ϕ almost all x.

For each $y \in Y$ it follows from 2.9.8, applied to the function mapping x onto $|f(x)-y|$, that $y \in T(x)$ for ϕ almost all x. Choosing a countable dense subset C of Y, we infer that $C \subset T(x)$ for ϕ almost all x. This suffices because $T(x)$ is closed for each x.

2.9.10. Combining 2.9.7, 2.9.8, 2.9.2 we see that, if $\psi \in M$, then

$$\mathbf{D}(\psi_\phi, \phi, V, x) = \mathbf{D}(\psi, \phi, V, x), \quad \mathbf{D}(\psi - \psi_\phi, \phi, V, x) = 0$$

for ϕ almost all x. Since ψ_ϕ was defined without reference to V, 2.9.7 implies that, if V_1 and V_2 are any two ϕ Vitali relations, then

$$\mathbf{D}(\psi, \phi, V_1, x) = \mathbf{D}(\psi, \phi, V_2, x) \text{ for } \phi \text{ almost all } x.$$

The following two observations relate the present theory with section 2.5:

If L is the lattice of all continuous, bounded, real valued functions on X with bounded support, then M is the class of all L regular measures associated with monotone Daniell integrals on L.

If $\psi \in M$, then $\mathbf{L}_\infty(\phi) \subset \mathbf{L}_\infty(\psi_\phi)$. In case $\psi(X) < \infty$ we can define a Daniell integral μ on $\mathbf{L}_\infty(\phi)$ by the formula

$$\mu(f) = \textstyle\int f\, d\psi_\phi \text{ whenever } f \in \mathbf{L}_\infty(\phi),$$

and obtain from 2.5.8 a function $k \in \mathbf{L}_1(\phi)$ such that

$$\mu(f) = \textstyle\int f k\, d\phi \text{ whenever } f \in \mathbf{L}_\infty(\phi).$$

Comparison with 2.9.7 shows that

$$k(x) = \mathbf{D}(\psi, \phi, V, x) \text{ for } \phi \text{ almost all } x.$$

However, while k was obtained (in the proof of 2.5.7) as a solution of a certain global extremal problem, the derivate has been constructed by a specific local limiting process, which is likely to be more useful in analytic and geometric applications. On the other hand the method of section 2.5 has the advantage of greater generality, because it does not depend on the existence of a ϕ Vitali relation.

2.9.11. Theorem. *If* $A \subset X$ *and*

$$P = \{x \colon (V) \lim_{S \to x} \phi(S \cap A)/\phi(S) = 1\},$$

$$Q = \{x \colon (V) \lim_{S \to x} \phi(S \sim A)/\phi(S) = 0\},$$

then P *and* Q *are* ϕ *measurable,*

$$\phi(A \sim P) = 0, \quad A \cup P \text{ is a } \phi \text{ hull of } A,$$

$$\phi(Q \sim A) = 0, \quad X \sim (A \cap Q) \text{ is a } \phi \text{ hull of } X \sim A.$$

Moreover, ϕ *measurability of* A *is equivalent to each of the two conditions:*

$$\phi(P \sim A) = 0; \quad \phi(A \sim Q) = 0.$$

Proof. If B is any ϕ hull of A, then

$$P = \{x \colon \mathbf{D}(\phi \llcorner B, \phi, V, x) = 1\},$$

hence 2.9.8 implies $\phi(B \sim P) + \phi(P \sim B) = 0$, and the assertions about P follow from the inclusions

$$A \sim P \subset B \sim P, \quad A \subset A \cup P \subset B \cup (P \sim B).$$

Similarly, if C is a ϕ hull of $X \sim A$, then

$$Q = \{x \colon \mathbf{D}(\phi \llcorner C, \phi, V, x) = 0\},$$

hence 2.9.8 implies $\phi(Q \cap C) + \phi[X \sim (C \cup Q)] = 0$, and the assertions about Q follow from the inclusions

$$Q \sim A \subset Q \cap C, \quad X \sim A \subset X \sim (A \cap Q) \subset C \cup [X \sim (C \cup Q)].$$

2.9.12. Whenever $A \subset X$ and $x \in X$, one calls

$$(V) \lim_{S \to x} \phi(S \cap A)/\phi(S)$$

the (ϕ, V) **density of** A **at** x. The preceding theorem shows that A is ϕ measurable if and only if the (ϕ, V) density function of A and the characteristic function of A are ϕ almost equal.

By means of such densities we define the following useful modifications of the concepts of limit and continuity: Suppose f is a function mapping a subset of X into a topological space Y, and $x \in X$. A point $y \in Y$ is called a (ϕ, V) **approximate limit of** f **at** x if and only if, *for every neighborhood* W *of* y *in* Y, *the set* $X \sim f^{-1}(W)$ *has density* 0 *at* x. In case Y is a Hausdorff space, there can be at most one such y; we denote it

$$(\phi, V) \operatorname{ap} \lim_{x} f, \quad \text{or} \quad (\phi, V) \operatorname{ap} \lim_{z \to x} f(z).$$

We say that f is (ϕ, V) **approximately continuous at** x if and only if $x \in \operatorname{dmn} f$ and

$$(\phi, V) \operatorname{ap} \lim_x f = f(x).$$

In case $Y = \overline{\mathbf{R}}$, we also define the (ϕ, V) **approximate upper limit of** f **at** x,

$$(\phi, V) \operatorname{ap} \limsup_x f,$$

as the *greatest lower bound of the set of all numbers t such that*

$$\{z\colon f(z) > t\} \text{ has density } 0 \text{ at } x.$$

The **approximate lower limit** is defined similarly.

In 2.3.5 we found a global connection between measurability and continuity. The next theorem localizes this relationship.

2.9.13. Theorem. *Suppose f is a function mapping ϕ almost all of X into a separable metric space. Then f is ϕ measurable if and only if f is (ϕ, V) approximately continuous at ϕ almost all points of X.*

Proof. If f is ϕ measurable it follows from 2.3.5 that ϕ almost all of X is the union of a countable family of closed sets C such that $f|C$ is continuous. Moreover 2.9.11 implies that, for ϕ almost all points x in each such set C,

$$(V) \lim_{S \to x} \phi(S \sim C)/\phi(S) = 0,$$

hence f is (V, ϕ) approximately continuous at x.

Conversely, if f is (ϕ, V) approximately continuous at ϕ almost all points of X, and if W is any open set in the range space, then 2.9.11 implies that $f^{-1}(W)$ is ϕ measurable because

$$(V) \lim_{S \to x} \phi[S \sim f^{-1}(W)]/\phi(S) = 0$$

whenever $x \in f^{-1}(W)$ and f is (ϕ, V) approximately continuous at x.

2.9.14. Suppose J is an open subset of $\{r\colon 0 < r < \infty\}$.
If α and β measure X, with $\beta \in \mathbf{M}$, then

$$\{x\colon \alpha\,\mathbf{B}(x,r) > \beta\,\mathbf{B}(x,r) \text{ for some } r \in J\} \textit{ is open.}$$

In fact, whenever $\alpha\,\mathbf{B}(a,r) > \beta\,\mathbf{B}(a,r)$ with $r \in J$, there exists a positive ε for which

$$r + \varepsilon \in J \quad \text{and} \quad \alpha\,\mathbf{B}(a,r) > \beta\,\mathbf{B}(a, r+2\varepsilon),$$

hence $x \in \mathbf{B}(a, \varepsilon)$ implies

$$\alpha\,\mathbf{B}(x, r+\varepsilon) \geq \alpha\,\mathbf{B}(a,r) > \beta\,\mathbf{B}(a, r+2\varepsilon) \geq \beta\,\mathbf{B}(x, r+\varepsilon).$$

Two corollaries follow:

If ψ *measures* X *and* $0<t<\infty$, *then*

$$\{x\colon \psi\mathbf{B}(x,r)>t\,\phi\mathbf{B}(x,r) \text{ for some } r\in J\} \text{ is open.}$$

If $\psi\in M$ *and* $0<t<\infty$, *then*

$$\{x\colon \psi\mathbf{B}(x,r)<t\,\phi\mathbf{B}(x,r) \text{ for some } r\in J\} \text{ is open.}$$

2.9.15. Theorem. *Suppose*

either $\limsup_{r\to 0+} \phi\mathbf{B}(x,5r)/\phi\mathbf{B}(x,r)<\infty$ *for every* $x\in X$,

or X *is the union of a countable family* K *such that* ρ *is directionally limited at each member of* K *and* X *is separable.*

If $\psi\in M$ *and*

$$W=\{x\colon \lim_{r\to 0+}\psi\mathbf{B}(x,r)/\phi\mathbf{B}(x,r)=\infty\},$$

$$P=\{x\colon \phi\mathbf{B}(x,r)=0 \text{ for some } r>0\},$$

then W *is a Borel set,* P *is open and*

$$\psi_\phi=\psi\llcorner[X\sim(W\cup P)].$$

Proof. According to either 2.8.17 or 2.8.18, the general theory of this section now applies with

$$V=\{(x,\mathbf{B}(x,r))\colon x\in X,\ 0<r<\infty\}.$$

In 2.9.5 we saw that $\phi(W\cup P)=0$, hence

$$\psi_\phi\le\psi\llcorner[X\sim(W\cup P)].$$

To prove the opposite inequality it will suffice to show that

$$\psi(T)=0 \text{ whenever } T\subset X\sim(W\cup P) \text{ with } \phi(T)=0.$$

For this purpose we may assume, after performing countable partitions that, for some $c<\infty$,

$$\liminf_{r\to 0+}\psi\mathbf{B}(x,r)/\phi\mathbf{B}(x,r)<c \text{ whenever } x\in T,$$

and also that *either*, for some $\lambda<\infty$,

$$\limsup_{r\to 0+}\phi\mathbf{B}(x,5r)/\phi\mathbf{B}(x,r)<\lambda \text{ whenever } x\in T,$$

or ρ is directionally ξ,η,ζ limited at T and X is separable.

Suppose E is an arbitrary open set containing T.

In the first alternative T is covered by the family F of all balls $\mathbf{B}(x,r)\subset E$ with

$$\psi\,\mathbf{B}(x,5r)<c\,\phi\,\mathbf{B}(x,5r)<c\,\lambda\,\phi\,\mathbf{B}(x,r).$$

Applying 2.8.4, 2.8.5, 2.8.8 with $\tau=2$, $\delta=\operatorname{diam}$ we obtain a disjointed subfamily G of F for which

$$T\subset\bigcup F\subset\bigcup\{\hat{S}\colon S\in G\}.$$

Since G is countable, because each member of G has positive ϕ measure, we find that

$$\psi(T)\le\sum_{S\in G}\psi(\hat{S})\le\sum_{S\in G}c\,\lambda\,\phi(S)\le c\,\lambda\,\phi(E).$$

In the second alternative we suppose $0<\mu<\xi/2$ and apply 2.8.15 to the family F' of all balls $\mathbf{B}(x,r)\subset E$ with

$$x\in T,\quad 0<r<\mu,\quad \psi\,\mathbf{B}(x,r)<c\,\phi\,\mathbf{B}(x,r).$$

We find that T is contained in the union of $2\zeta+1$ disjointed subfamilies $H_1,H_2,\ldots,H_{2\zeta+1}$ of F', which are countable because X is separable, and infer that

$$\psi(T)\le\sum_{j=1}^{2\zeta+1}\psi(\bigcup H_j)\le\sum_{j=1}^{2\zeta+1}c\,\phi(\bigcup H_j)\le(2\zeta+1)\,c\,\phi(E).$$

In both cases we conclude that $\psi(T)=0$ because $\phi(E)$ can be made arbitrarily small.

2.9.16. The following example shows that 2.9.15 cannot be generalized to arbitrary ϕ Vitali relations.

Taking $X=\mathbf{R}^n$ with a metric induced by some norm, we see from 2.8.17 and 2.8.8 that

$$V=\{(x,S)\colon S \text{ is a closed ball, } x\in S\}$$

is an $\mathscr{L}^n$ Vitali relation. Suppose $n\ge 2$, Y is a straight line in $\mathbf{R}^n$ and $\psi=\mathscr{H}^1 \llcorner Y$ (see 2.10.2). Then

$$(V)\liminf_{S\to x}\psi(S)/\mathscr{L}^n(S)=0 \quad\text{whenever } x\in\mathbf{R}^n,$$

the sets corresponding to W and P are empty, but

$$\psi_{\mathscr{L}^n}\neq\psi,\ \text{because } \mathscr{L}^n(Y)=0 \text{ and } \psi(Y)=\infty.$$

2.9.17. Theorem. *Suppose* $\lambda<\infty$, $\mu>0$ *and*

$$\phi\mathbf{B}(x,5r)<\lambda\phi\mathbf{B}(x,r) \text{ whenever } x\in X, \quad 0<r<\mu.$$

If ψ *measures* X, $\psi(T)=0$ *and* T *is* ϕ *measurable, then, for* ϕ *almost all* x *in* T,

$$\limsup_{r\to 0+} \psi\mathbf{B}(x,r)/\phi\mathbf{B}(x,r) \text{ equals either } 0 \text{ or } \infty.$$

Proof. In view of 2.2.3 we may assume that T is closed. Then 2.9.14 implies that

$$C_n=T\cap\{x\colon \psi\mathbf{B}(x,r)\le n\phi\mathbf{B}(x,r) \text{ for } 0<r<1/n\}$$

is closed whenever n is a positive integer, and since

$$T\cap\{x\colon \limsup_{r\to 0+}\psi\mathbf{B}(x,r)/\phi\mathbf{B}(x,r)<\infty\}=\bigcup_{n=1}^{\infty} C_n,$$

it will suffice to show that

$$\lim_{r\to 0+}\psi\mathbf{B}(x,r)/\phi\mathbf{B}(x,r)=0 \text{ for } \phi \text{ almost all } x \text{ in } C_n.$$

Suppose $x\in C_n$ and $0<r<\inf\{1/n,\mu\}/4$.

Then $\mathbf{B}(x,r)\sim C_n$ is covered by the family F of all closed balls $\mathbf{B}(y,s)$ with

$$y\in\mathbf{B}(x,r)\sim C_n \text{ and } s=\text{distance}(y,C_n)/2.$$

For each such (y,s), $s\le r/2$ and $\rho(y,z)<3s$ for some $z\in C_n$, hence

$$\mathbf{B}(y,s)\subset\mathbf{B}(x,2r), \quad 8s<1/n, \quad 5s<\mu,$$

$$\psi\mathbf{B}(y,5s)\le\psi\mathbf{B}(z,8s)\le n\phi\mathbf{B}(z,8s)\le n\phi\mathbf{B}(y,11s)<n\lambda^2\phi\mathbf{B}(y,s).$$

Applying 2.8.4, 2.8.5 with $\delta=\text{diam}$, $\tau=2$ we obtain a countable disjointed subfamily G of F with

$$\mathbf{B}(x,r)\sim C_n\subset\bigcup\{\hat{S}\colon S\in G\},$$

$$\psi\mathbf{B}(x,r)=\psi[\mathbf{B}(x,r)\sim C_n]\le\sum_{S\in G}\psi(\hat{S})$$

$$\le\sum_{S\in G}n\lambda^2\phi(S)\le n\lambda^2\phi[B(x,2r)\sim C_n],$$

$$\frac{\psi\mathbf{B}(x,r)}{\phi\mathbf{B}(x,r)}\le n\lambda^3\,\frac{\phi[B(x,2r)\sim C_n]}{\phi\mathbf{B}(x,2r)}.$$

Finally we observe that, as r approaches 0, the right fraction in the last inequality approaches 0 provided $X\sim C_n$ has density 0 at x, which is true for ϕ almost all x in C_n according to 2.8.17 and 2.9.11.

2.9.18. The following two examples show that the result obtained in 2.9.17 is rather precise.

(1) *In the conclusion,* lim sup *cannot be replaced by* lim. Suppose $X=\mathbf{R}$ with the usual metric, $\phi=\mathscr{L}^1$, and T is a nowhere dense compact set with positive $\mathscr{L}^1$ measure. We choose positive numbers c_j and finite sets S_j for which

$$S_j\subset\mathbf{R}\sim T \text{ and } T\subset\bigcup\{\mathbf{U}(a,c_j)\colon a\in S_j\},$$

subject to the inductive conditions $c_1=1$ and

$$j c_{j+1}<\text{distance}(S_j,T)=\inf\rho(S_j\times T)<c_j.$$

Defining the measure ψ over $\mathbf{R}$ by the formula

$$\psi(A)=\sup(\{0\}\cup\{c_j\colon A\cap S_j\neq\emptyset\}) \text{ for } A\subset X,$$

we see that $\psi(T)=0$ and, whenever $x\in T$,

$$\psi\mathbf{B}(x,r)/\mathscr{L}^1\mathbf{B}(x,r)=c_{j+1}/2r \text{ for } c_{j+1}\le r\le j c_{j+1},$$

hence the left ratio has lower limit 0, and upper limit $\ge\frac{1}{2}$, as r approaches 0; of course 2.9.17 implies that the upper limit equals ∞ for $\mathscr{L}^1$ almost all x in T.

(2) *The hypothesis* $\phi\mathbf{B}(x,5r)<\lambda\phi\mathbf{B}(x,r)$ *cannot be dropped for a set of points* x *with* ϕ *measure* 0, *even when* ρ *is directionally limited.* Suppose $X=\mathbf{R}^2$ with its usual metric, T is a straight line in $\mathbf{R}^2$, $\phi=\mathscr{H}^1\llcorner T$, $p\in\mathbf{O}^*(2,1)$ and

$$\psi(A)=\mathscr{L}^1[p(A\sim T)] \text{ whenever } A\subset\mathbf{R}^2.$$

Then $\phi(\mathbf{R}^2\sim T)=0$, $\psi(T)=0$ and, whenever $x\in T$ and $0<r<\infty$,

$$\phi\mathbf{B}(x,5r)/\phi\mathbf{B}(x,r)=5,\quad \psi\mathbf{B}(x,r)/\phi\mathbf{B}(x,r)=1.$$

2.9.19. Theorem. *If* $g\colon\mathbf{R}\to\mathbf{R}$ *with* (see 2.5.16)

$$\mathbf{V}_a^b g<\infty \text{ whenever } -\infty<a<b<\infty,$$

if ψ^+ *and* ψ^- *are the Radon measures over* $\mathbf{R}$ *induced by* g (see 2.5.17), *and if*

$$V=\{(x,S)\colon S \text{ is a compact nondegenerate interval}, x\in S\},$$

then g *has at* $\mathscr{L}^1$ *almost all points* x *in* $\mathbf{R}$ *the* **derivative**

$$g'(x)=\lim_{h\to 0}[g(x+h)-g(x)]/h\in\mathbf{R}$$

satisfying the two conditions:

(1) $|g'(x)|=\lim_{h\to 0}(\mathbf{V}_x^{x+h}g)/h.$

(2) *Either* $g'(x) = \mathbf{D}(\psi^+, \mathscr{L}^1, V, x)$ *and* $\mathbf{D}(\psi^-, \mathscr{L}^1, V, x) = 0$, *or* $g'(x) = -\mathbf{D}(\psi^-, \mathscr{L}^1, V, x)$ *and* $\mathbf{D}(\psi^+, \mathscr{L}^1, V, x) = 0$.

Moreover $\mathbf{V}_a^b g \geq \int_a^b |g'| \, d\mathscr{L}^1$ *whenever* $-\infty < a < b < \infty$.

Proof. From 2.8.17 we see that V is an $\mathscr{L}^1$ Vitali relation.

First we treat the *special case when g is nondecreasing*. We know from 2.9.5 that, for $\mathscr{L}^1$ almost all x,

$$\mathbf{D}(\psi^+, \mathscr{L}^1, V, x) \in \mathbf{R};$$

fixing such a point x and abbreviating the above derivate by λ, we can for each $\varepsilon > 0$ find a $\delta > 0$ such that

$$(\lambda - \varepsilon)(b - a) \leq \psi^+ \{t : a \leq t \leq b\} \leq (\lambda + \varepsilon)(b - a)$$

whenever $a \leq x \leq b$ and $0 < b - a < \delta$; therefore the inequality

$$(\lambda - \varepsilon)(b - a) \leq g(b) - g(a) \leq (\lambda + \varepsilon)(b - a)$$

holds for such a and b provided g is continuous at a and b; however, since the points of continuity of g are dense in $\mathbf{R}$, and since $g(b) - g(a)$ is nondecreasing with respect to b, nonincreasing with respect to a, the inequality holds also if g is discontinuous at a or b; consequently

$$\lambda - \varepsilon \leq [g(x+h) - g(x)]/h \leq \lambda + \varepsilon \text{ whenever } 0 < |h| < \delta.$$

We conclude that $g'(x) = \lambda$.

Next, in the *general case*, we define s as in 2.5.16, observe that s and $s - g$ are nondecreasing, and infer from the special case that, for $\mathscr{L}^1$ almost all x,

$$s'(x) \in \mathbf{R}, \quad (s-g)'(x) \in \mathbf{R}, \quad g'(x) = s'(x) - (s-g)'(x) \in \mathbf{R}.$$

Recalling (2.5.17) the relation between ψ^+, ψ^- and the increments

$$g(b) - g(a), \quad s(b) - s(a) = \mathbf{V}_a^b g,$$

we find that, for $\mathscr{L}^1$ almost all x,

$$\mathbf{D}(\psi^+, \mathscr{L}^1, V, x) + \mathbf{D}(\psi^-, \mathscr{L}^1, V, x) \leq s'(x),$$

$$\mathbf{D}(\psi^+, \mathscr{L}^1, V, x) - \mathbf{D}(\psi^-, \mathscr{L}^1, V, x) = g'(x),$$

and the last equation implies

$$|g'(x)| \leq \mathbf{D}(\psi^+, \mathscr{L}^1, V, x) + \mathbf{D}(\psi^-, \mathscr{L}^1, V, x).$$

Therefore, to show that g' satisfies the two conditions stated in the theorem, it remains to be proved only that

$$s'(x) \leq |g'(x)| \text{ for } \mathscr{L}^1 \text{ almost all } x.$$

We observe that

$$\{x\colon s'(x)>|g'(x)|\}\subset\bigcup_{m=1}^{\infty}C_m,$$

where C_m is the set of all $x\in\mathbf{R}$ for which

$$s(b)-s(a)-|g(b)-g(a)|>(b-a)/m$$

whenever $a\le x\le b$ and $0<b-a<1/m$, and will prove that each C_m has $\mathscr{L}^1$ measure 0.

Given $-\infty<p<q<\infty$ and $\varepsilon>0$, we can choose a finite sequence $p=t_1<t_2<\cdots<t_n<t_{n+1}=q$ with $t_{j+1}-t_j<1/m$ and

$$\mathbf{V}_p^q g-\sum_{j=1}^{n}|g(t_{j+1})-g(t_j)|<\varepsilon.$$

Letting $K=\{j\colon C_m\cap\{x\colon t_j\le x\le t_{j+1}\}\neq\varnothing\}$ we find that

$$\begin{aligned}\mathscr{L}^1(C_m\cap\{x\colon p\le x\le q\})&\le\sum_{j\in K}(t_{j+1}-t_j)\\&<\sum_{j\in K}m[s(t_{j+1})-s(t_j)-|g(t_{j+1})-g(t_j)|]<m\varepsilon.\end{aligned}$$

Finally we use 2.9.7 in verifying that, for $-\infty<a<b<\infty$,

$$\begin{aligned}\mathbf{V}_a^b g&\ge\psi^+\{x\colon a<x<b\}+\psi^-\{x\colon a<x<b\}\\&\ge\textstyle\int_a^b[\mathbf{D}(\psi^+,\mathscr{L}^1,V,x)+\mathbf{D}(\psi^-,\mathscr{L}^1,V,x)]\,d\mathscr{L}^1x\\&=\textstyle\int_a^b|g'(x)|\,d\mathscr{L}^1x.\end{aligned}$$

2.9.20. Corollary. *Whenever* $-\infty<a<b<\infty$ *the following four conditions are equivalent:*

(1) $g(x)-g(a)=\int_a^x g'\,d\mathscr{L}^1$ *for* $a\le x\le b$.

(2) $\mathbf{V}_a^b g=\int_a^b|g'|\,d\mathscr{L}^1$.

(3) *For each* $\varepsilon>0$ *there is a* $\delta>0$ *such that, for any finite sequences* $a\le u_1<v_1\le u_2<v_2\le\cdots\le u_n<v_n\le b$,

$$\sum_{j=1}^{n}(v_j-u_j)\le\delta\ \text{implies}\ \sum_{j=1}^{n}|g(v_j)-g(u_j)|\le\varepsilon.$$

(4) $g|\{x\colon a\le x\le b\}$ *is continuous and the measures*

$$\psi^+\llcorner\{x\colon a<x<b\},\qquad\psi^-\llcorner\{x\colon a<x<b\}$$

are absolutely continuous with respect to $\mathscr{L}^1$.

In case these conditions hold the function g is said to be **absolutely continuous** on $\{x\colon a\le x\le b\}$.

Proof. (1) implies (2), because if (1) holds and $a \le t_1 < t_2 < \cdots < t_n < t_{n+1} = b$, then

$$\sum_{j=1}^{n} |g(t_{j+1}) - g(t_j)| = \sum_{j=1}^{n} |\textstyle\int_{t_j}^{t_{j+1}} g' \, d\mathscr{L}^1| \le \int_a^b |g'| \, d\mathscr{L}^1.$$

Applying 2.4.11 with $\phi = \mathscr{L}^1 \llcorner \{x : a \le x \le b\}$ and $f = g'$ we similarly infer that (2) implies (3), because if (2) holds then

$$|g(v) - g(u)| \le \mathbf{V}_u^v g = \textstyle\int_u^v |g'| \, d\mathscr{L}^1 \text{ for } a \le u \le v \le b.$$

Next we observe that the implication in (3) has the three consequent implications:

For $a \le u_1 < v_1 \le u_2 < v_2 \le \cdots \le u_n < v_n \le b$,

$$\sum_{j=1}^{n} (v_j - u_j) \le \delta \text{ implies } \sum_{j=1}^{n} \mathbf{V}_{u_j}^{v_j} g \le \varepsilon.$$

For any open subset S of $\{x : a < x < b\}$,

$$\mathscr{L}^1(S) \le \delta \text{ implies } \psi^+(S) + \psi^-(S) \le \varepsilon.$$

For any subset S of $\{x : a < x < b\}$,

$$\mathscr{L}^1(S) = 0 \text{ implies } \psi^+(S) + \psi^-(S) = 0.$$

Therefore (3) implies (4).

Finally (4) implies (1), because if (4) holds and $a < u < v < b$, then

$$\begin{aligned} g(v) - g(u) &= \psi^+\{x : u < x < v\} - \psi^-\{x : u < x < v\} \\ &= \textstyle\int_u^v [\mathbf{D}(\psi^+, \mathscr{L}^1, V, x) - \mathbf{D}(\psi^-, \mathscr{L}^1, V, x)] \, d\mathscr{L}^1 x \\ &= \textstyle\int_u^v g'(x) \, d\mathscr{L}^1 x \end{aligned}$$

by virtue of 2.9.2 and 2.9.7, and use of 2.4.11 in letting u approach a, or v approach b, yields

$$g(v) - g(u) = \textstyle\int_u^v g' \, d\mathscr{L}^1 \text{ whenever } a \le u < v \le b.$$

2.9.21. Combining 2.9.19, 2.9.20 with 2.9.7, 2.4.10 we obtain the proposition: *If g is absolutely continuous on $\{x : a \le x \le b\}$, then*

$$\textstyle\int_a^b f \, dg = \int_a^b f g' \, d\mathscr{L}^1 \text{ whenever } f \in \mathscr{K}(\mathbf{R}).$$

In case g is also increasing, it follows from 2.5.18(2) that

$$\textstyle\int_a^b (f \circ g) \cdot g' \, d\mathscr{L}^1 = \int_{g(a)}^{g(b)} f \, d\mathscr{L}^1 \text{ whenever } f \in \mathscr{K}(\mathbf{R}).$$

However, much more general formulae for "change of variable of integration" will be proved in 3.2.6.

2.9.22. Here we shall see how *the results of* 2.9.19, 2.9.20 *can be extended to the case when*

$$g: \mathbf{R} \to Y, \qquad Y \text{ is a reflexive Banach space},$$
$$\mathbf{V}_a^b g < \infty \text{ whenever } -\infty < a < b < \infty.$$

In place of ψ^+, ψ^- we now consider the measures ψ_α^+, ψ_α^- induced by the real valued functions $\alpha \circ g$ corresponding to all α in the conjugate space Y^* of Y (see 2.5.12). With this substitution, the only difficulty encountered in the extension is to prove the existence, for $\mathscr{L}^1$ almost all x, of the limit

$$g'(x) = \lim_{h \to 0} [g(x+h) - g(x)]/h \in Y,$$

with respect to the norm topology of Y; then $g'(x)$ will be uniquely characterized by the condition

$$\langle g'(x), \alpha \rangle = (\alpha \circ g)'(x) \text{ whenever } \alpha \in Y^*.$$

Since $\operatorname{im} g$ is separable and every closed subspace of a reflexive space is reflexive, we may assume that Y is separable, hence $(Y^*)^*$ and Y^* are separable. Choosing a countable dense subset F of Y^* we observe that

$$|y - \eta| = \sup \{\langle y - \eta, \beta \rangle / |\beta| : 0 \neq \beta \in F\}$$

whenever $\eta \in Y$, and infer that a function u with values in Y is $\mathscr{L}^1$ measurable if and only if $\beta \circ u$ is $\mathscr{L}^1$ measurable for every $\beta \in F$.

Now we first treat the *special case when g is Lipschitzian.* For $\mathscr{L}^1$ almost all x it is true that

$$(\beta \circ g)'(x) \in \mathbf{R} \text{ whenever } \beta \in F;$$

moreover, if $\alpha \in Y^*$ and $\varepsilon > 0$, then there exist $\beta \in F$ and $\delta > 0$ such that $|\beta - \alpha| < \varepsilon$, and $0 < |h| < \delta$ implies

$$|\beta [g(x+h) - g(x)]/h - (\beta \circ g)'(x)| < \varepsilon,$$

hence

$$|\alpha [g(x+h) - g(x)]/h - (\beta \circ g)'(x)| < \varepsilon \operatorname{Lip}(g) + \varepsilon,$$

and Cauchy's criterion implies $(\alpha \circ g)'(x) \in \mathbf{R}$; the function mapping α onto $(\alpha \circ g)'(x)$ is an element of $(Y^*)^*$ with norm at most $\operatorname{Lip}(g)$, hence there exists a unique

$$v(x) \in Y \text{ with } |v(x)| \leq \operatorname{Lip}(g),$$
$$\langle v(x), \alpha \rangle = (\alpha \circ g)'(x) \text{ whenever } \alpha \in Y^*.$$

For $-\infty < a < b < \infty$ we see from 2.4.12 that

$$\alpha \left(\int_a^b v \, d\mathscr{L}^1 \right) = \int_a^b (\alpha \circ v) \, d\mathscr{L}^1 = \int_a^b (\alpha \circ g)' \, d\mathscr{L}^1 = \alpha [g(b) - g(a)]$$

whenever $\alpha \in Y^*$, because $\alpha \circ g$ is absolutely continuous, hence

$$\int_a^b v \, d\mathscr{L}^1 = g(b) - g(a).$$

Assuming that x belongs to the $(\mathscr{L}^1, V)$ Lebesgue set of v, we take $a \leq x \leq b$ and infer that

$$\begin{aligned} |[g(b)-g(a)]/(b-a) - v(x)| &= |\int_a^b [v(t)-v(x)] \, d\mathscr{L}^1 t/(b-a)| \\ &\leq \int_a^b |v(t)-v(x)| \, d\mathscr{L}^1 t/(b-a) \end{aligned}$$

approaches 0 with $b-a$, hence $g'(x) = v(x)$.

Next, in the *general case*, we factor $g = H \circ s$ as in 2.5.16. Applying the special case to H we find that

$$H'(z) \in Y \text{ for } \mathscr{L}^1 \text{ almost all } z.$$

Moreover the inequality

$$s(b) - s(a) \geq \int_a^b s' \, d\mathscr{L}^1 \quad \text{whenever } -\infty < a < b < \infty$$

implies that

$$\mathscr{L}^1(Z) \geq \int_{s^{-1}(Z)} s' \, d\mathscr{L}^1$$

if Z is an open interval, and hence if Z is an open subset of $\mathbf{R}$. We infer that, if $\mathscr{L}^1(Z) = 0$, then $s'(x) = 0$ for $\mathscr{L}^1$ almost all x in $s^{-1}(Z)$. Consequently, for $\mathscr{L}^1$ almost all x,

$$\text{either } s'(x) = 0 \text{ or } H'[s(x)] \in Y \text{ and } 0 < s'(x) < \infty;$$

in the first alternative

$$\lim_{h \to 0} |g(x+h) - g(x)|/|h| \leq s'(x) = 0, \qquad g'(x) = 0,$$

while in the second alternative

$$\frac{g(x+h)-g(x)}{h} = \frac{H[s(x+h)] - H[s(x)]}{s(x+h)-s(x)} \cdot \frac{s(x+h)-s(x)}{h}$$

approaches $H'[s(x)] \cdot s'(x)$ as h approaches 0.

2.9.23. The following example shows the need in 2.9.22 for some strong restriction on Y, like reflexivity – contrasting the freedom afforded in 2.4.12.

Suppose $g: \mathbf{R} \to \mathbf{L}_1(\mathscr{L}^1)$,

$$g(x) = \text{the characteristic function of } \{t: 0 < t < x\}$$

whenever $x \in \mathbf{R}$. Then $g(x) = 0$ whenever $x \leq 0$,

$$\mathscr{L}^1_{(1)}[g(u) - g(v)] = v - u \quad \text{whenever } 0 \leq u < v < \infty,$$

hence $\operatorname{Lip}(g)=1$. However

$$g'(x)\notin \mathbf{L}_1(\mathscr{L}^1) \text{ whenever } 0<x<\infty;$$

in fact the assumption $0<x\in \operatorname{dmn} g'$ would imply the existence of

$$\lim_{h\to 0}\int \frac{g(x+h)-g(x)}{h} f\,d\mathscr{L}^1 = \lim_{h\to 0}\int_x^{x+h} f\,d\mathscr{L}^1/h$$

for every $f\in \mathbf{L}_\infty(\mathscr{L}^1)$, which is absurd; not even the weak limit of the difference quotient exists!

2.9.24. *If $-\infty<a\le b<\infty$, f and g map $\mathbf{R}$ into $\mathbf{R}$, $\mathbf{V}_a^b f<\infty$ and g is absolutely continuous on $\{x\colon a\le x\le b\}$, then*

$$\int_a^b f\,g'\,d\mathscr{L}^1+\int_a^b g\,d f=f(b)\,g(b)-f(a)\,g(a).$$

Taking $t_{n,j}=a+(j-1)(b-a)/n$ we see from 2.9.20 and the proof of 2.5.18(1) that

$$\int_a^b f_n\,g'\,d\mathscr{L}^1+\sum_{j=1}^n g(t_{n,j})\,[f(t_{n,j+1})-f(t_{n,j})]=f(b)\,g(b)-f(a)\,g(a)$$

with $f_n(x)=f(t_{n,j+1})$ for $t_{n,j}<x\le t_{n,j+1}$. By 2.5.16 the functions f_n converge boundedly and $\mathscr{L}^1$ almost everywhere on $\{x\colon a\le x\le b\}$ to f as n approaches ∞, hence 2.4.9 implies the asserted formula for integration by parts.

2.9.25. Combining 2.6.2, 2.9.8 and 2.9.20 one obtains the following proposition about differentiation of an integral with respect to a parameter:

If ϕ measures S, $-\infty<a<b<\infty$, $I=\{t\colon a\le t\le b\}$,

$$u\colon S\times I\to\overline{\mathbf{R}},\qquad \int_{S\times I} u\,d(\phi\times\mathscr{L}^1)\in\mathbf{R},$$

$$g(t)=\int_S\int_a^t u(x,y)\,d\mathscr{L}^1 y\,d\phi\,x \text{ for } t\in I,$$

then g is absolutely continuous on I and

$$g'(t)=\int_S u(x,t)\,d\phi\,x \text{ for } \mathscr{L}^1 \text{ almost all } t \text{ in } I.$$

2.10. Carathéodory's construction

2.10.1. For each metric space X, each family F of subsets of X, and each function ζ such that

$$0\le\zeta(S)\le\infty \text{ whenever } S\in F,$$

we construct preliminary measures ϕ_δ corresponding to $0<\delta\le\infty$, and then a final measure ψ, as follows:

Whenever $A \subset X$, $\phi_\delta(A)$ *is the infimum of the set of numbers*

$$\sum_{S \in G} \zeta(S)$$

corresponding to all countable families G with

$$G \subset F \cap \{S\colon \operatorname{diam} S \le \delta\} \quad \text{and} \quad A \subset \bigcup G.$$

The fact that $\phi_\delta \ge \phi_\sigma$ *for* $0 < \delta < \sigma \le \infty$ *implies the existence of*

$$\psi(A) = \lim_{\delta \to 0+} \phi_\delta(A) = \sup_{\delta > 0} \phi_\delta(A) \quad \text{whenever} \quad A \subset X.$$

Clearly ϕ_δ and ψ measure X. From 2.3.2 (9) it follows that *all open subsets of X are ψ measurable;* in fact

$$\phi_\delta(A \cup B) \ge \phi_\delta(A) + \phi_\delta(B) \quad \text{whenever} \quad \operatorname{distance}(A, B) > \delta > 0,$$

because for each covering G of $A \cup B$, consisting of sets whose diameters do not exceed δ, the families

$$G \cap \{S\colon S \cap A \neq \varnothing\}, \quad G \cap \{S\colon S \cap B \neq \varnothing\}$$

are disjoint and cover A, B respectively. However, simple examples show that not all open sets need be ϕ_δ measurable.

If all members of F are Borel sets, then every subset of X is contained in a Borel set with equal ϕ_δ *measure, and* ψ *is a Borel regular measure.*

We shall call ψ the **result of Carathéodory's construction from ζ on F**, and refer to ϕ_δ as the **size δ approximating measure.**

Carathéodory's construction converts an arbitrary method ζ of estimation on F to a well behaved measure ψ over X. Usually ψ is not an extension of ζ, but reflects the properties of ζ and F more subtly. Appropriate choices of ζ and F yield measures ψ of basic geometric importance. Several such measures are defined in 2.10.2 – 2.10.5.

It was shown in [F 11, 2.5] that $\psi(A)$ can also be characterized as the infimum of the set of all numbers t with the following property: For every open covering H of A there exists a countable subfamily G of F such that each member of G is contained in some member of H, G covers A and

$$\sum_{S \in G} \zeta(S) < t.$$

This alternate description of ψ is perhaps theoretically preferable because it does not require X to be metrizable. For any topological space X, it remains clear that ψ measures X, and one sees from 2.3.2 (8) that every continuous map of X into a metric space is ψ measurable. In this book we do not adopt this more general alternate approach, because up to now it has been useful only in proving [F 11, 2.10, 3.3] which we will not treat here, and because we will make use of the measures ϕ_δ; of course such approximating measures could be defined for any uniform space X.

2.10.2. Here we apply Carathéodory's construction with

$$\zeta(S)=\alpha(m)\,2^{-m}(\operatorname{diam} S)^m \text{ whenever } \varnothing\neq S\subset X,$$

where m is a nonnegative integer and $\alpha(m)$ is defined in 2.7.16 (1). [More generally one can allow m to be any nonnegative real number, taking

$$\alpha(m)=\Gamma\left(\frac{1}{2}\right)^m\Big/\Gamma\left(\frac{m}{2}+1\right);$$

this choice is consistent by virtue of 3.2.13.]

(1) When F is the family of all *nonempty subsets of* X, the resulting measure ψ is called the m **dimensional Hausdorff measure** over X, denoted

$$\mathscr{H}^m \text{ (or } \mathscr{H}_\rho^m\text{, where } \rho \text{ is the metric of } X).$$

One obtains the same measure ψ by letting F be the family of all *nonempty closed subsets of* X, or the family of all *nonempty open subsets of* X (though in the latter case the size δ approximations can be different – consider a circle A with diameter δ, and $m=1$!). Therefore *$\mathscr{H}^m$ is Borel regular*. In case X is a normed vectorspace, the same measure ψ results when the members of F are restricted to be *convex*, because any set and its convex hull have equal diameters.

We observe that *$\mathscr{H}^0$ equals the counting measure over X.*

If $\mathscr{H}^m(A)<\infty$, then $\mathscr{H}^k(A)=0$ for $m<k<\infty$.

(2) When F is the family of all *closed balls* in X (see 2.8.1), the resulting measure ψ is called the m **dimensional spherical measure** over X, denoted

$$\mathscr{S}^m \text{ (or } \mathscr{S}_\rho^m\text{, where } \rho \text{ is the metric of } X).$$

In case $X=\mathbf{R}^n$ one obtains the same measure ψ by letting F be the family of all *open balls* in X.

Clearly $\mathscr{H}^m\leq\mathscr{S}^m\leq 2^m\mathscr{H}^m$; more precise inequalities, for the case of Euclidean space, may be found in 2.10.42, 2.10.6.

2.10.3. Next suppose $X=\mathbf{R}^n$, m is a positive integer, and

$$\zeta(S)=\alpha(m)\,2^{-m}\sup\{|(a_1-b_1)\wedge\cdots\wedge(a_m-b_m)|: a_1,b_1,\ldots,a_m,b_m\in S\}$$

whenever $\varnothing\neq S\subset\mathbf{R}^n$. The result of Carathéodory's construction from ζ on the family F of all nonempty subsets of $\mathbf{R}^n$ will be denoted

$$\mathscr{T}^m.$$

The same measure results when the members of F are required to be open, or compact, or convex; the function ζ assigns the same number to any set and its convex hull, because the map of $(\mathbf{R}^n)^{2m}$ into $\wedge_m(\mathbf{R}^n)$, which yields the above exterior product, is affine with respect to each of the $2m$ variables $a_1, b_1, \ldots, a_m, b_m$.

2.10.4. Assuming $X=\mathbf{R}^n$ and m is a positive integer, with $m \leq n$, we now take

$$\zeta(S)=\sup\{\mathscr{L}^m[p(S)]: p \in \mathbf{O}^*(n,m)\} \text{ for } S \subset \mathbf{R}^n.$$

(1) When F is the family of all *Borel subsets* of $\mathbf{R}^n$, the result of Carathéodory's construction is the m **dimensional Gross measure** over $\mathbf{R}^n$, denoted

$$\mathscr{G}^m.$$

(2) When F is the family of all *open convex subsets* of $\mathbf{R}^n$, the result of Carathéodory's construction is the m **dimensional Carathéodory measure** over $\mathbf{R}^n$, denoted

$$\mathscr{C}^m.$$

The same measure results when F is the family of all *closed convex subsets* of $\mathbf{R}^n$.

In case $m=1$, $\zeta(S)=\operatorname{diam}(S)$ whenever S is convex, hence

$$\mathscr{C}^1=\mathscr{H}^1.$$

2.10.5. Here we suppose $X=\mathbf{R}^n$, m is a positive integer, $m \leq n$, and define for $1 \leq t \leq \infty$ the function ζ_t on the class of all Suslin subsets of $\mathbf{R}^n$ as follows:

With each Suslin set S we associate the function f_S on $\mathbf{O}^*(n,m)$ such that

$$f_S(p)=\mathscr{L}^m[p(S)] \text{ whenever } p \in \mathbf{O}^*(n,m),$$

and let [see 2.4.12, 2.7.16 (6)]

$$\zeta_t(S)=[\theta^*_{n,m}]_{(t)}(f_S)/\beta_t(n,m);$$

the function f_S is $\theta^*_{n,m}$ measurable because

$$\{(x,y,p): x \in S \text{ and } y=p(x)\}$$

is a Suslin subset of $\mathbf{R}^n \times \mathbf{R}^m \times \mathbf{O}^*(n,m)$, hence 2.2.10, 2.2.12, 2.6.2 imply that

$$\{(y,p): y \in p(S)\} \text{ is } \mathscr{L}^m \times \theta^*_{n,m} \text{ measurable}.$$

In case S is bounded, so is f_S, and we infer from 2.4.17 that $\beta_t(n,m)\,\zeta_t(S)$ *is nondecreasing and continuous with respect to* t, and that $\zeta_t(S)$ *is continuous.*

(1) When F is the family of all *Borel subsets* of $\mathbf{R}^n$, the result of Carathéodory's construction from ζ_t on F is the m **dimensional integral-geometric measure with exponent** t over $\mathbf{R}^n$, denoted

$$\mathscr{I}_t^m.$$

(2) When F is the family of all *open convex subsets* of $\mathbf{R}^n$, the measure resulting by Carathéodory's construction from ζ_t on F will be denoted

$$\mathscr{Q}_t^m.$$

The same measure results when F is the family of all *closed convex subsets* of $\mathbf{R}^n$.

The function f_S associated with any bounded open convex set S is continuous (see 3.2.36). Therefore

$$\mathscr{Q}_\infty^m = \mathscr{C}^m.$$

On the other hand, $\mathscr{Q}_1^m$ is called the m **dimensional Gillespie measure** over $\mathbf{R}^n$.

For each $A \subset R^n$, $\beta_t(n,m)\,\mathscr{I}_t^m(A)$ and $\beta_t(n,m)\,\mathscr{Q}_t^m(A)$ are nondecreasing with respect to t; more precise results will be discussed in 3.2.45, 3.3.19, 3.3.16.

Other characterizations of $\mathscr{I}_1^m$ may be found in 2.10.15, 2.10.16.

We observe that $\mathscr{I}_t^m(A) = 0$ if and only if A is contained in a Borel set B with $\mathscr{L}^m[p(B)] = 0$ for $\theta_{n,m}^*$ almost all p in $\mathbf{O}^*(n,m)$. Thus all the measures $\mathscr{I}_t^m$ corresponding to $1 \le t \le \infty$ have the same null sets.

2.10.6. Anticipating some facts to be proved later on, we summarize here all known inequalities relating the various m dimensional measures over $\mathbf{R}^n$ constructed in 2.10.2–2.10.5. From the definitions and 2.10.34 we obtain (for $1 \le t \le \infty$):

$$\begin{array}{ccccccccccc} \mathscr{S}^m & \ge & \mathscr{H}^m & \ge & \mathscr{T}^m & \ge & \mathscr{C}^m & = & \mathscr{Q}_\infty^m & \ge & \beta_t(n,m)\,\mathscr{Q}_t^m \\ & & & & & & \text{\rotatebox{90}{$\le$}} & & \text{\rotatebox{90}{$\le$}} & & \text{\rotatebox{90}{$\le$}} \\ & & & & & & \mathscr{G}^m & \ge & \mathscr{I}_\infty^m & \ge & \beta_t(n,m)\,\mathscr{I}_t^m \end{array}$$

Moreover:

$$\mathscr{S}^m \le [2n/(n+1)]^{m/2}\,\mathscr{H}^m, \text{ by 2.10.42;}$$

$$\mathscr{S}^m \le n^{m/2}\,2^{m-1}\,\mathscr{T}^m, \text{ by 2.10.39;}$$

$$\mathscr{T}^m \le \alpha(m)\,2^{-m}\,m!\,\mathscr{Q}_t^m, \text{ by 2.10.37;}$$

$$\mathscr{C}^m \le \mathscr{Q}_1^m, \text{ by 3.2.45;}$$

$$\mathscr{I}_t^m \le \mathscr{I}_\infty^m, \text{ by 3.3.16.}$$

Therefore the ratios between any two of the measures $\mathscr{S}^m$, $\mathscr{H}^m$, $\mathscr{T}^m$, $\mathscr{Q}_t^m$ are bounded. However the measures $\mathscr{C}^m$, $\mathscr{G}^m$, $\mathscr{I}_\infty^m$ do not have bounded ratios, as shown by 3.3.19, 3.3.20. It is not known whether $\mathscr{I}_\infty^m = \mathscr{I}_1^m$ (see 3.3.16).

All these measures coincide with $\mathscr{L}^m$ in case $m=n$, according to 2.10.35. More generally they agree on the class of all m rectifiable subsets of $\mathbf{R}^n$, by 3.2.26. The comparative study of the behavior of some of these measures on the class of nonrectifiable sets gave much impetus to the development of the structure theory treated in 3.3.

Many particular (compact) sets have constructed to illustrate the differences between the above measures. We refer the reader to:

[B 1, § 47] for a set $A \subset \mathbf{R}^2$ with $\mathscr{S}^1(A) = 2/\sqrt{3}$, $\mathscr{H}^1(A) = 1$.

[FR 1] and [ME 2] for sets $A \subset \mathbf{R}^3$ with $0 < \mathscr{C}^2(A) < \mathscr{H}^2(A) < \infty$.

3.3.19 for a set $A \subset \mathbf{R}^2$ with $\mathscr{Q}_t^1(A) = 2\boldsymbol{\beta}_t(2,1)^{-1}$ for $1 \le t \le \infty$ and $\mathscr{G}^1(A) = 0$.

3.3.20 for a set $E \subset \mathbf{R}^2$ with $0 = \mathscr{I}_\infty^1(E) < \mathscr{G}^1(E) < \mathscr{H}^1(E) < \infty$.

Additional references are listed in 3.3.21.

2.10.7. Suppose $g: \mathbf{R} \to \mathbf{R}$ is nondecreasing, F is the family of all nonempty bounded open subintervals of $\mathbf{R}$, and

$$\zeta\{t: a<t<b\} = g(b) - g(a) \text{ whenever } -\infty < a < b < \infty.$$

Then the measure ψ over $\mathbf{R}$ resulting by Carathéodory's construction from ζ on F satisfies the condition

$$\psi\{t: a<t<b\} = g(b) - g(a)$$

provided g is continuous at a and b, with $a<b$; therefore *ψ equals the measure ψ^+ generated by g* (see 2.5.17).

To verify this condition we first observe that in this case all size δ approximation measures are equal, because if g is continuous at $t_1 < t_2 < \cdots < t_{n+1}$, then

$$g(t_{n+1}) - g(t_1) = \lim_{\varepsilon \to 0+} \sum_{j=1}^{n} [g(t_{j+1}+\varepsilon) - g(t_j - \varepsilon)].$$

Consequently $\psi\{t: a<t<b\} \le g(b) - g(a)$. Next we obtain the opposite inequality from the fact that, if G is any countable family of open intervals covering $\{t: a<t<b\}$, and if $\varepsilon > 0$, then $\{t: a+\varepsilon \le t \le b-\varepsilon\}$ is covered by some finite subfamily of G, consisting of intervals

$$\{t: u_1 < t < v_1\}, \ldots, \{t: u_n < t < v_n\}$$

with $u_1 < a+\varepsilon$, $v_n > b-\varepsilon$, $u_j < u_{j+1} < v_j$ for $j<n$, hence

$$\sum_{j=1}^{n} [g(v_j)-g(u_j)] \geq g(b-\varepsilon)-g(a+\varepsilon).$$

In particular, if $g(x)=x$ for all $x \in \mathbf{R}$, then $\psi = \mathscr{H}^1$, hence the 1 *dimensional Hausdorff measure over* $\mathbf{R}$ *equals the Lebesgue measure* $\mathscr{L}^1$.

2.10.8. Theorem. *Suppose* ψ *results by Carathéodory's construction from* ζ *on the family of all Borel subsets of a separable metric space* X, *with*

$$\zeta(A) \leq \sum_{B \in G} \zeta(B)$$

whenever G *is a countable family of Borel sets and* $A \subset \bigcup G$.

If A *is any Borel subset of* X, *then*

$$\psi(A) = \sup\{\sum_{B \in H} \zeta(B)\colon H \text{ is a Borel partition of } A\};$$

moreover, if $H_1, H_2, H_3, \ldots$ *are Borel partitions of* A, *then*

$$\lim_{j\to\infty} \sup\{\operatorname{diam} B\colon B \in H_j\} = 0 \quad \textit{implies} \quad \lim_{j\to\infty} \sum_{B \in H_j} \zeta(B) = \psi(A).$$

Proof. Clearly $\zeta(S) \leq \psi(S)$ for every Borel set S. Since all Borel sets are ψ measurable it follows that

$$\psi(A) = \sum_{B \in H} \psi(B) \geq \sum_{B \in H} \zeta(B)$$

whenever H is a Borel partition of A. Moreover, if the diameters of the members of the Borel partitions H_j of A approach 0, then

$$\liminf_{j\to\infty} \sum_{B \in H_j} \zeta(B) \geq \psi(A).$$

2.10.9. The preceding theorem applies to the construction of $\mathscr{G}^m$ and $\mathscr{I}_t^m$. It yields the equation

$$\mathscr{I}_t^m = \lim_{s\to t-} \mathscr{I}_s^m \quad \text{for } 1 < t \leq \infty.$$

The theorem does not apply to the construction of $\mathscr{H}^m$, $\mathscr{S}^m$, $\mathscr{T}^m$ or $\mathscr{Q}_t^m$. (However see 3.2.45.)

We shall use the theorem in 2.10.10, 2.10.13, 3.2.39 to integrate the **multiplicity**

$$N(f, y)$$

with which a function f assumes a value y, defined as the *number of elements* (possibly ∞) *of* $f^{-1}\{y\}$.

The following proposition is often employed in conjunction with 2.2.13.

2.10.10. Theorem. *Suppose X is a separable metric space, μ measures Y, f maps X into Y, and $f(A)$ is μ measurable whenever A is a Borel subset of X. If*

$$\zeta(S)=\mu[f(S)] \text{ for } S\subset X,$$

and ψ is the measure over X resulting by Carathéodory's construction from ζ on the family of all Borel subsets of X, then

$$\psi(A)=\int N(f|A,y)\,d\mu\,y \text{ for every Borel set } A\subset X.$$

Proof. We choose Borel partitions $H_1, H_2, H_3, \ldots$ of A such that each member of H_j is the union of some subfamily of H_{j+1}, and

$$\sup\{\operatorname{diam} S\colon S\in H_j\}\to 0 \text{ as } j\to\infty.$$

Letting c_S be the characteristic function of $f(S)$, we note that

$$\sum_{S\in H_j} c_S(y)\uparrow N(f|A,y) \text{ as } j\uparrow\infty$$

for each $y\in Y$, and use 2.10.8, 2.4.7 to obtain

$$\psi(A)=\lim_{j\to\infty}\sum_{S\in H_j}\mu[f(S)]=\lim_{j\to\infty}\int\sum_{S\in H_j}c_S\,d\mu=\int N(f|A,y)\,d\mu\,y.$$

2.10.11. Corollary. *If f is a Lipschitzian map of a complete separable metric space X into a metric space Y, $0\le m<\infty$, and A is a Borel subset of X, then*

$$(\operatorname{Lip} f)^m\cdot\mathscr{H}^m(A)\ge\int N(f|A,y)\,d\mathscr{H}^m y.$$

Proof. From 2.10.2 we see that

$$\zeta(S)=\mathscr{H}^m[f(S)]\le(\operatorname{Lip} f)^m\mathscr{H}^m(S) \text{ for } S\subset X.$$

(In this argument $\mathscr{H}^m$ can be replaced by $\mathscr{S}^m$.)

2.10.12. Corollary. *If X is a metric space, then*

$$\mathscr{H}^1(C)\ge\operatorname{diam} C \text{ for every connected set } C\subset X.$$

Proof. We may assume that $\mathscr{H}^1(C)<\infty$, X is separable and complete, and choose a Borel set B containing C with equal $\mathscr{H}^1$ measure.

Given $a,b\in C$ we define

$$f\colon X\to\mathbf{R},\qquad f(x)=\operatorname{distance}(a,x) \text{ for } x\in X,$$

and infer from 2.10.11, 2.10.7 that

$$\mathscr{H}^1(C)=\mathscr{H}^1(B)\ge\int N(f|B,y)\,d\mathscr{H}^1 y\ge\mathscr{H}^1[f(C)]\ge\operatorname{distance}(a,b)$$

because $0=f(a)$ and $f(b)$ belong to the interval $f(C)$.

2.10.13. Theorem. *If g is a continuous map of* $\mathbf{R}$ *into a metric space Y,* $-\infty<a<b<\infty$ *and* $A=\{t\colon a\le t\le b\}$, *then*

$$\mathbf{V}_a^b g=\int N(g|A,y)\,d\mathscr{H}^1 y.$$

Moreover, in case Y is a reflexive Banach space (see 2.9.22), *g is absolutely continuous on A if and only if* $\mathbf{V}_a^b g<\infty$ *and*

$$\mathscr{H}^1[g(T)]=0 \text{ whenever } T\subset A \text{ with } \mathscr{L}^1(T)=0.$$

Proof. For $a=t_1<t_2<\cdots<t_{n+1}=b$ it follows from 2.10.12 that

$$\sum_{j=1}^{n}\operatorname{distance}[g(t_j),g(t_{j+1})]\le\sum_{j=1}^{n}\mathscr{H}^1\{g(x)\colon t_j<x<t_{j+1}\}.$$

Consequently 2.10.10, 2.10.8 imply $\mathbf{V}_a^b g\le\int N(g|A,y)\,d\mathscr{H}^1 y$.

Henceforth we assume $\mathbf{V}_a^b g<\infty$ and factor $g=H\circ s$ as in 2.5.16. Then 2.10.7, 2.10.11 imply

$$\mathbf{V}_a^b g=s(b)-s(a)=\mathscr{H}^1[s(A)]\ge\int N[H|s(A),y]\,d\mathscr{H}^1 y=\int N(g|A,y)\,d\mathscr{H}^1 y$$

because $N[H|s(A),y]=N(g|A,y)$ unless $A\cap g^{-1}\{y\}$ contains a nondegenerate interval, and the class of all such points y is countable.

Having thus proved the first part of the theorem, we infer from it that the measure induced by the monotone function s equals the measure ψ constructed in 2.10.10 (with $f=g$), and deduce the equivalence of the following three properties (see 2.9.20): absolute continuity of g on A, absolute continuity of s on A, absolute continuity of $\psi\llcorner A$ with respect to $\mathscr{L}^1$.

2.10.14. The preceding theorem can be generalized by dropping the hypothesis that g be continuous, and changing the conclusion to:

$$\mathbf{V}_a^b g=\int N(g|A,y)\,d\mathscr{H}^1 y+\sum_{a\le x<b}\sigma_+(x)+\sum_{a<x\le b}\sigma_-(x)$$

where

$$\sigma_+(x)=\limsup_{t\to x+}\operatorname{distance}[g(t),g(x)],$$

$$\sigma_-(x)=\limsup_{t\to x-}\operatorname{distance}[g(t),g(x)].$$

A more significant extension may be found in [F 11, 3.3], which applies to any light continuous map of a locally compact separable metric space into a metric space.

2.10.15. Theorem. *If $n \geq m \geq 1$ are integers, A is a Borel subset of $\mathbf{R}^n$ and*

$$g(p) = \int N(p \mid A, y)\, d\mathscr{L}^m y \text{ for } p \in \mathbf{O}^*(n,m),$$

then

$$\mathscr{I}_t^m(A) \geq [\theta_{n,m}^*]_{(t)}(g)/\beta_t(n,m) \text{ for } 1 \leq t \leq \infty;$$

moreover equality holds in case $t=1$, that is

$$\mathscr{I}_1^m(A) = \iint N(p \mid A, y)\, d\mathscr{L}^m y\, d\theta_{n,m}^* p / \beta_1(n,m).$$

Proof. Choosing H_j as in the proof of 2.10.10, and recalling 2.10.5(1), we see that

$$\sum_{S \in H_j} f_S(p) \uparrow g(p) \text{ as } j \uparrow \infty \text{ for } p \in \mathbf{O}^*(n,m),$$

and use 2.4.7, 2.4.15, 2.10.8 to infer

$$\begin{aligned} [\theta_{n,m}^*]_{(t)}(g) &= \lim_{j \to \infty} [\theta_{n,m}^*]_{(t)}\Big(\sum_{S \in H_j} f_S\Big) \leq \lim_{j \to \infty} \sum_{S \in H_j} [\theta_{n,m}^*]_{(t)}(f_S) \\ &= \lim_{j \to \infty} \sum_{S \in H_j} \beta_t(n,m)\, \zeta_t(S) = \beta_t(n,m)\, \mathscr{I}_t^m(A); \end{aligned}$$

in case $t=1$ the equation 2.4.8 takes the place of Minkowski's inequality.

2.10.16. In view of 2.10.5 the proof of 2.10.15 also shows that $N(p \mid A, y)$ is $\mathscr{L}^m \times \theta_{n,m}^*$ measurable with respect to (y, p). Letting μ be the rigidly invariant integral over the space of all $n-m$ dimensional affine subspaces of $\mathbf{R}^n$, which was defined in 2.7.16(7) with m replaced by $n-m$, we find that

$$\mathscr{I}_1^m(A) = \beta_1(n,m)^{-1}\, \mu(v)$$

where $v(W)$ equals the number of points in $A \cap W$, for any $n-m$ dimensional affine subspace W of $\mathbf{R}^n$.

Similar arguments in [F 5, § 8] and [FR 1, 3.1] show that

$$\mathscr{I}_1^m(A) = \beta_1(j, j+m-n)\, \beta_1(n,m)^{-1} \int \mathscr{I}_1^{j+m-n}(A \cap Z)\, d\gamma\, Z$$

where $j > n-m$ and γ is our standard rigidly invariant measure over the space of all j dimensional affine subspaces of $\mathbf{R}^n$, and that

$$\mathscr{I}_1^m(A) = \beta_1(k,m)\, \beta_1(n,m)^{-1} \iint_{\mathbf{R}^k} N(p \mid A, y)\, d\mathscr{I}_1^m y\, d\theta_{n,k}^* p$$

whenever $m < k < n$.

2.10.17. Theorem. *Suppose X, F, ζ, ϕ_δ, ψ are as in 2.10.1, the covering relation*

$$C=\{(x,S)\colon x\in S\in F\}$$

is fine at each point of X, and μ measures X.

(1) *If $A\subset X$, $t>0$, $\delta>0$ and*

$$\mu(A\cap S)\le t\,\zeta(S) \text{ whenever } S\in F,\ \operatorname{diam} S\le\delta,$$

then $\mu(A)\le t\,\phi_\delta(A)$.

(2) *If μ is regular, $A\subset X$, $t>0$ and*

$$0\le(C)\limsup_{S\to x}\mu(A\cap S)/\zeta(S)<t \text{ whenever } x\in A,$$

then $\mu(A)\le t\,\psi(A)$.

(3) *If $A\subset X$ and $\phi_\delta(A)<\infty$ whenever $\delta>0$, and if S, $T\in F$ implies $S\cap T\in F$ with $\zeta(S\cap T)\le\zeta(T)$, then*

$$\psi\Big(A\cap\{x\colon 0\le(C)\limsup_{S\to x}\phi_\infty(A\cap S)/\zeta(S)<1\}\Big)=0.$$

Proof. (1) is evident from the definition of ϕ_δ. To verify (2) and (3) we let $B(\mu,t,\delta)$ be the set of all those points x of A for which

$$\mu(A\cap S)\le t\,\zeta(S) \text{ whenever } x\in S\in F,\ \operatorname{diam} S\le\delta,$$

and infer from (1) that

$$\mu B(\mu,t,\delta)\le t\,\phi_\delta B(\mu,t,\delta).$$

The hypothesis of (2) implies that

$$A=\bigcup_{n=1}^{\infty}B(\mu,t,1/n),\qquad \mu(A)=\lim_{n\to\infty}\mu B(\mu,t,1/n)\le t\,\psi(A).$$

The hypothesis of (3) implies that, for $0<\delta<1$,

$$\phi_\delta B(\phi_\delta,1-\delta,\delta)\le(1-\delta)\,\phi_\delta B(\phi_\delta,1-\delta,\delta)<\infty,$$

hence $\phi_\delta B(\phi_\delta,1-\delta,\delta)=0$; moreover

$$A\cap\{x\colon (C)\limsup_x(\phi_\infty\llcorner A)/\zeta<1\}=\bigcup_{n=1}^{\infty}B(\phi_\infty,1-1/n,1/n)$$

and $B(\phi_\infty,1-1/n,1/n)\subset B(\phi_\delta,1-\delta,\delta)$ for $0<\delta\le 1/n$, hence

$$\psi\,B(\phi_\infty,1-1/n,1/n)=0.$$

2.10.18. Theorem. *Suppose X, F, ζ, ϕ_δ, ψ, C, μ are as in 2.10.17, the members of F are closed and μ measurable, and there exists $\eta<\infty$ such that*

$$\zeta(\hat{S})<\eta\,\zeta(S)<\infty \text{ whenever } S\in F,$$

where $\hat{S}=\bigcup\{T\colon T\in F,\ T\cap S\neq\varnothing,\ \operatorname{diam} T\leq 2\operatorname{diam} S\}$.

(1) *If V is an open subset of X, $B\subset V$, $t>0$ and*

$$(C)\limsup_{S\to x}\mu(S)/\zeta(S)>t \text{ whenever } x\in B,$$

then $\mu(V)\geq t\,\psi(B)$.

(2) *If μ is Borel regular, $\mu(A)<\infty$ and A is μ measurable, then*

$$(C)\lim_{S\to x}\mu(A\cap S)/\zeta(S)=0 \text{ for } \psi \text{ almost all } x \text{ in } X\sim A.$$

(3) *If $A\subset X$ and $\psi(A)<\infty$, then*

$$0\leq(C)\limsup_{S\to x}\psi(A\cap S)/\zeta(S)\leq 1 \text{ for } \psi \text{ almost all } x.$$

Proof. To verify (1) we observe that for each $\delta>0$ the family

$$F\cap\{S\colon \mu(S)>t\,\zeta(S),\ \hat{S}\subset V,\ \operatorname{diam}\hat{S}<\delta\}$$

covers B finely, and has by 2.8.6 a disjointed subfamily G such that

$$B\sim\bigcup H\subset\bigcup\{\hat{S}\colon S\in G\sim H\} \text{ for every finite } H\subset G;$$

in case $\mu(V)<\infty$ the fact that $\mu(S)>0$ for $S\in G$ implies G is countable and, given $\varepsilon>0$, we can choose H so that

$$\sum_{S\in G\sim H}\mu(S)<\varepsilon,$$

hence

$$\begin{aligned}\phi_\delta(B)&\leq\sum_{S\in H}\zeta(S)+\sum_{S\in G\sim H}\zeta(\hat{S})\\&<t^{-1}\sum_{S\in G}\mu(S)+\eta\,t^{-1}\varepsilon\leq t^{-1}[\mu(V)+\eta\,\varepsilon].\end{aligned}$$

Next, to prove (2), suppose n is a positive integer and

$$B_n=(X\sim A)\cap\{x\colon (C)\limsup_x(\mu\llcorner A)/\zeta>1/n\}.$$

If $\psi(B_n)>0$ we could apply 2.2.3 to obtain a closed subset E of A with $\mu(A\sim E)<\psi(B_n)/n$, but then (1) with μ, V replaced by $\mu\llcorner A$, $X\sim E$ would imply the opposite inequality $\mu(A\sim E)\geq\psi(B_n)/n$.

In proving (3) we may assume, since every set of finite ψ measure has a Borel ψ hull, that A is a Borel set. If n is a positive integer and

$$B_n=\{x\colon (C)\limsup_x(\psi\llcorner A)/\zeta>1+1/n\},$$

then $\psi(A\cap B_n)=\inf\{\psi(A\cap V)\colon V$ is open, $B_n\subset V\}$, because $\psi \llcorner A$ is Borel regular and $(\psi\llcorner A)X<\infty$ (see 2.2.3, 2.2.2), and (1) with $\mu=\psi\llcorner A$ implies

$$\infty>\psi(A\cap B_n)\geq(1+1/n)\,\psi(B_n),\ \text{hence}\ \psi(B_n)=0.$$

2.10.19. Whenever μ measures a metric space X, $0\leq m<\infty$ and $a\in X$, one defines the m **dimensional upper and lower densities of** μ **at** a:

$$\Theta^{*m}(\mu,a)=\limsup_{r\to 0+}\alpha(m)^{-1}\,r^{-m}\,\mu\,\mathbf{B}(a,r),$$
$$\Theta_*^m(\mu,a)=\liminf_{r\to 0+}\alpha(m)^{-1}\,r^{-m}\,\mu\,\mathbf{B}(a,r).$$

In case the upper and lower densities are equal, their common value is the m **dimensional density** $\Theta^m(\mu,a)$.

The two preceding theorems imply relations between the m dimensional densities and the measures $\mathscr{H}^m$, $\mathscr{S}^m$ over X.

First we take ζ, F, ϕ_δ, $\psi=\mathscr{H}^m$ as in 2.10.2(1) and use 2.10.17 to obtain:

(1) *If* μ *is regular and* $\Theta^{*m}(\mu,x)<t$ *whenever* $x\in A$, *then*

$$\mu(A)\leq 2^m\,t\,\mathscr{H}^m(A).$$

(2) *If* $\phi_\delta(A)<\infty$ *whenever* $\delta>0$, *then*

$$\mathscr{H}^m(A\cap\{x\colon\Theta^{*m}(\phi_\infty\llcorner A,x)<2^{-m}\})=0.$$

In 3.3.19 we will give an example showing that 2^m cannot be replaced by 1 in (1) or (2).

Next we take ζ, F, ϕ_δ, $\psi=\mathscr{S}^m$ as in 2.10.2(2) and use 2.10.18 to obtain:

(3) *If all closed subsets of* X *are* μ *measurable,* V *is open,* $B\subset V$ *and* $\Theta^{*m}(\mu,x)>t$ *whenever* $x\in B$, *then* $\mu(V)\geq t\,\mathscr{S}^m(B)$.

(4) *If* μ *is Borel regular,* $\mu(A)<\infty$ *and* A *is* μ *measurable, then* $\Theta^m(\mu\llcorner A,x)=0$ *for* $\mathscr{S}^m$ *almost all* x *in* $X\sim A$.

(5) *If* $\mathscr{S}^m(A)<\infty$, *then* $\Theta^{*m}(\mathscr{S}^m\llcorner A,x)\leq 1$ *for* $\mathscr{S}^m$ *almost all* x *in* X.

2.10.20. Suppose X, F, ζ, ϕ_δ are as in 2.10.1.

If $0<\delta<\infty$, $A\subset X$ *and* $a\in X$, *then*

$$\phi_\delta(A)=\lim_{r\to\infty}\phi_\delta[A\cap\mathbf{B}(a,r)].$$

To prove this we abbreviate the above limit by s, assume $s<\infty$, suppose $\delta<t<\infty$, and let

$$B_j=A\cap\{x\colon(j-1)\,t\leq\operatorname{dist}(a,x)\leq j\,t\}\ \text{for}\ j=1,2,3,\ldots.$$

Since distance$(B_j, B_k) > \delta$ whenever $|j-k| \geq 2$, we readily see that

$$\sum_{k \in K} \phi_\delta(B_k) = \phi_\delta\Big(\bigcup_{k \in K} B_k\Big) \leq s$$

for every finite set K of positive integers whose elements are either all even or all odd, and hence that

$$\sum_{j=n}^{\infty} \phi_\delta(B_j) \to 0 \text{ as } n \to \infty, \quad \phi_\delta(A) \leq s.$$

The corresponding proposition about ψ is evident, because closed balls are ψ measurable. In 2.10.22 we will study the behavior of ϕ_δ on general increasing sequences of sets, though with hypotheses on X, F, ζ which may be unnecessarily restrictive.

The following two simple statements describe the behavior of the approximating measures on a *decreasing sequence* $C_1 \supset C_2 \supset C_3 \supset \cdots$ *of compact subsets of* X*:*

If the members of F *are open subsets of* X*, then*

$$\lim_{i \to \infty} \phi_\delta(C_i) = \phi_\delta\Big(\bigcap_{i=1}^{\infty} C_i\Big).$$

If $0 < \xi < \delta$ *and*

$$\zeta(S) = \inf\{\zeta(T) \colon T \in F,\ S \subset \operatorname{Int} T,\ \operatorname{diam} T \leq \delta\}$$

whenever $S \in F$ *with* diam $S \leq \xi$*, then*

$$\lim_{i \to \infty} \phi_\delta(C_i) \leq \phi_\xi\Big(\bigcap_{i=1}^{\infty} C_i\Big).$$

2.10.21. Assuming that X, Y are metrized by ρ, σ we consider the function space Y^X with the cartesian product topology, and observe:

If $0 \leq s < \infty$*, then*

$$\begin{aligned} E_s &= Y^X \cap \{f \colon \operatorname{Lip}(f) \leq s\} \\ &= \bigcap_{(u,v) \in X \times X} Y^X \cap \{f \colon \sigma[f(u), f(v)] \leq s\,\rho(u,v)\} \end{aligned}$$

is closed. Moreover *convergence of functions in* E_s, which by definition means convergence at each point of X, *is equivalent to uniform convergence on each compact subset of* X, because if Z is a finite subset of X, $\varepsilon > 0$, and $f, g \in E_s$, then

$$\begin{aligned} &\sup\Big\{\sigma[f(x), g(x)] \colon x \in \bigcup_{z \in Z} \mathbf{B}(z, \varepsilon)\Big\} \\ &\leq 2s\,\varepsilon + \sup\{\sigma[f(z), g(z)] \colon z \in Z\}. \end{aligned}$$

In case Y is boundedly compact, $a\in X$, $b\in Y$, $0\le r<\infty$, then

$$E_s\cap\{f:\sigma[f(a),b]\le r\} \text{ is compact},$$

because this set is closed and contained in the cartesian product of the compact balls $\mathbf{B}[b,r+s\,\rho(a,x)]$ corresponding to all $x\in X$.

Next we *suppose X is boundedly compact, let F be the family of all nonempty compact subsets of X*, and consider the map

$$\Omega\colon F\to E_1=\mathbf{R}^X\cap\{f\colon \operatorname{Lip}(f)\le 1\},$$

$$\Omega_C(x)=\operatorname{distance}(C,x)=\inf\rho(C\times\{x\}) \text{ for } C\in F,\ x\in X.$$

Since $C=\{x\colon \Omega_C(x)=0\}$, Ω is univalent. Also $C,D\in F$ implies

$$\sup\{\Omega_D(x)-\Omega_C(x)\colon x\in X\}=\sup\{\Omega_D(x)\colon x\in C\},$$

$$\sup\{|\Omega_D(x)-\Omega_C(x)|\colon x\in X\}=\sup[\Omega_D(C)\cup\Omega_C(D)];$$

the number $\sup\operatorname{im}|\Omega_D-\Omega_C|$ is called the **Hausdorff distance** between D and C. *If $a\in X$, $0<r<\infty$ and*

$$F_{a,r}=F\cap\{C\colon C\subset\mathbf{B}(a,r)\},$$

then $\Omega(F_{a,r})$ is a closed subset of $E_1\cap\{f\colon |f(a)|\le r\}$, hence compact; consequently $F_{a,r}$ is compact with respect to the Hausdorff distance. To verify that the set $\Omega(F_{a,r})$ is closed in E_1 we assume f belongs to its closure and let

$$D_\varepsilon=\{x\colon f(x)\le\varepsilon\} \text{ whenever } 0\le\varepsilon<\infty;$$

$D_\varepsilon\subset\mathbf{B}(a,r+\varepsilon)$ because $\Omega_C(x)>\varepsilon$ for $C\in F_{a,r}$ and $\rho(a,x)>r+\varepsilon$; if $\varepsilon>0$ there exists $C\in F_{a,r}$ with $f(x)<\Omega_C(x)+\varepsilon$ for $x\in\mathbf{B}(a,r+\varepsilon)$, hence $C\subset D_\varepsilon\in F$; therefore

$$\varnothing\neq\bigcap\{D_\varepsilon\colon \varepsilon>0\}=D_0\in F_{a,r};$$

if $\delta>0$, then $\{x\colon \Omega_{D_0}(x)<\delta\}$ is a neighborhood of D_0, hence contains D_ε for some $\varepsilon>0$, so that $\sup\Omega_{D_0}(D_\varepsilon)<\delta$, and for each $x\in X$ there exists

$$C\in F_{a,r} \text{ with } C\subset D_\varepsilon \text{ and } \Omega_C(x)<f(x)+\delta,$$

hence $\Omega_{D_0}-\Omega_C\le\sup\Omega_{D_0}(C)<\delta$, $\Omega_{D_0}(x)<f(x)+2\delta$; thus $\Omega_{D_0}\le f$, while $\Omega_{D_0}\ge f$ because $\operatorname{Lip}(f)\le 1$.

If $\varepsilon>0$, $Z\subset\mathbf{B}(a,r)\subset\bigcup\{\mathbf{B}(z,\varepsilon)\colon z\in Z\}$ and if $C\in F_{a,r}$, then the Hausdorff distance between C and

$$Z\cap\{z\colon \mathbf{B}(z,\varepsilon)\cap C\neq\varnothing\}$$

does not exceed ε. The nonempty finite sets are dense in $F_{a,r}$.

The function associating with each member of F its diameter is Lipschitzian relative to the Hausdorff distance, with Lipschitz constant 2.

If X is a finite dimensional normed vectorspace, *the class of all nonempty compact convex subsets of X is closed with respect to the Hausdorff distance;* hence *the class of all convex members* of $F_{a,r}$ *is compact.* This is true because a member C of F is convex if and only if Ω_C is a convex function, and the convex functions form a closed subset of $\mathbf{R}^X$. *The function mapping each member of F onto its convex hull is Lipschitzian relative to the Hausdorff distance, with Lipschitz constant 1. The convex hulls of nonempty finite sets are dense in the space of all convex subsets of X.*

2.10.22. Theorem. *Suppose X is a boundedly compact metric space, F is the family of all nonempty compact of X,*

$$\zeta: F \to \{y: 0 \leq y < \infty\}$$

is continuous relative to the Hausdorff distance, and

$$\zeta(C) > 0 \text{ whenever } C \in F \text{ with } \operatorname{diam} C > 0.$$

If $0 < \delta < \infty$, ϕ_δ is the size δ approximating measure occurring in Carathéodory's construction from ζ on F, and $A_1 \subset A_2 \subset A_3 \subset \cdots$ form an increasing sequence of subsets of X, then

$$\lim_{j \to \infty} \phi_\delta(A_j) = \phi_\delta\left(\bigcup_{j=1}^{\infty} A_j\right).$$

Proof. In view of 2.10.20 we may assume that, for some $a \in X$ and $r < \infty$,

$$B = \bigcup_{j=1}^{\infty} A_j \subset \mathbf{B}(a, r),$$

define $F_{a,r+\delta}$ as in 2.10.21 and choose

$$C_{j,k} \in F_{a,r+\delta} \quad \text{with} \quad \delta \geq \operatorname{diam} C_{j,k} \geq \operatorname{diam} C_{j,k+1},$$

$$A_j \subset \bigcup_{k=1}^{\infty} C_{j,k}, \qquad s = \lim_{j \to \infty} \phi_\delta(A_j) = \lim_{j \to \infty} \sum_{k=1}^{\infty} \zeta(C_{j,k}).$$

We note that $s \leq \phi_\delta \mathbf{B}(a, r) < \infty$ because $\mathbf{B}(a, r)$ is compact. Since $F_{a,r+\delta}$ is compact relative to the Hausdorff metric we may further assume, passing to a subsequence, that

$$\lim_{j \to \infty} C_{j,k} = D_k \in F_{a,r+\delta}$$

for each positive integer k (see [K, p. 238, D(a)]). Then

$$\sum_{k=1}^{n} \zeta(D_k) = \sum_{k=1}^{n} \lim_{j \to \infty} \zeta(C_{j,k}) = \lim_{j \to \infty} \sum_{k=1}^{n} \zeta(C_{j,k}) \leq s$$

for each positive integer n, hence

$$t=\sum_{k=1}^{\infty}\zeta(D_k)\le s,\qquad \lim_{n\to\infty}\lim_{j\to\infty}\sum_{k>n}\zeta(C_{j,k})=s-t.$$

Moreover $\delta\ge\operatorname{diam}D_k\ge\operatorname{diam}D_{k+1}$, $\zeta(D_k)\to 0$ as $k\to\infty$, and the compactness of $F_{a,r+\delta}$ implies $\operatorname{diam}D_k\to 0$ as $k\to\infty$, because ζ vanishes only at sets of diameter 0.

Now we shall prove that

$$\psi\Big(B\sim\bigcup_{k=1}^{\infty}V_k\Big)\le s-t$$

for any choice of open sets V_k containing D_k. Since

$$\psi\Big(B\sim\bigcup_{k=1}^{\infty}V_k\Big)=\lim_{i\to\infty}\psi\Big(A_i\sim\bigcup_{k=1}^{\infty}V_k\Big)$$

because ψ is Borel regular, it suffices to show that

$$\phi_\varepsilon\Big(A_i\sim\bigcup_{k=1}^{\infty}V_k\Big)\le s-t\ \text{ for } i=1,2,3,\dots \text{ and } \varepsilon>0.$$

Given i and ε, the inequality $\operatorname{diam}D_n\le\varepsilon/2$ holds provided n is sufficiently large; whenever j is sufficiently large it is also true that

$$A_i\subset A_j,\ \ C_{j,k}\subset V_k\ \text{ for } k\le n,\ \operatorname{diam}C_{j,n}\le\varepsilon,$$

hence

$$A_i\sim\bigcup_{k=1}^{\infty}V_k\subset A_j\sim\bigcup_{k\le n}V_k\subset\bigcup_{k>n}C_{j,k},\qquad \phi_\varepsilon\Big(A_i\sim\bigcup_{k=1}^{\infty}V_k\Big)\le\sum_{k>n}\zeta(C_{j,k});$$

moreover the last sum approaches $s-t$ as n and j approach ∞.

We infer that

$$\psi[B\sim(\bigcup_{k\le n}D_k\cup\bigcup_{k>n}V_k)]\le s-t$$

whenever n is a positive integer and V_k is a neighborhood of D_k for each $k>n$, because ψ is Borel regular and

$$B\sim(\bigcup_{k\le n}D_k\cup\bigcup_{k>n}V_k)=\bigcup_{m=1}^{\infty}[B\sim(\bigcup_{k\le n}\{x\colon\Omega_{D_k}(x)<1/m\}\cup\bigcup_{k>n}V_k)].$$

Next we verify that

$$\psi\Big(B\sim\bigcup_{k=1}^{\infty}D_k\Big)\le s-t.$$

Given $\varepsilon>0$ we choose n so that

$$\operatorname{diam}D_n<\varepsilon\quad\text{and}\quad\sum_{k>n}\zeta(D_k)<\varepsilon;$$

for each $k>n$ we select $S_k \in F$ with

$$D_k \subset \operatorname{Int} S_k, \qquad \operatorname{diam} S_k < \varepsilon, \qquad \sum_{k>n} \zeta(S_k) < \varepsilon;$$

then we find that

$$\phi_\varepsilon\Big(B \sim \bigcup_{k=1}^{\infty} D_k\Big) \le \phi_\varepsilon\big[(B \sim \bigcup_{k\le n} D_k) \sim \bigcup_{k>n} S_k\big] + \sum_{k>n} \zeta(S_k)$$
$$\le \psi\big[B \sim (\bigcup_{k\le n} D_k \cup \bigcup_{k>n} \operatorname{Int} S_k)\big] + \varepsilon \le s - t + \varepsilon.$$

Finally we reach the conclusion

$$\phi_\delta(B) \le \sum_{k=1}^{\infty} \zeta(D_k) + \phi_\delta\Big(B \sim \bigcup_{k=1}^{\infty} D_k\Big) \le t + s - t = s.$$

2.10.23. Corollary. *Suppose X, F, ζ satisfy the conditions of* 2.10.22, *ψ results by Carathéodory's construction from ζ on F, and S is a Suslin subset of X. Then*

$$\psi(S) = \sup\{\psi(C) \colon C \text{ is a compact subset of } S\}.$$

Proof. In view of 2.10.20 we may assume that X is compact.

For any positive numbers $t, \varepsilon, \delta, \xi$ such that

$$\phi_\delta(S) > t + \varepsilon \quad \text{and} \quad \delta > \xi,$$

we can perform the construction of 2.2.12 with γ replaced by the approximating measure ϕ_δ; more precisely, we use 2.10.22 to choose Z_i for $i>0$ so that

$$\phi_\delta[p(Z_{i-1})] - \phi_\delta[p(Z_i)] < \varepsilon\, 2^{-i},$$

and infer from the last statement of 2.10.20 that

$$\phi_\xi(C) \ge \lim_{i\to\infty} \phi_\delta[\text{Closure } p(Z_i)] > \phi_\delta(S) - \varepsilon > t.$$

2.10.24. Assuming the general conditions of 2.10.1 we let f be any nonnegative $\overline{\mathbf{R}}$ valued function on X. For $0<\delta\le\infty$ we define

$$\lambda_\delta(f)$$

as *the infimum of the set of numbers*

$$\sum_{S\in G} u(S)\cdot\zeta(S)$$

corresponding to all countable subfamilies G of F such that $\operatorname{diam} S \le \delta$ *whenever $S \in G$, and all nonnegative* $\mathbf{R}$ *valued functions u on G such that*

$$\sum_{x\in S\in G} u(S) \ge f(x) \text{ for } \psi \text{ almost all } x.$$

Clearly $\lambda_\delta(f) \geq \lambda_\sigma(f)$ for $0 < \delta < \sigma \leq \infty$, and

$$\lim_{\delta \to 0+} \lambda_\delta(f) \leq \int^* f \, d\psi.$$

The general problem whether or not the preceding inequality can always be replaced by the corresponding equation is unsolved. The following two propositions show that the answer is affirmative in at least some of the most interesting special cases:

(1) *equality holds provided* $\{x: f(x) > 0\}$ *is the union of a countable family of sets with finite* ψ *measure and* F, ζ *satisfy the conditions of* 2.10.18.

(2) *equality holds provided* X, F, ζ *satisfy the conditions of* 2.10.18 *and* 2.10.22.

We prove (1) by applying 2.10.18 (3) and the Borel regularity of ψ to obtain an increasing sequence of Borel sets B_j, and a sequence of positive numbers ε_j, such that

$$\psi\Big(\{x: f(x) > 0\} \sim \bigcup_{j=1}^{\infty} B_j\Big) = 0,$$

$$\psi(B_j \cap S) \leq (1 + j^{-1}) \zeta(S) \text{ whenever } S \in F, \; S \cap B_j \neq \varnothing, \; \operatorname{diam} S \leq \varepsilon_j.$$

We also choose countable families G_j, and nonnegative $\mathbf{R}$ valued functions u_j on G_j, such that

$$G_j \subset F \cap \{S: \operatorname{diam} S \leq \varepsilon_j\},$$

$$g_j(x) = \sum_{x \in S \in G_j} u_j(S) \geq f(x) \text{ for } \psi \text{ almost all } x,$$

$$\lim_{j \to \infty} \sum_{S \in G_j} u_j(S) \zeta(S) = \lim_{\delta \to 0+} \lambda_\delta(f).$$

Letting h_j be the product of g_j and the characteristic function of B_j we find that

$$\begin{aligned}
\int^* f \, d\psi &\leq \int \liminf_{j \to \infty} h_j \, d\psi \leq \liminf_{j \to \infty} \int h_j \, d\psi \\
&= \liminf_{j \to \infty} \sum_{S \in G_j} u_j(S) \psi(B_j \cap S) \\
&\leq \liminf_{j \to \infty} (1 + j^{-1}) \sum_{S \in G_j} u_j(S) \zeta(S) = \lim_{\delta \to 0+} \lambda_\delta(f).
\end{aligned}$$

Next we deduce (2) from (1) by showing that

$$\psi\{x: f(x) > t\} \leq t^{-1} \eta \lim_{\delta \to 0+} \lambda_\delta(f)$$

whenever $0<t<\infty$; here η is as in 2.10.18. For this purpose it suffices to verify that

$$\phi_{5\delta}\{x\colon \sum_{x\in S\in G} u(S)>t\}\le t^{-1}\eta\sum_{S\in G} u(S)\,\zeta(S)$$

for every countable $G\subset F\cap\{S\colon \operatorname{diam} S\le\delta\}$, and every positive $\mathbf{R}$ valued function u on G; moreover 2.10.22 reduces this verification to the case when G is finite, and then approximation of u by a function with rational values, and multiplication by a common denominator, allow a further reduction to the case when the values of u are positive integers. In this special case we let k be the least integer greater than or equal to t,

$$A=\{x\colon \sum_{x\in S\in G} u(S)\ge k\},$$

and we define functions $v_0, v_1, \ldots, v_k$ on G and subfamilies $H_1, \ldots, H_k$ of G by induction, starting with $v_0=u$, so that, for $1\le j\le k$,

$$H_j\subset G\cap\{S\colon v_{j-1}(S)\ge 1\},\ H_j \text{ is disjointed, } A\subset\bigcup\{\hat{S}\colon S\in H_j\},$$
$$v_j(S)=v_{j-1}(S)-1 \quad\text{whenever } S\in H_j,$$
$$v_j(S)=v_{j-1}(S) \quad\text{whenever } S\in G\sim H_j,$$
$$A\subset\{x\colon \sum_{x\in S\in G} v_j(S)\ge k-j\};$$

the existence of H_j is implied by 2.8.4. Then we find that

$$k\,\phi_{5\delta}(A)\le\sum_{j=1}^{k}\sum_{S\in H_j}\zeta(\hat{S})\le\eta\sum_{j=1}^{k}\sum_{G}(v_{j-1}-v_j)\,\zeta\le\eta\sum_{G}u\,\zeta.$$

The proposition (2) *applies to the construction* 2.10.2(1) *of the Hausdorff measures* $\mathscr{H}^m$ *over any boundedly compact metric space* X*;* the spherical measures $\mathscr{S}^m$ can be treated similarly because the closed balls in X form a closed set with respect to the Hausdorff distance. I do not know whether the desired equality holds for the construction of Hausdorff measures over an arbitrary metric space.

The equality can be proved for the constructions 2.10.4(1) and 2.10.5(1) of $\mathscr{G}^m$ and $\mathscr{I}_t^m$ by an argument using 2.10.8.

2.10.25. Theorem. *If $f\colon X\to Y$ is a Lipschitzian map of metric spaces, $A\subset X$, $0\le k<\infty$ and $0\le m<\infty$, then*

$$\int_Y^* \mathscr{H}^k(A\cap f^{-1}\{y\})\,d\mathscr{H}^m y\le(\operatorname{Lip} f)^m\,\frac{\alpha(k)\,\alpha(m)}{\alpha(k+m)}\,\mathscr{H}^{k+m}(A)$$

provided either $\{y\colon \mathscr{H}^k(A\cap f^{-1}\{y\})>0\}$ is the union of a countable family of sets with finite $\mathscr{H}^m$ measure, or Y is boundedly compact.

Proof. Letting $\zeta^k, \zeta^m, \zeta^{k+m}$ and $\phi_\delta^k, \phi_\delta^m, \phi_\delta^{k+m}$ be the auxiliary functions occurring in the construction of the Hausdorff measures $\mathscr{H}^k$ over X, $\mathscr{H}^m$ over Y, $\mathscr{H}^{k+m}$ over X, we note that

$$\zeta^k(S) \cdot \zeta^m[\operatorname{Clos} f(S)] \leq c\, \zeta^{k+m}(S) \text{ whenever } S \subset X,$$

where $c = \alpha(k) \cdot \alpha(m) \cdot (\operatorname{Lip} f)^m / \alpha(k+m)$. For each positive integer j we choose a countable family G_j of subsets of X such that

$$A \subset \bigcup G_j, \quad \operatorname{diam} S \leq 1/j \text{ whenever } S \in G_j, \quad \sum_{S \in G_j} \zeta^{k+m}(S) \leq \phi_{1/j}^{k+m}(A) + 1/j,$$

and let $H_j = \{\operatorname{Clos} f(S) \colon S \in G_j\}$,

$$u_j(T) = \sum_{S \in G_j,\ \operatorname{Clos} f(S) = T} \zeta^k(S) \text{ for } T \in H_j.$$

Since $\operatorname{diam} T \leq (\operatorname{Lip} f)/j$ whenever $T \in H_j$, and since

$$A \cap f^{-1}\{y\} \subset \bigcup \{S \colon S \in G_j,\ y \in \operatorname{Clos} f(S)\},$$
$$\phi_{1/i}^k(A \cap f^{-1}\{y\}) \leq \sum_{y \in T \in H_j} u_j(T)$$

whenever $y \in Y$ and $i \leq j$, we infer from 2.10.24 (1) or (2) that

$$\int_Y^* \phi_{1/i}^k(A \cap f^{-1}\{y\})\, d\mathscr{H}^m y \leq \liminf_{j \to \infty} \sum_{T \in H_j} u_j(T) \zeta^m(T)$$
$$= \liminf_{j \to \infty} \sum_{S \in G_j} \zeta^k(S) \zeta^m[\operatorname{Clos} f(S)] \leq c\, \mathscr{H}^{k+m}(A)$$

for every positive integer i.

2.10.26. I do not know whether in the preceding theorem the supplementary hypothesis "provided ... compact" is really necessary.

Another unsolved problem is to find very general conditions on X and A which imply that the integrand be $\mathscr{H}^m$ measurable. Would it be sufficient to assume that X is separable and complete, and that A is a Borel set? The need for some such condition is shown by the example where $A = X \subset Y = \mathbf{R}$, f is the inclusion map, $k = 0$, $m = 1$, and X is $\mathscr{L}^1$ nonmeasurable. For the purpose of this book the following result will be adequate:

If X is boundedly compact, A is $\mathscr{H}^{k+m}$ measurable and $\mathscr{H}^{k+m}(A) < \infty$, then

$$\mathscr{H}^k(A \cap f^{-1}\{y\}) \text{ is } \mathscr{H}^m \text{ measurable with respect to } y,$$

$$A \cap f^{-1}\{y\} \text{ is } \mathscr{H}^k \text{ measurable for } \mathscr{H}^m \text{ almost all } y.$$

Since these conclusions follow from 2.10.25 in case $\mathscr{H}^{k+m}(A)=0$, 2.2.3 reduces the proof to the case when A is compact. If A is compact and $t>0$, then

$$\{y:\ \mathscr{H}^k(A\cap f^{-1}\{y\})\le t\}=\bigcap_{j=1}^{\infty} V_j$$

where V_j consists of all points y such that $A\cap f^{-1}\{y\}$ has a finite open covering G with

$$\operatorname{diam} S\le 1/j \text{ for } S\in G, \qquad \sum_{S\in G}\zeta^k(S)<t+1/j;$$

moreover each set V_j is open in Y.

A special case of 2.10.25, in which $f(x)$ is the distance from x to some fixed point of X, was first proved as a lemma to show that, *if* $\mathscr{H}^{m+1}(X)=0$, *then the topological dimension of X does not exceed m;* the converse is false, but it is true that *every m dimensional separable metric space is homeomorphic to some subset of* $\mathbf{R}^{2m+1}$ *with $m+1$ dimensional Hausdorff measure* 0; an exposition of this theory may be found in [HW, Chapter 7]. Moreover it was shown in [F 5] that these statements remain valid when $\mathscr{H}^{m+1}$ is replaced by any one of the $m+1$ dimensional measures constructed in 2.10.2 to 2.10.6.

2.10.27. From 2.10.25 we deduce the corollary:

If Y, Z and $Y\times Z$ are metrized so that the projections

$$f:\ Y\times Z\to Y, \qquad g:\ Y\times Z\to Z$$

have the Lipschitz constant 1, *and if $A\subset Y\times Z$, $0\le k<\infty$, $0\le m<\infty$, then*

$$\int_Y^* \mathscr{H}^k\{z:\ (y,z)\in A\}\,d\mathscr{H}^m y\le\frac{\alpha(k)\,\alpha(m)}{\alpha(k+m)}\,\mathscr{H}^{k+m}(A)$$

provided either $\{y:\ \mathscr{H}^k\{z:\ (y,z)\in A\}>0\}$ is the union of a countable family of sets with finite $\mathscr{H}^m$ measure, or Y is boundedly compact.

Moreover, if $Y=\mathbf{R}^m$, $Z=\mathbf{R}^n$ and $Y\times Z$ is isometric with $\mathbf{R}^{m+n}$, then

$$\int^* \mathscr{S}^k\{z:\ (y,z)\in A\}\,d\mathscr{L}^m y\le\mathscr{S}^{k+m}(A).$$

To prove the last inequality (in which $\mathscr{S}^k$, $\mathscr{S}^{k+m}$ cannot always be replaced by $\mathscr{H}^k$, $\mathscr{H}^{k+m}$, as shown in [WA]), we note that if $u\in Y$, $v\in Z$, $0<r<\infty$, then

$$\{z:\ (y,z)\in\mathbf{B}[(u,v),r]\}=\mathbf{B}[v,(r^2-|y-u|^2)^{\frac{1}{2}}]$$

whenever $y \in f\,\mathbf{B}[(u,v),r] = \mathbf{B}(u,r)$, hence

$$\begin{aligned}&\int \zeta^k\{z\colon (y,z)\in \mathbf{B}[(u,v),r]\}\, d\mathscr{L}^m y\\ &= \int_{\mathbf{B}(u,r)} \alpha(k)\cdot(r^2-|y-u|^2)^{k/2}\, d\mathscr{L}^m y\\ &= \alpha(k+m)\, r^{k+m} = \zeta^{k+m}\,\mathbf{B}[(u,v),r].\end{aligned}$$

Therefore, if G is a countable family of closed balls in $Y\times Z$ covering A, then

$$H(y) = \{\{z\colon (y,z)\in S\}\colon S\in G,\ y\in f(S)\}$$

covers $\{z\colon (y,z)\in A\}$ whenever $y\in Y$, and

$$\sum_{S\in G}\zeta^{k+m}(S) = \sum_{S\in G}\int \zeta^k\{z\colon (y,z)\in S\}\, d\mathscr{L}^m y = \int \sum_{T\in H(y)} \zeta^k(T)\, d\mathscr{L}^m y.$$

We infer that $\mathscr{H}^m(Y)\,\mathscr{H}^k(Z) \le \alpha(m)\,\alpha(k)\,\alpha(m+k)^{-1}\,\mathscr{H}^{m+k}(Y\times Z)$ provided Y is countably $\mathscr{H}^m$ measurable or boundedly compact. Inequalities in the opposite direction are false in general (see 2.10.29), but true in important special cases (2.10.45, 3.2.23).

2.10.28. Here we let F be the family of all nonempty compact subintervals of $\mathbf{R}$, we let

$$\zeta(S) = h(\operatorname{diam} S) \text{ whenever } \varnothing \neq S\in F, \quad \zeta(\varnothing)=0,$$

where h *is continuous, increasing and strictly concave* ($-h$ is strictly convex) on $\{t\colon t\ge 0\}$, with $h(0)=0$, and we investigate how the measure ψ resulting by Carathéodory's construction from ζ on F behaves on certain particular subsets of $\mathbf{R}$. For instance the measure $\mathscr{H}^m$ over $\mathbf{R}$, with $0<m<1$, can be studied in this way by taking

$$h(t) = \alpha(m)\, 2^{-m}\, t^m \text{ whenever } 0\le t<\infty.$$

We shall say that H *is* **special** *for* J *if and only if*

$$J\in F,\quad H\subset F\cap\{S\colon S\subset J\},$$

$$H \textit{ is finite and disjointed, } \zeta(J) = \sum_{S\in H}\zeta(S),$$

$$\zeta(T) \ge \sum_{T\supset S\in H} \zeta(S) \textit{ whenever } T\in F.$$

This concept has the following properties:

(1) *If H is special for J, then*

$$\zeta(T) \ge \sum_{S\in H}\zeta(T\cap S) \textit{ whenever } T\in F.$$

(2) *If H is special for J, and $K(S)$ is special for S whenever $S \in H$, then*

$$\bigcup \{K(S)\colon S \in H\} \text{ is special for } J.$$

(3) *If $H_0, H_1, H_2, \ldots$ are finite disjointed subfamilies of F such that for $j \geq 1$ each member of H_j is contained in some member of H_{j-1} and*

$$H_j \cap \{T\colon T \subset S\} \text{ is special for } S \text{ whenever } S \in H_{j-1},$$

and if $\sup\{\operatorname{diam} S\colon S \in H_j\} \to 0$ *as* $j \to \infty$,

$$A = \bigcap_{j=0}^{\infty} \bigcup H_j,$$

then

$$\psi(A \cap T) \leq \zeta(T) \text{ whenever } T \in F,$$

$$\psi(A \cap T) = \zeta(T) \text{ whenever } T \in \bigcup_{j=0}^{\infty} H_j.$$

(4) *If $J \in F$, $2 \leq n \in \mathbf{Z}$ and $\Phi(J, n)$ consists of the intervals*

$$\{x\colon \inf J + (i-1)\,t \leq x \leq \inf J + (i-1)\,t + s\}$$

corresponding to $i = 1, \ldots, n$ where

$$h(s) = \zeta(J)/n, \qquad t = (\operatorname{diam} J - s)/(n-1),$$

then $0 < s < \operatorname{diam} J/n$, $t > s$, $\Phi(J, n)$ *is special for J and*

$$\sum_{S \in \Phi(J,n)} \zeta(T \cap S) \leq 3\,\zeta(J) \operatorname{diam} T/\operatorname{diam} J$$

whenever $T \in F$ with $\operatorname{diam} T \geq t$.

For the proof of (1) it suffices, since h is increasing, to consider the case when

$$\inf T \in V \in H, \qquad \sup T \in W \in H, \qquad V \cap W = \varnothing.$$

Also $h(a+r) - h(a) \geq h(b+r) - h(b)$ for $0 \leq a \leq b$ and $r \geq 0$, because h is concave, hence

$$\zeta(T) - \zeta(T \cap V) \geq \zeta(T \cup V) - \zeta(V),$$

$$\zeta(T \cup V) - \zeta(T \cap W) \geq \zeta(T \cup V \cup W) - \zeta(W),$$

and it follows that

$$\zeta(T) \geq \zeta(T \cup V \cup W) - \zeta(V) - \zeta(W) + \zeta(T \cap V) + \zeta(T \cap W)$$

$$\geq \sum_{\operatorname{Int} T \supset S \in H} \zeta(S) \quad + \zeta(T \cap V) + \zeta(T \cap W).$$

Now (2) and (3) become obvious. To verify (4) we first consider an interval

$$T=\{x:\ \inf J+(i-1)\,t\le x\le \inf J+(j-1)\,t+s\}$$

where $1\le i<j\le n$ are integers, hence $0\le u=(j-i)/(n-1)\le 1$ and

$$\begin{aligned}\zeta(T)&=h[(j-i)\,t+s]=h[u\operatorname{diam}J+(1-u)\,s]\\&\ge u\,\zeta(J)+(1-u)\,h(s)=(u\,n+1-u)\,h(s)\\&=(j-i+1)\,h(s)=\sum_{T\supset S\in H}\zeta(S).\end{aligned}$$

Next we consider $T\in F$ with $\operatorname{diam}T\ge t$, let k be the positive integer such that $k\,t\le\operatorname{diam}T\le(k+1)\,t$, then observe that T meets at most $k+2$ members of $\Phi(J,n)$ and

$$(k+2)\,h(s)=\frac{k+2}{k}\,\zeta(J)\,\frac{k\,t}{n\,t}\le 3\,\zeta(J)\,\frac{\operatorname{diam}T}{\operatorname{diam}J}.$$

Given $0<m<1$ and a sequence $n=(n_1,n_2,n_3,\ldots)$ of integers $n_j\ge 2$ we now take $h(t)=\alpha(m)\,2^{-m}\,t^m$ and apply (3) with $H_0=\{I\}$, where $I=\{t:\ 0\le t\le 1\}$, and

$$H_j=\bigcup\{\Phi(S,n_j):\ S\in H_{j-1}\}\ \text{ for } j\ge 1$$

to obtain a set $A(m,n)$ with $\mathscr{H}^m[A(m,n)]=\alpha(m)\,2^{-m}$; moreover the function f such that

$$f(x)=\mathscr{H}^m[A(m,n)\cap\{z:\ z\le x\}]\ \text{ for } x\in\mathbf{R}$$

satisfies the conditions

$$0\le f(b)-f(a)\le\alpha(m)\,2^{-m}(b-a)^m\ \text{ for } a\le b,$$

$$f[A(m,n)]=\operatorname{im}f=\{y:\ 0\le y\le\alpha(m)\,2^{-m}\},$$

$$f'(x)=0\ \text{ for } x\in\mathbf{R}\sim A(m,n),\ \text{hence for } \mathscr{L}^1 \text{ almost all } x.$$

For instance, if $m=\log(2)/\log(3)$ and $n_j=2$ for all j, then $A(m,n)$ is the Cantor set and f is the Cantor function.

2.10.29. With $A(m,n)$ defined as in 2.10.28 it is easy to verify that

$$\mathscr{H}^{2m}[A(m,n)\times A(m,n)]\le\alpha(2m)\,2^{-m}$$

because for each j the diameters of all intervals in H_j are equal. However *we will now construct sequences n and v of positive integers such that*

$$\mathscr{H}^k[A(m,n)\times A(m,v)]=\infty\ \textit{ for } 0\le k<m+1;$$

this is the worst behavior permitted by 2.10.45.

Since $(1-x^{-1/m})/(x-1)\to 0$ as $x\to\infty$ we can inductively define sequences n, s, t, v, σ, τ such that $s_0=\sigma_0=1$ and

$$2\le n_j\in\mathbf{Z},\quad s_j=s_{j-1}\,n_j^{-1/m},\quad t_j=(s_{j-1}-s_j)/(n_j-1),$$
$$t_j=s_{j-1}(1-n_j^{-1/m})/(n_j-1)\le\sigma_{j-1}^j,$$
$$2\le v_j\in\mathbf{Z},\quad \sigma_j=\sigma_{j-1}\,v_j^{-1/m},\quad \tau_j=(\sigma_{j-1}-\sigma_j)/(v_j-1),$$
$$\tau_j=\sigma_{j-1}\,(1-v_j^{-1/m})/(v_j-1)\le s_j^j$$

for $j=1,2,3,\ldots$. Using 2.10.28 (4) we obtain

$$\sigma_{j-1}\ge t_j\ge s_j\ge\tau_j\ge\sigma_j\ge t_{t+1}\to 0 \text{ as } j\to\infty.$$

We let $B=A(m,n)\times A(m,v)$ and observe that $(\mathscr{H}^m\times\mathscr{H}^m)\,B>0$. *We will prove $0<k<m+1$ implies $\mathscr{H}^k(B)=\infty$ by showing that if $\varepsilon>0$ and $c=6^{-1}\,\alpha(m)^{-2}\,2^{2m}$, then*

$$(\operatorname{diam} S)^k\ge c(\mathscr{H}^m\times\mathscr{H}^m)\,(B\cap S)/\varepsilon$$

whenever $S\subset\mathbf{R}^2$ with $\operatorname{diam} S\le\sup\{t_j\colon k+1/j\le m+1,\ \sigma_{j-1}^m\le\varepsilon\}$. Letting $d=\operatorname{diam} S$ we choose intervals U, V and a positive integer j such that

$$S\subset U\times V,\qquad \operatorname{diam} U=d=\operatorname{diam} V,$$
$$t_j\ge d\ge t_{j+1},\qquad k+1/j\le m+1,\qquad \sigma_{j-1}^m\le\varepsilon.$$

In case $t_j\ge d\ge\tau_j$ we infer from 2.10.28 (3) and (4) that

$$\mathscr{H}^m[A(m,n)\cap U]\le\alpha(m)\,2^{-m}\,d^m,$$
$$\mathscr{H}^m[A(m,v)\cap V]\le 6\alpha(m)\,2^{-m}\,\sigma_{j-1}^m\,d/\sigma_{j-1},$$

because $\sigma_{j-1}\ge d$ implies V meets at most two of the intervals of length σ_{j-1} occurring in the construction of $A(m,v)$, hence obtain

$$c(\mathscr{H}^m\times\mathscr{H}^m)\,(B\cap S)\le\varepsilon\,d^{m+1}/\sigma_{j-1}\le\varepsilon\,d^{k+1/j}/\sigma_{j-1}$$
$$\le\varepsilon\,d^k\,t_j^{1/j}/\sigma_{j-1}\le\varepsilon\,d^k.$$

In case $\tau_j\ge d\ge t_{j+1}$ we find similarly that

$$\mathscr{H}^m[A(m,n)\cap U]\le 6\alpha(m)\,2^{-m}\,s_j^m\,d/s_j,$$
$$\mathscr{H}^m[A(m,v)\cap V]\le\alpha(m)\,2^{-m}\,d^m,$$
$$c(\mathscr{H}^m\times\mathscr{H}^m)\,(B\cap S)\le\varepsilon\,d^{m+1}/s_j\le\varepsilon\,d^{k+1/j}/s_j$$
$$\le\varepsilon\,d^k\,\tau_j^{1/j}/s_j\le\varepsilon\,d^k.$$

It is now easy to see that for each positive integer q the subsets $A(m,n)^q$ and $A(m,v)^q$ of $\mathbf{R}^q$ have finite qm dimensional Hausdorff measure, yet

$$A(m,n)^q \times A(m,v)^q \simeq [A(m,n) \times A(m,v)]^q$$

has infinite k dimensional Hausdorff measure whenever $0 \leq k < qm+q$, by 2.10.27.

2.10.30. Steiner symmetrization with respect to an $n-1$ dimensional vector subspace V of $\mathbf{R}^n$ is the operation which associates with each bounded subset S of $\mathbf{R}^n$ the subset T of $\mathbf{R}^n$ such that, *for every straight line L perpendicular to V,*

either $L \cap S = \emptyset$ and $L \cap T = \emptyset$,

or $L \cap S \neq \emptyset$ and $L \cap T$ is a closed segment with center in V and

$$\mathscr{H}^1(L \cap T) = \mathscr{H}^1(L \cap S).$$

Using 2.10.7, 2.7.16(1) and the Fubini theorem one readily verifies:

If S is compact, then T is compact and $\mathscr{L}^n(S) = \mathscr{L}^n(T)$.

If S is convex, then T is convex.

Letting u be a unit vector orthogonal to V we define the **reflection**

$$\rho_u \in \mathbf{O}(n), \quad \rho_u(x) = x - 2(x \bullet u)\, u \text{ for } x \in \mathbf{R}^n,$$

and observe that $\rho_u(T) = T$. Moreover:

If S is a compact subset of $\mathbf{B}(0,r)$ and $Z = \operatorname{Bdry} \mathbf{B}(0,r)$ then $T \subset \mathbf{B}(0,r)$ and $(Z \sim S) \cup \rho_u(Z \sim S) \subset Z \sim T$.

2.10.31. Theorem. *If C is a nonempty class of nonempty compact subsets of $\mathbf{R}^n$, C is closed with respect to the Hausdorff distance, and Steiner symmetrization with respect to every $n-1$ dimensional vector subspace of $\mathbf{R}^n$ maps C into C, then either $\{0\} \in C$ or $\mathbf{B}(0,r) \in C$ for some $r > 0$.*

Proof. We define

$$r = \inf\{s \colon T \subset \mathbf{B}(0,s) \text{ for some } T \in C\}$$

and infer from the compactness of $C \cap F_{0,r+1}$ (see 2.10.21) that either $r=0$ and $\{0\} \in C$, or $r>0$ and there exists $S \in C$ with $S \subset \mathbf{B}(0,r)$.

Next we shall prove that $\mathbf{B}(0,r) \supset S \in C$ implies

$$Z = \operatorname{Bdry} \mathbf{B}(0,r) \subset S.$$

Otherwise $B(a,\varepsilon) \subset \mathbf{R}^n \sim S$ for some $a \in Z$ and $\varepsilon > 0$, we could choose distinct points $a = y_0, y_1, \ldots, y_k \in Z$ with

$$Z \subset \bigcup_{j=0}^{k} \mathbf{B}(y_j, \varepsilon),$$

and construct $S=T_0, T_1, \ldots, T_k \in C$ inductively so that, for $j \geq 1$, T_j is obtained from T_{j-1} by Steiner symmetrization with respect to

$$\mathbf{R}^n \cap \{v\colon v \bullet u_j = 0\}, \text{ where } u_j = (y_j - a)/|y_j - a|.$$

From the last statement in 2.10.30 it would follow that

$$\bigcup_{i=0}^{j} Z \cap \mathbf{B}(y_i, \varepsilon) \subset Z \sim T_j, \qquad \text{because } \rho_{u_j} \mathbf{B}(a, \varepsilon) = \mathbf{B}(y_j, \varepsilon),$$

hence $Z \subset Z \sim T_k$, $T_k \subset \mathbf{B}(a, s)$ for some $s < r$.

Finally we observe that $\mathbf{B}(0, r) \supset S \in C$ implies $\mathbf{B}(0, r) = S$, because otherwise there would exist a straight line L with

$$\mathscr{H}^1(L \cap S) < \mathscr{H}^1[L \cap \mathbf{B}(0, r)],$$

and Steiner symmetrization of S with respect to the vectorsubspace perpendicular to L would yield $T \in C$ with $T \subset \mathbf{B}(0, r)$ but

$$\emptyset \neq Z \cap L \subset Z \sim T.$$

2.10.32. Theorem. *If* $\emptyset \neq S \subset \mathbf{R}^m$, *then*

$$\mathscr{L}^m(S) \leq \boldsymbol{\alpha}(m)\, 2^{-m} \sup\{|(a_1 - b_1) \wedge \ldots \wedge (a_m - b_m)|\colon a_1, b_1, \ldots, a_m, b_m \in S\}.$$

Proof. We define ζ as in 2.10.3 with $n = m$, suppose λ, μ are positive numbers, and let C be the class of all those nonempty, compact, convex subsets S of $\mathbf{R}^m$ for which

$$\mathscr{L}^m(S) \geq \lambda \quad \text{and} \quad \zeta(S) \leq \mu.$$

Obviously C is closed with respect to the Hausdorff metric. Moreover we will show that if T is obtained from $S \in C$ by Steiner symmetrization with respect to

$$V = \mathbf{R}^n \cap \{v\colon v \bullet u = 0\}, \text{ where } u \in \mathbf{R}^n \text{ and } |u| = 1,$$

then $T \in C$. For any $a_1, b_1, \ldots, a_m, b_m \in T$ there exist $v_1, w_1, \ldots, v_m, w_m \in V$ and $\alpha_1, \beta_1, \ldots, \alpha_m, \beta_m \in \mathbf{R}$ with

$$a_i = v_i + \alpha_i u, \qquad |2\alpha_i| \leq \operatorname{diam}\{t\colon v_i + tu \in S\},$$
$$b_i = w_i + \beta_i u, \qquad |2\beta_i| \leq \operatorname{diam}\{t\colon w_i + tu \in S\},$$

and hence there exist $\gamma_1, \delta_1, \ldots, \gamma_m, \delta_m \in \mathbf{R}$ with

$$a_i' = v_i + \gamma_i u \in S, \qquad a_i'' = v_i + (\gamma_i + 2\alpha_i) u \in S,$$
$$b_i' = w_i + \delta_i u \in S, \qquad b_i'' = w_i + (\delta_i + 2\beta_i) u \in S.$$

It follows that

$$\bigwedge_{i=1}^{m} (a_i - b_i) = \bigwedge_{i=1}^{m} [(v_i - w_i) + (\alpha_i - \beta_i) u] = \sum_{j=1}^{m} (\alpha_j - \beta_j) u \wedge \xi_j$$

where $\xi_j=(-1)^{j-1}\bigwedge_{i\neq j}(v_i-w_i)\in\bigwedge_{m-1}(\mathbf{R}^m)$, and similarly

$$\bigwedge_{i=1}^{m}(a_i'-b_i')=\sum_{j=1}^{m}(\gamma_j-\delta_j)u\wedge\xi_j,$$

$$\bigwedge_{i=1}^{m}(a_i''-b_i'')=\sum_{j=1}^{m}[(\gamma_j-\delta_j)+2(\alpha_j-\beta_j)]u\wedge\xi_j,$$

hence $\left|\bigwedge_{i=1}^{m}(a_i-b_i)\right|=\frac{1}{2}\left|\bigwedge_{i=1}^{m}(a_i''-b_i'')-\bigwedge_{i=1}^{m}(a_i'-b_i')\right|\leq\mu.$

In case C is nonempty we infer from 2.10.31 that $\mathbf{B}(0,r)\in C$ for some $r>0$, and conclude

$$\lambda\leq\mathscr{L}^m\mathbf{B}(0,r)=\boldsymbol{\alpha}(m)r^m=\zeta\mathbf{B}(0,r)\leq\mu.$$

2.10.33. Corollary (Isodiametric inequality). *If* $\varnothing\neq S\subset\mathbf{R}^m$, *then*

$$\mathscr{L}^m(S)\leq\boldsymbol{\alpha}(m)\,2^{-m}(\operatorname{diam}S)^m.$$

Proof. From 1.7.5 we see that if $a_1,b_1,\ldots,a_m,b_m\in S$, then

$$|(a_1-b_1)\wedge\cdots\wedge(a_m-b_m)|\leq|a_1-b_1|\cdot\cdots\cdot|a_m-b_m|\leq(\operatorname{diam}S)^m.$$

2.10.34. Corollary. $\mathscr{C}^m(A)\leq\mathscr{T}^m(A)$ for $A\subset\mathbf{R}^n$.

Proof. Let ζ be defined as in 2.10.3. Whenever $S\subset\mathbf{R}^n$ and $p\in\mathbf{O}^*(n,m)$, Theorem 2.10.32 is applicable to $p(S)$, while 1.7.6 implies $\|\bigwedge_m p\|=1$, hence

$$\mathscr{L}^m[p(S)]\leq\zeta[p(S)]\leq\zeta(S).$$

2.10.35. Theorem. *If* $A\subset\mathbf{R}^m$ *and* $1\leq t\leq\infty$, *then*

$$\mathscr{L}^m(A)=\mathscr{S}^m(A)=\mathscr{H}^m(A)=\mathscr{T}^m(A)=\mathscr{Q}_t^m(A)=\mathscr{I}_t^m(A)=\mathscr{G}^m(A).$$

Proof. Suppose ψ is any one of the measures $\mathscr{H}^m$, $\mathscr{T}^m$, $\mathscr{Q}_t^m$, $\mathscr{I}_t^m$, $\mathscr{G}^m$ over $\mathbf{R}^m$. Using 2.10.32, the fact that

$$\mathscr{L}^m[p(S)]=\mathscr{L}^m(S)\text{ whenever }S\subset\mathbf{R}^m\text{ and }p\in\mathbf{O}(m),$$

and the equation $\boldsymbol{\beta}_t(m,m)=1$, we obtain the inequality

$$\mathscr{S}^m(A)\geq\psi(A)\geq\mathscr{L}^m(A).$$

Moreover, since $\Theta^m(\mathscr{L}^m,x)=1$ whenever $x\in A$, we infer from 2.10.19(3) that

$$\mathscr{L}^m(V)\geq\mathscr{S}^m(A)\text{ for every open set }V\text{ containing }A,$$

hence $\mathscr{L}^m(A)\geq\mathscr{S}^m(A)$.

2.10.36. Theorem. *If S is a convex subset of $\mathbf{R}^m$ and $a_1, b_1, \ldots, a_m, b_m \in S$, then*

$$|(a_1 - b_1) \wedge \cdots \wedge (a_m - b_m)| \leq m!\, \mathscr{L}^m(S).$$

Proof. The assertion is trivial in case $m = 1$. To apply induction with respect to m we assume $a_1 \neq b_1$ and choose

$$p \in \mathbf{O}^*(m, m-1) \text{ with } a_1 - b_1 \in \ker p.$$

Since $x - (p^* \circ p)x \in \ker p$ whenever $x \in \mathbf{R}^m$, and $\wedge_{m-1} p^*$ is an isometric injection, we obtain

$$\begin{aligned} \left| \bigwedge_{i=1}^{m} (a_i - b_i) \right| &= \left| (a_1 - b_1) \wedge \bigwedge_{i=2}^{m} [(p^* \circ p)(a_i - b_i)] \right| \\ &= |a_1 - b_1| \cdot \left| \bigwedge_{i=2}^{m} [p(a_i) - p(b_i)] \right| \\ &\leq |a_1 - b_1| \cdot (m-1)!\, \mathscr{L}^{m-1}[p(S)]. \end{aligned}$$

Subjecting S to Steiner symmetrization with respect to

$$\operatorname{im} p^* = \mathbf{R}^m \cap \{v \colon v \bullet (a_1 - b_1) = 0\},$$

and observing that the resulting set T contains the convex hull of

$$p^*[p(S)] \cup \{p^*[p(a_1)] + (a_1 - b_1)/2,\ p^*[p(a_1)] + (b_1 - a_1)/2\}$$

we also find that

$$\mathscr{L}^m(S) = \mathscr{L}^m(T) \geq |a_1 - b_1| \cdot \mathscr{L}^{m-1}[p(S)]/m.$$

2.10.37. Corollary. *If $A \subset \mathbf{R}^n$, m is a positive integer and $1 \leq t \leq \infty$, then*

$$\mathscr{T}^m(A) \leq \alpha(m)\, 2^{-m}\, m!\, \mathscr{Q}_t^m(A).$$

Proof. Assuming S is a convex subset of $\mathbf{R}^n$ and $a_1, b_1, \ldots, a_m, b_m \in S$, we let

$$g(p) = |p(a_1 - b_1) \wedge \cdots \wedge p(a_m - b_m)| \text{ for } p \in \mathbf{O}^*(n, m),$$

and define f_S, ζ_t as in 2.10.5. Then $g \leq m!\, f_S$, and we see from 2.7.16(6) that

$$\begin{aligned} |(a_1 - b_1) \wedge \cdots \wedge (a_m - b_m)| &= [\theta^*_{n,m}]_{(t)}(g)/\beta_t(n, m) \\ &\leq m!\, [\theta^*_{n,m}]_{(t)}(f_S)/\beta_t(n, m) = m!\, \zeta_t(S). \end{aligned}$$

2.10.38. Theorem. *For each compact $S \subset \mathbf{R}^n$ there exist $a_1, b_1, \ldots, a_n, b_n \in S$ and a rectangular box Q with side lengths $c_1 \geq c_2 \geq \cdots \geq c_n$ such that $S \subset Q \subset \mathbf{R}^n$ and*

$$|(a_1 - b_1) \wedge \cdots \wedge (a_m - b_m)| = c_1 \cdot \cdots \cdot c_m \text{ for } m = 1, \ldots, n.$$

Proof. The assertion is trivial in case $n=1$. Applying induction with respect to n we assume $n>1$ and $\operatorname{diam} S>0$, choose

$$a_1, b_1 \in S \text{ with } |a_1-b_1|=\operatorname{diam} S,$$
$$p\in \mathbf{O}^*(n, n-1), \quad a_1-b_1 \in \ker p,$$

then select $\alpha_2, \beta_2, \ldots, \alpha_n, \beta_n \in p(S)$ and a rectangular box W with side lengths $c_2 \geq c_3 \geq \cdots \geq c_n$ such that $p(S) \subset W \subset \mathbf{R}^{n-1}$ and

$$|(\alpha_2-\beta_2)\wedge \cdots \wedge (\alpha_m-\beta_m)| = c_2 \cdot \cdots \cdot c_m \text{ for } m=2, \ldots, n.$$

Finally we take $c_1 = |a_1-b_1|$,

$$Q=\mathbf{R}^n \cap \{x\colon 0 \leq (x-a_1)\bullet(b_1-a_1) \leq |a_1-b_1|^2,\ p(x)\in W\},$$
$$a_i, b_i \in S \text{ with } p(a_i)=\alpha_i,\ p(b_i)=\beta_i \text{ for } i=2, \ldots, n,$$

and compute

$$\left|\bigwedge_{i=1}^{m}(a_i-b_i)\right| = \left|(a_1-b_1)\wedge \bigwedge_{i=2}^{m}[(p^*\circ p)(a_i-b_i)]\right|$$
$$= |a_1-b_1| \cdot \left|\bigwedge_{i=2}^{m}(\alpha_i-\beta_i)\right| = c_1\cdot c_2 \cdot \cdots \cdot c_m.$$

2.10.39. Corollary. *$\mathscr{S}^m(A) \leq n^{m/2}\, 2^{m-1}\, \mathscr{T}^m(A)$ for $A \subset \mathbf{R}^n$.*

Proof. For each rectangular box Q with side lengths $c_1 \geq c_2 \geq \cdots \geq c_n > 0$ we can choose positive integers $k_1, \ldots, k_{m-1}$ such that

$$k_i/2 < c_i/c_m \leq k_i \text{ whenever } i=1, \ldots, m-1,$$

and express Q as the union of $k_1 \cdot \cdots \cdot k_{m-1}$ rectangular boxes with side lengths

$$c_1/k_1, \ldots, c_{m-1}/k_{m-1}, c_m, \ldots, c_n.$$

Each of these new boxes is contained in a cube with side length c_m, hence in a closed ball with diameter $n^{\frac{1}{2}} c_m$. The sum of the m'th powers of the diameters of the balls so obtained equals

$$k_1 \cdot \cdots \cdot k_{m-1} \cdot n^{m/2} \cdot (c_m)^m = n^{m/2}(k_1 c_m) \cdot \cdots \cdot (k_{m-1} c_m) \cdot c_m$$
$$< n^{m/2}(2c_1) \cdot \cdots \cdot (2c_{m-1}) \cdot c_m = n^{m/2}\, 2^{m-1} c_1 \cdot \cdots \cdot c_m.$$

2.10.40. Lemma. *If $\varnothing \neq P \subset \mathbf{R}^n \times \{r\colon 0<r<\infty\}$, P is compact, and*

$$Y_t = \{y\colon |y-a| \leq rt \text{ whenever } (a,r)\in P\}$$

for $0 \leq t < \infty$, then $c=\inf\{t\colon Y_t \neq \varnothing\} < \infty$, Y_c consists of a single point b, and b belongs to the convex hull of

$$A=\{a\colon \text{for some } r,\ (a,r)\in P \text{ and } |b-a|=rc\}.$$

Proof. Since each Y_t is compact, and

$$0 \in Y_t \text{ for } t \geq \sup\{|a|/r: (a,r) \in P\},$$

we see that $Y_c = \bigcap\{Y_t: c<t<\infty\} \neq \emptyset$. We define

$$\mu = \sup\{r: (a,r) \in P \text{ for some } a\}.$$

If $y, z \in Y_c$, then $(a,r) \in P$ implies

$$\begin{aligned}|(y+z)/2-a|^2 &= |y+z|^2/4+|a|^2-(y+z)\bullet a \\ &= |y|^2/2+|z|^2/2-|y-z|^2/4+|a|^2-y\bullet a - z\bullet a \\ &= (|y-a|^2+|z-a|^2)/2-|y-z|^2/4 \leq r^2c^2-r^2|y-z|^2/(4\mu^2),\end{aligned}$$

thus $(y+z)/2 \in Y_t$ with $t=[c^2-|y-z|^2/(4\mu^2)]^{\frac{1}{2}}$, hence $y=z$. Subjecting $\mathbf{R}^n$ to a translation, we henceforth assume $Y_c=\{0\}$.

If $u \in \mathbf{R}^n$ and $|u|=1$, then $\varepsilon>0$ implies $\varepsilon u \notin Y_c$, hence there exists $(a,r) \in P$ with

$$|a|^2 \leq r^2c^2 < |u\varepsilon - a|^2 = \varepsilon^2+|a|^2-2\varepsilon u\bullet a, \quad u\bullet a \leq \varepsilon/2;$$

since P is compact it follows that

$$P \cap \{(a,r): |a|=rc \text{ and } u\bullet a \leq 0\} \neq \emptyset, \quad A \cap \{a: u\bullet a \leq 0\} \neq \emptyset.$$

Thus no $n-1$ dimensional plane separates 0 from the compact set A.

2.10.41. Jung's theorem. *If* $S \subset \mathbf{R}^n$ *and* $0<\operatorname{diam} S<\infty$, *then* S *is contained in a unique closed ball with minimal diameter, which does not exceed*

$$[2n/(n+1)]^{\frac{1}{2}} \operatorname{diam} S.$$

Proof. We suppose S is compact and apply 2.10.40 with $P=S\times\{1\}$. Assuming $b=0$, hence $S \subset \mathbf{B}(0,c)$, we choose

$$a_1, \ldots, a_{n+1} \in A = S \cap \{a: |a|=c\}$$

and nonnegative numbers $\lambda_1, \ldots, \lambda_{n+1}$ with

$$0 = \sum_{i=1}^{n+1} \lambda_i a_i \text{ and } 1 = \sum_{i=1}^{n+1} \lambda_i.$$

We infer for each $j \in \{1, \ldots, n+1\}$ that

$$\begin{aligned}2c^2 &= \sum_{i=1}^{n+1} \lambda_i(2c^2-2a_i\bullet a_j) = \sum_{i=1}^{n+1} \lambda_i|a_i-a_j|^2 \\ &\leq \sum_{i\neq j} \lambda_i(\operatorname{diam} S)^2 = (1-\lambda_j)(\operatorname{diam} S)^2,\end{aligned}$$

and conclude by summation with respect to j that

$$(n+1)2c^2 \leq n(\operatorname{diam} S)^2.$$

2.10.42. Corollary. $\mathscr{S}^m(A) \leq [2n/(n+1)]^{m/2} \mathscr{H}^m(A)$ *for* $A \subset \mathbf{R}^n$.

2.10.43. Kirszbraun's theorem. *If* $S \subset \mathbf{R}^m$ *and* $f: S \to \mathbf{R}^n$ *is Lipschitzian, then* f *has a Lipschitzian extension* $g: \mathbf{R}^m \to \mathbf{R}^n$ *with* $\operatorname{Lip}(g) = \operatorname{Lip}(f)$.

Proof. We suppose $\operatorname{Lip}(f)=1$ and consider the class Ω of all those Lipschitzian extensions of f which map some subset of $\mathbf{R}^m$ into $\mathbf{R}^n$ and have the Lipschitz constant 1. By Hausdorff's maximal principle Ω has a maximal (with respect to inclusion) element $g: T \to \mathbf{R}^n$, where $T \subset \mathbf{R}^m$.

We will show that if $\xi \in \mathbf{R}^m \sim T$ there would exist $\eta \in \mathbf{R}^n$ with

$$|\eta - g(x)| \leq |\xi - x| \text{ whenever } x \in T,$$

hence $g \cup \{(\xi, \eta)\} \in \Omega$, and g would not be maximal in Ω. Thus we must prove that

$$\bigcap \{\mathbf{B}[g(x), |x-\xi|]: x \in T\} \neq \varnothing;$$

since these balls are compact it will suffice to verify that

$$\bigcap \{\mathbf{B}[g(x), |x-\xi|]: x \in F\} \neq \varnothing$$

for every finite subset F of T. For this purpose we apply 2.10.40 with

$$P = \{(g(x), |x-\xi|): x \in F\},$$

choose distinct points $x_1, \ldots, x_k \in F$ and positive numbers $\lambda_1, \ldots, \lambda_k$ with

$$g(x_i) \in A, \text{ hence } |b - g(x_i)| = |x_i - \xi| c \text{ for } i = 1, \ldots, k,$$

$$b = \sum_{i=1}^{k} \lambda_i g(x_i), \quad 1 = \sum_{i=1}^{k} \lambda_i,$$

and use the identity $2u \bullet v = |u|^2 + |v|^2 - |u-v|^2$ to obtain

$$\begin{aligned}
0 &= 2 |\sum_i \lambda_i [g(x_i) - b]|^2 = 2 \sum_{i,j} \lambda_i \lambda_j [g(x_i) - b] \bullet [g(x_j) - b] \\
&= \sum_{i,j} \lambda_i \lambda_j [|g(x_i) - b|^2 + |g(x_j) - b|^2 - |g(x_i) - g(x_j)|^2] \\
&\geq \sum_{i,j} \lambda_i \lambda_j [c^2 |x_i - \xi|^2 + c^2 |x_j - \xi|^2 - |x_i - x_j|^2] \\
&= \sum_{i,j} \lambda_i \lambda_j [2c(x_i - \xi) \bullet c(x_j - \xi) + (c^2 - 1) |x_i - x_j|^2] \\
&= 2 |c \sum_i \lambda_i (x_i - \xi)|^2 + (c^2 - 1) \sum_{i,j} \lambda_i \lambda_j |x_i - x_j|^2;
\end{aligned}$$

hence either $k=1$ and $c=0$ (because $\xi \neq x_1 \in T$), or $k>1$ and $c \leq 1$; we conclude that $c \leq 1$, $b \in Y_1$.

2.10.44. The conclusion of the preceding theorem may fail if either $\mathbf{R}^m$ or $\mathbf{R}^n$ is remetrized by some norm not induced by an inner product. For example if

$$S=\{(1,-1),(-1,1),(1,1)\}\subset\mathbf{R}^2, f: S\to\mathbf{R}^2,$$

$$f(1,-1)=(1,0),\quad f(-1,1)=(-1,0),\quad f(1,1)=(0,\sqrt{3}),$$

$$\mu(x)=\sup\{|x_1|,|x_2|\},\ \nu(x)=[(x_1)^2+(x_2)^2]^{\frac{1}{2}} \text{ for } x\in\mathbf{R}^2,$$

then $\mu(u-v)=2=\nu[f(u)-f(v)]$ for $u,v\in S$, and $\mu(u)=1$ for $u\in S$, but there exists no $\eta\in\mathbf{R}^2$ with $\nu[\eta-f(u)]\le 1$ for $u\in S$; thus f has no extension to $S\cup\{(0,0)\}$ with μ, ν Lipschitz constant 1.

On the other hand, if S is a subset of an arbitrary metric space X, then every Lipschitzian map $f: S\to\mathbf{R}$ has a Lipschitzian extension $g: X\to\mathbf{R}$ with $\operatorname{Lip}(g)=\operatorname{Lip}(f)$; in fact one may define

$$g(\xi)=\inf\{f(x)+\operatorname{Lip}(f)\cdot\operatorname{dist}(x,\xi): x\in S\} \text{ for } \xi\in X.$$

2.10.45. Theorem. *Suppose Z is a metric space and $\mathbf{R}^m\times Z$ is metrized so that*

$$\operatorname{dist}[(y,z),(u,v)]^2=\operatorname{dist}(y,u)^2+\operatorname{dist}(z,v)^2$$

whenever $y,u\in\mathbf{R}^m$ and $z,v\in Z$. For each $V\subset Z$ with $\mathscr{H}^k(V)<\infty$ there exists a number c such that

$$\alpha(m)^{-1}\le c\,\alpha(k)\,\alpha(k+m)^{-1}\le 2^{-m}(m+1)^{(k+m)/2},$$

$$\mathscr{H}^{m+k}(U\times V)=c\,\mathscr{L}^m(U)\,\mathscr{H}^k(V) \text{ for every } \mathscr{L}^m \text{ measurable set } U.$$

Proof. We let $\gamma(U)=\mathscr{H}^{k+m}(U\times V)$ for $U\subset\mathbf{R}^m$, and observe that γ measures $\mathbf{R}^m$, all closed subsets of $\mathbf{R}^m$ are γ measurable by 2.3.2(9), γ is invariant under translations, and

$$\mathscr{L}^m(U)\,\mathscr{H}^k(V)\,\alpha(k+m)\,\alpha(k)^{-1}\,\alpha(m)^{-1}\le\gamma(U) \text{ for } U\subset\mathbf{R}^m$$

by 2.10.27, 2.10.35. Letting $\xi=\alpha(k+m)\,\alpha(k)^{-1}\,2^{-m}(m+1)^{(k+m)/2}$ we will prove next that

$$\gamma(U)\le\mathscr{L}^m(U)\,\xi\,\mathscr{H}^k(V)$$

whenever U is an m dimensional cube in $\mathbf{R}^m$. Suppose u is the side length of U, $\delta>0$, and G is any countable covering of V such that $0<\operatorname{diam} T\le\delta$ for $T\in G$; with each $T\in G$ we associate the positive integer $j(T)$ such that

$$[j(T)-1]\operatorname{diam} T<u\le j(T)\operatorname{diam} T,$$

and we cover U by a family $H(T)$ consisting of $j(T)^m$ cubes whose side length equals diam T, hence

$$j(T)\operatorname{diam} T \le u + \operatorname{diam} T \le u + \delta,$$

$$\operatorname{diam}(S \times T) \le (m+1)^{\frac{1}{2}} \operatorname{diam} T \text{ whenever } S \in H(T);$$

then $\{S \times T\colon S \in H(T) \text{ for some } T \in G\}$ covers $U \times V$ and

$$\sum_{T\in G}\sum_{S\in H(T)} \alpha(k+m)\, 2^{-k-m} (\operatorname{diam} S\times T)^{k+m}$$
$$\le \sum_{T\in G} j(T)^m \alpha(k+m)\, 2^{-k-m} (m+1)^{(k+m)/2} (\operatorname{diam} T)^{k+m}$$
$$\le (u+\delta)^m\, \xi \sum_{T\in G} \alpha(k)\, 2^{-k} (\operatorname{diam} T)^k.$$

From these facts we conclude that, if $\mathscr{H}^k(V)=0$ then $\gamma=0$, while if $\mathscr{H}^k(V)>0$ then by 2.7.7 and 2.5.3 the restriction of γ to the class of Borel sets is a constant positive multiple of $\mathscr{L}^m$, satisfying the stated inequalities.

2.10.46. Concerning the dependence of c on V in the preceding theorem with $Z=\mathbf{R}^n$, we will prove in 3.2.23 that $c=1$ if V is m rectifiable, whereas [FR 1], [BM] discuss a compact set V for which $c>1$. I do not know whether $c<1$ for some V.

A study of the corresponding problem for the Carathéodory measures constructed in 2.10.4(2) has led to the general inequality

$$\mathscr{C}^{m+k}(U\times V) \le \mathscr{L}^m(U)\, \mathscr{C}^k(V) \text{ for } U\subset \mathbf{R}^m,\ V\subset \mathbf{R}^n,$$

and also to the construction in [FR 2] of a compact set V for which the inequality is strict. The general inequality (which was proved in [RJ 1] in case $n=2, m=k=1$) can be deduced from the following observation: Suppose S is a closed ball in $\mathbf{R}^m$ with radius r, and $T\subset \mathbf{R}^n$ with diam $T\le\delta$. For each orthogonal projection $f\colon \mathbf{R}^m\times\mathbf{R}^n\to\mathbf{R}^{m+k}$ there exist, by the last part of 1.7.4, $p\in\mathbf{O}^*(m+k,k)$ and $q\in\mathbf{O}^*(n,k)$ such that

$$p[f(x,y)] = q(y) \text{ whenever } (x,y)\in \mathbf{R}^m\times\mathbf{R}^n,$$

hence $p[f(S\times T)]=q(T)$; therefore Fubini's theorem and the isodiametric inequality imply

$$\mathscr{L}^{m+k}[f(S\times T)] \le \mathscr{L}^k[q(T)]\, \alpha(m)\, 2^{-m} (\operatorname{diam} S\times T)^m$$
$$\le \mathscr{L}^k[q(T)]\, \mathscr{L}^m(S)\, (1+\delta^2/4r^2)^{m/2}.$$

To apply this estimate one covers $U\times V$ by families of sets $S\times T$ such that r and δ/r are small.

The preceding argument applies equally to the Gross measures constructed in 2.10.4(1). However in this case the opposite inequality is also true, as one readily verifies with the help of 2.10.8, hence

$$\mathscr{G}^{m+k}(U \times V) = \mathscr{L}^m(U)\, \mathscr{G}^k(V) \text{ for } U \subset \mathbf{R}^m,\ V \subset \mathbf{R}^n.$$

The analogous equation for Gillespie measures (which was proved in [MR 1] in case $n=2$, $k=m=1$) will be derived in 3.2.45.

In [FR 2] it was shown that

$$\mathscr{I}_1^k(V) < \infty \text{ implies } \mathscr{I}_1^{m+k}(U \times V) = \mathscr{L}^m(U)\, \mathscr{I}_1^k(V),$$

provided U, V are Borel subsets of $\mathbf{R}^m$, $\mathbf{R}^n$. From 3.3.14, 3.2.23 one infers the same for $\mathscr{I}_\infty$.

2.10.47. Theorem. *If $0 \leq m < \infty$ and A is a compact subset of $\mathbf{R}^n$ with $\mathscr{H}^m(A) > 0$, then A contains a compact set B with $\infty > \mathscr{H}^m(B) > 0$.*

Proof. For each nonnegative integer j we choose a family H_j of compact n dimensional cubes with side length 2^{-j}, such that H_0 consists of a single cube C with $\mathscr{H}^m(C \cap A) > 0$, and such that, for $j \geq 1$, H_j consists of the cubes obtained by subdividing each member of H_{j-1} into 2^n congruent parts. Defining ζ as in 2.10.2 we let

$$\gamma_i(E) = \inf \Big\{ \sum_G \zeta :\ G \subset \bigcup_{j=i}^{\infty} H_j \text{ and } E \subset \bigcup G \Big\}$$

whenever i is a nonnegative integer and $E \subset \mathbf{R}^n$. The measures γ_i over $\mathbf{R}^n$ are related to the approximating measures ϕ_δ occurring in the construction 2.10.2(1) of $\mathscr{H}^m$ through the inequalities

$$\phi_{n^{\frac{1}{2}} 2^{-i}}(E) \leq \gamma_i(E) \leq 2^{n+m}\, n^{m/2}\, \phi_{2^{-i}}(E)$$

for $E \subset C$, because if $S \subset C$ with $2^{-j-1} \leq \operatorname{diam} S \leq 2^{-j}$, then S is contained in the union of at most 2^n members of H_j, and

$$2^n (n^{\frac{1}{2}} 2^{-j})^m \leq 2^n (n^{\frac{1}{2}} 2 \operatorname{diam} S)^m.$$

Applying induction with respect to n, we may assume that $\mathscr{H}^m(A \cap V) = 0$ *for every $n-1$ dimensional affine subspace V of* $\mathbf{R}^n$, which will allow us to deduce the three statements:

(1) *If $E \subset A \cap C$ and $k \leq i$, then*

$$\gamma_i(E) = \sum_{S \in H_k} \gamma_i(S \cap E).$$

(2) *If $E \subset A \cap C$ and $\gamma_{i+1}(S \cap E) \leq \zeta(S)$ for $S \in H_i$, then*

$$\gamma_{i+1}(E) = \gamma_i(E).$$

(3) *If E is a compact subset of $A \cap C$ and $0 \le y \le \gamma_i(E)$, then E contains a compact set D with $\gamma_i(D) = y$.*

The statement (1) holds because if $G \subset \bigcup \{H_j : i \le j\}$ with $E \subset \bigcup G$, then

$$\sum_{S \in H_k} \gamma_i(E \cap S) = \sum_{S \in H_k} \gamma_i(E \cap \operatorname{Int} S) \le \sum_{S \in H_k} \sum_{G \cap \{T : T \subset S\}} \zeta = \sum_G \zeta.$$

The hypothesis of (2) implies $\gamma_{i+1}(S \cap E) = \gamma_i(S \cap E)$ for $S \in H_i$, hence (1) yields

$$\gamma_{i+1}(E) = \sum_{S \in H_i} \gamma_{i+1}(S \cap E) = \sum_{S \in H_i} \gamma_i(S \cap E) = \gamma_i(E).$$

To obtain (3) we observe that $\gamma_i(E \cap \{x : x_1 \le t\})$ is continuous with respect to t, because every open covering of $E \cap \{x : x_1 = t\}$ also covers $E \cap \{x : t - \varepsilon < x_1 < t + \varepsilon\}$ for some $\varepsilon > 0$.

Next we choose h so that $\gamma_h(A \cap C) > 0$ and inductively define compact sets

$$A \cap C = E_h \supset E_{h+1} \supset E_{h+2} \supset E_{h+3} \supset \cdots$$

satisfying the condition

$$\gamma_j(S \cap E_j) = \gamma_{j-1}(S \cap E_{j-1}) \text{ for } h < j,\ S \in H_{j-1}.$$

To construct E_j from E_{j-1} we choose for each $S \in H_{j-1}$ a compact set

$$D_S \subset S \cap E_{j-1} \text{ with } \gamma_j(D_S) = \gamma_{j-1}(S \cap E_{j-1}),$$

which is possible by (3) because $\gamma_{j-1}(S \cap E_{j-1}) \le \gamma_j(S \cap E_{j-1})$, and we let $E_j = \bigcup \{D_S : S \in H_{j-1}\}$.

We infer from (1) that

$$\gamma_j(E_j) = \gamma_{j-1}(E_{j-1}) \text{ for } h < j,$$

and from (2) that

$$\gamma_i(E_j) = \gamma_{i+1}(E_j) \text{ for } h \le i < j,$$

because $\gamma_{i+1}(S \cap E_j) \le \gamma_{i+1}(S \cap E_{i+1}) = \gamma_i(S \cap E_i) \le \zeta(S)$ for $S \in H_i$. Consequently

$$\gamma_i(E_j) = \gamma_h(E_h) \text{ for } h \le i \le j.$$

Finally we define $B = \bigcap_{j=h}^{\infty} E_j$.

Clearly $\mathscr{H}^m(B) \le \lim_{i \to \infty} \gamma_i(B) \le \gamma_h(E_h) < \infty$. Applying the last assertion in 2.10.20 we also find that

$$\mathscr{H}^m(B) \ge \phi_{2^{-h-1}}(B) \ge \lim_{j \to \infty} \phi_{2^{-h}}(E_j) \ge 2^{-n-m} n^{-m/2} \gamma_h(E_h) > 0.$$

2.10.48. Corollary. *If S is a Suslin subset of $\mathbf{R}^n$ and $0 \le m < \infty$, then*

$$\mathscr{H}^m(S) = \sup\{\mathscr{H}^m(C): C \text{ is compact}, C \subset S, \mathscr{H}^m(C) < \infty\}.$$

Proof. Calling the above supremum y, we choose compact subsets $C_1 \subset C_2 \subset C_3 \subset \cdots$ of S with $\mathscr{H}^m(C_j) < \infty$ so that

$$y = \lim_{j \to \infty} \mathscr{H}^m(C_j) = \mathscr{H}^m\left(\bigcup_{j=1}^{\infty} C_j\right),$$

and let

$$T = S \sim \bigcup_{j=1}^{\infty} C_j.$$

If $y < \mathscr{H}^m(S)$, then $\mathscr{H}^m(T) > 0$, hence 2.10.23 and 2.10.47 would yield a compact subset B of T with $\infty > \mathscr{H}^m(B) > 0$, leading to the contradiction

$$\mathscr{H}^m(C_j \cup B) = \mathscr{H}^m(C_j) + \mathscr{H}^m(B) > y \text{ for large } j.$$

2.10.49. The problem of extending 2.10.47, 2.10.48 from $\mathbf{R}^n$ to more general metric spaces has not been solved. I do not know whether every complete metric space X, with $\mathscr{H}^m(X) > 0$, contains a set B with $\infty > \mathscr{H}^m(B) > 0$. The need for some assumption like completeness is demonstrated by certain subspaces of $\mathbf{R}^2$ constructed in [B 2], [S P 2], [E G 2] using the continuum hypothesis $2^{\aleph_0} = \aleph_1$.

We will show by an example in 3.3.20 that $\mathscr{H}^m$ cannot be replaced by $\mathscr{I}_\infty^m$ in 2.10.47.

CHAPTER THREE

Rectifiability

This chapter contains the basic facts concerning integration with respect to m dimensional measures over subsets of n dimensional Euclidean space. It centers about the tangential and rectifiability properties of sets, and the transformation formulae corresponding to Lipschitzian maps. Exterior algebra plays a useful role here, but the theory of integration of differential forms over oriented sets (which includes the boundary formulae of Gauss, Green and Stokes) will be treated in Chapter 4.

In 3.1 we prove some fundamental theorems on the differentiation of functions. Differentials of the first order are introduced in 3.1.1, differentials of higher order in 3.1.11, and analyticity in 3.1.24. Some generalizations of differentiation are defined in 3.1.2 and 3.1.22, tangent spaces of arbitrary subsets of $\mathbf{R}^n$ in 3.1.21, and differentiable submanifolds in 3.1.19. The main results of 3.1 are of the following four types:

Each of the Theorems 3.1.4, 3.1.6, 3.1.8, 3.1.9 (which were first proved in [SW1, 2], [RH], [F1, II], [SW1, 2]) shows that differentiability $\mathscr{L}^m$ almost everywhere is implied by some apparently weaker property.

Theorem 3.1.14 (which originates from [WH1] with some simplifications in the proof suggested by [HS], [GG]) tells us under what conditions a function defined on a closed subset of $\mathbf{R}^m$ has a k times continuously differentiable extension over $\mathbf{R}^m$.

The Theorems 3.1.15 and 3.1.16 (first obtained in [F1, II] for $k=0$, in [WH3] for $k>0$) determine when a function of class k coincides except on a set of small $\mathscr{L}^m$ measure with some function of class $k+1$.

Theorem 3.1.20 characterizes differentiable submanifolds of $\mathbf{R}^n$ as differentiable neighborhood retracts. (The "only if" part may be found in [WH4, p.121], with a different proof, while the "if" part is new.)

In 3.2 we first derive the Jacobian integral formulae which make it possible to compute the Hausdorff measures of sets defined by Lipschitzian functions. Our proof of the classical area formula 3.2.3 follows the spirit of [KA], but we employ a simple new device to show the irrelevance of the subset where the Jacobian has value 0. The transformation formulae 3.2.5 and 3.2.6 are obtained by the method of [F 2, 2.1]. Dual to the area formula is the coarea formula 3.2.11, which was discovered in [F 15, 3.1]. After defining rectifiable sets and approximate tangent planes in 3.1.14 and 3.2.16, we correspondingly generalize the calculus of area and coarea in 3.2.19 and 3.2.33. An alternate characterization of rectifiability may be found in 3.2.29. The new theorem 3.2.31 was motivated in part by [BJ 1, 10.3].

Further topics treated in 3.2 include Theorem 3.2.26 (originating mostly from independent work in [F 1, 3, 4] and [N 1]) which shows that all the m dimensional measures defined in 2.10 agree for m rectifiable sets, and Theorem 3.2.39 (first proved in [KM 1]) relating Hausdorff measure to Minkowski content. The classical proofs of the Brunn-Minkowski inequality and the isoperimetric inequality are presented in 3.2.41 and 3.2.43. We provide a sample of modern integralgeometry in Theorem 3.2.48 (taken from [BJ 1]) whose proof illustrates the use of the coarea formula; simpler illustrations occur in 3.2.13 and 3.2.28. In 3.2.46 it is indicated how the theory can be extended from Euclidean spaces to Riemannian manifolds.

In 3.3 we develop the key structure theorems 3.3.12 – 3.3.15 which characterize rectifiable sets by their projection properties. Such results were obtained first in [B 1, III] for the special case of 1 dimensional Hausdorff measure in the plane, then in [F 5] for general measures in n dimensional Euclidean space; our exposition includes also some simplifications and refinements (in particular 3.3.14) achieved in [M I]. The proofs of 3.3.6 and 3.3.17 use a method adapted from [MR 2]. The particular set A of 3.3.19 was studied in [B 1, I] and [MO] with regard to $\mathscr{H}^1$, while [MR 1] contains a statement of the fact that $\mathscr{Q}_1^1(A)=\pi$. The set E of 3.3.20 was discussed for the case $m=\frac{1}{2}$ in [GR 1].

In 3.4 we study some effects of high order smoothness of functions on the Hausdorff measures of certain associated sets. Theorem 3.4.3 and the examples constructed in 3.4.4 determine the largest possible dimension (with respect to Hausdorff measure) of the image of the set where a function of class k mapping m dimensional Euclidean space into a normed vectorspace Y has a differential with rank ν. Our Lemma 3.4.2 is a refinement of the key lemma in [M 1], where the problem was solved for the special case when $\nu=0$; partial results for $\nu>0$ were obtained in [SA 1, 3]. Theorem 3.4.8 describes some of the basic local properties of sets of zeros of real analytic functions. It shows that such a set has dimension m in the sense of local ideal theory if and only if its m dimensional Hausdorff measure is locally finite and positive. Similar results have also been announced in [BN] and [HR]. Much of the normalization procedure used is classical in algebraic geometry and in the theory of complex varieties (see [AB], [GUR], [H 3], [LP], [DR 2], [HEM],

[NR]), but some topological aspects (like the local connectedness properties first studied in [BC], and treated here somewhat differently) are more complicated for real varieties.

3.1. Differentials and tangents

3.1.1. Assuming X, Y are normed vectorspaces, $A \subset X$ and $f: A \to Y$, we say that f is **differentiable** at a if and only if $a \in \operatorname{Int} A$ and there exists a continuous linear map $L: X \to Y$ with

$$\lim_{x \to a} |f(x) - f(a) - L(x-a)|/|x-a| = 0;$$

then L is uniquely determined by the formula

$$\lim_{t \to 0} [f(a+t\,v) - f(a)]/t = L(v) \text{ for } v \in X,$$

and L is called the **differential** of f at a, denoted

$$D\,f(a);$$

moreover

$$\limsup_{x \to a} |f(x) - f(a)|/|x-a| = \sup\{|\langle v, D\,f(a)\rangle| : |v|=1\} = \|D\,f(a)\|,$$

$$\liminf_{x \to a} |f(x) - f(a)|/|x-a| = \inf\{|\langle v, D\,f(a)\rangle| : |v|=1\};$$

if $D\,f(a)$ has a continuous inverse, the last infimum equals $\|D\,f(a)^{-1}\|^{-1}$.

We note two special cases:

In case $X = \mathbf{R}$, differentiability of f at a is equivalent to existence of the derivative

$$f'(a) = \lim_{t \to 0} [f(a+t) - f(a)]/t \in Y,$$

and then $\langle v, D\,f(a)\rangle = v \cdot f'(a)$ whenever $v \in \mathbf{R}$.

In case $Y = \mathbf{R}$, X is an inner product space and $\dim X < \infty$ (more generally, if X is complete) the polarity corresponding to the inner product maps X isomorphically onto the space X^* of all continuous real valued linear functions on X; if f is differentiable at a, then $D\,f(a) \in X^*$ is the image of

$$\operatorname{grad} f(a) \in X,$$

the **gradient** of f at a, which is characterized by the condition

$$\langle v, D\,f(a)\rangle = v \bullet \operatorname{grad} f(a) \text{ whenever } v \in X.$$

Among the basic properties of differentials are the following:

(1) *If f maps a neighborhood of a homeomorphically onto a neighborhood of $f(a)$, and if $D\,f(a)$ is a linear homeomorphism of X onto Y, then*

$$D\,f^{-1}[f(a)] = [D\,f(a)]^{-1}.$$

(2) *If f is differentiable at a and g is differentiable at $f(a)$, where g maps a subset of Y into some normed vectorspace, then*

$$D(g\circ f)(a)=D g[f(a)]\circ D f(a).$$

(3) *If $f: X\to Y$ is linear and continuous, then $D f(a)=f$ whenever $a\in X$.*

(4) *If $f: X_1\times\cdots\times X_m\to Y$ is m linear and continuous, then*

$$\langle(v_1,\ldots,v_m), D f(a_1,\ldots,a_m)\rangle=\sum_{i=1}^{m} f(a_1,\ldots,a_{i-1},v_i,a_{i+1},\ldots,a_m)$$

whenever $a_i, v_i\in X_i$ for $i=1,\ldots,m$.

(5) *In case* $\dim Y<\infty$ *and $\omega_1,\ldots,\omega_n$ form a base of $\wedge^1(Y)$, then f is differentiable at a if and only if $\omega_1\circ f,\ldots,\omega_n\circ f$ are differentiable at a.*

A continuous function mapping some open subset of X into Y is said to be of **class** 0.

We say that f is of **class** 1 if and only if f is of class 0, f is differentiable at each point of its domain, and $D f$ is a continuous map with values in the normed vectorspace of all continuous linear maps of X into Y.

The composition of two maps of class 1 *is of class* 1.

If X is complete, f is of class 1 *and $D f(a)$ is a linear homeomorphism of X onto Y, then f maps some neighborhood of a homeomorphically onto a neighborhood of $f(a)$.* To prove this we may replace f by $D f(a)^{-1}\circ f$, hence assume $X=Y$, $D f(a)=\mathbf{1}_X$ and also $a=f(a)=0$. Then $g=f-\mathbf{1}_X$ is of class 1, and whenever $0<\varepsilon<1$ there exists $\delta>0$ with

$$\|D g(x)\|=\|D f(x)-\mathbf{1}_X\|\leq\varepsilon \text{ for } x\in\mathbf{B}(0,\delta),$$

hence (see 2.2.7) $u,v\in\mathbf{B}(0,\delta)$ implies

$$|g(u)-g(v)|\leq\varepsilon\,|u-v|,\qquad |f(u)-f(v)|\geq(1-\varepsilon)\,|u-v|.$$

Also, given $y\in\mathbf{B}[0,\delta(1-\varepsilon)]$ we inductively define

$$u_0=0,\qquad u_n=y-g(u_{n-1}) \text{ for } n=1,2,3,\ldots,$$

use the equation $u_{n+1}-u_n=g(u_n)-g(u_{n-1})$ to verify that

$$|u_{n+1}-u_n|\leq\varepsilon\,|u_n-u_{n-1}|\leq\varepsilon^n\,|y|,$$

$$|u_{n+1}|\leq(1+\varepsilon+\cdots+\varepsilon^n)\,|y|<\delta,$$

hence conclude the existence of

$$\lim_{n\to\infty} u_n=x\in\mathbf{B}(0,\delta) \text{ with } x=y-g(x),\ f(x)=y.$$

If f is a univalent function of class 1 *and $D f(a)$ maps X homeomorphically onto Y for every $a\in$* dmn *f, then f^{-1} is a function of class* 1.

If f is of class 0, $\dim X<\infty$ and Y is complete, then $\operatorname{dmn} Df$ *is a Borel set and Df is a Borel function.* Since $\operatorname{im} f$ is separable we may assume Y separable; in this case the space Z of all continuous linear maps of X into Y is a separable Banach space. We observe that

$$Df=\bigcap_{i=1}^{\infty}\bigcup_{j=1}^{\infty}C_{i,j}$$

where $C_{i,j}$ is the relatively closed subset of $(\operatorname{dmn} f)\times Z$ consisting of all pairs (a,L) such that

$$|f(a+h)-f(a)-L(h)|\le|h|/i \text{ for } |h|<1/j,$$

hence Df is a Borel set of $X\times Z$; since the projection $X\times Z\to X$ maps Df in one to one manner onto $\operatorname{dmn} Df$, the conclusions follow from 2.2.10.

In case $X=\mathbf{R}^m$ we shall also use the **partial derivatives** $D_1 f,\ldots,D_m f$ defined by the formulae

$$D_i f(a)=\lim_{t\to 0}[f(a_1,\ldots,a_{i-1},a_i+t,a_{i+1},\ldots,a_m)-f(a)]/t$$

for $a\in\mathbf{R}^m$. Letting $e_1,\ldots,e_m$ be the standard base of $\mathbf{R}^m$ we see that *if f is differentiable at a, then $D_i f(a)=\langle e_i, Df(a)\rangle$, hence*

$$\langle v, Df(a)\rangle=\sum_{i=1}^{m}v_i D_i f(a) \text{ whenever } v\in\mathbf{R}^m.$$

It can happen that $D_1 f(a),\ldots,D_m f(a)$ exist, but f is not differentiable at a. However, *if some neighborhood of a is contained in the domains of $D_1 f,\ldots, D_m f$ and if these partial derivatives are continuous at a, then f is differentiable at a.* To prove this we define

$$g(x)=f(x)-f(a)-\sum_{i=1}^{m}(x_i-a_i)D_i f(a) \text{ for } x\in\operatorname{dmn} f;$$

whenever $\varepsilon>0$ we choose $\delta>0$ so that

$$|D_i g(x)|=|D_i f(x)-D_i f(a)|\le\varepsilon \text{ for } x\in\mathbf{B}(a,\delta),\ i=1,\ldots,m,$$

and infer from 2.2.7 that $x\in\mathbf{B}(a,\delta)$ implies

$$|g(x)|\le\sum_{i=1}^{m}|g(x_1,\ldots,x_i,a_{i+1},\ldots,a_m)-g(x_1,\ldots,x_{i-1},a_i,\ldots,a_m)|$$
$$\le\sum_{i=1}^{m}\varepsilon|x_i-a_i|\le m\,\varepsilon\,|x-a|.$$

3.1.2. Recalling 2.9.12 we now generalize the local concepts described in 3.1.1 through replacement of limits by approximate limits. For this purpose we take $X=\mathbf{R}^m$ with the $\mathscr{L}^m$ Vitali relation

$$V=\{(a,\mathbf{B}(a,r)):\ a\in\mathbf{R}^m,\ 0<r<\infty\}$$

and abbreviate "$(\mathscr{L}^m, V)$ ap lim" by "ap lim".

Assuming that $A\subset\mathbf{R}^m$ and f maps A into a normed vectorspace Y, we say that f is **approximately differentiable** at a if and only if there exists a linear map $L\colon \mathbf{R}^m\to Y$ with

$$\operatorname*{ap\,lim}_{x\to a} |f(x)-f(a)-L(x-a)|/|x-a|=0;$$

this implies that $\mathbf{R}^m\sim A$ has density 0 at a; moreover L is unique, because the difference T of any two such linear maps satisfies the condition

$$\operatorname*{ap\,lim}_{v\to 0} |T(v)|/|v|=\operatorname*{ap\,lim}_{x\to a} |T(x-a)|/|x-a|=0,$$

hence if $0<\varepsilon<1$ there exists $r>0$ such that

$$\mathscr{L}^m[\mathbf{B}(0,r)\cap\{v\colon |T(v)|>\varepsilon|v|\}]<\varepsilon^m r^m\boldsymbol{\alpha}(m),$$

and then for every $u\in\mathbf{R}^m$ with $|u|=r-r\varepsilon$ there exists

$$v\in\mathbf{B}(u,\varepsilon r)\subset\mathbf{B}(0,r)\ \text{ with }\ |T(v)|\le\varepsilon|v|$$

which implies

$$|T(u)|\le|T(u-v)|+|T(v)|\le\|T\|\,\varepsilon r+\varepsilon r=|u|\,(\|T\|+1)\,\varepsilon/(1-\varepsilon),$$

hence $\|T\|\le(\|T\|+1)\,\varepsilon/(1-\varepsilon)$, and finally $\|T\|=0$; we call L the **approximate differential** of f at a and denote it

$$\operatorname{ap} Df(a).$$

We also use $\mathscr{L}^1$ to define the **approximate partial derivatives**

$$\operatorname{ap} D_i f(a)=\operatorname*{ap\,lim}_{t\to 0}[f(\ldots,a_{i-1},a_i+t,a_{i+1},\ldots)-f(a)]/t.$$

The simple relation between these concepts will be described in 3.1.4 and 3.1.7. Their utility for studying the more classical notions of 3.1.1 will appear in 3.1.5, 3.1.6, 3.1.9.

3.1.3. For use in the proof of the next theorem we observe:

(1) *If S is an $\mathscr{L}^m\times\mathscr{L}^k$ measurable subset of $\mathbf{R}^m\times\mathbf{R}^k$, $\varepsilon>0$, $\delta>0$, and T is the subset of $\mathbf{R}^m$ consisting of those points x for which*

$$\mathscr{L}^k\{z\colon (x,z)\in S,\ |z|\le r\}\le\varepsilon r^k\boldsymbol{\alpha}(k)$$

whenever $0<r<\delta$, then T is $\mathscr{L}^m$ measurable. In fact, for each r,

$$S_r=S\cap\{(x,z)\colon |z|\le r\}\ \text{ is }\ \mathscr{L}^m\times\mathscr{L}^k\ \text{measurable}$$

and Fubini's theorem implies

$$\mathscr{L}^k\{z\colon (x,z)\in S_r\} \text{ is } \mathscr{L}^m \text{ measurable with respect to } x;$$

moreover T equals the set of those points x for which

$$\mathscr{L}^k\{z\colon (x,z)\in S_r\}\le \varepsilon\, r^k \alpha(k)$$

whenever r is rational and $0<r<\delta$.

(2) *If σ is a real valued $\mathscr{L}^m\times\mathscr{L}^k$ measurable function, then*

$$\operatorname{ap}\limsup_{z\to 0}\sigma(x,z) \quad \text{and} \quad \operatorname{ap}\liminf_{z\to 0}\sigma(x,z)$$

are $\mathscr{L}^m$ measurable with respect to x. In fact, for each $t\in\mathbf{R}$, (1) is applicable to the sets

$$\{(x,z)\colon \sigma(x,z)>t\} \quad \text{and} \quad \{(x,z)\colon \sigma(x,z)<t\}.$$

(3) *If $\alpha\colon \mathbf{R}^n\to\mathbf{R}^m$ is a linear epimorphism and W is an $\mathscr{L}^m$ measurable set, then $\alpha^{-1}(W)$ is $\mathscr{L}^n$ measurable.* In fact there exists a linear isomorphism

$$\beta\colon \mathbf{R}^n\simeq\mathbf{R}^m\times\mathbf{R}^{n-m} \quad \text{with} \quad \beta[\alpha^{-1}(W)]=W\times\mathbf{R}^{n-m}.$$

(4) *If g is a real valued $\mathscr{L}^m$ measurable function and $k\le m$, then the $\mathscr{L}^k$ approximate limit*

$$\operatorname{ap}\lim_{z\to 0} g(a_1+z_1,\ldots,a_k+z_k,a_{k+1},\ldots,a_m)=g(a)$$

for $\mathscr{L}^m$ almost all a. In fact, identifying $\mathbf{R}^m\simeq\mathbf{R}^k\times\mathbf{R}^{m-k}$ and defining

$$\alpha\colon \mathbf{R}^k\times\mathbf{R}^{m-k}\times\mathbf{R}^k\to\mathbf{R}^k\times\mathbf{R}^{m-k}, \quad \alpha(u,v,z)=(u+z,v),$$

one sees from (3) that $g\circ\alpha$ is $\mathscr{L}^k\times\mathscr{L}^{m-k}\times\mathscr{L}^k$ measurable, hence from (2) that

$$E=\{(u,v)\colon \operatorname{ap}\lim_{z\to 0} g(u+z,v)=g(u,v)\}$$

is $\mathscr{L}^k\times\mathscr{L}^{m-k}$ measurable; from 2.6.2 and 2.9.13 one infers that, for $\mathscr{L}^{m-k}$ almost all v, $g(u,v)$ is $\mathscr{L}^k$ measurable with respect to u and

$$\mathscr{L}^k\{u\colon (u,v)\notin E\}=0;$$

by 2.6.2 the complement of E has $\mathscr{L}^k\times\mathscr{L}^{m-k}$ measure 0.

(5) *If A is an $\mathscr{L}^m$ measurable set and $k\le m$, then for $\mathscr{L}^m$ almost all a in A the set*

$$\mathbf{R}^k\cap\{u\colon (u_1,\ldots,u_k,a_{k+1},\ldots,a_m)\notin A\}$$

has $\mathscr{L}^k$ density 0 at $(a_1,\ldots,a_k)$. This results from (4) when g is the characteristic function of A.

3.1.4. Theorem. *If $f\colon \mathbf{R}^m \to \mathbf{R}^n$ is $\mathscr{L}^m$ measurable, then*

$$A_i = \operatorname{dmn} \operatorname{ap} D_i f \text{ is an } \mathscr{L}^m \text{ measurable set},$$

$$\operatorname{ap} D_i f \text{ is an } \mathscr{L}^m \llcorner A_i \text{ measurable function}$$

for $i=1,\ldots,m$, and f is approximately differentiable at $\mathscr{L}^m$ almost all points a in $\bigcap\{A_i\colon i=1,\ldots,m\}$, with

$$\langle v, \operatorname{ap} D f(a)\rangle = \sum_{i=1}^{m} v_i \cdot \operatorname{ap} D_i f(a) \text{ whenever } v \in \mathbf{R}^m.$$

Proof. We assume $n=1$, let $e_1,\ldots,e_m$ be the standard base of $\mathbf{R}^m$, and define

$$g_i(x) = \operatorname{ap} \limsup_{t\to 0} [f(x+t\,e_i)-f(x)]/t$$

whenever $x\in\mathbf{R}^m$. From 3.1.3 (3) [with $n=m+1$, $\alpha(x,t)=x+t\,e_i$] and (2) we see that g_i and the corresponding lower limit function are $\mathscr{L}^m$ measurable. Therefore A_i, the set where these upper and lower limits are finite and equal, is $\mathscr{L}^m$ measurable.

Whenever j is a positive integer, $x\in\mathbf{R}^m$ and $r>0$ we let

$$T(x,r,i,j)$$

be the set of all real numbers t for which

$$|t|<r \quad \text{and} \quad |f(x+t\,e_i)-f(x)-t\,g_i(x)|>|t|/j.$$

Whenever p is a positive integer we define

$$B(i,j,p) = A_i \cap \{x\colon \mathscr{L}^1\, T(x,r,i,j) \le r/j \text{ for } 0<r<1/p\}.$$

In case $i>1$ we let

$$Z(x,r,i,j,p)$$

be the subset of $\mathbf{R}^{i-1}$ consisting of all points z for which $|z|<r$ and

$$\text{either } x+z_1\,e_1+\cdots+z_{i-1}\,e_{i-1} \notin B(i,j,p)$$

$$\text{or } |g_i(x+z_1\,e_1+\cdots+z_{i-1}\,e_{i-1})-g_i(x)|>i/j.$$

Whenever q is a positive integer we define

$$C(i,j,p,q) = B(i,j,p) \cap \{x\colon \mathscr{L}^{i-1}\, Z(x,r,i,j,p) \le r^{i-1}/j \text{ for } 0<r<1/q\}.$$

We also take $C(1,j,p,q)=B(1,j,p)$ for all q.

From 3.1.3 we see that the sets $B(i,j,p)$ and $C(i,j,p,q)$ are $\mathscr{L}^m$ measurable with

$$A_i = \bigcup_{p=1}^{\infty} B(i,j,p) \text{ for each } (i,j),$$

$$\mathscr{L}^m\left[B(i,j,p) \sim \bigcup_{q=1}^{\infty} C(i,j,p,q)\right] = 0 \text{ for each } (i,j,p).$$

Moreover $B(i,j,p) \subset B(i,j,p+1)$ and $C(i,j,p,q) \subset C(i,j,p,q+1)$.

Given any $E \subset \bigcap\{A_i\colon i=1,\ldots,m\}$ with $\mathscr{L}^m(E) < \infty$, and any $\varepsilon > 0$, we can therefore choose sequences $p_1, p_2, p_3, \ldots$ and $q_1, q_2, q_3, \ldots$ of positive integers so that

$$\mathscr{L}^m[E \sim B(i,j,p_j)] < \varepsilon\, 2^{-j},$$

$$\mathscr{L}^m[E \cap B(i,j,p_j) \sim C(i,j,p_j,q_j)] < \varepsilon\, 2^{-j}$$

for $i=1,\ldots,m$ and every positive integer j, hence

$$\mathscr{L}^m(E \sim H) < 2m\,\varepsilon, \quad \text{where} \quad H = \bigcap_{i=1}^{m} \bigcap_{j=1}^{\infty} C(i,j,p_j,q_j).$$

We will show that f is (uniformly) approximately differentiable at the points of H.

Suppose $a \in H$, j is a positive integer and

$$0 < r < \inf\{1/p_j, 1/q_j\}.$$

For $i=1,\ldots,m$ we let S_i be the subset of $\mathbf{R}^m$ consisting of all points v such that $|v| \le r$ and

$$\text{either } i>1 \text{ and } (v_1,\ldots,v_{i-1}) \in Z(a,r,i,j,p_j)$$
$$\text{or } v_i \in T(a+v_1 e_1+\cdots+v_{i-1} e_{i-1}, r, i, j).$$

Since $\mathscr{L}^{i-1} Z(a,r,i,j,p_j) \le r^{i-1}/j$, and since

$$\mathscr{L}^1 T(x,r,i,j) \le r/j \quad \text{if} \quad x = a+v_1 e_1+\cdots+v_{i-1} e_{i-1} \in B(i,j,p_j),$$

we find that

$$\mathscr{L}^m(S_i) \le (r^{i-1}/j)\cdot\alpha(m-i+1)\, r^{m-i+1} + (r/j)\cdot\alpha(m-1)\, r^{m-1}$$
$$\le r^m\, 2^{m-i+1}/j + r^m\, 2^{m-1}/j \le 2^{m+1}\, r^m/j.$$

Furthermore, if $v \in \mathbf{B}(0,r) \sim S_i$, then

$$|f(a+v_1 e_1+\cdots+v_i e_i) - f(a+v_1 e_1+\cdots+v_{i-1} e_{i-1}) - v_i\, g_i(a)|$$
$$\le |v_i|/j + |v_i\, g_i(a+v_1 e_1+\cdots+v_{i-1} e_{i-1}) - v_i\, g_i(a)| \le 2\,|v_i|/j.$$

Letting $W = \bigcup \{S_i\colon i=1,\ldots,m\}$ we conclude that

$$\mathscr{L}^m(W)/r^m \leq m\, 2^{m+1}/j,$$

and that $v \in \mathbf{B}(0,r) \sim W$ implies

$$\left| f(a+v) - f(a) - \sum_{i=1}^{m} v_i\, g_i(a) \right| \leq \sum_{i=1}^{m} 2\,|v_i|/j \leq (2m/j)\cdot|v|.$$

3.1.5. Lemma. *Suppose* $C \subset B \subset \mathbf{R}^m$, $f\colon B \to \mathbf{R}^n$, $0<\eta$, $0<M<\infty$ *and*

$$z \in C \textit{ implies } \mathbf{U}(z,\eta) \subset B \textit{ with}$$

$$|f(x) - f(z)| \leq M \cdot |x-z| \text{ for } x \in \mathbf{U}(z,\eta).$$

If $a \in C$, $\mathbf{R}^m \sim C$ *has* $\mathscr{L}^m$ *density* 0 *at* a, *and* f *is approximately differentiable at* a, *then* f *is differentiable at* a.

Proof. Suppose $L = \operatorname{ap} D f(a)$, $0<\varepsilon<1$, $0<\delta\leq\eta$,

$$W = C \cap \{z\colon |f(z) - f(a) - L(z-a)| \leq \varepsilon\,|z-a|\},$$

and $\mathscr{L}^m[\mathbf{B}(a,r) \sim W] < \varepsilon^m\, r^m\, \boldsymbol{\alpha}(m)$ whenever $0<r<\delta$.

For $x \in \mathbf{U}(a, \delta - \varepsilon\,\delta)$ we take $r = |x-a|/(1-\varepsilon) < \delta$ and observe that

$$\mathbf{B}(x,\varepsilon\, r) \subset \mathbf{B}(a,r), \qquad \mathbf{B}(x,\varepsilon\, r) \cap W \neq \emptyset;$$

choosing $z \in \mathbf{B}(x,\varepsilon\, r) \cap W$ we infer $x \in \mathbf{B}(z,\varepsilon\, r) \subset \mathbf{U}(z,\eta)$,

$$\begin{aligned}
&|f(x) - f(a) - L(x-a)| \\
&\leq |f(z) - f(a) - L(z-a)| + |f(x) - f(z)| + |L(z-x)| \\
&\leq \varepsilon\,|z-a| + M\,|x-z| + \|L\|\cdot|x-z| \\
&\leq \varepsilon\,|x-a| + (\varepsilon + M + \|L\|)\,\varepsilon\, r \\
&= |x-a|\,\varepsilon[1 + (\varepsilon + M + \|L\|)/(1-\varepsilon)].
\end{aligned}$$

3.1.6. Rademacher's theorem. *If* $f\colon \mathbf{R}^m \to \mathbf{R}^n$ *is Lipschitzian, then* f *is differentiable at* $\mathscr{L}^m$ *almost all points of* $\mathbf{R}^m$.

Proof. For $i=1,\ldots,m$ we define A_i as in 3.1.4. Whenever $u \in \mathbf{R}^{m-1}$ the map

$$f_u\colon \mathbf{R} \to \mathbf{R}^n, \quad f_u(s) = f(u_1,\ldots,u_{i-1},s,u_i,\ldots,u_{m-1}) \text{ for } s \in \mathbf{R},$$

is Lipschitzian, hence 2.9.19 implies

$$D_i f(u_1,\ldots,u_{i-1},s,u_i,\ldots,u_{m-1}) = f_u'(s) \in \mathbf{R}^n$$

for $\mathscr{L}^1$ almost all s. Since A_i is $\mathscr{L}^m$ measurable it follows from Fubini's theorem that $\mathscr{L}^m(\mathbf{R}^m \sim A_i) = 0$.

Therefore $\mathscr{L}^m(\mathbf{R}^m \sim \bigcap\{A_i : i=1, \ldots, m\})=0$, and we obtain the desired conclusion from 3.1.4 and 3.1.5 [with $C=B=\mathbf{R}^m$, $\eta=1$, $M=\operatorname{Lip}(f)$].

3.1.7. Lemma. *If A is an $\mathscr{L}^m$ measurable set and $f: A \to \mathbf{R}^n$ is Lipschitzian, then f has $\mathscr{L}^m$ almost everywhere in A an approximate differential and approximate partial derivatives.*

Proof. By 2.10.43, f has a Lipschitzian extension $g: \mathbf{R}^m \to \mathbf{R}^n$. Now 3.1.6 implies that g is differentiable at $\mathscr{L}^m$ almost all points of $\mathbf{R}^m$, while 2.9.11 implies that $\mathbf{R}^m \sim A$ has density 0 at $\mathscr{L}^m$ almost all points of A; at those points where both conditions hold the differential of g is an approximate differential of f. Moreover $D_i g(a) \in \mathbf{R}^n$ for $\mathscr{L}^m$ almost all a in $\mathbf{R}^m$, hence 3.1.3 (5) implies $D_i g(a) = \operatorname{ap} D_i f(a)$ for $\mathscr{L}^m$ almost all a in A.

3.1.8. Theorem. *If $A \subset \mathbf{R}^m$, $f: A \to \mathbf{R}^n$ and*

$$\operatorname{ap}\limsup_{x \to a} |f(x)-f(a)|/|x-a| < \infty \quad \textit{whenever } a \in A,$$

then A is the union of a countable family of $\mathscr{L}^m$ measurable sets such that the restriction of f to each member of the family is Lipschitzian; moreover f is approximately differentiable at $\mathscr{L}^m$ almost all points of A.

Proof. The hypothesis implies that $\mathbf{R}^m \sim A$ has density 0 and f is approximately continuous at each point of A, hence A is $\mathscr{L}^m$ measurable and f is $\mathscr{L}^m \llcorner A$ measurable, by 2.9.11 and 2.9.13.

The ratio $\mathscr{L}^m[\mathbf{B}(u,|u-v|) \cap \mathbf{B}(v,|u-v|)]/|u-v|^m$ has a constant value ρ for all pairs (u, v) of distinct points of $\mathbf{R}^m$, because it is invariant under isometries and scalar multiplications of $\mathbf{R}^m$.

For each positive integer j we let

$$Q(u,r,j) = \mathbf{B}(u,r) \cap \{x : x \notin A \text{ or } |f(x)-f(u)| > j\,|x-u|\}$$

whenever $u \in A$, $r>0$; then we define

$$B_j = A \cap \{u : \mathscr{L}^m Q(u,r,j) < r^m \rho/2 \text{ for } 0<r<1/j\}.$$

Each set B_j is $\mathscr{L}^m$ measurable (see 3.1.3) and

$$A = \bigcup_{j=1}^{\infty} B_j.$$

Next we observe that

$$|f(u)-f(v)| \le 2j\,|u-v| \quad \text{whenever } u, v \in B_j,\ |u-v|<1/j,$$

because if $r=|u-v|$, then

$$\mathscr{L}^m[Q(u,r,j) \cup Q(v,r,j)] < r^m \rho = \mathscr{L}^m[\mathbf{B}(u,r) \cap \mathbf{B}(v,r)],$$

and choosing $x \in [\mathbf{B}(u,r) \cap \mathbf{B}(v,r)] \sim [Q(u,r,j) \cup Q(v,r,j)]$ we obtain

$$|f(u)-f(v)| \leq |f(x)-f(u)| + |f(x)-f(v)|$$
$$\leq j\,|x-u| + j\,|x-v| \leq 2jr = 2j\,|u-v|.$$

Finally we express B_j as the union of $\mathscr{L}^m$ measurable sets $B_{j,1}, B_{j,2}, B_{j,3}, \ldots$ with diameters less that $1/j$, note that $f \,|\, B_{j,k}$ is Lipschitzian, and infer from 3.1.7 that $f \,|\, B_{j,k}$ (hence also f) is approximately differentiable at $\mathscr{L}^m$ almost all points of $B_{j,k}$.

3.1.9. Stepanoff's theorem. *If $A \subset B \subset \mathbf{R}^m$, B is open, $f: B \to \mathbf{R}^n$ and*

$$\limsup_{x \to a} |f(x)-f(a)|/|x-a| < \infty \quad \textit{whenever} \quad a \in A,$$

then f is differentiable at $\mathscr{L}^m$ almost all points of A.

Proof. We observe that A is contained in the union of the sets

$$C_j = B \cap \{z\colon\ |f(x)-f(z)| \leq j\,|x-z| \ \text{ for } \ x \in U(z, 1/j)\}$$

corresponding to all positive integers j. Each C_j is closed; in fact suppose $C_j \ni \zeta_p \to z \in \mathbf{R}^m$ as $p \to \infty$, and $x \in \mathbf{U}(z, 1/j)$; if p is sufficiently large, then $\{z, x\} \subset \mathbf{U}(\zeta_p, 1/j) \subset B$,

$$|f(x)-f(z)| \leq |f(x)-f(\zeta_p)| + |f(z)-f(\zeta_p)|$$
$$\leq j\,|x-\zeta_p| + j\,|z-\zeta_p| \to j\,|x-z| \ \text{ as } \ p \to \infty,$$

hence $z \in C_j$. Next we express C_j as the union of closed sets $C_{j,1}, C_{j,2}, C_{j,3}, \ldots$ with diameters less that $1/j$, and note that $f \,|\, C_{j,k}$ is Lipschitzian. From 3.1.7 and 2.9.11 we infer that at $\mathscr{L}^m$ almost all points of $C_{j,k}$ the function $f \,|\, C_{j,k}$ is approximately differentiable and $\mathbf{R}^m \sim C_{j,k}$ has density 0, hence f is approximately differentiable and $\mathbf{R}^m \sim C_j$ has density 0, hence f is differentiable by 3.1.5.

3.1.10. In 3.1.8 we found that an approximate local growth condition on f suffices to guarantee that f is the union of a countable family of Lipschitzian functions. We shall obtain an even stronger result in 3.1.16, which shows that the same local condition implies the existence of a set S such that $\mathscr{L}^m(A \sim S) = 0$ and $f \,|\, S$ is contained in the union of a countable family of functions of class 1. Most of the technique developed for this purpose will apply also to the study of functions of class greater than 1 (which are defined in 3.1.11).

3.1.11. We will now inductively extend the concepts of 3.1.1 and discuss differentials of orders higher than 1. For this purpose we write

$$D^0 f = f, \quad D^1 f = Df,$$

we agree that "once differentiable" shall mean "differentiable", and we norm the vectorspace of all continuous m linear maps of X^m into Y by defining

$$\|\phi\| = \sup\{\phi(v_1, \ldots, v_m): v_i \in X \text{ and } |v_i| \leq 1 \text{ for } i = 1, \ldots, m\}.$$

In case $k \geq 2$ we say that f is k **times differentiable** *at a if and only if f is $k-1$ times differentiable at each point of some neighborhood of a and the function $D^{k-1} f$, with values in the space of continuous $k-1$ linear maps of X^{k-1} into Y, is differentiable at a; in this case the k-***th differential** *of f at a is the continuous k linear map*

$$D^k f(a): X^k \to Y$$

defined by the formula

$$\langle (v_1, \ldots, v_k), D^k f(a) \rangle = \langle (v_1, \ldots, v_{k-1}), \langle v_k, D D^{k-1} f(a) \rangle \rangle$$

whenever $v_1, \ldots, v_k \in X$.

By induction with respect to q one verifies that

$$\langle (v_1, \ldots, v_{p+q}), D^{p+q} f(a) \rangle = \langle (v_1, \ldots, v_p), \langle (v_{p+1}, \ldots, v_{p+q}), D^q D^p f(a) \rangle \rangle$$

whenever $v_1, \ldots, v_{p+q} \in X$.

Next we prove: *If f is twice differentiable at a, then*

$$\lim_{u, v \to 0} \frac{|f(a+u+v) - f(a+u) - f(a+v) + f(a) - \langle (u, v), D^2 f(a) \rangle|}{(|u| + |v|)^2} = 0.$$

We assume $a = 0$. Given $\varepsilon > 0$ we choose $\delta > 0$ so that

$$\|D f(x) - D f(0) - \langle x, D D f(0) \rangle\| \leq \varepsilon |x| \text{ for } x \in \mathbf{U}(0, 2\delta).$$

For $u, v \in \mathbf{B}(0, \delta)$ we define

$$g_v(u) = f(u+v) - f(u) - f(v) + f(0) - \langle u, \langle v, D D f(0) \rangle \rangle$$

and observe that

$$\begin{aligned} D g_v(u) &= D f(u+v) - D f(u) - \langle v, D D f(0) \rangle \\ &= [D f(u+v) - D f(0) - \langle u+v, D D f(0) \rangle] \\ &\quad - [D f(u) - D f(0) - \langle u, D D f(0) \rangle], \end{aligned}$$

$$\|D g_v(u)\| \leq \varepsilon |u+v| + \varepsilon |u| \leq 2\varepsilon (|u| + |v|);$$

therefore $\operatorname{Lip}[g_v | \mathbf{B}(0, |u|)] \leq 2\varepsilon(|u| + |v|)$, and since $g_v(0) = 0$ we conclude

$$|g_v(u)| \leq 2\varepsilon(|u| + |v|) \cdot |u| \leq 2\varepsilon(|u| + |v|)^2.$$

We deduce the corollary that *$D^2 f(a)$ is a symmetric bilinear form.*

By induction using the relation between D^k and $D^2\, D^{k-2}$ we obtain: *If f is k times differentiable at a, then $D^k f(a)$ is a symmetric k form.*

In case k is a positive integer, we say that a function f is of **class** k if and only if f is of class 0, f is k times differentiable at each point of its domain, and $D^k f$ is continuous.

If f is of class k, then f is of class j for $j=1, \ldots, k$.

We say f is of **class** ∞ if and only if f is of class k for every positive integer k.

The composition of two maps of class k is a map of class k. If a map of class k has an inverse of class 1, then the inverse is of class k. One readily verifies these assertions by induction with respect to k, using the formulae for the differentials of compositions and inverses (see 3.1.1), the bilinearity of the operation composing linear maps, and the infinite differentiability of the operation inverting linear maps: *If X is a Banach space and E is the Banach space of all continuous linear endomorphisms of X, then the set A of all linear automorphisms of X is open in E,*

$$(T+S)^{-1} = \sum_{k=0}^{\infty} (-T^{-1} \circ S)^k \circ T^{-1}$$

whenever $T \in A$, $S \in E$, $\|T^{-1} \circ S\| < 1$, and the k-th differential at T of the inversion operator maps $(S_1, \ldots, S_k) \in E^k$ onto the sum

$$\sum_p (-1)^k\, T^{-1} \circ S_{p(1)} \circ T^{-1} \circ \cdots \circ T^{-1} \circ S_{p(k)} \circ T^{-1}$$

over the set of all permutations of $\{1, \ldots, k\}$.

In the *special case where $X=\mathbf{R}$*, f is k times differentiable at a if and only if the **k-th derivative** $f^{(k)}(a) \in Y$, and then

$$\langle v, D^k f(a) \rangle = v \cdot f^{(k)}(a) \text{ for } v \in \mathbf{R} \simeq \odot_k \mathbf{R};$$

as customary we denote $f = f^{(0)}$ and $[f^{(k-1)}]' = f^{(k)}$ for $k=1,2,3,\ldots$. If f is of class $k \geq 1$ and $\{t\colon 0 \leq t \leq 1\} \subset \operatorname{dmn} f$, one verifies with the help of integration by parts that (see 2.5.17, 18)

$$f(1) - \sum_{i=0}^{k} f^{(i)}(0)/i! = \textstyle\int_0^1 [f^{(k)}(0) - f^{(k)}(t)]\, d_t(1-t)^k/k!$$

Considering again a general normed vectorspace X, we will next establish the general **Taylor formula:**

$$\begin{aligned} &f(x) - \sum_{i=0}^{k} \langle (x-a)^i/i!, D^i f(a) \rangle \\ &= \textstyle\int_0^1 \langle (x-a)^k/k!, D^k f(a) - D^k f[a+t(x-a)] \rangle\, d_t(1-t)^k \end{aligned}$$

provided f is of class $k\geq 1$ and $x+t(x-a)\in\operatorname{dmn} f$ for $0\leq t\leq 1$; the powers $(x-a)^i$ are computed in $\odot_* X$ (see 1.9.1). For the proof we assume $a=0$, let

$$g(t)=f(t\,x) \text{ for } t\in\mathbf{R}\cap\{t: t\,x\in\operatorname{dmn} f\},$$

verify by induction that

$$g^{(i)}(t)=\langle x^i, D^i f(t\,x)\rangle \text{ for } i=0,\ldots,k,$$

and apply to g the special Taylor formula from the preceding paragraph.

If f is of class k and $j<k$, then $D^j f$ is of class $k-j$ and

$$\langle\xi, D^i D^j f(a)\rangle=\xi \,\lrcorner\, D^{i+j} f(a) \text{ for } \xi\in\odot_i X,$$

hence the Taylor formula for $D^j f$ takes the form:

$$D^j f(x)-\sum_{i=0}^{k-j}(x-a)^i/i!\,\lrcorner\, D^{i+j} f(a)$$
$$=\int_0^1 (x-a)^{k-j}/(k-j)!\,\lrcorner\,\big(D^k f(a)-D^k f[a+t(x-a)]\big)\, d_t(1-t)^{k-j}.$$

If $\phi_m: X^m\to Y$ is a continuous symmetric m form whenever $m=0,\ldots,M$ and if the corresponding continuous polynomial functions P and S_i are constructed as in 1.10.4, then the formulae derived there imply

$$D^i P=S_i \text{ for } i=0,\ldots,M \text{ and } D^{M+1} P=0;$$

therefore P is of class ∞ and $D^i P(0)=\phi_i$ for $i=0,\ldots,M$; in the Taylor formula for P the integral (remainder) equals 0 if $k\geq M$.

Whenever f is k times differentiable at a we define the **k jet of f at a** to be the polynomial function P_a such that

$$P_a(x)=\sum_{i=0}^{k}\langle(x-a)^i/i!, D^i f(a)\rangle \text{ for } x\in X;$$

it follows that $D^j P_a$ is the $k-j$ jet of $D^j f$ at a for $j\leq k$. From the Taylor formula for $D^j f$ we see that, *if f is of class k, then*

$$\lim_{x\to a}\|D^j f(x)-D^j P_a(x)\|\cdot|x-a|^{j-k}\cdot(k-j)!=0$$

whenever $a\in A$ and $j=0,\ldots,k$; indeed the convergence is uniform for all points a in any compact subset of $\operatorname{dmn} f$. This shows that the hypothesis of 3.1.14 is essential.

The following corollary of the Taylor formula is sometimes useful for the computation of differentials: *If f is of class $k\leq M$, $\phi_i\in\odot^i(X,Y)$ are continuous for $i=0,\ldots,M$ and*

$$\lim_{t\to 0}\left|f(a+t\,v)-\sum_{i=0}^{M}\langle(t\,v)^i/i!,\phi_i\rangle\right|\Big/t^k=0$$

whenever $v\in X$, then $D^i f(a)=\phi_i$ for $i=0,\ldots,k$.

For example *suppose Y is a normed algebra, f and g are maps of class k and $a \in \operatorname{dmn} f \cap \operatorname{dmn} g$*; letting P_a and Q_a be the k jets of f and g at a we find that

$$\lim_{x \to a} |f(x)\, g(x) - P_a(x)\, Q_a(x)| / |x-a|^k = 0,$$

and use the isomorphism of the algebra of polynomial functions with the algebra $\odot^*(X, Y)$ to obtain the general **Leibnitz formula:**

$$D^i(f \cdot g)(a) = \sum_{j=0}^{i} D^{i-j} f(a) \odot D^j g(a) \textit{ for } i=0, \ldots, k.$$

For another example we shall derive the general formula for the differentials of a composition: *If X, Y, Z are normed vectorspaces, A and B are open subsets of X and Y, $f\colon A \to Y$ and $g\colon B \to Z$ are maps of class k, $a \in A$ and $f(a) = b \in B$, then*

$$D^p(g \circ f)(a) = \sum_{\alpha \in S(p)} D^{\Sigma \alpha} g(b) \circ [D^1 f(a)^{\alpha_1} \odot \cdots \odot D^k f(a)^{\alpha_k}] / \alpha!$$

for $p=1, \ldots, k$, where $S(p)$ is the set of all k termed sequences α of nonnegative integers such that

$$\sum_{i=1}^{k} i\, \alpha_i = p.$$

In this formula $D^j g(b)$ should be interpreted as a linear function on $\odot_j Y$, and $\circ$ is multiplication in $\odot^*(X, \odot_* Y)$. To prove the formula we assume $a=0$, $b=0$, $g(b)=0$, let P_0 and Q_0 be the k jets of f and g at 0 and 0, observe that

$$\lim_{x \to 0} |(g \circ f)(x) - (Q_0 \circ P_0)(x)| / |x|^k = 0,$$

hence $D^p(g \circ f)(0) = D^p(Q_0 \circ P_0)(0)$ for $p \le k$, abbreviate

$$\phi_i = D^i P_0(0), \quad \psi_i = D^i Q_0(0) \text{ for } i = 1, 2, 3, \ldots,$$

and compute

$$\begin{aligned} Q_0[P_0(x)] &= \sum_{j=1}^{k} \left\langle \left(\sum_{i=1}^{k} \langle x^i/i!, \phi_i \rangle \right)^j \Big/ j!, \psi_j \right\rangle \\ &= \sum_{j=1}^{k} \left\langle \sum_{\alpha \in \Xi(k,j)} \left(\prod_{i=1}^{k} \langle x^i/i!, \phi_i \rangle^{\alpha_i} \right) \Big/ \alpha!, \psi_j \right\rangle \\ &= \sum_{p=1}^{k^2} \sum_{\alpha \in S(p)} \langle \langle x^p/p!, [(\phi_1)^{\alpha_1} \odot \cdots \odot (\phi_k)^{\alpha_k}]/\alpha! \rangle, \psi_{\Sigma \alpha} \rangle. \end{aligned}$$

Applying the composition formula to the special case when $Y=Z=\mathbf{R}$ and $g(y)=1/y$ for $y\in\mathbf{R}$, we obtain

$$D^p\left(\frac{1}{f}\right)(a)=\sum_{\alpha\in S(p)}\frac{(\Sigma\alpha)!}{\alpha!}\cdot\frac{1}{f(a)}\cdot\left[\frac{D^1 f(a)}{-f(a)}\right]^{\alpha_1}\odot\cdots\odot\left[\frac{D^k f(a)}{-f(a)}\right]^{\alpha_k}.$$

In case $X=\mathbf{R}^m$ we also use the k**-th order partial derivatives** $D_v f$ defined by the recursion formula

$$D_v f=D_{v_1,\ldots,v_k} f=D_{v_1} D_{v_2,\ldots,v_k} f$$

for all sequences v: $\{1,\ldots,k\}\to\{1,\ldots,m\}$. Letting $e_1,\ldots,e_m$ and $\omega_1,\ldots,\omega_m$ be the standard dual bases of $\mathbf{R}^m$ and $\bigodot^1\mathbf{R}^m$, we see that *if f is k times differentiable at a, then*

$$D_v f(a)=\langle(e_{v_k},\ldots,e_{v_1}), D^k f(a)\rangle;$$

in this case we also define

$$D^\alpha f(a)=\langle e^\alpha, D^k f(a)\rangle \text{ for } \alpha\in\Xi(m,k),$$

and infer that

$$D^\alpha f(a)=D_v f(a) \text{ if } \alpha_i=\operatorname{card}\{j\colon v_j=i\} \text{ for } i=1,\ldots,m,$$

$$D^k f(a)=\sum_{\alpha\in\Xi(m,k)} D^\alpha f(a)\cdot\omega^\alpha/\alpha!.$$

Therefore *f is of class k if and only if $D^\alpha f$ is of class 1 whenever $\alpha\in\Xi(m,j)$ with $j<k$.*

We note that if f: $\mathbf{R}\to\mathbf{R}$,

$$f(x)=\exp(-x^{-2}) \text{ for } x>0, \qquad f(x)=0 \text{ for } x\le 0,$$

then f is of class ∞, and increasing on $\{x\colon x>0\}$. Consequently the function g defined by the formula

$$g(x)=f[f(1)-f(x)]/f[f(1)] \text{ for } x\in\mathbf{R}$$

is of class ∞ and nonincreasing on $\mathbf{R}$, with

$$g(x)=1 \text{ for } x\le 0, \qquad g(x)=0 \text{ for } x\ge 1.$$

3.1.12. Lemma. *If $S\subset U\subset\mathbf{R}^m$, h: $U\to\{r\colon 0<r<\infty\}$ is Lipschitzian, $\{\mathbf{B}[s,h(s)]\colon s\in S\}$ is disjointed, $\lambda\ge\operatorname{Lip}(h)$, $\alpha>0$, $\beta>0$, $\lambda\alpha<1$, $\lambda\beta<1$ and*

$$S_x=S\cap\{s\colon \mathbf{B}[x,\alpha h(x)]\cap\mathbf{B}[s,\beta h(s)]\neq\emptyset\}$$

for $x\in U$, then

$$(1-\lambda\beta)/(1+\lambda\alpha)\le h(x)/h(s)\le(1+\lambda\beta)/(1-\lambda\alpha)$$

whenever $s\in S_x$, and

$$\operatorname{card} S_x\le[\alpha+(\beta+1)(1+\lambda\alpha)(1-\lambda\beta)^{-1}]^m(1+\lambda\beta)^m(1-\lambda\alpha)^{-m}.$$

Proof. If $s \in S_x$, then

$$|h(x)-h(s)| \le \lambda |x-s| \le \lambda \alpha h(x) + \lambda \beta h(s),$$
$$(1-\lambda \alpha) h(x) \le (1+\lambda \beta) h(s), \quad (1-\lambda \beta) h(s) \le (1+\lambda \alpha) h(x),$$
$$|x-s| + h(s) \le \alpha h(c) + (\beta+1) h(s) \le \gamma h(x),$$

where $\gamma = \alpha + (\beta+1)(1+\lambda \alpha)(1-\lambda \beta)^{-1}$; thus

$$\mathbf{B}[s, h(s)] \subset \mathbf{B}[x, \gamma h(x)].$$

It follows that

$$(\operatorname{card} S_x) \cdot \boldsymbol{\alpha}(m) \cdot [(1-\lambda \alpha)(1+\lambda \beta)^{-1} h(x)]^m$$
$$\le \sum_{s \in S_x} \boldsymbol{\alpha}(m)\, h(s)^m \le \boldsymbol{\alpha}(m)\, [\gamma h(x)]^m.$$

3.1.13. The following construction is applicable to every family Φ of open subsets of $\mathbf{R}^m$:

Taking $\lambda = \frac{1}{20}$ we let h be the function on $\bigcup \Phi$ such that

$$h(x) = \lambda \sup \{\inf\{1, \operatorname{dist}(x, \mathbf{R}^m \sim T)\} : T \in \Phi\}$$

whenever $x \in \bigcup \Phi$, and apply 2.8.4, 5 with

$$F = \{\mathbf{B}[x, h(x)] : x \in \bigcup \Phi\}, \quad \delta = \operatorname{diam}, \quad \tau = 2$$

to choose $S \subset \bigcup \Phi$ so that

$$\{\mathbf{B}[s, h(s)] : s \in S\} \text{ is disjointed}, \quad \bigcup \{\mathbf{B}[s, 5h(s)] : s \in S\} = \bigcup \Phi.$$

Clearly S is countable. Letting $\alpha = \beta = 10$ we infer from 3.1.12 that $x \in \bigcup \Phi$ implies

$$\tfrac{1}{3} \le h(x)/h(s) \le 3 \text{ for } s \in S_x, \quad \operatorname{card} S_x \le (129)^m.$$

Using a map $\gamma : \mathbf{R} \to \{y : 0 \le y \le 1\}$ of class ∞ such that

$$\gamma(t) = 1 \text{ for } t \le 1, \quad \gamma(t) = 0 \text{ for } t \ge 2,$$

we define functions μ, u_s of class ∞ on $\mathbf{R}^m$ by the formulae

$$\mu(x) = \gamma(|x|) \text{ for } x \in \mathbf{R}^m,$$
$$u_s(x) = \mu([x-s]/[5h(s)]) \text{ for } s \in S,\ x \in \mathbf{R}^m,$$

and observe that

$$\operatorname{spt} u_s \subset \mathbf{B}[s, 10h(s)], \quad u_s(x) = 1 \text{ for } x \in \mathbf{B}[s, 5h(s)],$$
$$\|D^i u_s(x)\| \le G_i \cdot [5h(s)]^{-i} \le G_i \cdot h(x)^{-i} \text{ for } s \in S_x,$$

where $G_i = \sup \operatorname{im} \|D^i \mu\| < \infty$ for $i=0,1,2,\ldots$. Letting

$$\sigma(x) = \sum_{s\in S} u_s(x) = \sum_{s\in S_x} u_s(x) \text{ for } x\in \mathbf{R}^m,$$

we obtain

$$\sigma(x)\geq 1, \quad \|D^i \sigma(x)\| \leq (129)^m G_i \cdot h(x)^{-i} \text{ for } x\in \bigcup \Phi,$$

and infer that the functions $v_s = u_s/\sigma\colon \bigcup \Phi \to \{y\colon 0\leq y\leq 1\}$ satisfy the conditions

$$\sum_{s\in S} v_s(x) = 1, \qquad \sum_{s\in S} D^i v_s(x) = 0 \text{ if } i>0,$$

$$\|D^i v_s(x)\| \leq V_i \cdot h(x)^{-i} \text{ for } x\in \bigcup \Phi,$$

where V_i is determined by $m, G_0, \ldots, G_i$.

The functions v_s constitute a **partition of unity** on $\bigcup \Phi$ associated with the family Φ. They are of class ∞ and their supports from a locally finite refinement of the covering Φ of $\bigcup \Phi$. Each v_s has an extension of class ∞ over $\mathbf{R}^m$, with value 0 on $\mathbf{R}^m \sim \bigcup \Phi$.

3.1.14. Whitney's extension theorem. *Suppose Y is a normed vectorspace, k is a nonnegative integer, A is a closed subset of $\mathbf{R}^m$, and to each $a\in A$ corresponds a polynomial function*

$$P_a\colon \mathbf{R}^m \to Y \text{ with degree } P_a \leq k.$$

Whenever $C\subset A$ and $\delta>0$ let $\rho(C,\delta)$ be the supremum of the set of all numbers

$$\|D^i P_a(b) - D^i P_b(b)\| \cdot |a-b|^{i-k} \cdot (k-i)!$$

corresponding to $i=0,\ldots,k$ and $a,b\in C$ with $0<|a-b|\leq\delta$.

If $\rho(C,\delta)\to 0$ as $\delta\to 0+$ for each compact subset C of A, then there exists a map $g\colon \mathbf{R}^m\to Y$ of class k such that

$$D^i g(a) = D^i P_a(a) \text{ for } i=0,\ldots,k \text{ and } a\in A.$$

Proof. We let $U=\mathbf{R}^m \sim A$ and apply 3.1.13 with $\Phi=\{U\}$, so that

$$20\,h(x) = \inf\{1, \operatorname{dist}(x,A)\} \text{ for } x\in U.$$

For each $s\in S$ we choose

$$\xi(s)\in A \text{ with } |s-\xi(s)| = \operatorname{dist}(s,A).$$

Then we define $g\colon \mathbf{R}^m\to Y$ by the formulae

$$g(x) = P_x(x) \text{ for } x\in A, \qquad g(x) = \sum_{s\in S} v_s(x)\, P_{\xi(s)}(x) \text{ for } x\in U.$$

Clearly $g|U$ is of class ∞ with

$$D^i g(x) = \sum_{s\in S} \sum_{j=0}^{i} D^{i-j} v_s(x) \odot D^j P_{\xi(s)}(x)$$

for $x\in U$ and $i=0,1,2,\ldots$. We will show that g has the required behavior on and near A.

First, if $c,b\in C\subset A$, $x\in\mathbf{R}^m$ and $j\le k$, then

$$\|D^j P_c(x) - D^j P_b(x)\| = \left\| \sum_{i=0}^{k-j} (x-b)^i/i! \,\lrcorner\, [D^{j+i} P_c(b) - D^{j+i} P_b(b)] \right\|$$

$$\le \sum_{i=0}^{k-j} (|x-b|^i/i!)\cdot[|c-b|^{k-j-i}/(k-j-i)!]\,\rho(C,|c-b|)$$

$$=[(|x-b|+|c-b|)^{k-j}/(k-j)!]\,\rho(C,|c-b|).$$

Next, if $a\in A$, $C=A\cap\mathbf{B}(a,2)$ and $x\in U\cap\mathbf{B}(a,\frac{1}{3})$, then we choose $b\in A$ with $|x-b|=\operatorname{dist}(x,A)$ and observe that

$$|x-b|\le|x-a|\le\tfrac{1}{3},\qquad 20h(x)=|x-b|\le\tfrac{1}{3},$$

$$|b-a|\le|x-b|+|x-a|\le 2|x-a|<2,\ b\in C;$$

moreover $s\in S_x$ implies

$$20h(s)\le 60h(x)\le 1,\qquad 20h(s)=|s-\xi(s)|,$$

$$|s-x|\le 10h(s)+10h(x)\le 40h(x)\le\tfrac{2}{3},$$

$$|\xi(s)-a|\le|\xi(s)-s|+|s-x|+|x-a|\le 2,\ \xi(s)\in C,$$

$$|\xi(s)-b|\le|\xi(s)-s|+|s-x|+|x-b|\le 120h(x)=6|x-b|,$$

$$|x-b|+|\xi(s)-b|\le 140h(x)=7|x-b|;$$

for $i\le k$ we therefore obtain

$$\|D^i g(x) - D^i P_b(x)\| = \left\| \sum_{s\in S} \sum_{j=0}^{i} D^{i-j} v_s(x)\odot[D^j P_{\xi(s)}(x) - D^j P_b(x)] \right\|$$

$$\le \sum_{s\in S_x} \sum_{j=0}^{i} \binom{i}{j} V_{i-j}\, h(x)^{j-i}[140h(x)]^{k-j}\rho[C,120h(x)]$$

$$\le M_i\cdot|x-b|^{k-i}\rho(C,6|x-b|),$$

where M_i is determined by $m,k,V_0,\ldots,V_i$; it follows that

$$\|D^i g(x)-D^i P_a(x)\| \le \|D^i g(x)-D^i P_b(x)\| + \|D^i P_a(x)-D^i P_b(x)\|$$

$$\le M_i\cdot|x-b|^{k-i}\rho(C,6|x-b|)+(3|x-a|)^{k-i}\rho(C,2|x-a|)$$

$$\le(M_i+3^{k-i})|x-a|^{k-i}\rho(C,6|x-a|).$$

We now obtain the conclusion of the theorem by induction with respect to i. In fact, for any $i \le k$ the inductive assumption

$$D^i g(x) = D^i P_x(x) \text{ whenever } x \in A$$

implies by virtue of the preceding estimate and the hypothesis of the theorem that

$$\lim_{x \to a} \|D^i g(x) - D^i P_a(x)\| / |x-a|^{k-i} = 0$$

whenever $a \in A$, hence $D^i g$ is continuous at a and, in case $i < k$,

$$D\, D^i g(a) = D\, D^i P_a(a), \qquad D^{i+1} g(a) = D^{i+1} P_a(a).$$

3.1.15. Theorem. *If $A \subset B \subset \mathbf{R}^m$, $f: B \to \mathbf{R}^n$ is of class k and*

$$\limsup_{x \to a} \|D^k f(x) - D^k f(a)\| / |x-a| < \infty \text{ whenever } a \in A,$$

then for each $\varepsilon > 0$ there exists a map $g: \mathbf{R}^m \to \mathbf{R}^n$ of class $k+1$ such that

$$\mathscr{L}^m(A \sim \{x : f(x) = g(x)\}) < \varepsilon.$$

Proof. Assured by 3.1.9 that $\mathscr{L}^m(A \sim \operatorname{dmn} D\,D^k f) = 0$, and by 3.1.4 that $D\,D^k f$ is $\mathscr{L}^m \llcorner \operatorname{dmn} D\,D^k f$ measurable, we use 2.3.5 to obtain a closed set E such that

$$D\,D^k f | E \text{ is continuous and } \mathscr{L}^m(A \sim E) < \varepsilon.$$

For $q = 1, 2, 3, \ldots$ and $a \in E$ we let $\eta_q(a)$ be the supremum of the set of numbers

$$\|D^k f(x) - D^k f(a) - \langle x-a, D\,D^k f(a)\rangle\| / |x-a|$$

corresponding to all $x \in \mathbf{B}(a, 1/q) \sim \{a\}$. Since

$$\lim_{q \to \infty} \eta_q(a) = 0 \text{ whenever } a \in E,$$

and each η_q is a Borel function, 2.3.7 and 2.2.2 allows us to construct a closed subset F of E such that $\mathscr{L}^m(A \sim F) < \varepsilon$ and

$$\limsup_{q \to \infty} \{\eta_q(a) : a \in C\} = 0 \text{ for every compact } C \subset F.$$

For each $a \in F$ we let P_a be the $k+1$ jet of f at a, and observe that $P_a(a) = f(a)$. We will show that the hypotheses of 3.1.14 hold with Y, k, A replaced by $\mathbf{R}^n$, $k+1$, F.

Suppose C is a compact subset of F and $a, b \in C$. For $i = 0, \ldots, k-1$ and $0 < |b-a| \le 1/q$ we use the Taylor formulae for $D^i P_a$ and $D^i f$ to compute

$$D^i P_b(b) - D^i P_a(b) = D^i f(b) - \sum_{j=i}^{k+1} (b-a)^{j-i}/(j-i)! \,\lrcorner\, D^j f(a)$$

$$= \int_0^1 (b-a)^{k-i}/(k-i)! \,\lrcorner\, (D^k f(a) - D^k f[a+t(b-a)])\, d_t(1-t)^{k-i}$$

$$- (b-a)^{k+1-i}/(k+1-i)! \,\lrcorner\, D^{k+1} f(a)$$

$$= \int_0^1 (b-a)^{k-i}/(k-i)!$$

$$\lrcorner\, (D^k f(a) + \langle t(b-a), D D^k f(a)\rangle - D^k f[a+t(b-a)])\, d_t(1-t)^{k-i},$$

because $\int_0^1 t\, d_t(1-t)^{k-i} = -1/(k+1-i)$, hence obtain

$$\|D^i P_b(b) - D^i P_a(b)\| \cdot |b-a|^{i-k-1}(k+1-i)! \le \eta_q(a).$$

Next we compute

$$D^k P_b(b) - D^k P_a(b) = D^k f(b) - D^k f(a) - (b-a) \,\lrcorner\, D^{k+1} f(a)$$

$$= D^k f(b) - D^k f(a) - \langle b-a, D D^k f(a)\rangle,$$

and infer that $0 < |b-a| \le 1/q$ implies

$$\|D^k P_b(b) - D^k P_a(b)\| \cdot |b-a|^{-1} \le \eta_q(a).$$

Finally, since $D^{k+1} f$ is uniformly continuous on C, we observe that

$$D^{k+1} P_b(b) - D^{k+1} P_a(b) = D^{k+1} f(b) - D^{k+1} f(a)$$

is small for small $|b-a|$.

3.1.16. Theorem. *If $A \subset \mathbf{R}^m$, $f: A \to \mathbf{R}^n$ and*

$$\operatorname{ap}\limsup_{x \to a} |f(x) - f(a)|/|x-a| < \infty$$

for $\mathscr{L}^m$ almost all x in A, then for each $\varepsilon > 0$ there exists a map $g: \mathbf{R}^m \to \mathbf{R}^n$ of class 1 such that

$$\mathscr{L}^m(A \sim \{x: f(x) = g(x)\}) < \varepsilon.$$

Proof. From 3.1.8 we obtain compact sets $C_1, C_2, C_3, \ldots$ such that

$$C_i \subset \mathbf{U}(0,i) \sim \mathbf{U}(0,i-1),\ f \mid C_i \text{ is Lipschitzian},$$

$$\mathscr{L}^m(A \cap [\mathbf{U}(0,i) \sim \mathbf{U}(0,i-1)] \sim C_i) < 2^{-i}\varepsilon,$$

then use 2.10.43 repeatedly to construct $h: \mathbf{R}^m \to \mathbf{R}^n$ so that $h \mid \mathbf{U}(0,i)$ is a Lipschitzian extension of $h \mid \mathbf{U}(0,i-1) \cup f \mid C_i$. Observing that h is locally Lipschitzian and

$$\mathscr{L}^m(A \sim \{x: f(x) = h(x)\}) < \varepsilon,$$

we finally apply 3.1.15 with A, B, f, k replaced by $\mathbf{R}^m, \mathbf{R}^m, h, 0$.

3.1.17. Since the converse of Theorem 3.1.16 is clearly also true, this theorem achieves an optimal result. The corresponding problem for functions of higher class is unsolved: *Can* "lim sup" *be replaced by* "ap lim sup" *in* 3.1.15 *when* $k>0$?

3.1.18. A **diffeomorphism** of class k is a homeomorphism of class k whose inverse is also a map of class k; here it is understood that the domain and the image of the diffeomorphism are open subsets of (isomorphic) normed vectorspaces.

We now consider the problem of representing maps of class k locally as compositions of linear maps with diffeomorphisms. *Suppose A is an open subset of $\mathbf{R}^m$ and $f: A \to \mathbf{R}^n$ is of class $k \geq 1$.* For each nonnegative integer ν the set

$$\{x: \dim \operatorname{im} D f(x) \geq \nu\} = \{x: \wedge_\nu D f(x) \neq 0\}$$

is open, because $\langle \xi, \wedge_\nu D f(x) \rangle$ is continuous with respect to x whenever $\xi \in \wedge_\nu \mathbf{R}^m$. The set

$$\{x: \dim \operatorname{im} D f(x) = \nu\}$$

need not be open; its interior points are called **generic points of rank** ν with respect to f. We will study *the behavior of f near such a generic point a of rank ν.*

Clearly $\nu \leq \inf\{m, n\}$. If $\nu = 0$, then f maps a neighborhood of a in A onto $f(a)$. We *henceforth assume* $\nu > 0$ and choose a projection

$$p \in \mathbf{O}^*(n, \nu) \quad \text{with} \quad \operatorname{im} p^* = \operatorname{im} D f(a).$$

In case $\nu < m$ we also choose a projection

$$q \in \mathbf{O}^*(m, m-\nu) \quad \text{with} \quad \operatorname{im} q^* = \ker D f(a)$$

and define

$$g: A \to \mathbf{R}^\nu \times \mathbf{R}^{m-\nu}, \quad g(x) = ([p \circ f](x), q(x)) \text{ for } x \in A,$$
$$r: \mathbf{R}^\nu \times \mathbf{R}^{m-\nu} \to \mathbf{R}^\nu, \quad r(v, w) = v \text{ for } (v, w) \in \mathbf{R}^\nu \times \mathbf{R}^{m-\nu};$$

in case $\nu = m$ we take $g = p \circ f$, $r = \mathbf{1}_{\mathbf{R}^\nu}$. Since $p \mid \operatorname{im} D f(a)$ is univalent,

$$\ker D(p \circ f)\, a = \ker D f(a),$$

and $\ker q \cap \ker D f(a) = \ker q \cap \operatorname{im} q^* = \{0\}$ in case $\nu < m$, we see that $D g(a)$ is univalent and infer from 3.1.1, 11 that *a has a neighborhood U such that $g \mid U$ is a diffeomorphism of class k.* In fact we may suppose that $(p \circ f)\, a$ has a convex neighborhood V in $\mathbf{R}^\nu$, and in case $\nu < m$ that $q(a)$ has a convex neighborhood W in $\mathbf{R}^{m-\nu}$, such that

$$g(U) = V \times W \text{ if } \nu < m, \quad g(U) = V \text{ if } \nu = m.$$

We may also assume that $x \in U$ implies

$$\nu = \dim \operatorname{im} Df(x) \geq \dim \operatorname{im} D(p \circ f)(x) \geq \nu,$$

hence $\dim \ker Df(x) = m - \nu = \dim \ker D(p \circ f)(x)$,

$$\ker Df(x) = \ker D(p \circ f)(x);$$

applying $Dg(x)$ to this equation we find, since $p \circ f = r \circ g$, that

$$\ker D[f \circ (g|U)^{-1}]\, g(x) = \ker D[p \circ f \circ (g|U)^{-1}]\, g(x) = \ker r.$$

In case $\nu < m$ we infer that

$$\langle (0, z), D[f \circ (g|U)^{-1}](v, w) \rangle = 0 \text{ for } v \in V,\ w \in W,\ z \in \mathbf{R}^{m-\nu},$$

hence $f \circ (g|U)^{-1} = \psi \circ r$ where

$$\psi\colon V \to \mathbf{R}^n, \quad \psi(v) = [f \circ (g|U)^{-1}](v, q(a)) \text{ for } v \in V;$$

in case $\nu = m$ we take $\psi = f \circ (g|U)^{-1}$; in both cases we obtain

$$f|U = \psi \circ r \circ g|U = \psi \circ p \circ f|U$$

and note that $p \circ f|U = p \circ \psi \circ p \circ f|U$ with $(p \circ f)\, U = V$, hence

$$p \circ \psi = \mathbf{1}_V.$$

If follows that the map

$$h = \mathbf{1}_{\mathbf{R}^n} + (p^* \circ p) - (\psi \circ p)\colon p^{-1}(V) \to p^{-1}(V)$$

is a diffeomorphism of class k; in fact $p \circ h = p|p^{-1}(V)$ and

$$h^{-1} = \mathbf{1}_{\mathbf{R}^n} - (p^* \circ p) + (\psi \circ p)\colon p^{-1}(V) \to p^{-1}(V).$$

After verifying that

$$h \circ f|U = p^* \circ p \circ f|U = p^* \circ p \circ \psi \circ r \circ g|U = p^* \circ r \circ g|U$$

we obtain the conclusions

$$h[f(U)] = p^*(V) \subset \operatorname{im} Df(a), \quad f|U = h^{-1} \circ p^* \circ r \circ g|U.$$

Thus *$f|U$ is equivalent to the linear map $p^* \circ r$, modulo the diffeomorphisms $g|U$ and h.* We also observe that

$$D\psi[(p \circ f)\, a] = p^*$$

because $\operatorname{im} D\psi[(p \circ f)\, a] = \operatorname{im} Df(a) = \operatorname{im} p^*$ and $p \circ D\psi[(p \circ f)\, a] = \mathbf{1}_{\mathbf{R}^\nu}$; it follows that

$$Dh[f(a)] = \mathbf{1}_{\mathbf{R}^n}.$$

3.1.19. By a μ **dimensional submanifold of class** k of $\mathbf{R}^n$ we mean a subset B of $\mathbf{R}^n$ satisfying the condition:

(1) *For each* $b \in B$ *there exist a neighborhood* T *of* b *in* $\mathbf{R}^n$, *a diffeomorphism* $\sigma: T \to \mathbf{R}^n$ *of class* k, *and a* μ *dimensional vectorsubspace* Z *of* $\mathbf{R}^n$ *such that*

$$\sigma(B \cap T) = Z \cap \operatorname{im} \sigma.$$

In case $\mu \geq 1$ *and* $k \geq 1$, the condition (1) is equivalent to each of the following conditions (2) to (5):

(2) *For each* $b \in B$ *there exist a neighborhood* T *of* b *in* $\mathbf{R}^n$ *and a map* $f: T \to \mathbf{R}^i$ *of class* k, *with* $i \geq n - \mu$, *such that*

$$B \cap T = f^{-1}\{f(b)\}, \qquad \dim \operatorname{im} Df(x) = n - \mu \text{ whenever } x \in T.$$

(3) *For each* $b \in B$ *there exist an open subset* Q *of* $\mathbf{R}^m$, *with* $m \geq \mu$, *and a function* $f: Q \to \mathbf{R}^n$ *of class* k, *such that* f *maps each open subset of* Q *onto a relatively open subset of* B,

$$b \in \operatorname{im} f, \qquad \dim \operatorname{im} Df(x) = \mu \text{ whenever } x \in Q.$$

(4) *For each* $b \in B$ *there exist a neighborhood* T *of* b *in* $\mathbf{R}^n$, *a convex open subset* V *of* $\mathbf{R}^\mu$, *and maps* $\phi: T \to V$, $\psi: V \to T$ *of class* k *such that*

$$B \cap T = \operatorname{im} \psi, \qquad \phi \circ \psi = \mathbf{1}_V.$$

(5) *For each* $b \in B$ *there exist a neighborhood* T *of* b *in* $\mathbf{R}^n$ *and a projection* $p \in \mathbf{O}^*(n, \mu)$ *such that*

$$p|(B \cap T) \text{ is univalent}, \qquad p(B \cap T) = p(T) \text{ is convex},$$

$$[p|(B \cap T)]^{-1}: p(T) \to \mathbf{R}^n \text{ is of class } k, \qquad D[p|(B \cap T)]^{-1} p(b) = p^*.$$

The asserted equivalences may be verified as follows:

It is trivial that (1) implies (2), (1) implies (4), (5) implies (4).

From 3.1.18 one sees that (2) implies (1), (3) implies (1), (4) implies (5).

To prove (4) implies (3) one observes that $\psi(W) = B \cap \phi^{-1}(W)$ for $W \subset V$, $\operatorname{im} D\phi[\psi(x)] = \mathbf{R}^\mu$ for $x \in V$.

Next we note that every submanifold of $\mathbf{R}^n$ is locally compact. Whenever T, V, ϕ, ψ are as in (4), then ϕ maps $B \cap T$ homeomorphically onto V, and $\phi|(B \cap T) = \psi^{-1}$ is called a **coordinate system of class** k for B at b; moreover $\psi \circ \phi$ retracts T onto $T \cap B$.

A continuous map $f: C \to D$ is said to **retract** C onto D if and only if $D \subset C$ and $f|D = \mathbf{1}_D$.

Finally we observe that a subset of $\mathbf{R}^n$ is a 0 dimensional submanifold if and only if all of its points are isolated.

3.1.20. Theorem. *Suppose B is a connected subset of $\mathbf{R}^n$ and $k \geq 1$. Then B is a submanifold of class k of $\mathbf{R}^n$ if and only if there exists a map of class k retracting some open subset of $\mathbf{R}^n$ onto B.*

Proof. First, we assume B is a μ dimensional submanifold of class k, let Φ be the family of all open subsets T of $\mathbf{R}^n$ for which there exist V, ϕ, ψ with the properties described in 3.1.19 (4), apply the construction 3.1.13 to Φ, arrange the elements of the resulting set S in a univalent sequence $s_1, s_2, s_3, \ldots$ and denote

$$P_i = \mathbf{B}[s_i, 10h(s_i)], \; Q_i = \mathbf{U}[s_i, 5h(s_i)] \text{ for } i = 1, 2, 3, \ldots.$$

Starting with $W_0 = \varnothing, f_0 = \varnothing$ we inductively define open subsets W_i of $\mathbf{R}^n$ and maps f_i of class k which retract W_i onto $B \cap W_i$. To obtain W_i, f_i from W_{i-1}, f_{i-1} we choose T, V, ϕ, ψ with the properties described in 3.1.19 (4) and with $P_i \subset T$, we observe that

$$\rho = \psi \circ [(1 - u_{s_i}) \cdot (\phi \circ f_{i-1}) + u_{s_i} \cdot \phi] \colon T \cap f_{i-1}^{-1}(T) \to T$$

is a map of class k which retracts $T \cap f_{i-1}^{-1}(T)$ onto $B \cap T \cap W_{i-1}$ and for which

$$\rho(x) = f_{i-1}(x) \text{ whenever } x \in T \cap f_{i-1}^{-1}(T) \cap (W_{i-1} \sim P_i),$$

$$\rho(x) = (\psi \circ \phi)(x) \text{ whenever } x \in T \cap f_{i-1}^{-1}(T) \cap Q_i;$$

we take

$$W_i = (W_{i-1} \sim P_i) \cup [T \cap f_{i-1}^{-1}(T)] \cup Q_i,$$

$$f_i = [f_{i-1} | (W_{i-1} \sim P_i)] \cup \rho \cup (\psi \circ \phi | Q_i)$$

and note that $B \cap W_{i-1} \cap P_i \subset T \cap f_{i-1}^{-1}(T)$, hence

$$B \cap W_i = B \cap (W_{i-1} \cup Q_i).$$

Concluding therefore that

$$B = \bigcup_{j=1}^{\infty} (B \cap Q_j) \subset \bigcup_{j=1}^{\infty} \bigcap_{i=j}^{\infty} W_i = A,$$

we will verify that the set A so defined is open in $\mathbf{R}^n$, and that the functions f_i converge on A to a map of class k which retracts A onto B. In fact, for each positive integer c,

$$\operatorname{card}\{i \colon P_c \cap P_i \neq \varnothing\} \leq (129)^n,$$

hence there exists a positive integer d such that

$$P_c \cap P_i = \varnothing \text{ whenever } i > d;$$

consequently $Q_c \cap W_i = Q_c \cap W_d$ whenever $i > d$,

$$Q_c \cap A = \bigcup_{j=1}^{\infty} \bigcap_{i=j}^{\infty} (Q_c \cap W_i) = Q_c \cap W_d,$$

and $f_i | (Q_c \cap A) = f_d | (Q_c \cap W_d)$ whenever $i > d$.

Second, to prove the converse, we assume f is a map of class k which retracts an open subset A of $\mathbf{R}^n$ onto B, let

$$\mu = \sup\{\dim \operatorname{im} D f(x) \colon x \in A\}, \qquad G = \{x \colon \dim \operatorname{im} D f(x) = \mu\},$$

and note that G is open. In case $\mu = 0$, then f is constant on the component of B in A, card $B = 1$. From now on we assume $\mu \geq 1$. Since $f \circ f = f$ we see that

$$D f[f(x)] \circ D f(x) = D f(x), \qquad \operatorname{im} D f(x) \subset \operatorname{im} D f[f(x)]$$

whenever $x \in A$, hence $f(G) \subset B \cap G$; we also obtain

$$\langle v, D f(x) \rangle = v \text{ whenever } x \in B, \ v \in \operatorname{im} D f(x),$$

hence $B \cap G = B \cap \{x \colon \| \wedge_\mu D f(x) \| \geq 1\}$. We conclude that $B \cap G$ is nonempty, open and closed relative to B, hence $B \subset G$. It follows from 3.1.18 that every point b of B has a neighborhood U in $\mathbf{R}^n$ such that $f(U)$ is a μ dimensional submanifold of class k of $\mathbf{R}^n$; then $U \cap f(U) = U \cap B$ is both a submanifold of $\mathbf{R}^n$ and a neighborhood of b in B.

3.1.21. Whenever X is a normed vectorspace, $S \subset X$ and $a \in X$, we define the **tangent cone** of S at a, denoted

$$\operatorname{Tan}(S, a),$$

as the set of all $v \in X$ such that for every $\varepsilon > 0$ there exist

$$x \in S, \ 0 < r \in \mathbf{R} \text{ with } |x - a| < \varepsilon, \ |r(x-a) - v| < \varepsilon;$$

such vectors v are called **tangent vectors** of S at a. We observe that

$\operatorname{Tan}(S, a)$ is a closed subset of X;
$v \in \operatorname{Tan}(S, a)$, $0 \leq s \in \mathbf{R}$ implies $s\, v \in \operatorname{Tan}(S, a)$;
$0 \in \operatorname{Tan}(S, a)$ if and only if $a \in \operatorname{Clos} S$;
$\operatorname{Tan}(S, a) \cap \{v \colon |v| = 1\} = \bigcap_{\varepsilon > 0} \operatorname{Clos}\left\{\dfrac{x-a}{|x-a|} \colon a \neq x \in S \cap \mathbf{U}(a, \varepsilon)\right\}$;
$\operatorname{Tan}(S, a) = \operatorname{Tan}(\operatorname{Clos} S, a)$.

We also define

$$\operatorname{Tan}(S) = (S \times X) \cap \{(a, v) \colon v \in \operatorname{Tan}(S, a)\}.$$

If f maps X into another normed vector space Y and f is differentiable at a, then

$$Df(a)[\operatorname{Tan}(S,a)] \subset \operatorname{Tan}[f(S), f(a)];$$

moreover equality holds in case $f|S$ is univalent, $a \in S$, $(f|S)^{-1}$ is continuous at $f(a)$, and $Df(a)$ is a linear homeomorphism mapping X onto a closed vectorsubspace of Y. To prove this we assume $a=0$, $f(a)=0$, and abbreviate $Df(a)=L$. For each $v \in \operatorname{Tan}(S,0)$ there exist $x_i \in S$, $0<r_i \in \mathbf{R}$ with $x_i \to 0$, $r_i x_i \to v$ as $i \to \infty$, hence

$$L(v) - r_i f(x_i) = L(v - r_i x_i) + |r_i x_i| \cdot [L(x_i) - f(x_i)]/|x_i|$$

approaches 0 as $i \to \infty$, and $L(v) \in \operatorname{Tan}[f(S), 0]$. On the other hand, in case the additional hypotheses hold and $w \in \operatorname{Tan}[f(S), 0]$, there exist $x_i \in S$, $0<r_i \in \mathbf{R}$ *with* $f(x_i) \to 0$, $r_i f(x_i) \to w$ as $i \to \infty$, hence $x_i \to 0$ and

$$L(r_i x_i) - w = |r_i f(x_i)| \cdot [|x_i|/|f(x_i)|] \cdot [L(x_i) - f(x_i)]/|x_i| \\ + r_i f(x_i) - w \to 0 \text{ as } i \to \infty,$$

which implies that $r_i x_i$ converge to some v with $L(v)=w$, and $v \in \operatorname{Tan}(S,0)$.

Since the preceding proposition applies to diffeomorphisms of class 1, and since $\operatorname{Tan}(Z,z)=Z$ whenever Z is a closed vectorsubspace of X and $z \in Z$, we see from 3.1.19(1) that *if B is a μ dimensional submanifold of class $k \geq 1$ of $\mathbf{R}^n$ and $b \in B$, then* $\operatorname{Tan}(B,b)$ *is a μ dimensional vectorsubspace of $\mathbf{R}^n$; furthermore* $\operatorname{Tan}(B)$ *is a 2μ dimensional submanifold of class $k-1$ of $\mathbf{R}^n \times \mathbf{R}^n$* (called the **tangent bundle** of B).

Similarly we see from 3.1.18 that *if f is a function of class 1 mapping an open subset of $\mathbf{R}^m$ into $\mathbf{R}^n$, then every generic point a with respect to f has a neighborhood U in $\mathbf{R}^m$ such that*

$$\operatorname{Tan}[f^{-1}\{f(a)\}, a] = \ker Df(a), \qquad \operatorname{Tan}[f(U), f(a)] = \operatorname{im} Df(a).$$

In case X is an inner product space, $S \subset X$ and $a \in X$ we also define the **normal cone** of S at a,

$$\operatorname{Nor}(S,a) = X \cap \{u \colon u \bullet v \leq 0 \text{ for } v \in \operatorname{Tan}(S,a)\}.$$

Clearly $\operatorname{Nor}(S,a)$ is closed and convex; its elements are called **normal vectors** to S at a. We further let

$$\operatorname{Nor}(S) = (S \times X) \cap \{(a,u) \colon u \in \operatorname{Nor}(S,a)\}.$$

If B is a μ dimensional submanifold of class $k \geq 1$ of $\mathbf{R}^n$, then $\operatorname{Nor}(B)$ *is an n dimensional submanifold of class $k-1$ of $\mathbf{R}^n \times \mathbf{R}^n$* (called the **normal bundle** of B in $\mathbf{R}^n$).

Both tangent and normal cones play an important role in the study of *sets with positive reach* [F 15]; among these sets are all closed convex subsets of $\mathbf{R}^n$ and all closed submanifolds of class 2 of $\mathbf{R}^n$ (including manifolds with boundary).

3.1.22. Assuming that X, Y are normed vectorspaces, $S \subset X$ and $a \in \operatorname{Clos} S$, we discuss here the concept of differentiation relative to S at a.

First we observe that *if U is a neighborhood of a in X, h maps X into Y and h is differentiable at a, then*

$$h(S \cap U) = \{0\} \text{ implies } D\,h(a)[\operatorname{Tan}(S, a)] = \{0\}.$$

In fact, if $S \cap U \ni x_i \to a$ and $(x_i - a)/|x_i - a| \to v$ as $i \to \infty$, then $h(x_i) = 0$ for all i, hence $h(a) = 0$ and

$$\langle v, D\,h(a) \rangle = \lim_{i \to \infty} \langle x_i - a, D\,h(a) \rangle / |x_i - a| = 0.$$

This allows us to make the following definition:

Suppose f maps some subset of X into Y; we say that f is **differentiable relative to S at a**, and that L is the **differential of f relative to S at a**, if and only if *there exist a neighborhood U of a in X and a map $g: U \to Y$ such that g is differentiable at a with*

$$f|(S \cap U) = g|(\dot{S} \cap U) \quad \text{and} \quad L = D\,g(a)|\operatorname{Tan}(S, a).$$

Whenever $f: S \to Y$, $a \in \operatorname{Clos} S$ and f has the differential L relative to S at a we define

$$D\,f(a) = L.$$

We observe that *f is differentiable relative to S at a if and only if there exist $\eta \in Y$ and a continuous linear map $\zeta: X \to Y$ such that*

$$|f(x) - \eta - \zeta(x - a)|/|x - a| \to 0 \text{ as } S \ni x \to a;$$

in fact whenever this condition holds we can take

$$g(x) = f(x) \text{ for } x \in S, \quad g(x) = \eta + \zeta(x - a) \text{ for } x \in X \sim S,$$

and infer that $D\,g(a) = \zeta$, $D\,f(a) = \zeta|\operatorname{Tan}(S, a)$.

Of particular interest is differentiation relative to a *μ dimensional submanifold B of class* 1 of $\mathbf{R}^n$; whenever $f: B \to Y$ and $b \in B$, the following three conditions are equivalent:

(1) *L is a differential of f relative to B at b.*

(2) *For some neighborhood U of b in $\mathbf{R}^n$ and some map r of class* 1 *which retracts U onto $B \cap U$, $f \circ r$ is differentiable at b and*

$$L = D(f \circ r)(b)|\operatorname{Tan}(B, b).$$

(3) *For some coordinate system γ of class* 1 *for B at b, $f \circ \gamma^{-1}$ is differentiable at $\gamma(b)$ and*

$$L = D(f \circ \gamma^{-1})[\gamma(b)] \circ (D\,\gamma^{-1}[\gamma(b)])^{-1}.$$

This list of equivalents may be amplified by adjoining the conditions [2], [3] obtained through replacement of "some" by "every" in (2), (3).

Clearly [3] implies (3), [2] implies (2), (2) implies (1).

To prove (3) implies [2] we choose ϕ, ψ as in 3.1.19 (4) with $\gamma=\psi^{-1}$ and observe that $f\circ r$ agrees with $(f\circ\psi)\circ(\phi\circ r)$ near b, hence $f\circ r$ is differentiable at b; furthermore $(f\circ r)\circ\psi$ agrees with $f\circ\psi$ near $\gamma(b)$,

$$D(f\circ r)(b)\circ D\psi[\gamma(b)]=D(f\circ\psi)[\gamma(b)] \quad\text{and}\quad D\psi[\gamma(b)]\colon \mathbf{R}^{\mu}\simeq\operatorname{Tan}(B,b).$$

To prove (1) implies [3] we choose g according to the definition of (1), and also choose ϕ, ψ as above; then $f\circ\psi$ agrees with $g\circ\psi$ near $\gamma(b)$, hence $f\circ\psi$ is differentiable at $\gamma(b)$ and

$$D(f\circ\psi)[\gamma(b)]=D\,g(b)\circ D\psi[\gamma(b)].$$

Next we relativize the concept of function of class $k\geq 1$: We say that f is of **class** k **relative to** S if and only if *for each $a\in S$ there exist a neighborhood U of a in X and a map $g\colon U\to Y$ of class k with $f|(S\cap U)=g|(S\cap U)$.*

In case $X=\mathbf{R}^n$ it follows from 3.1.13 that *f is of class k relative to S if and only if there exist an open subset U of $\mathbf{R}^n$ and a map $g\colon U\to Y$ of class k with $S\subset U$ and $f|S=g|S$;* moreover, if S is closed, then one can take $U=\mathbf{R}^n$, by use of 3.1.14.

In case B is a submanifold of class k of $\mathbf{R}^n$, a function f mapping B into Y is of class k relative to B if and only if $f\circ\gamma^{-1}$ is of class k whenever γ is a coordinate system of class k for B.

3.1.23. Theorem. *Suppose B is a μ dimensional submanifold of class $k\geq 1$ of $\mathbf{R}^n$, $b\in B$, $0<t<1$ and*

$$A=\{b+v\colon v\in\operatorname{Tan}(B,b)\}.$$

For every sufficiently small positive number r there exists a diffeomorphism f of class k, mapping $\mathbf{R}^n$ onto $\mathbf{R}^n$, such that

$$\operatorname{Lip}(f)\leq t^{-1},\quad \operatorname{Lip}(f^{-1})\leq t^{-1},\quad f(x)=x \text{ whenever } x\in\mathbf{R}^n\sim\mathbf{U}(b,r),$$

$$\mathbf{U}(b,tr)\cap B=\mathbf{U}(b,tr)\cap f^{-1}(A).$$

Proof. We assume $b=0$, choose T and p according to 3.1.19 (5), abbreviate $[p|(B\cap T)]^{-1}=\psi$, select a function

$$\gamma\colon \mathbf{R}\to\{y\colon 0\leq y\leq 1\} \text{ of class } \infty$$

with

$$\gamma(s)=1 \text{ for } s\leq t,\qquad \gamma(s)=0 \text{ for } s\geq 1,$$

and let $\varepsilon=(1-t)/[\operatorname{Lip}(\gamma)+1]$.

For every sufficiently small positive number r it is true that $\mathbf{R}^n \cap \mathbf{B}(0,r) \subset T$ and

$$\|p^* - D\psi(z)\| < \varepsilon \text{ for } z \in \mathbf{R}^\mu \cap \mathbf{B}(0,r);$$

since $\psi(0)=0$ it follows that

$$|p^*(z) - \psi(z)| \leq \varepsilon\,|z| \text{ for } z \in \mathbf{R}^\mu \cap \mathbf{B}(0,r).$$

Under these conditions we define the maps

$$u\colon \mathbf{R}^n \to \{y\colon 0 \leq y \leq 1\}, \quad u(x) = \gamma(|x|/r) \text{ for } x \in \mathbf{R}^n,$$
$$f\colon \mathbf{R}^n \to \mathbf{R}^n, \quad f(x) = x \text{ for } x \in \mathbf{R}^n \sim T,$$
$$f(x) = x + u(x) \cdot [(p^* \circ p)\,x - (\psi \circ p)\,x] \text{ for } x \in T,$$

and observe that u is of class ∞,

$$u(x) = 0 \text{ and } f(x) = x \text{ for } x \in \mathbf{R}^n \sim \mathbf{U}(0,r),$$
$$f \text{ is of class } k, \; p \circ f = p,$$
$$u(x) = 1 \text{ and } x - \psi[p(x)] = f(x) - p^*[p(x)] \text{ for } x \in \mathbf{U}(0,t\,r),$$
$$\mathbf{U}(0,t\,r) \cap B = \mathbf{U}(0,t\,r) \cap f^{-1}(\operatorname{im} p^*),$$

and that, whenever $x \in \mathbf{R}^n \cap \mathbf{B}(0,r)$,

$$D f(x) - \mathbf{1}_{\mathbf{R}^n} = D\,u(x) \cdot (p^* - \psi)\,[p(x)] + u(x) \cdot (p^* - D\psi[p(x)]) \circ p,$$
$$\|D f(x) - \mathbf{1}_{\mathbf{R}^n}\| \leq [\operatorname{Lip}(\gamma)/r] \cdot \varepsilon\,|x| + \varepsilon \leq 1 - t.$$

We conclude

$$\operatorname{Lip}(f - \mathbf{1}_{\mathbf{R}^n}) \leq 1 - t,$$

hence $x, v \in \mathbf{R}^n$ implies

$$|f(x+v) - f(x) - v| \leq (1-t)\,|v| \leq (t^{-1} - 1)\,|v|,$$
$$t\,|v| = |v| - (1-t)\,|v| \leq |f(x+v) - f(x)| \leq |v| + (t^{-1} - 1)\,|v| = t^{-1}\,|v|.$$

Moreover the map f is open, proper and closed, hence $\operatorname{im} f = \mathbf{R}^n$.

3.1.24. Suppose X and Y are Banach spaces.

If $a \in A \subset X$ and $f\colon A \to Y$, then the following four conditions are equivalent; in case they hold we say that f is **analytic** at a:

(1) *There exist a neighborhood U of a in X and a finite number c such that $U \subset A$, $f|U$ is of class ∞ and*

$$\|D^m f(x)\| \leq c^m\, m!$$

whenever $x \in U$ and m is any positive integer.

(2) *There exist a neighborhood U of a in X and a finite number c such that $U \subset A$, $f|U$ is of class ∞ and*

$$|\langle v^m/m!, D^m f(x)\rangle| \leq c^m$$

whenever $x \in U$, $v \in X$, $|v| \leq 1$ and m is any positive integer.

(3) *There exists a neighborhood U of a in X such that $U \subset A$, $f|U$ is of class ∞ and*

$$\sup_{x\in U}\left| f(x) - \sum_{m=0}^{n} \langle (x-a)^m/m!, D^m f(a)\rangle \right| \to 0 \text{ as } n \to \infty.$$

(4) *There exist a neighborhood U of a in X and continuous homogeneous polynomial functions P_m: $X \to Y$ of degree m such that*

$$\sup_{x\in U}\left| f(x) - \sum_{m=0}^{n} P_m(x-a) \right| \to 0 \text{ as } n \to \infty.$$

Clearly (1) implies (2), and (3) implies (4).

To prove (2) implies (3) we choose $r>0$ so that $c\,r<1$ and $\mathbf{U}(a,r) \subset U$, then infer from Taylor's theorem that for $x \in \mathbf{U}(a,r)$ the norm of the difference in (3) does not exceed $(c\,r)^{n+1}$, hence (3) holds with U replaced by $\mathbf{U}(a,r)$.

To prove (4) implies (1) we choose $r>0$ so that $\mathbf{B}(a,r) \subset U$ and $|P_m(x-a)| \leq 1$ for every $x \in \mathbf{B}(a,r)$ and every positive integer m, hence (see 1.10.5)

$$r^m \, \|P_m\| \leq 1;$$

choosing $\phi_m \in \odot^m(X,Y)$ so that $P_m(v) = \langle v^m/m!, \phi_m\rangle$ whenever $v \in X$, we obtain

$$r^m \, \|\phi_m\| \leq m^m \leq e^m \, m!.$$

Next we suppose $0<s<r/(2e)$, $c = 2e/(r-2e\,s)$ and observe that, for $x \in \mathbf{U}(a,s)$ and every positive integer i,

$$\begin{aligned}
&\sum_{m=0}^{\infty} \|D^i P_m(x-a)\| = \sum_{m=i}^{\infty} \|(x-a)^{m-i}/(m-i)! \mathbin{\lrcorner} \phi_m\| \\
&\leq \sum_{m=i}^{\infty} s^{m-i} \, \|\phi_m\|/(m-i)! \leq \sum_{m=i}^{\infty} s^{m-i} r^{-m} e^m \, m!/(m-i)! \\
&\leq \sum_{m=i}^{\infty} s^{m-i} r^{-m} e^m \, 2^m \, i! = (2e/r)^i \, i!/[1-(2e\,s/r)] \\
&= (2e/r)^{i-1} \, c\, i! \leq c^i \, i!.
\end{aligned}$$

From this estimate it readily follows by induction with respect to i that

$$D^i f(x) = \sum_{m=0}^{\infty} D^i P_m(x-a) \quad \text{and} \quad \|D^i f(x)\| \le c^i\, i!$$

whenever $x \in \mathbf{U}(a,s)$ and i is any positive integer, hence (1) holds with U replaced by $\mathbf{U}(a,s)$.

By an **analytic function** we mean a function which is analytic at every point of its domain.

The composition of two analytic functions is analytic. In fact, if f is analytic at a, g is analytic at $f(a)$, with

$$\|D^m f(a)\| \le c^m\, m! \quad \text{and} \quad \|D^m g[f(a)]\| \le \gamma^m\, m!$$

for all positive integers m, then the formula for $D^p(g \circ f)$ derived in 3.1.11 shows that

$$\|D^p(g \circ f)(a)\| \le \sum_{\alpha \in S(p)} \gamma^{\Sigma\alpha}(\Sigma\alpha)!\, c^p\, p!/\alpha! = c^p \gamma(\gamma+1)^{p-1} p! \le [c(\gamma+1)]^p\, p!;$$

the above value for the sum over $S(p)$ may be obtained through consideration of the special case when

$$f(x) = c\,x/(1-c\,x) \text{ for } x \in \mathbf{R},\ |x| < 1/c,$$

$$g(y) = \gamma\, y/(1-\gamma\, y) \text{ for } y \in \mathbf{R},\ |y| < 1/\gamma,$$

hence $(g \circ f)x = c\,\gamma\, x/[1-c(\gamma+1)x]$ for small x in $\mathbf{R}$, the differentials of f, g, $g \circ f$ at 0 can be computed by the use of geometric series and $\|D^p(g \circ f)(0)\|$ equals the sum over $S(p)$.

If f is a map of class 1 *with an analytic inverse g, then f is analytic.* We see from 3.1.11 that f is of class ∞ and that, for $a \in \operatorname{dmn} f$ and $p \ge 2$,

$$D^p f(a) = -\sum_{\alpha \in T(p)} D^1 f(a) \circ D^{\Sigma\alpha} g[f(a)] \circ [D^1 f(a)^{\alpha_1} \odot \cdots \odot D^{p-1} f(a)^{\alpha_{p-1}}]/\alpha!$$

where $T(p)$ consists of all sequences $\alpha: \{1, \ldots, p-1\} \to \{0, \ldots, p-1\}$ with

$$\sum_{i=1}^{p-1} i\,\alpha_i = p \quad \text{and} \quad \sum_{i=1}^{p-1} \alpha_i > 1.$$

For any finite positive numbers β and γ we inductively define $\lambda_1, \lambda_2, \lambda_3, \ldots$ by the formulae $\lambda_1 = (\beta\,\gamma)^{-1}$,

$$\lambda_p = \sum_{\alpha \in T(p)} \lambda_1 \gamma^{\Sigma\alpha}(\Sigma\alpha)!/\alpha! \prod_{i=1}^{p-1} (\lambda_i)^{i\alpha_i} \text{ for } p \ge 2,$$

and observe that the hypotheses

$$\|D^1 f(a)\| \le (\beta\,\gamma)^{-1}, \quad \|D^m g[f(a)]\| \le \gamma^m\, m! \text{ for } m = 2, 3, 4, \ldots$$

imply that

$$\|D^p f(a)\| \le \lambda_p\, p! \text{ for } p=1,2,3,\ldots.$$

Moreover there exists a finite number c such that

$$\lambda_p \le c^p \text{ for } p=1,2,3,\ldots.$$

To prove this assertion we consider the special case when

$$g(y)=\beta\gamma y-\sum_{m=2}^{\infty}\gamma^m y^m=\beta\gamma y-\gamma^2 y^2(1-\gamma y)^{-1}$$

for $-1<\gamma y<1-(\beta+1)^{-\frac{1}{2}}$, invert g by solving a quadratic equation to obtain

$$f(x)=\left(x+\beta-[(x-\beta-2)^2-4(\beta+1)]^{\frac{1}{2}}\right)[2(\beta+1)\gamma]^{-1}$$

for $x<\beta+2-2(\beta+1)^{\frac{1}{2}}$, and note that this particular function f is analytic at 0 with $\|D^p f(0)\|=\lambda_p\, p!$ for all positive integers p.

If f is an analytic function whose domain is a connected open subset A of $\mathbf{R}^n$, then either $f^{-1}\{0\}=A$ or $\mathscr{L}^n[f^{-1}\{0\}]=0$. In fact

$$W=\{x: D^i f(x)=0 \text{ for } i=0,1,2,3,\ldots\}$$

is open and closed relative to A, hence either $W=A$ or $W=\varnothing$; moreover 2.9.11 implies that $\mathscr{L}^n$ almost all of $f^{-1}\{0\}$ is contained in

$$Z=A\cap\{x: \mathbf{R}^n\sim f^{-1}\{0\} \text{ has } \mathscr{L}^n \text{ density } 0 \text{ at } x\},$$

and approximate differentiation shows that $Z\subset W$, hence

$$\mathscr{L}^n[f^{-1}\{0\}]\le\mathscr{L}^n(W).$$

More precise results on the root sets of analytic functions will be obtained in 3.4.

Clearly maps of class k may be replaced by analytic maps in 3.1.18 and 3.1.19. This leads to the concept of **analytic submanifold** of $\mathbf{R}^n$. While the analytic analogue of 3.1.20 is true, a different proof is needed for the analytic case because there exist no analytic partitions of unity; one may use the analytic retraction which maps every point in a suitable neighborhood of an analytic submanifold B onto the unique nearest point in B (see [F 15, § 4]).

The concluding remarks of 1.10.4 and 1.10.5 show that in case $X=\mathbf{R}^n$ the condition (1) defining analyticity implies

$$\sum_{m=0}^{\infty}\langle (x-a)^m/m!, D^m f(a)\rangle=\sum_{m=0}^{\infty}\sum_{\alpha\in\Xi(n,m)}\left[\prod_{i=1}^{n}(x_i-a_i)^{\alpha_i}/\alpha_i!\right]D^\alpha f(a)$$

whenever $x\in\mathbf{R}^n$ with

$$\sum_{i=1}^{n}|x_i-a_i|<c^{-1},$$

and the norms of all terms in the repeated sum total at most

$$\sum_{m=0}^{\infty} c^m\, m!\left(\sum_{i=1}^{n}|x_i-a_i|\right)^m\Big/ m! < \infty .$$

Thus the coordinate free series representation (3) of f by its differentials may be replaced by a series of monomials in the coordinates with partial derivative coefficients; however the region of convergence of the second series is often smaller than that of the first (see [BOC], [KO]).

3.2. Area and coarea of Lipschitzian maps

3.2.1. Assuming that f maps a subset of $\mathbf{R}^m$ into $\mathbf{R}^n$, we recall 1.7.6 to define for each nonnegative integer k the k **dimensional Jacobian**

$$J_k f(a) = \|\wedge_k D f(a)\|$$

whenever f is differentiable at a; similarly we define

$$\operatorname{ap} J_k f(a) = \|\wedge_k \operatorname{ap} D f(a)\|$$

whenever f is approximately differentiable at a.

Usually taking $k=\inf\{m,n\}$ we will study the geometric significance of the Jacobian integral

$$\int_A J_k f(x)\, d\mathscr{L}^m x$$

for any Lipschitzian map $f\colon \mathbf{R}^m \to \mathbf{R}^n$ and any $\mathscr{L}^m$ measurable set A. In 3.2.3 we prove that in case $k=m\le n$ the Jacobian integral equals the m dimensional **Hausdorff area** of $f|A$ defined as

$$\int_{\mathbf{R}^n} N(f|A, y)\, d\mathscr{H}^m y$$

where N is the multiplicity function introduced in 2.10.9. Then we show in 3.2.11 that in case $m>n=k$ the Jacobian integral equals the **coarea** of $f|A$ defined as

$$\int_{\mathbf{R}^n} \mathscr{H}^{m-n}(A\cap f^{-1}\{y\})\, d\mathscr{L}^n y .$$

The area and coarea formulae lead to the Theorems 3.2.5 and 3.2.12 concerning transformations of integrals. In all these propositions the hypothesis that f be Lipschitzian may be replaced, in view of 3.1.8, by the weaker condition

$$\operatorname{ap}\limsup_{x\to a} |f(x)-f(a)|/|x-a| < \infty \quad \text{for } x\in A,$$

provided $J_k f$ is replaced by $\operatorname{ap} J_k f$ in the conclusions. Moreover it follows from 3.2.26, 3.2.15 that $\mathscr{H}^m$ and $\mathscr{H}^{m-n}$ may be replaced by any of the m and $m-n$ dimensional measures constructed in 2.10.2 to 2.10.5. Further generalizations will be obtained in 3.2.20, 3.2.22, 3.2.32, 3.2.46.

3.2.2. Lemma. *If $f: \mathbf{R}^m \to \mathbf{R}^n$ is continuous and $\lambda > 1$, then*

$$\{x: Df(x) \text{ is univalent}\}$$

has a countable covering G consisting of Borel sets E such that $f|E$ is univalent and there exists a linear automorphism s of $\mathbf{R}^m$ with

$$\operatorname{Lip}[(f|E)\circ s^{-1}] \le \lambda, \quad \operatorname{Lip}[s\circ(f|E)^{-1}] \le \lambda,$$

$$\lambda^{-1}|s(v)| \le |\langle v, Df(x)\rangle| \le \lambda |s(v)| \text{ for } x\in E,\ v\in\mathbf{R}^m,$$

$$\lambda^{-m}|\det(s)| \le J_m f(x) \le \lambda^m |\det(s)| \text{ for } x\in E.$$

Proof. We choose $\varepsilon > 0$ so that $\lambda^{-1}+\varepsilon < 1 < \lambda - \varepsilon$, and let S be a countable dense subset of $\mathbf{GL}(m, \mathbf{R})$. With each $s\in S$ and each positive integer i we associate the Borel set $Z(s, i)$ of $\mathbf{R}^m$ consisting of all points a such that

$$(\lambda^{-1}+\varepsilon)|s(v)| \le |Df(a)(v)| \le (\lambda-\varepsilon)|s(v)| \text{ for } v\in\mathbf{R}^m,$$

$$|f(b)-f(a)-Df(a)(b-a)| \le \varepsilon |s(b-a)| \text{ for } b\in\mathbf{B}(a, i^{-1});$$

the first of these two conditions implies

$$(\lambda^{-1}+\varepsilon)^m |\det(s)| \le J_m f(a) \le (\lambda-\varepsilon)^m |\det(s)|.$$

If $E\subset Z(s, i)$ with $\operatorname{diam} E \le i^{-1}$, then

$$|f(b)-f(a)| \le |Df(a)(b-a)| + \varepsilon |s(b-a)| \le \lambda |s(b)-s(a)|,$$

$$|f(b)-f(a)| \ge |Df(a)(b-a)| - \varepsilon |s(b-a)| \ge \lambda^{-1} |s(b)-s(a)|$$

whenever $a, b\in E$. Therefore $Z(s, i)$ has a countable covering consisting of Borel sets with the required properties.

To complete the proof we will show that each point a, for which $Df(a)$ is univalent, belongs to some $Z(s, i)$. Recalling 1.7.3 we factor

$$Df(a) = h\circ g \text{ with } g\in\mathbf{GL}(m, \mathbf{R}),\ h\in\mathbf{O}(m, n),$$

and infer that $|Df(a)(v)| = |g(v)|$ for $v\in\mathbf{R}^m$. Then we choose $s\in S$ so that

$$\|s\circ g^{-1}\| < (\lambda^{-1}+\varepsilon)^{-1} \quad \text{and} \quad \|g\circ s^{-1}\| < \lambda - \varepsilon,$$

hence $|s(v)| \le (\lambda^{-1}+\varepsilon)^{-1}|g(v)|$ and $|g(v)| \le (\lambda-\varepsilon)|s(v)|$ for $v\in\mathbf{R}^m$. Finally, since $|b-a| \le \|s^{-1}\|\cdot|s(b-a)|$, we obtain a suitable i from the definition of $Df(a)$.

3.2.3. Theorem. *Suppose $f: \mathbf{R}^m \to \mathbf{R}^n$ is Lipschitzian with $m \le n$.*
(1) *If A is an $\mathscr{L}^m$ measurable set, then*

$$\int_A J_m f(x)\, d\mathscr{L}^m x = \int_{\mathbf{R}^n} N(f|A, y)\, d\mathscr{H}^m y.$$

(2) *If u is an $\mathscr{L}^m$ integrable function, then*

$$\int_{\mathbf{R}^m} u(x)\, J_m f(x)\, d\mathscr{L}^m x = \int_{\mathbf{R}^n} \sum_{x \in f^{-1}\{y\}} u(x)\, d\mathscr{H}^m y.$$

Proof. Since (2) reduces to (1) in the special case when u is the characteristic function of A, one readily infers (2) from (1) by the usual method of approximation [2.3.3, 2.4.8, 2.4.4 (6)].

Turning to the proof of (1), we see from 2.10.11 and 2.10.35 that both integrals equal 0 in case $\mathscr{L}^m(A)=0$. In view of 3.1.6 we may henceforth assume that f is differentiable at each point of A, and that A is a Borel set with $\mathscr{L}^m(A) < \infty$.

First we consider the case when

$$A \subset \{x: Df(x) \text{ is univalent}\}.$$

For any $\lambda > 1$ we choose G according to 3.2.2 and construct a Borel partition H of A such that every member of H is contained in some member of G. If $B \in H$, $B \subset E \in G$ and s is as in 3.2.2, then

$$\lambda^{-m} |\det(s)| \cdot \mathscr{L}^m(B) \le \int_B J_m f\, d\mathscr{L}^m \le \lambda^m |\det(s)| \cdot \mathscr{L}^m(B),$$

while the equation $f(B) = [(f|B) \circ s^{-1}]\, s(B)$ implies

$$\lambda^{-m} \mathscr{H}^m[s(B)] \le \mathscr{H}^m[f(B)] \le \lambda^m \mathscr{H}^m[s(B)];$$

since $\mathscr{H}^m[s(B)] = \mathscr{L}^m[s(B)] = |\det(s)| \cdot \mathscr{L}^m(B)$ according to 2.10.35, 2.7.16 (1), it follows that

$$\lambda^{-2m} \mathscr{H}^m[f(B)] \le \int_B J_m f\, d\mathscr{L}^m \le \lambda^{2m} \mathscr{H}^m[f(B)].$$

Summing over H we obtain the inequality

$$\lambda^{-2m} \int N(f|A, y)\, d\mathscr{H}^m y \le \int_A J_m f\, d\mathscr{L}^m \le \lambda^{2m} \int N(f|A, y)\, d\mathscr{H}^m y.$$

From this we deduce the asserted equation by letting λ approach 1.

Second we consider the case when

$$A \subset \{x: \dim \ker Df(x) > 0\} = \{x: J_m f(x) = 0\}.$$

For any $\varepsilon > 0$ we factor $f = p \circ g$ where

$$g: \mathbf{R}^m \to \mathbf{R}^n \times \mathbf{R}^m, \quad g(x) = (f(x), \varepsilon x) \text{ for } x \in \mathbf{R}^m,$$

$$p: \mathbf{R}^n \times \mathbf{R}^m \to \mathbf{R}^n, \quad p(y, z) = y \text{ for } (y, z) \in \mathbf{R}^n \times \mathbf{R}^m,$$

and observe that $x \in A$ implies

$$\langle v, Dg(x)\rangle = (\langle v, Df(x)\rangle, \varepsilon v) \text{ for } v \in \mathbf{R}^m,$$
$$Dg(x) \text{ is univalent}, \quad \|Dg(x)\| \leq \operatorname{Lip}(f) + \varepsilon,$$
$$J_m g(x) \leq \varepsilon [\operatorname{Lip}(f) + \varepsilon]^{m-1}$$

because $|\langle v, Dg(x)\rangle| = \varepsilon |v|$ for $v \in \ker Df(x)$. We apply the first case to g and obtain

$$\mathscr{H}^m [f(A)] \leq \mathscr{H}^m [g(A)] = \int_A J_m g \, d\mathscr{L}^m \leq \varepsilon [\operatorname{Lip}(f) + \varepsilon]^{m-1} \mathscr{L}^m(A).$$

Letting ε approach 0 we conclude $\mathscr{H}^m [f(A)] = 0$. Thus both integrals in the asserted equation equal 0 in the second case.

3.2.4. Corollary. *For every $\mathscr{L}^m$ measurable set A there exists a Borel set*

$$B \subset A \cap \{x \colon J_m f(X) > 0\}$$

such that $f|B$ is univalent and $\mathscr{H}^m [f(A) \sim f(B)] = 0$.

Proof. In case $\mathscr{L}^m(A) = 0$ we take $B = \varnothing$.

In case A is a Borel set we use 3.2.2 to obtain Borel sets E_i such that $f|E_i$ is univalent for $i = 1, 2, 3, \ldots$ and

$$P = A \cap \{x \colon J_m f(x) > 0\} \subset \bigcup_{i=1}^{\infty} E_i,$$

then construct the Borel sets

$$F_i = P \cap E_i \sim \bigcup_{j=1}^{i-1} f^{-1}[f(P \cap E_j)] \text{ for } i = 1, 2, 3, \ldots,$$
$$B = \bigcup_{i=1}^{\infty} F_i,$$

and conclude that $f|B$ is univalent, $f(B) = f(P)$ and

$$\mathscr{H}^m [f(A) \sim f(B)] \leq \mathscr{H}^m [f(A \sim P)] \leq \int_{A \sim P} J_m f \, d\mathscr{L}^m = 0.$$

3.2.5. Theorem. *If $f \colon \mathbf{R}^m \to \mathbf{R}^n$ is Lipschitzian and $m \leq n$, then*

$$\int_A g[f(x)] \, J_m f(x) \, d\mathscr{L}^m x = \int_{\mathbf{R}^n} g(y) \, N(f|A, y) \, d\mathscr{H}^m y$$

whenever A is an $\mathscr{L}^m$ measurable set, $g \colon \mathbf{R}^n \to \overline{\mathbf{R}}$ and

either (1) *g is $\mathscr{H}^m$ measurable,*

or (2) *$N(f|A, y) < \infty$ for $\mathscr{H}^m$ almost all y,*

or (3) *$\alpha \cdot (g \circ f) \cdot J_m f$ is $\mathscr{L}^m$ measurable, where α is the characteristic function of A.*

Proof. In view of 3.2.3 we may assume that A is the union of a countable family of compact sets, hence $f(A)$ is a Borel set, and that

$$g(y)=0 \text{ for } y\in\mathbf{R}^n\sim f(A).$$

First we consider the case when g is the characteristic function of an $\mathscr{H}^m$ measurable subset T of $f(A)$. If T is a Borel set, then 3.2.3 is applicable to $A\cap f^{-1}(T)$, hence

$$\begin{aligned}\int_A (g\circ f)\, J_m f\, d\mathscr{L}^m &= \int_{A\cap f^{-1}(T)} J_m f\, d\mathscr{L}^m \\ &= \int N[f|A\cap f^{-1}(T), y]\, d\mathscr{H}^m y = \int g(y)\, N(f|A,y)\, d\mathscr{H}^m y.\end{aligned}$$

In particular, if T is a Borel set with $\mathscr{H}^m(T)=0$, then both integrals in the asserted equation equal 0. The Borel regularity of $\mathscr{H}^m$ and monotonicity imply that both integrals equal 0 if T is any subset of $f(A)$ with $\mathscr{H}^m(T)=0$. Moreover, since $f(A)$ is countably $\mathscr{H}^m$ measurable, every $\mathscr{H}^m$ measurable subset of $f(A)$ is the union of a Borel set and a set with $\mathscr{H}^m$ measure 0.

Now it is easy to verify by the usual approximation procedures [2.3.3, 2.4.8, 2.4.4 (6)] that (1) is sufficient for the validity of the asserted equation.

Next we observe that the conjunction of (2) with $\mathscr{H}^m$ measurability of the right integrand implies (1); in fact

$$g(y)=g(y)\, N(f|A,y)\cdot\sum_{i=1}^{\infty} h_i(y)/i$$

whenever $N(f|A,y)<\infty$, where h_i is the characteristic function of $\{y: N(f|A,y)=i\}$.

Finally we will prove that (3) implies (1). Choosing B as in 3.2.4, with the characteristic function β, we deduce from (3) that $\beta\cdot(g\circ f)$ is $\mathscr{L}^m$ measurable. Therefore, if V is any open subset of $\overline{\mathbf{R}}\sim\{0\}$, then

$$W=B\cap(g\circ f)^{-1}V \text{ is } \mathscr{L}^m \text{ measurable},$$

$$f(W)\subset g^{-1}(V)\subset f(W)\cup[f(A)\sim f(B)],$$

and it follows from 3.2.4 that $f(W)$ and $g^{-1}(V)$ are $\mathscr{H}^m$ measurable.

3.2.6. Theorem. *If $f:\mathbf{R}\to\mathbf{R}^n$ is absolutely continuous on every compact interval, then*

$$\int_A (g\circ f)\cdot|f'|\, d\mathscr{L}^1=\int_{\mathbf{R}^n} g(y)\, N(f|A,y)\, d\mathscr{H}^1 y$$

for every function $g:\mathbf{R}^n\to\overline{\mathbf{R}}$ and every bounded $\mathscr{L}^1$ measurable set A. Moreover, in case $n=1$, then

$$\int_a^b (g\circ f)\cdot f'\, d\mathscr{L}^1=\int_{f(a)}^{f(b)} g\, d\mathscr{L}^1$$

whenever $-\infty<a<b<\infty$ and $g(y)\,N(f|\{x: a\le x\le b\},y)$ is $\mathscr{L}^1$ summable with respect to y.

Proof. The first equation follows from 3.2.5(2) and 3.1.8 in case $A \subset \operatorname{dmn} f'$, and from 2.10.13 in case $\mathscr{L}^1(A)=0$.

To prove the second equation we assume $f(a)<f(b)$ and consider the three sets

$$A_1=\{x: a<x<b, f'(x)>0\}, \quad A_2=\{x: a<x<b, f'(x)<0\},$$
$$A_3=\{x: a\le x\le b\}\sim(A_1\cup A_2).$$

Applying the first equation to $g: \mathbf{R}\to\{1\}$, we find that $\mathscr{L}^1$ almost all real numbers y satisfy the conditions

$$N(f|A_1\cup A_2, y)<\infty \quad \text{and} \quad N(f|A_3, y)=0;$$

then we observe that under these conditions

$$N(f|A_1, y)-N(f|A_2, y)$$

equals 1 if $f(a)<y<f(b)$, and equals 0 otherwise. Finally we apply the first equation to any g such that $g(y)N(f|A_1\cup A_2, y)$ is $\mathscr{L}^1$ integrable with respect to y, and obtain

$$\begin{aligned}\textstyle\int_{f(a)}^{f(b)} g\,d\mathscr{L}^1 &= \textstyle\int g(y)N(f|A_1, y)\,d\mathscr{L}^1 y-\int g(y)N(f|A_2, y)\,d\mathscr{L}^1 y\\ &= \textstyle\int_{A_1}(g\circ f)\cdot f'\,d\mathscr{L}^1-\int_{A_2}(g\circ f)\cdot(-f')\,d\mathscr{L}^1\\ &= \textstyle\int_{A_1\cup A_2}(g\circ f)\cdot f'\,d\mathscr{L}^1=\int_a^b(g\circ f)\cdot f'\,d\mathscr{L}^1.\end{aligned}$$

3.2.7. The need for the alternative conditions (1), (2), (3) in 3.2.5 is shown by the example where $m=n=1$, $f=\sin$, $A=\mathbf{R}$, c is the characteristic function of an $\mathscr{L}^1$ nonmeasurable subset of $\{y: -1\le y\le 1\}$ and $g=1+c$; in this case the right integral equals ∞, but the left integrand is $\mathscr{L}^1$ nonmeasurable.

We also observe that $g\circ f$ may be $\mathscr{L}^m$ nonmeasurable even though $(g\circ f)\cdot J_m f$ is $\mathscr{L}^m$ measurable. Consider for instance the case when $m=n=2$, $A=\mathbf{R}^2$,

$$f(x)=(x_1, 0) \text{ and } g(x)=h(x_1) \text{ for } x\in\mathbf{R}^2,$$

where h is the characteristic function of an $\mathscr{L}^1$ nonmeasurable set; here both integrals equal 0.

The second equation in 3.2.6 does not hold for every $\mathscr{L}^1$ summable function g, as shown by the example where $a=0$, $b=1$ and

$$f(x)=|x|^{\frac{3}{2}}\sin(x^{-1}) \text{ for } 0\neq x\in\mathbf{R},\ f(0)=0,$$
$$g(y)=|y|^{-\frac{2}{3}} \text{ for } 0<|y|\le 1, \quad g(y)=0 \text{ for } |y|>1.$$

Finally we note that the conclusion of 3.2.3(2) fails when $m=n=1$, $f(x)=|x|$ and $u(x)=x$ for $x\in\mathbf{R}$.

3.2.8. Lemma. *If* $u: \mathbf{R}^m \to \mathbf{R}^m$, $v: \mathbf{R}^m \to \mathbf{R}^m$ *are Lipschitzian maps and* $C = \{x: v[u(x)] = x\}$, *then*

$$Dv[u(x)] \circ Du(x) = \mathbf{1}_{\mathbf{R}^m} \text{ for } \mathscr{L}^m \text{ almost all } x \text{ in } C.$$

Proof. If $S = C \cap \operatorname{dmn} Du \cap u^{-1}(\operatorname{dmn} Dv)$, then

$$C \sim S \subset (\mathbf{R}^m \sim \operatorname{dmn} Du) \cup v(\mathbf{R}^m \sim \operatorname{dmn} Dv),$$

hence $\mathscr{L}^m(C \sim S) = 0$ by 3.1.6, 2.10.35, 2.10.11; moreover

$$Dv[u(x)] \circ Du(x) = D(v \circ u)(x) \text{ for } x \in S.$$

On the other hand $v \circ u$ has the approximate differential $\mathbf{1}_{\mathbf{R}^m}$ wherever $\mathbf{R}^m \sim C$ has $\mathscr{L}^m$ density 0, which holds $\mathscr{L}^m$ almost everywhere in C by 2.9.11.

3.2.9. Lemma. *If* $f: \mathbf{R}^m \to \mathbf{R}^n$ *is continuous, then*

$$\{x: \operatorname{im} Df(x) = \mathbf{R}^n\}$$

has a countable covering G *consisting of Borel sets* E *such that there exists a projection* $p \in \mathbf{O}^*(m, m-n)$ *and Lipschitzian maps*

$$u: \mathbf{R}^m \to \mathbf{R}^n \times \mathbf{R}^{m-n} \quad \text{and} \quad v: \mathbf{R}^n \times \mathbf{R}^{m-n} \to \mathbf{R}^m$$

with

$$u(x) = (f(x), p(x)) \text{ and } v[u(x)] = x \text{ for } x \in E.$$

Proof. Recalling 1.7.4 we associate with each $\lambda \in \Lambda(m, m-n)$ the map

$$u_\lambda: \mathbf{R}^m \to \mathbf{R}^n \times \mathbf{R}^{m-n}, \quad u_\lambda(x) = (f(x), \mathbf{p}_\lambda(x)) \text{ for } x \in \mathbf{R}^m,$$

define $A_\lambda = \{x: Du_\lambda(x) \text{ is univalent}\}$, and note that

$$\ker Du_\lambda(x) = \ker Df(x) \cap \ker \mathbf{p}_\lambda$$

whenever f is differentiable at x. It follows from 1.7.4 that

$$\{x: \operatorname{im} Df(x) = \mathbf{R}^n\} = \{x: \dim \ker Df(x) = m-n\} = \bigcup_{\lambda \in \Lambda(m, m-n)} A_\lambda.$$

For each λ we apply 3.2.2 with f replaced by u_λ to cover A_λ by a countable family G_λ of Borel sets E such that $u_\lambda | E$ is univalent and $(u_\lambda | E)^{-1}$ is Lipschitzian, hence by 2.10.43 there exists a Lipschitzian map

$$v: \mathbf{R}^n \times \mathbf{R}^{m-n} \to \mathbf{R}^m \quad \text{with} \quad v | u_\lambda(E) = (u_\lambda | E)^{-1}.$$

3.2.10. Lemma. *If $f\colon \mathbf{R}^m \to \mathbf{R}^n$ is Lipschitzian, $m>n$, and E, p, u, v are as in 3.2.9 then*

$$\int_B J_n f(x)\, d\mathscr{L}^m x = \int_{\mathbf{R}^n} \mathscr{H}^{m-n}(B \cap f^{-1}\{y\})\, d\mathscr{L}^n y$$

for every $\mathscr{L}^m$ measurable subset B of E, and

$$E \cap f^{-1}\{y\} = v[\{y\} \times p(E \cap f^{-1}\{y\})] \text{ for } y \in \mathbf{R}^n.$$

Proof. We note that $v|u(E)$ is univalent and

$$u(E \cap f^{-1}\{y\}) = \{y\} \times p(E \cap f^{-1}\{y\}) \text{ for } y \in \mathbf{R}^n;$$

defining $v_y\colon \mathbf{R}^{m-n} \to \mathbf{R}^m$, $v_y(z) = v(y,z)$ for $z \in \mathbf{R}^{m-n}$, we see that v_y maps $p(E \cap f^{-1}\{y\})$ univalently onto $E \cap f^{-1}\{y\}$.

From 3.2.8 we know that, for $\mathscr{L}^m$ almost all x in E,

$$D u(x)^{-1} = D v[u(x)],$$

and we use the commutative diagram

$$\begin{array}{ccccc}
\mathbf{R}^{m-n} \simeq \{0\} \times \mathbf{R}^{m-n} & \xrightarrow{\subset} & \mathbf{R}^n \times \mathbf{R}^{m-n} & \xrightarrow{q} & \mathbf{R}^n \\
\downarrow{\scriptstyle L_x} & & {\scriptstyle Du(x)}\uparrow & \nearrow{\scriptstyle Df(x)} & \uparrow \\
\operatorname{im} L_x & \xrightarrow{\subset} & \mathbf{R}^m & \xleftarrow{\supset} & W
\end{array}$$

where $q(y,z) = y$ for $(y,z) \in \mathbf{R}^n \times \mathbf{R}^{m-n}$, $L_x = D\, v_{f(x)}[p(x)]$ and W is the orthogonal complement of

$$\operatorname{im} L_x = D u(x)^{-1}(\{0\} \times \mathbf{R}^{m-n}) = \ker D f(x),$$

to compute

$$J_m u(x) = \|\textstyle\bigwedge_{m-n}(L_x)\|^{-1}\, \|\textstyle\bigwedge_n [Df(x)|W]\| = J_{m-n}\, v_{f(x)}[p(x)]^{-1} J_n f(x).$$

Applying 3.2.5, 2.6.2, 3.2.3 we find that

$$\begin{aligned}
\int_B J_n f(x)\, d\mathscr{L}^m x &= \int_B J_{m-n}\, v_{f(x)}[p(x)]\, J_m u(x)\, d\mathscr{L}^m x \\
&= \int_{u(B)} J_{m-n}\, v_y(z)\, d(\mathscr{L}^n \times \mathscr{L}^{m-n})(y,z) \\
&= \int_{\mathbf{R}^n} \int_{p(B \cap f^{-1}\{y\})} J_{m-n}\, v_y(z)\, d\mathscr{L}^{n-m} z\, d\mathscr{L}^n y \\
&= \int_{\mathbf{R}^n} \mathscr{H}^{m-n}(B \cap f^{-1}\{y\})\, d\mathscr{L}^n y.
\end{aligned}$$

3.2.11. Theorem. *If $f\colon \mathbf{R}^m \to \mathbf{R}^n$ is Lipschitzian and $m>n$, then*

$$\int_A J_n f(x)\, d\mathscr{L}^m x = \int_{\mathbf{R}^n} H^{m-n}(A \cap f^{-1}\{y\})\, d\mathscr{L}^n y$$

for every $\mathscr{L}^m$ measurable set A.

Proof. From 2.10.25, 2.10.35 we know that both integrals equal 0 in case $\mathscr{L}^m(A)=0$. In view of 3.1.6 we may henceforth assume that f is differentiable at each point of A, and that A is a Borel set with $\mathscr{L}^m(A)<\infty$.

First we consider the case when

$$A\subset\{x\colon \operatorname{im} Df(x)=\mathbf{R}^n\}.$$

Choosing G according to 3.2.9 we construct a Borel partition H of A such that every member of H is contained in some member of G, we apply 3.2.10 to each $B\in H$, and sum over H.

Second we consider the case when

$$A\subset\{x\colon \dim\ker \textstyle\bigwedge^1 Df(x)>0\}=\{x\colon J_n f(x)=0\}.$$

For any $\varepsilon>0$ we define the maps

$$g\colon \mathbf{R}^m\times\mathbf{R}^n\to\mathbf{R}^n \quad\text{and}\quad p\colon \mathbf{R}^m\times\mathbf{R}^n\to\mathbf{R}^n,$$

$$g(x,z)=f(x)+\varepsilon z \text{ and } p(x,z)=z \text{ for } (x,z)\in\mathbf{R}^m\times\mathbf{R}^n,$$

and observe that $(x,z)\in A\times\mathbf{R}^n$ implies

$$\langle(v,w),Dg(x,z)\rangle=\langle v,Df(x)\rangle+\varepsilon w \text{ for } (v,w)\in\mathbf{R}^m\times\mathbf{R}^n,$$

$$\operatorname{im} Dg(x,z)=\mathbf{R}^n,\quad \|Dg(x,z)\|\le \operatorname{Lip}(f)+\varepsilon,$$

$$J_n g(x,z)=\|\textstyle\bigwedge^n Dg(x,z)\|\le\varepsilon[\operatorname{Lip}(f)+\varepsilon]^{n-1}$$

because $|\alpha\circ Dg(x,z)|=\varepsilon|\alpha|$ for $\alpha\in\ker\bigwedge^1 Df(x)$. Applying the first case with f, A replaced by g, $Q=A\times\mathbf{B}(0,1)\subset\mathbf{R}^m\times\mathbf{R}^n$ and using 2.10.25, 2.10.26, 3.1.3(3), 2.6.2 we obtain

$$\begin{aligned}
\varepsilon[\operatorname{Lip}(f)+\varepsilon]^{n-1}\mathscr{L}^n(A)\,\boldsymbol{\alpha}(n)&\ge\textstyle\int_Q J_n g\, d(\mathscr{L}^m\times\mathscr{L}^n)\\
&=\textstyle\int_{\mathbf{R}^n}\mathscr{H}^m(Q\cap g^{-1}\{y\})\,d\mathscr{L}^n y\\
&\ge\textstyle\int_{\mathbf{R}^n} c\int_{\mathbf{R}^n}\mathscr{H}^{m-n}(Q\cap g^{-1}\{y\}\cap p^{-1}\{w\})\,d\mathscr{L}^n w\,d\mathscr{L}^n y\\
&=c\textstyle\int_{\mathbf{R}^n}\int_{\mathbf{B}(0,1)}\mathscr{H}^{m-n}(A\cap f^{-1}\{y-\varepsilon w\})\,d\mathscr{L}^n w\,d\mathscr{L}^n y\\
&=c\textstyle\int_{\mathbf{B}(0,1)}\int_{\mathbf{R}^n}\mathscr{H}^{m-n}(A\cap f^{-1}\{y-\varepsilon w\})\,d\mathscr{L}^n y\,d\mathscr{L}^n w\\
&=c\,\boldsymbol{\alpha}(n)\textstyle\int_{\mathbf{R}^n}\mathscr{H}^{m-n}(A\cap f^{-1}\{y\})\,d\mathscr{L}^n y
\end{aligned}$$

where $c=\boldsymbol{\alpha}(m)\,\boldsymbol{\alpha}(m-n)^{-1}\,\boldsymbol{\alpha}(n)^{-1}$. Letting ε approach 0 we conclude that the right integral in the asserted equation equals 0 in the second case, as does the left integral.

3.2.12. Theorem. *If $f\colon\mathbf{R}^m\to\mathbf{R}^n$ is Lipschitzian and $m>n$, then*

$$\int_{\mathbf{R}^m} g(x)\,J_n f(x)\,d\mathscr{L}^m x=\int_{\mathbf{R}^n}\int_{f^{-1}\{y\}} g(x)\,d\mathscr{H}^{m-n}x\,d\mathscr{L}^n y$$

for every $\mathscr{L}^m$ integrable $\overline{\mathbf{R}}$ valued function g.

Proof. By the usual approximation procedures [2.3.3, 2.4.8, 2.4.4 (6)] the problem is easily reduced to the special case when g is the characteristic function of an $\mathscr{L}^m$ measurable set A, but then the asserted equation is clearly equivalent to the equation proved in 3.2.11.

3.2.13. Applying 3.2.12 to the special case when

$$f: \mathbf{R}^m \to \mathbf{R}, \quad f(x) = |x| \text{ for } x \in \mathbf{R}^m,$$

$$\langle v, Df(x) \rangle = v \bullet x/|x|, \; J_1 f(x) = 1 \text{ for } 0 \neq x \in \mathbf{R}^m, \; v \in \mathbf{R}^m,$$

we obtain the formula

$$\int g \, d\mathscr{L}^m = \int_0^\infty \int_{\{x: |x| = r\}} g \, d\mathscr{H}^{m-1} \, d\mathscr{L}^1 r$$
$$= \int_0^\infty r^{m-1} \int_{\mathbf{S}^{m-1}} g(r u) \, d\mathscr{H}^{m-1} u \, d\mathscr{L}^1 r$$

where

$$\mathbf{S}^{m-1} = \mathbf{R}^m \cap \{u: |u| = 1\},$$

because the map sending u into $r u$ changes $\mathscr{H}^{m-1}$ by the factor r^{m-1}. In particular, letting g be the characteristic function of $\mathbf{B}(0,1)$, we find that

$$\alpha(m) = \int_0^1 r^{m-1} \mathscr{H}^{m-1}(\mathbf{S}^{m-1}) \, d\mathscr{L}^1 r = \mathscr{H}^{m-1}(\mathbf{S}^{m-1})/m.$$

Next we derive some properties of Euler's function Γ, which may be defined by the formula

$$\Gamma(s) = \int_0^\infty \exp(-x) \, x^{s-1} \, d\mathscr{L}^1 x \text{ for } 0 < s < \infty.$$

Integration by parts shows that $\Gamma(s+1) = s\,\Gamma(s)$. Transforming the integral by the function mapping $y \in \mathbf{R}$ onto y^2, one obtains

$$\Gamma(s) = \int_{-\infty}^\infty \exp(-y^2) \, |y|^{2s-1} \, d\mathscr{L}^1 y.$$

Whenever $z \in \mathbf{R}^m$ with $z_j > 0$ for $j = 1, \ldots, m$ it follows from Fubini's theorem and the result of the preceding paragraph that

$$\prod_{j=1}^m \Gamma(z_j) = \int_{\mathbf{R}^m} \exp(-|x|^2) \prod_{j=1}^m |x_j|^{2z_j - 1} \, d\mathscr{L}^m x$$
$$= \int_0^\infty \exp(-r^2) \, r^{2\Sigma z - 1} \, 2B(z) \, d\mathscr{L}^1 r,$$

where B is the function defined by the formula

$$2B(z) = \int_{\mathbf{S}^{m-1}} \prod_{j=1}^m |u_j|^{2z_j - 1} \, d\mathscr{H}^{m-1} u;$$

we conclude that

$$\prod_{j=1}^m \Gamma(z_j) = \Gamma\Big(\sum_{j=1}^m z_j\Big) \cdot B(z).$$

In particular we obtain

$$\mathscr{H}^{m-1}(\mathbf{S}^{m-1})=2B(\tfrac{1}{2},\ldots,\tfrac{1}{2})=2\Gamma(\tfrac{1}{2})^m/\Gamma(m/2),$$
$$\alpha(m)=\Gamma(\tfrac{1}{2})^m/[(m/2)\,\Gamma(m/2)]=\Gamma(\tfrac{1}{2})^m/\Gamma[(m/2)+1],$$
$$\int_{\mathbf{S}^{m-1}}|u_1|^t\,d\mathscr{H}^{m-1}u=2B[(t+1)/2,\tfrac{1}{2},\ldots,\tfrac{1}{2}]$$
$$=2\Gamma[(t+1)/2]\,\Gamma(\tfrac{1}{2})^{m-1}/\Gamma[(t+m)/2]$$

for $t>-1$. To compute $\beta_t(m,1)$ in case $1\le t<\infty$ [see 2.7.16 (6)] we use the fact that $\mathbf{S}^{m-1}$ is a homogeneous space of $\mathbf{O}(m)$; choosing $c\in\mathbf{S}^{m-1}$ we let

$$A^c(g)=g(c)\ \text{for}\ g\in\mathbf{O}(m),$$

and infer from 2.7.11 (2) that the measure $\mathscr{H}^{m-1}$ over $\mathbf{S}^{m-1}$ is a constant multiple of $A^c_{\#}(\theta_m)$, because both are Radon measures invariant under the action of $\mathbf{O}(m)$; defining $q\in\mathbf{O}^*(m,1)$ by the formula $q(u)=u_1$ for $u\in\mathbf{R}^m$, we conclude

$$\int_{\mathbf{S}^{m-1}}|u_1|^t\,d\mathscr{H}^{m-1}u=\mathscr{H}^{m-1}(\mathbf{S}^{m-1})\int_{\mathbf{O}(m)}|q[A^c(g)]|^t\,d\theta_m g$$
$$=\mathscr{H}^{m-1}(\mathbf{S}^{m-1})\cdot\beta_t(m,1)^t,$$

hence

$$\beta_t(m,1)^t=\frac{\Gamma[(t+1)/2]\cdot\Gamma[m/2]}{\Gamma[(t+m)/2]\cdot\Gamma[\tfrac{1}{2}]}.$$

Furthermore we see from 2.7.16 (6) that

$$\beta_t(n,m)/\beta_t(n-1,m)=\beta_t(n,1)/\beta_t(n-m,1)$$

whenever $n>m\ge 1$. Since $\beta_t(m,m)=1$ one readily verifies by induction with respect to n that

$$\beta_t(n,m)^t=\prod_{j=m+1}^{n}\frac{\Gamma[j/2]\cdot\Gamma[(t+j-m)/2]}{\Gamma[(j-m)/2]\cdot\Gamma[(t+j)/2]}$$

for all integers $n\ge m\ge 1$. In particular

$$\beta_1(n,m)=\frac{\Gamma[(m+1)/2]\cdot\Gamma[(n-m+1)/2]}{\Gamma[(n+1)/2]\cdot\Gamma[\tfrac{1}{2}]}.$$

3.2.14. We shall use the following terminology when E is a subset of a metric space X and m is a positive integer:

(1) E is m **rectifiable** if and only if there exists a Lipschitzian function mapping some bounded subset of $\mathbf{R}^m$ onto E.

(2) E is **countably** m **rectifiable** if and only if E equals the union of some countable family whose members are m rectifiable.

(3) E is **countably** (ϕ,m) **rectifiable** if and only if ϕ measures X and there exists a countably m rectifiable set containing ϕ almost all of E.

(4) E is (ϕ, m) **rectifiable** if and only if E is countably (ϕ, m) rectifiable and $\phi(E) < \infty$.

We observe that if S and T are m rectifiable sets, so are Clos S and $S \cup T$. In case all closed subsets of X are ϕ measurable and $\phi(E) < \infty$, E is (ϕ, m) rectifiable if and only if for each $\varepsilon > 0$ there exists an m rectifiable set F with $\phi(E \sim F) < \varepsilon$.

These notions are particularly useful when $X = \mathbf{R}^n$ and $\phi = \mathscr{H}^m$. We will show in 3.2.19 that the tangential properties of $(\mathscr{H}^m, m)$ rectifiable sets generalize those of m dimensional submanifolds of class 1 (see also 3.2.29). In 3.2.20, 22, 31, 32 we will obtain transformation formulae for integrals over Hausdorff rectifiable sets.

We say that E is **purely** (ϕ, m) **unrectifiable** if and only if ϕ measures X and E contains no m rectifiable set F with $\phi(F) > 0$.

If all closed subsets of X are ϕ measurable and $\phi(A) < \infty$, then there exists a countably m rectifiable Borel set B such that $A \sim B$ is purely (ϕ, m) unrectifiable. In fact, such a set B can be constructed by maximizing $\phi \llcorner A$ on the class of all countably m rectifiable Borel subsets of X.

The tangential and projection properties of purely (ϕ, m) unrectifiable subsets of $\mathbf{R}^n$ will be studied in 3.3.

A useful extension of these notions to the case $m = 0$ is obtained through replacement of (1) by the convention that "0 rectifiable set" means "finite set".

3.2.15. Theorem. *If $f: \mathbf{R}^m \to \mathbf{R}^n$ is Lipschitzian and $m > n$, then $f^{-1}\{y\}$ is countably $(\mathscr{H}^{m-n}, m-n)$ rectifiable for $\mathscr{L}^n$ almost all y.*

Proof. Letting $P = \{x: \operatorname{im} Df(x) = \mathbf{R}^n\} = \{x: J_n f(x) > 0\}$ we see from 3.2.9, 3.2.10 that $P \cap f^{-1}\{y\}$ is countably $m-n$ rectifiable for all $y \in \mathbf{R}^n$, and from 3.2.11 that

$$\mathscr{H}^{m-n}[(\mathbf{R}^m \sim P) \cap f^{-1}\{y\}] = 0 \text{ for } \mathscr{L}^n \text{ almost all } y.$$

3.2.16. Assuming ϕ measures a normed vectorspace X, m is a positive integer and $a \in X$, we define the cone

$$\operatorname{Tan}^m(\phi, a) = \bigcap \{\operatorname{Tan}(S, a): S \subset X,\ \Theta^m(\phi \llcorner X \sim S, a) = 0\},$$

whose elements are called (ϕ, m) **approximate tangent vectors** at a. Letting

$$\mathbf{E}(a, v, \varepsilon) = X \cap \{x: |r(x-a) - v| < \varepsilon \text{ for some } r > 0\}$$

whenever $v \in X$ and $\varepsilon > 0$, one readily verifies that $v \in \operatorname{Tan}^m(\phi, a)$ if and only if

$$\Theta^{*m}[\phi \llcorner \mathbf{E}(a, v, \varepsilon), a] > 0 \text{ for every } \varepsilon > 0.$$

If C is any compact subset of $X \sim \operatorname{Tan}^m(\phi, a)$ and

$$T = \{a + r\,v \colon r > 0,\ v \in C\},$$

then $\Theta^m(\phi \llcorner T, a) = 0$; in fact $\{a + v \colon v \in C\}$ can be covered by a finite family of sets $\mathbf{E}(a, v, \varepsilon)$ with $\Theta^m[\phi \llcorner \mathbf{E}(a, v, \varepsilon), a] = 0$, and this family also covers T.

In case X is an inner product space we also define the cone

$$\operatorname{Nor}^m(\phi, a) = X \cap \{u \colon u \bullet v \leq 0 \text{ for } v \in \operatorname{Tan}^m(\phi, a)\},$$

whose elements are called **(ϕ, m) approximate normal vectors** at a.

Now suppose f maps a subset of X into another normed vectorspace Y. We say that f is **(ϕ, m) approximately differentiable** at a if and only if there exists a neighborhood U of a in X and a map $g \colon U \to Y$ such that g is differentiable at a and

$$\Theta^m[\phi \llcorner \{x \colon f(x) \neq g(x)\}, a] = 0.$$

This implies that f is differentiable relative to $\{x \colon f(x) = g(x)\}$ at a, and that f determines the restriction of $D\,g(a)$ to

$$\operatorname{Tan}[\{x \colon f(x) = g(x)\}, a] \supset \operatorname{Tan}^m(\phi, a);$$

we then call

$$D\,g(a) | \operatorname{Tan}^m(\phi, a)$$

the **(ϕ, m) approximate differential** of f at a, denoted

$$(\phi, m) \operatorname{ap} D\,f(a),$$

or more briefly $\operatorname{ap} D\,f(a)$ when ϕ and m are clear from context.

We note that *f is (ϕ, m) approximately differentiable at a if and only if there exist $\eta \in Y$ and a continuous linear map $\zeta \colon X \to Y$ such that*

$$\Theta^m[\phi \llcorner X \sim \{x \colon |f(x) - \eta - \zeta(x-a)| \leq \varepsilon\,|x-a|\}, a] = 0$$

for every $\varepsilon > 0$. In fact if this condition holds and

$$S_i = \{x \colon |f(x) - \eta - \zeta(x-a)| \leq |x-a|/i\} \text{ for } i = 1, 2, 3, \ldots,$$

one can choose $\delta_i > 0$ so that

$$\phi[\mathbf{B}(a, r) \sim S_i] \leq r^m\, 2^{-i} \text{ whenever } 0 < r \leq \delta_i,$$

and so that $\delta_{i+1} < \delta_i$; letting

$$T = \bigcup_{i=1}^{\infty} [S_i \cap \mathbf{B}(a, \delta_i) \sim \mathbf{B}(a, \delta_{i+1})]$$

one readily verifies that $\Theta^m(\phi \llcorner X \sim T, a) = 0$ and f is differentiable relative to T at a.

Therefore, if $X=\mathbf{R}^m$, then $(\mathscr{L}^m, m)$ approximate differentiability coincides with the concept introduced in 3.1.2. However, in the following theorems we will study the case when $X=\mathbf{R}^n$, $n>m$ and $\phi=\mathscr{H}^m \llcorner W$ for some $(\mathscr{H}^m, m)$ rectifiable subset W of $\mathbf{R}^n$.

3.2.17. Lemma. *If $\psi\colon \mathbf{R}^m \to \mathbf{R}^n$, $a\in K\subset \mathbf{R}^m$, $\mathbf{R}^m \sim K$ has $\mathscr{L}^m$ density 0 at a, ψ has a univalent differential at a, $\psi|K$ is univalent, $1<\lambda<\infty$ and λ is a Lipschitz constant for $\psi|K$ and $(\psi|K)^{-1}$, then*

$$\lambda^{-2m} \le \Theta^m_*[\mathscr{H}^m \llcorner \psi(K), \psi(a)] \le \Theta^{*m}[\mathscr{H}^m \llcorner \psi(K), \psi(a)] \le \lambda^{2m},$$
$$\operatorname{Tan}^m[\mathscr{H}^m \llcorner \psi(K), \psi(a)] = \operatorname{Tan}[\psi(K), \psi(a)] = \operatorname{im} D\psi(a).$$

Moreover, if $F\colon \mathbf{R}^n \to \mathbf{R}^\nu$ and $F\circ\psi$ is differentiable at a, then F is differentiable relative to $\psi(K)$ at $\psi(a)$ with

$$D[F|\psi(K)][\psi(a)] \circ D\psi(a) = D(F\circ\psi)(a).$$

Proof. We assume $a=0$, $\psi(a)=0$ and abbreviate $D\psi(a)=L$. First we observe that

$$K\cap \mathbf{B}(0,\delta/\lambda) \subset \psi^{-1}[\psi(K)\cap \mathbf{B}(0,\delta)], \quad \psi(K)\cap \mathbf{B}(0,\delta) \subset \psi[K\cap \mathbf{B}(0,\delta\lambda)],$$

hence

$$\lambda^{-2m}\frac{\mathscr{L}^m[K\cap \mathbf{B}(0,\delta/\lambda)]}{\alpha(m)(\delta/\lambda)^m} \le \frac{\mathscr{H}^m[\psi(K)\cap \mathbf{B}(0,\delta)]}{\alpha(m)\,\delta^m}$$
$$\le \lambda^{2m}\frac{\mathscr{L}^m[K\cap \mathbf{B}(0,\delta\lambda)]}{\alpha(m)(\delta\lambda)^m}$$

whenever $0<\delta<\infty$, and infer the first conclusion because K has $\mathscr{L}^m$ density 1 at a. Next we see from 3.2.16 and 3.1.21 that

$$\operatorname{Tan}^m[\mathscr{H}^m \llcorner \psi(K), 0] \subset \operatorname{Tan}[\psi(K), 0] \subset \operatorname{im} L.$$

To prove the opposite inclusions we suppose $v\in\mathbf{R}^m$ and $\varepsilon>0$; choosing

$$\eta>0 \quad\text{and}\quad \zeta>0 \quad\text{with}\quad (|v|+\eta)\cdot\zeta + \|L\|\cdot\eta \le \varepsilon,$$

we note that the condition

$$|\psi(x)-L(x)| \le \zeta\,|x| \ \text{ for } x\in\mathbf{R}^m\cap \mathbf{B}(0,\delta/\lambda)$$

holds whenever δ is a sufficiently small positive number; it implies

$$\psi[\mathbf{E}(0,v,\eta)\cap \mathbf{B}(0,\delta/\lambda)] \subset \mathbf{E}[0, L(v), \varepsilon]$$

because, if $x\in\mathbf{R}^m$, $r>0$, $|x|\le\delta/\lambda$ and $|r\,x-v|<\eta$, then

$$|r\,\psi(x)-L(v)|\le r\,|\psi(x)-L(x)|+|L(r\,x-v)|$$
$$\le|r\,x|\,\zeta+\|L\|\cdot|r\,x-v|<(|v|+\eta)\,\zeta+\|L\|\,\eta\le\varepsilon;$$

therefore

$$\mathscr{H}^m[\psi(K)\cap\mathbf{E}[0,L(v),\varepsilon]\cap\mathbf{B}(0,\delta)]/[\alpha(m)\,\delta^m]$$
$$\ge\lambda^{-2m}\,\mathscr{H}^m[K\cap\mathbf{E}(0,v,\eta)\cap\mathbf{B}(0,\delta/\lambda)]/[\alpha(m)(\delta/\lambda)^m],$$

which approaches $\lambda^{-2m}\,\mathscr{L}^m[\mathbf{E}(0,v,\eta)\cap\mathbf{B}(0,1)]/\alpha(m)$ as δ approaches 0, and we conclude that

$$\Theta^m_*[\mathscr{H}^m\llcorner\psi(K)\cap\mathbf{E}[0,L(v),\varepsilon],0]>0,$$

hence $L(v)\in\operatorname{Tan}^m[\mathscr{H}^m\llcorner\psi(K),0]$.

Moreover, to prove the differentiability of F relative to $\psi(K)$ at $\psi(0)=0$, we assume $F(0)=0$, recall 3.1.21 and 3.1.22, choose a linear function

$$\zeta:\mathbf{R}^n\to\mathbf{R}^\nu\quad\text{with}\quad\zeta\circ L=D(F\circ\psi)(0)$$

and observe that

$$(F[\psi(x)]-\zeta[\psi(x)])/|\psi(x)|$$
$$=\frac{|x|}{|\psi(x)|}\left(\frac{[F\circ\psi](x)-\langle x,D(F\circ\psi)(0)\rangle}{|x|}-\zeta\left[\frac{\psi(x)-L(x)}{|x|}\right]\right)$$

approaches 0 as x approaches 0, hence

$$[F(y)-\zeta(y)]/|y|\to0\ \text{ as }\ \psi(K)\ni y\to0.$$

3.2.18. Lemma. *If W is an $(\mathscr{H}^m,m)$ rectifiable and $\mathscr{H}^m$ measurable subset of $\mathbf{R}^n$, and if $1<\lambda<\infty$, then there exist compact subsets $K_1,K_2,K_3,\ldots$ of $\mathbf{R}^m$ and Lipschitzian maps $\psi_1,\psi_2,\psi_3,\ldots$ of $\mathbf{R}^m$ into $\mathbf{R}^n$ such that $\psi_1(K_1)$, $\psi_2(K_2)$, $\psi_3(K_3),\ldots$ are disjoint subsets of W with*

$$\mathscr{H}^m\left[W\sim\bigcup_{i=1}^{\infty}\psi_i(K_i)\right]=0$$

and, for each positive integer i,

$$\operatorname{Lip}(\psi_i)\le\lambda,\qquad\psi_i|K_i\ \text{is univalent},\qquad\operatorname{Lip}[(\psi_i|K_i)^{-1}]\le\lambda,$$
$$\lambda^{-1}|v|\le|\langle v,D\,\psi_i(a)\rangle|\le\lambda\,|v|\ \text{for}\ a\in K_i,\ v\in\mathbf{R}^m.$$

Proof. We combine 3.2.14, 3.2.4, 3.2.2, 2.10.43 and 2.2.3.

3.2.19. Theorem. *If W is an $(\mathscr{H}^m, m)$ rectifiable and $\mathscr{H}^m$ measurable subset of $\mathbf{R}^n$, then, for $\mathscr{H}^m$ almost all w in W,*

$$\Theta^m(\mathscr{H}^m \llcorner W, w) = 1$$

and $\operatorname{Tan}^m(\mathscr{H}^m \llcorner W, w)$ *is an m dimensional vectorsubspace of $\mathbf{R}^n$.*

Moreover, if $f\colon W \to \mathbf{R}^\nu$ is Lipschitzian, then f has at $\mathscr{H}^m$ almost all points w in W an $(\mathscr{H}^m \llcorner W, m)$ approximate differential

$$\operatorname{ap} D f(w)\colon \operatorname{Tan}^m(\mathscr{H}^m \llcorner W, w) \to \mathbf{R}^\nu.$$

Proof. Assuming $1 < \lambda < \infty$ we choose K_i, ψ_i as in 3.2.18. Moreover we let $F\colon \mathbf{R}^n \to \mathbf{R}^\nu$ be a Lipschitzian extension of f. From 2.10.19 (5), (4) we know that

$$\Theta^{*m}(\mathscr{H}^m \llcorner W, w) \le 1, \qquad \Theta^m[\mathscr{H}^m \llcorner W \sim \psi_i(K_i), w] = 0$$

for $\mathscr{H}^m$ almost all w in $\psi_i(K_i)$. Furthermore 2.9.11, 3.1.6, 2.10.11 imply that for $\mathscr{H}^m$ almost all points w in $\psi_i(K_i)$ the hypotheses of 3.2.17 hold with $a = (\psi_i | K_i)^{-1} w$. We conclude that

$$\lambda^{-2m} \le \Theta^m_*(\mathscr{H}^m \llcorner W, w) \le \Theta^{*m}(\mathscr{H}^m \llcorner W, w) \le 1,$$
$$\operatorname{Tan}^m(\mathscr{H}^m \llcorner W, w) = \operatorname{im} D\psi_i[(\psi_i | K_i)^{-1} w]$$

and f has the $(\mathscr{H}^m \llcorner W, m)$ approximate differential

$$\operatorname{ap} D f(w) = D[f | \psi_i(K_i)](w) \text{ with}$$
$$\operatorname{ap} D f(w) \circ D\psi_i[(\psi_i | K_i)^{-1} w] = D(F \circ \psi_i)[(\psi_i | K_i)^{-1} w]$$

at $\mathscr{H}^m$ almost all points w in $\psi_i(K_i)$.

We deduce the theorem by applying the above argument for each positive integer k with $\lambda = 1 + 1/k$.

3.2.20. Corollary. *If $m \le \nu$ and the $(\mathscr{H}^m \llcorner W, m)$ approximate m dimensional Jacobian of f is defined by the formula*

$$\operatorname{ap} J_m f(w) = \|\textstyle\bigwedge_m \operatorname{ap} D f(w)\|$$

whenever f is $(\mathscr{H}^m \llcorner W, m)$ approximately differentiable at w, then

$$\int_W (g \circ f) \cdot \operatorname{ap} J_m f \, d\mathscr{H}^m = \int_{\mathbf{R}^\nu} g(z)\, N(f, z)\, d\mathscr{H}^m z$$

for every function $g\colon \mathbf{R}^\nu \to \overline{\mathbf{R}}$.

Proof. We observe that, for each positive integer i,

$$\operatorname{ap} J_m f[\psi_i(x)] \cdot J_m \psi_i(x) = J_m(F \circ \psi_i)(x)$$

for $\mathscr{L}^m$ almost all x in K_i. Inasmuch as

$$N[f \mid \psi_i(K_i), z] = N(F \circ \psi_i \mid K_i, z) < \infty$$

for $\mathscr{H}^m$ almost all z in $\mathbf{R}^\nu$, by 3.2.3, we infer from 3.2.5 (2) that

$$\begin{aligned}\int_{\mathbf{R}^\nu} g(z)\, N[f \mid \psi_i(K_i), z]\, d\mathscr{H}^m z &= \int_{K_i} (g \circ F \circ \psi_i) \cdot J_m(F \circ \psi_i)\, d\mathscr{L}^m \\ &= \int_{K_i} (g \circ F \circ \psi_i) \cdot [(\operatorname{ap} J_m f) \circ \psi_i] \cdot J_m \psi_i\, d\mathscr{L}^m \\ &= \int_{\psi_i(K_i)} (g \circ F) \cdot (\operatorname{ap} J_m f)\, d\mathscr{H}^m.\end{aligned}$$

Then we sum with respect to i.

3.2.21. Lemma. *If $v\colon \mathbf{R}^m \to \mathbf{R}^\nu$ is Lipschitzian, μ is a positive integer, and T is a purely $(\mathscr{H}^\mu, \mu)$ unrectifiable Borel subset of $\mathbf{R}^\nu$, then*

$$\dim \operatorname{im} D\, v(x) < \mu \quad \textit{for } \mathscr{L}^m \textit{ almost all } x \textit{ in } v^{-1}(T).$$

Proof. In the special case when $\mu = m$ we obtain the conclusion by applying 3.2.3 with $f = v$ and $A = v^{-1}(T)$.

We now assume $\mu < m$ and identify $\mathbf{R}^m$, $\mathscr{L}^m$ with $\mathbf{R}^\mu \times \mathbf{R}^{m-\mu}$, $\mathscr{L}^\mu \times \mathscr{L}^{m-\mu}$. If the conclusion were false we could choose $g \in \mathbf{O}(m)$ so that $\mathscr{L}^m(A) > 0$ with

$$\begin{aligned}A &= v^{-1}(T) \cap \{x\colon \wedge_\mu [D\, v(x) \mid g(\mathbf{R}^\mu \times \{0\})] \neq 0\} \\ &= g\left[(v \circ g)^{-1}(T) \cap \{z\colon \wedge_\mu [D(v \circ g)(z) \mid \mathbf{R}^\mu \times \{0\}] \neq 0\}\right],\end{aligned}$$

then select $\eta \in \mathbf{R}^{m-\mu}$ so that $\mathscr{L}^\mu(B) > 0$ with

$$B = \mathbf{R}^\mu \cap \{\xi\colon (\xi, \eta) \in g^{-1}(A)\},$$

define $f\colon \mathbf{R}^\mu \to \mathbf{R}^\nu$, $f(\xi) = (v \circ g)(\xi, \eta)$ for $\xi \in \mathbf{R}^\mu$, note that $B \subset f^{-1}(T)$, and infer from the preceding special case that

$$J_\mu f(\xi) = 0 \quad \text{for } \mathscr{L}^\mu \text{ almost all } \xi \text{ in } B;$$

since $J_\mu f(\xi) = \|\wedge_\mu [D(v \circ g)(\xi, \eta) \mid \mathbf{R}^\mu \times \{0\}]\| > 0$ for $\xi \in B$, we would obtain $\mathscr{L}^m(B) = 0$, contrary to our choice of η.

3.2.22. Theorem. *If* $W \subset \mathbf{R}^n$, $Z \subset \mathbf{R}^\nu$, $m \geq \mu$

$$W \text{ is } (\mathscr{H}^m, m) \text{ rectifiable and } \mathscr{H}^m \text{ measurable},$$

$$Z \text{ is } (\mathscr{H}^\mu, \mu) \text{ rectifiable and } \mathscr{H}^\mu \text{ measurable},$$

$$f\colon W \to Z \text{ is a Lipschitzian map},$$

and if the $(\mathscr{H}^m \llcorner W, m)$ *approximate* μ *dimensional Jacobian of* f *is defined by the formula*

$$\operatorname{ap} J_\mu f(w) = \|\wedge_\mu \operatorname{ap} D f(w)\|$$

whenever f *is* $(\mathscr{H}^m \llcorner W, m)$ *approximately differentiable at* w, *then:*

(1) *For* $\mathscr{H}^m$ *almost all* w *in* W, *either* $\operatorname{ap} J_\mu f(w) = 0$ *or*

$$\operatorname{im} \operatorname{ap} D f(w) = \operatorname{Tan}^\mu[\mathscr{H}^\mu \llcorner Z, f(w)] \text{ is a } \mu \text{ dimensional vectorspace.}$$

(2) *For* $\mathscr{H}^\mu$ *almost all* z *in* Z, $f^{-1}\{z\}$ *is* $(\mathscr{H}^{m-\mu}, m-\mu)$ *rectifiable and* $\mathscr{H}^{m-\mu}$ *measurable.*

(3) *For every* $\mathscr{H}^m \llcorner W$ *integrable* $\overline{\mathbf{R}}$ *valued function* g *on* W,

$$\int_W g \cdot \operatorname{ap} J_\mu f \, d\mathscr{H}^m = \int_Z \int_{f^{-1}\{z\}} g \, d\mathscr{H}^{m-\mu} \, d\mathscr{H}^\mu z.$$

Proof. Assuming $1 < \lambda < \infty$, we choose K_i and ψ_i related to W as in 3.2.18; we also choose compact subsets C_j of $\mathbf{R}^\mu$ and Lipschitzian maps γ_j of $\mathbf{R}^\mu$ into $\mathbf{R}^\nu$ which are similarly related to Z. Defining

$$P_{i,j} = \psi_i(K_i) \cap f^{-1}[\gamma_j(C_j)]$$

whenever i, j are positive integers, and

$$Q = Z \sim \bigcup_{j=1}^{\infty} \gamma_j(C_j),$$

we obtain the decomposition

$$W = f^{-1}(Q) \cup \bigcup_{i=1}^{\infty} \bigcup_{j=1}^{\infty} P_{i,j}.$$

Moreover we use 2.10.43 to secure Lipschitzian maps

$$F\colon \mathbf{R}^n \to \mathbf{R}^\nu \quad \text{with} \quad F|W = f,$$

$$G_j\colon \mathbf{R}^\nu \to \mathbf{R}^\mu \quad \text{with} \quad G_j|\gamma_j(C_j) = (\gamma_j|C_j)^{-1}, \quad \operatorname{Lip}(G_j) \leq \lambda.$$

Since $\mathscr{H}^\mu(Q) = 0$ we infer from 3.2.21 that

$$\dim \operatorname{im} D(F \circ \psi_i)(x) < \mu$$

for $\mathscr{L}^m$ almost all x in $(F \circ \psi_i)^{-1} Q$, and hence see from the proof of 3.2.19 that

$$\dim \operatorname{im} \operatorname{ap} D f(w) < \mu$$

for $\mathscr{H}^m$ almost all w in $\psi_i(K_i) \cap f^{-1}(Q)$. It follows that

$$\operatorname{ap} J_\mu f(w) = 0 \text{ for } \mathscr{H}^m \text{ almost all } w \text{ in } f^{-1}(Q).$$

Furthermore 2.10.25 implies

$$\mathscr{H}^{m-\mu}[f^{-1}\{z\}] = 0 \text{ for } \mathscr{H}^\mu \text{ almost all } z \text{ in } Q.$$

On account of 2.10.26 we therefore obtain the equations

$$\int_W \operatorname{ap} J_\mu f \, d\mathscr{H}^m = \sum_{i=1}^\infty \sum_{j=1}^\infty a_{i,j} \quad \text{with} \quad a_{i,j} = \int_{P_{i,j}} \operatorname{ap} J_\mu f \, d\mathscr{H}^m;$$

$$\int_Z \mathscr{H}^{m-\mu}(f^{-1}\{z\}) \, d\mathscr{H}^\mu z = \sum_{i=1}^\infty \sum_{j=1}^\infty b_{i,j}$$

$$\text{with } b_{i,j} = \int_{\gamma_j(C_j)} \mathscr{H}^{m-\mu}(P_{i,j} \cap f^{-1}\{z\}) \, d\mathscr{H}^\mu z.$$

Next we recall the proof of 3.2.19 and observe that

$$\operatorname{im} \operatorname{ap} D f[\psi_i(x)] = \operatorname{im} D(F \circ \psi_i)(x), \qquad \lambda^{-m} \le J_m \psi_i(x) \le \lambda^m,$$

$$\lambda^{-\mu} \operatorname{ap} J_\mu f[\psi_i(x)] \le J_\mu (F \circ \psi_i)(x) \le \lambda^\mu \operatorname{ap} J_\mu f[\psi_i(x)],$$

for $\mathscr{L}^m$ almost all x in K_i. Defining

$$\phi_{i,j} = G_j \circ F \circ \psi_i \quad \text{and} \quad A_{i,j} = K_i \cap \psi_i^{-1}(P_{i,j})$$

we note that $F \circ \psi_i | A_{i,j} = \gamma_j \circ \phi_{i,j} | A_{i,j}$ and $C_j \subset \operatorname{dmn} D\gamma_j$, hence

$$D(F \circ \psi_i)(x) = D\gamma_j[\phi_{i,j}(x)] \circ D\phi_{i,j}(x),$$

$$\lambda^{-\mu} J_\mu \phi_{i,j}(x) \le J_\mu (F \circ \psi_i)(x) \le \lambda^\mu J_\mu \phi_{i,j}(x)$$

for $\mathscr{L}^m$ almost all x in $A_{i,j}$. Moreover, inasmuch as

$$\mathscr{L}^\mu(C_j \cap \{y : \operatorname{im} D\gamma_j(y) \neq \operatorname{Tan}^\mu[\mathscr{H}^\mu \llcorner Z, \gamma_j(y)]\}) = 0,$$

we infer from 3.2.21 that

$$\text{either } J_\mu \phi_{i,j}(x) = 0 \text{ and } \wedge_\mu D(F \circ \psi_i)(x) = 0$$

$$\text{or } \operatorname{im} D(F \circ \psi_i)(x) = \operatorname{Tan}^\mu[\mathscr{H}^\mu \llcorner Z, (F \circ \psi_i)(x)]$$

for $\mathscr{L}^m$ almost all x in $A_{i,j}$. Since 3.2.5 implies

$$a_{i,j} = \int_{A_{i,j}} [(\operatorname{ap} J_\mu f) \circ \psi_i] \cdot J_m \psi_i \, d\mathscr{L}^m,$$

we conclude from the above inequalities that

$$\lambda^{-2\mu-m} a_{i,j} \le \int_{A_{i,j}} J_\mu \phi_{i,j} \, d\mathscr{L}^m \le \lambda^{2\mu+m} a_{i,j}.$$

Furthermore 3.2.11 and 3.2.5 yield the equations

$$\int_{A_{i,j}} J_\mu \phi_{i,j}\, d\mathscr{L}^m = \int_{C_j} \mathscr{H}^{m-\mu}(A_{i,j} \cap \phi_{i,j}^{-1}\{y\})\, d\mathscr{L}^\mu y,$$

$$b_{i,j} = \int_{C_j} \mathscr{H}^{m-\mu}[P_{i,j} \cap f^{-1}\{\gamma_j(y)\}] \cdot J_\mu \gamma_j(y)\, d\mathscr{L}^\mu y.$$

Inasmuch as $\lambda^{-\mu} \leq J_\mu \gamma_j(y) \leq \lambda^\mu$ and

$$P_{i,j} \cap f^{-1}\{\gamma_j(y)\} = \psi_j(A_{i,j} \cap \phi_{i,j}^{-1}\{y\})$$

for $y \in C_j$, and since λ is a Lipschitz constant for ψ_j and $(\psi_j | A_{i,j})^{-1}$, it follows that

$$\lambda^{-2\mu-2m} a_{i,j} \leq b_{i,j} \leq \lambda^{2\mu+2m} a_{i,j}.$$

Summing with respect to i and j, then noting that λ can be taken arbitrarily close to 1, we find that

$$\int_W \operatorname{ap} J_\mu f\, d\mathscr{H}^m = \int_Z \mathscr{H}^{m-\mu}(W \cap f^{-1}\{z\})\, d\mathscr{H}^m z.$$

The same argument is of course applicable with W replaced by any $\mathscr{H}^m$ measurable subset of W, hence (3) results by the standard approximation procedure.

3.2.23. Theorem. *If W is an m rectifiable Borel subset of $\mathbf{R}^n$ and Z is an $(\mathscr{H}^\mu, \mu)$ rectifiable Borel subset of $\mathbf{R}^\nu$, then $W \times Z$ is an $(\mathscr{H}^{m+\mu}, m+\mu)$ rectifiable subset of $\mathbf{R}^n \times \mathbf{R}^\nu$ and*

$$\mathscr{H}^{m+\mu} \llcorner (W \times Z) = (\mathscr{H}^m \llcorner W) \times (\mathscr{H}^\mu \llcorner Z).$$

Proof. Suppose $f: \mathbf{R}^m \to \mathbf{R}^n$ is Lipschitzian with $W \subset \operatorname{im} f$.

If $\mathscr{H}^\mu(Z) = 0$, then $\mathscr{H}^{m+\mu}(\mathbf{R}^m \times Z) = 0$ according to 2.10.45, hence $\mathscr{H}^{m+\mu}(W \times Z) \leq \mathscr{H}^{m+\mu}[\operatorname{im}(f \times \mathbf{1}_Z)] = 0$.

If $Z \subset \operatorname{im} g$ for some Lipschitzian $g: \mathbf{R}^\mu \to \mathbf{R}^\nu$, then $W \times Z \subset \operatorname{im}(f \times g)$, and 3.2.3 implies

$$\mathscr{H}^{m+\mu}[f(A) \times g(B)] = \int_{A \times B} J_{m+\mu}(f \times g)\, d(\mathscr{L}^m \times \mathscr{L}^\mu)$$

$$= \int_A \int_B J_m f(x)\, J_\mu g(y)\, d\mathscr{L}^\mu y\, d\mathscr{L}^m x = \mathscr{H}^m[f(A)] \cdot \mathscr{H}^\mu[g(B)]$$

whenever $A \subset \mathbf{R}^m$, $B \subset \mathbf{R}^\mu$ are Borel sets with $f|A$, $g|B$ univalent; use of 3.2.4, 2.6.4 and the Borel regularity of Hausdorff measures yields the conclusion.

3.2.24. The conclusions of 3.2.23 need not hold in case W is merely $(\mathscr{H}^m, m)$ rectifiable. In fact 2.10.29 provides examples where $\mathscr{H}^m(W) = 0$, $\mathscr{H}^\mu(Z) = 0$, $\mathscr{H}^{m+\mu}(W \times Z) = \infty$.

3.2.25. Lemma. *For every $(\mathscr{H}^m, m)$ rectifiable subset W of $\mathbf{R}^n$ there exists an $\mathscr{H}^m \llcorner W$ measurable function T with values in*

$$\wedge_m \mathbf{R}^n \cap \{\xi: \xi \text{ is simple and } |\xi|=1\}$$

such that, for $\mathscr{H}^m$ almost all w in W, $\operatorname{Tan}^m(H^m \llcorner W, w)$ is the m dimensional vectorsubspace of $\mathbf{R}^n$ associated with $T(w)$.

Proof. Assuming W is $\mathscr{H}^m$ measurable we apply 3.2.18, choose a nonzero m vector η of $\mathbf{R}^m$, define

$$S(w)=\langle \eta, \wedge_m D\psi_i[(\psi_i|K_i)^{-1} w]\rangle, \qquad T(w)=S(w)/|S(w)|$$

whenever $w \in \psi_i(K_i)$, and recall the proof of 3.2.19.

3.2.26. Theorem. *If W is an $(\mathscr{H}^m, m)$ rectifiable subset of $\mathbf{R}^n$ and $1 \le t \le \infty$, then*

$$\mathscr{S}^m(W)=\mathscr{H}^m(W)=\mathscr{T}^m(W)=\mathscr{G}^m(W)=\mathscr{Q}_t^m(W)=\mathscr{I}_t^m(W).$$

Proof. In view of the general inequalities summarized in 2.10.6 we need only show that our present hypothesis has the following three consequences:

(1) $\mathscr{H}^m(W) \ge \mathscr{S}^m(W)$. (2) $\mathscr{H}^m(W) \ge \mathscr{Q}_t^m(W)$. (3) $\mathscr{H}^m(W) \le \mathscr{I}_t^m(W)$.

To prove (1) we apply 2.10.19(3) with $\mu=\mathscr{H}^m \llcorner W$, $V=\mathbf{R}^n$ and $B=\{w: \Theta^m(\mu, w)=1\}$, knowing from 3.2.19 that $\mathscr{S}^m(W \sim B)=0$.

In proving (2) we may, by virtue of 3.2.18 and 3.2.17, restrict our discussion to the special case when, for every $w \in W$, $\operatorname{Tan}(W, w)$ is contained in an m dimensional vectorsubspace $\sigma(w)$ of $\mathbf{R}^n$. We define ζ_t as in 2.10.15, choose an m dimensional vectorsubspace Y of $\mathbf{R}^n$, and observe that

$$\begin{aligned} g(\varepsilon) &= \zeta_t[\mathbf{R}^n \cap \{x: |x| \le 1 \text{ and } \operatorname{dist}(x, Y) \le \varepsilon\}] \\ &\to \zeta_t[\mathbf{B}(0,1) \cap Y] = \alpha(m) \text{ as } \varepsilon \to 0+. \end{aligned}$$

Given $\delta>0$ we select $\varepsilon>0$ with $g(\varepsilon)<(1+\delta)\,\alpha(m)$, and consider the family F of all closed balls $\mathbf{B}(w,r)$ such that $w \in W$, $0<2r<\delta$,

$$\begin{gathered} (1+\delta)\,\mathscr{H}^m[W \cap \mathbf{B}(w,r)] \ge \alpha(m)\, r^m \text{ and} \\ W \cap \mathbf{B}(w,r) \subset C(w,r) = \mathbf{B}(w,r) \cap \{x: \operatorname{dist}[x-a, \sigma(w)] \le \varepsilon r\}. \end{gathered}$$

For $\mathscr{H}^m$ almost all w in W it is true that $\mathbf{B}(w,r)\in F$ whenever r is sufficiently small, hence 2.8.15 allows us to choose a sequence of disjoint balls $\mathbf{B}(w_i,r_i)\in F$ whose union contains $\mathscr{H}^m$ almost all of W. Letting φ_δ be the size δ approximating measure occurring in the construction of $\mathscr{Q}_t^m$ we conclude that

$$\begin{aligned}\varphi_\delta(W) &\le \sum_{i=1}^{\infty} \zeta_t[C(w_i,r_i)] = \sum_{i=1}^{\infty} g(\varepsilon)\cdot(r_i)^m \le \sum_{i=1}^{\infty}(1+\delta)\,\alpha(m)\cdot(r_i)^m \\ &\le \sum_{i=1}^{\infty}(1+\delta)^2\,\mathscr{H}^m[W\cap\mathbf{B}(w_i,r_i)] = (1+\delta)^2\,\mathscr{H}^m(W).\end{aligned}$$

In proving (3) we assume W is a Borel set and choose T according to 3.2.25. Given $\delta>0$ we can find a Borel partition H of W such that for each $S\in H$ there exists a simple m vector ξ_S of $\mathbf{R}^n$ with $|\xi_S|=1$ and

$$|T(w)-\xi_S|\le\delta \text{ for } \mathscr{H}^m \text{ almost all } w \text{ in } S.$$

Whenever $S\in H$ and $p\in\mathbf{O}^*(n,m)$ we define

$$g_S(p)=\int N(p|S,y)\,d\mathscr{L}^m y$$

and infer with the help of 2.10.20 that

$$\begin{aligned}g_S(p) &= \int_S \operatorname{ap} J_m(p|W)\,d\mathscr{H}^m = \int_S |\langle T(w),\textstyle\bigwedge_m p\rangle|\,d\mathscr{H}^m w \\ &\ge (|\langle\xi_S,\textstyle\bigwedge_m p\rangle|-\delta)\cdot\mathscr{H}^m(S);\end{aligned}$$

recalling 2.7.16 (6) and 2.10.15 we obtain the inequalities

$$\begin{aligned}[\theta^*_{n,m}]_{(t)}(g_S) &\ge [\boldsymbol{\beta}_t(n,m)-\delta]\cdot\mathscr{H}^m(S), \\ \mathscr{I}_t^m(S) &\ge [1-\delta/\boldsymbol{\beta}_t(n,m)]\cdot\mathscr{H}^m(S).\end{aligned}$$

Finally we sum over H to conclude

$$\mathscr{I}_t^m(W)\ge[1-\delta/\boldsymbol{\beta}_t(n,m)]\cdot\mathscr{H}^m(W).$$

3.2.27. Theorem. *If W is an $(\mathscr{H}^m,m)$ rectifiable subset of $\mathbf{R}^n$, $m\le n$ and*

$$a_\lambda=\int N(\mathbf{p}_\lambda|W,y)\,d\mathscr{L}^m y \text{ for } \lambda\in\Lambda(n,m),$$

then

$$\Big[\sum_{\lambda\in\Lambda(n,m)}(a_\lambda)^2\Big]^{\frac12}\le\mathscr{H}^m(W)\le\sum_{\lambda\in\Lambda(n,m)}a_\lambda.$$

Proof. Choosing T according to 3.2.25 and recalling 1.7.4 we define

$$\xi\colon \operatorname{dmn} T \to \wedge_m \mathbf{R}^n, \qquad \xi_\lambda\colon \operatorname{dmn} T \to \mathbf{R}$$

so that $\langle \xi(w), \omega_\lambda \rangle = |\langle T(w), \wedge_m \mathbf{p}_\lambda \rangle| = \xi_\lambda(w)$ whenever $w \in \operatorname{dmn} T$ and $\lambda \in \Lambda(n, m)$. Then

$$\mathscr{H}^m(W) = \int_W |T| \, d\mathscr{H}^m = \int_W |\xi| \, d\mathscr{H}^m,$$

while 3.2.20 and 2.4.12 imply

$$a_\lambda = \int_W \operatorname{ap} J_m(\mathbf{p}_\lambda | W) \, d\mathscr{H}^m = \int_W \xi_\lambda \, d\mathscr{H}^m = \langle \textstyle\int \xi \, d\mathscr{H}^m, \omega_\lambda \rangle$$

for each $\lambda \in \Lambda(n, m)$, hence

$$\left[\sum_\lambda (a_\lambda)^2\right]^{\frac{1}{2}} = \left|\int_W \xi \, d\mathscr{H}^m\right| \leq \int_W |\xi| \, d\mathscr{H}^m \leq \int_W \sum_\lambda \xi_\lambda \, d\mathscr{H}^m = \sum_\lambda a_\lambda.$$

3.2.28. By studying certain particular maps we will now exhibit the spaces $\mathbf{O}(n, m)$, $\mathbf{G}(n, m)$, $\mathbf{G}_0(n, m)$ as submanifolds (of class ∞) of suitable Euclidean spaces, and compute the Hausdorff measures of these manifolds.

(1) Letting Y_m be the vectorspace of all linear symmetric endomorphisms of $\mathbf{R}^m$, and assuming $m \leq n$, we consider the map

$$P\colon \operatorname{Hom}(\mathbf{R}^m, \mathbf{R}^n) \to Y_m, \qquad P(g) = g^* \circ g \text{ for } g \in \operatorname{Hom}(\mathbf{R}^m, \mathbf{R}^n).$$

We find that

$$\langle h, DP(g) \rangle = h^* \circ g + g^* \circ h \text{ for } h \in \operatorname{Hom}(\mathbf{R}^m, \mathbf{R}^n),$$

$$\operatorname{im} DP(g) = Y_m \text{ in case } g \text{ is univalent},$$

because in this case every $s \in Y_m$ has a factorization $s = t \circ g$ with $t \in \operatorname{Hom}(\mathbf{R}^n, \mathbf{R}^m)$, hence $h = t^*/2 \in \operatorname{Hom}(\mathbf{R}^m, \mathbf{R}^n)$ and

$$h^* \circ g + g^* \circ h = (t \circ g + g^* \circ t^*)/2 = (s + s^*)/2 = s.$$

Therefore, in case g is univalent,

$$\begin{aligned} \dim \ker DP(g) &= \dim \operatorname{Hom}(\mathbf{R}^m, \mathbf{R}^n) - \dim Y_m \\ &= m\,n - m(m+1)/2 = m(2n - m - 1)/2. \end{aligned}$$

Using 3.1.19 and 3.1.21 we conclude

$$\mathbf{O}(n, m) = P^{-1}\{\mathbf{1}_{\mathbf{R}^m}\}$$

is an $m(2n - m - 1)/2$ *dimensional submanifold of class* ∞ *of* $\operatorname{Hom}(\mathbf{R}^m, \mathbf{R}^n)$ *with*

$$\operatorname{Tan}[\mathbf{O}(n, m), g] = \operatorname{Hom}(\mathbf{R}^m, \mathbf{R}^n) \cap \{h\colon h^* \circ g + g^* \circ h = 0\}$$

for $g \in \mathbf{O}(n, m)$.

(2) Choosing $\zeta \in \wedge_m \mathbf{R}^m$ with $|\zeta|=1$, and recalling 1.7.9, we observe that the linear isomorphism

$$\wedge_m \mathbf{R}^n \simeq \operatorname{Hom}(\wedge_m \mathbf{R}^m, \wedge_m \mathbf{R}^n),$$

under which $\langle \zeta, \phi \rangle$ corresponds to ϕ, is an isometry; therefore 1.6.2 implies

$$\mathbf{G}_0(n,m) \simeq \wedge_m \mathbf{R}^n \cap \{\xi : \xi \text{ is simple, } |\xi|=1\} \simeq \{\wedge_m g : g \in \mathbf{O}(n,m)\}.$$

We may thus identify $\mathbf{G}_0(n,m)$ with the image of the map

$$Q: \mathbf{O}(n,m) \to \operatorname{Hom}(\wedge_m \mathbf{R}^m, \wedge_m \mathbf{R}^n), \qquad Q(g) = \wedge_m g \text{ for } g \in \mathbf{O}(n,m).$$

If $g \in \mathbf{O}(n,m)$, then

$$Z(g) = \operatorname{Hom}(\mathbf{R}^m, \mathbf{R}^n) \cap \{h : h^* \circ g = 0\} \subset \operatorname{Tan}[\mathbf{O}(n,m), g]$$

with $\dim Z(g) = m(n-m)$ because $\dim(\mathbf{R}^n/\operatorname{im} g) = n-m$. For $h \in Z(g)$ we see from 1.7.9 that

$$|\wedge_m(g+h) - \wedge_m g|^2 = \det(\mathbf{1}_{\mathbf{R}^m} + h^* \circ h) - 1$$

and use 1.4.5 to compute

$$\begin{aligned} |\langle h, DQ(g) \rangle|^2 &= \lim_{t \to 0} |\wedge_m(g+t\,h) - \wedge_m g|^2/t^2 \\ &= \lim_{t \to 0} [\det(\mathbf{1}_{\mathbf{R}^m} + t^2\, h^* \circ h) - 1]/t^2 = \operatorname{trace}(h^* \circ h) = |h|^2. \end{aligned}$$

Thus $DQ(g)|Z(g)$ is an isometric injection. On the other hand the map

$$R_g : \mathbf{SO}(m) = \mathbf{O}(m) \cap \{f : \det(f) = 1\} \to \mathbf{O}(n,m)$$
$$R_g(f) = g \circ f \text{ for } f \in \mathbf{SO}(m),$$

is differentiable, and $\operatorname{im}(Q \circ R_g) = \{\wedge_m g\}$, hence

$$\ker DQ(g) \supset \operatorname{im} DR_g(\mathbf{1}_{\mathbf{R}^m}) = \{g \circ u : u \in \operatorname{Tan}[\mathbf{O}(m), \mathbf{1}_{\mathbf{R}^m}]\}.$$

If $h \in Z(g)$ and $u \in \operatorname{Tan}[\mathbf{O}(m), \mathbf{1}_{\mathbf{R}^m}]$, then

$$h \bullet (g \circ u) = \operatorname{trace}(h^* \circ g \circ u) = 0.$$

Moreover it is true for any $h \in \operatorname{Tan}[\mathbf{O}(n,m), g]$ that

$$h - g \circ g^* \circ h \in Z(g), \qquad g^* \circ h \in \operatorname{Tan}[\mathbf{O}(m), \mathbf{1}_{\mathbf{R}^m}].$$

We conclude that $\ker DQ(g)$ is the orthogonal complement of $Z(g)$ in $\operatorname{Tan}[\mathbf{O}(n,m), g]$, and that

$$\dim \operatorname{im} DQ(g) = \dim Z(g) = m(n-m).$$

Next we observe that Q is an open map; in fact the topologies of $\mathbf{O}(n, m)$ and $\operatorname{im} Q$ are uniquely determined by the transitive actions of $\mathbf{O}(n)$, because $\mathbf{O}(n)$ is compact, and all maps A^x corresponding to such actions A are open (see 2.7.1). Consequently 3.1.19(3) implies $\operatorname{im} Q \simeq \mathbf{G}_0(n, m)$ *is an* $m(n-m)$ *dimensional submanifold of class* ∞ *of* $\operatorname{Hom}(\wedge_m \mathbf{R}^m, \wedge_m \mathbf{R}^n) \simeq \wedge_m \mathbf{R}^n$. Finally we note that

$$J_{m(n-m)} Q(g) = 1, \qquad Q^{-1}\{Q(g)\} = \operatorname{im} R_g$$

and R_g is an isometry whenever $g \in \mathbf{O}(n, m)$, hence

$$\mathscr{H}^{m(m-1)/2}[Q^{-1}(y)] = 2^{-1} \mathscr{H}^{m(m-1)/2}[\mathbf{O}(m)]$$

whenever $y \in \operatorname{im} Q$, and use 3.2.22 to obtain the formula

$$2\mathscr{H}^{m(2n-m-1)/2}[\mathbf{O}(n, m)] = \mathscr{H}^{m(m-1)/2}[\mathbf{O}(m)] \cdot \mathscr{H}^{m(n-m)}[\mathbf{G}_0(n, m)].$$

(3) Assuming $n \geq m \geq 2$ we now choose $\alpha \in \mathbf{O}(n, m-1)$ and consider the commutative diagram

$$\begin{array}{ccc}
\mathbf{O}(n, m) \subset \operatorname{Hom}(\mathbf{R}^m, \mathbf{R}^n) & \simeq & \wedge^1 \mathbf{R}^m \otimes \mathbf{R}^n \\
S \big\downarrow & & \big\downarrow \wedge^1 \alpha \otimes 1_{\mathbf{R}^n} \\
\mathbf{O}(n, m-1) \subset \operatorname{Hom}(\mathbf{R}^{m-1}, \mathbf{R}^n) & \simeq & \wedge^1 \mathbf{R}^{m-1} \otimes \mathbf{R}^n
\end{array}$$

where $S(g) = g \circ \alpha$ for $g \in \mathbf{O}(n, m)$. In order to compute $DS(g)$ we choose dual orthonormal bases $e_1, \ldots, e_m$ and $\omega_1, \ldots, \omega_m$ of $\mathbf{R}^m$ and $\wedge^1 \mathbf{R}^m$ so that $e_1, \ldots, e_{m-1}$ span $\operatorname{im} \alpha$, as well as an orthonormal base $v_1, \ldots, v_n$ of $\mathbf{R}^n$ with

$$g(e_i) = v_i \text{ for } i = 1, \ldots, m.$$

From (1) we know that $\operatorname{Tan}[\mathbf{O}(n, m), g]$ is the set of all $h \in \operatorname{Hom}(\mathbf{R}^m, \mathbf{R}^n)$ such that

$$h(e_i) \bullet v_j + v_i \bullet h(e_j) = e_i \bullet (h^* \circ g + g^* \circ h)\, e_j = 0 \text{ for } 1 \leq i \leq j \leq m.$$

It readily follows that the image of $\operatorname{Tan}[\mathbf{O}(n, m), g]$ in $\wedge^1 \mathbf{R}^m \otimes \mathbf{R}^n$ has an orthonormal base consisting of the elements

$$a_{i,j} = 2^{-\frac{1}{2}}(\omega_i \otimes v_j - \omega_j \otimes v_i) \text{ with } 1 \leq i < j \leq m,$$

and

$$b_{i,j} = \omega_i \otimes v_j \text{ with } 1 \leq i \leq m < j \leq n.$$

Similarly the image of $\operatorname{Tan}[\mathbf{O}(n, m-1), S(g)]$ in $\wedge^1 \mathbf{R}^{m-1} \otimes \mathbf{R}^n$ has an orthonormal base consisting of the elements

$$c_{i,j} = 2^{-\frac{1}{2}}[(\omega_i \circ \alpha) \otimes v_j - (\omega_j \circ \alpha) \otimes v_i] \text{ with } 1 \leq i < j \leq m-1,$$

and

$$d_{i,j}=(\omega_i\circ\alpha)\otimes v_j \text{ with } 1\le i\le m-1<j\le n.$$

Moreover $DS(g)$ is equivalent to the restriction σ of $\bigwedge^1\alpha\otimes\mathbf{1}_{\mathbf{R}^m}$ to the image of $\operatorname{Tan}[\mathbf{O}(n,m),g]$, and for σ we find the formulae

$$\sigma(a_{i,j})=c_{i,j} \text{ for } 1\le i<j<m,$$
$$\sigma(a_{i,m})=2^{-\frac{1}{2}}d_{i,m} \text{ for } 1\le i<m,$$
$$\sigma(b_{i,j})=d_{i,j} \text{ for } 1\le i<m<j\le n,$$
$$\sigma(b_{m,j})=0 \text{ for } m<j\le n;$$

consequently

$$J_{(m-1)(2n-m)/2}\,S(g)=2^{(1-m)/2}.$$

Furthermore we can map

$$S^{-1}\{S(g)\}=\mathbf{O}(n,m)\cap\{f\colon f(e_i)=v_i \text{ for } 1\le i<m\}$$

isometrically onto

$$\mathbf{R}^n\cap\{u\colon |u|=1,\,u\bullet v_i=0 \text{ for } 1\le i<m\},$$

by letting $f(e_m)$ correspond to f. Therefore

$$\mathscr{H}^{n-m}(S^{-1}\{y\})=\mathscr{H}^{n-m}(\mathbf{S}^{n-m}) \text{ for } y\in\mathbf{O}(n,m-1),$$

and we use 3.2.22 to conclude

$$\begin{aligned}&\mathscr{H}^{m(2n-m-1)/2}[\mathbf{O}(n,m)]\\&=2^{(m-1)/2}\mathscr{H}^{n-m}(\mathbf{S}^{n-m})\cdot\mathscr{H}^{(m-1)(2n-m)/2}[\mathbf{O}(n,m-1)].\end{aligned}$$

(4) Assuming that X is a finite dimensional inner product space we now consider the polynomial map

$$T\colon X\to\bigodot_2 X,\qquad T(x)=x^2/2 \text{ for } x\in X,$$

and compute

$$\langle h,DT(x)\rangle=x\,h \text{ for } x,h\in X.$$

We define $U=X\cap\{x\colon |x|=1\}$ and see from 1.10.5 that

$$|\langle h,DT(x)\rangle|=|x\,h|=|h|$$

whenever $x\in U$ and $h\in X\cap\{y\colon y\bullet x=0\}=\operatorname{Tan}(U,x)$; therefore $D(T|U)(x)$ is an isometric injection of $\operatorname{Tan}(U,x)$ into $\bigodot_2 X$.

If $x, y \in X$, then $x^2 \bullet y^2 = 2(x \bullet y)^2$ and

$$|x^2 - y^2|^2 = 2\,|x|^4 + 2\,|y|^4 - 4(x \bullet y)^2.$$

Consequently $x, y \in U$ implies

$$|T(x) - T(y)|^2 = 1 - (x \bullet y)^2 = (1 - x \bullet y)(1 + x \bullet y)$$
$$= |x - y|^2 (1 + x \bullet y)/2,$$

hence $T(x) = T(y)$ if and only if $x = \pm y$. It also follows that

$$T(U) \cap \mathbf{U}[T(x), r] \subset T[U \cap \mathbf{U}(x, r 2^{\frac{1}{2}})]$$

whenever $x \in U$ and $r > 0$, hence $T|U$ is an open map of U onto $T(U)$.

If B is any μ dimensional submanifold of class $k \geq 1$ of U which satisfies the condition

$$b \in B \text{ if and only if } -b \in B,$$

then $T|B$ is an open map of B onto $T(B)$ and, for each $b \in B$, $D(T|B)(b)$ is an isometric injection of $\operatorname{Tan}(B, b)$ into $\odot_2 X$, hence $J_\mu(T|B)(b) = 1$; applying 3.1.19 (3) and 3.2.20 we conclude that *$T(B)$ is a μ dimensional submanifold of class k of $\odot_2 X$ and*

$$\mathscr{H}^\mu(B) = 2\mathscr{H}^\mu[T(B)].$$

By virtue of (2) we can in particular take

$$X = \wedge_m \mathbf{R}^n, \quad B = \wedge_m \mathbf{R}^n \cap \{\xi : \xi \text{ is simple}, |\xi| = 1\}.$$

In this case 1.6.2 implies $B \simeq \mathbf{G}_0(n, m)$ and $T(B) \simeq \mathbf{G}(n, m)$. We may therefore *identify $\mathbf{G}(n, m)$ with an $m(n-m)$ dimensional submanifold of class ∞ of $\odot_2 \wedge_m \mathbf{R}^n$*; moreover

$$\mathscr{H}^{m(n-m)}[\mathbf{G}_0(n, m)] = 2\mathscr{H}^{m(n-m)}[\mathbf{G}(n, m)].$$

(5) Combining (3), (4), (2) with 3.2.13 we see by induction, starting from $\mathbf{O}(n, 1) \simeq \mathbf{S}^{n-1}$, that

$$\mathscr{H}^{m(2n-m-1)/2}[\mathbf{O}(n, m)] = \frac{2^{m(m+3)/4}\,\Gamma(\tfrac{1}{2})^{m(2n-m+1)/2}}{\prod_{j=1}^{m} \Gamma[(n-j+1)/2]},$$

$$\mathscr{H}^{m(n-m)}[\mathbf{G}(n, m)] = \Gamma(\tfrac{1}{2})^{m(n-m)} \prod_{j=1}^{m} \frac{\Gamma(j/2)}{\Gamma[(n-j+1)/2]}.$$

3.2.29. Theorem. *A subset W of $\mathbf{R}^n$ is countably $(\mathscr{H}^m, m)$ rectifiable if and only if $\mathscr{H}^m$ almost all of W is contained in the union of some countable family of m dimensional submanifolds of class 1 of $\mathbf{R}^n$.*

Proof. Whenever $F\colon \mathbf{R}^m \to \mathbf{R}^n$ is Lipschitzian and $\varepsilon > 0$, we can use 3.1.16 to obtain a map $f\colon \mathbf{R}^m \to \mathbf{R}^n$ of class 1 such that

$$\mathscr{L}^m(\mathbf{R}^m \sim C) < \varepsilon \quad \text{with} \quad C = \{x\colon f(x) = F(x)\}.$$

From 3.1.18 we see that the open set

$$P = \{x\colon \dim \operatorname{im} Df(x) = m\} = \{x\colon J_m f(x) > 0\}$$

is the union of a countable family G of open sets U such that f maps U homeomorphically onto $f(U)$. For each $U \in G$ it follows from 3.1.19 (3) that $f(U)$ is an m dimensional submanifold of class 1 of $\mathbf{R}^n$. Furthermore

$$(\operatorname{im} F) \sim \bigcup \{f(U)\colon U \in G\} = F(\mathbf{R}^m \sim C) \cup f(C \sim P),$$

$$\mathscr{H}^m[F(\mathbf{R}^m \sim C)] < \varepsilon \cdot (\operatorname{Lip} F)^m, \qquad \mathscr{H}^m[f(C \sim P)] = 0$$

according to 3.2.3.

3.2.30. *If $f\colon X \to Y$ is a linear map of Euclidean spaces, $A \subset X$ and m is a positive integer, then*

$$\mathscr{T}^m[f(A)] \le \|\wedge_m f\| \, \mathscr{T}^m(A).$$

This assertion obviously follows from 2.10.3; in view of 2.10.39 it implies the inequality:

$$\mathscr{H}^m[f(A)] \le (\dim Y)^{m/2} \, 2^{m-1} \, \|\wedge_m f\| \, \mathscr{H}^m(A).$$

3.2.31. Theorem. *If W is an $(\mathscr{H}^m, m)$ rectifiable and $\mathscr{H}^m$ measurable subset of $\mathbf{R}^n$, $f\colon W \to \mathbf{R}^\nu$ is a Lipschitzian map, μ is an integer, $0 \le \mu \le m$, $\lambda > 0$ and*

$$Z = \mathbf{R}^\nu \cap \{y\colon \mathscr{H}^{m-\mu}(f^{-1}\{y\}) \ge \lambda\},$$

then Z is $(\mathscr{H}^\mu, \mu)$ rectifiable.

Proof. In view of 3.2.18 and 2.10.25 we consider only the case when $m = n$ and W is compact, hence Z is a Borel set according to 2.10.26; we also assume $\mu \ge 1$. Using 2.10.43 we extend f to a Lipschitzian map $g\colon \mathbf{R}^m \to \mathbf{R}^\nu$ and define

$$V = W \cap \{x\colon \dim \operatorname{im} Dg(x) < \mu\}.$$

Since $\mathscr{H}^\mu(Z) < \infty$ by 2.10.25, we can choose a countably μ rectifiable Borel subset R of $\mathbf{R}^\nu$ such that $Z \sim R$ is purely $(\mathscr{H}^\mu, \mu)$ unrectifiable, infer from 3.2.21 that

$$\mathscr{L}^m[f^{-1}(Z \sim R) \sim V] = 0,$$

apply 2.10.25 again to obtain

$$\mathscr{H}^{m-\mu}(f^{-1}\{z\}\sim V)=0 \text{ for } \mathscr{H}^{\mu} \text{ almost all } z \text{ in } Z\sim R,$$

hence

$$\mathscr{H}^{m-\mu}(f^{-1}\{z\}\cap V)\geq\lambda \text{ for } \mathscr{H}^{\mu} \text{ almost all } z \text{ in } Z\sim R.$$

Next, given any $\varepsilon>0$, we let T be a countable, dense subset of the set of all linear symmetric automorphisms t of $\mathbf{R}^{\nu}$ with $\|\wedge_{\mu} t\|<\varepsilon$. With each $t\in T$ and any positive integers i, j we associate the subset $U(t,i,j)$ of V consisting of all points a such that

$$\|t^{-1}\circ Dg(a)\|\leq 1-i^{-1}$$

and

$$|t^{-1}[g(b)-g(a)-\langle b-a, Dg(a)\rangle]|\leq i^{-1}|b-a| \text{ for } b\in\mathbf{B}(a,j^{-1}).$$

If $E\subset U(t,i,j)$ with $\operatorname{diam} E\leq 1/j$, then

$$\begin{aligned}|(t^{-1}\circ g)b-(t^{-1}\circ g)a|&\leq|\langle b-a, t^{-1}\circ Dg(a)\rangle|+i^{-1}|b-a|\\&\leq(1-i^{-1})|b-a|+i^{-1}|b-a|=|b-a|\end{aligned}$$

whenever $a,b\in E$; thus $\operatorname{Lip}(t^{-1}\circ f|E)\leq 1$ and it follows from 3.2.30, 2.10.25 that

$$\begin{aligned}\int^{*}\mathscr{H}^{m-\mu}(E\cap f^{-1}\{z\})\,d\mathscr{H}^{\mu}z&=\int^{*}\mathscr{H}^{m-\mu}[E\cap(t^{-1}\circ f)^{-1}\{t^{-1}(z)\}]\,d\mathscr{H}^{\mu}z\\&\leq\nu^{\mu/2}\,2^{\mu-1}\,\|\wedge_{\mu}t\|\int^{*}\mathscr{H}^{m-\mu}[E\cap(t^{-1}\circ f)\{y\}]\,d\mathscr{H}^{\mu}y\\&\leq c\,\varepsilon\,\mathscr{L}^{m}(E), \text{ with } c=\nu^{\mu/2}\,2^{\mu-1}\,\alpha(m-\mu)\,\alpha(\mu)/\alpha(m).\end{aligned}$$

Moreover V equals the union of all the sets $U(t,i,j)$; in fact if $a\in V$ we apply 1.7.3 to factor $Dg(a)=s\circ p$, where p is an orthogonal projection of $\mathbf{R}^{m}$ onto $\operatorname{im} Dg(a)$, and s is a symmetric endomorphism of $\mathbf{R}^{\nu}$ with $\wedge_{\mu}s=0$; then we choose $t\in T$ with $\|t^{-1}\circ Dg(a)\|=\|t^{-1}\circ s\|<1$, and select suitable integers i,j. We can therefore represent V as the union of a countable disjointed family F consisting of $\mathscr{L}^{m}$ measurable sets E for which

$$\int\mathscr{H}^{m-\mu}(E\cap f^{-1}\{z\})\,d\mathscr{H}^{\mu}z\leq c\,\varepsilon\,\mathscr{L}^{m}(E).$$

Summing over F we conclude that

$$\lambda\,\mathscr{H}^{\mu}(Z\sim R)\leq\int\mathscr{H}^{m-\mu}(V\cap f^{-1}\{z\})\,d\mathscr{H}^{\mu}z\leq c\,\varepsilon\,\mathscr{L}^{m}(V).$$

The arbitrariness of ε implies $\mathscr{H}^{\mu}(Z\sim R)=0$.

3.2.32. Corollary. *If*

$$Y=\mathbf{R}^{\nu}\cap\{y\colon \mathscr{H}^{m-\mu}(f^{-1}\{y\})>0\},\qquad X=W\cap\{x\colon \dim\operatorname{im}\operatorname{ap}Df(x)\leq\mu\},$$

then $\mathscr{H}^{m}[f^{-1}(Y)\sim X]=0$ *and*

$$\int_{f^{-1}(Y)}\operatorname{ap}J_{\mu}f\,d\mathscr{H}^{m}=\int_{\mathbf{R}^{\nu}}\mathscr{H}^{m-\mu}(f^{-1}\{y\})\,d\mathscr{H}^{\mu}y.$$

Proof. For each positive integer k we apply 3.2.31 with $\lambda=1/k$, and 3.2.22 with W replaced by $f^{-1}(Z)$; then we let k approach ∞.

3.2.33. When $\mu<m$ the conditions of 3.2.32 allow the possibility that

$$\int_{X\sim f^{-1}(Y)} J_\mu f \, d\mathscr{H}^m>0,$$

even in case f is a map of class ∞, $\nu=\mu+1$ and

$$W=\mathbf{R}^m\cap\{x\colon 0<x_i<1 \text{ for } i=1,\ldots,m\}.$$

To construct such an example we let $T=\{t\colon 0<t<1\}$, choose a function $\phi\colon \mathbf{R}\to\mathbf{R}$ of class ∞ with

$$\phi(t)>0 \text{ for } t\in T, \quad \phi(t)=0 \text{ for } t\in\mathbf{R}\sim T,$$

select nonempty, open subintervals $U_1, U_2, U_3, \ldots$ of T so that

$$V=\bigcup_{j=1}^{\infty} U_j \text{ is dense in } T,\ \mathscr{L}^1(T\sim V)>0,$$

then define maps ψ, g, f of class ∞ by the formulae

$$\psi(t)=\sum_{j=1}^{\infty}(\operatorname{diam} U_j)^j\,\phi[(t-\inf U_j)/\operatorname{diam} U_j] \text{ for } t\in T,$$

$$g(t)=\int_0^t \psi \, d\mathscr{L}^1 \text{ for } t\in T,$$

$$f(x)=(x_1,\ldots,x_\mu, g(x_{\mu+1}))\in\mathbf{R}^{\mu+1} \text{ for } x\in W;$$

if $\mu=0$ we take $f(x)=g(x_1)$. It follows that g is increasing, $f^{-1}\{y\}$ is an $m-\mu-1$ dimensional cube whenever $y\in\operatorname{im} f$, hence $Y=\varnothing$; on the other hand

$$\{t\colon g'(t)=0\}=T\sim V, \quad X=W\cap\{x\colon x_{\mu+1}\in T\sim V\},$$

$$\int_{X\sim f^{-1}(Y)} J_\mu f\, d\mathscr{H}^m\geq\mathscr{L}^m(X)=\mathscr{L}^1(T\sim V)>0.$$

We also observe that such behavior is impossible when W is a connected analytic m dimensional submanifold of $\mathbf{R}^n$ and the map f is analytic; in fact the functions h_i defined by the formula (see 1.7.9)

$$h_i(x)=\textstyle\bigwedge_i Df(x)\bullet\bigwedge_i Df(x) \text{ for } x\in W$$

are analytic, hence the set

$$S=\{x\colon \dim\operatorname{im} Df(x)=\mu\}=\{x\colon h_\mu(x)>0=h_{\mu+1}(x)\}$$

has the property that either $\mathscr{H}^m(S)=0$ or S is an open set containing $\mathscr{H}^m$ almost all of W, and 3.1.24 shows that the second alternative implies $S\subset f^{-1}(Y)$.

3.2.34. Lemma. *If S is a nonempty closed subset of $\mathbf{R}^n$,*

$$f(x)=\operatorname{distance}(x,S) \text{ for } x\in\mathbf{R}^n,$$

and $0\leq a<b$, then

$$\mathscr{L}^n\{x: a<f(x)<b\}=\int_a^b \mathscr{H}^{n-1}(f^{-1}\{y\})\, d\mathscr{L}^1 y.$$

Proof. Clearly $\operatorname{Lip}(f)\leq 1$. For any $x\in\mathbf{R}^n\sim S$ we can choose $s\in S$ with $f(x)=|s-x|$ and observe that

$$f[x+t(s-x)]=f(x)-t|s-x| \text{ for } 0\leq t\leq 1;$$

in case f is differentiable at x we therefore obtain

$$\langle s-x, Df(x)\rangle=-|s-x|, \quad J_1 f(x)=\|Df(x)\|=1.$$

Then we apply 3.2.11 with $A=\{x: a<f(x)<b\}$.

3.2.35. Theorem. *If S is a nonempty bounded convex subset of $\mathbf{R}^n$,*

$$\zeta^m(S)=\int_{\mathbf{O}^*(n,m)} \mathscr{L}^m[p(S)]\, d\theta^*_{n,m}\, p/\beta_1(n,m)$$

for $m=1,\ldots,n$ and $\zeta^0(S)=1$, then

$$\mathscr{L}^n\{x: \operatorname{dist}(x,S)<r\}=\sum_{m=0}^{n}\zeta^m(S)\,\alpha(n-m)\, r^{n-m}$$

for $0<r<\infty$. Moreover, if S is open, then Bdry S *is $n-1$ rectifiable and*

$$\mathscr{H}^{n-1}(\text{Bdry } S)=2\zeta^{n-1}(S).$$

Proof. First we prove the second assertion. Assuming that S is open and $0\in S$, we define

$$g(x)=\inf\{t: 0<t<\infty \text{ and } x/t\in S\} \text{ for } x\in\mathbf{R}^n,$$

and verify that g is a positively homogeneous convex function; in fact if $x, y\in\mathbf{R}^n$ and a, b, α, β are positive numbers with x/a, $y/b\in S$, and if $c=\alpha a+\beta b$, then

$$(\alpha x+\beta y)/c=(\alpha a/c)(x/a)+(\beta b/c)(y/b)\in S.$$

Therefore Bdry $S=\{g(x)x: x\in\mathbf{S}^{n-1}\}$ is $n-1$ rectifiable, and it follows from 3.2.26, 2.10.15 that

$$\beta_1(n,n-1)\,\mathscr{H}^{n-1}(\text{Bdry } S)=\iint N(p|\text{Bdry } S, y)\, d\mathscr{L}^{n-1} y\, d\theta^*_{n,n-1} p.$$

Furthermore, if $p \in \mathbf{O}^*(n, n-1)$, then

$$p(\operatorname{Bdry} S) = p(S) \cup \operatorname{Bdry} p(S), \quad N(p|\operatorname{Bdry} S, y) = 2 \text{ for } y \in p(S),$$
$$\operatorname{Bdry} p(S) \text{ is } n-2 \text{ rectifiable}, \quad \mathscr{L}^{n-1}[\operatorname{Bdry} p(S)] = 0.$$

Next we prove the first assertion of the theorem by induction with respect to n. In case $n=1$ it is trivial. Assuming $n>1$ we define f as in 3.2.34. If $0<t<\infty$, then $\{x: f(x)<t\}$ is a convex open set with boundary $f^{-1}\{t\}$, and

$$p\{x: f(x)<t\} = \mathbf{R}^{n-1} \cap \{y: \operatorname{dist}[y, p(S)] < t\}$$

whenever $p \in \mathbf{O}^*(n, n-1)$. We may therefore inductively compute

$$\begin{aligned}
&\mathscr{L}^n\{x: f(x)<r\} - \zeta^n(S) \\
&= \mathscr{L}^n\{x: 0<f(x)<r\} = \textstyle\int_0^r \mathscr{H}^{n-1}(f^{-1}\{t\})\, d\mathscr{L}^1 t \\
&= \textstyle\int_0^r 2 \int \mathscr{L}^{n-1}[p\{x: f(x)<t\}]\, d\theta^*_{n,n-1}\, p\, d\mathscr{L}^1 t / \beta_1(n, n-1) \\
&= \sum_{m=0}^{n-1} \gamma_m \textstyle\int_0^r t^{n-1-m} \iint \mathscr{L}^m[(q \circ p) S]\, d\theta^*_{n-1,m}\, q\, d\theta^*_{n,n-1}\, p\, d\mathscr{L}^1 t,
\end{aligned}$$

where $\gamma_m = 2\alpha(n-1-m)/[\beta_1(n-1, m)\, \beta_1(n, n-1)]$; applying 2.7.11 (2) and 3.2.13 we find that the m-th term of this sum equals

$$\gamma_m r^{n-m} (n-m)^{-1} \beta_1(n, m)\, \zeta^m(S) = r^{n-m} \alpha(n-m)\, \zeta^m(S).$$

3.2.36. From 3.2.35 and 2.10.33 we see that the restriction of $\mathscr{L}^n$ to the class C of all nonempty compact convex subsets of $\mathbf{R}^n$ is continuous with respect to the Hausdorff distance. It follows that $\mathscr{L}^m[p(S)]$ is continuous with respect to (p, S) on $\mathbf{O}^*(n, m) \times C$, and the function ζ^m defined in 3.2.35 is continuous on C for $m = 1, \ldots, n$.

The first conclusion of 3.2.35 is called **Steiner's formula**, the second **Cauchy's formula**. An extension of Steiner's formula from convex sets to sets with positive reach may be found in [F 15, Theorem 5.6].

If P and Q are nonempty bounded convex subsets of $\mathbf{R}^\mu$ and $\mathbf{R}^\nu$, then

$$\zeta^m(P \times Q) = \sum_{i+j=m} \zeta^i(P)\, \zeta^j(Q)$$

for $m = 0, \ldots, \mu+\nu$. To prove this we define

$$f(y, z) = \operatorname{dist}(y, P) \text{ for } (y, z) \in \mathbf{R}^\mu \times \mathbf{R}^\nu,$$

note that $J_1 f(y,z)=1$ whenever f is differentiable at (y,z) with $y\notin P$, and apply 3.2.11, 3.2.23, 3.2.35 to compute

$$\begin{aligned}
&\mathscr{L}^{\mu+\nu}\{(y,z)\colon \operatorname{dist}[(y,z),P\times Q]<r\}\\
&=\mathscr{L}^{\mu+\nu}(f^{-1}\{0\}\cap\{(y,z)\colon \operatorname{dist}[(y,z),P\times Q]<r\})\\
&\quad+\textstyle\int_0^r \mathscr{H}^{\mu+\nu-1}(f^{-1}\{s\}\cap\{(y,z)\colon \operatorname{dist}[(y,z),P\times Q]<r\})\,d\mathscr{L}^1 s\\
&=\mathscr{L}^{\mu}(P)\cdot\mathscr{L}^{\nu}\{z\colon \operatorname{dist}(z,Q)<r\}+\textstyle\int_0^r \mathscr{H}^{\mu-1}\{y\colon \operatorname{dist}(y,P)=s\}\\
&\qquad\qquad\cdot\mathscr{L}^{\nu}\{z\colon \operatorname{dist}(z,Q)<(r^2-s^2)^{\frac{1}{2}}\}\,d\mathscr{L}^1 s\\
&=\zeta^{\mu}(P)\cdot\sum_{j=0}^{\nu}\zeta^{j}(Q)\,\alpha(\nu-j)\,r^{\nu-j}+\textstyle\int_0^r\sum_{i=0}^{\mu-1}\zeta^{i}(P)\,\alpha(\mu-i)(\mu-i)\,s^{\mu-i-1}\\
&\qquad\qquad\cdot\sum_{j=0}^{\nu}\zeta^{j}(Q)\,\alpha(\nu-j)(r^2-s^2)^{(\nu-j)/2}\,d\mathscr{L}^1 s\\
&=\sum_{i=0}^{\mu}\sum_{j=0}^{\nu}\zeta^{i}(P)\,\zeta^{j}(Q)\,c(\mu-i,\nu-j)\,r^{\mu+\nu-i-j}
\end{aligned}$$

for $0<r<\infty$, where $c(\xi,\eta)$ is defined whenever ξ and η are nonnegative integers by the formulae $c(0,\eta)=\alpha(\eta)$,

$$c(\xi,\eta)=\alpha(\xi)\,\xi\,\alpha(\eta)\textstyle\int_0^1 t^{\xi-1}(1-t^2)^{\eta/2}\,d\mathscr{L}^1 t$$

in case $\xi>0$; moreover one readily sees, either from 3.2.13 or more quickly by considering the special case when P and Q consist of single points, that $c(\xi,\eta)=\alpha(\xi+\eta)$.

From the preceding proposition we see that, if S is a rectangular parallelepiped, then $\zeta^m(S)$ is obtained by applying the m-the elementary symmetric function to the lengths of the edges of S. Also, letting P consist of a single point, we find that the functions ζ^m are invariant under isometric injections; hence the same holds for the measures $\mathscr{Q}_1^m$.

3.2.37. For $S\subset\mathbf{R}^n$ and each integer m with $0\le m\le n$ we define the m **dimensional upper Minkowski content**

$$\mathscr{M}^{*m}(S)=\limsup_{r\to 0+}\mathscr{L}^n\{x\colon \operatorname{dist}(x,S)<r\}/[\alpha(n-m)\,r^{n-m}]$$

and the m **dimensional lower Minkowski content**

$$\mathscr{M}_*^m(S)=\liminf_{r\to 0+}\mathscr{L}^n\{x\colon \operatorname{dist}(x,S)<r\}/[\alpha(n-m)\,r^{n-m}];$$

in case these upper and lower Minkowski contents are equal, their common value is called the m **dimensional Minkowski content** $\mathscr{M}^m(S)$.

The functions $\mathscr{M}^{*m}$ and $\mathscr{M}_*^m$ are not measures. Each assigns the same number to S and Clos S. Using Fubini's theorem we see that

$$\mathscr{M}_*^m(S) \geq \mathscr{L}^m[p(S)] \quad \text{for } p \in \mathbf{O}^*(n,m).$$

It follows that

$$\mathscr{M}_*^m(A) \geq \mathscr{G}^m(A) \quad \text{for } A \subset \mathbf{R}^n,$$

because, if A is a Borel set, then $\mathscr{G}^m(A)$ equals (see 2.10.8) the supremum of the set of all numbers

$$\sum_{S \in H} \mathscr{L}^m[p_S(S)]$$

where H is a finite family of disjoint compact subsets of A and $p_S \in \mathbf{O}^*(n,m)$ for $S \in H$.

3.2.38. Lemma. *If $f: \mathbf{R}^m \to \mathbf{R}^n$ is Lipschitzian, $m \leq n$ and C is a compact subset of $\mathbf{R}^m$, then*

$$\mathscr{M}^{*m}[f(C)] \leq \int_C J_m f \, d\mathscr{L}^m.$$

Proof. Given any positive integer ν we choose $\delta > 0$ so that

$$\mathscr{L}^m\{x: 0 < \operatorname{dist}(x, C) < \delta\} + \mathscr{L}^m(C \sim E) < \nu^{-1},$$

where E is the set of those points $a \in C$ which satisfy the two conditions:

$$|f(x) - f(a) - \langle x - a, Df(a) \rangle| \leq \nu^{-2} |x - a| \quad \text{for } x \in \mathbf{B}(a, \delta),$$
$$\int_K |J_m f(x) - J_m f(a)| \, d\mathscr{L}^m x \leq \nu^{-1} \mathscr{L}^m(K)$$

for every cube K with $a \in K$ and diam $K < \delta$ (see 2.9.9).

Assuming $0 < r < \delta/(m\,\nu)$ we cover C by a finite family F of compact cubes with disjoint interiors, each with side length $\nu\, r$ and meeting C. We also define

$$G = F \cap \{K: K \cap E \neq \varnothing\},$$

and let H be the family of cubes with side length r obtained by dividing every member of $F \sim G$ into ν^m subcubes. Then

$$C \subset \bigcup(G \cup H), \qquad \mathscr{L}^m[(\bigcup G) \sim C] + \mathscr{L}^m(\bigcup H) < \nu^{-1}.$$

For each $K \in G \cup H$ we will estimate the ratio

$$\rho(K) = \mathscr{L}^n\{z: \operatorname{dist}[z, f(K)] < r\}/[\alpha(n-m)\, r^{n-m}].$$

In case $K \in G$ we choose $a \in E \cap K$ and consider the affine map

$$g: \mathbf{R}^m \to \mathbf{R}^n, \qquad g(x) = f(a) + \langle x - a, Df(a) \rangle \quad \text{for } x \in \mathbf{R}^m.$$

We note that diam $K \leq m\,\nu\, r$, hence

$$\{z: \operatorname{dist}[z, f(K)] < r\} \subset \{z: \operatorname{dist}[z, g(K)] < r + m\,\nu^{-1} r\},$$

and infer from 3.2.35 that $\rho(K)$ does not exceed

$$\sum_{j=1}^{n} \zeta^j[g(K)]\,\alpha(n-j)(1+m\,v^{-1})^{n-j}\,r^{m-j}\,\alpha(n-m)^{-1}.$$

Inasmuch as $\zeta^j[g(K)]=0$ for $j>m$,

$$\zeta^m[g(K)]=J_m f(a)\,\mathscr{L}^m(K)\le\int_K J_m f\,d\mathscr{L}^m+v^{-1}\,\mathscr{L}^m(K),$$

$$\begin{aligned}\zeta^j[g(K)]\,r^{m-j}&\le\alpha(j)\,2^{-j}[\operatorname{diam} g(K)]^j\,\beta_1(n,j)^{-1}\,r^{m-j}\\&\le\alpha(j)\,2^{-j}\,\beta_1(n,j)^{-1}\,m^j[\operatorname{Lip}(f)]^j\,\mathscr{L}^m(K)\,v^{j-m}\end{aligned}$$

for $j<m$, we find that

$$\rho(K)\le\int_K J_m f\,d\mathscr{L}^m+\xi\,v^{-1}\,\mathscr{L}^m(K),$$

where ξ is a number determined by n, m and $\operatorname{Lip}(f)$.

In case $K\in H$, then $\operatorname{diam} f(K)\le\operatorname{Lip}(f)\,m\,r$, hence

$$\rho(K)\le\alpha(n)[r+\operatorname{Lip}(f)\,m\,r]^n\,\alpha(n-m)^{-1}\,r^{m-n}=\eta\,\mathscr{L}^m(K),$$

with $\eta=\alpha(n)[1+\operatorname{Lip}(f)\,m]^n\,\alpha(n-m)^{-1}$.

We conclude that

$$\begin{aligned}\mathscr{L}^n\{z\colon \operatorname{dist}[z,f(C)]\}/[\alpha(n-m)\,r^{n-m}]&\le\sum_{K\in G}\rho(K)+\sum_{K\in H}\rho(K)\\&\le\int_{\bigcup G} J_m f\,d\mathscr{L}^m+\xi\,v^{-1}\,\mathscr{L}^m(\textstyle\bigcup G)+\eta\,\mathscr{L}^m(\textstyle\bigcup H)\\&\le\int_C J_m f\,d\mathscr{L}^m+([\operatorname{Lip}(f)]^m+\xi\,v^{-1})\,\mathscr{L}^m[(\textstyle\bigcup G)\sim C]\\&\qquad+\xi\,v^{-1}\mathscr{L}^m(C)+\eta\,\mathscr{L}^m(\textstyle\bigcup H)\\&\le\int_C J_m f\,d\mathscr{L}^m+v^{-1}([\operatorname{Lip}(f)]^m+\xi\,v^{-1}+\xi\,\mathscr{L}^m(C)+\eta).\end{aligned}$$

3.2.39. Theorem. *If W is a closed m rectifiable subset of $\mathbf{R}^n$, then*

$$\mathscr{M}^m(W)=\mathscr{H}^m(W).$$

Proof. We choose a Lipschitzian map $f\colon \mathbf{R}^m\to\mathbf{R}^n$ and a compact subset K of $\mathbf{R}^m$ with $W\subset f(K)$, hence

$$A=K\cap f^{-1}(W)\ \text{is compact and}\ f(A)=W.$$

Given any $\varepsilon>0$, we apply 3.2.3, 2.10.10, 2.10.8 to obtain a finite family G of closed subsets of A such that

$$\sum_{C\in G}\int_C J_m f\,d\mathscr{L}^m=\int_A J_m f\,d\mathscr{L}^m,\qquad \sum_{C\in G}\mathscr{H}^m[f(C)]\ge\int_A J_m f\,d\mathscr{L}^m-\varepsilon.$$

For $j=1,2,\ldots,v=\operatorname{card} G$ we define the set

$$S_j=\{y\colon \operatorname{card}(G\cap\{C\colon y\in f(C)\})\ge j\}$$

with the characteristic function s_j. Letting t_C be the characteristic function of $f(C)$, we observe that

$$\sum_{j=1}^{\nu} s_j = \sum_{C\in G} t_C, \qquad \sum_{j=1}^{\nu} \mathscr{H}^m(S_j) = \sum_{C\in G} \mathscr{H}^m[f(C)].$$

Furthermore, for $r>0$, we let $s_{j,r}$ and $t_{C,r}$ be the characteristic functions of the sets

$$\{z\colon \operatorname{dist}(z,S_j)<r\} \quad \text{and} \quad \{z\colon \operatorname{dist}[z,f(C)]<r\},$$

and note that

$$\sum_{j=1}^{\nu} s_{j,r} \le \sum_{C\in G} t_{C,r}.$$

Integrating this inequality with respect to $\mathscr{L}^n$, dividing by $\alpha(n-m)\,r^{n-m}$, and letting r approach 0 we obtain

$$\mathscr{M}^{*m}(S_1) + \sum_{j=2}^{\nu} \mathscr{M}_*^m(S_j) \le \sum_{C\in G} \mathscr{M}^{*m}[f(C)].$$

Next we apply 3.2.38 to infer that

$$\mathscr{M}^{*m}(S_1) - \mathscr{H}^m(S_1) + \sum_{j=2}^{\nu} [\mathscr{M}_*^m(S_j) - \mathscr{H}^m(S_j)]$$

$$\le \sum_{C\in G} \left(\textstyle\int_C J_m f\, d\mathscr{L}^m - \mathscr{H}^m[f(C)]\right) < \varepsilon.$$

Since $\mathscr{M}_*^m(S_j) \ge \mathscr{G}^m(S_j) = \mathscr{H}^m(S_j)$ for $j=1,2,\ldots,\nu$ according to 3.2.37 and 3.2.26, and since $S_1 = W$, we conclude

$$\mathscr{M}^{*m}(W) < \mathscr{H}^m(W) + \varepsilon \le \mathscr{M}_*^m(W) + \varepsilon.$$

3.2.40. Applying 2.10.28 with $0<m\le\frac{1}{2}$ and a sequence n such that $n_j \to \infty$ as $j \to \infty$, one obtains a compact set

$$E = A(m,n) \times A(m,n) \subset \mathbf{R}^2$$

for which $\mathscr{M}^{*1}(E) = \infty$, $\mathscr{M}_*^1(E) = 0$ if $m<\frac{1}{2}$, $0<\mathscr{M}_*^1(E)<\infty$ if $m=\frac{1}{2}$. Since $\mathscr{H}^1(E)=0$ if $m<\frac{1}{2}$, this example shows the need for the strong hypothesis in 3.2.38. A detailed discussion of a similar example may be found in [GR 2, pp. 184–187].

We refer to [KM 1, p. 387] for the construction of a compact subset S of $\mathbf{R}^3$ such that S is the union of a countable family of triangles and there exists an orthogonal projection of $\mathbf{R}^3$ onto $\mathbf{R}^2$ which maps S homeomorphically onto a square, yet $\mathscr{M}^2(S) = \infty > \mathscr{H}^2(S)$.

3.2.41. Brunn-Minkowski theorem. *If A and B are nonempty subsets of $\mathbf{R}^n$, then*

$$\mathscr{L}^n(A+B)^{1/n} \geq \mathscr{L}^n(A)^{1/n} + \mathscr{L}^n(B)^{1/n}$$

where $A+B=\{x+y: x\in A,\ y\in B\}$.

Proof. Let F be the family of all rectangular boxes

$$P_1 \times \cdots \times P_n$$

where $P_1, \ldots, P_n$ are nonempty, bounded, open subintervals of $\mathbf{R}$.

If $A = P_1 \times \cdots \times P_n \in F$ and $B = Q_1 \times \cdots \times Q_n \in F$, then

$$A+B=(P_1+Q_1)\times \cdots \times (P_n+Q_n);$$

taking $u_j = \mathscr{L}^1(P_j)/\mathscr{L}^1(P_j+Q_j)$ and $v_j = \mathscr{L}^1(Q_j)/\mathscr{L}^1(P_j+Q_j)$ for $j=1,\ldots,n$ we recall 2.4.13 to obtain

$$[\mathscr{L}^n(A)^{1/n}+\mathscr{L}^n(B)^{1/n}]\,\mathscr{L}^n(A+B)^{-1/n}$$

$$= \prod_{j=1}^n u_j^{1/n} + \prod_{j=1}^n v_j^{1/n} \leq \sum_{j=1}^n u_j/n + \sum_{j=1}^n v_j/n = 1$$

because $u_j + v_j = 1$.

Next we treat the case when $A = \bigcup G$ and $B = \bigcup H$ for some finite disjointed subfamilies G and H of F, by applying induction with respect to $\operatorname{card}(G)+\operatorname{card}(H)$.

If $\operatorname{card}(G)>1$ we choose $i\in\{1,\ldots,n\}$ and $a\in\mathbf{R}$ so that each of the two sets

$$A_1 = A \cap \{x: x_i < a\}, \quad A_2 = A\cap\{x: x_i > a\}$$

contains some member of G, and also choose $b\in\mathbf{R}$ so that the sets

$$B_1 = B\cap\{x: x_i<b\}, \quad B_2 = B\cap \{x: x_i > b\}$$

satisfy the equations

$$\mathscr{L}^n(A_k)/\mathscr{L}^n(A) = \mathscr{L}^n(B_k)/\mathscr{L}^n(B) \text{ for } k=1,2;$$

defining

$$G_k = \{U\cap A_k: U\in G,\ U\cap A_k \neq \varnothing\}, \quad H_k = \{V\cap B_k: V\in H,\ V\cap B_k\neq\varnothing\},$$

we see that $A_k = \bigcup G_k$ and $B_k = \bigcup H_k$ with

$$\operatorname{card}(G_k) < \operatorname{card}(G) \quad \text{and} \quad \operatorname{card}(H_k) \leq \operatorname{card}(H);$$

since A_1+B_1 and A_2+B_2 are separated by $\{x: x_i = a+b\}$, induction yields

$$\begin{aligned}\mathscr{L}^n(A+B) &\geq \mathscr{L}^n(A_1+B_1)+\mathscr{L}^n(A_2+B_2)\\ &\geq [\mathscr{L}^n(A_1)^{1/n}+\mathscr{L}^n(B_1)^{1/n}]^n + [\mathscr{L}^n(A_2)^{1/n}+\mathscr{L}^n(B_2)^{1/n}]^n\\ &= [\mathscr{L}^n(A)^{1/n}+\mathscr{L}^n(B)^{1/n}]^n.\end{aligned}$$

From the preceding elementary case one readily infers by approximation that the conclusion of the theorem holds when A and B are compact, hence when A and B are $\mathscr{L}^n$ measurable.

Finally, to treat the quite general case, we associate with each subset S of $\mathbf{R}^n$ its $\mathscr{L}^n$ hull (see 2.9.11)

$$S^* = \mathbf{R}^n \cap \{x: \lim_{r \to 0+} \mathscr{L}^n[S \cap B(x,r)]/[\alpha(n)\, r^n] = 1\}$$

and observe that $A^* + B^* \subset (A+B)^*$.

3.2.42. Corollary. *If* $S \subset \mathbf{R}^n$, $\mathscr{L}^n(\operatorname{Clos} S) < \infty$ *and* $0 < r < \infty$, *then*

$$\mathscr{L}^n\{x: 0 < \operatorname{dist}(x,S) < r\}/r \geq n\,\alpha(n)^{1/n} \mathscr{L}^n(\operatorname{Clos} S)^{(n-1)/n},$$

$$\mathscr{L}^n\{x: 0 < \operatorname{dist}(x, \mathbf{R}^n \sim S) < r\}/r$$
$$\geq n\,\alpha(n)^{1/n} \mathscr{L}^n\{x: \operatorname{dist}(x, \mathbf{R}^n \sim S) \geq r\}^{(n-1)/n}.$$

Proof. $\{x: \operatorname{dist}(x,S) < r\} = (\operatorname{Clos} S) + \mathbf{U}(0,r)$, hence

$$\mathscr{L}^n\{x: \operatorname{dist}(x,S) < r\} \geq [\mathscr{L}^n(\operatorname{Clos} S)^{1/n} + \alpha(n)^{1/n} r]^n$$
$$\geq \mathscr{L}^n(\operatorname{Clos} S) + n\,\mathscr{L}^n(\operatorname{Clos} S)^{(n-1)/n} \alpha(n)^{1/n} r.$$

Moreover, if $T = \{x: \operatorname{dist}(x, \mathbf{R}^n \sim S) \geq r\}$, then $\operatorname{Int} S \supset T + \mathbf{U}(0,r)$,

$$\mathscr{L}^n(\operatorname{Int} S) \geq \mathscr{L}^n(T) + n\,\mathscr{L}^n(T)^{(n-1)/n} \alpha(n)^{1/n} r.$$

3.2.43. Isoperimetric inequality. *If* $S \subset \mathbf{R}^n$ *and* $\mathscr{L}^n(\operatorname{Clos} S) < \infty$, *then*

$$\mathscr{M}_*^{n-1}(\operatorname{Bdry} S) \geq n\,\alpha(n)^{1/n}\, \mathscr{L}^n(\operatorname{Clos} S)^{(n-1)/n}.$$

Proof. We apply 3.2.42 observing that, for $0 < r < \infty$,

$$\{x: 0 < \operatorname{dist}(x,S) < r\} \quad \text{and} \quad \{x: 0 < \operatorname{dist}(x, \mathbf{R}^n \sim S) < r\}$$

are disjoint subsets of $\{x: \operatorname{dist}(x, \operatorname{Bdry} S) < r\}$. Furthermore

$$\mathscr{L}^n\{x: \operatorname{dist}(x, \mathbf{R}^n \sim S) \geq r\} \uparrow \mathscr{L}^n(\operatorname{Int} S) \text{ as } r \downarrow 0,$$

and $\mathscr{M}_*^{n-1}(\operatorname{Bdry} S) < \infty$ implies $\mathscr{L}^n(\operatorname{Bdry} S) = 0$.

3.2.44. According to 3.2.39, $\mathscr{M}_*^{n-1}(\operatorname{Bdry} S)$ may be replaced by $\mathscr{H}^{n-1}(\operatorname{Bdry} S)$ in 3.2.43, provided $\operatorname{Bdry} S$ is $n-1$ rectifiable; however this restriction will be removed in 4.5.9(31).

3.2.45. Assuming $n \geq m \geq 1$ are integers and $A \subset \mathbf{R}^n$ we will now prove that $\mathcal{Q}_t^m(A)$ *is continuous with respect to t on* $\{t: 1 \leq t \leq \infty\}$. We define ζ_t^m as in 2.10.5 and observe that there exists a finite positive number γ such that

$$\zeta_\infty^m(S) \leq \gamma\, \zeta_1^m(S) \text{ for every bounded convex set } S \subset \mathbf{R}^n;$$

in fact 2.10.32 and the proof of 2.10.37 imply this inequality with $\gamma = \alpha(m)\, 2^{-m}\, m!$; a more refined argument outlined below shows that it holds with $\gamma = 1$. Letting

$$h_S(u) = \log[\beta_{1/u}(n,m)\, \zeta_{1/u}^m(S)] \text{ for } 0 \leq u \leq 1,$$

where $1/0 = \infty$, we see from 2.4.17 that the function h_S is convex and nonincreasing, hence

$$0 \leq [h_S(u) - h_S(v)]/(v-u) \leq h_S(0) - h_S(1) \leq \log[\gamma/\beta_1(n,m)]$$

whenever $0 \leq u < v \leq 1$. We conclude that

$$1 \leq \frac{\beta_{1/u}(n,m)\, \mathcal{Q}_{1/u}^m(A)}{\beta_{1/v}(n,m)\, \mathcal{Q}_{1/v}^m(A)} \leq [\gamma/\beta_1(n,m)]^{v-u} \text{ for } 0 \leq u < v \leq 1.$$

A relevant example will be constructed in 3.3.19.

Next we derive the inequality

$$\int_{\mathbf{R}^m} \zeta_1^k [\mathbf{R}^{n-m} \cap \{z: (y,z) \in C\}]\, d\mathcal{L}^m y \leq \zeta_1^{m+k}(C)$$

for every nonempty bounded convex subset C of $\mathbf{R}^m \times \mathbf{R}^{n-m}$ and $k = 0, \ldots, n-m$. Letting

$$T = (\mathbf{R}^m \times \mathbf{R}^{n-m}) \cap \{(y,z): y = 0, |z| < 1\}$$

we apply Fubini's theorem and 3.2.35 to compute

$$(\mathcal{L}^m \times \mathcal{L}^{n-m})(C + rT) = \int \mathcal{L}^{n-m}\{w: \operatorname{dist}(w, \{z: (y,z) \in C\}) < r\}\, d\mathcal{L}^m y$$
$$= \int \sum_{k=0}^{n-m} \zeta_1^k\{z: (y,z) \in C\}\, \alpha(n-m-k)\, r^{n-m-k}\, d\mathcal{L}^m y$$

whenever $0 < r < \infty$. On the other hand 3.2.35 tells us that

$$(\mathcal{L}^m \times \mathcal{L}^{n-m})[C + r\mathbf{U}(0,1)] = \sum_{j=0}^{n} \zeta_1^j(C)\, \alpha(n-j)\, r^{n-j},$$

and since $T \subset \mathbf{U}(0,1)$ we infer from a fundamental property of mixed volumes of convex sets (see [BF, p. 41, Proposition 5] or [EG, p. 86, Theorem 42]) that each coefficient in the first polynomial is less than or equal to the coefficient of the same power of r in the second polynomial. Thus we obtain the asserted inequality.

Applying the case $k=0$ we recall that $\zeta_1^0(E)=1$ for every nonempty bounded convex set E, and obtain

$$\mathscr{L}^m\{y\colon (y,z)\in C \text{ for some } z\}\le\zeta_1^m(C).$$

Thus $\mathscr{L}^m[p(S)]\le\zeta_1^m(S)$ for every bounded convex subset S of $\mathbf{R}^n$ and every $p\in\mathbf{O}^*(n,m)$, hence $\zeta_\infty^m(S)\le\zeta_1^m(S)$, and we conclude

$$\mathscr{Q}_\infty^m\le\mathscr{Q}_1^m.$$

For all $k\in\{0,\ldots,n-m\}$ it readily follows from our inequality that

$$\int_{\mathbf{R}^m}^* \mathscr{Q}_1^k[\mathbf{R}^{n-m}\cap\{z\colon (y,z)\in B\}]\,d\mathscr{L}^m y\le\mathscr{Q}_1^{m+k}(B)$$

whenever $B\subset\mathbf{R}^m\times\mathbf{R}^{n-m}$. In particular we see that the left member cannot exceed the right member of the equation

$$\mathscr{L}^m(U)\cdot\mathscr{Q}_1^k(V)=\mathscr{Q}_1^{m+k}(U\times V),$$

when U is an $\mathscr{L}^m$ measurable set and $V\subset\mathbf{R}^{n-m}$. To prove that the right member does not exceed the left we note that, for every nonempty bounded convex subset Q of $\mathbf{R}^{n-m}$,

$$\beta_1(n-m,j)\,\zeta_1^j(Q)\le\beta_1(n-m,k)\,\zeta_1^k(Q)\,(\operatorname{diam} Q)^{j-k}$$

whenever $j\ge k$, hence infer from 3.2.36 that, for every m dimensional cube P with side length s contained in $\mathbf{R}^m$,

$$\begin{aligned}\zeta_1^{m+k}(P\times Q)&=\sum_i\binom{m}{i}s^i\,\zeta_1^{m+k-i}(Q)\le\sum_{i=0}^m\binom{m}{i}s^i\,\zeta_1^k(Q)\,(M\operatorname{diam} Q)^{m-i}\\&=\mathscr{L}^m(P)\,\zeta_1^k(Q)\,[1+M(\operatorname{diam} Q)/s]^m\end{aligned}$$

where M is a suitable finite constant depending only on m,n,k; then we cover $U\times V$ by families of sets $P\times Q$ such that s and $(\operatorname{diam} Q)/s$ are small.

The above results, which depend on references to the theory of mixed volumes, will not be used later in this book.

3.2.46. We will now describe a simple procedure whereby much of the theory of Hausdorff measure may be extended from Euclidean spaces to arbitrary Riemannian manifolds (as defined in [HE, p. 47] or [KN, p. 154] or [ST, p. 85, 173]). Since measure theoretic problems are easily localized, we consider here only the case where the whole manifold is the domain of a single coordinate system; thus we will study an open subset of Euclidean space with a new metric.

Suppose A is an open subset of $\mathbf{R}^n$ and

$$g: A \to \odot^2(\mathbf{R}^n)$$

is a continuous map such that $g(x)$ is an inner product for each $x \in A$. Corresponding to this Riemannian tensor g we define the distance $\rho(x,y)$ between $x, y \in A$ as the infimum of the set of numbers

$$\int_K \langle C'(t) \odot C'(t), g[C(t)] \rangle^{\frac{1}{2}} \, d\mathscr{L}^1 t$$

corresponding to all Lipschitzian functions C mapping some compact interval K into A with $C(\inf K) = x$, $C(\sup K) = y$. Clearly ρ is a symmetric function satisfying the triangle inequality; the next paragraph shows that it is a metric inducing the usual topology of A.

For each $a \in A$ we apply 1.7.3 to obtain orthonormal base vectors $e_1, \ldots, e_n$ of $\mathbf{R}^n$ such that

$$\langle e_i \odot e_j, g(a) \rangle = 0 \text{ whenever } i \neq j,$$

let L be the symmetric linear automorphism of $\mathbf{R}^n$ such that

$$L(e_i) = \langle e_i \odot e_i, g(a) \rangle^{\frac{1}{2}} e_i \text{ for } i = 1, \ldots, n,$$

and note that

$$\langle v \odot v, g(a) \rangle^{\frac{1}{2}} = |L(v)| \text{ whenever } v \in \mathbf{R}^n.$$

Given $1 < \lambda < \infty$ we use the continuity of g at a to choose $\delta > 0$ so that

$$\lambda^{-1} |L(v)| \le \langle v \odot v, g(x) \rangle^{\frac{1}{2}} \le \lambda |L(v)|$$

whenever $v \in \mathbf{R}^n$ and $|x - a| \le \delta$. Comparing each of the above integrals with the Euclidean curve length of $L \circ C$ we find that

$$\lambda^{-1} |L(x-y)| \le \rho(x,y) \le \lambda |L(x-y)|$$

whenever $|L(x-a)| \le \delta/2$ and $|L(y-a)| \le \delta/2$. It follows that

$$\lambda^{-m} \mathscr{H}^m[L(S)] \le \mathscr{H}^m_\rho(S) \le \lambda^m \mathscr{H}^m[L(S)]$$

whenever $S \subset \{x: |L(x-a)| \le \delta/2\}$ and $0 \le m < \infty$, where $\mathscr{H}^m_\rho$ and $\mathscr{H}^m$ are the m dimensional Hausdorff measures corresponding to ρ and the ordinary metric of $\mathbf{R}^n$; in particular (see 1.7.10)

$$\lambda^{-n} [\operatorname{discr} g(a)]^{\frac{1}{2}} \mathscr{L}^n(S) \le \mathscr{H}^n_\rho(S) \le \lambda^n [\operatorname{discr} g(a)]^{\frac{1}{2}} \mathscr{L}^n(S)$$

because $\det(L) = [\operatorname{discr} g(a)]^{\frac{1}{2}}$. Consequently the derivate of $\mathscr{H}^n_\rho$ with respect to $\mathscr{L}^n$ at a equals $[\operatorname{discr} g(a)]^{\frac{1}{2}}$.

We conclude that

$$\mathscr{H}_\rho^n(B) = \int_B (\operatorname{discr} g)^{\frac{1}{2}} \, d\mathscr{L}^n$$

for every $\mathscr{L}^n$ measurable subset B of A. Using linear maps L as above (through which the usual inner product of $\mathbf{R}^n$ becomes tangent to g at a in the sense of [CE, p. 86]) one also verifies easily that the basic transformation formulae 3.2.5, 3.2.12 remain true for Hausdorff measures induced by Riemannian metrics, provided the Jacobians are computed with respect to the Riemannian inner products.

3.2.47. In the next theorem we will derive an integralgeometric formula concerning the Hausdorff measures of rectifiable subsets of $\mathbf{S}^{n-1}$, by studying $\mathbf{S}^{n-1}$ as a homogeneous space of $\mathbf{O}(n)$. To prepare for this we fix here a point $c \in \mathbf{S}^{n-1}$ and consider the $\mathbf{O}(n)$ left covariant map

$$\Phi: \mathbf{O}(n) \to \mathbf{S}^{n-1}, \qquad \Phi(g) = g(c) \text{ for } g \in \mathbf{O}(n),$$

with the isotropy group $I = \Phi^{-1}\{c\} \simeq \mathbf{O}(n-1)$. We also choose an orthonormal base $e_1, \ldots, e_n$ of $\mathbf{R}^n$ such that $e_1 = c$, let

$$U = \mathbf{S}^{n-1} \cap \{u \colon u \wedge e_2 \wedge \cdots \wedge e_n \neq 0\},$$

associate with each $u \in U$ inductively the vectors $f_1(u) = u$ and

$$f_i(u) = \Big[e_i - \sum_{j<i} [e_i \bullet f_j(u)] \, f_j(u)\Big] \Big/ \Big|e_i - \sum_{j<i} [e_i \bullet f_j(u)] \, f_j(u)\Big|$$

corresponding to $i = 2, \ldots, n$, then define $F(u) \in \mathbf{O}(n)$ so that

$$\langle e_i, F(u) \rangle = f_i(u) \text{ for } i = 1, \ldots, n.$$

Clearly U is a neighborhood of c relative to $\mathbf{S}^{n-1}$, the functions $f_1, \ldots, f_n$, F are differentiable, and $\Phi \circ F = \mathbf{1}_U$. Whenever $\xi \in \mathbf{O}(n)$ we let

$$H_\xi \colon \Phi^{-1}[\xi(U)] = \xi \circ \Phi^{-1}(U) \to I,$$

$$H_\xi(g) = [(F \circ \Phi)(\xi^{-1} \circ g)]^{-1} \circ \xi^{-1} \circ g \text{ for } g \in \Phi^{-1}[\xi(U)],$$

and construct a diffeomorphism

$$\Phi^{-1}[\xi(U)] \simeq \xi(U) \times I$$

by mapping g onto $(\Phi(g), H_\xi(g))$. We abbreviate

$$\nu = n(n-1)/2, \qquad \mu = \nu - n + 1, \qquad \mathbf{1} = \mathbf{1}_{\mathbf{R}^n}.$$

Recalling 3.2.28 one now readily verifies that, for each $g \in \mathbf{O}(n)$,

$$\operatorname{Tan}[\mathbf{O}(n), g] = g(\operatorname{Tan}[\mathbf{O}(n), 1]), \qquad \ker D\Phi(g) = g[\operatorname{Tan}(I, 1)],$$

and $D\Phi(g)$ maps

$$W(g) = \operatorname{Tan}[\mathbf{O}(n), g] \cap \{w: w \bullet z = 0 \text{ for } z \in \ker D\Phi(g)\}$$

onto $\operatorname{Tan}[\mathbf{S}^{n-1}, \Phi(g)]$ with

$$|\langle w, D\Phi(g)\rangle| = 2^{-\frac{1}{2}} |w| \text{ whenever } w \in W(g);$$

in fact when base vectors $a_{i,j}$ are chosen as in 3.2.28 (3) with $m = n$ and $e_1 = c$, hence $v_1 = \Phi(g)$, then $D\Phi(g)$ is represented by the function mapping $a_{i,j}$ onto

$$2^{-\frac{1}{2}}[\omega_i(c)\, v_j - \omega_j(c)\, v_i],$$

which equals 0 in case $i > 1$, and equals $2^{-\frac{1}{2}} v_j$ in case $i = 1$.

The following three statements hold when $E \subset \mathbf{S}^{n-1}$ and r is a nonnegative integer:

(1) *If E is r rectifiable, then $\Phi^{-1}(E)$ is $r+\mu$ rectifiable.*

(2) *If $\mathscr{H}^r(E) = 0$, then $\mathscr{H}^{r+\mu}[\Phi^{-1}(E)] = 0$.*

(3) *If $\Phi^{-1}(E)$ is $(\mathscr{H}^{r+\mu}, r+\mu)$ rectifiable, then E is $(\mathscr{H}^r, r)$ rectifiable and*

$$\mathscr{H}^{r+\mu}[\Phi^{-1}(E)] = 2^{r/2} \mathscr{H}^\mu(I)\, \mathscr{H}^r(E).$$

Since the sets $\xi(U)$ form an open covering of $\mathbf{S}^{n-1}$, the proof is easily reduced to the special case when $E \subset \xi(U)$ for some $\xi \in \mathbf{O}(n)$, hence the above diffeomorphism maps $\Phi^{-1}(E)$ onto $E \times I$. Noting that I is a μ dimensional submanifold of class ∞ of $\operatorname{Hom}(\mathbf{R}^n, \mathbf{R}^n)$, we find (1) obvious and deduce (2) from 2.10.45. Applying 3.2.22 (2) with f replaced by $H_\xi | \Phi^{-1}(E)$ we obtain the first conclusion of (3) because

$$\Phi[\Phi^{-1}(E) \cap H_\xi^{-1}\{h\}] = E \text{ whenever } h \in I.$$

Finally we derive the second conclusion of (3) by applying 3.2.22 (3) with f replaced by $\Phi | \Phi^{-1}(E)$; in view of 3.2.29 and (2) we may assume that E is an r dimensional submanifold of class 1 of $\mathbf{R}^n$, hence

$$\ker D\Phi(g) \subset \operatorname{Tan}[\Phi^{-1}(E), g],$$

$$\dim(W(g) \cap \operatorname{Tan}[\Phi^{-1}(E), g]) = r, \quad J_r[\Phi | \Phi^{-1}(E)](g) = 2^{-r/2}$$

whenever $g \in \Phi^{-1}(E)$; moreover $\Phi^{-1}\{z\}$ is isometric with I for each $z \in \mathbf{S}^{n-1}$.

3.2.48. Theorem. *If A and B are subsets of $\mathbf{S}^{n-1}$, α and β are real valued functions on $\mathbf{S}^{n-1}$,*

$$A \text{ is } (\mathscr{H}^k, k) \text{ rectifiable and } \mathscr{H}^k \text{ measurable},$$

$$B \text{ is } m \text{ rectifiable and } \mathscr{H}^m \text{ measurable},$$

$$\alpha \text{ is } \mathscr{H}^k \llcorner A \text{ summable}, \qquad \beta \text{ is } \mathscr{H}^m \llcorner B \text{ summable},$$

and $\lambda = k+m-n+1 \geq 0$, then $A \cap g(B)$ is $(\mathscr{H}^\lambda, \lambda)$ rectifiable for $\boldsymbol{\theta}_n$ almost all g in $\mathbf{O}(n)$ and

$$\int_{\mathbf{O}(n)} \int_{A \cap g(B)} \alpha \cdot (\beta \circ g^{-1}) \, d\mathscr{H}^\lambda \, d\boldsymbol{\theta}_n g$$
$$= \frac{\Gamma[(k+1)/2] \, \Gamma[(m+1)/2]}{2\Gamma[\frac{1}{2}]^n \Gamma[(k+m-n+2)/2]} \cdot \int_A \alpha \, d\mathscr{H}^k \cdot \int_B \beta \, d\mathscr{H}^m.$$

Proof. Readopting the conventions of 3.2.47 we see that

$$X = \Phi^{-1}(A) \text{ is } (\mathscr{H}^{k+\mu}, k+\mu) \text{ rectifiable},$$

$$Y = \Phi^{-1}(B) \text{ is } m+\mu \text{ rectifiable},$$

$$X \times Y \text{ is } (\mathscr{H}^{k+m+2\mu}, k+m+2\mu) \text{ rectifiable}$$

according to 3.2.23. We now define

$$\Psi: X \times Y \to \mathbf{O}(n), \qquad \Psi(x,y) = x \circ y^{-1} \text{ for } (x,y) \in X \times Y,$$

and apply 3.2.22 with f, m, μ, $m-\mu$ replaced by Ψ, $k+m+2\mu$, ν, $\lambda+\mu$. Whenever $g \in \mathbf{O}(n)$,

$$\Psi^{-1}\{g\} = \{(x, g^{-1} \circ x) : x \in \Phi^{-1}[A \cap g(B)]\}$$

and the function mapping x onto $(x, g^{-1} \circ x)$ magnifies all distances by the factor $\sqrt{2}$; moreover, for $\mathscr{H}^\nu$ almost all g in $\mathbf{O}(n)$, $\Psi^{-1}\{g\}$ is $(\mathscr{H}^{\lambda+\mu}, \lambda+\mu)$ rectifiable, hence $A \cap g(B)$ is $(\mathscr{H}^\lambda, \lambda)$ rectifiable, and

$$\int_{\Psi^{-1}\{g\}} \alpha[\Phi(x)] \cdot \beta[\Phi(y)] \, d\mathscr{H}^{\lambda+\mu}(x,y)$$
$$= 2^{(\lambda+\mu)/2} \int_{\Phi^{-1}[A \cap g(B)]} \alpha[\Phi(x)] \cdot \beta[\Phi(g^{-1} \circ x)] \, d\mathscr{H}^{\lambda+\mu} x$$
$$= 2^{\mu/2+\lambda} \mathscr{H}^\mu(I) \int_{A \cap g(B)} \alpha(a) \cdot \beta[g^{-1}(a)] \, d\mathscr{H}^\lambda a.$$

Abbreviating

$$\Omega = \int_{X \times Y} \alpha[\Phi(x)] \cdot \beta[\Phi(y)] \cdot \operatorname{ap} J_\nu \Psi(x,y) \, d\mathscr{H}^{k+m+2\mu}(x,y)$$

we infer that

$$\Omega = 2^{\mu/2+\lambda} \mathscr{H}^\mu(I) \int_{\mathbf{O}(n)} \int_{A \cap g(B)} \alpha \cdot (\beta \circ g^{-1}) \, d\mathscr{H}^\lambda \, d\mathscr{H}^\nu g$$
$$= 2^{\nu/2+\lambda} [\mathscr{H}^\mu(I)]^2 \cdot \mathscr{H}^{n-1}(\mathbf{S}^{n-1}) \cdot \int_{\mathbf{O}(n)} \int_{A \cap g(B)} \alpha \cdot (\beta \circ g^{-1}) \, d\mathscr{H}^\lambda \, d\boldsymbol{\theta}^n g$$

because $\mathscr{H}^\nu[\mathbf{O}(n)] = 2^{(n-1)/2} \mathscr{H}^\mu(I) \cdot \mathscr{H}^{n-1}(\mathbf{S}^{n-1})$ and $\mathscr{H}^\nu \llcorner \mathbf{O}(n)$ is a Haar measure of $\mathbf{O}(n)$.

If $\mathscr{H}^k(A)=0$ or $\mathscr{H}^m(B)=0$, then $\Omega=0$.

In view of 3.2.29 we may henceforth assume that A and B are k and m dimensional submanifolds of class 1 of $\mathbf{R}^n$; this allows us to use tangent spaces rather than approximate tangent spaces in the following computation of the Jacobian of Ψ.

For $x\in X$ and $y\in Y$ we let

$$S(x)=x^{-1}(\operatorname{Tan}[A,\Phi(x)]), \qquad S'(x)=W(1)\cap[x^{-1}\circ\operatorname{Tan}(X,x)],$$
$$T(y)=y^{-1}(\operatorname{Tan}[B,\Phi(y)]), \qquad T'(y)=W(1)\cap[y^{-1}\circ\operatorname{Tan}(Y,y)];$$

we also choose $\sigma(x)\in\bigwedge_k\operatorname{Tan}(\mathbf{S}^{n-1},c)$ and $\tau(y)\in\bigwedge_m\operatorname{Tan}(\mathbf{S}^{n-1},c)$ so that $\sigma(x)$ and $\tau(y)$ are simple, $|\sigma(x)|=1=|\tau(y)|$,

$$S(x) \text{ is associated with } \sigma(x),\ T(y) \text{ is associated with } \tau(y).$$

Observing that Ψ is constructed by composition from a bilinear map and inversion, we see from 3.1.1 and 3.1.11 that

$$\begin{aligned}\langle(u,v),D\Psi(x,y)\rangle&=u\circ y^{-1}-x\circ y^{-1}\circ v\circ y^{-1}\\&=x\circ(x^{-1}\circ u-y^{-1}\circ v)\circ y^{-1}\end{aligned}$$

whenever $u\in\operatorname{Tan}(X,x)$ and $v\in\operatorname{Tan}(Y,y)$, hence obtain the commutative diagram:

$$\begin{array}{ccc}
\operatorname{Tan}[X\times Y,(x,y)] & \xrightarrow{D\Psi(x,y)} & \operatorname{Tan}[\mathbf{O}(n),x\circ y^{-1}]\\
\wr\| & & \\
\operatorname{Tan}(X,x)\times\operatorname{Tan}(Y,y) & & \Big\uparrow{\scriptstyle x\circ\,|\,\circ y^{-1}}\\
\Big\downarrow{\scriptstyle x^{-1}\circ\,|\,y^{-1}\circ} & & \\
[x^{-1}\circ\operatorname{Tan}(X,x)]\times[y^{-1}\circ\operatorname{Tan}(Y,y)] & \xrightarrow{-} & \operatorname{Tan}[\mathbf{O}(n),1]\\
\| & & \|\\
[\operatorname{Tan}(I,1)\oplus S'(x)]\times[\operatorname{Tan}(I,1)\oplus T'(y)] & \xrightarrow{-} & \operatorname{Tan}(I,1)\oplus W(1)
\end{array}$$

Using vectorsubtraction to define

$$f_1\colon \operatorname{Tan}(I,1)\times\operatorname{Tan}(I,1)\to\operatorname{Tan}(I,1)$$

as well as the functions f_2 and f_3 in the commutative diagram

$$\begin{array}{ccc}
S'(x)\times T'(y) & \xrightarrow{f_2} & W(1)\\
\downarrow & & \downarrow\\
S(x)\times T(y) & \xrightarrow{f_3} & \operatorname{Tan}(\mathbf{S}^{n-1},c),
\end{array}$$

whose vertical maps are induced by $D\Phi(1)$ and multiply norms by $2^{-\frac{1}{2}}$, we apply 1.7.11 twice to compute

$$J_\nu \Psi(x,y) = \|\wedge_\mu f_1\| \cdot \|\wedge_{n-1} f_2\| = 2^{\mu/2} \cdot \|\wedge_{n-1} f_3\|$$
$$= 2^{(\mu+\lambda)/2}\, |{*\sigma(x)} \wedge {*\tau(y)}|$$

where $*$ is defined relative to $\operatorname{Tan}(\mathbf{S}^{n-1}, c)$.

Next we choose a Haar measure γ of I with $\gamma(I)=1$, and observe that $\{h|\operatorname{Tan}(\mathbf{S}^{n-1}, c) : h \in I\}$ equals the orthogonal group of $\operatorname{Tan}(\mathbf{S}^{n-1}, c)$. Since

$$\Phi(y \circ h^{-1}) = \Phi(y), \qquad T(y \circ h^{-1}) = h[T(y)],$$
$$\tau(y \circ h^{-1}) = \pm \langle \tau(y), \wedge_m [h|\operatorname{Tan}(\mathbf{S}^{n-1}, c)] \rangle$$

whenever $h \in I$, we infer with the help of 2.7.16(6) that

$$\int_I |{*\sigma(x)} \wedge {*\tau(y \circ h^{-1})}|\, d\gamma\, h$$
$$= \int_I |{*\sigma(x)} \wedge \langle {*\tau(y)}, \wedge_m [h|\operatorname{Tan}(\mathbf{S}^{n-1}, c)] \rangle|\, d\gamma\, h$$
$$= \boldsymbol{\beta}_1(n-1, n-1-m) / \boldsymbol{\beta}_1(k, n-1-m).$$

Making use of the fact that the metric of $\mathbf{O}(n)$ and the function Φ are invariant under right multiplication by members of I, we now conclude

$$\Omega = \int_I \int_{X \times Y} \alpha[\Phi(x)] \cdot \beta[\Phi(y)] \cdot J_\nu \Psi(x, y \circ h^{-1})\, d\mathscr{H}^{k+m+2\mu}(x,y)\, d\gamma\, h$$
$$= \int_X \int_Y 2^{(\mu+\lambda)/2}\, \boldsymbol{\beta}_1(n-1, n-1-m) \cdot \boldsymbol{\beta}_1(k, n-1-m)^{-1}$$
$$\cdot\, \alpha[\Phi(x)] \cdot \beta[\Phi(y)]\, d\mathscr{H}^{k+\mu} x\, d\mathscr{H}^{m+\mu} y$$
$$= 2^{\nu/2+\lambda} [\mathscr{H}^\mu(I)]^2\, \boldsymbol{\beta}_1(n-1, n-1-m) \cdot \boldsymbol{\beta}_1(k, n-1-m)^{-1}$$
$$\cdot \int_A \alpha\, d\mathscr{H}^k \cdot \int_B \beta\, d\mathscr{H}^m.$$

Equating this value of Ω with the value obtained earlier in the proof, and using 3.2.13 to compute the constants, one obtains the equation asserted in the theorem.

3.2.49. The preceding theorem is a special case of the theory developed in [BJ1] for general Riemannian manifolds with a transitive Lie group of isometries. As long as the isotropy group at each point is represented by the full orthogonal group of the tangent space at that point, the argument remains essentially the same as that given here. However if the group is tangentially less transitive, then the choice of sets A and B must be correspondingly restricted; the modular function of the group may also appear in the formula. For example in studying the action of the unitary group $\mathbf{U}(n)$ on $\mathbf{S}^{2n-1}$ one assumes that the tangent spaces of A and B are complex vectorsubspaces of $\mathbf{C}^n$. Earlier work in [F 3, 10] on integralgeometric theorems for subsets of $\mathbf{R}^n$ was simplified by the fact that the group of all isometries of $\mathbf{R}^n$ is the semidirect product of the isotropy group $\mathbf{O}(n)$ and the group of translations. Similar methods were used in [F 15] to prove the principal kinematic formula concerning the curvature measures associated with subsets of $\mathbf{R}^n$ which have positive reach. Very likely the kinematic formula could also be extended to more general Riemannian homogeneous spaces.

I do not know whether the hypothesis "B is m rectifiable" could be replaced in 3.2.48 by the weaker assumption "B is $(\mathscr{H}^m, m)$ rectifiable". This problem may be restated as follows: *Do the conditions*

$$A \subset \mathbf{S}^{n-1}, \quad \mathscr{H}^k(A)=0, \quad B \subset \mathbf{S}^{n-1}, \quad \mathscr{H}^m(B)=0$$

imply that $\mathscr{H}^{k+m-n+1}[A \cap g(B)]=0$ *for* θ_n *almost all* g *in* $\mathbf{O}(n)$?

3.3. Structure theory

3.3.1 Assuming throughout this section that *ϕ measures $\mathbf{R}^n$ and all closed subsets of $\mathbf{R}^n$ are ϕ measurable*, we will develop criteria for the (ϕ, m) rectifiability of subsets of $\mathbf{R}^n$. We will study sets locally, near any point of $\mathbf{R}^n$, by examining their position relative to $n-m$ dimensional affine subspaces of $\mathbf{R}^n$ through that point.

For $a \in \mathbf{R}^n$, $0 < r \leq \infty$, $V \in \mathbf{G}(n, n-m)$, $0 < s < 1$ we define

$$\mathbf{X}(a, r, V, s) = \mathbf{R}^n \cap \{x\colon s^{-1} \operatorname{dist}(x-a, V) < |x-a| < r\}.$$

We note that, if $p \in \mathbf{O}^*(n, m)$, then $\operatorname{dist}(x, \ker p) = |p(x)|$ whenever $x \in \mathbf{R}^n$, hence

$$\mathbf{X}(a, r, \ker p, s) = \mathbf{R}^n \cap \{x\colon s^{-1}|p(x-a)| < |x-a| < r\}.$$

The map associating $\ker p \in \mathbf{G}(n, n-m)$ with $p \in \mathbf{O}^*(n, m)$ is continuous and open, and it carries $\theta^*_{n,m}$ onto $\gamma_{n,n-m}$ [see 2.7.16(6)].

3.3.2. Lemma. *If $A \subset \mathbf{R}^n$ and $\delta > 0$, then the function mapping (a, p) onto*

$$\limsup_{s \to 0+} \sup_{0 < r < \delta} \phi[A \cap \mathbf{X}(a, r, \ker p, s)]\, r^{-m} s^{-m}$$

is a Borel function on $\mathbf{R}^n \times \mathbf{O}^*(n, m)$.

Proof. Replacing ϕ by $\phi \llcorner A$, we may assume $A = \mathbf{R}^n$.

Suppose $\lambda \in \mathbf{R}$. First we observe that

$$\{(a, r, p, s)\colon C \subset \mathbf{X}(a, r, \ker p, s)\}$$

is open in $\mathbf{R}^n \times \mathbf{R} \times \mathbf{O}^*(n, m) \times \mathbf{R}$ whenever C is a compact subset of $\mathbf{R}^n$. Since each $\mathbf{X}(a, r, \ker p, s)$ is the union of an increasing sequence of compact subsets, it follows from 2.1.3(4) that

$$S=\{(a,r,p,s)\colon\ \phi[\mathbf{X}(a,r,\ker p,s)]\,r^{-m}s^{-m}>\lambda\}$$

is open, hence for each positive integer i the set

$$T_i=\{(a,p)\colon\ (a,r,p,s)\in S \text{ for some } r<\delta,\ s<1/i\}$$

is open in $\mathbf{R}^n\times\mathbf{O}^*(n,m)$. Moreover the set where our function has values greater than λ equals $\bigcap\{T_i\colon i=1,2,3,\ldots\}$.

3.3.3. Lemma. *If A is a Suslin subset of $\mathbf{R}^n$, then*

$$\{(a,p)\colon\ a\in\operatorname{Closure}\{x\colon\ x-a\in\ker p,\ a\neq x\in A\}\}$$

is a Suslin subset of $\mathbf{R}^n\times\mathbf{O}^*(n,m)$.

Proof. For each positive integer i the set

$$S_i=\{(x,a,p)\colon\ p(x)=p(a),\ 0<|x-a|<1/i,\ x\in A\}$$

is a Suslin subset of $\mathbf{R}^n\times\mathbf{R}^n\times\mathbf{O}^*(n,m)$. Consequently

$$\bigcap_{i=1}^{\infty}\{(a,p)\colon\ (x,a,p)\in S_i \text{ for some } x\}$$

is a Suslin subset of $\mathbf{R}^n\times\mathbf{O}^*(n,m)$.

3.3.4. Theorem. *If A is a Suslin subset of $\mathbf{R}^n$ and $a\in\mathbf{R}^n$, then $\gamma_{n,n-m}$ almost every V in $\mathbf{G}(n,n-m)$ satisfies one of the following three conditions:*

(1) *For some $\delta>0$,*

$$\lim_{s\to 0+}\ \sup_{0<r<\delta}\ \phi[A\cap\mathbf{X}(a,r,V,s)]\,r^{-m}s^{-m}=0.$$

(2) *For all $\delta>0$,*

$$\limsup_{s\to 0+}\ \sup_{0<r<\delta}\ \phi[A\cap\mathbf{X}(a,r,V,s)]\,r^{-m}s^{-m}=\infty.$$

(3) $a\in\operatorname{Closure}\{x\colon\ x-a\in V,\ a\neq x\in A\}$.

Proof. We assume $a=0\notin A$, choose an orthonormal base $e_1,\ldots,e_n$ of $\mathbf{R}^n$, and associate with each $g\in\mathbf{O}(n)$ the vectors

$$g_1=g(e_1),\ \ldots,\ g_n=g(e_n)$$

as well as the $n-m$ dimensional vectorspace

$$v(g)=\mathbf{R}^n\cap\{x\colon\ x\bullet g_i=0 \text{ for } i=1,\ldots,m\}.$$

Recalling 2.7.16(6) we note that $v_{\#}(\theta_n)=\gamma_{n,n-m}$, because $v(g)=g[v(\mathbf{1}_{\mathbf{R}^n})]$ for $g\in\mathbf{O}(n)$.

If $g \in \mathbf{O}(n)$ and $0 < s < 1$, then

$$\mathbf{X}[0, \infty, v(g), s] = \mathbf{R}^n \cap \left\{x: s^{-2} \sum_{i=1}^{m} (x \bullet g_i)^2 < |x|^2\right\}$$

$$= \mathbf{R}^n \cap \left\{x: (s^{-2} - 1) \sum_{i=1}^{m} (x \bullet g_i)^2 < \sum_{j=m+1}^{n} (x \bullet g_j)^2\right\}.$$

For $j = m+1, \ldots, n$ and $0 < r < \infty$, $0 < t < 1$ we define

$$Y_j(r, g, t) = A \cap \mathbf{U}(0, r) \cap \left\{x: (t^{-2} - 1) \sum_{i=1}^{m} (x \bullet g_i)^2 < (x \bullet g_j)^2\right\}.$$

Letting

$$\sigma(s) = s(n-m)^{\frac{1}{2}} [1 + (n-m-1) s^2]^{-\frac{1}{2}}$$

we use the equation

$$[\sigma(s)]^{-2} - 1 = (s^{-2} - 1)(n-m)^{-1}$$

to infer that

$$\bigcup_{j=m+1}^{n} Y_j(r, g, s) \subset A \cap \mathbf{X}[0, r, v(g), s] \subset \bigcup_{j=m+1}^{n} Y_j[r, g, \sigma(s)].$$

Corresponding to $j = m+1, \ldots, n$ we now consider:

The set B_j of all $g \in \mathbf{O}(n)$ such that, for some $\delta > 0$,

$$\lim_{t \to 0+} \sup_{0 < r < \delta} \phi[Y_j(r, g, t)]\, r^{-m} t^{-m} = 0.$$

The set C_j of all $g \in \mathbf{O}(n)$ such that, for all $\delta > 0$,

$$\limsup_{t \to 0+} \sup_{0 < r < \delta} \phi[Y_j(r, g, t)]\, r^{-m} t^{-m} = \infty.$$

Observing that

$$\sigma(s)/s \to (n-m)^{\frac{1}{2}} \quad \text{as} \quad s \to 0+,$$

we obtain the following two implications:

If $g \in \bigcap_{j=m+1}^{n} B_j$, then (1) holds with $V = v(g)$.

If $g \in \bigcup_{j=m+1}^{n} C_j$, then (2) holds with $V = v(g)$.

We also define

$$D = \mathbf{O}(n) \cap \{g: 0 \in \operatorname{Closure}[A \cap v(g)]\}.$$

Inasmuch as

$$\mathbf{O}(n) \sim \left[\left(\bigcap_{j=m+1}^{n} B_j\right) \cup \left(\bigcup_{j=m+1}^{n} C_j\right) \cup D\right] \subset \bigcup_{j=m+1}^{n} [\mathbf{O}(n) \sim (B_j \cup C_j \cup D)],$$

we plan to complete the proof by showing that

$$\theta_n[\mathbf{O}(n) \sim (B_j \cup C_j \cup D)] = 0 \text{ for } j = m+1, \ldots, n.$$

We henceforth fix j, let ξ be the characteristic function of $\mathbf{O}(n) \sim (B_j \cup C_j \cup D)$, and consider the subgroup

$$H = \mathbf{O}(n) \cap \{h\colon h(e_i) = e_i \text{ for } m < i \neq j\}$$

with the Haar measure α such that $\alpha(H) = 1$. Observing that D, B_j, C_j are Suslin sets, which follows from 3.3.3 and an obvious modification of 3.3.2, we use 2.2.12, 2.6.2 to obtain

$$\theta_n[\mathbf{O}(n) \sim (B_j \cup C_j \cup D)] = \int \xi \, d\theta_n$$
$$= \iint \xi(g \circ h) \, d\theta_n \, g \, d\alpha \, h = \iint \xi(g \circ h) \, d\alpha \, h \, d\theta_n \, g.$$

It will therefore suffice to prove that

$$\int \xi(g \circ h) \, d\alpha \, h = 0 \text{ whenever } g \in \mathbf{O}(n).$$

From now on we consider a fixed $g \in \mathbf{O}(n)$ and note that the corresponding inner automorphism of $\mathbf{O}(n)$, which associates $g \circ h \circ g^{-1}$ with h, maps H onto the subgroup

$$K = \mathbf{O}(n) \cap \{k\colon k(g_i) = g_i \text{ for } m < i \neq j\}$$

and carries α onto the Haar measure β of K with $\beta(K) = 1$, hence

$$\int \xi(g \circ h) \, d\alpha \, h = \int \xi(k \circ g) \, d\beta \, k.$$

Letting X be the $m+1$ dimensional vectorspace generated by $g_j, g_1, \ldots, g_m$ we consider the commutative diagram

$$\begin{array}{ccccccc}
\mathbf{R}^n & \xrightarrow{q} & X & & X & \xrightarrow{T} & \odot_2 X \\
\cup & & \cup & & \cup & & \cup \\
\mathbf{R}^n \sim \ker q & \longrightarrow & X \sim \{0\} & \xrightarrow{w} & U & \longrightarrow & T(U)
\end{array}$$

where U, T are defined as in 3.2.28 (4) and

$$q(x) = (x \bullet g_j) \, g_j + \sum_{i=1}^{m} (x \bullet g_i) \, g_i \text{ for } x \in \mathbf{R}^n,$$

$$w(x) = x/|x| \text{ for } x \in X.$$

We also let

$$f = T \circ w \circ q | \mathbf{R}^n \sim \ker q\colon \mathbf{R}^n \sim \ker q \to T(U),$$

$$F\colon K \to T(U), \quad F(k) = T[k(g_j)] \text{ for } k \in K.$$

In view of the restriction isomorphism mapping K onto the orthogonal group of X, K acts isometrically and transitively on $T(U)$. Since $T(U)$ is a compact m dimensional submanifold of class 1 of $\odot_2 X$, we see that $\mathscr{H}^m \llcorner T(U)$ is a Radon measure invariant under the action of K, and infer from 2.7.11 (2) that

$$\mathscr{H}^m \llcorner T(U) = \mathscr{H}^m[T(U)] \cdot F_\#(\beta).$$

Next we observe that the ratios

$$\mathscr{H}^m[T(U)\cap \mathbf{B}(z,t)]/[\alpha(m)\,t^m],$$

which correspond to $z\in T(U)$ and $0<t<\infty$, are really independent of z, because of the action of K, and approach 1 as t approaches 0, according to 3.2.19. It follows that 2.9.17 will be applicable with ϕ replaced by $\mathscr{H}^m \llcorner T(U)$.

For each positive integer ν we consider the measure ψ_ν over $T(U)$ given by the formula

$$\psi_\nu(S)=\sup_{0<r<1/\nu}\phi[A\cap \mathbf{U}(0,r)\cap f^{-1}(S)]\,r^{-m}$$

whenever $S\subset T(U)$. Letting

$$P_\nu=T(U)\cap\{z:\lim_{t\to 0+}\psi_\nu[\mathbf{U}(z,t)]\,t^{-m}=0\},$$

$$Q_\nu=T(U)\cap\{z:\limsup_{t\to 0+}\psi_\nu[\mathbf{U}(z,t)]\,t^{-m}=\infty\},$$

$$R_\nu=f[A\cap\mathbf{U}(0,1/\nu)\sim\ker q],$$

we note that $\psi_\nu[T(U)\sim R_\nu]=0$ and infer from 2.2.13, 2.9.17 that

$$\mathscr{H}^m[T(U)\sim(R_\nu\cup P_\nu\cup Q_\nu)]=0.$$

Observing that $\psi_\nu\geq\psi_{\nu+1}$, $P_\nu\subset P_{\nu+1}$, $Q_\nu\supset Q_{\nu+1}$, $R_\nu\supset R_{\nu+1}$ we also let

$$P=\bigcup_{\nu=1}^{\infty}P_\nu,\qquad Q=\bigcap_{\nu=1}^{\infty}Q_\nu,\qquad R=\bigcap_{\nu=1}^{\infty}R_\nu$$

and verify that

$$\bigcap_{\nu=1}^{\infty}(P_\nu\cup Q_\nu\cup R_\nu)\subset P\cup Q\cup R,$$

hence $\mathscr{H}^m[T(U)\sim(P\cup Q\cup R)]=0$, which implies

$$\beta[K\sim F^{-1}(P\cup Q\cup R)]=0.$$

Furthermore we see from 3.2.28 (4) that

$$U\cap T^{-1}\mathbf{U}[F(k),t]=U\cap\{x:1-[x\bullet k(g_j)]^2<t^2\}$$

whenever $k\in K$ and $0<t<\infty$; it follows that

$$A\cap\mathbf{U}(0,r)\cap f^{-1}\mathbf{U}[F(k),t]=Y_j(r,k\circ g,t)$$

whenever $0<r<\infty$ because for each $x\in A\cap\mathbf{U}(0,r)$ the three inequalities

$$1-[|q(x)|^{-1}q(x)\bullet k(g_j)]^2<t^2,\qquad |q(x)|^2-[x\bullet(k\circ g)_j]^2<t^2\,|q(x)|^2,$$

$$\sum_{i=1}^{m}[x\bullet(k\circ g)_i]^2<t^2\Big([x\bullet(k\circ g)_j]^2+\sum_{i=1}^{m}[x\bullet(k\circ g)_i]^2\Big)$$

are equivalent; consequently

$$\psi_\nu \mathbf{U}[F(k),t] = \sup_{0<r<1/\nu} \phi[Y_j(r, k\circ g, t)]\, r^{-m}$$

for each positive integer ν. We infer that

$$F^{-1}(P) = K \cap \{k: k\circ g \in B_j\}, \quad F^{-1}(Q) = K \cap \{k: k\circ g \in C_j\}.$$

Next we observe that

$$U \cap T^{-1}\{F(k)\} = \{k(g_j), -k(g_j)\},$$
$$f^{-1}\{F(k)\} = \{x: q(x)/|q(x)| = \pm(k\circ g)_j\} \subset v(k\circ g),$$
$$F(k)\in R_\nu \text{ implies } A \cap \mathbf{U}(0,1/\nu) \cap v(k\circ g) \neq \varnothing$$

whenever $k\in K$ and ν is a positive integer; therefore

$$F^{-1}(R) \subset K \cap \{k: k\circ g \in D\}.$$

Finally we conclude

$$\int \xi(k\circ g)\, d\beta\, k = \beta[K \sim \{k: k\circ g \in (B_j \cup C_j \cup D)\}]$$
$$\leq \beta[K \sim F^{-1}(P \cup Q \cup R)] = 0.$$

3.3.5. Lemma. *If* $E\subset \mathbf{R}^n$, $p\in \mathbf{O}^*(n,m)$, $r>0$, $0<s<1$ *and*

$$E \cap \mathbf{X}(a, r, \ker p, s) = \varnothing \text{ whenever } a\in E,$$

then every subset of E *with diameter less than* r *is contained in the image of some Lipschitzian map*

$$f: \mathbf{R}^m \to \mathbf{R}^n \text{ with } \operatorname{Lip}(f) \leq s^{-1} \text{ and } p\circ f = \mathbf{1}_{\mathbf{R}^m};$$

therefore E *is countably* m *rectifiable.*

Proof. We suppose $S\subset E$ with $\operatorname{diam} S<r$, and choose $q\in \mathbf{O}^*(n,n-m)$ so that $\operatorname{im} q^* = \ker p$. If $a,b\in S$, then $|b-a|<r$ but $b\notin \mathbf{X}(a,r,\ker p,s)$, hence

$$s^{-1}|p(b-a)| \geq |b-a|, \quad (s^{-2}-1)|p(b-a)|^2 \geq |q(b-a)|^2$$

because $|x|^2 = |p(x)|^2 + |q(x)|^2$ for $x\in\mathbf{R}^n$. Thus $p|S$ is univalent,

$$\operatorname{Lip}[(p|S)^{-1}] \leq s^{-1}, \quad \operatorname{Lip}[q\circ (p|S)^{-1}] \leq (s^{-2}-1)^{\frac{1}{2}}.$$

Using 2.10.43 we extend $q\circ(p|S)^{-1}$ to a map

$$g: \mathbf{R}^m \to \mathbf{R}^{n-m} \text{ with } \operatorname{Lip}(g) \leq (s^{-2}-1)^{\frac{1}{2}},$$

and easily verify that $f = p^* + (q^*\circ g)$ is an extension of $(p|S)^{-1}$ with the required properties.

3.3.6. Lemma. *If A is a purely (ϕ, m) unrectifiable subset of $\mathbf{R}^n$, $p \in \mathbf{O}^*(n,m)$, $0<s<1$, $\lambda>0$, $\delta>0$ and*

$$\phi[A \cap \mathbf{X}(x, r, \ker p, s)] \le \lambda\, \alpha(m)\, r^m s^m$$

whenever $x \in A$ and $0<r\le\delta$, then

$$\phi(A \cap \mathbf{B}(a, 4\rho/s) \cap p^{-1}\mathbf{B}[p(a), \rho]) \le 2(84)^m \lambda\, \alpha(m)\, \rho^m$$

whenever $a \in \mathbf{R}^n$ and $0<\rho\le s\,\delta/24$; consequently

$$\Theta^{*m}(\phi \llcorner A, a) \le 2(84)^m \lambda.$$

Proof. Assuming $a \in \mathbf{R}^n$, $0<\rho\le s\,\delta/24$ we let $\varepsilon = s/4$,

$$B = A \cap \mathbf{B}(a, \rho/\varepsilon) \cap p^{-1}\mathbf{B}[p(a), \rho],$$

$$C = B \cap \{x : B \cap \mathbf{X}(x, \infty, \ker p, \varepsilon) \ne \varnothing\}$$

and infer from 3.3.5 that $B \sim C$ is countably m rectifiable, hence

$$\phi(B \sim C) = 0.$$

For each $x \in C$ we define

$$h(x) = \sup\{|y-x| : y \in B \cap \mathbf{X}(x, \infty, \ker p, \varepsilon)\}$$

and observe that $0<h(x)\le \operatorname{diam} B \le 2\rho/\varepsilon$. Then

$$p(C) \subset \bigcup_{x \in C} \mathbf{B}[p(x), \varepsilon\, h(x)/5] \subset \mathbf{B}[p(a), 7\rho/5],$$

hence 2.8.5 provides a countable subset D of C such that

$$\{\mathbf{B}[p(x), \varepsilon\, h(x)/5] : x \in D\} \text{ is disjointed,} \quad p(C) \subset \bigcup_{x \in D} \mathbf{B}[p(x), \varepsilon\, h(x)]$$

and $p|D$ is univalent. Inasmuch as

$$\sum_{x \in D} \alpha(m)\,[\varepsilon\, h(x)]^m \le 7^m \alpha(m)\, \rho^m,$$

it will be sufficient to show that

$$\phi(C \cap p^{-1}\mathbf{B}[p(x), \varepsilon\, h(x)]) \le 2(12)^m \lambda\, \alpha(m)\, [\varepsilon\, h(x)]^m$$

for each $x \in D$.

For this purpose we now fix $x \in D$, choose

$$y \in B \cap \mathbf{X}(x, \infty, \ker p, \varepsilon) \quad \text{with} \quad 4|y-x| > 3h(x),$$

and note that $|p(y-x)| < \varepsilon\,|y-x| \le \varepsilon\, h(x)$. We assert:

$$C \cap p^{-1}\mathbf{B}[p(x), \varepsilon\, h(x)] \subset \mathbf{X}[x, 3h(x), \ker p, s] \cup \mathbf{X}[y, 3h(x), \ker p, s].$$

In fact, if $z \in C$ and $|p(z-x)| \leq \varepsilon\, h(x)$, then

either $|z-x| \leq h(x)$

or $z \notin \mathbf{X}(x, \infty, \ker p, \varepsilon)$, $|z-x| \leq \varepsilon^{-1} |p(z-x)| \leq h(x)$,

hence always

$$|z-x| \leq h(x), \qquad |z-y| \leq |z-x| + |y-x| \leq 2h(x);$$

moreover

$$\begin{aligned} &|p(z-x)| + |p(z-y)| \leq 2\,|p(z-x)| + |p(x-y)| \leq 3\varepsilon\, h(x) \\ &< 4\varepsilon\, |y-x| = s\,|y-x| \leq s\,|z-x| + s\,|z-y|, \end{aligned}$$

hence either $|p(z-x)| < s\,|z-x|$ or $|p(z-y)| < s\,|z-y|$, and z belongs to the above union. Having thus verified our assertion we apply the hypothesis of the theorem twice with

$$r = 3h(x) \leq 6\rho/\varepsilon = 24\rho/s \leq \delta$$

to infer that the ϕ measure of the union does not exceed

$$2\lambda\, \boldsymbol{\alpha}(m)\, [3h(x)]^m\, s^m = 2(12)^m\, \lambda\, \boldsymbol{\alpha}(m)\, [\varepsilon\, h(x)]^m.$$

The second conclusion of the theorem is a consequence of the first conclusion because $A \cap \mathbf{B}(a, \rho) \subset B$.

3.3.7. Corollary. *If A is a purely (ϕ, m) unrectifiable subset of $\mathbf{R}^n$, $p \in \mathbf{O}^*(n, m)$, $0 < \sigma < 1$, $\lambda > 0$ and*

$$\phi[A \cap \mathbf{X}(x, r, \ker p, s)] \leq \lambda\, \boldsymbol{\alpha}(m)\, r^m s^m$$

whenever $x \in A$, $r > 0$ and $0 < s \leq \sigma$, then

$$\phi(A) \leq 2(84)^m\, \lambda\, \mathscr{L}^m[p(A)].$$

Proof. Suppose A is bounded. Whenever $0 < \rho \leq (\sigma \operatorname{diam} A)/4$ the hypothesis of 3.3.6 holds with $s = 4\rho/\operatorname{diam} A \leq \sigma$ and $\delta = \infty$; for each $a \in A$ it is also true that $A \subset \mathbf{B}(a, \operatorname{diam} A) = \mathbf{B}(a, 4\rho/s)$, hence 3.3.6 implies

$$p_\# (\phi \llcorner A)\, \mathbf{B}[p(a), \rho] \leq 2(84)^m\, \lambda\, \boldsymbol{\alpha}(m)\, \rho^m.$$

Defining the Radon measure ψ over $\mathbf{R}^m$ by the formula

$$\psi(S) = \inf\{p_\# (\phi \llcorner A)(T) \colon S \subset T,\ T \text{ is a Borel set}\}$$

for $S \subset \mathbf{R}^m$, we see that

$$\psi[\mathbf{B}(z, \rho)] \leq 2(84)^m\, \lambda\, \mathscr{L}^m[\mathbf{B}(z, \rho)]$$

whenever $z \in p(A)$ and $0 < \rho < (\sigma \operatorname{diam} A)/4$, then use 2.9.15 and 2.9.7 to conclude

$$\phi(A) = \psi[p(A)] \leq 2(84)^m\, \lambda\, \mathscr{L}^m[p(A)].$$

3.3.8. Lemma. *If B is a purely (ϕ, m) unrectifiable subset of $\mathbf{R}^n$, $p \in \mathbf{O}^*(n,m)$ and for each $x \in B$ there exists a $\delta > 0$ such that*

$$\lim_{s \to 0+} \sup_{0<r<\delta} \phi[B \cap \mathbf{X}(x, r, \ker p, s)]\, r^{-m} s^{-m} = 0,$$

then $\phi(B) = 0$.

Proof. For each positive integer i we let B_i be the set of those points x in B for which

$$\lim_{s \to 0+} \sup_{0<r<1/i} \phi[B \cap \mathbf{X}(x, r, \ker p, s)]\, r^{-m} s^{-m} = 0.$$

We also choose a countable family F_i such that $\bigcup F_i = B_i$ and each member of F_i has diameter less than $1/i$.

Since B is the union of all the sets B_i, we need only show that each member of each family F_i has ϕ measure 0.

Given $E \in F_i$ and $\lambda > 0$, we associate with each positive integer j the subset C_j of $\mathbf{R}^n$ consisting of all points x for which

$$\phi[E \cap \mathbf{X}(x, r, \ker p, s)] \leq \lambda\, \alpha(m)\, r^m s^m$$

whenever $0 < r < 1/i$ and $0 < s < 1/j$; we see from the proof of 3.3.2 that C_j is closed. For each $x \in E \cap C_j$ the above inequality holds whenever $r > 0$ and $0 < s < 1/j$, because diam $E < 1/i$. Consequently 3.3.7 implies

$$\phi(E \cap C_j) \leq 2(84)^m \lambda \mathscr{L}^m[p(E \cap C_j)].$$

Since the sets C_j form an increasing sequence whose union contains E, we can apply 2.1.3(4) to $\phi \llcorner E$ and infer

$$\phi(E) = \lim_{j \to \infty} \phi(E \cap C_j) \leq 2(84)^m \lambda \mathscr{L}^m[p(E)].$$

Finally we observe that $\mathscr{L}^m[p(E)] < \infty$ and λ can be chosen arbitrarily small, hence conclude $\phi(E) = 0$.

3.3.9. Lemma. *If $A \subset \mathbf{R}^n$, $\phi(A) < \infty$, $p \in \mathbf{O}^*(n,m)$, $B \subset \mathbf{R}^n$ and*

$$\limsup_{s \to 0+} \sup_{0<r<\delta} \phi[A \cap \mathbf{X}(x, r, \ker p, s)]\, r^{-m} s^{-m} = \infty$$

whenever $x \in B$ and $\delta > 0$, then $\mathscr{L}^m[p(B)] = 0$.

Proof. Since $\mathbf{X}(x, r, \ker p, s) \subset p^{-1}\mathbf{U}[p(x), r\,s]$ whenever $x \in \mathbf{R}^n$ and r, s are positive numbers, our hypothesis implies

$$\limsup_{(r,s) \to (0,0)} p_\#(\phi \llcorner A)\, \mathbf{U}[p(x), r\,s]/(r\,s)^m = \infty$$

whenever $x \in B$. Defining the Radon measure ψ over $\mathbf{R}^m$ by the formula

$$\psi(S) = \inf\{p_\#(\phi \llcorner A)(T)\colon S \subset T, T \text{ is a Borel set}\}$$

whenever $S \subset \mathbf{R}^m$, we see that

$$\limsup_{\rho \to 0+} \psi[\mathbf{B}(z,\rho)]/\mathscr{L}^m[\mathbf{B}(z,\rho)] = \infty$$

whenever $z \in p(B)$, and use 2.9.5 to conclude $\mathscr{L}^m[p(B)]=0$.

3.3.10. Theorem. *If A is a purely (ϕ, m) unrectifiable Suslin subset of $\mathbf{R}^n$ with $\phi(A) < \infty$ and, for $i \in \{1, 2, 3\}$,*

$$S_i \text{ is the set of all } (a, V) \in A \times \mathbf{G}(n, n-m)$$
$$\text{which satisfy the condition } (i) \text{ of } 3.3.4,$$

then

$$(\phi \times \gamma_{n,n-m})(S_1) = 0, \quad (\phi \times \gamma_{n,n-m})[A \times \mathbf{G}(n, n-m) \sim (S_2 \cup S_3)] = 0$$

and, for every $p \in \mathbf{O}^*(n,m)$,

$$\phi\{a: (a, \ker p) \in S_1\} = 0, \quad \mathscr{L}^m[p\{a: (a, \ker p) \in S_2\}] = 0,$$
$$p\{a: (a, \ker p) \in S_3\} \subset \{y: A \cap p^{-1}\{y\} \text{ is infinite}\};$$

moreover, for $\theta^*_{n,m}$ almost all p in $\mathbf{O}^*(n,m)$,

$$\phi[A \cap \{a: (a, \ker p) \notin S_2 \cup S_3\}] = 0.$$

Proof. For every $p \in \mathbf{O}^*(n,m)$ we apply

$$3.3.8 \text{ with } B = \{a: (a, \ker p) \in S_1\},$$
$$3.3.9 \text{ with } B = \{a: (a, \ker p) \in S_2\},$$

and observe that

$$(a, \ker p) \in S_3 \text{ if and only if } a \text{ is a cluster point of } A \cap p^{-1}\{p(a)\}.$$

Assured by 3.3.2, 3.3.3 that S_1, S_2, S_3 are $\phi \times \gamma_{n,n-m}$ measurable sets, we use 2.6.2 and 3.3.4 to compute

$$(\phi \times \gamma_{n,n-m})(S_1) = \int \phi\{a: (a, V) \in S_1\}\, d\gamma_{n,n-m} V = 0,$$
$$(\phi \times \gamma_{n,n-m})[A \times \mathbf{G}(n, n-m) \sim (S_1 \cup S_2 \cup S_3)]$$
$$= \int_A \gamma_{n,n-m}\{V: (a, V) \notin S_1 \cup S_2 \cup S_3\}\, d\phi\, a = 0,$$

hence conclude that

$$0 = (\phi \times \gamma_{n,n-m})[A \times G(n, n-m) \sim (S_2 \cup S_3)]$$
$$= \int \phi[A \cap \{a: (a, V) \notin S_2 \cup S_3\}]\, d\gamma_{n,n-m} V.$$

3.3.11. A particular example illustrating 3.3.10 with

$$(\phi \times \gamma_{n,n-m})[A \times \mathbf{G}(n, n-m) \sim S_2] = 0$$

may be found in 3.3.19. However I do not know any example in which

$$(\phi \times \gamma_{n,n-m}) S_3 > 0.$$

3.3.12. Theorem. *Suppose E is a Suslin subset of $\mathbf{R}^n$ with $\phi(E) < \infty$, and $\theta^*_{n,m}$ almost all p in $\mathbf{O}^*(n,m)$ have the following two properties:*

(I) $\mathscr{L}^m[p(B)] = 0$ *whenever $B \subset E$ with $\phi(B) = 0$.*

(II) $E \cap p^{-1}\{y\}$ *is finite for $\mathscr{L}^m$ almost all y in $\mathbf{R}^m$.*

Then there exists a countably m rectifiable Borel subset R of $\mathbf{R}^n$ such that $E \sim R$ is purely (ϕ, m) unrectifiable and

$$\mathscr{L}^m[p(E \sim R)] = 0 \textit{ for } \theta^*_{n,m} \textit{ almost all } p \textit{ in } \mathbf{O}^*(n,m).$$

Proof. According to 3.2.14 there exists a countably m rectifiable Borel set R such that $E \sim R$ is purely (ϕ, m) unrectifiable. Letting $A = E \sim R$ we see from 3.3.10 that for $\theta^*_{n,m}$ almost all p in $\mathbf{O}^*(n,m)$ the property (I) implies

$$\mathscr{L}^m[p(A \cap \{a: (a, \ker p) \notin S_2 \cup S_3\})] = 0,$$

and the property (II) implies

$$\mathscr{L}^m[p\{a: (a, \ker p) \in S_3\}] = 0;$$

since we also know that

$$\mathscr{L}^m[p\{a: (a, \ker p) \in S_2\}] = 0,$$

we conclude $\mathscr{L}^m[p(A)] = 0$.

3.3.13. Theorem. *If $E \subset \mathbf{R}^n$ with $\mathscr{H}^m(E) < \infty$, and $1 \le t \le \infty$, then there exists a countably m rectifiable Borel subset R of $\mathbf{R}^n$ such that $E \sim R$ is purely $(\mathscr{H}^m, m)$ unrectifiable and*

$$\mathscr{I}^m_t(E \sim R) = 0;$$

moreover, $\mathscr{H}^m(E) = \mathscr{I}^m_t(E)$ if and only if E is $(\mathscr{H}^m, m)$ rectifiable.

Proof. Since $\mathscr{H}^m$ and $\mathscr{I}^m_t$ are Borel regular we need only consider the case when E is a Borel set.

Every $p \in \mathbf{O}^*(n,m)$ has the properties (I) and (II) of 3.3.12 because 2.10.11 implies

$$\infty > \mathscr{H}^m(B) \ge \int N(p|B, y)\, d\mathscr{L}^m y$$

for every Borel subset B of E. Accordingly the first conclusion of the theorem follows from 3.3.12 and 2.10.5. Moreover

$$\mathscr{H}^m(E)=\mathscr{I}_t^m(E\cap R)+\mathscr{H}^m(E\sim R)=\mathscr{I}_t^m(E)+\mathscr{H}^m(E\sim R)$$

by virtue of 3.2.26, whence the second conclusion follows.

3.3.14. Theorem. *If $E\subset\mathbf{R}^n$ with $\mathscr{I}_\infty^m(E)<\infty$, then E is $(\mathscr{I}_\infty^m, m)$ rectifiable.*

Proof. Since $\mathscr{I}_\infty^m$ is Borel regular we need only consider the case when E is a Borel set.

From 2.10.15 we see that $\theta_{n,m}^*$ almost all $p\in\mathbf{O}^*(n,m)$ satisfy the condition

$$\infty>\mathscr{I}_\infty^m(E)\geq\int N(p|E,y)\,d\mathscr{L}^m y,$$

hence have the property (II) of 3.3.12.

Defining ζ_∞ as in 2.10.5 we choose Borel partitions $H_1, H_2, H_3, \dots$ of E as in 2.10.8 (with A replaced by E), and so that each member of H_{j+1} is contained in some member of H_j. We let

$$P=\bigcup_{j=1}^{\infty}\bigcup_{S\in H_j}\mathbf{O}^*(n,m)\cap\{p\colon \mathscr{L}^m[p(S)]>\zeta_\infty(S)\},$$

note that $\theta_{n,m}^*(P)=0$, and will show that every $p\in\mathbf{O}^*(n,m)\sim P$ has the property (I) of 3.3.12.

Whenever $p\in\mathbf{O}^*(n,m)\sim P$, $B\subset E$, $\mathscr{I}_\infty^m(B)=0$ and $\varepsilon>0$, there exists an open subset U of $\mathbf{R}^n$ with

$$B\subset E\cap U \quad\text{and}\quad \mathscr{I}_\infty^m(E\cap U)<\varepsilon,$$

because $\mathscr{I}_\infty^m \llcorner E$ is a Radon measure. For each positive integer j we consider the family

$$G_j=H_j\cap\{S\colon S\subset U\}$$

and use 2.10.8 to infer

$$\mathscr{L}^m[p(\bigcup G_j)]\leq\sum_{S\in G_j}\mathscr{L}^m[p(S)]\leq\sum_{S\in G_j}\zeta_\infty(S)\leq\mathscr{I}_\infty^m(E\cap U)<\varepsilon.$$

Since the sets $\bigcup G_j$ form a nondecreasing sequence with union $E\cap U$, it follows that

$$\mathscr{L}^m[p(B)]\leq\mathscr{L}^m[p(E\cap U)]=\lim_{j\to\infty}\mathscr{L}^m[p(\bigcup G_j)]<\varepsilon.$$

Accordingly 3.3.12 yields a countably m rectifiable set R such that $\mathscr{I}_\infty^m(E\sim R)=0$.

3.3.15. Theorem. *If* $W \subset \mathbf{R}^n$, $\phi(W) < \infty$ *and*

$$\Theta^{*m}(\phi \llcorner W, x) > 0 \text{ for } \phi \text{ almost all } x \text{ in } W,$$

then there exists a countably (ϕ, m) *rectifiable and* ϕ *measurable set* Q *such that* $W \sim Q$ *is purely* (ϕ, m) *unrectifiable and*

$$\mathscr{L}^m[p(W \sim Q)] = 0 \text{ for } \theta^*_{n,m} \text{ almost all } p \text{ in } \mathbf{O}^*(n,m).$$

Proof. We define the Radon measure μ over $\mathbf{R}^n$ by the formula

$$\mu(S) = \inf\{\phi(W \cap T)\colon S \subset T, T \text{ is a Borel set}\}$$

whenever $S \subset \mathbf{R}^n$, and consider the Borel sets

$$E_i = \mathbf{R}^n \cap \{x\colon \Theta^{*m}(\mu, x) > i^{-1}\}$$

corresponding to all positive integers i. Using 2.10.19 (3) and 2.10.11 we see that

$$\infty > i\,\mu(B) \geq \mathscr{S}^m(B) \geq \int N(p|B, y)\, d\mathscr{L}^m y$$

for every Borel set $B \subset E_i$ and every $p \in \mathbf{O}^*(n,m)$, hence infer that the hypotheses of 3.3.12 hold with ϕ, E replaced by μ, E_i. Accordingly E_i contains a countably m rectifiable Borel set R_i such that $E_i \sim R_i$ is purely (μ, m) unrectifiable and

$$\mathscr{L}^m[p(E_i \sim R_i)] = 0 \text{ for } \theta^*_{n,m} \text{ almost all } p \text{ in } \mathbf{O}^*(n,m).$$

Letting

$$F = W \sim \bigcup_{i=1}^{\infty} E_i, \quad Q = F \cup \bigcup_{i=1}^{\infty} R_i,$$

we complete the proof by observing that $\phi(F) = 0$,

$$W \sim Q \subset \bigcup_{i=1}^{\infty} (W \cap E_i \sim R_i),$$

and each $W \cap E_i \sim R_i$ is purely (ϕ, m) unrectifiable.

3.3.16. *In* 3.3.13 *one may replace* $\mathscr{H}^m$ *by any one of the measures* $\mathscr{S}^m$, $\mathscr{T}^m$, $\mathscr{Q}^m_t$, $\mathscr{G}^m$; appropriate modifications in the proof are readily made with the help of 2.10.6, 2.10.8, 2.10.10.

Combining 3.3.14 with 3.2.26 one obtains the proposition:

If $E \subset \mathbf{R}^n$ *with* $\mathscr{I}^m_\infty(E) < \infty$, *then*

$$\mathscr{I}^m_t(E) = \mathscr{I}^m_\infty(E) \text{ for } 1 \leq t \leq \infty.$$

However it is not known whether the hypothesis $\mathscr{I}^m_\infty(E) < \infty$ is really necessary, because *it is not known whether* $\mathscr{I}^m_\infty$ *may be replaced in* 3.3.14

by $\mathscr{I}_t^m$ *with* $1 \le t < \infty$; in view of 3.3.15 and 3.2.19 a negative solution of this problem would be equivalent to the construction of a compact subset E of $\mathbf{R}^n$ such that

$$\mathscr{I}_t^m(E) > 0 \quad \text{and} \quad \Theta^m(\mathscr{I}_t^m \llcorner E, x) = 0 \text{ for } x \in E.$$

In [BJ 2] our theory has been generalized from $\mathbf{R}^n$ to a large class of homogeneous spaces.

3.3.17. Theorem. *If* $E \subset \mathbf{R}^n$, $\phi(E) < \infty$, $c = 2^{-1}(84)^{-m}$ *and for* ϕ *almost all* a *in* E *there exist* $V \in \mathbf{G}(n, n-m)$, $0 < s < 1$ *such that*

$$\Theta^{*m}[\phi \llcorner E \cap \mathbf{X}(a, \infty, V, s), a] < c\, s^m\, \Theta^{*m}(\phi \llcorner E, a),$$

then E *is* (ϕ, m) *rectifiable.*

Proof. We define the Radon measure ψ over $\mathbf{R}^n$ by the formula

$$\psi(S) = \inf\{\phi(E \cap T) \colon S \subset T, \ T \text{ is a Borel set}\}$$

whenever $S \subset \mathbf{R}^n$, and choose a countably m rectifiable Borel set R such that $\mathbf{R}^n \sim R$ is purely (ψ, m) unrectifiable.

With any $p \in \mathbf{O}^*(n, m)$, $0 < s < 1$, $\lambda > 0$, $\delta > 0$ we associate the set

$$B(p, s, \lambda, \delta)$$

consisting of those points a in $\mathbf{R}^n$ which satisfy the condition

$$\psi[\mathbf{X}(a, r, \ker p, s)]/[\alpha(m)\, r^m\, s^m] \le \lambda < c\, \Theta^{*m}(\psi, a)$$

whenever $0 < r \le \delta$. Our hypothesis implies that ϕ almost all of E is contained in the union of all the sets $B(p, s, \lambda, \delta)$. Moreover this union equals the union of the countably many sets $B(q, t, \mu, \varepsilon)$ corresponding to $t, \mu, \varepsilon \in \mathbf{Q}$ and $q \in D$, where D is a countable dense subset of $\mathbf{O}^*(n, m)$; in fact, given $a \in B(p, s, \lambda, \delta)$ we can choose ε, μ, t, q so that

$$0 < \varepsilon < \delta, \quad \lambda < \mu < c\, \Theta^{*m}(\psi, a), \quad 0 < t < s \quad \text{with} \quad \lambda\, s^m < \mu\, t^m,$$

$$\|q - p\| < s - t, \quad \text{hence} \quad \mathbf{X}(a, \infty, \ker q; t) \subset \mathbf{X}(a, \infty, \ker p, s),$$

and infer $a \in \mathbf{B}(q, t, \mu, \varepsilon)$.

We will complete the proof by verifying the statement:

If $p \in \mathbf{O}^*(n, m)$, $0 < s < 1$, $\lambda > 0$, $\delta > 0$ *and*

$$A = B(p, s, \lambda, \delta) \sim R,$$

then $\psi(A) = 0$. First we apply 3.3.6 to obtain

$$c\, \Theta^{*m}(\psi \llcorner A, a) \le \lambda \quad \text{whenever } a \in \mathbf{R}^n.$$

Next we infer from 2.8.9, 2.8.18, 2.9.11 that ψ almost all points a in A satisfy the equation

$$\lim_{r\to 0+} \psi[\mathbf{B}(a,r)\cap A]/\psi[\mathbf{B}(a,r)]=1,$$

hence

$$c\,\Theta^{*m}(\psi,a)=c\,\Theta^{*m}(\psi \llcorner A,a)\le\lambda.$$

Since $c\,\Theta^{*m}(\psi,a)>\lambda$ whenever $a\in A$, we conclude $\psi(A)=0$.

3.3.18. The converse of Theorem 3.3.17 is false (as may be seen from simple examples in which spt ϕ is countable), but the following proposition holds:

If E is a (ϕ,m) rectifiable subset of $\mathbf{R}^n$, then

$$\Theta^{*m}(\phi\llcorner E,a)>0 \text{ for } \phi \text{ almost all } a \text{ in } E;$$

moreover, in case $\Theta^{*m}(\phi\llcorner E,a)<\infty$ for ϕ almost all a in E, then ϕ almost all points a of E have the property that $\operatorname{Tan}^m(\phi\llcorner E,a)$ *is an m dimensional vectorspace and*

$$\Theta^m(\phi\llcorner E\cap\mathbf{X}[a,\infty,\operatorname{Nor}^m(\phi\llcorner E,a),s],a)=0$$

whenever $0<s<1$.

To verify this we define ψ as in the proof of 3.3.17 and let R be any compact m rectifiable subset of $\mathbf{R}^n$. Since $\mathscr{H}^m(R)<\infty$ it follows from 2.10.19(1) that

$$\psi(R\cap\{a\colon \Theta^m(\psi,a)=0\})=0.$$

Moreover we consider the sets

$$R_i=R\cap\{a\colon i^{-1}<\Theta^{*m}(\psi,a)<i\}$$

corresponding to all positive integers i, and infer from 2.10.19(3), (1), (4) that

$$i^{-1}\mathscr{H}^m(S)\le\psi(S)\le 2^m\, i\,\mathscr{H}^m(S) \text{ whenever } S\subset R_i,$$

$$\Theta^m(\psi\llcorner \mathbf{R}^n\sim R_i,a)=0 \text{ for } \psi \text{ almost all } a \text{ in } R_i.$$

Applying 3.2.19 we conclude that, for ψ almost all a in R_i,

$$\operatorname{Tan}^m(\psi,a)=\operatorname{Tan}^m(\psi\llcorner R_i,a)=\operatorname{Tan}^m(\mathscr{H}^m\llcorner R_i,a)$$

is an m dimensional vectorspace; for every such a the final assertion of the proposition is implied by 3.2.16 because

$$\mathbf{R}^n\cap\{v\colon s^{-1}\operatorname{dist}[v,\operatorname{Nor}^m(\psi,a)]\le|v|=1\}$$

is a compact subset of $\mathbf{R}^n\sim\operatorname{Tan}^m(\psi,a)$ whenever $0<s<1$.

3.3.19. Here we study a particular example which illustrates several features of the theory of purely $(\mathscr{H}^1, 1)$ unrectifiable subsets of $\mathbf{R}^2$.

With each closed circular disc $D = \mathbf{B}(z, r) \subset \mathbf{R}^2 = \mathbf{C}$ and each positive integer n we associate the family $F_n(D)$ consisting of the n discs

$$\mathbf{B}[z + (1 - n^{-1}) r \exp(k n^{-1} 2\pi \mathbf{i}), n^{-1} r]$$

corresponding to $k = 1, 2, \ldots, n$. We note that the minimum distance between the centers of distinct members of $F_n(D)$ equals

$$2(1 - n^{-1}) \sin(n^{-1} \pi) r$$

and that no halfplane bounded by a straight line through the center of D meets more than $n/2 + 1$ members of $F_n(D)$.

We inductively construct families $G_4, G_5, G_6, \ldots$ of closed circular discs, taking

$$G_4 = F_4[\mathbf{B}(0, 1)], \quad G_n = \bigcup \{F_n(D) : D \in G_{n-1}\} \text{ for } n \geq 5.$$

Thus, for any integer $n \geq 4$, G_n consists of $n!/3!$ discs with radius

$$\rho_n = 3!/n!,$$

and the minimum distance between distinct members of G_n equals

$$\begin{aligned} &2(1 - n^{-1}) \sin(n^{-1} \pi)\, n\, \rho_n - 2\rho_n = 2\rho_n[(n-1) \sin(n^{-1} \pi) - 1] \\ &> 2\rho_n[(n-1)\, 2n^{-1} - 1] = 2\rho_n(1 - 2n^{-1}) \geq \rho_n, \end{aligned}$$

because $\sin(x) > 2x/\pi$ for $0 < x < \pi/2$. Then we consider the compact set

$$A = \bigcap_{n=4}^{\infty} \bigcup G_n.$$

We see that $\mathscr{H}^1(A) \leq 2$ because

$$\sum_{B \in G_n} \operatorname{diam} B = 2 \text{ for } n = 4, 5, 6, \ldots.$$

Next we will prove that $\mathscr{Q}_1^1(A) \geq \pi$.

Given $0 < \lambda < \pi$ we can choose $\delta > 0$ so that, for any positive integer $n \geq 4$, the condition $\rho_n \leq \delta$ implies the inequalities

$$\begin{gathered} n \geq 5, \quad (n-1) \sin(n^{-1} \pi) > \lambda \quad \text{and} \\ (n-1) \sin(n^{-1} \pi) - 1 > \lambda[2^{-1} + (n+1)^{-1}]. \end{gathered}$$

Defining ζ_1 as in 2.10.5 with $m = 1$, we will show that

$$\sum_{S \in H} \zeta_1(S) \geq \lambda$$

for every countable covering H of A consisting of convex open sets S with $\operatorname{diam} S \leq \delta$. Since A and all $\bigcup G_n$ are compact, we may assume H to be finite and choose an integer v such that each member of G_v is contained in some member of H; we may further assume that each member of H contains some member of G_v. For every integer $n \geq 4$ we define the measure ψ_n over $\mathbf{R}^2$ by the formula

$$\psi_n(T) = \rho_n \operatorname{card}(G_n \cap \{B: B \cap T \neq \varnothing\})$$

whenever $T \subset \mathbf{R}^2$, and observe that $\psi_n \geq \psi_{n+1}$.

If $S \in H$, then either S meets a single member D of G_v, hence $D \subset S$ and 3.2.13 implies

$$\begin{aligned}\zeta_1(S) &\geq \zeta_1(D) = 2\rho_v/\boldsymbol{\beta}_1(2,1) = \boldsymbol{\Gamma}(\tfrac{1}{2})^2 \rho_v \\ &= \boldsymbol{\alpha}(2)\rho_v = \pi\rho_v = \pi\psi_v(S) > \lambda\psi_{v+1}(S),\end{aligned}$$

or S meets at least two members of G_v, hence there exists a smallest integer n such that $4 \leq n \leq v$ and S meets at least two members of G_n. Henceforth considering the second alternative, we infer that $\rho_n \leq \delta$, $n \geq 5$ and S meets a single member $D = \mathbf{B}(z, \rho_{n-1})$ of G_{n-1}. We will use the radial retraction (see 4.1.16)

$$\begin{gathered} f: \mathbf{R}^2 \to E = \mathbf{B}(z, \rho_{n-1} - \rho_n) \quad \text{with} \quad \operatorname{Lip}(f) = 1, \quad f(x) = x \text{ for } x \in E, \\ f(x) = z + (\rho_{n-1} - \rho_n)(x - z)/|x - z| \text{ for } x \in \mathbf{R}^2 \sim E. \end{gathered}$$

For each $B \in G_n$ we let $c(B)$ be the center of B; whenever B meets S we can choose

$$\xi(B) \in f^{-1}(B) \cap \operatorname{Bdry} S,$$

subject to the further requirement

$$f[\xi(B)] = c(B) \text{ in case } c(B) \in S;$$

it follows that

$$\lambda\psi_{n+1}(B \cap S) \leq (1 - n^{-1}) \sin(n^{-1}\pi)\rho_{n-1} - |f[\xi(B)] - c(B)|$$

because either $c(B) \in S$, $f[\xi(B)] = c(B)$ and

$$\lambda\psi_{n+1}(B \cap S) \leq \lambda\psi_n(B) = \lambda\rho_n < (n-1)\sin(n^{-1}\pi)\rho_n,$$

or $c(B) \notin S$, $B \cap S$ is contained in a halfplane bounded by a straight line through $c(B)$ and

$$\begin{aligned}\lambda\psi_{n+1}(B \cap S) &\leq \lambda[(n+1)/2 + 1]\rho_{n+1} \\ &= \lambda[2^{-1} + (n+1)^{-1}]\rho_n < [(n-1)\sin(n^{-1}\pi) - 1]\rho_n \\ &\leq (n-1)\sin(n^{-1}\pi)\rho_n - |f[\xi(B)] - c(B)|.\end{aligned}$$

Since Bdry S is a simple closed curve we can arrange all those members of G_n which meet S into a univalent sequence $B_1, \ldots, B_m$ (with $m \geq 2$) so that the corresponding points $\xi(B_1), \ldots, \xi(B_m)$ occur in a natural order along Bdry S. Letting $B_0 = B_m$ we infer from 3.2.35 and 2.10.13 that

$$2\zeta_1(S) = \mathscr{H}^1(\text{Bdry } S) \geq \sum_{j=1}^{m} |\xi(B_j) - \xi(B_{j-1})|$$

$$\geq \sum_{j=1}^{m} |f[\xi(B_j)] - f[\xi(B_{j-1})]|$$

$$\geq \sum_{j=1}^{m} (|c(B_j) - c(B_{j-1})| - |f[\xi(B_j)] - c(B_j)| - |f[\xi(B_{j-1})] - c(B_{j-1})|)$$

$$\geq 2 \sum_{j=1}^{m} ((1 - n^{-1}) \sin(n^{-1}\pi)\, \rho_{n-1} - |f[\xi(B_j)] - c(B_j)|)$$

$$\geq 2 \sum_{j=1}^{m} \lambda\, \psi_{n+1}(B_j \cap S) = 2\lambda\, \psi_{n+1}(S) \geq 2\lambda\, \psi_{\nu+1}(S).$$

Accordingly we find that

$$\sum_{S \in H} \zeta_1(S) \geq \sum_{S \in H} \lambda\, \psi_{\nu+1}(S) \geq \lambda\, \psi_{\nu+1}(A) = \lambda,$$

as asserted. Therefore $\mathscr{Q}_1^1(A) \geq \pi$.

Since $\beta_t(2,1)\, \mathscr{Q}_t^1(A)$ is nondecreasing with respect to t for $1 \leq t \leq \infty$, whereas we have shown that

$$\beta_\infty(2,1)\, \mathscr{Q}_\infty^1(A) = \mathscr{H}^1(A) \leq 2 = \beta_1(2,1)\, \pi \leq \beta_1(2,1)\, \mathscr{Q}_1^1(A),$$

we conclude

$$\beta_t(2,1)\, \mathscr{Q}_t^1(A) = 2 \quad \text{for} \quad 1 \leq t \leq \infty;$$

in particular $\mathscr{H}^1(A) = 2$ and $\mathscr{Q}_1^1(A) = \pi$. We infer that, for each integer $n \geq 4$,

$$\mathscr{H}^1(A \cap B) = 2\rho_n \quad \text{and} \quad \mathscr{Q}_1^1(A \cap B) = \pi\, \rho_n \quad \text{whenever} \quad B \in G_n,$$

because the $1/\rho_n$ sets $A \cap B$ corresponding to $B \in G_n$ are congruent and form a partition of A. Since $\{A \cap B : B \in G_n \text{ for some } n \geq 4\}$ is a basis for the relative topology of A, it follows that

$$(\mathscr{H}^1 \llcorner A)/2 = (\mathscr{Q}_1^1 \llcorner A)/\pi.$$

Accordingly 3.2.26 implies A is purely $(\mathscr{H}^1, 1)$ unrectifiable.

For $n \geq 5$ we define $\eta_n: \bigcup G_n \to \mathbf{S}^1$ by the formula

$$\eta_n(x) = (x-z)/|x-z| \text{ whenever } x \in \bigcup G_n,\ x \in \mathbf{B}(z, \rho_{n-1}) \in G_{n-1}.$$

Then $\mathscr{H}^1$ almost all points a of A have the property that

$$\{\eta_n(a): n > \nu\} \text{ is dense in } \mathbf{S}^1$$

for every integer $\nu \geq 5$. In fact, for any closed proper subarc J of $\mathbf{S}^1$ we consider the families $K_\nu = G_\nu$ and

$$K_n = \bigcup_{D \in K_{n-1}} F_n(D) \cap \{B: \eta_n(B) \cap (\mathbf{S}^1 \sim J) \neq \varnothing\}$$

corresponding to $n > \nu$; since $\sigma = \mathscr{H}^1(\mathbf{S}^1 \sim J)/2\pi < 1$ and

$$\mathscr{H}^1(A \cap \bigcup K_n) \leq \mathscr{H}^1(A \cap \bigcup K_{n-1}) \cdot (\sigma + 2/n)$$

for $n > \nu$, we find that the set

$$A \cap \bigcap_{n > \nu} \{a: \eta_n(a) \notin J\} \subset \bigcap_{n > \nu} (A \cap \bigcup K_n)$$

has $\mathscr{H}^1$ measure 0.

For $\mathscr{H}^1$ almost all points a of A it is true that

$$\Theta^{*1}[\mathscr{H}^1 \llcorner A \cap \mathbf{X}(a, \infty, V, s)] \geq 1/(2\pi)$$

whenever $V \in \mathbf{G}(2,1)$ and $0 < s < 1$. In view of the preceding paragraph it suffices to show that for every sufficiently large integer n there exists a positive number $t \leq 2\rho_{n-1}$ such that the ratio

$$\mathscr{H}^1[A \cap \mathbf{X}(a, t, \{x: x \bullet \eta_n(a) = 0\}, s)]/(2t)$$

is not much smaller than $1/(2\pi)$. For this purpose we define

$$S(z, r, u, \gamma) = \mathbf{B}(z, r) \cap \{x: (x-z) \bullet u \geq r\cos(\gamma)\}$$

whenever $z \in \mathbf{R}^2$, $r > 0$, $u \in \mathbf{S}^1$, $0 \leq \gamma \leq \pi$, and observe that

$$\mathscr{H}^1[A \cap S(z, \rho_{n-1}, u, \gamma)]/(2\rho_{n-1}) \text{ is close to } \gamma/\pi$$

in case $\mathbf{B}(z, \rho_{n-1}) \in G_{n-1}$ and n is large. Assuming

$$a \in A, \quad a \in \mathbf{B}(z, \rho_{n-1}) \in G_{n-1}, \quad V = \{x: x \bullet \eta_n(a) = 0\},$$
$$0 < \alpha < \pi/2, \quad \sin(\alpha) = s, \quad 0 < \beta < \pi, \quad \cos(\beta) = 1 - 2/n$$

one readily sees that

$$S[z, \rho_{n-1}, \eta_n(a)\exp(i\alpha), \alpha] \sim S[z, \rho_{n-1}, \eta_n(a), \beta] \subset \mathbf{X}(a, 2\rho_{n-1} s, V, s);$$

if n is large, then β is small and

$$\mathscr{H}^1[A \cap \mathbf{X}(a, 2\rho_{n-1} s, V, s)]/(4\rho_{n-1} s)$$

is not much smaller than $\alpha/(2\pi s) \geq 1/(2\pi)$.

Readopting now the notation of 3.3.10, we infer from the above inequality that

$$\mathscr{H}^1\Big(A \sim \bigcap_{V \in \mathbf{G}(2,1)} \{a\colon (a, V) \in S_2\}\Big) = 0,$$

hence $\mathscr{L}^1[p(A)] = \mathscr{L}^1[p\{a\colon (a, \ker p) \in S_2\}] = 0$ for every $p \in \mathbf{O}^*(2,1)$; thus $\mathscr{G}^1(A) = 0$.

It may also be shown that

$$\Theta^{*1}(\mathscr{H}^1 \llcorner A, a) = \tfrac{1}{2} \text{ for } \mathscr{H}^1 \text{ almost all } a \text{ in } A.$$

3.3.20. Assuming $\frac{1}{2} \le m < 1$ we now apply 2.10.28, with $n_j = 2$ for every positive integer j, to construct the compact set

$$E = A(m, n) \times A(m, n) \subset \mathbf{R}^2.$$

We see from 2.10.27 and the initial remark of 2.10.29 that

$$\alpha(2m)\, 2^{-2m} \le \mathscr{H}^{2m}(E) \le \alpha(2m)\, 2^{-m}.$$

Since $\mathscr{L}^1[A(m,n)] = 0$, we infer from 3.2.27 that *E is purely $(\mathscr{H}^1, 1)$ unrectifiable*, hence purely $(\mathscr{H}^{2m}, 1)$ unrectifiable. From 3.3.14 and 3.2.26 it follows that *E has no subset with finite and positive $\mathscr{I}^1_\infty$ measure*. Therefore the projection properties proved below imply

$$\mathscr{I}^1_\infty(E) = \infty \text{ in case } m > \tfrac{1}{2}.$$

This behavior of $\mathscr{I}^1_\infty$ contrasts the more supple manner of $\mathscr{H}^1$ guaranteed by 2.10.48. We will also show that

$$\mathscr{G}^1(E) = 3/\sqrt{5} \text{ and } \mathscr{H}^1(E) = \sqrt{2} \text{ in case } m = \tfrac{1}{2}.$$

From 2.10.28 (4) we see that, for $J \in F$, $\Phi(J, 2)$ consists of the intervals

$$\{\xi\colon \inf J \le \xi \le (\inf J) + s\}, \qquad \{\xi\colon (\sup J) - s \le \xi \le \sup J\}$$

with $s = \operatorname{diam} J / 2^{1/m}$. We now let

$$\Psi(J \times K) = \{P \times Q\colon P \in \Phi(J, 2), Q \in \Phi(K, 2)\}$$

whenever $J, K \in F$, and inductively define the families

$$G_0 = \{I \times I\}, \quad \text{where} \quad I = \{\xi\colon 0 \le \xi \le 1\},$$

$$G_j = \bigcup \{\Psi(T)\colon T \in G_{j-1}\} \text{ for } j = 1, 2, 3, \ldots.$$

Recalling the definition of $A(m, n)$ we find that

$$E = \bigcap_{j=0}^{\infty} \bigcup G_j.$$

With each $u \in \mathbf{S}^1$ we associate

$$p_u \in \mathbf{O}^*(2,1), \quad p_u(x) = u \bullet x \text{ for } x \in \mathbf{R}^2,$$

and observe that there exists a positive number $c_u \leq 1$ such that

$$\mathscr{L}^1[p_u(\bigcup \Psi\, T)] = c_u\, \mathscr{L}^1[p_u(T)]$$

for all squares $T = J \times K$ corresponding to $J, K \in F$ with $\operatorname{diam} J = \operatorname{diam} K$; it follows that

$$\mathscr{L}^1[p_u(E)] = 0 \text{ whenever } c_u < 1,$$

$$\mathscr{L}^1[p_u(E)] = \mathscr{L}^1[p_u(I \times I)] = |u_1| + |u_2| \text{ whenever } c_u = 1.$$

If $u \in \mathbf{S}^1$ with $u_1 \geq u_2 \geq 0$, then p_u maps the four squares in $\Psi(I \times I)$ onto the intervals

$$\{\xi: 0 \leq \xi \leq s(u_1 + u_2)\}, \quad \{\xi: (1-s)\, u_2 \leq \xi \leq s\, u_1 + u_2\},$$

$$\{\xi: (1-s)\, u_1 \leq \xi \leq u_1 + s\, u_2\}, \quad \{\xi: (1-s)(u_1 + u_2) \leq \xi \leq u_1 + u_2\}$$

with $s = 2^{-1/m}$; in order that the union of these intervals be equal to

$$p_u(I \times I) = \{\xi: 0 \leq \xi \leq u_1 + u_2\}$$

it is necessary and sufficient that

$$1 - 2s \leq u_2/u_1 \leq s/(1-2s).$$

Making use of the symmetries of $I \times I$ we conclude that

$$U = \mathbf{S}^1 \cap \{u: c_u = 1\}$$

consists of those $u \in \mathbf{S}^1$ *for which either* $|u_2/u_1|$ or $|u_1/u_2|$ *belongs to the compact interval with endpoints*

$$1 - 2^{1-1/m} \text{ and } \inf\{1, (2^{1/m} - 2)^{-1}\}.$$

In case $m > \frac{1}{2}$, this interval is nondegenerate, hence

$$0 < \mathscr{H}^1(U) = 2\pi\, \theta^*_{2,1}\{p_u: u \in U\} \text{ and } \mathscr{I}^1_\infty(E) > 0.$$

In case $m = \frac{1}{2}$, this interval consists of the single number $\frac{1}{2}$, hence

$$U = \mathbf{S}^1 \cap \{u: |u_1| = 2\,|u_2| \text{ or } |u_2| = 2\,|u_1|\}$$

and $\mathscr{L}^1[p_u(E)] = 3/\sqrt{5}$ whenever $u \in U$; since the families $\{T \cap E: T \in G_j\}$ form Borel partitions of E with

$$\mathscr{L}^1[p_u(T \cap E)] = \mathscr{L}^1[p_u(E)]/\operatorname{card} G_j$$

whenever $T \in G_j$ and $u \in U$, it follows from 2.10.8 that $\mathscr{G}^1(E) = 3/\sqrt{5}$.

In case $m=\frac{1}{2}$, we know that $\mathscr{H}^1(E)\leq\alpha(1)\,2^{-\frac{1}{2}}=\sqrt{2}$. Defining the measures ψ_j over $\mathbf{R}^2$ by the formula

$$\psi_j(S)=(\sqrt{2}/4^j)\operatorname{card}(G_j\cap\{T\colon T\cap S\neq\varnothing\})$$

whenever $S\subset\mathbf{R}^2$, we observe that $\psi_j(E)=\sqrt{2}$ and $\psi_j\geq\psi_{j+1}$, hence the inequality $\mathscr{H}^1(E)\geq\sqrt{2}$ follows from the following proposition:

If $S\subset\mathbf{R}^2$, *then* $\operatorname{diam}(S)\geq\psi_j(S)$ *for some* j.

An obvious similarity transformation reduces the proof of this proposition to the special case when S meets at least two members of G_1, hence $\operatorname{diam} S\geq\frac{1}{2}$. Defining

$$R_j=G_j\cap\{T\colon T\cap S=\varnothing\},\qquad r_j=\operatorname{card}R_j=4^j[1-\psi_j(S)/\sqrt{2}],$$
$$\sigma=1-(\operatorname{diam}S)/\sqrt{2},$$

and assuming the conclusion of the proposition to fail, we infer that

$$r_j<4^j\sigma \text{ for every positive integer } j.$$

To prove this impossible we consider the class Ω of all those quadruples (j,λ,μ,δ) for which there exist a family

$$K\subset R_j \text{ with } \operatorname{card}K=\lambda,$$

and 2μ distinct squares $T_1,T_2,\ldots,T_{2\mu-1},T_{2\mu}\in G_j\sim K$ with

$$\operatorname{distance}(T_{2i-1},T_{2i})\geq\delta \text{ for } i=1,\ldots,\mu.$$

We observe:

If $(j,\lambda,\mu,\delta)\in\Omega$ and $r_j<\lambda+\mu$, then $\sigma\leq1-\delta/\sqrt{2}$; in fact

$$\operatorname{card}\{i\colon T_{2i-1}\in R_j \text{ or } T_{2i}\in R_j\}\leq r_j-\operatorname{card}K<\mu,$$

hence $\operatorname{diam}S\geq\delta$.

First we note that $r_1\leq2$ and $\sigma\leq1-1/(2\sqrt{2})<11/16$, hence $r_2<11$.
In case $r_1\leq1$, the fact that $(1,0,2,\sqrt{2}/2)\in\Omega$ implies $\sigma\leq\frac{1}{2}$.
In case $r_1=2$, then $(2,8,4,11/16)\in\Omega$ and $r_2<11$, hence

$$\sigma\leq1-11/(16\sqrt{2}),\ r_2<16-11/\sqrt{2}<9;$$

moreover $(2,8,1,7/8)\in\Omega$, hence $\sigma\leq1-7/(8\sqrt{2})<7/16$.

Thus assured that $0<\sigma\leq2/4$, we choose a positive integer k for which

$$2/4^{k+1}<\sigma\leq2/4^k.$$

Since $r_{k+1}<4^{k+1}\sigma\leq8$ and

$$(k+1,0,8,\sqrt{2}-5\sqrt{2}/4^{k+1})\in\Omega,$$

as one verifies by using the 16 squares of G_{k+1} nearest the corners of $I\times I$, judiciously pairing squares near diagonally opposite corners, one infers $\sigma \leq 5/4^{k+1}$, hence $r_{k+2} < 20$. Examining next the 40 squares of G_{k+2} nearest the corners of $I \times I$ one sees that

$$(k+2, 0, 20, \sqrt{2} - 14\sqrt{2}/4^{k+2}) \in \Omega,$$

hence $\sigma \leq 14/4^{k+2}$ and $r_{k+2} < 14$; similarly

$$(k+2, 0, 14, \sqrt{2} - (19/\sqrt{2})\, 4^{k+2}) \in \Omega,$$

hence $\sigma \leq (19/2)/4^{k+2}$ and $r_{k+2} < 10$; finally

$$(k+2, 0, 10, \sqrt{2} - 2\sqrt{2}/4^{k+1}) \in \Omega,$$

hence $\sigma \leq 2/4^{k+1}$, contrary to our choice of k.

3.3.21. Applying 2.10.28 with $0 < m < 1$ and a sequence n such that $n_j \to \infty$ as $j \to \infty$, one obtains a compact set

$$E = A(m,n) \times A(m,n) \subset \mathbf{R}^2$$

for which $0 < \mathscr{H}^{2m}(E) < \infty$ and $\Theta^m_*(\mathscr{H}^{2m} \llcorner E, x) = 0$ for $\mathscr{H}^{2m}$ almost all x.

Other instructive examples of unrectifiable sets are discussed in [MR 2, p. 285], [MR 1, p. 411], [DD], [WG], [MO], [B 1, pp. 435, 456, 459], [BU], [BEM], [RJ 2], [B 3], [SP 1, p. 184], [FR 1, p. 262], [FR 2], [ME 2, p. 600], [WA], [S 2], [MAS].

3.3.22. The development of the theory of Hausdorff measure was greatly stimulated by the problem of proving the following statement (essentially the converse of the first implication in 3.2.19):

If $W \subset \mathbf{R}^n$, m is a positive integer and

$$\Theta^m(\mathscr{H}^m \llcorner W, w) = 1 \text{ for } \mathscr{H}^m \text{ almost all } w \text{ in } W,$$

then W is countably $(\mathscr{H}^m, m)$ rectifiable.

This statement was proved for $m=1$ and $n=2$ in [B 1]; for $m=1$ and all n in [ME 1]; for $m=2$ and $n=3$ in [MJ 4]. Moreover it seems that the method of [MJ 4] could be modified so as to apply to all positive integers m and n.

However the above statement appears to be a special case of the following still conjectural proposition:

If ϕ is a Borel regular measure over $\mathbf{R}^n$, m is a positive integer and

$$0 < \Theta^m(\phi, a) < \infty \text{ for } \phi \text{ almost all } a \text{ in } \mathbf{R}^n,$$

then $\mathbf{R}^n$ is countably (ϕ, m) rectifiable.

This conjecture is true when $m=1$, as proved for $n=2$ in [MR 2], then for all n in [ME 1]; actually these authors show the sufficiency of the hypothesis

$$\Theta^{*1}(\phi, a) < (1.01)\, \Theta^1_*(\phi, a) \text{ for } \phi \text{ almost all } a.$$

For $m>1$ only partial results have been obtained, in [MJ 5]. One reason for the difficulty is the fact that, while every connected subset C of $\mathbf{R}^n$ with $\mathscr{H}^1(C)<\infty$ is 1 rectifiable (by [EH, Theorem 2] and 2.5.16), no comparable assertion is valid when $m>1$; we will see in 4.2.25 that every compact totally disconnected subset A of $\mathbf{R}^3$ is contained in a set B homeomorphic with $\mathbf{S}^2$, such that $B \sim A$ is a 2 dimensional submanifold of class ∞ of $\mathbf{R}^3$ with $\mathscr{H}^2(B\sim A)<\infty$.

The results described above will be neither proved nor used in this book.

3.4. Some properties of highly differentiable functions

3.4.1. In this section we will first solve the following problem: *Given integers $k\geq 1$ and $m>\nu\geq 0$, find the least number s such that*

$$\mathscr{H}^s[f\{x\colon \dim \operatorname{im} Df(x)\leq \nu\}]=0$$

for every function f of class k mapping an open subset of $\mathbf{R}^m$ into some normed vectorspace Y. It is shown by 3.4.3 and 3.4.4 that $s=\nu+(m-\nu)/k$.

From 3.2.3 we already know that $s\leq m$ if $\nu=m-1$ and $\dim Y<\infty$; however the present argument will be independent of 3.2.3.

3.4.2. Lemma. *Suppose:*

(1) *Y is a normed vectorspace.*

(2) *Z is a vectorsubspace of $\mathbf{R}^m$, $\nu=m-\dim Z$.*

(3) *Whenever $\mu\geq\nu$ and $k\geq 1$ are integers,*

$$\Omega(\mu, k)$$

is the set of those μ dimensional submanifolds S of class k of $\mathbf{R}^m$ for which either $\mu=0$ or there exists $p\in\mathbf{O}^*(m,\mu)$ such that

$$\ker p\subset Z, \quad p(S) \text{ is convex and open in } \mathbf{R}^\mu,$$

$$p|S \text{ is univalent}, \quad (p|S)^{-1} \text{ is a map of class } k.$$

(4) *Whenever $C\subset S\in\Omega(\mu,1)$, $f\colon S\to Y$, $0\leq\lambda<\infty$ and $\delta>0$,*

$$\eta(S, C, f, \lambda, \delta)$$

is the supremum of the set consisting of 0 and the numbers

$$|f(x)-f(c)|/|x-c|^\lambda$$

corresponding to all $x \in S$, $c \in C$ *with* $x - c \in Z$, $0 < |x - c| \leq \delta$; *moreover, in case* f *is of class* 1,

$$\zeta(S, C, f, \lambda, \delta)$$

is the supremum of the set consisting of 0 *and the numbers*

$$|\langle z, Df(x) \rangle| / |x - c|^{\lambda}$$

corresponding to all $x \in S$, $c \in C$, $z \in Z \cap \operatorname{Tan}(S, x)$ *with*

$$x - c \in Z, \quad 0 < |x - c| \leq \delta, \quad |z| \leq 1.$$

(5) *Whenever* $C \subset S \in \Omega(\mu, 1)$ *and* k *is a positive integer,*

$$\Phi(S, C, k)$$

is the set of all maps $f: S \to Y$ *of class* k *such that* $f(x) = 0$ *for* $x \in C$; *moreover*

$$\Psi(S, C, k)$$

is the set of all pairs (T, G) *such that* $T \in \Omega(\gamma, 1)$ *for some* $\gamma \leq \mu$, $T \subset S$, G *is a compact subset of* $T \cap C$ *and*

$$\eta(T, G, f | T, k, \delta) \to 0 \text{ as } \delta \to 0+ \text{ for every } f \in \Phi(S, C, k).$$

Then the following four conclusions hold:

(6) *If* $S \in \Omega(\mu, 1)$, C *is a compact subset of* S,

$$f: S \to Y \text{ is of class } 1,\ 1 \leq \lambda < \infty,\ \zeta(S, C, f, \lambda - 1, \delta) \to 0 \text{ as } \delta \to 0+,$$

then $\eta(S, C, f, \lambda, \delta) \to 0$ *as* $\delta \to 0+$.

(7) *If* $S \in \Omega(\mu, k)$ *and* C *is a relatively closed subset of* S, *then there exists a sequence of pairs* $(T_j, G_j) \in \Psi(S, C, k)$ *such that*

$$C = \bigcup_{j=1}^{\infty} G_j.$$

(8) *If* $S \in \Omega(\mu, 1)$, C *is a compact subset of* S,

$$f: S \to Y \text{ is of class } 1,\ 0 < \lambda < \infty,\ \eta(S, C, f, \lambda, \delta) \to 0 \text{ as } \delta \to 0+,$$

and $\mu > \nu$, $\sigma = \nu + (\mu - \nu)/\lambda$, *then* $\mathscr{H}^{\sigma}[f(C)] = 0$.

(9) *If* $S \in \Omega(\mu, k)$, $f: S \to Y$ *is of class* $k \geq 1$,

$$C = S \cap \{x : \langle z, Df(x) \rangle = 0 \text{ for } z \in Z \cap \operatorname{Tan}(S, x)\},$$

and $\mu > \nu$, $\tau = \nu + (\mu - \nu)/k$, *then* $\mathscr{H}^{\tau}[f(C)] = 0$.

Proof of (6). Choosing p according to (3) we observe that

$$r=(p|S)^{-1}\circ p$$

is a map of class 1 retracting $U=p^{-1}[p(S)]$ onto S, with

$$r(x)-x\in\ker p\subset Z,\ \langle z, D\,r(x)\rangle\in Z \text{ for } x\in U,\ z\in Z.$$

For every sufficiently small $\delta>0$ it is true that

$$V_\delta=\{x\colon \operatorname{dist}(x,C)\le\delta\}\subset U,\ M_\delta=\operatorname{Lip}(r|V_\delta)<\infty.$$

If $x\in S$, $c\in C$ with $x-c\in Z$, $|x-c|\le\delta$, then

$$g(t)=r[c+t(x-c)]\in S,\ g(t)-c\in Z \text{ and}$$

$$|g(t)-c|\le M_\delta|x-c|\le M_\delta\,\delta \text{ for } 0\le t\le 1;$$

$$g'(t)=\langle x-c, D\,r[c+t(x-c)]\rangle\in Z\cap\operatorname{Tan}[S,g(t)] \quad\text{and}$$

$$\begin{aligned}|(f\circ g)'(t)|&=|\langle g'(t), D\,f[g(t)]\rangle|\\ &\le|g'(t)|\,\zeta(S,C,f,\lambda-1,M_\delta\,\delta)[M_\delta|x-c|]^{\lambda-1}\\ &\le M_\delta^\lambda\,\zeta(S,C,f,\lambda-1,M_\delta\,\delta)\,|x-c|^\lambda\end{aligned}$$

for $0<t<1$; consequently

$$|f(x)-f(c)|=|(f\circ g)\,1-(f\circ g)\,0|\le M_\delta^\lambda\,\zeta(S,C,f,\lambda-1,M_\delta\,\delta)\,|x-c|^\lambda.$$

Proof of (7). We proceed by simultaneous induction with respect to $\mu(\ge\nu)$ and $k(\ge 1)$.

Since C is the union of a countable family of compact sets, we will consider only the case when C is compact.

In case $\mu=\nu>0$ we choose p according to (3) and observe that

$$\dim\ker p=m-\mu=m-\nu=\dim Z,$$

hence $\ker p=Z$ and there exist no distinct points $x,c\in S$ with $x-c\in Z$; consequently

$$\eta(S,C,f,k,\delta)=0 \text{ for } f\in\Phi(S,C,k),\ \delta>0.$$

In case $\mu=\nu=0$ the same is true when δ is sufficiently small, because S is discrete.

In case $\mu>\nu$ we consider the closed set

$$B=C\cap\{x\colon \langle z, D\,f(x)\rangle=0 \text{ for } f\in\Phi(S,C,k),\ z\in Z\cap\operatorname{Tan}(S,x)\},$$

and will first construct a sequence of pairs $(U_j,H_j)\in\Psi(S,C,k)$ such that

$$B=\bigcup_{j=1}^{\infty}H_j.$$

In the subcase $k=1$ we note that, for $f\in\Phi(S,C,1)$,

$$\zeta(S,B,f,0,\delta)\to 0 \quad\text{as}\quad \delta\to 0+$$

because Df is continuous and

$$\{(x,z):\ x\in S,\ \operatorname{dist}(x,B)\le\delta,\ z\in Z\cap\operatorname{Tan}(S,x),\ |z|\le 1\}$$

is compact for every sufficiently small $\delta>0$, hence (6) implies

$$\eta(S,B,f,1,\delta)\to 0 \quad\text{as}\quad \delta\to 0+;$$

accordingly $(S,B)\in\Psi(S,C,1)$ and we take $U_j=S$, $H_j=B$ for all positive integers j.

In the subcase $k\ge 2$ we inductively apply (7) with k and C replaced by $k-1$ and B to obtain a sequence of pairs $(U_j,H_j)\in\Psi(S,B,k-1)$ such that

$$B=\bigcup_{j=1}^{\infty} H_j.$$

We also choose p according to (3), define r as in the proof of (6), let $e_1,\ldots,e_{m-\nu}$ be an orthonormal basis of Z, and associate with every function $f\colon S\to Y$ of class k the functions $f_1,\ldots,f_{m-\nu}\colon S\to Y$ of class $k-1$ by the formula

$$f_i(x)=\langle\langle e_i, D\,r(x)\rangle, D\,f(x)\rangle \ \text{ for }\ x\in S.$$

If $f\in\Phi(S,C,k)$, then $f_i\in\Phi(S,B,k-1)$ for $i=1,\ldots,m-\nu$ because

$$\langle e_i, D\,r(x)\rangle\in Z\cap\operatorname{Tan}(S,x) \ \text{ for }\ x\in S,$$

hence

$$\eta(U_j,H_j,f_i|U_j,k-1,\delta)\to 0 \quad\text{as}\quad \delta\to 0+$$

for each positive integer j; moreover, if $x\in U_j$ and $z\in Z\cap\operatorname{Tan}(U_j,x)$, then $z\in\operatorname{Tan}(S,x)$ and

$$\langle z, D\,f(x)\rangle=\langle\langle z, D\,r(x)\rangle, D\,f(x)\rangle=\sum_{i=1}^{m-\nu}(z\bullet e_i)\,f_i(x);$$

consequently

$$\zeta(U_j,H_j,f|U_j,k-1,\delta)\le\sum_{i=1}^{m-\nu}\eta(U_j,H_j,f_i|U_j,k-1,\delta)\to 0 \quad\text{as}\quad \delta\to 0+$$

and (6) implies

$$\eta(U_j,H_j,f|U_j,k,\delta)\to 0 \quad\text{as}\quad \delta\to 0+.$$

Thus $(U_j,H_j)\in\Psi(S,C,k)$.

Next we will show that for each $\xi\in C\sim B$ there exists a manifold $V\in\Omega(\mu-1,k)$ such that $V\cap C$ is a neighborhood of ξ relative to C.

For this purpose we choose $f \in \Phi(S, C, k)$ and $z \in Z \cap \operatorname{Tan}(S, \xi)$ with $\langle z, Df(\xi)\rangle \neq 0$, then use the Hahn-Banach theorem to obtain a continuous linear map

$$\alpha: Y \to \mathbf{R} \quad \text{with} \quad \langle\langle z, Df(\xi)\rangle, \alpha\rangle \neq 0,$$

and observe that

$$F = \alpha \circ f \circ (p|S)^{-1}: p(S) \to \mathbf{R}$$

is a function of class k for which

$$F \circ p|S = \alpha \circ f, \qquad \langle p(z), DF[p(\xi)]\rangle \neq 0.$$

In the subcase $\mu > 1$ we also choose

$$q \in \mathbf{O}^*(\mu, \mu-1) \quad \text{with} \quad \ker q = \{t\, p(z): t \in \mathbf{R}\},$$

and define the map $g: p(S) \to \mathbf{R} \times \mathbf{R}^{\mu-1}$ of class k,

$$g(w) = (F(w), q(w)) \text{ for } w \in p(S).$$

Since $\ker Dg[p(\xi)] = \ker DF[p(\xi)] \cap \ker q = \{0\}$, $p(\xi)$ has a neighborhood W in $p(S)$ such that $g|W$ is a diffeomorphism, $g(W) = F(W) \times q(W)$, $F(W)$ is an open interval, and $q(W)$ is a convex, open subset of $\mathbf{R}^{\mu-1}$. Inasmuch as $C \subset f^{-1}\{0\}$, the set

$$V = S \cap p^{-1}(W \cap F^{-1}\{0\})$$

is a neighborhood of ξ relative to C, the orthogonal projection $q \circ p$ maps V homeomorphically onto $q(W)$, $[(q \circ p)|V]^{-1}$ is a map of class k, and

$$\ker(q \circ p) = p^{-1}(\ker q) \subset p^{-1}[p(Z)] = Z,$$

hence $V \in \Omega(\mu-1, k)$.

In the subcase $\mu = 1$ the point ξ is isolated in C and we take $V = \{\xi\}$.

To complete the proof we observe that $C \sim B$ is separable, choose a sequence of manifolds $V_i \in \Omega(\mu-1, k)$ such that

$$C \sim B = \bigcup_{i=1}^{\infty} (V_i \cap C),$$

and inductively apply (7) for each i with μ, S, C replaced by $\mu-1, V_i, V_i \cap C$ to obtain a sequence of pairs $(W_{i,j}, K_{i,j}) \in \Psi(V_i, V_i \cap C, k)$ such that

$$V_i \cap C = \bigcup_{j=1}^{\infty} K_{i,j},$$

hence $(W_{i,j}, K_{i,j}) \in \Psi(S, C, k)$ and

$$C \sim B = \bigcup_{i=1}^{\infty} \bigcup_{j=1}^{\infty} K_{i,j}.$$

Proof of (8). We choose p according to (3) and let

$$Q = \mathbf{R}^\mu \cap \{v\colon v \bullet u = 0 \text{ for } u \in p(Z)\};$$

then $\mathbf{R}^\mu = p(Z) \oplus Q$ and

$$\dim p(Z) = \dim Z - \dim \ker p = \mu - \nu, \quad \dim Q = \nu.$$

Assuming that $r > 0$, A is a compact $\mu - \nu$ dimensional cube with side length r contained in $p(Z)$, B is a compact ν dimensional cube with side length r contained in Q if $\nu > 0$, $B = \{0\}$ if $\nu = 0$, and that

$$A + B = \{u + v\colon u \in A, v \in B\} \subset p(S),$$

we will show that

$$\mathscr{H}^\sigma(f[C \cap p^{-1}(A+B)]) = 0.$$

First we choose a common Lipschitz constant M for the restrictions of $(p|S)^{-1}$ and $f \circ (p|S)^{-1}$ to $A + B$, and let $N = M\,\mu + M^\lambda\,\mu^\lambda$.

Next, given $\varepsilon > 0$, we choose $\delta > 0$ so that

$$\eta(S, C, f, \lambda, \delta) \leq \varepsilon,$$

select positive integers α and β for which

$$M\,\mu\, r/\alpha \leq \delta \quad \text{and} \quad \varepsilon(r/\alpha)^\lambda \leq r/\beta \leq 2\varepsilon(r/\alpha)^\lambda,$$

represent A as the union of $\alpha^{\mu-\nu}$ compact $\mu - \nu$ dimensional cubes A_i with side length r/α, and represent B as the union of β^ν compact ν dimensional cubes B_j with side length r/β; in case $\nu = 0$ we use a single cube $B_j = B = \{0\}$.

For each pair (i, j) such that

$$C \cap p^{-1}(A_i + B_j) \neq \varnothing$$

we can choose $c \in C$, $a \in A_i$, $b \in B_j$ with $p(c) = a + b$. Whenever

$$s \in S \cap p^{-1}(A_i + B_j)$$

we similarly find $u \in A_i$, $v \in B_j$ with $p(s) = u + v$; then

$$u + b \in A_i + B_j, \quad x = (p|S)^{-1}(u+b) \in S,$$

$$p(x - c) = u + b - (a + b) = u - a \in p(Z), \quad x - c \in Z,$$

$$|x - c| \leq M\,|u - a| \leq M \operatorname{diam} A_i \leq M(\mu - \nu)\, r/\alpha \leq \delta,$$

$$|f(x) - f(c)| \leq \varepsilon\, |x - c|^\lambda \leq \varepsilon(M\,\mu\, r/\alpha)^\lambda,$$

$$p(s - x) = u + v - (u + b) = v - b,$$

$$|f(s) - f(x)| \leq M\,|v - b| \leq M \operatorname{diam} B_j \leq M\,\nu\, r/\beta,$$

$$|f(s) - f(c)| \leq M\,\nu\, r/\beta + \varepsilon(M\,\mu\, r/\alpha)^\lambda \leq N\, r/\beta.$$

We infer that

$$\operatorname{diam} f[S \cap p^{-1}(A_i + B_j)] \leq 2N\, r/\beta.$$

Accordingly the size $2Nr/\beta$ approximating measure occurring in the construction of $\mathscr{H}^\sigma$ assigns to the set

$$f[C\cap p^{-1}(A+B)]$$

a number not exceeding

$$\begin{aligned}\alpha^{\mu-\nu}\beta^\nu\boldsymbol{\alpha}(\sigma)\,2^{-\sigma}(2Nr/\beta)^\sigma &\le \alpha^{\mu-\nu}\beta^\nu\boldsymbol{\alpha}(\sigma)\,N^\sigma(r/\beta)^\nu[2\varepsilon(r/\alpha)^\lambda]^{(\mu-\nu)/\lambda}\\ &=\boldsymbol{\alpha}(\sigma)\,N^\sigma r^\mu(2\varepsilon)^{(\mu-\nu)/\lambda}.\end{aligned}$$

Proof of (9). *In case* $k=1$ we note that, for each compact subset B of C,

$$\zeta(S,B,f,0,\delta)\to 0 \quad\text{as}\quad \delta\to 0+,$$

hence (6) implies

$$\eta(S,B,f,1,\delta)\to 0 \quad\text{as}\quad \delta\to 0+,$$

and (8) shows that $\mathscr{H}^\mu[f(B)]=0$.

In case $k\ge 2$ we apply (7) with k replaced by $k-1$ to obtain a sequence of pairs $(T_j,G_j)\in\Psi(S,C,k-1)$ such that

$$C=\bigcup_{j=1}^{\infty}G_j.$$

Defining f_i for $i=1,\ldots,m-\nu$ as in the proof of (7), we see that $f_i\in\Phi(S,C,k-1)$, hence

$$\begin{gathered}\eta(T_j,G_j,f_i|T_j,k-1,\delta)\to 0 \quad\text{as}\quad \delta\to 0+,\\ \zeta(T_j,G_j,f|T_j,k-1,\delta)\to 0 \quad\text{as}\quad \delta\to 0+,\\ \eta(T_j,G_j,f|T_j,k,\delta)\to 0 \quad\text{as}\quad \delta\to 0+\end{gathered}$$

by (6), for each positive integer j; moreover

$$T_j\in\Omega(\gamma_j,1) \text{ with } \nu\le\gamma_j\le\mu, \quad \sigma_j=\nu+(\gamma_j-\nu)/k\le\nu+(\mu-\nu)/k=\tau.$$

If $\gamma_j>\nu$, then (8) implies

$$\mathscr{H}^{\sigma_j}[f(G_j)]=0, \quad\text{hence}\quad \mathscr{H}^\tau[f(G_j)]=0.$$

If $\gamma_j=\nu$, then

$$\mathscr{H}^\nu[f(G_j)]<\infty \quad\text{and}\quad \mathscr{H}^\tau[f(G_j)]=0,$$

because $f(G_j)$ is ν rectifiable and $\tau>\nu$.

3.4.3. Theorem. *If* $m>\nu\ge 0$ *and* $k\ge 1$ *are integers, A is an open subset of* $\mathbf{R}^m$, $B\subset A$, *Y is a normed vectorspace, and*

$$f\colon A\to Y \textit{ is a map of class } k, \qquad \dim\operatorname{im} Df(x)\le\nu \textit{ for } x\in B,$$

then

$$\mathscr{H}^{\nu+(m-\nu)/k}[f(B)]=0.$$

Proof. Assuming $a \in A$ with $\dim \operatorname{im} Df(a) = \nu$, we will show that a has a neighborhood U in A for which

$$\mathscr{H}^{\nu+(m-\nu)/k}[f(B \cap U)] = 0.$$

First we use the Hahn-Banach theorem to obtain a continuous linear map $L\colon Y \to \mathbf{R}^\nu$ whose restriction to $\operatorname{im} Df(a)$ is an isomorphism. We also choose

$$q \in \mathbf{O}^*(m, m-\nu) \quad \text{with} \quad \operatorname{im} q^* = \ker Df(a)$$

and define the map $g\colon A \to \mathbf{R}^\nu \times \mathbf{R}^{m-\nu}$ of class k,

$$g(x) = (L[f(x)], q(x)) \text{ for } x \in A.$$

Since $\ker Dg(a) = \ker L \circ Df(a) \cap \ker q = \{0\}$, the point a has a neighborhood U in A such that $g|U$ is a diffeomorphism and $S = g(U)$ is convex. It follows that

$$F = f \circ (g|U)^{-1}\colon S \to Y \text{ is of class } k, \quad L[F(u,v)] = u \text{ for } (u,v) \in S,$$

$$L[\langle(\alpha,\beta), DF(u,v)\rangle] = \alpha \text{ for } (u,v) \in S,\ \alpha \in \mathbf{R}^\nu,\ \beta \in \mathbf{R}^{m-\nu},$$

and to each $(u,v) \in S$ corresponds the commutative diagram:

$$\begin{array}{ccccc}
\{0\} \times \mathbf{R}^{m-\nu} & \subset & \mathbf{R}^\nu \times \mathbf{R}^{m-\nu} & \supset & \mathbf{R}^\nu \times \{0\} \\
\downarrow & & \downarrow{\scriptstyle DF(u,v)} & & \wr\| \\
\ker L & \subset & Y & \xrightarrow{\ L\ } & \mathbf{R}^\nu.
\end{array}$$

Defining $C = S \cap \{(u,v)\colon DF(u,v)|\{0\} \times \mathbf{R}^{m-\nu} = 0\}$ we infer that

$$g(U \cap B) \subset S \cap \{(u,v)\colon \dim \operatorname{im} DF(u,v) \le \nu\} = C,$$

hence $f(U \cap B) \subset F(C)$. On the other hand 3.4.2(9) is applicable with μ, f, Z replaced by $m, F, \{0\} \times \mathbf{R}^{m-\nu}$ and yields the equation

$$\mathscr{H}^{\nu+(m-\nu)/k}[F(C)] = 0.$$

3.4.4. We will now construct examples showing that the result of the preceding theorem is optimal; for any positive number $\gamma < \nu + (m-\nu)/k$ it can happen that $\mathscr{H}^\gamma[f(B)] > 0$.

Given any positive number $\beta < 1/k$, we choose a positive number $\alpha < 1$ such that $\beta < \alpha/k$. Then we construct the sets

$$A(\alpha, n) \quad \text{and} \quad A(\beta, n)$$

by the method of 2.10.28, with m replaced by α and β, taking $n_j = 2$ for every positive integer j. Letting $H_{\alpha,j}$ and $H_{\beta,j}$ be the families of intervals occurring in the constructions of $A(\alpha, n)$ and $A(\beta, n)$, we observe that the obvious order preserving correspondences between $H_{\alpha,j}$ and $H_{\beta,j}$ induce a homeomorphism

$$\phi\colon A(\alpha, n) \simeq A(\beta, n).$$

For any two distinct points $u, v \in A(\alpha, n)$ there exists a least integer j such that u and v belong to distinct intervals in $H_{\alpha,j+1}$; then u and v belong to the same interval in $H_{\alpha,j}$, hence $\phi(u)$ and $\phi(v)$ belong to the same interval in $H_{\beta,j}$, and we obtain

$$|u-v| \geq 2^{-j/\alpha} - 2\cdot 2^{-(j+1)/\alpha}, \quad |\phi(u)-\phi(v)| \leq 2^{-j/\beta},$$
$$|\phi(u)-\phi(v)|/|u-v|^{\alpha/\beta} \leq (1-2^{1-1/\alpha})^{-\alpha/\beta};$$

since $k < \alpha/\beta$ it follows that $|\phi(u)-\phi(v)|/|u-v|^k$ is small when $|u-v|$ is small. Applying 3.1.14 with

$$P_a(x) = \phi(a) \text{ for } a \in A(\alpha, n) \text{ and } x \in \mathbf{R},$$

we obtain a map $g\colon \mathbf{R} \to \mathbf{R}$ of class k such that

$$g(a) = \phi(a) \quad \text{and} \quad g^{(i)}(a) = 0 \text{ for } i = 1, \ldots, k$$

whenever $a \in A(\alpha, n)$.

We let $f\colon \mathbf{R}^m \to \mathbf{R}^m$ be the map of class k such that $f(x) = y$ if and only if

$$y_i = x_i \text{ for } i = 1, \ldots, \nu \quad \text{and} \quad y_i = g(x_i) \text{ for } i = \nu+1, \ldots, m.$$

Taking $B = \mathbf{R}^m \cap \{x\colon x_i \in A(\alpha, n) \text{ for } i = \nu+1, \ldots, m\}$ we see that

$$\dim \operatorname{im} Df(x) = \nu \text{ for } x \in B,$$

and $f(B) = \mathbf{R}^m \cap \{y\colon y_i = A(\beta, n) \text{ for } i = \nu+1, \ldots, m\}$, hence 2.10.27 and 2.10.28 imply

$$\mathscr{H}^{\nu+(m-\nu)\beta}[f(B)] > 0.$$

Moreover $\nu + (m-\nu)\beta$ can be made arbitrarily close to $\nu + (m-\nu)/k$ by appropriate choice of $\beta < 1/k$.

Additionally, the reader may consult [W 2] or [CG] for examples where $\nu = 0$, $k < m$, $Y = \mathbf{R}$, B is an arc and $f(B)$ is a nondegenerate interval. Related problems are studied in [SI], [SA 2] and [KP].

3.4.5. In the second part of this section we will derive some of the fundamental local algebraic and measure theoretic properties of sets defined by real analytic equations. The main results are contained in Theorem 3.4.8. Here we discuss the basic concepts.

Suppose X is a Banach space and $a \in X$.

In the study of real valued functions which are analytic at a, one considers two such functions f and g equivalent if and only if there exists a neighborhood U of a in X such that

$$U \subset \operatorname{dmn} f \cap \operatorname{dmn} g \quad \text{and} \quad f \mid U = g \mid U .$$

The resulting equivalence classes, called **germs of real valued analytic functions at** a, form the commutative ring

$$\mathcal{O}_a(X)$$

whose operations are induced by functional addition and multiplication, and which has no proper divisors of 0 (as seen from the Leibnitz formula). The equivalence class of f is termed the **germ of** f **at** a and denoted

$$\gamma_a(f);$$

we observe that, for each nonnegative integer m, $D^m f(a)$ is determined by $\gamma_a(f)$, and we agree to write

$$D^m \sigma = D^m f(a) \quad \text{whenever} \quad \sigma = \gamma_a(f).$$

We metrize $\mathcal{O}_a(X)$ by defining (see 1.10.5)

$$\operatorname{distance}(\sigma, \tau) = \sum_{m=0}^{\infty} 2^{-m} \frac{\|D^m \sigma - D^m \tau\|}{1 + \|D^m \sigma - D^m \tau\|}$$

whenever $\sigma, \tau \in \mathcal{O}_a(X)$, and we observe that $\mathcal{O}_a(X)$ is the union of the closed subsets

$$\{\sigma \colon \|D^m \sigma\| \leq s\, t^m\, m! \text{ for } m = 0, 1, 2, \ldots\}$$

corresponding to all positive integers s and t. In case $\dim X < \infty$, each of these subsets is compact, hence $\mathcal{O}_a(X)$ is separable. The ring operations are continuous.

Each analytic map u of some neighborhood of a into another Banach space Y induces a continuous ring homomorphism

$$\mathcal{O}_a(u) \colon \mathcal{O}_{u(a)}(Y) \to \mathcal{O}_a(X)$$

such that $\langle \gamma_{u(a)}(h), \mathcal{O}_a(u) \rangle = \gamma_a(h \circ u)$ for every real valued function h which is analytic at $u(a)$.

One uses a similar concept of germ in studying the behavior of subsets of X near a; in this context one considers two subsets S and T of X equivalent if and only if

$$S \cap U = T \cap U \text{ for some neighborhood } U \text{ of } a \text{ in } X;$$

the resulting equivalence classes are called **set germs of** X **at** a. Since the Boolean operations $\cup$, $\cap$, $\sim$ are compatible with this equivalence relation, they induce like operations applicable to set germs. We will designate the germ of a set S at a by $\gamma_a(S)$.

Whenever $S \subset X$ and $a \in X$, one defines the **ideal of S at a** as the subset of $\mathcal{O}_a(X)$ consisting of the germs of those real valued analytic functions f for which there exists a neighborhood U of a in X with

$$U \subset \operatorname{dmn} f \quad \text{and} \quad f(x) = 0 \text{ for } x \in S \cap U;$$

observing that this ideal is uniquely determined by the germ of S at a, we denote it

$$\operatorname{ideal} \gamma_a(S).$$

For each open subset U of X and each finite class F of real valued analytic functions whose domains contain U, we define

$$\operatorname{var}(U, F) = U \cap \{x : f(x) = 0 \text{ whenever } f \in F\},$$

the **analytic subvariety of U** associated with F. We note that if G is another finite class of real valued analytic functions on U, then

$$\operatorname{var}(U, F) \cup \operatorname{var}(U, G) = \operatorname{var}(U, \{f \cdot g : f \in F, g \in G\}),$$

$$\operatorname{var}(U, F) \cap \operatorname{var}(U, G) = \operatorname{var}(U, F \cup G);$$

moreover

$$\operatorname{var}(U, F) = \operatorname{var}(U, \{h\}) \quad \text{with} \quad h = \sum_{f \in F} (f)^2.$$

The germs at a of the analytic subvarieties of neighborhoods of a in X are called **analytic variety germs of X at a**; they form the class $\mathscr{V}_a(X)$.

With every finite generated ideal I of the ring $\mathcal{O}_a(X)$ one associates the analytic variety germ

$$\operatorname{var}(I) \in \mathscr{V}_a(X)$$

characterized by the property: *If F is any finite class of real valued analytic functions and U is any neighborhood of a in X such that*

$$U \subset \bigcap \{\operatorname{dmn} f : f \in F\}, \qquad \{\gamma_a(f) : f \in F\} \text{ generates } I,$$

then $\operatorname{var}(I) = \gamma_a[\operatorname{var}(U, F)]$. Clearly

$$I \subset \operatorname{ideal} \operatorname{var}(I);$$

equality holds in case $I = \operatorname{ideal} \gamma_a(S)$ for some $S \subset X$.

If $\alpha \in \mathscr{V}_a(X)$ and ideal α is finitely generated, then

$$\alpha = \operatorname{var}(\operatorname{ideal} \alpha).$$

A germ $\alpha \in \mathscr{V}_a(X)$ is called **irreducible** if and only if there exists no decomposition

$$\alpha = \alpha_1 \cup \alpha_2 \quad \text{with} \quad \alpha_1, \alpha_2 \in \mathscr{V}_a(X), \ \alpha_1 \neq \alpha \neq \alpha_2.$$

One readily verifies that α *is irreducible if and only if* ideal α *is prime.*

If the ring $\mathcal{O}_a(X)$ is Nötherian (according to 3.4.7 this condition holds when $\dim X < \infty$), then every nonempty subfamily H of $\mathscr{V}_a(X)$ has a member which contains no other member of H; it follows that *every germ $\alpha \in \mathscr{V}_a(X)$ is the union of a unique finite subfamily Γ of $\mathscr{V}_a(X)$ such that each member of Γ is irreducible and contains no other member of Γ;* the members of Γ are the maximal irreducible germs contained in α, and are called the **irreducible components** of α; we note that

$$\text{ideal}\,\alpha = \bigcap \{\text{ideal}\,\beta \colon \beta \in \Gamma\}.$$

When U is an open subset of X, the Boolean operations $\cup$, $\cap$, $\sim$ lead from the class of all analytic subvarieties of U to the class of sets

$$\bigcup_{i=1}^{\nu} (S_i \sim T_i)$$

corresponding to all finite sequences of analytic subvarieties $S_1, T_1, \ldots, S_\nu, T_\nu$ of U. Moreover the (topological) components of such sets are relevant for the study of analytic inequalities; in case $\dim X < \infty$ it follows from 3.4.8 (11) that no compact subset of U meets an infinite number of components of such a set. *We will use the following special types of sets when $X = \mathbf{R}^n$:*

We call S an m **dimensional functional analytic submanifold of** U if and only if there exist real valued analytic functions $f_{m+1}, \ldots, f_n$, g on U such that

$$S = \operatorname{var}(U, \{f_{m+1}, \ldots, f_n\}) \sim \operatorname{var}(U, \{g\})$$

and, for each $x \in S$, the sequence $D f_{m+1}(x), \ldots, D f_n(x)$ is linearly independent [in case $n = m$ we mean simply that $S = U \sim \operatorname{Var}(U, \{g\})$]. We refer to the components of such a set S as m **dimensional analytic blocks of** U.

When S and A are subsets of X, one terms S **analytic in** A if and only if the germs of S at all points of A are analytic variety germs; this holds if and only if A has a covering consisting of open sets U such that $U \cap S$ is an analytic subvariety of U. If A is open, then every analytic subvariety of A is analytic in A, but a subset of A can be analytic in A without being an analytic subvariety of A, as shown by examples in [CH, pp. 97–99].

3.4.6. Weierstrass preparation theorem. *If X is a Banach space, Φ is a real valued function which is analytic at 0 in X, k is a positive integer, $v \in X$ and*

$$\langle v^i, D^i \Phi(0) \rangle = 0 \textit{ for } 0 \le i < k, \quad \langle v^k, D^k \Phi(0) \rangle \neq 0,$$

then: (1) *For every real valued function* Ψ *which is analytic at* 0 *in* X, *there exist real valued analytic functions* Q *and* R *satisfying the equations*

$$\Psi(x)=Q(x)\,\Phi(x)+R(x) \text{ for } x \text{ near } 0 \text{ in } X,$$

$$\langle v^i, D^i R(x)\rangle = 0 \text{ for } i\geq k \text{ and } x \text{ near } 0 \text{ in } X;$$

the germs of such functions Q *and* R *at* 0 *are uniquely determined by* v *and the germs of* Φ *and* Ψ.

(2) *For every continuous real valued linear function* ξ *on* X *with* $\xi(v)=1$, *there exist real valued functions* Q, W *analytic at* 0 *in* X, *and* $W_0, W_1, \ldots, W_{k-1}$ *analytic at* 0 *in* $\ker \xi$, *which satisfy the equation*

$$Q(x)\,\Phi(x)=W(x)=\xi(x)^k+\sum_{i=0}^{k-1}\xi(x)^i\,W_i[x-\xi(x)\,v]$$

for x *near* 0 *in* X; *the germs of such functions* $Q, W, W_0, \ldots, W_{k-1}$ *at* 0 *are uniquely determined by* v, ξ *and the germ of* Φ; *moreover*

$$Q(0)\neq 0, \qquad W_i(0)=0 \text{ for } 0\leq i<k.$$

(One calls W a **Weierstrass polynomial** with respect to v and ξ at 0.)

Proof. We assume $|v|=1$ and $\langle v^k/k!, D^k\Phi(0)\rangle=1$. Letting

$$\eta\colon X\to X, \qquad \eta(x)=x-\xi(x)\,v \text{ for } x\in X,$$

we see that $\operatorname{im}\eta=\ker\xi$ and $X=\{t\,v\colon t\in\mathbf{R}\}\oplus\operatorname{im}\eta$.

For every real valued function f which is analytic at 0, and for all nonnegative integers i and j, we define the homogeneous polynomial function $f_{i,j}$ of degree j on $\operatorname{im}\eta$ by the formula

$$f_{i,j}(z)=\langle (v^i/i!)\odot(z^j/j!), D^{i+j}f(0)\rangle \text{ for } z\in\operatorname{im}\eta;$$

using the binomial theorem we find that, for x near 0 and each nonnegative integer s,

$$\langle v^s/s!, D^s f(x)\rangle=\sum_{m=0}^{\infty}\langle v^s/s!, x^m/m! \,\lrcorner\, D^{m+s}f(0)\rangle$$

$$=\sum_{m=0}^{\infty}\langle v^s/s!\odot[\xi(x)\,v+\eta(x)]^m/m!, D^{m+s}f(0)\rangle$$

$$=\sum_{m=0}^{\infty}\sum_{i=0}^{m}\xi(x)^i\langle v^s/s!\odot v^i/i!\odot\eta(x)^{m-i}/(m-i)!, D^{m+s}f(0)\rangle$$

$$=\sum_{m=0}^{\infty}\sum_{i=0}^{m}\binom{s+i}{i}\xi(x)^i f_{s+i,m-i}[\eta(x)]$$

$$=\sum_{i=0}^{\infty}\binom{s+i}{i}\xi(x)^i\sum_{j=0}^{\infty}(f_{s+i,j}\circ\eta)(x);$$

to justify the last step in the preceding computation we observe that if b, c are finite positive numbers such that $\|D^m f(0)\| \le b(c/2)^m m!$ for all nonnegative integers m, then

$$\|f_{i,j}\| \le b\binom{i+j}{i}(c/2)^{i+j} \le b\, c^{i+j},$$

and make use of the following fact:

If α, β, γ are finite positive numbers and $g_{i,j}$ are homogeneous polynomial functions of degree j on $\operatorname{im}\eta$ with

$$\|g_{i,j}\| \le \alpha\, \beta^i\, \gamma^j$$

for all nonnegative integers i and j, then, for each nonnegative integer s,

$$\begin{aligned}
&\sum_{i=0}^{\infty}\sum_{j=0}^{\infty}\binom{s+i}{i}|\xi(x)^i(g_{s+i,j}\circ\eta)(x)| \\
&\le \sum_{i=0}^{\infty}\sum_{j=0}^{\infty} 2^{s+i}|\xi(x)|^i\,\alpha\,\beta^{s+i}\,|\gamma\,\eta(x)|^j \\
&= \alpha[2\beta]^s[1-2\beta|\xi(x)|]^{-1}[1-\gamma|\eta(x)|]^{-1}
\end{aligned}$$

whenever $2\beta|\xi(x)|<1$ and $\gamma|\eta(x)|<1$. Taking $s=0$ and recalling 3.1.24(4) we also see that the function

$$\sum_{i=0}^{\infty}\xi^i\sum_{j=0}^{\infty}(g_{i,j}\circ\eta) = \sum_{m=0}^{\infty}\sum_{i=0}^{m}\xi^i\cdot(g_{i,m-i}\circ\eta)$$

is analytic at 0; this function has germ 0 at 0 if and only if $g_{i,j}=0$ for all i and j, because it maps $t\,v+z$ onto

$$\sum_{i=0}^{\infty}t^i\sum_{j=0}^{\infty}g_{i,j}(z)$$

whenever $t\in\mathbf{R}$, $z\in\operatorname{im}\eta$ and $|t|, |z|$ are sufficiently small.

Thus the members of $\mathcal{O}_0(X)$ are uniquely represented by certain double sequences of polynomial functions on $\operatorname{im}\eta$.

Our assumptions concerning Φ may be restated as

$$\Phi_{i,0}=0 \text{ for } 0\le i<k, \quad\text{and}\quad \Phi_{k,0}=1.$$

The equations in (1) are equivalent to the conditions

$$\Psi_{i,j}=R_{i,j}+\sum_{\mu=0}^{i}\sum_{\nu=0}^{j}Q_{\mu,\nu}\,\Phi_{i-\mu,j-\nu} \text{ for all } i \text{ and } j,$$

$$R_{i,j}=0 \text{ for } i\ge k \text{ and all } j;$$

the resulting equations

$$Q_{i,j} = \Psi_{i+k,j} - \sum_{\mu=0}^{i-1} Q_{\mu,j} \Phi_{i+k-\mu,0} - \sum_{\mu=0}^{i+k} \sum_{\nu=0}^{j-1} Q_{\mu,\nu} \Phi_{i+k-\mu,j-\nu}$$

determine the polynomials $Q_{i,j}$ for all nonnegative integers i and j, by induction with respect to $i+(k+1)j$, hence characterize the germ of Q at 0, while the germ of R equals the germ of $\Psi - Q\,\Phi$. To prove the existential part of (1) we choose finite positive numbers $b, c, \alpha, \beta, \gamma, \delta$ so that

$$\|D^m \Phi(0)\| \leq b(c/2)^m\, m! \quad \text{and} \quad \|D^m \Psi(0)\| \leq b(c/2)^m\, m!$$

for all nonnegative integers m, $\alpha > b\,c^k$, $\beta > c$, $\gamma > c$,

$$\delta = \frac{b\,c^k}{\alpha} + \frac{b\,c^{k+1}}{\beta - c} + \frac{b\,c\,\beta^{k+1}}{(\beta-c)(\gamma-c)} < 1,$$

and we verify that the polynomials $Q_{i,j}$ defined by the above recursion formulae satisfy the conditions

$$\|Q_{i,j}\| \leq \alpha\,\beta^i\,\gamma^j$$

for all nonnegative integers i and j; in fact we inductively estimate

$$\|Q_{i,j}\| \leq b\,c^{i+j+k} + \sum_{\mu=0}^{i-1} \alpha\,\beta^\mu\,\gamma^j\,b\,c^{i+k-\mu} + \sum_{\mu=0}^{i+k} \sum_{\nu=0}^{j-1} \alpha\,\beta^\mu\,\gamma^\nu\,b\,c^{i+k-\mu+j-\nu}$$

$$= b\,c^{i+j+k} + \alpha\,b\,c^{k+1}\,\frac{\beta^i - c^i}{\beta - c}\,\gamma^j + \alpha\,b\,c\,\frac{\beta^{i+k+1} - c^{i+k+1}}{\beta - c} \cdot \frac{\gamma^j - c^j}{\gamma - c}$$

$$< \alpha\,\beta^i\,\gamma^j\,\delta < \alpha\,\beta^i\,\gamma^j.$$

To prove (2) we apply (1) with $\Psi = \xi^k$, note that

$$0 = \Psi_{i,0} = R_{i,0} + \sum_{\mu=0}^{i} Q_{\mu,0}\,\Phi_{i-\mu,0} = R_{i,0} \quad \text{for } i < k,$$

and take

$$W = \xi^k - R = \xi^k - \sum_{i=0}^{k-1} \xi^i \sum_{j=1}^{\infty} (R_{i,j} \circ \eta).$$

3.4.7. If Φ is a real valued function analytic at $a \in \mathbf{R}^n$ with $\gamma_a(\Phi) \neq 0$, and if

$$0 < k = \inf\{i\colon D^i\,\Phi(a) \neq 0\} < \infty,$$

then 2.6.5 and 3.2.13 imply

$$\mathscr{L}^n(\mathbf{R}^n \cap \{v\colon \langle v^k, D^k\,\Phi(a)\rangle = 0\}) = 0,$$

$$\mathscr{H}^{n-1}(\mathbf{S}^{n-1} \cap \{v\colon \langle v^k, D^k\,\Phi(a)\rangle = 0\}) = 0;$$

thus the hypothesis of 3.4.6 holds for almost all vectors v.

We now deduce, by induction with respect to n, that $\mathcal{O}_a(\mathbf{R}^n)$ *is a Nötherian ring and a unique factorization domain:*

Assuming $a=0$, I is an ideal of $\mathcal{O}_0(\mathbf{R}^n)$, $0 \neq \gamma_0(\Phi) \in I$ and $\Phi(0)=0$, we choose v so that 3.4.6 is applicable and define η as in the proof of 3.4.6. Observing that $\mathcal{O}_0(\eta)$ is a ring monomorphism making $\mathcal{O}_0(\mathbf{R}^n)$ a module over

$$\operatorname{im} \mathcal{O}_0(\eta) \simeq \mathcal{O}_0(\operatorname{im} \eta) \simeq \mathcal{O}_0(\mathbf{R}^{n-1}),$$

we let M be the submodule generated by

$$\{\gamma_0(\xi^i): i=0,1,\ldots,k-1\}$$

and inductively infer that the $\mathcal{O}_0(\mathbf{R}^{n-1})$ module $M \cap I$ is finitely generated (see [ZS, p. 158]). Using 3.4.6(1) we see that

$$I = \mathcal{O}_0(\mathbf{R}^n) \cdot \Phi + (M \cap I),$$

and conclude that the $\mathcal{O}_0(\mathbf{R}^n)$ module I is finitely generated. (In case $n=1$, $\mathcal{O}_0(\mathbf{R}^{n-1})$ should be replaced by $\mathbf{R}$.)

To obtain the prime factorization of $\gamma_0(\Phi)$ we apply 3.4.6(2). We observe that $\gamma_0(Q)$ is a unit because $Q(0) \neq 0$, hence the problem is reduced to the prime factorization of $\gamma_0(W)$. Now $\gamma_0(W)$ belongs to the subring P generated by $\gamma_0(\xi)$ and $\operatorname{im} \mathcal{O}_0(\eta)$, which is a simple transcendental extension of $\operatorname{im} \mathcal{O}_0(\eta) \simeq \mathcal{O}_0(\mathbf{R}^{n-1})$. Using the Gauss lemma (see [ZS, p. 32]) we inductively infer that P is a unique factorization domain. Moreover, in case there exists a factorization

$$\gamma_0(W) = \gamma_0(f) \cdot \gamma_0(g)$$

where f and g are real valued analytic functions on some neighborhood of 0 in $\mathbf{R}^n$ with $f(0)=0$ and $g(0)=0$, we see from the equation

$$f(t\,v)\,g(t\,v) = W(t\,v) = t^k \text{ for small } t \in \mathbf{R}$$

that 3.4.6 is applicable with Φ, k replaced by f, μ and by g, ν where μ and ν are positive integers such that $\mu+\nu=k$, hence obtain Weierstrass polynomials F and G of degrees μ and ν, as well as real valued analytic functions M and N, such that

$$\gamma_0(M\,f) = \gamma_0(F) \quad \text{and} \quad \gamma_0(N\,g) = \gamma_0(G);$$

then $\gamma_0(M\,N\,W) = \gamma_0(F\,G)$ and $F\,G$ is a Weierstrass polynomial of degree k, hence the uniqueness assertion in 3.4.6(2) implies $\gamma_0(F\,G) = \gamma_0(W)$, a nontrivial factorization of $\gamma_0(W)$ in P.

3.4.8. Theorem. *Suppose:*

(1) $e_1, \ldots, e_n$ *and* $\omega_1, \ldots, \omega_n$ *are dual orthonormal bases of* $\mathbf{R}^n$ *and* $\wedge_1 \mathbf{R}^n$. *For* $j=1, \ldots, n$,

$$p_j \in \mathbf{O}^*(n,j), \quad p_j(x)=(\omega_1(x), \ldots, \omega_j(x)) \text{ for } x \in \mathbf{R}^n,$$

and $\Omega_j = \operatorname{im} \mathcal{O}_0(p_j)$; *also* $\Omega_0 = \mathbf{R}$.

(2) *Whenever* I *is an ideal of* $\mathcal{O}_0(\mathbf{R}^n)$ *and* $I \neq \mathcal{O}_0(\mathbf{R}^n)$,

$$I_g = \{\langle \sigma, \mathcal{O}_0(g) \rangle : \sigma \in I\} \text{ for } g \in \mathbf{O}(n).$$

For $j=1, \ldots, n$,

$$A_j(I) = \mathbf{O}(n) \cap \{g : I_g \cap \Omega_j \neq \{0\}\},$$

$$B_j(I) = \mathbf{O}(n) \cap \{g : \langle (e_j)^k, D^k \phi \rangle \neq 0 \text{ for some } \phi \in I_g \cap \Omega_j \text{ and some positive integer } k\};$$

also $A_0(I) = \varnothing$. *For* $m=0, \ldots, n-1$,

$$C_m(I) = \mathbf{O}(n) \cap \{g : \text{for } j=m+1, \ldots, n \text{ there exist } \delta_j \in \Omega_m \sim I_g \text{ and } \tau_j \in \Omega_{m+1} \text{ with } \delta_j \gamma_0(\omega_j) - \tau_j \in I_g\},$$

$$E_m(I) = C_m(I) \cap \bigcap_{j=m+1}^{n} B_j(I) \sim A_m(I);$$

also $E_n(I) = \mathbf{O}(n) \sim A_n(I)$. *In case* $I \neq \{0\}$,

$$\operatorname{depth}(I) = \inf \left\{ m : \bigcap_{j=m+1}^{n} B_j(I) \neq \varnothing \right\};$$

also $\operatorname{depth}(\{0\}) = n$.

(3) *Whenever* $\gamma_0(\varnothing) \neq \alpha \in \mathscr{V}_0(\mathbf{R}^n)$, *the dimension of* α *is defined as*

$$\dim \alpha = \operatorname{depth}(\operatorname{ideal} \alpha).$$

(4) μ_j *is a Haar measure of* $H_j = \mathbf{O}(n) \cap \{h : h(e_i) = e_i \text{ for } i > j\}$;
ν_m *is a Haar measure of* $Q_m = \mathbf{O}(n) \cap \{q : q(e_i) = e_i \text{ for } i \leq m\}$.

Then the following twelve conclusions hold:

(5) $A_j(I)$, $B_j(I)$, $C_m(I)$, $E_m(I)$ *are Borel subsets of* $\mathbf{O}(n)$.

(6) *If* $g \in A_j(I)$, *then* $g \circ h \in B_j(I)$ *for* μ_j *almost all* h *in* H_j.

(7) $\theta_n[A_j(I) \sim B_j(I)] = 0$ *for* $j=1, \ldots, n$.

(8) *If* I *is a prime ideal and* $g \in \bigcap_{j=m+1}^{n} B_j(I)$ *with* $m<n$, *then*

$$g \circ q \in C_m(I) \cap \bigcap_{j=m+1}^{n} B_j(I) \text{ for } \nu_m \text{ almost all } q \text{ in } Q_m.$$

(9) *If I is a prime ideal, then*

$$\theta_n\left[\bigcap_{j=m+1}^{n} B_j(I)\sim C_m(I)\right]=0 \quad \text{for } m=0,\ldots,n-1.$$

(10) *If I is a prime ideal and $g\in E_m(I)$ with $0<m<n$, then there exist a neighborhood Y of 0 in $\mathbf{R}^m$, a positive integer k, real valued analytic functions Δ, $b_{i,j}$ on Y corresponding to $i=0,\ldots,k-1$ and $j=m+1,\ldots,n$, and functions f_j on $p_m^{-1}(Y)$ such that*

$$f_{m+1}=(\omega_{m+1})^k+\sum_{i=0}^{k-1}(\omega_{m+1})^i(b_{i,m+1}\circ p_m),$$

$$f_j=(\Delta\circ p_m)\cdot\omega_j-\sum_{i=0}^{k-1}(\omega_{m+1})^i(b_{i,j}\circ p_m) \quad \text{for } j>m+1,$$

$$\operatorname{var}(I_g)=\operatorname{var}(J)\cup\gamma_0(Z),$$

where J is the ideal of $\mathcal{O}_0(\mathbf{R}^n)$ generated by $I_g\cup\{\gamma_0(\Delta\circ p_m)\}$,

$$Z=\bigcap_{j=m+1}^{n}\{x\colon f_j(x)=0\}\sim\{x\colon \Delta[p_m(x)]=0\};$$

moreover $b_{i,m+1}(0)=0$, $Z\subset\{x\colon \langle e_{m+1}, Df_{m+1}(x)\rangle\neq 0\}$, Z is an m dimensional functional analytic submanifold of $p_m^{-1}(Y)$; p_m maps each component of Z homeomorphically onto some component of $Y\cap\{y\colon \Delta(y)\neq 0\}$;

$$\gamma_0(\Delta\circ p_m)\notin I_g,\quad \gamma_0(f_j)\in I_g \text{ for } j=m+1,\ldots,n,$$

and $K\cap\Omega_m\neq\{0\}$ for every ideal K of $\mathcal{O}_0(\mathbf{R}^n)$ which properly contains I_g.

In case $I=\operatorname{ideal}\operatorname{var}(I)$, then $\gamma_0(Z)\neq\gamma_0(\varnothing)$.

(11) *If $\alpha,\beta\in\mathscr{V}_0(\mathbf{R}^n)$, $\dim\alpha=m$, V_0 is a neighborhood of 0 in $\mathbf{R}^n$ and S_0 is an analytic subvariety of V_0 with $\gamma_0(S_0)=\beta$, then there exist neighborhoods $V_1,\ldots,V_\mu$ of 0 in V_0 and sets $S_1,\ldots,S_\mu$ such that*

$$\alpha\sim\beta=\gamma_0\left(\bigcup\{S_i\colon 1\le i\le\mu\}\right)$$

and, for $1\le i\le\mu$, S_i is a functional analytic submanifold of V_i with dimension at most m, S_i has only finitely many (topological) components,

$$S_i\cap\bigcup\{S_j\colon 0\le j<i\}=\varnothing,\quad V_i\cap\operatorname{Clos}S_i\subset\bigcup\{S_j\colon 0\le j\le i\}.$$

(12) *If $\alpha\in\mathscr{V}_0(\mathbf{R}^n)$ and U is any neighborhood of 0 in $\mathbf{R}^n$, then there exists a set S such that $\gamma_0(S)=\alpha$ and, for every analytic subvariety T of U,*

$$\alpha\subset\gamma_0(T) \text{ implies } S\subset T.$$

(13) *If* $\gamma_0(\varnothing)\neq\alpha\in\mathscr{V}_0(\mathbf{R}^n)$ *and* α *is irreducible, then*

$$E_m(\text{ideal }\alpha)\neq\varnothing \text{ if and only if } m=\dim\alpha;$$

moreover, if $m=\dim\alpha$, *then*

$$\theta_n[\mathbf{O}(n)\sim E_m(\text{ideal }\alpha)]=0$$

and there exists an $(\mathscr{H}^m, m)$ *rectifiable set* S *such that* $\gamma_0(S)=\alpha$ *and*

$$\sup\{\mathscr{H}^m(T)/(\operatorname{diam} T)^m\colon \varnothing\neq T\subset S\}<\infty.$$

(14) *If* $\gamma_0(\varnothing)\neq\alpha\in\mathscr{V}_0(\mathbf{R}^n)$, *then*

$$\dim\alpha=\sup\{\dim\beta\colon \beta \text{ is an irreducible component of } \alpha\}.$$

(15) *If* $\alpha,\beta\in\mathscr{V}_0(\mathbf{R}^n)$, $\beta\subset\alpha$, $\gamma_0(\varnothing)\neq\beta\neq\alpha$ *and* α *is irreducible, then*

$$\dim\beta<\dim\alpha.$$

(16) *If* S *is an analytic subvariety of some neighborhood of* 0 *in* $\mathbf{R}^n$, T *is a relatively open subset of* S, u *is a nonnegative integer and* $\mathscr{H}^{u+1}(T)=0$, *then*

$$\dim\operatorname{var}[\text{ideal }\gamma_0(T)]\leq u.$$

Proof of (5). From 3.4.5 we know that $\mathscr{O}_0(\mathbf{R}^n)$ is the union of a countable family of compact subsets; the same holds for I because, if I is generated by a set with r elements, there exists a continuous map of $\mathscr{O}_0(\mathbf{R}^n)^r$ onto I. Moreover Ω_j equals the closed subset of $\mathscr{O}_0(\mathbf{R}^n)$ consisting of all germs σ such that

$$e_i \,\lrcorner\, D^s\sigma=0 \text{ whenever } i>j \text{ and } s\geq 1.$$

Since the function mapping (σ, g) onto $\langle\sigma, \mathscr{O}_0(g^{-1})\rangle$ is continuous we find that

$$P=[\Omega_j\times\mathbf{O}(n)]\cap\{(\sigma,g)\colon \langle\sigma,\mathscr{O}_0(g^{-1})\rangle\in I\sim\{0\}\}$$

is the union of a countable family of compact sets. Then we observe that the standard projection of $\mathscr{O}_0(\mathbf{R}^n)\times\mathbf{O}(n)$ onto $\mathbf{O}(n)$ maps P onto $A_j(I)$.

Similarly one verifies that $B_j(I)$, $C_m(I)$, $E_m(I)$ are unions of countable families of compact subsets of $\mathbf{O}(n)$.

Proof of (6) **and** (7). Observing that $H_j\simeq\mathbf{O}(j)$ acts transitively on

$$S=\mathbf{R}^n\cap\{v\colon |v|=1 \text{ and } v\bullet e_i=0 \text{ for } i>j\}\simeq\mathbf{S}^{j-1},$$

we define $V\colon H_j\to S$, $V(h)=h(e_j)$ for $h\in H_j$, and note that $V_\#(\mu_j)$ is a constant multiple of $\mathscr{H}^{j-1}\llcorner S$.

Whenever $g \in A_j(I)$ we can choose $\phi \in I_g \cap \Omega_j$ and a positive integer k with $D^k \phi \neq 0$, and infer as in 3.4.7 that

$$\langle v^k, D^k \phi \rangle \neq 0 \text{ for } \mathscr{H}^{j-1} \text{ almost all } v \text{ in } S;$$

consequently μ_j almost all h in H_j satisfy the condition

$$0 \neq \langle V(h)^k, D^k \phi \rangle = \langle (e_j)^k, D^k \langle \phi, \mathscr{O}_0(h) \rangle \rangle$$

with $\langle \phi, \mathscr{O}_0(h) \rangle \in I_{g \circ h} \cap \Omega_j$, hence $g \circ h \in B_j(I)$.

It follows that, for every $g \in \mathbf{O}(n)$,

$$\text{either } (g H_j) \cap A_j(I) = \varnothing \text{ or } \mu_j \{h \colon g \circ h \notin B_j(I)\} = 0.$$

Letting c be the characteristic function of $A_j(I) \sim B_j(I)$ we conclude

$$\mu_j(H_j) \int c \, d\theta_n = \iint c(g \circ h) \, d\mu_j \, h \, d\theta_n \, g = 0.$$

Proof of (8) **and** (9). Observing that $Q_m \simeq \mathbf{O}(n-m)$ acts transitively on

$$S = \mathbf{R}^n \cap \{v \colon |v| = 1 \text{ and } v \bullet e_i = 0 \text{ for } i \leq m\} \simeq \mathbf{S}^{n-m-1},$$

we define $V \colon Q_m \to S$, $V(q) = q(e_{m+1})$ for $q \in Q_m$, and note that $V_\# (v_m)$ is a constant multiple of $\mathscr{H}^{n-m-1} \llcorner S$.

Whenever $g \in B_j(I)$ for $j = m+1, \ldots, n$ we apply 3.4.6(2) to obtain positive integers $r(j)$ and Weierstrass polynomials

$$W_j = (\omega_j)^{r(j)} + \sum_{i=0}^{r(j)-1} (\omega_j)^i (W_{i,j} \circ p_{j-1})$$

with respect to e_j and ω_j, where the functions $W_{i,j}$ are analytic at 0 in $\mathbf{R}^{j-1}$ with $W_{i,j}(0) = 0$, and

$$\gamma_0(W_j) \in I_g \cap \Omega_j;$$

in case $m = 0$ we use $W_{i,1} \in \mathbf{R}$ in place of $W_{i,1} \circ p_0$. We let

$$\rho \colon \mathscr{O}_0(\mathbf{R}^n) \to \mathscr{O}_0(\mathbf{R}^n)/I_g$$

be the canonical ring homomorphism, abbreviate $\zeta_j = \rho[\gamma_0(\omega_j)]$, and observe that $\rho(\Omega_j)$ is the ring generated by $\rho(\Omega_{j-1})$ and ζ_j; in fact 3.4.6(1) applied with $\Phi = W_j$, $v = e_j$ and $\gamma_0(\Psi) \in \Omega_j$ shows that $\rho[\gamma_0(\Psi)] = \rho[\gamma_0(R)]$ and $\gamma_0(R)$ belongs to the ring generated by Ω_{j-1} and $\gamma_0(\omega_j)$. Moreover the equation $\rho[\gamma_0(W_j)] = 0$ tells us that ζ_j is integral over $\rho(\Omega_{j-1})$. Recalling the transitivity of integral dependence (see [ZS, p. 256]) we conclude:

$$\mathscr{O}_0(\mathbf{R}^n)/I_g \text{ is an integral ring extension of } \rho(\Omega_m),$$

$$\text{generated by } \rho(\Omega_m) \text{ and } \zeta_{m+1}, \ldots, \zeta_n.$$

We also consider the field L of fractions of the integral domain $\mathscr{O}_0(\mathbf{R}^n)/I_g$, and the subfield M generated by $\rho(\Omega_m)$. The classical reasoning regarding the simplicity of finite separable algebraic field extensions (see [ZS, pp. 84, 85]) shows that, for $\mathscr{H}^{n-m-1}$ almost all v in S,

$$L \text{ is the ring generated by } M \text{ and } \sum_{j=m+1}^{n} \omega_j(v)\,\zeta_j.$$

Inasmuch as, for every $q \in Q_m$,

$$\begin{aligned}\sum_{j=m+1}^{n} \omega_j[V(q)]\,\omega_j &= \sum_{j=m+1}^{n} [q(e_{m+1})\bullet e_j]\,\omega_j \\ &= \sum_{j=m+1}^{n} [e_{m+1}\bullet q^{-1}(e_j)]\,\omega_j = \omega_{m+1}\circ q^{-1},\end{aligned}$$

we infer that, for v_m almost all q in Q_m,

$$L \text{ is the ring generated by } M \text{ and } \rho[\gamma_0(\omega_{m+1}\circ q^{-1})].$$

Whenever q has this property and $\sigma \in \mathscr{O}_0(\mathbf{R}^n)$, there exist $\delta \in \Omega_m \sim I_g$ and $\tau \in \Omega_{m+1}$ with

$$\rho\langle\sigma, \mathscr{O}_0(q^{-1})\rangle = \rho\langle\tau, \mathscr{O}_0(q^{-1})\rangle/\rho(\delta),$$

hence $\langle\delta\sigma-\tau, \mathscr{O}_0(q^{-1})\rangle \in I_g$, $\delta\sigma-\tau \in I_{g\circ q}$; there also exist a positive integer u and $\beta_0, \ldots, \beta_{u-1} \in \Omega_m$ with

$$\langle\sigma, \mathscr{O}_0(q^{-1})\rangle^u + \sum_{i=0}^{u-1} \langle\sigma, \mathscr{O}_0(q^{-1})\rangle^i\,\beta_i \in I_g, \quad \text{hence} \quad \sigma^u + \sum_{i=0}^{u-1} \sigma^i\,\beta_i \in I_{g\circ q}.$$

Taking $\sigma = \gamma_0(\omega_j)$ we find that $g\circ q \in C_m(I)$ and also that $g\circ q \in B_j(I)$ for $j = m+1, \ldots, n$.

Letting c be the characteristic function of $\bigcap_{j=m+1}^{n} B_j(I) \sim C_m(I)$ we conclude

$$v_m(Q_m)\int c\,d\theta_n = \iint c(g\circ q)\,dv_m\,q\,d\theta_n\,g = 0.$$

Proof of (10). We readopt the notation used in the proof of (8). Now the additional assumption $g \in C_m(I) \sim A_m(I)$ implies L is the field generated by $\rho(\Omega_m)$ and ζ_{m+1}, and that $\rho|\Omega_m$ is a monomorphism. Therefore $\rho(\Omega_m)$ is integrally closed in M, and the monic minimal polynomial of ζ_{m+1} over M has coefficients in $\rho(\Omega_m)$ (see [ZS, pp. 260, 261]). Letting P be the ring generated by Ω_m and $\gamma_0(\omega_{m+1})$, we may identify this minimal polynomial with a generator of $P \cap I_g$; in fact we may choose W_{m+1} so that

$$P \cap I_g = P \cdot \gamma_0(W_{m+1}).$$

We take $k = r(m+1)$ and let Γ be the Ω_m module generated by $\{\gamma_0(\omega_{m+1})^i : i = 0, \ldots, k-1\}$. Defining

$$W'_{m+1}(x) = \langle e_{m+1}, D\,W_{m+1}(x)\rangle \text{ for } x \in \operatorname{dmn} W_{m+1},$$

we see that $\gamma_0(W'_{m+1}) \in \Gamma \sim \{0\} \subset P \sim I_g$.

For any $\sigma\in\mathcal{O}_0(\mathbf{R}^n)\sim I_g$, $\rho(\sigma)$ has an inverse in L, hence there exist $\delta\in\Omega_m\sim I_g$ and $\tau\in\Gamma$ with

$$\rho(\sigma)\cdot\rho(\tau)/\rho(\delta)=1,\qquad \sigma\tau-\delta\in I_g.$$

Therefore every ideal of $\mathcal{O}_0(\mathbf{R}^n)$, which properly contains I_g, meets $\Omega_m\sim I_g$. In particular we choose $\delta_{m+1}\in\Omega_m\sim I_g$ and $\tau_{m+1}\in\Gamma$ with

$$\gamma_0(W'_{m+1})\cdot\tau_{m+1}-\delta_{m+1}\in I_g\cap P.$$

On the other hand we select $\delta_j\in\Omega_m\sim I_g$ and $\tau_j\in\Gamma$ so that

$$\delta_j\,\gamma_0(\omega_j)-\tau_j\in I_g\ \text{ for } j=m+2,\dots,n.$$

Replacing $\delta_{m+1},\dots,\delta_n$ by a common multiple and changing $\tau_{m+1},\dots,\tau_n$ accordingly, we may assume that all δ_j equal the same $\delta\in\Omega_m\sim I_g$. Then we choose Δ with $\gamma_0(\Delta\circ p_m)=\delta$, define f_j so that $f_{m+1}=W_{m+1}$ and

$$\gamma_0(f_j)=\delta\,\gamma_0(\omega_j)-\tau_j\ \text{ for } j=m+2,\dots,n,$$

and also select real valued analytic functions t_{m+1}, s_{m+1} with

$$\gamma_0(t_{m+1})=\tau_{m+1}\in\Gamma,\quad \gamma_0(s_{m+1})\in P,\quad W'_{m+1}\cdot t_{m+1}-(\Delta\circ p_m)=s_{m+1}\cdot W_{m+1}.$$

Since the germs of $\Delta\circ p_m$, t_{m+1}, s_{m+1} belong to P, we may assume that $\Delta\circ p_m$, t_{m+1}, s_{m+1} and f_j have domain $p_m^{-1}(Y)$, where Y is a neighborhood of 0 in $\mathbf{R}^m$.

Let F be the ideal of $\mathcal{O}_0(\mathbf{R}^n)$ generated by $\gamma_0(f_{m+1}),\dots,\gamma_0(f_n)$. Clearly $F\subset I_g$. We will prove that

$$\sigma\in I_g\ \text{ implies }\ \delta^u\sigma\in F\ \text{ for some positive integer } u.$$

Assuming $\sigma\in I_g\cap\Omega_j$ we use induction with respect to j. First we apply 3.4.6(1) with $\Phi=W_j$, $v=e_j$ to reduce the problem to the case when σ belongs to the Ω_{j-1} module generated by $\{\gamma_0(\omega_j)^i\colon i=0,\dots,r(j)\}$; for such special σ we observe that, if $j\ge m+2$ then

$$\delta^{r(j)}\sigma\in I_g\cap[(\Gamma+F)\cdot\Omega_{j-1}]\subset(I_g\cap\Omega_{j-1})+F$$

because $\delta\,\gamma_0(\omega_j)\in\Gamma+F$ and $\Gamma\subset\Omega_{j-1}$, while if $j=m+1$ then

$$\sigma\in P\cap I_g=P\cdot\gamma_0(f_{m+1})\subset F.$$

Using the preceding result one readily verifies that

$$\operatorname{var}(I_g)=\operatorname{var}(J)\cup\gamma_0(Z).$$

Moreover Z is an m dimensional functional analytic submanifold of $p_m^{-1}(Y)$ because, for each $x\in Z$, $W_{m+1}(x)=0$ and $\Delta[p_m(x)]\neq 0$, hence $W'_{m+1}(x)\neq 0$,

$$\langle e_{m+1}\wedge\cdots\wedge e_n, D f_{m+1}(x)\wedge\cdots\wedge D f_n(x)\rangle=W'_{m+1}(x)\cdot\Delta[p_m(x)]^{n-m-1}\neq 0$$

and $Df_{m+1}(x), \ldots, Df_n(x)$ are linearly independent; also x has a neighborhood U in $\mathbf{R}^n$ such that $p_m|Z \cap U$ is an analytic coordinate system for Z at x.

Next we verify that $Z \cap \{x: p_m(x) \in C\}$ is compact for every compact subset C of $Y \cap \{y: \Delta(y) \neq 0\}$. Clearly the set $Z \cap p_m^{-1}(C)$ is closed. To prove it bounded we choose $M \in \mathbf{R}$ so that

$$|\Delta(y)| \geq 1/M \quad \text{and} \quad \sum_{i=0}^{k-1} |b_{i,j}(y)| \leq M \text{ for } j = m+1, \ldots, n$$

whenever $y \in C$; if $x \in Z \cap p_m^{-1}(C)$, then either $|\omega_{m+1}(x)| \leq 1$ or

$$|\omega_{m+1}(x)|^k \leq M\, |\omega_{m+1}(x)|^{k-1} \quad \text{because} \quad f_{m+1}(x) = 0,$$

hence $|\omega_{m+1}(x)| \leq \sup\{1, M\}$ and it follows for $j > m+1$ that

$$M^{-1}\, |\omega_j(x)| \leq M \sup\{1, M^{k-1}\} \quad \text{because} \quad f_j(x) = 0.$$

We observe that $Y \cap \{y: \Delta(y) \neq 0\}$ is the union of the sets

$$Y_\nu = Y \cap \{y: \Delta(y) \neq 0, \operatorname{card}(Z \cap p_m^{-1}\{y\}) = \nu\}$$

corresponding to $\nu = 0, 1, \ldots, k$. If $\eta \in Y_\nu$, then there exist disjoint open subsets $U_1, \ldots, U_\nu$ of $\mathbf{R}^n$ such that p_m maps each of the sets $Z \cap U_1, \ldots, Z \cap U_\nu$ homeomorphically onto a neighborhood of η in $\mathbf{R}^m$; for every sufficiently small positive number ε the set

$$D_\varepsilon = Z \cap \{x: |p_m(x) - \eta| \leq \varepsilon\} \sim (U_1 \cup \cdots \cup U_\nu)$$

is compact, and the intersection of all such sets D_ε is empty, hence $D_\varepsilon = \varnothing$ for some $\varepsilon > 0$. This shows that Y_ν is open; in case $\nu > 0$ it also implies the continuity of the functions $\psi_1, \ldots, \psi_\nu$ on Y_ν characterized by the conditions

$$\psi_\lambda(y) \in Z \cap p_m^{-1}\{y\} \text{ for } y \in Y \text{ and } \lambda = 1, \ldots, \nu,$$

$$\omega_{m+1}[\psi_\lambda(y)] < \omega_{m+1}[\psi_{\lambda+1}(y)] \text{ if } \lambda < \nu.$$

Thus $Z \cap p_m^{-1}(Y_\nu)$ is the union of its relatively open subsets $\operatorname{im}\psi_1, \ldots, \operatorname{im}\psi_\nu$ and p_m maps each of these sets homeomorphically onto Y_ν.

Since $Y_0, \ldots, Y_k$ are open and disjoint, each component of $Y \cap \{y: \Delta(y) \neq 0\}$ is a component of Y_ν for some ν.

Finally, if $\gamma_0(Z) = \gamma_0(\varnothing)$, then $\operatorname{var}(I_g) = \operatorname{var}(J)$, hence

$$\delta \in J \sim I_g \subset \operatorname{ideal} \operatorname{var}(I_g) \sim I_g.$$

Proof of (11). We will use induction with respect to m, simultaneously for all n.

In case $m=0$ we choose $g\in\mathbf{O}(n)$ and real valued analytic functions Φ_j such that $\gamma_0(\Phi_j)\in I_g\cap\Omega_j$ with $I=\text{ideal}\,\alpha$,

$$\Phi_j(t\,e_j)\neq 0 \text{ for small } t\in\mathbf{R}\sim\{0\} \text{ and } j=1,\ldots,n,$$

hence infer $\operatorname{var}(I_g)=\gamma_0(\{0\})$.

To treat $m>0$ we first observe that if $\alpha_1,\ldots,\alpha_s$ are the irreducible components of α, then

$$\alpha\sim\beta=\bigcup_{i=1}^{s}\left[\alpha_i\sim\left(\beta\cup\bigcup_{j=1}^{i-1}\alpha_j\right)\right]$$

with $\text{ideal}\,\alpha_i\supset\text{ideal}\,\alpha$, hence $\dim\alpha_i\le\dim\alpha$. Hereafter we assume that α is irreducible with $\alpha\not\subset\beta$, and let $I=\text{ideal}\,\alpha$.

In case $n=m$, then $B_m(I)=\varnothing$, $A_m(I)=\varnothing$ by (6), hence $I=\{0\}$ and $\alpha=\gamma_0(\mathbf{R}^m)$. We choose a real valued function Φ which is analytic at 0 in $\mathbf{R}^m$ with $0\neq\gamma_0(\Phi)\in\text{ideal}\,\beta$, let K be the ideal of $\mathcal{O}_0(\mathbf{R}^m)$ generated by $\gamma_0(\Phi)$, and note that

$$\alpha\sim\beta=[\alpha\sim\operatorname{var}(K)]\cup[\operatorname{var}(K)\sim\beta] \quad\text{with}\quad \dim\operatorname{var}(K)\le\operatorname{depth}(K)\le m-1$$

by (6), hence induction is applicable to $\operatorname{var}(K)\sim\beta$. We will also show that the open set

$$R=\mathbf{R}^m\cap\{y\colon |y|<\varepsilon \text{ and } \Phi(y)\neq 0\}$$

has only finitely many components, provided $0<\varepsilon<\infty$ and $\mathbf{B}(0,\varepsilon)\subset\operatorname{dmn}\Phi$. For this purpose we define

$$\Psi(y)=[\varepsilon^2-|y|^2]\,\Phi(y)^2 \text{ for } y\in\operatorname{dmn}\Phi,\qquad H=\{y\colon \|D\,\Psi(y)\|^2=0\},$$

and suppose Φ is not constant on $\mathbf{B}(0,\varepsilon)$. For each $a\in\mathbf{B}(0,\varepsilon)$, Ψ is not constant on any neighborhood of a, hence

$$\gamma_a(\|D\,\Psi\|^2)\neq 0,\qquad \dim\gamma_a(H)\le m-1$$

by (6), and induction implies that some neighborhood of a meets only finitely many components of H. Since $\mathbf{B}(0,\varepsilon)$ is compact, $\mathbf{B}(0,\varepsilon)$ meets only finitely many components of H. Moreover each component C of R contains some component of H, because

$$\Psi(y)=0 \text{ for } y\in\operatorname{Bdry}C,\qquad \Psi(y)>0 \text{ for } y\in C,$$

hence Ψ has a relative maximum at some point $\eta\in C\cap H$; the component of η in H cannot meet Bdry C, because Ψ is constant on every connected analytic manifold contained in H, hence on every component of H.

In case $n>m>0$ we use (8) to choose

$$g\in C_m(I)\cap\bigcap_{j=m+1}^{n}B_j(I).$$

Since $\bigcap_{j=m}^{n}B_j(I)=\varnothing$ we see from (6) that $g\notin A_m(I)$, hence $g\in E_m(I)$ and the results of (10) are at our disposal. Observing that

$$\operatorname{ideal} g^{-1}(\alpha\cap\beta)\quad\text{properly contains}\quad \operatorname{ideal} g^{-1}(\alpha)=I_g$$

because $\alpha\not\subset\alpha\cap\beta$, we choose a real valued function χ which is analytic at 0 in $\mathbf{R}^m$ with

$$0\neq\gamma_0(\chi\circ p_m)\in\operatorname{ideal} g^{-1}(\alpha\cap\beta).$$

Next we define $\Phi=\chi\cdot\Delta$, let K be the ideal of $\mathcal{O}_0(\mathbf{R}^n)$ generated by $I_g\cup\{\gamma_0(\Phi\circ p_m)\}$, and note that

$$g^{-1}(\alpha\sim\beta)=[g^{-1}(\alpha)\sim\operatorname{var}(K)]\cup[\operatorname{var}(K)\sim g^{-1}(\beta)]$$
$$\text{with}\quad \dim\operatorname{var}(K)\leq\operatorname{depth}(K)\leq m-1,$$

by (6), hence induction is applicable to $\operatorname{var}K\sim g^{-1}(\beta)$. Assuming $\operatorname{dmn}\chi=Y$ we let

$$D=Z\cap\{x\colon \chi[p_m(x)]\neq 0\}$$

and infer

$$g^{-1}(\alpha)\sim\operatorname{var}(K)=\gamma_0(Z)\sim\operatorname{var}(K)=\gamma_0(D)$$

because $g^{-1}(\alpha)=\operatorname{var}(I_g)=\gamma_0(Z)\cup\operatorname{var}(J)$ and

$$\operatorname{var}(J)\subset\operatorname{var}(K)=\operatorname{var}(I_g)\cap\gamma_0\{x\colon \Phi[p_m(x)]=0\}.$$

Furthermore, *given a neighborhood U of 0 in $\mathbf{R}^n$ and an analytic subvariety G of U with $\gamma_0(G)=\operatorname{var}(K)$*, we recall the Weierstrass polynomials W_j from the proof of (8) and successively select positive numbers ε_n, $\varepsilon_{n-1},\ldots,\varepsilon_{m+1},\varepsilon_m$ so that $\varepsilon_n<1$,

$$\mathbf{R}^m\cap\{y\colon |y|\leq\varepsilon_n\}\subset Y,\qquad \mathbf{R}^n\cap\{x\colon |x|\leq\varepsilon_n\}\subset U,$$
$$\mathbf{R}^n\cap\{x\colon |x|\leq\varepsilon_n\}\cap\operatorname{Clos}Z\subset Z\cup G,$$
$$Z\cap\{x\colon |x|\leq\varepsilon_n\}\subset\bigcap_{j=m+1}^{n}\{x\colon W_j(x)=0\},$$

and so that, for $j=n,\ldots,m+1$, $\varepsilon_{j-1}<\varepsilon_j/2$ and

$$\sum_{i=0}^{r(j)-1}|W_{i,j}(w)|<(\varepsilon_j/2)^{r(j)}\ \text{ whenever } w\in\mathbf{R}^{j-1}\text{ and }|w|\leq\varepsilon_{j-1}.$$

Finally we define

$$V=\mathbf{R}^n\cap\{x:\ |p_j(x)|<\varepsilon_j \text{ for } j=m,\dots,n\},$$
$$S=D\cap V,\qquad T=D\cap\{x:\ |p_m(x)|<\varepsilon_m\},$$

and note that $\operatorname{Clos} V\subset U$, $S=T\cap V$ is open relative to T. Moreover S is closed relative to T, because if $x\in T\cap\operatorname{Clos} S$, then $x\in Z$, $|p_m(x)|<\varepsilon_m$ and $|p_j(x)|\le\varepsilon_j$ for $j=m+1,\dots,n$, hence

$$W_j(x)=0 \quad\text{and}\quad |p_{j-1}(x)|\le\varepsilon_{j-1},$$
$$|\omega_j(x)|^{r(j)}<(\varepsilon_j/2)^{r(j)}\sup\{1,|\omega_j(x)|^{r(j)-1}\},$$
$$|\omega_j(x)|<1 \quad\text{because}\quad \varepsilon_j/2<1,$$
$$|\omega_j(x)|<\varepsilon_j/2,\qquad |p_j(x)|<\varepsilon_{j-1}+\varepsilon_j/2<\varepsilon_j,$$

and $x\in V$. We infer that every component of S is a component of T. On the other hand T has only finitely many components, because p_m maps each component of T homeomorphically onto some component of

$$R=\mathbf{R}^m\cap\{y:\ |y|<\varepsilon_m \text{ and } \Phi(y)\neq 0\},$$

the preceding special case shows that R has only finitely many components, and the $p_m|T$ counterimage of each component of R is the union of at most k components of T. Thus S has only finitely many components, and S is an m dimensional functional analytic submanifold of $V\subset U$, with

$$g^{-1}(\alpha)\sim\operatorname{var}(K)=\gamma_0(S),\qquad S\cap G=\varnothing,\qquad V\cap\operatorname{Clos} S\subset S\cup G.$$

Proof of (12). Applying (11) with $V_0=U$, $\beta=\gamma_0(\varnothing)$ and $S_0=\varnothing$, we let S be the union of all sets C such that $0\in\operatorname{Clos} C$ and C is a component of S_i for some $i\in\{1,\dots,\mu\}$. If Φ is any real valued analytic function on U for which $\gamma_0(\Phi)\in\operatorname{ideal}\alpha$, then $\Phi(x)=0$ whenever $x\in S$.

Proof of (13). Let $I=\operatorname{ideal}\alpha$. From (6) and (8) we see [as in the proof of (11)] that $E_{\dim\alpha}(I)\neq\varnothing$.

Observing next that

$$\mathbf{O}(n)\sim E_n(I)=\bigcup_{m=0}^{n-1}\Big[\bigcap_{j=m+1}^{n}A_j(I)\sim A_m(I)\Big]$$

we infer from (7) and (9) that

$$\theta_n\Big[\mathbf{O}(n)\sim\bigcup_{m=0}^{n}E_m(I)\Big]=0.$$

If $E_n(I) \neq \varnothing$, then $I = \{0\}$, hence $E_n(I) = \mathbf{O}(n)$ and $\alpha = \gamma_0(\mathbf{R}^n)$. Using this fact in conjunction with (11) and (10) we see that for $m = 0, \ldots, n$ the assumption $E_m(I) \neq \varnothing$ implies

$$\mathscr{H}^{m+1}(S) = 0 \text{ for some set } S \text{ with } \gamma_0(S) = \alpha,$$

$$\mathscr{H}^m(S) > 0 \text{ for every set with } \gamma_0(S) = \alpha,$$

because $U \cap Z \neq \varnothing$, hence $\mathscr{H}^m(U \cap Z) > 0$ for every neighborhood U of 0 in $\mathbf{R}^n$. However there can be only one integer m with the above two properties; for this m we conclude that $E_m(I)$ contains θ_n almost all of $\mathbf{O}(n)$, and that $m = \dim \alpha$.

We now assume $m = \dim \alpha < n$. For each $\lambda \in \Lambda(n, m)$ we choose $\psi_\lambda \in \mathbf{O}(n)$ with $p_m = \mathbf{p}_\lambda \circ \psi_\lambda$. Then we select $g \in \mathbf{O}(n)$ so that

$$g \circ \psi_\lambda \in E_m(I) \text{ whenever } \lambda \in \Lambda(n, m).$$

Applying (10) with g replaced by $g \circ \psi_\lambda$, as well as (11), we obtain integers k_λ, m dimensional analytic manifolds Z_λ and closed sets S_λ such that

$$\operatorname{var}(I_{g \circ \psi_\lambda}) = \gamma_0(S_\lambda), \qquad \mathscr{H}^m(S_\lambda \sim Z_\lambda) = 0,$$

$$N(p_m | Z_\lambda, y) \leq k_\lambda \text{ whenever } y \in \mathbf{R}^m.$$

Letting $S = \bigcap \{\psi_\lambda(S_\lambda) \colon \lambda \in \Lambda(n, m)\}$ we find that $\operatorname{var}(I_g) = \gamma_0(S)$, S is countably $(\mathscr{H}^m, m)$ rectifiable and

$$N(\mathbf{p}_\lambda | S, y) \leq N(p_m | S_\lambda, y) \leq k_\lambda$$

for $\mathscr{L}^m$ almost all y in $\mathbf{R}^m$. Using 3.2.27 we conclude

$$\mathscr{H}^m(T) \leq \sum_{\lambda \in \Lambda(n,m)} k_\lambda \mathscr{L}^m[\mathbf{p}_\lambda(T)] \leq \sum_{\lambda \in \Lambda(n,m)} k_\lambda \boldsymbol{\alpha}(m) (\operatorname{diam} T)^m$$

for every nonempty Borel subset T of S.

Proof of (14). We let Γ be the family of all irreducible components of α, define $m = \sup\{\dim \beta \colon \beta \in \Gamma\}$, and use (13) to select

$$g \in \bigcap_{\beta \in \Gamma} E_{\dim \beta}(\text{ideal } \beta).$$

When $j \in \{m+1, \ldots, n\}$ we choose for each $\beta \in \Gamma$ a germ $\phi_\beta \in \Omega_j \cap (\text{ideal } \beta)_g$ such that $\langle (e_j)^k, D^k \phi_\beta \rangle \neq 0$ for some positive integer k; then

$$\phi = \prod_{\beta \in \Gamma} \phi_\beta \in \Omega_j \cap (\text{ideal } \alpha)_g$$

and $\langle (e_j)^k, D^k \phi \rangle \neq 0$ for some positive integer k, hence $g \in B_j(\text{ideal } \alpha)$. We conclude that $\dim \alpha \leq m$. The opposite inequality is trivial.

Proof of (15). In view of (14) we assume β irreducible. Using (13) we choose

$$g \in E_{\dim \alpha}(\text{ideal } \alpha) \cap E_{\dim \beta}(\text{ideal } \beta),$$

note that $(\text{ideal } \beta)_g$ properly contains $(\text{ideal } \alpha)_g$, and infer from (10) that $g \in A_{\dim \alpha}(\text{ideal } \beta)$, hence $g \notin E_{\dim \alpha}(\text{ideal } \beta)$. Therefore $\dim \beta \neq \dim \alpha$.

Proof of (16). We abbreviate $v = \text{var}[\text{ideal } \gamma_0(T)]$, observe that v is the minimal member of $\mathscr{V}_0(\mathbf{R}^n)$ containing $\gamma_0(T)$, and consider any irreducible component α of v with $\dim \alpha = m$. Applying (13) and (10) we find an m dimensional analytic submanifold Z of $\mathbf{R}^n$ and germs $\beta, w \in \mathscr{V}_0(\mathbf{R}^n)$ such that

$$\alpha = \gamma_0(Z) \cup \beta, \quad \alpha \not\subset \beta, \quad v = \alpha \cup w, \quad \alpha \not\subset w.$$

It follows that $\gamma_0(T \cap Z) \neq \gamma_0(\varnothing)$, because otherwise

$$\gamma_0(T) \subset [\alpha \cap \gamma_0(T)] \cup w \subset \beta \cup w, \qquad v \subset \beta \cup w,$$

and $\alpha = \beta \cup (\alpha \cap w)$ would be reducible. Noting that $v \subset \gamma_0(S)$, we choose a neighborhood U of 0 in $\mathbf{R}^n$ with $Z \cap U \subset S$, a point $z \in T \cap Z \cap U$, and a neighborhood V of z in U with $S \cap V \subset T$. Then

$$Z \cap V = Z \cap U \cap V \subset S \cap V \subset T,$$

hence $\mathscr{H}^{u+1}(Z \cap V) = 0$. Since also $\mathscr{H}^m(Z \cap V) > 0$, we conclude that $m \leq u$. Now reference to (14) completes this proof.

3.4.9. We see from 3.4.8(11) that *the family of all (topological) components of the difference of two analytic subvarieties of an open subset U of $\mathbf{R}^n$ is locally finite in U*. It follows that in the conclusion of 3.4.8(11) the sets V_i and S_i may be replaced by $\mathbf{U}(0,r)$ and $S_i \cap \mathbf{U}(0,r)$ for every sufficiently small positive number r, because $S_i \cap \{x \colon |x|^2 - r^2 \neq 0\}$ is a functional analytic submanifold of V_i when $i \geq 1$.

3.4.10. Whenever $S \subset U \subset \mathbf{R}^n$, U is open and S is analytic in U one defines

$$\dim S = \sup \{\dim \gamma_a(S) \colon a \in U\}.$$

In case $\dim S = m$ the **regular part** of S is the subset R consisting of those points of S which have a neighborhood V in U such that $S \cap V$ is an m dimensional analytic submanifold of $\mathbf{R}^n$. From 3.4.8 it follows that

$$\mathscr{H}^m(C \cap S) < \infty \text{ for every compact } C \subset U,$$

and that $S \sim R$ can be covered by a family F of analytic submanifolds of $\mathbf{R}^n$ with dimensions less than m such that each compact subset of U meets only finitely many members of F, and

$$\mathscr{H}^{m-1}(C \cap S \sim R) < \infty \text{ for every compact } C \subset U.$$

However $S \sim R$ need not be analytic in U, as shown by an example in [WB, §11(c), p. 159]. There and in [CH], [LS 1, 2], [MB 1, 2], [NR] other aspects of the theory of analytic sets have also been studied.

3.4.11. *If* $S \subset U \subset \mathbf{R}^n$, *$U$ is open, S is analytic in U and* $a \in U$ *with* $\gamma_a(S) \neq \gamma_a(\varnothing)$, *then*

$$\dim \operatorname{var} \operatorname{ideal} \gamma_0[\operatorname{Tan}(S, a)] \leq \dim \gamma_a(S).$$

To prove this we assume $a=0$ and define

$$I = \operatorname{ideal} \gamma_0(S), \qquad J = \operatorname{ideal} \gamma_0[\operatorname{Tan}(S, 0)].$$

If ϕ is a real valued analytic function with $0 \neq \gamma_0(\phi) \in I$, k is the least integer such that $D^k \phi(0) \neq 0$, and P is the homogeneous polynomial function corresponding to $D^k \phi(0)$, then

$$|x|^{-k} |P(x) - \phi(x)| \to 0 \quad \text{as} \quad x \to 0 \text{ in } \mathbf{R}^n,$$

$$P(|x|^{-1} x) \to 0 \quad \text{as} \quad x \to 0 \text{ in } S,$$

hence $P(v)=0$ for $v \in \operatorname{Tan}(S, 0)$, and $\gamma_0(P) \in J$. We infer by the argument used in the proof of 3.4.8 (6) that

$$g \in A_j(I) \quad \text{implies} \quad g \circ h \in B_j(J) \text{ for } \mu_j \text{ almost all } h \text{ in } H_j,$$

and we conclude that $\operatorname{depth}(I) \geq \operatorname{depth}(J)$.

The preceding construction also yields a finite set F of real valued homogeneous polynomial functions on $\mathbf{R}^n$ such that

$$\operatorname{Tan}(S, a) \subset \operatorname{var}(\mathbf{R}^n, F) \quad \text{and} \quad \dim \operatorname{var}(\mathbf{R}^n, F) \leq \dim \gamma_a(S).$$

However $\operatorname{Tan}(S, a)$ need not be analytic in $\mathbf{R}^n$, as shown by the example where $U = \mathbf{R}^2$,

$$S = \mathbf{R}^2 \cap \{x: (x_1)^3 = (x_2)^2\}, \qquad \operatorname{Tan}(S, 0) = \mathbf{R}^2 \cap \{x: x_1 \geq 0, x_2 = 0\}.$$

3.4.12. In the theory developed in 3.4.5−11 one can replace the real field $\mathbf{R}$ by the complex field $\mathbf{C}$, making only relatively minor modifications which we will discuss here. The complex case is actually simpler and more classical; extensive expositions of it may be found in [AB], [GUR], [NR]. We will use the term "holomorphic" as synonymous with "complex analytic"; the single word "analytic" will continue to mean "real analytic".

To define the concept of a **holomorphic function** we *require in* 3.1.24 *additionally that X, Y are complex Banach spaces and $Df(x)$ is* $\mathbf{C}$ *linear for each* $x \in A$; this implies that $D^m f(x)$ is complex m linear for all positive integers m.

We say that B is a **complex m dimensional holomorphic submanifold of $\mathbf{C}^n$** if and only if for each $b \in B$ there exists a neighborhood T of b in $\mathbf{C}^n$, a holomorphic diffeomorphism σ mapping T onto an open subset of $\mathbf{C}^n$, and a $\mathbf{C}$ vectorsubspace Z of $\mathbf{C}^n$ such that

$$\dim_{\mathbf{C}}(Z) = m, \qquad \sigma(B \cap T) = Z \cap \operatorname{im} \sigma .$$

It follows that B is a $2m$ dimensional analytic submanifold of $\mathbf{C}^n \simeq \mathbf{R}^{2n}$; the real dimension of B equals twice its complex dimension.

Replacing real valued analytic functions by complex valued holomorphic functions throughout 3.4.5 we define

$$\mathscr{O}_a^{\mathbf{C}}(X), \quad \mathscr{O}_a^{\mathbf{C}}(u), \quad \operatorname{ideal}^{\mathbf{C}} \gamma_a(S), \quad \mathscr{V}_a^{\mathbf{C}}(X)$$

and the notions of **holomorphic ideal, subvariety, variety germ, irreducibility, component, functional submanifold** and **block** of an open subset U of $\mathbf{C}^n$.

In 3.4.6, 7 the change from $\mathbf{R}$ to $\mathbf{C}$ requires only formal modifications.

To adapt 3.4.8 we replace $\mathbf{R}^n$, $\mathbf{O}(n)$ and θ_n by $\mathbf{C}^n$, the unitary group $\mathbf{U}(n)$ consisting of all $\mathbf{C}$ linear isometries of $\mathbf{C}^n$, and a Haar measure of $\mathbf{U}(n)$. We use dual $\mathbf{C}$ bases for $\mathbf{C}^n$ and $\bigwedge^1_{\mathbf{C}}(\mathbf{C}^n, \mathbf{C})$. Instead of $\dim \alpha$ we employ the complex dimension

$$\dim_{\mathbf{C}} \alpha = \operatorname{depth}(\operatorname{ideal}^{\mathbf{C}} \alpha).$$

In the proof of (10), use of the function ψ_λ defined by inequalities can be replaced by lifting curves from Y_ν to Z; we note also that in the complex case $Y_\nu = \varnothing$ for $\nu \neq k$, because $\mathbf{C}$ is algebraically closed. In the proof of (11) for the complex case one needs a different argument to show that R has only finitely many components; one can either apply the result of the real case to the analytic function $|\Phi|^2$, or verify by an elementary inductive argument that in the complex case R is connected. *The last part of the conclusion of* (13) *for the complex case should refer to an* $(\mathscr{H}^{2m}, 2m)$ *rectifiable set* S *such that* $\gamma_0(S) = \alpha$ *and*

$$\sup\{\mathscr{H}^{2m}(T)/(\operatorname{diam} T)^{2m} : \varnothing \neq T \subset S\} < \infty ;$$

to prove this inequality one can either apply the corresponding result for the real case, because it is by now clear that

$$\dim_{\mathbf{C}} \alpha = 2 \dim \alpha \ \text{ for } \ \alpha \in \mathscr{V}_0^{\mathbf{C}}(\mathbf{C}^n) \subset \mathscr{V}_0(\mathbf{R}^{2n}),$$

or modify 3.2.27 using Wirtinger's inequality and the fact that $\operatorname{Tan}(S, x)$ is a complex m dimensional vectorsubspace of $\mathbf{C}^n$ for $\mathscr{H}^{2m}$ almost all x in S (a version of the second argument will appear in 5.4.19). *The last part of the hypothesis of* (16) *for the complex case should require that* $\mathscr{H}^{2u+2}(T) = 0$.

Proceeding as in 3.4.10 we find that if $S \subset U \subset \mathbf{C}^n$, S is holomorphic in U and

$$m = \dim_{\mathbf{C}} S = \sup \{\dim_{\mathbf{C}} \gamma_a(S) \colon a \in U\},$$

then the regular part R of S is a complex m dimensional holomorphic submanifold of $\mathbf{C}^n$ and

$$\mathscr{H}^{2m}(C \cap S) < \infty, \quad \mathscr{H}^{2m-2}(C \cap S \sim R) < \infty$$

for every compact $C \subset U$; moreover S is analytic in U with $\dim S = 2m$. It can also be shown that $S \sim R$ is holomorphic in U with $\dim_{\mathbf{C}}(S \sim R) \leq m-1$, and that every point of S has a neighborhood V in U such that $V \cap R$ is connected; however these additional facts will not be needed in this book.

In 3.4.11 the change from **R** to **C** is trivial, except for the example in the last paragraph. (It can be shown that $\operatorname{Tan}(S, a)$ is holomorphic whenever S is holomorphic at a; see [WH 6, 7] where many interesting facts about tangent cones are discussed.) We will study tangential properties of oriented holomorphic varieties in 4.3.19 and 5.4.19.

CHAPTER FOUR

Homological integration theory

In this chapter we make a measure theoretic study of certain groups of chains, with a boundary operator, which are appropriate for both topological and analytic uses. They yield the standard homology groups with coefficients in **Z** or $\mathbf{Z}_\nu$ or **R**. They have compactness properties implying the existence of solutions of variational problems. They obey approximation theorems, representation theorems and isoperimetric inequalities of simultaneous geometric and analytic interest.

We employ the concepts of distribution and current introduced in [SCH] and [DR 1], and defined here in 4.1.1 and 4.1.7. Much of our work involves the functorial properties of differential forms and currents, which transform with opposite variance under maps of their underlying spaces. Therefore we must distinguish consciously – not just logically – between a differential form ϕ of degree m on $\mathbf{R}^n$ and the corresponding $n-m$ dimensional current $\mathbf{E}^n \llcorner \phi$. We must follow the practice of topologists in maintaining a separation between chains and cochains, and we cannot adopt the viewpoint of those analysts who think of distributions and currents mainly as generalized functions and differential forms.

The principal objects of our investigation are the normal, rectifiable and integral currents introduced in [FF] and defined here in 4.1.7 and 4.1.24, the flat chains introduced in [WH 4] and defined here in 4.1.12, and the integral flat chains defined in 4.1.24. We follow the lead of [WH 4] in using Lipschitzian maps and the notion of mass, but the equivalence of Whitney's concepts of flat chain with ours is not obvious (the isomorphism follows from 4.1.23 and 4.2.23), and the two books have very different aims; Whitney's book is directed to (real) cohomology with general cochains, ours to (real and integral) homology with general chains.

In 4.1.5 we derive basic properties of the variation measure $\|T\|$ corresponding to a distribution T representable by integration. In 4.1.6 we use dual procedures to define the exterior derivative of a differential form of degree m and the divergence of a p vectorfield. In 4.1.8 and 4.1.11 we discuss cartesian products and joins of currents, and we construct oriented simplexes as joins of point masses at the vertexes. This leads in 4.1.22, 4.1.27, 4.1.32 to simple proofs of the elementary properties of polyhedral chains. In 4.1.28 we obtain various geometric characterizations of rectifiable currents, which we relate in 4.1.31 to the concept of oriented manifold. In 4.1.13 – 4.1.15 and 4.1.25, 4.1.30 we discuss mapping properties of flat chains and rectifiable currents.

In 4.1.18 we prove two new propositions concerning the representation of m dimensional flat chains by pairs of $\mathscr{L}^n$ summable m and $m+1$ vectorfields. In 4.1.20 and 4.1.21 we establish the important relation of such chains to m dimensional integralgeometric measure.

The pivotal results of this chapter are the Deformation Theorem 4.2.9, the Closure Theorem 4.2.16, the Compactness Theorem 4.2.17 and the Approximation Theorem 4.2.20, all of which originated in [FF]. However the proof of 4.2.16 given here differs greatly from the corresponding argument in [FF, 8.10 – 8.12]. The present method has the advantage of applying also to flat chains modulo ν, which we treat in 4.2.26. The theory of flat chains modulo ν was first developed, by methods different from ours, for special dimensions and $\nu=2$ in [Z 1], then for general dimensions and $\nu \geq 2$ in [FL 4]. The discovery of 4.2.17 (2) in [FF] was preceded by treatment of the special case $m=n$ in [DG 1, 2], and the subcase $m=3=n$ in [FY 2].

In 4.2.1 we slice normal currents by Lipschitzian maps as in [FF, 3.9], and in Section 4.3 we slice flat chains by Lipschitzian maps into $\mathbf{R}^n$ using the method of [F 19]. The proof of 4.3.10 combines the techniques of [F 19, 3.15] and [MCB 1, 2.1]. In 4.3.20 we indicate a new approach to intersection theory by way of slicing.

Analytic and holomorphic chains are discussed in 4.2.28, 4.3.18 and 4.2.29, 4.3.19.

In 4.4.1 we construct the integral homology groups of local Lipschitz neighborhood retracts in $\mathbf{R}^n$ (see 4.1.29) by using complexes of integral flat chains. The axioms of Eilenberg and Steenrod are verified with particular ease in our context. The reader need not have any previous knowledge of homology theory. Those facts actually used in our book follow readily from our definitions and theorems. However we do not duplicate the deduction of all the propositions of classical homology theory from the axioms, which may be found in standard treatises like [ES] or [SE]. The most significant additional features distinguishing our treatment of homology theory are the isoperimetric inequalities proved in 4.4.2, 4.4.3, 4.2.10, 4.5.14, 4.5.2, 4.5.9 (18), (31), which originated in [FF], [F 18], and the compactness properties of homology classes established in 4.4.4, which were obtained in [FF] for the special case $B=\emptyset$.

The proposition 4.5.4 was first proved in [GW]; here we use the method of [F 16]. A lemma similar to 4.5.3 occurs in [DG 3].

We present an elementary version of the Gauss-Green theorem in 4.1.31, and perhaps optimal results in 4.5.6, 4.5.11. Research on the problem of finding the most natural and general form of this theorem has contributed greatly to the development of geometric measure theory. The concept of exterior normal defined in 4.5.5 was introduced in [F 2] and used in [F 2, 3, 4] to establish the validity of the Gauss-Green theorem for every open subset A of $\mathbf{R}^n$ with $\mathscr{H}^{n-1}(\operatorname{Bdry} A)<\infty$. Next it was discovered in [DG 1, 2] that if A is a Borel subset of $\mathbf{R}^n$ such that the current $T=\partial(\mathbf{E}^n \llcorner A)$ has finite mass, then $\mathbf{R}^n$ is $(\|T\|, n-1)$ rectifiable and $*\,\mathbf{n}(A,b)=\vec{T}(b)$ for $\|T\|$ almost all b. This and some additional observations in [F 14] led to 4.5.6. The proposition 4.5.11 is new.

In 4.5.7–4.5.13 we establish the fundamental properties of those real valued functions which correspond to locally normal n dimensional currents in $\mathbf{R}^n$ or, equivalently by 4.5.16, to distributions of type $\mathbf{R}$ whose first order partial derivatives are representable by integration. Classes of such function have been studied for several decades, beginning long before the concept of distribution was explicitly defined, in numerous papers on potential theory, area theory and calculus of variations, for instance in [EGC], [TL], [CL 1], [MCB 1], [F 11], [KK], [FL 2]. Regarding our theorem 4.5.9 we note that the propositions (27) and (30), without (30 II), may be considered classical; part of (29) was proved in [GC 2]; (26 II) was proved in [CZ]; (14) was proved in [Z 3] under the conditions of (29); (13) was proved in [FLR]; (5) was proved for continuous f in [F 4] for $n=2$, and in [F 16] for all n. However our present method involving $\lambda, \mu, F, G, S, C, E$ is quite different from earlier work, and the general results obtained in 4.5.9 (5), (14)–(22), (24), (25), (26 I, IV), (29) are new.

The results of 4.5.15 for $1<\xi<n$, which we deduce from our isoperimetric inequality 4.5.9 (31), were first derived from potential theory in [SOB]. In case $\xi=1$ our method yields a stronger result. We follow [MCB 1] in proving the Hölder condition for $\xi>n$.

4.1. Differential forms and currents

4.1.1. Assuming that X, Y are Banach spaces with $\dim X<\infty$ and that U is an open subset of X we let

$$\mathscr{E}(U,Y)$$

be the vectorspace of all functions of class ∞ mapping U into Y. For each nonnegative integer i and each compact subset K of U we define the seminorm

$$\nu_K^i(\phi)=\sup\{\|D^j\phi(x)\|: 0\leq j\leq i \text{ and } x\in K\}$$

whenever $\phi \in \mathscr{E}(U, Y)$. The family of all seminorms ν_K^i induces a locally convex, translation invariant Hausdorff topology on $\mathscr{E}(U, Y)$; basic neighborhood of any $\psi \in \mathscr{E}(U, Y)$ are the sets

$$\mathscr{E}(U, Y) \cap \{\phi : \nu_K^i(\phi - \psi) < r\}$$

corresponding to all i, K and all $r > 0$. We let

$$\mathscr{E}'(U, Y)$$

be the vectorspace of all continuous real valued linear functions on $\mathscr{E}(U, Y)$, and we endow $\mathscr{E}'(U, Y)$ with the weak topology generated by the sets

$$\mathscr{E}'(U, Y) \cap \{T : a < T(\phi) < b\}$$

corresponding to all $\phi \in \mathscr{E}(U, Y)$ and all $a, b \in \mathbf{R}$. Defining

$$\operatorname{spt} \phi = U \cap \operatorname{Clos}\{x : \phi(x) \neq 0\} \text{ for } \phi \in \mathscr{E}(U, Y),$$

$$\operatorname{spt} T = U \sim \bigcup \{W : W \text{ is open, } T(\phi) = 0 \text{ whenever } \phi \in \mathscr{E}(U, Y) \text{ and } \operatorname{spt} \phi \subset W\}$$

for $T \in \mathscr{E}'(U, Y)$, we observe that $\operatorname{spt} T$ is compact because

$$T \leq M \cdot \nu_K^i \text{ for some } i, K \text{ and some } M < \infty.$$

Thus we find that $\mathscr{E}'(U, Y)$ is the union of its closed subsets

$$\mathscr{E}'_K(U, Y) = \mathscr{E}'(U, Y) \cap \{T : \operatorname{spt} T \subset K\}$$

corresponding to all compact subsets K of U. It may also be shown that all members of any convergent sequence in $\mathscr{E}'(U, Y)$ belong to some single set $\mathscr{E}'_K(U, Y)$.

For each compact subset K of U we define

$$\mathscr{D}_K(U, Y) = \mathscr{E}(U, Y) \cap \{\phi : \operatorname{spt} \phi \subset K\}$$

and observe that $\mathscr{D}_K(U, Y)$ is closed in $\mathscr{E}(U, Y)$. Then we consider the vectorspace

$$\mathscr{D}(U, Y) = \bigcup \{\mathscr{D}_K(U, Y) : K \text{ is a compact subset of } U\}$$

with the largest topology such that the inclusion maps from all the sets $\mathscr{D}_K(U, Y)$ are continuous; accordingly a subset of $\mathscr{D}(U, Y)$ is open if and only if its intersection with each $\mathscr{D}_K(U, Y)$ belongs to the relative topology of $\mathscr{D}_K(U, Y)$ in $\mathscr{E}(U, Y)$. It follows that the inclusion map of $\mathscr{D}(U, Y)$ into $\mathscr{E}(U, Y)$ is continuous; this map is not a homeomorphic embedding unless $U = \varnothing$ or $Y = \{0\}$, but the topologies of $\mathscr{E}(U, Y)$ and $\mathscr{D}(U, Y)$ induce the same relative topology on each $\mathscr{D}_K(U, Y)$. We let

$$\mathscr{D}'(U, Y)$$

be the vectorspace of all continuous real valued linear functions on $\mathscr{D}(U,Y)$, and we endow $\mathscr{D}'(U,Y)$ with the weak topology generated by the sets

$$\mathscr{D}'(U,Y)\cap\{T\colon a<T(\phi)<b\}$$

corresponding to all $\phi\in\mathscr{D}(U,Y)$ and all $a,b\in\mathbf{R}$. Defining supports as before we see that the support of each member of $\mathscr{D}(U,Y)$ is compact, but the support of a member of $\mathscr{D}'(U,Y)$ need not be compact. *A real valued linear function T on $\mathscr{D}(U,Y)$ belongs to $\mathscr{D}'(U,Y)$ if and only if for each compact subset K of U there exist nonnegative integers i and M such that*

$$T(\phi)\le M\,\nu_K^i(\phi)\ \text{whenever}\ \phi\in\mathscr{D}_K(U,Y).$$

The members of $\mathscr{D}'(U,Y)$ are called **distributions in U of type Y.**

Functional restriction from $\mathscr{E}(U,Y)$ to $\mathscr{D}(U,Y)$ yields a univalent continuous linear map

$$\mathscr{E}'(U,Y)\to\mathscr{D}'(U,Y)$$

whose image consists of all distributions with compact support in U of type Y; this map is usually not a homeomorphic embedding, but it does embed $\mathscr{E}_K'(U,Y)$ for each compact subset K of U. Often one fails to distinguish notationally between a member of $\mathscr{E}'(U,Y)$ and its image in $\mathscr{D}'(U,Y)$.

For every open subset V of X such that $U\subset V$ we construct a continuous linear monomorphism

$$\mathscr{D}(U,Y)\to\mathscr{D}(V,Y)$$

by associating $\phi\cup[(V\sim U)\times\{0\}]\in\mathscr{D}(V,Y)$ with $\phi\in\mathscr{D}(U,Y)$; this function maps $\mathscr{D}_K(U,Y)$ homeomorphically onto $\mathscr{D}_K(V,Y)$ whenever K is a compact subset of U. The dual linear map

$$\mathscr{D}'(V,Y)\to\mathscr{D}'(U,Y)$$

is continuous; it is usually not surjective, but its image always contains

$$\mathscr{D}'(U,Y)\cap\{T\colon \operatorname{spt}T \text{ is a compact subset of } U\}.$$

If $T\in\mathscr{D}'(U,Y)$, Φ is an open covering of U and the functions v_s form an associated partition of unity on U (see 3.1.13), then

$$T=\sum_{s\in S}T\llcorner v_s$$

where $(T\llcorner v_s)\phi=T(v_s\cdot\phi)$ for $s\in S$ and $\phi\in\mathscr{D}(U,Y)$; we mean that T is the limit in $\mathscr{D}'(U,Y)$ of the partial sums corresponding to any exhausting sequence of finite subsets of S. It follows that

$$T(\phi)=0\ \text{whenever}\ \phi\in D(U,Y)\ \text{with}\ \operatorname{spt}T\cap\operatorname{spt}\phi=\varnothing.$$

Sometimes we write $T_x\phi(x)$ in place of $T(\phi)$.

Now suppse $X=\mathbf{R}^n$ with the standard base $e_1,\ldots,e_n$. For each distribution T we define the distributions $TD_1,\ldots,TD_n$ by the formula

$$(TD_j)\phi=T(D_j\phi) \text{ whenever } \phi\in\mathscr{D}(U,Y);$$

similarly we define TD^α for $\alpha\in\Xi(n,i)$. In case $U=\mathbf{R}^n$ it follows from Taylor's theorem that for each $\phi\in\mathscr{D}(\mathbf{R}^n,Y)$ the functions mapping $x\in\mathbf{R}^n$ onto

$$[\phi(x+h\,e_j)-\phi(x)]/h-D_j\phi(x)\in Y$$

converge to 0 in $\mathscr{D}(\mathbf{R}^n,Y)$ as h approaches 0 in $\mathbf{R}$, hence

$$(TD_j)\phi=\lim_{h\to 0}[T_x\phi(x+h\,e_j)-T_x\phi(x)]/h$$

for $T\in\mathscr{D}'(\mathbf{R}^n,Y)$.

Next suppose Y separable and let Y^* be the Banach space of all continuous linear functions on Y. *To each $\mathscr{L}^n$ measurable function ξ with values in the weakly topologized space Y^** (see 2.5.12) *such that*

$$\textstyle\int_K \|\xi\|\,d\mathscr{L}^n<\infty \text{ for every compact subset } K \text{ of } \mathbf{R}^n,$$

there corresponds the distribution $T\in\mathscr{D}'(\mathbf{R}^n,Y)$ defined by the formula

$$T(\phi)=\textstyle\int\langle\phi(x),\xi(x)\rangle\,d\mathscr{L}^n x \text{ for } \phi\in\mathscr{D}(\mathbf{R}^n,Y).$$

We abbreviate the integrand function by $\langle\phi,\xi\rangle$. If ξ is of class 1, then $D_j\langle\phi,\xi\rangle=\langle D_j\phi,\xi\rangle+\langle\phi,D_j\xi\rangle$ with $\int D_j\langle\phi,\xi\rangle\,d\mathscr{L}^n=0$ because $\operatorname{spt}\langle\phi,\xi\rangle$ is compact, hence

$$-T(D_j\phi)=\textstyle\int\langle\phi,D_j\xi\rangle\,d\mathscr{L}^n;$$

thus $-TD_j$ is the distribution corresponding to the function $D_j\xi$.

The preceding observation leads to the following definitions valid for arbitrary $T\in\mathscr{D}'(U,Y)$:

$$D_jT=-TD_j \text{ whenever } j\in\{1,\ldots,n\},$$
$$D^\alpha T=(-1)^i\,TD^\alpha \text{ whenever } \alpha\in\Xi(n,i).$$

Each of these partial differential operators is a continuous endomorphism of $\mathscr{D}'(U,Y)$.

4.1.2. Here we derive some facts about **smoothing** (also called **regularization**). For this purpose we choose a nonnegative function Φ of class ∞ on $\mathbf{R}^n$ with

$$\operatorname{spt}\Phi\subset\mathbf{R}^n\cap\{x\colon |x|<1\} \quad\text{and}\quad \textstyle\int\Phi\,d\mathscr{L}^n=1.$$

Letting $\Phi_\varepsilon(x)=\varepsilon^{-n}\Phi(\varepsilon^{-1}x)$ for $0<\varepsilon<\infty$ and $x\in\mathbf{R}^n$, we find that

$$\operatorname{spt}\Phi_\varepsilon\subset\mathbf{R}^n\cap\{x\colon |x|<\varepsilon\},\quad \textstyle\int\Phi_\varepsilon\,d\mathscr{L}^n=1,$$
$$\sup\operatorname{im}\|D^i\Phi_\varepsilon\|=\varepsilon^{-n-i}\sup\operatorname{im}\|D^i\Phi\|<\infty$$

for all nonnegative integers i. Whenever f is an $\mathscr{L}^n$ measurable function with values in a separable Banach space Y such that

$$\int_K \|f\| \, d\mathscr{L}^n < \infty \text{ for every compact subset } K \text{ of } \mathbf{R}^n,$$

we consider the convolution $\Phi_\varepsilon * f$ defined by the formula

$$(\Phi_\varepsilon * f)(x) = \int \Phi_\varepsilon(x-z) f(z) \, d\mathscr{L}^n z = \int \Phi_\varepsilon(z) f(x-z) \, d\mathscr{L}^n z \text{ for } x \in \mathbf{R}^n.$$

Letting $e_1, \ldots, e_n$ be the standard base of $\mathbf{R}^n$ we compute

$$\begin{aligned}(\Phi_\varepsilon * f)(x + h e_j) - (\Phi_\varepsilon * f)(x) &= \int \int_0^h D_j \Phi_\varepsilon(x - z + t e_j) \, d\mathscr{L}^1 t \, f(z) \, d\mathscr{L}^n z \\ &= \int_0^h \int D_j \Phi_\varepsilon(x - z + t e_j) f(z) \, d\mathscr{L}^n z \, d\mathscr{L}^1 t\end{aligned}$$

for $j = 1, \ldots, n$ and $h \in \mathbf{R}$, hence

$$D_j(\Phi_\varepsilon * f)(x) = \int D_j \Phi_\varepsilon(x-z) f(z) \, d\mathscr{L}^n z.$$

We infer that $\Phi_\varepsilon * f$ is a function of class ∞ with

$$D^\alpha(\Phi_\varepsilon * f) = (D^\alpha \Phi_\varepsilon) * f \text{ for } \alpha \in \Xi(n, i).$$

Similarly we see that $D_j(\Phi_\varepsilon * f) = \Phi_\varepsilon * D_j f$ if f is of class 1 (more generally if f is locally Lipschitzian and Y is reflexive), hence

$$D^i(\Phi_\varepsilon * f) = \Phi_\varepsilon * D^i f \text{ if } f \text{ is of class } i.$$

We also observe that

$$\begin{aligned}\|(\Phi_\varepsilon * f)(x) - f(x)\| &= \|\int \Phi_\varepsilon(x-z)[f(z) - f(x)] \, d\mathscr{L}^n z\| \\ &\leq (\sup \operatorname{im} \Phi) \cdot \varepsilon^{-n} \int_{\mathbf{B}(x,\varepsilon)} \|f(x) - f(z)\| \, d\mathscr{L}^n z \to 0\end{aligned}$$

as $\varepsilon \to 0+$ whenever x belongs to the $\mathscr{L}^n$ Lebesgue set of f (see 2.9.9); in case f is continuous the convergence is locally uniform with respect to x. We conclude:

If $f \in \mathscr{D}(\mathbf{R}^n, Y)$, then $\Phi_\varepsilon * f \to f$ in $\mathscr{D}(\mathbf{R}^n, Y)$ as $\varepsilon \to 0+$.

If $T \in \mathscr{D}'(\mathbf{R}^n, Y)$, then $T_\varepsilon \to T$ in $\mathscr{D}'(\mathbf{R}^n, Y)$ as $\varepsilon \to 0+$, where

$$T_\varepsilon(f) = T(\Phi_\varepsilon * f) \text{ for } f \in \mathscr{D}(\mathbf{R}^n, Y);$$

moreover T_ε corresponds to the function $g_\varepsilon \in \mathscr{E}(\mathbf{R}^n, Y^*)$ such that

$$\langle y, g_\varepsilon(z) \rangle = T_x[\Phi_\varepsilon(x-z) y] \text{ for } y \in Y \text{ and } z \in \mathbf{R}^n,$$

because $D^i \Phi_\varepsilon(x-z) y$ is continuous with respect to (x, z, y) for each nonnegative integer i and

$$\begin{aligned}\int \langle f(z), g_\varepsilon(z) \rangle \, d\mathscr{L}^n z &= \int T_x[\Phi_\varepsilon(x-z) f(z)] \, d\mathscr{L}^n z \\ &= T_x \int \Phi_\varepsilon(x-z) f(z) \, d\mathscr{L}^n z = T(\Phi_\varepsilon * f) = T_\varepsilon(f)\end{aligned}$$

for $f \in \mathscr{D}(\mathbf{R}^n, Y)$; to justify the interchange of $\int$ and T we approximate $(\Phi_\varepsilon * f)(x)$ by

$$\psi(x) = \sum_{k=1}^{\nu} \mathscr{L}^n(A_k)\, \Phi_\varepsilon(x - a_k)\, f(a_k)$$

where $A_1, \ldots, A_\nu$ form a Borel partition of $\operatorname{spt} f$ and $a_k \in A_k$, noting that, for each $\alpha \in \Xi(n, i)$,

$$D^\alpha \psi(x) = \sum_{k=1}^{\nu} \mathscr{L}^n(A_k)\, D^\alpha \Phi_\varepsilon(x - a_k)\, f(a_k)$$

is uniformly close to $D^\alpha(\Phi_\varepsilon * f)(x)$ provided $A_1, \ldots, A_\nu$ have small diameters.

In order to extend smoothing from $\mathscr{D}'(\mathbf{R}^n, Y)$ to $\mathscr{D}'(U, Y)$, where U is any open subset of $\mathbf{R}^n$, we choose a sequence of functions $w_j \in \mathscr{D}(U, \mathbf{R})$ such that, for each compact subset K of U there exists a positive integer ν with $w_j(x) = 1$ whenever $j \geq \nu$ and $x \in K$. If $T \in \mathscr{D}'(U, Y)$, then each $T \llcorner w_j$ is the restriction of some member of $\mathscr{D}'(\mathbf{R}^n, Y)$, hence $(T \llcorner w_j)_\varepsilon$ may be defined as before, and approximates T in $\mathscr{D}'(U, Y)$ when j is large and ε is small.

4.1.3. Applying 4.1.2 with a function Φ such that

$$\Phi(x) = \gamma(x_1) \cdot \gamma(x_2) \cdot \cdots \cdot \gamma(x_n) \text{ whenever } x \in \mathbf{R}^n,$$

for a suitable $\gamma \in \mathscr{D}(\mathbf{R}, \mathbf{R})$, and recalling 1.1.3, we find that the resulting functions ψ belong to *the image of* $[\bigotimes_n \mathscr{D}(\mathbf{R}, \mathbf{R})] \otimes Y$ *in* $\mathscr{D}(\mathbf{R}^n, Y)$; accordingly *this image is dense in* $\mathscr{D}(\mathbf{R}^n, Y)$.

4.1.4. *If U is an open subset of* $\mathbf{R}^n$, $T \in \mathscr{D}'(U, Y)$, *A is a connected open subset of U and*

$$\operatorname{spt} D_j T \subset U \sim A \quad \text{for } j = 1, \ldots, n,$$

then there exists a continuous linear function α on Y such that

$$T(f) = \int_U (\alpha \circ f)\, d\mathscr{L}^n$$

whenever $f \in D(U, Y)$ with $\operatorname{spt} f \subset A$. It will suffice to prove this in case $\operatorname{Clos} A$ is a compact subset of U; we may then assume that $\operatorname{spt} T$ is a compact subset of U, hence also that $U = \mathbf{R}^n$. We choose $a \in A$ and, for each $\varepsilon > 0$, let C_ε be the component of a in

$$A \cap \{z : \operatorname{dist}(z, \mathbf{R}^n \sim A) > \varepsilon\}.$$

Applying 4.1.2 we find that $D_j g_\varepsilon(z)=0$ for $z\in C_\varepsilon$ and $j=1,\dots,n$, hence g_ε maps C_ε onto a single point $\alpha_\varepsilon\in Y^*$, and

$$T(f)=\lim_{\varepsilon\to 0+}\int\langle f(x),\alpha_\varepsilon\rangle\, d\mathscr{L}^n x$$

whenever $f\in\mathscr{D}(\mathbf{R}^n, Y)$ with $\operatorname{spt} f\subset A$. Moreover, for all $y\in Y$ and small $\delta>0$,

$$\begin{aligned}\langle y,\alpha_\delta\rangle=\langle y, g_\delta(a)\rangle &= T_x[\Phi_\delta(x-a)\,y]\\ &=\lim_{\varepsilon\to 0+}\int\langle\Phi_\delta(x-a)\,y,\alpha_\varepsilon\rangle\, d\mathscr{L}^n x=\lim_{\varepsilon\to 0+}\langle y,\alpha_\varepsilon\rangle;\end{aligned}$$

thus α_δ is independent of δ.

4.1.5. Suppose U is an open subset of $\mathbf{R}^n$, Y is a separable Banach space and Y^* is the Banach space of all continuous real valued linear functions on Y. For each $S\in\mathscr{D}'(U, Y)$ we define the function $\|S\|$ on $\mathscr{K}(U)^+$ (see 2.5.14) by the formula

$$\|S\|(f)=\sup\{S(\phi)\colon \phi\in\mathscr{D}(U,Y) \text{ and } \|\phi\|\le f\}$$

whenever $f\in\mathscr{K}(U)^+$. We will prove that *the condition*

$$\|S\|(f)<\infty \quad \textit{whenever } f\in\mathscr{K}(U)^+$$

is necessary and sufficient for the existence of a Radon measure ρ over U and a ρ measurable function ξ with values in the weakly topologized space Y^* such that

$$\int_K\|\xi\|\,d\rho<\infty \quad \textit{for every compact subset } K \textit{ of } U,$$

$$S(\phi)=\int\langle\phi(x),\xi(x)\rangle\, d\rho\, x \quad \textit{whenever } \phi\in\mathscr{D}(U,Y).$$

The condition is clearly necessary. To prove its sufficiency we apply 2.5.12 taking $L=\mathscr{K}(U)$, letting Ω be the vectorspace of all continuous maps of U into Y with compact support, and extending S to a linear function T on Ω by the formula (see 4.1.2)

$$T(\omega)=\lim_{\varepsilon\to 0+} S(\Phi_\varepsilon * \omega) \quad \text{for } \omega\in\Omega.$$

In order to verify that this limit exists and does not exceed $\|S\|(\|\omega\|)$ we first select

$$g\in\mathscr{K}(U)^+ \quad \text{with} \quad \operatorname{spt}\omega\subset\operatorname{Int}\{x\colon g(x)=1\};$$

for every positive number η it is true that

$$\|\Phi_\varepsilon*\omega-\Phi_\delta*\omega\|\le\eta\, g,$$

hence

$$|S(\Phi_\varepsilon*\omega)-S(\Phi_\delta*\omega)|\le\|S\|\,(\eta\, g)=\eta\,\|S\|\,(g)$$

whenever ε and δ are sufficiently small positive numbers; moreover we can choose

$$h \in \mathscr{K}(U)^+ \quad \text{with} \quad h \cdot \|\omega\| = (\|\omega\| - \eta\, g)^+,$$

observe that

$$\|\Phi_\varepsilon * (h\,\omega)\| \leq \|\omega\| \quad \text{and} \quad \|\Phi_\varepsilon * (\omega - h\,\omega)\| \leq \eta\, g$$

for small $\varepsilon > 0$, hence infer

$$T(\omega) = T(h\,\omega) + T(\omega - h\,\omega) \leq \|S\|(\|\omega\|) + \eta\, \|S\|(g).$$

It follows that $\|S\|$ equals the Daniell integral λ defined in the first half of 2.5.12 (5); the second half is here a consequence of the first, because the argument of 2.5.13 can be applied with $g - h_n, c, \mu$ replaced by $\|\xi_n\|, \lambda(f), T$.

In case the above condition holds we call S a **distribution representable by integration;** the Radon measure associated with the Daniell integral $\|S\|$ will usually be denoted by the same symbol $\|S\|$, so that

$$S(\phi) = \int \langle \phi, k \rangle \, d\|S\| \quad \text{for } \phi \in \mathscr{D}(U, Y),$$

where k is a Y^* valued weakly $\|S\|$ measurable function with $\|k(x)\| = 1$ for $\|S\|$ almost all x in U. Moreover the symbol $S(\phi)$ will often be employed to designate the integral on the right even when $\phi \notin \mathscr{D}(U, Y)$, for example when $\phi \in \mathbf{L}_1(\|S\|, Y)$; by 4.1.2 this extension of S is uniquely determined by its continuity with respect to $\|S\|_{(1)}$, because $\mathscr{D}(U, Y)$ is $\|S\|_{(1)}$ dense in $\mathbf{L}_1(\|S\|, Y)$. From 2.8.18 and 2.9.8 we see that for $\|S\|$ almost all x in U the linear function $k(x)$ is given by the formula

$$\langle y, k(x) \rangle = \lim_{r \to 0+} S(b_{x,r}\, y) / \|S\|\, \mathbf{B}(x, r)$$

for $y \in Y$, where $b_{x,r}$ is the characteristic function of $\mathbf{B}(x, r)$; in fact, since both sides of this equation depend continuously on y, it suffices to verify the equation for all y in some countable dense subset of Y.

If μ is a Daniell integral on $\mathscr{K}(U)$ and $S = \mu | \mathscr{D}(U, \mathbf{R})$, then $S \in \mathscr{D}'(U, \mathbf{R})$ and $\|S\| = \mu^+ + \mu^-$ (see 2.5.6).

If $S_i \to S$ in $\mathscr{D}'(U, Y)$ as $i \to \infty$, then

$$\liminf_{i \to \infty} \|S_i\|(f) \geq \|S\|(f) \quad \text{for } f \in \mathscr{K}(U)^+,$$

$$\liminf_{i \to \infty} \|S_i\|(E) \geq \|S\|(E) \quad \text{for every open set } E \subset U.$$

If $M: \mathscr{K}(U)^+ \to \{t: 0 \leq t < \infty\}$ is monotone, then

$$\begin{aligned} &\mathscr{D}'(U, Y) \cap \{S: \|S\|(f) \leq M(f) \text{ whenever } f \in \mathscr{K}(U)^+\} \\ &= \bigcap_{\phi \in \mathscr{D}(U, Y)} \mathscr{D}'(U, Y) \cap \{S: S(\phi) \leq M(\|\phi\|)\} \end{aligned}$$

is compact.

4.1.6. A function mapping an open subset U of $\mathbf{R}^n$ into $\wedge^m(\mathbf{R}^n, W)$ is called a **differential form of degree** m **on** U **with coefficients in** W (see 1.4.1); in case $W=\mathbf{R}$ one speaks simply of a differential form of degree m on U.

If W is any vectorspace, so is $\wedge^m(\mathbf{R}^n, W)$, hence the set of all differential forms of degree m with coefficients in W is a vectorspace with respect to pointwise addition and scalar multiplication; in case W is an algebra we similarly use the alternating product of $\wedge^*(\mathbf{R}^n, W)$ constructed in 1.4.2 to define, for any two differential forms ϕ and ψ of degrees p and q on U with coefficients in W, the differential form $\phi \wedge \psi$ of degree $p+q$ with coefficients in W by the formula

$$(\phi \wedge \psi)(x) = \phi(x) \wedge \psi(x) \text{ whenever } x \in U.$$

We also write $E \wedge \psi = \phi \wedge \psi$ when $E \subset U$ and ϕ is the characteristic function of E.

Differential forms which are Baire functions are called **Baire forms.**

Functions mapping U into $\wedge_m \mathbf{R}^n$ are called m**-vectorfields on** U. If ξ is an m-vectorfield on U, ϕ is a differential form of degree k on U with coefficients in W, and $m \leq k$, then $\xi \,\lrcorner\, \phi$ is the differential form of degree $k-m$ on U with coefficients in W such that (see 1.5.1)

$$(\xi \,\lrcorner\, \phi)(x) = \xi(x) \,\lrcorner\, \phi(x) \text{ whenever } x \in U;$$

in case $m=k$ one usually denotes $\xi \,\lrcorner\, \phi$ by $\langle \xi, \phi \rangle$. When $m \geq k$ and $W=\mathbf{R}$ one similarly defines the $(m-k)$-vectorfield $\xi \,\llcorner\, \phi$ on U. In the same manner we use the inner products constructed in 1.7.5 to associate a real valued function with a pair of m-vectorfields on U, or a pair of differential forms of degree k, as well as to define the functions $|\xi|$ and $|\phi|$; we also apply 1.8.1 to obtain the functions $\|\xi\|$ and $\|\phi\|$.

The standard coordinate functions $X_1, \ldots, X_n$ on $\mathbf{R}^n$ [defined by $X_i(x) = x_i$ for $x = (x_1, \ldots, x_n) \in \mathbf{R}^n$] form a base of $\wedge^1 \mathbf{R}^n$ which is dual to the standard base $e_1, \ldots, e_n$ of $\mathbf{R}^n$, hence the products

$$X_\lambda = X_{\lambda(1)} \wedge \cdots \wedge X_{\lambda(m)} \quad \text{and} \quad e_\lambda = e_{\lambda(1)} \wedge \cdots \wedge e_{\lambda(m)}$$

corresponding to all $\lambda \in \Lambda(n, m)$ form dual bases of $\wedge^m \mathbf{R}^n$ and $\wedge_m \mathbf{R}^n$. We note that $DX_1, \ldots, DX_n$ are differential forms of degree 1 on $\mathbf{R}^n$ with

$$DX_i(x) = X_i \text{ whenever } x \in \mathbf{R}^n,$$

hence the differential forms

$$(DX)_\lambda = DX_{\lambda(1)} \wedge \cdots \wedge DX_{\lambda(m)}$$

of degree m on $\mathbf{R}^n$ satisfy the equations

$$(DX)_\lambda(x) = X_\lambda \text{ whenever } x \in \mathbf{R}^n.$$

We also define the m vectorfields $\mathbf{e}_\lambda$ on $\mathbf{R}^n$ by the formula

$$\mathbf{e}_\lambda(x) = e_\lambda \text{ whenever } x \in \mathbf{R}^n.$$

Then we find that, for every differential form ϕ of degree m on an open subset U of $\mathbf{R}^n$,

$$\phi = \sum_{\lambda \in \Lambda(n,m)} \langle \mathbf{e}_\lambda, \phi \rangle \wedge (DX)_\lambda.$$

This representation of ϕ (which historically motivated the term "differential form") remains valid when ϕ has coefficients in any algebra W with a unit (whose real multiples form a subalgebra isomorphic with $\mathbf{R}$); moreover every vectorspace W can be embedded in such an algebra, for example in $\bigotimes_* W$.

Recalling 1.5.2 we associate with every p-vectorfield ξ on U the differential form

$$\mathbf{D}_p\,\xi = \xi \,\lrcorner\, (DX_1 \wedge \cdots \wedge DX_n)$$

of degree $n-p$, and with every differential form ϕ of degree p on U the $(n-p)$-vectorfield

$$\mathbf{D}^p\,\phi = (\mathbf{e}_1 \wedge \cdots \wedge \mathbf{e}_n) \,\llcorner\, \phi.$$

Similarly we define $*\xi$ and $*\phi$ by applying 1.7.8 with $E = e_1 \wedge \cdots \wedge e_n$ to $\xi(x)$ and $\phi(x)$ for each $x \in U$.

We now assume that W is a normed vectorspace and use the induced norm on $\bigwedge^m(\mathbf{R}^n, W)$ constructed by the method of 3.1.11. For every differential form ϕ of degree m on U with coefficients in W, such that ϕ is differentiable at each point of U, we define the **exterior derivative**

$$d\phi$$

as the differential form of degree $m+1$ on U with coefficients in W given by the formula

$$\langle v_1 \wedge \cdots \wedge v_{m+1}, d\phi(x) \rangle = \sum_{i=1}^{m+1} \langle v^{(i)}, \langle v_i, D\phi(x) \rangle \rangle$$

with

$$v^{(i)} = (-1)^{i-1} v_1 \wedge \cdots \wedge v_{i-1} \wedge v_{i+1} \wedge \cdots \wedge v_{m+1}$$

whenever $x \in U$ and $v_1, \ldots, v_{m+1} \in \mathbf{R}^n$; we justify this definition by observing that for each $x \in U$ the above sum varies linearly with each v_j, and alternates upon interchange of v_j and v_{j+1}. The formula should be interpreted so that

$$d\phi = D\phi \textit{ if } \phi \textit{ has degree } 0.$$

For any differential forms ϕ and ψ on U, which are differentiable on U, the following four equations hold:

$$d\phi = \sum_{j=1}^{n} DX_j \wedge D_j \phi;$$

$$d(\phi+\psi) = d\phi + d\psi \text{ if } \phi \text{ and } \psi \text{ have equal degree;}$$

$$d(\phi \wedge \psi) = (d\phi) \wedge \psi + (-1)^{\text{degree}\,\phi}\, \phi \wedge d\psi;$$

$$d(d\phi) = 0 \text{ if } \phi \text{ is twice differentiable on } U.$$

We deduce the first equation from the definition of $d\phi$ by observing that

$$\langle v_1 \wedge \cdots \wedge v_{m+1}, d\phi(x)\rangle = \sum_{i=1}^{m+1} \Big\langle v^{(i)}, \sum_{j=1}^{n} \langle v_i, X_j\rangle D_j \phi(x)\Big\rangle$$

$$= \sum_{j=1}^{n} \sum_{i=1}^{m+1} \langle v_i, X_j\rangle \langle v^{(i)}, D_j \phi(x)\rangle = \sum_{j=1}^{n} \langle v_1 \wedge \cdots \wedge v_{m+1}, X_j \wedge D_j \phi(x)\rangle$$

because the shuffle of type $(1, m)$ mapping 1 onto i has index $(-1)^{i-1}$. Verification of the second equation is trivial. We derive the third and fourth equations from the first by computing

$$d(\phi \wedge \psi) = \sum_{j=1}^{n} DX_j \wedge D_j(\phi \wedge \psi) = \sum_{j=1}^{n} DX_j \wedge (D_j \phi \wedge \psi + \phi \wedge D_j \psi)$$

$$= \sum_{j=1}^{n} [DX_j \wedge D_j \phi \wedge \psi + (-1)^{\text{degree}\,\phi}\, \phi \wedge DX_j \wedge D_j \psi],$$

$$d(d\phi) = \sum_{k=1}^{n} DX_k \wedge D_k \sum_{j=1}^{n} DX_j \wedge D_j \phi = \sum_{k=1}^{n} \sum_{j=1}^{n} DX_k \wedge DX_j \wedge D_{k,j} \phi = 0$$

because $DX_k \wedge DX_j = -DX_j \wedge DX_k$ and $D_{k,j}\phi = D_{j,k}\phi$. Applying the first equation we also obtain the representation

$$d\phi = \sum_{j=1}^{n} DX_j \wedge \sum_{\lambda \in \Lambda(n,\, \text{degree}\,\phi)} D_j \langle \mathbf{e}_\lambda, \phi\rangle \wedge (DX)_\lambda$$

$$= \sum_{\lambda \in \Lambda(n,\, \text{degree}\,\phi)} D\langle \mathbf{e}_\lambda, \phi\rangle \wedge (DX)_\lambda .$$

We observe that the exterior derivative $d\phi$ may also be characterized by the condition that, for $x \in U$,

$d\phi(x)$ is the image of $D\phi(x)$ under the composition

$$\operatorname{Hom}(\mathbf{R}^n, \wedge^m \mathbf{R}^n) \simeq \wedge^1 \mathbf{R}^n \otimes \wedge^m \mathbf{R}^n \to \wedge^{1+m} \mathbf{R}^n$$

of the isomorphism described in 1.1.4 with the linear map induced by exterior multiplication.

Dualizing this procedure we associate with each differentiable p-vectorfield ξ on U the $(p-1)$-vectorfield

$$\operatorname{div}\xi,$$

called the **divergence** (or **interior derivative**) of ξ, by requiring that, for each $x \in U$,

$$\operatorname{div}\xi(x) \text{ is the image of } D\,\xi(x) \text{ under the composition}$$

$$\operatorname{Hom}(\mathbf{R}^n, \wedge_p \mathbf{R}^n) \simeq \wedge_p \mathbf{R}^n \otimes \wedge^1 \mathbf{R}^n \to \wedge_{p-1} \mathbf{R}^n$$

of an isomorphism obtained from 1.1.4 and 1.1.2 with the linear map induced by interior multiplication; thus we compute

$$\operatorname{div}\xi = \sum_{j=1}^{n} D_j\,\xi \llcorner D X_j.$$

We see that the operator $\mathbf{D}$ links exterior and interior differentiation by the equation

$$d(\mathbf{D}_p\,\xi) = (-1)^{n-p}\,\mathbf{D}_{p-1}(\operatorname{div}\xi)$$

because the last assertion in 1.5.2 implies

$$\sum_{j=1}^{n} \mathbf{D}_p(D_j\,\xi) \wedge D X_j = \sum_{j=1}^{n} \mathbf{D}_{p-1}(D_j\,\xi \llcorner D X_j).$$

Whenever f is a function mapping an open subset A of $\mathbf{R}^\nu$ into $\mathbf{R}^n$, f is differentiable at each point of A and the image of f is contained in the domain of the differential form ϕ of degree m with coefficients in W we define

$$f^{\#}\,\phi$$

as the differential form of degree m on A with values in W such that, for each $a \in A$, $(f^{\#}\,\phi)(a)$ is the image of $\phi[f(a)]$ under the linear map (see 1.4.1)

$$\wedge^m[D f(a), W]\colon \wedge^m(\mathbf{R}^n, W) \to \wedge^m(\mathbf{R}^\nu, W);$$

this implies in particular that

$$f^{\#}\,\phi = \phi \circ f \text{ if } \phi \text{ has degree } 0.$$

The operation $f^{\#}$ commutes with addition and alternating multiplication, because $\wedge^* D f(a)$ does for each $a \in A$. Furthermore

$$d(f^{\#}\,\phi) = f^{\#}(d\phi) \text{ if } f \text{ is twice differentiable.}$$

In case ϕ has degree 0 this equation merely reasserts that

$$D(\phi\circ f)(a)=D\phi[f(a)]\circ Df(a) \text{ whenever } a\in A;$$

in case $\phi=D\psi$ for some ψ of degree 0 we prove the equation by computing

$$d(f^{\#}d\psi)=d(df^{\#}\psi)=0=f^{\#}(d\,d\psi);$$

through addition and alternating multiplication an arbitrary form ϕ can be obtained from forms of the preceding two special types, namely $\langle \mathbf{e}_\lambda, \phi\rangle$ and DX_i; moreover we see from the formulae for the exterior derivatives of sums and products that the class of all forms ϕ satisfying the asserted equation is closed to addition and multiplication.

If A, B, C are open subsets of Euclidean spaces and $f: A\to B$, $g: B\to C$ are maps of class ∞, then

$$(g\circ f)^{\#}\phi=f^{\#}(g^{\#}\phi)$$

for every differential form ϕ on C.

4.1.7. For each open subset U of $\mathbf{R}^n$ and each nonnegative integer m we define

$$\mathscr{E}^m(U)=\mathscr{E}(U,\textstyle\bigwedge^m\mathbf{R}^n),\qquad \mathscr{E}_m(U)=\mathscr{E}'(U,\textstyle\bigwedge^m\mathbf{R}^n),$$

$$\mathscr{D}^m(U)=\mathscr{D}(U,\textstyle\bigwedge^m\mathbf{R}^n),\qquad \mathscr{D}_m(U)=\mathscr{D}'(U,\textstyle\bigwedge^m\mathbf{R}^n).$$

Thus $\mathscr{E}^m(U)$ is the vector space of all differential forms of degree m and class ∞ on U with real coefficients, and $\mathscr{D}^m(U)$ is the vector subspace consisting of those forms whose support is a compact subset of U. The members of $\mathscr{D}_m(U)$ are called m **dimensional currents in** U, and the image of $\mathscr{E}_m(U)$ in $\mathscr{D}_m(U)$ consists of all m dimensional currents with compact support in U.

In addition to the concepts applicable to distributions of arbitrary type discussed in 4.1.1–4.1.5, the following special operations for currents are constructed by dualizing some of the notions on differential forms and vectorfields described in 4.1.6:

Suppose $T\in\mathscr{D}_m(U)$. For $\phi\in\mathscr{E}^k(U)$ with $k\leq m$ we define

$$T\llcorner\phi\in\mathscr{D}_{m-k}(U),\quad (T\llcorner\phi)(\psi)=T(\phi\wedge\psi) \text{ whenever } \psi\in\mathscr{D}^{m-k}(U).$$

For each p-vectorfield ξ of class ∞ on U we define

$$T\wedge\xi\in\mathscr{D}_{m+p}(U),\quad (T\wedge\xi)(\psi)=T(\xi\,\lrcorner\,\psi) \text{ whenever } \psi\in\mathscr{D}^{m+p}(U).$$

The restriction that ϕ and ξ be functions of class ∞ is often needlessly restrictive; for instance in case T is representable by integration it suffices to assume that ϕ and ξ be $\|T\|$ summable over every compact subset of U, and in particular we let

$$T \llcorner A = T \llcorner (\text{characteristic function of } A) \in \mathscr{E}_m(U)$$

for every $\|T\|$ measurable set A. If $m \geq 1$ we define the **boundary** of T,

$$\partial T \in \mathscr{D}_{m-1}(U), \quad (\partial T)(\psi) = T(d\psi) \text{ whenever } \psi \in \mathscr{D}^{m-1}(U).$$

Assuming ϕ and ξ of class ∞ on U one readily verifies the eight equations:

$$\partial(\partial T) = 0 \text{ if } \dim T \geq 2;$$

$$(\partial T) \llcorner \phi = T \llcorner d\phi + (-1)^{\text{degree}\,\phi}\, \partial(T \llcorner \phi);$$

$$\partial T = -\sum_{j=1}^{n} (D_j T) \llcorner D X_j \text{ if } \dim T \geq 1;$$

$$T = \sum_{\lambda \in \Lambda(n,m)} [T \llcorner (DX)_\lambda] \wedge e_\lambda;$$

$$D_j(T \llcorner \phi) = (D_j T) \llcorner \phi + T \llcorner D_j \phi; \quad D_j(T \wedge \xi) = (D_j T) \wedge \xi + T \wedge D_j \xi;$$

$$(T \wedge \xi) \llcorner \phi = T \wedge (\xi \llcorner \phi) \text{ if } \dim T = 0 \text{ and degree } \phi \leq \dim \xi = p;$$

$$\partial(T \wedge \xi) = -T \wedge \operatorname{div} \xi - \sum_{j=1}^{n} D_j T \wedge (\xi \llcorner D X_j) \text{ if } \dim T = 0 \leq \dim \xi = p.$$

We will often use the same symbol to designate a Radon measure over U, the associated Daniell integral over $\mathscr{K}(U)$, and the 0 dimensional current obtained by restriction to $\mathscr{D}^0(U)$.

To each $\mathscr{L}^n$ measurable p-vectorfield ξ, such that

$$\textstyle\int_K \|\xi\|\, d\mathscr{L}^n < \infty \text{ for every compact subset } K \text{ of } \mathbf{R}^n,$$

there corresponds [as in 4.1.1 because $(\wedge^p \mathbf{R}^n)^* \simeq \wedge_p \mathbf{R}^n$] *the current* $\mathscr{L}^n \wedge \xi \in \mathscr{D}_p(\mathbf{R}^n)$ *defined by the formula*

$$(\mathscr{L}^n \wedge \xi)(\psi) = \int \langle \xi, \psi \rangle \, d\mathscr{L}^n \text{ for } \psi \in \mathscr{D}^p(\mathbf{R}^n).$$

Observing that $D_j \mathscr{L}^n = 0$ we find that, if ξ is of class 1, then

$$D_j(\mathscr{L}^n \wedge \xi) = \mathscr{L}^n \wedge D_j \xi, \quad \partial(\mathscr{L}^n \wedge \xi) = -\mathscr{L}^n \wedge \operatorname{div} \xi.$$

Moreover, if $\phi \in \mathscr{D}^{n-p}(\mathbf{R}^n)$ and $\psi \in \mathscr{D}^p(\mathbf{R}^n)$, then

$$(\mathscr{L}^n \wedge \mathbf{D}^{n-p} \phi)(\psi) = \int \langle \mathbf{D}^{n-p} \phi, \psi \rangle \, d\mathscr{L}^n = \int \langle e_{(1,\ldots,n)}, \phi \wedge \psi \rangle \, d\mathscr{L}^n.$$

We also note the commutative diagram:

$$\begin{array}{ccccc}
\mathscr{E}^{n-p}(\mathbf{R}^n) & \xrightarrow{\mathbf{D}^{n-p}} & \mathscr{E}(\mathbf{R}^n, \wedge_p \mathbf{R}^n) & \xrightarrow{\mathscr{L}^n \wedge} & \mathscr{D}_p(\mathbf{R}^n) \\
\big\downarrow {\scriptstyle (-1)^{n-p} d} & & \big\downarrow {\scriptstyle \operatorname{div}} & & \big\downarrow {\scriptstyle -\partial} \\
\mathscr{E}^{n-p+1}(\mathbf{R}^n) & \xrightarrow{\mathbf{D}^{n-p+1}} & \mathscr{E}(\mathbf{R}^n, \wedge_{p-1} \mathbf{R}^n) & \xrightarrow{\mathscr{L}^n \wedge} & \mathscr{D}_{p-1}(\mathbf{R}^n).
\end{array}$$

We abbreviate

$$\mathbf{E}^n = \mathscr{L}^n \wedge \mathbf{e}_1 \wedge \cdots \wedge \mathbf{e}_n \in \mathscr{D}_n(\mathbf{R}^n),$$

so that $\mathbf{E}^n(\phi) = \int \langle e_1 \wedge \cdots \wedge e_n, \phi(x) \rangle \, d\mathscr{L}^n x$ for $\phi \in \mathscr{D}^n(\mathbf{R}^n)$; clearly

$$D_j \mathbf{E}^n = 0 \text{ for } j = 1, \ldots, n \text{ and } \partial \mathbf{E}^n = 0.$$

We note that, since $\bigwedge^{n+1} \mathbf{R}^n = \{0\}$, 1.5.3 implies

$$(\partial T) \wedge \mathbf{e}_j = (-1)^n D_j T \text{ for } T \in \mathscr{D}_n(U) \text{ and } j = 1, \ldots, n.$$

Applying 4.1.4 with $Y = \bigwedge^n \mathbf{R}^n$, and observing that for each $\alpha \in Y^*$ there exists a $c \in \mathbf{R}$ such that

$$\langle y, \alpha \rangle = c \langle e_1 \wedge \cdots \wedge e_n, y \rangle \text{ whenever } y \in \textstyle\bigwedge^n \mathbf{R}^n,$$

we obtain the following **constancy theorem**:

If $T \in \mathscr{D}_n(U)$, A is a connected open subset of U and

$$\operatorname{spt} \partial T \subset U \sim A,$$

then there exists a real number c such that

$$\operatorname{spt}(T - c\,\mathbf{E}^n \llcorner U) \subset U \sim A.$$

In dealing with any p dimensional current S in U which is representable by integration, we will apply the concepts of 4.1.5 with $\bigwedge^m \mathbf{R}^n = Y$ and $\bigwedge_m R^n \simeq Y^*$ normed by mass and comass according to 1.8.1. We let

$$\vec{S}$$

be the $\|S\|$ measurable function with values in $\bigwedge_m \mathbf{R}^n$ defined by the formula

$$\langle \vec{S}(x), y \rangle = \lim_{r \to 0+} S(b_{x,r}\, y) / \|S\| \, \mathbf{B}(x, r)$$

for all $y \in \bigwedge^m \mathbf{R}^n$, where $b_{x,r}$ is the characteristic function of $\mathbf{B}(x, r)$; thus $x \in \operatorname{dmn} \vec{S}$ if and only if the above limit exists for each $y \in \bigwedge^m \mathbf{R}^n$. Then

$$\|\vec{S}(x)\| = 1 \text{ for } \|S\| \text{ almost all } x \text{ in } U,$$

$$S(\psi) = \int \langle \vec{S}, \psi \rangle \, d\|S\| = \|S\| \, (\vec{S} \lrcorner \psi)$$

for $\psi \in \mathscr{D}^m(U)$, hence

$$S = \|S\| \wedge \vec{S}.$$

It follows that

$$\|\gamma \wedge \xi\| = \gamma \wedge \|\xi\|$$

whenever γ is a Radon measure over U and ξ is a γ measurable p-vectorfield such that $\int_K \|\xi\| \, d\gamma < \infty$ for every compact subset K of U; in fact 2.5.9 implies the existence of a bounded nonnegative γ measurable function k such that

$$\|\gamma \wedge \xi\| = \gamma \wedge \|\xi\| \wedge k,$$

hence $\gamma \wedge \xi = \gamma \wedge \|\xi\| \wedge k \wedge \overrightarrow{(\gamma \wedge \xi)}$ and, for γ almost all x,

$$\xi(x) = \|\xi(x)\| \, k(x) \overrightarrow{(\gamma \wedge \xi)}(x), \qquad \|\xi(x)\| = \|\xi(x)\| \, k(x).$$

For example $\|\mathbf{E}^n\| = \mathscr{L}^n$ and $\overrightarrow{\mathbf{E}^n} = \mathbf{e}_1 \wedge \cdots \wedge \mathbf{e}_n$.

We define the **comass** of a differential form ϕ of degree m on U as

$$\mathbf{M}(\phi) = \sup\{\|\phi(x)\| : x \in U\},$$

and the **mass** of an m dimensional current T in U as

$$\mathbf{M}(T) = \sup\{T(\phi) : \phi \in \mathscr{D}^m(U) \text{ and } \mathbf{M}(\phi) \leq 1\}.$$

If $\mathbf{M}(T) < \infty$, then T is representable by integration. If T is representable by integration, then $\mathbf{M}(T) = \|T\|(U)$.

Assuming $T \in \mathscr{D}_m(U)$, we call T **locally normal** if and only if T is representable by integration and either ∂T is representable by integration or $m = 0$. Furthermore call T **normal** if and only if T is locally normal and spt T is compact. We also define

$$\mathbf{N}(T) = \mathbf{M}(T) + \mathbf{M}(\partial T) \text{ in case } m > 0, \qquad \mathbf{N}(T) = \mathbf{M}(T) \text{ in case } m = 0.$$

If $\mathbf{N}(T) < \infty$, then T is locally normal. If T is normal, then $\mathbf{N}(T) < \infty$. We let

$$\mathbf{N}_m(U) \quad \text{and} \quad \mathbf{N}_m^{\mathrm{loc}}(U)$$

be the vectorspaces of all m dimensional normal and locally normal currents in U, respectively. For each compact subset K of U we define

$$\mathbf{N}_{m,K}(U) = \mathbf{N}_m(U) \cap \{T : \operatorname{spt} T \subset K\}.$$

Clearly $T \in \mathbf{N}_m^{\mathrm{loc}}(U)$ if and only if $T \llcorner \gamma \in \mathbf{N}_m(U)$ for every $\gamma \in \mathscr{D}^0(U)$, and also if and only if for every $x \in U$ there exists an $S \in \mathbf{N}_m(U)$ with $x \notin \operatorname{spt}(T - S)$.

We note that $(\partial T)(\psi) = T(d\psi)$ in case $T \in \mathbf{N}_m(U)$, $m > 0$ and ψ is a differential form of degree $m-1$ and class 1 on U, because exterior differentiation commutes with smoothing.

Whenever U, V are open subsets of Euclidean spaces and

$$f: U \to V \text{ is of class } \infty, \quad T \in \mathscr{D}_m(U), \; f|\operatorname{spt} T \text{ is proper},$$

we define

$$f_\# T \in \mathscr{D}_m(V)$$

so as to satisfy the equation

$$(f_\# T)(\phi) = T[\gamma \wedge (f^\# \phi)]$$

for all $\phi \in \mathscr{D}^m(V)$ and $\gamma \in \mathscr{D}^0(U)$ with

$$(\operatorname{spt} T) \cap f^{-1}(\operatorname{spt} \phi) \subset \operatorname{Int}\{x: \gamma(x) = 1\}.$$

Under these conditions it follows that

$$\operatorname{spt}(f_\# T) \subset f(\operatorname{spt} T), \quad \partial(f_\# T) = f_\#(\partial T),$$

$$f_\#(T \llcorner f^\# \psi) = (f_\# T) \llcorner \psi \text{ for } \psi \in \mathscr{E}^k(V) \text{ with } m \geq k.$$

If, also, T is representable by integration, then $f_\#(\|T\| \llcorner \|(\wedge_m Df)\vec{T}\|)$ is a Radon measure over V and $f_\# T$ is representable by integration with

$$\|f_\# T\| \leq f_\#(\|T\| \llcorner \|(\wedge_m Df)\vec{T}\|),$$

because $T(f^\# \phi) = \|T\| \langle (\wedge_m Df)\vec{T}, \phi \circ f \rangle$ for $\phi \in \mathscr{D}^m(V)$. Therefore $f_\# T$ is (locally) normal whenever T is (locally) normal (and $f|\operatorname{spt} T$ is proper); in this case $f_\#(T \llcorner f^\# \psi) = (f_\# T) \llcorner \psi$ for every bounded Baire form ψ of degree $k \leq m$ on V, hence

$$f_\#[T \llcorner f^{-1}(B)] = (f_\# T) \llcorner B \text{ for every Borel subset } B \text{ of } V.$$

From the last paragraph of 4.1.6 we see that

$$(g \circ f)_\# T = g_\#(f_\# T) \text{ if } g \circ f|\operatorname{spt} T \text{ is proper.}$$

The definition of $f_\# T$ will be extended in 4.1.14 and 4.2.2. Here we illustrate the difficulty in weakening the hypothesis that $f|\operatorname{spt} T$ be proper by the *example when f is the inclusion map of $U = \{x: 0 < x < \infty\}$ into* $\mathbf{R}$, $0 < b < \infty$ *and*

$$T \in \mathscr{D}_1(U), \quad T(\phi) = \int_0^b \langle 1, \phi(x) \rangle \, d\mathscr{L}^1 x \text{ for } \phi \in \mathscr{D}^1(U);$$

recalling 2.5.19 we find that

$$\partial T = \delta_b^U \in \mathscr{D}_0(U), \quad f_\#(\delta_b^U) = \delta_b^{\mathbf{R}} \in \mathscr{D}_0(\mathbf{R})$$

and the only $S \in D_1(\mathbf{R})$ with $\partial S = f_\#(\partial T)$ is given by the formula

$$S(\psi) = \int_{-\infty}^b \langle 1, \psi(x) \rangle \, d\mathscr{L}^1 x \text{ for } \psi \in \mathscr{D}^1(\mathbf{R});$$

inasmuch as $\operatorname{spt} S \not\subset f(\operatorname{spt} T)$, no satisfactory definition of $f_\# T$ exists in this case, even though $\mathbf{N}(T) < \infty$ and f is Lipschitzian.

4.1.8. Suppose A and B are open subsets of $\mathbf{R}^m$ and $\mathbf{R}^n$,

$$p\colon A\times B\to A \text{ and } q\colon A\times B\to B,$$
$$p(a,b)=a \text{ and } q(a,b)=b \text{ whenever } (a,b)\in A\times B.$$

We will prove that *for any* $S\in\mathscr{D}_i(A)$ *and* $T\in\mathscr{D}_j(B)$ *there exists one and only one current*

$$S\times T\in\mathscr{D}_{i+j}(A\times B),$$

called the **cartesian product of** S **and** T, *satisfying the condition: If* $\alpha\in\mathscr{D}^k(A)$ *and* $\beta\in\mathscr{D}^{i+j-k}(B)$, *then*

$$(S\times T)(p^{\#}\alpha\wedge q^{\#}\beta)=S(\alpha)\,T(\beta) \text{ in case } k=i,$$
$$(S\times T)(p^{\#}\alpha\wedge q^{\#}\beta)=0 \text{ in case } k\neq i.$$

Observing that

$$\textstyle\bigwedge^{i+j}(\mathbf{R}^m\times\mathbf{R}^n)=\bigoplus_{k=0}^{i+j}(\operatorname{im}\bigwedge^k p)\wedge(\operatorname{im}\bigwedge^{i+j-k} q)$$

and recalling 4.1.3, we see that the above differential forms $p^{\#}\alpha\wedge q^{\#}\beta$ generate a dense vectorsubspace of $\mathscr{D}^{i+j}(A\times B)$, hence $S\times T$ is uniquely characterized by our condition.

To prove the existence of $S\times T$ we use the maps

$$P\colon \mathbf{R}^m\to\mathbf{R}^m\times\mathbf{R}^n,\quad P(a)=(a,0) \text{ for } a\in\mathbf{R}^m,$$
$$Q\colon \mathbf{R}^n\to\mathbf{R}^m\times\mathbf{R}^n,\quad Q(b)=(0,b) \text{ for } b\in\mathbf{R}^n.$$

Whenever $\phi\in\mathscr{D}^{i+j}(A\times B)$ we define $\phi_S\in\mathscr{D}^j(B)$ so that

$$\langle\eta,\phi_S(b)\rangle=S_a(\textstyle\bigwedge^i P)\,[(\bigwedge_j Q)\,\eta\;\lrcorner\;\phi(a,b)]$$

for $b\in B$ and $\eta\in\bigwedge_j\mathbf{R}^n$, then let

$$(S\times T)(\phi)=T(\phi_S);$$

observing that $D\,p(a)=P^*$ for $a\in A$, $D\,q(b)=Q^*$ for $b\in B$, $P^*\circ P$ and $Q^*\circ Q$ are identity maps (see 1.7.4), one readily checks that $S\times T$ satisfies the required condition.

From the characterizing condition one infers

$$\operatorname{spt}(S\times T)=\operatorname{spt}S\times\operatorname{spt}T,$$
$$\partial(S\times T)=(\partial S)\times T+(-1)^i S\times\partial T \text{ if } i>0<j,$$
$$\partial(S\times T)=(\partial S)\times T \text{ if } i>j=0,\qquad \partial(S\times T)=S\times\partial T \text{ if } i=0<j.$$

The cartesian product of currents is a bilinear operation, and is associative (compare 2.6.5). Moreover it is anticommutative in the sense that

$$r_\#(T\times S)=(-1)^{ij}S\times T$$

where $r\colon B\times A\to A\times B$, $r(b,a)=(a,b)$ for $(b,a)\in B\times A$; in fact

$$(-1)^{ij}r_\#(T\times S)(p^\#\alpha\wedge q^\#\beta)=(T\times S)[(q\circ r)^\#\beta\wedge(p\circ r)^\#\alpha]=T(\beta)\,S(\alpha)$$

whenever $\alpha\in\mathscr{D}^i(A)$ and $\beta\in\mathscr{D}^j(B)$.

In case $\operatorname{spt} S$ is compact, then

$$q_\#(S\times T)=0 \text{ if } i>0,\qquad q_\#(S\times T)=S(1)\,T \text{ if } i=0.$$

If S and T are representable by integration, then

$$(S\times T)(\phi)=\int\langle\zeta,\phi\rangle\,d(\|S\|\times\|T\|)$$

whenever $\phi\in\mathscr{D}^{i+j}(A\times B)$, where

$$\zeta(a,b)=(\textstyle\bigwedge_i P)\,\vec{S}(a)\wedge(\textstyle\bigwedge_j Q)\,\vec{T}(b)$$

for $\|S\|\times\|T\|$ almost all (a,b), with $\|\zeta(a,b)\|\le 1$ according to 1.8.1. We infer that $S\times T$ is representable by integration and

$$\|S\times T\|\le\|S\|\times\|T\|.$$

In case either $\vec{S}(a)$ or $\vec{T}(b)$ is simple for $\|S\|\times\|T\|$ almost all (a,b), hence $\|\zeta(a,b)\|=1$ according to 1.8.4, we obtain the equations

$$\|S\times T\|=\|S\|\times\|T\|,$$

$$\overrightarrow{S\times T}(a,b)=\zeta(a,b) \text{ for } \|S\times T\| \text{ almost all } (a,b).$$

I do not know whether these equations are always true (without simplicity assumptions on $\vec{S}$ or $\vec{T}$); this problem is equivalent to the problem at the end of 1.8.4. However we are at least assured that always

$$\binom{i+j}{i}\|S\times T\|\ge\|S\|\times\|T\|,$$

because

$$\|p^\#\alpha\wedge q^\#\beta\|\le\binom{i+j}{i}(\|\alpha\|\circ p)(\|\beta\|\circ q)$$

whenever $\alpha\in\mathscr{D}^i(A)$ and $\beta\in\mathscr{D}^j(B)$, by 1.8.1.

Whenever A contains the line segment with endpoints u and v we define

$$[u,v]\in\mathscr{D}_1(A),$$
$$[u,v](\alpha)=\int_0^1\langle v-u,\alpha[(1-t)u+tv]\rangle\, d\mathscr{L}^1 t$$

for $\alpha\in\mathscr{D}^1(A)$. Clearly $[v,u]=-[u,v]$. Recalling 2.5.19 we compute

$$[u,v](D\psi)=\psi(v)-\psi(u)=\boldsymbol{\delta}_v(\psi)-\boldsymbol{\delta}_u(\psi)$$

for $\psi\in\mathscr{D}^0(A)$, hence

$$\partial[u,v]=\boldsymbol{\delta}_v-\boldsymbol{\delta}_u.$$

We infer that, for $T\in\mathscr{D}_j(B)$,

$$\partial([u,v]\times T)=\boldsymbol{\delta}_v\times T-\boldsymbol{\delta}_u\times T-[u,v]\times\partial T \text{ if } j>0,$$

$$\partial([u,v]\times T)=\boldsymbol{\delta}_v\times T-\boldsymbol{\delta}_u\times T \text{ if } j=0.$$

We also note that, whenever $z\in A$,

$$\boldsymbol{\delta}_z\times T=\varepsilon_{z\#}T \quad\text{with}\quad \varepsilon_z(b)=(z,b) \text{ for } b\in B,$$

because $\phi_{\boldsymbol{\delta}_z}=\varepsilon_z^{\#}\phi$ for $\phi\in\mathscr{D}^j(A\times B)$. Furthermore

$$\langle\eta,\phi_{[u,v]}(b)\rangle=\int_0^1\langle P(v-u)\wedge(\textstyle\bigwedge_j Q)\eta,\phi[(1-t)u+tv,b]\rangle\, d\mathscr{L}^1 t$$

whenever $\phi\in\mathscr{D}^{1+j}(A\times B)$, $b\in B$ and $\eta\in\bigwedge_j\mathbf{R}^n$.

From the above boundary formulae we infer by induction with respect to ν that

$$\partial([u_1,v_1]\times\cdots\times[u_\nu,v_\nu])=\sum_{k=1}^{\nu}(-1)^k[u_1,v_1]\times\cdots$$
$$\times[u_{k-1},v_{k-1}]\times(\boldsymbol{\delta}_{u_k}-\boldsymbol{\delta}_{v_k})\times[u_{k+1},v_{k+1}]\times\cdots\times[u_\nu,v_\nu]$$

whenever, for each $k\in\{1,\ldots,\nu\}$, u_k and v_k belong to the same Euclidean space; this result is Stokes' Theorem for the case of an **oriented rectangular parallelepiped.**

We note that if $-\infty<u_k<v_k<\infty$ for $k=1,\ldots,\nu$, then

$$[u_1,v_1]\times\cdots\times[u_\nu,v_\nu]=\mathbf{E}^\nu \llcorner \{x\colon u_k<x_k<v_k \text{ for } k=1,\ldots,\nu\};$$

this follows from the general facts on cartesian products because

$$[u_k,v_k](\alpha)=\int_{u_k}^{v_k}\langle 1,\alpha(s)\rangle\, d\mathscr{L}^1 s \text{ for } \alpha\in\mathscr{D}^1(\mathbf{R}),$$

hence $[u_k,v_k]=\mathbf{E}^1\llcorner\{s\colon u_k<s<v_k\}$ for $k=1,\ldots,\nu$.

4.1.9. Suppose U is an open subset of $\mathbf{R}^n$, V is an open subset of $\mathbf{R}^\nu$, f and g are functions mapping U into V. A **homotopy of class r from f to g** is a map

$$h: A \times U \to V$$

of class r such that A is an open subinterval of $\mathbf{R}$, $0 \in A$, $1 \in A$,

$$h(0,x)=f(x) \quad \text{and} \quad h(1,x)=g(x) \text{ for } x \in U.$$

Whenever $t \in A$ we define

$$h_t: U \to V, \quad h_t(x)=h(t,x) \text{ for } x \in U,$$

hence $h_0=f$ and $h_1=g$; in case $r \geq 1$ we also define

$$\dot{h}_t: U \to \mathbf{R}^\nu, \quad \dot{h}_t(x)=\langle (1,0), Dh(t,x)\rangle \text{ for } x \in U,$$

hence $\langle (v,w), Dh(t,x)\rangle = v\,\dot{h}_t(x)+\langle w, D h_t(x)\rangle$ for $t \in A$, $x \in U$, $v \in \mathbf{R}$, $w \in \mathbf{R}^n$.

Assuming $r=\infty$, $T \in \mathscr{D}_j(U)$ *and* $h|\{t: 0 \leq t \leq 1\} \times \operatorname{spt} T$ *is proper* we use 4.1.7 and 4.1.8 to define the h **deformation chain of T** as

$$h_\#([0,1] \times T) \in \mathscr{D}_{1+j}(V),$$

and compute its boundary to obtain the **homotopy formulae for currents:**

$$g_\# T - f_\# T = \partial h_\#([0,1] \times T) + h_\#([0,1] \times \partial T) \text{ if } j>0;$$

$$g_\# T - f_\# T = \partial h_\#([0,1] \times T) \text{ if } j=0.$$

Recalling the existence proof in 4.1.8 we may restate the preceding formulae by saying that, whenever $\psi \in \mathscr{D}^j(V)$,

$$T(g^\# \psi - f^\# \psi) = T([h^\#(d\psi)]_{[0,1]} + d[(h^\# \psi)_{[0,1]}])$$

if $j>0$; the second summand on the right must be omitted if $j=0$. Applying these equations for each $x \in U$ and each $\eta \in \wedge_j R^n$ to the particular current T such that $T(\beta)=\langle \eta, \beta(x)\rangle$ for $\beta \in \mathscr{D}^j(U)$, we obtain the **homotopy formulae for differential forms:**

$$g^\# \psi - f^\# \psi = [h^\#(d\psi)]_{[0,1]} + d[(h^\# \psi)_{[0,1]}] \text{ if } j>0,$$

$$g^\# \psi - f^\# \psi = [h^\#(d\psi)]_{[0,1]} \text{ if } j=0.$$

If T is representable by integration, then

$$[\wedge_{1+j} Dh(t,x)]\overrightarrow{[0,1] \times T}(t,x) = \dot{h}_t(x) \wedge [\wedge_j D h_t(x)]\,\vec{T}(x)$$

for $\mathscr{L}^1 \times \|T\|$ almost all (t,x) in $\{t: 0 \leq t \leq 1\} \times U$, whence it follows that

$$\|h_\#([0,1] \times T)\| \leq \int_0^1 h_{t\,\#}(\|T\| \llcorner \|\dot{h}_t \wedge (\wedge_j D h_t)\,\vec{T}\|)\, d\mathscr{L}^1 t.$$

We will often use the **affine homotopy** h **from** f **to** g defined by the formula

$$h(t,x)=(1-t)f(x)+t\,g(x) \text{ for } (t,x)\in A\times U.$$

In this case $\dot{h}_t=g-f$ and $h_t=(1-t)f+t\,g$ for $t\in A$. If T is representable by integration, then

$$\begin{aligned}\mathbf{M}[h_\#([0,1]\times T)] &\le \textstyle\int_0^1 \|T\|(|g-f|\cdot[(1-t)\|Df\|+t\|Dg\|]^j)\,d\mathscr{L}^1 t\\ &=\|T\|\Big(|g-f|\sum_{k=0}^{j}\|Df\|^k\|Dg\|^{j-k}/(j+1)\Big)\\ &\le \|T\|(|g-f|\sup\{\|Df\|^j,\|Dg\|^j\}).\end{aligned}$$

4.1.10. Here we consider for each $u\in\mathbf{R}^n$ *the affine homotopy* h *from the constant map* f *with value* u *to the identity map* g *of* $\mathbf{R}^n$. *Whenever* $\psi\in\mathscr{D}_j(\mathbf{R}^n)$ *with* $j>0$ we find that $g^\#\psi=\psi$ and $f^\#\psi=0$; therefore

$$d\psi=0 \quad \textit{implies} \quad \psi=d[(h^\#\psi)_{[0,1]}].$$

Whenever $T\in\mathscr{D}_j(\mathbf{R}^n)$ *and* spt T *is compact* we compute $g_\# T=T$, $f_\# T=0$ if $j>0$ and $f_\# T=T(1)\,\boldsymbol{\delta}_u$ if $j=0$; since $\mathscr{D}_{n+1}(\mathbf{R}^n)=0$ we conclude:

$$\partial T=0 \text{ implies } T=\partial h_\#([0,1]\times T) \text{ if } 0<j<n;$$
$$\partial T=0 \text{ implies } T=0 \text{ if } j=n;$$
$$T=T(1)\,\boldsymbol{\delta}_u+\partial h_\#([0,1]\times T) \text{ if } j=0.$$

The preceding argument remains valid with $\mathbf{R}^n$ replaced by any open subset U such that $u\in U$ *and, for every* $v\in U$, U *contains the line segment from* u *to* v. However we will see that no similar conclusions need to hold for an arbitrary open subset U of $\mathbf{R}^n$; one measures the extent of this failure by the homology groups of U (defined in 4.4).

4.1.11. Here we use the map $F:\mathbf{R}^n\times\mathbf{R}\times\mathbf{R}^n\to\mathbf{R}^n$ such that

$$F(x,t,y)=(1-t)\,x+t\,y \text{ whenever } x,y\in\mathbf{R}^n \text{ and } t\in\mathbf{R}$$

to define, for any currents $S\in\mathscr{D}_i(\mathbf{R}^n)$ and $T\in\mathscr{D}_j(\mathbf{R}^n)$ with compact supports, the **join of** S **and** T as

$$S \times\!\!\!\times\, T=F_\#(S\times[0,1]\times T)\in\mathscr{D}_{i+1+j}(\mathbf{R}^n).$$

Since $S \times \delta_t \times T = G_{t\#}(S \times T)$ with $G_t(x, y) = (x, t, y)$, and since $F \circ G_0$ and $F \circ G_1$ are the standard projections of $\mathbf{R}^n \times \mathbf{R}^n$ onto $\mathbf{R}^n$, we find that

$$\partial(S \times T) = (\partial S) \times T - (-1)^i S \times \partial T \quad \text{if } i > 0 < j,$$
$$\partial(S \times T) = (\partial S) \times T - (-1)^i T(1) S \quad \text{if } i > 0 = j,$$
$$\partial(S \times T) = S(1) T - S \times \partial T \quad \text{if } i = 0 < j,$$
$$\partial(S \times T) = S(1) T - T(1) S \quad \text{if } i = 0 = j.$$

The anticommutativity of the cartesian product implies

$$T \times S = (-1)^{(i+1)(j+1)} S \times T;$$

in fact, if $\rho(x, t, y) = (y, 1-t, x)$ for $x, y \in \mathbf{R}^n$ and $t \in \mathbf{R}$, then $F = F \circ \rho$ and

$$T \times [0, 1] \times S = (-1)^{ij+i+j+1} \rho_\#(S \times [0, 1] \times T).$$

We also observe that (see 2.7.16)

$$A_\#(S \times T) = A_\#(S) \times A_\#(T) \text{ for every affine map } A,$$

because $A \circ F = F \circ (A \times \mathbf{1}_{\mathbf{R}} \times A)$.

If S and T are representable by integration, then

$$\begin{aligned}\mathbf{M}(S \times T) &\le \int\int_0^1\int \|(1-t)^i \vec{S}(x) \wedge (y-x) \wedge t^j \vec{T}(y)\| \, d\|S\| \, x \, d\mathscr{L}^1 \, t \, d\|T\| \, y \\ &= [i!\,j!/(i+j+1)!] \, \|S\|_x \, \|T\|_y \, \|\vec{S}(x) \wedge (y-x) \wedge \vec{T}(y)\|.\end{aligned}$$

If $u \in \mathbf{R}^n$, then

$$\delta_u \times T = h_\#([0, 1] \times T)$$

where h is the affine homotopy from the constant map with value u to the identity map of $\mathbf{R}^n$, because $F(u, t, y) = h(t, y)$ for $t \in \mathbf{R}$ and $y \in \mathbf{R}^n$; in case T is representable by integration, then

$$\mathbf{M}(\delta_u \times T) \le \|T\|_y \, \|(y-u) \wedge \vec{T}(y)\|/(j+1).$$

Using the alternate notation

$$[u] = \delta_u \text{ whenever } u \in \mathbf{R}^n,$$

and observing that

$$[u, v] = \delta_u \times \delta_v \text{ whenever } u, v \in \mathbf{R}^n,$$

we inductively define the m dimensional **oriented simplex**

$$[u_0, \ldots, u_m] = [u_0] \times [u_1, \ldots, u_m] \in \mathscr{D}_m(\mathbf{R}^n)$$

for any $m+1$ termed sequence of points $u_0, \ldots, u_m \in \mathbf{R}^n$. We compute

$$\begin{aligned}\partial[u_0, \ldots, u_m] &= [u_1, \ldots, u_m] - [u_0] \times \partial[u_1, \ldots, u_m] \\ &= \sum_{k=0}^{m} (-1)^k [u_0, \ldots, u_{k-1}, u_{k+1}, \ldots, u_m].\end{aligned}$$

We also note that

$$A_\#[u_0, \ldots, u_m] = [A(u_0), \ldots, A(u_m)]$$

for every affine map A. Therefore every m dimensional oriented simplex in any Euclidean space is representable as affine image of any nondegenerate m dimensional oriented simplex in $\mathbf{R}^m$. Next we will prove that

$$[u_0, u_1, u_2, \ldots, u_m] + [u_1, u_0, u_2, \ldots, u_m] = 0.$$

For $m=1$ we know this from 4.1.8. For $m>1$ we inductively verify that the above sum has boundary 0; we infer from 4.1.10 that the sum equals zero in case $u_0, \ldots, u_m$ belong to $\mathbf{R}^m$; then we use an affine map to show that the sum equals zero in case $u_0, \ldots, u_m$ belong to any Euclidean space. It follows from the above equation that

$$[u_{\sigma(0)}, \ldots, u_{\sigma(m)}] = \operatorname{index}(\sigma)[u_0, \ldots, u_m]$$

for every permutation σ of $\{0, \ldots, m\}$. By a similar inductive argument (applying ∂ and using 4.1.10) one obtains the formula for the simplicial decomposition of an **oriented simplicial prism:**

$$[a, b] \times [u_0, \ldots, u_m] = \sum_{k=0}^{m} (-1)^k [(a, u_0), \ldots, (a, u_k), (b, u_k), \ldots, (b, u_m)].$$

Through repeated application of this formula one can represent any m dimensional oriented rectangular parallelepiped as sum of $m!$ oriented simplexes, whose vertexes are among the vertexes of the parallelepiped.

The associative law

$$(R \times S) \times T = R \times (S \times T)$$

holds whenever R, S, T are currents with compact support in $\mathbf{R}^n$. To prove this we define

$$H: \mathbf{R}^n \times \mathbf{R}^n \times \mathbf{R}^n \times \mathbf{R}^3 \to \mathbf{R}^n, \quad g: \mathbf{R}^2 \to \mathbf{R}^3, \quad f: \mathbf{R}^2 \to \mathbf{R}^3,$$

$$H(x, y, z, w) = w_1 x + w_2 y + w_3 z, \quad g(t) = (1-t_1, t_1(1-t_2), t_1 t_2)$$

$$f(t) = ((1-t_1)(1-t_2), t_1(1-t_2), t_2),$$

for $x, y, z \in \mathbf{R}^n$, $w \in \mathbf{R}^3$, $t \in \mathbf{R}^2$ and verify that

$$F[F(x, t_1, y), t_2, z] = H[x, y, z, f(t)], \quad F[x, t_1, F(y, t_2, z)] = H[x, y, z, g(t)];$$

we also let v_1, v_2, v_3 be the standard base of $\mathbf{R}^3$ and observe that

$$f_\#([0, 1] \times [0, 1]) = [v_1, v_2] \times [v_3] = [v_3, v_1, v_2]$$

$$= [v_1, v_2, v_3] = [v_1] \times [v_2, v_3] = g_\#([0, 1] \times [0, 1]).$$

4.1.12. Suppose U is an open subset of $\mathbf{R}^n$. For each compact subset K of U we recall from 4.1.1 that

$$\nu_K^0(\phi)=\sup\{\|\phi(x)\|: x\in K\}$$

whenever ϕ is a differential form of class ∞ on U, and define the **flat seminorm**

$$\mathbf{F}_K(\phi)=\sup\{\nu_K^0(\phi), \nu_K^0(d\phi)\}.$$

For every real valued linear function T on $\mathscr{D}^m(U)$ we let

$$\mathbf{F}_K(T)=\sup\{T(\phi): \phi\in\mathscr{D}^m(U) \text{ and } \mathbf{F}_K(\phi)\le 1\}$$

be the dual flat seminorm, and observe:

If $\mathbf{F}_K(T)<\infty$, *then* $T\in\mathscr{D}_m(U)$ *with* spt $T\subset K$.

$\mathscr{D}_m(U)\cap\{T: \mathbf{F}_K(T)<\infty\}$ *is* $\mathbf{F}_K$ *complete.*

If $T\in\mathscr{D}_m(U)$, *then* $\mathbf{F}_K(\partial T)\le\mathbf{F}_K(T)$; *moreover*

$$\mathbf{F}_{K\cap \operatorname{spt}\gamma}(T\llcorner\gamma)\le[\nu_K^0(\gamma)+\nu_K^0(d\gamma)]\,\mathbf{F}_K(T) \text{ whenever } \gamma\in\mathscr{D}^0(U).$$

Next we prove that, *for every* $T\in\mathscr{D}_m(U)$ *with* spt $T\subset K$, $\mathbf{F}_K(T)$ *is the least member* (possibly ∞) *of the set*

$$\{\mathbf{M}(T-\partial S)+\mathbf{M}(S): S\in\mathscr{D}_{m+1}(U) \text{ with spt } S\subset K\}.$$

For any $S\in\mathscr{D}_{m+1}(U)$ with spt $S\subset K$

$$T(\phi)=(T-\partial S)(\phi)+S(d\phi)\le\mathbf{M}(T-\partial S)+\mathbf{M}(S)$$

whenever $\phi\in\mathscr{D}^m(U)$ with $\mathbf{F}_K(\phi)\le 1$, hence

$$\mathbf{F}_K(T)\le\mathbf{M}(T-\partial S)+\mathbf{M}(S).$$

In order to show that equality holds for some S we assume $\mathbf{F}_K(T)<\infty$, endow the vectorspace $P=\mathscr{D}^m(U)\times\mathscr{D}^{m+1}(U)$ with the seminorm

$$\nu(\phi,\psi)=\sup\{\nu_K^0(\phi), \nu_K^0(\psi)\} \text{ for } (\phi,\psi)\in P,$$

and consider the linear monomorphism

$$Q: \mathscr{D}^m(U)\to P, \quad Q(\phi)=(\phi, d\phi) \text{ for } \phi\in\mathscr{D}^m(U).$$

Since $\nu[Q(\phi)]=\mathbf{F}_K(\phi)$ for $\phi\in\mathscr{D}^m(U)$, we see that

$$T\circ Q^{-1}\le\mathbf{F}_K(T)\,\nu|\operatorname{im} Q$$

and use the Hahn-Banach theorem to obtain a real valued linear function L on P such that

$$L|\operatorname{im} Q=T\circ Q^{-1} \quad\text{and}\quad L\le\mathbf{F}_K(T)\,\nu.$$

Then we define the linear functions R and S by the formulae

$$R(\phi)=L(\phi,0) \text{ for } \phi\in\mathscr{D}^m(U), \qquad S(\psi)=L(0,\psi) \text{ for } \psi\in\mathscr{D}^{m+1}(U),$$

observe that $\nu(\phi,\psi)\leq 1$ implies $R(\phi)+S(\psi)=L(\phi,\psi)\leq\mathbf{F}_K(T)$, hence $R\in\mathscr{D}_m(U)$, $S\in\mathscr{D}_{m+1}(U)$,

$$\mathbf{M}(R)+\mathbf{M}(S)\leq\mathbf{F}_K(T) \quad\text{with}\quad \operatorname{spt} R\cup\operatorname{spt} S\subset K,$$

and that $T(\phi)=L(\phi,d\phi)=R(\phi)+S(d\phi)$ for every $\phi\in\mathscr{D}^m(U)$, hence $T=R+\partial S$. (We also note that *if* $\mathbf{M}(T)<\infty$, *then* S *is normal.*)

With each compact subset K of U we associate the vectorspace

$$\mathbf{F}_{m,K}(U)=\text{the } \mathbf{F}_K \text{ closure of } \mathbf{N}_{m,K}(U) \text{ in } \mathscr{D}_m(U),$$

which is complete relative to $\mathbf{F}_K$. Then we define

$$\mathbf{F}_m(U)$$

as the union of the vectorspaces $\mathbf{F}_{m,K}(U)$ corresponding to all compact subsets K of U. The members of $\mathbf{F}_m(U)$ are called m dimensional **flat chains** in U. We also let

$$\mathbf{F}_m^{\text{loc}}(U)$$

be the subset of $\mathscr{D}_m(U)$ consisting of all T such that

$$T\llcorner\gamma\in\mathbf{F}_m(U) \text{ for every } \gamma\in\mathscr{D}^0(U);$$

such currents are termed **locally flat**. We note that $T\in\mathbf{F}_m^{\text{loc}}(U)$ if and only if for every $x\in U$ there exists an $S\in\mathbf{F}_m(U)$ with $x\notin\operatorname{spt}(T-S)$; this may be verified with the aid of a suitable partition of unity on U.

If $T\in\mathbf{F}_{m,K}(U)$ *with* $m>0$, *then* $\partial T\in\mathbf{F}_{m-1,K}(U)$.

If $T\in\mathbf{F}_{m,K}(U)$ *and* $\gamma\in\mathscr{D}^0(U)$, *then*

$$T\llcorner\gamma\in\mathbf{F}_{m,K\cap\operatorname{spt}\gamma}(U).$$

For every neighborhood Z *of* spt T *in* U we may choose γ so that $\operatorname{spt}\gamma\subset Z$ and $\operatorname{spt} T\subset\operatorname{Int}\{x:\gamma(x)=1\}$, hence $T=T\llcorner\gamma$, and infer that

$$T\in\mathbf{F}_{m,K\cap\operatorname{Clos} Z}(U).$$

However *it can happen that* $T\notin\mathbf{F}_{m,\operatorname{spt} T}(U)$, *for instance when*

$$T=\partial\sum_{k=1}^{\infty}[2^{-2k},2^{1-2k}]\in\mathbf{F}_0(\mathbf{R});$$

anticipating 4.1.18 we see that the support of this particular current T does not contain the support of any nonzero 1 dimensional normal current, hence $\mathbf{F}_{0,\operatorname{spt} T}(\mathbf{R})$ equals the $\mathbf{M}$ closure of $\mathbf{N}_{0,\operatorname{spt} T}(\mathbf{R})$, whereas $\mathbf{M}(T)=\infty$.

An example in 4.1.15 will show that not all 1 dimensional currents with compact support and finite mass are flat chains.

Next we recall 4.1.8 and assume $S\in\mathscr{D}_i(A)$, $T\in\mathscr{D}_j(B)$, C is a compact subset of A, D is a compact subset of B. One readily verifies that

$$\mathbf{F}_{C\times D}(S\times T)\le\mathbf{N}(S)\,\mathbf{F}_D(T) \text{ if } \operatorname{spt} S\subset C,$$

$$\mathbf{F}_C(S)\,\mathbf{F}_D(T)\le\binom{i+j+2}{i+1}\mathbf{F}_{C\times D}(S\times T).$$

If $S\in\mathbf{N}_{i,C}(A)$ and $T\in\mathbf{F}_{j,D}(B)$, then $S\times T\in\mathbf{F}_{i+j,C\times D}(A\times B)$. However the cartesian product of two flat chains need not be flat, as shown by the example where

$$S=\sum_{k=1}^{\infty} k[u_k,v_k]\in\mathbf{F}_1(\mathbf{R}),\qquad T=\partial S\in\mathbf{F}_0(\mathbf{R}),$$

$$u_1=0,\quad u_{k+1}=u_k+4k^{-3} \text{ and } v_k=u_k+2k^{-3} \text{ for } k=1,2,3,\ldots.$$

Clearly

$$\operatorname{spt} S\subset\left\{x\colon 0\le x\le\sum_{k=1}^{\infty}4k^{-3}\right\},\qquad \mathbf{M}(S)=2\sum_{k=1}^{\infty}k^{-2}<\infty.$$

We choose $\phi_k\in\mathscr{D}^0(\mathbf{R})$ with $\operatorname{spt}\phi_k\subset\{x\colon u_k<x<u_{k+1}\}$,

$$\mathbf{M}(\phi_k)=\phi_k(v_k)=1 \quad\text{and}\quad \mathbf{M}(d\phi_k)\le k^3,$$

define $\psi_m\in\mathscr{D}^0(\mathbf{R}^2)$ for $m=1,2,3,\ldots$ by the formula

$$\psi_m(x,y)=\sum_{k=1}^{m}k^{-3}\,\phi_k(x)\,\phi_k(y) \text{ for } (x,y)\in\mathbf{R}^2,$$

then observe that $\mathbf{M}(\psi_m)\le 1$ and $\mathbf{M}(d\psi_m)\le 2$. Since

$$(T\times T)(\psi_m)=\sum_{k=1}^{m}k^{-3}\,T(\phi_k)^2=\sum_{k=1}^{m}k^{-1}\to\infty \text{ as } m\to\infty,$$

we infer $T\times T$ is not a flat chain. Finally the equation $\partial(S\times T)=T\times T$ shows that $S\times T$ is not a flat chain.

4.1.13. The importance of the flat seminorms for the theory of normal currents stems mainly from the following proposition: *If $T\in\mathbf{N}_m(U)$, f and g are functions of class ∞ mapping U into V with*

$$C=\{(1-t)\,f(x)+t\,g(x)\colon x\in\operatorname{spt} T,\ 0\le t\le 1\}\subset V,$$

and if $\rho(x)=\sup\{\|Df(x)\|,\ \|Dg(x)\|\}$ for $x\in U$, then

$$\mathbf{F}_C(g_\# T-f_\# T)\le\|T\|\,(|g-f|\,\rho^m)+\|\partial T\|\,(|g-f|\,\rho^{m-1})$$

in case $m>0$, and $\mathbf{F}_C(g_\# T-f_\# T)\le\|T\|\,(|g-f|)$ in case $m=0$. To prove this we replace U by a suitable neighborhood of $\operatorname{spt} T$, so that the affine homotopy from $f|U$ to $g|U$ can be defined with values in V, recall the homotopy formulae for currents, observe that C contains the supports of the deformation chains of T and ∂T, and estimate the masses of both deformation chains by means of inequalities at the end of 4.1.9.

4.1.14. Here we assume U and V are open subsets of Euclidean spaces and $f: U \to V$ is a *locally Lipschitzian* map. We will define

$$f_\# T \in \mathbf{F}_{m,f(K)}(V) \text{ for } T \in \mathbf{F}_{m,K}(U),$$

where K is any compact subset of U, so as to satisfy the convergence condition

$$\lim_{i\to\infty} \mathbf{F}_E(f_\# T - g_{i\,\#} S) = 0$$

whenever E is a compact subset of V with $f(K) \subset \operatorname{Int} E$, Z is a neighborhood of K in U, ζ is the inclusion map of Z into U,

$$S \in \mathbf{F}_{m,K}(Z) \quad \textit{with} \quad \zeta_\# S = T,$$

and g_i: $Z \to V$ are maps of class ∞ with

$$c = \sup \bigcup_{i=1}^{\infty} \{\|D g_i(x)\| : x \in K\} < \infty,$$

$$\lim_{i\to\infty} \sup \{|f(x) - g_i(x)| : x \in K\} = 0.$$

In case $\operatorname{Clos} Z$ is a compact subset of U, such approximating functions g_i can be constructed by smoothing f, because the restriction of f to some neighborhood of $\operatorname{Clos} Z$ is Lipschitzian, hence the maps $(\Phi_\varepsilon * f)|Z$ corresponding to all sufficiently small positive numbers ε have a common Lipschitz constant. Therefore $f_\# T$ is uniquely characterized by the above convergence condition. We establish the existence of $f_\# T$ in two steps as follows:

First we suppose $T \in \mathbf{N}_{m,K}(U)$. For any approximating functions g_i we can apply 4.1.13 with f, g replaced by g_i, g_j and with

$$\{t\, g_i(x) + (1-t)\, g_j(x) : x \in K,\ 0 \le t \le 1\} \subset E$$

when i, j are large, to infer that the currents $g_{i\,\#} S$ form an $\mathbf{F}_E$ Cauchy sequence, which is $\mathbf{N}$ bounded because the differentials $D g_i$ are uniformly bounded on $\operatorname{spt} T$; accordingly the sequence of currents $g_{i\,\#} S$ has an $\mathbf{F}_E$ limit in $\mathbf{N}_{m,E}(V)$. This limit is independent of all the choices involved in its construction, because any two neighborhoods of K may be replaced by their intersection and any two sequences of approximating functions may be jointly rearranged into a single approximating sequence. Recalling 2.10.43 we see that

$$\mathbf{M}(f_\# T) \le \operatorname{Lip}(f|K)^m \mathbf{M}(T).$$

Next we observe that the operator $f_\#$ constructed in the preceding paragraph defines a linear map of $\mathbf{N}_{m,K}(U)$ into $\mathbf{N}_{m,f(K)}(V)$ which satisfies, as a consequence of 4.1.12, the flat boundedness condition

$$\mathbf{F}_{f(K)}(f_\# T) \leq \sup\{\mathrm{Lip}(f|K)^m, \mathrm{Lip}(f|K)^{m+1}\}\, \mathbf{F}_K(T)$$

for all $T \in \mathbf{N}_{m,K}(U)$. This map can be uniquely extended to a linear map of $\mathbf{F}_{m,K}(U)$ into $\mathbf{F}_{m,f(K)}(V)$, also denoted $f_\#$, which satisfies the same flat boundedness condition for all $T \in \mathbf{F}_{m,K}(U)$. To verify that the convergence condition holds when $T \in \mathbf{F}_{m,K}(U)$ we note that for every $\varepsilon > 0$ there exists

$$R \in \mathbf{N}_{m,K}(Z) \quad \text{with} \quad \mathbf{F}_K(S-R) \leq \varepsilon,$$

hence $\mathbf{F}_E(f_\# \zeta_\# R - g_{i\#} R) \to 0$ as $i \to \infty$, and the upper limit of the $\mathbf{F}_E$ norm of

$$f_\# T - g_{i\#} S = f_\# \zeta_\# (S-R) + (f_\# \zeta_\# R - g_{i\#} R) + g_{i\#}(R-S)$$

does not exceed

$$\varepsilon \sup\{\mathrm{Lip}(f|K)^m, \mathrm{Lip}(f|K)^{m+1}\} + \varepsilon \sup\{c^m, c^{m+1}\}.$$

We also define

$$f_\# T \in \mathbf{F}_m^{\mathrm{loc}}(V) \text{ whenever } T \in \mathbf{F}_m^{\mathrm{loc}}(U) \text{ and } f|\mathrm{spt}\, T \text{ is proper,}$$

so as to satisfy the equation

$$(f_\# T)(\phi) = [f_\#(T \llcorner \gamma)](\phi)$$

for all $\phi \in \mathscr{D}^m(V)$ and $\gamma \in \mathscr{D}^0(U)$ with

$$\mathrm{spt}\, T \cap f^{-1}(\mathrm{spt}\, \phi) \subset \mathrm{Int}\{x : \gamma(x) = 1\}.$$

One readily verifies that *the basic formal properties of $f_\# T$ generalize from infinitely differentiable to locally Lipschitzian maps f, provided T is locally flat and $f|\mathrm{spt}\, T$ is proper.* In particular $f_\#$ is a linear operator,

$$\mathrm{spt}(f_\# T) \subset f(\mathrm{spt}\, T), \qquad \partial(f_\# T) = f_\#(\partial T),$$

the homotopy formula holds for every locally Lipschitzian homotopy whose restriction to $\{t : 0 \leq t \leq 1\} \times \mathrm{spt}\, T$ is proper, and $(g \circ f)_\# = g_\# \circ f_\#$ when g is a locally Lipschitzian function on V. Using the fact that differentiation of f commutes with smoothing one also obtains the proposition:

If $T \in \mathbf{F}_m(U)$, $\mathbf{M}(T) < \infty$, $U \subset \mathbf{R}^n$ and σ is a continuous function on U with $\|Df(x)\| \leq \sigma(x)$ for $\mathscr{L}^n$ almost all x in U, then

$$\|f_\# T\| \leq f_\#(\|T\| \llcorner \sigma^m);$$

if also $g: U \to V$ *is locally Lipschitzian with* $\|Dg(x)\| \le \sigma(x)$ *for* $\mathscr{L}^n$ *almost all* x *in* U,

$$\{(1-t)f(x)+t\,g(x): 0 \le t \le 1,\ x \in \operatorname{spt} T\} \subset V$$

and h *is the affine homotopy from* f *to* g, *then*

$$\mathbf{M}[h_\# (\llbracket 0,1 \rrbracket \times T)] \le \|T\| (|g-f|\,\sigma^m);$$

in case T *is normal the conclusions of* 4.1.13 *hold with* ρ *replaced by* σ.

We note that $(f_\# T)(\phi) = T(f^\# \phi)$ in case f is of class 1, $T \in \mathbf{F}_m(U)$, $\mathbf{M}(T) < \infty$ and ϕ is a continuous differential form of degree m on V.

4.1.15. If $0 < r < \infty$ and $\Psi_r: \mathbf{R}^\nu \to \mathbf{R}^\nu$,

$$\Psi_r(y) = 0 \text{ for } |y| < r, \qquad \Psi_r(y) = (1 - r/|y|)\,y \text{ for } |y| \ge r,$$

then $|\Psi_r(y) - y| \le r$ for $y \in \mathbf{R}^\nu$ and $\operatorname{Lip}(\Psi_r) \le 1$. In fact, for any a and b in $\mathbf{R}^\nu$ with $|a| \ge r$ and $|b| \ge r$,

$$|\Psi_r(a) - \Psi_r(b)|^2 = (r-|a|)^2 + (r-|b|)^2 - 2(1-r/|a|)(1-r/|b|)\,a \bullet b,$$

hence

$$\begin{aligned} &|\Psi_r(a) - \Psi_r(b)|^2 - |a-b|^2 \\ &= 2r^2 - 2r\,|a| - 2r\,|b| - 2(r^2 - r\,|a| - r\,|b|)\,\frac{a}{|a|} \bullet \frac{b}{|b|} \\ &= 2r(r-|a|-|b|)\left(1 - \frac{a}{|a|} \bullet \frac{b}{|b|}\right) \le 0. \end{aligned}$$

If f and g are locally Lipschitzian functions mapping an open subset U of $\mathbf{R}^n$ into an open subset V of $\mathbf{R}^\nu$, then

$$G_r = f + \Psi_r \circ (g-f): U \to \mathbf{R}^\nu$$

with $|G_r(x) - g(x)| \le r$ for $x \in U$, $G_r(x) = f(x)$ in case $|g(x) - f(x)| \le r$. Smoothing f and G_r we find that $\Phi_\varepsilon * f$ and $\Phi_\varepsilon * G_r$ agree on the open set

$$Z_{r,\varepsilon} = U \cap \{z: |f(x) - g(x)| < r \text{ for } x \in \mathbf{B}(z,\varepsilon)\}.$$

Moreover $\Phi_\varepsilon * f$ and $\Phi_\varepsilon * G_r$ approximate f and g uniformly on each compact subset K of U, provided ε and r are sufficiently small, and the differentials of these functions are uniformly bounded on K. We conclude:

If $T \in \mathbf{F}_{m,K}(U)$, *then*

$$f \,|\, \operatorname{spt} T = g \,|\, \operatorname{spt} T \textit{ implies } f_\# T = g_\# T.$$

In fact for every $r > 0$ it is true that $\operatorname{spt} T \subset Z_{r,\varepsilon}$ provided ε is sufficiently small, hence $(\Phi_\varepsilon * f)_\# T = (\Phi_\varepsilon * G_r)_\# T$.

We obtain the corollary: *If there exists a Lipschitzian map retracting K onto a subset C, then*

$$\mathbf{F}_{m,K}(U) \cap \{T: \operatorname{spt} T \subset C\} = \mathbf{F}_{m,C}(U).$$

In fact by 2.10.43 the given retraction can be extended to a Lipschitzian function f mapping a neighborhood Z of K in U into U, which agrees on C with the inclusion map g of Z into U. In particular, *if $u \in U$, then*

$$T = 0 \text{ whenever } m > 0,\ T \in \mathbf{F}_m(U) \text{ and } \operatorname{spt} T \subset \{u\},$$

because $f_\# T = 0$ for the retraction f of U onto $\{u\}$; recalling 4.1.10 we also find that

$$T = T(1)\,\delta_u \text{ whenever } T \in \mathbf{F}_0(U) \text{ and } \operatorname{spt} T \subset \{u\}.$$

Another corollary is the following proposition: *If $f: U \to V$ and $g: V \to U$ are locally Lipschitzian with $g \circ f = \mathbf{1}_U$, then*

$$\begin{aligned}\mathbf{F}_m(V) \cap \{S: \operatorname{spt} S \subset \operatorname{im} f\} &= \mathbf{F}_m(V) \cap \{S: S = f_\#(g_\# S)\}\\ &= \{f_\# T: T \in \mathbf{F}_m(U)\},\end{aligned}$$

because $f \circ g$ retracts V onto $\operatorname{im} f$. This implies that no $m-1$ dimensional affine subspace of $\mathbf{R}^n$ contains the support of a nonzero m dimensional flat chain; however a much more precise result will be obtained in 4.1.20.

The above properties of flat chains are not shared by all currents of finite mass, as shown by the example where

$$T = \delta_0 \wedge \mathbf{e}_1 \in \mathscr{D}_1(\mathbf{R}), \qquad T(\phi) = \langle 1, \phi(0) \rangle \text{ for } \phi \in \mathscr{D}^1(\mathbf{R}),$$

$$f(x) = 0 \text{ and } g(x) = x \text{ for } x \in \mathbf{R},$$

hence $\operatorname{spt} T = \{0\}$, $\mathbf{M}(T) = 1$, $(\partial T)\psi = \psi'(0)$ for $\psi \in \mathscr{D}^0(\mathbf{R})$, $\mathbf{M}(\partial T) = \infty$, $f_\# T = 0$ and $g_\# T = T$. We also observe that this particular current T does not satisfy the convergence condition of 4.1.14, because if $\psi_i \in \mathscr{D}^0(\mathbf{R})$ converge to 0 uniformly on $\mathbf{R}$ with uniformly bounded differentials, but the numbers $\psi_i'(0)$ do not converge, then the currents $\psi_{i\#} T$ do not converge in $\mathscr{D}_0(\mathbf{R})$.

4.1.16. *If C is a nonempty closed convex subset of R^n, then*

$$f = \{(x, y): x \in \mathbf{R}^n,\ y \in C,\ |x - y| = \operatorname{dist}(x, C)\}$$

is a Lipschitzian function retracting $\mathbf{R}^n$ onto C, with $\operatorname{Lip}(f) \le 1$. In fact, if $(x, y) \in f$ and $(u, v) \in f$, then

$$\begin{aligned}|x-y|^2 &\le |x - [(1-t)y + t v]|^2\\ &= |x-y|^2 + 2t(x-y) \bullet (y-v) + t^2 |y-v|^2\end{aligned}$$

whenever $0 \le t \le 1$, hence

$$(x-y) \bullet (y-v) \ge 0; \quad \text{similarly} \quad (u-v) \bullet (v-y) \ge 0;$$

adding these two inequalities we conclude

$$(x-u) \bullet (y-v) \ge (y-v) \bullet (y-v), \qquad |x-u| \ge |y-v|.$$

(In [F 15, 4.8] it is shown how this property of convex sets generalizes to sets with positive reach.)

Applying 4.1.14, 4.1.15 we infer that *if* $T \in \mathbf{F}_m(\mathbf{R}^n)$ *and* C *is the convex hull of* spt T, *then* $T \in \mathbf{F}_{m,C}(\mathbf{R}^n)$ *and*

$$\mathbf{F}_K(T) = \mathbf{F}_C(T) \text{ for every compact } K \text{ containing } C.$$

4.1.17. Here we observe that

$$\mathbf{F}_{m,K}(U) \cap \{T\colon \mathbf{M}(T) < \infty\} = \text{the } \mathbf{M} \text{ closure of } \mathbf{N}_{m,K}(U) \text{ in } \mathscr{D}_m(U).$$

In fact, whenever $T \in \mathbf{F}_{m,K}(U)$, $\mathbf{M}(T) < \infty$ and $\varepsilon > 0$, we can successively choose

$$Q \in \mathbf{N}_{m,K}(U) \quad \text{and} \quad S \in \mathscr{D}_{m+1}(U) \quad \text{with} \quad \operatorname{spt} S \subset K,$$

$$\varepsilon > \mathbf{F}_K(T-Q) = \mathbf{M}(T-Q-\partial S) + \mathbf{M}(S),$$

hence infer $Q - \partial S \in \mathbf{N}_{m,K}(U)$ with $\mathbf{M}(T-Q-\partial S) < \varepsilon$.

Also, *if* $T \in \mathbf{F}_{m,K}(U)$, $\mathbf{M}(T) < \infty$ *and* $m \ge i \ge 0$, *then* $T \llcorner \phi \in \mathbf{F}_{m-i,K}(U)$ *for every* $\|T\|$ *summable differential form* ϕ *of degree* i *on* U, because $\mathscr{D}^i(U)$ is $\|T\|_{(1)}$ dense in $\mathbf{L}_1(\|T\|, \wedge^i \mathbf{R}^n)$.

4.1.18. Recalling 4.1.2 and 2.7.16 we see that

$$\Phi_\varepsilon * \phi = \int \Phi_\varepsilon(-z)\, \tau_z^\# \phi \, d\mathscr{L}^n z, \qquad T_\varepsilon(\phi) = \int \Phi_\varepsilon(-z) (\tau_{z\#} T) \phi \, d\mathscr{L}^n z$$

whenever $\phi \in \mathscr{D}^m(\mathbf{R}^n)$ and $T \in \mathscr{D}_m(\mathbf{R}^n)$. This motivates us to associate with each $z \in \mathbf{R}^n$ the affine homotopy

$$H^z\colon \mathbf{R} \times \mathbf{R}^n \to \mathbf{R}^n, \qquad H^z(t,x) = x + t z \text{ for } (t,x) \in \mathbf{R} \times \mathbf{R}^n,$$

from the identity map of $\mathbf{R}^n$ to τ_z, and to define the **smoothing deformation chain** $H_\varepsilon T \in \mathscr{D}_{m+1}(\mathbf{R}^n)$ so that

$$(H_\varepsilon T) \psi = \int \Phi_\varepsilon(-z) [H^z_\# ([0,1] \times T)] \psi \, d\mathscr{L}^n z$$

whenever $\psi \in \mathscr{D}^{m+1}(\mathbf{R}^n)$. We infer the **smoothing homotopy formulae for currents:**

$$T_\varepsilon - T = \partial(H_\varepsilon T) + H_\varepsilon(\partial T) \text{ if } 0 < m < n,$$

$$T_\varepsilon - T = \partial(H_\varepsilon T) \text{ if } m = 0, \qquad T_\varepsilon - T = H_\varepsilon(\partial T) \text{ if } m = n.$$

Moreover $\mathbf{M}(T_\varepsilon) \le \mathbf{M}(T)$, $\mathbf{M}(H_\varepsilon T) \le \varepsilon\, \mathbf{M}(T)$, $\partial T_\varepsilon = (\partial T)_\varepsilon$,

$$\operatorname{spt} T_\varepsilon \cup \operatorname{spt} H_\varepsilon T \subset C(\varepsilon) = \{x \colon \operatorname{dist}(x, \operatorname{spt} T) \le \varepsilon\},$$

$$\mathbf{F}_{C(\varepsilon)}(T_\varepsilon - T) \le \varepsilon\, \mathbf{N}(T) \text{ if } \operatorname{spt} T \text{ is compact}.$$

In case T is representable by integration the Daniell integrals $\|T_\varepsilon\|$ on $\mathscr{K}(\mathbf{R}^n)$ converge weakly (see 2.5.9) *to* $\|T\|$, because

$$\|T_\varepsilon\| \le \int \Phi_\varepsilon(-z)\, \tau_{z\,\#} \|T\|\, d\mathscr{L}^n z;$$

given $f \in \mathscr{K}(\mathbf{R}^n)^+$ and $\delta > 0$ we can choose $g \in \mathscr{K}(\mathbf{R}^n)^+$ so that

$$\|T\|(g) < \|T\|(f) + \delta \quad \text{and} \quad f \circ \tau_z < g \text{ for } z \text{ near } 0 \text{ in } \mathbf{R}^n,$$

and infer $\|T_\varepsilon\|(f) \le \|T\|(g)$ for small ε, hence

$$\|T\| f \le \liminf_{\varepsilon \to 0+} \|T_\varepsilon\| f \le \limsup_{\varepsilon \to 0+} \|T_\varepsilon\| f \le \delta + \|T\| f.$$

Similarly one can derive smoothing homotopy formulae for differential forms (compare 4.1.9), to represent $\Phi_\varepsilon * \phi - \phi$.

Next we suppose U is an open subset of $\mathbf{R}^n$ and let Ω_m be the class of all m-vectorfields ξ on U such that $\operatorname{spt} \xi$ *is a compact subset of U and ξ is $\mathscr{L}^n$ summable over U.* Assigning to each (ξ, η) in the vectorspace $\Omega_m \times \Omega_{m+1}$ the norm

$$\int_U (\|\xi\| + \|\eta\|)\, d\mathscr{L}^n,$$

we exhibit $\mathbf{F}_m(U)$ as a quotient space of $\Omega_m \times \Omega_{m+1}$ through the following two propositions:

If $\xi \in \Omega_m$, $\eta \in \Omega_{m+1}$, $T = \mathscr{L}^n \wedge \xi + \partial(\mathscr{L}^n \wedge \eta) \in \mathscr{D}_m(U)$, $0 < \delta < \infty$ *and* $K = \{x \colon \operatorname{dist}(x, \operatorname{spt} \xi \cup \operatorname{spt} \eta) \le \delta\} \subset U$, *then*

$$T \in \mathbf{F}_{m,K}(U) \quad \textit{and} \quad \mathbf{F}_K(T) \le \int_U (\|\xi\| + \|\eta\|)\, d\mathscr{L}^n.$$

If K is a compact subset of U, $T \in \mathbf{F}_{m,K}(U)$, $0 < \delta < \infty$ and

$$E = \{x \colon \operatorname{dist}(x, K) \le \delta\} \subset U,$$

then there exist $\xi \in \Omega_m$ and $\eta \in \Omega_{m+1}$ such that

$$T = \mathscr{L}^n \wedge \xi + \partial(\mathscr{L}^n \wedge \eta), \qquad \operatorname{spt} \xi \cup \operatorname{spt} \eta \subset E$$

and

$$\int_U (\|\xi\| + \|\eta\|)\, d\mathscr{L}^n \le \mathbf{F}_K(T) + \delta.$$

The second conclusion of the first proposition is trivial, and its first conclusion is clear in case ξ and η are of class ∞; moreover (see 2.7.18)

$$\int_U (\|\Phi_\varepsilon * \xi - \xi\| + \|\Phi_\varepsilon * \eta - \eta\|)\, d\mathscr{L}^n \to 0 \quad \text{as} \quad \varepsilon \to 0+.$$

Observing that the second proposition is trivial when $m>n$, we prove it by *downward induction with respect to m*. We let $T_0=T$, $K_0=K$ and choose $R_i, S_i, \varepsilon_i, K_i, \xi_i, T_i$ for every positive integer i so that

$$R_i\in\mathbf{N}_{m,K_{i-1}}(U),\quad S_i\in\mathbf{N}_{m+1,K_{i-1}}(U),\quad \mathbf{F}_{K_{i-1}}(T_{i-1}-R_i-\partial S_i)<2^{-i-3}\delta,$$
$$\mathbf{M}(R_i)+\mathbf{M}(S_i)<\mathbf{F}_{K_{i-1}}(T_{i-1})+2^{-i-2}\delta,\ 0<\varepsilon_i\le 2^{-i-1}\delta,\ \varepsilon_i\mathbf{N}(R_i)<2^{-i-3}\delta,$$
$$K_i=\mathbf{R}^n\cap\{x:\operatorname{dist}(x,K_{i-1})\le 2^{-i-1}\delta\}\subset U,$$
$$(R_i)_{\varepsilon_i}\text{ is obtained by smoothing } R_i,$$
$$(R_i)_{\varepsilon_i}=\mathscr{L}^n\wedge\xi_i,\quad \xi_i\in\mathscr{D}(U,\wedge_m\mathbf{R}^n),\quad \operatorname{spt}\xi_i\subset K_i,$$
$$T_i=T_{i-1}-\mathscr{L}^n\wedge\xi_i-\partial S_i\in\mathbf{F}_{m,K_i}(U),$$

hence $\int_U\|\xi_i\|\,d\mathscr{L}^n\le\mathbf{M}(R_i)$ and

$$\begin{aligned}\mathbf{F}_{K_i}(T_i)&\le\mathbf{F}_{K_i}(T_{i-1}-R_i-\partial S_i)+\mathbf{F}_{K_i}(R_i-\mathscr{L}^n\wedge\xi_i)\\&<2^{-i-3}\delta+\varepsilon_i\mathbf{N}(R_i)<2^{-i-2}\delta.\end{aligned}$$

Letting $C=\mathbf{R}^n\cap\{x:\operatorname{dist}(x,K)\le\delta/2\}\subset U$ we infer

$$\sum_{i=1}^{\infty}[\textstyle\int_U\|\xi_i\|\,d\mathscr{L}^n+\mathbf{M}(S_i)]\le\sum_{i=1}^{\infty}[\mathbf{M}(R_i)+\mathbf{M}(S_i)]$$
$$\le\sum_{i=1}^{\infty}[\mathbf{F}_{K_{i-1}}(T_{i-1})+2^{-i-2}\delta]\le\mathbf{F}_K(T)+2\sum_{i=1}^{\infty}2^{-i-2}\delta=\mathbf{F}_K(T)+\delta/2,$$
$$\xi=\sum_{i=1}^{\infty}\xi_i\in\Omega_m,\quad \operatorname{spt}\xi\subset C,\quad S=\sum_{i=1}^{\infty}S_i\in\mathbf{F}_{m+1,C}(U),$$
$$\textstyle\int_U\|\xi\|\,d\mathscr{L}^n+\mathbf{M}(S)\le\mathbf{F}_K(T)+\delta/2,$$
$$T=T_i+\sum_{j=1}^{i}(\mathscr{L}^n\wedge\xi_j+\partial S_j)\text{ for every } i,\quad T=\mathscr{L}^n\wedge\xi+\partial S.$$

Moreover, by downward induction, there exist $\eta\in\Omega_{m+1}$ and $\zeta\in\Omega_{m+2}$ such that

$$S=\mathscr{L}^n\wedge\eta+\partial(\mathscr{L}^n\wedge\zeta),\quad \operatorname{spt}\eta\cup\operatorname{spt}\zeta\subset E$$

and

$$\textstyle\int_U(\|\eta\|+\|\zeta\|)\,d\mathscr{L}^n\le\mathbf{M}(S)+\delta/2.$$

Since $\partial S=\partial(\mathscr{L}^n\wedge\eta)$ the required representation of T results.

In the special case when $m=n$ the above propositions show that $\mathbf{F}_n(U)$ *consists of the currents* $\mathscr{L}^n\wedge\xi$ *corresponding to all* $\mathscr{L}^n$ *summable n-vectorfields* ξ *on U with compact support.*

Returning to the general case we note that

$$(\Omega_m \times \Omega_{m+1}) \cap \{(\xi, \eta) : \mathscr{L}^n \wedge \xi + \partial(\mathscr{L}^n \wedge \eta) = 0\}$$
$$= \text{Closure}\,\{(\operatorname{div} \eta, \eta) : \eta \in \mathscr{D}(U, \wedge_{m+1} \mathbf{R}^n)\},$$

as may be verified by smoothing.

4.1.19. Suppose U is an open subset of $\mathbf{R}^n$. We say that α is an m dimensional **flat cochain** of U if and only if α is a real valued linear function on $\mathbf{F}_m(U)$ and there exists a number $c < \infty$ such that

$$\alpha(T) \le c\, \mathbf{F}_K(T)$$

whenever K is a compact subset of U and $T \in \mathbf{F}_{m,K}(U)$. The least such number c is called the **flat norm** of α, denoted $\mathbf{F}(\alpha)$. We also define $d\alpha$, the **coboundary** of α, as the $m+1$ dimensional flat cochain of U such that

$$(d\alpha)\, S = \alpha(\partial S) \quad \text{whenever} \quad S \in \mathbf{F}_{m+1}(U).$$

Using 4.1.18 *we represent flat cochains of* U *by bounded* $\mathscr{L}^n \llcorner U$ *measurable differential forms on* U as follows: To each m dimensional flat cochain α of U corresponds the linear function β on $\Omega_m \times \Omega_{m+1}$ defined by

$$\beta(\xi, \eta) = \alpha[\mathscr{L}^n \wedge \xi + \partial(\mathscr{L}^n \wedge \eta)] \quad \text{for } \xi \in \Omega_m,\ \eta \in \Omega_{m+1};$$

it follows that

$$\mathbf{F}(\alpha) = \sup\{\beta(\xi, \eta) : \textstyle\int_U (\|\xi\| + \|\eta\|)\, d\mathscr{L}^n \le 1\}.$$

According to 2.5.9, 12 there exist bounded $\mathscr{L}^n \llcorner U$ measurable differential forms ϕ and ψ of degrees m and $m+1$ on U satisfying the four conditions:

$$\alpha(\mathscr{L}^n \wedge \xi) = \textstyle\int_U \langle \xi, \phi \rangle\, d\mathscr{L}^n \quad \text{for } \xi \in \Omega_m,$$
$$\alpha[\partial(\mathscr{L}^n \wedge \eta)] = \textstyle\int_U \langle \eta, \psi \rangle\, d\mathscr{L}^n \quad \text{for } \eta \in \Omega_{m+1},$$
$$\mathbf{F}(\alpha) = \sup\{\mathbf{M}(\phi), \mathbf{M}(\psi)\},$$
$$\textstyle\int_U (\langle \operatorname{div} \eta, \phi \rangle + \langle \eta, \psi \rangle)\, d\mathscr{L}^n = 0 \quad \text{for } \eta \in \mathscr{D}(U, \wedge_{m+1} \mathbf{R}^n);$$

moreover ϕ and ψ are $\mathscr{L}^n \llcorner U$ almost unique. Conversely, any two bounded $\mathscr{L}^n \llcorner U$ measurable differential forms ϕ and ψ of degrees m and $m+1$ on U satisfying the fourth condition are associated with a unique m dimensional flat cochain α of U such that the first three conditions hold. (In case ϕ is of class 1 the fourth condition holds if and only if $d\phi = \psi$; in general it holds if and only if $d(\Phi_\varepsilon * \phi) = \Phi_\varepsilon * \psi$ for every $\varepsilon > 0$.) If α is represented by ϕ and ψ, then $d\alpha$ is represented by ψ and 0.

From 4.1.23 and 4.2.23 it follows that our concept of flat cochain is equivalent to the notion bearing that name in [WH 4]. The preceding paragraph shows a simple new approach to the main results in Chapter 9 of Whitney's book. However, while flat chains play an important role in the present book, flat cochains occur here only rarely.

4.1.20. Theorem. *If* $T \in \mathbf{F}_m(U)$ *with* $m>0$, *then*

$$\mathscr{I}_1^m(\operatorname{spt} T)=0 \quad \textit{implies} \quad T=0.$$

Proof. Recalling 2.10.5 and 1.7.4 we choose $g \in \mathbf{O}(n)$ so that

$$\mathscr{L}^m[(\mathbf{p}_\lambda \circ g)\operatorname{spt} T]=0 \text{ for all } \lambda \in \Lambda(n,m).$$

Whenever $\phi \in \mathscr{D}^m(U)$ we see from 4.1.6 that

$$\phi = \sum_{\lambda \in \Lambda(n,m)} \phi_\lambda \wedge D X_{\lambda_1} \wedge \cdots \wedge D X_{\lambda_m} = \sum_{\lambda \in \Lambda(n,m)} \phi_\lambda \wedge \mathbf{p}_\lambda^\# (D Y_1 \wedge \cdots \wedge D Y_m),$$

where $\phi_\lambda = \langle \mathbf{e}_\lambda, \phi \rangle$ and $Y_1, \ldots, Y_m$ are the standard coordinate functions on $\mathbf{R}^m$, hence

$$\begin{aligned}(g_\# T)\phi &= \sum_{\lambda \in \Lambda(n,m)} [(g_\# T) \llcorner \phi_\lambda]\, \mathbf{p}_\lambda^\# (D Y_1 \wedge \cdots \wedge D Y_m) \\ &= \sum_{\lambda \in \Lambda(n,m)} \mathbf{p}_{\lambda\#} [(g_\# T) \llcorner \phi_\lambda](D Y_1 \wedge \cdots \wedge D Y_m)=0\end{aligned}$$

by 4.1.18, because each $\mathbf{p}_{\lambda\#}[(g_\# T)\llcorner \phi_\lambda]$ is an m dimensional flat chain in $\mathbf{R}^m$ with support contained in the set

$$\mathbf{p}_\lambda[\operatorname{spt}(g_\# T)] = (\mathbf{p}_\lambda \circ g)\operatorname{spt} T$$

with $\mathscr{L}^m$ measure 0. Thus $g_\# T=0$, $T=0$.

4.1.21. The preceding theorem has two important corollaries:

If $T \in \mathbf{F}_m(U)$, $m>0$, $B \subset U$ *and* $(\operatorname{spt} T) \sim B$ *is closed relative to* U, *then*

$$\mathscr{I}_1^m(B \cap \operatorname{spt} T)=0 \quad \textit{implies} \quad \operatorname{spt} T \subset U \sim B.$$

In fact, for each $x \in B$ we can choose $\gamma \in \mathscr{D}^0(U)$ so that

$$x \notin \operatorname{spt}(T - T \llcorner \gamma) \quad \text{and} \quad [(\operatorname{spt} T) \sim B] \cap \operatorname{spt} \gamma = \varnothing,$$

hence $\operatorname{spt}(T \llcorner \gamma) \subset B \cap \operatorname{spt} T$, $T \llcorner \gamma = 0$, $x \notin \operatorname{spt} T$.

If $T \in \mathbf{F}_m(U)$, $m>0$, $\mathbf{M}(T)<\infty$ *and* $E \subset U$, *then*

$$\mathscr{I}_1^m(E)=0 \quad \textit{implies} \quad \|T\|(E)=0.$$

Since $\mathscr{I}_1^m$ is Borel regular and $\|T\|$ is a Radon measure we may assume that E is compact, hence

$$T \llcorner E \in \mathbf{F}_m(\mathbf{R}^n) \quad \text{with} \quad \operatorname{spt}(T \llcorner E) \subset E,$$

and conclude $T \llcorner E = 0$.

4.1.22. Whenever U is an open subset of a Euclidean space and K is a compact subset of U we define

$$\mathscr{P}_{m,K}(U)$$

as the additive subgroup of $\mathscr{D}_m(U)$ generated by all m dimensional oriented simplexes $[u_0, \ldots, u_m]$ such that K contains the convex hull of $\{u_0, \ldots, u_m\}$. We also let

$$\mathbf{P}_{m,K}(U)$$

be the vectorspace generated by $\mathscr{P}_{m,K}(U)$.

We define the abelian group

$$\mathscr{P}_m(U),$$

whose members are called m dimensional **integral polyhedral chains** in U, as the union of the groups $\mathscr{P}_{m,K}(U)$ corresponding to all compact subsets K of U. Similarly the vectorspace

$$\mathbf{P}_m(U),$$

whose members are called m dimensional **polyhedral chains** in U, is the union of all the vectorspaces $\mathbf{P}_{m,K}(U)$.

4.1.23. Theorem. *Whenever U is an open subset of $\mathbf{R}^n$, K and C are compact subsets of U, $K \subset \operatorname{Int} C$, $T \in \mathbf{F}_{m,K}(U)$ and $\varepsilon > 0$, there exists a $P \in \mathbf{P}_{m,C}(U)$ with*

$$\mathbf{F}_C(T-P) \leq \varepsilon \quad \textit{and} \quad \mathbf{M}(P) \leq \mathbf{M}(T) + \varepsilon.$$

Proof. We reduce the problem:

first, applying 4.1.17, to the case when $T \in \mathbf{N}_{m,K}(U)$;

second, approximating T by T_ε as in 4.1.18, to the case when $T = \mathscr{L}^n \wedge \xi$ with $\xi \in \mathscr{D}(U, \bigwedge_m \mathbf{R}^n)$ and $\operatorname{spt} \xi \subset \operatorname{Int} C$;

third, approximating ξ by step functions relative to the norm $\mathscr{L}^n_{(1)}$, to the case when

$$T(\phi) = \int_A \langle \eta, \phi(x) \rangle \, d\mathscr{L}^n x \text{ for } \phi \in \mathscr{D}^m(U),$$

where A is an $\mathscr{L}^n$ measurable subset of $\operatorname{Int} C$ and $\eta \in \bigwedge_m(\mathbf{R}^n)$;

fourth, using the second characterization of mass in 1.8.1, to the subcase when η is a simple m-vector with $|\eta| = 1$;

fifth, rotating U and computing $\mathscr{L}^n$ by suitable coverings, to the further subcase when

$$A = \mathbf{R}^n \cap \{x \colon u_i \leq x_i \leq v_i \text{ for } i = 1, \ldots, n\}, \qquad \eta = e_1 \wedge \cdots \wedge e_m,$$

where $-\infty < u_i < v_i < \infty$ and e_i are the standard base vectors of $\mathbf{R}^n$.

Identifying $\mathbf{R}^n$ with $\mathbf{R}^m \times \mathbf{R}^{n-m}$ and recalling 4.1.8 we reformulate the last subcase by expressing $T = R \times S$ with

$$R = [u_1, v_1] \times \cdots \times [u_m, v_m] \in \mathscr{D}_m(\mathbf{R}^m), \qquad S = \mathscr{L}^{n-m} \llcorner B \in \mathscr{D}_0(\mathbf{R}^{n-m}),$$
$$B = \mathbf{R}^{n-m} \cap \{z \colon u_{m+j} \le z_j \le v_{m+j} \text{ for } j = 1, \ldots, n-m\}.$$

Then we represent B as the union of disjoint Borel sets $B_1, \ldots, B_\nu$ with $\operatorname{diam} B_k \le \varepsilon / [\mathbf{N}(R)\, \mathscr{L}^{n-m}(B)]$, select points $b_k \in B_k$, define

$$Q = \sum_{k=1}^{\nu} \mathscr{L}^{n-m}(B_k)[b_k] \in \mathscr{D}_0(\mathbf{R}^{n-m}),$$

infer from 4.1.10 and 4.1.9 that

$$\begin{aligned} \mathbf{F}_B(S - Q) &\le \sum_{k=1}^{\nu} \mathbf{F}_B\big((\mathscr{L}^{n-m} \llcorner B_k) - \mathscr{L}^{n-m}(B_k)[b_k]\big) \\ &\le \sum_{k=1}^{\nu} (\operatorname{diam} B_k)\, \mathbf{M}(\mathscr{L}^{n-m} \llcorner B_k) \le \varepsilon / \mathbf{N}(R), \end{aligned}$$

let $P = R \times Q$ and conclude $\mathbf{F}_A(T - P) \le \mathbf{N}(R)\, \mathbf{F}_B(S - Q) \le \varepsilon$ with $\mathbf{M}(P) = \mathbf{M}(R)\,\mathbf{M}(Q) = \mathbf{M}(R)\,\mathbf{M}(S) = \mathbf{M}(T)$. Moreover we see from 4.1.11 that $[b_k] \times R \in \mathscr{P}_m(\mathbf{R}^n)$ for each k, hence $P \in \mathbf{P}_m(\mathbf{R}^n)$.

(We will show in 4.2.24 that $\mathbf{M}$ may be replaced by $\mathbf{N}$ in the conclusion of 4.1.23.)

4.1.24. Whenever U is an open subset of $\mathbf{R}^n$ and K is a compact subset of U we define

$$\mathscr{R}_{m,K}(U)$$

as the class of all m dimensional currents T in U with the following property: *For every $\varepsilon > 0$ there exist an open subset Z of some Euclidean space, a compact subset C of Z, a Lipschitzian map $f \colon Z \to U$ with $f(C) \subset K$, and an integral polyhedral chain*

$$P \in \mathscr{P}_{m,C}(Z) \quad \textit{with} \quad \mathbf{M}(T - f_\# P) < \varepsilon.$$

Clearly $\mathscr{R}_{m,K}(U)$ is an additive subgroup of $\mathbf{F}_{m,K}(U)$, and

$$g_\# \mathscr{R}_{m,K}(U) \subset \mathscr{R}_{m,g(K)}(V)$$

for every locally Lipschitzian map $g \colon U \to V$. We also define the abelian group

$$\mathscr{R}_m(U),$$

whose members are called m dimensional **rectifiable currents** in U, as the union of the groups $\mathscr{R}_{m,K}(U)$ corresponding to all compact subsets K of U. We note that

$$\mathscr{R}_0(U) = \mathscr{P}_0(U).$$

Alternate characterizations of rectifiable currents may be found in 4.1.28.

For each compact subset K of U we let

$$\mathbf{I}_{m,K}(U) = \{T\colon T \in \mathscr{R}_{m,K}(U),\ \partial T \in \mathscr{R}_{m-1,K}(U)\}$$

in case $m > 0$, and $\mathbf{I}_{0,K}(U) = \mathscr{R}_{0,K}(U)$. We also define the abelian group

$$\mathbf{I}_m(U),$$

whose members are called m dimensional **integral currents** in U, as the union of the groups $\mathbf{I}_{m,K}(U)$ corresponding to all compact subsets K of U.

For each compact subset K of U we let

$$\mathscr{F}_{m,K}(U) = \{R + \partial S\colon R \in \mathscr{R}_{m,K}(U),\ S \in \mathscr{R}_{m+1,K}(U)\}.$$

We also define the abelian group

$$\mathscr{F}_m(U),$$

whose members are called m dimensional **integral flat chains** in U, as the union of the groups $\mathscr{F}_{m,K}(U)$ corresponding to all compact subsets K of U.

The following diagram of inclusions shows the classes of currents which are most important in this book:

$$\begin{array}{ccccccc}
\mathscr{P}_m(U) \subset \mathbf{I}_m(U) & \subset & \mathscr{R}_m(U) & \subset & \mathscr{F}_m(U) \\
\cap \qquad\quad \cap & & \cap & & \cap \\
\mathbf{P}_m(U) \subset \mathbf{N}_m(U) \subset & & \mathbf{F}_m(U) \cap \{T\colon \mathbf{M}(T) < \infty\} & \subset & \mathbf{F}_m(U).
\end{array}$$

Each of these concepts is capable of localization. In particular, we call T an m dimensional **locally rectifiable current** in U if and only if $T \in \mathscr{D}_m(U)$ and for each $x \in U$ there exists an $S \in \mathscr{R}_m(U)$ with $x \notin \operatorname{spt}(T - S)$; the class of all such currents T is denoted

$$\mathscr{R}_m^{\mathrm{loc}}(U).$$

Similarly we define the classes

$$\mathbf{I}_m^{\mathrm{loc}}(U) \quad \text{and} \quad \mathscr{F}_m^{\mathrm{loc}}(U)$$

of **locally integral currents** and **locally integral flat chains.**

For each $T \in \mathscr{F}_{m,K}(U)$ we define

$$\mathscr{F}_K(T)$$

as *the infimum of the set of numbers* $\mathbf{M}(R)+\mathbf{M}(S)$ *corresponding to all* $R \in \mathscr{R}_{m,K}(U)$ *and* $S \in \mathscr{R}_{m+1,K}(U)$ *with* $T = R + \partial S$.

Observing that

$$\mathscr{F}_K(T_1+T_2) \leq \mathscr{F}_K(T_1)+\mathscr{F}_K(T_2) \text{ for } T_1, T_2 \in \mathscr{F}_{m,K}(U),$$

we metrize $\mathscr{F}_{m,K}(U)$ by letting the distance between T_1 and T_2 equal $\mathscr{F}_K(T_1-T_2)$. We note that

$$\mathbf{I}_{m,K}(U) \text{ is } \mathscr{F}_K \text{ dense in } \mathscr{F}_{m,K}(U),$$

because $\mathbf{I}_{m,K}(U)$ and $\mathbf{I}_{m+1,K}(U)$ are $\mathbf{M}$ dense in $\mathscr{R}_{m,K}(U)$ and $\mathscr{R}_{m+1,K}(U)$, respectively, and that

$$\mathscr{F}_{m,K}(U) \text{ is } \mathscr{F}_K \text{ complete},$$

because $\mathscr{R}_{m,K}(U)$ and $\mathscr{R}_{m+1,K}(U)$ are $\mathbf{M}$ complete; in fact whenever $T_i \in \mathscr{F}_{m,K}(U)$ for $i=1,2,3,\ldots$ with

$$\sum_{i=2}^{\infty} \mathscr{F}_K(T_i - T_{i-1}) < \infty,$$

we can choose $R_i \in \mathscr{R}_{m,K}(U)$ and $S_i \in \mathscr{R}_{m+1,K}(U)$ so that

$$T_i - T_{i-1} = R_i + \partial S_i \text{ for } i=2,3,4,\ldots \quad \text{and} \quad \sum_{i=2}^{\infty} [\mathbf{M}(R_i)+\mathbf{M}(S_i)] < \infty,$$

hence conclude

$$T = T_1 + \sum_{i=2}^{\infty} R_i + \partial \sum_{i=2}^{\infty} S_i \in \mathscr{F}_{m,K}(U)$$

and

$$\mathscr{F}_K(T-T_j) \leq \sum_{i=j+1}^{\infty} [\mathbf{M}(R_i)+\mathbf{M}(S_i)] \to 0 \text{ as } j \to \infty.$$

A similar argument shows that

$$\mathbf{F}_{m,K}(U) = \{R+\partial S : R \in \mathbf{F}_{m,K}(U), \mathbf{M}(R)<\infty, S \in \mathbf{F}_{m+1,K}(U), \mathbf{M}(S)<\infty\}.$$

We infer that for each $T \in \mathbf{F}_{m,K}(U)$ there exists

$$S \in \mathbf{F}_{m+1,K}(U) \quad \text{with} \quad \mathbf{M}(T-\partial S)+\mathbf{M}(S) = \mathbf{F}_K(T),$$

by recalling 4.1.12 and observing that if R_1, R_2, S_1, S_2 are currents with finite mass and $R_1 + \partial S_1 = R_2 + \partial S_2$, then the current $S_1 - S_2$ is normal. The analogous proposition about $\mathscr{F}_K$ will be proved in 4.2.18.

4.1.25. Lemma. *If ξ is an $\mathscr{L}^m$ summable m-vectorfield with compact support and $f: \mathbf{R}^m \to \mathbf{R}^n$ is Lipschitzian, then*

$$[f_\# (\mathscr{L}^m \wedge \xi)]\, \phi = \int \langle [\textstyle\bigwedge_m D f(x)]\, \xi(x), \phi[f(x)] \rangle \, d\mathscr{L}^m x$$

whenever $\phi \in \mathscr{D}^m(\mathbf{R}^n)$; moreover

$$f_\# (\mathscr{L}^m \wedge \xi) = \mathscr{H}^m \wedge \eta$$

where, for $\mathscr{H}^m$ almost all y in $\mathbf{R}^n$,

$$\eta(y) = \sum_{x \in f^{-1}\{y\}} [\textstyle\bigwedge_m D f(x)]\, \xi(x) / J_m f(x)$$

is a simple m-vector of $\mathbf{R}^n$ such that either $\eta(y) = 0$ or

$$\operatorname{Tan}^m [\mathscr{H}^m \llcorner f(\operatorname{spt} \xi), y] \text{ is associated with } \eta(y)$$

and $[\bigwedge_m D f(x)]\, \xi(x) / J_m f(x) = \pm |\xi(x)|\, \eta(y) / |\eta(y)|$ for every $x \in f^{-1}\{y\}$.

Proof. The integral formula for $[f_\# (\mathscr{L}^m \wedge \xi)]\, \phi$ clearly holds in case f is a map of class ∞; it remains valid for every Lipschitzian function f because 4.1.2 shows that $D(\Phi_\varepsilon * f)$ converges boundedly and $\mathscr{L}^m$ almost everywhere to Df as $\varepsilon \to 0+$. Letting

$$C = \operatorname{spt} \xi \cap \{x \colon D f(x) \text{ is univalent}\},$$

$$u(x) = \langle [\textstyle\bigwedge_m D f(x)]\, \xi(x) / J_m f(x), \Phi[f(x)] \rangle \text{ for } x \in C,$$

we use 3.2.3 to express the integral as

$$\begin{aligned} \int_C u(x)\, J_m f(x)\, d\mathscr{L}^m x &= \int_{\mathbf{R}^n} \sum_{x \in C \cap f^{-1}\{y\}} u(x)\, d\mathscr{H}^m y \\ &= \int_{\mathbf{R}^n} \langle \eta(y), \phi(y) \rangle \, d\mathscr{H}^m y = (\mathscr{H}^m \wedge \eta)\, \phi, \end{aligned}$$

because $\mathscr{H}^m [f(\operatorname{spt} \xi \sim C)] = 0$. Moreover 3.2.2, 3.2.17, 2.10.19 (4) imply (compare 3.2.19) that, for $\mathscr{H}^m$ almost all y in $f(\operatorname{spt} \xi)$,

$$\operatorname{im} D f(x) = \operatorname{Tan}^m [\mathscr{H}^m \llcorner f(\operatorname{spt} \xi), y]$$

whenever $x \in \operatorname{spt} \xi \cap f^{-1}\{y\}$.

4.1.26. Corollary. *If $f: \mathbf{R}^m \to \mathbf{R}^m$ is Lipschitzian and A is a bounded $\mathscr{L}^m$ measurable set, then*

$$f_\# (\mathbf{E}^m \llcorner A) = \mathbf{E}^m \llcorner (\text{degree } f | A)$$

where, for $\mathscr{L}^m$ almost all y in $\mathbf{R}^m$,

$$(\text{degree } f | A)(y) = \sum_{x \in A \cap f^{-1}\{y\}} \text{sign det } D f(x).$$

Moreover degree $f|A$ *is $\mathscr{L}^m$ almost constant on each component of $\mathbf{R}^m \sim f(\operatorname{Bdry} A)$.*

In case A is connected and $f(A) \cap f(\text{Bdry}\, A) = \varnothing$, then degree $f|A$ *is $\mathscr{L}^m$ almost constant on $f(A)$; if in addition $f|A$ is univalent, then* sign det Df *is $\mathscr{L}^m$ almost constant on $A \cap \{x\colon J_m f(x) > 0\}$.*

Proof. We apply 4.1.25 with $\xi = A \wedge \mathbf{e}_1 \wedge \cdots \wedge \mathbf{e}_m$, observe that

$$\partial \mathbf{E}^m = 0, \quad \text{hence} \quad \partial(\mathbf{E}^m \llcorner A) = -\partial(\mathbf{E}^m \llcorner \mathbf{R}^m \sim A),$$

$$\operatorname{spt} \partial(\mathbf{E}^m \llcorner A) \subset \operatorname{Clos} A \cap \operatorname{Clos}(\mathbf{R}^m \sim A) = \operatorname{Bdry} A,$$

$$\operatorname{spt} \partial[\mathbf{E}^m \llcorner (\operatorname{degree} f|A)] \subset f(\operatorname{Bdry} A),$$

and use the constancy theorem of 4.1.7, as well as 3.2.3 (1).

4.1.27. Assuming $u_0, \ldots, u_m \in \mathbf{R}^n$ we see from 4.1.11 and 4.1.8 that

$$[u_0, \ldots, u_m] = f_\#(\mathbf{E}^m \llcorner C) \text{ where } f\colon \mathbf{R}^m \to \mathbf{R}^n,$$

$$f(x) = u_0 + \sum_{i=1}^m \Big(\prod_{j=1}^i x_j\Big)(u_i - u_{i-1}) \text{ for } x \in \mathbf{R}^m,$$

$$C = \mathbf{R}^m \cap \{x\colon 0 < x_i < 1 \text{ for } i = 1, \ldots, m\},$$

$$f(C) = \Big\{\sum_{i=0}^m t_i u_i\colon \sum_{i=0}^m t_i = 1,\ t_i > 0 \text{ for } i = 0, \ldots, m\Big\}.$$

Moreover $f|C$ is univalent provided no $m-1$ dimensional affine subspace of $\mathbf{R}^n$ contains $\{u_0, \ldots, u_m\}$; letting $e_1, \ldots, e_m$ be the standard base vectors of $\mathbf{R}^m$, and

$$\zeta = (u_1 - u_0) \wedge (u_2 - u_1) \wedge \cdots \wedge (u_m - u_{m-1}),$$

we compute

$$\langle e_1, Df(x)\rangle \wedge \cdots \wedge \langle e_m, wf(x)\rangle = \Big(\prod_{i=1}^{m-1} x_i^{m-i}\Big)\zeta$$

for $x \in \mathbf{R}^m$, and infer from 4.1.25 that

$$[u_0, \ldots, u_m] = \mathscr{H}^m \wedge \eta$$

with

$$\eta(y) = \zeta/|\zeta| \text{ for } y \in f(C), \qquad \eta(y) = 0 \text{ for } y \in \mathbf{R}^n \sim f(C).$$

4.1.28. Theorem. *Whenever T is an m dimensional current with compact support in an open subset U of $\mathbf{R}^n$, the following five conditions are equivalent:*

(1) *T is a rectifiable current.*

(2) *$T \in \mathscr{R}_{m,K}(U)$ for every compact subset K of U such that* $\operatorname{spt} T \subset \operatorname{Int} K$.

(3) *For every* $\varepsilon>0$ *there exist an open subset* Z *of* $\mathbf{R}^m$, *a compact subset* A *of* Z *and a Lipschitzian map* $f: Z \to U$ *such that*

$$\mathbf{M}[T-f_{\#}(\mathbf{E}^m \llcorner A)]<\varepsilon.$$

(4) *There exist an* $\mathscr{H}^m$ *measurable and* $(\mathscr{H}^m, m)$ *rectifiable subset* B *of* $\operatorname{spt} T$ *and an* $\mathscr{H}^m \llcorner B$ *summable m-vectorfield* η *such that*

$$T=(\mathscr{H}^m \llcorner B) \wedge \eta$$

and, for $\mathscr{H}^m$ *almost all* x *in* B,

$$\eta(x) \text{ is simple}, \quad |\eta(x)| \text{ is a positive integer},$$

$$\operatorname{Tan}^m(\mathscr{H}^m \llcorner B, x) \text{ is associated with } \eta(x).$$

(5) $\mathbf{M}(T)<\infty$, U *is* $(\|T\|, m)$ *rectifiable and, for* $\|T\|$ *almost all* x *in* U,

$$\Theta^m(\|T\|, x) \text{ is a positive integer}, \quad \vec{T}(x) \text{ is simple},$$

$$\operatorname{Tan}^m(\|T\|, x) \text{ is associated with } \vec{T}(x);$$

moreover $\|T\|=\mathscr{H}^m \llcorner \Theta^m(\|T\|, \cdot)$.

Proof. (1) implies (3) by 4.1.27 and 2.10.43, because every m dimensional oriented simplex in $\mathbf{R}^n$ is representable as Lipschitzian image of any m dimensional oriented cube in $\mathbf{R}^m$, hence every m dimensional integral polyhedral chain in $\mathbf{R}^n$ is representable as Lipschitzian image of a finite sum of such oriented cubes with disjoint supports.

(3) implies (2), because $A \cap f^{-1}(\operatorname{spt} T)$ can be $\mathscr{L}^m$ approximated by the union of a finite disjointed family of rectangular parallelepipeds contained in $f^{-1}(\operatorname{Int} K)$ and, according to 4.1.11, each oriented parallelepiped is representable as finite sum of oriented simplexes.

Clearly (2) implies (1).

(3) implies (4) by 4.1.25 and 2.10.19 (4). In fact (3) allows the inductive selection for $j=1, 2, 3, \ldots$ of compact subsets A_j of $\mathbf{R}^m$, Lipschitzian maps $f_j: \mathbf{R}^m \to \mathbf{R}^n$ and $\mathscr{H}^m$ measurable m-vectorfields η_j on $\mathbf{R}^n$ such that

$$f_{j\#}(\mathbf{E}^m \llcorner A_j)=\mathscr{H}^m \wedge \eta_j, \quad \mathbf{M}\Big(T-\sum_{i=1}^{j} \mathscr{H}^m \wedge \eta_i\Big)<2^{-j-2}$$

and, for $\mathscr{H}^m$ almost all y in $\mathbf{R}^n$, $|\eta_j(y)|$ is an integer, $\eta_j(y)$ is simple and either $\eta_j(y)=0$ or

$$\operatorname{Tan}^m[\mathscr{H}^m \llcorner f(A_j), y] \text{ is associated with } \eta_j(y).$$

Consequently $B_j=\{y\colon \eta_j(y)\neq 0\}$ is $\mathscr{H}^m$ almost contained in $f(A_j)$,

$$\mathscr{H}^m(B_j)\leq \int_{B_j}|\eta_j|\,d\mathscr{H}^m=\mathbf{M}(\mathscr{H}^m\wedge\eta_j)<2^{-j},$$

$$B=\bigcup_{j=1}^{\infty}B_j \text{ is } (\mathscr{H}^m,m) \text{ rectifiable}, \qquad T=\mathscr{H}^m\wedge\eta \text{ with } \eta=\sum_{j=1}^{\infty}\eta_j$$

and, for $\mathscr{H}^m$ almost all y in each B_j,

$$\operatorname{Tan}^m(\mathscr{H}^m \llcorner B,y)=\operatorname{Tan}^m(\mathscr{H}^m \llcorner B_j,y)=\operatorname{Tan}^m[\mathscr{H}^m \llcorner f_j(A_j),y].$$

(4) implies (3) by 3.2.4 and 4.1.25. Observing first that (4) implies

$$T=\sum_{j=1}^{\infty}j[(\mathscr{H}^m \llcorner B\cap\{y\colon |\eta(y)|=j\})\wedge\eta/j],$$

we reduce the problem to the *special case when* $|\eta(y)|=1$ *for* $y\in B$. Then we choose a compact subset C of $\mathbf{R}^m$ and a Lipschitzian map $f\colon \mathbf{R}^m\to\mathbf{R}^n$ such that $f|C$ is univalent, $f(C)\subset B$ and

$$\mathbf{M}(T-[\mathscr{H}^m \llcorner f(C)]\wedge\eta)=\mathscr{H}^m[B\sim f(C)]<\varepsilon,$$

associate with $i\in\{1,-1\}$ the set C_i consisting of all points $x\in C$ for which

$$\langle e_1\wedge\cdots\wedge e_m, \textstyle\bigwedge_m Df(x)\rangle/J_m f(x)=i\,\eta[f(x)],$$

and conclude

$$[\mathscr{H}^m \llcorner f(C)]\wedge\eta=f_\#(\mathbf{E}^m \llcorner C_1)-f_\#(\mathbf{E}^m \llcorner C_{-1}).$$

(4) implies (5) by 3.2.19, 2.8.18 (or 2.8.17) and 2.9.8, because

$$\|T\|=(\mathscr{H}^m \llcorner B)\wedge\|\eta\|, \qquad \|T\|(\mathbf{R}^n\sim B)=0,$$

$$\Theta^m(\|T\|,x)=|\eta(x)| \quad\text{and}\quad \vec{T}(x)=\eta(x)/|\eta(x)|$$

for $\mathscr{H}^m$ almost all x in B.

(5) implies (4) with $B=\{x\colon \Theta^m(\|T\|,x)>0\}$ and $\eta=\Theta^m(\|T\|,\cdot)\vec{T}$.

4.1.29. We say that A is a **(local) Lipschitz neighborhood retract** in U if and only if $A\subset U$ and there exists a (locally) Lipschitzian map which retracts some neighborhood of A in U onto A. (An alternate characterization of such subsets of Euclidean spaces may be found in [AF 1].)

We see from the preceding theorem that

$$\mathscr{R}_{m,K}(U)=\mathscr{R}_m(U)\cap\{T\colon \operatorname{spt}T\subset K\}$$

whenever K is a compact Lipschitz neighborhood retract in U. This equation does not hold for arbitrary compact sets K; for example if $K\subset\mathbf{R}$ with $\mathscr{L}^1(K)>0$ and $\operatorname{Int}K=\varnothing$, then

$$0\neq\mathbf{E}^1 \llcorner K\in\mathscr{R}_1(\mathbf{R}) \quad\text{and}\quad \mathscr{R}_{1,K}(\mathbf{R})=\{0\}.$$

It readily follows from 4.1.18 (or 4.1.23, or 4.2.8) that *the metric space* $\mathbf{F}_{m,K}(U)$ *is separable in case K is a compact Lipschitz neighborhood retract in U.*

Now we will prove the following separation property:

If $C \subset W \subset \mathbf{R}^n$, *C is closed and W is open, then there exists a closed local Lipschitz neighborhood retract D such that* $C \subset \operatorname{Int} D$ *and* $D \subset W$.

From 3.1.14 we obtain a function $\psi: \mathbf{R}^n \to \mathbf{R}$ of class n such that

$$\psi(x)=1 \text{ for } x \in C, \quad \psi(x)=0 \text{ for } x \in \mathbf{R}^n \sim W.$$

Letting $Z=\{x: D\psi(x)=0\}$ we apply 3.4.3 with ν, m, k replaced by $0, n, n$ to infer $\mathscr{H}^1[\psi(Z)]=0$, and choose a number r for which $0<r<1$ and $r \notin \psi(Z)$. We infer from 3.1.19(2) that $\psi^{-1}\{r\}$ is an $n-1$ dimensional submanifold of class n of $\mathbf{R}^n$, and use 3.1.20 to secure a map f of class n which retracts an open subset V of $\mathbf{R}^n$ onto $\psi^{-1}\{r\}$. Then

$$D=\{x: \psi(x) \geq r\} \subset W, \ D \cup V \text{ is open},$$

and the retraction $g: \mathbf{1}_D \cup f|(V \sim D): D \cup V \to D$ is locally Lipschitzian; in fact, for every compact $K \subset D \cup V$,

$$L=K \cap \{x: \psi(x) \leq r\} \subset V, \quad 2\delta=\operatorname{dist}(L, \mathbf{R}^n \sim V)>0,$$
$$M=\{x: \operatorname{dist}(x, L) \leq \delta\} \subset V, \quad \mu=\operatorname{Lip}(f|M)<\infty,$$

and for $a \in K \sim D$, $b \in D$ with $|a-b| \leq \delta$ there is a point $u \in \psi^{-1}\{r\} \cap M$ on the line segment from a to b, hence

$$\begin{aligned}|g(a)-g(b)| &= |f(a)-f(u)+u-b| \\ &\leq \mu|a-u|+|u-b| \leq \sup\{\mu, 1\}\,|a-b|.\end{aligned}$$

(One can prove this separation property also in more elementary fashion, constructing D and a neighborhood of D as unions of suitable families of cubes, and defining the retraction by means of central projections similar to the map σ_{n-1} in 4.2.5.)

4.1.30. We will now generalize 4.1.25 to the following proposition: *If U is an open subset of* $\mathbf{R}^\nu$, *K is a compact subset of U*, ξ *is an* $\mathscr{H}^m$ *summable m-vectorfield,*

$W=\{x: \xi(x) \neq 0\}$ *is an* $(\mathscr{H}^m, m)$ *rectifiable subset of K,*

$\xi(x)$ *is simple and* $\operatorname{Tan}^m(\mathscr{H}^m \llcorner W, x)$ *is the vectorsubspace of* $\mathbf{R}^\nu$ *associated with* $\xi(x)$ *for* $\mathscr{H}^m$ *almost all x in W,*

$G: U \to \mathbf{R}^\mu$ *is a locally Lipschitzian map and* $g=G|W$, *then*

$$[G_\#(\mathscr{H}^m \wedge \xi)]\,\phi=\int_W \langle [\textstyle\bigwedge_m \operatorname{ap} D g(x)]\,\xi(x), \phi[g(x)] \rangle \, d\mathscr{H}^m x$$

whenever $\phi \in \mathscr{D}^m(\mathbf{R}^\mu)$; *moreover*

$$G_\#(\mathscr{H}^m \wedge \xi) = \mathscr{H}^m \wedge \eta$$

where, for $\mathscr{H}^m$ *almost all* y *in* $\mathbf{R}^\mu$,

$$\eta(y) = \sum_{x \in g^{-1}\{y\}} [\wedge_m \operatorname{ap} D g(x)]\, \xi(x)/\operatorname{ap} J_m g(x)$$

is a simple m-vector of $\mathbf{R}^\mu$ *such that either* $\eta(y) = 0$ *or*

$$\operatorname{Tan}^m[\mathscr{H}^m \llcorner g(W), y] \text{ is associated with } \eta(y)$$

and

$$[\wedge_m \operatorname{ap} D g(x)]\, \xi(x)/\operatorname{ap} J_m g(x) = \pm |\xi(x)|\, \eta(y)/|\eta(y)| \text{ for every } x \in g^{-1}\{y\}.$$

[Here $\operatorname{ap} D g(x)$ and $\operatorname{ap} J_m g(x) = \| \wedge_m \operatorname{ap} D g(x) \|$ are $(\mathscr{H}^m \llcorner W, m)$ approximate differentials and Jacobians, as in 3.2.19, 3.2.20.]

Using 3.2.18 and 4.1.25 we reduce the proposition to the special case when there exists a Lipschitzian map $f: \mathbf{R}^m \to \mathbf{R}^\nu$ and an $\mathscr{L}^m$ summable m-vectorfield ζ on $\mathbf{R}^m$ with compact support such that

$$f(\operatorname{spt} \zeta) = W, \quad f | \operatorname{spt} \zeta \text{ is univalent}, \quad f_\#(\mathscr{L}^m \wedge \zeta) = \mathscr{H}^m \wedge \xi.$$

Then we see with the help of 3.2.17 that the equations

$$D(G \circ f)(z) = \operatorname{ap} D g[f(z)] \circ D f(z), \qquad [\wedge_m D f(z)]\, \zeta(z) = J_m f(z) \cdot \xi[f(z)],$$
$$[\wedge_m D(G \circ f)(z)]\, \zeta(z) = J_m f(z) \cdot (\wedge_m \operatorname{ap} D g[f(z)])\, \xi[f(z)]$$

hold for $\mathscr{L}^m$ almost all z in $\operatorname{spt} \zeta$, and that

$$\begin{aligned}
[G_\#(\mathscr{H}^m \wedge \xi)]\, \phi &= [(G \circ f)_\#(\mathscr{L}^m \wedge \zeta)]\, \phi \\
&= \int \langle [\wedge_m D(G \circ f)]\, \zeta, \phi \circ G \circ f \rangle \, d\mathscr{L}^m \\
&= \int_{\operatorname{spt} \zeta} \langle [(\wedge_m \operatorname{ap} D g) \circ f]\, (\xi \circ f), \phi \circ G \circ f \rangle \, J_m f \, d\mathscr{L}^m \\
&= \int_W \langle [\wedge_m \operatorname{ap} D G]\, \xi, \phi \circ G \rangle \, d\mathscr{H}^m
\end{aligned}$$

whenever $\phi \in \mathscr{D}^m(\mathbf{R}^\mu)$, by 3.2.5. Moreover for $\mathscr{H}^m$ almost all y in $\mathbf{R}^\mu$ it is true that

$$\begin{aligned}
&[\wedge_m D(G \circ f)(z)]\, \zeta(z)/J_m(G \circ f)(z) \\
&= (\wedge_m \operatorname{ap} D g[f(z)])\, \xi[f(z)]/\operatorname{ap} J_m g[f(z)]
\end{aligned}$$

whenever $z \in (\operatorname{spt} \xi) \cap (G \circ f)^{-1}\{y\} = (\operatorname{spt} \xi) \cap f^{-1}(g^{-1}\{y\})$.

4.1.31. Suppose U is an open subset of $\mathbf{R}^n$ and B is an m-dimensional submanifold of class 1 of U. Assuming $m>0$, one terms B **orientable** *if and only if there exists a continuous m vectorfield ζ on B such that, for every $b\in B$, $\zeta(b)$ is simple, $|\zeta(b)|=1$ and* $\operatorname{Tan}(B,b)$ *is associated with $\zeta(b)$*; such a function ζ is called an m **vectorfield orienting** B. Clearly, if ζ orients B, so does $-\zeta$; in case B is connected there exist at most two m vectorfields orienting B.

(1) *If B is orientable, ζ is an m-vectorfield orienting B and $\mathscr{H}^m(B\cap K)<\infty$ for every compact subset K of U, then*

$$(\mathscr{H}^m \llcorner B)\wedge\zeta\in\mathscr{R}_m^{\mathrm{loc}}(U),\qquad \operatorname{spt}\partial[(\mathscr{H}^m\llcorner B)\wedge\zeta]\subset U\sim B.$$

(2) *If $\Gamma\in\mathbf{F}_m^{\mathrm{loc}}(U)$,*

$$(\operatorname{spt}\Gamma)\sim B \text{ is closed relative to } U,\qquad \operatorname{spt}\partial\Gamma\subset U\sim B$$

and C is a component of B with $C\cap\operatorname{spt}\Gamma\neq\varnothing$, then C is orientable and for every m vectorfield ζ orienting C there exists a real number r such that

$$\operatorname{spt}[\Gamma-r(\mathscr{H}^m\llcorner C)\wedge\zeta]\subset U\sim C.$$

To prove these assertions we apply 3.1.19(4) for each $b\in B$, requiring that $T\cap\operatorname{spt}\Gamma\subset B$, that $\operatorname{Clos}T$ be a compact subset of U, and that there exist a neighborhood Z of $\operatorname{Clos}V$ in $\mathbf{R}^m$ as well as Lipschitzian maps

$$\Phi\colon U\to Z,\qquad \Psi\colon Z\to U\quad\text{with}\quad \Phi|T=\phi,\qquad \Psi|V=\psi.$$

We let ξ be a (constant) m vectorfield orienting $\mathbf{R}^m$ and infer from 4.1.25, 28 that

$$\Psi_\#[(\mathscr{L}^m\llcorner V)\wedge\xi]=(\mathscr{H}^m\llcorner B\cap T)\wedge\eta\in\mathscr{R}_m(U),$$

where η is an m vectorfield orienting $B\cap T$.

In case the hypothesis of (1) holds we can choose ξ so that $\eta=\zeta|(B\cap T)$ and conclude

$$\begin{aligned}T\cap\operatorname{spt}\partial[(\mathscr{H}^m\llcorner B)\wedge\zeta]&=T\cap\operatorname{spt}\Psi_\#\,\partial[(\mathscr{L}^m\llcorner V)\wedge\xi]\\&\subset T\cap\Psi(\operatorname{Bdry}V)=\Psi(V)\cap\Psi(\operatorname{Bdry}V)=\varnothing,\end{aligned}$$

because $\Psi|\operatorname{Clos}V$ is univalent.

In case the hypothesis of (2) holds and $b\in C$ we select $\gamma\in\mathscr{D}^0(U)$ with $\operatorname{spt}\gamma\subset T$, $b\in W=\operatorname{Int}\{x\colon\gamma(x)=1\}$ and we find that

$$\Gamma\llcorner\gamma\in\mathbf{F}_m(U),\qquad \operatorname{spt}\partial(\Gamma\llcorner\gamma)\subset B\cap T\sim W,$$
$$\Phi_\#(\Gamma\llcorner\gamma)\in\mathbf{F}_m(Z),\qquad \operatorname{spt}\partial\Phi_\#(\Gamma\llcorner\gamma)\subset V\sim\phi(B\cap W).$$

Letting Q be the component of b in $B \cap W$, and possibly replacing ξ by $-\xi$, we infer from the constancy theorem in 4.1.7 the existence of a nonnegative number ρ with

$$\operatorname{spt}[\Phi_\#(\Gamma \llcorner \gamma) - \rho(\mathscr{L}^m \llcorner V) \wedge \xi] \subset V \sim \phi(Q).$$

Since $\Psi[\Phi(x)] = x$ for $x \in \operatorname{spt}(\Gamma \llcorner \gamma)$, it follows from 4.1.15 that

$$\operatorname{spt}[(\Gamma \llcorner \gamma) - \rho(\mathscr{H}^m \llcorner B \cap T) \wedge \eta] \subset B \cap T \sim Q,$$

$$Q \cap \operatorname{spt}[\Gamma - \rho(\mathscr{H}^m \llcorner Q) \wedge \eta] = \varnothing.$$

To complete the proof of (2) we consider the class Ω of all such triples (Q, ρ, η). Clearly

$$C = \bigcup \{Q\colon (Q, \rho, \eta) \in \Omega \text{ for some } \rho \text{ and } \eta\}.$$

If $(Q_1, \rho_1, \eta_1) \in \Omega$ and $(Q_2, \rho_2, \eta_2) \in \Omega$, then

$$\rho_1 \eta_1 | Q_1 \cap Q_2 = \rho_2 \eta_2 | Q_1 \cap Q_2,$$

because $\mathscr{H}^m(A) > 0$ for every nonempty relatively open subset A of $Q_1 \cap Q_2$; in case $Q_1 \cap Q_2 \neq \varnothing$ and $\rho_1 \neq 0$ it follows that

$$\rho_1 = \rho_2 \quad \text{and} \quad \eta_1 | Q_1 \cap Q_2 = \eta_2 | Q_1 \cap Q_2.$$

Since $C \cap \operatorname{spt} \Gamma \neq \varnothing$ we can select a positive number r such that

$$\Upsilon = \Omega \cap \{(Q, \rho, \eta)\colon \rho = r\} \neq \varnothing,$$

and infer that $\bigcup \{Q\colon (Q, r, \eta) \in \Omega \text{ for some } \eta\}$ is a nonempty, relatively open and relatively closed subset of C, hence equals C. We conclude that $\Omega = \Upsilon$ and C is oriented by

$$\bigcup \{\eta | Q\colon (Q, r, \eta) \in \Omega \text{ for some } Q \text{ and } \eta\}.$$

As postscript to (2) we observe that *if Γ is representable by integration, then $\vec{\Gamma} | C$ orients C.* We also see from 4.1.28 that *r is an integer in case $\Gamma \in \mathscr{R}_m^{\text{loc}}(U)$*; this result will be sharpened in 4.2.27.

In (2) *the assumption that Γ be locally flat cannot be omitted*, as shown by the example where

$$U = \mathbf{R}^2, \quad B = \mathbf{R}^2 \cap \{x\colon x_2 = 0\}, \quad \Gamma \in \mathscr{D}_1(\mathbf{R}^2),$$

$$\Gamma(\alpha) = \textstyle\int_0^1 \langle e_1, D_2 \alpha(t, 0) \rangle \, d\mathscr{L}^1 t - \langle e_2, \alpha(1, 0) \rangle + \langle e_2, \alpha(0, 0) \rangle$$

for $\alpha \in \mathscr{D}^1(\mathbf{R}^2)$; in this case one finds that

$$\partial \Gamma = 0, \quad \operatorname{spt} \Gamma = \mathbf{R}^2 \cap \{x\colon 0 \leq x_1 \leq 1, \, x_2 = 0\}.$$

If α is a simple m vector of $\mathbf{R}^n$ such that $|\alpha| = 1$, and if B is the m dimensional vectorsubspace of $\mathbf{R}^n$ associated with α, then B is oriented by the *constant m vectorfield mapping B onto α*; the concept of orientation used here is thus consistent with 1.6.2.

In particular, we usually orient $\mathbf{R}^m$ by the m vectorfield $\mathbf{e}_1 \wedge \cdots \wedge \mathbf{e}_m$, where $e_1, \ldots, e_m$ are the standard base vectors of $\mathbf{R}^m$.

It is clear from the preceding discussion (or from 4.1.30) that *if f is a diffeomorphism of class 1 mapping U onto an open subset of $\mathbf{R}^n$, and if ζ is an m-vectorfield orienting B, then*

$$f_\# [(\mathscr{H}^m \llcorner B) \wedge \zeta] = [\mathscr{H}^m \llcorner f(B)] \wedge \chi$$

for some m-vectorfield χ orienting $f(B)$.

Next we derive a simple version of the **Gauss-Green Theorem** (for the optimal version see 4.5.6):

If A is a nonempty bounded open subset of $\mathbf{R}^n$ and $B = \operatorname{Bdry} A$ is a connected $n-1$ dimensional submanifold of class 1 of $\mathbf{R}^n$, then

$$\partial(\mathbf{E}^n \llcorner A) = (\mathscr{H}^{n-1} \llcorner B) \wedge \zeta$$

for some $n-1$ vectorfield ζ orienting B.

Since $\partial(\mathbf{E}^n \llcorner A) \neq 0$ by 4.1.10 and $\operatorname{spt} \partial(\mathbf{E}^n \llcorner A) \subset B$ by 4.1.7, we infer from (2) that $\partial(\mathbf{E}^n \llcorner A) = r(\mathscr{H}^{n-1} \llcorner B) \wedge \zeta$ where $0 \neq r \in \mathbf{R}$ and ζ is an $n-1$ vectorfield orienting B. To simplify the problem of determining r we subject $\mathbf{R}^n$ to a suitable diffeomorphism; in view of 3.1.23 we may then assume that

$$B \cap W = W \cap \{x : x_1 = 0\} \quad \text{with} \quad W = \{x : |x_i| < 1 \text{ for } i = 1, \ldots, n\}.$$

We observe that A contains at least one of the two regions

$$W^+ = W \cap \{x : x_1 > 0\} \quad \text{and} \quad W^- = W \cap \{x : x_1 < 0\}$$

because $B \cap W \subset \operatorname{Bdry} A$, but not both because this would imply $\mathscr{L}^n(W \sim A) = 0$ and $W \cap \operatorname{spt} \partial(\mathbf{E}^n \llcorner A) = \varnothing$. In case $W^+ \subset A$ and $W^- \subset \mathbf{R}^n \sim A$ we use 4.1.8 to compute

$$\begin{aligned} \partial(\mathbf{E}^n \llcorner A) \llcorner W &= \partial(\mathbf{E}^n \llcorner W^+) \llcorner W \\ &= -[\mathscr{H}^{n-1} \llcorner (B \cap W)] \wedge \mathbf{e}_2 \wedge \cdots \wedge \mathbf{e}_n = \pm [(\mathscr{H}^{n-1} \llcorner B) \wedge \zeta] \llcorner W, \end{aligned}$$

because $\mathbf{e}_2 \wedge \cdots \wedge \mathbf{e}_n$ orients $B \cap W$, hence conclude $r = \pm 1$.

Whenever B is orientable and a specific m-vectorfield ζ orienting B has been selected – or is clear from context – it is customary to use the notation

$$\textstyle\int_B \phi = [(\mathscr{H}^m \llcorner B) \wedge \zeta](\phi) = \int_B \langle \zeta, \phi \rangle \, d\mathscr{H}^m$$

for $\phi \in \mathscr{D}^m(U)$. In particular, when B is an open subset of $\mathbf{R}^m$, one often writes

$$\textstyle\int_B \phi = [\mathbf{E}^m \llcorner B](\phi) \text{ for } \phi \in \mathscr{D}^m(\mathbf{R}^m).$$

In case A and B are $m+1$ and m dimensional submanifolds of class 1 of U, with specified orienting $m+1$ and m vectorfields η and ζ, which satisfy the condition

$$\partial[(\mathscr{H}^{m+1} \llcorner A) \wedge \eta] = (\mathscr{H}^m \llcorner B) \wedge \zeta,$$

then it is customary to denote $B = \partial A$; translating the above equation into the language described in the preceding paragraph one obtains the formula

$$\textstyle\int_A d\psi = \int_{\partial A} \psi \text{ for } \psi \in \mathscr{D}^m(U).$$

4.1.32. Here we establish some useful facts about polyhedral chains.

If $T \in \mathbf{F}_m(\mathbf{R}^n)$, $\partial T \in \mathscr{P}_{m-1}(\mathbf{R}^n)$ *and* spt T *is contained in an m dimensional affine subspace of* $\mathbf{R}^n$, *then* $T \in \mathscr{P}_m(\mathbf{R}^n)$. By means of the second corollary in 4.1.15 we reduce the problem to the case when $n = m$. Then we choose $u \in \mathbf{R}^m$, infer from 4.1.11 that

$$\boldsymbol{\delta}_u \times \partial T \in \mathscr{P}_m(\mathbf{R}^n) \quad \text{with} \quad \partial(T - \boldsymbol{\delta}_u \times \partial T) = 0,$$

and use 4.1.10 to conclude $T - \boldsymbol{\delta}_u \times \partial T = 0$.

If U *is an open subset of* $\mathbf{R}^\nu$, $f\colon U \to \mathbf{R}^n$ *is locally Lipschitzian and* A *is a convex subset of* U *with*

$$f\{(1-s)u + s v\colon 0 \le s \le 1\} \subset \{(1-t)f(u) + t f(v)\colon 0 \le t \le 1\}$$

whenever $u \in A$ *and* $v \in A$, *then*

$$f_\#[u_0, \ldots, u_m] = [f(u_0), \ldots, f(u_m)]$$

whenever $u_0, \ldots, u_m \in A$. Since the conclusion is trivial in case $m = 0$, we may inductively assume that the current

$$Q = f_\#[u_0, \ldots, u_m] - [f(u_0), \ldots, f(u_m)] \in \mathbf{F}_m(\mathbf{R}^n)$$

satisfies the condition $\partial Q = 0$. Moreover f maps the convex hull of $\{u_0, \ldots, u_m\}$ into the convex hull H of $\{f(u_0), \ldots, f(u_m)\}$, hence spt $Q \subset H$ and we conclude as above that $Q = 0$.

A map f with the preceding property will be called **projective on** A. Of course every affine map is projective on its domain, but so is every monotone map of $\mathbf{R}$ into $\mathbf{R}$, and every central projection

$$f\colon U = \mathbf{R}^{n+1} \cap \{x\colon x_{n+1} > 0\} \to \mathbf{R}^n,$$

$$f(x) = (x_1/x_{n+1}, \ldots, x_n/x_{n+1}) \text{ for } x \in U.$$

If two locally Lipschitzian maps f *and* g *of* U *into* $\mathbf{R}^n$ *are projective on the convex subset* A *of* U, h *is the affine homotopy from* f *to* g *and if, for all m,*

$$h(\{t\colon 0 \le t \le 1\} \times E)$$

is contained in an $m+1$ dimensional affine subspace of $\mathbf{R}^n$ whenever E is an m dimensional simplex contained in A, then

$$h_{\#}([0,1]\times[u_0,\ldots,u_m])\in\mathscr{P}_{m+1}(\mathbf{R}^n)$$

whenever $u_0,\ldots,u_m\in A$. Observing that the conclusion is trivial in case $m=0$, we inductively infer from the homotopy formula for currents that the deformation chain of $[u_0,\ldots,u_m]$ has an integral polyhedral boundary. Since the support of this $m+1$ dimensional flat chain is contained in an $m+1$ dimensional affine subspace of $\mathbf{R}^n$, it must be an integral polyhedral chain.

If $\alpha\in\bigwedge^1\mathbf{R}^n$, $c\in\mathbf{R}$, $A=\mathbf{R}^n\cap\{x\colon \alpha(x)\le c\}$ and K is a compact subset of $\mathbf{R}^n$, then

$$T\in\mathscr{P}_{m,K}(\mathbf{R}^n)\quad\textit{implies}\quad T\llcorner A\in\mathscr{P}_{m,K\cap A}(\mathbf{R}^n).$$

In case $m=0$ this implication is trivial. In case $m>0$ we assume that $T=[u_0,\ldots,u_m]$, K is the convex hull of $\{u_0,\ldots,u_m\}$ and there exists a point

$$v\in K\cap\{x\colon \alpha(x)=c\};$$

then $T=\boldsymbol{\delta}_v\times\partial T=\boldsymbol{\delta}_v\times[(\partial T)\llcorner A+(\partial T)\llcorner(K\sim A)]$, hence

$$T\llcorner A=\boldsymbol{\delta}_v\times[(\partial T)\llcorner A]$$

by virtue of 4.1.11, 4.1.9 because the affine homotopy h from the constant map with value v to the identity map of $\mathbf{R}^n$ satisfies the conditions

$$h_t(A)\subset A,\quad h_t(K\sim A)\subset K\sim A\ \text{for}\ 0<t<1;$$

accordingly the inductive hypothesis $(\partial T)\llcorner A\in\mathscr{P}_{m-1,K\cap A}(\mathbf{R}^n)$ yields the conclusion $T\llcorner A\in\mathscr{P}_{m,K\cap A}(\mathbf{R}^n)$.

The last result implies that

$$\mathbf{E}^m\llcorner W\in\mathscr{P}_{m,W}(\mathbf{R}^n)$$

for every compact subset W of $\mathbf{R}^m$ which can be defined by finitely many linear inequalities; the images of such currents under affine maps of $\mathbf{R}^m$ into $\mathbf{R}^n$ are called m dimensional **oriented convex cells** in $\mathbf{R}^n$.

It follows that *every $T\in\mathscr{P}_m(\mathbf{R}^n)$ has a representation*

$$T=\sum_{i=1}^{k} r_i S_i\quad\textit{with}\quad \mathbf{M}(T)=\sum_{i=1}^{k} r_i\,\mathbf{M}(S_i)$$

where $r_1,\ldots,r_k$ are positive integers and $S_1,\ldots,S_k$ are m dimensional oriented convex cells such that, for $i\neq j$, $\operatorname{spt} S_i\cap\operatorname{spt} S_j$ *is contained in an $m-1$ dimensional affine subspace of $\mathbf{R}^n$.*

In the above six propositions and their proofs, *integral polyhedral chains may be replaced by polyhedral chains;* of course $r_1,\ldots,r_k$ must then be permitted to be arbitrary positive numbers.

4.1.33. We observe that

$$T(\mathbf{D}_{n-m}\eta)=\|T\|(\eta\bullet *\vec{T})$$

whenever $T\in D_m(\mathbf{R}^n)$, *T is representable by integration and* $\eta\in\mathscr{D}(\mathbf{R}^n,\wedge_{n-m}\mathbf{R}^n)$; in fact 1.7.8 implies

$$\langle\vec{T},\mathbf{D}_{n-m}\eta\rangle=\langle *\vec{T},*\mathbf{D}_{n-m}\eta\rangle=\langle *\vec{T},\gamma_{n-m}\eta\rangle=(*\vec{T})\bullet\eta.$$

Recalling 4.1.6 we infer that

$$\|\partial S\|(\xi\bullet *\overrightarrow{\partial S})=(-1)^{m-1}\|S\|(\operatorname{div}\xi\bullet *\vec{S})$$

whenever $S\in\mathbf{N}_m^{\mathrm{loc}}(\mathbf{R}^n)$ *and* $\xi\in\mathscr{D}(\mathbf{R}^n,\wedge_{n-m+1}\mathbf{R}^n)$, because

$$(\partial S)\,\mathbf{D}_{n-m+1}\xi=S(d\mathbf{D}_{n-m+1}\xi)=(-1)^{m-1}S(\mathbf{D}_{n-m}\operatorname{div}\xi).$$

4.1.34. The **Lie product** of two vectorfields α and β of class 1 on an open subset U of $\mathbf{R}^n$ is the vectorfield

$$[\alpha,\beta]=\langle\alpha,D\beta\rangle-\langle\beta,D\alpha\rangle.$$

Since $\langle\alpha,D\langle\beta,Dg\rangle\rangle=\langle\langle\alpha,D\beta\rangle,Dg\rangle+\langle\alpha\odot\beta,D^2g\rangle$ for every function g of class 2 on U, it follows that

$$\langle\alpha,D\langle\beta,Dg\rangle\rangle-\langle\beta,D\langle\alpha,Dg\rangle\rangle=\langle[\alpha,\beta],Dg\rangle.$$

Associating with each vectorfield α the differential operator mapping g onto $\langle\alpha,Dg\rangle$, we see that the operator associated with $[\alpha,\beta]$ measures the extent by which the operators associated with α and β fail to commute.

The following proposition relates Lie products with exterior differentiation: If ψ is a differential form of degree p and class 1 on U, and $\xi_1,\ldots,\xi_{p+1}$ are vectorfields of class 1 on U, then

$$\begin{aligned}\langle\xi_1\wedge\cdots\wedge\xi_{p+1},d\psi\rangle=&\sum_{i=1}^{p+1}(-1)^{i-1}\langle\xi_i,D\langle\xi^{(i)},\psi\rangle\rangle\\&+\sum_{i=1}^{p+1}\sum_{j=i+1}^{p+1}(-1)^{i+j}\langle[\xi_i,\xi_j]\wedge\xi^{(i,j)},\psi\rangle\end{aligned}$$

where $\xi^{(i)}=\xi_1\wedge\cdots\wedge\xi_{i-1}\wedge\xi_{i+1}\wedge\cdots\wedge\xi_{p+1}$ and

$$\xi^{(i,j)}=\xi_1\wedge\cdots\wedge\xi_{i-1}\wedge\xi_{i+1}\wedge\cdots\wedge\xi_{j-1}\wedge\xi_{j+1}\wedge\cdots\wedge\xi_{p+1}.$$

In case all ξ_i are constant this equation reduces to the definition of $d\psi$; one readily verifies that replacement of one ξ_i by $\phi\,\xi_i$ multiplies both sides of the equation by ϕ, for any $\phi\colon U\to\mathbf{R}$ of class 1; then the general assertion follows from multilinearity.

4.2. Deformations and compactness

4.2.1. Whenever $T\in\mathbf{N}_m(\mathbf{R}^n)$, $m\geq 1$, $u:\mathbf{R}^n\to\mathbf{R}$ is Lipschitzian and $r\in\mathbf{R}$ we define

$$\begin{aligned}\langle T,u,r+\rangle&=(\partial T)\llcorner\{x\colon u(x)>r\}-\partial(T\llcorner\{x\colon u(x)>r\})\\&=\partial(T\llcorner\{x\colon u(x)\leq r\})-(\partial T)\llcorner\{x\colon u(x)\leq r\}\end{aligned}$$

and

$$\begin{aligned}\langle T,u,r-\rangle&=\partial(T\llcorner\{x\colon u(x)<r\})-(\partial T)\llcorner\{x\colon u(x)<r\}\\&=(\partial T)\llcorner\{x\colon u(x)\geq r\}-\partial(T\llcorner\{x\colon u(x)\geq r\}).\end{aligned}$$

For all but countably many real numbers r it is also true that

$$[\|T\|+\|\partial T\|]\{x\colon u(x)=r\}=0,\quad\text{hence}\quad\langle T,u,r+\rangle=\langle T,u,r-\rangle.$$

We will now prove that, for every $r\in\mathbf{R}$,

$$\operatorname{spt}\langle T,u,r+\rangle\subset u^{-1}\{r\}\cap\operatorname{spt}T$$

and

$$\mathbf{M}\langle T,u,r+\rangle\leq\operatorname{Lip}(u)\liminf_{h\to 0+}\|T\|\{x\colon r<u(x)<r+h\}/h.$$

Letting $\gamma_h(t)=(|t-r|-|t-r-h|+h)/(2h)$ for $h>0$ and $t\in\mathbf{R}$, we observe that $\gamma_h\circ u$ is between the characteristic functions of the sets $\{x\colon u(x)\geq r+h\}$ and $\{x\colon u(x)>r\}$, hence

$$\begin{aligned}&\mathbf{F}_{\operatorname{spt}T}((\partial T)\llcorner(\gamma_h\circ u)-\partial[T\llcorner(\gamma_h\circ u)]-\langle T,u,r+\rangle)\\&\leq[\|\partial T\|+\|T\|]\{x\colon r<u(x)<r+h\}\to 0\quad\text{as}\quad h\to 0+.\end{aligned}$$

For each $h>0$ we uniformly approximate $\gamma_h\circ u$ by a sequence of functions

$$\chi_i\in\mathscr{E}^0(\mathbf{R}^n)\quad\text{with}\quad\operatorname{spt}d\chi_i\subset\{x\colon r<u(x)<r+h\},$$
$$\operatorname{Lip}(\chi_i)\to\operatorname{Lip}(\gamma_h\circ u)\leq\operatorname{Lip}(u)/h\quad\text{as}\quad i\to\infty,$$

and infer that $(\partial T)\llcorner(\gamma_h\circ u)-\partial[T\llcorner(\gamma_h\circ u)]$ is the $\mathbf{F}_{\operatorname{spt}T}$ limit of the sequence of currents

$$(\partial T)\llcorner\chi_i-\partial(T\llcorner\chi_i)=T\llcorner d\chi_i$$

with

$$\mathbf{M}(T\llcorner d\chi_i)\leq\operatorname{Lip}(\chi_i)\,\|T\|\{x\colon r<u(x)<r+h\}.$$

Twice applying the lowersemicontinuity of $\mathbf{M}$, we obtain the asserted estimate for $\mathbf{M}\langle T,u,r+\rangle$.

Since the function mapping $r\in\mathbf{R}$ onto $\|T\|\{x\colon u(x)>r\}$ is nonincreasing, it follows from 2.9.19 that

$$\int_a^{*b}\mathbf{M}\langle T,u,r+\rangle\,d\mathscr{L}^1r\leq\operatorname{Lip}(u)\,\|T\|\{x\colon a<u(x)<b\}$$

whenever $-\infty \leq a < b \leq \infty$. Replacing T by ∂T and observing that

$$\partial\langle T, u, r+\rangle = -\langle \partial T, u, r+\rangle \text{ for } r \in \mathbf{R},$$

we conclude

$$\langle T, u, r+\rangle \in \mathbf{N}_{m-1}(\mathbf{R}^n) \text{ for } \mathscr{L}^1 \text{ almost all } r.$$

Moreover, if $T \in \mathbf{N}_{m,K}(\mathbf{R}^n)$, where K is a compact subset of $\mathbf{R}^n$, then

$$\int^* \mathbf{F}_{K \cap \{x: u(x)=r\}} \langle T, u, r+\rangle \, d\mathscr{L}^1 r \leq \operatorname{Lip}(u) \, \mathbf{F}_K(T),$$

$$\int_a^{*b} \mathbf{F}_{K \cap \{x: u(x) \leq r\}} [T \llcorner \{x: u(x) \leq r\}] \, d\mathscr{L}^1 r \leq [b - a + \operatorname{Lip}(u)] \, \mathbf{F}_K(T)$$

whenever $-\infty < a < b < \infty$, because, for all $S \in \mathbf{N}_{m+1,K}(\mathbf{R}^n)$ and $r \in \mathbf{R}$,

$$\langle T, u, r+\rangle = \langle T - \partial S, u, r+\rangle - \partial\langle S, u, r+\rangle,$$

$$T \llcorner \{x: u(x) \leq r\} = (T - \partial S) \llcorner \{x: u(x) \leq r\}$$
$$+ \partial[S \llcorner \{x: u(x) \leq r\}] - \langle S, u, r+\rangle.$$

We also note that, if $r < s$ and $E(r, s) = \{x: r < u(x) \leq s\}$, then

$$\langle T, u, r+\rangle - \langle T, u, s+\rangle = (\partial T) \llcorner E(r, s) - \partial[T \llcorner E(r, s)],$$

$$\mathbf{F}_K[\langle T, u, r+\rangle - \langle T, u, s+\rangle] \leq \operatorname{Lip}(u) [\|\partial T\| + \|T\|] \, E(r, s),$$

hence the $\mathbf{F}_K$ length (see 2.5.16) of the function mapping $r \in \mathbf{R}$ onto $\langle T, u, r+\rangle \in \mathbf{F}_{m,K}(\mathbf{R}^n)$ does not exceed $\operatorname{Lip}(u) \, \mathbf{N}(T)$.

As a corollary of the above results we obtain the inequality

$$\int_a^{*b} \mathbf{M}(\partial[X \llcorner \{x: u(x) > r\}]) \, d\mathscr{L}^1 r \leq \operatorname{Lip}(u) \, \|X\| \, \{x: a < u(x) < b\}$$

whenever $X \in \mathscr{D}_m(\mathbf{R}^n)$, $m \geq 1$, $\sup u(\operatorname{spt} \partial X) < a < b$, $\operatorname{spt} X$ is compact and $\mathbf{M}(X) < \infty$; for this purpose we choose $\alpha \in \mathscr{E}^0(\mathbf{R}^n)$ with

$$(\operatorname{spt} \partial X) \cap (\operatorname{spt} \alpha) = \varnothing, \quad \alpha(x) = 1 \text{ and } d\alpha(x) = 0 \text{ if } u(x) \geq a,$$

infer that $T = X \llcorner \alpha$ and $\partial T = -X \llcorner d\alpha$ have finite mass, and observe that $r \geq a$ implies

$$\langle T, u, r+\rangle = -\partial[T \llcorner \{x: u(x) > r\}] = -\partial[X \llcorner \{x: u(x) > r\}].$$

From 4.3.4 it may be seen how the construction used here fits into the general theory of slicing developed in 4.3.

4.2.2. We recall that $f_\# T$ has been defined only in case f is proper on $\operatorname{spt} T$. Here we show how this restriction can be replaced by a different condition, which allows f to have essential singularities of a certain kind on $\operatorname{spt} T$.

Assuming that $u\colon \mathbf{R}^n \to \{r\colon 0 \le r < \infty\}$ is Lipschitzian, we call f a u **admissible function** if and only if, for some k,

$$f\colon \mathbf{R}^n \cap \{x\colon u(x) > 0\} \to \mathbf{R}^k \text{ is locally Lipschitzian,}$$

$$f \text{ maps bounded sets onto bounded sets and}$$

$$\|Df(x)\| \le 1/u(x) \text{ for } \mathscr{L}^n \text{ almost all } x \text{ in } \operatorname{dmn} f.$$

We say that T is an m dimensional u **admissible current** if and only if $T \in \mathbf{N}_m(\mathbf{R}^n)$,

$$\|T\|\{x\colon u(x)=0\}=0, \quad \|T\|(u^{-m})<\infty$$

and either $m=0$ or

$$\|\partial T\|\{x\colon u(x)=0\}=0, \quad \|\partial T\|(u^{1-m})<\infty.$$

Under these conditions we can define

$$f_{\# u} T = \lim_{r \to 0+} f_{\#}(T \llcorner U_r) \in \mathbf{D}_m(\mathbf{R}^k) \quad \text{with} \quad \mathbf{M}(f_{\# u} T) \le \|T\|(u^{-m}),$$

where $U_r = \{x\colon u(x) > r\}$ for $r > 0$; in fact $\mathscr{D}_m(\mathbf{R}^k)$ is $\mathbf{M}$ complete and, whenever $r > s > 0$,

$$\begin{aligned}\mathbf{M}[f_{\#}(T \llcorner U_s) - f_{\#}(T \llcorner U_r)] &= \mathbf{M} f_{\#}[T \llcorner (U_s \sim U_r)] \\ &\le \int_{\{x\colon u(x) < r\}} u^{-m}\, d\|T\| \to 0 \quad \text{as} \quad r \to 0.\end{aligned}$$

If $m > 0$, then ∂T is also u admissible, hence

$$f_{\# u}(\partial T) = \lim_{r \to 0+} f_{\#}[(\partial T) \llcorner U_r] \in \mathscr{D}_{m-1}(\mathbf{R}^k)$$

with

$$\mathbf{M}[f_{\# u}(\partial T)] \le \|\partial T\|(u^{1-m}).$$

In case g is another u admissible function with values in $\mathbf{R}^k$ we let h be the affine homotopy from f to g and define the u deformation chain

$$H_u(f,g)\, T = \lim_{r \to 0+} h_{\#}([0,1] \times (T \llcorner U_r)) \in \mathscr{D}_{m+1}(\mathbf{R}^k)$$

with

$$\mathbf{M}[H_u(f,g)\, T] \le \|T\|(|g-f|\, u^{-m});$$

in fact, whenever $r > s > 0$,

$$\begin{aligned}&\mathbf{M}\big[h_{\#}([0,1] \times (T \llcorner U_s)) - h_{\#}([0,1] \times (T \llcorner U_r))\big] \\ &\le \int_{\{x\colon u(x) < r\}} |g-f|\, u^{-m}\, d\|T\| \to 0 \quad \text{as} \quad r \to 0+.\end{aligned}$$

Next we will verify the equations

$$f_{\# u}(\partial T) = \partial(f_{\# u} T) \quad \text{and} \quad g_{\# u} T - f_{\# u} T = \partial[H_u(f,g)\, T] + H_u(f,g)(\partial T)$$

if $m>0$; $g_{\#u}T-f_{\#u}T=\partial[H_u(f,g)\,T]$ if $m=0$. We observe that, in case $m>0$,

$$f_{\#u}(\partial T)-\partial(f_{\#u}T)=\lim_{r\to 0+} f_{\#}\langle T,u,r+\rangle$$

with

$$\mathbf{M}[f_{\#}\langle T,u,r+\rangle]\le r^{1-m}\,\mathbf{M}\langle T,u,r+\rangle,$$

because $\operatorname{spt}\langle T,f,r+\rangle\subset\{x\colon u(x)=r\}$, and

$$\begin{aligned}
&H_u(f,g)(\partial T)+\partial[H_u(f,g)\,T]-g_{\#u}T+f_{\#u}T\\
&=\lim_{r\to 0+}\Big(h_{\#}([0,1]\times[(\partial T)\llcorner U_r])\\
&\qquad\qquad+\partial h_{\#}([0,1]\times(T\llcorner U_r))-g_{\#}(T\llcorner U_r)+f_{\#}(T\llcorner U_r)\Big)\\
&=\lim_{r\to 0+} h_{\#}([0,1]\times\langle T,u,r+\rangle),
\end{aligned}$$

with

$$\mathbf{M}\,h_{\#}([0,1]\times\langle T,u,r+\rangle)\le\|\langle T,u,r+\rangle\|(|g-f|\,r^{1-m})\le\xi\, r^{1-m}\,\mathbf{M}\langle T,u,r+\rangle,$$

where $\xi=\sup\{|g(x)-f(x)|\colon x\in\operatorname{spt}T,\ u(x)>0\}<\infty$. Inasmuch as

$$\begin{aligned}
\delta^{-1}\textstyle\int_{\delta}^{*2\delta} r^{1-m}\,\mathbf{M}\langle T,u,r+\rangle\,d\mathscr{L}^1 r&\le\delta^{-m}\operatorname{Lip}(u)\,\|T\|\{x\colon\delta<u(x)<2\delta\}\\
&\le 2^m\operatorname{Lip}(u)\textstyle\int_{\{x\colon u(x)<2\delta\}}u^{-m}\,d\|T\|\to 0
\end{aligned}$$

as $\delta\to 0+$, we find that

$$\liminf_{r\to 0+} r^{1-m}\,\mathbf{M}\langle T,u,r+\rangle=0$$

and infer the stated equations. In case $m=0$ the corresponding argument is trivial. We note that

$$f_{\#u}T\in\mathbf{N}_m(\mathbf{R}^k)\quad\text{and}\quad H_u(f,g)\,T\in\mathbf{N}_{m+1}(\mathbf{R}^k).$$

If T is rectifiable, so are $f_{\#u}T$ and $H_u(f,g)\,T$.

If f,g,T are both u and v admissible, then

$$f_{\#u}T=f_{\#v}T\quad\text{and}\quad H_u(f,g)\,T=H_v(f,g)\,T.$$

This fact is obvious in the special case when $u\ge v$, to which we reduce the general case replacing v by $\inf\{u,v\}$.

4.2.3. Here we will give alternate proofs of three propositions which are simple consequences of 4.1.20 and 4.1.31. However the present method has the advantage of generalizing to flat chains modulo ν (see 4.2.26), because it makes no use of the constancy theorem derived in 4.1.4 and 4.1.7.

If $X \in \mathbf{F}_m(\mathbf{R}^n)$, $\mathbf{M}(X) < \infty$, $\mathscr{H}^m(\operatorname{spt} X) = 0$ *and* spt X *is contained in the union of a finite family of* m *dimensional affine subspaces of* $\mathbf{R}^n$, *then* $X = 0$. In fact we see from 4.1.18 and the second corollary in 4.1.15 that $X \llcorner E = 0$ for every m dimensional affine subspace E of $\mathbf{R}^n$.

If $-\infty < a_i < b_i < \infty$ *for* $i = 1, \ldots, m$,

$$W = \mathbf{R}^m \cap \{x\colon a_i < x_i < b_i \text{ for } i = 1, \ldots, m\},$$

$X \in \mathbf{F}_m(\mathbf{R}^m)$, $\mathbf{M}(X) < \infty$ *and* spt $\partial X \subset$ Bdry W, *then there exists a real number* r *such that*

$$X = r[a_1, b_1] \times \cdots \times [a_m, b_m].$$

Letting $u(x) = \operatorname{dist}(x, \operatorname{Bdry} W)$ for $x \in \mathbf{R}^m$, and choosing $w \in W$, we apply 4.2.1 to obtain a number s such that

$$0 < s < u(w), \quad \mathbf{M}[\partial(X \llcorner H)] < \infty \quad \text{with} \quad H = \{x\colon u(x) > s\}.$$

We also secure a locally Lipschitzian retraction ψ of $\mathbf{R}^m \sim \{w\}$ onto Bdry W (see the proof of 3.2.35), note that

$$\partial(X \llcorner H) = \partial X - \partial(X \llcorner \mathbf{R}^m \sim H), \quad \operatorname{spt} \partial(X \llcorner H) \subset \mathbf{R}^m \sim H,$$

$\psi_\#(X \llcorner \mathbf{R}^m \sim H) = 0$ because $\operatorname{im} \psi = \operatorname{Bdry} W$ is contained in the union of a finite family of $m-1$ dimensional affine subspaces of $\mathbf{R}^m$, and use 4.1.15 to infer

$$\partial X = \psi_\#(\partial X) = \psi_\# \, \partial(X \llcorner H), \quad \mathbf{M}(\partial X) < \infty.$$

Next we define $A = \mathbf{R}^m \cap \{x\colon x_1 = a_1\}$, $B = \mathbf{R}^m \cap \{x\colon x_1 = b_1\}$,

$$f(x) = (a_1, x_2, \ldots, x_m) \text{ and } g(x) = x \text{ for } x \in \mathbf{R}^m,$$

$$h(t, x) = (1-t) f(x) + t\, g(x) = ((1-t)\, a_1 + t\, x_1, x_2, \ldots, x_m)$$

for $(t, x) \in \mathbf{R} \times \mathbf{R}^m$, and we see from 4.1.9 that

$$X = h_\#([0,1] \times [(\partial X) \llcorner B]);$$

in fact $g_\# X = X$, $f_\# X = 0$ because $\operatorname{im} f \subset A$, and h maps $\mathbf{R} \times [(\operatorname{Bdry} W) \sim B]$ into the union of a finite family of $m-1$ dimensional affine subspaces of $\mathbf{R}^m$.

In case $m = 1$, then $B = \{b_1\}$, hence $\partial X \llcorner B = r[b_1]$ for some real number r, and

$$X = r\, h_\#([0,1] \times [b_1]) = r\, h_\#[(0, b_1), (1, b_1)] = r[a_1, b_1].$$

In case $m>1$ we let

$$Z=\mathbf{R}^{m-1}\cap\{y\colon a_{j+1}<y_j<b_{j+1} \text{ for } j=1,\ldots,m-1\},$$

$$p\colon \mathbf{R}^m\to\mathbf{R}^{m-1},\quad p(x)=(x_2,\ldots,x_m) \text{ for } x\in\mathbf{R}^m,$$

$$\xi=(p|B)^{-1},\quad \xi(y)=(b_1,y_1,\ldots,y_{m-1}) \text{ for } y\in\mathbf{R}^{m-1},$$

observe that $\partial(\partial X\llcorner B)=-\partial[\partial X\llcorner(\operatorname{Bdry} W)\sim B]$,

$$\operatorname{spt}\partial(\partial X\llcorner B)\subset B\cap\operatorname{Clos}[(\operatorname{Bdry} W)\sim B]=\xi(\operatorname{Bdry} Z),$$

$$\operatorname{spt}\partial p_\#(\partial X\llcorner B)\subset\operatorname{Bdry} Z,$$

hence inductively obtain a real number r with

$$p_\#(\partial X\llcorner B)=r[a_2,b_2]\times\cdots\times[a_m,b_m]$$

and conclude

$$\begin{aligned}X&=r\,h_\#([0,1]\times\xi_\#([a_2,b_2]\times\cdots\times[a_m,b_m]))\\&=r[a_1,b_1]\times[a_2,b_2]\times\cdots\times[a_m,b_m]\end{aligned}$$

because $h(t,\xi(y))=((1-t)a_1+t\,b_1,y_1,\ldots,y_{m-1})$ for $(t,y)\in\mathbf{R}\times\mathbf{R}^{m-1}$.

Suppose W is as above and $L_1,\ldots,L_\mu$ are univalent affine maps of $\mathbf{R}^m$ into $\mathbf{R}^n$ such that $L_j(W)\cap L_k(\operatorname{Clos} W)=\varnothing$ whenever $j\neq k$. If $T\in\mathbf{F}_m(\mathbf{R}^n)$, $\mathbf{M}(T)<\infty$ and

$$\operatorname{spt} T\subset\bigcup\{L_j(\operatorname{Clos} W)\colon j=1,\ldots,\mu\},$$

$$\operatorname{spt}\partial T\subset\bigcup\{L_j(\operatorname{Bdry} W)\colon j=1,\ldots,\mu\},$$

then there exist real numbers $r_1,\ldots,r_\mu$ such that

$$T=\sum_{j=1}^{\mu} r_j\,L_{j\#}([a_1,b_1]\times\cdots\times[a_m,b_m]);$$

if also $T\in\mathscr{R}_m(\mathbf{R}^n)$, then $r_1,\ldots,r_\mu$ are integers. In fact

$$T=\sum_{j=1}^{\mu} T\llcorner L_j(W)$$

because $\operatorname{spt} T\sim\bigcup\{L_j(W)\colon j=1,\ldots,\mu\}$ is contained in the union of a finite family of $m-1$ dimensional affine subspaces of $\mathbf{R}^n$; for each j we note that

$$\begin{aligned}&\operatorname{spt}\partial[T\llcorner L_j(W)]\\&\subset L_j(\operatorname{Clos} W)\cap[(\operatorname{spt}\partial T)\cup\bigcup\{L_k(\operatorname{Clos} W)\colon k\neq j\}]\subset L_j(\operatorname{Bdry} W),\end{aligned}$$

hence infer from the preceding proposition and 4.1.15 that

$$T\llcorner L_j(W)=r_j\,L_{j\#}([a_1,b_1]\times\cdots\times[a_m,b_m])$$

for some real number r_j; the postscript follows from 4.1.28.

4.2.4. Whenever V and Y are disjoint affine subspaces of $\mathbf{R}^n$ such that $\dim V = n-m-1$, $\dim Y = m < n$ and the affine space generated by $V \cup Y$ equals $\mathbf{R}^n$, we define the **projection with center V onto Y** as the class f of all ordered pairs (x, y) such that $x \in \mathbf{R}^n \sim V$, $y \in Y$ and $V \cup \{x\} \cup \{y\}$ is contained in an $n-m$ dimensional affine subspace of $\mathbf{R}^n$.

Choosing $c \in V$ and letting W be the affine subspace of $\mathbf{R}^n$ generated by $Y \cup \{c\}$, we observe that

$$\mathbf{R}^n = \tau_{-c}(V) \oplus \tau_{-c}(W)$$

and there exist unique linear functions

$$\alpha\colon \mathbf{R}^n \to \mathbf{R} \quad \text{with} \quad \tau_{-c}(V) \subset \ker \alpha, \quad \tau_{-c}(Y) = \tau_{-c}(W) \cap \{x\colon \alpha(x) = 1\},$$
$$\beta\colon \mathbf{R}^n \to \tau_{-c}(W), \quad \beta(x) - x \in \tau_{-c}(V) \text{ for } x \in \mathbf{R}^n,$$

hence conclude $\operatorname{dmn} f = \tau_c(\mathbf{R}^n \sim \ker \alpha)$,

$$f(x) = c + [\alpha(x-c)]^{-1} \beta(x-c) \text{ for } x \in \operatorname{dmn} f.$$

In particular, if

$$V = \mathbf{R}^n \cap \{x\colon x_i = 0 \text{ for } i = 1, \ldots, m+1\},$$
$$Y = \mathbf{R}^n \cap \{x\colon x_i = 1 \text{ for } i = m+1, \ldots, n\},$$
$$c_i = 0 \text{ for } i = 1, \ldots, m+1 \quad \text{and} \quad c_i = 1 \text{ for } i = m+2, \ldots, n,$$

then $\alpha(x) = x_{m+1}$ and $\beta(x) = (x_1, \ldots, x_{m+1}, 0, \ldots, 0)$ for $x \in \mathbf{R}^n$,

$$f(x) = (x_1/x_{m+1}, \ldots, x_m/x_{m+1}, 1, \ldots, 1) \text{ in case } x_{m+1} \neq 0;$$

moreover $Df(x)$ maps $h \in \mathbf{R}^n$ onto the point whose i-th coordinate equals $(h_i - h_{m+1} x_i/x_{m+1})/x_{m+1}$ for $i = 1, \ldots, m$ and whose other coordinates equal 0, hence

$$\|Df(x)\| \le \Big(1 + \sum_{i=1}^{m} |x_i/x_{m+1}|\Big) \Big/ |x_{m+1}|.$$

4.2.5. For $j = 0, 1, \ldots, n$ we define $\mathbf{Z}_j^n$ as the set of all n termed sequences $z = (z_1, \ldots, z_n)$ such that

$$\operatorname{card}\{i\colon z_i \text{ is an even integer}\} = j$$

and

$$\operatorname{card}\{i\colon z_i \text{ is an odd integer}\} = n - j.$$

With each $z \in \mathbf{Z}_j^n$ we associate the j dimensional cube

$$\mathbf{W}'(z) = \mathbf{R}^n \cap \{x\colon |x_i - z_i| < 1 \text{ if } z_i \text{ is even}$$
$$\text{and } x_i = z_i \text{ if } z_i \text{ is odd}\}$$

and also the $n-j$ dimensional cube

$$\mathbf{W}''(z) = \mathbf{R}^n \cap \{x\colon x_i = z_i \text{ if } z_i \text{ is even}$$
$$\text{and } |x_i - z_i| < 1 \text{ if } z_i \text{ is odd}\}.$$

Clearly $\mathbf{Z}^n = \bigcup \{\mathbf{Z}_j^n\colon j = 0, 1, \ldots, n\}$. The families

$$\{\mathbf{W}'(z)\colon z \in \mathbf{Z}^n\} \quad \text{and} \quad \{\mathbf{W}''(z)\colon z \in \mathbf{Z}^n\}$$

are locally finite partitions of $\mathbf{R}^n$, which we call the **standard dual cubical subdivisions** of $\mathbf{R}^n$; we define their k dimensional skeletons as the closed sets

$$\begin{aligned}\mathbf{W}'_k &= \bigcup \{\mathbf{W}'(z)\colon z \in \mathbf{Z}_j^n \text{ with } j \le k\}\\ &= \mathbf{R}^n \cap \{x\colon \operatorname{card}\{i\colon x_i \text{ is an odd integer}\} \ge n-k\},\\ \mathbf{W}''_k &= \bigcup \{\mathbf{W}''(z)\colon z \in \mathbf{Z}_j^n \text{ with } n-j \le k\}\\ &= \mathbf{R}^n \cap \{x\colon \operatorname{card}\{i\colon x_i \text{ is an even integer}\} \ge n-k\}.\end{aligned}$$

We observe that $\mathbf{W}'_m \sim \mathbf{W}'_{m-1}$ is an m dimensional submanifold of $\mathbf{R}^n$ whose components are the cubes $\mathbf{W}'(z)$ corresponding to all $z \in \mathbf{Z}_m^n$. From 4.2.3 (or from 4.1.31 and 4.1.20) we infer that, in case $m > 0$,

$$\mathbf{N}_m(\mathbf{R}^n) \cap \{T\colon \operatorname{spt} T \subset \mathbf{W}'_m,\ \operatorname{spt} \partial T \subset \mathbf{W}'_{m-1}\}$$

is the vectorspace generated by the oriented m dimensional cubes

$$[\mathscr{H}^m \llcorner \mathbf{W}'(z)] \wedge \mathbf{e}_\lambda$$

with $z \in \mathbf{Z}_m^n$, $\lambda \in \Lambda(n,m)$ *and* $\operatorname{im} \lambda = \{i\colon z_i \text{ is even}\}$, because $\mathbf{e}_\lambda | \mathbf{W}'(z)$ orients $\mathbf{W}'(z)$; moreover

$$\mathbf{I}_m(\mathbf{R}^n) \cap \{T\colon \operatorname{spt} T \subset \mathbf{W}'_m,\ \operatorname{spt} \partial T \subset \mathbf{W}'_{m-1}\}$$

is the additive group generated by these oriented cubes. We also see from 4.1.15 that

$$\mathbf{N}_0(R^n) \cap \{T\colon \operatorname{spt} T \subset \mathbf{W}'_0\} \quad \text{and} \quad \mathbf{I}_0(\mathbf{R}^n) \cap \{T\colon \operatorname{spt} T \subset \mathbf{W}'_0\}$$

are the vectorspace and the additive group generated by the currents $\boldsymbol{\delta}_z$ *corresponding to all* $z \in \mathbf{Z}_0^n$.

4.2.6. Here we will construct retractions σ_m of $\mathbf{R}^n \sim \mathbf{W}''_{n-m-1}$ onto $\mathbf{W}'_m$ such that

$$\sigma_m = \sigma_m \circ \sigma_{m+1} \text{ for } m < n$$

and $\sigma_m | \mathbf{W}'_{m+1} \sim \mathbf{W}''_{n-m-1}$ has the geometric description:

If $c \in \mathbf{Z}_{m+1}^n$, then $\mathbf{W}'(c) \cap \mathbf{W}''_{n-m-1} = \{c\}$; for any $x \in W'(c) \sim \{c\}$, $\sigma_m(x) \in [\operatorname{Clos} \mathbf{W}'(c)] \sim \mathbf{W}'(c)$ and x lies on the line segment from c to $\sigma_m(x)$.

This description is helpful in visualizing σ_m but, since estimates for $\|D\sigma_m\|$ will be of crucial importance, we give our formal definition of σ_m in direct and purely algebraic terms as follows:

We let $u_n(x)=1$ and $\sigma_n(x)=x$ for $x\in\mathbf{R}^n$.

For $m=0,1,\ldots,n-1$ we define θ_m: $\mathbf{R}^n\to\mathbf{R}$ so that

$$\operatorname{card}\{i\colon |x_i|\leq\theta_m(x)\}\geq m+1 \quad\text{and}\quad \operatorname{card}\{i\colon |x_i|\geq\theta_m(x)\}\geq n-m$$

whenever $x\in\mathbf{R}^n$. We further let

$$u_m\colon \mathbf{R}^n\to\mathbf{R},\quad \sigma_m\colon \mathbf{R}^n\sim\mathbf{W}''_{n-m-1}\to\mathbf{W}'_m$$

be the functions characterized by the condition:

If $x\in\mathbf{R}^n$, $z\in\mathbf{Z}^n_n$ and $|x_i-z_i|\leq 1$ for $i=1,\ldots,n$, then

$$0\leq u_m(x)=\theta_m(x-z)\leq 1;$$

if also $x\notin\mathbf{W}''_{n-m-1}$, then $u_m(x)>0$ because

$$\operatorname{card}\{x\colon x_i \text{ is an even integer}\}<m+1,$$

and $\sigma_m(x)$ is the point $y\in\mathbf{W}'_m$ such that

$$y_i-z_i=(x_i-z_i)/u_m(x) \text{ whenever } |x_i-z_i|\leq u_m(x),$$
$$y_i-z_i=\operatorname{sign}(x_i-z_i) \quad \text{whenever } |x_i-z_i|\geq u_m(x).$$

This condition implies no inconsistency because if also $\zeta\in\mathbf{Z}^n_n$ and $|x_i-\zeta_i|\leq 1$ for $i=1,\ldots,n$, then for each i

$$\text{either } z_i=\zeta_i \text{ or } x_i-z_i=\zeta_i-x_i=\pm 1,$$

hence $\theta_m(x-z)=\theta_m(x-\zeta)$.

In case $x\in\mathbf{W}'_m$, then $\operatorname{card}\{i\colon x_i$ is an odd integer$\}\geq n-m$, hence $u_m(x)=1$ and $\sigma_m(x)=x$.

Clearly $\mathbf{W}''_{n-m-1}=\{x\colon u_m(x)=0\}$.

One readily verifies that $\operatorname{Lip}(u_m)\leq 1$ and σ_m is locally Lipschitzian. Moreover

$$|\sigma_m(x)-x|\leq n^{\frac{1}{2}} \text{ whenever } x\in\mathbf{R}^n.$$

We define $B=\mathbf{R}^n\cap\{x\colon 0\leq x_1\leq x_2\leq\cdots\leq x_n\leq 1\}$,

$$A=\mathbf{R}^n\cap\{x\colon |x_i|\leq 1 \text{ for } i=1,\ldots,n\},$$

and let Ξ be the group of all isometries ξ of $\mathbf{R}^n$ such that

$$u_m\circ\xi=u_m \text{ and } \sigma_m\circ\xi=\xi\circ\sigma_m \text{ for } m=0,1,\ldots,n.$$

Since all those linear automorphisms of $\mathbf{R}^n$ which permute the set consisting of the standard base vectors and their negatives belong to Ξ, and since $\tau_z \in \Xi$ whenever $z \in \mathbf{Z}^n_n$, we find that A is the union of an $\mathscr{L}^n$ almost disjointed family of $2^n n!$ images of B under transformations in Ξ, and that

$$\mathbf{R}^n = \bigcup \{\xi(B) \colon \xi \in \Xi\}.$$

If $x \in B$, then $u_m(x) = x_{m+1}$ and

$$\sigma_m(x) = (x_1/x_{m+1}, \ldots, x_m/x_{m+1}, 1, \ldots, 1) \in B$$

in case $x_{m+1} > 0$. Observing that σ_m agrees on B with the particular central projection discussed in 4.2.4, we obtain

$$\|D\,\sigma_m(x)\| \le (1+m)/u_m(x)$$

for $\mathscr{L}^n$ almost all x in B; since Ξ is countable we infer that this inequality holds for $\mathscr{L}^n$ almost all x in $\mathbf{R}^n$. Similarly we verify that

$$u_m \le u_{m+1} \quad \text{and} \quad \sigma_m = \sigma_m \circ \sigma_{m+1}.$$

Recalling 4.1.32 we see that σ_m is projective on $\xi(B \sim \mathbf{W}''_{n-m-1})$ whenever $\xi \in \Xi$; moreover

$$B \cap \mathbf{W}'_{m+1} = B \cap \{x \colon x_i = 1 \text{ for } i = m+2, \ldots, n\},$$

$$B \cap \mathbf{W}''_{n-m-1} = B \cap \{x \colon x_i = 0 \text{ for } i = 1, \ldots, m+1\},$$

$$B \cap \mathbf{W}'_{m+1} \cap \mathbf{W}''_{n-m-1} \text{ consists of a single point } c,$$

$$x = (1 - x_{m+1})\,c + x_{m+1}\,\sigma_m(x) \text{ for } x \in B \cap \mathbf{W}'_{m+1} \sim \mathbf{W}''_{n-m-1}$$

(which confirms the initial geometric description of σ_m); replacing x by $\sigma_{m+1}(x)$ we conclude that, for every j dimensional simplex $E \subset \xi(B \sim \mathbf{W}''_{n-m-1})$, $\sigma_m(E)$ is a simplex with dimension at most j, and $\sigma_{m+1}(E)$ is contained in the convex hull of $\{\xi(c)\} \cup \sigma_m(E)$.

4.2.7. Lemma. *Let u_m, A be as in 4.2.6, and $0^{-m} = \infty$.*

(1) *If ρ is a Radon measure over $\mathbf{R}^n$, then*

$$\textstyle\int_A \int (u_m \circ \tau_a)^{-m}\, d\rho\, d\mathscr{L}^n a = \binom{n}{m} \rho(\mathbf{R}^n)\, \mathscr{L}^n(A).$$

(2) *If ρ_i are Radon measures over $\mathbf{R}^n$ and $0 \le m_i \le n$ are integers for $i = 1, \ldots k$, then the set of all points a in A such that*

$$\textstyle\int (u_{m_i} \circ \tau_a)^{-m_i}\, d\rho_i \le k \binom{n}{m_i} \rho_i(\mathbf{R}^n) \text{ for } i = 1, \ldots, k$$

has positive $\mathscr{L}^n$ measure.

Proof. Assuming $m<n$ we define B as in 4.2.6, let

$$B_t=\mathbf{R}^{n-1}\cap\{y: 0\le y_1\le\cdots\le y_m\le t\le y_{m+1}\le\cdots\le y_{n-1}\le 1\}$$

whenever $t\in\mathbf{R}$ and compute

$$\int_B (x_{m+1})^{-m}\,d\mathscr{L}^n x=\int t^{-m}\mathscr{L}^{n-1}(B_t)\,d\mathscr{L}^1 t$$
$$=\int_0^1 t^{-m}[t^m/m!]\,[(1-t)^{n-1-m}/(n-1-m)!]\,d\mathscr{L}^1 t=1/[m!\,(n-m)!],$$

$$\int_A (u_m)^{-m}\,d\mathscr{L}^n=2^n n!\int_B (u_m)^{-m}\,d\mathscr{L}^n=2^n\binom{n}{m}.$$

Since $u_m\circ\tau_z=u_m$ for $z\in\mathbf{Z}^n_n$, and the characteristic function α of A satisfies the condition

$$\sum_{z\in\mathbf{Z}^n_n}\alpha(x-z)=1 \text{ for } \mathscr{L}^n \text{ almost all } x \text{ in } \mathbf{R}^n,$$

we infer that, whenever $w\in\mathbf{R}^n$,

$$\int_A [u_m(w+a)]^{-m}\,d\mathscr{L}^n a=\int [u_m(w+x)]^{-m}\alpha(x)\,d\mathscr{L}^n x$$
$$=\int [u_m(x)]^{-m}\alpha(x-w)\sum_{z\in\mathbf{Z}^n_n}\alpha(x-z)\,d\mathscr{L}^n x$$
$$=\sum_{z\in\mathbf{Z}^n_n}\int [u_m(x)]^{-m}\alpha(x+z-w)\,\alpha(x)\,d\mathscr{L}^n x$$
$$=\int [u_m(x)]^{-m}\alpha(x)\,d\mathscr{L}^n x=2^n\binom{n}{m}.$$

Applying Fubini's theorem we conclude

$$\mathscr{L}^n(A)\binom{n}{m}\rho(\mathbf{R}^n)=\int_{\mathbf{R}^n}2^n\binom{n}{m}\,d\rho\, w=\int_A\int_{\mathbf{R}^n}[u_m(w+a)]^{-m}\,d\rho\, w\,d\mathscr{L}^n a,$$

as asserted in (1). The denial of (2) would imply

$$\sum_{i=1}^k\left[k\binom{n}{m_i}\rho_i(\mathbf{R}^n)\right]^{-1}\int (u_{m_i}\circ\tau_a)^{-m_i}\,d\rho_i>1$$

for $\mathscr{L}^n$ almost all a in A, and $\mathscr{L}^n$ integration over A would yield a strict inequality contradicting (1).

4.2.8. Whenever $0<\varepsilon<\infty$ we define

$$\mu_\varepsilon:\ \mathbf{R}^n\to\mathbf{R}^n,\qquad \mu_\varepsilon(x)=\varepsilon\, x \text{ for } x\in\mathbf{R}^n,$$

and observe that $\|\mu_{\varepsilon\,\#}T\|=\varepsilon^m\,\mu_{\varepsilon\,\#}\,\|T\|$ for all those m dimensional currents T in $\mathbf{R}^n$ which are representable by integration.

4.2.9. Deformation theorem. *Whenever* $T \in \mathbf{N}_m(\mathbf{R}^n)$ *and* $\varepsilon > 0$ *there exist* $P \in \mathbf{N}_m(\mathbf{R}^n)$, $Q \in \mathbf{N}_m(\mathbf{R}^n)$, $S \in \mathbf{N}_{m+1}(\mathbf{R}^n)$ *such that the following conditions hold with* $\gamma = 2n^{2m+2}$:

(1) $T = P + Q + \partial S$.

(2) *In case* $m > 0$, *then*

$$\mathbf{M}(P)/\varepsilon^m \le \gamma[\mathbf{M}(T)/\varepsilon^m + \mathbf{M}(\partial T)/\varepsilon^{m-1}], \quad \mathbf{M}(\partial P)/\varepsilon^{m-1} \le \gamma \mathbf{M}(\partial T)/\varepsilon^{m-1},$$

$$\mathbf{M}(Q)/\varepsilon^m \le \gamma \mathbf{M}(\partial T)/\varepsilon^{m-1}, \quad \mathbf{M}(S)/\varepsilon^{m+1} \le \gamma \mathbf{M}(T)/\varepsilon^m.$$

(3) *In case* $m = 0$, *then* $\mathbf{M}(P) \le \mathbf{M}(T)$, $Q = 0$, $\mathbf{M}(S)/\varepsilon \le \gamma \mathbf{M}(T)$.

(4) $\operatorname{spt} P \cup \operatorname{spt} S \subset \{x: \operatorname{dist}(x, \operatorname{spt} T) \le 2n\varepsilon\}$;

$\operatorname{spt} \partial P \cup \operatorname{spt} Q \subset \{x: \operatorname{dist}(x, \operatorname{spt} \partial T) \le 2n\varepsilon\}$ *if* $m > 0$.

(5) $\operatorname{spt} P \subset \mu_\varepsilon(\mathbf{W}'_m)$, $\operatorname{spt} \partial P \subset \mu_\varepsilon(\mathbf{W}'_{m-1})$ *if* $m > 0$, *hence P is a polyhedral chain.*

(6) *If T is an integral current, then P is an integral polyhedral chain and Q, S are integral currents.*

(7) *If* ∂T *is a rectifiable current, then Q is an integral current.*

(8) *If* ∂T *is a (integral) polyhedral chain, so is Q.*

(9) *If T is a (integral) polyhedral chain, so is S.*

Proof. Since the general theorem is easily reduced to the special case $\varepsilon = 1$ by means of the map μ_ε, we assume that $\varepsilon = 1$.

In case $m > 0$ we apply 4.2.7 to choose $a \in A$ so that

$$\int (u_m \circ \tau_a)^{-m} \, d\|T\| \le 2\binom{n}{m} \mathbf{M}(T),$$

$$\int (u_{m-1} \circ \tau_a)^{1-m} \, d\|\partial T\| \le 2\binom{n}{m-1} \mathbf{M}(\partial T)$$

with $0^{-m} = \infty$, hence

$$\|T\| \{x: u_m[\tau_a(x)] = 0\} = 0, \qquad \|\partial T\| \{x: u_{m-1}[\tau_a(x)] = 0\} = 0.$$

From 4.2.6 we see that $\mathscr{L}^n$ almost all points x in $\mathbf{R}^n$ satisfy the conditions

$$\|D\sigma_i(x)\| \le (i+1)/u_i(x) \le n/u_j(x) \text{ for } j \le i < n.$$

Letting $v = (u_m \circ \tau_a)/n$, $w = (u_{m-1} \circ \tau_a)/n$ and recalling 4.2.2 we find that

$$\sigma_i \circ \tau_a \text{ and } T \text{ are } v \text{ admissible for } m \le i \le n,$$

$$\sigma_i \circ \tau_a \text{ and } \partial T \text{ are } w \text{ admissible for } m-1 \le i \le n.$$

We define

$$S_i = H_v(\sigma_i \circ \tau_a, \sigma_{i+1} \circ \tau_a)\, T \text{ for } m \le i < n, \quad S_n = H_v(\sigma_n \circ \tau_a, \sigma_n)\, T,$$

$$Q_i = H_w(\sigma_i \circ \tau_a, \sigma_{i+1} \circ \tau_a)\, \partial T \text{ for } m-1 \le i < n, \quad Q_n = H_w(\sigma_n \circ \tau_a, \sigma_n)\, \partial T,$$

$$S = \sum_{i=m}^{n} S_i, \quad Q = \sum_{i=m-1}^{n} Q_i, \quad P = [(\sigma_m \circ \tau_a)_{\#\, v} T] - Q_{m-1}$$

and infer from the homotopy formulae involving H_v and H_w that

$$\partial S + Q - Q_{m-1} = T - (\sigma_m \circ \tau_a)_{\#\, v} T, \quad \partial Q = \partial T - (\sigma_{m-1} \circ \tau_a)_{\#\, w} \partial T,$$

hence (1) holds and

$$\partial P = \partial T - \partial Q = (\sigma_{m-1} \circ \tau_a)_{\#\, w} \partial T.$$

Observing that $\operatorname{spt} Q_{m-1} \subset \mathbf{W}'_m$, because $\mathbf{W}'_m$ contains the line segment from $(\sigma_{m-1} \circ \tau_a)(x)$ to $(\sigma_m \circ \tau_a)(x)$ whenever $w(x) > 0$, we deduce (5) and (6) with the help of 4.2.5. Now (7) and (4) are evident. Furthermore (2) follows readily from the estimates

$$\mathbf{M}(S_i) \le \|T\|\,(n\, v^{-m}) \le n^{1+m}\, 2 \binom{n}{m} \mathbf{M}(T),$$

$$\mathbf{M}(Q_i) \le \|\partial T\|\,(n\, w^{1-m}) \le n^m\, 2 \binom{n}{m-1} \mathbf{M}(\partial T),$$

$$\mathbf{M}(\sigma_m \circ \tau_a)_{\#\, v} T \le \|T\|\,(v^{-m}) \le n^m\, 2 \binom{n}{m} \mathbf{M}(T),$$

$$\mathbf{M}(\sigma_{m-1} \circ \tau_a)_{\#\, w} (\partial T) \le \|\partial T\|\,(w^{-m}) \le n^{m-1}\, 2 \binom{n}{m-1} \mathbf{M}(\partial T).$$

If ∂T is a (integral) polyhedral chain, then $\mathbf{W}''_{n-m} \times \operatorname{spt} \partial T$ is contained in the union of a countable family of $n-1$ dimensional affine subspaces of $\mathbf{R}^n \times \mathbf{R}^n$, hence

$$\mathbf{R}^n \cap \{\eta : \mathbf{W}''_{n-m} \cap \tau_\eta(\operatorname{spt} \partial T) \ne \varnothing\} = \{x - y : (x,y) \in \mathbf{W}''_{n-m} \times \operatorname{spt} \partial T\}$$

has $\mathscr{L}^n$ measure 0, and we can choose a so that

$$\tau_a(\operatorname{spt} \partial T) \subset \mathbf{R}^n \sim \mathbf{W}''_{n-m}.$$

Under these conditions $\tau_a(\partial T)$ is expressible as finite sum of polyhedral chains, each with support contained in $\xi(B) \sim \mathbf{W}''_{n-m}$ for some $\xi \in \Xi$, hence with (integral) polyhedral images under the maps $\sigma_{i\,\#}$ corresponding to $m-1 \le i < n$, and with (integral) polyhedral affine deformation chains from σ_i to σ_{i+1}. Thus (8) is proved. Similarly one verifies (9), choosing a so that

$$\tau_a(\operatorname{spt} T) \subset \mathbf{R}^n \sim \mathbf{W}''_{n-m-1} \text{ if } T \text{ is polyhedral}.$$

In case $m=0$ we apply 4.2.7 to choose $a\in A$ so that

$$\|T\|\{x: u_0[\tau_a(x)]=0\}=0,$$

let $v=(u_0\circ\tau_a)/n$, $P=(u_0\circ\tau_a)_{\#\,v}T$, $Q=0$, define S as in the previous case, and verify the relevant conclusions by a much simplified version of the previous argument.

4.2.10. As a corollary of the preceding theorem we obtain the following isoperimetric inequality:

If $T\in\mathbf{I}_m(R^n)$ with $\partial T=0$, then there exists

$$S\in\mathbf{I}_{m+1}(\mathbf{R}^n) \text{ with } \partial S=T \text{ and } \mathbf{M}(S)^{m/(m+1)}\leq\gamma\mathbf{M}(T).$$

In fact, choosing ε so that $\gamma\mathbf{M}(T)=\varepsilon^m$, we find that $Q=0$ because $\partial T=0$, $P=0$ because $\mathbf{M}(P)\leq\varepsilon^m$ while 4.2.5 shows that $\mathbf{M}(P)$ is an integral multiple of $(2\varepsilon)^m$, and

$$\mathbf{M}(S)\leq\varepsilon\gamma\mathbf{M}(T)=\varepsilon^{m+1}=[\gamma\mathbf{M}(T)]^{(m+1)/m}.$$

Of course the constant γ used here is much larger than necessary. For the special cases $m=n-1$ and $m=1$ the optimal constants will be obtained in 4.5.9 (31) and 4.5.14. For other values of m the optimal constants are unknown.

A much more general isoperimetric inequality will be proved in 4.4.2.

A rather trivial corollary of 4.2.9 is the proposition that $\mathscr{R}_m(\mathbf{R}^m)$ *equals the* $\mathbf{M}$ *closure of* $\mathscr{P}_m(\mathbf{R}^m)$; clearly $\mathscr{R}_m(\mathbf{R}^m)$ equals the $\mathbf{M}$ closure of $\mathbf{I}_m(\mathbf{R}^m)$, and applying the theorem with $T\in\mathbf{I}_m(\mathbf{R}^m)$ we obtain $S=0$, $\mathbf{M}(T-P)=\mathbf{M}(Q)\leq\varepsilon\gamma\mathbf{M}(\partial T)$. (This proposition also follows easily from 4.1.28.) Recalling 4.1.15 we find that *every m dimensional rectifiable current in* $\mathbf{R}^n$, *whose support is contained in some m dimensional affine subspace of* $\mathbf{R}^n$, *belongs to the* $\mathbf{M}$ *closure of* $\mathscr{P}_m(\mathbf{R}^n)$.

In the following 4.2.11–4.2.14 we present an alternate proof of a proposition derived previously and more simply in 4.1.21. However the new method has the advantage of generalizing to flat chains modulo v (see 4.2.26), and Lemma 4.2.11 is of intrinsic interest.

4.2.11. Lemma. *If $E\subset\mathbf{R}^n$ and $\mathscr{I}_1^m(E)=0$ with $0<m<n$, then almost all* [see 2.7.16 (7)] *$n-m-1$ dimensional affine subspaces V of* $\mathbf{R}^n$ *have the property that*

$$\mathscr{H}^m[f(E\cap\operatorname{dmn}f)]=0$$

for every projection f with center V (see 4.2.4).

Proof. We let G be the isometry group of $\mathbf{R}^n$ with the Haar measure ϕ,

$$U=\mathbf{R}^n\cap\{u\colon u_i=0 \text{ for } i=1,\ldots,m+1\},$$

$$S=\mathbf{R}^n\cap\{s\colon |s|=1,\ s\bullet u=0 \text{ for } u\in U\},$$

$$\zeta(s)=\text{the vectorspace generated by } U\cup\{s\} \text{ whenever } s\in S,$$

and assume that E is a Borel set, hence

$$(G\times S)\cap\{(g,s)\colon g[\zeta(s)]\cap E\neq\emptyset\}$$

is a Suslin set. From 2.10.16 and Fubini's Theorem we infer

$$\phi\{g\colon g[\zeta(s)]\cap E\neq\emptyset\}=0 \text{ for } s\in S,$$

$$\mathscr{H}^m\{s\colon g[\zeta(s)]\cap E\neq\emptyset\}=0 \text{ for } \phi \text{ almost all } g.$$

We consider henceforth a fixed $g\in G$ satisfying the last equation, and will show that $V=g(U)$ has the property asserted in the lemma. Letting Y and f be as in 4.2.4 we define the locally Lipschitzian map

$$\psi=f\circ g|S\colon S\cap g^{-1}(\operatorname{dmn} f)\to Y$$

and observe that, for each $y\in Y$, $f^{-1}\{y\}$ is an $n-m$ dimensional affine subspace of $\mathbf{R}^n$ containing $g(U)$, hence

$$f^{-1}\{y\}=g[\zeta(s)] \text{ for some } s\in S,$$

which implies $y=\psi(s)$; consequently

$$\{y\colon f^{-1}\{y\}\cap E\neq\emptyset\}=\{\psi(s)\colon g[\zeta(s)]\cap E\neq\emptyset\}$$

has $\mathscr{H}^m$ measure 0.

4.2.12. Corollary. $\mathscr{H}^m(\sigma_m[g(E)\cap\operatorname{dmn}\sigma_m])=0$ *for almost all isometries* g *of* $\mathbf{R}^n$, *where* σ_m *is as in* 4.2.6.

Proof. We recall that $\sigma_m|B=\rho|B$, where ρ is a projection with center U, and observe that $g^{-1}\circ\rho\circ g$ is a projection with center $g^{-1}(U)$ whenever $g\in G$. Letting

$$\Omega=G\cap\{g\colon \mathscr{H}^m[(\rho\circ g)(E\cap\operatorname{dmn}\rho\circ g)]=0\}$$

we infer from the lemma that $G\sim\Omega$ has Haar measure 0, hence almost all isometries g of $\mathbf{R}^n$ satisfy the condition

$$\xi^{-1}\circ g\in\Omega \text{ for all } \xi\in\Xi;$$

since $\sigma_m|\xi(B)=\xi\circ\rho\circ\xi^{-1}|\xi(B)$ this condition implies that the sets

$$\sigma_m[\xi(B)\cap g(E)\cap\operatorname{dmn}\sigma_m]\subset(\xi\circ\rho\circ\xi^{-1}\circ g)[E\cap\operatorname{dmn}\rho\circ\xi^{-1}\circ g]$$

have $\mathscr{H}^m$ measure 0 for all $\xi\in\Xi$.

4.2.13. Lemma. *If* $T \in \mathbf{N}_m(\mathbf{R}^n)$, $0<m<n$, $\varepsilon>0$ *and* E *is a compact subset of* $\mathbf{R}^n$ *with* $\mathscr{I}_1^m(E)=0$, *then there exist* $R \in \mathbf{N}_m(\mathbf{R}^n)$ *and* $S \in \mathbf{N}_{m+1}(\mathbf{R}^n)$ *such that*

$$T = R + \partial S, \quad \mathbf{M}(S)/\varepsilon^{m+1} \le 6 n^{2m+1} \mathbf{M}(T)/\varepsilon^m,$$

$$\mathbf{M}(R)/\varepsilon^m \le 3 n^{2m} [\|T\|(\mathbf{R}^n \sim E)/\varepsilon^m + \mathbf{M}(\partial T)/\varepsilon^{m-1}].$$

Proof. We will discuss only the special case when $\varepsilon=1$, to which the general case is easily reduced by means of the map μ_ε. First we apply 4.2.12 and 4.2.7 to choose $t \in \mathbf{O}(n)$ and $a \in A$ so that

$$\mathscr{H}^m[\sigma_m(\tau_a[t(E)] \cap \operatorname{dmn} \sigma_m)] = 0, \quad \int (u_m \circ \tau_a)^{-m} \, d\|t_\# T\| \le 3\binom{n}{m} \mathbf{M}(T),$$

$$\int_{\mathbf{R}^n \sim t(E)} (u_m \circ \tau_a)^{-m} \, d\|t_\# T\| \le 3\binom{n}{m} \|T\|(\mathbf{R}^n \sim E),$$

$$\int (u_{m-1} \circ \tau_a)^{1-m} \, d\|\partial t_\# T\| \le 3\binom{n}{m-1} \mathbf{M}(\partial T),$$

note that $\sigma_m \circ \tau_a$, σ_n and $t_\# T$ are $v = (u_m \circ \tau_a)/n$ admissible, define

$$R = t_\#^{-1}[(\sigma_m \circ \tau_a)_{\# v}(t_\# T) + H_v(\sigma_m \circ \tau_a, \sigma_n)(\partial t_\# T)],$$

$$S = t_\#^{-1}[H_v(\sigma_m \circ \tau_a, \sigma_n)(t_\# T)]$$

and infer the equation $T = R + \partial S$ from the homotopy formula for H_v.

For each $r>0$ the first proposition in 4.2.3 implies

$$X = (\sigma_m \circ \tau_a)_\# [(t_\# T) \llcorner \{x: v(x) > r\} \cap t(E)] = 0,$$

because $\operatorname{spt} X \subset (\sigma_m \circ \tau_a)[\{x: v(x) \ge r\} \cap t(E)]$, which is a compact subset of $\mathbf{W}_m'$ with $\mathscr{H}^m$ measure 0; accordingly

$$\mathbf{M}((\sigma_m \circ \tau_a)_\# [(t_\# T) \llcorner \{x: v(x) > r\}]) \le \int_{\mathbf{R}^n \sim t(E)} v^{-m} \, d\|t_\# T\|.$$

Since $|\sigma_m[\tau_a(x)] - x| \le 2n$ for $x \in \operatorname{dmn} \sigma_m$, we conclude

$$\mathbf{M}(R) \le \int_{\mathbf{R}^n \sim t(E)} v^{-m} \, d\|t_\# T\| + \int 2n \, v^{1-m} \, d\|\partial t_\# T\|$$

$$\le n^m \, 3\binom{n}{m} \|T\|(\mathbf{R}^n \sim E) + 6 n^m \binom{n}{m-1} \mathbf{M}(\partial T),$$

$$\mathbf{M}(S) \le \int 2n \, v^{-m} \, d\|t_\# T\| \le 6 n^{1+m} \binom{n}{m} \mathbf{M}(T).$$

4.2.14. Theorem. *If* $T \in \mathbf{F}_m(\mathbf{R}^n)$, $m>0$, $\mathbf{M}(T)<\infty$ *and* $E \subset \mathbf{R}^n$, *then*

$$\mathscr{I}_1^m(E) = 0 \quad \textit{implies} \quad \|T\|(E) = 0.$$

Proof. Since $\mathscr{I}_1^m$ is Borel regular and $\|T\|$ is a Radon measure we may assume that E is compact and $\mathscr{I}_1^m(E)=0$. Then Fubini's Theorem implies $\mathscr{L}^n(E)=0$ and the representation of T derived in 4.1.18 reduces our problem to the case when $T=\partial(\mathscr{L}^n \wedge \eta)$. Thus we may also assume that $\partial T=0$ and $m<n$. [In case $T\in\mathscr{F}_m(\mathbf{R}^n)$ this reduction could be accomplished by use of 4.1.28 (4) and 3.2.26 in place of 4.1.18.]

Letting $u(x)=\operatorname{dist}(x,E)$ whenever $x\in\mathbf{R}^n$, we see from 4.2.1 that, for $\mathscr{L}^1$ almost all positive numbers r,

$$T_r=T\llcorner\{x\colon u(x)\leq r\}\in\mathbf{N}_m(\mathbf{R}^n)$$

because $\partial T_r=\langle T,u,r+\rangle$, and infer from 4.2.13 that

$$\mathbf{F}_C(T_r)\leq 6n^{2m+1}[\|T_r\|(\mathbf{R}^n\sim E)+\varepsilon\,\mathbf{M}(\partial T_r)+\varepsilon\,\mathbf{M}(T_r)]$$

for every $\varepsilon>0$, where C is the convex hull of spt T (see 4.1.16), hence

$$\mathbf{F}_C(T_r)\leq 6n^{2m+1}\|T\|\{x\colon 0<u(x)\leq r\}.$$

We conclude

$$\mathbf{F}_C(T\llcorner E)=\lim_{r\to 0+}\mathbf{F}_C(T_r)=0,\qquad T\llcorner E=0.$$

4.2.15. Lemma. *If $T\in\mathbf{N}_m(\mathbf{R}^n)$, $m>0$, $\partial T=0$ and if, for each $a\in\mathbf{R}^n$,*

$$\partial[T\llcorner\mathbf{B}(a,r)]\in\mathscr{R}_{m-1}(\mathbf{R}^n)$$

for $\mathscr{L}^1$ almost all positive numbers r, then $T\in\mathscr{R}_m(\mathbf{R}^n)$.

Proof. Choosing positive numbers ξ,η with

$$2m\,n^{2m}[\boldsymbol{\alpha}(m)\,\eta\,\xi]^{1/m}<\xi<1,$$

we will first show that the set

$$C=\mathbf{R}^n\cap\{a\colon \Theta^{*m}(\|T\|,a)<\eta\}$$

has $\|T\|$ measure 0.

If $a\in(\operatorname{spt}T)\cap C$, then $f(r)=\|T\|\,\mathbf{B}(a,r)>0$ for $r>0$,

$$\delta^{-m}f(\delta)<\boldsymbol{\alpha}(m)\,\eta\quad\text{for small }\delta>0.$$

Since $f^{1/m}$ is a nondecreasing function it follows from 2.9.19 that

$$\delta^{-1}\textstyle\int_0^\delta(f^{1/m})'\,d\mathscr{L}^1\leq\delta^{-1}f(\delta)^{1/m}<[\boldsymbol{\alpha}(m)\,\eta]^{1/m},$$

hence there exists a positive number $r<\delta$ such that

$$m^{-1}f(r)^{(1-m)/m}f'(r)=(f^{1/m})'(r)<[\boldsymbol{\alpha}(m)\,\eta]^{1/m}$$

and $\partial[T \llcorner \mathbf{B}(a,r)] \in \mathscr{R}_{m-1}(\mathbf{R}^n)$; we also see from 4.2.1, with $u(x)=|x-a|$ for $x \in \mathbf{R}^n$, that

$$\mathbf{M}(\partial[T \llcorner \mathbf{B}(a,r)]) = \mathbf{M}\langle T, u, r+\rangle \leq f'(r).$$

In case $m>1$ we use 4.2.10 with m replaced by $m-1$ to obtain

$$S \in \mathbf{I}_m(\mathbf{R}^n) \quad \text{with} \quad \partial S = \partial[T \llcorner \mathbf{B}(a,r)], \quad \operatorname{spt} S \subset \mathbf{B}(a,r),$$

$$\mathbf{M}(S)^{(m-1)/m} \leq 2n^{2m} f'(r) < [\xi f(r)]^{(m-1)/m}, \text{ hence } \mathbf{M}(S) < \xi \, \|T\| \, \mathbf{B}(a,r).$$

In case $m=1$, then $f'(r) < \boldsymbol{\alpha}(1)\,\eta < 1$, $\partial[T \llcorner \mathbf{B}(a,r)] = 0$ and we take $S=0$. In both cases 4.1.10, 11 imply

$$S - T \llcorner \mathbf{B}(a,r) = \partial(\boldsymbol{\delta}_a \times [S - T \llcorner \mathbf{B}(a,r)]),$$

$$\mathbf{F}_{\mathbf{B}(a,r)}[S - T \llcorner \mathbf{B}(a,r)] < r \, \|T\| \, \mathbf{B}(a,r).$$

If $\|T\| \, C > 0$ we could choose a neighborhood V of C in $\mathbf{R}^n$ such that $\xi \, \|T\| \, V < \|T\| \, C$, and for every $\rho > 0$ we could apply 2.8.18 to cover $\|T\|$ almost all of C by a sequence of disjoint balls $\mathbf{B}(a_i, r_i) \subset V$ such that $r_i < \rho$ and there exist

$$S_i \in \mathbf{I}_m(\mathbf{R}^n) \quad \text{with} \quad \mathbf{M}(S_i) < \xi \, \|T\| \, \mathbf{B}(a_i, r_i),$$

$$\mathbf{F}_K[S_i - T \llcorner \mathbf{B}(a_i, r_i)] < \rho \, \|T\| \, \mathbf{B}(a_i, r_i)$$

where $K = \operatorname{spt} T \cup \operatorname{Clos} V$, hence the current

$$R_\rho = T + \sum_{i=1}^{\infty} [S_i - T \llcorner \mathbf{B}(a_i, r_i)]$$

would satisfy the conditions

$$\mathbf{M}(R_\rho) \leq \|T\| (\mathbf{R}^n \sim C) + \xi \, \|T\| \, V, \quad \mathbf{F}_K(T - R_\rho) \leq \rho \, \|T\| \, V.$$

Using the lowersemicontinuity of $\mathbf{M}$ we could conclude

$$\mathbf{M}(T) \leq \|T\| (\mathbf{R}^n \sim C) + \xi \, \|T\| \, V < \mathbf{M}(T).$$

Since $\|T\| \, C = 0$ we infer from 3.3.15 and 4.2.14 that $\mathbf{R}^n$ is $\|T\|$ rectifiable. Applying 3.2.18 and 4.2.14 we cover $\|T\|$ almost all of $\mathbf{R}^n$ by a sequence of disjoint sets $\psi_j(K_j)$ where

$$K_j \subset \mathbf{R}^m, \quad K_j \text{ is compact}, \quad \psi_j \colon \mathbf{R}^m \to \mathbf{R}^n, \quad \operatorname{Lip}(\psi_j) < 2,$$

$$\psi_j | K_j \text{ is univalent}, \quad \operatorname{Lip}[(\psi_j | K_j)^{-1}] < 2.$$

Therefore

$$T = \sum_{j=1}^{\infty} [T \llcorner \psi_j(K_j)]$$

and it will suffice to prove that each summand is a rectifiable current.

Using 2.10.43 we extend $(\psi_j|K_j)^{-1}$ to a map

$$g: \mathbf{R}^n \to \mathbf{R}^m \quad \text{with} \quad \operatorname{Lip}(g)<2.$$

For every $\rho>0$ we apply 2.8.18 and 2.9.11 to cover $\|T\|$ almost all of $\psi_j(K_j)$ by a sequence of disjoint balls $\mathbf{B}(a_i,r_i)$ such that

$$\|T\|\,[\mathbf{B}(a_i,r_i)\sim\psi_j(K_j)]<\rho\,\|T\|\,\mathbf{B}(a_i,r_i),$$
$$\partial T_i\in\mathscr{R}_{m-1}(\mathbf{R}^n) \quad \text{with} \quad T_i=T\llcorner\mathbf{B}(a_i,r_i).$$

Since $\partial(g_\# T_i)\in\mathscr{R}_{m-1}(\mathbf{R}^m)$ we see from 4.2.10 and 4.1.10 that

$$g_\# T_i\in\mathbf{I}_m(\mathbf{R}^m), \qquad (\psi_j\circ g)_\# T_i\in\mathbf{I}_m(\mathbf{R}^n).$$

Moreover $\psi_j[g(x)]=x$ for $x\in\psi_j(K_j)$, hence 4.1.15 implies

$$(\psi_j\circ g)_\# T_i=T_i\llcorner\psi_j(K_j)+(\psi_j\circ g)_\#[T\llcorner\mathbf{B}(a_i,r_i)\sim\psi_j(K_j)].$$

Inasmuch as $\operatorname{Lip}(\psi_j\circ g)<4$ we infer

$$Q=\sum_{i=1}^\infty(\psi_j\circ g)_\# T_i\in\mathbf{I}_m(\mathbf{R}^n),$$

$$\mathbf{M}[Q-T\llcorner\psi_j(K_j)]\le\sum_{i=1}^\infty 4^m\rho\,\|T\|\,\mathbf{B}(a_i,r_i)\le 4\rho\,\mathbf{M}(T).$$

4.2.16. Closure theorem. *If K is a compact Lipschitz neighborhood retract in $\mathbf{R}^n$ and m is a nonnegative integer, then:*

(1) *$\mathbf{I}_{m,K}(\mathbf{R}^n)$ is $\mathbf{F}_K$ closed in $\mathbf{N}_{m,K}(\mathbf{R}^n)$.*

(2) *$\mathscr{R}_{m+1,K}(\mathbf{R}^n)\cap\{S:\mathbf{M}(\partial S)<\infty\}=\mathbf{I}_{m+1,K}(\mathbf{R}^n)$.*

(3) *$\mathscr{F}_{m,K}(\mathbf{R}^n)\cap\{T:\mathbf{M}(T)<\infty\}=\mathscr{R}_{m,K}(\mathbf{R}^n)$.*

Proof. For each m, clearly (2) implies (3); moreover (1) implies (2), because if $S\in\mathscr{R}_{m+1,K}(\mathbf{R}^n)$, then S belongs to the $\mathbf{M}$ closure of $\mathbf{I}_{m+1,K}(\mathbf{R}^n)$, hence ∂S belongs to the $\mathbf{F}_K$ closure of $\mathbf{I}_{m,K}(\mathbf{R}^n)$, and in case $\mathbf{M}(\partial S)<\infty$ it follows from (1) that $\partial S\in\mathbf{I}_{m,K}(\mathbf{R}^n)$.

To prove (1) we suppose $T\in\mathbf{N}_{m,K}(\mathbf{R}^n)$, $Q_i\in\mathbf{I}_{m,K}(\mathbf{R}^n)$ and

$$\sum_{i=1}^\infty\mathbf{F}_K(T-Q_i)<\infty.$$

For every nonempty closed subset E of $\mathbf{R}^n$ we apply 4.2.1, with $u(x)=\operatorname{dist}(x,E)$ for $x\in\mathbf{R}^n$, to infer

$$\int_0^{*b}\sum_{i=1}^\infty\mathbf{F}_K[(T-Q_i)\llcorner E_r]\,d\mathscr{L}^1 r<\infty \quad \text{for } 0<b<\infty,$$

where $E_r=\{x:\operatorname{dist}(x,E)\le r\}$, hence

$$\lim_{i\to\infty}\mathbf{F}_K[(T-Q_i)\llcorner E_r]=0 \quad \text{for } \mathscr{L}^1 \text{ almost all } r.$$

In case $m=0$, then $Q_i \in \mathscr{P}_0(\mathbf{R}^n)$ and we see that

$$(T \llcorner E_r)(1) = \lim_{i\to\infty} (Q_i \llcorner E_r)(1) \in \mathbf{Z} \text{ for } \mathscr{L}^1 \text{ almost all } r,$$

$$(T \llcorner E)(1) = \lim_{r\to 0+} (T \llcorner E_r)(1) \in \mathbf{Z}$$

for every nonempty closed subset E of $\mathbf{R}^n$, hence $T \in \mathscr{P}_0(\mathbf{R}^n)$.

In case $m>0$ we first observe that

$$\mathbf{F}_K(\partial T - \partial Q_i) \to 0 \quad \text{as} \quad i \to \infty,$$

inductively use (1) with m replaced by $m-1$ to infer $\partial T \in \mathbf{I}_{m-1,K}(\mathbf{R}^n)$, and apply 4.2.10 to select

$$R \in \mathbf{I}_m(\mathbf{R}^n) \quad \text{with} \quad \partial R = \partial T, \quad \text{hence} \quad \partial(T-R)=0.$$

In view of 4.1.29 it will suffice to prove $T-R \in \mathscr{R}_m(\mathbf{R}^n)$. Replacing T, Q_i and K by $T-R$, $Q_i - R$ and the convex hull of $K \cup \operatorname{spt} R$, we assume from now on that $\partial T = 0$.

Whenever $a \in \mathbf{R}^n$ we take $E=\{a\}$ to obtain

$$\begin{aligned} &\mathbf{F}_K(\partial[T \llcorner \mathbf{B}(a,r)] - \partial[Q_i \llcorner \mathbf{B}(a,r)]) \\ &\leq \mathbf{F}_K[T \llcorner \mathbf{B}(a,r) - Q_i \llcorner \mathbf{B}(a,r)] \to 0 \quad \text{as} \quad i \to \infty \end{aligned}$$

for $\mathscr{L}^1$ almost all r. Moreover 4.2.1 implies $T \llcorner \mathbf{B}(a,r)$ and $Q_i \llcorner \mathbf{B}(a,r)$ are normal currents for $\mathscr{L}^1$ almost all r. Whenever r satisfies these conditions we inductively use (2) and (1) with m replaced by $m-1$ to infer

$$\partial[Q_i \llcorner \mathbf{B}(a,r)] \in \mathbf{I}_{m-1}(\mathbf{R}^n) \text{ for } i=1,2,3,\ldots,$$

$$\partial[T \llcorner \mathbf{B}(a,r)] \in \mathbf{I}_{m-1}(\mathbf{R}^n).$$

Applying 4.2.15 we conclude $T \in \mathbf{I}_m(\mathbf{R}^n)$.

4.2.17. Compactness theorem. *If K is a compact Lipschitz neighborhood retract in $\mathbf{R}^n$, m is a nonnegative integer and $0 \leq c < \infty$, then:*

(1) $\mathbf{N}_{m,K}(\mathbf{R}^n) \cap \{T\colon \mathbf{N}(T) \leq c\}$ *is $\mathbf{F}_K$ compact.*

(2) $\mathbf{I}_{m,K}(\mathbf{R}^n) \cap \{T\colon \mathbf{N}(T) \leq c\}$ *is $\mathscr{F}_K$ compact.*

Proof. Since N is a lowersemicontinuous function, the set in (1) is $\mathbf{F}_K$ complete. From 4.1.24 and 4.2.16 we see that the set in (2) is $\mathscr{F}_K$ complete. We will show that these sets are totally bounded.

Let f be a Lipschitzian function retracting some bounded neighborhood of K in $\mathbf{R}^n$ onto K. Whenever

$$0<\varepsilon<1 \quad \text{with} \quad \{x\colon \operatorname{dist}(x,K)\leq 2n\,\varepsilon\}\subset \operatorname{dmn} f$$

we consider the sets

$$\Phi_1=\mathbf{P}_m(\mathbf{R}^n)\cap\{P\colon \operatorname{spt} P\subset \boldsymbol{\mu}_\varepsilon(\mathbf{W}'_m)\cap \operatorname{dmn} f, m=0 \text{ or}$$
$$\operatorname{spt}\partial P\subset \boldsymbol{\mu}_\varepsilon(\mathbf{W}'_{m-1}), \mathbf{N}(P)\leq 4n^{2m+2}c\},$$
$$\Phi_2=\mathscr{P}_m(\mathbf{R}^n)\cap\Phi_1, \quad \Psi_i=\{f_\# P\colon P\in\Phi_i\} \text{ for } i=1,2.$$

Since $\operatorname{dmn} f$ contains only finitely many cubes $\boldsymbol{\mu}_\varepsilon[\mathbf{W}'(z)]$, Φ_1 is $\mathbf{N}$ compact, Ψ_1 is $\mathbf{N}$ and $\mathbf{F}_K$ compact, Φ_2 and Ψ_2 are finite. Therefore Ψ_1 and Ψ_2 are totally bounded relative to $\mathbf{F}_K$ and $\mathscr{F}_K$, respectively.

For each member T of the set in (1) we apply 4.2.9 and infer $P\in\Phi_1$, $f_\# P\in\Psi_1$,

$$T=f_\# T=f_\# P+f_\# Q+\partial f_\# S,$$
$$\mathbf{F}_K(T-f_\# P)\leq \mathbf{M}(f_\# Q)+\mathbf{M}(f_\# S)\leq[\operatorname{Lip}(f)^m+\operatorname{Lip}(f)^{m+1}]\,2n^{2m+2}c\,\varepsilon.$$

Similarly we treat (2) by using Φ_2, Ψ_2 and $\mathscr{F}_K$.

4.2.18. Corollary. *If $T\in\mathscr{F}_{m,K}(\mathbf{R}^n)$, then $\mathscr{F}_K(T)$ is the least number in the set*

$$\{\mathbf{M}(T-\partial S)+\mathbf{M}(S)\colon S\in\mathscr{R}_{m+1,K}(\mathbf{R}^n), T-\partial S\in\mathscr{R}_{m,K}(\mathbf{R}^n)\}.$$

Proof. Whenever $\mathscr{F}_K(T)<c<\infty$ the set

$$\Omega=\mathscr{R}_{m+1,K}(\mathbf{R}^n)\cap\{S\colon T-\partial S\in\mathscr{R}_{m,K}(\mathbf{R}^n), \mathbf{M}(T-\partial S)+\mathbf{M}(S)\leq c\}$$

is nonempty and $\mathscr{F}_K$ compact, because for each $Q\in\Omega$ the set $\{S-Q\colon S\in\Omega\}$ is an $\mathscr{F}_K$ closed subset of

$$\mathbf{I}_{m+1,K}(\mathbf{R}^n)\cap\{X\colon \mathbf{N}(X)\leq 2c\},$$

and translation by Q is an $\mathscr{F}_K$ isometry.

4.2.19. Lemma. *If $R\in\mathbf{I}_m(\mathbf{R}^n)$, $\delta>0$ and either $m=0$ or $\partial R\in\mathscr{P}_{m-1}(\mathbf{R}^n)$, then there exist $S\in\mathbf{I}_{m+1}(\mathbf{R}^n)$ and a diffeomorphism g of class 1 mapping $\mathbf{R}^n$ onto $\mathbf{R}^n$ such that*

$$g_\# R-\partial S\in\mathscr{P}_m(\mathbf{R}^n), \quad \mathbf{N}(S)<\delta, \quad \operatorname{spt} S\subset\{x\colon \operatorname{dist}(x,\operatorname{spt} R)\leq\delta\},$$
$$\operatorname{Lip}(g)\leq 1+\delta, \quad \operatorname{Lip}(g^{-1})\leq 1+\delta, \quad |g(x)-x|\leq\delta \text{ for } x\in\mathbf{R}^n,$$
$$g(x)=x \text{ whenever } x\in\operatorname{spt}\partial R \text{ or } \operatorname{dist}(x,\operatorname{spt} R)\geq\delta.$$

Proof. In case $m=0$ we take $S=0$ and $g=\mathbf{1}_{\mathbf{R}^n}$.

In case $m>0$ we let $\gamma=2n^{m+2}$, choose a number t for which

$$0<t<1, \quad t^{-1}<1+\delta, \quad 6\gamma(t^{-m}-1)\,\mathbf{M}(R)<\delta/2,$$

and recall 4.1.28. Our hypothesis implies

$$\mathscr{H}^m(\operatorname{spt}\partial R)=0, \quad \text{hence} \quad \|R\|(\operatorname{spt}\partial R)=0.$$

Moreover $\mathbf{R}^n$ is $(\|R\|, m)$ rectifiable and 3.2.29 enables us to cover $\|R\|$ almost all of $\mathbf{R}^n$ by a countable family G of m dimensional submanifolds of class 1 of $\mathbf{R}^n$. Using 2.10.19 (4) we infer that for $\|R\|$ almost all points b in $\mathbf{R}^n$ there exist

$$B\in G \quad \text{with} \quad b\in B\sim\operatorname{spt}\partial R$$

and

$$\Theta^m(\|R\|\llcorner\mathbf{R}^n\sim B, b)=0<\Theta^m(\|R\|, b)<\infty.$$

Abbreviating $C_r=\mathbf{B}(b, t\,r)\cap B$ and observing that

$$\|R\|\,\mathbf{B}(b,t\,r)/[\alpha(m)\,r^m]\to t^m\,\Theta^m(\|R\|,b) \quad \text{as} \quad r\to 0+,$$

$$\|R\|\,[\mathbf{B}(b,r)\sim C_r]/\|R\|\,B(b,r)\to 1-t^m \quad \text{as} \quad r\to 0+,$$

we obtain arbitrarily small positive numbers r such that

$$\|R\|\,[\mathbf{B}(b,r)\sim C_r]<2(1-t^m)\,\|R\|\,\mathbf{B}(b,r)$$

and there exists a diffeomorphism f satisfying the conclusion of 3.1.23. Since $f(C_r)$ is contained in an m dimensional affine subspace of $\mathbf{R}^n$, we see from 4.2.10 that the rectifiable current $f_\#(R\llcorner C_r)$ can be $\mathbf{M}$ approximated by an m dimensional integral polyhedral chain Ψ so that

$$\mathbf{M}(\Psi-f_\#[R\llcorner\mathbf{B}(b,r)])\le\mathbf{M}[\Psi-f_\#(R\llcorner C_r)]+t^{-m}\,\|R\|\,[\mathbf{B}(b,r)\sim C_r]$$
$$<2(t^{-m}-1)\,\|R\|\,\mathbf{B}(b,r)$$

and $\operatorname{spt}\Psi\subset\mathbf{B}(b,r)$.

Accordingly 2.8.18 allows us to cover $\|R\|$ almost all of $\mathbf{R}^n$ by a sequence of disjoint balls $\mathbf{B}(b_i,r_i)\subset\mathbf{R}^n\sim\operatorname{spt}\partial R$, with $4r_i<\delta$, for which there exist $\Psi_i\in\mathscr{P}_m(\mathbf{R}^n)$ and diffeomorphisms f_i mapping $\mathbf{R}^n$ onto $\mathbf{R}^n$ such that

$$\operatorname{Lip}(f_i)\le t^{-1}, \quad \operatorname{Lip}(f_i^{-1})\le t^{-1}, \quad f_i(x)=x \text{ whenever } x\in\mathbf{R}^n\sim\mathbf{B}(b_i,r_i),$$

$$\mathbf{M}(\Psi_i-f_{i\#}[R\llcorner\mathbf{B}(b_i,r_i)])<2(t^{-m}-1)\,\|R\|\,\mathbf{B}(b_i,r_i)$$

and $\operatorname{spt}\Psi_i\subset\mathbf{B}(b_i,r_i)$. We choose a positive integer k for which

$$\sum_{i=k+1}^{\infty}\|R\|\,\mathbf{B}(b_i,r_i)<(t^{-m}-1)\,\mathbf{M}(R),$$

let $g=f_1\circ f_2\circ\cdots\circ f_k$ and observe that the integral current

$$T=g_\# R-\sum_{i=1}^{k}\Psi_i=\sum_{i=1}^{k}(f_{i\#}[R\llcorner B(b_i,r_i)]-\Psi_i)+\sum_{i=k+1}^{\infty}R\llcorner \mathbf{B}(b_i,r_i)$$

satisfies the conditions

$$\partial T=\partial R-\sum_{i=1}^{k}\partial\Psi_i\in\mathscr{P}_{m-1}(\mathbf{R}^n),$$

$$\operatorname{spt} T\subset\{x:\operatorname{dist}(x,\operatorname{spt} R)\le\delta/2\},\quad \mathbf{M}(T)<3(t^{-m}-1)\,\mathbf{M}(R).$$

Applying 4.2.9 with a number ε such that

$$0<2n\,\varepsilon<\delta/2,\quad 2\gamma\,\varepsilon\,\mathbf{N}(T)<\delta/2$$

we obtain $P,Q\in\mathscr{P}_m(\mathbf{R}^n)$ and $S\in\mathbf{I}_{m+1}(\mathbf{R}^n)$ with

$$g_\# R-\partial S=P+Q+\sum_{i=1}^{k}\Psi_i\in\mathscr{P}_m(\mathbf{R}^n),$$

$$\mathbf{N}(S)\le\mathbf{M}(S)+\mathbf{M}(Q)+\mathbf{M}(P)+\mathbf{M}(T)\le 2\gamma\,\varepsilon\,\mathbf{N}(T)+2\gamma\,\mathbf{M}(T)<\delta.$$

4.2.20. Approximation theorem. *Whenever* $T\in\mathbf{I}_m(\mathbf{R}^n)$ *and* $\varepsilon>0$ *there exist*

$$P\in\mathscr{P}_m(\mathbf{R}^n)\ \textit{with}\ \operatorname{spt} P\subset\{x:\operatorname{dist}(x,\operatorname{spt} T)\le\varepsilon\}$$

and a diffeomorphism f of class 1 mapping $\mathbf{R}^n$ onto $\mathbf{R}^n$ such that

$$\mathbf{N}(P-f_\# T)\le\varepsilon,\quad \operatorname{Lip}(f)\le 1+\varepsilon,\quad \operatorname{Lip}(f^{-1})\le 1+\varepsilon,$$

$$|f(x)-x|\le\varepsilon\ \textit{for}\ x\in\mathbf{R}^n,\quad f(x)=x\ \textit{if}\ \operatorname{dist}(x,\operatorname{spt} T)\ge\varepsilon.$$

Proof. In case $m=0$ we take $P=T$ and $f=\mathbf{1}_{\mathbf{R}^n}$.

In case $m>0$ we choose a positive number δ for which

$$(1+\delta)^m\,\delta+\delta\le\varepsilon,\quad\text{hence}\quad (1+\delta)^2\le 1+\varepsilon,\quad 2\delta\le\varepsilon,$$

and apply 4.2.19 twice, first with $R_1=\partial T$ to obtain $S_1\in\mathbf{I}_m(\mathbf{R}^n)$ and a diffeomorphism g_1 such that

$$\partial(g_{1\#}T-S_1)=g_{1\#}\,\partial T-\partial S_1\in\mathscr{P}_{m-1}(\mathbf{R}^n),$$

second with $R_2=g_{1\#}T-S_1$ to obtain $S_2\in\mathbf{I}_{m+1}(\mathbf{R}^n)$ and a diffeomorphism g_2 such that

$$P=g_{2\#}(g_{1\#}T-S_1)-\partial S_2\in\mathscr{P}_m(\mathbf{R}^n).$$

Letting $f=g_2\circ g_1$ we conclude

$$\mathbf{N}(P-f_\# T)\le\mathbf{N}(g_{2\#}S_1)+\mathbf{N}(S_2)\le(1+\delta)^m\,\delta+\delta\le\varepsilon.$$

4.2.21. Corollary. *If* $T \in \mathbf{I}_m(\mathbf{R}^n)$, $\rho > 0$, K *is a compact subset of* $\mathbf{R}^n$ *and* $\operatorname{spt} T \subset \operatorname{Int} K$, *then there exists*

$$P \in \mathscr{P}_m(R^n) \quad \text{with} \quad \mathscr{F}_K(P-T) \le \rho, \quad \mathbf{N}(P) \le \mathbf{N}(T) + \rho.$$

Proof. Assuming $\{x: \operatorname{dist}(x, \operatorname{spt} T) \le \varepsilon\} \subset K$, and letting h be the affine homotopy from $\mathbf{1}_{\mathbf{R}^n}$ to f, we find that

$$\mathbf{N}(P) \le \mathbf{N}(f_\# T) + \varepsilon \le (1+\varepsilon)^m \mathbf{N}(T) + \varepsilon,$$

$$P - T = P - f_\# T + \partial h_\#([0,1] \times T) + h_\#([0,1] \times \partial T),$$

$$\mathscr{F}_K(P-T) \le \varepsilon + \varepsilon(1+\varepsilon)^m \mathbf{N}(T).$$

4.2.22. Theorem. *If* $T \in \mathscr{F}_m(\mathbf{R}^n)$ *and* K *is a compact subset of* $\mathbf{R}^n$ *with* $\operatorname{spt} T \subset \operatorname{Int} K$, *then* $T \in \mathscr{F}_{m,K}(\mathbf{R}^n)$ *and for each* $\varepsilon > 0$ *there exists*

$$P \in \mathscr{P}_m(\mathbf{R}^n) \quad \text{with} \quad \mathscr{F}_K(P-T) \le \varepsilon, \quad \mathbf{M}(P) \le \mathbf{M}(T) + \varepsilon.$$

Proof. Letting $u(x) = \operatorname{dist}(x, \operatorname{spt} T)$ for $x \in \mathbf{R}^n$,

$$\delta > 0 \quad \text{with} \quad \{x: u(x) \le 2\delta\} \subset K,$$

$$\alpha \in \mathscr{E}^0(\mathbf{R}^n) \quad \text{with} \quad \operatorname{spt} \alpha \cap \operatorname{spt} T = \varnothing, \quad \alpha(x) = 1 \text{ if } u(x) \ge \delta,$$

$$R \in \mathscr{R}_m(\mathbf{R}^n), \quad S \in \mathscr{R}_{m+1}(\mathbf{R}^n) \quad \text{with} \quad T = R + \partial S,$$

we find that

$$\partial(S \llcorner \alpha) = (\partial S) \llcorner \alpha - S \llcorner d\alpha = -R \llcorner \alpha - S \llcorner d\alpha,$$

hence $S \llcorner \alpha$ is normal. Applying 4.2.1 we choose r so that

$$\delta < r < 2\delta \quad \text{and} \quad \mathbf{M}\langle S \llcorner \alpha, u, r+\rangle < \infty,$$

abbreviate $W = \{x: u(x) > r\}$ and infer

$$(S \llcorner \alpha) \llcorner W = S \llcorner W \in \mathscr{R}_{m+1}(\mathbf{R}^n), \quad [\partial(S \llcorner \alpha)] \llcorner W = -R \llcorner W \in \mathscr{R}_m(\mathbf{R}^n),$$

$$\langle S \llcorner \alpha, u, r+\rangle \in R_m(\mathbf{R}^n) \text{ by 4.2.16(3)},$$

$$\partial(S \llcorner W) + \langle S \llcorner \alpha, u, r+\rangle + R \llcorner W = 0,$$

$$\partial[S \llcorner (\mathbf{R}^n \sim W)] - \langle S \llcorner \alpha, u, r+\rangle + R \llcorner (\mathbf{R}^n \sim W) = T$$

with $(\mathbf{R}^n \sim W) \cup u^{-1}\{r\} = \{x: u(x) \le r\} \subset \operatorname{Int} K$.

In case $\mathbf{M}(T) < \infty$, then $T \in \mathscr{R}_{m,K}(\mathbf{R}^n)$ by 4.2.16(3), and since $\mathbf{I}_{m,K}(\mathbf{R}^n)$ is $\mathbf{M}$ dense in $\mathscr{R}_{m,K}(\mathbf{R}^n)$ the second conclusion follows from 4.2.21.

In case $\mathbf{M}(T) = \infty$ the second conclusion follows from 4.2.9.

4.2.23. Lemma. *If* $C \subset V \subset \mathbf{R}^n$, *$C$ is compact, V is open,* $X \in \mathbf{P}_m(\mathbf{R}^n)$ *and* $\operatorname{spt} X \subset C$, *then*

$$\inf\{\mathbf{M}(X-\partial Y)+\mathbf{M}(Y)\colon Y \in \mathbf{P}_{m+1}(\mathbf{R}^n), \operatorname{spt} Y \subset V\}$$

does not exceed $\mathbf{F}_C(X)$.

Proof. Letting $G(X)$ be the above infimum we will first show that

$$G(X) \leq \gamma\, \mathbf{F}_C(X) \quad \text{with} \quad \gamma = 4n^{2m+4}.$$

For this purpose we choose

$$N \in \mathbf{N}_{m+1,C}(\mathbf{R}^n) \quad \text{with} \quad \mathbf{F}_C(X) = \mathbf{M}(X-\partial N)+\mathbf{M}(N),$$

note that $\partial(X-\partial N) = \partial X \in \mathbf{P}_{m-1}(\mathbf{R}^n)$ or $m=0$, apply 4.2.9 with $T = X - \partial N$ and small ε to express

$$X - \partial N = R_1 + \partial S_1$$

where

$$R_1 \in \mathbf{P}_m(\mathbf{R}^n), \quad S_1 \in \mathbf{N}_{m+1}(\mathbf{R}^n), \quad \operatorname{spt} R_1 \cup \operatorname{spt} S_1 \subset V,$$

$$\mathbf{M}(R_1) \leq \gamma[\mathbf{M}(X-\partial N)+\varepsilon\,\mathbf{M}(\partial X)], \quad \mathbf{M}(S_1) \leq \varepsilon\,\gamma\,\mathbf{M}(X-\partial N),$$

note that $\partial(N+S_1) = X - R_1 \in \mathbf{P}_m(\mathbf{R}^n)$, apply 4.2.9 with $T = N + S_1$ to express

$$N + S_1 = R_2 + \partial S_2$$

where

$$R_2 \in \mathbf{P}_{m+1}(\mathbf{R}^n), \quad S_2 \in \mathbf{N}_{m+2}(\mathbf{R}^n), \quad \operatorname{spt} R_2 \cup \operatorname{spt} S_2 \subset V,$$

$$\mathbf{M}(R_2) \leq \gamma[\mathbf{M}(N+S_1)+\varepsilon\,\mathbf{M}(X-R_1)],$$

hence infer $X = R_1 + \partial(N+S_1) = R_1 + \partial R_2$ with

$$\mathbf{M}(R_1)+\mathbf{M}(R_2) \leq \gamma(1+\varepsilon\gamma)[\mathbf{F}_C(X)+\varepsilon\,\mathbf{N}(X)],$$

which implies the preliminary estimate.

Next we suppose $\rho > 0$, choose N as before, select a compact set $K \subset V$ for which $C \subset \operatorname{Int} K$, use 4.1.23 twice to obtain

$$P_1 \in \mathbf{P}_m(\mathbf{R}^n), \quad P_2 \in \mathbf{P}_{m+1}(\mathbf{R}^n) \quad \text{with} \quad \operatorname{spt} P_1 \cup \operatorname{spt} P_2 \subset K,$$

$$\mathbf{F}_K(X - \partial N - P_1) \leq \rho, \quad \mathbf{F}_K(N - P_2) \leq \rho,$$

$$\mathbf{M}(P_1) \leq \mathbf{M}(X-\partial N)+\rho, \quad \mathbf{M}(P_2) \leq \mathbf{M}(N)+\rho,$$

note that $X - P_1 - \partial P_2 \in \mathbf{P}_m(\mathbf{R}^n)$ with

$$\mathbf{F}_K(X-P_1-\partial P_2) \leq \mathbf{F}_K(X-\partial N-P_1)+\mathbf{F}_K(N-P_2) \leq 2\rho,$$

and apply the preliminary estimate with C, X replaced by $K, X - P_1 - \partial P_2$ to conclude

$$\begin{aligned} G(X) &\leq G(P_1+\partial P_2)+G(X-P_1-\partial P_2) \\ &\leq \mathbf{M}(P_1)+\mathbf{M}(P_2)+\gamma\, 2\rho \leq \mathbf{F}_C(X)+(1+\gamma)\, 2\rho. \end{aligned}$$

4.2.24. Theorem. *If* $T \in \mathbf{N}_m(\mathbf{R}^n)$, $\rho > 0$, *K is a compact subset of* $\mathbf{R}^n$ *and* spt $T \subset \operatorname{Int} K$, *then there exists*

$$P \in \mathbf{P}_m(\mathbf{R}^n) \quad \text{with} \quad \mathbf{F}_K(P-T) \le \rho, \quad \mathbf{N}(P) \le \mathbf{N}(T) + \rho.$$

Proof. Assuming $m > 0$ we choose a compact set $C \subset \operatorname{Int} K$ for which spt $T \subset \operatorname{Int} C$, apply 4.1.23 twice to obtain

$$P_1 \in \mathbf{P}_m(\mathbf{R}^n), \quad P_2 \in \mathbf{P}_{m-1}(\mathbf{R}^n) \quad \text{with} \quad \operatorname{spt} P_1 \cup \operatorname{spt} P_2 \subset C,$$
$$\mathbf{F}_C(T - P_1) < \rho/4, \quad \mathbf{F}_C(\partial T - P_2) < \rho/4$$
$$\mathbf{M}(P_1) < \mathbf{M}(T) + \rho/4, \quad \mathbf{M}(P_2) < \mathbf{M}(\partial T) + \rho/4$$

note that $P_2 - \partial P_1 \in \mathbf{P}_{m-1}(\mathbf{R}^n)$ with

$$\mathbf{F}_C(P_2 - \partial P_1) \le \mathbf{F}_C(P_2 - \partial T) + \mathbf{F}_C(T - P_1) < \rho/2,$$

use 4.2.23 to secure $Y \in \mathbf{P}_m(\mathbf{R}^n)$ with spt $Y \subset \operatorname{Int} K$ and

$$\mathbf{M}(P_2 - \partial P_1 - \partial Y) + \mathbf{M}(Y) < \rho/2,$$

let $P = P_1 + Y$ and conclude

$$\mathbf{F}_K(P-T) \le \mathbf{F}_K(P - P_1) + \mathbf{M}(Y) < \rho,$$
$$\mathbf{N}(P) \le \mathbf{M}(P_1) + \mathbf{M}(Y) + \mathbf{M}(\partial P_1 + \partial Y - P_2) + \mathbf{M}(P_2) < \mathbf{N}(T) + \rho.$$

4.2.25. *A current* $T \in \mathbf{I}_m(\mathbf{R}^n)$ *is called* **indecomposable** *if and only if there exists no* $R \in \mathbf{I}_m(\mathbf{R}^n)$ *with*

$$R \neq 0 \neq T - R \quad \text{and} \quad \mathbf{N}(T) = \mathbf{N}(R) + \mathbf{N}(T-R).$$

It may be shown that *for every* $T \in \mathbf{I}_m(\mathbf{R}_n)$ *there exists a sequence of indecomposable currents* $T_i \in \mathbf{I}_m(\mathbf{R}^n)$ *such that*

$$T = \sum_{i=1}^{\infty} T_i \quad \text{and} \quad \mathbf{N}(T) = \sum_{i=1}^{\infty} \mathbf{N}(T_i).$$

A key to the proof is the observation:

If $R \in \mathbf{I}_{m,K}(\mathbf{R}^n)$, *$K$ is convex and* $\mathbf{N}(R) < 1$, *then*

$$\mathscr{F}_K(R) \le c\, \mathbf{N}(R)^{(m+1)/m}$$

where $c = 9(2n^{2m+2})^4$. We verify this inequality in case $m > 1$ by applying 4.2.10 twice to choose

$$X \in \mathbf{I}_{m,K}(\mathbf{R}^n) \text{ with } \partial X = \partial R,\ \mathbf{M}(X) \le \gamma^{\alpha} \mathbf{M}(\partial R)^{\alpha},$$
$$Y \in \mathbf{I}_{m+1,K}(\mathbf{R}^n) \text{ with } \partial Y = R - X,\ \mathbf{M}(Y) \le \gamma^{\beta} \mathbf{M}(R-X)^{\beta},$$

where $\gamma = 2n^{2m+2}$, $\alpha = m/(m-1) > \beta = (m+1)/m > 1$, and by estimating

$$\mathscr{F}_K(R) \leq \mathbf{M}(X) + \mathbf{M}(Y) \leq \gamma^\alpha \mathbf{M}(\partial R)^\alpha + 2^\beta \gamma^\beta \mathbf{M}(R)^\beta + 2^\beta \gamma^{\beta+\alpha\beta} \mathbf{M}(\partial R)^{\alpha\beta}$$
$$\leq c\, \mathbf{N}(R)^\beta.$$

In case $m=1$ the hypothesis implies $\partial R = 0$, we take $X=0$ and estimate $\mathscr{F}_K(R) \leq \mathbf{M}(Y) \leq \gamma^\beta \mathbf{N}(R)^\beta$. Our observation has the following consequence:

If $T \in \mathbf{I}_{m,K}(\mathbf{R}^n)$, K is convex and $\mathscr{F}_K(T) > c\,\mathbf{N}(T)\,\varepsilon^{1/m}$, then there exists no sequence of currents $R_i \in \mathbf{I}_{m,K}(\mathbf{R}^n)$ such that

$$T = \sum_{i=1}^{\infty} R_i, \quad \mathbf{N}(T) = \sum_{i=1}^{\infty} \mathbf{N}(R_i), \quad \mathbf{N}(R_i) \leq \varepsilon \text{ for all } i.$$

For this reason the obvious procedure for decomposing T is effective.

The structure of 1 dimensional integral currents is relatively simple: *For every indecomposable $T \in \mathbf{I}_1(\mathbf{R}^n)$ there exists a function*

$$f\colon \mathbf{R} \to \mathbf{R}^n \text{ such that } \operatorname{Lip}(f) \leq 1, \quad f_\# [0, \mathbf{M}(T)] = T,$$
$$f \mid \{t\colon 0 \leq t < \mathbf{M}(T)\} \text{ is univalent;}$$

moreover $\partial T = 0$ if and only if $f(0) = f[\mathbf{M}(T)]$. (Such a current T is called an **oriented simple curve with finite length; a closed** curve if $\partial T = 0$.) This assertion may be verified in the special case when $T \in \mathscr{P}_1(\mathbf{R}^n)$ by a simple combinatorial argument, and extended to the general case by use of 4.2.19, 2.10.21 and the preceding paragraph.

We see from 4.1.31 and 4.2.16 that *if B is a connected m dimensional submanifold of class 1 of $\mathbf{R}^n$, ζ is an m-vectorfield orienting B and the current $(\mathscr{H}^m \llcorner B) \wedge \zeta$ is normal, then this current is integral and indecomposable.*

In case $1 < m < n$ there exists an indecomposable m dimensional integral current in $\mathbf{R}^n$ whose support is not countably $(\mathscr{H}^m, m)$ rectifiable. For example, given any totally disconnected (see [HW, pp. 22 and 48]) compact nonempty subset A of $\mathbf{R}^3$, we will indicate the construction of an indecomposable current $T \in \mathbf{I}_2(\mathbf{R}^3)$ with $\partial T = 0$ and $A \subset \operatorname{spt} T$.

First we inductively select finite open coverings G_j of A such that G_1 consists of a single open cube E and the following two conditions hold for $j > 1$:

If $U \in G_j$, then U is the union of a finite family of cubes, $U \cap A \neq \varnothing$, $\operatorname{diam} U < 2^{-j}$ and $\operatorname{Clos} U \subset V$ for some $V \in G_{j-1}$.

If $V \in G_{j-1}$, then $F = \{\operatorname{Clos} U\colon U \in G_j, \operatorname{Clos} U \subset V\}$ is disjointed and $V \sim \bigcup F$ is connected.

Next we choose $\phi, \psi \in \mathscr{E}^0(\mathbf{R})$ so that

$$\phi(t)=0 \text{ for } t\le 1, \quad \phi'(t)>0 \text{ for } t>1, \quad \phi(2)=1,$$
$$\psi(t)=1 \text{ for } t\le 1, \quad \psi(t)=0 \text{ for } t>4, \quad \psi'(t)<0 \text{ for } 1<t<4,$$

consider the sets

$$\Phi=\mathbf{R}^3\cap\{x\colon \phi[(x_1)^2+(x_2)^2]+(x_3)^2<1\},$$
$$\Psi=\mathbf{R}^3\cap\{x\colon 0<x_3<\psi[(x_1)^2+(x_2)^2]\},$$
$$\Delta=\mathbf{R}^3\cap\{x\colon x_3=0,\ (x_1)^2+(x_2)^2<4\},$$
$$\Gamma=\mathbf{R}^3\cap\{x\colon x_3=1,\ (x_1)^2+(x_2)^2<1\},$$

and note that $\Delta\cup\Gamma\subset \operatorname{Bdry}\Psi$, $\Gamma\subset\operatorname{Bdry}\Phi$, $\operatorname{Bdry}\Phi$ is diffeomorphic with $\mathbf{S}^2$. We also let F be the set of all those diffeomorphisms of class ∞ mapping a neighborhood of $\operatorname{Clos}\Psi$ in $\mathbf{R}^3$ onto an open subset of $\mathbf{R}^3$, whose restriction to each of the discs Δ and Γ is affine.

Assuming $\operatorname{Clos}\Phi\cap\operatorname{Clos}E=\varnothing$ we construct $f_E\in F$ so that

$$f_E(\operatorname{Clos}\Delta)\subset\Gamma, \quad f_E(\Gamma)\subset\operatorname{Bdry}E,$$
$$f_E[\operatorname{Clos}\Psi\sim\operatorname{Clos}(\Delta\cup\Gamma)]\subset\mathbf{R}^3\sim\operatorname{Clos}(\Phi\cup E),$$

and for $U\in G_j$ with $j>1$ and $\operatorname{Clos}U\subset V\in G_{j-1}$ we construct $f_U\in F$ inductively so that

$$f_U(\operatorname{Clos}\Delta)\subset f_V(\Gamma), \quad f_U(\Gamma)\subset\operatorname{Bdry}U,$$
$$f_U[\operatorname{Clos}\Psi\sim\operatorname{Clos}(\Delta\cup\Gamma)]\subset V\sim\operatorname{Clos}\bigcup G_j,$$
$$\mathscr{H}^2[f_U(\operatorname{Bdry}\Psi)]<2^{-j}/\operatorname{card}G_j;$$

we also require that $f_U(\operatorname{Clos}\Psi)\cap f_W(\operatorname{Clos}\Psi)=\varnothing$ whenever U and W are distinct members of G_j. (The f_U image of the vertical axis of Ψ is an arc joining $f_V(\Gamma)$ to $\operatorname{Bdry}U$ in $V\sim\operatorname{Clos}\bigcup G_j$, and $f_U(\Psi)$ is obtained by thickening this arc.) Defining

$$\Omega_j=\Phi\cup\bigcup_{i=1}^{j}\{f_U(\Psi\cup\Delta)\colon U\in G_i\}, \qquad T_j=\partial(\mathbf{E}^3\llcorner\Omega_j)$$

and recalling 4.1.25 one infers that $\operatorname{spt}T_j=\operatorname{Bdry}\Omega_j$ is diffeomorphic with $\mathbf{S}^2$ and

$$\mathbf{M}(T_j-T_{j-1})\le\sum_{U\in G_j}\mathscr{H}^2[f_U(\operatorname{Bdry}\Psi)]<2^{-j}$$

for $j>1$. The sequence of rectifiable currents T_j has the $\mathbf{M}$ limit

$$T=\partial(\mathbf{E}^3\llcorner\Omega) \text{ with } \Omega=\bigcup_{j=1}^{\infty}\Omega_j,$$

and one finds that $A \subset \operatorname{Bdry} \Omega = \operatorname{spt} T$, $X = (\operatorname{spt} T) \sim A$ is a connected 2 dimensional submanifold of class ∞ of $\mathbf{R}^3$,

$$T = (\mathscr{H}^2 \llcorner X) \wedge \xi$$

for some 2-vectorfield ξ orienting X, and $\operatorname{spt} T$ is homeomorphic but not diffeomorphic with $\mathbf{S}^2$. (An account of a similar polyhedral construction may be found in [N 2, § 1].)

Choosing A in particular so that $\mathscr{L}^3(A) > 0$, one infers not only that $\mathscr{L}^3(\operatorname{spt} T) > 0$, but one is also led to consider the current

$$S = \mathbf{E}^3 \llcorner A \in \mathscr{R}_3(\mathbf{R}^3) \text{ with } \operatorname{spt} S \subset A,$$

whose boundary ∂S is a nonzero 2 dimensional flat cycle with support in the proper subset A of the topological 2-sphere $\operatorname{spt} T$. This shows the need for the assumption in 4.1.31 that B have class 1.

Alternately, it is instructive to choose A purely $(\mathscr{H}^2, 2)$ unrectifiable with $0 < \mathscr{H}^2(A) < \infty$.

It would be extremely difficult to determine the structure of arbitrary m dimensional indecomposable integral currents in $\mathbf{R}^n$, when $1 < m < n$. Using 4.5.9 (13) one can show that the irreducible members of $\mathbf{I}_n(\mathbf{R}^n)$ are the currents $\mathbf{E}^n \llcorner A$ corresponding to certain $\mathscr{L}^n$ measurable sets A.

4.2.26. In view of the results obtained so far in this section, *flat chains* and *integral flat chains* appropriately generalize elementary notions on polyhedral chains with *real* and *integral* coefficients. For the coefficient group $\mathbf{Z}/\nu\,\mathbf{Z}$, a *cyclic group of order* ν, the analogous purpose is served by the *flat chains modulo* ν which will be considered here.

Whenever ν is a nonnegative integer, U is an open subset of $\mathbf{R}^n$, K is a compact subset of U and $T \in \mathscr{F}_{m,K}(U)$ we define

$$\mathscr{F}_K^\nu(T)$$

as *the infimum of the set of numbers* $\mathbf{M}(R) + \mathbf{M}(S)$ *corresponding to all* $R \in \mathscr{R}_{m,K}(U)$, $S \in \mathscr{R}_{m+1,K}(U)$ and $Q \in \mathscr{F}_{m,K}(U)$ with

$$T = R + \partial S + \nu Q.$$

Since $\mathbf{I}_{m,K}(U)$ is $\mathscr{F}_K$ dense in $\mathscr{F}_{m,K}(U)$ one may require that $Q \in \mathbf{I}_{m,K}(U)$; in case $T \in \mathbf{I}_{m,K}(U)$ such a choice of Q implies $R \in \mathbf{I}_{m,K}(U)$ and $S \in \mathbf{I}_{m+1,K}(U)$. Observing that

$$\mathscr{F}_K^\nu(T_1 + T_2) \le \mathscr{F}_K^\nu(T_1) + \mathscr{F}_K^\nu(T_2) \text{ for } T_1, T_2 \in \mathscr{F}_{m,K}^\nu(U),$$

we define the $\mathscr{F}_K^\nu$ pseudodistance between T_1 and T_2 as $\mathscr{F}_K^\nu(T_1 - T_2)$. We note that

$$\mathscr{F}_{m,K}(U) \cap \{T : \mathscr{F}_K^\nu(T) = 0\}$$

is an additive subgroup of $\mathscr{F}_{m,K}(U)$, and equals the $\mathscr{F}_K$ closure of $\nu\mathscr{F}_{m,K}(U)$; it is not known whether $\nu\mathscr{F}_{m,K}(U)$ is always $\mathscr{F}_K$ closed. The above pseudometric induces a metric, also denoted $\mathscr{F}_K^\nu$, on the factorgroup

$$\mathscr{F}^\nu_{m,K}(U)=\mathscr{F}_{m,K}(U)/[\mathscr{F}_{m,K}(U)\cap\{T:\ \mathscr{F}_K^\nu(T)=0\}];$$

reasoning as in 4.1.24 we find that

$$\mathscr{F}^\nu_{m,K}(U)\ \text{is}\ \mathscr{F}_K^\nu\ \text{complete}.$$

If $T\in\mathscr{F}_{m,K}(U)$, then

$$\mathscr{F}^\nu_{f(K)}(f_\# T)\le\sup\{\operatorname{Lip}(f|K)^m,\operatorname{Lip}(f|K)^{m+1}\}\,\mathscr{F}_K^\nu(T)$$

for every locally Lipschitzian map f of U into an open subset V of another Euclidean space, and

$$\mathscr{F}_K^\nu(\partial T)\le\mathscr{F}_K^\nu(T)\ \text{in case}\ m>0;$$

therefore $f_\#$ and ∂ induce continuous homomorphisms

$$f_\#:\ \mathscr{F}^\nu_{m,K}(U)\to\mathscr{F}^\nu_{m,f(K)}(V),\qquad \partial:\ \mathscr{F}^\nu_{m,K}(U)\to\mathscr{F}^\nu_{m-1,K}(U)\ \text{in case}\ m>0.$$

Applying the method of 4.2.1 in conjunction with 4.2.16 and 4.1.28 one readily verifies that

$$\int_a^{*b}\mathscr{F}^\nu_{C(r)}[T\llcorner\{x:\ u(x)\le r\}]\,d\mathscr{L}^1 r\le[b-a+\operatorname{Lip}(u)]\,\mathscr{F}_K^\nu(T)$$

whenever $T\in\mathbf{I}_{m,K}(U)$, $u: U\to\mathbf{R}$ *is Lipschitzian,* $-\infty<a<b<\infty$ *and* $C(r)$ *is a compact subset of* U *with*

$$K\cap\{x:\ u(x)\le r\}\subset\operatorname{Int}C(r)\ \textit{for}\ a<r<b.$$

In particular, assuming that G, H, K are compact subsets of U, we obtain the inequality

$$\mathscr{F}_H^\nu(T)\le[1+1/\operatorname{dist}(U\sim H,G)]\,\mathscr{F}_K^\nu(T)$$
$$\textit{whenever}\ T\in\mathscr{F}_{m,K}(U)\ \textit{with}\ \operatorname{spt}T\subset G\subset\operatorname{Int}H;$$

to this end we observe that $\mathbf{I}_{m,K}(U)$ is $\mathscr{F}_K$ dense in $\mathscr{F}_{m,K}(U)$ and take $u(x)=\operatorname{dist}(x,G)$ for $x\in U$, $C(r)=H$ for $a=0<r<\operatorname{dist}(U\sim H,G)=b$.

We say that two integral flat chains $T_1, T_2\in\mathscr{F}_m(U)$ are **congruent modulo** ν, and write $T_1\equiv T_2 \bmod \nu$, *if and only if* $\mathscr{F}_K^\nu(T_1-T_2)=0$ *for some compact subset* K *of* U. The resulting congruence classes are called m dimensional **flat chains modulo** ν; they are the cosets in the factorgroup

$$\mathscr{F}_m^\nu(U)=\mathscr{F}_m(U)/[\mathscr{F}_m(U)\cap\{T:\ T\equiv 0 \bmod \nu\}].$$

Whenever $T \in \mathscr{F}_m(U)$ we define

$$\operatorname{spt}^\nu(T) = \bigcap \{\operatorname{spt} R \colon R \in \mathscr{F}_m(U),\ R \equiv T \bmod \nu\}$$

and assert that, *for every neighborhood V of* $\operatorname{spt}^\nu(T)$ *in U, T is congruent modulo ν to some integral flat chain with support in V.* To make this obvious we will show that *if $T \equiv R \bmod \nu$ and V is any neighborhood of* $\operatorname{spt} T \cap \operatorname{spt} R$ *in U, then $T \equiv S \bmod \nu$ for some S with* $\operatorname{spt} S \subset V$. We let

$$A_\varepsilon = \{x \colon \operatorname{dist}(x, \operatorname{spt} R) \le \varepsilon\}, \quad B_\varepsilon = \{x \colon \operatorname{dist}(x, \operatorname{spt} T) \le \varepsilon\}$$

for $\varepsilon > 0$, secure a compact subset K of U with

$$\operatorname{spt} R \cup \operatorname{spt} T \subset \operatorname{Int} K, \quad \mathscr{F}_K^\nu(R - T) = 0,$$

and choose ξ so that $0 < \xi \le 1$, $A_{2\xi} \subset \operatorname{Int} K$, $H = A_{3\xi} \cap B_{3\xi} \subset V$. Whenever $\xi > \delta > 0$ we apply 4.2.22 to obtain

$$R_\delta, T_\delta \in \mathbf{I}_{m,K}(U) \text{ with } \operatorname{spt} R_\delta \subset A_\delta,\ \operatorname{spt} T_\delta \subset B_\delta,$$

$$\begin{aligned} \delta^3 &> \mathscr{F}_K(R_\delta - R) + \mathscr{F}_K(T_\delta - T) \ge \mathscr{F}_K^\nu(R_\delta - T_\delta) \\ &\ge 2^{-1} \textstyle\int_\delta^{*2\delta} \mathscr{F}_K^\nu[(R_\delta - T_\delta) \llcorner A_r]\, d\mathscr{L}^1 r, \end{aligned}$$

where the last inequality results from the preceding paragraph with $u(x) = \operatorname{dist}(x, \operatorname{spt} R)$ for $x \in U$; then we choose r so that

$$\delta < r < 2\delta, \quad \mathscr{F}_K^\nu(R_\delta - S_\delta) < 2\delta^2 \text{ with } S_\delta = T_\delta \llcorner A_r,$$

hence $\mathscr{F}_K^\nu(T - S_\delta) < 3\delta^2$ and $\operatorname{spt} S_\delta \subset G = A_{2\xi} \cap B_{2\xi}$. Using the second inequality in the preceding paragraph we infer that

$$\mathscr{F}_H^\nu(S_\delta - S_\varepsilon) \le (1 + \xi^{-1}) \mathscr{F}_K^\nu(S_\delta - S_\varepsilon) < (1 + \xi^{-1})\, 3(\delta^2 + \varepsilon^2) \le 12\delta$$

whenever $\xi > \delta > \varepsilon > 0$, because $\operatorname{dist}(G, U \sim H) \ge \xi$, and conclude the existence of

$$S \in \mathscr{F}_{m,H}(U) \text{ with } \lim_{\delta \to 0+} \mathscr{F}_H^\nu(S_\delta - S) = 0,$$

hence $\mathscr{F}_{K \cup H}^\nu(T - S) = 0$, $T \equiv S \bmod \nu$. *It is not known whether there always exists an $S \equiv T \bmod \nu$ with* $\operatorname{spt} S = \operatorname{spt}^\nu(T)$.

To each compact subset K of U corresponds the commutative diagram of additive homomorphisms:

$$\begin{array}{ccc} \mathscr{F}_{m,K}(U) & \xrightarrow{\subset} & \mathscr{F}_m(U) \\ \downarrow & & \downarrow \\ \mathscr{F}_{m,K}^\nu(U) & \longrightarrow & \mathscr{F}_m^\nu(U) \end{array}$$

In case K is a Lipschitz neighborhood retract, the map represented by the lower horizontal arrow is a monomorphism; it is not known whether this is true for all K. Whenever $T \in \mathscr{F}_{m,K}(U)$ we let

$$(T)^\nu_K \in \mathscr{F}^\nu_{m,K}(U) \quad \text{and} \quad (T)^\nu \in \mathscr{F}^\nu_m(U)$$

be the cosets of T. We define

$$\mathscr{R}^\nu_m(U) = \{(T)^\nu\colon\ T \in \mathscr{R}_m(U)\}.$$

For $m > 0$, ∂ induces a homomorphism of $\mathscr{F}^\nu_m(U)$ into $\mathscr{F}^\nu_{m-1}(U)$, which we denote by the same symbol ∂, and we define

$$\mathbf{I}^\nu_m(U) = \mathscr{R}^\nu_m(U) \cap \{\tau\colon\ \partial\tau \in \mathscr{R}^\nu_{m-1}(U)\};$$

we also define $\mathbf{I}^\nu_0(U) = \mathscr{R}^\nu_0(U)$. Similarly we construct groups $\mathscr{R}^\nu_{m,K}(U)$ and $\mathbf{I}^\nu_{m,K}(U)$. Furthermore we let

$$\mathscr{P}^\nu_m(U) = \{(T)^\nu\colon\ T \in \mathscr{P}_m(U)\}.$$

It is important to realize that usually

$$\mathbf{I}^\nu_m(U) \neq \{(T)^\nu\colon\ T \in \mathbf{I}_m(U)\}.$$

For example if S is the sum of an infinite series of disjoint real projective planes in a compact subset of $\mathbf{R}^6$, such that the sum of the areas is finite, but the sum of the lengths of the bounding projective lines is infinite, then $(S)^2 \in \mathbf{I}^2_2(\mathbf{R}^6)$, but no member of $\mathbf{I}_2(\mathbf{R}^6)$ is congruent to S modulo 2; to be specific we recall 3.2.28 (4) and let

$$S = \sum_{j=1}^{\infty} f_\# R_j$$

where $R_j \in \mathbf{I}_2(\mathbf{R}^3)$ is an oriented hemisphere with center 0 and radius $j^{-\frac{1}{2}}$, and

$$f\colon \mathbf{R}^3 \to \odot_2 \mathbf{R}^3 \simeq \mathbf{R}^6, \qquad f(x) = x^2/2 \text{ for } x \in \mathbf{R}^3.$$

Whenever $T \in \mathscr{F}_m(U)$ we define

$$\mathbf{M}^\nu(T)$$

as *the least* $t \in \overline{\mathbf{R}}$ *such that for every* $\varepsilon > 0$ *there exists a compact subset* K *of* U *and a rectifiable current*

$$R \in \mathscr{R}_{m,K}(U) \text{ with } \mathscr{F}^\nu_K(T - R) \le \varepsilon \text{ and } \mathbf{M}(R) \le t + \varepsilon;$$

we note that if $\delta > 0$ and $H = \{x\colon \operatorname{dist}(x, \operatorname{spt} T) \le 2\delta\} \subset U$, then

$$\delta^{-1} \int_\delta^{*2\delta} \mathscr{F}^\nu_H [T - R \llcorner \{x\colon \operatorname{dist}(x, \operatorname{spt} T) \le r\}]\, d\mathscr{L}^1 r \le (1 + \delta^{-1})\, \varepsilon;$$

moreover $\mathbf{I}_{m,H}(U)$ is $\mathbf{M}$ dense in $\mathscr{R}_{m,H}(U)$. It follows that *for every compact subset* K *of* U *with* $\operatorname{spt} T \subset \operatorname{Int} K$ *there exists a sequence of currents* $R_i \in \mathbf{I}_{m,K}(U)$ *with*

$$\sum_{i=0}^{\infty} \mathscr{F}^\nu_K(R_{i+1} - R_i) < \infty, \qquad \lim_{i\to\infty} \mathscr{F}^\nu_K(R_i - T) = 0, \qquad \lim_{i\to\infty} \mathbf{M}(R_i) = \mathbf{M}^\nu(T).$$

Clearly $\mathbf{M}^\nu(T_1+T_2)\le\mathbf{M}^\nu(T_1)+\mathbf{M}^\nu(T_2)$ for $T_1, T_2\in\mathscr{F}_m(U)$,

$$\mathbf{M}^\nu(T_1)=\mathbf{M}^\nu(T_2) \text{ if } T_1\equiv T_2 \bmod \nu,$$

$$\mathscr{F}_K^\nu(T)\le\mathbf{M}^\nu(T) \text{ if } \operatorname{spt} T\subset\operatorname{Int} K,$$

and $\mathbf{M}^\nu$ is lowersemicontinuous on $\mathscr{F}_{m,K}(U)$ relative to $\mathscr{F}_K^\nu$. We will now prove that if $\mathbf{M}^\nu(T)<\infty$, then for every approximating sequence of currents R_i as above *the corresponding sequence of Daniell integrals* $\|R_i\|$ *on* $\mathscr{K}(U)$ *converges weakly* (see 2.5.19) *to a Daniell integral*

$$\|T\|^\nu,$$

which is necessarily the same for all such approximating sequences, depends only on the congruence class of T modulo ν, and satisfies the equations

$$\|T\|^\nu(U)=\mathbf{M}^\nu(T), \quad \operatorname{spt}\|T\|^\nu=\operatorname{spt}^\nu(T).$$

Since $\|R_i\|(U)\to\mathbf{M}^\nu(T)<\infty$ it suffices to show that for every Lipschitzian real valued function u on U the sequence of numbers $\|R_i\|(u)$ converges. Choosing a compact subset C of U with $K\subset\operatorname{Int} C$, abbreviating

$$E_r=\{x\colon u(x)\le r\} \text{ for } r\in\mathbf{R},$$

and taking $-\infty<a<\inf u(K)\le\sup u(K)<b<\infty$, we see by addition of the inequalities

$$\textstyle\int_a^{*b}\mathscr{F}_C^\nu[(R_{i+1}-R_i)\llcorner E_r]\,d\mathscr{L}^1 r\le[b-a+\operatorname{Lip}(u)]\,\mathscr{F}_K^\nu(R_{i+1}-R_i)$$

that $\mathscr{L}^1$ almost all real numbers belong to the set

$$\Gamma=\Big\{r\colon \sum_{i=1}^\infty\mathscr{F}_C^\nu[(R_{i+1}-R_i)\llcorner E_r]<\infty\Big\}.$$

For each $r\in\Gamma$ we select a current

$$S_r\in\mathscr{F}_{m,C}(U) \text{ with } \lim_{i\to\infty}\mathscr{F}_C^\nu(S_r-R_i\llcorner E_r)=0.$$

Letting $g_i(r)=\mathbf{M}(R_i\llcorner E_r)=\mathbf{M}(R_i)-\mathbf{M}(R_i-R_i\llcorner E_r)$ for $r\in\mathbf{R}$, we find that, for $r\in\Gamma$,

$$\begin{aligned}\liminf_{i\to\infty} g_i(r)&\ge\mathbf{M}^\nu(S_r)\ge\mathbf{M}^\nu(T)-\mathbf{M}^\nu(T-S_r)\\&\ge\mathbf{M}^\nu(T)-\liminf_{i\to\infty}\mathbf{M}(R_i-R_i\llcorner E_r)=\limsup_{i\to\infty} g_i(r),\end{aligned}$$

hence $g_i(r)\to\mathbf{M}^\nu(S_r)$ as $i\to\infty$. Then we apply 2.4.18, 2.6.7 with $\alpha=\mathscr{L}^1\llcorner\{r\colon r\ge a\}$ and $\beta=u_\#\|R_i\|$, and 2.4.9, to conclude

$$\begin{aligned}\textstyle\int u\,d\|R_i\|&=\textstyle\int r\,d(u_\#\|R_i\|)\,r=b\,g_i(b)-\int_a^b g_i\,d\mathscr{L}^1\\&\to b\,\mathbf{M}^\nu(T)-\textstyle\int_a^b\mathbf{M}^\nu(S_r)\,d\mathscr{L}^1 r \text{ as } i\to\infty.\end{aligned}$$

We note that, if $T_j \in \mathscr{F}_m(U)$ with $\mathbf{M}^\nu(T_j) < \infty$ for $j=1,2$, then

$$\|T_1+T_2\|^\nu \le \|T_1\|^\nu + \|T_2\|^\nu.$$

Assuming $A \subset U$, $T \in \mathscr{F}_m(U)$ *and* $\mathbf{M}^\nu(T) < \infty$, we say that A **splits** T **modulo** ν *if and only if there exists* $X \in \mathscr{F}_m(U)$ *with* $\mathbf{M}^\nu(X) < \infty$ *and*

$$\|T\|^\nu = \|X\|^\nu + \|T-X\|^\nu, \qquad \|X\|^\nu(U \sim A) = 0 = \|T-X\|^\nu(A);$$

we note that if these conditions are satisfied by two currents X and X', then

$$\|X-X'\|^\nu(U\sim A) \le [\|X\|^\nu + \|X'\|^\nu](U\sim A) = 0,$$
$$\|X-X'\|^\nu(A) \le [\|X-T\|^\nu + \|T-X'\|^\nu](A) = 0,$$

hence $\mathbf{M}^\nu(X-X')=0$; thus A and the congruence class of T modulo ν determine the congruence class of X modulo ν, which we denote

$$(T)^\nu \llcorner A,$$

because $X = T \llcorner A$ in case $\nu = 0$. Returning to the construction discussed in the preceding paragraph we observe that $\mathscr{L}^1$ almost all real numbers belong to the set

$$\Delta = \Gamma \cap \{r: \ \|T\|^\nu u^{-1}\{r\} = 0\},$$

and that $r \in \Delta$ implies

$$\|T\|^\nu = \|S_r\|^\nu + \|T-S_r\|^\nu, \qquad \|S_r\|^\nu(U\sim E_r) = 0 = \|T-S_r\|^\nu(E_r)$$

because $\|R_i \llcorner E_r\|$ and $\|R_i\| - \|R_i \llcorner E_r\| = \|R_i - R_i \llcorner E_r\|$ converge weakly to $\|S_r\|^\nu$ and $\|T-S_r\|^\nu$; thus

$$(T)^\nu \llcorner E_r = (S_r)^\nu \text{ for } r \in \Delta.$$

Moreover E_t splits T for every real number t, because if $r, s \in \Delta$ with $r > s > t$, then

$$\begin{aligned}\mathbf{M}^\nu(S_r - S_s) &\le \liminf_{i\to\infty} \mathbf{M}[R_i \llcorner (E_r \sim E_s)] \\ &= \lim_{i\to\infty} [g_i(r) - g_i(s)] = \mathbf{M}^\nu(S_r) - \mathbf{M}^\nu(S_s) \\ &= \|T\|^\nu(E_r \sim E_s) \le \|T\|^\nu\{x: r \ge u(x) > t\}\end{aligned}$$

is small for r close to t, hence there exists $X \in \mathscr{F}_{m,C}(U)$ such that

$$\lim_{\Delta \ni r \to t+} \mathbf{M}^\nu(X - S_r) = 0$$

and X satisfies the splitting conditions for $A = E_t$. We infer *that every relatively closed subset A of U splits T modulo ν*, by taking $u(x) = \operatorname{dist}(x, A)$ for $x \in U$, and $t=0$. Furthermore one readily verifies that the class

$$\{A: \ A \subset U, A \text{ splits every } T \in \mathscr{F}_m(U) \text{ with } \mathbf{M}^\nu(T) < \infty\}$$

is a Borel family with respect to U; for instance if A and B belong to this class one can choose X as above, choose $Y \in \mathscr{F}_m(U)$ with $\mathbf{M}^\nu(Y) < \infty$ and

$$\|T-X\|^\nu = \|Y\|^\nu + \|T-X-Y\|^\nu, \qquad \|Y\|^\nu(U \sim B) = 0 = \|T-X-Y\|^\nu(B),$$

hence infer $X+Y \in \mathscr{F}_m(U)$ with $\mathbf{M}^\nu(X+Y) < \infty$ and

$$\|T\|^\nu = \|X\|^\nu + \|Y\|^\nu + \|T-X-Y\|^\nu \geq \|X+Y\|^\nu + \|T-X-Y\|^\nu \geq \|T\|^\nu,$$

$$\|X+Y\|^\nu [U \sim (A \cup B)] \leq \|X\|^\nu (U \sim A) + \|Y\|^\nu (U \sim B) = 0,$$

$$\begin{aligned} \|T-X-Y\|^\nu (A \cup B) &= \|T-X-Y\|^\nu (A \sim B) \\ &\leq \|T-X\|^\nu (A) + \|Y\|^\nu (A \sim B) = 0. \end{aligned}$$

It follows that *every Borel subset of U splits every $T \in \mathscr{F}_m(U)$ with* $\mathbf{M}^\nu(T) < \infty$.

Whenever $T \in \mathscr{F}_m(U)$ we define

$$\mathbf{N}^\nu(T) = \mathbf{M}^\nu(T) + \mathbf{M}^\nu(\partial T) \text{ if } m > 0, \qquad \mathbf{N}^\nu(T) = \mathbf{M}^\nu(T) \text{ if } m = 0.$$

Since the functions $\mathbf{M}^\nu$ and $\mathbf{N}^\nu$ are constant on each congruence class modulo ν, they induce functions on $\mathscr{F}_m^\nu(U)$, which will be denoted by the same symbols. In case $\mathbf{N}^\nu(T) < \infty$ with $\operatorname{spt} T \subset \operatorname{Int} K$ and $m > 0$ we can approximate T by currents R_i as in the two preceding paragraphs, and similarly approximate ∂T by currents $Q_i \in \mathbf{I}_{m-1,K}(U)$ with

$$\sum_{i=1}^{\infty} \mathscr{F}_K^\nu(Q_{i+1} - Q_i) < \infty, \qquad \lim_{i\to\infty} \mathscr{F}_K^\nu(Q_i - \partial T) = 0, \qquad \lim_{i\to\infty} \mathbf{M}(Q_i) = \mathbf{M}^\nu(\partial T);$$

inferring that $\mathscr{F}_K^\nu(Q_i - \partial R_i) \to 0$ as $i \to \infty$ we may also require, after passage to subsequences, that

$$\sum_{i=1}^{\infty} \mathscr{F}_K^\nu(Q_i - \partial R_i) < \infty.$$

Taking u, E_r, C as above we then see that, for $\mathscr{L}^1$ almost all r,

$$\mathscr{F}_C^\nu[(Q_i - \partial R_i) \llcorner E_r] \to 0 \text{ as } i \to \infty$$

and $\langle (T)^\nu, u, r+ \rangle = \partial[(T)^\nu \llcorner E_r] - (\partial T)^\nu \llcorner E_r$ equals the congruence class modulo ν of the $\mathscr{F}_C^\nu$ limit of the currents

$$\partial(R_i \llcorner E_r) - Q_i \llcorner E_r = \langle R_i, u, r+ \rangle + (\partial R_i - Q_i) \llcorner E_r,$$

hence there exists a current $Z_r \in \mathscr{F}_{m,C}(U)$ with

$$(Z_r)^\nu = \langle (T)^\nu, u, r+ \rangle \quad \text{and} \quad \lim_{i\to\infty} \mathscr{F}_C^\nu(Z_r - \langle R_i, u, r+ \rangle) = 0.$$

We conclude that, whenever $-\infty \leq a < b \leq \infty$,

$$\int_a^{*b} \mathbf{M}^\nu \langle (T)^\nu, u, r+ \rangle \, d\mathscr{L}^1 r \leq \operatorname{Lip}(u) \, \|T\|^\nu \{x : a < x < b\}.$$

From here on we assume that the integer ν is positive.

A current $T \in \mathscr{R}_m(U)$ is called **representative modulo** ν if and only if

$$\|T\|(A) \le (\nu/2)\,\mathscr{H}^m(A) \text{ for every subset } A \text{ of } U;$$

from 4.1.28 we see that this is equivalent to

$$\Theta^m(\|T\|, x) \le \nu/2 \text{ for } \mathscr{H}^m \text{ almost all } x \text{ in } U.$$

Since every integer is congruent modulo ν to some integer with absolute value at most $\nu/2$, *every rectifiable current is expressible as a sum $T + \nu\, Q$ where T, Q are rectifiable currents and T is representative modulo ν.*

We will now show that

$$\mathbf{M}^\nu(T) = \mathbf{M}(T), \quad \textit{hence} \quad \|T\|^\nu = \|T\|,$$

whenever $T \in \mathscr{R}_m(\mathbf{R}^n)$ and T is representative modulo ν. First we treat the case when $m = n$; for every $\varepsilon > 0$ we choose $R, Q \in \mathscr{R}_m(\mathbf{R}^m)$ so that

$$\mathbf{M}(T - R - \nu\, Q) \le \varepsilon \quad \text{and} \quad \mathbf{M}(R) \le \mathbf{M}^\nu(T) + \varepsilon,$$

and represent $T = \mathbf{E}^m \llcorner t$, $R = \mathbf{E}^m \llcorner r$, $Q = \mathbf{E}^m \llcorner q$ where t, r, q are integer valued $\mathscr{L}^m$ summable functions, with $|t(x)| \le \nu/2$ for $x \in \mathbf{R}^n$ because T is representative modulo ν, hence infer

$$\mathbf{M}(T) = \int |t|\, d\mathscr{L}^m \le \int |t - \nu\, q|\, d\mathscr{L}^m = \mathbf{M}(T - \nu\, Q) \le \mathbf{M}(R) + \varepsilon \le \mathbf{M}^\nu(T) + 2\varepsilon.$$

Recalling 4.1.15 we observe next that the assertion holds in case spt T can be covered by a finite family of m dimensional affine subspaces of $\mathbf{R}^n$. Finally we treat the general case by applying 4.2.20; for every $\varepsilon > 0$ there exists a current $X \in \mathscr{R}_m(\mathbf{R}^n)$, which is representative modulo ν, and a diffeomorphism f mapping $\mathbf{R}^n$ onto $\mathbf{R}^n$ such that

$$\mathbf{M}(T - X) \le \varepsilon, \quad \operatorname{Lip}(f) \le 1 + \varepsilon, \quad \operatorname{Lip}(f^{-1}) \le 1 + \varepsilon$$

and $f(\operatorname{spt} X)$ can be covered by a finite family of m dimensional affine subspaces of $\mathbf{R}^n$; since $f_\# X$ is representative modulo ν, as seen from the proof of 4.1.28 or from 4.1.30, we infer

$$\begin{aligned} \mathbf{M}(T) - \varepsilon \le \mathbf{M}(X) &\le (1+\varepsilon)^m\, \mathbf{M}(f_\# X) = (1+\varepsilon)^m\, \mathbf{M}^\nu(f_\# X) \\ &\le (1+\varepsilon)^{2m}\, \mathbf{M}^\nu(X) \le (1+\varepsilon)^{2m} [\mathbf{M}^\nu(T) + \varepsilon]. \end{aligned}$$

From these facts it follows that

$$\mathscr{R}_m(U) \cap \{T\colon T \equiv 0 \bmod \nu\} = \nu\, \mathscr{R}_m(U).$$

We observe that

$$\mathscr{H}^m(W) \le \nu^{-1} \|S\|(W) + \|S\|^\nu(W)$$

whenever $S \in \mathscr{R}_m(\mathbf{R}^n)$ and $W \subset \{x : \Theta^m(\|S\|, x) \ge 1\}$; to prove this inequality we express $S = T + \nu Q$ where T, Q are rectifiable currents and $\|S\|^\nu = \|T\|$,, let

$$A = W \cap \{x : \Theta^m(\|T\|, x) = 0, \ \Theta^m(\|Q\|, x) \text{ is an integer}\}$$

and $B = W \sim A$, then use 4.1.28 to infer

$$\Theta^m(\|S\|, x) = \nu\, \Theta^m(\|Q\|, x) \ge \nu \text{ for } x \in A, \quad \nu\, \mathscr{H}^m(A) \le \|S\|(A),$$
$$\Theta^m(\|T\|, x) \ge 1 \text{ for } \mathscr{H}^m \text{ almost all } x \text{ in } B, \quad \mathscr{H}^m(B) \le \|T\|(B).$$

Now, having established the basic concepts of the theory of flat chains modulo $\nu > 0$, one can apply to these chains the same geometric constructions which were used for integral flat chains (the case $\nu = 0$) in 4.2.2 to 4.2.22, and 4.2.25. Since our treatment of the integral case was carefully planned so as to facilitate this extension, we may leave the verification of details to the reader, and merely summarize the principal results in their modified form (expressed in terms of congruence classes modulo ν):

(4.2.9)$^\nu$ *Whenever* $T \in \mathscr{F}_m^\nu(\mathbf{R}^n)$, $\mathbf{N}^\nu(T) < \infty$ *and* $\varepsilon > 0$ *there exist* $P \in P_m^\nu(\mathbf{R}^n)$, $Q \in \mathscr{F}_m^\nu(\mathbf{R}^n)$, $S \in F_{m+1}^\nu(\mathbf{R}^n)$ *such that the conditions* (1)–(5) *are satisfied with* $\mathbf{M}$, spt *replaced by* $\mathbf{M}^\nu$, spt^ν *and the following implications hold:*

(6)$^\nu$ *If* $T \in \mathbf{I}_m^\nu(\mathbf{R}^n)$, *then* $Q \in \mathbf{I}_m^\nu(\mathbf{R}^n)$ *and* $S \in \mathbf{I}_{m+1}^\nu(\mathbf{R}^n)$.

(7)$^\nu$ *If* $\partial T \in \mathscr{R}_{m-1}^\nu(\mathbf{R}^n)$, *then* $Q \in \mathbf{I}_m^\nu(\mathbf{R}^n)$.

(8)$^\nu$ *If* $\partial T \in \mathscr{P}_{m-1}^\nu(\mathbf{R}^n)$, *then* $Q \in \mathscr{P}_m^\nu(\mathbf{R}^n)$.

(9)$^\nu$ *If* $T \in \mathscr{P}_m^\nu(\mathbf{R}^n)$, *then* $S \in \mathscr{P}_{m+1}^\nu(\mathbf{R}^n)$.

(4.2.10)$^\nu$ *If* $T \in \mathbf{I}_m^\nu(\mathbf{R}^n)$ *with* $\partial T = 0$; *then there exists*

$$S \in \mathbf{I}_{m+1}^\nu(\mathbf{R}^n) \quad \text{with} \quad \partial S = T \quad \text{and} \quad \mathbf{M}^\nu(S)^{m/(m+1)} \le \gamma\, \mathbf{M}^\nu(T).$$

(4.2.14)$^\nu$ *If* $T \in \mathscr{F}_m^\nu(\mathbf{R}^n)$, $m > 0$, $\mathbf{M}^\nu(T) < \infty$ *and* $E \subset \mathbf{R}^n$, *then*

$$\mathbf{I}_1^m(E) = 0 \text{ implies } \|T\|^\nu(E) = 0.$$

(4.2.16)$^\nu$ *If K is a compact Lipschitz neighborhood retract in* $\mathbf{R}^n$ *and m is a nonnegative integer, then*

$$\mathscr{R}_{m+1,K}^\nu(\mathbf{R}^n) \cap \{S : \mathbf{M}^\nu(\partial S) < \infty\} = \mathbf{I}_{m+1,K}^\nu(\mathbf{R}^n),$$
$$\mathscr{F}_{m,K}^\nu(\mathbf{R}^n) \cap \{T : \mathbf{M}^\nu(T) < \infty\} = \mathscr{R}_{m,K}^\nu(\mathbf{R}^n).$$

$(4.2.17)^\nu$ *If K is a compact Lipschitz neighborhood retract in* $\mathbf{R}^n$, *m is a nonnegative integer and* $0 \le c < \infty$, *then*

$$\mathbf{I}^\nu_{m,K}(\mathbf{R}^n) \cap \{T\colon \mathbf{N}^\nu(T) \le c\} \text{ is } \mathscr{F}^\nu_K \text{ compact.}$$

$(4.2.20)^\nu$ *One may replace* $\mathbf{I}_m$, $\mathscr{P}_m$, spt, $\mathbf{N}$ *in* 4.2.20 *by* $\mathbf{I}^\nu_m$, $\mathscr{P}^\nu_m$, spt^ν, $\mathbf{N}^\nu$.

We note that $(4.2.16)^\nu$ implies

$$\mathscr{F}^\nu_m(\mathbf{R}^n) \cap \{T\colon \mathbf{N}^\nu(T) < \infty\} = \mathbf{I}^\nu_m(\mathbf{R}^n);$$

however this equation is a highly nontrivial result and may not be used in the earlier part of the theory outlined here.

We see from $(4.2.16)^\nu$ and our discussion of representative currents that for every $T \in \mathscr{F}_m(\mathbf{R}^n)$ with $\mathbf{M}^\nu(T) < \infty$ there exists $R \in \mathscr{R}_m(\mathbf{R}^n)$ such that $T \equiv R \mod \nu$ and R is representative modulo ν, hence $\|T\|^\nu = \|R\| \le (\nu/2)\mathscr{H}^m$. However it is not known whether there exists currents $T \in \mathscr{F}_m(\mathbf{R}^n)$ with $\mathbf{M}^\nu(T) = \infty$ and $\mathscr{H}^m[\mathrm{spt}^\nu(T)] = 0$.

In case ν is odd, distinct representative rectifiable currents cannot be congruent modulo ν, hence 4.1.31 (2) may be generalized to flat chains modulo ν. In case ν is even, the example of a Möbius band shows the impossibility of such a generalization.

4.2.27. Here we localize 4.2.16 (3) by proving the following proposition: *If $X \in \mathscr{F}_m(\mathbf{R}^n)$, $a \in \mathbf{R}^n$ and there exists*

$$Y \in \mathscr{D}_m(\mathbf{R}^n) \quad \text{with} \quad \mathbf{M}(Y) < \infty \quad \text{and} \quad a \notin \mathrm{spt}(X - Y),$$

then there exist $Z \in \mathscr{R}_m(\mathbf{R}^n)$ with $a \notin \mathrm{spt}(X - Z)$.

Choosing $\delta > 0$ and $\alpha \in \mathscr{D}^0(\mathbf{R}^n)$ so that

$$\mathbf{B}(a, \delta) \subset \{x\colon \alpha(x) = 1\} \quad \text{and} \quad X \llcorner \alpha = Y \llcorner \alpha,$$

and representing $X = R + \partial S$ with $R \in \mathscr{R}_m(\mathbf{R}^n)$, $S \in \mathscr{R}_{m+1}(\mathbf{R}^n)$ we find that $\partial(S \llcorner \alpha) = (\partial S) \llcorner \alpha - S \llcorner d\alpha = Y \llcorner \alpha - R \llcorner \alpha - S \llcorner d\alpha$ has finite mass, hence $S \llcorner \alpha \in \mathbf{N}_{m+1}(\mathbf{R}^n)$. We apply 4.2.1 with $u(x) = |x - a|$ for $x \in \mathbf{R}^n$ to obtain a number r such that

$$0 < r < \delta \quad \text{and} \quad \mathbf{M}\langle S \llcorner \alpha, u, r+\rangle < \infty,$$

note that $(S \llcorner \alpha) \llcorner \mathbf{B}(a, r) = S \llcorner \mathbf{B}(a, r) \in \mathscr{R}_{m+1}(\mathbf{R}^n)$ and

$$\partial[S \llcorner \mathbf{B}(a, r)] = \langle S \llcorner \alpha, u, r+\rangle + [\partial(S \llcorner \alpha)] \llcorner \mathbf{B}(a, r)$$

has finite mass, infer from 4.2.16 (3) that $S \llcorner \mathbf{B}(a, r) \in \mathbf{I}_{m+1}(\mathbf{R}^n)$, and conclude

$$Z = R + \partial[S \llcorner \mathbf{B}(a, r)] \in \mathscr{R}_m(\mathbf{R}^n) \quad \text{with} \quad \mathrm{spt}(X - Z) \subset \mathbf{R}^n \sim \mathbf{U}(a, r).$$

Combining this proposition with 4.1.28 we can now *supplement* 4.1.31 (2) *by the statement that r is an integer in case $\Gamma \in \mathscr{F}^{\mathrm{loc}}_m(U)$.*

4.2.28. Assuming that W is an open subset of $\mathbf{R}^n$ we call T an m dimensional **analytic chain** in W if and only if $T \in \mathscr{F}_m^{\text{loc}}(W)$ and (see 3.4.5, 3.4.8)

$$\dim \operatorname{var}[\operatorname{ideal} \gamma_a(\operatorname{spt} T)] \leq m \text{ for } a \in \operatorname{spt} T,$$

$$\dim \operatorname{var}[\operatorname{ideal} \gamma_a(\operatorname{spt} \partial T)] \leq m-1 \text{ for } a \in \operatorname{spt} \partial T;$$

in case $m > 0$ it follows that ∂T is an $m-1$ dimensional analytic chain in W.

The m dimensional analytic chains in W form a subgroup of $\mathscr{F}_m^{\text{loc}}(W)$, because

$$\dim(\alpha \cup \beta) = \sup\{\dim \alpha, \dim \beta\} \text{ for } \alpha, \beta \in \mathscr{V}_a(\mathbf{R}^n).$$

Next we consider an *m dimensional analytic block B of any open subset U of W.* Choosing $f_{m+1}, \ldots, f_n, g$ according to the definition in 3.4.5 and letting

$$\xi = \mathbf{D}^{n-m}(D f_{m+1} \wedge \cdots \wedge D f_n), \qquad \zeta = \xi / |\xi|$$

we see that *$\zeta | B$ is an m-vectorfield orienting B.* For each $a \in U \cap \operatorname{Clos} B$ it follows from 3.4.8 (16), since $\mathscr{H}^{m+1}(B) = 0$ and B is relatively open in $\operatorname{Var}(U, \{f_{m+1}, \ldots, f_n\})$, that

$$\dim v \leq m \quad \text{with} \quad v = \operatorname{var} \operatorname{ideal}[\gamma_a(B)],$$

and from 3.4.8 (13) that

$$\mathscr{H}^m(B \cap Q) < \infty \text{ for some neighborhood } Q \text{ of } a;$$

moreover 3.4.8 (14), (15) imply

$$\dim(v \cap w) \leq m-1 \quad \text{with} \quad w = \gamma_a\{x : g(x) = 0\},$$

because $\gamma_a(B) \cap w = \gamma_a(\varnothing)$, hence $v \cap w$ contains no irreducible component of v. Using 4.1.31 (1) we conclude that *if* $\operatorname{Clos} B$ *is a compact subset of* U, *then the two currents*

$$\pm(\mathscr{H}^m \llcorner B) \wedge \zeta \in \mathscr{R}_m(W)$$

are m dimensional analytic chains in W; we will refer to such currents as m dimensional **oriented analytic blocks** with compact support in W.

We see from 3.4.8 (11) that if R is an oriented analytic block with compact support in W and h is any real valued analytic function on W, then $R \llcorner \{x : h(x) > 0\}$ is the sum of finitely many oriented analytic blocks.

Now one readily verifies the following basic proposition: *Every m dimensional analytic chain T in W is locally representable as finite sum of oriented analytic blocks with compact support in W, hence T is a locally rectifiable current; similarly ∂T is an $m-1$ dimensional locally rectifiable current if $m > 0$. The group of all m dimensional analytic chains with compact support in W is generated by the m dimensional oriented analytic* *blocks with compact support in W, and consists of integral currents.* In fact if $0 \in \operatorname{spt} T$ one can apply 3.4.8 (11) with

$$\alpha = \operatorname{var}[\operatorname{ideal} \gamma_0(\operatorname{spt} T)] \quad \text{and} \quad \beta = \operatorname{var}[\operatorname{ideal} \gamma_0(\operatorname{spt} \partial T)],$$

choosing $V_0 = V_1 = \cdots = V_\mu = \mathbf{U}(0,r)$ as explained in 3.4.9, and use downward induction to construct m dimensional currents T_i for $i = \mu, \ldots, 1$ so that

$$\mathbf{U}(0,r) \cap \operatorname{spt}[T - (T_\mu + \cdots + T_i)] \subset (S_{i-1} \cup \cdots \cup S_0)$$

and either $\dim S_i = m$ and T_i is an integral linear combination of oriented components of S_i, in accordance with 4.1.31 (2) and 4.2.27, or $\dim S_i < m$ and $T_i = 0$, in accordance with 4.1.21; since $\dim \beta < m$ we may assume $\dim S_0 < m$ and conclude

$$\mathbf{U}(0,r) \cap \operatorname{spt}[T - (T_\mu + \cdots + T_1)] = \varnothing;$$

moreover $T_i \llcorner \mathbf{U}(0,s)$ is an integral linear combination of finitely many analytic blocks with compact support in W whenever $0 < s < r$.

4.2.29. Here we suppose W is an open subset of $\mathbf{C}^n$ and modify 4.2.28 in accordance with 3.4.12.

We say that T is a **complex m dimensional holomorphic chain of W in U** if and only if U is an open subset of W and

$$T \in \mathscr{R}_{2m}^{\mathrm{loc}}(W), \qquad U \cap \operatorname{spt} \partial T = \varnothing,$$

$$\dim_{\mathbf{C}} \operatorname{var}[\operatorname{ideal}^{\mathbf{C}} \gamma_a(\operatorname{spt} T)] \le m \ \text{ for } a \in U \cap \operatorname{spt} T.$$

We call T **positive** in U if and only if for $\|T\|$ almost all x in U the simple $2m$ vector $\vec{T}(x)$ is complex and positive (see 1.6.6).

Whenever T_1 and T_2 are complex m dimensional holomorphic chains of W in U, so are $T_1 + T_2$ and $T_1 - T_2$; in case T_1 and T_2 are positive in U, so is $T_1 + T_2$.

The most important examples of holomorphic chains are constructed as follows: *If $S \subset A \subset W$,*

$$A \textit{ is open, } S \textit{ is holomorphic in } A, \ m = \dim_{\mathbf{C}} S$$

and R is the regular part of S, then R is oriented by a unique vectorfield ζ such that for every $z \in R$ the simple $2m$ vector $\zeta(z)$ is complex and positive; if furthermore

$$U \subset A, \ U \textit{ is open, } W \cap \operatorname{Clos} U \subset A$$

and B is the union of some family of components of $U \cap R$, then the current

$$Q = (\mathscr{H}^{2m} \llcorner B) \wedge \zeta$$

is a positive complex m dimensional holomorphic chain of W in U. This is implied by 4.1.31 (1) and 4.1.20 because

$$\operatorname{spt} Q \subset \operatorname{Clos} B \subset S, \quad U \cap \operatorname{spt} \partial Q \subset S \sim R,$$
$$\mathscr{H}^{2m-1}(U \cap \operatorname{spt} \partial Q) \leq \mathscr{H}^{2m-1}(S \sim R) = 0.$$

In case $B = U \cap R$ the resulting current Q satisfies the equation $\|Q\| = \mathscr{H}^{2m} \llcorner (U \cap S)$. In case $U \cap R$ is a functional holomorphic submanifold of U and B is a component of $U \cap R$, hence B is a holomorphic block of U, we call the corresponding current Q a **positive holomorphic block of U**.

Applying the complex version of 3.4.8 we can decompose any holomorphic chain locally into multiples of holomorphic blocks: *If T is a complex m dimensional holomorphic chain of W in U, then U is the union of open sets V for which there exist integers $r_1, \ldots, r_s$ and positive complex m dimensional holomorphic blocks $Q_1, \ldots, Q_s$ of V such that*

$$T \llcorner V = \sum_{j=1}^{s} r_j Q_j \quad \textit{and} \quad \|T\| \llcorner V = \sum_{j=1}^{s} \|r_j Q_j\|;$$

in case T is positive in U, then $r_1, \ldots, r_s$ are positive.

4.3. Slicing

4.3.1. In this section we assume that *U is an open subset of some Euclidean space, $f: U \to \mathbf{R}^n$ is a locally Lipschitzian map, K is a compact subset of U and $T \in \mathbf{F}_{m,K}(U)$ with $m \geq n$.* For $\mathscr{L}^n$ almost all y in $\mathbf{R}^n$ we will construct an $m-n$ dimensional current $\langle T, f, y \rangle$, to be considered the **slice of T in $f^{-1}\{y\}$**, as a limit of currents $T \llcorner f^{\#} \phi$ corresponding to suitable differential forms ϕ of degree n on $\mathbf{R}^n$ whose supports tend to y and whose integrals over $\mathbf{R}^n$ equal 1.

First we observe that, if f is of class ∞ and $\phi \in \mathscr{D}^n(\mathbf{R}^n)$, then

$$[T \llcorner f^{\#} \phi](\psi) = (-1)^{n(m-n)} [f_{\#}(T \llcorner \psi)](\phi)$$

whenever $\psi \in \mathscr{D}^{m-n}(U)$. However *the right side of the above equation remains meaningful for any locally Lipschitzian map f and any bounded Baire form ϕ of degree n on $\mathbf{R}^n$*, because 4.1.18 implies that

$$f_{\#}(T \llcorner \psi) = \mathscr{L}^n \wedge \xi_{\psi}$$

for some $\mathscr{L}^n$ summable n-vectorfield ξ_{ψ} on $\mathbf{R}^n$, and *for such general f and ϕ we define $T \llcorner f^{\#} \phi$ by the above equation;* thus $T \llcorner f^{\#} \phi$ is the linear function on $\mathscr{D}^{m-n}(U)$ given by the formula

$$[T \llcorner f^{\#} \phi](\psi) = (-1)^{n(m-n)} \int \langle \xi_{\psi}, \phi \rangle \, d\mathscr{L}^n$$

whenever $\psi \in \mathscr{D}^{m-n}(U)$. *We will prove that $T \llcorner f^{\#} \phi$ is a flat chain;* clearly

$$\operatorname{spt}(T \llcorner f^{\#} \phi) \subset f^{-1}(\operatorname{spt} \phi) \cap \operatorname{spt} T.$$

If $m>n$ and $\sigma\in\mathscr{D}^{m-n-1}(U)$; then

$$f_{\#}(T\llcorner d\sigma)=f_{\#}[(\partial T)\llcorner\sigma],$$

because $f_{\#}(T\llcorner\sigma)\in\mathscr{D}_{n+1}(\mathbf{R}^n)=\{0\}$; consequently

$$\partial(T\llcorner f^{\#}\phi)=(-1)^n(\partial T)\llcorner f^{\#}\phi.$$

If $\mathbf{M}(T)<\infty$ we see from 4.1.14 and 1.8.1, by approximation starting with the case when f and ϕ are of class ∞, that

$$[T\llcorner f^{\#}\phi](\psi)\le \operatorname{Lip}(f|K)^n\,\|T\|\,[\|\phi\|\circ f)\cdot\|\psi\|],$$

hence $T\llcorner f^{\#}\phi\in\mathscr{D}_{m-n}(U)$ with $\mathbf{M}(T\llcorner f^{\#}\phi)<\infty$ and

$$\|T\llcorner f^{\#}\phi\|\le \operatorname{Lip}(f|K)^n\,\|T\|\llcorner(\|\phi\|\circ f).$$

For any $T\in\mathbf{F}_{m,K}(U)$ we recall the last paragraph of 4.1.24 to choose

$$S\in\mathbf{F}_{m+1,K}(U)\quad\text{with}\quad \mathbf{M}(T-\partial S)+\mathbf{M}(S)=\mathbf{F}_K(T)<\infty,$$

and infer $T\llcorner f^{\#}\phi=(T-\partial S)\llcorner f^{\#}\phi+(-1)^n\partial(S\llcorner f^{\#}\phi)$, hence

$$\begin{aligned}\mathbf{F}_K(T\llcorner f^{\#}\phi)&\le\mathbf{M}[(T-\partial S)\llcorner f^{\#}\phi]+\mathbf{M}(S\llcorner f^{\#}\phi)\\&\le\operatorname{Lip}(f|K)^n[\|T-\partial S\|+\|S\|](\|\phi\|\circ f)\le\operatorname{Lip}(f|K)^n\,\mathbf{F}_K(T)\,\mathbf{M}(\phi).\end{aligned}$$

We also observe that if $g\colon U\to\mathbf{R}^n$ is another locally Lipschitzian map, h is the affine homotopy from f to g, σ is as in 4.1.14 and T is normal, then

$$g_{\#}(T\llcorner\psi)-f_{\#}(T\llcorner\psi)=h_{\#}([0,1]\times\partial(T\llcorner\psi)),$$

$$\begin{aligned}\mathbf{M}[g_{\#}(T\llcorner\psi)-f_{\#}(T\llcorner\psi)]&\le\|\partial(T\llcorner\psi)\|\,(|g-f|\,\sigma^{n-1})\\&\le 2^m\,\mathbf{F}_K(\psi)\,[\|T\|+\|\partial T\|](|g-f|\,\sigma^{n-1})\end{aligned}$$

whenever $\psi\in\mathscr{D}^{m-n}(U)$, because $\mathscr{D}_{n+1}(\mathbf{R}^n)=\{0\}$, hence

$$\mathbf{F}_K(T\llcorner g^{\#}\phi-T\llcorner f^{\#}\phi)\le 2^m\,\mathbf{M}(\phi)\,[\|T\|+\|\partial T\|](|g-f|\,\sigma^{n-1}).$$

From these estimates we conclude:

If ϕ_j are Baire forms of degree n on $\mathbf{R}^n$ such that

$$\phi_j(y)\to\phi(y)\ \text{as}\ j\to\infty\ \text{whenever}\ y\in f(K),$$

$$\sup\{\|\phi_j(y)\|: y\in f(K)\ \text{and}\ j=1,2,3,\ldots\}<\infty,$$

then $\mathbf{F}_K(T\llcorner f^{\#}\phi_j-T\llcorner f^{\#}\phi)\to 0$ *as* $j\to\infty$.

If $T_j\in\mathbf{F}_{m,K}(U)$ *and* $\mathbf{F}_K(T_j-T)\to 0$ *as* $j\to\infty$, *then*

$$\mathbf{F}_K(T_j\llcorner f^{\#}\phi-T\llcorner f^{\#}\phi)\to 0\ \text{as}\ j\to\infty.$$

If $T\in\mathbf{N}_{m,K}(U)$, *then* $T\llcorner f^{\#}\phi\in\mathbf{N}_{m-n,K}(U)$.

If $T\in\mathbf{F}_{m,K}(U)$, *then* $T\llcorner f^{\#}\phi\in\mathbf{F}_{m-n,K}(U)$.

If $f_j\colon U\to\mathbf{R}^n$ are maps of class ∞ such that

$$\sup\{|f_j(x)-f(x)|\colon x\in K\}\to 0 \text{ as } j\to\infty,$$

$$\sup\{\|Df_j(x)\|\colon x\in K \text{ and } j=1,2,3,\ldots\}<\infty,$$

then $\mathbf{F}_K(T\llcorner f_j^{\#}\phi - T\llcorner f^{\#}\phi)\to 0$ *as* $j\to\infty$.

Letting $Y_1,\ldots,Y_n$ be the standard coordinate functions on $\mathbf{R}^n$ we abbreviate

$$\Omega = DY_1\wedge\cdots\wedge DY_n\in\mathscr{E}^n(\mathbf{R}^n)$$

and will now prove that *for $\mathscr{L}^n$ almost all y in $\mathbf{R}^n$ there exists a current $\langle T,f,y\rangle\in\mathscr{D}_{m-n}(U)$ defined by the formula*

$$\langle T,f,y\rangle(\psi)=\lim_{\rho\to 0+}(T\llcorner f^{\#}[\mathbf{B}(y,\rho)\wedge\Omega]/[\alpha(n)\rho^n])(\psi)$$

whenever $\psi\in\mathscr{D}^{m-n}(U)$. For each ψ the above limit equals

$$\lim_{\rho\to 0+}(-1)^{n(m-n)}\int_{\mathbf{B}(y,\rho)}\langle\xi_\psi,\Omega\rangle\,d\mathscr{L}^n/[\alpha(n)\rho^n]$$
$$=(-1)^{n(m-n)}\langle\xi_\psi(y),Y_1\wedge\cdots\wedge Y_n\rangle\in\mathbf{R}$$

for $\mathscr{L}^n$ almost all y, according to 2.9.8. Choosing S as in the preceding paragraph and taking $\mu=f_{\#}(\|T-\partial S\|+\|S\|)$, we also see from 2.9.7 that

$$\int\Theta^n(\mu,y)\,d\mathscr{L}^n\le\mu(\mathbf{R}^n)=\mathbf{M}(T-\partial S)+\mathbf{M}(S)=\mathbf{F}_K(T).$$

Using 4.1.2 we select a countable $\mathbf{F}_K$ dense subset C of $\mathscr{D}^{m-n}(U)$. Then $\mathscr{L}^n$ almost all points y have the property that the above limit exists for every $\psi\in C$, and also $\Theta^n(\mu,y)<\infty$; for each such y we apply the estimate

$$\mathbf{F}_K(T\llcorner f^{\#}[\mathbf{B}(y,\rho)\wedge\Omega]/[\alpha(n)\rho^n])\le\operatorname{Lip}(f|K)^n\,\mu\,\mathbf{B}(y,\rho)/[\alpha(n)\rho^n]$$

to infer that the limit in question exists for every $\psi\in\mathscr{D}^{n-m}(U)$, and that

$$\mathbf{F}_K\langle T,f,y\rangle\le\operatorname{Lip}(f|K)^n\,\Theta^n(\mu,y).$$

By integration we obtain the inequality

$$\int^*\mathbf{F}_K\langle T,f,y\rangle\,d\mathscr{L}^n\,y\le\operatorname{Lip}(f|K)^n\,\mathbf{F}_K(T).$$

Clearly, whenever $\langle T,f,y\rangle\in\mathscr{D}_{m-n}(U)$,

$$\operatorname{spt}\langle T,f,y\rangle\subset f^{-1}\{y\}\cap\operatorname{spt}T,$$

$$\partial\langle T,f,y\rangle=(-1)^n\langle\partial T,f,y\rangle \text{ in case } m>n,$$

$$\langle T\llcorner\beta,f,y\rangle=(-1)^{nk}\langle T,f,y\rangle\llcorner\beta \text{ for } \beta\in\mathscr{D}^k(U) \text{ with } k\le m-n.$$

The last equation allows us to define

$$\langle R,f,y\rangle\in\mathscr{D}_{m-n}(U) \text{ whenever } R\in\mathbf{F}_m^{\mathrm{loc}}(U),$$

for $\mathscr{L}^n$ almost all y, by requiring that

$$\langle R,f,y\rangle\llcorner\beta=\langle R\llcorner\beta,f,y\rangle \text{ for all } \beta\in\mathscr{D}^0(U).$$

4.3.2. Theorem. *Under the conditions of* 4.3.1 *the following statements hold for every bounded real valued Baire function Φ on $\mathbf{R}^n$:*

(1) *For every $\psi \in \mathscr{D}^{m-n}(U)$,*

$$\int \Phi(y)\langle T, f, y\rangle(\psi)\, d\mathscr{L}^n y = [T \llcorner f^{\#}(\Phi \wedge \Omega)](\psi).$$

(2) *If $\mathbf{M}(T) < \infty$, then $T \llcorner f^{\#}(\Phi \wedge \Omega) = (T \llcorner f^{\#}\Omega) \llcorner (\Phi \circ f)$ and*

$$\int |\Phi(y)|\, \mathbf{M}\langle T, f, y\rangle\, d\mathscr{L}^n y = \mathbf{M}[T \llcorner f^{\#}(\Phi \wedge \Omega)]$$
$$\leq [\operatorname{Lip}(f|K)]^n\, \|T\|\,(|\Phi| \circ f);$$

moreover $\int \|\langle T, f, y\rangle\|(v)\, d\mathscr{L}^n y = \|T \llcorner f^{\#}\Omega\|(v)$ for every bounded real valued Baire function v on U.

(3) *If $T \in \mathbf{N}_{m,K}(U)$, then $\langle T, f, y\rangle \in \mathbf{N}_{m-n,K}(U)$ for $\mathscr{L}^n$ almost all y.*

(4) *$\langle T, f, y\rangle \in \mathbf{F}_{m-n,K}(U)$ for $\mathscr{L}^n$ almost all y.*

(5) *If K is a Lipschitz neighborhood retract in U, then the function mapping y onto $\langle T, f, y\rangle$ is $\mathscr{L}^n$ summable, with respect to the norm $\mathbf{F}_K$ on $\mathbf{F}_{m-n,K}(U)$, and*

$$\mathbf{F}_K(\langle T, f, y\rangle - T \llcorner f^{\#}[\mathbf{B}(y,\rho) \wedge \Omega]/[\alpha(n)\rho^n])$$
$$\leq \int_{\mathbf{B}(y,\rho)} \mathbf{F}_K[\langle T, f, y\rangle - \langle T, f, z\rangle]\, d\mathscr{L}^n z/[\alpha(n)\rho^n] \to 0$$

as $\rho \to 0+$ for $\mathscr{L}^n$ almost all y.

(6) *If $g: \mathbf{R}^n \to \mathbf{R}^n$ is locally Lipschitzian, then*

$$\langle T, g \circ f, z\rangle = \sum_{y \in g^{-1}\{z\}} \operatorname{sign}[\det D g(y)] \cdot \langle T, f, y\rangle \text{ for } \mathscr{L}^n \text{ almost all } z.$$

(7) *If $G: U \to V$ and $H: V \to \mathbf{R}^n$ are locally Lipschitzian, where V is an open subset of some Euclidean space, then*

$$G_{\#}[T \llcorner (H \circ G)^{\#}(\Phi \wedge \Omega)] = (G_{\#} T) \llcorner H^{\#}(\Phi \wedge \Omega),$$
$$G_{\#}\langle T, H \circ G, y\rangle = \langle G_{\#} T, H, y\rangle \text{ for } \mathscr{L}^n \text{ almost all } y.$$

Proof. From 4.3.1 we see that (1) holds because

$$[T \llcorner f^{\#}(\Phi \wedge \Omega)](\psi) = \int (-1)^{n(m-n)} \langle \xi_\psi, \Phi \wedge \Omega\rangle\, d\mathscr{L}^n y$$
$$= \int \Phi(y)(-1)^{n(m-n)} \langle \xi_\psi(y), Y_1 \wedge \cdots \wedge Y_n\rangle\, d\mathscr{L}^n y = \int \Phi(y)\langle T, f, y\rangle(\psi)\, d\mathscr{L}^n y.$$

The first conclusion of (2) clearly holds in case f and Φ are of class ∞; assuming $\mathbf{M}(T) < \infty$ we readily extend it to general f and Φ by approximation; in particular

$$T \llcorner f^{\#}[\mathbf{B}(y,\rho) \wedge \Omega] = (T \llcorner f^{\#}\Omega) \llcorner f^{-1}\mathbf{B}(y,\rho),$$
$$\mathbf{M}(T \llcorner f^{\#}[\mathbf{B}(y,\rho) \wedge \Omega]) \leq [f_{\#}\|T \llcorner f^{\#}\Omega\|]\,\mathbf{B}(y,\rho)$$

whenever $y \in \mathbf{R}^n$ and $0<\rho<\infty$, hence for any Φ

$$\begin{aligned}&\int^* |\Phi(y)|\, \mathbf{M}\langle T, f, y\rangle \, d\mathscr{L}^n y \le \int |\Phi(y)|\, \Theta^n(f_\# \|T \llcorner f^* \Omega\|, y)\, d\mathscr{L}^n y\\ &\le \int |\Phi|\, d f_\# \|T \llcorner f^* \Omega\| = \|T \llcorner f^* \Omega\| \, (|\Phi| \circ f)\\ &= \mathbf{M}[(T \llcorner f^* \Omega) \llcorner (\Phi \circ f)] = \mathbf{M}[T \llcorner f^* (\Phi \wedge \Omega)]\end{aligned}$$

by virtue of 2.9.7 and 2.4.18; on the other hand (1) implies

$$[T \llcorner f^* (\Phi \wedge \Omega)](\psi) \le \int_* |\Phi(y)|\, \mathbf{M}\langle T, f, y\rangle \, d\mathscr{L}^n y$$

for every $\psi \in \mathscr{D}^{m-n}(U)$ with $\mathbf{M}(\psi) \le 1$; combining these inequalities we obtain the second conclusion of (2). Since $\|\Phi \wedge \Omega\| = |\Phi|$, the third conclusion of (2) follows from 4.3.1. We deduce the fourth conclusion in case $0 \le v \in \mathscr{D}^0(U)$ through replacement of T by $T \llcorner v$, then in general through linearity and approximation.

We prove (3) by applying (2) to T and ∂T.

To deduce (4) from (3) we choose currents $T_j \in \mathbf{N}_{m,K}(U)$ so that $\mathbf{F}_K(T - T_j) \to 0$ as $j \to \infty$, then use 2.4.6 and 4.3.1 to infer

$$\begin{aligned}&\int^* \liminf_{j\to\infty} \mathbf{F}_K(\langle T, f, y\rangle - \langle T_j, f, y\rangle)\, d\mathscr{L}^n y\\ &\le \liminf_{j\to\infty} \int^* \mathbf{F}_K \langle T - T_j, f, y\rangle \, d\mathscr{L}^n y \le \liminf_{j\to\infty} \operatorname{Lip}(f|K)^n \, \mathbf{F}_K(T - T_j) = 0.\end{aligned}$$

The hypothesis of (5) implies that $\mathbf{F}_{m-n,K}(U)$ is $\mathbf{F}_K$ separable (see 4.1.29), hence $\mathscr{L}^n$ measurability of the function $\langle T, f, \cdot\rangle$ follows from the fact that, whenever $Q \in \mathbf{F}_{m-n,K}(U)$,

$$\begin{aligned}&\mathbf{F}_K(Q - \langle T, f, y\rangle)\\ &= \sup\{Q(\psi) - (-1)^{n(m-n)} \langle \xi_\psi(y), Y_1 \wedge \cdots \wedge Y_n\rangle : \psi \in C,\ \mathbf{F}_K(\psi) \le 1\}\end{aligned}$$

for $\mathscr{L}^n$ almost all y, where C and the $\mathscr{L}^n$ measurable functions ξ_ψ are chosen as in 4.3.1. To complete the proof of (5) we apply 2.9.9 with Y and f replaced by $\mathbf{F}_{m-n,K}(U)$ and $\langle T, f, \cdot\rangle$, noting that (1) and 2.4.12 imply

$$T \llcorner f^* [\mathbf{B}(y, \rho) \wedge \Omega] = \int_{\mathbf{B}(y,\rho)} \langle T, f, z\rangle \, d\mathscr{L}^n z.$$

We deduce (6) from 4.1.25; whenever $\psi \in \mathscr{D}^{m-n}(U)$ we find that

$$(g \circ f)_\# (T \llcorner \psi) = g_\# (\mathscr{L}^n \wedge \xi_\psi) = \mathscr{L}^n \wedge \eta_\psi$$

with

$$\eta_\psi(z) = \sum_{y \in g^{-1}\{z\}} \operatorname{sign}[\det D g(y)] \cdot \xi_\psi(y)$$

for $\mathscr{L}^n$ almost all z, hence

$$\begin{aligned}&[T \llcorner (g \circ f)^* (E \wedge \Omega)](\psi) = (-1)^{n(m-n)} \int_E \langle \eta_\psi, \Omega\rangle \, d\mathscr{L}^n\\ &= \int_E \sum_{y \in g^{-1}\{z\}} \operatorname{sign}[\det D g(y)] \langle T, f, y\rangle(\psi)\, d\mathscr{L}^n z\end{aligned}$$

for every Borel subset E of $\mathbf{R}^n$.

The first conclusion of (7) is trivial in case G, H, Φ are of class ∞, and may be extended to the general case by approximation. The second conclusions holds because (5) implies that $\mathscr{L}^n$ almost all points y in $\mathbf{R}^n$ satisfy the condition

$$\mathbf{F}_K[T \llcorner (H \circ G)^\# \phi_\rho - \langle T, H \circ G, y \rangle] \to 0 \text{ as } \rho \to 0+$$

with $\phi_\rho = \mathbf{B}(y, \rho) \wedge \Omega / [\alpha(n)\, \rho^n]$, hence

$$\mathbf{F}_{G(K)}[(G_\# T) \llcorner H^\# \phi_\rho - G_\# \langle T, H \circ G, y \rangle] \to 0 \text{ as } \rho \to 0+ .$$

4.3.3. Applying 4.3.1 to the *special case when U is an open subset of $\mathbf{R}^n$, $f\colon U \to \mathbf{R}^n$ is the inclusion map, $m = n$ and $T = \mathbf{E}^n \llcorner \tau$ where τ is an $\mathscr{L}^n$ summable function whose support is a compact subset of U*, we find

$$\langle \mathbf{E}^n \llcorner \tau, f, x \rangle = \tau(x)\, \boldsymbol{\delta}_x \in \mathbf{F}_0(U)$$

whenever x belongs to the $\mathscr{L}^n$ Lebesgue set of τ. For any locally Lipschitzian map $g\colon \mathbf{R}^n \to \mathbf{R}^n$ we infer from 4.3.2(6) that

$$\langle \mathbf{E}^n \llcorner \tau, g, y \rangle = \sum_{x \in g^{-1}\{y\}} \operatorname{sign}[\det D g(x)]\, \tau(x)\, \boldsymbol{\delta}_x$$

for $\mathscr{L}^n$ almost all y.

As an illustration we consider an open set $U \subset \mathbf{R}^2 = \mathbf{C}$ and a *(complex) holomorphic function* $g\colon U \to \mathbf{C}$. Letting τ be the characteristic function of an open set A with compact closure in U, we obtain

$$(\mathbf{E}^2 \llcorner A) \llcorner g^\# (B \wedge \Omega) = \int_B \sum_{x \in A \cap g^{-1}\{y\}} \boldsymbol{\delta}_x \, d\mathscr{L}^2 y$$

for every Borel subset B of $\mathbf{C}$, because

$$\det D g(x) = |g'(x)|^2 \geq 0 \text{ for } x \in U;$$

recalling the basic facts about the local behavior of holomorphic maps (see [AL, p. 131] or [RU, p. 216]) we infer that

$$\langle \mathbf{E}^2 \llcorner A, g, y \rangle = \sum_{x \in A} \mathbf{o}(g - y, x)\, \boldsymbol{\delta}_x$$

for every $y \in \mathbf{C} \sim g(\operatorname{Bdry} A)$, where $\mathbf{o}(g-y, x)$ is the order of the function $g - y$ at the point x.

4.3.4. Here we consider the *special case when $n = 1$ and $T \in \mathbf{N}_m(U)$*. Recalling 4.2.1 we will show that in this case

$$\langle T, f, r \rangle = [\langle T, f, r+ \rangle + \langle T, f, r- \rangle]/2 \in \mathbf{F}_{m-1}(U)$$

for every real number r.

First we observe that

$$T \llcorner f^{*}(D\gamma) = (\partial T) \llcorner (\gamma \circ f) - \partial[T \llcorner (\gamma \circ f)]$$

for every Lipschitzian function $\gamma\colon \mathbf{R} \to \mathbf{R}$; this equation follows from 4.1.7 when f and γ are of class ∞, because then $f^{*}(D\gamma) = D(\gamma \circ f)$, and is readily extended to general f and γ by smoothing. [Since $\mathscr{L}^1(\mathbf{R} \sim \operatorname{dmn} D\gamma) = 0$ and $\|D\gamma\| \leq \operatorname{Lip}(\gamma)$, the current $T \llcorner f^{*}(D\gamma)$ is well defined by the method of 4.3.1.]

Whenever $r \in \mathbf{R}$ and $h > 0$ we apply the above equation to the function γ_h defined in 4.2.1; since

$$\gamma_h'(t) = 1/h \text{ if } r < t < r+h, \qquad \gamma_h'(t) = 0 \text{ if } t < r \text{ or } t > r+h,$$

and $D\gamma_h = \gamma_h' \wedge \Omega$, we obtain the formula

$$T \llcorner f^{*}[\{t\colon r < t < r+h\} \wedge \Omega]/h = (\partial T) \llcorner (\gamma_h \circ f) - \partial[T \llcorner (\gamma_h \circ f)],$$

and infer from 4.2.1 that

$$\mathbf{F}_K(T \llcorner f^{*}[\{t\colon r < t < r+h\} \wedge \Omega]/h - \langle T, f, r+\rangle) \to 0$$

as $h \to 0+$, where $K = \operatorname{spt} T$. Similarly one verifies that

$$\mathbf{F}_K(T \llcorner f^{*}[\{t\colon r-h < t < r\} \wedge \Omega]/h - \langle T, f, r-\rangle) \to 0$$

as $h \to 0+$. Adding and dividing by 2, we conclude

$$\mathbf{F}_K(T \llcorner f^{*}[\mathbf{B}(r,h) \wedge \Omega]/(2h) - [\langle T, f, r+\rangle + \langle T, f, r-\rangle]/2) \to 0$$

as $h \to 0+$, hence $\langle T, f, r\rangle = [\langle T, f, r+\rangle + \langle T, f, r-\rangle]/2$.

Now we also see from 4.2.1 that the function $\langle T, f, \cdot\rangle$ has finite $\mathbf{F}_K$ length, and that $\langle T, f, \cdot\rangle$ is $\mathbf{F}_K$ continuous at every point y for which $[\|T\| + \|\partial T\|]\, f^{-1}\{y\} = 0$.

4.3.5. Theorem. *If $f\colon U \to \mathbf{R}^n$, $g\colon U \to \mathbf{R}^\nu$, $h\colon U \to \mathbf{R}^n \times \mathbf{R}^\nu$ are locally Lipschitzian maps such that*

$$h(x) = (f(x), g(x)) \text{ for } x \in U,$$

and if $T \in \mathbf{F}_{m,K}(U)$, where $m \geq n + \nu$ and K is a compact Lipschitz neighborhood retract in U, then

$$\langle T, h, (y,z)\rangle = \langle\langle T, f, y\rangle, g, z\rangle$$

for $\mathscr{L}^n \times \mathscr{L}^\nu$ almost all (y,z) in $\mathbf{R}^n \times \mathbf{R}^\nu$.

Proof. From 4.3.2 we see that $\langle T, h, \cdot\rangle$ is an $\mathscr{L}^n \times \mathscr{L}^\nu$ summable function with values in $\mathbf{F}_{m-n-\nu,K}(U)$, and from 2.6.2 we infer the same for the functions M_σ, $N_{\rho,\sigma}$ defined whenever $0<\rho<\infty$, $0<\sigma<\infty$ by the formulae

$$M_\sigma(y,z)=\int_{\mathbf{B}(0,\sigma)}\langle T,h,(y,z+v)\rangle\, d\mathscr{L}^\nu v/[\alpha(\nu)\,\sigma^\nu],$$

$$N_{\rho,\sigma}(y,z)=\int_{\mathbf{B}(0,\rho)} M_\sigma(y+u,z)\, d\mathscr{L}^n u/[\alpha(n)\,\rho^n]$$

$$=\int_{\mathbf{B}(y,\rho)\times\mathbf{B}(z,\sigma)}\langle T,h,\cdot\rangle\, d(\mathscr{L}^n\times\mathscr{L}^\nu)/[\alpha(n)\,\rho^n\,\alpha(\nu)\,\sigma^\nu]$$

for $(y,z)\in\mathbf{R}^n\times\mathbf{R}^\nu$. Moreover 2.9.9 implies that

$$\mathbf{F}_K[M_\sigma(y,z)-\langle T,h,(y,z)\rangle]\to 0 \quad\text{as}\quad \sigma\to 0+$$

for $\mathscr{L}^n\times\mathscr{L}^\nu$ almost all (y,z), and also that, for each σ,

$$\mathbf{F}_K[N_{\rho,\sigma}(y,z)-M_\sigma(y,z)]\to 0 \quad\text{as}\quad \rho\to 0+$$

for $\mathscr{L}^n\times\mathscr{L}^\nu$ almost all (y,z).

Defining $p(y,z)=y$ and $q(y,z)=z$ for $(y,z)\in\mathbf{R}^n\times\mathbf{R}^\nu$, we observe next that

$$T\llcorner h^\#(p^\#\eta\wedge q^\#\zeta)=(T\llcorner f^\#\eta)\llcorner g^\#\zeta$$

for all bounded Baire functions

$$\eta\colon \mathbf{R}^n\to\textstyle\bigwedge^n\mathbf{R}^n \quad\text{and}\quad \zeta\colon\mathbf{R}^\nu\to\textstyle\bigwedge^\nu\mathbf{R}^\nu;$$

in fact this equation holds in case f, g, η, ζ are of class ∞, because $p\circ h=f$ and $q\circ h=g$, and is readily extended to the general case by approximation. Letting $Y_1,\ldots,Y_n$ and $Z_1,\ldots,Z_\nu$ be the standard coordinate functions on $\mathbf{R}^n$ and $\mathbf{R}^\nu$, we apply the above equation to the differential forms

$$\eta_{y,\rho}=\mathbf{B}(y,\rho)\wedge DY_1\wedge\cdots\wedge DY_n/[\alpha(n)\,\rho^n],$$

$$\zeta_{z,\sigma}=\mathbf{B}(z,\sigma)\wedge DZ_1\wedge\cdots\wedge DZ_\nu/[\alpha(\nu)\,\sigma^\nu]$$

and use 4.3.2 (1) with f replaced by h, to infer that

$$N_{\rho,\sigma}(y,z)=(T\llcorner f^\#\eta_{y,\rho})\llcorner g^\#\zeta_{z,\sigma}$$

whenever $y\in\mathbf{R}^n$, $z\in\mathbf{R}^\nu$, $0<\rho<\infty$, $0<\sigma<\infty$.

We know from 4.3.2 (5) that

$$\mathbf{F}_K[T\llcorner f^\#\eta_{y,\rho}-\langle T,f,y\rangle]\to 0 \quad\text{as}\quad \rho\to 0+$$

for $\mathscr{L}^n$ almost all y, and for each such y we see from 4.3.1 that

$$\mathbf{F}_K[N_{\rho,\sigma}(y,z)-\langle T,f,y\rangle\llcorner g^\#\zeta_{z,\sigma}]\to 0 \quad\text{as}\quad \rho\to 0+$$

whenever $z\in\mathbf{R}^\nu$ and $0<\sigma<\infty$. We infer that $\mathscr{L}^n\times\mathscr{L}^\nu$ almost all (y,z) satisfy the equation

$$M_\sigma(y,z)=\langle T,f,y\rangle\llcorner g^\#\zeta_{z,\sigma}$$

for every positive rational number σ, hence by continuity whenever $0<\sigma<\infty$. Consequently

$$\mathbf{F}_K[\langle T, f, y\rangle \llcorner g^* \zeta_{z,\sigma} - \langle T, h, (y,z)\rangle] \to 0 \quad \text{as} \quad \sigma \to 0+$$

for $\mathscr{L}^n \times \mathscr{L}^\nu$ almost all (y, z), and for each such (y, z) we conclude $\langle\langle T, f, y\rangle, g, z\rangle = \langle T, h, (y, z)\rangle$.

4.3.6. From 4.3.4, 4.2.16, 4.3.5 and 4.3.2 (2) it follows by induction with respect to n that the following three propositions hold whenever $f: U \to \mathbf{R}^n$ is locally Lipschitzian and $m \geq n$:

If $T \in \mathbf{I}_m(U)$, *then* $\langle T, f, y\rangle \in \mathbf{I}_{m-n}(U)$ *for* $\mathscr{L}^n$ *almost all* y.
If $T \in \mathscr{R}_m(U)$, *then* $\langle T, f, y\rangle \in \mathscr{R}_{m-n}(U)$ *for* $\mathscr{L}^n$ *almost all* y.
If $T \in \mathscr{F}_m(U)$, *then* $\langle T, f, y\rangle \in \mathscr{F}_{m-n}(U)$ *for* $\mathscr{L}^n$ *almost all* y.

However, more precise information about the slices of rectifiable currents will be obtained in 4.3.8 through use of the coarea theorem 3.2.22.

We also observe that *if* K *is a compact Lipschitz neighborhood retract in* U *and* $T \in \mathscr{F}_{m,K}(U)$, *then*

$$\int^* \mathscr{F}_K \langle T, f, y\rangle \, d\mathscr{L}^n y \leq \operatorname{Lip}(f|K)^n \mathscr{F}_K(T),$$

by virtue of 4.3.2 (2); however the flat chains

$$T \llcorner f^* [B(y,\rho) \wedge \Omega] / [\alpha(n) \rho^n]$$

are usually not integral flat chains, hence $\mathbf{F}_K$ cannot be replaced by $\mathscr{F}_K$ in 4.3.2. (5).

4.3.7. With every locally Lipschitzian map $F: U \to \mathbf{R}^n$ we associate the commutative diagram

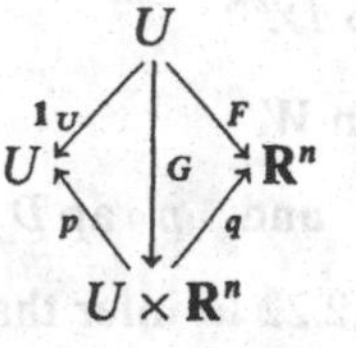

where $G(x) = (x, F(x))$, $p(x, y) = x$, $q(x, y) = y$ for $x \in U$, $y \in \mathbf{R}^n$. Assuming $T \in \mathbf{F}_m(U)$ with $m \geq n$ we infer from 4.3.2 (7) that

$$G_\#(T \llcorner F^* \phi) = (G_\# T) \llcorner q^* \phi, \qquad T \llcorner F^* \phi = p_\# [(G_\# T) \llcorner q^* \phi]$$

for every bounded Baire form ϕ of degree n on $\mathbf{R}^n$, and that

$$G_\# \langle T, F, y\rangle = \langle G_\# T, q, y\rangle, \qquad \langle T, F, y\rangle = p_\# \langle G_\# T, q, y\rangle$$

for $\mathscr{L}^n$ almost all y. Thus slicing T by F corresponds to slicing $G_\# T$ by the map q of class ∞.

4.3.8. Theorem. *If U is an open subset of R^ν, K is a compact subset of U, ξ is an $\mathscr{H}^m$ summable m-vectorfield,*

$$W=\{x\colon \xi(x)\neq 0\} \text{ is an } (\mathscr{H}^m, m) \text{ rectifiable subset of } K,$$

$$\xi(x) \text{ is simple and } \operatorname{Tan}^m(\mathscr{H}^m \llcorner W, x) \text{ is the vectorsubspace}$$

$$\text{of } \mathbf{R}^\nu \text{ associated with } \xi(x) \text{ for } \mathscr{H}^m \text{ almost all } x \text{ in } W,$$

$F\colon U\to \mathbf{R}^n$ is a locally Lipschitzian map, $f=F|W$, $m\geq n$ and ζ is the $m-n$ vectorfield such that (see 3.2.16, 22)

$$\zeta(x)=\xi(x)\llcorner\langle Y_1\wedge\cdots\wedge Y_n, \textstyle\bigwedge^n \operatorname{ap} Df(x)\rangle/\operatorname{ap} J_n f(x)$$

whenever $\operatorname{Tan}^m(\mathscr{H}^m\llcorner W,x)$ *is an m dimensional vectorspace associated with $\xi(x)$ and* $\operatorname{ap} J_n f(x)>0$ [here we identify $\xi(x)$ and $\zeta(x)$ with the corresponding m and $m-n$ vectors of $\operatorname{Tan}^m(\mathscr{H}^m\llcorner W,x)$], $\zeta(x)=0$ *for all other x in U, then:*

(1) $(\mathscr{H}^m\wedge\xi)\llcorner F^{\#}\phi=\mathscr{H}^m\wedge(\xi\llcorner\langle\phi\circ f, \bigwedge^n \operatorname{ap} Df\rangle)$ *for every bounded Baire form ϕ of degree n on* $\mathbf{R}^n$.

(2) $\langle\mathscr{H}^m\wedge\xi, F, y\rangle=(\mathscr{H}^{m-n}\llcorner f^{-1}\{y\})\wedge\zeta$ *for $\mathscr{L}^n$ almost all y.*

(3) $|\zeta(x)|=|\xi(x)|$, $\zeta(x)$ *is simple and* $\ker \operatorname{ap} Df(x)$ *is the vectorsubspace of* $\mathbf{R}^\nu$ *associated with $\zeta(x)$ whenever $\zeta(x)\neq 0$.*

Proof. First we deduce (1) from 4.3.7 and 4.1.30; assuming $\psi\in\mathscr{D}^{m-n}(U)$ and letting $g=G|W$ we compute

$$(\mathscr{H}^m\wedge\xi)\llcorner F^{\#}\phi](\psi)=(p_{\#}[G_{\#}(\mathscr{H}^m\wedge\xi)\llcorner q^{\#}\phi])(\psi)$$
$$=G_{\#}(\mathscr{H}^m\wedge\xi)[q^{\#}\phi\wedge p^{\#}\psi]=\textstyle\int_W\langle[\bigwedge_m \operatorname{ap} Dg]\,\xi, [q^{\#}\phi\wedge p^{\#}\psi]\circ g\rangle\, d\mathscr{H}^m$$
$$=\textstyle\int_W\langle\xi, [\bigwedge^n \operatorname{ap} Df](\phi\circ f)\wedge\psi\rangle\, d\mathscr{H}^m$$

because, for $\mathscr{H}^m$ almost all x in W,

$$q\circ\operatorname{ap} Dg(x)=\operatorname{ap} Df(x) \quad\text{and}\quad p\circ\operatorname{ap} Dg(x)=\mathbf{1}_{\operatorname{Tan}^m(\mathscr{H}^m\llcorner W,x)}.$$

Next we use 4.3.2(1) and 3.2.22 to infer that, for every Borel subset E of $\mathbf{R}^n$,

$$\textstyle\int_E\langle\mathscr{H}^m\wedge\xi, F, y\rangle(\psi)\, d\mathscr{L}^n y=[(\mathscr{H}^m\wedge\xi)\llcorner F^{\#}(E\wedge\Omega)](\psi)$$
$$=\textstyle\int_{f^{-1}(E)}\langle\xi(x)\llcorner\langle Y_1\wedge\cdots\wedge Y_n, \bigwedge^n \operatorname{ap} Df(x)\rangle, \psi(x)\rangle\, d\mathscr{H}^m x$$
$$=\textstyle\int_{f^{-1}(E)}\langle\zeta(x),\psi(x)\rangle \operatorname{ap} J_n f(x)\, d\mathscr{H}^m x$$
$$=\textstyle\int_E\int_{f^{-1}\{y\}}\langle\zeta(x),\psi(x)\rangle\, d\mathscr{H}^{m-n} x\, d\mathscr{L}^n y$$
$$=\textstyle\int_E[(\mathscr{H}^{m-n}\llcorner f^{-1}\{y\})\wedge\zeta](\psi)\, d\mathscr{L}^n y,$$

whence we conclude (2).

Whenever $\zeta(x)\neq 0$ we choose $c\in\mathbf{R}$ and orthonormal base vectors $v_1,\ldots,v_m$ of $\operatorname{Tan}^m(\mathscr{H}^m \llcorner W, x)$ so that $v_1,\ldots,v_{m-n}$ span $\ker \operatorname{ap} Df(x)$ and $\xi(x)=c\, v_1\wedge\cdots\wedge v_m$, then compute

$$|\langle v_{m-n+1}\wedge\cdots\wedge v_n, \langle Y_1\wedge\cdots\wedge Y_n, \textstyle\bigwedge^n \operatorname{ap} Df(x)\rangle\rangle| = \operatorname{ap} J_n f(x),$$

hence $\zeta(x)=\pm c\, v_1\wedge\cdots\wedge v_{m-n}$.

4.3.9. Theorem. *Whenever $T\in\mathbf{F}_{m,K}(U)$ with $m\geq n$,*

$$h\colon \mathbf{R}\times U\to\mathbf{R}^n \text{ is a locally Lipschitzian map},$$

$$h_t(x)=h(t,x), \quad \dot h_t(x)=\langle(1,0), Dh(t,x)\rangle, \quad q(t,x)=x$$

for $(t,x)\in\mathbf{R}\times U$, and $f=h_0$, $g=h_1$ the following statements hold:

(1) *For $\mathscr{L}^n$ almost all y in $\mathbf{R}^n$,*

$$\langle T,g,y\rangle-\langle T,f,y\rangle = q_\#[\langle[0,1]\times\partial T,h,y\rangle+(-1)^n\partial\langle[0,1]\times T,h,y\rangle].$$

(2) *If $\mathbf{M}(T)<\infty$, h is of class ∞, ϕ is a bounded Baire form of degree n on $\mathbf{R}^n$ and $\gamma\in\mathscr{K}(U)^+$, then*

$$\begin{aligned}&\|q_\#[([0,1]\times T)\llcorner h^\#\phi]\|(\gamma)\\ &\leq \textstyle\int_0^1\int \|\vec T\llcorner\langle \dot h_t \lrcorner(\phi\circ h_t), \bigwedge^{n-1} Dh_t\rangle\|\cdot\gamma\, d\|T\|\, d\mathscr{L}^1 t\\ &\leq \textstyle\int_0^1\int |\dot h_t|\cdot\|\phi\circ h_t\|\cdot\|Dh_t\|^{n-1}\cdot\gamma\, d\|T\|\, d\mathscr{L}^1 t.\end{aligned}$$

(3) *If $T\in\mathbf{N}_{m,K}(U)$, h is the affine homotopy from f to g, E is a Borel subset of $\mathbf{R}^n$ and $\lambda=\sup\{\operatorname{Lip}(f|K),\operatorname{Lip}(g|K)\}$, then*

$$\begin{aligned}&\textstyle\int_E \mathbf{F}_K[\langle T,g,y\rangle-\langle T,f,y\rangle]\, d\mathscr{L}^n y\\ &\leq \lambda^{n-1}\textstyle\int_0^1\int_{h_t^{-1}(E)}|g-f|\, d(\|\partial T\|+\|T\|)\, d\mathscr{L}^1 t.\end{aligned}$$

Proof. Defining $\varepsilon_t(x)=(t,x)$ whenever $(t,x)\in\mathbf{R}\times U$ we see from 4.3.2(7) that, for each $t\in\mathbf{R}$,

$$\langle T,h_t,y\rangle = q_\#\,\varepsilon_{t\#}\langle T,h\circ\varepsilon_t,y\rangle = q_\#\langle\varepsilon_{t\#}T,h,y\rangle$$

for $\mathscr{L}^n$ almost all y, then infer from 4.1.8 and 4.3.1 that

$$\begin{aligned}\langle T,g,y\rangle-\langle T,f,y\rangle &= q_\#\langle\varepsilon_{1\#}T-\varepsilon_{0\#}T,h,y\rangle\\ &= q_\#\langle[0,1]\times(\partial T)+\partial([0,1]\times T),h,y\rangle\\ &= q_\#[\langle[0,1]\times\partial T,h,y\rangle+(-1)^n\partial\langle[0,1]\times T,h,y\rangle]\end{aligned}$$

for $\mathscr{L}^n$ almost all y, as asserted in (1).

To verify (2) we suppose $\psi \in \mathscr{D}^{1+m-n}(U)$ and estimate

$$[([0,1] \times T) \llcorner h^\# \phi]\, q^\# \psi = \int_0^1 \int \langle \overrightarrow{[0,1] \times T}, h^\# \phi \wedge q^\# \psi \rangle \, d\|T\| \, d\mathscr{L}^1 t$$

by observing that for $\mathscr{L}^1 \times \|T\|$ almost all (t, x) the integrand equals

$$\begin{aligned} &\langle (1,0) \wedge [\textstyle\bigwedge_m \varepsilon_0]\, \vec{T}(x), [h^\# \phi \wedge q^\# \psi](t,x) \rangle \\ &= (-1)^{n-1} \langle [\textstyle\bigwedge_m \varepsilon_0]\, \vec{T}(x), [(1,0) \lrcorner (h^\# \phi)(t,x)] \wedge (q^\# \psi)(t,x) \rangle \\ &= (-1)^{n-1} \langle \vec{T}(x), \langle \dot{h}_t(x) \lrcorner \phi[h_t(x)], \textstyle\bigwedge^{n-1} D h_t(x) \rangle \wedge \psi(x) \rangle, \end{aligned}$$

because $(1,0) \in \ker q$, $(1,0) \lrcorner (q^\# \psi)(t,x) = 0$,

$$(1,0) \lrcorner \langle \phi[h(t,x)], \textstyle\bigwedge^n D h(t,x) \rangle = \langle \dot{h}_t(x) \lrcorner \phi[h(t,x)], \textstyle\bigwedge^{n-1} D h(t,x) \rangle$$

and $D h(t,x) \circ \varepsilon_0 = D h_t(x)$; moreover the absolute value of integrand does not exceed

$$\begin{aligned} &\|\vec{T}(x) \llcorner \langle [\dot{h}_t(x) \lrcorner \phi[h_t(x)], \textstyle\bigwedge^{n-1} D h_t(x) \rangle\| \cdot \|\psi(x)\| \\ &\leq |\dot{h}_t(x)| \cdot \|\phi[h_t(x)]\| \cdot \|D h_t(x)\|^{n-1} \cdot \|\psi(x)\| \end{aligned}$$

because $\dot{h}_t(x) \lrcorner \phi[h_t(x)]$ is a simple $n-1$ covector with the comass $|\dot{h}_t(x)| \cdot \|\phi[h_t(x)]\|$.

Under the hypothesis of (3) we apply (2) to obtain (first in case f, g have class ∞ and then in general by approximation) the inequality

$$\mathbf{M}(q_\# [([0,1] \times T) \llcorner h^\# \phi]) \leq \mu(\|\phi\|),$$

where μ is the monotone Daniell integral on $\mathscr{K}(\mathbf{R}^n)$ such that

$$\mu(\alpha) = \int_0^1 \int |g - f| (\alpha \circ h_t)\, \lambda^{n-1} \, d\|T\| \, d\mathscr{L}^1 t$$

whenever $\alpha \in \mathscr{K}(\mathbf{R}^n)$. Taking $\phi = \mathbf{B}(y, \rho) \wedge \Omega / [\boldsymbol{\alpha}(n) \rho^n]$ we infer

$$\mathbf{M}[q_\# \langle [0,1] \times T, h, y \rangle] \leq \Theta^n(\mu, y)$$

for $\mathscr{L}^n$ almost all y, hence use 2.9.7 to conclude

$$\int_E \mathbf{M}[q_\# \langle [0,1] \times T, h, y \rangle] \, d\mathscr{L}^n y \leq \mu(E) = \lambda^{n-1} \int_0^1 \int_{h_t^{-1}(E)} |g - f| \, d\|T\| \, d\mathscr{L}^1 t.$$

Similarly we estimate

$$\int_E \mathbf{M}[q_\# \langle [0,1] \times \partial T, h, y \rangle] \, d\mathscr{L}^n y \leq \lambda^{n-1} \int_0^1 \int_{h_t^{-1}(E)} |g - f| \, d\|\partial T\| \, d\mathscr{L}^1 t.$$

Adding these estimates, we deduce (3) from (1).

4.3.10. Theorem. *If* $T \in \mathbf{N}_{m,K}(U)$, $m \geq n$, $f: U \to \mathbf{R}^n$ *is locally Lipschitzian,* $\lambda = \operatorname{Lip}(f|K)$, $\mu = f_{\#}(\|T\| + \|\partial T\|)$, V *is an open subset of* $\mathbf{R}^n$, $\gamma < \infty$, $0 < \delta \leq 1$,

$$\mu\, \mathbf{B}(y, \rho) \leq \boldsymbol{\alpha}(n)\, \gamma\, \rho^{n-1+\delta} \quad \textit{whenever} \quad \mathbf{B}(y, \rho) \subset V,$$

$$\mu \llcorner V \textit{ is absolutely continuous with respect to } \mathscr{L}^n,$$

then $\langle T, f, y\rangle \in \mathbf{F}_{m-n,K}(U)$ *for all y in V, and*

$$\mathbf{F}_K[\langle T, f, a\rangle - \langle T, f, b\rangle] \leq 2^{2-\delta}\, \lambda^{n-1}\, \gamma\, \delta^{-1}\, |a-b|^\delta$$

whenever $a \in V$, $b \in V$ *with* $\mathbf{B}[(a+b)/2, |a-b|/2] \subset V$.

Proof. We abbreviate $c = (a+b)/2$, $r = |a-b|/2$. Whenever

$$0 < \rho < \infty, \quad \mathbf{B}(c, r+\rho) \subset V, \quad z \in \mathbf{B}(c, r)$$

we apply 4.3.9 (3) with $h(t, x) = f(x) + t(a-z)$ for $(t, x) \in \mathbf{R} \times U$, so that $\langle T, g, y\rangle = \langle T, f, y+z-a\rangle$ for $y \in \mathbf{R}^n$, and infer

$$\begin{aligned}
&\mathbf{F}_K(T \llcorner f^{\#}[\mathbf{B}(z,\rho) \wedge \Omega] - T \llcorner f^{\#}[\mathbf{B}(a,\rho) \wedge \Omega]) \\
&\leq \textstyle\int_{\mathbf{B}(a,\rho)} \mathbf{F}_K[\langle T, f, y+z-a\rangle - \langle T, f, y\rangle]\, d\mathscr{L}^n y \\
&\leq \lambda^{n-1} \textstyle\int_0^1 \int_{f^{-1}\mathbf{B}[a+t(z-a),\rho]} |a-z|\, d(\|T\| + \|\partial T\|)\, d\mathscr{L}^1 t \\
&\leq \lambda^{n-1}\, 2r \textstyle\int_0^1 \mu\, \mathbf{B}[a+t(z-a), \rho]\, d\mathscr{L}^1 t \\
&= \lambda^{n-1}\, 2r \textstyle\int_0^1 \int_{\mathbf{B}[a+t(z-a),\rho]} \Theta^n(\mu, w)\, d\mathscr{L}^n w\, d\mathscr{L}^1 t \\
&= \lambda^{n-1}\, 2r \textstyle\int_0^1 \int_{\mathbf{B}(a,\rho)} \Theta^n[\mu, y+t(z-a)]\, d\mathscr{L}^n y\, d\mathscr{L}^1 t
\end{aligned}$$

by virtue of 4.3.2 (1) and 2.9.7, because $\mathbf{B}(c, r)$ contains the line segment from a to z, hence $\mathbf{B}[a+t(z-a), \rho] \subset \mathbf{B}(c, r+\rho) \subset V$ for $0 \leq t \leq 1$. Consequently

$$\begin{aligned}
&\textstyle\int_{\mathbf{B}(c,r)} \mathbf{F}_K(T \llcorner f^{\#}[B(z,\rho) \wedge \Omega] - T \llcorner f^{\#}[B(a,\rho) \wedge \Omega])\, d\mathscr{L}^n z \\
&\leq \lambda^{n-1}\, 2r \textstyle\int_0^1 \int_{\mathbf{B}(a,\rho)} \int_{\mathbf{B}(c,r)} \Theta^n[\mu, y+t(z-a)]\, d\mathscr{L}^n z\, d\mathscr{L}^n y\, d\mathscr{L}^1 t \\
&= \lambda^{n-1}\, 2r \textstyle\int_0^1 \int_{\mathbf{B}(a,\rho)} \int_{\mathbf{B}[y+t(c-a),tr]} \Theta^n(\mu, u)\, t^{-n}\, d\mathscr{L}^n u\, d\mathscr{L}^n y\, d\mathscr{L}^1 t \\
&= \lambda^{n-1}\, 2r \textstyle\int_0^1 \int_{\mathbf{B}(a,\rho)} t^{-n}\, \mu\, \mathbf{B}[y+t(c-a), tr]\, d\mathscr{L}^n y\, d\mathscr{L}^1 t \\
&\leq \lambda^{n-1}\, 2r \textstyle\int_0^1 \int_{\mathbf{B}(a,\rho)} t^{-n}\, \boldsymbol{\alpha}(n)\, \gamma (t\, r)^{n-1+\delta}\, d\mathscr{L}^n y\, d\mathscr{L}^1 t \\
&= \lambda^{n-1}\, 2 r^{n+\delta}\, \boldsymbol{\alpha}(n)^2\, \rho^n\, \gamma \textstyle\int_0^1 t^{\delta-1}\, d\mathscr{L}^1 t.
\end{aligned}$$

The same holds with a replaced by b. Adding the two estimates and dividing by $\alpha(n)^2 \rho^n r^n$ we obtain

$$\mathbf{F}_K(T \llcorner f^{\#}[\mathbf{B}(a,\rho) \wedge \Omega] - T \llcorner f^{\#}[\mathbf{B}(b,\rho) \wedge \Omega])/[\alpha(n)\rho^n]$$
$$\leq \lambda^{n-1} 4 r^\delta \gamma \delta^{-1} = \lambda^{n-1} 2^{2-\delta} |a-b|^\delta \gamma \delta^{-1},$$

and conclude that, if $\langle T, f, a\rangle$ and $\langle T, f, b\rangle$ exist, then

$$\mathbf{F}_K[\langle T, f, a\rangle - \langle T, f, b\rangle] \leq \lambda^{n-1} 2^{2-\delta} \gamma \delta^{-1} |a-b|^\delta.$$

It follows that the function $\langle T, f, \cdot\rangle$ is uniformly continuous on $C \cap \operatorname{dmn}\langle T, f, \cdot\rangle$ for every compact subset C of V, hence from 4.3.2(1) that $V \subset \operatorname{dmn}\langle T, f, \cdot\rangle$.

4.3.11. The importance of $\|\partial T\|$ in 4.3.9(3) and 4.3.10 is illustrated by the simple example where $U = \mathbf{R}$, $K = \{x: 0 \leq x \leq 2\}$, $m = n = 1$, $f = 1_{\mathbf{R}}$, k is a positive integer,

$$T = \sum_{i=1}^{k} [(2i-1)/k, 2i/k], \quad \int_0^2 \mathbf{F}_K[\langle T, f, y+1/k\rangle - \langle T, f, y\rangle]\, d\mathscr{L}^1 y = 2,$$

$\|T\| \leq \mathscr{L}^1$, $\mathbf{M}(\partial T) = 2k$ and $\langle T, f, \cdot\rangle$ is discontinuous at j/k for $j = 1, 2, \ldots, 2k$.

It is also instructive to consider the example where $U = \mathbf{R}^2$, $f \in \mathbf{O}^*(2,1)$ and T is an oriented square; the hypothesis of 4.3.10 holds with $V = \mathbf{R}$ and $\delta = 1$ in case no edge of the square is parallel to $\ker f$; however in the alternate case the function $\langle T, f, \cdot\rangle$ is discontinuous at the image points of the edges parallel to $\ker f$.

In 4.5.15 the reader will find an important application of 4.3.10 and an example showing the need for the hypothesis $\delta > 0$.

If $T \in \mathbf{I}_{m,K}(U)$, then $\mathbf{F}_K$ may be replaced by $\mathscr{F}_K$ in 4.3.9(3) and 4.3.10.

The following situation illustrates both 4.3.8 and 4.3.10 (with $\delta = 1$): *Suppose* $R \in \mathbf{I}_m(U)$, $m \geq n$,

$$\alpha: U \to \mathbf{R} \textit{ and } f: U \to \mathbf{R}^n \textit{ are maps of class } 1,$$

$$\sup\{\Theta^{*m}(\|R\|, x) + \Theta^{*m-1}(\|\partial R\|, x): x \in U\} < \infty,$$

spt R *is contained in an* m *dimensional submanifold* A *of class* 1 *of* U *with*

$$\operatorname{im}[Df(x)|\operatorname{Tan}(A,x)] = \mathbf{R}^n \textit{ for } x \in \operatorname{spt}\alpha \cap \operatorname{spt} R,$$

and $\operatorname{spt}\alpha \cap \operatorname{spt}\partial R$ *has a neighborhood* H *in* U *such that* $H \cap \operatorname{spt}\partial R$ *is covered by a finite family* Λ *of* $m-1$ *dimensional submanifolds* L *of class* 1 *of* A *with* $H \cap \operatorname{Clos} L \subset L$,

$$\operatorname{im}[Df(x)|\operatorname{Tan}(L,x)] = \mathbf{R}^n \textit{ for } x \in L \cap \operatorname{spt}\alpha \cap \operatorname{spt}\partial R.$$

If $z \in \mathbf{R}^n$, *then* $\mathscr{H}^{m-n}(f^{-1}\{z\} \cap \operatorname{spt} \alpha \cap \operatorname{spt} \partial R) = 0$ *and*

$$\langle R \llcorner \alpha, f, z \rangle = (\mathscr{H}^{m-n} \llcorner f^{-1}\{z\} \cap \operatorname{spt} \alpha \cap \operatorname{spt} R) \wedge \gamma \in \mathbf{I}_{m-n}(U)$$

with

$$\gamma = \alpha\, \Theta^m(\|R\|, \cdot)\, |\vec{R} \llcorner f^{\#} \Omega|^{-1}\, \vec{R} \llcorner f^{\#} \Omega .$$

Using suitable coordinate systems and a partition of unity one can derive this proposition from the simple special case where

$$A = U = I^m \subset \mathbf{R}^m \text{ for some open interval } I \subset \mathbf{R},$$

$$f(x) = (x_1, \ldots, x_n) \text{ whenever } x = (x_1, \ldots, x_m) \in I^m,$$

and there exists a finite set Φ of maps $\phi\colon I^{m-1} \to I$ of class 1 such that

$$\operatorname{spt} \partial R \subset \bigcup_{\phi \in \Phi} I^m \cap \{x\colon \phi(x_1, \ldots, x_{m-1}) = x_m\}.$$

4.3.12. Here we show how slicing may be used to formulate the idea that *the roots of a polynomial depend continuously on its coefficients.* With each $a \in \mathbf{R}^n$ we associate the polynomial function P_a such that

$$P_a(s) = s^n + \sum_{i=1}^{n} a_i\, s^{n-i} \text{ for } s \in \mathbf{R},$$

and we observe that the set

$$W = (\mathbf{R}^n \times \mathbf{R}) \cap \{(a, s)\colon P_a(s) = 0\}$$

is an n dimensional orientable analytic submanifold of $\mathbf{R}^n \times \mathbf{R}$; in fact the functions

$$G\colon \mathbf{R}^{n-1} \times \mathbf{R} \to \mathbf{R}^n \times \mathbf{R}, \quad p\colon \mathbf{R}^n \times \mathbf{R} \to \mathbf{R}^{n-1} \times \mathbf{R}$$

$$G(\alpha, s) = \left(\left(\alpha_1, \ldots, \alpha_{n-1}, -s^n - \sum_{i=1}^{n-1} \alpha_i\, s^{n-i}\right), s\right) \text{ for } (\alpha, s) \in \mathbf{R}^{n-1} \times \mathbf{R},$$

$$p(a, s) = ((a_1, \ldots, a_{n-1}), s) \text{ for } (a, s) \in \mathbf{R}^n \times \mathbf{R},$$

satisfy the conditions $\operatorname{im} G = W$ and $p \circ G = \mathbf{1}_{\mathbf{R}^{n-1} \times \mathbf{R}}$; letting ξ be the standard n vectorfield orienting $\mathbf{R}^{n-1} \times \mathbf{R}$, we orient W by the n vectorfield η such that

$$G_{\#}(\mathscr{H}^n \wedge \xi) = (\mathscr{H}^n \llcorner W) \wedge \eta .$$

We will slice the current $(\mathscr{H}^n \llcorner W) \wedge \eta$ by the function

$$q\colon \mathbf{R}^n \times \mathbf{R} \to \mathbf{R}^n, \quad q(a, s) = a \text{ for } (a, s) \in \mathbf{R}^n \times \mathbf{R},$$

noting that $W \cap q^{-1}\{a\} = \{a\} \times \{s\colon P_a(s) = 0\}$ for $a \in \mathbf{R}^n$.

Assuming $1 \le r < \infty$ we see that

$$W \cap q^{-1}\mathbf{U}(0,r) \subset \mathbf{U}(0,r) \times \{s\colon |s| < n\,r\},$$

because $P_a(s)=0$ implies

$$\text{either } |s|<1 \text{ or } |s|^n \le \sum_{i=1}^{n} |a_i\, s^{n-i}| \le n\,|a| \cdot |s|^{n-1},$$

hence $|s| \le \sup\{1, n\,|a|\}$. Next we will prove that whenever $b \in \mathbf{R}^n$, $\rho > 0$ and $|b|+\rho<r$ there exist linear intervals $S_1, \ldots, S_n$ with diameters at most $2n\,r\,\rho^{1/n}$ such that

$$W \cap q^{-1}\mathbf{B}(b,\rho) \subset \bigcup \{\mathbf{B}(b,\rho) \times S_i\colon i=1,\ldots,n\};$$

for this purpose we consider the complex roots $u_1, \ldots, u_n$ of P_b, so that

$$P_b(s) = \prod_{i=1}^{n} (s-u_i) \text{ for all } s \in \mathbf{R},$$

and infer for any $(a,s) \in W$ with $|a-b| \le \rho$ that $a \in \mathbf{U}(0,r)$,

$$\prod_{i=1}^{n} |s-u_i| = |P_b(s)| = |P_b(s) - P_a(s)|$$

$$\le \sum_{i=1}^{n} |a_i - b_i| \cdot |s|^{n-i} \le n\,|a-b|\,(n\,r)^{n-1} \le (n\,r)^n\,\rho,$$

hence $|s-u_i| \le n\,r\,\rho^{1/n}$ for some i; it follows that

$$(q \circ G)^{-1}\mathbf{B}(b,\rho) = p[W \cap q^{-1}\mathbf{B}(b,\rho)]$$

$$\subset \bigcup \{\mathbf{B}[(b_1,\ldots,b_{n-1}),\rho] \times S_i\colon i=1,\ldots,n\},$$

$$\mathscr{H}^n[(q \circ G)^{-1}\mathbf{B}(b,\rho)] \le n\,\boldsymbol{\alpha}(n-1)\,\rho^{n-1}\,2n\,r\,\rho^{1/n} = \boldsymbol{\alpha}(n)\,\gamma\,\rho^{n-1+1/n}$$

where $\gamma = 2n^2\,r\,\boldsymbol{\alpha}(n-1)/\boldsymbol{\alpha}(n)$. Taking

$$K = (\mathbf{R}^{n-1} \times \mathbf{R}) \cap \{(\alpha,s)\colon |\alpha| \le r,\ |s| \le n\,r\}$$

we estimate $\operatorname{Lip}(G|K) \le 3\,n^{n+1}\,r^n$, and observe that

$$\det D(q \circ G)(\alpha,s) = -P'_{(q \circ G)(\alpha,s)}(s) \text{ for } (\alpha,s) \in \mathbf{R}^{n-1} \times \mathbf{R},$$

$$\det D(q \circ G)(\alpha,s) \ne 0 \text{ for } \mathscr{H}^n \text{ almost all } (\alpha,s),$$

hence $(q \circ G)_{\#}\,\mathscr{H}^n$ is absolutely continuous with respect to $\mathscr{L}^n$, by 3.2.3. Applying 4.3.10 with $T = (\mathscr{H}^n \llcorner K) \wedge \xi$ we obtain the Hölder condition

$$\mathbf{F}_K[\langle \mathscr{H}^n \wedge \xi, q \circ G, a\rangle - \langle \mathscr{H}^n \wedge \xi, q \circ G, b\rangle] \le \Gamma\,|a-b|^{1/n}$$

with $\Gamma = 2^{2-1/n}(3\,n^{n+1}\,r^n)^{n-1}\,\gamma\,n$, whenever $a \in \mathbf{R}^n$, $b \in \mathbf{R}^n$ and

$$\mathbf{B}[(a+b)/2, |a-b|/2] \subset \mathbf{U}(0,r);$$

from 4.3.2(7) we see that the slices

$$\langle (\mathscr{H}^n \llcorner W) \wedge \eta, q, a\rangle = G_{\#}\,\langle \mathscr{H}^n \wedge \xi, q \circ G, a\rangle$$

satisfy a similar Hölder condition, with Γ replaced by $3n^{n+1}r^n\Gamma$. Recalling 4.3.3 we also find that

$$\langle(\mathscr{H}^n \llcorner W)\wedge\eta, q, a\rangle = \sum_{s\in P_a^{-1}\{0\}} \operatorname{sign}[-P_a'(s)]\cdot\delta_{(a,s)}$$

for $\mathscr{L}^n$ almost all a. We finally conclude

$$\langle(\mathscr{H}^n \llcorner W)\wedge\eta, q, b\rangle = \sum_{t\in\mathbf{R}} \varepsilon(b,t)\,\delta_{(b,t)}$$

for every $b\in\mathbf{R}^n$, where

$$\varepsilon(b,t)=0 \text{ if } \mathbf{o}(P_b,t)=\inf\{j\colon P_b^{(j)}(t)\neq 0\} \text{ is even},$$

$$\varepsilon(b,t)=\operatorname{sign}[-P_b^{(k)}(t)] \text{ if } k=\mathbf{o}(P_b,t) \text{ is odd},$$

because $\langle(\mathscr{H}^m \llcorner W)\wedge\eta, q,\cdot\rangle$ is continuous at b and each $t\in\mathbf{R}$ belongs to an open interval X such that

$$\begin{aligned}\varepsilon(b,t)=&\operatorname{card}\{s\colon s\in X,\ P_a(s)=0,\ P_a'(s)<0\}\\ &-\operatorname{card}\{s\colon s\in X,\ P_a(s)=0,\ P_a'(s)>0\}\end{aligned}$$

whenever a is close to b in $\mathbf{R}^n$ and all roots of P_a are simple.

A similar treatment may be applied to the *complex roots of polynomials with complex coefficients;* to each $a\in\mathbf{C}^n$ corresponds in this case the slice

$$\sum_{z\in P_a^{-1}\{0\}} \mathbf{o}(P_a,z)\,\delta_{(a,z)}\in\mathscr{D}_0(\mathbf{C}^n\times\mathbf{C}).$$

Here the continuity of the slice function does not follow from 4.3.10, but is a consequence of Rouché's Theorem [RU, p. 218] or of [F 19, Theorem 3.13].

4.3.13. It is easy to generalize the theory of slicing from functions with values in $\mathbf{R}^n$ to functions with values in any *oriented n dimensional submanifold B of class* 1 of any Euclidean space. Assuming that β is an n-vectorfield orienting B one replaces $DY_1\wedge\cdots\wedge DY_n$ by a continuous differential form Ω of degree n, whose domain is some neighborhood of B and which satisfies the condition

$$\langle\beta(y),\Omega(y)\rangle=1 \text{ for } y\in B.$$

Using $\mathscr{H}^n \llcorner B$ instead of $\mathscr{L}^n$ one extends all the basic results of this section; for example the generalization of 4.3.2 (1) is the proposition that

$$\textstyle\int_B \Phi(y)\,\langle T,f,y\rangle(\psi)\,d\mathscr{H}^n y=[T\llcorner f^\#(\Phi\wedge\Omega)](\psi)$$

whenever $T\in\mathbf{F}_m(U)$, $m\geq n$, $f\colon U\to B$ is a locally Lipschitzian map, Φ is a bounded real valued Baire function on B, and $\psi\in\mathscr{D}^{m-n}(U)$.

Alternately the slices of a map into B may be constructed by slicing the compositions of this map with those coordinate systems on B which relate to β a positive multiple of $\mathbf{e}_1 \wedge \cdots \wedge \mathbf{e}_n$; the consistency of this procedure is guaranteed by 4.3.2(6), which shows that if g is a diffeomorphism of class 1, then

$$\langle T, g \circ f, g(y) \rangle = \operatorname{sign}[\det D g(y)] \langle T, f, y \rangle$$

for $\mathscr{L}^n$ almost all y.

An application of slicing to the integralgeometry of homogeneous spaces was made in [BJ 1].

4.3.14. We call C an m dimensional **oriented cone** in $\mathbf{R}^n$ if and only if $C \in \mathscr{D}_m(\mathbf{R}^n)$ and

$$\mu_{r\#} C = C \text{ whenever } 0 < r < \infty.$$

Clearly $\boldsymbol{\delta}_0$ is a 0 dimensional oriented cone.

If $m \geq 1$, $S \in \mathscr{D}_m(\mathbf{R}^n)$ *and* $\operatorname{spt} S$ *is a compact subset of* $\mathbf{R}^n \sim \{0\}$, *then the m dimensional current*

$$C = \int_0^\infty t^{-1} \mu_{t\#} S \, d\mathscr{L}^1 t$$

is an oriented cone because, for $\phi \in \mathscr{D}^m(\mathbf{R}^n)$,

$$\mu_r^\# \phi = r^m \phi \circ \mu_r \text{ whenever } 0 < r < \infty, \quad C(\phi) = \int_0^\infty t^{m-1} S(\phi \circ \mu_t) \, d\mathscr{L}^1 t,$$

$$\alpha = \sup\{|x| : x \in \operatorname{spt} \phi\} < \infty, \quad \beta = \inf\{|x| : x \in \operatorname{spt} S\} > 0,$$

$$\{t : t \geq 0, \operatorname{spt}(\phi \circ \mu_t) \cap \operatorname{spt} S \neq \varnothing\} \subset \{t : 0 \leq t \leq \alpha/\beta\},$$

$$\mu_{r\#} C = \int_0^\infty t^{-1} \mu_{r\#} \mu_{t\#} S \, d\mathscr{L}^1 t = \int_0^\infty (r t)^{-1} \mu_{(rt)\#} S \, r \, d\mathscr{L}^1 t = C.$$

We will neither prove nor use the converse proposition, that all oriented cones are representable in the foregoing manner. However we will derive a stronger result concerning the structure of locally rectifiable oriented cones:

If $m \geq 1$, $C \in \mathscr{R}_m^{\text{loc}}(\mathbf{R}^n)$ *and* C *is an oriented cone, then*

$$\langle C, f, 1 \rangle \in \mathscr{R}_{m-1}(\mathbf{R}^n), \quad \operatorname{spt} \langle C, f, 1 \rangle \subset \mathbf{S}^{n-1},$$

$$C = h_\# [(\mathbf{E}^1 \llcorner \{t : t > 0\}) \times \langle C, f, 1 \rangle]$$

$$= \int_0^\infty t^{-1} \mu_{t\#} [(-1)^{m-1} \langle C, f, 1 \rangle \wedge \operatorname{grad} f] \, d\mathscr{L}^1 t$$

where $h: \mathbf{R} \times \mathbf{R}^n \to \mathbf{R}^n$, $h(t, x) = t x$ *for* $(t, x) \in \mathbf{R} \times \mathbf{R}^n$,

$$f: \mathbf{R}^n \to \mathbf{R}, \ f(x) = |x| \text{ for } x \in \mathbf{R}^n.$$

Assuming $0<\gamma<1$ we see from 4.1.28 that $\|C\|$ almost all of $\mathbf{R}^n$ equals the union of the sets

$$B_j=\{x: \ \|C\|\,\mathbf{U}(x,\rho)>\gamma\,\alpha(m)\,\rho^m \text{ for } 0<\rho<|x|/j\}$$

corresponding to all positive integers j, and we note that

$$\|C\|=\|\mu_{r\#}\,C\|=r^m\,\mu_{r\#}\,\|C\|, \qquad \mu_r(B_j)=B_j$$

whenever $0<r<\infty$. If $b\in B_j$, then $b\in\operatorname{Tan}^m(\|C\|,b)$, because (see 3.2.16)

$$\mathbf{U}(b+s\,b,s\,\varepsilon)\subset\mathbf{E}(b,b,\varepsilon)\cap\mathbf{U}(b,s\,|b|+s\,\varepsilon),$$
$$\gamma\,\alpha(m)\,s^m\,\varepsilon^m\leq\|C\|\,[\mathbf{E}(b,b,\varepsilon)\cap\mathbf{U}(b,s\,|b|+s\,\varepsilon)]$$

whenever $0<\varepsilon<\infty$ and $0<s\,\varepsilon<|b+s\,b|/j$, hence

$$\Theta^{*m}[\|C\|\llcorner\mathbf{E}(b,b,\varepsilon),b]\geq\gamma\,\varepsilon^m(|b|+\varepsilon)^{-m};$$

in case $\operatorname{Tan}^m(\|C\|,b)$ is associated with $\vec{C}(b)$ and $b\neq 0$ we infer $b\wedge\vec{C}(b)=0$ and

$$\vec{C}(b)=[\operatorname{grad} f]\,(b)\wedge[\vec{C}(b)\llcorner Df(b)]$$

by 1.5.3 because $[\operatorname{grad} f]\,(b)=b/|b|$. Using 4.3.2(1) we compute

$$C=(-1)^{m-1}\,\|C\|\wedge(\vec{C}\llcorner Df)\wedge\operatorname{grad} f=(-1)^{m-1}(C\llcorner Df)\wedge\operatorname{grad} f$$
$$=\textstyle\int_0^\infty[(-1)^{m-1}\langle C,f,t\rangle\wedge\operatorname{grad} f]\,d\mathscr{L}^1\,t.$$

If $0<t<\infty$, then $f\circ\mu_t=g_t\circ f$ with $g_t(s)=t\,s$ for $s\in\mathbf{R}$, and we infer that

$$C\llcorner f^{\#}\,\Psi=(\mu_{t\#}\,C)\llcorner f^{\#}\,\Psi=\mu_{t\#}\,[C\llcorner\mu_t^{\#}\,(f^{\#}\,\Psi)]=\mu_{t\#}\,[C\llcorner f^{\#}\,(g_t^{\#}\,\Psi)]$$

for every bounded Baire form Ψ of degree 1 on $\mathbf{R}$; taking

$$\Psi=\mathbf{B}(t,t\,\rho)\wedge DY_1, \text{ hence } g_t^{\#}\,\Psi=t\,\mathbf{B}(1,\rho)\wedge DY_1,$$

dividing by $2t\,\rho$ and letting $\rho\to 0+$ we obtain

$$\langle C,f,t\rangle=\mu_{t\#}\,\langle C,f,1\rangle;$$

from this equation it follows that

$$\mu_{t\#}\,(\langle C,f,1\rangle\wedge\operatorname{grad} f)=t\,\langle C,f,t\rangle\wedge\operatorname{grad} f$$

because $(\operatorname{grad} f)\circ\mu_t=\operatorname{grad} f$ and

$$\operatorname{grad} f\,\lrcorner\,\mu_t^{\#}\,\phi=t^m\operatorname{grad} f\,\lrcorner\,(\phi\circ\mu_t)$$
$$=t^m(\operatorname{grad} f\,\lrcorner\,\phi)\circ\mu_t=t\,\mu_t^{\#}\,(\operatorname{grad} f\,\lrcorner\,\phi)$$

for $\phi \in \mathscr{D}^m(\mathbf{R}^n)$. Abbreviating

$$P = \mathbf{E}^1 \llcorner \{t\colon t>0\}, \qquad Q = \langle C, f, 1\rangle$$

and recalling 4.1.9 we also compute

$$\dot{h}_t(x) = x, \ \langle w, D\, h_t(x)\rangle = t\, w \ \text{ for } w \in \mathbf{R}^n,$$

$$[\textstyle\bigwedge_m D\, h(t,x)]\, \overrightarrow{P \times Q}(t,x) = x \wedge t^{m-1}\, \vec{Q}(x) = (-1)^{m-1}\, t^{m-1} [\vec{Q} \wedge \operatorname{grad} f](x)$$

for $t>0$ and $\|Q\|$ almost all x, since $\operatorname{spt} Q \subset \mathbf{S}^{n-1}$; we conclude

$$\begin{aligned}(-1)^{m-1}[h_\#(P \times Q)]\,\phi &= \textstyle\int_0^\infty \int \langle t^{m-1} [\vec{Q} \wedge \operatorname{grad} f](x), \phi(t\,x)\rangle\, d\,\|Q\|\, x\, d\mathscr{L}^1\, t \\ &= \textstyle\int_0^\infty \int \langle \vec{Q}, t^{m-1} \operatorname{grad} f \,\lrcorner\, \phi \circ \mu_t\rangle\, d\,\|Q\|\, d\mathscr{L}^1\, t \\ &= \textstyle\int_0^\infty Q[\mu_t^\# (\operatorname{grad} f \,\lrcorner\, \phi)]\, d\mathscr{L}_1\, t \\ &= \textstyle\int_0^\infty \langle C, f, t\rangle (\operatorname{grad} f \,\lrcorner\, \phi)\, d\mathscr{L}^1\, t = (-1)^{m-1}\, C(\phi).\end{aligned}$$

Next we note that *in case* $C \in \mathbf{I}_m^{\mathrm{loc}}(\mathbf{R}^n)$ *with* $m \geq 1$, C *is an oriented cone if and only if*

$$x \wedge \vec{C}(x) = 0 \ \textit{for}\ \|C\| \ \textit{almost all}\ x$$

$$\textit{and either}\ m=1,\ \operatorname{spt} \partial C \subset \{0\}$$

$$\textit{or}\ m \geq 2,\ x \wedge \overrightarrow{\partial C}(x) = 0\ \textit{for}\ \|\partial C\|\ \textit{almost all}\ x;$$

the sufficiency of this condition may be verified inductively by applying the homotopy formula to the affine homotopy H from μ_1 to μ_r, because

$$\dot{H}_t(x) \wedge [\textstyle\bigwedge_m DH_t(x)]\, \vec{C}(x) = (r-1)\, x \wedge (1-t+t\,r)^m\, \vec{C}(x)$$

for $t \in \mathbf{R}$ and $\|C\|$ almost all x.

Finally we observe that *if* $m \geq 1$ *and*

$$Q \in \mathscr{D}_{m-1}(\mathbf{R}^n), \quad \operatorname{spt} Q \subset \mathbf{S}^{n-1}, \quad 0 \times Q \in \mathscr{R}_m(\mathbf{R}^n),$$

then there exist an oriented cone $C \in \mathscr{R}_m^{\mathrm{loc}}(\mathbf{R}^n)$ *with*

$$C \llcorner \mathbf{U}(0,1) = 0 \times Q;$$

in fact if h, P are as above and $C = h_\#(P \times Q)$, then

$$h_\#([0,r] \times Q) = \mu_{r\#}\, h_\#([0,1] \times Q) = \mu_{r\#}(0 \times Q) \in \mathscr{R}_m(\mathbf{R}^n),$$

$$\operatorname{spt}[C - h_\#([0,r] \times Q)] \subset h(\{t\colon t \geq r\} \times \mathbf{S}^{n-1}) = \mathbf{R}^n \sim \mathbf{U}(0,r)$$

whenever $0 < r < \infty$.

4.3.15. We call C an m dimensional **oriented cylinder with direction** u in $\mathbf{R}^n$ if and only if $C \in \mathscr{D}_m(R^n)$, $u \in \mathbf{S}^{n-1}$ and

$$\tau_{tu\#}\, C = C \quad \text{whenever} \quad t \in \mathbf{R}.$$

Such cylinders may be studied by the method applied to cones in 4.3.14. Fixing $u \in \mathbf{S}^{n-1}$ and using the maps

$$h\colon \mathbf{R} \times \mathbf{R}^n \to \mathbf{R}^n, \quad h(t, x) = t\,u + x \text{ for } (t, x) \in \mathbf{R} \times \mathbf{R}^n,$$

$$f\colon \mathbf{R}^n \to \mathbf{R}, \quad f(x) = u \bullet x \text{ for } x \in \mathbf{R}^n,$$

$$g\colon \mathbf{R}^n \to \mathbf{R}^n, \quad g(x) = x - (u \bullet x)\, x \text{ for } x \in \mathbf{R}^n,$$

one thus readily verifies the following propositions:

If $S \in \mathscr{D}_m(\mathbf{R}^n)$ and $f(\operatorname{spt} S)$ is bounded, then

$$h_\# (\mathscr{L}^1 \times S) = \int \tau_{tu\#}\, S \, d\mathscr{L}^1 t$$

is an m dimensional oriented cylinder with direction u in $\mathbf{R}^n$.

If C is an m dimensional oriented cylinder with direction u in $\mathbf{R}^n$, $\psi \in \mathscr{D}^0(\mathbf{R})$ and $\int \psi \, d\mathscr{L}^1 = 1$, then

$$C = h_\# [\mathscr{L}^1 \times (C \llcorner \psi \circ f)];$$

in case $m = 0$, then also $C = h_\# [\mathscr{L}^1 \times g_\# (C \llcorner \psi \circ f)]$.

If $C \in \mathscr{R}_m^{\mathrm{loc}}(\mathbf{R}^n)$ and C is an oriented cylinder with direction u, then $m \geq 1$ and

$$\langle C, f, 0 \rangle \in \mathscr{R}_{m-1}^{\mathrm{loc}}(\mathbf{R}^n), \qquad \operatorname{spt} \langle C, f, 0 \rangle \subset \ker f,$$

$$C = h_\# [\mathbf{E}^1 \times \langle C, f, 0 \rangle] = (-1)^{m-1}\, h_\# [\mathscr{L}^1 \times (\langle C, f, 0 \rangle \wedge \operatorname{grad} f)].$$

In case $C \in \mathbf{I}_m^{\mathrm{loc}}(\mathbf{R}^n)$ and $m \geq 1$, C is an oriented cylinder with direction u if and only if

$$u \wedge \vec{C}(x) = 0 \text{ for } \|C\| \text{ almost all } x$$

$$\text{and either } m = 1,\ \partial C = 0$$

$$\text{or } m \geq 2,\ u \wedge \overrightarrow{\partial C}(x) = 0 \text{ for } \|\partial C\| \text{ almost all } x.$$

We also observe that a *current C in $\mathbf{R}^n$ is a cylinder with direction e_j*, the j'th standard base vector of $\mathbf{R}^n$, *if and only if $D_j C = 0$.*

4.3.16. We now topologize the abelian group $\mathscr{F}_m^{\mathrm{loc}}(U)$ by constructing *basic neighborhoods of* 0 as follows: With each pair (W, δ) such that

$$W \text{ is open, } \operatorname{Clos} W \text{ is a compact subset of } U,\ \delta > 0$$

we associate *the subset of $\mathscr{F}_m^{\mathrm{loc}}(U)$ consisting of those currents T for which there exist $R \in \mathscr{R}_m(U)$ and $S \in \mathscr{R}_{m+1}(U)$ satisfying the conditions*

$$\operatorname{spt}(T - R - \partial S) \subset U \sim W, \qquad \mathbf{M}(R) + \mathbf{M}(S) < \delta;$$

we may add the requirement that

$$\operatorname{spt} R \cup \operatorname{spt} S \subset \operatorname{Clos} W,$$

because R and S can be replaced by $R \llcorner W$ and $S \llcorner W$.

If f is a locally Lipschitzian proper map of U into V, then the induced homomorphism $f_\#$ maps $\mathscr{F}_m^{\text{loc}}(U)$ continuously into $\mathscr{F}_m^{\text{loc}}(V)$.

In case $m \geq 1$ the boundary operator ∂ maps $\mathscr{F}_m^{\text{loc}}(U)$ continuously into $\mathscr{F}_{m-1}^{\text{loc}}(U)$.

Assuming $T \in \mathscr{F}_m^{\text{loc}}(\mathbf{R}^n)$ and $b \in \mathbf{R}^n$ we say that C is an **oriented tangent cone of T at b** if and only if $C \in \mathscr{F}_m^{\text{loc}}(\mathbf{R}^n)$, *$C$ is an oriented cone and there exists a sequence of positive numbers r_j such that*

$$\lim_{j\to\infty} r_j = \infty \quad \textit{and} \quad \lim_{j\to\infty} (\mu_{r_j} \circ \tau_{-b})_\# T = C;$$

clearly this implies $\operatorname{spt} C \subset \operatorname{Tan}(\operatorname{spt} T, b)$.

Simple examples show that T may have several distinct oriented tangent cones at b, corresponding to different sequences of numbers approaching ∞. However, if r_j, C are as above, then *every sequence of positive numbers s_j satisfying the relative growth condition*

$$0 < \liminf_{j\to\infty} s_j/r_j \leq \limsup_{j\to\infty} s_j/r_j < \infty$$

yields the same tangent cone C because

$$C - \lim_{j\to\infty} (\mu_{s_j} \circ \tau_{-b})_\# T = \lim_{j\to\infty} \mu_{s_j/r_j\#} [C - (\mu_{r_j} \circ \tau_{-b})_\# T] = 0.$$

In the *special case where T is an oriented cone and $0 \neq b \in \mathbf{R}^n$, every oriented tangent cone of T at b is an oriented cylinder with direction $b/|b|$*; taking C, r_j as above, $t \in \mathbf{R}$ and $s_j = r_j - t$ we find here that

$$\begin{aligned}\tau_{tb\#} C &= \lim_{j\to\infty} (\tau_{tb} \circ \mu_{r_j} \circ \tau_{-b})_\# T \\ &= \lim_{j\to\infty} (\mu_{s_j} \circ \tau_{-b} \circ \mu_{r_j/s_j})_\# T = \lim_{j\to\infty} (\mu_{s_j} \circ \tau_{-b})_\# T\end{aligned}$$

equals C, because $s_j/r_j \to 1$ as $j \to \infty$.

If C is an oriented tangent cone of T at b, f is a locally Lipschitzian proper map of $\mathbf{R}^n$ into $\mathbf{R}^\nu$, f is differentiable at b and there exist positive numbers α, β such that

$$|f(x) - f(b)| \geq \alpha |x - b| \quad \textit{whenever} \quad |f(x) - f(b)| \leq \beta,$$

then $Df(b)_\# C$ is an oriented tangent cone of $f_\# T$ at $f(b)$. To verify this we assume $b = 0$, $f(b) = 0$ and consider the sequence of maps

$$g_j \colon \mathbf{R}^n \to \mathbf{R}^\nu, \quad g_j(x) = r_j f(x/r_j) \text{ for } x \in \mathbf{R}^n,$$

whose restrictions to each compact subset of $\mathbf{R}^n$ converge uniformly to $Df(0)$, with a common Lipschitz constant. We observe that

$$|g_j(x)| \geq \alpha |x| \quad \text{whenever} \quad |g_j(x)| \leq r_j \beta,$$

hence for $0 < \sigma < \infty$ the conditions

$$r_j \beta \geq \sigma \quad \text{and} \quad g_j^{-1} \mathbf{B}(0, \sigma) \subset \mathbf{B}(0, \sigma/\alpha)$$

hold when j is large, and we compute

$$\lim_{j\to\infty} \mu_{r_j\#}(f_\# T) = \lim_{j\to\infty} g_{j\#}(\mu_{r_j\#} T) = \lim_{j\to\infty} [g_{j\#}(\mu_{r_j\#} T - C) + g_{j\#} C]$$
$$= Df(0)_\# C.$$

We will prove that in many important cases (4.3.17, 4.3.18, 4.5.5) the currents $(\mu_{r_j} \circ \tau_{-b})_\# T$ have the same limit for all sequences of positive numbers r_j approaching ∞; if

$$(\mu_r \circ \tau_{-b})_\# T \to C \text{ in } \mathscr{F}_m^{\mathrm{loc}}(\mathbf{R}^n) \text{ as } \mathbf{R} \ni r \to \infty,$$

then C is an oriented cone, because

$$\mu_{s\#} [\lim_{r\to\infty} (\mu_r \circ \tau_{-b})_\# T] = \lim_{r\to\infty} (\mu_{sr} \circ \tau_{-b})_\# T$$

whenever $0 < s < \infty$, hence C is the unique tangent cone of T at b. On the other hand, the use of particular sequences of numbers r_j will allow us to apply 4.2.17(2) in the proof of 5.4.3.

4.3.17. *If* $T \in \mathscr{R}_m^{\mathrm{loc}}(\mathbf{R}^n)$ *and* $E = \{x: \Theta^{*m}(\|T\|, x) > 0\}$, *then:*

(1) *For* $\mathscr{H}^m$ *almost all* b *in* E,

$$\lim_{r\to\infty} (\mu_r \circ \tau_{-b})_\# T = [\mathscr{H}^m \llcorner \operatorname{Tan}^m(\|T\|, b)] \wedge \zeta$$
$$\textit{with } \zeta(x) = \Theta^m(\|T\|, b)\, \vec{T}(b) \textit{ for } x \in \mathbf{R}^n.$$

(2) *For* $b \in \mathbf{R}^n \sim E$, $\quad \lim_{r\to\infty} (\mu_r \circ \tau_{-b})_\# T = 0.$

Proof. Using 4.3.16 we immediately obtain (2), and we verify (1) first in the special case where $T \in \mathscr{P}_m(\mathbf{R}^n)$, next in the special case where $T = f_\# P$ for some $P \in \mathscr{P}_m(\mathbf{R}^n)$ and some diffeomorphism f of class 1 mapping $\mathbf{R}^n$ onto $\mathbf{R}^n$; for this purpose we recall 4.1.27 and 4.1.30.

To prove (1) in case $T \in \mathscr{R}_m(\mathbf{R}^n)$ we employ 4.2.20 for each $\varepsilon > 0$ to secure $P \in \mathscr{P}_m(\mathbf{R}^n)$ and a diffeomorphism f of class 1 mapping $\mathbf{R}^n$ onto $\mathbf{R}^n$ such that

$$\varepsilon \geq \mathbf{M}(T - f_\# P) = \int \Theta^m(\|T - f_\# P\|, \cdot)\, d\mathscr{H}^m \geq \mathscr{H}^m(G),$$

where $G=\{x\colon \Theta^{*m}(\|T-f_{\#}P\|,x)>0\}$, and apply (2) with T, E replaced by $T-f_{\#}P, G$ in verifying that $\mathscr{H}^m$ almost all points b in $E\sim G$ satisfy the equations

$$\lim_{r\to\infty}(\mu_r\circ\tau_{-b})_{\#}(T-f_{\#}P)=0,\qquad 0<\Theta^m(\|T\|,b)=\Theta^m(\|f_{\#}P\|,b)<\infty,$$
$$\vec{T}(b)=\overrightarrow{f_{\#}P}(b),\qquad \operatorname{Tan}^m(\|T\|,b)=\operatorname{Tan}^m(\|f_{\#}P\|,b).$$

4.3.18. Combining the concepts of 4.2.28 and 4.3.14, 15 we say that C is an **analytic oriented cone (cylinder)** in $\mathbf{R}^n$ if and only if C is both an analytic chain and an oriented cone (cylinder) in $\mathbf{R}^n$. Recalling 4.3.16 we will prove:

Whenever T is an m dimensional analytic chain in $\mathbf{R}^n$ and $b\in\mathbf{R}^n$, there exists an m dimensional analytic cone C such that

$$(\mu_r\circ\tau_{-b})_{\#}T\to C \text{ in } \mathscr{F}_m^{\mathrm{loc}}(\mathbf{R}^n) \text{ as } r\to\infty.$$

It suffices to consider the case when T is an oriented analytic block with compact support in $\mathbf{R}^n$, say

$$T=(\mathscr{H}^m \llcorner B)\wedge(\xi/|\xi|),$$

B is a component of $V=\operatorname{var}(U,\{f_{m+1},\ldots,f_n\})\sim\operatorname{var}(U,\{g\})$, U is an open subset of $\mathbf{R}^n$, $\operatorname{Clos} B$ is a compact subset of U and $f_{m+1},\ldots,f_n$, g are real valued analytic functions on U,

$$Df_{m+1}(x)\wedge\cdots\wedge Df_n(x)\neq 0 \text{ for } x\in V,$$

$\xi=\mathbf{D}^{n-m}(Df_{m+1}\wedge\cdots\wedge Df_n)$ and $b=0\in(\operatorname{Clos} B)\sim V$. Defining

$$p\colon \mathbf{R}\times\mathbf{R}^n\to\mathbf{R},\qquad q\colon \mathbf{R}\times\mathbf{R}^n\to\mathbf{R}^n,\qquad h\colon \mathbf{R}\times\mathbf{R}^n\to\mathbf{R}^n,$$
$$p(t,x)=t,\quad q(t,x)=x,\quad h(t,x)=tx \text{ for } (t,x)\in\mathbf{R}\times\mathbf{R}^n,$$
$$F_j=f_j\circ h,\quad G=g\circ h,\quad Z=(\mathbf{R}\times\mathbf{R}^n)\cap\{(t,x)\colon t>0\},$$
$$k\colon Z\to Z,\quad k(t,x)=(t,t^{-1}x) \text{ for } (t,x)\in Z,$$
$$A=h^{-1}(B)\cap\{(t,x)\colon t>0\}=k(\{t\colon t>0\}\times B),$$
$$\eta(t,x)=[\textstyle\bigwedge_{m+1}Dk(t,tx)]\big((1,0)\wedge[\bigwedge_m q^*]\,\xi(tx)\big)$$

for $(t,x)\in A$, we find that

$$h^{-1}(V)=\operatorname{var}[h^{-1}(U),\ \{F_{m+1},\ldots,F_n\}]\sim\operatorname{var}[h^{-1}(U),\ \{G\}],$$
$$DF_{m+1}(t,x)\wedge\cdots\wedge DF_n(t,x)\neq 0 \text{ for } (t,x)\in h^{-1}(V),$$

A is relatively open and closed in $h^{-1}(V)$, A is connected, A is an analytic block of $h^{-1}(U)$,

$$\operatorname{Clos} A\subset h^{-1}(\operatorname{Clos} B)\subset h^{-1}(U),$$

the $m+1$ vectorfield $\eta/|\eta|$ orients A, the current

$$S=(\mathscr{H}^{m+1}\llcorner A)\wedge(\eta/|\eta|)$$

is an $m+1$ dimensional analytic chain in $\mathbf{R}\times\mathbf{R}^n$, hence

$$(\partial S)\llcorner\ker p=(\partial S)-(\partial S)\llcorner Z$$

is an m dimensional analytic chain in $\mathbf{R}\times\mathbf{R}^n$ and

$$C=-q_\#[(\partial S)\llcorner\ker p]$$

is an m dimensional analytic chain in $\mathbf{R}^n$; moreover C is an oriented cone because for $0<r<\infty$ the linear automorphism α of $\mathbf{R}\times\mathbf{R}^n$ mapping (t,x) onto $(r^{-1}t, rx)$ satisfies the equations

$$h\circ\alpha=h,\quad \alpha(A)=A,\quad \alpha\circ k=k\circ(\beta\times\mathbf{1}_{\mathbf{R}^n})\ \text{ with }\ \beta(t)=r^{-1}t \text{ for } t\in\mathbf{R},$$
$$(\textstyle\bigwedge_{m+1}\alpha)\circ\eta=r^{-1}\eta\circ\alpha,\ \alpha_\# S=S,\ \alpha(\ker p)=\ker p,\ \boldsymbol{\mu}_r\circ q=q\circ\alpha,\ \boldsymbol{\mu}_{r\#}C=C.$$

From 4.3.4, 6, 8 and 4.2.1 we infer that $\mathscr{L}^1$ almost all positive numbers δ satisfy the conditions

$$\langle S,|q|,\delta-\rangle\in\mathbf{I}_m^{\mathrm{loc}}(\mathbf{R}\times\mathbf{R}^n)$$

and

$$\langle S,|q|,\delta-\rangle=(\mathscr{H}^m\llcorner A\cap\{(t,x)\colon |x|=\delta\})\wedge(\gamma/|\gamma|)$$

with $\gamma=\eta\llcorner D|q|$, hence $\|\langle S,|q|,\delta-\rangle\|\ker p=0$; we fix such a δ and note that

$$R=S\llcorner\{(t,x)\colon |x|<\delta\}\in\mathbf{I}_{m+1}^{\mathrm{loc}}(\mathbf{R}\times\mathbf{R}^n),$$
$$q_\#[(\partial R)\llcorner\ker p]=q_\#[(\partial S)\llcorner\{(t,x)\colon t=0,|x|<\delta\}]$$
$$=q_\#([(\partial S)\llcorner\ker p]\llcorner q^{-1}\mathbf{U}(0,\delta))=-C\llcorner\mathbf{U}(0,\delta).$$

Next we see similarly that $\mathscr{L}^1$ almost all positive numbers ε satisfy the condition

$$\langle R,p,\varepsilon-\rangle=(\mathscr{H}^m\llcorner E_\varepsilon)\wedge(\zeta/|\zeta|)\in\mathbf{I}_m(\mathbf{R}\times\mathbf{R}^n)$$

with $E_\varepsilon=A\cap\{(t,x)\colon t=\varepsilon,|x|<\delta\}$ and $\zeta=\eta\llcorner Dp$; since

$$q(E_\varepsilon)=\mu_{1/\varepsilon}(B)\cap\mathbf{U}(0,\delta),$$
$$(\textstyle\bigwedge_m q)\circ\zeta|E_\varepsilon=\varepsilon^{-m}\xi\circ\boldsymbol{\mu}_\varepsilon\circ q|E_\varepsilon,\quad q_\#\langle R,p,\varepsilon-\rangle=(\boldsymbol{\mu}_{1/\varepsilon\#}T)\llcorner\mathbf{U}(0,\delta),$$
$$\langle R,p,\varepsilon-\rangle+(\partial R)\llcorner\ker p=\partial[R\llcorner\{(t,x)\colon t<\varepsilon\}]-(\partial R)\llcorner\{(t,x)\colon 0<t<\varepsilon\},$$

we conclude that

$$\mathscr{F}_{\mathbf{B}(0,\delta)}[(\mu_{1/\varepsilon\#}T)\llcorner\mathbf{U}(0,\delta)-C\llcorner\mathbf{U}(0,\delta)]$$
$$\le\mathscr{F}_{\{(t,x)\colon 0\le t\le\varepsilon,|x|\le\delta\}}[\langle R,p,\varepsilon-\rangle+(\partial R)\llcorner\ker p]$$
$$\le[\|R\|+\|\partial R\|]\{(t,x)\colon 0<t<\varepsilon\}\to 0 \text{ as } \varepsilon\to 0+.$$

4.3.19. Recalling 4.2.29 we say that C is a **holomorphic oriented cone** in $\mathbf{C}^n$ if and only if C is both a holomorphic chain of $\mathbf{C}^n$ in $\mathbf{C}^n$ and an oriented cone.

Whenever T is a complex m dimensional holomorphic chain of $\mathbf{C}^n$ in U, and $b \in U$, there exists a complex m dimensional holomorphic oriented cone C in $\mathbf{C}^n$ such that

$$(\mu_r \circ \tau_{-b})_\# T \to C \quad \textit{in} \quad \mathscr{F}_{2m}^{\mathrm{loc}}(\mathbf{C}^n) \ \textit{as} \ r \to \infty .$$

In fact, if $\mathbf{B}(b, \varepsilon) \subset U$, then $T \llcorner \mathbf{U}(b, \varepsilon)$ is a $2m$ dimensional analytic chain in $\mathbf{C}^n \simeq \mathbf{R}^{2n}$, hence 4.3.18 implies the existence of a limit cone C. Moreover $\partial C = 0$, because $U \cap \operatorname{spt} \partial C = \varnothing$. On the other hand the complex version of 3.4.11 implies that $\operatorname{Tan}(\operatorname{spt} T, b)$ is contained in a complex m dimensional holomorphic subvariety of $\mathbf{C}^n$. Since $\operatorname{spt} C \subset \operatorname{Tan}(\operatorname{spt} T, b)$ it follows that C is a complex m dimensional holomorphic chain.

4.3.20. To construct **intersections of flat chains** in $\mathbf{R}^n$ we slice their cartesian products by means of the subtraction map

$$f \colon \mathbf{R}^n \times \mathbf{R}^n \to \mathbf{R}^n, \qquad f(x, y) = x - y \text{ for } (x, y) \in \mathbf{R}^n \times \mathbf{R}^n,$$

and also use the diagonal map

$$g \colon \mathbf{R}^n \to \mathbf{R}^n \times \mathbf{R}^n, \qquad g(x) = (x, x) \text{ for } x \in \mathbf{R}^n,$$

as follows: *If* $S \in \mathbf{F}_k^{\mathrm{loc}}(\mathbf{R}^n)$, $T \in \mathbf{F}_l^{\mathrm{loc}}(\mathbf{R}^n)$, $S \times T \in \mathbf{F}_{k+l}^{\mathrm{loc}}(\mathbf{R}^n \times \mathbf{R}^n)$, $k + l \geq n$ *and*

$$\langle S \times T, f, 0 \rangle \in \mathbf{F}_{k+l-n}^{\mathrm{loc}}(\mathbf{R}^n \times \mathbf{R}^n),$$

then $\operatorname{spt} \langle S \times T, f, 0 \rangle \subset \ker f = \operatorname{im} g$ *and* 4.1.15 *implies the existence of a unique current*

$$S \cap T \in \mathbf{F}_{k+l-n}^{\mathrm{loc}}(\mathbf{R}^n) \quad \textit{with} \quad g_\#(S \cap T) = (-1)^{(n-k)l} \langle S \times T, f, 0 \rangle .$$

Letting $r \colon \mathbf{R}^n \times \mathbf{R}^n \to \mathbf{R}^n$, $r(x, y) = (y, x)$ for $(x, y) \in \mathbf{R}^n \times \mathbf{R}^n$, we see that $f \circ r = -f$, $(f \circ r)^\# \Omega = (-1)^n f^\# \Omega$ and $r \circ g = g$, hence

$$(-1)^{kl + (n-k)l} g_\#(S \cap T) = (-1)^{kl} \langle S \times T, f, 0 \rangle = \langle r_\#(T \times S), f, 0 \rangle$$
$$= r_\# \langle T \times S, f \circ r, 0 \rangle = (-1)^n r_\# \langle T \times S, f, 0 \rangle = (-1)^{n + (n-l)k} g_\#(T \cap S),$$

and we obtain the formula

$$S \cap T = (-1)^{(n-k)(n-l)} T \cap S .$$

One verifies readily that if $S \cap T$ exists and $k + l > n$, $k > 0$, $l > 0$, then

$$\partial(S \cap T) - S \cap (\partial T) = (-1)^{n-l} (\partial S) \cap T .$$

The theory of intersections is complicated by the fact that, while $\langle S\times T, f, z\rangle \in \mathbf{F}^{\text{loc}}_{k+l-n}(\mathbf{R}^n)$ for $\mathscr{L}^n$ almost all z in $\mathbf{R}^n$, this may fail when $z=0$. However in many important special cases, like those discussed below, one can prove continuity with respect to z at 0. We observe also that, in any event,

$$g_\#[(\tau_{a\#} S)\cap(\tau_{b\#} T)]=(-1)^{(n-k)l}\,\tau_{(a,b)\#}\langle S\times T, f, b-a\rangle$$

for $\mathscr{L}^n\times\mathscr{L}^n$ almost all (a,b), because $f\circ\tau_{(a,b)}=\tau_{a-b}\circ f$.

If $S\in\mathbf{F}_k(\mathbf{R}^n)$, $\psi\in\mathscr{D}^{n-l}(\mathbf{R}^n)$ and $k+l\geq n$, then

$$S\cap(\mathbf{E}^n\llcorner\psi)=S\llcorner\psi.$$

In fact for $\xi\in\mathscr{D}(\mathbf{R}^n,\wedge_k\mathbf{R}^n)$ and $\eta\in\mathscr{D}(\mathbf{R}^n,\wedge_l\mathbf{R}^n)$ we see from 4.1.8 and 1.6.5 that

$$[(\mathscr{L}^n\wedge\xi)\times(\mathscr{L}^n\wedge\eta)]\llcorner f^\#\Omega=(\mathscr{L}^n\times\mathscr{L}^n)\wedge\zeta,$$

where ζ is the $k+l-n$ vectorfield on $\mathbf{R}^n\times\mathbf{R}^n$ mapping (x,y) onto

$$\begin{aligned}&[\langle\xi(x),\wedge_k P\rangle\wedge\langle\eta(y),\wedge_l Q\rangle]\llcorner\langle Y_1\wedge\cdots\wedge Y_n,\wedge^n f\rangle\\ &=(-1)^{(n-k)l}\langle\xi(x)\llcorner\mathbf{D}_l\,\eta(y),\wedge_{k+l-n}g\rangle,\end{aligned}$$

hence infer for $\alpha\in\mathscr{D}^0(\mathbf{R}^n)$ and $\beta\in\mathscr{D}^{k+l-n}(\mathbf{R}^n\times\mathbf{R}^n)$ that

$$\begin{aligned}&([(\mathscr{L}^n\wedge\xi)\times(\mathscr{L}^n\wedge\eta)]\llcorner f^\#(\alpha\,\Omega))(\beta)\\ &=\textstyle\iint\langle\alpha(x-y)\,\zeta(x,y),\beta(x,y)\rangle\,d\mathscr{L}^n y\,d\mathscr{L}^n x\\ &=\textstyle\iint\langle\alpha(z)\,\zeta(x,x-z),\beta(x,x-z)\rangle\,d\mathscr{L}^n z\,d\mathscr{L}^n x\\ &=\textstyle\int\alpha(z)(-1)^{(n-k)l}\langle\langle\xi(x)\llcorner\mathbf{D}_l\,\eta(x-z),\wedge_{k+l-n}g\rangle,\beta(x,x-z)\rangle\,d\mathscr{L}^n x\,d\mathscr{L}^n z\\ &=\textstyle\int\alpha(z)(-1)^{(n-k)l}\,g_{z\#}[(\mathscr{L}^n\wedge\xi)\llcorner\mathbf{D}_l(\eta\circ\tau_{-z})](\beta)\,d\mathscr{L}^n z\end{aligned}$$

with $g_z(x)=(x,x-z)$ for $z,x\in\mathbf{R}^n$. Through regularization we approximate S by such currents $\mathscr{L}^n\wedge\xi$ and obtain

$$\begin{aligned}&([S\times(\mathscr{L}^n\wedge\eta)]\llcorner f^\#(\alpha\,\Omega))(\beta)\\ &=\textstyle\int\alpha(z)(-1)^{(n-k)l}\,g_{z\#}[S\llcorner\mathbf{D}_l(\eta\circ\tau_{-z})](\beta)\,d\mathscr{L}^n z.\end{aligned}$$

We conclude that, for every $z\in\mathbf{R}^n$,

$$\langle S\times(\mathscr{L}^n\wedge\eta), f, z\rangle=(-1)^{(n-k)l}\,g_{z\#}[S\llcorner\mathbf{D}_l(\eta\circ\tau_{-z})].$$

Taking $z=0$ and $\eta=\mathbf{D}^{n-l}\,\psi$ we deduce the asserted formula. In particular we find that

$$(\mathbf{E}^n\llcorner\phi)\cap(\mathbf{E}^n\llcorner\psi)=\mathbf{E}^n\llcorner(\phi\wedge\psi)\ \text{ for }\ \phi\in\mathscr{D}^{n-k}(\mathbf{R}^n),\ \psi\in\mathscr{D}^{n-l}(\mathbf{R}^n).$$

The preceding results show that the intersection product generalizes interior and exterior multiplication. However the more significant part of intersection theory concerns the case when neither factor corresponds to a differential form, in particular when, S, T and $S \cap T$ are integral currents. For example the following situation (manifolds in general relative position) occurs often in classical geometry:

Suppose $S \in \mathbf{I}_k(\mathbf{R}^n)$, $T \in \mathbf{I}_l(\mathbf{R}^n)$, $k+l \geq n$,

$$\sup\{\Theta^{*k}(\|S\|, x) + \Theta^{*k-1}(\|\partial S\|, x) \colon x \in \mathbf{R}^n\} < \infty,$$
$$\sup\{\Theta^{*l}(\|T\|, x) + \Theta^{*l-1}(\|\partial T\|, x) \colon x \in \mathbf{R}^n\} < \infty,$$

spt S *and* spt T *are contained in k and l dimensional submanifolds M and N of class* 1 *of* $\mathbf{R}^n$ *with*

$$\dim[\operatorname{Tan}(M, x) \cap \operatorname{Tan}(N, x)] = k+l-n \text{ for } x \in M \cap N,$$

either $k=0$ *or* spt $\partial S \cap$ spt T *has a neighborhood E in* $\mathbf{R}^n$ *such that* $E \cap$ spt ∂S *is covered by a finite family Γ of $k-1$ dimensional submanifolds G of class* 1 *of M with* $E \cap \operatorname{Clos} G \subset G$,

$$\dim[\operatorname{Tan}(G, x) \cap \operatorname{Tan}(N, x)] = k-1+l-n \text{ for } x \in G \cap N,$$

and either $l=0$ *or* spt $S \cap$ spt ∂T *has a neighborhood F in* $\mathbf{R}^n$ *such that* $F \cap$ spt ∂T *is covered by a finite family Δ of $l-1$ dimensional submanifolds D of class* 1 *of N with* $F \cap \operatorname{Clos} D \subset F$,

$$\dim[\operatorname{Tan}(M, x) \cap \operatorname{Tan}(D, x)] = k+l-1-n \text{ for } x \in M \cap D.$$

Then $\zeta(x) = \mathbf{D}^{2n-k-l}[\mathbf{D}_k \vec{S}(x) \wedge \mathbf{D}_l \vec{T}(x)]$ *is a nonzero $k+l-n$ vector whenever* $x \in (\text{spt } S \cap \text{spt } T) \sim (\text{spt } \partial S \cup \text{spt } \partial T)$, *and*

$$\mathscr{H}^{k+l-n}[\text{spt } S \cap \text{spt } T \cap (\text{spt } \partial S \cup \text{spt } \partial T)] = 0,$$
$$S \cap T = [\mathscr{H}^{k+l-n} \llcorner (\text{spt } S \cap \text{spt } T)] \wedge \Xi \in \mathbf{I}_{k+l-n}(\mathbf{R}^n)$$

with

$$\Xi = \Theta^k(\|S\|, \cdot)\, \Theta^l(\|T\|, \cdot)\, |\zeta|^{-1} \zeta.$$

This assertion may be deduced from 1.6.3, 5 and the proposition in 4.3.11, applied with $U = \mathbf{R}^n \times \mathbf{R}^n$, $R = S \times T$, $A = M \times N$,

$$\Lambda = \{G \times N \colon G \in \Gamma\} \cup \{M \times D \colon D \in \Delta\}, \quad \alpha = \beta \circ f,$$

where $\beta \in \mathscr{D}^0(\mathbf{R}^n)$, $0 \notin \operatorname{spt}(\beta - 1)$ and spt β is contained in a sufficiently small neighborhood of 0 in $\mathbf{R}^n$.

It appears that the hypotheses of the preceding proposition are unduly restrictive, and that future research should yield better criteria for the existence of intersections of some types of integral currents. For instance it is a reasonable conjecture that $S \cap T$ exists, and depends continuously on S and T, whenever S and T are k and l dimensional analytic chains (see 4.2.28) for which there exist analytic sets A and B with (see 3.4.10)

$$\operatorname{spt} S \cap \operatorname{spt} T \subset A, \qquad \dim A \leq k+l-n$$

and

$$(\operatorname{spt} \partial S \cap \operatorname{spt} T) \cup (\operatorname{spt} S \cap \operatorname{spt} \partial T) \subset B, \qquad \dim B < k+l-n.$$

The analogous conjecture about holomorphic chains (see 4.2.29) is even more plausible, because for the special case of complex algebraic cycles it follows from [F 19, § 4.5–4.8] that the algebraic intersection product (see [S A M, § 6]) can be constructed alternately by our method of slicing.

Another problem requiring further study is to find broad conditions which imply associativity of the intersection product of flat chains. Perhaps one should modify 4.3.1 through replacement of balls by more general sets in the definition of slices.

If $S \in \mathbf{F}_k(\mathbf{R}^n)$ and $T \in \mathbf{F}_{n-k}(\mathbf{R}^n)$ with $S \times T \in \mathbf{F}_n(\mathbf{R}^n \times \mathbf{R}^n)$,

$$k=0 \text{ or } \operatorname{spt} \partial S \cap \operatorname{spt} T = \varnothing, \qquad k=n \text{ or } \operatorname{spt} S \cap \operatorname{spt} \partial T = \varnothing,$$

then $f_\#(S \times T) \in \mathbf{F}_n(\mathbf{R}^n)$ with $0 \notin \operatorname{spt} \partial f_\#(S \times T)$, hence there exists a unique real number c such that

$$0 \notin \operatorname{spt}[(-1)^{n-k} f_\#(S \times T) - c\,\mathbf{E}^n];$$

in case S and T are integral currents, c is an integer. One calls c the **Kronecker index** of S and T. It follows that

$$((-1)^{n-k}(S \times T) \llcorner f^\#[\mathbf{B}(0,\rho) \wedge \Omega])(\chi \circ f) = c \int_{\mathbf{B}(0,\rho)} \chi \, d\mathscr{L}^n$$

whenever $\chi \in \mathscr{E}^0(\mathbf{R}^n)$ and ρ is a sufficiently small positive number. Taking $\chi(z)=1$ for $z \in \mathbf{R}^n$, we infer that *if* $S \cap T$ *exists, then* $(S \cap T)(1)$ *equals the Kronecker index of* S *and* T. We also observe that if $\partial S = 0$ and $\partial T = 0$, then $\partial f_\#(S \times T) = 0$, hence $f_\#(S \times T) = 0$ and the Kronecker index of S and T equals 0.

4.4. Homology groups

4.4.1. Here we study the homology theory of the **local Lipschitz category**; an object of this category is a pair (A, B) such that, for some n, A and B are local Lipschitz neighborhood retracts in $\mathbf{R}^n$ with $B \subset A$; a morphism of (A, B) into another object (A', B') is a locally Lipschitzian map

$$f\colon (A, B) \to (A', B'),$$

which means that f is a locally Lipschitzian function such that $\operatorname{dmn} f = A$, $\operatorname{im} f \subset A'$ and $f(B) \subset B'$.

For each nonnegative integer m we define the group of m dimensional **integral flat cycles**

$$\mathscr{Z}_m(A,B)=\{T\colon T\in\mathscr{F}_m(\mathbf{R}^n),\ \operatorname{spt} T\subset A,\ \operatorname{spt}\partial T\subset B \text{ or } m=0\},$$

the subgroup of m dimensional **integral flat boundaries**

$$\mathscr{B}_m(A,B)=\{T+\partial S\colon T\in\mathscr{F}_m(\mathbf{R}^n),\ \operatorname{spt} T\subset B,\ S\in\mathscr{F}_{m+1}(\mathbf{R}^n),\ \operatorname{spt} S\subset A\},$$

and the m dimensional **integral homology group**

$$\mathbf{H}_m(A,B)=\mathscr{Z}_m(A,B)/\mathscr{B}_m(A,B).$$

Since the currents in $\mathscr{F}_m(\mathbf{R}^n)$ have compact supports, replacement of $\mathbf{R}^n$ by any neighborhood of A yields isomorphic groups. We also observe that these homology groups are isomorphic to the homology groups of the chain complex (see [MCL, p. 39] or [ES, p. 125] with the chain groups

$$K_m=\{T\colon T\in\mathscr{F}_m(\mathbf{R}^n),\ \operatorname{spt} T\subset A\}/\{T\colon \operatorname{spt} T\subset B\}$$

for $m\geq 0$, $K_m=\{0\}$ for $m<0$, and with the boundary homomorphisms $\partial_m\colon K_m\to K_{m-1}$ induced by ∂ for $m>0$.

The cosets in $\mathbf{H}_m(A,B)$ are called m dimensional **integral homology classes** of (A,B).

For each locally Lipschitzian map $f\colon (A,B)\to(A',B')$ and each nonnegative integer m we define a homomorphism

$$\mathbf{H}_m(f)\colon \mathbf{H}_m(A,B)\to\mathbf{H}_m(A',B');$$

selecting a locally Lipschitzian retraction r of some neighborhood of A onto A we observe that $(f\circ r)_\#$ maps $\mathscr{Z}_m(A,B)$ into $\mathscr{Z}_m(A',B')$, and $\mathscr{B}_m(A,B)$ into $\mathscr{B}_m(A',B')$, hence induces a homomorphism of the homology groups; from 4.1.15 we see that this homomorphism does not depend on the choice of r.

Whenever $C\subset B\subset A$ are local Lipschitz neighborhood retracts in the same Euclidean space and m is a positive integer, the boundary operator ∂ maps $\mathscr{Z}_m(A,B)$ into $\mathscr{Z}_{m-1}(B,C)$, and $\mathscr{B}_m(A,B)$ into $\mathscr{B}_{m-1}(B,C)$, hence induces a homomorphism

$$\partial\colon \mathbf{H}_m(A,B)\to\mathbf{H}_{m-1}(B,C).$$

We abbreviate

$$\mathscr{Z}_m(A,\varnothing)=\mathscr{Z}_m(A),\quad \mathscr{B}_m(A,\varnothing)=\mathscr{B}_m(A),\quad \mathbf{H}_m(A,\varnothing)=\mathbf{H}_m(A).$$

The operations $\mathbf{H}_m$ and ∂ defined above, which lead from the local Lipschitz category to the category of abelian groups, satisfy the **axioms of Eilenberg and Steenrod** for a homology theory with coefficient group $\mathbf{Z}$ (see [ES, p. 10]) in the following form:

(1) *If f is the identity map of (A,B), then $\mathbf{H}_m(f)$ is the identity map of* $\mathbf{H}_m(A,B)$.

(2) *If $f: (A,B)\to(A',B')$ and $g: (A',B')\to(A'',B'')$, then*

$$\mathbf{H}_m(g\circ f)=\mathbf{H}_m(g)\circ\mathbf{H}_m(f).$$

(3) *If $C\subset B\subset A$, $C'\subset B'\subset A'$, $f: (A,B)\to(A',B')$, $f|B: (B,C)\to(B',C')$ and $m>0$, then*

$$\mathbf{H}_{m-1}(f|B)\circ\partial=\partial\circ\mathbf{H}_m(f).$$

(4) *If $i: (B,C)\to(A,C)$ and $j: (A,C)\to(A,B)$ are inclusion maps, then the sequence of homomorphisms*

$$\mathbf{H}_m(B,C)\xrightarrow{\mathbf{H}_m(i)}\mathbf{H}_m(A,C)\xrightarrow{\mathbf{H}_m(j)}\mathbf{H}_m(A,B)$$
$$\xrightarrow{\partial}\mathbf{H}_{m-1}(B,C)\xrightarrow{\mathbf{H}_{m-1}(i)}\mathbf{H}_{m-1}(A,C)\xrightarrow{\mathbf{H}_{m-1}(j)}\mathbf{H}_{m-1}(A,B)$$

is exact (which means that the image of each homomorphism equals the kernel of the next) *for $m>0$, and* $\operatorname{im}\mathbf{H}_0(j)=\mathbf{H}_0(A,B)$.

(5) *If $I=\{t: 0\le t\le 1\}$, $h: (I\times A, I\times B)\to(A',B')$ and*

$$h_t: (A,B)\to(A',B'),\quad h_t(x)=h(t,x)\ \textit{for}\ t\in I,\ x\in A,$$

then $\mathbf{H}_m(h_0)=\mathbf{H}_m(h_1)$.

(6) *If $f: (A,B)\to(A',B')$ is an inclusion map such that*

$$A\cap B'=B\quad\textit{and}\quad\operatorname{Clos}(A'\sim B')\cap\operatorname{Clos}(A'\sim A)=\varnothing,$$

then $\mathbf{H}_m(f)$ is an isomorphism.

(7) *If $a\in\mathbf{R}^n$, then $\mathbf{H}_0(\{a\})\simeq\mathbf{Z}$, $\mathbf{H}_m(\{a\})=\{0\}$ for $m>0$.*

Using the basic properties of currents established in 4.1 one readily verifies the conditions (1), (2), (3), (4), (5) and (7). Moreover it follows from 4.2.22 and 4.1.15 that

$$\{T: T\in\mathscr{F}_m(\mathbf{R}^n),\ \operatorname{spt}T\subset W\}$$
$$=\{R+\partial S: R\in\mathscr{R}_m(\mathbf{R}^n),\ S\in\mathscr{R}_{m+1}(\mathbf{R}^n),\ \operatorname{spt}R\cup\operatorname{spt}S\subset W\}$$

whenever W is a local Lipschitz neighborhood retract in $\mathbf{R}^n$. Therefore, to prove (6) we observe that the set

$$E=A'\cap\{x: \operatorname{dist}(x,A'\sim B')\le\operatorname{dist}(x,A'\sim A)\}$$

is closed relative to A', with $E\subset A$ and $A'\cap\operatorname{Clos}(A'\sim E)\subset B'$, and we reason as follows:

If $Y\in\mathscr{R}_m(\mathbf{R}^n)\cap\mathscr{Z}_m(A',B')$ and $X=Y\llcorner E$, then

$$\operatorname{spt}X\subset E,\quad\operatorname{spt}(Y-X)\subset B',\quad\operatorname{spt}\partial X\subset E\cap B'\subset B,$$

hence $Y-X\in\mathscr{B}_m(A',B')$ and $X\in\mathscr{Z}_m(A,B)$; thus $\mathbf{H}_m(f)$ is an epimorphism.

If $X \in \mathscr{R}_m(\mathbf{R}^n) \cap \mathscr{Z}_m(A, B) \cap \mathscr{B}_m(A', B')$, then there exist $R \in \mathscr{R}_m(\mathbf{R}^n)$, $S \in \mathscr{R}_{m+1}(\mathbf{R}^n)$ such that

$$X = R + \partial S, \quad \operatorname{spt} R \subset B', \quad \operatorname{spt} S \subset A',$$

hence $X = T + \partial(S \llcorner E)$ with $T = R + \partial[S \llcorner (A' \sim E)]$, which implies that $\operatorname{spt} T \subset A \cap B' = B$ and $X \in \mathscr{B}_m(A, B)$; thus $\mathbf{H}_m(f)$ is a monomorphism.

4.4.2. General isoperimetric theorem. *If $\mathbf{R}^n \supset A \supset B$, U and V are neighborhoods of A and B in $\mathbf{R}^n$, f and g are locally Lipschitzian retractions of U and V onto A and B,*

$$W = V \cap \{x \colon (1-t)\, g(x) + t\, x \in U \text{ for } 0 \le t \le 1\},$$

$$K \subset A, \; K \text{ and } B \cap K \text{ are compact}, \; \alpha > 0, \; \beta > 0,$$

$$L = \{x \colon \operatorname{dist}(x, K) \le \alpha\} \subset U, \quad \lambda = \sup\{1, \operatorname{Lip}(f \mid L)\},$$

$$M = \{x \colon \operatorname{dist}(x, B \cap K) \le \beta\} \subset V, \quad \mu = \sup\{1, \operatorname{Lip}(g \mid M)\},$$

$$\gamma = 2 n^{2m+2}, \quad \delta = \inf\{\alpha/\mu, \beta\}/(3n),$$

$$\rho = \lambda^m [(\mu+1)\, 3n\, \mu^m + 1]^{m/(m+1)}\, 2\gamma + \mu^m\, 2\gamma,$$

then W is a neighborhood of B in $U \cap V$ and the following five statements hold:

(1) *If $m > 0$ and $X \in \mathscr{R}_m(\mathbf{R}^n)$ with*

$$\operatorname{spt} X \subset K, \quad \operatorname{spt} \partial X \subset B, \quad \mathbf{M}(X) \le \delta^m/(2\gamma),$$

then there exists $Y \in \mathbf{I}_{m+1}(\mathbf{R}^n)$ with

$$\operatorname{spt} Y \subset f(L), \quad \operatorname{spt}(X - \partial Y) \subset g(M),$$

$$\mathbf{M}(Y)^{m/(m+1)} + \mathbf{M}(X - \partial Y) \le \rho\, \mathbf{M}(X).$$

(2) *There exists a finite positive number σ with the property: If*

$$X \in \mathscr{R}_m(\mathbf{R}^n) \cap \mathscr{Z}_m(A, B) \cap \mathscr{B}_m(U, W) \quad \text{with} \quad \operatorname{spt} X \subset K,$$

then there exists $Y \in \mathbf{I}_{m+1}(\mathbf{R}^n)$ with

$$\operatorname{spt} Y \subset A, \quad \operatorname{spt}(X - \partial Y) \subset B, \quad \mathbf{M}(Y)^{m/(m+1)} + \mathbf{M}(X - \partial Y) \le \sigma\, \mathbf{M}(X).$$

(3) *If $i \colon (A, B) \to (U, W)$ is the inclusion map, then*

$$\mathbf{H}_m(i) \colon \mathbf{H}_m(A, B) \to \mathbf{H}_m(U, W)$$

is a monomorphism, whose image is a direct summand of $\mathbf{H}_m(U, W)$.

(4) *The subgroup of* $\mathbf{H}_m(A, B)$, *consisting of those homology classes which meet* $\mathscr{R}_m(\mathbf{R}^n) \cap \{X: \operatorname{spt} X \subset K\}$, *is finitely generated.*

(5) *If* $0<c<\infty$, *then only finitely many integral homology classes of* (A, B) *meet* $\{X: \operatorname{spt} X \subset K, \mathbf{M}(X) \leq c\}$.

Proof. Assuming $B \cap K \neq \emptyset$ and $m>0$ we define

$$u(x) = \operatorname{dist}(x, B \cap K) \text{ for } x \in \mathbf{R}^n.$$

Whenever $0<\varepsilon \leq \delta$ and $X \in \mathscr{R}_m(\mathbf{R}^n)$, $\operatorname{spt} X \subset K$, $\operatorname{spt} \partial X \subset B$ we see from the corollary in 4.2.1 that

$$\textstyle\int_0^{*\varepsilon} \mathbf{M}(\partial[X \llcorner \{x: u(x)>r\}])\, d\mathscr{L}^1 r \leq \mathbf{M}(X)$$

because $\operatorname{spt} \partial X \subset \{x: u(x)=0\}$, choose r and T so that

$$0<r<\varepsilon, \quad T = X \llcorner \{x: u(x)>r\}, \quad \varepsilon\, \mathbf{M}(\partial T) \leq \mathbf{M}(X),$$

and infer from 4.2.16 (2) that $T \in \mathbf{I}_m(\mathbf{R}^n)$. We also recall 4.2.5 to define

$$\Omega = \mathbf{I}_m(\mathbf{R}^n) \cap \{R: \operatorname{spt} R \subset L \cap \mu_\varepsilon(\mathbf{W}'_m), \operatorname{spt} \partial R \subset \mu_\varepsilon(\mathbf{W}'_{m-1})\}$$

and let h be the affine homotopy from g to $\mathbf{1}_U$. We observe that for each $x \in M$ there exists $b \in B \cap K$ with $u(x)=|x-b|$, hence

$$u[g(x)] \leq |g(x)-b| = |g(x)-g(b)| \leq \mu\, u(x),$$
$$|g(x)-x| \leq |g(x)-b| + |b-x| \leq (\mu+1)\, u(x).$$

Applying 4.2.9 we express

$$T = P + Q + \partial S \quad \text{with} \quad P \in \Omega, \quad Q \in \mathbf{I}_m(\mathbf{R}^n), \quad S \in \mathbf{I}_{m+1}(\mathbf{R}^n),$$
$$\mathbf{M}(P) \leq 2\gamma\, \mathbf{M}(X), \quad \mathbf{M}(Q) \leq \gamma\, \mathbf{M}(X), \quad \mathbf{M}(S) \leq \varepsilon\, \gamma\, \mathbf{M}(X),$$
$$\operatorname{spt} \partial T \subset \{x: u(x)=r\}, \quad \operatorname{spt} Q \subset \{x: u(x) \leq 3n\varepsilon\},$$
$$h(\{t: 0 \leq t \leq 1\} \times [\operatorname{spt} Q \cup \operatorname{spt}(X-T)]) \cup \operatorname{spt} S \subset L$$

because $3n\varepsilon \leq \beta$ and $3n\varepsilon\mu \leq \alpha$. Letting

$$Z = h_\#([0,1] \times (Q+X-T)) + S$$

we see that $\operatorname{spt} Z \subset L$ and

$$\partial Z = X - P - g_\#(Q+X-T) + h_\#([0,1] \times \partial P)$$

because $\partial(Q+X-T) = \partial X - \partial P$ with $g(x)=x$ for $x \in \operatorname{spt} \partial X$, hence $Z \in \mathbf{I}_{m+1}(\mathbf{R}^n)$, and we estimate

$$\mathbf{M}(f_\# Z) \leq \lambda^{m+1}[(\mu+1)\, 3n\varepsilon\, \mu^m\, 2\gamma\, \mathbf{M}(X) + \varepsilon\, \gamma\, \mathbf{M}(X)]$$
$$\leq \lambda^{m+1}[(\mu+1)\, 3n\, \mu^m + 1]\, \varepsilon\, 2\gamma\, \mathbf{M}(X).$$

To prove (1) *we suppose* $0<2\gamma\,\mathbf{M}(X)\le\delta^m$ *and choose* ε *so that* $\varepsilon^m=2\gamma\,\mathbf{M}(X)$, *hence* $\mathbf{M}(P)\le\varepsilon^m$; since $\mathbf{M}(P)$ is an integral multiple of $(2\varepsilon)^m$, it follows that $P=0$. Taking $Y=f_\# Z$ we infer the conclusions of (1) because

$$X-\partial Y=f_\#(X-\partial Z)=g_\#(Q+X-T),$$

$$[\varepsilon\,2\gamma\,\mathbf{M}(X)]^{m/(m+1)}=2\gamma\,\mathbf{M}(X),\qquad \mathbf{M}(X-\partial Y)\le\mu^m\,2\gamma\,\mathbf{M}(X).$$

From here on we assume that $\alpha<\infty$, *hence* L *is compact and* Ω *is a finitely generated free abelian group.*

To prove (2) *we take* $\varepsilon=\delta$ *and consider the subgroup*

$$\Phi=\Omega\cap\mathscr{B}_m(U,W).$$

From 4.2.22 we see that for each $R\in\Phi$ there exist

$$F\in\mathscr{R}_{m+1}(\mathbf{R}^n)\ \text{ with }\ \operatorname{spt}F\subset U,\quad \operatorname{spt}(R-\partial F)\subset W,$$

$$G\in\mathscr{R}_m(\mathbf{R}^n),\quad H\in\mathscr{R}_{m+1}(\mathbf{R}^n)\ \text{ with }\ R-\partial F=G+\partial H,\quad \operatorname{spt}G\cup\operatorname{spt}H\subset W,$$

hence $F+H\in\mathbf{I}_{m+1}(\mathbf{R}^n)$, $\operatorname{spt}(F+H)\subset U$, $\operatorname{spt}[R-\partial(F+H)]\subset W$. Since Ω and Φ are free abelian groups, generated over $\mathbf{Z}$ by $\mathbf{R}$ independent finite subsets, we can therefore construct a homomorphism

$$\Gamma\colon\Phi\to\mathbf{I}_{m+1}(\mathbf{R}^n)$$

such that $\operatorname{spt}\Gamma(R)\subset U$ and $\operatorname{spt}[R-\partial\Gamma(R)]\subset W$ for $R\in\Phi$. Observing that Φ generates a finite dimensional vectorsubspace of $\mathbf{P}_m(\mathbf{R}^n)$ normed by $\mathbf{M}$, and that Γ has a linear extension mapping this vectorsubspace into $\mathbf{N}_{m+1}(\mathbf{R}^n)$, we obtain a finite positive number ξ such that

$$\mathbf{M}\,f_\#\,[\Gamma(R)+h_\#([0,1]\times[R-\partial\Gamma(R)])]\le\xi\,\mathbf{M}(R)$$

$$\text{and }\mathbf{M}\,g_\#\,[R-\partial\Gamma(R)]\le\xi\,\mathbf{M}(R)\ \text{ for }R\in\Phi.$$

We now suppose X *satisfies the hypothesis of* (2) *and also the condition*

$$2\gamma\,\mathbf{M}(X)>\delta^m,$$

because in the alternate case a suitable current Y would be furnished by (1). We find that

$$P=X-(Q+X-T)-\partial S\in\Phi,$$

$$\begin{aligned}&X-\partial[Z+\Gamma(P)+h_\#([0,1]\times[P-\partial\Gamma(P)])]\\&=X-\partial Z-\partial\Gamma(P)+h_\#([0,1]\times\partial P)-P+\partial\Gamma(P)+g_\#[P-\partial\Gamma(P)]\\&=g_\#(Q+X-T)+g_\#[P-\partial\Gamma(P)].\end{aligned}$$

Taking $Y=f_\#[Z+\Gamma(P)+h_\#([0,1]\times[P-\partial\Gamma(P)])]$ we infer the conclusions of (2) because

$$\mathbf{M}(X-\partial Y)\le\mu^m\,\mathbf{M}(Q+X-T)+\xi\,\mathbf{M}(P)\le(\mu^m+\xi)\,2\gamma\,\mathbf{M}(X),$$
$$\mathbf{M}(Y)\le\mathbf{M}(f_\#Z)+\xi\,\mathbf{M}(P)\le\eta\,2\gamma\,\mathbf{M}(X)$$

where $\eta=\lambda^{m+1}[(\mu+1)\,3n\,\mu^m+1]\,\delta+\xi$, hence

$$\mathbf{M}(Y)^m\,\delta^m\le\eta^m[2\gamma\,\mathbf{M}(X)]^{m+1},\quad \mathbf{M}(Y)^{m/(m+1)}\le(\eta/\delta)^{m/(m+1)}\,2\gamma\,\mathbf{M}(X).$$

To prove (3) *we define the homomorphism* $E\colon \mathscr{Z}_m(U,W)\to\mathscr{Z}_m(A,B)$,

$$E(R)=f_\#[R-h_\#([0,1]\times\partial R)]\quad\textit{for}\quad R\in\mathscr{Z}_m(U,W),$$

which maps $\mathscr{B}_m(U,W)$ *into* $\mathscr{B}_m(A,B)$; in fact the conditions

$$F\in\mathscr{F}_m(\mathbf{R}^n),\quad \operatorname{spt}F\subset W,\quad G\in\mathscr{F}_{m+1}(\mathbf{R}^n),\quad \operatorname{spt}G\subset U$$

imply $\operatorname{spt}g_\#F\subset B$, $\operatorname{spt}f_\#[h_\#([0,1]\times F)+G]\subset A$ and

$$E(F+\partial G)=g_\#F+\partial f_\#[h_\#([0,1]\times F)+G].$$

Accordingly E induces a homomorphism $E_*\colon \mathbf{H}_m(U,W)\to\mathbf{H}_m(A,B)$. Since $E(R)=R$ for $R\in\mathscr{Z}_m(A,B)$, we conclude that

$$E_*\circ\mathbf{H}_m(i)\text{ is the identity map of }\mathbf{H}_m(A,B).$$

To prove (4) *we take* $\varepsilon=\delta$ *and consider the group*

$$\Psi=\Omega\cap\mathscr{Z}_m(U,W).$$

Since Ψ is finitely generated, so is the image of Ψ under the composed homomorphism

$$\mathscr{Z}_m(U,W)\xrightarrow{E}\mathscr{Z}_m(A,B)\to\mathbf{H}_m(A,B).$$

For each $X\in\mathscr{R}_m(\mathbf{R}^n)$ with $\operatorname{spt}X\subset K$ and $\operatorname{spt}\partial X\subset B$, the corresponding current P satisfies the conditions

$$P\in\Psi,\quad P-X\in\mathscr{B}_m(U,W),\quad E(P)-X\in\mathscr{B}_m(A,B),$$

hence the homology class of X belongs to image of Ψ.

Similarly we prove (5) by observing that

$$\Psi_c=\Psi\cap\{R\colon \mathbf{M}(R)\le2\gamma c\}\text{ is finite,}$$

and that the additional hypothesis $\mathbf{M}(X)\le c$ implies $P\in\Psi_c$, hence the homology class of X belongs to the image of Ψ_c.

The preceding arguments are easily adapted to the case when $B\cap K=\varnothing$, which is actually much simpler; then one can take $\delta=\alpha/3n$, $T=X$, $Q=0$, $Z=S$. Similarly one verifies (2), (3), (4), (5) in case $m=0$.

4.4.3. A corollary of 4.4.2 (2) is the proposition:

If $T \in \mathbf{I}_{m+1}^{\mathrm{loc}}(\mathbf{R}^n)$ *with* $\operatorname{spt} T \subset A$ *and* $\operatorname{spt} \partial T \subset K \cup B$, *then there exists* $S \in \mathbf{I}_{m+1}^{\mathrm{loc}}(\mathbf{R}^n) \cap \mathscr{Z}_{m+1}(A, B)$ *with*

$$\mathbf{M}(S-T)^{m/(m+1)} + \|\partial S - \partial T\|(B) \le \sigma \, \|\partial T\|(A \sim B).$$

To verify this we consider the rectifiable current

$$X = (\partial T) \llcorner (A \sim B) = \partial T - (\partial T) \llcorner B \in \mathscr{B}_m(A, B),$$

choose Y according to 4.4.2 (2), take $S = T - Y$ and note that

$$\partial S = X - \partial Y + (\partial T) \llcorner B, \ \operatorname{spt} \partial S \subset B.$$

4.4.4. Here we assume, in addition to the hypotheses of Theorem 4.4.2, that K *is a Lipschitz neighborhood retract and* $0 < c < \infty$, and we study compactness properties of the set

$$\Delta = \mathscr{Z}_m(A, B) \cap \{X \colon \operatorname{spt} X \subset K, \mathbf{M}(X) \le c\}.$$

First we observe:

If $B = \emptyset$, *then* Δ *is* $\mathscr{F}_K$ *compact, and* $\Delta \cap \chi$ *is* $\mathscr{F}_K$ *compact for* $\chi \in \mathbf{H}_m(A)$.

In fact the first conclusion is a consequence of 4.2.16 (3) and 4.2.17 (2), because $\partial X = 0$ for $X \in \Delta$ in case $B = \emptyset$. Since the sets $\Delta \cap \chi$ form a partition of Δ, the second conclusion follows from 4.4.2 (1), which implies that each set $\Delta \cap \chi$ is $\mathscr{F}_K$ open relative to Δ.

In case $B \ne \emptyset$ the set Δ need not be $\mathscr{F}_K$ compact, as shown by the example when $m = 1$, $c = \pi/2$,

$$A = K = \mathbf{R}^3 \cap \{x \colon 0 \le x_i \le 1 \text{ for } i = 1, 2, 3\}, \quad B = A \cap \{x \colon x_3 = 0\};$$

letting $R_i \in \Delta$ be the sum of i^2 oriented semicircles from $(j\,i^{-1}, k\,i^{-1}, 0)$ to $(j\,i^{-1} + i^{-2}, k\,i^{-1}, 0)$ corresponding to $j, k \in \{0, \ldots, i-1\}$, one readily verifies that

$$\lim_{i \to \infty} R_i(\phi) = \int_B \langle (1,0,0), \phi(x) \rangle \, d\mathscr{H}^2 x$$

for $\phi \in \mathscr{D}^1(\mathbf{R}^3)$; since 4.2.16 (3) and 4.1.28 imply that the limit current defined by the above integral formula does not belong to $\mathscr{F}_1(\mathbf{R}^3)$, the sequence of currents R_i has no $\mathscr{F}_K$ convergent subsequence.

To obtain positive results when $K \cap B \ne \emptyset$ we use the function $\mathscr{F}_{K,B}$ defined by the formula

$$\mathscr{F}_{K,B}(R) = \int_0^{*\infty} \mathscr{F}_K[R \llcorner \{x \colon \operatorname{dist}(x, K \cap B) > r\}] \, d\mathscr{L}^1 r$$

whenever $R \in \mathscr{R}_{m,K}(\mathbf{R}^n)$. Letting $\mathscr{F}_{K,B}(R_1 - R_2)$ be the pseudodistance between two elements R_1 and R_2 of Δ, we will prove:

$$\Delta \text{ is } \mathscr{F}_{K,B} \text{ compact, and } \Delta \cap \chi \text{ is } \mathscr{F}_{K,B} \text{ compact for } \chi \in \mathbf{H}_m(A, B).$$

We consider any sequence $R_1, R_2, R_3, \ldots \in \Delta$. Abbreviating

$$H(r) = \{x \colon \operatorname{dist}(x, K \cap B) > r\} \text{ for } r \in \mathbf{R},$$

we see from the corollary in 4.2.1 that

$$\int_0^{*\infty} \mathbf{M}(\partial[R_i \llcorner H(r)])\, d\mathscr{L}^1 r \le \mathbf{M}(R_i) \le c,$$

because spt $\partial R_i \subset \mathbf{R}^n \sim H(0)$. For all positive integers i and j we can therefore choose $r_{i,j}$ and $R_{i,j}$ so that

$$0 < r_{i,j} < 1/j, \quad R_{i,j} = R_i \llcorner H(r_{i,j}), \quad \mathbf{M}(\partial R_{i,j}) \le j\,c,$$

hence $\mathbf{N}(R_{i,j}) \le (1+j)\,c$ and $R_{i,j} \in \mathbf{I}_{m,K}(\mathbf{R}^n)$, by 4.2.16(2). Applying 4.2.17(2) for each j and using Cantor's diagonal process, we replace the given sequence by a subsequence (again denoted $R_1, R_2, R_3, \ldots$) such that for every j there exists

$$G_j \in \mathbf{I}_{m,K}(\mathbf{R}^n) \quad \text{with} \quad \lim_{i \to \infty} \mathscr{F}_K(G_j - R_{i,j}) = 0;$$

since $R_i \llcorner H(r) = R_{i,j} \llcorner H(r)$ whenever $r > 1/j$, we also see from 4.2.1 (or 4.2.26) that

$$\int_{1/j}^{*\infty} \mathscr{F}_K[(G_j - R_i) \llcorner H(r)]\, d\mathscr{L}^1 r \le [(\operatorname{diam} K) + 1]\, \mathscr{F}_K(G_j - R_{i,j}).$$

For any two positive integers $j < k$ we infer, by adding the corresponding inequalities and letting i approach ∞, that

$$\int_{1/j}^{*\infty} \mathscr{F}_K[(G_j - G_k) \llcorner H(r)]\, d\mathscr{L}^1 r = 0,$$

hence $G_j \llcorner H(1/j) = G_k \llcorner H(1/j)$. Accordingly

$$\mathbf{M}[G_1 \llcorner H(1)] + \sum_{j=2}^{k} \mathbf{M}[G_j \llcorner H(1/j) - G_{j-1} \llcorner H(1/(j-1))] \\ + \mathbf{M}[G_k - G_k \llcorner H(1/k)] = \mathbf{M}(G_k) \le c$$

for every positive integer k, and we conclude that the sequence of currents $G_j \llcorner H(1/j)$ has the $\mathbf{M}$ limit

$$G = G_1 \llcorner H(1) + \sum_{j=2}^{\infty} [G_j \llcorner H(1/j) - G_{j-1} \llcorner H(1/(j-1))].$$

Since $\operatorname{spt} \partial[G_j \llcorner H(1/j)] \subset \{x: \operatorname{dist}(x, K \cap B) = 1/j\}$ we find that $G \in \Delta$. Moreover $\mathscr{F}_{K,B}(G - R_i)$ does not exceed

$$[\mathbf{M}(G) + \mathbf{M}(R_i)]/j + \int_{1/j}^{*\infty} \mathscr{F}_K[(G_j - R_i) \llcorner H(r)] \, d\mathscr{L}^1 r,$$

because $G \llcorner H(r) = G_j \llcorner H(r)$ for $r > 1/j$, hence

$$\lim_{i \to \infty} \mathscr{F}_{K,B}(G - R_i) = 0.$$

Assuming from now on that $\chi \in \mathbf{H}_m(A, B)$ and $R_i \in \chi$ for all i, we will show that $G \in \chi$. From 4.1.29 we obtain a closed local Lipschitz neighborhood retract D with $K \cap B \subset \operatorname{Int} D$ and $D \subset W$. Applying 4.4.2(1) with A, B replaced by U, D we secure a positive number ζ such that

$$\mathscr{R}_{m,K}(\mathbf{R}^n) \cap \{X: \operatorname{spt} \partial X \subset D, \mathbf{M}(X) \leq \zeta\} \subset \mathscr{B}_m(U, D).$$

We choose j so that $1/j < \operatorname{dist}(K \cap B, \mathbf{R}^n \sim D)$, hence

$$\operatorname{spt}(G - G_j) \subset D, \quad \operatorname{spt}(R_{i,j} - R_i) \subset D \text{ for all } i;$$

then we choose i so large that

$$\mathscr{F}_K(G_j - R_{i,j}) \leq \zeta, \quad \text{hence} \quad G_j - R_{i,j} \in \mathscr{B}_m(U, D),$$

and infer $G - R_i \in \mathscr{B}_m(U, D) \subset \mathscr{B}_m(U, W)$; using 4.4.2(3) we conclude $G \in \chi$.

4.4.5. One may use integral currents in place of integral flat chains to compute the integral homology groups of a pair (A, B) in the local Lipschitz category; reasoning as in 4.4.2 one readily obtains a natural isomorphism

$$\mathbf{H}_m(A, B) \simeq [\mathbf{I}_m(\mathbf{R}^n) \cap \mathscr{Z}_m(A, B)]/[\mathbf{I}_m(\mathbf{R}^n) \cap \mathscr{B}_m(A, B)].$$

For consistency with the conventions introduced in 4.4.6, we denote

$$\mathbf{H}_m(A, B; \mathbf{Z}) = \mathbf{H}_m(A, B).$$

4.4.6. Modifying 4.4.1 in the spirit of 4.2.26 we construct for each positive integer ν the **homology theory with coefficient group $\mathbf{Z}_\nu = \mathbf{Z}/\nu\,\mathbf{Z}$** on the local Lipschitz category by considering the groups

$$\mathscr{Z}_m^\nu(A, B) = \{T: T \in \mathscr{F}_m^\nu(\mathbf{R}^n), \operatorname{spt}^\nu T \subset A, \operatorname{spt}^\nu \partial T \subset B \text{ or } m = 0\},$$

$$\mathscr{B}_m^\nu(A, B) = \{T + \partial S: T \in \mathscr{F}_m^\nu(\mathbf{R}^n), \operatorname{spt}^\nu T \subset B, S \in \mathscr{F}_{m+1}^\nu(\mathbf{R}^n), \operatorname{spt}^\nu S \subset A\},$$

whose elements are called **flat cycles modulo ν** and **flat boundaries modulo ν**, and by defining

$$\mathbf{H}_m(A, B; \mathbf{Z}_\nu) = \mathscr{Z}_m^\nu(A, B)/\mathscr{B}_m^\nu(A, B).$$

All concepts and results of 4.4.1 – 4.4.5 extend immediately to this modification, with $\mathbf{Z}$ replaced by $\mathbf{Z}_\nu$ in 4.4.1(7).

We also construct the **homology theory with coefficient group R** on the local Lipschitz category by considering the groups

$$\mathbf{Z}_m(A,B)=\{T:\ T\in\mathbf{F}_m(\mathbf{R}^n),\ \operatorname{spt}T\subset A,\ \operatorname{spt}\partial T\subset B \text{ or } m=0\},$$

$$\mathbf{B}_m(A,B)=\{T+\partial S:\ T\in\mathbf{F}_m(\mathbf{R}^n),\ \operatorname{spt}T\subset B,\ S\in\mathbf{F}_{m+1}(\mathbf{R}^n),\ \operatorname{spt}S\subset A\},$$

whose elements are called **real flat cycles** and **real flat boundaries**, and by defining

$$\mathbf{H}_m(A,B;\mathbf{R})=\mathbf{Z}_m(A,B)/\mathbf{B}_m(A,B).$$

In this case the Eilenberg-Steenrod axioms hold with $\mathbf{Z}$ replaced by $\mathbf{R}$ in 4.4.1 (7). The analogue of 4.4.2 (3) is true, and 4.4.2 (4) is replaced by the proposition: *The vectorsubspace of* $\mathbf{H}_m(A,B;\mathbf{R})$, *consisting of those homology classes which meet* $\mathbf{F}_m(\mathbf{R}^n)\cap\{X:\ \operatorname{spt}X\subset K\}$, *has finite dimension.* However the analogues of 4.4.2 (1), (2), (5) are false (see 4.5.13). Modifying 4.4.5 one finds that normal currents may be used in place of flat chains to compute homology groups with real coefficients; moreover 4.1.18 shows that arbitrary currents with compact supports may be used in place of flat chains if A and B are open.

4.4.7. Using 4.2.3, 4.2.9, 4.2.26 and 4.4.2 (3) one readily sees that

$$\mathbf{H}_n(\mathbf{R}^n,\mathbf{R}^n\sim C;G)\simeq G,\quad \mathbf{H}_m(\mathbf{R}^n,\mathbf{R}^n\sim C;G)=\{0\}\ \text{for}\ m<n$$

whenever C is an open n dimensional cube in $\mathbf{R}^n$ and G equals either $\mathbf{Z}$ or $\mathbf{Z}_\nu$ or $\mathbf{R}$. [This will hardly surprise readers familiar with algebraic topology; in fact a classical theorem of Eilenberg and Steenrod implies that a homology theory is determined up to a natural equivalence by its coefficient group, on the subcategory of the Lipschitz category consisting of those pairs (A,B) for which A and B are compact.]

Next we prove: *If $T\in\mathscr{D}_n(\mathbf{R}^n)$ and ∂T is representable by integration, then T is representable by integration.* Letting C be any open n dimensional cube in $\mathbf{R}^n$ we choose

$$\alpha\in\mathscr{D}^0(\mathbf{R}^n)\quad\text{with}\quad \operatorname{Clos}C\subset\operatorname{Int}\{x:\ \alpha(x)=1\}$$

and observe that $\operatorname{spt}d\alpha\subset\mathbf{R}^n\sim C$,

$$\partial(T\llcorner\alpha)+T\llcorner d\alpha=(\partial T)\llcorner\alpha\in\mathbf{Z}_{n-1}(\mathbf{R}^n,\mathbf{R}^n\sim C);$$

since $\mathbf{H}_{n-1}(\mathbf{R}^n,\mathbf{R}^n\sim C;\mathbf{R})=\{0\}$ there exists

$$S\in\mathbf{F}_n(\mathbf{R}^n)\quad\text{with}\quad \operatorname{spt}[\partial S-\partial(T\llcorner\alpha)]\subset\mathbf{R}^n\sim C,$$

hence the constancy theorem in 4.1.7 yields

$$c\in\mathbf{R}\quad\text{with}\quad \operatorname{spt}(c\,\mathbf{E}^n-S-T\llcorner\alpha)\subset\mathbf{R}^n\sim C;$$

we conclude $\|T\|\,C\le|c|\,\mathscr{L}^n(C)+\mathbf{M}(S)<\infty$.

4.4.8. The homology groups of a compact Lipschitz neighborhood retract A in $\mathbf{R}^n$ are related to the homotopy groups of the abelian group spaces (with the $\mathscr{F}_A$ topology)

$$\tilde{\mathscr{Z}}_0(A)=\{T\colon\ T\in\mathscr{R}_0(\mathbf{R}^n),\ \operatorname{spt} T\subset A,\ T(1)=0\},$$

$$\tilde{\mathscr{Z}}_k(A)=\{T\colon\ T\in\mathbf{I}_k(\mathbf{R}^n),\ \operatorname{spt} T\subset A,\ \partial T=0\}\ \text{ for } k>0,$$

by natural isomorphisms

$$\mathbf{H}_{j+k}(A)\simeq\pi_j[\tilde{\mathscr{Z}}_k(A),0]$$

constructed in [AF 1]. In case $\alpha\in\mathbf{H}_{j+k}(A)$ and $T\in\alpha\cap\mathscr{P}_{j+k,A}(\mathbf{R}^n)$ one can choose $p\in\mathbf{O}^*(n,j)$ so that the map

$$\langle T,p,\cdot\rangle\colon\ \mathbf{R}^j\to\tilde{\mathscr{Z}}_k(A)\ \text{ is continuous},$$

with compact support, and belongs to the homotopy class corresponding to α. This isomorphism was used in [AF 3] for a partial extension of the theory of M. Morse to multidimensional variational problems.

4.4.9. For pairs in the local Lipschitz category, the cohomology groups with *real coefficients* can be obtained by use of flat cochains. Such generalizations of the classical theorem of de Rham may be found in [WH 4, IV 29, VII 12]. On the other hand the connection between cochains with *integral coefficients* and integral flat chains is not yet fully understood.

4.5. Normal currents of dimension n in $\mathbf{R}^n$

4.5.1. In the first part of this section, and again in 4.5.11, we study $\mathscr{L}^n$ measurable sets A satisfying the condition

$$\mathbf{E}^n\llcorner A\in\mathbf{I}_n^{\mathrm{loc}}(\mathbf{R}^n),$$

which are called **sets with locally finite perimeter,** because the Radon measure $\|\partial(\mathbf{E}^n\llcorner A)\|$ and the exterior normal defined in 4.5.5 embody the main geometric and analytic properties of the perimeter of such a set A. Using these concepts we prove an essentially optimal version of the Gauss-Green theorem in 4.5.6. Later in 4.5.11, 4.5.12 we will derive geometric criteria for the analytic condition that $\partial(\mathbf{E}^n\llcorner A)$ be representable by integration.

It is evident from 4.1.28 that the group $\mathscr{F}_n^{\mathrm{loc}}(\mathbf{R}^n)$ consists of the currents $\mathbf{E}^n\llcorner f$ corresponding to all *integer valued $\mathscr{L}^n$ measurable functions f such that*

$$\textstyle\int_K|f|\,d\mathscr{L}^n<\infty\ \text{ for every compact } K\subset\mathbf{R}^n.$$

Topologizing $\mathscr{F}_n^{\mathrm{loc}}(\mathbf{R}^n)$ according to 4.3.16 one readily verifies that, for any functions f and $f_1, f_2, f_3, \ldots$ of this kind,

$$\mathbf{E}^n \llcorner f_j \to \mathbf{E}^n \llcorner f \quad \text{in} \quad \mathscr{F}_n^{\mathrm{loc}}(\mathbf{R}^n) \quad \text{as} \quad j \to \infty$$

if and only if

$$\lim_{j\to\infty} \textstyle\int_K |f_j - f| \, d\mathscr{L}^n = 0 \text{ for every compact } K \subset \mathbf{R}^n.$$

We will use the induced topology on the subgroup $\mathbf{I}_n^{\mathrm{loc}}(\mathbf{R}^n)$ of $\mathscr{F}_n^{\mathrm{loc}}(\mathbf{R}^n)$.

4.5.2. Lemma. *If R is a connected, open, bounded subset of $\mathbf{R}^n$, $n > 1$ and* Bdry R *is a Lipschitz neighborhood retract, then there exists a finite positive number σ with the following properties:*

(1) *For each $T \in \mathbf{I}_n^{\mathrm{loc}}(\mathbf{R}^n)$ there exists an integer c such that*

$$\|c\,\mathbf{E}^n - T\|(R) \le [\sigma \, \|\partial T\|(R)]^{n/(n-1)}.$$

(2) *If P is an $\mathscr{L}^n$ measurable set and $\partial(\mathbf{E}^n \llcorner P)$ is representable by integration, then*

$$\inf\{\mathscr{L}^n(R \cap P),\ \mathscr{L}^n(R \sim P)\} \le [\sigma \, \|\partial(\mathbf{E}^n \llcorner P)\|(R)]^{n/(n-1)}.$$

(3) *If P and Q are $\mathscr{L}^n$ measurable sets, $\partial(\mathbf{E}^n \llcorner P)$ and $\partial(\mathbf{E}^n \llcorner Q)$ are representable by integration, and if*

$$\delta = \inf\{\mathbf{M}(X) + \mathbf{M}(Y)\colon X \in \mathscr{R}_{n-1}(\mathbf{R}^n) \text{ and } Y \in \mathscr{R}_n(\mathbf{R}^n) \text{ with}$$
$$\operatorname{spt}[\partial(\mathbf{E}^n \llcorner Q) - \partial(\mathbf{E}^n \llcorner P) - X - \partial Y] \subset \mathbf{R}^n \sim R\},$$

then

$$\inf\{\mathscr{L}^n(R \cap P) + \mathscr{L}^n(R \sim Q),\ \mathscr{L}^n(R \sim P) + \mathscr{L}^n(R \cap Q),$$
$$\|(\mathbf{E}^n \llcorner Q) - (\mathbf{E}^n \llcorner P)\|(R)\} \le (\sigma\,\delta)^{n/(n-1)} + \delta.$$

Proof. Choosing a neighborhood H of Bdry R in $\mathbf{R}^n$ and a Lipschitzian retraction h of H onto Bdry R, we construct a Lipschitzian retraction g of $H \cup (\mathbf{R}^n \sim R)$ onto $\mathbf{R}^n \sim R$ by defining

$$g(x) = x \text{ for } x \in \mathbf{R}^n \sim R, \qquad g(x) = h(x) \text{ for } x \in H \cap R;$$

in fact whenever $a \in \mathbf{R}^n \sim R$ and $b \in H \cap R$ the line segment from a to b meets Bdry R at some point z, hence

$$|g(a) - g(b)| = |a - z + h(z) - g(b)|$$
$$\le |a - z| + |z - b| \operatorname{Lip}(h) \le |a - b| \sup\{1, \operatorname{Lip}(h)\}.$$

We let σ be the number provided by 4.4.3 with $A = \mathbf{R}^n$, $B = \mathbf{R}^n \sim R$, $K = \operatorname{Clos} R$ and $m = n - 1$.

For each $T \in \mathbf{I}_n^{\text{loc}}(\mathbf{R}^n)$ we choose S according to 4.4.3, note that $\operatorname{spt} \partial S \subset \mathbf{R}^n \sim R$, then use the constancy theorem of 4.1.7 in conjunction with 4.1.28 to secure an integer c such that

$$\operatorname{spt}(c\,\mathbf{E}^n - S) \subset \mathbf{R}^n \sim R, \text{ hence}$$

$$\|c\,\mathbf{E}^n - T\|(R) = \|S - T\|(R) \le [\sigma\, \|\partial T\|(R)]^{n/(n-1)}.$$

To obtain (2) we take $T = \mathbf{E}^n \llcorner P$ and observe that

$$\|c\,\mathbf{E}^n - \mathbf{E}^n \llcorner P\|(R) = |c-1|\,\mathscr{L}^n(P \cap R) + |c|\,\mathscr{L}^n(R \sim P)$$

with $\sup\{|c-1|, |c|\} \ge 1$.

To verify (3) we suppose $\varepsilon > \delta$, choose

$$X \in \mathscr{R}_{n-1}(\mathbf{R}^n),\ Y \in \mathscr{R}_n(\mathbf{R}^n) \text{ with } \mathbf{M}(X) + \mathbf{M}(Y) < \varepsilon$$
$$\text{and } \operatorname{spt}[\partial(\mathbf{E}^n \llcorner Q) - \partial(\mathbf{E}^n \llcorner P) - X - \partial Y] \subset \mathbf{R}^n \sim R,$$

then use (1) with $T = \mathbf{E}^n \llcorner Q - \mathbf{E}^n \llcorner P - Y$, which implies $(\partial T) \llcorner R = X \llcorner R$, to infer

$$\begin{aligned}
&(\sigma\,\varepsilon)^{n/(n-1)} + \varepsilon > [\sigma\, \|(\partial T)\|(R)]^{n/(n-1)} + \mathbf{M}(Y) \\
&\ge \|c\,\mathbf{E}^n - T\|(R) + \mathbf{M}(Y) \ge \|c\,\mathbf{E}^n - \mathbf{E}^n \llcorner Q + \mathbf{E}^n \llcorner P\|(R) \\
&\qquad = \textstyle\int_R |c - q(x) + p(x)|\, d\mathscr{L}^n x
\end{aligned}$$

where p, q are the characteristic functions of P, Q; moreover the above integral equals

$$c\,\mathscr{L}^n(R) - \mathscr{L}^n(R \cap Q) + \mathscr{L}^n(R \cap P) \ge \mathscr{L}^n(R \sim Q) + \mathscr{L}^n(R \cap P) \text{ if } c \ge 1,$$

$$-c\,\mathscr{L}^n(R) - \mathscr{L}^n(R \cap P) + \mathscr{L}^n(R \cap Q) \ge \mathscr{L}^n(R \sim P) + \mathscr{L}^n(R \cap Q) \text{ if } c \le -1,$$

$$\|\mathbf{E}^n \llcorner Q - \mathbf{E}^n \llcorner P\|(R) \text{ if } c = 0.$$

4.5.3. Corollary. *For each integer $n > 1$ there exists a finite positive number σ such that*

$$\inf\{\mathscr{L}^n[\mathbf{U}(b,\rho) \cap P],\ \mathscr{L}^n[\mathbf{U}(b,\rho) \sim P]\} \le [\sigma\, \|\partial(\mathbf{E}^n \llcorner P)\|\, \mathbf{U}(b,\rho)]^{n/(n-1)}$$

whenever $b \in \mathbf{R}^n$, $0 < \rho < \infty$, P is an $\mathscr{L}^n$ measurable set and $\partial(\mathbf{E}^n \llcorner P)$ is representable by integration.

Proof. Using $\mu_{1/\rho} \circ \tau_{-b}$ we reduce the problem to the special case where $b = 0$ and $\rho = 1$. Then we apply 4.5.2(2) with $R = \mathbf{U}(0, 1)$.

4.5.4. Corollary. *Suppose n and σ are as in 4.5.3. If P is an $\mathscr{L}^n$ measurable set, $\mathscr{L}^n(P)<\infty$, $0<\tau<\frac{1}{2}$ and*

$$\Theta^{*n}(\mathscr{L}^n \llcorner P, b)>\tau \text{ whenever } b\in P,$$

then P can be covered by a sequence of balls $\mathbf{B}(b_i, r_i)$ with $b_i\in P$ and

$$\sum_{i=1}^{\infty}(r_i)^{n-1}\le\sigma[\tau\,\boldsymbol{\alpha}(n)]^{(1-n)/n}\,5^{n-1}\,\mathbf{M}[\partial(\mathbf{E}^n\llcorner P)].$$

Proof. Whenever $b\in P$ there exists $\rho>0$ such that

$$\mathscr{L}^n[\mathbf{B}(b,\rho)\cap P]/[\boldsymbol{\alpha}(n)\,\rho^n]=\tau,$$

because this ratio depends continuously on ρ, exceeds τ for some ρ near 0, and approaches 0 as ρ gets large; since $\tau<\frac{1}{2}$ it follows from 4.5.3 and the above equation that

$$[\tau\,\boldsymbol{\alpha}(n)\,\rho^n]^{(n-1)/n}\le\sigma\,\|\partial(\mathbf{E}^n\llcorner P)\|\,\mathbf{U}(b,\rho).$$

Applying 2.8.5 to the family F of all such balls $\mathbf{B}(b,\rho)$ we obtain a sequence of disjoint balls $\mathbf{B}(b_i,\rho_i)$ such that the enlarged balls $\mathbf{B}(b_i, 5\rho_i)$ cover P, and conclude

$$[\tau\,\boldsymbol{\alpha}(n)]^{(n-1)/n}\sum_{i=1}^{\infty}(5\rho_i)^{n-1}\le 5^{n-1}\,\sigma\,\mathbf{M}[\partial(\mathbf{E}^n\llcorner P)].$$

4.5.5. Assuming $A\subset\mathbf{R}^n$ and $b\in\mathbf{R}^n$ we say that u is an **exterior normal of A at b** if and only if $u\in\mathbf{S}^{n-1}$

$$\Theta^n[\mathscr{L}^n\llcorner\{x\colon (x-b)\bullet u>0\}\cap A, b]=0$$
$$\text{and } \Theta^n[\mathscr{L}^n\llcorner\{x\colon (x-b)\bullet u<0\}\sim A, b]=0.$$

We note that *if u and v are exterior normals of A at b, then $u=v$*, because otherwise $u-v$ would belong to the open cone

$$W=\mathbf{R}^n\cap\{w\colon w\bullet u>0 \text{ and } w\bullet v<0\},$$

hence $W\neq\emptyset$ and $\Theta^n[\mathscr{L}^n\llcorner\tau_b(W), b]=\mathscr{L}^n[\mathbf{U}(0,1)\cap W]>0$, whereas

$$\tau_b(W)\subset[\{x\colon (x-b)\bullet u>0\}\cap A]\cup[\{x\colon (x-b)\bullet v<0\}\sim A]$$

has $\mathscr{L}^n$ density 0 at b. We define

$$\mathbf{n}(A,b)=u \text{ if } u \text{ is an exterior normal of } A \text{ at } b,$$
$$\mathbf{n}(A,b)=0 \text{ if there exists no exterior normal of } A \text{ at } b.$$

Clearly $\mathbf{n}(\mathbf{R}^n\sim A, b)=-\mathbf{n}(A,b)$.

Recalling 4.3.16 and 4.5.5 we compute

$$\mathbf{M}\big([(\mu_{1/\rho}\circ\tau_{-b})_\#(\mathbf{E}^n \llcorner A)-\mathbf{E}^n\llcorner\{w:\ w\bullet u<0\}]\llcorner\mathbf{U}(0,\delta)\big)$$
$$=\rho^{-n}\mathscr{L}^n[\mathbf{U}(b,\rho\,\delta)\cap\{x:\ (x-b)\bullet u>0\}\cap A]$$
$$+\rho^{-n}\mathscr{L}^n[\mathbf{U}(b,\rho\,\delta)\cap\{x:\ (x-b)\bullet u<0\}\sim A]$$

for $u\in\mathbf{S}^{n-1}$, $0<\delta<\infty$, $0<\rho<\infty$ and infer that u *is an exterior normal of* A *at* b *if and only if* $u\in\mathbf{S}^{n-1}$ *and*

$$(\mu_{1/\rho}\circ\tau_{-b})_\#(\mathbf{E}^n\llcorner A)\to\mathbf{E}^n\llcorner\{w:\ w\bullet u<0\}\ \text{ as }\ \rho\to 0+.$$

We also see from 3.2.16 that u *is an exterior normal of* A *at* b *if and only if* $u\in\mathbf{S}^{n-1}$ *and*

$$\operatorname{Tan}^n(\mathscr{L}^n\llcorner A,b)=\mathbf{R}^n\cap\{w:\ w\bullet u\le 0\};$$

this condition implies $\operatorname{Nor}^n(\mathscr{L}^n\llcorner A,b)=\{t\,u:\ 0\le t<\infty\}$.

4.5.6. Gauss-Green theorem. *If* A *is an* $\mathscr{L}^n$ *measurable set,*

$$B=\mathbf{R}^n\cap\{b:\ \mathbf{n}(A,b)\in\mathbf{S}^{n-1}\}$$

and $T=\partial(\mathbf{E}^n\llcorner A)$ *is representable by integration, then:*

(1) $T\in\mathscr{R}^{\mathrm{loc}}_{n-1}(\mathbf{R}^n)$ *and* $\|T\|=\mathscr{H}^{n-1}\llcorner B$.

(2) *For* $\mathscr{H}^{n-1}$ *almost all* b *in* B,

$$\Theta^{n-1}(\|T\|,b)=1\quad\text{and}\quad *\mathbf{n}(A,b)=\vec{T}(b).$$

(3) *For* $\mathscr{H}^{n-1}$ *almost all* b *in* $\mathbf{R}^n\sim B$, $\Theta^{n-1}(\|T\|,b)=0$ *and*

$$\text{either }\ \Theta^n(\mathscr{L}^n\llcorner A,b)=0\ \text{ or }\ \Theta^n(\mathscr{L}^n\llcorner\mathbf{R}^n\sim A,b)=0.$$

(4) $T=\mathscr{H}^{n-1}\wedge *\mathbf{n}(A,\cdot)$.

(5) *For every Lipschitzian* 1*-vectorfield* ξ *on* $\mathbf{R}^n$ *with compact support,*

$$\int\xi(x)\bullet\mathbf{n}(A,x)\,d\mathscr{H}^{n-1}x=\int_A\operatorname{div}\xi(x)\,d\mathscr{L}^n x.$$

Proof. Assured by 4.2.27 that T is locally rectifiable we define E as in 4.3.17 with $m=n-1$ and recall 4.1.28. For $\mathscr{H}^{n-1}$ almost all b in E we find that $\Theta^{n-1}(\|T\|,b)$ is a positive integer and

$$\partial[\mathbf{E}^n\llcorner(\mu_r\circ\tau_{-b})\,A]=(\mu_r\circ\tau_{-b})_\#\,T\to C\ \text{ as }\ r\to\infty$$
$$\text{with }\ C=[\mathscr{H}^{n-1}\llcorner\operatorname{Tan}^{n-1}(\|T\|,b)]\wedge\zeta,$$
$$\zeta(x)=\Theta^{n-1}(\|T\|,b)\,\vec{T}(b)\ \text{ for }\ x\in\mathbf{R}^n,$$
$$\operatorname{Tan}^{n-1}(\|T\|,b)=\mathbf{R}^n\cap\{x:\ x\wedge\vec{T}(b)=0\};$$

letting $u=(-1)^{n-1}*\vec{T}(b)\in\mathbf{S}^{n-1}$, hence $*u=\vec{T}(b)$ by 1.7.8, and noting that

$$C=\partial[\Theta^{n-1}(\|T\|,b)\,\mathbf{E}^n\llcorner P]\ \text{ with }\ P=\{x:\ x\bullet u<0\},$$

we use 4.5.2 (3) to infer

$$\mathbf{E}^n \llcorner (\mu_r \circ \tau_{-b}) A \to \Theta^{n-1}(\|T\|, b)\, \mathbf{E}^n \llcorner P \text{ as } r \to \infty;$$

we also conclude, since $P \cap \mathbf{U}(0,1)$ is open, that

$$\begin{aligned}&\Theta^{n-1}(\|T\|, b)\, \|\mathbf{E}^n \llcorner P\|\, [P \cap \mathbf{U}(0,1)]\\ &\le \liminf_{r\to\infty} \|\mathbf{E}^n \llcorner (\mu_r \circ \tau_{-b}) A\|\, [P \cap \mathbf{U}(0,1)]\\ &\le \mathscr{L}^n [P \cap \mathbf{U}(0,1)] = \|\mathbf{E}^n \llcorner P\|\, [P \cap \mathbf{U}(0,1)],\end{aligned}$$

hence $\Theta^{n-1}(\|T\|, b) = 1$ and u is an exterior normal of A at b. Thus

$$\|T\| = \mathscr{H}^{n-1} \llcorner \Theta^{n-1}(\|T\|, \cdot) = \mathscr{H}^{n-1} \llcorner E$$

and $\mathscr{H}^{n-1}(E \sim B) = 0$.

Next we consider any point $b \in \mathbf{R}^n$ such that

$$\inf\{\Theta^{*n}(\mathscr{L}^n \llcorner A, b),\ \Theta^{*n}(\mathscr{L}^n \llcorner \mathbf{R}^n \sim A, b)\} > \delta$$

with $0 < \delta < \frac{1}{2}$; defining the continuous function ψ by the formula

$$\psi(\rho) = \frac{\mathscr{L}^n [\mathbf{U}(b,\rho) \cap A]}{[\alpha(n)\, \rho^n]} = 1 - \frac{\mathscr{L}^n [\mathbf{U}(b,\rho) \sim A]}{[\alpha(n)\, \rho^n]}$$

for $0 < \rho < \infty$, we see that

$$\begin{aligned}&\limsup_{\rho\to 0+} \psi(\rho) = \Theta^{*n}(\mathscr{L}^n \llcorner A, b) > \delta,\\ &\liminf_{\rho\to 0+}\ \psi(\rho) = 1 - \Theta^{*n}(\mathscr{L}^n \llcorner \mathbf{R}^n \sim A, b) < 1 - \delta,\end{aligned}$$

with $\delta < 1 - \delta$, hence there exist arbitrarily small positive numbers ρ satisfying the condition

$$\delta < \psi(\rho) < 1 - \delta,$$

and for every such ρ we infer from 4.5.3 that

$$\delta\, \alpha(n)\, \rho^n \le [\sigma\, \|T\|\, \mathbf{U}(b,\rho)]^{n/(n-1)};$$

accordingly $\delta\, \alpha(n) \le [\alpha(n-1)\, \sigma\, \Theta^{*n-1}(\|T\|, b)]^{n/(n-1)}$ and $b \in E$.

Now we conclude that $B \subset E$, the assertions (1), (2), (3) are true, and (4) holds because $T = \|T\| \wedge \vec{T} = (\mathscr{H}^{n-1} \llcorner B) \wedge *\mathbf{n}(A, \cdot)$.

We verify (5) first in case ξ if of class ∞, applying 4.1.33 with $S = \mathbf{E}^n \llcorner A$ to compute

$$\begin{aligned}&\int \xi \bullet n(A,\cdot)\, d\mathscr{H}^{n-1} = \|T\|\, [*\xi \bullet (-1)^{n-1} \vec{T}]\\ &= \|\mathbf{E}^n \llcorner A\|\, (\operatorname{div} \xi \bullet 1) = \int_A \operatorname{div} \xi\, d\mathscr{L}^n,\end{aligned}$$

then in the general case by smoothing ξ.

In case $n=1$ we modify the preceding argument, replacing reference to 4.5.3 by the following observation: If U is an open interval for which

$$\mathscr{L}^1(U\cap A)>0 \quad \text{and} \quad \mathscr{L}^1(U\sim A)>0,$$

then $U\cap \operatorname{spt} T \neq \varnothing$, hence $\|T\|(U)\geq 1$. Consequently A is $\mathscr{L}^1$ almost equal to the union of a locally finite family of intervals.

4.5.7. We see from 4.1.2, 7, 18 that $\mathbf{F}_n^{\mathrm{loc}}(\mathbf{R}^n)$ consists of the currents $\mathbf{E}^n \llcorner f$ corresponding to all real valued $\mathscr{L}^n$ measurable functions f satisfying the condition

$$\textstyle\int_K |f|\, d\mathscr{L}^n < \infty \text{ for every compact } K\subset \mathbf{R}^n,$$

and that $\mathbf{E}^n \llcorner f \in \mathbf{N}_n^{\mathrm{loc}}(\mathbf{R}^n)$ *if and only if there exists a sequence of functions* $f_j \in \mathscr{E}^0(\mathbf{R}^n)$ *with*

$$\lim_{j\to\infty} \textstyle\int_K |f_j - f|\, d\mathscr{L}^n = 0 \quad \textit{and} \quad \liminf_{j\to\infty} \textstyle\int_K \|D f_j\|\, d\mathscr{L}^n < \infty$$

for every compact $K\subset \mathbf{R}^n$; moreover

$$\partial(\mathbf{E}^n \llcorner f) = -\mathbf{E}^n \llcorner Df \text{ in case } f \text{ is locally Lipschitzian.}$$

Recalling 2.5.1 we find that

$$\{f\colon \mathbf{E}^n \llcorner f \in \mathbf{N}_n^{\mathrm{loc}}(\mathbf{R}^n),\ f\colon \mathbf{R}^n \to \mathbf{R}\} \text{ is a lattice of functions on } \mathbf{R}^n;$$

if f and g are locally Lipschitzian real valued functions on $\mathbf{R}^n$, so is $h=\inf\{f,g\}$, and approximate differentiation with the help of 3.1.6, 2.9.11 yields

$$Dh(x) = Df(x) \text{ for } \mathscr{L}^n \text{ almost all } x \text{ such that } f(x)\leq g(x),$$

$$Dh(x) = Dg(x) \text{ for } \mathscr{L}^n \text{ almost all } x \text{ such that } f(x)\geq g(x).$$

The structure of n dimensional locally normal currents in $\mathbf{R}^n$ will be analyzed in 4.5.9 through detailed study of certain sets with locally finite perimeter in $\mathbf{R}^{n+1}$ and $\mathbf{R}^n$. The allied general proposition 4.5.8 relates n dimensional locally normal currents in $\mathbf{R}^{n+1}$ and $\mathbf{R}^n$. For use in this work we fix the standard dual bases

$$e_1, \ldots, e_n \text{ and } X_1, \ldots, X_n \text{ of } \mathbf{R}^n \text{ and } \textstyle\bigwedge^1 \mathbf{R}^n,$$

$$\varepsilon_1, \ldots, \varepsilon_{n+1} \text{ and } Y_1, \ldots, Y_{n+1} \text{ of } \mathbf{R}^{n+1} \text{ and } \textstyle\bigwedge^1 \mathbf{R}^{n+1},$$

as well as the orthogonal projection

$$p\colon \mathbf{R}^{n+1} \to \mathbf{R}^n,\ p(y) = (y_1, \ldots, y_n) \text{ for } y\in \mathbf{R}^{n+1};$$

accordingly $p(\varepsilon_i)=e_i$ and $X_i \circ p = Y_i$ for $i=1,\ldots,n$.

4.5.8. Theorem. *If $S \in \mathbf{N}_n^{\text{loc}}(\mathbf{R}^{n+1})$ and the $p_\#$ images of the measures*

$$\|S \llcorner DY_1 \wedge \cdots \wedge DY_n \wedge Y_{n+1}\|, \quad \|S \llcorner DY_{n+1}\|, \quad \|(\partial S) \llcorner Y_{n+1}\|$$

are Radon measures over $\mathbf{R}^n$, then there exists a unique $T \in \mathbf{N}_n^{\text{loc}}(\mathbf{R}^n)$ such that

$$T(\phi) = S(Y_{n+1} \wedge p^\# \phi) \text{ for } \phi \in \mathscr{D}^n(\mathbf{R}^n);$$

furthermore $(\partial T)\psi = (\partial S)(Y_{n+1} \wedge p^\# \psi) - S(DY_{n+1} \wedge p^\# \psi)$ for $\psi \in \mathscr{D}^{n-1}(\mathbf{R}^n)$,

$$T = \mathbf{E}^n \llcorner \gamma \quad \text{with} \quad \gamma(x) = \langle S, p, x \rangle (Y_{n+1}) \text{ for } \mathscr{L}^n \text{ almost all } x.$$

Proof. In case $Y_{n+1}(\operatorname{spt} S)$ is bounded, hence $p | \operatorname{spt} S$ is proper, we take $T = p_\#(S \llcorner Y_{n+1})$ and infer from 4.1.7 that

$$\partial T = p_\# [(\partial S) \llcorner Y_{n+1} - S \llcorner DY_{n+1}].$$

To treat the general case we observe first that if μ is any Radon measure over $\mathbf{R}^{n+1}$ such that $p_\# \mu$ is a Radon measure, then

$$\mu[p^{-1}(K) \cap \{y : |y_{n+1}| \geq j\}] \to 0 \quad \text{as} \quad j \to \infty$$

for every compact $K \subset \mathbf{R}^n$, because $\mu[p^{-1}(K)] < \infty$. We choose $\alpha \in \mathscr{E}^0(\mathbf{R})$ with

$$\alpha(t) = 1 \text{ for } t \leq 0, \quad \alpha(t) = 0 \text{ for } t \geq 1,$$

$$1 > \alpha(t) > 0 \text{ and } 0 > \alpha'(t) > -2 \text{ for } 0 < t < 1,$$

and define $\beta_j \in \mathscr{D}^0(\mathbf{R})$ for $j = 1, 2, 3, \ldots$ so that

$$\beta_j(r) = \alpha[\log(|r|/j)] \text{ for } 0 \neq r \in \mathbf{R},$$

hence

$$\beta_j(r) = 1 \text{ for } |r| \leq j, \quad \beta_j(r) = 0 \text{ for } |r| \geq j\,e,$$

$$0 \leq \beta_j(r) \leq 1 \text{ and } |r\,\beta_j'(r)| \leq 2 \text{ for } r \in \mathbf{R}.$$

Letting $S_j = S \llcorner (\beta_j \circ Y_{n+1})$ we find that

$$\|(S_j - S) \llcorner DY_1 \wedge \cdots \wedge DY_n \wedge Y_{n+1}\|\, p^{-1}(K) \to 0$$

and

$$[\|(\partial S_j - \partial S) \llcorner Y_{n+1}\| + \|(S_j - S) \llcorner DY_{n+1}\|]\, p^{-1}(K) \to 0$$

as $j \to \infty$ for every compact $K \subset \mathbf{R}^n$, because

$$\partial S_j = (\partial S) \llcorner (\beta_j \circ Y_{n+1}) - S \llcorner (\beta_j' \circ Y_{n+1})\, DY_{n+1},$$

$$\|(\partial S_j - \partial S) \llcorner Y_{n+1}\| \leq [\|(\partial S) \llcorner Y_{n+1}\| + 2\|S \llcorner DY_{n+1}\|] \llcorner \{y : |y_{n+1}| \geq j\}.$$

Noting that $Y_{n+1}(\operatorname{spt} S_j)$ is bounded and taking $T_j = p_\#(S_j \llcorner Y_{n+1})$ we infer the existence of

$$T(\phi) = \lim_{j \to \infty} T_j(\phi) = (S \llcorner Y_{n+1})\, p^\# \phi$$

for $\phi \in \mathcal{D}^n(\mathbf{R}^n)$, and

$$(\partial T)\psi = \lim_{j\to\infty}(\partial T_j)\psi = [(\partial S) \llcorner Y_{n+1} - S \llcorner DY_{n+1}]\, p^\# \psi$$

for $\psi \in \mathcal{D}^{n-1}(\mathbf{R}^n)$; clearly T is a locally normal current. We also see from 4.3.2(1) that, for $\phi \in \mathcal{D}^n(\mathbf{R}^n)$,

$$T(\phi) = [S \llcorner p^\# \phi](Y_{n+1}) = \int \langle e_1, \ldots, e_n, \phi(x)\rangle \cdot \langle S, p, x\rangle (Y_{n+1})\, d\mathscr{L}^n x.$$

4.5.9. Theorem. *If f is a real valued $\mathscr{L}^n$ measurable function such that*

$$T = \mathbf{E}^n \llcorner f \in \mathbf{N}_n^{\mathrm{loc}}(\mathbf{R}^n)$$

and if

$$\lambda(x) = (\mathscr{L}^n) \operatorname{ap} \liminf_{z\to x} f(z) \in \overline{R} \;\; \text{for} \;\; x \in \mathbf{R}^n,$$

$$\mu(x) = (\mathscr{L}^n) \operatorname{ap} \limsup_{z\to x} f(z) \in \overline{R} \;\; \text{for} \;\; x \in \mathbf{R}^n,$$

$$F(x) = [\lambda(x) + \mu(x)]/2 \;\; \text{for} \;\; x \in \mathbf{R}^n,$$

$$G = \mathbf{R}^{n+1} \cap \{y : \mu(y_1, \ldots, y_n) \geq y_{n+1}\}, \quad S = (-1)^n\, \partial(\mathbf{E}^{n+1} \llcorner G),$$

$$C = \mathbf{R}^{n+1} \cap \{y : \lambda(y_1, \ldots, y_n) \leq y_{n+1} \leq \mu(y_1, \ldots, y_n)\},$$

$$E = \mathbf{R}^n \cap \{x : \lambda(x) < \mu(x)\},$$

then $\lambda, \mu, F, G, S, C, E$ are uniquely determined by T (because f is $\mathscr{L}^n$ almost determined by T) and the following thirty-one statements hold:

(1) *If $f: \mathbf{R}^n \to \mathbf{R}$ is locally Lipschitzian and*

$$g: \mathbf{R}^n \to \mathbf{R}^{n+1}, \quad g(x) = (x_1, \ldots, x_n, f(x)) \;\; \text{for} \;\; x \in \mathbf{R}^n,$$

then g is locally Lipschitzian and

$$S = g_\# \mathbf{E}^n, \quad \|S\| = g_\# [\mathscr{L}^n \llcorner (1 + |Df|^2)^{\frac{1}{2}}].$$

(2) *λ, μ, F are Borel functions.*

(3) *For $\mathscr{H}^{n-1}$ almost all x in $\mathbf{R}^n$, $-\infty < \lambda(x) \leq \mu(x) < \infty$.*

(4) *$S \in \mathscr{R}_n^{\mathrm{loc}}(\mathbf{R}^{n+1})$.*

(5) *$\|S\| = \mathscr{H}^n \llcorner C$.*

(6) *For $\phi \in \mathcal{D}^n(\mathbf{R}^n)$, $T(\phi) = S(Y_{n+1} \wedge p^\# \phi)$ and $\mathbf{E}^n(\phi) = S(p^\# \phi)$.*

(7) *For $\psi \in \mathcal{D}^{n-1}(\mathbf{R}^n)$, $(\partial T)\psi = -S(DY_{n+1} \wedge p^\# \psi)$.*

(8) *$\|T\| = p_\# \|S \llcorner DY_1 \wedge \cdots \wedge DY_n \wedge Y_{n+1}\|$.*

(9) *$\mathscr{L}^n = p_\# \|S \llcorner DY_1 \wedge \cdots \wedge DY_n\|$.*

(10) *$\|\partial T\| = p_\# \|S \llcorner DY_{n+1}\|$ and, for $i \in \{1, \ldots, n\}$,*

$$\|(\partial T) \llcorner DX_1 \wedge \cdots \wedge DX_{i-1} \wedge DX_{i+1} \wedge \cdots \wedge DX_n\|$$
$$= p_\# \|S \llcorner DY_1 \wedge \cdots \wedge DY_{i-1} \wedge DY_{i+1} \wedge \cdots \wedge DY_{n+1}\|.$$

(11) *$\mathscr{L}^n + \|\partial T\| \geq p_\# \|S\|$.*

(12) *For $\mathscr{L}^1$ almost all real numbers s,*

$$p_\# \langle S, Y_{n+1}, s\rangle = -\partial[\mathbf{E}^n \llcorner \{x\colon f(x)\geq s\}] \in \mathscr{R}^{\mathrm{loc}}_{n-1}(\mathbf{R}^n).$$

(13) $\partial T = \int \partial[\mathbf{E}^n \llcorner \{x\colon f(x)\geq s\}]\, d\mathscr{L}^1 s$ *and*

$$\|\partial T\| = \int \|\partial[\mathbf{E}^n \llcorner \{x\colon f(x)\geq s\}]\|\, d\mathscr{L}^1 s.$$

(14) *For every $\|\partial T\|$ integrable $\overline{R}$ valued function k,*

$$\|\partial T\|(k) = \iint_{\{x\colon \lambda(x)\leq s\leq \mu(x)\}} k\, d\mathscr{H}^{n-1}\, d\mathscr{L}^1 s.$$

(15) *For every Borel subset W of E,*

$$\|\partial T\|(W) = \mathscr{H}^n[C\cap p^{-1}(W)] = \int_W (\mu-\lambda)\, d\mathscr{H}^{n-1}.$$

(16) *E is countably $(\mathscr{H}^{n-1}, n-1)$ rectifiable.*

(17) *For $\mathscr{H}^{n-1}$ almost all b in E there exists $u\in\mathbf{S}^{n-1}$ such that*

$$\mathbf{n}[\{x\colon f(x)\geq s\}, b] = u \quad \text{and} \quad \mathbf{n}[G, (b_1,\ldots,b_n,s)] = (u_1,\ldots,u_n,0)$$

whenever $\lambda(b)<s<\mu(b)$.

(18) *If $n>1$, R and σ are as in 4.5.2, $t\in\mathbf{R}$,*

$$\mathscr{L}^n[R\cap\{x\colon f(x)>t\}]\leq \mathscr{L}^n(R)/2, \qquad \mathscr{L}^n[R\cap\{x\colon f(x)<t\}]\leq \mathscr{L}^n(R)/2,$$

and $\beta = n/(n-1)$, then

$$\left(\int_R |f(x)-t|^\beta\, d\mathscr{L}^n x\right)^{1/\beta} \leq \sigma\, \|\partial T\|(R).$$

(19) *If $n>1$, σ is as in 4.5.3, $b\in\mathbf{R}^n$, $0<\rho<\infty$, $t\in\mathbf{R}$,*

$$\mathscr{L}^n[\mathbf{U}(b,\rho)\cap\{x\colon f(x)>t\}]\leq \boldsymbol{\alpha}(n)\,\rho^n/2,$$
$$\mathscr{L}^n[\mathbf{U}(b,\rho)\cap\{x\colon f(x)<t\}]\leq \boldsymbol{\alpha}(n)\,\rho^n/2,$$

and $\beta = n/(n-1)$, then

$$\left(\rho^{-n}\int_{\mathbf{U}(b,\rho)} |f(x)-t|^\beta\, d\mathscr{L}^n x\right)^{1/\beta} \leq \sigma\, \rho^{1-n}\, \|\partial T\|\, \mathbf{U}(b,\rho).$$

(20) *If $\lambda(b)=\mu(b)\in\mathbf{R}$, $n>1$, $\beta=n/(n-1)$ and σ is as in 4.5.3, then*

$$\limsup_{\rho\to 0+}\left(\rho^{-n}\int_{\mathbf{U}(b,\rho)} |f(x)-F(b)|^\beta\, d\mathscr{L}^n x\right)^{1/\beta} \leq \sigma\,\boldsymbol{\alpha}(n-1)\,\Theta^{*n-1}(\|\partial T\|, b).$$

(21) *If $n>1$, then, for $\mathscr{H}^{n-1}$ almost all b in $\mathbf{R}^n\sim E$,*

$$\lim_{\rho\to 0+} \rho^{-n}\int_{\mathbf{U}(b,\rho)} |f(x)-F(b)|^{n/(n-1)}\, d\mathscr{L}^n x = 0.$$

(22) *If $n>1$, then for $\mathscr{H}^{n-1}$ almost all b in E the vector u characterized by* (17) *satisfies also the conditions*

$$\lim_{\rho\to 0+} \rho^{-n} \int_{Q^+(b,\rho)} |f(x)-\lambda(b)|^{n/(n-1)}\, d\mathscr{L}^n x = 0$$

with $Q^+(b,\rho)=\mathbf{U}(b,\rho)\cap\{x\colon (x-b)\bullet u>0\}$, *and*

$$\lim_{\rho\to 0+} \rho^{-n} \int_{Q^-(b,\rho)} |f(x)-\mu(b)|^{n/(n-1)}\, d\mathscr{L}^n x = 0$$

with $Q^-(b,\rho)=\mathbf{U}(b,\rho)\cap\{x\colon (x-b)\bullet u<0\}$.

(23) *If $n=1$, U is an open interval and*

$$r=\inf\{s\colon \mathscr{L}^1[U\cap\{x\colon f(x)<s\}]>0\},$$

$$t=\sup\{s\colon \mathscr{L}^1[U\cap\{x\colon f(x)>s\}]>0\},$$

then $t-r\le \|\partial T\|(U)$; *consequently*

$$\mathbf{V}_a^b F \le \|\partial T\|\{x\colon a\le x\le b\} \text{ for } -\infty<a<b<\infty,$$

$$\{\lambda(b),\mu(b)\}=\{\lim_{x\to b-} F(x),\ \lim_{x\to b+} F(x)\} \text{ for } b\in\mathbf{R}.$$

(24) *For $\mathscr{H}^{n-1}$ almost all b in* $\mathbf{R}^n$,

$$F(b)=\lim_{\varepsilon\to 0+}\int f(b+\varepsilon z)\,\psi(|z|)\, d\mathscr{L}^n z = \lim_{\varepsilon\to 0+}\int f(x)\,\varepsilon^{-n}\,\psi(\varepsilon^{-1}|b-x|)\, d\mathscr{L}^n x$$

whenever ψ is a real valued $\mathscr{L}^1$ measurable function with compact support such that

$$\mathscr{H}^{n-1}(\mathbf{S}^{n-1})\int_0^\infty r^{n-1}\,\psi(r)\, d\mathscr{L}^1 r=1 \quad\text{and}\quad \int_0^\infty r^{n-1}|\psi(r)|^n\, d\mathscr{L}^1 r<\infty;$$

furthermore $\langle T, \mathbf{1}_{\mathbf{R}^n}, b\rangle = F(b)\,\boldsymbol{\delta}_b$.

(25) *For $\|\partial T\|$ almost all b in* $\mathbf{R}^n\sim E$,

$$\overrightarrow{\partial T}(b)=*\mathbf{n}[\{x\colon f(x)\ge F(b)\}, b]=-(\wedge_{n-1} p)\,\eta/|\eta|$$

where $\eta=\vec{S}(b_1,\ldots,b_n,F(b))\llcorner Y_{n+1}$.

(26) *For $\mathscr{L}^n$ almost all b the function f has an $\mathscr{L}^n$ approximate differential L at b such that*

(I) *either* $\Theta^n(\|\partial T\|,b)=0$ *and* $L=0$,
or $L=-\mathbf{D}_{n-1}[\Theta^n(\|\partial T\|,b)\,\overrightarrow{\partial T}(b)]$;

(II) *in case $n>1$ and* $\beta=n/(n-1)$,

$$\lim_{\rho\to 0+}\rho^{-n}\int_{\mathbf{U}(b,\rho)}\left|\frac{f(x)-f(b)-L(x-b)}{|x-b|}\right|^\beta d\mathscr{L}^n x=0;$$

(III) *in case $n=1$, L is the differential of F at b;*

(IV) $\vec{S}(b_1,\ldots,b_n,f(b))=(-1)^n\,\mathbf{D}^1(M/|M|)$ *with*

$$M=Y_{n+1}-(L\circ p)\in\wedge^1\mathbf{R}^{n+1}.$$

(27) *If* $n>1$, $i\in\{1,2,\ldots,n\}$,

$$\Omega_i=(-1)^i DX_1\wedge\cdots\wedge DX_{i-1}\wedge DX_{i+1}\wedge\cdots\wedge DX_n,$$

$$q_i(x)=(x_1,\ldots,x_{i-1},x_{i+1},\ldots,x_n)\in\mathbf{R}^{n-1} \text{ for } x\in\mathbf{R}^n,$$

$$p_i(y)=(y_1,\ldots,y_{i-1},y_{i+1},\ldots,y_{n+1})\in\mathbf{R}^n \text{ for } y\in\mathbf{R}^{n+1},$$

$$\chi_{i,z}(t)=(z_1,\ldots,z_{i-1},t,z_i,\ldots,z_{n-1})\in\mathbf{R}^n \text{ for } z\in\mathbf{R}^{n-1},\ t\in\mathbf{R},$$

and if W is a Borel subset of $\mathbf{R}^n$, Z is a Borel subset of $\mathbf{R}^{n-1}$, $-\infty<\alpha<\beta<\infty$, $\gamma\in\mathscr{D}^0(\mathbf{R}^n)$, $\operatorname{spt}\gamma\subset\{x:\ \alpha<x_i<\beta\}$, then

$$\|(\partial T)\llcorner\Omega_i\|(W)=\int N[p_i|C\cap p^{-1}(W),v]\,d\mathscr{L}^n v,$$

$$\|(\partial T)\llcorner\Omega_i\|\{x:\ q_i(x)\in Z,\alpha<x_i<\beta\}=\int_Z \lim_{\delta\to 0+}\mathbf{V}_{\alpha+\delta}^{\beta-\delta}(F\circ\chi_{i,z})\,d\mathscr{L}^{n-1}z,$$

$$(\partial T)(\gamma\wedge\Omega_i)=\int\int_\alpha^\beta(\gamma\circ\chi_{i,z})\,d(F\circ\chi_{i,z})\,d\mathscr{L}^{n-1}z.$$

(28) *If $n=1$, W is a Borel subset of $\mathbf{R}$, $-\infty<\alpha<\beta<\infty$, $\gamma\in\mathscr{D}^0(\mathbf{R})$, $\operatorname{spt}\gamma\subset\{x:\ \alpha<x<\beta\}$, then*

$$\|\partial T\|(W)=\int N[Y_2|C\cap p^{-1}(W),v]\,d\mathscr{L}^1 v,$$

$$\|\partial T\|\{x:\ \alpha<x<\beta\}=\lim_{\delta\to 0+}\mathbf{V}_{\alpha+\delta}^{\beta-\delta}F,\quad (\partial T)(-\gamma)=\int_\alpha^\beta\gamma\,dF.$$

(29) *In case $n>1$ the following three conditions are equivalent:*

(I) *For $W\subset\mathbf{R}^n$, $\mathscr{H}^{n-1}(W)<\infty$ implies $\|\partial T\|(W)=0$.*

(II) *F is ($\mathscr{L}^n$) approximately continuous at $\mathscr{H}^{n-1}$ almost all points of $\mathbf{R}^n$.*

(III) *For $i=1,2,\ldots,n$ the functions $F\circ\chi_{i,z}$ corresponding to $\mathscr{L}^{n-1}$ almost all z in $\mathbf{R}^{n-1}$ are continuous on $\mathbf{R}$.*

In case $n=1$ the equivalence holds provided (III) *is replaced by the condition that F be a continuous function.*

(30) *In case $n>1$ the following six conditions are equivalent:*

(I) *For $W\subset\mathbf{R}^n$, $\mathscr{L}^n(W)=0$ implies $\|\partial T\|(W)=0$.*

(II) *For $W\subset\mathbf{R}^n$, $\mathscr{L}^n(W)=0$ implies $\mathscr{H}^n[C\cap p^{-1}(W)]=0$.*

(III) *$(\partial T)\llcorner\Omega_i=\mathscr{L}^n\llcorner D_i F$ for $i=1,2,\ldots,n$.*

(IV) *$\|(\partial T)\llcorner\Omega_i\|=\mathscr{L}^n\llcorner|D_i F|$ for $i=1,2,\ldots,n$.*

(V) *For $i=1,2,\ldots,n$ the functions $F\circ\chi_{i,z}$ corresponding to $\mathscr{L}^{n-1}$ almost all z in $\mathbf{R}^{n-1}$ are absolutely continuous on $\mathbf{R}$.*

(VI) *There exists a sequence of functions $f_j\in\mathscr{E}^0(\mathbf{R}^n)$ such that, for every compact $K\subset\mathbf{R}^n$,*

$$\lim_{j\to\infty}\int_K|f_j-f|\,d\mathscr{L}^n=0 \quad\text{and}\quad \lim_{(j,k)\to(\infty,\infty)}\int_K\|D f_j-D f_k\|\,d\mathscr{L}^n=0.$$

Furthermore (VI) *implies, for every compact* $K \subset \mathbf{R}^n$,

$$\lim_{j\to\infty} \int_K \|D f_j - \operatorname{ap} D f\| \, d\mathscr{L}^n = 0.$$

In case $n=1$ *the equivalence holds provided* (V) *is replaced by the condition that* F *be an absolutely continuous function, and* Ω_1 *is replaced by* -1.

(31) *If* $n>1$, $\beta = n/(n-1)$ *and* $\mathbf{M}(\partial T) < \infty$, *then there exists a unique* $c \in \mathbf{R}$ *such that*

$$\left[\int |f(x) - c|^\beta \, d\mathscr{L}^n x\right]^{1/\beta} \le n^{-1} \alpha(n)^{-1/n} \mathbf{M}(\partial T).$$

In case $\operatorname{spt} f$ *is compact, then* $c=0$.

In case f *is integer valued, then* c *is an integer and*

$$\mathbf{M}(T - c\,\mathbf{E}^n)^{1/\beta} \le n^{-1} \alpha(n)^{-1/n} \mathbf{M}(\partial T).$$

Proof of (1). In this case we define the locally Lipschitzian homeomorphism h of $\mathbf{R}^{n+1}$ onto $\mathbf{R}^{n+1}$ by the formula

$$h(y) = (y_1, \ldots, y_n, y_{n+1} + f(y_1, \ldots, y_n)) \text{ for } y \in \mathbf{R}^{n+1},$$

observe that $G = h\{y\colon y_{n+1} \le 0\}$, $g = h \circ p^*$, and use 4.1.26, 4.1.8 to compute

$$\begin{aligned} S &= (-1)^n \partial h_\#(\mathbf{E}^{n+1} \llcorner \{y\colon y_{n+1} \le 0\}) \\ &= h_\#[(-1)^n \partial(\mathbf{E}^{n+1} \llcorner \{y\colon y_{n+1} \le 0\})] = h_\# p^*_\# \mathbf{E}^n = g_\# \mathbf{E}^n, \end{aligned}$$

because $\partial[\mathbf{E}^n \times (\mathbf{E}^1 \llcorner \{t\colon t \le 0\})] = (-1)^n \mathbf{E}^n \times \delta_0$. From 4.1.25 and 3.2.3 we see that

$$\|S\| = \mathscr{H}^n \llcorner \operatorname{im} g = g_\#(\mathscr{L}^n \llcorner J_n g),$$

since g is univalent, and for $\mathscr{L}^n$ almost all x we obtain

$$D_i g(x) = \varepsilon_i + D_i f(x)\, \varepsilon_{n+1} \text{ for } i = 1, \ldots, n,$$

$$\begin{aligned} \langle e_1 \wedge \cdots \wedge e_n, \textstyle\bigwedge_n D g(x)\rangle &= D_1 g(x) \wedge \cdots \wedge D_n g(x) \\ &= \varepsilon_1 \wedge \cdots \wedge \varepsilon_n + \sum_{i=1}^n D_i f(x)\, \varepsilon_1 \wedge \cdots \wedge \varepsilon_{i-1} \wedge \varepsilon_{n+1} \wedge \varepsilon_{i+1} \wedge \cdots \wedge \varepsilon_n, \end{aligned}$$

$$[J_n g(x)]^2 = 1 + \sum_{i=1}^n [D_i f(x)]^2 = 1 + |D f(x)|^2.$$

Thus we have verified (1). However we proceed to amplify our discussion of the locally Lipschitzian case by some observations which will be useful in the sequel. Noting that $p \circ g = \mathbf{1}_{\mathbf{R}^n}$ we obtain

$$p_\# \|S\| = \mathscr{L}^n \llcorner (1 + \|D f\|^2)^{\frac{1}{2}} \le \mathscr{L}^n \llcorner (1 + \|D f\|) = \mathscr{L}^n + \|\partial T\|.$$

Since $Y_i \circ g = X_i$ for $i = 1, \ldots, n$, and $Y_{n+1} \circ g = f$, we see from 4.1.25 that, whenever $\gamma \in \mathscr{D}^0(\mathbf{R}^{n+1})$ and $\Psi \in \mathscr{D}^{n-1}(\mathbf{R}^{n+1})$,

$$\begin{aligned}
&[S \llcorner DY_1 \wedge \cdots \wedge DY_n \wedge Y_{n+1}](\gamma)\\
&= \mathbf{E}^n[DX_1 \wedge \cdots \wedge DX_n \wedge (Y_{n+1} \circ g) \wedge (\gamma \circ g)] = \mathscr{L}^n[f \cdot (\gamma \circ g)],\\
&\qquad [S \llcorner DY_1 \wedge \cdots \wedge DY_n](\gamma) = \mathscr{L}^n(\gamma \circ g),\\
&|\langle\langle e_1 \wedge \cdots \wedge e_n, \textstyle\bigwedge_n Dg(x)\rangle, Y_{n+1} \wedge \Psi[g(x)]\rangle|\\
&= \left|\sum_{i=1}^n (-1)^{i-1} D_i f(x) \langle \varepsilon_1 \wedge \cdots \wedge \varepsilon_{i-1} \wedge \varepsilon_{i+1} \wedge \cdots \wedge \varepsilon_n, \Psi[g(x)]\rangle\right|\\
&\le \|Df(x)\| \cdot \|\Psi[g(x)]\| \text{ for } \mathscr{L}^n \text{ almost all } x,\\
&\qquad [S \llcorner DY_{n+1}](\Psi) \le \mathscr{L}^n[\|Df\| \cdot (\|\Psi\| \circ g)],\\
&[S \llcorner DY_1 \wedge \cdots \wedge DY_{i-1} \wedge DY_{i+1} \wedge \cdots \wedge DY_{n+1}](\gamma)\\
&= \mathbf{E}^n[DX_1 \wedge \cdots \wedge DX_{i-1} \wedge DX_{i+1} \wedge \cdots \wedge DX_n \wedge Df \wedge (\gamma \circ g)]\\
&= (-1)^{n-1}[(\partial T) \llcorner DX_1 \wedge \cdots \wedge DX_{i-1} \wedge DX_{i+1} \wedge \cdots \wedge DX_n](\gamma \circ g);
\end{aligned}$$

it follows that

$$\begin{gathered}
\|S \llcorner DY_1 \wedge \cdots \wedge DY_n \wedge Y_{n+1}\| \le g_\#(\mathscr{L}^n \llcorner |f|) = g_\# \|T\|,\\
\|S \llcorner DY_1 \wedge \cdots \wedge DY_n\| \le g_\# \mathscr{L}^n, \quad \|S \llcorner DY_{n+1}\| \le g_\#(\mathscr{L}^n \llcorner \|Df\|) = g_\# \|\partial T\|,\\
\|S \llcorner DY_1 \wedge \cdots \wedge DY_{i-1} \wedge DY_{i+1} \wedge \cdots \wedge DY_{n+1}\|\\
\le g_\# \|(\partial T) \llcorner DX_1 \wedge \cdots \wedge DX_{i-1} \wedge DX_{i+1} \wedge \cdots \wedge DX_n\|,
\end{gathered}$$

hence

$$\begin{gathered}
p_\# \|S \llcorner DY_1 \wedge \cdots \wedge DY_n \wedge Y_{n+1}\| \le \|T\|,\\
p_\# \|S \llcorner DY_1 \wedge \cdots \wedge DY_n\| \le \mathscr{L}^n, \quad p_\# \|S \llcorner DY_{n+1}\| \le \|\partial T\|,\\
p_\# \|S \llcorner DY_1 \wedge \cdots \wedge DY_{i-1} \wedge DY_{i+1} \wedge \cdots \wedge DY_{n+1}\|\\
\le \|(\partial T) \llcorner DX_1 \wedge \cdots \wedge DX_{i-1} \wedge DX_{i+1} \wedge \cdots \wedge DX_n\|.
\end{gathered}$$

Finally we observe that, for every $k \in \mathscr{E}^0(\mathbf{R})$,

$$\begin{aligned}
p_\#[S \llcorner (k \circ Y_{n+1})] &= p_\#[(g_\# \mathbf{E}^n) \llcorner (k \circ f \circ p)]\\
&= (p_\# g_\# \mathbf{E}^n) \llcorner (k \circ f) = \mathbf{E}^n(k \circ f).
\end{aligned}$$

Proof of (2). Regarding λ we note that $\lambda(x) \ge t \in \mathbf{R}$ if and only if

$$\lim_{j \to \infty} j^n \mathscr{L}^n[\mathbf{B}(x, j^{-1}) \cap \{z \colon f(z) < t - i^{-1}\}] = 0$$

for every positive integer i.

Proof of (4), (6), (7), (8), (9), (10), (11). Recalling 4.1.2, 4.1.18 we regularize f and T by means of a function Φ such that

$$\Phi(-z)=\Phi(z) \text{ for } z\in\mathbf{R}^n;$$

since $\tau_{z\#}T=\tau_{z\#}(\mathbf{E}^n\llcorner f)=\mathbf{E}^n\llcorner\tau_{-z}^{\#}f$ we obtain

$$T_\varepsilon=\int\Phi_\varepsilon(-z)(\mathbf{E}^n\llcorner\tau_{-z}^{\#}f)\,d\mathscr{L}^n z=\mathbf{E}^n\llcorner\int\Phi_\varepsilon(-z)\,\tau_z^{\#}f\,d\mathscr{L}^n z=\mathbf{E}^n\llcorner(\Phi_\varepsilon*f)$$

whenever $0<\varepsilon<\infty$. Defining

$$G_\varepsilon=\mathbf{R}^{n+1}\cap\{y\colon(\Phi_\varepsilon*f)(y_1,\ldots,y_n)\geq y_{n+1}\},\qquad S_\varepsilon=(-1)^n\,\partial(\mathbf{E}^{n+1}\llcorner G_\varepsilon)$$

we see with the help of 2.9.13 that, for every compact subset K of $\mathbf{R}^n$,

$$\begin{aligned}&\mathbf{M}([\mathbf{E}^{n+1}\llcorner G_\varepsilon-\mathbf{E}^{n+1}\llcorner G]\llcorner p^{-1}(K))\\&=\mathscr{L}^{n+1}([(G_\varepsilon\sim G)\cup(G\sim G_\varepsilon)]\cap p^{-1}(K))\\&=\int_K|(\Phi_\varepsilon*f)-f|\,d\mathscr{L}^n=\|T_\varepsilon-T\|\,(K)\\&=\|H_\varepsilon(\partial T)\|\,(K)\leq\varepsilon\,\|\partial T\|\,\{x\colon\operatorname{dist}(x,K)\leq\varepsilon\}\to 0\end{aligned}$$

as $\varepsilon\to 0+$, hence

$$\begin{gathered}\mathbf{E}^{n+1}\llcorner G_\varepsilon\to\mathbf{E}^{n+1}\llcorner G \text{ in } \mathscr{F}_{n+1}^{\mathrm{loc}}(\mathbf{R}^{n+1}) \text{ as } \varepsilon\to 0+,\\ S_\varepsilon\to S \text{ in } \mathscr{F}_n^{\mathrm{loc}}(\mathbf{R}^{n+1}) \text{ as } \varepsilon\to 0+.\end{gathered}$$

Next we apply the observations following the proof of (1) with f, T replaced by $\Phi_\varepsilon*f$, T_ε to infer that

$$\begin{aligned}\|S\|\,(\gamma\circ p)&\leq\liminf_{\varepsilon\to 0+}\|S_\varepsilon\|\,(\gamma\circ p)\\&\leq\liminf_{\varepsilon\to 0+}[\mathscr{L}^n+\|\partial T_\varepsilon\|]\,(\gamma)=[\mathscr{L}^n+\|\partial T\|]\,(\gamma)\end{aligned}$$

whenever $\gamma\in\mathscr{K}(\mathbf{R}^n)^+$, hence S is representable by integration, (11) holds, and (4) follows from 4.2.27; similarly we obtain the inequalities

$$\begin{gathered}p_\#\,\|S\llcorner DY_1\wedge\cdots\wedge DY_n\wedge Y_{n+1}\|\leq\|T\|,\\ p_\#\,\|S\llcorner DY_1\wedge\cdots\wedge DY_n\|\leq\mathscr{L}^n,\qquad p_\#\,\|S\llcorner DY_{n+1}\|\leq\|\partial T\|,\\ p_\#\,\|S\llcorner DY_1\wedge\cdots\wedge DY_{i-1}\wedge DY_{i+1}\wedge\cdots\wedge DY_{n+1}\|\\ \leq\|(\partial T)\llcorner DX_1\wedge\cdots\wedge DX_{i-1}\wedge DX_{i+1}\wedge\cdots\wedge DX_n\|.\end{gathered}$$

We also define β_j for $j=1,2,3,\ldots$ as in the proof of 4.5.8 and infer that

$$\begin{aligned}&p_\#\,[S_\varepsilon\llcorner(\beta_j\circ Y_{n+1})\wedge Y_{n+1}]\\&=\mathbf{E}^n\llcorner[\beta_j\circ(\Phi_\varepsilon*f)]\wedge(\Phi_\varepsilon*f)=T_\varepsilon\llcorner[\beta_j\circ(\Phi_\varepsilon*f)]\end{aligned}$$

whenever $0<\varepsilon<\infty$, by taking $k(r)=\beta_j(r)\cdot r$ for $r\in\mathbf{R}$; since

$$(\beta_j\circ Y_{n+1})\wedge Y_{n+1}\wedge p^*\,\phi\in\mathscr{D}^n(\mathbf{R}^{n+1})\ \text{ for }\ \phi\in\mathscr{D}^n(\mathbf{R}^n),$$

we may let ε approach 0 to infer

$$p_\#\,[S\llcorner(\beta_j\circ Y_{n+1})\wedge Y_{n+1}]=T\llcorner[\beta_j\circ f];$$

then we let j approach ∞ to obtain the first conclusion of (6). Similarly we verify the second conclusion of (6) by taking $k=\beta_j$. Having thus proved (6), we deduce (7) from 4.5.8. Since (6) and (7) clearly imply

$$\|T\|\le p_\#\,\|S\llcorner DY_1\wedge\cdots\wedge DY_n\wedge Y_{n+1}\|,$$

$$\mathscr{L}^n=\|E^n\|\le p_\#\,\|S\llcorner DY_1\wedge\cdots\wedge DY_n\|,\qquad \|\partial T\|\le p_\#\,\|S\llcorner DY_{n+1}\|,$$

$$\|(\partial T)\llcorner DX_1\wedge\cdots\wedge DX_{i-1}\wedge DX_{i+1}\wedge\cdots\wedge DX_n\|$$
$$\le p_\#\,\|S\llcorner DY_1\wedge\cdots\wedge DY_{i-1}\wedge DY_{i+1}\wedge\cdots\wedge DY_{n+1}\|,$$

and we have previously obtained the opposite inequalities, we now deduce (8), (9), (10).

Proof of (12) **and** (13). Inasmuch as

$$[(-1^n)\,\varepsilon_1\wedge\cdots\wedge\varepsilon_{n+1}]\llcorner Y_{n+1}=\varepsilon_1\wedge\cdots\wedge\varepsilon_n$$

we see from 4.3.8 that, for $\mathscr{L}^1$ almost all real numbers s,

$$\langle(-1)^n\,\mathbf{E}^{n+1}\llcorner G,\,Y_{n+1},s\rangle=(\mathscr{H}^n\llcorner G\cap\{y\colon\ y_{n+1}=s\})\wedge\varepsilon_1\wedge\cdots\wedge\varepsilon_n$$
$$=\Delta_{s\#}\,[(\mathscr{L}^n\llcorner\{x\colon\ \mu(x)\ge s\})\wedge\mathbf{e}_1\wedge\cdots\wedge\mathbf{e}_n]=\Delta_{s\#}\,[\mathbf{E}^n\llcorner\{x\colon\ f(x)\ge s\}]$$

where $\Delta_s\colon\mathbf{R}^n\to\mathbf{R}^{n+1}$, $\Delta_s(x)=(x_1,\ldots,x_n,s)$ for $x\in\mathbf{R}^n$; noting that $p\,|\operatorname{im}\Delta_s$ is proper and $p\circ\Delta_s=\mathbf{1}_{\mathbf{R}^n}$, we apply $p_\#\,\partial$ to the above equation and obtain (12) with the help of 4.3.1, 4.2.27.

For $\phi\in\mathscr{D}^{n-1}(\mathbf{R}^n)$ we use (7), (12) and 4.3.2(1) to compute

$$(\partial T)(\phi)=-(S\llcorner DY_{n+1})(p^*\,\phi)=-\textstyle\int\langle S,Y_{n+1},s\rangle\,(p^*\,\phi)\,d\mathscr{L}^1 s$$
$$=\textstyle\int(\partial\,[\mathbf{E}^n\llcorner\{x\colon\ f(x)\ge s\}])(\phi)\,d\mathscr{L}^1 s,$$

and for $\gamma\in\mathscr{D}^0(\mathbf{R}^n)$ we apply (10), (12) and 4.3.2(2), 4.3.1 to obtain

$$\|\partial T\|(\gamma)=\|S\llcorner DY_{n+1}\|(\gamma\circ p)=\mathbf{M}\,[(S\llcorner\gamma\circ p)\llcorner DY_{n+1}]$$
$$=\textstyle\int\mathbf{M}\,\langle S\llcorner\gamma\circ p,\,Y_{n+1},s\rangle\,d\mathscr{L}^1 s=\int\|\langle S,Y_{n+1},s\rangle\|\,(\gamma\circ p)\,d\mathscr{L}^1 s$$
$$=\textstyle\int\|\partial\,[\mathbf{E}^n\llcorner\{x\colon\ f(x)\ge s\}]\|\,(\gamma)\,d\mathscr{L}^1 s$$

because Δ_s is an isometric embedding.

Proof of (16) **and** (17). We define

$$Q_s=\{x\colon \mu(x)\ge s\},\qquad B_s=\{x\colon \mathbf{n}(Q_s,x)\in\mathbf{S}^{n-1}\},$$

$$Z_s=\{x\colon \Theta^{*n}(\mathscr{L}^n\llcorner Q_s,x)>0 \text{ and } \Theta^{*n}(\mathscr{L}^n\llcorner \mathbf{R}^n\sim Q_s,x)>0\}$$

for $s\in\mathbf{R}$ and apply (12) to choose a countable dense subset Ξ of $\mathbf{R}$ such that

$$\partial(\mathbf{E}^n\llcorner Q_r)\in\mathscr{R}^{\text{loc}}_{n-1}(\mathbf{R}^n) \text{ whenever } r\in\Xi.$$

From 4.5.6, 4.1.28 we see that $\bigcup\{B_r\colon r\in\Xi\}$ is countably $(\mathscr{H}^{n-1},n-1)$ rectifiable and

$$\mathscr{H}^{n-1}(\bigcup\{Z_r\sim B_r\colon r\in\Xi\})=0.$$

Observing that

$$\{x\colon \lambda(x)<s<\mu(x)\}\subset Z_s \text{ for } s\in\mathbf{R},$$

hence $E\subset\bigcup\{Z_r\colon r\in\Xi\}$, $\mathscr{H}^{n-1}(E\sim\bigcup\{B_r\colon r\in\Xi\})=0$, we obtain (16).

Next we will show that the conclusions of (17) hold for each point

$$b\in E\sim\bigcup\{Z_r\sim B_r\colon r\in\Xi\}.$$

Suppose $r\in\Xi$, $t\in\Xi$ with $\lambda(b)<r<t<\mu(b)$; then $b\in Z_r\cap Z_t$,

$$u=\mathbf{n}(Q_r,b)\in\mathbf{S}^{n-1},\qquad v=\mathbf{n}(Q_t,b)\in\mathbf{S}^{n-1},$$

$$\Theta^n(\mathscr{L}^n\llcorner Q_r,b)=\tfrac{1}{2}=\Theta^n(\mathscr{L}^n\llcorner Q_t,b)$$

and $Q_t\subset Q_r$, $\Theta^n(\mathscr{L}^n\llcorner Q_r\sim Q_t,b)=0$, hence $u=v$. For all numbers s such that $r<s<t$ we infer

$$Q_t\subset Q_s\subset Q_r,\qquad \Theta^n(\mathscr{L}^n\llcorner Q_s\sim Q_t,b)=0,$$

hence $\mathbf{n}(Q_s,b)=u$; letting $a=(b_1,\ldots,b_n,s)$ we find also that

$$\begin{aligned}&\{y\colon (y-a)\bullet p^*(u)>0 \text{ and } y_{n+1}\ge r\}\cap G\\ &=\{y\colon [p(y)-b]\bullet u>0 \text{ and } \mu[p(y)]\ge y_{n+1}\ge r\}\\ &\subset p^{-1}[\{x\colon (x-b)\bullet u>0\}\cap Q_r],\\ &\Theta^{n+1}[\mathscr{L}^{n+1}\llcorner\{y\colon (y-a)\bullet p^*(u)>0\}\cap G,a]=0\end{aligned}$$

because $\Theta^n[\mathscr{L}^n\llcorner\{x\colon (x-b)\bullet u>0\}\cap Q_r,b]=0$ and $s>r$,

$$\begin{aligned}&\{y\colon (y-a)\bullet p^*(u)<0 \text{ and } y_{n+1}\le t\}\sim G\\ &=\{y\colon [p(y)-b]\bullet u<0 \text{ and } \mu[p(y)]<y_{n+1}\le t\}\\ &\subset p^{-1}[\{x\colon (x-b)\bullet u<0\}\sim Q_t],\\ &\Theta^{n+1}[\mathscr{L}^{n+1}\llcorner\{y\colon (y-a)\bullet p^*(u)<0\}\sim G,a]=0\end{aligned}$$

because $\Theta^n[\mathscr{L}^n\llcorner\{x\colon (x-b)\bullet u<0\}\sim Q_t,b]=0$ and $s<t$, whence we conclude $\mathbf{n}(G,a)=p^*(u)$.

Proof of (5). Noting that 4.5.6(1) implies

$$\|S\| = \mathscr{H}^n \llcorner B \text{ with } B = \mathbf{R}^{n+1} \cap \{y\colon \mathbf{n}(G, y) \in \mathbf{S}^n\},$$

we will verify (5) by showing that $B \subset C$ and $\mathscr{H}^n(C \sim B) = 0$.

If $a \in \mathbf{R}^{n+1} \sim C$, then

$$\text{either } a_{n+1} > \mu[p(a)] \text{ or } a_{n+1} < \lambda[p(a)];$$

in the first case we choose s with $a_{n+1} > s > \mu[p(a)]$, note that $\Theta^n[\mathscr{L}^n \llcorner \{x\colon f(x) \geq s\}, p(a)] = 0$ and

$$\mathbf{U}(a, \rho) \cap G \subset p^{-1}(\mathbf{U}[p(a), \rho] \cap \{x\colon \mu(x) \geq s\})$$

for $0 < \rho < a_{n+1} - s$, hence infer

$$\Theta^{n+1}(\mathscr{L}^{n+1} \llcorner G, a) = 0;$$

in the second case we similarly obtain

$$\Theta^{n+1}(\mathscr{L}^{n+1} \llcorner \mathbf{R}^{n+1} \sim G, a) = 0;$$

in both cases we conclude $a \notin B$. Thus $B \subset C$.

Next we let ϕ_δ be the size δ approximating measure occurring in the construction of $\mathscr{H}^n$ over $\mathbf{R}^{n+1}$, define

$$D = \mathbf{R}^{n+1} \cap \{y\colon y_{n+1} = \lambda[p(y)] \text{ or } y_{n+1} = \mu[p(y)]\},$$

$$L = \mathbf{R}^{n+1} \cap \{y\colon \Theta^{*n}(\phi_\infty \llcorner D, y) > 0\}, \quad M = \mathbf{R}^{n+1} \cap \{y\colon \Theta^{*n}(\|S\|, y) > 0\},$$

and let N be the subset of E consisting of those points b for which the conclusion of (17) fails. Noting that

$$(C \sim D) \sim p^{-1}(N) \subset B, \quad \mathscr{H}^{n-1}(N) = 0$$

and $p^{-1}(N)$ is isometric with $N \times \mathbf{R}$, we use 2.10.45 to infer

$$\mathscr{H}^n[p^{-1}(N)] = 0, \quad \mathscr{H}^n[(C \sim D) \sim B] = 0.$$

We also see from 2.10.19(2) and 4.5.6(3) that

$$\mathscr{H}^n(D \sim L) = 0 \quad \text{and} \quad \mathscr{H}^n(M \sim B) = 0.$$

Thus $\mathscr{H}^n(C \sim B) \leq \mathscr{H}^n(D \sim B) \leq \mathscr{H}^n(L \sim B) \leq \mathscr{H}^n(L \sim M)$, and we will complete the proof of (5) by showing that $L \subset M$.

Suppose $a \in L$ and $0 < \rho < \infty$. Assured by 4.2.1 that

$$\textstyle\int_\rho^{2\rho} \mathbf{M}(\partial[S \llcorner \mathbf{U}(a, r)])\, d\mathscr{L}^1 r \leq \|S\|\, \mathbf{U}(a, 2\rho)$$

because $\partial S = 0$, we choose a number r for which

$$\rho < r < 2\rho \quad \text{and} \quad \rho\, \mathbf{M}(\partial[S \llcorner \mathbf{U}(a, r)]) \leq \|S\|\, \mathbf{U}(a, 2\rho).$$

Since $\mathscr{H}^n(E)=0$ we infer from 4.1.30 that

$$p_\# [S \llcorner \mathbf{U}(a,r) \cap p^{-1}(E)] = 0.$$

We note that $S = S \llcorner C$ because $B \subset C$, that

$$C \cap \mathbf{U}(a,r) \sim p^{-1}(E) = C \cap p^{-1}(p[C \cap \mathbf{U}(a,r)] \sim E)$$

because $p|[C \sim p^{-1}(E)]$ is univalent, and that

$$p[C \cap \mathbf{U}(a,r)] \sim E \subset p[D \cap \mathbf{U}(a,r)] \sim N \subset p[C \cap \mathbf{U}(a,r)].$$

Using these facts and the second equation of (6) we obtain

$$\begin{aligned} p_\# [S \llcorner \mathbf{U}(a,r)] &= p_\# [S \llcorner p^{-1}(p[C \cap \mathbf{U}(a,r)] \sim E)] \\ &= \mathbf{E}^n \llcorner (p[C \cap \mathbf{U}(a,r)] \sim E) = \mathbf{E}^n \llcorner P \end{aligned}$$

with $P = p[D \cap \mathbf{U}(a,r)] \sim N$, hence

$$\rho\, \mathbf{M}[\partial(\mathbf{E}^n \llcorner P)] \le \|S\|\, \mathbf{U}(a, 2\rho).$$

Now we observe that

$$\Theta^n_*(\mathscr{L}^n \llcorner P, b) \ge \tfrac{1}{2} \text{ for } b \in P,$$

because if $y \in D \cap \mathbf{U}(a,r)$ with $p(y) = b \notin N$, and

$$W_\varepsilon = \{x\colon |f(x) - y_{n+1}| < \varepsilon\} \text{ whenever } \varepsilon > 0,$$

then either $b \notin E$, $y_{n+1} = \lambda(b) = \mu(b)$ and

$$\Theta^n(\mathscr{L}^n \llcorner W_\varepsilon, b) = 1 \text{ for every } \varepsilon > 0,$$

or $b \in E \sim N$, $\lambda(b) < \mu(b)$, $y_{n+1} \in \{\lambda(b), \mu(b)\}$ and the conclusion of (17) implies

$$\Theta^n(\mathscr{L}^n \llcorner W_\varepsilon, b) = \tfrac{1}{2} \text{ for } 0 < \varepsilon < \mu(b) - \lambda(b).$$

In case $n=1$ we find that $P \neq \emptyset$, because $a \in L$ and $N = \emptyset$, hence infer

$$0 \neq \mathbf{E}^1 \llcorner P \in \mathbf{I}_1(R), \quad \mathbf{M}[\partial(\mathbf{E}^1 \llcorner P)] \ge 1, \quad \|S\|\, \mathbf{U}(a, 2\rho) \ge \rho,$$

and we conclude $\Theta^1_*(\|S\|, a) \ge \frac{1}{2}$, $a \in M$.

In case $n > 1$ we choose σ according to 4.5.3 and let

$$c = \sigma [3^{-1} \alpha(n)]^{(1-n)/n}\, 5^{n-1}.$$

From 4.5.4 we infer that P can be covered by a sequence of balls $\mathbf{B}(b_i, r_i)$ with $b_i \in P$, $0 < r_i < 2\rho$ and

$$\sum_{i=1}^{\infty} (r_i)^{n-1} \le c\, \mathbf{M}[\partial(\mathbf{E}^n \llcorner P)].$$

Letting k_i be the integer such that

$$(k_i - 1) r_i < \rho \le k_i r_i, \text{ hence } k_i r_i < 3\rho,$$

we cover $p^{-1}[\mathbf{B}(b_i, r_i)] \cap \mathbf{U}(a, \rho)$ by k_i sets with diameter $2r_i 2^{\frac{1}{2}}$ and estimate

$$\phi_\infty[p^{-1}(P) \cap \mathbf{U}(a, \rho)] \le \sum_{i=1}^{\infty} k_i \alpha(n) (r_i 2^{\frac{1}{2}})^n$$
$$\le \sum_{i=1}^{\infty} 2^{2+n} \rho\, \alpha(n) (r_i)^{n-1} \le 2^{2+n} \alpha(n)\, c\, \|S\|\, \mathbf{U}(a, 2\rho).$$

Since $\phi_\infty[p^{-1}(N)] \le \mathscr{H}^n[p^{-1}(N)] = 0$ and

$$D \cap \mathbf{U}(a, \rho) \subset [p^{-1}(P) \cap \mathbf{U}(a, \rho)] \cup p^{-1}(N),$$

we obtain the inequality

$$\phi_\infty[D \cap \mathbf{U}(a, \rho)] \le 2^{2+n} \alpha(n)\, c\, \|S\|\, \mathbf{U}(a, 2\rho).$$

Consequently the hypothesis $a \in L$ implies that $a \in M$.

Proof of (14). Since $S \llcorner D Y_{n+1} = \|S\| \wedge (\vec{S} \llcorner D Y_{n+1})$ we see from (5) that

$$\|S \llcorner D Y_{n+1}\| = (\mathscr{H}^n \llcorner C) \wedge \|\vec{S} \llcorner D Y_{n+1}\|.$$

In view of (4), (5) and 4.3.8, $\operatorname{Tan}^n(\mathscr{H}^n \llcorner C, y)$ is the n dimensional vectorsubspace of $\mathbf{R}^{n+1}$ associated with $\vec{S}(y)$ and

$$\operatorname{ap} J_1(Y_{n+1} | C)(y) = \|\vec{S}(y) \llcorner Y_{n+1}\|$$

for $\mathscr{H}^n$ almost all y in C. Using (10) and 3.2.22 we compute

$$\|\partial T\|(k) = \|S \llcorner D Y_{n+1}\|(k \circ p) = \int_C (k \circ p) \cdot \operatorname{ap} J_1(Y_{n+1} | C)\, d\mathscr{H}^n$$
$$= \iint_{C \cap \{y:\, y_{n+1} = s\}} (k \circ p)\, d\mathscr{H}^{n-1}\, d\mathscr{L}^1 s = \iint_{\{x:\, \lambda(x) \le s \le \mu(x)\}} k\, d\mathscr{H}^{n-1}\, d\mathscr{L}^1 s$$

because, for each $s \in \mathbf{R}$, p maps $C \cap \{y:\, y_{n+1} = s\}$ isometrically onto $\{x:\, \lambda(x) \le s \le \mu(x)\}$.

Proof of (15). From (16) and 3.2.23 we see that

$$\mathscr{H}^n \llcorner (E \times \mathbf{R}) = (\mathscr{H}^{n-1} \llcorner E) \times \mathscr{L}^1;$$

since the standard isometry $\mathbf{R}^{n+1} \simeq \mathbf{R}^n \times \mathbf{R}$ maps $p^{-1}(E)$ onto $E \times \mathbf{R}$, we infer from 2.6.2 (3) that

$$\mathscr{H}^n(V) = \int \mathscr{H}^{n-1}\{x:\, (x_1, \ldots, x_n, s) \in V\}\, d\mathscr{L}^1 s$$
$$= \int \mathscr{L}^1\{s:\, (x_1, \ldots, x_n, s) \in V\}\, d\mathscr{H}^{n-1} x$$

for every Borel subset V of $p^{-1}(E)$. Taking

$$V = C \cap p^{-1}(W) = \{(x_1, \ldots, x_n, s):\, x \in W, \lambda(x) \le s \le \mu(x)\}$$

we see from (14) that the first of the two above integrals equals $\|\partial T\|(W)$, and we evaluate the integrand of the second integral as $\mu(x) - \lambda(x)$.

Proof of (3). It will suffice to discuss the special case when spt T is compact. We define

$$P_s = \{x: \lambda(x) > s\} \text{ for } s \in \mathbf{R},$$

infer from (13) that

$$\int \mathbf{M}[\partial(\mathbf{E}^n \llcorner P_s)]\, d\mathscr{L}^1 s = \mathbf{M}(\partial T) < \infty,$$

and note that

$$\Theta^n(\mathscr{L}^n \llcorner P_s, b) = 1 \text{ whenever } b \in P_s.$$

Taking $0 < \tau < \frac{1}{2}$, σ as in 4.5.3, $c = \sigma[\tau\,\boldsymbol{\alpha}(n)]^{(1-n)/n}\, 5^{n-1}$ we apply 4.5.4 to each set P_s and conclude

$$\begin{aligned}\mathscr{H}^{n-1}\{x: \lambda(x) = \infty\} &= \mathscr{H}^{n-1}[\bigcap\{P_s: s \in \mathbf{R}\}] \\ &\le \liminf_{s\to\infty} \boldsymbol{\alpha}(n-1)\, c\, \mathbf{M}[\partial(\mathbf{E}^n \llcorner P_s)] = 0.\end{aligned}$$

Replacing f by $-f$ we find that also

$$\mathscr{H}^{n-1}\{x: \mu(x) = -\infty\} = 0.$$

Finally we observe that (15) implies

$$\mathscr{H}^{n-1}\{x: \mu(x) - \lambda(x) = \infty\} = 0.$$

Proof of (18) **and** (19). Replacement of T and f by $T - t\,\mathbf{E}^n$ and $f - t$ reduces the problem to the special case when $t = 0$. To verify (18) in this case we define

$$V_s = \{x: f(x) > s\}, \quad f_s = \inf\{f, s\},$$

$$\psi(s) = \left(\int_{R \cap V_0} |f_s|^\beta\, d\mathscr{L}^n\right)^{1/\beta} \text{ for } 0 \le s < \infty,$$

and use Minkowski's inequality to estimate

$$\begin{aligned}\psi(s+h) - \psi(s) &\le \left(\int_R |f_{s+h} - f_s|^\beta\, d\mathscr{L}^n\right)^{1/\beta} \\ &\le \left(\int_{R \cap V_s} h^\beta\, d\mathscr{L}^n\right)^{1/\beta} = h[\mathscr{L}^n(R \cap V_s)]^{1/\beta}\end{aligned}$$

for $0 < h < \infty$. Thus $\operatorname{Lip}(\psi) \le [\mathscr{L}^n(R)]^{1/\beta} < \infty$, we see from (12) and 4.5.2(2) that

$$\psi'(s) \le [\mathscr{L}^n(R \cap V_s)]^{1/\beta} \le \sigma\, \|\partial(\mathbf{E}^n \llcorner V_s)\|(R)$$

for $\mathscr{L}^1$ almost all positive numbers s, because $V_s \subset V_0$,

$$2\mathscr{L}^n(R \cap V_s) \le \mathscr{L}^n(R), \qquad \mathscr{L}^n(R \cap V_s) \le \mathscr{L}^n(R \sim V_s),$$

and we conclude

$$\begin{aligned}\left(\int_{R \cap \{x: f(x) > 0\}} |f|^\beta\, d\mathscr{L}^n\right)^{1/\beta} &= \lim_{s\to\infty} \psi(s) = \int_0^\infty \psi'\, d\mathscr{L}^1 \\ &\le \sigma \int_0^\infty \|\partial[\mathbf{E}^n \llcorner \{x: f(x) > s\}]\|(R)\, d\mathscr{L}^1 s.\end{aligned}$$

Replacing f by $-f$ we find that also

$$\left(\int_{R\cap\{x:\,f(x)<0\}} |f|^\beta \, d\mathscr{L}^n\right)^{1/\beta} \le \sigma \int_{-\infty}^{0} \|\partial[\mathbf{E}^n \llcorner \{x:\, f(x)<s\}]\|(R)\, d\mathscr{L}^1 s.$$

Adding these two results, applying Minkowski's inequality, noting that

$$\begin{aligned}\mathbf{E}^n \llcorner \{x:\, f(x)>s\} &= \mathbf{E}^n \llcorner \{x:\, f(x)\ge s\}\\ &= \mathbf{E}^n - \mathbf{E}^n \llcorner \{x:\, f(x)<s\} \text{ for } \mathscr{L}^1 \text{ almost all } s,\end{aligned}$$

and applying (13) we obtain the conclusion of (18).

To prove (19) we simply take $R=\mathbf{U}(b,\rho)$.

Proof of (20). For $0<\rho<\infty$ we apply (19) with t replaced by

$$t_\rho = \inf\{s:\, \mathscr{L}^n[\mathbf{U}(b,\rho)\cap\{x:\, f(x)>s\}] \le \alpha(n)\,\rho^n/2\},$$

in conjunction with Minkowski's inequality, to infer

$$\begin{aligned}&\left(\rho^{-n}\int_{\mathbf{U}(b,\rho)} |f(x)-F(b)|^\beta \, d\mathscr{L}^n x\right)^{1/\beta}\\ &\le \sigma\,\rho^{1-n}\,\|\partial T\|\,\mathbf{U}(b,\rho) + \alpha(n)^{1/\beta}\,|t_\rho - F(b)|.\end{aligned}$$

Moreover $t_\rho \to F(b)$ as $\rho\to 0+$, because $\lambda(b)=\mu(b)\in\mathbf{R}$.

Proof of (21) **and** (22). In the special case when f is bounded it is obvious that the conclusion of (21) holds for every $b\in\mathbf{R}^n\sim E$, and that (17) implies (22).

To treat the general case we consider the sequence of bounded functions f_j such that

$$f_j(x)=f(x) \text{ if } |f(x)|\le j,\; f_j(x)=j \operatorname{sign} f(x) \text{ if } |f(x)|>j,$$

let $W_j=\{b:\, -j\le\lambda(b)\le\mu(b)\le j\}$, and observe that for every positive integer j the following three assertions are true, with $\beta=n/(n-1)$ and σ as in 4.5.3:

If $b\in W_j\sim E$, then

$$\lim_{\rho\to 0+} \rho^{-n}\int_{\mathbf{U}(b,\rho)} |f_j(x)-F(b)|^\beta \, d\mathscr{L}_n x = 0.$$

For $\mathscr{H}^{n-1}$ almost all b in $W_j\cap E$,

$$\lim_{\rho\to 0+} \rho^{-n}\int_{Q^+(b,\rho)} |f_j(x)-\lambda(b)|^\beta \, d\mathscr{L}^n x = 0$$

and

$$\lim_{\rho\to 0+} \rho^{-n}\int_{Q^-(b,\rho)} |f_j(x)-\mu(b)|^\beta \, d\mathscr{L}^n x = 0.$$

If $b \in W_j$, then

$$\limsup_{\rho \to 0+} [\rho^{-n} \int_{\mathbf{U}(b,\rho)} |f(x)-f_j(x)|^\beta \, d\mathscr{L}^n x]^{1/\beta}$$
$$\le \sigma\, \alpha(n-1)\, \Theta^{*n-1}(\|\partial[\mathbf{E}^n \llcorner (f-f_j)]\|, b).$$

The third assertion follows from (20) because $f-f_j$ has the $\mathscr{L}^n$ approximate limit 0 at each point of W_j.

Noting that $W_j \subset W_{j+1}$ and, by (3),

$$\mathscr{H}^{n-1}\left(\mathbf{R}^n \sim \bigcup_{j=1}^{\infty} W_j\right)=0,$$

we will complete the proof of (21) and (22) by showing that, for every $\varepsilon>0$,

$$\mathscr{H}^{n-1}\left(\mathbf{R}^n \sim \bigcap_{i=1}^{\infty} \bigcup_{j=i}^{\infty} Z_j\right)=0$$

with

$$Z_j=\{b\colon \Theta^{*n-1}(\|\partial[\mathbf{E}^n \llcorner (f-f_j)]\|, b) \le \varepsilon\}.$$

For each bounded open subset V of $\mathbf{R}^n$ we use 2.10.19(3) and the statement (13) of the present theorem to estimate

$$\begin{aligned}
\varepsilon\, \mathscr{H}^{n-1}(V \sim Z_j) &\le \|\partial[\mathbf{E}^n \llcorner (f-f_j)]\|(V)\\
&= \int_0^\infty \|\partial[\mathbf{E}^n \llcorner \{x\colon f(x)-f_j(x) \ge s\}]\|(V)\, d\mathscr{L}^1 s\\
&\quad + \int_{-\infty}^0 \|\partial[\mathbf{E}^n \llcorner \{x\colon f(x)-f_j(x) < s\}]\|(V)\, d\mathscr{L}^1 s\\
&= \int_0^\infty \|\partial[\mathbf{E}^n \llcorner \{x\colon f(x) \ge j+s\}]\|(V)\, d\mathscr{L}^1 s\\
&\quad + \int_{-\infty}^0 \|\partial[\mathbf{E}^n \llcorner \{x\colon f(x) < -j+s\}]\|(V)\, d\mathscr{L}^1 s\\
&= \int_{\{s\colon |s|>j\}} \|\partial[\mathbf{E}^n \llcorner \{x\colon f(x) \ge s\}]\|(V)\, d\mathscr{L}^1 s;
\end{aligned}$$

since $\|\partial T\|(V)<\infty$, the last integral approaches 0 as j tends to ∞, and we conclude

$$\mathscr{H}^{n-1}\left[\bigcup_{i=1}^{\infty} \bigcap_{j=i}^{\infty} (V \sim Z_j)\right]=0.$$

Proof of (23). Abbreviating $P_s=\{x\colon f(x) \ge s\}$ we note that, for $\mathscr{L}^1$ almost all numbers s between r and t,

$$\mathscr{L}^1(U \cap P_s)>0 \quad \text{and} \quad \mathscr{L}^1(U \sim P_s)>0,$$
$$U \cap \operatorname{spt} \partial(\mathbf{E}^1 \llcorner P_s) \neq \varnothing, \qquad \|\partial(\mathbf{E}^1 \llcorner P_s)\|(U) \ge 1$$

according to (12), and we apply (13) to conclude

$$\|\partial T\|(U)=\int \|\partial(\mathbf{E}^1 \llcorner P_s)\|(U)\, d\mathscr{L}^1 s \ge t-r.$$

Proof of (24). Abbreviating

$$\beta=n/(n-1), \quad \gamma=\mathscr{H}^{n-1}(\mathbf{S}^{n-1}), \quad \delta=\sup \operatorname{spt}\psi$$

we consider a point b for which either the conclusion of (21) or the conclusion of (22) holds.

In case $b\in\mathbf{R}^n\sim E$ we define

$$g(x)=f(x)-F(b) \text{ whenever } x\in\mathbf{R}^n,$$

and in case $b\in E$ we let

$$g(x)=f(x)-\lambda(b) \text{ if } (x-b)\bullet u>0,$$
$$g(x)=f(x)-\mu(b) \text{ if } (x-b)\bullet u<0;$$

in both cases it is true that

$$\lim_{\rho\to 0+}\rho^{-n}\int_{\mathbf{U}(b,\rho)}|g|^\beta\,d\mathscr{L}^n=0.$$

For $0<\varepsilon<\infty$ we use 3.2.13 to compute

$$\int[f(b+\varepsilon z)-g(b+\varepsilon z)]\,\psi(|z|)\,d\mathscr{L}^n z=\int_0^\infty r^{n-1}F(b)\,\gamma\,\psi(r)\,d\mathscr{L}^1 r=F(b),$$

and we apply Hölder's inequality to see that

$$\begin{aligned}&\int|g(b+\varepsilon z)\,\psi(|z|)|\,d\mathscr{L}^n z\\ &\le\Big[\int_{\mathbf{U}(0,\delta)}|g(b+\varepsilon z)|^\beta\,d\mathscr{L}^n z\Big]^{1/\beta}\cdot\Big[\int|\psi(|z|)|^n\,d\mathscr{L}^n z\Big]^{1/n}\\ &=\Big[\varepsilon^{-n}\int_{\mathbf{U}(b,\varepsilon\delta)}|g|^\beta\,d\mathscr{L}^n\Big]^{1/\beta}\cdot\Big[\int_0^\infty r^{n-1}|\psi(r)|^n\,\gamma\,d\mathscr{L}^1 r\Big]^{1/n}\end{aligned}$$

approaches 0 as ε tends to 0. Furthermore

$$\langle T,\mathbf{1}_{\mathbf{R}^n},b\rangle(\phi)=\lim_{\rho\to 0+}\int_{\mathbf{U}(b,\rho)}f\,\phi\,d\mathscr{L}^n/[\alpha(n)\,\rho^n]=F(b)\,\phi(b)$$

for $\phi\in\mathscr{D}^0(\mathbf{R}^n)$, because the continuity and boundedness of ϕ imply the equations

$$\lim_{\rho\to 0+}\int_{\mathbf{U}(b,\rho)}(f-g)\,\phi\,d\mathscr{L}^n/[\alpha(n)\,\rho^n]=F(b)\,\phi(b),$$
$$\lim_{\rho\to 0+}\int_{\mathbf{U}(b,\rho)}|g\,\phi|\,d\mathscr{L}^n/[\alpha(n)\,\rho^n]=0.$$

In case $n=1$ we use (23) in place of (21) and (22).

Proof of (25). We define $\eta(y)=\vec{S}(y) \llcorner Y_{n+1}$ for $\|S\|$ almost all y in $\mathbf{R}^{n+1}$. Whenever $\psi\in\mathscr{D}^{n-1}(\mathbf{R}^n)$ we apply (10) and (7), respectively, to obtain the formulae

$$(\partial T)\,\psi=\int\langle\overrightarrow{\partial T},\psi\rangle\,d\|\partial T\|=\int\langle(\overrightarrow{\partial T})\circ p,\psi\circ p\rangle\,d\|S\llcorner D\,Y_{n+1}\|$$
$$=\int\langle(\overrightarrow{\partial T})\circ p,\psi\circ p\rangle\,|\eta|\,d\|S\|,$$
$$(\partial T)\,\psi=-\int\langle\vec{S}(y),\,Y_{n+1}\wedge(\wedge^n p)\,\psi[p(y)]\rangle\,d\|S\|\,y$$
$$=-\int\langle(\wedge_n p)\,\eta(y),\psi[p(y)]\rangle\,d\|S\|\,y;$$

the resulting equation

$$\int\langle|\eta|\cdot[(\overrightarrow{\partial T})\circ p]+(\wedge_n p)\,\eta,\psi\circ p\rangle\,d\|S\|=0$$

remains valid for all bounded Baire forms ψ of degree $n-1$ with compact support on $\mathbf{R}^n$; using (5) and the fact that $p|[C\sim p^{-1}(E)]$ is univalent we infer

$$|\eta(y)|\,\overrightarrow{\partial T}[p(y)]=-(\wedge_n p)\,\eta(y)$$

for $\|S\|$ almost all y in $C\sim p^{-1}(E)$. Since the above formulae also imply

$$\|\partial T\|\,[p\{y:\ \eta(y)=0\}\sim E]=0,$$

it follows from (10) that

$$\overrightarrow{\partial T}(b)=-(\wedge_{n-1}p)\,\eta(b_1,\ldots,b_n,F(b))/|\eta(b_1,\ldots,b_n,F(b))|$$

for $\|\partial T\|$ almost all b in $\mathbf{R}^n\sim E$.

For $\mathscr{L}^1$ almost all s we see from (5) and 4.3.8 that

$$\langle S,Y_{n+1},s\rangle=(\mathscr{H}^{n-1}\llcorner C\cap\{y:\ y_{n+1}=s\})\wedge(\eta/|\eta|),$$

from (12) and the preceding paragraph that

$$(\partial[\mathbf{E}^n\llcorner\{x:\ f(x)\geq s\}])\llcorner(\mathbf{R}^n\sim E)=[\mathscr{H}^{n-1}\llcorner\{x:\ F(x)=s\}\sim E]\wedge\overrightarrow{\partial T},$$

hence from 4.5.6(2) that $*\mathbf{n}[\{x:\ f(x)\geq s\},b]=\overrightarrow{\partial T}(b)$ for $\mathscr{H}^{n-1}$ almost all b in $\{x:\ F(x)=s\}\sim E$. Letting k be the characteristic function of the subset of $\mathbf{R}^n\sim E$ consisting of those points b for which the first conclusion of (25) fails, we use (14) to infer $\|\partial T\|\,(k)=0$.

Proof of (26). Defining

$$Q=\{x: \Theta^n(\|\partial T\|, x)\in\mathbf{R}\}, \qquad Z=\{x: \Theta^n(\|\partial T\|, x)=0\}$$

we see from 2.9.15, 2.9.7, 2.9.10, 2.9.11 that

$$\mathscr{L}^n(\mathbf{R}^n\sim Q)=0, \qquad \|\partial T\|\llcorner Q=\mathscr{L}^n\llcorner\Theta^n(\|\partial T\|,\cdot),$$

$$\Theta^n(\|\partial T\|\llcorner\mathbf{R}^n\sim Q, x)=0 \text{ for } \mathscr{L}^n \text{ almost all } x \text{ in } Q,$$

$$\mathscr{L}^n[Q\sim(Z\cup\operatorname{dmn}\overrightarrow{\partial T})]=0 \quad\text{because}\quad \|\partial T\|(Q\sim\operatorname{dmn}\overrightarrow{\partial T})=0.$$

We let ξ be the $\mathscr{L}^n$ measurable $n-1$ vectorfield such that

$$\xi(x)=\Theta^n(\|\partial T\|,x)\,\overrightarrow{\partial T}(x) \text{ for } x\in Q\cap\operatorname{dmn}\overrightarrow{\partial T}, \qquad \xi(x)=0 \text{ for } x\in Z,$$

and infer $(\partial T)\llcorner Q=(\|\partial T\|\llcorner Q)\wedge\overrightarrow{\partial T}=\mathscr{L}^n\wedge\xi$. Moreover it follows from (4), (5) and 4.5.6 that the set

$$\Upsilon=C\cap\{z: *\mathbf{n}(G,z)\neq(-1)^n\,\vec{S}(z)\}$$

has $\mathscr{H}^n$ measure 0, hence $\mathscr{L}^n[p(\Upsilon)]=0$.

Recalling 2.9.8 we consider henceforth a point b in the $\mathscr{L}^n$ Lebesgue set of ξ, with $\lambda(b)=\mu(b)=f(b)\in\mathbf{R}$,

$$\Theta^n(\|\partial T-\mathscr{L}^n\wedge\xi\|,b)=\Theta^n(\|\partial T\|\llcorner\mathbf{R}^n\sim Q,b)=0$$

and $b\notin p(\Upsilon)$. We use 1.5.2 to define $L=-\mathbf{D}_{n-1}[\xi(b)]$ and

$$\zeta(x)=\xi(b)=-\mathbf{D}^1(L) \text{ for } x\in\mathbf{R}^n.$$

Since $DL(x)=L$ for $x\in\mathbf{R}^n$, we see from the commutative diagram in 4.1.7 that

$$\partial(\mathbf{E}^n\llcorner L)=-\mathscr{L}^n\wedge\mathbf{D}^1(dL)=\mathscr{L}^n\wedge\zeta,$$

and we infer

$$\Theta^n(\|\partial(T-\mathbf{E}^n\llcorner L)\|,b)=\Theta^n(\|\mathscr{L}^n\wedge(\xi-\zeta)\|,b)=0.$$

We also observe that $T-\mathbf{E}^n\llcorner L=\mathbf{E}^n\llcorner(f-L)$.

In case $n>1$ we apply (19) with T replaced by $\mathbf{E}^n\llcorner(f-L)$ to obtain the inequality

$$[\rho^{-n-\beta}\textstyle\int_{\mathbf{U}(b,\rho)}|f(x)-L(x)-t_\rho|^\beta\,d\mathscr{L}^n x]^{1/\beta}\le\sigma\rho^{-n}\,\|\partial(T-\mathbf{E}^n\llcorner L)\|\,\mathbf{U}(b,\rho)$$

for $0<\rho<\infty$, where t_ρ is the infimum of the set of all real numbers s satisfying the condition

$$\mathscr{L}^n[\mathbf{U}(b,\rho)\cap\{x: f(x)-L(x)>s\}]\le\alpha(n)\,\rho^n/2.$$

Inasmuch as

$$\lim_{\rho\to0+}t_\rho=(\mathscr{L}^n)\operatorname{ap}\lim_{x\to b}[f(x)-L(x)]=f(b)-L(b),$$

we see with the help of Minkowski's inequality that

$$\lim_{\rho\to 0+} \rho^{-n-\beta} \int_{\mathbf{U}(b,\rho)} g^\beta \, d\mathscr{L}^n = 0$$

with $g(x) = |f(x) - L(x) - f(b) + L(b)|$ for $x \in \mathbf{R}^n$, then use the estimate

$$\begin{aligned}
&\rho^{-n} \int_{\mathbf{U}(b,\rho)} [|x-b|^{-1} g(x)]^\beta \, d\mathscr{L}^n x \\
&\le \rho^{-n} \sum_{i=1}^\infty \int_{\mathbf{U}(b,2^{1-i}\rho) \sim \mathbf{U}(b,2^{-i}\rho)} (2^i \rho^{-1} g)^\beta \, d\mathscr{L}^n \\
&\le 2^{n+\beta} \sum_{i=1}^\infty 2^{-ni} (2^{1-i}\rho)^{-n-\beta} \int_{\mathbf{U}(b,2^{1-i}\rho)} g^\beta \, d\mathscr{L}^n
\end{aligned}$$

to deduce the conclusion of (II).

In case $n=1$ we infer from (23) that

$$\rho^{-1} \mathbf{V}_{b-\rho}^{b+\rho}(F-L) \le \rho^{-1} \|\partial(T - \mathbf{E}^1 \llcorner L)\| \, \mathbf{U}(b,\rho) \to 0$$

as $\rho \to 0+$, hence obtain the conclusion of (III).

In both cases it follows that L is an ($\mathscr{L}^n$) approximate differential of f at b.

To verify (IV) we abbreviate $a = (b_1, \ldots, b_n, f(b))$. Since

$$M(y-a) = y_{n+1} - f(b) - L[p(y) - b] \quad \text{for } y \in \mathbf{R}^{n+1},$$

we find that

$$\begin{aligned}
&\rho^{-n-1}(\mathscr{L}^{n+1}[\{y: p(y) \in \mathbf{U}(b,\rho), M(y-a) \ge 0\} \cap G] \\
&\qquad\qquad + \mathscr{L}^{n+1}[\{y: p(y) \in \mathbf{U}(b,\rho), M(y-a) \le 0\} \sim G]) \\
&= \rho^{-n-1} \int_{\mathbf{U}(b,\rho)} |f(x) - f(b) - L(x-b)| \, d\mathscr{L}^n x \\
&\le \rho^{-n-1} [\textstyle\int_{\mathbf{U}(b,\rho)} g^\beta \, d\mathscr{L}^n]^{1/\beta} [\alpha(n) \rho^n]^{1/n} \\
&= \alpha(n)^{1/n} [\rho^{-n-\beta} \textstyle\int_{\mathbf{U}(b,\rho)} g^\beta \, d\mathscr{L}^n]^{1/\beta} \to 0
\end{aligned}$$

as $\rho \to 0+$. Recalling 1.7.5, 8 we infer $\mathbf{n}(G,a) = \gamma_1(M/|M|)$ and deduce (IV) because $a \in C \sim \Upsilon$.

Proof of (27). Using (7), (10), (4), (5) we compute

$$\begin{aligned}
&\|(\partial T) \llcorner \Omega_i\| (W) = \|S \llcorner (DY_{n+1} \wedge p^\# \Omega_i)\| \, p^{-1}(W) \\
&= \int_{C \cap p^{-1}(W)} |\langle \vec{S}, DY_{n+1} \wedge p^\# \Omega_i \rangle| \, d\mathscr{H}^n = \int_{C \cap p^{-1}(W)} \operatorname{ap} J_n(p_i | C) \, d\mathscr{H}^n
\end{aligned}$$

and apply 3.2.20 to obtain the first conclusion of (27).

Next we observe that

$$[\partial(\mathbf{E}^n \llcorner f)](\gamma \wedge \Omega_i) = -\int f \cdot D_i \gamma \, d\mathscr{L}^n,$$

which equals $\int \gamma D_i f \, d\mathscr{L}^n$ in case $f \in \mathscr{E}^0(\mathbf{R}^n)$.

We regularize f and T by means of a function Φ such that

$$\Phi(x) = \psi(|x|) \text{ for } x \in \mathbf{R}^n,$$

where ψ is a nonnegative function of class ∞ satisfying the conditions of (24). Suppose $\zeta \in \mathscr{K}(\mathbf{R}^{n-1})^+$ and ξ is the characteristic function of $\{t: \alpha < t < \beta\}$. For each $\delta > 0$ we choose $\xi_\delta \in \mathscr{K}(\mathbf{R})^+$ so that

$$\xi_\delta \leq \xi, \; \xi_\delta(t) = 1 \text{ whenever } \alpha - \delta \leq t \leq \beta - \delta,$$

and use the fact that $\|(\partial T) \llcorner \Omega_i\|$ is the weak limit of the measures $\|[(\partial T) \llcorner \Omega_i]_\varepsilon\| = \|(\partial T_\varepsilon) \llcorner \Omega_i\|$ to obtain

$$\begin{aligned}
&\|(\partial T) \llcorner \Omega_i\| \, [(\zeta \circ q_i) \cdot (\xi_\delta \circ X_i)] = \lim_{\varepsilon \to 0+} \|(\partial T_\varepsilon) \llcorner \Omega_i\| \, [(\zeta \circ q_i) \cdot (\xi_\delta \circ X_i)] \\
&= \lim_{\varepsilon \to 0+} \int (\zeta \circ q_i) \cdot (\xi_\delta \circ X_i) \cdot |D_i(f * \Phi_\varepsilon)| \, d\mathscr{L}^n \\
&\geq \liminf_{\varepsilon \to 0+} \int \zeta(z) \, \mathbf{V}_{\alpha+\delta}^{\beta-\delta} [(f * \Phi_\varepsilon) \circ \chi_{i,z}] \, d\mathscr{L}^{n-1} z \\
&\geq \int \zeta(z) \liminf_{\varepsilon \to 0+} \mathbf{V}_{\alpha+\delta}^{\beta-\delta} [(f * \Phi_\varepsilon) \circ \chi_{i,z}] \, d\mathscr{L}^{n-1} z \\
&\geq \int^* \zeta(x) \, \mathbf{V}_{\alpha+\delta}^{\beta-\delta} (F \circ \chi_{i,z}) \, d\mathscr{L}^{n-1} z
\end{aligned}$$

because q_i maps the set of those points b in $\mathbf{R}^n$, for which the conclusion of (24) fails, onto a set with $\mathscr{L}^{n-1}$ measure 0. We infer the inequality

$$\int^* \zeta(z) \lim_{\delta \to 0+} \mathbf{V}_{\alpha+\delta}^{\beta-\delta} (F \circ \chi_{i,z}) \, d\mathscr{L}^{n-1} z \leq \|(\partial T) \llcorner \Omega_i\| \, [(\zeta \circ q_i) \cdot (\xi \circ X_i)] < \infty.$$

Applying 2.9.24 we find that

$$-\int_\alpha^\beta (F \circ \chi_{i,z}) \cdot (\gamma \circ \chi_{i,z})' \, d\mathscr{L}^1 t = \int_\alpha^\beta (\gamma \circ \chi_{i,z}) \, d(F \circ \chi_{i,z})$$

for $\mathscr{L}^{n-1}$ almost all z in $\mathbf{R}^{n-1}$, hence obtain the third conclusion of (27). From it we deduce the inequality

$$\|(\partial T) \llcorner \Omega_i\| \, [(\zeta \circ q_i) \cdot (\xi \circ X_i)] \leq \int_* \zeta(z) \lim_{\delta \to 0+} \mathbf{V}_{\alpha+\delta}^{\beta-\delta} (F \circ \chi_{i,z}) \, d\mathscr{L}^{n-1} z.$$

Combining these inequalities we obtain the second conclusion of (27).

The same argument, with obvious simplifications, yields a proof of (28).

Proof of (29). If (I) holds, then $\|\partial T\|(E)=0$ by (16), hence $\mathscr{H}^{n-1}(E)=0$ by (15), and (II) follows in view of (3).

If (II) holds, then $\mathscr{H}^{n-1}E)=0$; for each Borel subset W of $\mathbf{R}^n$ with $\mathscr{H}^{n-1}(W)<\infty$, (15) implies $\|\partial T\|(W\cap E)=0$, and (14) implies

$$\|\partial T\|(W\sim E)=\int \mathscr{H}^{n-1}[(W\sim E)\cap\{x\colon F(x)=s\}]\,d\mathscr{L}^1 s=0$$

because the integrand is positive for at most countably many real numbers s. Thus (I) follows from (II).

Whenever Z is a compact subset of $\mathbf{R}^{n-1}$, $-\infty<\alpha<\beta<\infty$ and

$$H=\mathbf{R}^n\cap\{x\colon q_i(x)\in Z,\ \alpha<x_i<\beta\},$$

we see from (16) and 2.10.11 that, for $\mathscr{L}^{n-1}$ almost all z in Z,

$$E\cap q_i^{-1}\{z\}\ \text{is countable},$$

hence, for all but countably many real numbers s,

$$\begin{aligned}N[F\circ\chi_{i,z}|\{t\colon\alpha<t<\beta\},s]&=\operatorname{card}\{x\colon x\in H\sim E,\ q_i(x)=z,\ F(x)=s\}\\&=N[p_i|C\cap p^{-1}(H\sim E),(z_1,\ldots,z_{n-1},s)];\end{aligned}$$

consequently (27) implies

$$\begin{aligned}&\|(\partial T)\llcorner\Omega_i\|(H\cap E)=\|(\partial T)\llcorner\Omega_i\|(H)-\|(\partial T)\llcorner\Omega_i\|(H\sim E)\\&=\int_Z\Big[\lim_{\delta\to 0+}\mathbf{V}_{\alpha+\delta}^{\beta-\delta}(F\circ\chi_{i,z})-\int N(F\circ\chi_{i,z}|\{t\colon\alpha<t<\beta\},s)\,d\mathscr{L}^1 s\Big]\,d\mathscr{L}^{n-1}z\end{aligned}$$

and we infer from 2.10.14 [or alternatively from (23) and (28)] that $\|(\partial T)\llcorner\Omega_i\|(H\cap E)=0$ if and only if the functions $F\circ\chi_{i,z}|\{t\colon\alpha<t<\beta\}$ corresponding to $\mathscr{L}^{n-1}$ almost all z in Z are continuous. Combining this result with the observations that

$$\|\partial T\|\le\sum_{i=1}^n\|(\partial T)\llcorner\Omega_i\|\le n\,\|\partial T\|,$$

$$\|\partial T\|(E)=0\ \text{if and only if}\ \mathscr{H}^{n-1}(E)=0,\ \text{by (15)},$$

we deduce the equivalence of (III) and (II).

Proof of (30). (I) and (II) are equivalent by (5), (10), (11).

If (I) holds, then $(\partial T)\llcorner\Omega_i=\mathscr{L}^n\wedge\langle\xi,\Omega_i\rangle$, where ξ is the $n-1$ vectorfield defined in the proof of (26), and

$$\xi(x)=-\mathbf{D}^1[\operatorname{ap}DF(x)]\ \text{for}\ \mathscr{L}^n\ \text{almost all}\ x;$$

on the other hand (27), 2.9.19 and 3.1.4 imply that the partial derivatives of F exist and

$$\operatorname{ap}DF(x)=\sum_{i=1}^n D_iF(x)X_i\ \text{for}\ \mathscr{L}^n\ \text{almost all}\ x;$$

we infer that, for $\mathscr{L}^n$ almost all x,

$$\xi(x)=\sum_{i=1}^{n}(-1)^i D_i F(x)\, X_1\wedge\cdots\wedge X_{i-1}\wedge X_{i+1}\wedge\cdots\wedge X_n,$$

hence $\langle\xi(x),\Omega_i(x)\rangle=D_i F(x)$. Thus (III) follows from (I).

Clearly (III) implies (IV).

If (IV) holds and $-\infty<\alpha\le\beta<\infty$, then (27) implies

$$\int_Z\int_\alpha^\beta|(F\circ\chi_{i,z})'|\,d\mathscr{L}^1\,d\mathscr{L}^{n-1}z=\int_Z\lim_{\delta\to 0+}\mathbf{V}_{\alpha+\delta}^{\beta-\delta}(F\circ\chi_{i,z})\,d\mathscr{L}^{n-1}z$$

for every Borel subset Z of $\mathbf{R}^{n-1}$, hence the integrands are equal for $\mathscr{L}^{n-1}$ almost all z, and (V) follows by virtue of 2.9.20.

If (V) holds, then regularization of F yields

$$D_i(\Phi_\varepsilon * F)=\Phi_\varepsilon * D_i F \text{ for } 0<\varepsilon<\infty \text{ and } i=1,\ldots,n,$$

because for each $x\in\mathbf{R}^n$ the functions $F\circ\chi_{i,q_i(x-w)}$ corresponding to $\mathscr{L}^n$ almost all w absolutely continuous, hence

$$(\Phi_\varepsilon * F)(x+h\,e_i)-(\Phi_\varepsilon * F)(x)=\int\Phi_\varepsilon(w)\,[F(x+h\,e_i-w)-F(x-w)]\,d\mathscr{L}^n w$$
$$=\int\Phi_\varepsilon(w)\int_0^h D_i F(x+t\,e_i-w)\,d\mathscr{L}^1 t\,d\mathscr{L}^n w=\int_0^h(\Phi_\varepsilon * D_i F)(x+t\,e_i)\,d\mathscr{L}^1 t$$

whenever $h\in\mathbf{R}$; we infer that, for each compact $K\subset\mathbf{R}^n$,

$$\int_K|D_i(\Phi_\varepsilon * F)(x)-D_i F(x)|\,d\mathscr{L}^n x$$
$$=\int_K\left|\int\Phi_\varepsilon(w)\,[D_i F(x-w)-D_i F(x)]\,d\mathscr{L}^n w\right|\,d\mathscr{L}^n x$$
$$\le\sup\{\int_K|D_i F(x-w)-D_i F(x)|\,d\mathscr{L}^n x\colon w\in\operatorname{spt}\Phi_\varepsilon\}$$

approaches 0 as ε tends to 0, by 2.7.18. Thus (VI) follows from (V).

If (VI) holds, then 2.4.12 yields real valued $\mathscr{L}^n$ measurable functions g_i such that

$$\lim_{j\to\infty}\int_K|D_i f_j-g_i|\,d\mathscr{L}^n$$

for every compact $K\subset\mathbf{R}^n$, hence

$$[\partial(\mathbf{E}^n\llcorner f_j)]\llcorner\Omega_i=\mathscr{L}^n\llcorner D_i f_j\to\mathscr{L}^n\llcorner g_i$$

in $\mathscr{D}_0(\mathbf{R}^n)$ as $j\to\infty$; since $\mathbf{E}^n\llcorner f_j\to T$ in $\mathscr{D}_n(\mathbf{R}^n)$ as $j\to\infty$, and the operations ∂ and $\llcorner\Omega_i$ are continuous, we infer

$$(\partial T)\llcorner\Omega_i=\mathscr{L}^n\llcorner g_i.$$

Thus (VI) implies (I) and, by (III), that

$$g_i(x)=D_i F(x) \text{ for } \mathscr{L}^n \text{ almost all } x.$$

Proof of (31). For $0<\rho<\infty$ we define t_ρ as in the proof of (20) and infer from (19) that

$$[\int_{\mathbf{U}(0,\rho)} |f(x)-t_\rho|^\beta \, d\mathscr{L}^n x]^{1/\beta} \le \sigma \mathbf{M}(\partial T).$$

Using Minkowski's inequality we obtain

$$[|t_\rho - t_r|^\beta \alpha(n) \rho^n]^{1/\beta} \le 2\sigma \mathbf{M}(\partial T)$$

whenever $0<\rho<r<\infty$, and conclude the existence of

$$c = \lim_{\rho\to\infty} t_\rho \in \mathbf{R} \quad \text{with} \quad [\int |f(x)-c|^\beta \, d\mathscr{L}^n x]^{1/\beta} \le \sigma \mathbf{M}(\partial T)$$

by 2.4.6. In case f is integer valued, then the numbers t_ρ and c are integers.

We will show that σ may be replaced by $\tau = n^{-1}\alpha(n)^{-1/n}$ in the last inequality. Replacing f by $f-c$ we assume henceforth that $f\in \mathbf{L}_\beta(\mathscr{L}^n)$.

First we consider the special case when f is the characteristic function of an $\mathscr{L}^n$ measurable set A with finite $\mathscr{L}^n$ measure. Assured by 4.2.1 that

$$\int \mathbf{M}(\partial[T \llcorner \mathbf{B}(0,r)] - (\partial T) \llcorner \mathbf{B}(0,r)) \, d\mathscr{L}^1 r \le \mathbf{M}(T) = \mathscr{L}^n(A) < \infty,$$

hence

$$\liminf_{r\to\infty} \mathbf{M}(\partial[T \llcorner \mathbf{B}(0,r)]) = \mathbf{M}(\partial T),$$

we may replace A by $A \cap U(0,r)$, and assume from now on that A is bounded. Whenever $0<\delta<1$ and $0<\varepsilon<\infty$ we see by regularization and use of 3.2.11 that

$$\begin{aligned}\int \mathscr{H}^{n-1}\{x\colon (\Phi_\varepsilon * f)(x)=s\} \, d\mathscr{L}^1 s &= \int \|D(\Phi_\varepsilon * f)\| \, d\mathscr{L}^n \\ &= M(\partial[\mathbf{E}^n \llcorner (\Phi_\varepsilon * f)]) \le \mathbf{M}(\partial T),\end{aligned}$$

apply 3.4.3 to choose a number s for which

$$0<s<\delta, \qquad \mathscr{H}^{n-1}\{x\colon (\Phi_\varepsilon * f)(x)=s\} \le \mathbf{M}(\partial T)/\delta$$

and $\{x\colon (\Phi_\varepsilon * f)(x)=s\} = \operatorname{Bdry}\{x\colon (\Phi_\varepsilon * f)(x)>s\}$ is a compact $n-1$ dimensional submanifold of class ∞ of $\mathbf{R}^n$, then infer from 3.2.43, 44 that

$$\begin{aligned}\mathscr{L}^n\{x\colon (\Phi_\varepsilon * f)(x)>\delta\}^{1/\beta} &\le \mathscr{L}^n\{x\colon (\Phi_\varepsilon * f)(x)>s\}^{1/\beta} \\ &\le \tau \mathscr{H}^{n-1}\{x\colon (\Phi_\varepsilon * f)(x)=s\} \le \tau \mathbf{M}(\partial T)/\delta.\end{aligned}$$

We let $\varepsilon\to 0+$ to conclude $\mathscr{L}^n(A)^{1/\beta} \le \tau \mathbf{M}(\partial T)/\delta$, and finally let $\delta\to 1-$.

Next we treat the case when $f \in \mathbf{L}_\beta(\mathscr{L}^n)$ by modifying the proof of (18). We define $V_s, f_s, \psi(s)$ as in (18), with R replaced by $\mathbf{R}^n$, note that

$$\mathscr{L}^n(V_s) \le s^{-\beta} \int |f|^\beta \, d\mathscr{L}^n < \infty \text{ for } 0 < s < \infty,$$

hence ψ is absolutely continuous, infer from the preceding paragraph that

$$\mathscr{L}^n(V_s)^\beta \le \tau \mathbf{M}[\partial(\mathbf{E}^n \llcorner V_s)] \text{ for } 0 < s < \infty,$$

and proceed as in (18) to conclude $(\mathscr{L}^n)_{(\beta)}(f) \le \tau \mathbf{M}(\partial T)$.

If f is integer valued, then $|f| \le |f|^\beta$, hence

$$\mathbf{M}(T) \le \int |f|^\beta \, d\mathscr{L}^n.$$

4.5.10. Whenever f is an $\mathscr{L}^1$ measurable function with values in a metric space and $-\infty < \alpha < \beta < \infty$ we define the **essential variation**

$$\operatorname{ess} \mathbf{V}_\alpha^\beta f$$

as the supremum of the set of numbers

$$\sum_{j=1}^{\nu} \operatorname{distance}[f(t_j), f(t_{j+1})]$$

corresponding to all finite sequences of points $t_1, t_2, \ldots, t_\nu, t_{\nu+1}$ of $\mathscr{L}^1$ approximate continuity of f with $\alpha < t_1 \le t_2 \le \cdots \le t_\nu \le t_{\nu+1} < \beta$.

We observe that, *in case f is a real valued $\mathscr{L}^1$ measurable function, $\mathbf{E}^1 \llcorner f \in \mathbf{N}_1^{\mathrm{loc}}(\mathbf{R})$ if and only if*

$$\int_\alpha^\beta |f| \, d\mathscr{L}^1 + \operatorname{ess} \mathbf{V}_\alpha^\beta f < \infty \textit{ whenever } -\infty < \alpha < \beta < \infty,$$

and that this condition implies

$$\|\partial(\mathbf{E}^1 \llcorner f)\| \{x : \alpha < x < \beta\} = \operatorname{ess} \mathbf{V}_\alpha^\beta f;$$

smoothing f by means of a function Φ with $\operatorname{spt} \Phi \subset \{x : |x| \le 1\}$ we find that

$$\sum_{j=1}^{\nu} |(\Phi_\varepsilon * f)(t_j) - (\Phi_\varepsilon * f)(t_{j+1})|$$

$$\le \int_{-\varepsilon}^{\varepsilon} \Phi_\varepsilon(w) \sum_{j=1}^{\nu} |f(t_j - w) - f(t_{j+1} - w)| \, d\mathscr{L}^1 w \le \operatorname{ess} \mathbf{V}_\alpha^\beta f$$

whenever $\alpha + \varepsilon = t_1 \le t_2 \le \cdots \le t_\nu \le t_{\nu+1} = \beta - \varepsilon$, because f is $\mathscr{L}^1$ approximately continuous at $t_j - w$ for $\mathscr{L}^1$ almost all w, hence that

$$\int_{\alpha+\varepsilon}^{\beta-\varepsilon} \|D(\Phi_\varepsilon * f)\| \, d\mathscr{L}^1 = \mathbf{V}_{\alpha+\varepsilon}^{\beta-\varepsilon}(\Phi_\varepsilon * f) \le \operatorname{ess} \mathbf{V}_\alpha^\beta f$$

whenever $0 < 2\varepsilon < \beta - \alpha$; from this fact and 4.5.7, 4.5.9 (28) our assertions follow.

Combining the preceding result with 4.5.9 (27) we infer that, *in case f is a real valued $\mathscr{L}^n$ measurable function with $n>1$, $\mathbf{E}^n \llcorner f \in \mathbf{N}_n^{\mathrm{loc}}(\mathbf{R}^n)$ if and only if f satisfies the conditions*

$$\int_K |f| \, d\mathscr{L}^n < \infty \text{ for every compact } K \subset \mathbf{R}^n,$$
$$\int_* z \operatorname{ess} \mathbf{V}_\alpha^\beta (f \circ \chi_{i,z}) \, d\mathscr{L}^{n-1} z < \infty$$

for every compact $Z \subset \mathbf{R}^{n-1}$, $-\infty < \alpha < \beta < \infty$, $i \in \{1, \ldots, n\}$; in fact

$$\begin{aligned}[\partial(\mathbf{E}^n \llcorner f)](\gamma \wedge \Omega_i) &= -\int f \cdot D_i \gamma \, d\mathscr{L}^n \\ &= -\int_Z (\mathbf{E}^1 \llcorner (f \circ \chi_{i,z})) [d(\gamma \circ \chi_{i,z})] \, d\mathscr{L}^{n-1} z \\ &\le \mathbf{M}(\gamma) \int_* z \operatorname{ess} \mathbf{V}_\alpha^\beta (f \circ \chi_{i,z}) \, d\mathscr{L}^{n-1} z\end{aligned}$$

whenever $\gamma \in \mathscr{D}^0(\mathbf{R}^n)$ with $\operatorname{spt} \gamma \subset \{x : q_i(x) \in Z, \alpha < x_i < \beta\}$.

4.5.11. Theorem. *If $A \subset \mathbf{R}^n$ and*

$$Q = \{x : \Theta^n[\mathscr{L}^n \llcorner (\mathbf{R}^n \sim A), x] = 0\}, \quad R = \{x : \Theta^n[\mathscr{L}^n \llcorner A, x] = 0\},$$

then the following two conditions are equivalent:

(I) *A is $\mathscr{L}^n$ measurable and $\partial(\mathbf{E}^n \llcorner A)$ is representable by integration.*

(II) *$\mathscr{I}_1^{n-1}[K \sim (Q \cup R)] < \infty$ for every compact $K \subset \mathbf{R}^n$.*

Proof. If (I) holds, then 4.5.6 implies

$$\mathscr{H}^{n-1}[(\mathbf{R}^n \sim B) \sim (Q \cup R)] = 0,$$

hence $\mathscr{H}^{n-1}[K \sim (Q \cup R)] \le \mathscr{H}^{n-1}(K \cap B) < \infty$ for every compact subset K of $\mathbf{R}^n$, and (II) follows.

If (II) holds, then $\mathscr{L}^n[\mathbf{R}^n \sim (Q \cup R)] = 0$ by 2.10.15; since Q and R are disjoint Borel sets, while 2.9.11 implies

$$\mathscr{L}^n(Q \sim A) = 0 \quad \text{and} \quad \mathscr{L}^n(R \cap A) = 0,$$

it follows that $\mathscr{L}^n[(\mathbf{R}^n \sim A) \sim R] = 0$ and $\mathscr{L}^n(A \sim Q) = 0$, hence A is $\mathscr{L}^n$ measurable. We define q_i and $\chi_{i,z}$ as in 4.5.9 (27), use 2.10.15 to obtain $g \in \mathbf{O}(n)$ with $\det(g) = 1$ and

$$\int N(q_i \circ g \,|\, [\mathbf{B}(0,j) \sim (Q \cup R)], z) \, d\mathscr{L}^{n-1} z < \infty$$

for $i = 1, \ldots, n$ and all positive integers j. Noting that

$$\mathbf{E}^n \llcorner A = \mathbf{E}^n \llcorner Q = g_\#^{-1}[\mathbf{E}^n \llcorner g(Q)],$$

we will complete the proof of (I) by applying 4.5.10 to the characteristic function f of $g(Q)$. For this purpose we consider a fixed $i \in \{1, \ldots, n\}$ and let

$$G(k) = \mathbf{R}^n \cap \{x \colon \mathscr{L}^n[g(R) \cap \mathbf{B}(x,\rho)] \le 3^{-n-1} \alpha(n-1) \rho^n \text{ for } 0<\rho<3/k\},$$
$$H(k) = \mathbf{R}^n \cap \{x \colon \mathscr{L}^n[g(Q) \cap \mathbf{B}(x,\rho)] \le 3^{-n-1} \alpha(n-1) \rho^n \text{ for } 0<\rho<3/k\},$$
$$G^+(k,m) = G(k) \cap \{x \colon x + s\, e_i \in g(R) \text{ for } 0<s<3/m\},$$
$$G^-(k,m) = G(k) \cap \{x \colon x - s\, e_i \in g(R) \text{ for } 0<s<3/m\},$$
$$H^+(k,m) = H(k) \cap \{x \colon x + s\, e_i \in g(Q) \text{ for } 0<s<3/m\},$$
$$H^-(k,m) = H(k) \cap \{x \colon x - s\, e_i \in g(Q) \text{ for } 0<s<3/m\}$$

whenever k and m are positive integers. First we prove that

$$\mathscr{L}^{n-1}(q_i[G^+(k,m)]) = 0,$$

by expressing $G^+(k,m)$ as union of the sets

$$W_\nu = G^+(k,m) \cap \{x \colon (\nu-1)/m < x_i \le \nu/m\}$$

corresponding to all integers ν, and showing that

$$\Theta^{n-1}[\mathscr{L}^{n-1} \llcorner q_i(W_\nu), z] \le \tfrac{1}{3} \text{ for } z \in \mathbf{R}^{n-1};$$

if $0<r<\inf\{1/k, 1/m\}$, $\mathbf{B}(z,r) \cap q_i(W_\nu) \neq \varnothing$ and $\varepsilon>0$, then there exists $b \in W_\nu$ with $q_i(b) \in \mathbf{B}(z,r)$ and

$$b_i + \varepsilon r > \sup\{x_i \colon x \in W_\nu,\ q_i(x) \in \mathbf{B}(z,r)\},$$

hence the set

$$\{y \colon q_i(y) \in q_i(W_\nu) \cap \mathbf{B}(z,r),\ b_i + \varepsilon r \le y_i \le b_i + r\}$$

is contained in $g(R) \cap \mathbf{B}(b, 3r)$ and

$$\mathscr{L}^{n-1}[q_i(W_\nu) \cap \mathbf{B}(z,r)] \cdot (r - \varepsilon r) \le 3^{-n-1} \alpha(n-1)(3r)^n,$$
$$\mathscr{L}^{n-1}[q_i(W_\nu) \cap \mathbf{B}(z,r)] / [\alpha(n-1)\, r^{n-1}] \le 1/(3-3\varepsilon).$$

Similarly we see that

$$\mathscr{L}^{n-1}(q_i[G^-(k,m) \cup H^+(k,m) \cup H^-(k,m)]) = 0.$$

Next we prove that if

$$z \in \mathbf{R}^{n-1} \sim \bigcup_{k=1}^{\infty} \bigcup_{m=1}^{\infty} q_i[G^+(k,m) \cup G^-(k,m) \cup H^+(k,m) \cup H^-(k,m)]$$

and $-\infty<\alpha<\beta<\infty$, then

$$\begin{aligned}\operatorname{ess} \mathbf{V}_\alpha^\beta(f\circ\chi_{i,z}) &\le \operatorname{card}\{t\colon \alpha<t<\beta,\ \chi_{i,z}(t)\notin g(Q\cup R)\}\\ &= N[q_i\,|\,\{x\colon \alpha<x_i<\beta\}\sim g(Q\cup R), z],\end{aligned}$$

by showing that whenever u and v are real numbers with

$$\chi_{i,z}(u)\in g(Q) \quad\text{and}\quad \chi_{i,z}(v)\in g(R)$$

there exists a number t between u and v with $\chi_{i,z}(t)\notin g(Q\cup R)$; otherwise the fact that

$$g(Q)\subset\bigcup_{k=1}^{\infty}G(k), \qquad g(R)\subset\bigcup_{k=1}^{\infty}H(k),$$

$$G(k) \text{ and } H(k) \text{ are closed and disjoint for each } k,$$

would permit the inductive selection of positive integers k_ν and real numbers u_ν, v_ν such that

$$\chi_{i,z}(u_\nu)\in G(k_\nu) \quad\text{and}\quad \chi_{i,z}(v_\nu)\in H(k_\nu)$$

for all nonnegative integers ν, $u_0=u$, $v_0=v$, and

$$u_\nu \text{ and } v_\nu \text{ are strictly between } u_{\nu-1} \text{ and } v_{\nu-1},$$

$$\chi_{i,z}(t)\notin G(k_\nu)\cup H(k_\nu) \text{ for all } t \text{ strictly between } u_\nu \text{ and } v_\nu$$

in case $\nu\ge 1$; for every number t in the intersection of the nested sequence of intervals with endpoints u_ν and v_ν it would follow that $\chi_{i,z}(t)\notin g(R\cup Q)$. We conclude that, for every positive integer j,

$$\begin{aligned}&\int^{*}_{\mathbf{R}^{n-1}\cap\mathbf{B}(0,j)} \operatorname{ess}\mathbf{V}_{-j}^{j}(f\circ\chi_{i,z})\,d\mathscr{L}^{n-1}z\\ &\le\int N[q_i\,|\,\mathbf{R}^n\cap\mathbf{B}(0,2j)\sim g(Q\cup R)]\,d\mathscr{L}^{n-1}z<\infty.\end{aligned}$$

4.5.12. Whenever $A\subset\mathbf{R}^n$ and Q, R are defined as in 4.5.11 it is reasonable to consider Q the essential interior of A, R the essential interior of $\mathbf{R}^n\sim A$, hence $\mathbf{R}^n\sim(Q\cup R)$ the essential boundary of A. Noting that $\operatorname{Bdry} A$, the ordinary boundary of A, contains the essential boundary of A we obtain the following corollary of 4.5.11:

If $A\subset\mathbf{R}^n$ and $\mathscr{I}_1^{n-1}(K\cap\operatorname{Bdry}A)<\infty$ for every compact subset K of $\mathbf{R}^n$, then A is $\mathscr{L}^n$ measurable and $\partial(\mathbf{E}^n\llcorner A)$ is representable by integration.

We now combine the theorems 4.5.9 and 4.5.11 to establish a geometric criterion telling us which locally bounded functions occur as density functions of n dimensional locally normal currents in $\mathbf{R}^n$. *If f maps $\mathscr{L}^n$ almost all of $\mathbf{R}^n$ into $\mathbf{R}$, $f(K)$ is bounded for every compact $K\subset\mathbf{R}^n$, and λ, μ, C are defined as in 4.5.9, then the following two conditions are equivalent:*

(I) *f is $\mathscr{L}^n$ measurable and $\mathbf{E}^n\llcorner f\in\mathbf{N}_n^{\mathrm{loc}}(\mathbf{R}^n)$.*

(II) *$\mathscr{I}_1^n(M\cap C)<\infty$ for every compact $M\subset\mathbf{R}^{n+1}$.*

To prove this we let G, S be as in 4.5.9. In case (I) holds, then (II) follows from 4.5.9 (5) and (4). In case (II) holds, then 2.10.15 and 2.6.2 imply that $\mathscr{L}^{n+1}(C)=0$, and that $\mu(x)=\lambda(x)$ for $\mathscr{L}^n$ almost all x, hence 2.9.13 implies f is $\mathscr{L}^n$ measurable; moreover the reasoning in the second paragraph of the proof of 4.5.9 (5) shows that C contains the essential boundary of G, hence 4.5.11 implies S is locally normal; since f is locally bounded, $p|\operatorname{spt} S$ is proper, hence $p_\#(S \llcorner Y_{n+1})$ is locally normal, and we will deduce (I) by verifying the equation

$$p_\#(S \llcorner Y_{n+1}) = \mathbf{E}^n \llcorner f;$$

given $\gamma \in \mathscr{D}^0(\mathbf{R}^n)$ we choose $\alpha \in \mathscr{D}^0(\mathbf{R})$ so that

$$\alpha(y_{n+1}) = 1 \text{ whenever } y \in C \cap p^{-1}(\operatorname{spt} \gamma),$$

let $\beta(t) = t\,\alpha(t)$ for $t \in \mathbf{R}$, and use 4.5.6 to compute

$$\begin{aligned}
&(S \llcorner Y_{n+1})\, p^\#(\gamma \wedge DX_1 \wedge \cdots \wedge DX_n) = S[(\beta \circ Y_{n+1}) \wedge (\gamma \circ p) \wedge DY_1 \wedge \cdots \wedge DY_n]\\
&= (\mathbf{E}^{n+1} \llcorner G)[(\beta' \circ Y_{n+1}) \wedge (\gamma \circ p) \wedge DY_1 \wedge \cdots \wedge DY_{n+1}]\\
&= \textstyle\int_G (\beta' \circ Y_{n+1}) \cdot (\gamma \circ p)\, d\mathscr{L}^{n+1} = \int \gamma(x) \int_{-\infty}^{f(x)} \beta'(t)\, d\mathscr{L}^1 t\, d\mathscr{L}^n x\\
&= \textstyle\int \gamma(x) f(x)\, d\mathscr{L}^n x = (\mathbf{E}^n \llcorner f)(\gamma \wedge DX_1 \wedge \cdots \wedge DX_n).
\end{aligned}$$

4.5.13. The following example shows that the third conclusion of 4.5.9 (31) may fail in case f is not integer valued, even when $\operatorname{spt} f$ is compact. Assuming $n-1<\gamma<n$ and $1<\nu<\infty$ we let f be the function on $\mathbf{R}^n$ such that

$$f(x)=1 \text{ for } |x|<1, \quad f(x)=|x|^{-\gamma} \text{ for } 1 \le |x| < \nu,$$

$$f(x)=\nu^{-\gamma}(\nu+1-|x|) \text{ for } \nu \le |x| < \nu+1, \quad f(x)=0 \text{ for } |x| \ge \nu+1,$$

and use 3.2.13 to estimate

$$\begin{aligned}
[n\,\alpha(n)]^{-1} \textstyle\int |f|\, d\mathscr{L}^n &\ge \textstyle\int_1^\nu r^{n-1-\gamma}\, d\mathscr{L}^1 r = (\nu^{n-\gamma}-1)/(n-\gamma),\\
[n\,\alpha(n)]^{-1} \textstyle\int \|Df\|\, d\mathscr{L}^n &= \textstyle\int_1^\nu \gamma\, r^{n-2-\gamma}\, d\mathscr{L}^1 r + \int_\nu^{\nu+1} r^{n-1} \nu^{-\gamma}\, d\mathscr{L}^1 r\\
&\le \gamma/(\gamma+1-n) + 2^n \nu^{n-1-\gamma}.
\end{aligned}$$

By taking ν large we can therefore make the isoperimetric ratio

$$\mathbf{M}(\mathbf{E}^n \llcorner f)^{(n-1)/n} / \mathbf{M}[\partial(\mathbf{E}^n \llcorner f)]$$

arbitrarily large. Moreover this ratio does not change when f is replaced by the function $f \circ \mu_{2\nu/\varepsilon}$ with support in $\mathbf{U}(0, \varepsilon)$.

4.5.14. We now derive the optimal isoperimetric inequality for 1 dimensional integral cycles in $\mathbf{R}^n$:

If $R \in \mathbf{I}_1(\mathbf{R}^n)$ *with* $\partial R = 0$, *then there exists*

$$S \in \mathbf{I}_2(\mathbf{R}^n) \text{ with } \partial S = R \text{ and } [4\pi \mathbf{M}(S)]^{\frac{1}{2}} \le \mathbf{M}(R).$$

In view of 4.2.20 and 4.2.17 the problem is easily reduced to the special case when R is a simple closed polygon. Assuming that

$$R = \sum_{i=1}^{\nu} [a_{i-1}, a_i] \quad \text{and} \quad \mathbf{M}(R) = \sum_{i=1}^{\nu} |a_i - a_{i-1}|$$

with $a_0, a_1, \ldots, a_{\nu-1}, a_\nu = a_0 \in \mathbf{R}^n$, we note that

$$R = \partial(a_0 \times R) \quad \text{and} \quad a_0 \times R = \sum_{i=1}^{\nu-1} [a_0, a_{i-1}, a_i].$$

We let $b_0 = 0$ and for $i = 1, 2, \ldots, \nu$ we inductively choose $b_i \in \mathbf{R}^2$ so that the triangle with vertices $0, b_{i-1}, b_i$ is congruent to the triangle with vertices a_0, a_{i-1}, a_i and

$$\langle b_i \wedge b_{i-1}, X_1 \wedge X_2 \rangle \ge 0,$$

where X_1, X_2 are usual coordinate functions on $\mathbf{R}^2$. Applying 4.5.9 (31) with n replaced by 2 and

$$T = \sum_{i=1}^{\nu-1} [b_0, b_{i-1}, b_i], \qquad \partial T = \sum_{i=1}^{\nu} [b_{i-1}, b_i]$$

we conclude

$$\mathbf{M}(a_0 \times R)^{\frac{1}{2}} \le \mathbf{M}(T)^{\frac{1}{2}} \le 2^{-1} \pi^{-\frac{1}{2}} \mathbf{M}(\partial T)$$
$$\le (4\pi)^{-\frac{1}{2}} \sum_{i=1}^{\nu} |b_i - b_{i-1}| = (4\pi)^{-\frac{1}{2}} \mathbf{M}(R).$$

4.5.15. *Here we consider an* $\mathscr{L}^n$ *measurable real valued function* f *with compact support which satisfies the conditions of* 4.5.9 (30), and study consequences of $\mathscr{L}^n$ summability of powers of $\|\operatorname{ap} Df\|$.

If $1 \le \xi < n$ *and* $\eta = n\xi/(n-\xi)$, *then*

$$\left(\int |f|^\eta \, d\mathscr{L}^n\right)^{1/\eta} \le c \left(\int \|\operatorname{ap} Df\|^\xi \, d\mathscr{L}^n\right)^{1/\xi}$$

with $c = \alpha(n)^{-1/n} \xi(n-1)/[n(n-\xi)]$.

This assertion holds in case $f \in \mathscr{D}^0(\mathbf{R}^n)$ because 4.5.9 (31), applied with f replaced by $|f|^{\eta/\beta}$, and Hölder's inequality imply

$$\left(\int |f|^\eta \, d\mathscr{L}^n\right)^{1/\beta} \le c \int |f|^{(\eta/\beta)-1} \|Df\| \, d\mathscr{L}^n$$
$$\le c \left(\int |f|^\eta \, d\mathscr{L}^n\right)^{1-1/\xi} \left(\int \|Df\|^\xi \, d\mathscr{L}^n\right)^{1/\xi}.$$

The assertion follows in the general case by regularization.

Since $n\xi/(n-\xi)\to\infty$ as $\xi\to n-$, *the hypothesis that* $\|\operatorname{ap} Df\|^n$ *be* $\mathscr{L}^n$ *summable implies that* $|f|^\zeta$ *is* $\mathscr{L}^n$ *summable whenever* $1\le\zeta<\infty$, but in case $n>1$ this hypothesis does not imply that f is $\mathscr{L}^n$ essentially bounded, as shown by the example at the end of our discussion.

If $n<\xi<\infty$, $\int\|\operatorname{ap} Df\|^\xi\,d\mathscr{L}^n<\infty$ *and* $\delta=1-n/\xi$, *then the function* F *defined in* 4.5.9 *satisfies the Hölder condition*

$$|F(a)-F(b)|\le 2^{2-\delta}\gamma\,\delta^{-1}|a-b|^\delta \text{ for } a,b\in\mathbf{R}^n$$

with $\gamma=[\int(|f|+\|\operatorname{ap} Df\|)^\xi\,d\mathscr{L}^n/\alpha(n)]^{1/\xi}<\infty$.

The fact that $\gamma<\infty$ follows from the preceding paragraph (applied with ξ,η replaced by $n\xi/(n+\xi),\xi$) and Minkowski's inequality. Letting

$$T=\mathbf{E}^n\llcorner f,\quad g=|f|+\|\operatorname{ap} Df\|,\quad\text{hence}\quad \|T\|+\|\partial T\|=\mathscr{L}^n\llcorner g,$$

we use Hölder's inequality to estimate

$$\int_{\mathbf{B}(y,\rho)}|g|\,d\mathscr{L}^n\le\left(\int|g|^\xi\,d\mathscr{L}^n\right)^{1/\xi}[\alpha(n)\,\rho^n]^{1-1/\xi}=\gamma\,\alpha(n)\,\rho^{n-1+\delta}$$

for $y\in\mathbf{R}^n$ and $0<\rho<\infty$. Then we apply 4.3.10 with f, λ replaced by $\mathbf{1}_{\mathbf{R}^n}$, 1 and recall 4.5.9 (24) to complete the proof.

The following example shows the need for the hypothesis $\xi>n$ in the preceding paragraph, and for the hypothesis $\delta>0$ in 4.3.10. Assuming $n>1$ we let

$$h(r)=\log[\log(\sup\{1+r^{-1},e\})] \text{ for } 0<r<\infty,$$
$$f(x)=h(|x|) \text{ for } x\in\mathbf{R}^n\sim\{0\},\quad T=\mathbf{E}^n\llcorner f$$

and note that $\operatorname{spt} f=\mathbf{B}(0,k)$ with $k=1/(e-1)$,

$$\begin{aligned}[n\,\alpha(n)]^{-1}\int\|Df\|^n\,dL^n&=\int_0^k r^{n-1}|h'(r)|^n\,d\mathscr{L}^1 r\\ &=\int_0^k r^{-1}[\log(1+r^{-1})\cdot(r+1)]^{-n}\,d\mathscr{L}^1 r\\ &\le\int_0^k r^{-1}[-\log(r)]^{-n}\,d\mathscr{L}^1 r=[\log(e-1)]^{1/n}/(n-1),\end{aligned}$$

hence $\gamma=[\int(|f|+\|Df\|)^n\,d\mathscr{L}^n/\alpha(n)]^{1/n}<\infty$,

$$[\|T\|+\|\partial T\|]\,\mathbf{B}(y,\rho)\le\gamma\,\alpha(n)\,\rho^{n-1} \text{ for } y\in\mathbf{R}^n \text{ and } 0<\rho<\infty,$$

but f is $\mathscr{L}^n$ essentially unbounded because $f(x)\to\infty$ as $x\to 0$.

4.5.16. We see from 4.1.7 that the isomorphism

$$\mathscr{D}_n(\mathbf{R}^n)\simeq\mathscr{D}_0(\mathbf{R}^n),$$

which maps $T\in\mathscr{D}_n(\mathbf{R}^n)$ onto $T\llcorner(DX_1\wedge\cdots\wedge DX_n)$, and whose inverse maps $S\in\mathscr{D}_0(\mathbf{R}^n)$ onto $S\wedge\mathbf{e}_1\wedge\cdots\wedge\mathbf{e}_n$, relates the operators ∂ and D_i by the equations

$$D_i[T \llcorner (DX_1 \wedge \cdots \wedge DX_n)]$$
$$=(-1)^i (\partial T) \llcorner (DX_1 \wedge \cdots \wedge DX_{i-1} \wedge DX_{i+1} \wedge \cdots \wedge DX_n)$$

for $i=1,\ldots,n$ and

$$\partial(S \wedge \mathbf{e}_1 \wedge \cdots \wedge \mathbf{e}_n) = -\sum_{i=1}^{n} (D_i S \wedge \mathbf{e}_1 \wedge \cdots \wedge \mathbf{e}_n) \llcorner DX_i .$$

Recalling 4.4.7 we infer that *this isomorphism maps*

$$\mathbf{N}_n^{\mathrm{loc}}(\mathbf{R}^n) = \mathscr{D}_n(\mathbf{R}^n) \cap \{T \colon \partial T \text{ is representable by integration}\}$$

onto the space of all those distributions S *of type* $\mathbf{R}$ *whose partial derivatives* $D_1 S, \ldots, D_n S$ *are representable by integration.* Every such distribution S corresponds to a locally $\mathscr{L}^n$ summable function, because our isomorphism maps $\mathbf{E}^n \llcorner f$ onto $\mathscr{L}^n \llcorner f$. Moreover 4.5.9 (30) shows that

$$\mathbf{N}_n^{\mathrm{loc}}(\mathbf{R}^n) \cap \{T \colon \|\partial T\| \text{ is absolutely continuous with respect to } \mathscr{L}^n\}$$

is mapped onto the subspace of those distributions whose partial derivatives correspond to locally $\mathscr{L}^n$ summable functions.

4.5.17. *If* $R \in \mathscr{R}_{n-1}(\mathbf{R}^n)$ *with* $\partial R = 0$, *then there exist* $\mathscr{L}^n$ *measurable sets* M_i *corresponding to all integers* i *such that* $M_i \subset M_{i-1}$ *and*

$$R = \sum_{i \in \mathbf{Z}} \partial(\mathbf{E}^n \llcorner M_i), \qquad \|R\| = \sum_{i \in \mathbf{Z}} \|\partial(\mathbf{E}^n \llcorner M_i)\| .$$

In fact $R = \partial T$ with $T \in \mathbf{I}_n(\mathbf{R}^n)$, and $T = \mathbf{E}^n \llcorner f$ for some integer valued $\mathscr{L}^n$ summable function f, hence 4.5.9 (13) implies the assertion with $M_i = \{x \colon f(x) \geq i\}$.

CHAPTER FIVE

Applications to the calculus of variations

The theory developed in this chapter is based on the concept of the integral

$$\langle \Phi, T\rangle = \int_T \Phi = \int \Phi[z, \vec{T}(z)]\, d\|T\|\, z$$

of a parametric integrand Φ of degree m over an m dimensional rectifiable current T defined in 5.1.1, on the concept of ellipticity defined in 5.1.2, and on the notions of minimizing current defined in 5.1.6. For example, a current $S \in \mathscr{R}_m(\mathbf{R}^n)$ is absolutely Φ minimizing with respect to $\mathbf{R}^n$ if and only if

$$\langle \Phi, S\rangle \leq \langle \Phi, T\rangle \text{ for all } T \in \mathscr{R}_m(\mathbf{R}^n) \text{ with } \partial S = \partial T.$$

After proving the lowersemicontinuity theorem 5.1.5, we deduce in 5.1.6 two powerful propositions on the existence of Φ minimizing currents from the compactness properties established in 4.4.4.

Most of our effort will be directed to the problem of interior regularity of a Φ minimizing current S. Combining the general properties of rectifiable currents with an isoperimetric density estimate, we see quickly in 5.1.6 that $\operatorname{Tan}(\operatorname{spt} S, z)$ is an m dimensional vectorspace for $\|S\|$ almost all z in $\operatorname{spt} S \sim \operatorname{spt} \partial S$. The problem is to show that these tangent spaces vary continuously on a large set of points z, and that a large part of $\operatorname{spt} S \sim \operatorname{spt} \partial S$ is an m dimensional differentiable manifold. Even though $\operatorname{spt} S \sim \operatorname{spt} \partial S$ can have a nonempty singular subset (see 5.4.19), we may reasonably conjecture that the singular subset cannot be too large, and we will indeed prove several results of this type (see 5.3.17).

Using methods originating from [AF 6] we present in Section 5.3 all the general theorems discovered up to now concerning interior regularity properties of Φ minimizing currents for an arbitrary positive elliptic integrand Φ of class 3. While Almgren treated a problem different from the one studied here, because he minimized the integral

$$\int_{\{z:\, \Theta^m(\|T\|, z) \geq 1\}} \Phi[z, \vec{T}(z)]\, d\mathscr{H}^m\, z$$

in place of the integral defined in 5.1.1, we carry over from his paper in modified form all the principal ideas for the key lemmas 5.3.4, 5.3.7, 5.3.8, 5.3.10 and the main theorems 5.3.13, 5.3.14, 5.3.16.

Regularity theorems were obtained earlier for the special case of the area integrand of degree m on $\mathbf{R}^n$, for $m=2=n-1$ in [FL 3], for $m=n-1$ in [DG 6], and for all $m<n$ as a consequence of [R 3]. In fact the special case of 5.3.16, when Φ is the area integrand, can be deduced from the epiperimetric inequality of [R 3] by methods of [FF] and [FL 3, 4]. An exposition of this deduction was planned for inclusion in our book, but has now been displaced by adoption of the superior general approach of [AF 6].

There are still some very interesting special results about the regularity of area minimizing currents, which are not implied by the general theory of Section 5.3, and which we present in Section 5.4. Additional techniques used in this special case depend on the monotonicity of density ratios, the existence of tangent cones, and formulae from differential geometry. The propositions 5.4.3 (1) and 5.4.5 (3) were discovered in [FL 3]; 5.4.3 (3) and its modification in 5.4.4 originate from [FF]; 5.4.3 (7) and 5.4.8 were proved for $m=n-1$ in [TD 1, 3]. The lemma 5.4.14 was obtained first for $n=3$ in [FL 3], next for $n=4$ in [AF 5], then for $3\leq n\leq 7$ in [SJ], where the present method including 5.4.13 was developed. This method uses certain facts from differential geometry, which we derive in 5.4.11 and 5.4.12 so as to keep our treatment selfcontained; an extensive exposition placing these facts in the general theory of connections may be found in [KN, vol. 2]. The passage from Lemma 5.4.14 to Theorem 5.4.15, and its partial extension in 5.4.16, follows a pattern designed in [FL 3] and [TD 1, 2, 3]. The result of 5.4.17 is new. Limit cones were first used in the manner of 5.4.18 for a new proof of the classical ($n=3$) Bernstein theorem in [FL 3]. The proposition in 5.4.19 originates from [F 19]; earlier a special case, in which spt $T\sim$ spt ∂T is a complex μ dimensional holomorphic submanifold of $\mathbf{C}^\nu$, was treated in [WW], but our main interest attaches to the case when the singular subset is nonempty.

Variations associated with isotopic deformations of the ambient space play an important role in the study of Φ minimizing currents. These concepts are defined in 5.1.7 and applied in 5.1.8 – 5.1.11, 5.2.18, 5.3.7, 5.3.8, 5.4.3 (1), 5.4.12, 5.4.14.

It is very useful for geometric constructions that the concept of elliptic integrand, as formulated in 5.1.2 following [AF 6], has the invariance property 5.1.4. It is equally useful for analytic considerations that, by virtue of 5.1.9 and 5.2.17, the second differentials of the nonparametric integrand $\Phi^{\S}$ associated with an elliptic parametric integrand Φ of degree m satisfy a strong ellipticity condition in the sense of 5.2.3. This nonparametric Legendre condition implies by way of 5.2.5 some crucial estimates in the proof of 5.3.8, and allows the use of 5.2.15 in 5.2.18 to prove higher order smoothness of all Φ minimizing currents whose supports lie on m dimensional submanifolds of class 1.

Our treatment of strongly elliptic systems of second order partial differential equations in Section 5.2 is not intended as a comprehensive account of the subject, such as may be found in [MCB 4], but it contains a complete exposition of those results which we need for the geometric applications in Section 5.3. We prove no existence theorem for the Dirichlet problem, though such a theorem follows readily from 1.7.13 and the Gårding inequality in 5.2.3 (see [NI 1]). We usually require the weak solutions of the differential equations to be functions of class 1, and often also that their differentials satisfy Hölder conditions. The Fourier transformation is not used in our work.

After proving a basic a priori estimate for strongly elliptic bilinear forms in 5.2.3, following the scheme of [NI 2, Lecture 4], we deduce inequalities of Cauchy type for solutions of homogeneous equations with constant coefficients in 5.2.5. Next we construct fundamental solutions of equations with constant coefficients in 5.2.7–5.2.11 by the method of [J], as extended to systems in [MCB 2]. Then we prove the Theorems 5.2.14 and 5.2.15 implying the higher differentiability of solutions of linear strongly elliptic equations and of nonlinear equations arising from elliptic variational problems. For this purpose we use part of the technique of [MCB 2], where such results for general dimensions were first obtained; the special case $m=n-1$ had been treated earlier in [HOE 2].

In 5.2.18 we prove the classical maximum principle originating from [HOE 1] and also a second proposition recently discovered in [AF 7], which has the advantage of applying to arbitrary elliptic variational problems with $m=n-1$.

5.1. Integrands and minimizing currents

5.1.1. Suppose Z is an open subset of $\mathbf{R}^n$. By a **parametric integrand of degree** m on Z we mean a continuous map

$$\Phi\colon Z \times \bigwedge_m \mathbf{R}^n \to \mathbf{R}$$

satisfying the condition

$$\Phi(z, r\,\alpha) = r\,\Phi(z,\alpha) \text{ for } z\in Z,\ \alpha\in\bigwedge_m \mathbf{R}^n,\ 0<r\in\mathbf{R}.$$

We term Φ an **integrand of class** k **(analytic integrand)** if and only if the function $\Phi|[Z\times(\bigwedge_m \mathbf{R}^n \sim \{0\})]$ is of class k (analytic). We say that Φ is a **positive integrand** if and only if

$$\Phi(z,\alpha)>0 \text{ for } z\in Z,\ 0\neq\alpha\in\bigwedge_m \mathbf{R}^n.$$

Whenever $T\in\mathscr{R}_m(Z)$ we call

$$\int \Phi[z, \vec{T}(z)]\, d\,\|T\|\, z$$

the **integral of Φ over T** and denote it

$$\textstyle\int_T \Phi \text{ or } \langle \Phi, T\rangle.$$

This amounts to an extension of the linear function T from the space of differential forms to the space of parametric integrands; in fact to each continuous differential form ϕ of degree m on Z corresponds the parametric integrand Φ defined by the formula

$$\Phi(z,\alpha)=\langle \alpha, \phi(z)\rangle \text{ for } z\in Z,\ \alpha\in \wedge_m \mathbf{R}^n,$$

and in this case the integral of Φ over T equals $T(\phi)$, according to 4.1.7.

We construct the simplest and most important positive integrand from the inner product of $\wedge_m \mathbf{R}^n$ by letting

$$\Phi(z,\alpha)=|\alpha|=(\alpha\bullet\alpha)^{\frac{1}{2}} \text{ for } z\in \mathbf{R}^n,\ \alpha\in\wedge_m \mathbf{R}^n;$$

in this case $\langle \Phi, T\rangle = \mathbf{M}(T)$ whenever $T\in\mathscr{R}_m(\mathbf{R}^n)$, because 4.1.28 implies that $|\vec{T}(z)|=\|\vec{T}(z)\|=1$ for $\|T\|$ almost all x; this particular function Φ is called the **parametric area integrand of degree** m **on** $\mathbf{R}^n$.

Whenever Ψ is a parametric integrand of degree m on an open subset V of some Euclidean space and $G: Z\to V$ is a map of class 1 we define the parametric integrand $G^{\#}\Psi$ of degree m on Z by the formula

$$(G^{\#}\Psi)(z,\alpha)=\Psi[G(z), \langle \alpha, \wedge_m DG(z)\rangle]$$

for $z\in Z$, $\alpha\in\wedge_m\mathbf{R}^n$. *If* $T\in\mathscr{R}_m(Z)$ *and* $G|\operatorname{spt} T$ *is univalent, then*

$$\langle G^{\#}\Psi, T\rangle = \langle \Psi, G_{\#} T\rangle;$$

we prove this by applying 4.1.30, with $T=\mathscr{H}^m \wedge \xi$, and 3.2.20 to compute

$$\begin{aligned}\langle G^{\#}\Psi, T\rangle &= \textstyle\int \Psi[G(z), \langle \vec{T}(z), \wedge_m DG(z)\rangle]\, d\|T\|\, z\\ &= \textstyle\int_W \Psi[G(z), \langle \xi(z), \wedge_m DG(z)\rangle]\, d\mathscr{H}^m z\\ &= \textstyle\int_W \Psi[G(z), \eta[G(z)]]\ \operatorname{ap} J_m g(z)\, d\mathscr{H}^m z\\ &= \textstyle\int_{G(W)} \Psi[v, \eta(v)]\, d\mathscr{H}^m v\\ &= \textstyle\int \Psi[v, \overrightarrow{G_{\#} T}(v)]\, d\|G_{\#} T\|\, v = \langle \Psi, G_{\#} T\rangle.\end{aligned}$$

The need for the univalence hypothesis is shown by the example where

$$G: \mathbf{R}\to\mathbf{R}, \quad G(z)=z^2 \text{ for } z\in\mathbf{R}, \quad T=[-1,1],$$

$$\Psi(v,\beta)=|\beta| \text{ for } v\in\mathbf{R},\ \beta\in\wedge_1\mathbf{R},$$

$$\langle G^{\#}\Psi, T\rangle = \textstyle\int_{-1}^{1} |2z|\, d\mathscr{L}^1 z = 2 \quad\text{and}\quad G_{\#} T=0.$$

If Φ is a positive parametric integrand of degree m on Z and if $T_1, T_2 \in \mathscr{R}_m(Z)$, then

$$\langle \Phi, T_1+T_2\rangle \le \langle \Phi, T_1\rangle + \langle \Phi, T_2\rangle;$$

recalling 4.1.28 we express $T_1 = \mathscr{H}^m \wedge \xi_1$, $T_2 = \mathscr{H}^m \wedge \xi_2$ and observe that for $\mathscr{H}^m$ almost all z in Z the m-vectors $\xi_1(z)$ and $\xi_2(z)$ are linearly dependent, hence

$$\Phi[z, \xi_1(z)+\xi_2(z)] \le \Phi[z, \xi_1(z)] + \Phi[z, \xi_2(z)].$$

5.1.2. Assuming that Φ is a parametric integrand of degree m on U we associate with each $a \in Z$ the map

$$\Phi_a\colon \wedge_m \mathbf{R}^n \to \mathbf{R}, \quad \Phi_a(a) = \Phi(a, \alpha) \text{ for } \alpha \in \wedge_m \mathbf{R}^n.$$

We say that Φ is **elliptic** at a if and only if *there exists a positive number c such that the inequality*

$$\|R\|(\Phi_a \circ \vec{R}) - \|S\|(\Phi_a \circ \vec{S}) \ge c[\mathbf{M}(R) - \mathbf{M}(S)]$$

holds whenever $R \in \mathscr{R}_m(\mathbf{R}^n)$, $S \in \mathscr{R}_m(\mathbf{R}^n)$, $\partial R = \partial S$, spt S *is contained in the vectorsubspace of $\mathbf{R}^n$ associated with a simple m vector γ of $\mathbf{R}^n$, and*

$$\vec{S}(z) = \gamma \text{ for } \|S\| \text{ almost all } z;$$

we will refer to such a positive number c as an **ellipticity bound** for Φ at a.

We call Φ an **elliptic integrand** if and only if Φ is elliptic at each point of Z and for every compact subset K of Z there exists a common ellipticity bound for Φ at all points of K.

Whenever R, S, γ are as above, then

$$\int \vec{R}\, d\|R\| = \int \vec{S}\, d\|S\| = \mathbf{M}(S)\, \gamma$$

because $\partial(R-S)=0$, $R-S = \partial[\delta_0 \times (R-S)]$ and for every real valued linear function χ on $\wedge_m \mathbf{R}^n$ the constant differential form ϕ mapping $\mathbf{R}^n$ onto χ has exterior derivative 0, hence

$$\chi\left(\int \vec{R}\, d\|R\| - \int \vec{S}\, d\|S\|\right) = \int (\chi \circ \vec{R})\, d\|R\| - \int (\chi \circ \vec{S})\, d\|S\|$$
$$= (R-S)\,\phi = [\delta_0 \times (R-S)]\, d\phi = 0.$$

Assuming $0 < c \in \mathbf{R}$ we consider the function F defined by the formula

$$F(\alpha) = \Phi_a(\alpha) - c|\alpha| \text{ for } \alpha \in \wedge_m \mathbf{R}^n,$$

and note that $F(t\alpha) = t\, F(\alpha)$ for $0 \le t \in \mathbf{R}$, hence F is convex if and only if $F(\alpha+\beta) \le F(\alpha) + F(\beta)$ for all $\alpha, \beta \in \wedge_m \mathbf{R}^n$. We verify that *convexity of F implies ellipticity of Φ at a with bound c*, by using 2.4.19 to infer

$$\int (\Phi_a \circ \vec{S})\, d\|S\| - c\,\mathbf{M}(S) = F[\mathbf{M}(S)\,\gamma] = F\left(\int \vec{R}\, d\|R\|\right)$$
$$\le \int (F \circ \vec{R})\, d\|R\| = \int (\Phi_a \circ \vec{R})\, d\|R\| - c\,\mathbf{M}(R).$$

The notion of **semielliptic integrand** is obtained through replacement above of c by 0; one finds that *convexity of Φ_a implies semiellipticity of Φ at a.*

5.1.3. Assuming X is a normed vectorspace, $F: X \to \mathbf{R}$ is continuous and $F|X \sim \{0\}$ is of class 2, we observe that F is a convex function if and only if for any two linearly independent vectors α and β the function on $\mathbf{R}$ mapping t onto $F(\alpha + t\beta)$ is convex; computing the second derivative of the latter function we find that convexity of F is equivalent to the condition

$$\langle \beta \odot \beta, D^2 F(\alpha) \rangle \geq 0 \text{ for } 0 \neq \alpha \in X,\ \beta \in X.$$

In case X is an inner product space, $c \in \mathbf{R}$ and

$$F(\alpha) = \Omega(\alpha) - c|\alpha| \text{ for } \alpha \in X,$$

we infer that *convexity of F is equivalent to the condition*

$$\langle \beta \odot \beta, D^2 \Omega(\alpha) \rangle \geq \frac{c}{|\alpha|} \left[|\beta|^2 - \left(\beta \bullet \frac{\alpha}{|\alpha|} \right)^2 \right]$$

$$\textit{whenever } 0 \neq \alpha \in X \textit{ and } \beta \in X;$$

we also note that $|\beta|^2 - (\beta \bullet \alpha/|\alpha|)^2 = |\beta \wedge \alpha|^2/|\alpha|^2$.

In particular, if L is a continuous linear automorphism of X and $\Omega(\alpha) = |L(\alpha)|$ for $\alpha \in X$, then

$$\begin{aligned} \langle \beta \odot \beta, D^2 \Omega(\alpha) \rangle &= |L(\alpha)|^{-3} |L(\beta) \wedge L(\alpha)|^2 \\ &\geq \|L\|^{-3} |\alpha|^{-3} \|L^{-1}\|^{-4} |\beta \wedge \alpha|^2, \end{aligned}$$

hence F is convex provided $c \leq \|L\|^{-3} \|L^{-1}\|^{-4}$.

When Φ is a parametric integrand of degree m and class 2, we can take $X = \wedge_m \mathbf{R}^n$ and $\Omega = \Phi_a$ to obtain a differential criterion for convexity of the function F considered in 5.1.2. This condition on $D^2 \Phi_a$ is called the **parametric Legendre condition** for Φ at a with bound c. If it holds with $c > 0$, then Φ is elliptic at a with bound c.

For every *linear automorphism* f of $\mathbf{R}^n$ one can apply the above reasoning with $L = \wedge_m f$ to see that

$$\begin{aligned} \mathbf{M}(f_\# R) - \mathbf{M}(f_\# S) &= \|R\| (|L| \circ \vec{R}) - \|S\| (|L| \circ \vec{S}) \\ &\geq \| \wedge_m f \|^{-3} \| \cdot \| \wedge_m f^{-1} \|^{-4} [\mathbf{M}(R) - \mathbf{M}(S)] \end{aligned}$$

whenever R and S are as in 5.1.2. This method will be used in the proof of the following more general theorem, which establishes the invariance of ellipticity under diffeomorphisms.

5.1.4. Theorem. *Suppose Z and V are open subsets of $\mathbf{R}^n$ and $\mathbf{R}^\nu$, $G: Z \to V$ is a map of class 1 and $DG(a)$ is univalent for each $a \in Z$. If Ψ is an elliptic integrand of degree m on V, then $G^\# \Psi$ is an elliptic integrand on Z; in fact, if $a \in Z$ and δ is an ellipticity bound for Ψ at $G(a)$, then*

$$\delta \, \|\wedge_m DG(a)\|^{-3} \, \|\wedge_m DG(a)^{-1}\|^{-4}$$

is an ellipticity bound for $G^\# \Psi$ at a. Similarly, if Ψ is a semielliptic integrand, so is $G^\# \Psi$.

Proof. Observing that the assertion is trivial in case G is an isometric injection, and recalling 3.1.18, we will discuss only the case when $n = \nu$ and G is a diffeomorphism.

Suppose $a \in Z$, δ is an ellipticity bound for Ψ at $G(a)$, $f = DG(a)$, $L = \wedge_m f$ and $c = \|L\|^{-3} \|L^{-1}\|^{-4}$. If R and S satisfy the requirements of 5.1.2, so do $f_\# R$ and $f_\# S$; using the method of 5.1.1 to transform integrals by the linear automorphism f of $\mathbf{R}^n$, and noting that 5.1.3 with $\Omega = |L|$ permits us to apply the reasoning of 5.1.2 with Φ_a replaced by $|L|$, we estimate

$$\begin{aligned}
&\|R\| \, [(G^\# \Psi)_a \circ \vec{R}] - \|S\| \, [(G^\# \Psi)_a \circ \vec{S}] \\
&= \|R\| \, [\Psi_{G(a)} \circ L \circ \vec{R}] - \|S\| \, [\Psi_{G(a)} \circ L \circ \vec{S}] \\
&= \|f_\# R\| \, [\Psi_{G(a)} \circ \overrightarrow{f_\# R}] - \|f_\# S\| [\Psi_{G(a)} \circ \overrightarrow{f_\# S}] \geq \delta [\mathbf{M}(f_\# R) - \mathbf{M}(f_\# S)] \\
&= \delta [\|R\| (|L| \circ \vec{R}) - \|S\| (|L| \circ \vec{S})] \geq \delta c [\mathbf{M}(R) - \mathbf{M}(S)].
\end{aligned}$$

A similar but simpler argument, with δ replaced by 0, applies in case Ψ is semielliptic.

5.1.5. Theorem. *If Φ is a positive semielliptic integrand of degree m on Z and K is a compact subset of Z, then the function on $\mathscr{R}_{m,K}(Z)$ mapping T onto $\langle \Phi, T \rangle$ is lowersemicontinuous with respect to the $\mathscr{F}_K$ topology.*

Proof. Replacing Z by a suitable neighborhood of K, we assume the existence of a finite, positive number λ such that

$$\lambda^{-1} |\alpha| \leq \Phi(z, \alpha) \leq \lambda |\alpha| \text{ for } z \in Z, \ \alpha \in \wedge_m \mathbf{R}^n.$$

We suppose that the sequence $T_1, T_2, T_3, \ldots$ is $\mathscr{F}_K$ convergent to Q in $\mathscr{R}_{m,K}(Z)$, and will show that

$$\langle \Phi, Q \rangle \leq \liminf_{i \to \infty} \langle \Phi, T_i \rangle.$$

First we discuss the *special case when Q is an integral polyhedral chain. Given* $\varepsilon>0$, we consider the family F of all balls $\mathbf{B}(a,r)$ contained in Z such that

$$|\Phi(z,\alpha)-\Phi(a,\alpha)|\le\varepsilon|\alpha| \text{ for } z\in\mathbf{B}(a,r),\ \alpha\in\wedge_m\mathbf{R}^n,$$

$\mathbf{B}(a,r)\cap\operatorname{spt}Q$ is contained in some m dimensional affine subspace of $\mathbf{R}^n$ and $\mathbf{B}(a,r)\cap\operatorname{spt}\partial Q=\varnothing$, hence $Q\llcorner\mathbf{B}(a,r)$ is an integral multiple of an oriented m dimensional ball. We verify that

$$\langle\Phi,Q\llcorner\mathbf{B}(a,r)\rangle\le\varepsilon\,\|Q\|\,\mathbf{B}(a,r)+(1+\varepsilon\,\lambda)\liminf_{i\to\infty}\langle\Phi,T_i\llcorner\mathbf{B}(a,r)\rangle$$

whenever $\mathbf{B}(a,r)\in F$; assuming $0<\rho<r$ we choose

$$X_i\in\mathscr{R}_{m,K}(Z),\qquad Y_i\in\mathscr{R}_{m+1,K}(Z) \text{ with } Q-T_i=X_i+\partial Y_i$$

so that $\mathbf{M}(X_i)+\mathbf{M}(Y_i)\to 0$ as $i\to\infty$, apply 4.2.1 with $u(z)=|z-a|$ for $z\in\mathbf{R}^n$ to obtain numbers s_i such that

$$\rho<s_i<r,\qquad (r-\rho)\,\mathbf{M}\langle Y_i,u,s_i+\rangle\le\mathbf{M}(Y_i),$$

hence $Q\llcorner\mathbf{B}(a,s_i)=S_i+\partial[Y_i\llcorner\mathbf{B}(a,s_i)]$ with

$$S_i=(T_i+X_i)\llcorner\mathbf{B}(a,s_i)-\langle Y_i,u,s_i+\rangle\in\mathscr{R}_m(Z)$$

by 4.2.16(3); using the semiellipticity of Φ at a we estimate

$$\begin{aligned}
&\langle\Phi,Q\llcorner\mathbf{B}(a,\rho)\rangle-\varepsilon\,\|Q\|\,\mathbf{B}(a,r)\le\|Q\llcorner\mathbf{B}(a,s_i)\|\,(\Phi_a\circ\vec{Q})\le\|S_i\|\,(\Phi_a\circ\vec{S}_i)\\
&\le\|T_i\llcorner\mathbf{B}(a,s_i)\|\,(\Phi_a\circ\vec{T}_i)+\lambda[\mathbf{M}(X_i)+\mathbf{M}\langle Y_i,u,s_i+\rangle]\\
&\le\langle\Phi,T_i\llcorner\mathbf{B}(a,r)\rangle+\varepsilon\,\|T_i\|\,\mathbf{B}(a,r)+\lambda[\mathbf{M}(X_i)+\mathbf{M}(Y_i)/(r-\rho)]\\
&\le(1+\lambda\,\varepsilon)\langle\Phi,T_i\llcorner\mathbf{B}(a,r)\rangle+\lambda[\mathbf{M}(X_i)+\mathbf{M}(Y_i)/(r-\rho)]
\end{aligned}$$

and let $i\to\infty$, then $\rho\to r-$. Recalling 2.8.18 we cover $\|Q\|$ almost all of Z by a countable disjointed subfamily of F and conclude

$$\langle\Phi,Q\rangle\le\varepsilon\,\mathbf{M}(Q)+(1+\varepsilon\,\lambda)\liminf_{i\to\infty}\langle\Phi,T_i\rangle.$$

Next we treat the *general case*. Given $\varepsilon>0$, we choose a compact subset C of Z with $K\subset\operatorname{Int}C$, and apply 4.2.20 to obtain $P\in\mathscr{P}_{m,C}(Z)$ and a diffeomorphism f of class 1 mapping Z onto Z such that

$$\mathbf{M}(P-f_\#Q)\le\varepsilon,\qquad \operatorname{Lip}(f^{-1})\le 2,\qquad f(K)\subset C.$$

Then $\Psi=f^{-1\#}\Phi$ is a positive semielliptic integrand, by 5.1.4, with

$$\Psi(z,\alpha)\le\lambda\,2^m|\alpha| \text{ for } z\in Z,\ \alpha\in\wedge_m\mathbf{R}^n,$$

and the sequence of currents $P+f_\#(T_i-Q)$ is $\mathscr{F}_C$ convergent to P in $\mathscr{R}_{m,C}(Z)$. Using 5.1.1 and the result of the preceding special case we conclude

$$\begin{aligned}&\langle\Phi,Q\rangle-\lambda 2^m\varepsilon\leq\langle\Psi,f_\# Q\rangle-\langle\Psi,f_\# Q-P\rangle\\&\leq\langle\Psi,P\rangle\leq\liminf_{i\to\infty}\langle\Psi,P+f_\#(T_i-Q)\rangle\\&\leq\langle\Psi,P-f_\# Q\rangle+\liminf_{i\to\infty}\langle\Psi,f_\# T_i\rangle\leq\lambda 2^m\varepsilon+\liminf_{i\to\infty}\langle\Phi,T_i\rangle.\end{aligned}$$

5.1.6. *Here we assume that Φ is a positive semielliptic integrand of degree m on Z and $B\subset A$ are compact Lipschitz neighborhood retracts in Z.* We will use the following terminology:

A current $S\in\mathscr{R}_{m,A}(Z)$ is called **absolutely Φ minimizing with respect to** (A,B) if and only if

$$\langle\Phi,S\rangle\leq\langle\Phi,S+X\rangle \text{ whenever } X\in\mathscr{Z}_m(A,B)\cap\mathscr{R}_m(Z).$$

A current $S\in\mathscr{R}_{m,A}(Z)$ is called **homologically Φ minimizing with respect to** (A,B) if and only if

$$\langle\Phi,S\rangle\leq\langle\Phi,S+X\rangle \text{ whenever } X\in\mathscr{B}_m(A,B)\cap\mathscr{R}_m(Z).$$

A current $S\in\mathscr{R}_{m,A}(Z)$ is called **locally Φ minimizing with respect to** (A,B) if and only if A has a covering consisting of open sets N such that

$$\langle\Phi,S\rangle\leq\langle\Phi,S+X\rangle \text{ whenever } X\in\mathscr{Z}_m(A\cap N,B\cap N)\cap\mathscr{R}_m(Z).$$

Clearly *every absolutely Φ minimizing current is homologically Φ minimizing.* Furthermore *every homologically Φ minimizing current is locally Φ minimizing;* to prove this we assume f,g,U,V,W as in 4.4.2 with $V\subset U\subset Z$, we cover A by the family

$$F=\{W\}\cup\{\mathbf{U}(a,r)\colon \mathbf{U}(a,r)\subset U\sim B\},$$

and we verify that

$$\mathscr{Z}_m(A\cap N,B\cap N)\subset\mathscr{B}_m(A,B) \text{ for } N\in F;$$

in case $N=W$ the above assertion follows from 4.4.2(3), and in case $N=\mathbf{U}(a,r)\subset U\sim B$ it holds because

$$X=f_\#\partial(\delta_a \times X) \text{ whenever } X\in\mathscr{Z}_m(A\cap N).$$

The existence of certain Φ minimizing currents is guaranteed by the following two propositions:

(1) *If $R\in\mathscr{B}_{m-1}(A,B)$, then there exists an absolutely Φ minimizing current $S\in\mathscr{R}_{m,A}(Z)$ such that* $\operatorname{spt}(R-\partial S)\subset B$.

(2) *If $T\in\mathscr{R}_{m,A}(Z)$ and $\chi\in\mathscr{H}_m(A,B)$, then there exists a homologically Φ minimizing current $S\in\mathscr{R}_{m,A}(Z)$ such that $T-S\in\chi$.*

To prove (1) we define

$$\Upsilon=\mathscr{R}_{m,A}(Z)\cap\{Q:\ \operatorname{spt}(R-\partial Q)\subset B\},\qquad \mu=\inf\{\langle\Phi,Q\rangle:\ Q\in\Upsilon\},$$

note that $\Upsilon\neq\emptyset$, $0\leq\mu<\infty$, and choose a sequence of currents $Q_i\in\Upsilon$ such that

$$\lim_{i\to\infty}\langle\Phi,Q_i\rangle=\mu.$$

Since there exists a finite, positive number λ with

$$\Phi(x,\alpha)\geq\lambda^{-1}|\alpha|\ \text{ for }\ x\in A,\ \alpha\in\wedge_m\mathbf{R}^n,$$

we see that

$$\limsup_{i\to\infty}\mathbf{M}(Q_i)\leq\lambda\,\mu.$$

From 4.4.4 we infer that the $\mathbf{M}$ bounded sequence of currents

$$Q_i-Q_1\in\mathscr{Z}_m(A,B)$$

has a subsequence which is $\mathscr{F}_{A,B}$ convergent in $\mathscr{Z}_m(A,B)$. Thus we obtain a current

$$P\in\mathscr{Z}_m(A,B)\quad\text{with}\quad\liminf_{i\to\infty}\mathscr{F}_{A,B}(Q_i-Q_1-P)=0,$$

and letting $S=(Q_1+P)\llcorner(Z\sim B)$ we find that

$$\liminf_{i\to\infty}\int_0^{*\infty}\mathscr{F}_A[(Q_i-S)\llcorner H(r)]\,d\mathscr{L}^1 r=0$$

where $H(r)=Z\cap\{z:\ \operatorname{dist}(z,B)>r\}$. It follows from 2.4.6 that there exist arbitrarily small positive numbers r satisfying the condition

$$\liminf_{i\to\infty}\mathscr{F}_A[(Q_i-S)\llcorner H(r)]=0,$$

and 5.1.5 implies for each such r that

$$\langle\Phi,S\llcorner H(r)\rangle\leq\liminf_{i\to\infty}\langle\Phi,Q_i\llcorner H(r)\rangle\leq\mu;$$

since $\|S\|\,B=0$ we infer

$$\langle\Phi,S\rangle=\liminf_{r\to0+}\langle\Phi,S\llcorner H(r)\rangle\leq\mu.$$

Observing that $S+X\in\Upsilon$ for every $X\in\mathscr{Z}_m(A,B)$, we conclude $\langle\Phi,S\rangle=\mu$ and S is absolutely Φ minimizing with respect to (A,B).

The proposition (2) may be proved by a similar argument, in which $\mathscr{R}_{m,A}(Z)\cap\{Q:\ T-Q\in X\}$ plays the role analogous to Υ.

Next we establish a basic density property of minimizing currents:

If $S\in\mathscr{R}_{m,A}(Z)$ *is absolutely* Φ *minimizing with respect to* (A,B), $0<\lambda<\infty$,

$$\lambda^{-1}|\alpha|\leq\Phi(z,\alpha)\leq\lambda|\alpha|\ \textit{for}\ (z,\alpha)\in Z\times\wedge_m\mathbf{R}^n,$$

and if $a \in \operatorname{spt} S$, $0 < \rho \le \operatorname{dist}(a, \operatorname{spt} \partial S)$, *then*

$$\rho^{-m} \|S\| \mathbf{U}(a, \rho) \ge \lambda^{2-2m} \sigma^{-m} m^{-m}$$

where σ *is the isoperimetric constant of* (A, B) *obtained by applying* 4.4.2 (2) *with* m *replaced by* $m-1$.

Defining $S_r = S \llcorner \mathbf{U}(a, r)$ and $f(r) = \mathbf{M}(S_r)$ for $r \in \mathbf{R}$, we use 4.2.1 with $u(z) = |z-a|$ for $z \in Z$ to infer that $\mathscr{L}^1$ almost all numbers r between 0 and ρ satisfy the condition

$$\mathbf{M}(\partial S_r) = \mathbf{M}\langle S, u, r- \rangle \le f'(r),$$

because $\operatorname{spt} \partial S \subset Z \sim \mathbf{U}(a, \rho)$; then 4.4.2 (2) yields a current $Y_r \in \mathscr{R}_{m,A}(Z)$ such that

$$\operatorname{spt}(\partial S_r - \partial Y_r) \subset B \quad \text{and} \quad \mathbf{M}(Y_r)^{(m-1)/m} \le \sigma f'(r),$$

hence $Y_r - S_r \in \mathscr{Z}_m(A, B)$ and

$$\langle \Phi, S_r \rangle + \langle \Phi, S - S_r \rangle = \langle \Phi, S \rangle \le \langle \Phi, S + Y_r - S_r \rangle \le \langle \Phi, Y_r \rangle + \langle \Phi, S - S_r \rangle,$$

$$\lambda^{-1} \mathbf{M}(S_r) \le \langle \Phi, S_r \rangle \le \langle \Phi, Y_r \rangle \le \lambda \mathbf{M}(Y_r),$$

$$[\lambda^{-2} f(r)]^{(m-1)/m} \le \sigma f'(r), \quad \lambda^{2/m-2} \le \sigma m (f^{1/m})'(r);$$

to justify the last step we observe that $f(r) > 0$ for $r > 0$, because $a \in \operatorname{spt} S$. Applying 2.9.19 to the nondecreasing function $f^{1/m}$ we conclude

$$[f(\rho)]^{1/m} \ge \textstyle\int_0^\rho (f^{1/m})' \, d\mathscr{L}^1 \ge \lambda^{2/m-2} \sigma^{-1} m^{-1} \rho.$$

We deduce the corollary: *If S is absolutely* Φ *minimizing with respect to* (A, B) *and* $a \in (\operatorname{spt} S) \sim (\operatorname{spt} \partial S)$, *then*

$$\Theta_*^m(\|S\|, a) \ge [\lambda^{2m-2} \sigma^m m^m \alpha(m)]^{-1} \quad \textit{and} \quad \operatorname{Tan}^m(\|S\|, a) = \operatorname{Tan}(\operatorname{spt} S, a).$$

Recalling 3.1.16 we suppose $v \in \operatorname{Tan}(\operatorname{spt} S, a)$, $|v| = 1$ and $0 < \varepsilon < \frac{1}{2}$. Whenever $0 < 2\delta \le \operatorname{dist}(a, \operatorname{spt} \partial S)$ there exists

$$x \in \operatorname{spt} S, r > 0 \quad \text{with} \quad |r(x-a) - v| < \varepsilon, \quad |x-a| < \delta,$$

and it follows that

$$\mathbf{U}(x, \varepsilon/r) \subset \mathbf{E}(a, v, 2\varepsilon) \cap \mathbf{U}(a, |x-a| + \varepsilon/r),$$

$$\varepsilon < 1 - \varepsilon = |v| - \varepsilon < r|x-a| < r\delta, \quad |x-a| + \varepsilon/r < 2\delta,$$

$$1 + \varepsilon > r|x-a|, \quad (1+2\varepsilon)\varepsilon/r > \varepsilon(|x-a| + \varepsilon/r),$$

$$\|S\| [\mathbf{E}(a, v, 2\varepsilon) \cap \mathbf{U}(a, |x-a| + \varepsilon/r) \ge \|S\| \mathbf{U}(x, \varepsilon/r)$$

$$\ge \gamma(\varepsilon/r)^m \ge \gamma \varepsilon^m (1+2\varepsilon)^{-m} (|x-a| + \varepsilon/r)^m,$$

where $\gamma = (\lambda^{2m-2} \sigma^m m^m)^{-1}$; accordingly

$$\Theta_*^m [\|S\| \llcorner \mathbf{E}(a, v, 2\varepsilon), a] \ge \gamma \varepsilon^m (1+2\varepsilon)^{-m} \alpha(m)^{-1} > 0.$$

Recalling 4.2.25 we observe that *if T is absolutely (homologically) Φ minimizing with respect to (A, B), so is each of the irreducible currents T_i*, because

$$\langle \Phi, T \rangle = \sum_{i=1}^{\infty} \langle \Phi, T_i \rangle .$$

The definitions of minimizing currents of various types remain meaningful and useful *in case A or B fails to be compact*, and this extension does not affect local properties, but the existence theorems (1) and (2) cannot be generalized, as shown by the impossibility of minimizing the length of 1 dimensional cycles on the surface of a cone whose vertex has been removed.

We extend the concepts of minimizing currents also to *the case when S is locally rectifiable and Φ is a positive integrand*, by substituting the requirement that

$$\langle \Phi, S \llcorner K \rangle \le \langle \Phi, (S \llcorner K) + X \rangle$$

for all compact subsets K of A (and all X as before).

5.1.7. Assuming $W \subset Z$ are open subsets of $\mathbf{R}^n$ we consider an **isotopic deformation** h of W in Z of class $k+1 \ge 3$; this means that

$$h: I \times W \to Z$$

is a map of class $k+1$, I is an open interval, $0 \in I$, h_0 is the inclusion map of W into Z and, for each $t \in I$, h_t is a diffeomorphism mapping W onto an open subset of Z; we define $h_t(z) = h(t, z)$ and

$$h_t^{(i)}(z) = \langle (1, 0)^i, D^i h(t, z) \rangle \text{ for } (t, z) \in I \times W,$$

and also denote $h_t^{(1)} = \dot{h}_t$, $h_t^{(2)} = \ddot{h}_t$.

If $T \in \mathscr{R}_m(W)$, Φ is a parametric integrand of degree m and class k on Z, and

$$J(t) = \langle \Phi, h_{t\#} T \rangle \text{ for } t \in I,$$

then the function J is of class k. In fact 5.1.1 implies

$$J(t) = \langle h_t^{\#} \Phi, T \rangle = \int \phi_z(t) \, d \|T\| z$$

with $\phi_z(t) = \Phi[h_t(z), \langle \vec{T}(z), \textstyle\bigwedge_m D h_t(z) \rangle]$ for $\|T\|$ almost all z. The functions ϕ_z are of class k, and their first k derivatives have bounds independent of z on every compact subinterval of I, because the function mapping (t, z) onto $\bigwedge_m D h_t(z)$ is of class k; the differentials of this latter function may be computed explicitly in terms of the differentials of h by use of the formula

$$\Gamma[\textstyle\bigwedge_m D h_t(z)] = \Gamma[D h_t(z)]^m / m!$$

obtained in 1.4.5. One concludes with the aid of 2.4.9 that

$$J^{(i)}(t) = \int \phi_z^{(i)}(t) \, d \|T\| z \text{ for } t \in I, \ i = 0, 1, \ldots, k.$$

The number $J^{(i)}(0)$ is called the **i-th variation associated with T, Φ, h** and denoted

$$\delta^{(i)}(T, \Phi, h).$$

For $\|T\|$ almost all z we can choose an orthonormal sequence $w_1, \ldots, w_m$ such that $\vec{T}(z) = w_1 \wedge \cdots \wedge w_m$, hence express

$$\langle \vec{T}(z), \wedge_m D\,h_t(z)\rangle = \langle w_1, D\,h_t(z)\rangle \wedge \cdots \wedge \langle w_m, D\,h_t(z)\rangle;$$

differentiating with respect to t at 0 we obtain the m-vector

$$\sum_{i=1}^{m} w_1 \wedge \cdots \wedge w_{i-1} \wedge \langle w_i, D\,\dot{h}_0(z)\rangle \wedge w_{i+1} \wedge \cdots \wedge w_m$$

whose norm does not exceed $m\,\|D\,\dot{h}_0(z)\|$. We conclude

$$|\delta^{(1)}(T, \Phi, h)| \le \int \|D\Phi[z, \vec{T}(z)]\| \cdot [|\dot{h}_0(z)| + m\,\|D\,\dot{h}_0(z)\|]\, d\,\|T\|\, z.$$

If $B \subset A \subset Z$, T is homologically Φ minimizing with respect to (A, B) and $h(I \times \operatorname{spt} T) \subset A$, $h[I \times (B \cap \operatorname{spt} \partial T)] \subset B$,

$$h(t, z) = z \text{ for } t \in I \text{ and } z \in (\operatorname{spt} \partial T) \sim B,$$

then $\delta^{(1)}(T, \Phi, h) = 0$ and $\delta^{(2)}(T, \Phi, h) \ge 0$. We will prove that the function J has a minimum at 0, hence $J'(0) = 0$ and $J''(0) \ge 0$. Assuming $s \in I$ we verify first that

$$\operatorname{spt} h_\#([0, s] \times \partial T) \subset B;$$

we define $H(t, z) = z$ for $(t, z) \in I \times Z$, observe that

$$H_\#([0, s] \times Q) = 0 \text{ for } Q \in \mathscr{D}_{m-1}(Z),$$

and let K be the compact interval with endpoints 0 and s; if V is any neighborhood of B, then the set

$$U = \{z\colon h(K \times \{z\}) \subset V\}$$

is a neighborhood of B and there exists a function

$$\phi \in \mathscr{D}^0(Z) \text{ with } \operatorname{spt} \phi \subset Z \sim B, \quad (\operatorname{spt} \partial T) \sim U \subset \operatorname{Int}\{x\colon \phi(x) = 1\};$$

taking $Q = (\partial T) \llcorner \phi$ we use 4.1.15 to infer

$$h_\#([0, s] \times Q) = H_\#([0, s] \times Q) = 0,$$

$$\operatorname{spt}(\partial T - Q) \subset U, \quad \operatorname{spt} h_\#([0, s] \times \partial T) \subset h(K \times U) \subset V.$$

We conclude that

$$h_{s\#} T - T = \partial h_\#([0, s] \times T) + h_\#([0, s] \times \partial T)$$

belongs to $\mathscr{B}_m(A, B)$ and $J(0) = \langle \Phi, T\rangle \le \langle \Phi, h_{s\#} T\rangle = J(s)$.

If ζ is any vectorfield of class $j \ge 2$ on Z and Clos W is a compact subset of Z, then there exist isotopic deformations h of W in Z of class j

which satisfy the condition $\dot{h}_0 = \zeta \mid W$. An example is the **affine deformation** h defined by the formula

$$h(t,z) = z + t\,\zeta(z) \text{ for } z \in W \text{ and } -\delta < t < \delta,$$

where δ is a positive number such that

$$\delta \cdot \operatorname{Lip}(\zeta \mid W) < 1 \text{ and } \delta\,|\zeta(z)| < \operatorname{dist}(z, \mathbf{R}^n \sim Z) \text{ for } z \in W;$$

in this case the equation $\dot{h}_t = \zeta$ holds whenever $|t| < \delta$. Another example is the **steady flow** h of W in Z with the velocity field ζ, constructed by solving the differential equation $\dot{h}_t = \zeta \circ h_t$ with the initial condition that h_0 be the inclusion map of W into Z.

5.1.8. *If Φ is the parametric area integrand of degree m on $\mathbf{R}^n$, $T \in \mathscr{R}_m(\mathbf{R}^n)$ and h is an isotopic deformation of class 3 of a neighborhood of* spt T, *then*

$$\delta^{(1)}(T, \Phi, h) = \int \operatorname{trace}(M_z)\, d\,\|T\|\, z,$$

$$\delta^{(2)}(T, \Phi, h) = \int [(\operatorname{trace} M_z)^2 + \operatorname{trace}(N_z^* \circ N_z + 2\Xi_z - M_z \circ M_z)]\, d\,\|T\|\, z$$

where, for $\|T\|$ almost all z in $\mathbf{R}^n$,

$$\rho_z \colon \operatorname{Tan}^m(\|T\|, z) \to \mathbf{R}^n \textit{ is the inclusion map},$$

$$M_z = \rho_z^* \circ D\,\dot{h}_0(z) \circ \rho_z, \quad \Xi_z = \rho_z^* \circ D\,\ddot{h}_0(z) \circ \rho_z, \quad N_z = D\,\dot{h}_0(z) \circ \rho_z - \rho_z \circ M_z.$$

Writing $\approx$ between expressions which describe functions with equal 2 jets at 0 in $\mathbf{R}$, we obtain

$$D\,h_t(z) \circ \rho_z \approx \rho_z + t\,\rho_z \circ M_z + t\,N_z + t^2\, D\,\ddot{h}_0(z) \circ \rho_z,$$

$$[D\,h_t(z) \circ \rho_z]^* \circ [D\,h_t(z) \circ \rho_z]$$

$$\approx \mathbf{1} + t[M_z + M_z^* + t(M_z^* \circ M_z + N_z^* \circ N_z + \Xi_z + \Xi_z^*)],$$

because $\rho_z^* \circ \rho_z$ is the identity map of $\operatorname{Tan}^m(\|T\|, z)$ and $\rho_z^* \circ N_z = 0$; we define ϕ_z as in 5.1.7 and use 1.7.9, 1.4.5, 1.7.12 to compute

$$\begin{aligned}[\phi_z(t)]^2 &= |\wedge_m [D\,h_t(z) \circ \rho_z]|^2 = \det([D\,h_t(z) \circ \rho_z]^* \circ [D\,h_t(z) \circ \rho_z]) \\ &\approx 1 + t \operatorname{trace}[M_z + M_z^* + t(M_z^* \circ M_z + N_z^* \circ N_z + \Xi_z + \Xi_z^*)] \\ &\qquad + t^2 \operatorname{trace}[\wedge_2(M_z + M_z^*)] \\ &= 1 + t\,2 \operatorname{trace}(M_z) \\ &\qquad + t^2[2(\operatorname{trace} M_z)^2 + \operatorname{trace}(N_z^* \circ N_z + 2\Xi_z - M_z \circ M_z)],\end{aligned}$$

$$\begin{aligned}\phi_z(t) &\approx 1 + t \operatorname{trace}(M_z) \\ &\qquad + (t^2/2)[(\operatorname{trace} M_z)^2 + \operatorname{trace}(N_z^* \circ N_z + 2\Xi_z - M_z \circ M_z)].\end{aligned}$$

5.1.9. Whenever Φ is a parametric integrand of degree m and class k on the open subset Z of $\mathbf{R}^n$ we define

$$\tilde{\Phi}: Z \times \operatorname{Hom}(\mathbf{R}^m, \mathbf{R}^n) \to \mathbf{R},$$
$$\tilde{\Phi}(z, \sigma) = \Phi[z, (\wedge_m \sigma)(e_1 \wedge \cdots \wedge e_m)]$$

for $z \in Z$ and $\sigma \in \operatorname{Hom}(\mathbf{R}^m, \mathbf{R}^n)$, where $e_1, \ldots, e_m$ are the standard base vectors of $\mathbf{R}^m$; we note that

$$\tilde{\Phi}(z, \sigma \circ g) = \tilde{\Phi}(z, \sigma) \cdot \det(g)$$

for $g \in \operatorname{Hom}(\mathbf{R}^m, \mathbf{R}^m)$ with $\det(g) > 0$. We also define

$$\tilde{\Phi}_a(\sigma) = \tilde{\Phi}(a, \sigma) \text{ for } a \in Z, \ \sigma \in \operatorname{Hom}(\mathbf{R}^m, \mathbf{R}^n).$$

If $K \subset Q \subset \mathbf{R}^m$, K is compact, Q is open, $F: Q \to \mathbf{R}^n$ is of class 1 and $F|K$ is univalent, then

$$\langle \Phi, F_\# (\mathbf{E}^m \llcorner K) \rangle = \langle F^\# \Phi, \mathbf{E}^m \llcorner K \rangle = \int_K \tilde{\Phi}[F(x), DF(x)] \, d\mathscr{L}^m x.$$

If Ψ is a parametric integrand of degree m on an open subset V of $\mathbf{R}^\nu$ and $G: Z \to V$ is of class 1, then

$$\widetilde{G^\# \Psi}(z, \sigma) = \tilde{\Psi}[G(z), DG(z) \circ \sigma]$$

for $z \in Z$ and $\sigma \in \operatorname{Hom}(\mathbf{R}^m, \mathbf{R}^n)$.

Assuming Φ, Z, m, n, k as above and representing $\mathbf{R}^n \simeq \mathbf{R}^m \times \mathbf{R}^{n-m}$ by the projections

$$\mathbf{p} \in \mathbf{O}^*(n, m), \quad \mathbf{p}(z) = (z_1, \ldots, z_m) \text{ for } z \in \mathbf{R}^n,$$
$$\mathbf{q} \in \mathbf{O}^*(n, n-m), \quad \mathbf{q}(z) = (z_{m+1}, \ldots, z_n) \text{ for } z \in \mathbf{R}^n,$$

we define the **nonparametric integrand $\Phi^\S$ associated with Φ** as the map

$$\Phi^\S: Z \times \operatorname{Hom}(\mathbf{R}^m, \mathbf{R}^{n-m}) \to \mathbf{R} \text{ of class } k,$$
$$\Phi^\S(z, \tau) = \tilde{\Phi}(z, \mathbf{p}^* + \mathbf{q}^* \circ \tau)$$

for $z \in Z$ and $\tau \in \operatorname{Hom}(\mathbf{R}^m, \mathbf{R}^{n-m})$; we observe that

$$\mathbf{p} \circ (\mathbf{p}^* + \mathbf{q}^* \circ \tau) = \mathbf{1}_{\mathbf{R}^m}, \quad \wedge_m(\mathbf{p}^* + \mathbf{q}^* \circ \tau) \neq 0.$$

In case $x \in \mathbf{R}^m$, $y \in \mathbf{R}^{n-m}$ and

$$z = \mathbf{p}^*(x) + \mathbf{q}^*(y) = (x_1, \ldots, x_m, y_1, \ldots, y_{n-m}) \in \mathbf{R}^n$$

we sometimes write (x, y) in place of z, and $\Phi^\S(x, y, \tau)$ in place of $\Phi^\S(z, \tau)$. We also define

$$\Phi^\S_a(\tau) = \Phi^\S(a, \tau) \text{ for } a \in Z, \ \tau \in \operatorname{Hom}(\mathbf{R}^m, \mathbf{R}^{n-m}).$$

If Q is an open subset of $\mathbf{R}^m$ and $f\colon Q \to \mathbf{R}^{n-m}$ is of class 1, then $F = \mathbf{p}^* + \mathbf{q}^* \circ f\colon Q \to \mathbf{R}^n$ is a univalent map of class 1 and

$$DF(x) = \mathbf{p}^* + \mathbf{q}^* \circ Df(x) \text{ for } x \in Q;$$

if K is a compact subset of Q with $F(K) \subset Z$, then

$$\langle \Phi, F_\# (\mathbf{E}^m \llcorner K) \rangle = \int_K \Phi^\S [x, f(x), Df(x)] \, d\mathscr{L}^m x.$$

Given any $\theta\colon \mathbf{R}^m \to \mathbf{R}^{n-m}$ of class 1 we choose a neighborhood W of $F(K)$ in Z and a positive number δ such that

$$\delta \, |\theta[\mathbf{p}(z)]| < \operatorname{dist}(z, \mathbf{R}^n \sim Z) \text{ for } z \in W,$$

and we define a **nonparametric isotopic deformation** h of W in Z by the formula

$$h(t, z) = z + t[\mathbf{q}^* \circ \theta \circ \mathbf{p}](z) \text{ for } z \in W, \; -\delta < t < \delta.$$

Observing that $h_t \circ F = \mathbf{p}^* + \mathbf{q}^* \circ (f + t\,\theta)$ we compute

$$\langle \Phi, h_{t\#} F_\# (\mathbf{E}^m \llcorner K) \rangle = \int_K \Phi^\S [x, f(x) + t\,\theta(x), Df(x) + t\,D\theta(x)] \, d\mathscr{L}^m x$$

for $-\delta < t < \delta$, and infer by differentiation with respect to t that

$$\delta^{(i)} [F_\# (\mathbf{E}^m \llcorner K), \Phi, h] = \int_K \langle [0, \theta(x), D\theta(x)]^i, D^i \Phi^\S [x, f(x), Df(x)] \rangle \, d\mathscr{L}^m x$$

for $i = 1, \ldots, k$; the above i-th power is computed in the symmetric algebra of the vectorspace $\mathbf{R}^m \times \mathbf{R}^{n-m} \times \operatorname{Hom}(\mathbf{R}^m, \mathbf{R}^{n-m})$.

Recalling 1.7.9 we see that

$$(\mathbf{p}^* + \mathbf{q}^* \circ \tau)^* \circ (\mathbf{p}^* + \mathbf{q}^* \circ \tau) = \mathbf{1}_{\mathbf{R}^m} + \tau^* \circ \tau,$$
$$|\textstyle\bigwedge_m (\mathbf{p}^* + \mathbf{q}^* \circ \tau)|^2 = \det(\mathbf{1}_{\mathbf{R}^m} + \tau^* \circ \tau)$$

for $\tau \in \operatorname{Hom}(\mathbf{R}^m, \mathbf{R}^{n-m})$, and infer from 1.4.5 that *the nonparametric integrand $\Phi^\S$ associated with the parametric area integrand Φ of degree m* is given by the formula

$$\Phi^\S(z, \tau) = \Big(\sum_{i=0}^m \operatorname{trace}[\textstyle\bigwedge_i (\tau^* \circ \tau)] \Big)^{\frac{1}{2}} = \Big(\sum_{i=0}^m |\textstyle\bigwedge_i \tau|^2 \Big)^{\frac{1}{2}}$$

for $z \in \mathbf{R}^n$; it follows that $\Phi^\S_z(0) = 1$, $D\Phi^\S_z(0) = 0$ and

$$\langle \tau^2, D^2 \Phi^\S_z(0) \rangle = \operatorname{trace}(\tau^* \circ \tau) = |\tau|^2.$$

If Ψ is a parametric integrand of degree m on an open subset V of $\mathbf{R}^\nu$ and $G\colon Z \to V$ is of class 1, then

$$(G^\# \Psi)^\S (z, \tau) = \tilde{\Psi} [G(z), DG(z) \circ \mathbf{p}^* + DG(z) \circ \mathbf{q}^* \circ \tau]$$

for $z \in Z$ and $\tau \in \operatorname{Hom}(\mathbf{R}^m, \mathbf{R}^{n-m})$.

5.1.10. *If Φ is a parametric integrand of degree m and class 2 on $Z \subset \mathbf{R}^n$, and if Φ is elliptic at a with ellipticity bound c, then*

$$\int (\langle [D\theta(x)]^2, D^2 \Phi_a^{\S}(0)\rangle - c\,|D\theta(x)|^2)\, d\mathscr{L}^m x \geq 0$$

for every function θ: $\mathbf{R}^m \to \mathbf{R}^{n-m}$ of class 1 with compact support. To verify this we define $\mathbf{p}$ and $\mathbf{q}$ as in 5.1.9, let

$$h(t,z) = z + t[\mathbf{q}^* \circ \theta \circ \mathbf{p}](z) \text{ for } (t,z) \in \mathbf{R} \times \mathbf{R}^n,$$

$$S = \mathbf{p}^*_{\#}(\mathbf{E}^m \llcorner \operatorname{spt} \theta),$$

$$\Psi(z,\alpha) = \Phi(a,\alpha) - c\,|\alpha| \text{ for } (z,\alpha) \in \mathbf{R}^n \times \wedge_m \mathbf{R}^n,$$

note that $h_t(z) = z$ whenever $z \in \operatorname{spt} \partial S \subset \mathbf{p}^*\{x: \theta(x) = 0\}$, hence $\partial h_{t\#} S = \partial S$, infer from the ellipticity of Φ that $\langle h_{t\#} S, \Psi\rangle$ is least when $t=0$, and use 5.1.9 to conclude

$$\begin{aligned} 0 \leq \delta^{(2)}(S,\Psi,h) &= \int \langle [0,\theta(x), D\theta(x)]^2, D^2 \Psi^{\S}[x,0,0]\rangle\, d\mathscr{L}^m x \\ &= \int (\langle [D\theta(x)]^2, D^2 \Phi_a^{\S}(0)\rangle - c\,|D\theta(x)|^2)\, d\mathscr{L}^m x. \end{aligned}$$

The above proposition may be extended to Lipschitzian functions θ by regularization. As a corollary we will derive the **nonparametric Legendre condition**

$$\langle (\xi y)^2, D^2 \Phi_a^{\S}(0)\rangle \geq c\,|\xi|^2\,|y|^2 \text{ for } \xi \in \wedge^1 \mathbf{R}^m,\ y \in \mathbf{R}^{n-m};$$

here $\xi y \in \operatorname{Hom}(\mathbf{R}^n, \mathbf{R}^{n-m})$ maps $x \in \mathbf{R}^m$ onto $\xi(x)\, y \in \mathbf{R}^{n-m}$. We consider the case when $|y|=1$, and $\xi(x) = x_1$ for $x \in \mathbf{R}^m$. First we choose $\lambda < \infty$ so that

$$|\langle \zeta, D^2 \Phi_a^{\S}(0)\rangle| \leq \lambda\,|\zeta| \text{ for } \zeta \in \bigodot_2 \operatorname{Hom}(\mathbf{R}^m, \mathbf{R}^{n-m}).$$

Assuming $m \leq \rho < \infty$ we then define

$$\phi(t) = \sup\{1-|t|, 0\} \text{ for } t \in \mathbf{R},$$

$$\psi(x) = \phi(x_2)\cdot\phi(x_3)\cdot \dots \cdot \phi(x_m) \text{ for } x \in \mathbf{R}^m,$$

$$\theta(x) = \rho^{-1}\phi(\rho^2 x_1)\,\psi(x)\, y \text{ for } x \in \mathbf{R}^m,$$

and compute

$$\int \phi^2\, d\mathscr{L}^1 = \tfrac{2}{3}, \qquad \mathscr{L}^m(\operatorname{spt}\theta) = 2^m \rho^{-2},$$

$$\int_{\operatorname{spt}\theta} \psi^2\, d\mathscr{L}^m = 2\rho^{-2}\left(\int \phi^2\, d\mathscr{L}^1\right)^{m-1} = 2^m\, 3^{1-m}\, \rho^{-2},$$

$$D_1\theta(x) = \pm \rho\,\psi(x)\, y, \quad |D_i\theta(x)| \leq \rho^{-1} \text{ for } i = 2,\dots,m,$$

$$\begin{aligned} |[D\theta(x)]^2 - [\rho\,\psi(x)\,\xi y]^2| &= |[D\theta(x) - \rho\,\psi(x)\,\xi y] \odot [D\theta(x) + \rho\,\psi(x)\,\xi y]| \\ &\leq \sqrt{2(m-1)}\,\rho^{-1}[(m-1)\rho^{-1} + 2\rho] \leq 6m \end{aligned}$$

for $\mathscr{L}^m$ almost all x in spt θ (see 1.10.5), hence infer

$$\begin{aligned}
&2^m 3^{1-m}(\langle(\xi y)^2, D^2 \Phi_a^{\S}(0)\rangle - c|\xi|^2|y|^2) \\
&= \int_{\operatorname{spt}\theta}(\langle[\rho\,\psi(x)\,\xi\, y]^2, D^2 \Phi_a^{\S}(0)\rangle - c|\rho\,\psi(x)\,\xi\, y|^2)\, d\mathscr{L}^m x \\
&\geq \int_{\operatorname{spt}\theta}(\langle[D\,\theta(x)]^2, D^2 \Phi_a^{\S}(0)\rangle - \lambda\, 6m - c|D\theta(x)|^2 - c\,6m)\, d\mathscr{L}^m x \\
&\geq -(\lambda+c)\,6m\,\mathscr{L}^m(\operatorname{spt}\theta) = -(\lambda+c)\,6m\,2^m\rho^{-2}.
\end{aligned}$$

Letting ρ approach ∞ we obtain the corollary.

Recalling 5.1.3 we infer that *the parametric Legendre condition implies the nonparametric Legendre condition.* One can also verify this directly by choosing orthonormal base vectors $u_1, \ldots, u_m$ of $\mathbf{R}^m$ with $\xi(u_i)=0$ for $i=2, \ldots, m$ and observing that

$$\langle u_1 \wedge \cdots \wedge u_m, \textstyle\bigwedge_m[\mathbf{p}^* + \mathbf{q}^* \circ (t\,\xi\, y)]\rangle = \alpha + t\,\beta$$

for $t \in \mathbf{R}$, where

$$\alpha = \mathbf{p}^*(u_1) \wedge \cdots \wedge \mathbf{p}^*(u_m), \qquad \beta = \xi(u_1)\,\mathbf{q}^*(y) \wedge \mathbf{p}^*(u_2) \wedge \cdots \wedge \mathbf{p}^*(u_m),$$

hence $|\alpha|=1$, $|\beta| = |\xi| \cdot |y|$, $\alpha \bullet \beta = 0$ and

$$\langle(\xi\, y)^2, D^2 \Phi_a^{\S}(0)\rangle = \langle\beta^2, D^2 \Phi_a(\alpha)\rangle.$$

On the other hand, *convexity of Φ_a does not imply convexity of $\Phi_a^{\S}$*, as shown by the example of the area integrand of degree 2 on $\mathbf{R}^4$.

5.1.11. In computations with functions whose domain is a subset of $\mathbf{R}^m \times \mathbf{R}^{n-m} \times \operatorname{Hom}(\mathbf{R}^m, \mathbf{R}^{n-m})$, we will sometimes use the following notational conventions:

Employing the standard base vectors and coordinate functions

$$e_1, \ldots, e_m \text{ and } X_1, \ldots, X_m \text{ of } \mathbf{R}^m,$$

$$v_1, \ldots, v_{n-m} \text{ and } Y_1, \ldots, Y_{n-m} \text{ of } \mathbf{R}^{n-m},$$

and defining

$$\mathbf{p}(x, y) = x, \quad \mathbf{q}(x, y) = y \text{ for } (x, y) \in \mathbf{R}^m \times \mathbf{R}^{n-m},$$

we consider $\mathbf{R}^m \times \mathbf{R}^{n-m} \simeq \mathbf{R}^n$ with the coordinate functions

$$Z_1 = X_1 \circ \mathbf{p}, \ldots, Z_m = X_m \circ \mathbf{p}, \; Z_{m+1} = Y_1 \circ \mathbf{q}, \ldots, Z_n = Y_{n-m} \circ \mathbf{q}.$$

Then the functions $X_i\, v_j$ form a basis of $\operatorname{Hom}(\mathbf{R}^m, \mathbf{R}^{n-m})$; in fact

$$\tau = \sum_{i=1}^{m} \sum_{j=1}^{n-m} Y_j[\tau(e_i)]\, X_i\, v_j \text{ for } \tau \in \operatorname{Hom}(\mathbf{R}^m, \mathbf{R}^{n-m}).$$

Whenever ϕ is a function of class 1 on a neighborhood of (x, y, τ) in $\mathbf{R}^m \times \mathbf{R}^{n-m} \times \operatorname{Hom}(\mathbf{R}^m, \mathbf{R}^{n-m})$ we denote the first order partial derivatives of ϕ by

$$D_i\,\phi(x, y, \tau) = \langle (e_i, 0, 0), D\,\phi(x, y, \tau)\rangle,$$
$$D_{m+j}\,\phi(x, y, \tau) = \langle (0, v_j, 0), D\,\phi(x, y, \tau)\rangle,$$
$$\phi_{(i,j)}(x, y, \tau) = \langle (0, 0, X_i\, v_j), D\,\phi(x, y, \tau)\rangle$$

for $i=1, \ldots, m$ and $j=1, \ldots, n-m$. In case ϕ is of class 2 we similarly denote the second order partial derivatives of ϕ by

$$D_{i,k}\,\phi(x, y, \tau) = D_i\, D_k\,\phi(x, y, \tau) = \langle (e_i, 0, 0) \odot (e_k, 0, 0), D^2\,\phi(x, y, \tau)\rangle,$$
$$D_{i,m+l}\,\phi(x, y, \tau) = D_i\, D_{m+l}\,\phi(x, y, \tau) = \langle (e_i, 0, 0) \odot (0, v_l, 0), D^2\,\phi(x, y, \tau)\rangle,$$
$$D_{m+j,m+l}\,\phi(x, y, \tau) = D_{m+j}\, D_{m+l}\,\phi(x, y, \tau)$$
$$= \langle (0, v_j, 0) \odot (0, v_l, 0), D^2\,\phi(x, y, \tau)\rangle,$$
$$D_i\,\phi_{(k,l)}(x, y, \tau) = (D_i\,\phi)_{(k,l)}(x, y, \tau)$$
$$= \langle (e_i, 0, 0) \odot (0, 0, X_k\, v_l), D^2\,\phi(x, y, \tau)\rangle,$$
$$D_{m+j}\,\phi_{(k,l)}(x, y, \tau) = (D_{m+j}\,\phi)_{(k,l)}(x, y, \tau)$$
$$= \langle (0, v_j, 0) \odot (0, 0, X_k\, v_l), D^2\,\phi(x, y, \tau)\rangle,$$
$$\phi_{(i,j;k,l)}(x, y, \tau) = (\phi_{(i,j)})_{(k,l)}(x, y, \tau)$$
$$= \langle (0, 0, X_i\, v_j) \odot (0, 0, X_k\, v_l), D^2\,\phi(x, y, \tau)\rangle$$

for $\{i, k\} \subset \{1, \ldots, m\}$ and $\{j, l\} \subset \{1, \ldots, n-m\}$.

Applying this notation with $\phi = \Phi^\S$ we express the first variation considered in 5.1.9 as

$$\delta^{(1)}[F_\#(\mathbf{E}^m \llcorner K), \Phi, h] = \int_K \langle [0, \theta(x), D\,\theta(x)], D\,\Phi^\S[x, f(x), D\,f(x)]\rangle\, d\mathscr{L}^m\, x$$
$$= \int_K \Big(\sum_{j=1}^{n-m} \theta_j(x)\, D_{m+j}\,\Phi^\S[x, f(x), D\,f(x)]$$
$$+ \sum_{i=1}^{m} \sum_{j=1}^{n-m} D_i\,\theta_j(x)\, \Phi^\S_{(i,j)}[x, f(x), D\,f(x)] \Big)\, d\mathscr{L}^m\, x$$

where $\theta_j = Y_j \circ \theta$. *In case Φ and f are of class 2 and* spt $\theta \subset K$, integration by parts yields the **Euler-Lagrange formula**

$$\delta^{(1)}[F_\#(\mathbf{E}^m \llcorner K), \Phi, h] = \int_K \sum_{j=1}^{n-m} \theta_j \cdot \Big[(D_{m+j}\,\Phi^\S) \circ \psi - \sum_{i=1}^{m} D_i(\Phi^\S_{(i,j)} \circ \psi) \Big]\, d\mathscr{L}^m$$

with $\psi(x)=[x, f(x), Df(x)]$ for $x\in\mathbf{R}^m$. *In order that this first variation equal* 0 *for all* $\theta\in\mathscr{D}(\mathbf{R}^m,\mathbf{R}^{n-m})$ *with* $\operatorname{spt}\theta\subset\operatorname{Int}K$ *it is necessary and sufficient that the functions*

$$(D_{m+j}\Phi^\S)\circ\psi-\sum_{i=1}^m D_i(\Phi^\S_{(i,j)}\circ\psi)$$

$$=(D_{m+j}\Phi^\S)\circ\psi-\sum_{i=1}^m(D_i\Phi^\S_{(i,j)})\circ\psi-\sum_{i=1}^m\sum_{l=1}^{n-m}[(D_{m+l}\Phi^\S_{(i,j)})\circ\psi]\,D_i f_l$$

$$-\sum_{i=1}^m\sum_{k=1}^m\sum_{l=1}^{n-m}[\Phi^\S_{(i,j;k,l)}\circ\psi]\cdot D_{i,k}f_l$$

corresponding to all $j\in\{1,\ldots,n-m\}$ *vanish on* $\operatorname{Int}K$; here $f_l=Y_l\circ f$.

5.2. Regularity of solutions of certain differential equations

5.2.1. For use throughout Section 5.2 we fix two positive integers $m<n$ and readopt the notational conventions of 5.1.11.

Assuming that U is an open subset of $\mathbf{R}^m$ we denote

$$(f,g)_U=\int_U[f(x)\bullet g(x)]\,d\mathscr{L}^m x,\qquad |f|_U=\left(\int_U|f(x)|^2\,d\mathscr{L}^m x\right)^{\frac12}=(f,f)_U^{\frac12}$$

whenever f and g are $\mathscr{L}^m\llcorner U$ measurable functions with values in some Hilbert space (see 1.7.1, 1.7.13, 2.4.12, 2.4.14). In case $U=\mathbf{R}^m$ the subscript U will be omitted. In case $b\in\mathbf{R}^m$ and $0<r\in\mathbf{R}$ we abbreviate

$$(f,g)_{\mathbf{U}(b,r)}=(f,g)_{b,r},\qquad |f|_{\mathbf{U}(b,r)}=|f|_{b,r}.$$

We observe that if f is $\mathscr{L}^m$ measurable, then

$$\int|f|_{b,r}^2\,d\mathscr{L}^m b=\alpha(m)\,r^m\,|f|^2.$$

Whenever $0<\delta\le 1$ and f is a map whose domain and range are metric spaces we define

$$\mathbf{h}_\delta(f)$$

as the smallest t such that $0\le t\le\infty$ and

$$\operatorname{dist}[f(x),f(u)]\le t[\operatorname{dist}(x,u)]^\delta\quad\text{for } x,u\in\operatorname{dmn}f.$$

In case $\mathbf{h}_\delta(f)<\infty$ the function f is said to satisfy a **Hölder condition with exponent** δ.

Each function f of class p mapping U into $\mathbf{R}^{n-m}$ has the p-th differential $D^p f$ mapping U into the inner product space $\bigodot^p(\mathbf{R}^m,\mathbf{R}^{n-m})$ described in 1.10.5, 1.10.6. When f is a weak solution of a strongly elliptic differential equation, in the sense explained in 5.2.3, one can study f by estimating the numbers

$$|D^p f|_{b,r} \quad \text{and} \quad \mathbf{h}_\delta[D^p f|\mathbf{B}(b,r)]$$

corresponding to $b \in U$ and $0 < r < \operatorname{dist}(b, \mathbf{R}^m \sim U)$. Thus we will derive results about certain linear equations in 5.2.5, 5.2.14 and about nonlinear equations arising from elliptic variational problems in 5.2.15, 5.2.18.

5.2.2. *If $T \in \mathscr{D}_0(U)$, $0 < \delta \leq 1$, $\beta < \infty$, $\gamma < \infty$ and there exists a sequence of real valued functions g_ν of class 1 on U such that*

$$\mathbf{h}_\delta(D\,g_\nu) \leq \beta \quad \text{and} \quad |D\,g_\nu|_U \leq \gamma \ \text{for} \ \nu = 0, 1, 2, \ldots,$$

$$T(\phi) = \lim_{\nu \to \infty} \textstyle\int_U g_\nu\, \phi\, d\mathscr{L}^m \ \text{for} \ \phi \in \mathscr{D}^0(U),$$

then there exists a real valued function g of class 1 on U such that

$$\mathbf{h}_\delta(D\,g) \leq \beta, \quad |D\,g|_U \leq \gamma, \quad T(\phi) = \textstyle\int_U g\, \phi\, d\mathscr{L}^m \ \text{for} \ \phi \in \mathscr{D}^0(U);$$

moreover g_ν and $D\,g_\nu$ converge to g and $D\,g$ uniformly on each compact subset of U. To prove this we observe that

$$\textstyle\int_U \liminf_{\nu \to \infty} |D\,g_\nu(x)|^2\, d\mathscr{L}^m\, x \leq \gamma$$

and pass to a subsequence (without changing notation) such that the differential forms $D\,g_\nu$ converge, uniformly on each compact subset of U, to a continuous differential form ψ of degree 1 on U (compare 2.10.21). Then the restrictions of the functions g_ν to each compact subset of U have a common Lipschitz constant. For each $b \in U$ the set of all numbers $g_\nu(b)$ is bounded, because otherwise we could select another subsequence such that for some $\rho > 0$ the functions $g_\nu|\mathbf{B}(b,\rho)$ converge uniformly to ∞ or $-\infty$, choose a nonnegative function $\phi \in \mathscr{D}^0(U) \sim \{0\}$ with $\operatorname{spt}\phi \subset \mathbf{U}(b,\rho)$, and infer that $T(\phi) = \pm\infty$. Consequently we can pass to another subsequence such that the functions g_ν converge, uniformly on each compact subset of U, to a continuous real valued function g. Since

$$[a,b](\psi) = \lim_{\nu \to \infty} [a,b](D\,g_\nu) = \lim_{\nu \to \infty} [g_\nu(b) - g_\nu(a)] = g(b) - g(a)$$

whenever U contains the line segment from a to b, it follows that $\psi = D\,g$. Moreover g is uniquely determined by T, hence all convergent subsequences of the original sequence have the same limit, and the original sequence converges.

5.2.3. A form $\Upsilon \in \bigodot^2 \operatorname{Hom}(\mathbf{R}^m, \mathbf{R}^{n-m})$ is termed **strongly elliptic** if and only if there exists a positive number c such that

$$\int \langle D\theta(x) \odot D\theta(x), \Upsilon \rangle \, d\mathscr{L}^m x \geq c\, |D\theta|^2$$

whenever $\theta \in \mathscr{D}(\mathbf{R}^m, \mathbf{R}^{n-m})$; one calls c an **ellipticity bound** for Υ. [Here $\odot$ means multiplication in $\bigodot_* \operatorname{Hom}(\mathbf{R}^m, \mathbf{R}^{n-m})$; see 1.9.1.] For example we have seen in 5.1.10 that this condition is satisfied by the form $D^2 \Phi_a^\S(0)$ associated with an elliptic parametric integrand Φ of degree m and class 2. Reasoning as in the derivation of the nonparametric Legendre condition one finds also that strong ellipticity of Υ with bound c implies the inequality

$$\Upsilon(\xi\, y, \xi\, y) \geq c\, |\xi|^2\, |y|^2 \quad \text{for } \xi \in \bigodot^1 \mathbf{R}^m,\ y \in \mathbf{R}^{n-m}.$$

[The converse implication is easily obtained by Fourier's transformation (see [VH 1], [H 2, § 1.7]), but will not be used in this book.]

We now consider *a continuous map*

$$A \colon U \to \bigodot^2 \operatorname{Hom}(\mathbf{R}^m, \mathbf{R}^{n-m})$$

for which there exist numbers $0 < c \leq M < \infty$ *such that* $A(x)$ *is strongly elliptic with ellipticity bound* c *and* $\|A(x)\| \leq M$ *whenever* $x \in U$; in accordance with 1.10.5 the last condition means that

$$\langle \sigma \odot \tau, A(x) \rangle \leq M\, |\sigma| \cdot |\tau| \quad \text{for } \sigma, \tau \in \operatorname{Hom}(\mathbf{R}^m, \mathbf{R}^{n-m}).$$

We define the bilinear symmetric function B by the formula

$$B(\eta, \zeta) = \int_U \langle \eta(x) \odot \zeta(x), A(x) \rangle \, d\mathscr{L}^m x$$

whenever η, ζ are $\mathscr{L}^m \llcorner U$ measurable functions with values in $\operatorname{Hom}(\mathbf{R}^m, \mathbf{R}^{n-m})$; if $|\eta| \cdot |\zeta|$ is $\mathscr{L}^m \llcorner U$ summable, then the above integral exists and

$$|B(\eta, \zeta)| \leq M(|\eta|, |\zeta|)_U \leq M\, |\eta|_U\, |\zeta|_U,$$

$$B(\phi\, \eta, \zeta) = B(\eta, \phi\, \zeta) \quad \text{for } \phi \in \mathscr{D}^0(U).$$

Taking $\theta \in \mathscr{D}(U, \mathbf{R}^{n-m})$, $b \in U$ and comparing $B(D\theta, D\theta)$ with the integral obtained through replacement of Υ by $A(b)$ in the preceding paragraph, one might hope to majorize $|D\theta|_U^2$ by a constant multiple of $B(D\theta, D\theta)$; however we will see that in general one must add a multiple of $|\theta|_U^2$ to obtain a valid estimate.

Applying 4.1.2 with n replaced by m and $\Phi^{\ddagger} \in \mathscr{D}^0(\mathbf{R}^m)$, hence

$$N = \sup\{\sup \operatorname{im} \Phi^{\ddagger}, \sup \operatorname{im} |D\Phi^{\ddagger}|\} < \infty,$$

we first prove a version of **Gårding's inequality**:

If $0<d<\inf\{N^2\alpha(m), c(c+M+1)^{-1}\}$, $P=N^4\alpha(m)\,d^{-1}$,

$$\theta\in\mathscr{D}(\mathbf{R}^m,\mathbf{R}^{n-m}),\qquad 0<\varepsilon<\operatorname{dist}(\operatorname{spt}\theta,\mathbf{R}^m\sim U),$$

$$\|A(x)-A(b)\|\le d \text{ whenever } x\in\operatorname{spt}\theta \text{ and } b\in\mathbf{B}(x,\varepsilon),$$

then

$$B(D\theta,D\theta)+(c+M)\,2\alpha(m)\,P\,\varepsilon^{-2}\,|\theta|_U^2\ge[c-(c+M+1)\,d]\,|D\theta|_U^2.$$

Letting $\psi_b(x)=\Phi_\varepsilon^{\frac12}(x-b)\,\theta(x)$ for $b,x\in\mathbf{R}^m$ and using the norm $\|\ \|$ (see 1.10.5) on $\bigodot_2\operatorname{Hom}(\mathbf{R}^m,\mathbf{R}^{n-m})$ we see that

$$\begin{aligned}
&\int\|[D\psi_b(x)]^2-\Phi_\varepsilon(x-b)[D\theta(x)]^2\|\,d\mathscr{L}^m x\\
&=\int\|[D\Phi_\varepsilon^{\frac12}(x-b)\,\theta(x)]^2+2D\Phi_\varepsilon^{\frac12}(x-b)\,\theta(x)\odot\Phi_\varepsilon^{\frac12}(x-b)\,D\theta(x)\|\,d\mathscr{L}^m x\\
&\le N^2\varepsilon^{-m-2}\,|\theta|_{b,\varepsilon}^2+2N^2\varepsilon^{-m-1}\,|\theta|_{b,\varepsilon}\,|D\theta|_{b,\varepsilon}\\
&\le\varepsilon^{-m}[P\varepsilon^{-2}\,|\theta|_{b,\varepsilon}^2+2P^{\frac12}\varepsilon^{-1}\,|\theta|_{b,\varepsilon}\,\alpha(m)^{-\frac12}\,d^{\frac12}\,|D\theta|_{b,\varepsilon}]\\
&\le\varepsilon^{-m}[2P\varepsilon^{-2}\,|\theta|_{b,\varepsilon}^2+\alpha(m)^{-1}\,d\,|D\theta|_{b,\varepsilon}^2]
\end{aligned}$$

and $\int\langle[D\psi_b(x)]^2,A(b)\rangle\,d\mathscr{L}^m x\ge c\,|D\psi_b|^2$, hence

$$\begin{aligned}
&\int\langle\Phi_\varepsilon(x-b)[D\theta(x)]^2,A(b)\rangle\,d\mathscr{L}^m x\\
&\ge c\,|D\psi_b|^2-M\varepsilon^{-m}[2P\varepsilon^{-2}\,|\theta|_{b,\varepsilon}^2+\alpha(m)^{-1}\,d\,|D\theta|_{b,\varepsilon}^2]\\
&\ge c\int\Phi_\varepsilon(x-b)|D\theta(x)|^2\,d\mathscr{L}^m x-(c+M)\varepsilon^{-m}[2P\varepsilon^{-2}\,|\theta|_{b,\varepsilon}^2+\alpha(m)^{-1}d|D\theta|_{b,\varepsilon}^2]
\end{aligned}$$

and conclude

$$\begin{aligned}
B(D\theta,D\theta)+d\,|D\theta|^2&\ge\int\langle[D\theta(x)]^2,\int\Phi_\varepsilon(x-b)A(b)\,d\mathscr{L}^m b\rangle\,d\mathscr{L}^m x\\
&=\int\int\langle\Phi_\varepsilon(x-b)[D\theta(x)]^2,A(b)\rangle\,d\mathscr{L}^m x\,d\mathscr{L}^m b\\
&\ge c\,|D\theta|^2-(c+M)[2\alpha(m)\,P\,\varepsilon^{-2}\,|\theta|^2+d\,|D\theta|^2].
\end{aligned}$$

In some special circumstances it is possible to sharpen Gårding's inequality by omission of the summand involving $|\theta|_U$. Evidently:

If $\|A(x)-A(b)\|\le d$ *whenever* $x,b\in U$, *then*

$$B(D\theta,D\theta)\ge(c-d)\,|D\theta|_U^2 \text{ for } \theta\in\mathscr{D}(U,\mathbf{R}^{n-m}).$$

Also, if $m=1$ *or* $n-m=1$, *then* $B(\eta,\eta)\ge c\,|\eta|_U^2$ *for every* $\mathscr{L}^m\llcorner U$ *measurable function with values in* $\operatorname{Hom}(\mathbf{R}^m,\mathbf{R}^{n-m})$, because in these cases the extension of the nonparametric Legendre condition to each form $A(x)$ implies

$$\langle\sigma\odot\sigma,A(x)\rangle\ge c\,|\sigma|^2 \text{ for } \sigma\in\operatorname{Hom}(\mathbf{R}^m,\mathbf{R}^{n-m}).$$

To construct an example showing that *such sharpening is generally impossible when $m \geq 2$ and $n-m \geq 2$* we choose

$$\alpha \in \mathscr{D}(\mathbf{R}^m, \mathbf{R}^{n-m}) \quad \text{so that} \quad |\beta|>0$$

with

$$\beta(x) = \langle D_1 \alpha(x) \wedge D_2 \alpha(x), Y_1 \wedge Y_2 \rangle \text{ for } x \in \mathbf{R}^m,$$

select a number $\gamma > |D\alpha|^2\, 2^{-1} |\beta|^{-2}$, define $A(x)$ as the function mapping (σ, τ) onto

$$\sigma \bullet \tau - \gamma\, \beta(x) \langle \sigma(e_1) \wedge \tau(e_2) - \sigma(e_2) \wedge \tau(e_1), Y_1 \wedge Y_2 \rangle,$$

and verify that $\int \langle [D\theta(x)]^2, A(b) \rangle \, d\mathscr{L}^m x = |D\theta|^2$ whenever $b \in \mathbf{R}^m$ and $\theta \in \mathscr{D}(\mathbf{R}^m, \mathbf{R}^{n-m})$, but $B(D\alpha, D\alpha) = |D\alpha|^2 - 2\gamma |\beta|^2 < 0$.

Next we establish a basic estimate concerning weak solutions of certain elliptic differential equations:

Suppose $\lambda \in \mathbf{R}$, $\chi \in \mathbf{L}_2(\mathscr{L}^m \llcorner U, \mathbf{R}^{n-m})$, $\Omega \in \mathbf{L}_2[\mathscr{L}^m \llcorner U, \operatorname{Hom}(\mathbf{R}^m, \mathbf{R}^{n-m})]$ and $f: U \to \mathbf{R}^{n-m}$ is a map of class 1 such that

$$B(Df, D\theta) + \lambda (f, \theta)_U = (\Omega, D\theta)_U + (\chi, \theta)_U$$

for all $\theta \in \mathscr{D}(U, \mathbf{R}^{n-m})$. If $0 \leq \mu \leq M$ and

$$B(D\theta, D\theta) + \lambda |\theta|_U^2 \geq \mu |D\theta|^2$$

for all $\theta \in \mathscr{D}(U, \mathbf{R}^{n-m})$, then

$$\mu |Df|_{b,r} \leq |\Omega|_{b,\rho} + (\rho - r) |\chi|_{b,\rho} + \mu^{\frac{1}{2}} M^{\frac{1}{2}} (\rho - r)^{-1} |f|_{b,\rho}$$

whenever $b \in U$, $0 < r < \rho < \infty$ and $\mathbf{U}(b, \rho) \subset U$.

Regularization shows that the hypotheses remain valid for every function θ of class 1 mapping U into $\mathbf{R}^{n-m}$ such that $\operatorname{spt} \theta$ is a compact subset of U. Whenever $\rho - r < t < \infty$ we choose

$$\phi \in \mathscr{D}^0(\mathbf{R}^m) \quad \text{with} \quad 0 \leq \phi(x) \leq 1 \text{ and } |D\phi(x)| < t^{-1} \text{ for } x \in \mathbf{R}^m,$$

$$\phi(x) = 1 \text{ for } x \in \mathbf{B}(b, r), \qquad \operatorname{spt} \phi \subset \mathbf{U}(b, \rho),$$

infer that

$$\begin{aligned} \mu |D(\phi f)|_U^2 &\leq B[(D\phi) f + \phi\, Df, (D\phi) f + \phi\, Df] + \lambda |\phi f|_U^2 \\ &= B[(D\phi) f, (D\phi) f] + B[Df, D(\phi^2 f)] + \lambda (f, \phi^2 f)_U \\ &= B[(D\phi) f, (D\phi) f] + (\Omega, D(\phi^2 f))_U + (\chi, \phi^2 f)_U \\ &= B[(D\phi) f, (D\phi) f] + (\phi\,\Omega, (D\phi) f)_U + (\phi\,\Omega, D(\phi f))_U + (\phi\,\chi, \phi f)_U \\ &\leq M t^{-2} |f|_{b,\rho}^2 + |\Omega|_{b,\rho}\, t^{-1} |f|_{b,\rho} + |\Omega|_{b,\rho} |D(\phi f)|_U + |\chi|_{b,\rho} |f|_{b,\rho} \end{aligned}$$

and conclude

$$\begin{aligned}&(\mu\,|D(\phi f)|_U-2^{-1}\,|\Omega|_{b,\rho})^2\\&\leq 2^{-2}\,|\Omega|_{b,\rho}^2+\mu M t^{-2}\,|f|_{b,\rho}^2+\mu(|\Omega|_{b,\rho}+t\,|\chi|_{b,\rho})\,t^{-1}\,|f|_{b,\rho}\\&\leq(2^{-1}\,|\Omega|_{b,\rho}+t\,|\chi|_{b,\rho}+\mu^{\frac{1}{2}} M^{\frac{1}{2}} t^{-1}\,|f|_{b,\rho})^2\end{aligned}$$

with $|Df|_{b,r}\leq|D(\phi f)|_U$.

Finally we explain why *the above function f of class 1 is considered a weak solution of a strongly elliptic system of second order partial differential equations.* Letting

$$A_{i,j;k,l}(x)=\langle X_i v_j \odot X_k v_l, A(x)\rangle,$$
$$\Omega_{i,j}(x)=\langle\langle e_i,\Omega(x)\rangle, Y_j\rangle,\qquad g_j(x)=\langle g(x), Y_j\rangle$$

whenever $\{i,k\}\subset\{1,\dots,m\}$, $\{j,l\}\subset\{1,\dots,n-m\}$, $x\in U$ and g maps U into $\mathbf{R}^{n-m}$, we find that our hypothesis concerning f is equivalent to the condition

$$\begin{aligned}\int_U\Bigg(&\sum_{i=1}^{m}\sum_{j=1}^{n-m}\sum_{k=1}^{m}\sum_{l=1}^{n-m}A_{i,j;k,l}\,D_i f_j\,D_k\theta_l+\lambda\sum_{l=1}^{n-m}f_l\theta_l\\&-\sum_{k=1}^{m}\sum_{l=1}^{n-m}\Omega_{k,l}\,D_k\theta_l-\sum_{l=1}^{n-m}\chi_l\theta_l\Bigg)\,d\mathscr{L}^m=0\end{aligned}$$

for all $\theta\in\mathscr{D}(U,\mathbf{R}^{n-m})$; integrating by parts we infer that, *in case f is of class 2, the condition holds if and only if the functions $f_1,\dots,f_{n-m}$ satisfy the equations*

$$-\sum_{i=1}^{m}\sum_{j=1}^{n-m}\sum_{k=1}^{m}D_k(A_{i,j;k,l}\,D_i f_j)+\lambda f_l+\sum_{k=1}^{m}D_k\,\Omega_{k,l}-\chi_l=0$$

for $l=1,\dots,n-m$. These equations do not constitute the most general second order strongly elliptic linear system, because additional linear combinations of $D_i f_j$ and f_j with prescribed coefficient functions could be included, and treated similarly. However the more general systems will not be needed in this book. On the other hand we will replace the functions f_j by distributions whose first partial derivatives correspond to $\mathscr{L}^m$ square-summable functions in 5.2.6, for later use in 5.3.8.

5.2.4. Sobolev's inequality. *If f is a function of class $N>m/2$ mapping $\mathbf{R}^m\cap\mathbf{U}(b,\rho)$ into a Hilbert space, then*

$$|f(b)|\leq\sum_{i=0}^{N}2N\,[i!\,\alpha(m)]^{-\frac{1}{2}}\,\rho^{i-m/2}\,|D^i f|_{b,\rho}.$$

Proof. We compute $f(b)$ by Taylor's expansion at each point x of $\mathbf{U}(b,\rho)$, and integrate to obtain

$$\alpha(m)\,\rho^m f(b)=\int_{\mathbf{U}(b,\rho)}\left[\sum_{i=0}^{N-1}\langle (b-x)^i/i!,\,D^i f(x)\rangle+R(x)\right]d\mathscr{L}^m x$$

with $R(x)=\int_0^1\langle (b-x)/N!,\,D^N f[x+t(b-x)]\rangle\,N(1-t)^{N-1}\,d\mathscr{L}^1 t$. Noting that $\int_{\mathbf{U}(b,\rho)}|b-x|^s\,d\mathscr{L}^m x\le\rho^{m+s}\alpha(m)$ whenever $s\ge 0$, we estimate

$$\int_{\mathbf{U}(b,\rho)}|\langle (b-x)^i/i!,\,D^i f(x)\rangle|\,d\mathscr{L}^m x$$
$$\le\int_{\mathbf{U}(b,\rho)} i!^{-\frac{1}{2}}\,|b-x|^i\,|D^i f(x)|\,d\mathscr{L}^m x\le i!^{-\frac{1}{2}}[\rho^{m+2i}\alpha(m)]^{\frac{1}{2}}\,|D^i f(x)|_{b,\rho}$$

and

$$\int_{\mathbf{U}(b,\rho)}|R(x)|\,dL^m x$$
$$\le\int_0^1\int_{\mathbf{U}(b,\rho)}|\langle (b-x)^N/N!,\,D^N f[t\,b+(1-t)\,x]\rangle|\,d\mathscr{L}^m x\,N(1-t)^{N-1}\,d\mathscr{L}^1 t$$
$$\le\int_0^1 N!^{-\frac{1}{2}}[\rho^{m+2N}\alpha(m)]^{\frac{1}{2}}$$
$$\cdot[\textstyle\int_{\mathbf{U}(b,\rho)}|(D^N f)\circ\tau_{tb}\circ\mu_{1-t}|^2\,d\mathscr{L}^m]^{\frac{1}{2}}\,N(1-t)^{N-1}\,d\mathscr{L}^1 t$$
$$\le\int_0^1 N!^{-\frac{1}{2}}\,\rho^{m/2+N}\alpha(m)^{\frac{1}{2}}\,|D^N f|_{b,\rho}(1-t)^{-m/2+N-1}\,N\,d\mathscr{L}^1 t$$
$$=N!^{-\frac{1}{2}}\,\rho^{m/2+N}\alpha(m)^{\frac{1}{2}}\,|D^N f|_{b,\rho}\,N/(N-m/2)$$

with $2N-m\ge 1$.

5.2.5. Theorem. *Suppose* $0<c\le M<\infty$, $\Upsilon\in\odot^2\operatorname{Hom}(\mathbf{R}^m,\mathbf{R}^{n-m})$, Υ *is strongly elliptic with ellipticity bound* c, $\|\Upsilon\|\le M$, N *is the least integer exceeding* $m/2$, *and* $0<\delta\le 1$.

If $b\in\mathbf{R}^m$, $0<\rho<\infty$, $f\in\mathscr{E}[\mathbf{U}(b,\rho),\mathbf{R}^{n-m}]$ *with*

$$\int_{\mathbf{U}(b,\rho)}\langle Df(x)\odot D\theta(x),\Upsilon\rangle\,d\mathscr{L}^m x=0$$

for all $\theta\in\mathscr{D}[\mathbf{U}(b,\rho),\mathbf{R}^{n-m}]$ (such a function f is called Υ **harmonic**), $0\le p\le q$ *are integers and* $0<r<\rho$, *then*

$$q!^{\frac{1}{2}}\,|D^q f|_{b,r}\le\left[\left(\frac{M}{c}\right)^{\frac{1}{2}}\frac{q-p}{\rho-r}\right]^{q-p} p!^{\frac{1}{2}}\,|D^p f|_{b,\rho},$$

$$q!^{\frac{1}{2}}\,|D^q f(b)|\le 2N\,e\,\alpha(m)^{-\frac{1}{2}}(N+q-p)^{N+q-p}$$
$$\cdot(M/c)^{(N+q-p)/2}(2/\rho)^{m/2+q-p}\,p!^{\frac{1}{2}}\,|D^p f|_{b,\rho},$$

$$\mathbf{h}_\delta[D^p f\,|\,\mathbf{B}(b,r)]\le 4(N+1)^{N+2}\,e\,\alpha(m)^{-\frac{1}{2}}\,2^{m/2}$$
$$\cdot(M/c)^{(N+1)/2}(\rho-r)^{-m/2-\delta}\,|D^p f|_{b,\rho}$$

and f *is an analytic function.*

Proof. Replacing ρ by a slightly smaller number we may assume that $|D^i f|_{b,\rho} < \infty$ for every nonnegative integer i.

From 1.10.6 we see that

$$i!\,|D^i f(x)|^2 = \sum_{s\in\mathscr{S}(m,i)} |D_s f(x)|^2,$$

$$(i+j)!\,|D^{i+j} f(x)|^2 = i!\,j!\,|D^i D^j f(x)|^2$$

whenever $x\in\mathbf{U}(b,\rho)$ and i,j are nonnegative integers.

If $k\in\{1,\ldots,m\}$, then $\theta\in\mathscr{D}[\mathbf{U}(b,\rho),\mathbf{R}^{n-m}]$ implies $D_k\theta\in\mathscr{D}[\mathbf{U}(b,\rho),\mathbf{R}^{n-m}]$ and integration by parts yields

$$\int_{\mathbf{U}(b,\rho)} \langle D D_k f(x)\odot D\theta(x), \Upsilon\rangle\, d\mathscr{L}^m x$$
$$= -\int_{\mathbf{U}(b,\rho)} \langle D f(x)\odot D D_k\theta(x), \Upsilon\rangle\, d\mathscr{L}^m x = 0,$$

hence $D_k f$ is Υ harmonic. By induction we infer that $D_s f$ is Υ harmonic for every finite sequence s of nonnegative integers, and we use the basic estimate in 5.2.3 to obtain the inequality

$$|D D_s f|_{b,r} \le (M/c)^{\frac{1}{2}}(R-r)^{-1}\,|D_s f|_{b,R}$$

for $0<r<R\le\rho$; squaring and summing with respect to s over $\mathscr{S}(m,i)$ we find that

$$(i+1)!\,|D^{i+1} f|^2_{b,r} \le M\,c^{-1}(R-r)^{-2}\,i!\,|D^i f|^2_{b,R}.$$

From the resulting inequalities

$$(p+j+1)!\,|D^{p+j+1} f|^2_{b,\,\rho-(j+1)(\rho-r)/(q-p)}$$
$$\le M\,c^{-1}(\rho-r)^{-2}(q-p)^2(p+j)!\,|D^{p+j}|^2_{b,\,\rho-j(\rho-r)/(q-p)}$$

for $j\in\{0,\ldots,q-p-1\}$ we deduce the first conclusion of the theorem.

Applying 5.2.4 with f,ρ replaced by $D^q f,\rho/2$ we obtain

$$|D^q f(b)| \le \sum_{i=0}^{N} 2N\,[i!\,\alpha(m)]^{-\frac{1}{2}}(\rho/2)^{i-m/2}\binom{i+q}{i}^{\frac{1}{2}}|D^{i+q} f|_{b,\rho}$$
$$\le \sum_{i=0}^{N} 2N\,i!^{-1}\,\alpha(m)^{-\frac{1}{2}}\,q!^{-\frac{1}{2}}(2/\rho)^{m/2+q-p}$$
$$\cdot (M/c)^{(i+q-p)/2}(i+q-p)^{i+q-p}\,p!^{\frac{1}{2}}\,|D^p f|_{b,\rho}$$

and infer the second conclusion of the theorem.

Whenever $x\in\mathbf{B}(b,r)$ we apply the second conclusion with b,ρ replaced by $x,\rho-r$ and infer

$$|D^p f(x)| \le \alpha(\rho-r)^{-m/2}, \quad |D^{p+1} f(x)| \le 2\alpha(\rho-r)^{-m/2-1}$$

where $\alpha = 2(N+1)^{N+2}\,e\,\alpha(m)^{-\frac{1}{2}}(M/c)^{(N+1)/2}\,2^{m/2}\,|D^p f|_{b,\rho}$.

If $x, u \in \mathbf{B}(b,r)$, then either $|x-u| \geq \rho - r$ and

$$|D^p f(x) - D^p f(u)| \leq 2\alpha(\rho-r)^{-m/2} \leq 2\alpha(\rho-r)^{-m/2-\delta}|x-u|^\delta,$$

or $|x-u| \leq \rho - r$ and

$$|D^p f(x) - D^p f(u)| \leq 2\alpha(\rho-r)^{-m/2-1}|x-u| \leq 2\alpha(\rho-r)^{-m/2-\delta}|x-u|^\delta.$$

Accordingly the third conclusions holds.

Applying the second conclusion once more, as above, in conjunction with the fact that

$$(N+q)^{N+q} \leq (N+q)!\, e^{N+q} \leq N!\, q!\, (2e)^{N+q}$$

for all nonnegative integers q, we obtain for $x \in \mathbf{B}(b,r)$ the estimate

$$q!^{\frac{1}{2}}\, |D^q f(x)| \leq N\, \alpha(m)^{-\frac{1}{2}}\, N!\, q!\, (2e)^{1+N+q} (M/c)^{(N+q)/2} \cdot [2/(\rho-r)]^{m/2+q}\, |f|_{b,\rho},$$

and deduce the analyticity of f from 3.1.24(1) and 1.10.6.

5.2.6. Corollary. *If $H_1, \ldots, H_{n-m} \in \mathscr{D}_0[\mathbf{R}^m \cap \mathbf{U}(b,\rho)]$ with*

$$\sum_{i=1}^{m} \sum_{j=1}^{n-m} \sum_{k=1}^{m} \sum_{l=1}^{n-m} \Upsilon_{i,j;k,l}\, D_k H_l(D_i\, \theta_j) = 0$$

for all $\theta \in \mathscr{D}[\mathbf{U}(b,\rho), \mathbf{R}^{n-m}]$, where

$$\Upsilon_{i,j,k,l} = \Upsilon(X_i\, v_j;\, X_k\, v_l) \quad \text{and} \quad \theta_j = Y_j \circ \theta,$$

and if $0 \leq \mu < \infty$ with

$$\sum_{i=1}^{m} \sum_{j=1}^{n-m} D_i\, H_j(w_{i,j}) \leq \mu \Big(\sum_{i=1}^{m} \sum_{j=1}^{n-m} |w_{i,j}|_{b,\rho}^2 \Big)^{\frac{1}{2}}$$

whenever $w_{i,j} \in \mathscr{D}^0[\mathbf{U}(b,\rho)]$, then there exists a Υ harmonic function g such that $|D\, g|_{b,\rho} \leq \mu$ and

$$H_j(\phi) = \int_{\mathbf{U}(b,\rho)} \phi \cdot (Y_j \circ g)\, d\mathscr{L}^m$$

for $j = 1, \ldots, n-m$ and $\phi \in \mathscr{D}^0[\mathbf{U}(b,\rho)]$.

Proof. Recalling 4.1.2 we regularize H_l by means of the functions and distributions

$$g_{l,\varepsilon} \in \mathscr{E}^0[\mathbf{U}(b,\rho-\varepsilon)] \quad \text{and} \quad H_{l,\varepsilon} \in \mathscr{D}_0[\mathbf{U}(b,\rho-\varepsilon)]$$

satisfying the equations

$$g_{l,\varepsilon}(u) = (H_l)_x\, \Phi_\varepsilon(x-u) \text{ for } u \in \mathbf{U}(b,\rho-\varepsilon),$$

$$H_{l,\varepsilon}(\phi) = \int_{\mathbf{U}(b,\rho-\varepsilon)} \phi\, g_{l,\varepsilon}\, d\mathscr{L}^m = H_l(\Phi_\varepsilon * \phi)$$

for $\phi \in \mathscr{D}^0[\mathbf{U}(b,\rho-\varepsilon)]$; to compute $\Phi_\varepsilon * \phi$ we extend ϕ by mapping $\mathbf{U}(b,\rho) \sim U(b,\rho-\varepsilon)$ onto 0; we note that

$$\int_{\mathbf{U}(b,\rho-\varepsilon)} \phi \, D_k g_{i,\varepsilon} \, d\mathscr{L}^m = D_k H_i(\Phi_\varepsilon * \phi).$$

We also let $g_\varepsilon(u) = (g_{1,\varepsilon}(u), \ldots, g_{n-m,\varepsilon}(u))$ for $u \in \mathbf{U}(b,\rho-\varepsilon)$.

If $\theta \in \mathscr{D}[\mathbf{U}(b,\rho-\varepsilon), \mathbf{R}^{n-m}]$ and $\theta_j = Y_j \circ \theta$, then

$$\begin{aligned}
&\int_{\mathbf{U}(b,\rho-\varepsilon)} \langle D\theta(x) \odot D g_\varepsilon(x), \Upsilon \rangle \, d\mathscr{L}^m x \\
&= \int_{\mathbf{U}(b,\rho-\varepsilon)} \sum_{i=1}^{m} \sum_{j=1}^{n-m} \sum_{k=1}^{m} \sum_{l=1}^{n-m} \Upsilon_{i,j;k,l} \, D_i \theta_j \, D_k g_{l,\varepsilon} \, d\mathscr{L}^m \\
&= \sum_{i=1}^{m} \sum_{j=1}^{n-m} \sum_{k=1}^{m} \sum_{l=1}^{n-m} \Upsilon_{i,j;k,l} \, D_k H_l[D_i(\Phi_\varepsilon * \theta_j)] = 0.
\end{aligned}$$

Thus g_ε is a Υ harmonic function.

If $w_{i,j} \in \mathscr{D}^0[\mathbf{U}(b,\rho-\varepsilon)]$, then

$$\begin{aligned}
&\sum_{i=1}^{m} \sum_{j=1}^{n-m} \int_{\mathbf{U}(b,\rho)} w_{i,j} \, D_i g_{j,\varepsilon} \, d\mathscr{L}^m = \sum_{i=1}^{m} \sum_{j=1}^{n-m} D_i H_j(\Phi_\varepsilon * w_{i,j}) \\
&\le \mu \left(\int_{\mathbf{U}(b,\rho)} \sum_{i=1}^{m} \sum_{j=1}^{n-m} |\Phi_\varepsilon * w_{i,j}|^2 \, d\mathscr{L}^m \right)^{\frac{1}{2}} \\
&\le \mu \left(\sum_{i=1}^{m} \sum_{j=1}^{n-m} \int_{\mathbf{U}(b,\rho-\varepsilon)} |w_{i,j}|^2 \, d\mathscr{L}^m \right)^{\frac{1}{2}}.
\end{aligned}$$

It follows that $|D g_\varepsilon|_{\mathbf{U}(b,\rho-\varepsilon)} \le \mu$.

Applying 5.2.5 with $f = g_\varepsilon$, we see that if $0 < r < \rho$ and q is a positive integer, then the set of all numbers

$$|D^q g_\varepsilon|_{b,r} \quad \text{and} \quad \mathbf{h}_1[D^q g_\varepsilon | \mathbf{B}(b,r)]$$

corresponding to $0 < \varepsilon \le (\rho - r)/2$ is bounded. Then we use 5.2.2, with T replaced by the distributions H_l and their partial derivatives, to conclude that as ε approaches 0 the functions g_ε and their partial derivatives converge to a harmonic function g and its partial derivatives, uniformly on $\mathbf{U}(b,r)$ for $0 < r < \rho$.

5.2.7. A function K mapping $\mathbf{R}^m \sim \{0\}$ into a normed vectorspace V is called **essentially homogeneous of degree** λ if and only if $\lambda \in \mathbf{R}$ and

$$K(x) = \log(|x|)\, P(x) + Q(x) \quad \text{for } x \in \mathbf{R}^m \sim \{0\},$$

where $Q: \mathbf{R}^m \sim \{0\} \to V$ with $Q(t\,x) = t^\lambda Q(x)$ for $0 < t \in \mathbf{R}$, and either $P = 0$ or λ is a nonnegative integer and $P: \mathbf{R}^m \to V$ is a homogeneous polynomial function of degree λ.

Assuming that K is essentially homogeneous of degree λ, and of class ∞, we see that $D^\alpha K$ is essentially homogeneous of degree $\lambda - p$ whenever $\alpha \in \Xi(m, p)$.

In case $\lambda > -m$, $V = \operatorname{Hom}(\mathbf{R}^{n-m}, \mathbf{R}^{n-m})$ and $\chi: \mathbf{R}^m \to \mathbf{R}^{n-m}$ is a bounded $\mathscr{L}^m$ measurable function with compact support, we define the convolution $\chi * K$ by the formula

$$(\chi * K)(x) = \int \langle \chi(x-w), K(w) \rangle \, d\mathscr{L}^m w$$

for $x \in \mathbf{R}^m$; this integral exists because, for $0 < \rho < \infty$,

$$\begin{aligned} \int_{\mathbf{B}(0,\rho)} |K| \, d\mathscr{L}^m &= \int_0^\rho \int_{\{w: |w|=r\}} |K| \, d\mathscr{H}^{m-1} \, d\mathscr{L}^1 r \\ &\leq \int_0^\rho r^{m-1+\lambda} [\beta |\log(r)| + \gamma] \, d\mathscr{L}^1 r < \infty \end{aligned}$$

with $\beta = \int_{\mathbf{S}^{m-1}} |P| \, d\mathscr{H}^{m-1}$ and $\gamma = \int_{\mathbf{S}^{m-1}} |Q| \, d\mathscr{H}^{m-1}$. Since

$$\begin{aligned} |(\chi * K)(x+h) - (\chi * K)(x)| &= |\int \langle \chi(x-w), K(w+h) - K(w) \rangle \, d\mathscr{L}^m w| \\ &\leq (\sup \operatorname{im} |\chi|) \int_{\mathbf{B}(0,\rho)} \|K(w+h) - K(w)\| \, d\mathscr{L}^m w \end{aligned}$$

whenever $x, h \in \mathbf{R}^m$ and $\operatorname{spt} \chi \subset \mathbf{B}(x, \rho)$, we see that the function $\chi * K$ is continuous. Expressing

$$(\chi * K)(x) = \int \langle \chi(u), K(x-u) \rangle \, d\mathscr{L}^m u$$

one readily verifies (*compare* 4.1.2) *that if p is a nonnegative integer and $\lambda - p > -m$, then $\chi * K$ is of class p and*

$$D^\alpha(\chi * K) = \chi * D^\alpha K \quad \textit{for } \alpha \in \Xi(m, p).$$

Questions concerning partial derivatives of $\chi * K$ of order $p \geq \lambda + m$ are much more subtle, but may be studied by means of the singular integrals discussed in 5.2.8, in case χ satisfies a Hölder condition.

Clearly, *if $\lambda > -m$ and χ is of class p, then $\chi * K$ is of class p and*

$$D^\alpha(\chi * K) = (D^\alpha \chi) * K \quad \textit{for } \alpha \in \Xi(m, p).$$

5.2.8. Assuming here that $\lambda \in \mathbf{R}$,

$$K: \mathbf{R}^m \sim \{0\} \to \operatorname{Hom}(\mathbf{R}^{n-m}, \mathbf{R}^{n-m}) \text{ is of class } \infty,$$

$$K(t\,x) = t^\lambda K(x) \text{ for } x \in \mathbf{R}^m \sim \{0\} \text{ and } 0 < t \in \mathbf{R},$$

$$\chi: \mathbf{R}^m \to \mathbf{R}^{n-m}, \quad |\chi(x)| \leq \mu < \infty \text{ for } x \in \mathbf{R}^m,$$

$$0 < \delta < 1 \quad \text{and} \quad \mathbf{h}_\delta(\chi) \leq \beta < \infty,$$

we suppose $0 < \rho < \infty$ and define

$$g_\varepsilon(x) = \int_{\mathbf{B}(0,\rho) \sim \mathbf{B}(0,\varepsilon)} \langle \chi(x-w), K(w) \rangle \, d\mathscr{L}^m w$$

for $0 \leq \varepsilon \leq \rho$ and $x \in \mathbf{R}^m$. The following five propositions relate to the existence, continuity and differentiation of the **singular integral**

$$g(x) = \lim_{\varepsilon \to 0+} g_\varepsilon(x).$$

We abbreviate $\kappa_p = \int_{\mathbf{S}^{m-1}} \|D^p K\| \, d\mathscr{H}^{m-1}$ for $p \in \{0, 1, 2\}$.

(1) *If* $i \in \{1, \ldots, m\}$, $0 < \varepsilon < \rho$, $x \in \mathbf{R}^m$ *and* $y \in \mathbf{R}^{n-m}$, *then*

$$\begin{aligned} D_i g_\varepsilon(x) = & \int_{\mathbf{B}(0,\rho) \sim \mathbf{B}(0,\varepsilon)} \langle \chi(x-w) - y, D_i K(w) \rangle \, d\mathscr{L}^m w \\ & - \rho^{m-1+\lambda} \int_{\mathbf{S}^{m-1}} \langle \chi(x - \rho u) - y, u_i K(u) \rangle \, d\mathscr{H}^{m-1} u \\ & + \varepsilon^{m-1+\lambda} \int_{\mathbf{S}^{m-1}} \langle \chi(x - \varepsilon u) - y, u_i K(u) \rangle \, d\mathscr{H}^{m-1} u. \end{aligned}$$

In case $\lambda + \delta \neq 1 - m$,

$$|D_i g_\varepsilon(x)| \leq [\kappa_0 + \kappa_1 |m + \lambda + \delta - 1|^{-1}] \, \beta (\rho^{m+\lambda+\delta-1} + \varepsilon^{m+\lambda+\delta-1}).$$

(2) *If* $\lambda > -m$, *then*

$$g(x) = \int_{\mathbf{B}(0,\rho)} \langle \chi(x-w), K(w) \rangle \, d\mathscr{L}^m w$$

$$\text{and } |g(x)| \leq \kappa_0 \, \mu \, \rho^{m+\lambda} (m+\lambda)^{-1} \text{ for } x \in \mathbf{R}^m,$$

$$\mathbf{h}_\delta(g) \leq \kappa_0 \, \beta \, \rho^{m+\lambda} (m+\lambda)^{-1}.$$

(3) *If* $\lambda = -m$ *and* $\int_{\mathbf{S}^{m-1}} K \, d\mathscr{H}^{m-1} = 0$, *then*

$$g(x) = \int_{\mathbf{B}(0,\rho)} \langle \chi(x-w) - \chi(x), K(w) \rangle \, d\mathscr{L}^m w$$

$$\text{and } |g(x)| \leq \kappa_0 \, \beta \, \rho^\delta \, \delta^{-1} \text{ for } x \in \mathbf{R}^m,$$

$$\mathbf{h}_\delta(g) \leq 2m [\kappa_0 (1 + \delta^{-1}) + \kappa_1 (1 - \delta)^{-1}] \, \beta.$$

(4) *If* $\lambda > 1 - m$, $i \in \{1, \ldots, m\}$ *and* $x \in \mathbf{R}^m$, *then*

$$\begin{aligned} D_i g(x) = & \int_{\mathbf{B}(0,\rho)} \langle \chi(x-w), D_i K(w) \rangle \, d\mathscr{L}^m w \\ & - \rho^{m-1+\lambda} \int_{\mathbf{S}^{m-1}} \langle \chi(x - \rho u), u_i K(u) \rangle \, d\mathscr{H}^{m-1} u. \end{aligned}$$

(5) *If* $\lambda = 1 - m$ *and* $i \in \{1, \ldots, m\}$, *then*

$$\int_{\mathbf{S}^{m-1}} D_i K \, d\mathscr{H}^{m-1} = 0,$$

$$\begin{aligned} D_i g(x) = & \int_{\mathbf{B}(0,\rho)} \langle \chi(x-w) - \chi(x), D_i K(w) \rangle \, d\mathscr{L}^m w \\ & - \int_{\mathbf{S}^{m-1}} \langle \chi(x - \rho u) - \chi(x), u_i K(u) \rangle \, d\mathscr{H}^{m-1} u \end{aligned}$$

$$\text{and } |D_i g(x)| \leq (\kappa_1 \delta^{-1} + \kappa_0) \, \beta \, \rho^\delta \text{ for } x \in \mathbf{R}^m,$$

$$\mathbf{h}_\delta(D_i g) \leq 2m [\kappa_1 (1 + \delta^{-1}) + \kappa_2 (1 - \delta)^{-1} + \kappa_0] \, \beta.$$

We verify (1) in case $\chi \in \mathscr{E}(\mathbf{R}^m, \mathbf{R}^{n-m})$, observing that the general case follows easily by regularization. In the special case we compute

$$D_i g_\varepsilon(x) = \int_{\mathbf{B}(0,\rho) \sim \mathbf{B}(0,\varepsilon)} \langle D_i \chi(x-w), K(w) \rangle \, d\mathscr{L}^m w,$$

let $\theta(w) = \langle \chi(x-w) - y, K(w) \rangle$ for $w \in \mathbf{R}^m \sim \{0\}$, apply 4.5.6 with $\xi = (Y_j \circ \theta)\, e_i$ for $j \in \{1, \ldots, n-m\}$ to obtain

$$\begin{aligned} &\int_{\{w: |w| = \rho\}} |w|^{-1} w_i \theta(w) \, d\mathscr{H}^{m-1} w - \int_{\{w: |w| = \varepsilon\}} |w|^{-1} w_i \theta(w) \, d\mathscr{H}^{m-1} w \\ &= \int_{\mathbf{B}(0,\rho) \sim \mathbf{B}(0,\varepsilon)} D_i \theta(w) \, d\mathscr{L}^m w \\ &= -D_i g_\varepsilon(x) + \int_{\mathbf{B}(0,\rho) \sim \mathbf{B}(0,\varepsilon)} \langle \chi(x-w) - y, D_i K(w) \rangle \, d\mathscr{L}^m w, \end{aligned}$$

then transform the $\mathscr{H}^{m-1}$ integrals by the maps μ_ρ and μ_ε. Taking $y = \chi(x)$ and recalling 3.2.13 we estimate

$$|D_i g_\varepsilon(x)| \le \int_\varepsilon^\rho r^{m-1} \beta r^\delta \kappa_1 r^{\lambda-1} \, d\mathscr{L}^1 r + \rho^{m-1+\lambda} \beta \rho^\delta \kappa_0 + \varepsilon^{m-1+\lambda} \beta \varepsilon^\delta \kappa_0,$$

since $D_i K(t\,u) = t^{\lambda-1} D_i K(u)$ for $u \in \mathbf{R}^m \sim \{0\}$ and $0 < t \in \mathbf{R}$.

The assertion (2) is trivial (see 5.2.7).

If the hypothesis of (3) holds, then

$$\begin{aligned} &\int_{\mathbf{B}(0,\rho) \sim \mathbf{B}(0,\varepsilon)} K \, d\mathscr{L}^m = \int_\varepsilon^\rho r^{m-1} 0 \, d\mathscr{L}^1 r = 0, \\ &\int_{\mathbf{B}(0,\varepsilon)} |\langle \chi(x-w) - \chi(x), K(w) \rangle| \, d\mathscr{L}^m w \\ &\le \int_0^\varepsilon r^{m-1} \beta r^\delta \kappa_0 r^{-m} \, d\mathscr{L}^1 r = \beta \kappa_0 \varepsilon^\delta \delta^{-1}, \end{aligned}$$

whence the first two conclusions follow. Moreover (1) implies

$$\begin{aligned} &|D_i g_\varepsilon(x)| \le [\kappa_0 + \kappa_1 (1-\delta)^{-1}] \beta \, 2\varepsilon^{\delta-1} \quad \text{for } x \in \mathbf{R}^m, \\ &|g_\varepsilon(x+h) - g_\varepsilon(x)| \le m [\kappa_0 + \kappa_1 (1-\delta)^{-1}] \beta \, 2\varepsilon^{\delta-1} |h| \end{aligned}$$

for $x, h \in \mathbf{R}^m$; letting $\varepsilon = \inf\{|h|, \rho\}$ and observing that

$$|g(x+h) - g(x)| \le |g_\varepsilon(x+h) - g_\varepsilon(x)| + 2\beta \kappa_0 \varepsilon^\delta \delta^{-1}$$

we infer the third conclusion of (3).

Taking $\chi = 0$ and $\lambda = 1 - m$ in (1) we obtain

$$\begin{aligned} 0 &= \int_{\mathbf{B}(0,\rho) \sim \mathbf{B}(0,\varepsilon)} \langle -y, D_i K(w) \rangle \, d\mathscr{L}^m w \\ &= -\int_\varepsilon^\rho r^{-1} \int_{\mathbf{S}^{m-1}} \langle y, D_i K(w) \rangle \, d\mathscr{H}^{m-1} w \, d\mathscr{L}^1 r \end{aligned}$$

and deduce the first conclusion of (5).

To prove (4) and the remainder of (5) we use (1) with $y = 0$ and $y = \chi(x)$, and apply (2) and (3) with $K, \lambda, \kappa_0, \kappa_1$ replaced by $D_i K, \lambda - 1, \kappa_1, \kappa_2$; in both cases $D_i g_\varepsilon$ converges uniformly to $D_i g$ as ε approaches 0 (see 5.2.2).

Combining the results of 5.2.7 with the above proposition (5) we find that *if p is a positive integer, K is of class ∞ and essentially homogeneous of degree $p - m$, χ has compact support and satisfies a Hölder condition, then $\chi * K$ is of class p.*

5.2.9. *If* $-1<\lambda\in\mathbf{R}$, $\alpha\in\mathbf{R}$, $\beta\in\mathbf{R}$,

$$\phi(s)=|s|^\lambda(\alpha+\beta\log|s|)\ \textit{for}\ s\in\mathbf{R}\sim\{0\},$$

and T is a function of class ∞ *mapping* $\mathbf{R}^m\sim\{0\}$ *into a Banach space, then there exists a function K of class* ∞ *on* $\mathbf{R}^m\sim\{0\}$ *such that*

$$K(x)=\int_{\mathbf{S}^{m-1}}\phi(x\bullet u)\,T(u)\,d\mathscr{H}^{m-1}u\ \text{ for }\ x\in\mathbf{R}^m\sim\{0\}.$$

In case either $\beta=0$ *or* λ *is an even integer, K is essentially homogeneous of degree* λ.

To prove this we construct (see 3.2.47) for each $b\in\mathbf{R}^m\sim\{0\}$ a neighborhood V of b in $\mathbf{R}^m$ and an analytic map

$$R\colon V\to\mathbf{O}(m)\quad\text{with}\quad\langle e_1,R(x)\rangle=|x|^{-1}x\ \text{ for } x\in V,$$

note that $x\bullet\langle w,R(x)\rangle=|x|\,w_1$ for $w\in\mathbf{R}^m$, and compute

$$K(x)=\int_{\mathbf{S}^{m-1}}\phi(|x|\,w_1)\,T(\langle w,R(x)\rangle)\,d\mathscr{H}^{m-1}w$$
$$=\int_{\mathbf{S}^{m-1}}|x|^\lambda[\phi(w_1)+\log(|x|)\,\psi(w_1)]\,T(\langle w,R(x)\rangle)\,d\mathscr{H}^{m-1}w$$

where $\psi(s)=\beta\,|s|^\lambda$ for $s\in\mathbf{R}$. Since ψ and ϕ are $\mathscr{L}^1$ summable over every compact interval, $\psi\circ X_1$ and $\phi\circ X_1$ are $\mathscr{L}^m$ summable over $\mathbf{B}(0,r)$, hence $\mathscr{H}^{m-1}$ summable over $\{w\colon |w|=r\}$ for $\mathscr{L}^1$ almost all r; using the equation

$$\phi\circ X_1\circ\mu_r=r^\lambda[(\phi\circ X_1)+\log(r)(\psi\circ X_1)]$$

we infer that $\psi\circ X_1$ and $\phi\circ X_1$ are $\mathscr{H}^{m-1}$ summable over $\mathbf{S}^{m-1}$. Then we conclude that K is of class ∞ on V. In case $\beta=0$ or λ is an even integer the conditions of 5.2.7 hold with

$$P(x)=\int_{\mathbf{S}^{m-1}}\beta\,|x\bullet u|^\lambda\,T(u)\,d\mathscr{H}^{m-1}u,$$
$$Q(x)=\int_{\mathbf{S}^{m-1}}|x\bullet u|^\lambda(\alpha+\beta\log\big||x|^{-1}x\bullet u\big|)\,T(u)\,d\mathscr{H}^{m-1}u.$$

5.2.10. The **Laplacian** differential operator is defined by the formula

$$\operatorname{Lap}f=\sum_{i=1}^m D_{ii}f$$

for every function f of class 2 on an open subset of $\mathbf{R}^m$.

Letting $\mathbf{r}(x)=|x|$ for $x\in\mathbf{R}^m$ we compute

$$\operatorname{Lap}\mathbf{r}^\lambda=\lambda(m-2+\lambda)\,\mathbf{r}^{\lambda-2}\ \text{ for }\ \lambda\in\mathbf{R}\sim\{0\},\quad \operatorname{Lap}(\log\circ\mathbf{r})=(m-2)\,\mathbf{r}^{-2}.$$

If m is odd, then

$$\operatorname{Lap}^{(m-1)/2}\mathbf{r}=(-1)^{(m-1)/2}(m-1)!\,\mathbf{r}^{2-m}/(2-m).$$

If m is even and $m>2$, then

$$\operatorname{Lap}^{(m-2)/2}(\log\circ\mathbf{r})=(-1)^{(m-2)/2}\,2^{m-2}\,[(m-2)/2]!^2\,\mathbf{r}^{2-m}/(2-m).$$

Whenever $\chi\colon \mathbf{R}^m \to \mathbf{R}^{n-m}$ *satisfies a Hölder condition and* $\operatorname{spt}\chi$ *is compact* we see from 5.2.7 and 5.2.8 that the convolution $\mathbf{r}^{2-m} * \chi$, defined by the formula

$$(\mathbf{r}^{2-m} * \chi)(x) = \int |w|^{2-m}\, \chi(x-w)\, d\mathscr{L}^m w$$

for $x \in \mathbf{R}^m$, is a function of class 2 on $\mathbf{R}^m$ with

$$D_i(\mathbf{r}^{2-m} * \chi)(x) = \int (2-m)\, w_i\, |w|^{-m}\, \chi(x-w)\, d\mathscr{L}^m w,$$

$$\operatorname{Lap}(\mathbf{r}^{2-m} * \chi)(x) = (2-m)\, \mathscr{H}^{m-1}(\mathbf{S}^{m-1})\, \chi(x).$$

Similarly we find that

$$\operatorname{Lap}[(\log \circ \mathbf{r}) * \chi] = \mathscr{H}^1(\mathbf{S}^1)\, \chi \quad \text{in case} \quad m = 2.$$

Proceeding in the above manner we conclude:

If m is odd, then

$$\operatorname{Lap}^{(m+1)/2}(\mathbf{r} * \chi) = (-1)^{(m-1)/2}(m-1)!\, \mathscr{H}^{m-1}(\mathbf{S}^{m-1})\, \chi.$$

If m is even, then

$$\operatorname{Lap}^{m/2}[(\log \circ \mathbf{r}) * \chi] = (-1)^{(m-2)/2}\, 2^{m-2}\, [(m-2)/2]!^2\, \mathscr{H}^{m-1}(\mathbf{S}^{m-1})\, \chi.$$

From 1.7.4 and 2.7.16(6) we see that

$$\mathbf{O}^*(m,1) = (\odot^1 \mathbf{R}^m) \cap \{\xi\colon |\xi| = 1\},$$

$$\int_{\mathbf{O}^*(m,1)} |\xi(w)|\, d\mathscr{H}^{m-1}\xi = \beta_1(m,1)\, \mathscr{H}^{m-1}(\mathbf{S}^{m-1})\, |w|$$

whenever $w \in \mathbf{R}^m$, and for each w there exists $g \in \mathbf{O}(m)$ with $g(w) = |w|\, e_1$, hence

$$\int_{\mathbf{O}^*(m,1)} \log|\xi(w)|\, d\mathscr{H}^{m-1}\xi = \int_{\mathbf{O}^*(m,1)} \log|\xi[g(w)]|\, d\mathscr{H}^{m-1}\xi$$

$$= \mathscr{H}^{m-1}(\mathbf{S}^{m-1}) \log|w| + \int_{\mathbf{O}^*(m,1)} \log|\xi(e_1)|\, d\mathscr{H}^{m-1}\xi.$$

Combining these facts we obtain **John's formulae:**

If m is odd, $\mathbf{c}(m) = (-1)^{(m-1)/2}(m-1)!\, \mathscr{H}^{m-1}(\mathbf{S}^{m-1})^2\, \beta_1(m,1)$,

$$F(x) = \iint_{\mathbf{O}^*(m,1)} |\xi(w)|\, \chi(x-w)\, d\mathscr{H}^{m-1}\xi\, d\mathscr{L}^m w \text{ for } x \in \mathbf{R}^m,$$

then $\operatorname{Lap}^{(m+1)/2} F = \mathbf{c}(m)\, \chi$.

If m is even, $\mathbf{c}(m) = (-1)^{(m-2)/2}\, 2^{m-2}\, [(m-2)/2]!^2\, \mathscr{H}^{m-1}(\mathbf{S}^{m-1})^2$,

$$F(x) = \iint_{\mathbf{O}^*(m,1)} \log|\xi(w)|\, \chi(x-w)\, d\mathscr{H}^{m-1}\xi\, d\mathscr{L}^m w \text{ for } x \in \mathbf{R}^m,$$

then $\operatorname{Lap}^{m/2} F = \mathbf{c}(m)\, \chi$.

5.2.11. With $\Upsilon \in \odot^2 \operatorname{Hom}(\mathbf{R}^m, \mathbf{R}^{n-m})$ we associate the linear map

$$S\colon \odot^2(\mathbf{R}^m, \mathbf{R}^{n-m}) \simeq [\odot^2 \mathbf{R}^m] \otimes \mathbf{R}^{n-m} \to \mathbf{R}^{n-m}$$

characterized by the condition

$$S[(\xi \circ \psi)\, y] \bullet v = \Upsilon(\xi\, y, \psi\, v) + \Upsilon(\psi\, y, \xi\, v)$$

for $\xi, \psi \in \odot^1 \mathbf{R}^m$ and $y, v \in \mathbf{R}^{n-m}$. Moreover, to each $\xi \in \odot^1 \mathbf{R}^m$ corresponds the linear function

$$S_\xi\colon \mathbf{R}^{n-m} \to \mathbf{R}^{n-m}, \quad S_\xi(y) = S[(\xi \circ \xi/2)\, y] \text{ for } y \in \mathbf{R}^{n-m};$$

the map carrying ξ onto S_ξ is a homogeneous polynomial function of degree 2, called the **symbol** of Υ. *Assuming that* $\|\Upsilon\| \le M$ *and* Υ *is elliptic with ellipticity bound* c *we infer*

$$S_\xi(y) \bullet v = \Upsilon(\xi\, y, \xi\, v) \le M |\xi|^2\, |y| \cdot |v|, \quad |S_\xi(y)| \cdot |y| \ge \Upsilon(\xi\, y, \xi\, y) \ge c |\xi|^2 |y|^2$$

for $y, v \in \mathbf{R}^{n-m}$, hence $|\xi| = 1$ implies that

$$\|S_\xi\| \le M, \ S_\xi \text{ is univalent}, \ \|S_\xi^{-1}\| \le c^{-1}.$$

Letting $\Upsilon_{i,j;k,l} = \Upsilon(X_i\, v_j, X_k\, v_l)$ we find that

$$S(\phi) = \sum_{i=1}^{m} \sum_{j=1}^{n-m} \sum_{k=1}^{m} \sum_{l=1}^{n-m} \Upsilon_{i,j;k,l} [\phi(e_i, e_k) \bullet v_j]\, v_l$$

whenever $\phi \in \odot^2(\mathbf{R}^m, \mathbf{R}^{n-m})$.

If $f\colon \mathbf{R}^m \to \mathbf{R}^{n-m}$ is of class 2 and $\theta \in \mathscr{D}(\mathbf{R}^m, \mathbf{R}^{n-m})$, then

$$\begin{aligned}\int \Upsilon[D f(x), D\theta(x)]\, d\mathscr{L}^m x &= \int \sum_{i=1}^{m} \sum_{j=1}^{n-m} \sum_{k=1}^{m} \sum_{l=1}^{n-m} \Upsilon_{i,j;k,l} (D_i f_j) \cdot (D_k \theta_l)\, d\mathscr{L}^m \\ &= -\int \sum_{i=1}^{m} \sum_{j=1}^{n-m} \sum_{k=1}^{m} \sum_{l=1}^{n-m} \Upsilon_{i,j;k,l} (D_{i,k} f_j) \cdot \theta_l\, d\mathscr{L}^m \\ &= -\int S[D^2 f(x)] \bullet \theta(x)\, d\mathscr{L}^m x\end{aligned}$$

where $f_j = Y_j \circ f$ and $\theta_l = Y_l \circ \theta$. To the form Υ thus corresponds the linear differential operator L (with constant coefficients) defined by the formula

$$L f = S \circ D^2 f.$$

For example, to the standard inner product of $\operatorname{Hom}(\mathbf{R}^m, \mathbf{R}^{n-m})$ corresponds the Laplace operator.

For $G \in \mathscr{E}^0(\mathbf{R} \sim \{0\})$, $\xi \in \odot^1 \mathbf{R}^m$, $y \in \mathbf{R}^{n-m}$ we compute

$$D^2[(G \circ \xi)\, y](x) = G''[\xi(x)](\xi \odot \xi/2)\, y,$$

$$L[(G \circ \xi)\, y](x) = G''[\xi(x)]\, S_\xi(y), \qquad \operatorname{Lap}(G \circ \xi)(x) = G''[\xi(x)]\, |\xi|^2$$

whenever $x \in \mathbf{R}^m \sim \ker \xi$. In particular we choose G so that, for $s \in \mathbf{R} \sim \{0\}$,

$$G(s) = |s|^3/6,\ G''(s) = |s| \text{ if } m \text{ is odd},$$

$$G(s) = s^2[\log(s^2) - 3]/4,\ G''(s) = \log|s| \text{ if } m \text{ is even}.$$

Recalling 5.2.9 we consider the maps

$$H: \mathbf{R}^m \sim \{0\} \to \operatorname{Hom}(\mathbf{R}^{n-m}, \mathbf{R}^{n-m}), \qquad H_v: \mathbf{R}^m \sim \{0\} \to \mathbf{R}^{n-m},$$

$$\langle v, H(x) \rangle = H_v(x) = \int_{\mathbf{O}^*(m,1)} G[\xi(x)]\, S_\xi^{-1}(v)\, d\mathscr{H}^{m-1}\xi$$

for $v \in \mathbf{R}^{n-m}$ and $x \in \mathbf{R}^m \sim \{0\}$; we observe that H is essentially homogeneous of degree 3 or 2, and class ∞, with $H(-x) = H(x)$,

$$(L H_v)(x) = \int_{\mathbf{O}^*(m,1)} L[(G \circ \xi)\, S_\xi^{-1}(v)](x)\, d\mathscr{H}^{m-1}\xi$$
$$= \int_{\mathbf{O}^*(m,1)} G''[\xi(x)]\, v\, d\mathscr{H}^{m-1}\xi,$$

$$(\operatorname{Lap} H_v)(x) = \int_{\mathbf{O}^*(m,1)} G''[\xi(x)]\, S_\xi^{-1}(v)\, d\mathscr{H}^{m-1}\xi.$$

We define the **elementary solution** E associated with Υ as

$$E = \mathbf{c}(m)^{-1} \operatorname{Lap}^{(m+1)/2} H \text{ in case } m \text{ is odd},$$

$$E = \mathbf{c}(m)^{-1} \operatorname{Lap}^{m/2} H \text{ in case } m \text{ is even},$$

and note that E is of class ∞ on $\mathbf{R}^m \sim \{0\}$ with $E(-x) = E(x)$,

$$E(t\,x) = t^{2-m} E(x) \text{ if } m \neq 2,$$

$$E(t\,x) = E(x) + \log(t) \int_{\mathbf{O}^*(2,1)} S_\xi^{-1}\, d\mathscr{H}^1 \xi \text{ if } m = 2,$$

$$D E(t\,x) = t^{1-m} D E(x)$$

for $x \in \mathbf{R}^m \sim \{0\}$ and $0 < t \in \mathbf{R}$. The following proposition shows that convolution with E provides a well behaved solution f of the equation $Lf = \chi$.

If $\chi: \mathbf{R}^m \to \mathbf{R}^{n-m}$ *satisfies a Hölder condition and has compact support, then* $\chi * E$ *is of class* 2 *and*

$$L(\chi * E) = \chi.$$

To prove this we first recall 5.2.7 and compute

$$L(\chi * H)(x) = \int [L H_{\chi(w)}](x - w)\, d\mathscr{L}^m w$$
$$= \iint_{\mathbf{O}^*(m,1)} G''[\xi(x - w)]\, \chi(w)\, d\mathscr{H}^{m-1}\xi\, d\mathscr{L}^m w$$

for $x \in \mathbf{R}^m$. Next we apply 5.2.8 and 5.2.10 (John's formulae). In case m is odd we find that $\chi * H$ is of class $3+m$ and

$$\begin{aligned} \mathbf{c}(m)\,\chi &= \operatorname{Lap}^{(m+1)/2}[L(\chi * H)] \\ &= L[\operatorname{Lap}^{(m+1)/2}(\chi * H)] = L[\chi * \operatorname{Lap}^{(m+1)/2} H]. \end{aligned}$$

In case m is even we find that $\chi * H$ is of class $2+m$ and

$$\mathbf{c}(m)\,\chi = \operatorname{Lap}^{m/2}[L(\chi * H)] = L[\operatorname{Lap}^{m/2}(\chi * H)] = L(\chi * \operatorname{Lap}^{m/2} H).$$

If $\Omega\colon \mathbf{R}^m \to \operatorname{Hom}(\mathbf{R}^m, \mathbf{R}^{n-m})$ satisfies a Hölder condition and has compact support, then the convolution

$$\Omega * DE = \sum_{i=1}^{m} (\mathbf{e}_i \lrcorner \Omega) * D_i E$$

is of class 1 *and has the property that*

$$\int \langle D(\Omega * DE)(x) \odot D\theta(x), \Upsilon \rangle \, d\mathscr{L}^m x = (\Omega, D\theta)$$

for all $\theta \in \mathscr{D}(\mathbf{R}^m, \mathbf{R}^{n-m})$; moreover

$$c\,|D(\Omega * DE)| \le |\Omega|.$$

In case $\Omega \in \mathscr{D}[\mathbf{R}^m, \operatorname{Hom}(\mathbf{R}^m, \mathbf{R}^{n-m})]$ the preceding proposition is applicable with

$$\chi = \sum_{i=1}^{m} D_i(\mathbf{e}_i \lrcorner \Omega),$$

and the asserted property of

$$\Omega * DE = \sum_{i=1}^{m} D_i[(\mathbf{e}_i \lrcorner \Omega) * E] = \chi * E$$

holds because

$$\begin{aligned} \int \langle D(\chi * E)(x) \odot D\theta(x), \Upsilon \rangle \, d\mathscr{L}^m x &= -(L(\chi * E), \theta) = -(\chi, \theta) \\ &= \sum_{i=1}^{m} (\mathbf{e}_i \lrcorner \Omega, D_i \theta) = (\Omega, D\theta); \end{aligned}$$

in the general case this property follows by regularization using 5.2.8. Next we prove that

$$|\Omega * DE| < \infty \text{ if } m > 1,$$

by choosing finite, positive numbers s, γ, κ with $\operatorname{spt} \Omega \subset \mathbf{B}(0, s)$,

$$|\Omega(w)| \le \gamma \text{ for } w \in \mathbf{R}^m, \qquad |DE(u)| \le \kappa \text{ for } u \in \mathbf{S}^{m-1},$$

and observing that

$$\begin{aligned} |(\Omega * DE)(x)| &\le \int |\Omega(w)| \cdot |DE(x-w)| \, d\mathscr{L}^m w \\ &\le \int_{\mathbf{B}(0,s)} \gamma \kappa |x-w|^{1-m} \, d\mathscr{L}^m w \le \alpha(m)\, s^m \gamma \kappa (|x|/2)^{1-m} \end{aligned}$$

whenever $x \in \mathbf{R}^m \sim \mathbf{B}(0, 2s)$; then we apply the basic estimate in 5.2.3 with $\rho = 2r$ to infer that

$$c\,|D(\Omega * DE)|_{0,r} \leq |\Omega| + c^{\frac{1}{2}} M^{\frac{1}{2}} r^{-1} |\Omega * DE|$$

whenever $0 < r < \infty$, and let r approach ∞ to obtain the inequality $c\,|D(\Omega * DE)| \leq |\Omega|$.

If $m=1$, then $E(x) = 2^{-1}|x|\,S_{X_1}^{-1}$ for $x \in \mathbf{R} \sim \{0\}$, and

$$(\Omega * DE)' = S_{X_1}^{-1} \circ (\mathbf{e}_1 \lrcorner \Omega).$$

5.2.12. *If* $b \in \mathbf{R}^m$, $0 < \rho < \infty$, $0 < \delta \leq 1$, V *is a normed vectorspace,* $\omega: \mathbf{B}(b, \rho) \to V$, $0 \in \operatorname{im} \omega$ *and* $\mathbf{h}_\delta(\omega) < \infty$, *then* ω *has an extension* $\Omega: \mathbf{R}^m \to V$ *with* spt $\Omega \subset \mathbf{B}(0, 3\rho/2)$,

$$\mathbf{h}_\delta(\Omega) \leq 9\,\mathbf{h}_\delta(\omega) \quad \textit{and} \quad |\Omega| \leq 2^m\,|\omega|_{b,\rho}.$$

To prove this in case $b=0$ we define $\phi: \mathbf{R}^m \to \mathbf{B}(0, \rho)$,

$$\phi(x) = x \text{ if } |x| \leq \rho, \quad \phi(x) = 0 \text{ if } |x| \geq 2\rho,$$
$$\phi(x) = (2\rho\,|x|^{-1} - 1)\,x \text{ if } \rho < |x| < 2\rho,$$

and note that if $\rho < |x| < 2\rho$, then

$$\langle u, D\phi(x)\rangle = (2\rho\,|x|^{-1} - 1)\,u \text{ for } u \in \mathbf{R}^m \text{ with } u \bullet x = 0,$$
$$\langle x, D\phi(x)\rangle = -x, \quad \|D\phi(x)\| = 1, \quad J_m\phi(x) = (2\rho\,|x|^{-1} - 1)^{m-1}.$$

We infer $\operatorname{Lip}(\phi) = 1$, $\mathbf{h}_\delta(\omega \circ \phi) = \mathbf{h}_\delta(\omega)$ and

$$3^{1-m} \int_{\{x:\, \rho < |x| < 3\rho/2\}} |\omega \circ \phi|^2 \, d\mathscr{L}^m$$
$$\leq \int_{\{x:\, \rho < |x| < 3\rho/2\}} |\omega \circ \phi|^2 \, J_m\phi \, d\mathscr{L}^m \leq |\omega|_{0,\rho}^2,$$

hence $|\omega \circ \phi|_{0, 3\rho/2} \leq (1 + 3^{(m-1)/2})\,|\omega|_{0,\rho} \leq 2^m\,|\omega|_{0,\rho}$. We further define $\psi: \mathbf{R}^m \to \{t: 0 \leq t \leq 1\}$,

$$\psi(x) = 1 \text{ if } |x| \leq \rho, \quad \psi(x) = 0 \text{ if } |x| \geq 3\rho/2,$$
$$\psi(x) = 3 - 2\,|x|/\rho \text{ if } \rho < |x| < 3\rho/2,$$

and let $\Omega = \psi \cdot (\omega \circ \phi)$. To estimate $\mathbf{h}_\delta(\Omega)$ we suppose $x, u \in \mathbf{R}^m$ with $|x| \leq 3\rho/2$, infer from the hypothesis $0 \in \omega[\mathbf{B}(0, \rho)]$ that

$$|\Omega(x)| \leq |\omega[\phi(x)]| \leq \mathbf{h}_\delta(\omega)\,(2\rho)^\delta,$$

and observe that either $|u| > 2\rho$, hence $|x - u| > \rho/2$ and

$$|\Omega(x) - \Omega(u)| = |\Omega(x)| \leq \mathbf{h}_\delta(\omega)\,4^\delta\,|x - u|^\delta,$$

or $|u| \le 2\rho$, hence $|x-u| \le 4\rho$ and

$$\begin{aligned}|\Omega(x)-\Omega(u)| &\le |\psi(x)-\psi(u)| \cdot |\omega[\phi(x)]| + \psi(u)\,|\omega[\phi(x)]-\omega[\phi(u)]| \\ &\le 2\rho^{-1}\,|x-u|\,\mathbf{h}_\delta(\omega)\,(2\rho)^\delta + \mathbf{h}_\delta(\omega)\,|x-u|^\delta \\ &\le [2\rho^{-1}(4\rho)^{1-\delta}(2\rho)^\delta + 1]\,\mathbf{h}_\delta(\omega)\,|x-u|^\delta \le 9\,\mathbf{h}_\delta(\omega)\,|x-u|^\delta.\end{aligned}$$

5.2.13. Lemma. *To any numbers $0<c\le M<\infty$ and $0<\delta<1$ corresponds a number $\gamma<\infty$ with the following property:*

If $\Upsilon \in \bigodot^2 \operatorname{Hom}(\mathbf{R}^m, \mathbf{R}^{n-m})$, Υ is strongly elliptic with ellipticity bound c, $\|\Upsilon\| \le M$, $b \in \mathbf{R}^m$, $0<\rho<\infty$ and

$$\eta:\ \mathbf{B}(b,\rho) \to \operatorname{Hom}(\mathbf{R}^m, \mathbf{R}^{n-m}) \text{ with } \mathbf{h}_\delta(\eta) < \infty,$$

then there exists $F: \mathbf{R}^m \to \mathbf{R}^{n-m}$ of class 1 with

$$\mathbf{h}_\delta(DF) \le \gamma\, \mathbf{h}_\delta(\eta), \qquad |DF| \le 2^{m+1} c^{-1} |\eta|_{b,\rho}$$

and satisfying the condition

$$\int_{\mathbf{U}(b,\rho)} \langle DF(x) \odot D\theta(x), \Upsilon \rangle \, d\mathscr{L}^m x = (\eta, D\theta)_{b,\rho}$$

for all $\theta \in \mathscr{D}[\mathbf{U}(b,\rho), \mathbf{R}^{n-m}]$.

Proof. Analyzing the constructions performed in 5.2.11 and 5.2.9 one deduces the existence of a finite number κ (determined by m, n, c, M) such that for every strongly elliptic $\Upsilon \in \bigodot^2 \operatorname{Hom}(\mathbf{R}^m, \mathbf{R}^{n-m})$ with ellipticity bound c and $\|\Upsilon\| \le M$ the corresponding elementary solution E satisfies the inequalities

$$\int_{\mathbf{S}^{m-1}} \|D^p E\| \, d\mathscr{H}^{m-1} \le \kappa \text{ for } p \in \{1,2,3\};$$

in fact one readily estimates the differentials of the function mapping ξ onto S_ξ^{-1}, and of functions R suitable for use in 5.2.9. We will verify that the number

$$\gamma = 18\,m^3\,[2+\delta^{-1}+(1-\delta)^{-1}]\,\kappa$$

has the asserted property. Observing that

$$\begin{aligned}&\int_{\mathbf{U}(b,\rho)} \int_{\mathbf{U}(b,\rho)} |\eta(x)-\eta(u)|^2 \, d\mathscr{L}^m x \, d\mathscr{L}^m u \\ &\le \int_{\mathbf{U}(b,\rho)} \int_{\mathbf{U}(b,\rho)} 2\,[|\eta(x)|^2 + |\eta(u)|^2] \, d\mathscr{L}^m x \, d\mathscr{L}^m u \\ &= \int_{\mathbf{U}(b,\rho)} 4\,|\eta|_{b,\rho}^2 \, d\mathscr{L}^m u,\end{aligned}$$

we choose $u \in \mathbf{B}(b,\rho)$ so that $|\omega|_{b,\rho} \le 2\,|\eta|_{b,\rho}$ with

$$\omega(x) = \eta(x) - \eta(u) \text{ for } x \in \mathbf{B}(b,\rho),$$

extend ω to a function $\Omega: \mathbf{R}^m \to \operatorname{Hom}(\mathbf{R}^m, \mathbf{R}^{n-m})$ as in 5.2.12, and use 5.2.11 to construct

$$F = \Omega * DE = \sum_{k=1}^m (\mathbf{e}_k \,\lrcorner\, \Omega) * D_k E.$$

We find that

$$\textstyle\int_{\mathbf{U}(b,\rho)} \langle DF(x) \odot D\theta(x), \Upsilon \rangle \, d\mathscr{L}^m x = (\Omega, D\theta)_{b,\rho} = (\eta, D\theta)_{b,\rho}$$

for all $\theta \in \mathscr{D}[\mathbf{U}(b,\rho), \mathbf{R}^{n-m}]$ because $\Omega - \eta$ is constant on $\mathbf{U}(b,\rho)$, note that

$$c\,|DF| \le |\Omega| \le 2^m\,|\omega|_{b,\rho} \le 2^{m+1}\,|\eta|, \qquad \mathbf{h}_\delta(\Omega) \le 9\,\mathbf{h}_\delta(\omega) = 9\,\mathbf{h}_\delta(\eta),$$

and apply 5.2.8 (5) with $K = D_k E$ to estimate

$$\mathbf{h}_\delta(D_k F) \le m\,2m[(1+\delta^{-1}) + (1-\delta)^{-1} + 1]\,\kappa\,\mathbf{h}_\delta(\Omega).$$

5.2.14. Theorem. *To any numbers $0 < c \le M < \infty$ and $0 < \delta < 1$ correspond numbers $\varepsilon > 0$ and $\Gamma < \infty$ with the following property:*

If $b \in \mathbf{R}^m$, $0 < R < \infty$,

$$f\colon \mathbf{U}(b,R) \to \mathbf{R}^{n-m} \text{ is of class } 1, \qquad \mathbf{h}_\delta(Df) < \infty,$$

$$\Omega\colon \mathbf{B}(b,R) \to \operatorname{Hom}(\mathbf{R}^m, \mathbf{R}^{n-m}), \qquad \mathbf{h}_\delta(\Omega) < \infty,$$

$$A\colon \mathbf{B}(b,R) \to \odot^2 \operatorname{Hom}(\mathbf{R}^m, \mathbf{R}^{n-m}), \qquad R^\delta\,\mathbf{h}_\delta(A) \le \varepsilon,$$

$$A(b) \text{ is strongly elliptic with ellipticity bound } c, \qquad \|A(b)\| \le M,$$

$$\textstyle\int_{\mathbf{U}(b,R)} \langle Df(x) \odot D\theta(x), A(x) \rangle \, d\mathscr{L}^m x = (\Omega, D\theta)_{b,R}$$

for all $\theta \in \mathscr{D}[\mathbf{U}(b,R), \mathbf{R}^{n-m}]$, then

$$(R-r)^{m/2+\delta}\,\mathbf{h}_\delta[Df \,|\, \mathbf{B}(b,r)] \le \Gamma[|Df|_{b,R} + R^{m/2+\delta}\,\mathbf{h}_\delta(\Omega)]$$

whenever $0 < r < R$.

Proof. We choose γ according to 5.2.13, let

$$\Delta_1 = 2^{m/2}\,\boldsymbol{\alpha}(m)^{-\frac{1}{2}}, \qquad \Delta_2 = 2^{m/2} + 1, \qquad \varepsilon = (2^{1+m/2+\delta}\,\gamma\,\Delta_2)^{-1},$$

$$\Delta_3 = 4(m+1)^{m+2}\,e\,\boldsymbol{\alpha}(m)^{-\frac{1}{2}}\,2^{m/2}\,(M/c)^{(m+1)/2},$$

$$\Delta_4 = \Delta_3(1 + 2^{m+1}\,c^{-1}\,\varepsilon), \qquad \Delta_5 = \Delta_3\,2^{m+1}\,c^{-1}\,\boldsymbol{\alpha}(m)^{\frac{1}{2}},$$

$$\Delta_6 = \Delta_1 + 3^{m/2+\delta}\,\Delta_4, \qquad \Delta_7 = \gamma + 3^{m/2+\delta}\,\Delta_5, \qquad \Gamma = 2\sup\{\Delta_6, \Delta_7\},$$

and suppose b, R, f, Ω, A satisfy the hypothesis of the asserted property. Since subtraction of a constant vector from Ω changes neither $\mathbf{h}_\delta(\Omega)$ nor $(\Omega, D\theta)_{b,R}$, we may also assume that $\Omega(b) = 0$, which implies

$$|\Omega|_{b,R} \le \boldsymbol{\alpha}(m)^{\frac{1}{2}}\,R^{m/2+\delta}\,\mathbf{h}_\delta(\Omega).$$

We define

$$\mu = \sup\{(R-r)^{m/2+\delta}\,\mathbf{h}_\delta[Df \,|\, \mathbf{B}(b,r)] : 0 < r < R\}$$

and note that $0 \le \mu \le R^{m/2+\delta}\,\mathbf{h}_\delta(Df) < \infty$.

Considering now some particular number r between 0 and R we let

$$d=(R-r)/3, \quad \rho=r+d=R-2d, \quad \zeta: \mathbf{B}(b,\rho)\to \mathrm{Hom}(\mathbf{R}^m, \mathbf{R}^{n-m}),$$
$$\zeta(x)\bullet\sigma=\langle Df(x)\odot\sigma, A(b)-A(x)\rangle$$

whenever $x\in\mathbf{B}(b,\rho)$ and $\sigma\in\mathrm{Hom}(\mathbf{R}^m,\mathbf{R}^{n-m})$.

If $x\in\mathbf{B}(b,\rho)$ and $u\in\mathbf{B}(b,\rho)$, then

$$\begin{aligned}\alpha(m)^{\frac{1}{2}}\, d^{m/2}\,|Df(x)| &\le |Df|_{x,d}+\left[\textstyle\int_{\mathbf{B}(x,d)}|Df(x)-Df(w)|^2\, d\mathscr{L}^m w\right]^{\frac{1}{2}}\\ &\le |Df|_{b,R}+d^{-m/2-\delta}\,\mu\, d^{\delta}\,\alpha(m)^{\frac{1}{2}}\, d^{m/2}\end{aligned}$$

because $\mathbf{B}(x,d)\subset\mathbf{B}(b,R-d)$ and $d^{m/2+\delta}\,\mathbf{h}_\delta[Df\,|\,\mathbf{B}(b,R-d)]\le\mu$,

$$\begin{aligned}[\zeta(x)-\zeta(u)]\bullet\sigma &= \langle Df(x)\odot\sigma, A(u)-A(x)\rangle\\ &\quad+\langle[Df(x)-Df(u)]\odot\sigma, A(b)-A(u)\rangle\end{aligned}$$

whenever $\sigma\in\mathrm{Hom}(\mathbf{R}^m,\mathbf{R}^{n-m})$, hence

$$\begin{aligned}|\zeta(x)-\zeta(u)| &\le |Df(x)|\cdot\|A(u)-A(x)\|+|Df(x)-Df(u)|\cdot\|A(b)-A(u)\|\\ &\le d^{-m/2}\left[\alpha(m)^{-\frac{1}{2}}\,|Df|_{b,R}+\mu\right]\mathbf{h}_\delta(A)\,|u-x|^\delta\\ &\quad+(2d)^{-m/2-\delta}\,\mu\,|x-u|^\delta\,\mathbf{h}_\delta(A)\,|b-u|^\delta\\ &\le (2d)^{-m/2-\delta}\left[\Delta_1\,|Df|_{b,R}+\Delta_2\,\mu\right]R^\delta\,\mathbf{h}_\delta(A)\,|x-u|^\delta.\end{aligned}$$

We infer that

$$\mathbf{h}_\delta(\zeta)\le(2d)^{-m/2-\delta}\left[\Delta_1\,|Df|_{b,R}+\Delta_2\,\mu\right]\varepsilon,$$
$$|\zeta|_{b,\rho}\le|Df|_{b,\rho}\,\mathbf{h}_\delta(A)\,R^\delta\le|Df|_{b,R}\,\varepsilon.$$

Applying 5.2.13 with $\Upsilon=A(b)$ and $\eta=\zeta+\Omega$ we find that the resulting function F of class 1 satisfies the conditions

$$\mathbf{h}_\delta(DF)\le\gamma\left[\mathbf{h}_\delta(\zeta)+\mathbf{h}_\delta(\Omega)\right], \quad |DF|_{b,\rho}\le 2^{m+1}c^{-1}\left[|\zeta|_{b,\rho}+|\Omega|_{b,\rho}\right],$$
$$\begin{aligned}&\textstyle\int_{\mathbf{U}(b,\rho)}\langle DF(x)\odot D\theta(x), A(b)\rangle\, d\mathscr{L}^m x=(\eta, D\theta)_{b,\rho}\\ &=\textstyle\int_{\mathbf{U}(b,\rho)}\langle Df(x)\odot D\theta(x), A(b)-A(x)\rangle\, d\mathscr{L}^m x+(\Omega, D\theta)_{b,\rho}\\ &=\textstyle\int_{\mathbf{U}(b,\rho)}\langle Df(x)\odot D\theta(x), A(b)\rangle\, d\mathscr{L}^m x\end{aligned}$$

for all $\theta\in\mathscr{D}[\mathbf{U}(b,\rho),\mathbf{R}^{n-m}]$, hence $g=(f-F)|\mathbf{U}(b,\rho)$ is a function of class 1 such that

$$\int_{\mathbf{U}(b,\rho)}\langle Dg(x)\odot D\theta(x), A(b)\rangle\, d\mathscr{L}^m x=0$$

for all $\theta\in\mathscr{D}[\mathbf{U}(b,\rho),\mathbf{R}^{n-m}]$, and we see from 5.2.5, 5.2.6 that g is an $A(b)$ harmonic function with

$$\mathbf{h}_\delta[Dg\,|\,\mathbf{B}(b,r)]\le\Delta_3\, d^{-m/2-\delta}\,|Dg|_{b,\rho}\le\Delta_3\, d^{-m/2-\delta}\left[|Df|_{b,R}+|DF|_{b,\rho}\right].$$

Combining our estimates we obtain

$$\begin{aligned}\mathbf{h}_\delta[Df\,|\,\mathbf{B}(b,r)] &\le \mathbf{h}_\delta(DF)+\mathbf{h}_\delta[Dg\,|\,\mathbf{B}(b,r)]\\ &\le \gamma(2d)^{-m/2-\delta}[\Delta_1\,|Df|_{b,R}+\Delta_2\,\mu]\,\varepsilon+\gamma\,\mathbf{h}_\delta(\Omega)\\ &\quad +d^{-m/2-\delta}[\Delta_4\,|Df|_{b,R}+\Delta_5\,R^{m/2+\delta}\,\mathbf{h}_\delta(\Omega)],\end{aligned}$$

and conclude

$$(R-r)^{m/2+\delta}\,\mathbf{h}_\delta[Df\,|\,\mathbf{B}(b,r)]\le\mu/2+\Delta_6\,|Df|_{b,R}+\Delta_7\,R^{m/2+\delta}\,\mathbf{h}_\delta(\Omega).$$

Since this inequality holds for all numbers r between 0 and R, it follows that

$$\mu\le\mu/2+\Delta_6\,|Df|_{b,R}+\Delta_7\,R^{m/2+\delta}\,\mathbf{h}_\delta(\Omega).$$

5.2.15. Theorem. *Suppose:*

(1) *q is an integer, $q\ge 2$.*

(2) *V is an open subset of $\mathbf{R}^m\times\mathbf{R}^{n-m}\times\operatorname{Hom}(\mathbf{R}^m,\mathbf{R}^{n-m})$,*

$$G\colon V\to R \text{ is of class } q+1.$$

(3) *U is an open subset of $\mathbf{R}^m$,*

$$f\colon U\to\mathbf{R}^{n-m} \text{ is of class 1, } 0<\delta<1,\ \mathbf{h}_\delta(Df)<\infty,$$

$$\psi(x)=[x,f(x),Df(x)]\in V \text{ whenever } x\in U.$$

(4) *$\int_U\langle[0,\theta(x),D\theta(x)],DG[\psi(x)]\rangle\,d\mathscr{L}^m x=0$ whenever $\theta\in\mathscr{D}(U,\mathbf{R}^{n-m})$.*

(5) *$A\colon U\to\odot^2\operatorname{Hom}(\mathbf{R}^m,\mathbf{R}^{n-m})$,*

$$\langle\sigma\odot\tau,A(x)\rangle=\langle(0,0,\sigma)\odot(0,0,\tau),D^2\,G[\psi(x)]\rangle$$

whenever $x\in U$ and $\sigma,\tau\in\operatorname{Hom}(\mathbf{R}^m,\mathbf{R}^{n-m})$.

(6) *$A(x)$ is strongly elliptic whenever $x\in U$.*

Then the following four statements hold:

(7) *f is of class q, $\mathbf{h}_\delta(D^q f\,|\,K)<\infty$ for every compact $K\subset U$.*

(8) *To each $s\in\mathscr{S}(m,p)$ with $p\in\{1,\ldots,q-1\}$ corresponds*

$$\Omega_s\colon U\to\operatorname{Hom}(\mathbf{R}^m,\mathbf{R}^{n-m}) \text{ of class } q-p \text{ such that}$$

$$\mathbf{h}_\delta(D^{q-p}\,\Omega_s\,|\,K)<\infty \text{ for every compact } K\subset U,$$

$$\int_U\langle DD_s f(x)\odot D\theta(x),A(x)\rangle\,d\mathscr{L}^m x=(\Omega_s,D\theta)_U \text{ whenever } \theta\in\mathscr{D}(U,\mathbf{R}^{n-m}).$$

(9) *$\Omega_{(i)}(x)\bullet\sigma=\langle[0,\sigma(e_i),0],DG[\psi(x)]\rangle-\langle[0,0,\sigma]\odot[e_i,D_i f(x),0],$ $D^2\,G[\psi(x)]\rangle$ whenever $i\in\{1,\ldots,m\}$, $x\in U$ and $\sigma\in\operatorname{Hom}(\mathbf{R}^m,\mathbf{R}^{n-m})$.*

(10) *$\Omega_{(s_1,\ldots,s_p,i)}(x)\bullet\sigma=D_i\,\Omega_s(x)\bullet\sigma-\langle DD_s f(x)\odot\sigma,D_i A(x)\rangle$ whenever $s\in\mathscr{S}(m,p)$, $p\in\{1,\ldots,q-2\}$, $i\in\{1,\ldots,m\}$, $x\in U$ and $\sigma\in\operatorname{Hom}(\mathbf{R}^m,\mathbf{R}^{n-m})$.*

Proof. Given $b \in U$ we will show that the conclusion holds with U replaced by $\mathbf{U}(b, \rho)$ when ρ is a sufficiently small positive number. First we choose $M < \infty$ and $c > 0$ so that $A(b)$ belongs to the convex open set

$$\Delta = [\odot^2 \operatorname{Hom}(\mathbf{R}^m, \mathbf{R}^{n-m})] \cap \{\Upsilon \colon \|\Upsilon\| < M \text{ and } \Upsilon \text{ is strongly}$$
$$\text{elliptic with some ellipticity bound greater than } c\},$$

and choose a convex neighborhood W of $\psi(b)$ in V such that

$$\alpha_p = \sup\{\|D^p G(x, y, \lambda)\| \colon (x, y, \lambda) \in W\} < \infty \text{ for } p \in \{0, \ldots, q+1\},$$
$$D^2 G_{(x,y)}(\lambda) \in \Delta \text{ for } (x, y, \lambda) \in W, \text{ where}$$
$$\langle \sigma \odot \tau, D^2 G_{(x,y)}(\lambda) \rangle = \langle (0, 0, \sigma) \odot (0, 0, \tau), D^2 G(x, y, \lambda) \rangle$$

for $\sigma, \tau \in \operatorname{Hom}(\mathbf{R}^m, \mathbf{R}^{n-m})$. Then we choose ε and Γ according to 5.2.14. We observe that, for small $\rho > 0$,

$$\mathbf{B}(b, \rho) \subset \psi^{-1}(W), \quad \mathbf{h}_\delta[\psi | \mathbf{B}(b, \rho)] \le 1 + \operatorname{Lip}[f | \mathbf{B}(b, \rho)] + \mathbf{h}_\delta(D f),$$
$$\operatorname{Lip}[f | \mathbf{B}(b, \rho)] \le \|D f(b)\| + \rho^\delta \mathbf{h}_\delta(D f),$$

and choose ρ so that also

$$\rho^\delta \alpha_3 \mathbf{h}_\delta[\psi | \mathbf{B}(b, \rho)] < \inf\{\varepsilon, c/2\}.$$

We now consider the special case when $q = 2$.

Fixing $i \in \{1, \ldots, m\}$ we suppose $0 < r < \rho$, $d = (\rho - r)/3$ and define for each integer $\nu > 1/d$ the maps

$$f_\nu \colon \mathbf{B}(b, \rho - d) \to \mathbf{R}^{n-m}, \quad A_\nu \colon \mathbf{B}(b, \rho - d) \to \Delta,$$
$$\psi_\nu \colon \mathbf{B}(b, \rho - d) \to \mathbf{R}^m \times \mathbf{R}^{n-m} \times \operatorname{Hom}(\mathbf{R}^m, \mathbf{R}^{n-m}),$$
$$\phi_\nu \colon \{t \colon 0 \le t \le 1\} \times \mathbf{B}(b, \rho - d) \to W,$$
$$P_\nu, Q_\nu \colon \mathbf{B}(b, \rho - d) \to \operatorname{Hom}(\mathbf{R}^m, \mathbf{R}^{n-m})$$

by the formulae

$$f_\nu(x) = \nu[f(x + \nu^{-1} e_i) - f(x)] = \textstyle\int_0^1 D_i f(x + t \nu^{-1} e_i) \, d\mathscr{L}^1 t,$$
$$\psi_\nu(x) = \nu[\psi(x + \nu^{-1} e_i) - \psi(x)] = [e_i, f_\nu(x), D f_\nu(x)],$$
$$\phi_\nu(t, x) = (1 - t) \psi(x) + t \psi(x + \nu^{-1} e_i),$$
$$\langle \sigma \odot \tau, A_\nu(x) \rangle = \textstyle\int_0^1 \langle (0, 0, \sigma) \odot (0, 0, \tau), D^2 G[\phi_\nu(t, x)] \rangle \, d\mathscr{L}^1 t,$$
$$\sigma \bullet P_\nu(x) = \textstyle\int_0^1 \langle [0, \sigma(e_i), 0], D G[\psi(x + t \nu^{-1} e_i)] \rangle \, d\mathscr{L}^1 t,$$
$$\sigma \bullet Q_\nu(x) = \textstyle\int_0^1 \langle [0, 0, \sigma] \odot [e_i, f_\nu(x), 0], D^2 G[\phi_\nu(t, x)] \rangle \, d\mathscr{L}^1 t$$

whenever $x \in \mathbf{B}(b, \rho-d)$, $0 \le t \le 1$ and $\sigma, \tau \in \operatorname{Hom}(\mathbf{R}^m, \mathbf{R}^{n-m})$; we also note the equation

$$v(DG[\psi(x+v^{-1}e_i)] - DG[\psi(x)]) = \int_0^1 \langle \psi_v(x), DDG[\phi_v(t,x)] \rangle \, d\mathscr{L}^1 t.$$

We see that f_v is of class 1, $\mathbf{h}_\delta(Df_v) \le 2v\,\mathbf{h}_\delta(Df)$,

$$\mathbf{h}_\delta(f_v) \le \mathbf{h}_\delta(D_i f), \quad |f_v|_{b,\rho-d} \le |D_i f|_{b,\rho},$$

$$\operatorname{Lip}(D^p G|W) \le \alpha_{p+1} \text{ for } p \in \{0, \ldots, q\},$$

$$\mathbf{h}_\delta(P_v) \le \alpha_2\,\mathbf{h}_\delta[\psi|B(b,\rho)], \quad |P_v|_{b,\rho-d} \le |(DG)\circ\psi|_{b,\rho},$$

$$\mathbf{h}_\delta(Q_v) \le \mathbf{h}_\delta(D_i f)\,\alpha_2 + \operatorname{Lip}[Df|\mathbf{B}(b,\rho)]\,\alpha_3\,\mathbf{h}_\delta[\psi|\mathbf{B}(b,\rho)],$$

$$|Q_v|_{b,\rho-d} \le |D_i f|_{b,\rho}\,\alpha_2,$$

$$\mathbf{h}_\delta(A_v) \le \alpha_3\,\mathbf{h}_\delta[\psi|\mathbf{B}(b,\rho)], \quad \rho^\delta\,\mathbf{h}_\delta(A_v) < \inf\{\varepsilon, c/2\}.$$

Recalling 5.2.3 we infer the inequality

$$\int_{\mathbf{U}(b,\rho-d)} \langle [D\theta(x)]^2, A_v(x) \rangle \, d\mathscr{L}^m x \ge (c/2)\,|D\theta|^2_{b,\rho-d}$$

for all $\theta \in \mathscr{D}[\mathbf{U}(b,\rho-d), \mathbf{R}^{n-m}]$.

If $\theta \in \mathscr{D}(\mathbf{R}^m, \mathbf{R}^{n-m})$ with $\operatorname{spt}\theta \subset \mathbf{U}(b,\rho-d)$, then

$$\int_{\mathbf{U}(b,\rho-d)} \langle Df_v(x) \odot D\theta(x), A_v(x) \rangle \, d\mathscr{L}^m x = (P_v - Q_v, D\theta)_{b,\rho-d}$$

because (4) is applicable with θ replaced by the function $\theta_v \in \mathscr{D}(\mathbf{R}^m, \mathbf{R}^{n-m})$ with $\operatorname{spt}\theta_v \subset \mathbf{U}(b,\rho)$ such that

$$\theta_v(x) = v[\theta(x - v^{-1}e_i) - \theta(x)] = -\int_0^1 D_i\theta(x - t v^{-1} e_i)\, d\mathscr{L}^1 t$$

whenever $x \in \mathbf{R}^m$, and because

$$\begin{aligned}
&\int_U \langle [0, \theta_v(x), 0], DG[\psi(x)] \rangle \, d\mathscr{L}^m x \\
&= -\int_0^1 \int_{\mathbf{U}(b,\rho)} \langle [0, D_i\theta(x - t v^{-1} e_i), 0], DG[\psi(x)] \rangle \, d\mathscr{L}^m x \, d\mathscr{L}^1 t \\
&= -\int_0^1 \int_{\mathbf{U}(b,\rho-d)} \langle [0, D_i\theta(x), 0], DG[\psi(x + t v^{-1} e_i)] \rangle \, d\mathscr{L}^m x \, d\mathscr{L}^1 t \\
&= -(D\theta, P_v)_{b,\rho-d}
\end{aligned}$$

while

$$\begin{aligned}
&\int_U \langle [0, 0, D\theta_v(x)], DG[\psi(x)] \rangle \, d\mathscr{L}^m x \\
&= \int_{\mathbf{U}(b,\rho)} v \langle [0, 0, D\theta(x - v^{-1}e_i) - D\theta(x)], DG[\psi(x)] \rangle \, d\mathscr{L}^m x \\
&= \int_{\mathbf{U}(b,\rho-d)} v \langle [0, 0, D\theta(x)], DG[\psi(x + v^{-1}e_i)] - DG[\psi(x)] \rangle \, d\mathscr{L}^m x \\
&= \int_{\mathbf{U}(b,\rho-d)} \int_0^1 \langle [0, 0, D\theta(x)] \odot \psi_v(x), D^2 G[\phi_v(t,x)] \rangle \, d\mathscr{L}^1 t \, d\mathscr{L}^m x \\
&= \int_{\mathbf{U}(b,\rho-d)} \langle D\theta(x) \odot Df_v(x), A_v(x) \rangle \, d\mathscr{L}^m x + (D\theta, Q_v)_{b,\rho-d}.
\end{aligned}$$

Using the basic estimate in 5.2.3 we find that

$$\begin{aligned}(c/2)\,|D f_\nu|_{b,r+d} &\le |P_\nu - Q_\nu|_{b,\rho-d} + (c\,M/2)^{\frac12}\, d^{-1}\, |f_\nu|_{b,\rho-d}\\ &\le |(D\,G)\circ\psi|_{b,\rho-d} + [\alpha_2 + (c\,M/2)^{\frac12}\, d^{-1}]\,|D_i f|_{b,\rho}\end{aligned}$$

and applying 5.2.14 we obtain

$$d^{m/2+\delta}\,\mathbf{h}_\delta[D f_\nu|\mathbf{B}(b,r)] \le \Gamma[|D f_\nu|_{b,r+d} + (r+d)^{m/2+\delta}\,\mathbf{h}_\delta(P_\nu - Q_\nu)];$$

moreover $\mathbf{h}_\delta(P_\nu - Q_\nu)$ does not exceed

$$\mathbf{h}_\delta(D_i f)\,\alpha_2 + \big(\alpha_2 + \operatorname{Lip}[D f|\mathbf{B}(b,\rho)]\,\alpha_3\big)\,\mathbf{h}_\delta[\psi|\mathbf{B}(b,\rho)].$$

Accordingly $\mathbf{h}_\delta[D f_\nu|\mathbf{B}(b,r)]$ and $|D f_\nu|_{b,r+d}$ have bounds independent of ν. Since the functions f_ν converge to $D_i f$ uniformly on $\mathbf{B}(b,\rho-d)$ as ν approaches ∞, we infer from 5.2.2 that $D_i f\,|\,\mathbf{U}(b,r)$ is of class 1,

$$\mathbf{h}_\delta[D\,D_i f\,|\,\mathbf{U}(b,r)] < \infty$$

and the functions $D f_\nu$ converge to $D\,D_i f$ uniformly on each compact subset of $\mathbf{U}(b,r)$. Defining $\Omega_{(i)}$ by (9) and observing that A_ν, $P_\nu - Q_\nu$ converge to $A, \Omega_{(i)}$ uniformly on $\mathbf{B}(b,r)$, we conclude that (7) and (8) hold in case $q=2$.

Proceeding by induction with respect to q, *we assume next that* $q>2$ *and the conclusions hold with* q *replaced by* $q-1$.

Fixing $s\in\mathscr{S}(m,q-2)$ and $i\in\{1,\ldots,m\}$ we take r, d, ν, f_ν as before and also define

$$\begin{gathered}\eta_\nu, \zeta_\nu\colon \mathbf{B}(b,\rho-d)\to \operatorname{Hom}(\mathbf{R}^m, \mathbf{R}^{n-m}),\\ \eta_\nu(x)\bullet\sigma = \nu\langle D\,D_s f(x+\nu^{-1} e_i)\odot\sigma,\ A(x+\nu^{-1} e_i) - A(x)\rangle,\\ \zeta_\nu(x) = \nu[\Omega_s(x+\nu^{-1} e_i) - \Omega_s(x)]\end{gathered}$$

whenever $x\in\mathbf{B}(b,\rho-d)$ and $\sigma\in\operatorname{Hom}(\mathbf{R}^m,\mathbf{R}^{n-m})$. We see that $D_s f_\nu$, Ω_s, A are functions of class 1, whose differentials satisfy Hölder conditions with exponent δ on $\mathbf{B}(b,\rho)$, and that

$$|D_s f_\nu|_{b,\rho-d} \le |D_i\, D_s f|_{b,\rho},$$

$$\mathbf{h}_\delta(\zeta_\nu) \le \mathbf{h}_\delta[D_i\,\Omega_s|\mathbf{B}(b,\rho)], \qquad |\zeta_\nu|_{b,\rho-d} \le |D_i\,\Omega_s|_{b,\rho},$$

$$\begin{aligned}\mathbf{h}_\delta(\eta_\nu) \le\ & \mathbf{h}_\delta[D^{q-1} f\,|\mathbf{B}(b,\rho)]\operatorname{Lip}[D_i A|\mathbf{B}(b,\rho)]\\ &+ \sup\{|D^{q-1} f(x)|\colon x\in\mathbf{B}(b,\rho)\}\,\mathbf{h}_\delta[A|\mathbf{B}(b,\rho)],\\ |\eta_\nu|_{b,\rho-d} \le\ & |D\,D_s f|_{b,\rho}\operatorname{Lip}[A|\mathbf{B}(b,\rho)].\end{aligned}$$

If $\theta \in \mathscr{D}(\mathbf{R}^m, \mathbf{R}^{n-m})$ with spt $\theta \subset \mathbf{U}(b, \rho - d)$, then

$$\int_{\mathbf{U}(b,\rho-d)} \langle D D_s f_\nu(x) \odot D\theta(x), A(x)\rangle \, d\mathscr{L}^m x = (\zeta_\nu - \eta_\nu)_{b,\rho-d}$$

because the last conclusion of (8) is applicable with θ replaced by θ_ν, defined as before, and because

$$\begin{aligned}
&\int_U D D_s f(x) \odot D\theta_\nu(x), A(x)\rangle \, d\mathscr{L}^m x \\
&= \int_{\mathbf{U}(b,\rho)} \nu \langle D D_s f(x) \odot [D\theta(x - \nu^{-1} e_i) - D\theta(x)], A(x)\rangle \, d\mathscr{L}^m x \\
&= \int_{\mathbf{U}(b,\rho-d)} \nu [\langle D D_s f(x + \nu^{-1} e_i) \odot D\theta(x), A(x + \nu^{-1} e_i)\rangle \\
&\qquad\qquad - \langle D D_s f(x) \odot D\theta(x), A(x)\rangle] \, d\mathscr{L}^m x \\
&= \int_{\mathbf{U}(b,\rho-d)} \langle D D_s f_\nu(x) \odot D\theta(x), A(x)\rangle \, d\mathscr{L}^m x + (\eta_\nu, D\theta)_{b,\rho-d}
\end{aligned}$$

while $(\Omega_s, D\theta_\nu)_U = (\zeta_\nu, D\theta)_{b,\rho-d}$.

Using the basic estimate in 5.2.3 we find that

$$(c/2)\, |D D_s f_\nu|_{b,r+d} \le |\zeta_\nu - \eta_\nu|_{b,\rho-d} + (c M/2)^{\frac{1}{2}} d^{-1} |D_s f_\nu|_{b,\rho-d}$$

and applying 5.2.14 we obtain

$$d^{m/2+d}\, \mathbf{h}_\delta [D D_s f | \mathbf{B}(b,r)] \le \Gamma [|D D_s f_\nu|_{b,r+d} + (r+d)^{m/2+\delta} \mathbf{h}_\delta(\zeta_\nu - \eta_\nu)].$$

Inasmuch as $\mathbf{h}_\delta(\zeta_\nu - \eta_\nu)$, $|\zeta_\nu - \eta_\nu|_{b,\rho-d}$ and $|D_s f_\nu|_{b,\rho-d}$ have bounds independent of ν, we infer that $\mathbf{h}_\delta [D D_s f_\nu | \mathbf{B}(b,r)]$ and $|D D_s f_\nu|_{b,r+d}$ have bounds independent of ν. Since the functions $D_s f_\nu$ converge to $D_i D_s f$ uniformly on $\mathbf{B}(b, \rho - d)$ as ν approaches ∞, we infer from 5.2.2 that $D_i D_s f | \mathbf{U}(b,r)$ is of class 1,

$$\mathbf{h}_\delta [D D_i D_s f | \mathbf{U}(b,r)] < \infty$$

and the functions $D D_s f_\nu$ converge to $D D_i D_s f$ uniformly on each compact subset of $\mathbf{U}(b,r)$. Observing that the functions $\zeta_\nu - \eta_\nu$ converge uniformly on $U(b,r)$ to the function

$$\Omega_{(s_1, \ldots, s_{q-2}, i)}$$

defined by (10), we deduce (7) and (8); we verify the class $q - p$ and Hölder condition asserted in (8) by induction with respect to p using (9) and (10).

5.2.16. While the regularity theorem 5.2.15 suffices for applications in this book (5.3.14, 5.3.16, 5.3.18, 5.3.19 and 5.4.15), we call attention to the following additional results in the literature:

If G is an analytic function, then f is an analytic function; this follows from [PE] or [MCB 3] or [FRA] or [MCB 4, § 6.7].

The hypothesis $\mathbf{h}_\delta(Df)<\infty$ in (3) is redundant; only the assumption that f be of class 1 occurs in [MCB 2]. Our proof of 5.2.14, 5.2.15 uses much of Morrey's method, but is greatly simplified by the Hölder condition on Df.

For special dimensions and special types of integrands the conditions on f can be further weakened in the spirit of 5.2.6; such results are summarized in [MCB 4, §1.10, 1.11]. However in case $m=n-m\geq 3$ one cannot replace the real valued functions $Y_j\circ f$ of class 1 by distributions whose first order partial derivatives correspond to locally square integrable functions, even when G is analytic and quasilinear; a counterexample was constructed in [GM 1].

5.2.17. *If Φ is an elliptic integrand of degree m and class 2 on the open subset W of $\mathbf{R}^n$, then*

$$D^2\,\Phi^{\S}_w(\lambda) \text{ is strongly elliptic for } (w,\lambda)\in W\times\operatorname{Hom}(\mathbf{R}^m,\mathbf{R}^{n-m}).$$

To prove this we define $\gamma=\mathbf{1}_{\mathbf{R}^n}+\mathbf{q}^*\circ\lambda\circ\mathbf{p}$ and $z=\gamma^{-1}(w)$, infer from the last paragraph in 5.1.9 that

$$(\gamma^{\#}\Phi)^{\S}_z(\tau)=\Phi^{\S}_w(\lambda+\tau) \text{ for } \tau\in\operatorname{Hom}(\mathbf{R}^m,\mathbf{R}^{n-m}),$$
$$\text{hence } D^2(\gamma^{\#}\Phi^{\S}_z)(0)=D^2\,\Phi^{\S}_w(\lambda),$$

observe that $\gamma^{\#}\Phi$ is an elliptic integrand according to 5.1.4, and apply 5.1.10 with Φ, a replaced by $\gamma^{\#}\Phi$, z.

5.2.18. Theorem. *If q is an integer, $q\geq 2$, Φ is an elliptic parametric integrand of degree m and class $q+1$ on the open subset W of $\mathbf{R}^m\times\mathbf{R}^{n-m}$,*

$$U \text{ is an open subset of } \mathbf{R}^m, \quad f\colon U\to\mathbf{R}^{n-m} \text{ is of class } 1,$$
$$0<\delta<1, \quad \mathbf{h}_\delta(Df)<\infty, \quad \operatorname{im}(\mathbf{p}^*+\mathbf{q}^*\circ f)\subset W,$$

and $(\mathbf{p}^*+\mathbf{q}^*\circ f)_{\#}(\mathbf{E}^m\llcorner U)$ is absolutely Φ minimizing with respect to W, then f is of class q and $\mathbf{h}_\delta(D^q f\,|\,K)<\infty$ for every compact subset K of U.

Proof. Recalling 5.1.9 we apply 5.2.15 with $G=\Phi^{\S}$. To verify (4) we consider the nonparametric isotopic deformation h corresponding to θ and compute

$$\int_U\langle[0,\theta(x),D\theta(x)],D\Phi^{\S}[\psi(x)]\rangle\,d\mathscr{L}^m x$$
$$=\delta^{(1)}[(\mathbf{p}^*+\mathbf{q}^*\circ f)_{\#}(\mathbf{E}^m\llcorner\operatorname{spt}\theta),\Phi,h]=0$$

by 5.1.7. We deduce (6) from 5.2.17.

5.2.19. Here we derive two propositions concerning the special case $n=m+1$. First we prove the **maximum principle:**

If U is an open subset of $\mathbf{R}^m$, $\phi: U \to \mathbf{R}$ is of class 2,

$$A: U \to \odot^2 \operatorname{Hom}(\mathbf{R}^m, \mathbf{R}) \text{ is of class } 1,$$

$$A(x) \text{ is strongly elliptic for } x \in U,$$

$$\int_U \langle D\phi(x) \odot D\theta(x), A(x)\rangle \, d\mathscr{L}^m x = 0 \text{ for } \theta \in \mathscr{D}(U, \mathbf{R}),$$

then $C = \{x: \phi(x) = \sup \operatorname{im} \phi\}$ is open. Letting

$$A_{i,k}(x) = \langle X_i \odot X_k, A(x)\rangle$$

for $x \in U$ and $\{i, k\} \subset \{1, \ldots, m\}$ we define

$$L\psi = \sum_{i=1}^{m} \sum_{k=1}^{m} D_k(A_{i,k} D_i \psi)$$

for every map $\psi: U \to \mathbf{R}$ of class 2, and we integrate by parts as in 5.2.3 to infer $L\phi = 0$. If C were not open, we could choose $b, r, s, t, \gamma, h, \varepsilon, \Phi, u$ so that

$$b \in U \sim C, \quad 0 < r < s < t < \infty, \quad \mathbf{B}(b, t) \subset U,$$

$$C \cap \mathbf{B}(b, s) \neq \varnothing, \quad C \cap \mathbf{U}(b, s) = \varnothing, \quad -\infty < \gamma < 0,$$

$$h(x) = \exp(\gamma |x-b|^2) - \exp(\gamma s^2) \text{ for } x \in U,$$

$$(Lh)(x) = \exp(\gamma |x-b|^2) \Big(\sum_{i=1}^{m} 2\gamma A_{ii}(x) + \sum_{i=1}^{m} \sum_{k=1}^{m}$$
$$[4\gamma^2 (x_i - b_i)(x_k - b_k) A_{i,k}(x) + 2\gamma (x_i - b_i) D_k A_{i,k}(x)] \Big) > 0$$

for all x in the compact set $\mathbf{B}(b, t) \sim \mathbf{U}(b, r)$, which holds when $|\gamma|$ is large because $A(x)$ is an inner product,

$$0 < \varepsilon < (\sup \operatorname{im} \phi) - \sup \phi[\mathbf{B}(b, r)], \quad \Phi = \phi + \varepsilon h,$$

$$u \in \mathbf{B}(b, s) \sim \mathbf{B}(b, r), \quad \Phi(u) = \sup \operatorname{im} \phi,$$

hence $D\Phi(u) = 0$; such a point u would exist because

$$\Phi(x) = \phi(x) \text{ if } |x-b| = s, \quad \sup \operatorname{im} \Phi \geq \sup \operatorname{im} \phi,$$

$$\Phi(x) < \phi(x) \text{ if } |x-b| > s, \quad \Phi(x) < \sup \operatorname{im} \phi \text{ if } |x-b| \leq r.$$

We could then conclude that

$$0 < \varepsilon (Lh)(u) = (L\Phi)(u) = \sum_{i=1}^{m} \sum_{k=1}^{m} A_{i,k}(u) D_{i,k} \Phi(u) = \langle D^2 \Phi(u), A(u)\rangle \leq 0,$$

because $\langle \sigma \odot \sigma, A(u) \rangle \geq 0$ for $\sigma \in \bigodot^1 \mathbf{R}^m$ and 1.7.3 yields a representation

$$D^2 \Phi(u) = \sum_{i=1}^{m} \lambda_i (\sigma_i \odot \sigma_i) \text{ with } \lambda_i \leq 0, \ \sigma_i \in \bigodot^1 \mathbf{R}^m.$$

We deduce a corollary related to certain variational problems (see 5.3.18):

If $f: U \to \mathbf{R}$ and $g: U \to \mathbf{R}$ are of class 2,

$$f(x) \leq g(x) \text{ for } x \in U, \quad H: U \times \operatorname{Hom}(\mathbf{R}^m, \mathbf{R}) \to \mathbf{R} \text{ is of class } 3,$$

$$\int_U \langle [0, D\theta(x)], DH[x, Df(x)] \rangle \, d\mathscr{L}^m x$$
$$= \int_U \langle [0, D\theta(x)], DH[x, Dg(x)] \rangle \, d\mathscr{L}^m x$$

for $\theta \in \mathscr{D}(U, \mathbf{R})$, and the form mapping $(\sigma, \tau) \in \operatorname{Hom}(\mathbf{R}^m, \mathbf{R}) \times \operatorname{Hom}(\mathbf{R}^m, \mathbf{R})$ onto

$$\langle (0, \sigma) \odot (0, \tau), D^2 H(x, \lambda) \rangle$$

is strongly elliptic for $(x, \lambda) \in U \times \operatorname{Hom}(\mathbf{R}^m, \mathbf{R})$, then

$$\{x: f(x) = g(x)\} \text{ is open.}$$

We apply the maximum principle with $\phi(x) = f(x) - g(x)$ and

$$\langle \sigma \odot \tau, A(x) \rangle = \int_0^1 \langle (0, \sigma) \odot (0, \tau), D^2 H[x, (1-t) Dg(x) + t Df(x)] \rangle \, d\mathscr{L}^1 t$$

for $x \in U$ and $\sigma, \tau \in \operatorname{Hom}(\mathbf{R}^m, \mathbf{R})$.

Next we will prove the second proposition:

If U is a connected open subset of $\mathbf{R}^m$,

$$\Xi: U \to \bigodot^2 [\mathbf{R} \times \operatorname{Hom}(\mathbf{R}^m, \mathbf{R})] \text{ is continuous,}$$

$$\langle (0, \sigma) \odot (0, \sigma), \Xi(x) \rangle > 0 \text{ for } x \in U \text{ and } 0 \neq \sigma \in \operatorname{Hom}(\mathbf{R}^m, \mathbf{R}),$$

$$\phi: U \to \mathbf{R} \cap \{y: y \leq 0\} \text{ is of class } 1, \quad C = \{x: \phi(x) = 0\},$$

$$\int_U \langle [\phi(x), D\phi(x)] \odot [\theta(x), D\theta(x)], \Xi(x) \rangle \, d\mathscr{L}^m x = 0 \text{ for } \theta \in \mathscr{D}(U, \mathbf{R}),$$

then either $C = U$ or $\mathscr{H}^{m-1}(C) = 0$.

We assume $C \neq U$ and let γ be the characteristic function of C. Given any connected open set W such that $\operatorname{Clos} W$ is a compact subset of U and $W \not\subset C$, we choose

$$\psi \in \mathscr{D}^0(U) \text{ with } \psi(x) \geq 0 \text{ for } x \in U, \ \psi(x) = 1 \text{ for } x \in W,$$

$$0 < \lambda \leq \mu < \infty \text{ with } \|\Xi(x)\| \leq \mu \text{ for } x \in \operatorname{spt} \psi,$$

$$\langle (0, \sigma) \odot (0, \sigma), \Xi(x) \rangle \geq \lambda |\sigma|^2 \text{ for } x \in \operatorname{spt} \psi \text{ and } \sigma \in \operatorname{Hom}(\mathbf{R}^m, \mathbf{R}).$$

Clearly our hypothesis remains valid for every Lipschitzian real valued function θ on U such that $\operatorname{spt}\theta$ is a compact subset of U. We define

$$\theta_j=\sup\{\psi+j\,\phi,0\},\qquad R_j=\{x\colon \theta_j(x)>0\},\qquad \alpha_j=|\psi|_{R_j}+|D\,\psi|_{R_j}$$

for each positive integer j, observe that

$$\begin{aligned}&[0,D\,\theta_j(x)]\odot[0,D\,\theta_j(x)]-j\,[\phi(x),D\,\phi(x)]\odot[\theta_j(x),D\,\theta_j(x)]\\&=[-j\,\phi(x),D\,\psi(x)]\odot[\theta_j(x),D\,\theta_j(x)]-[0,D\,\theta_j(x)]\odot[\theta_j(x),0]\end{aligned}$$

with $0\le -j\,\phi(x)<\psi(x)$ and $0<\theta_j(x)\le\psi(x)$ for $x\in R_j$, and that

$$D\,\theta_j(x)=0 \text{ for } \mathscr{L}^m \text{ almost all } x \text{ in } U\sim R_j,$$

hence estimate

$$\begin{aligned}&\lambda\,|D\,\theta_j|_U^2\le\mu\,\alpha_j(\alpha_j+|D\,\theta_j|_U)+\mu\,|D\,\theta_j|_U\,\alpha_j,\\&|D\,\theta_j|_U^2-2\lambda^{-1}\,\mu\,\alpha_j\,|D\,\theta_j|_U\le(\lambda^{-1}\,\mu\,\alpha_j)^2,\qquad |D\,\theta_j|_U\le 3\,\lambda^{-1}\,\mu\,\alpha_j.\end{aligned}$$

We also note that $\theta_j(x)\to\gamma(x)\,\psi(x)$ as $j\to\infty$ for $x\in U$. Extending γ,ψ,θ_j to $\mathbf{R}^m$ by mapping $\mathbf{R}^m\sim U$ onto 0, we infer that

$$\mathbf{E}^m\llcorner\theta_j\to\mathbf{E}^m\llcorner\gamma\,\psi \text{ in } \mathscr{D}_m(\mathbf{R}^m) \text{ as } j\to\infty.$$

If $\beta\in\mathscr{D}^{m-1}(\mathbf{R}^m)$, then

$$\begin{aligned}&|\langle\beta,\partial(\mathbf{E}^m\llcorner\theta_j)\rangle|=|\mathbf{E}^m[(D\,\theta_j)\wedge\beta]|\le 3\,\lambda^{-1}\,\mu\,\alpha_j\,|\beta| \text{ for all } j,\\&|\langle\beta,\partial(\mathbf{E}^m\llcorner\gamma\,\psi)\rangle|\le 3\,\lambda^{-1}\,\mu(|\psi|+|D\,\psi|)\,|\beta|.\end{aligned}$$

Consequently $\|\partial(\mathbf{E}^m\llcorner\gamma\,\psi)\|$ is absolutely continuous with respect to $\mathscr{L}^m$. Moreover $(\mathbf{E}^m\llcorner\gamma\,\psi)\llcorner W=(\mathbf{E}^m\llcorner C)\llcorner W$ and for each $a\in W$ it follows from 4.2.1, 4.2.17, 4.1.28 that $\mathscr{L}^1$ almost all numbers ρ between 0 and $\operatorname{dist}(a,\mathbf{R}^m\sim W)$ satisfy the conditions

$$T_\rho=(\mathbf{E}^m\llcorner C)\llcorner\mathbf{U}(a,\rho)\in\mathbf{I}_m(\mathbf{R}^m),\qquad \|\partial T_\rho\|=\|\partial T_\rho\|\llcorner B_\rho$$

with $B_\rho=\{x\colon\Theta^{m-1}(\|\partial T_\rho\|,x)\ge 1\}$ and $\mathscr{H}^{m-1}(B_\rho)<\infty$, hence

$$\|\partial(\mathbf{E}^m\llcorner C)\|\llcorner\mathbf{U}(a,\rho)=\|\partial T_\rho\|\llcorner\mathbf{U}(a,\rho)\le\|\partial T_\rho\|\llcorner B_\rho$$

and $\|\partial(\mathbf{E}^m\llcorner C)\|\,\mathbf{U}(a,\rho)\le\|\partial(\mathbf{E}^m\llcorner C)\|\,B_\rho=0$, because $\mathscr{L}^m(B_\rho)=0$. Thus we infer

$$W\cap\operatorname{spt}\partial(\mathbf{E}^m\llcorner C)=\varnothing,\qquad \mathscr{L}^m(W\cap C)=0,$$

since the alternative $\mathscr{L}^m(W\sim C)=0$ would imply $W\subset C$.

Expressing U as the union of such sets W, we conclude $\mathscr{L}^m(C)=0$.

Applying our construction again to a particular set W, we now observe that $R_j \supset R_{j+1}$ for all positive integers j, and

$$\bigcap_{j=1}^{\infty} R_j \subset C, \text{ hence } \lim_{j\to\infty} \mathscr{L}^m(R_j)=0, \ \lim_{j\to\infty} \alpha_j = 0.$$

We infer with the help of 4.5.9(13) that

$$\begin{aligned}&\textstyle\int_0^1 \mathbf{M}[\partial(\mathbf{E}^m \llcorner \{x\colon \theta_j(x)>s\})]\, d\mathscr{L}^1 s \le \mathbf{M}[\partial(\mathbf{E}^m \llcorner \theta_j)]\\ &= \textstyle\int_{R_j} |D\theta_j|\, d\mathscr{L}^m \le 3\lambda^{-1}\mu\,\alpha_j [\mathscr{L}^m(R_j)]^{\frac{1}{2}} \to 0 \text{ as } j\to\infty,\end{aligned}$$

choose numbers s_j between 0 and 1 so that

$$\mathbf{M}[\partial(\mathbf{E}^m \llcorner P_j)] \to 0 \text{ as } j\to\infty \text{ with } P_j=\{x\colon \theta_j(x)>s_j\},$$

and apply 4.5.4 to cover each of the open sets P_j by a countable family F_j of closed balls so that

$$\sum_{S\in F_j} (\operatorname{diam} S)^{m-1} \to 0 \text{ as } j\to\infty.$$

Since $C\cap W \subset P_j$ for all j, we conclude $\mathscr{H}^{m-1}(C\cap W)=0$.

We derive the following corollary (for use in 5.3.19):

If U and V are connected open subsets of $\mathbf{R}^m$ and $\mathbf{R}$,

$$f\colon U\to V \text{ and } g\colon U\to V \textit{ are of class } 1,$$

$$f(x)\le g(x) \textit{ for } x\in U, \quad C=U\cap\{x\colon f(x)=g(x)\},$$

$$G\colon U\times V\times \operatorname{Hom}(\mathbf{R}^m,\mathbf{R})\to\mathbf{R} \textit{ is of class } 2,$$

$$\begin{aligned}&\textstyle\int_U \langle [0,\theta(x),D\theta(x)], DG[x,f(x),Df(x)]\rangle\, d\mathscr{L}^m x\\ &= \textstyle\int_U \langle [0,\theta(x),D\theta(x)], DG[x,g(x),Dg(x)]\rangle\, d\mathscr{L}^m x\end{aligned}$$

for $\theta\in\mathscr{D}(U,\mathbf{R})$, and the form mapping $(\sigma,\tau)\in\operatorname{Hom}(\mathbf{R}^m,\mathbf{R})\times\operatorname{Hom}(\mathbf{R}^m,\mathbf{R})$ onto $\langle (0,0,\sigma)\odot(0,0,\tau), D^2G(x,y,\lambda)\rangle$ is strongly elliptic for $(x,y,\lambda)\in U\times V\times\operatorname{Hom}(\mathbf{R}^m,\mathbf{R})$, then either $C=U$ or $\mathscr{H}^{m-1}(C)=0$.

We apply the second proposition with $\phi(x)=f(x)-g(x)$ and

$$\begin{aligned}&\langle (r,\sigma)\odot(s,\tau), \Xi(x)\rangle\\ &= \textstyle\int_0^1 \langle (0,r,\sigma)\odot(0,s,\tau), D^2G[x, g(x)+t\,\phi(x), Dg(x)+t\,D\phi(x)]\rangle\, d\mathscr{L}^1 t\end{aligned}$$

for $x\in U$, $r\in\mathbf{R}$, $\sigma\in\operatorname{Hom}(\mathbf{R}^m,\mathbf{R})$, $s\in\mathbf{R}$, $\tau\in\operatorname{Hom}(\mathbf{R}^m,\mathbf{R})$.

5.2.20. In case $m=1$ we will use the uniqueness theorem for ordinary differential equations: *If Ω is an open subset of a normed vectorspace E, $H: \Omega \to E$ is Lipschitzian, U is an open interval and*

$$\alpha: U \to \Omega,\ \beta: U \to \Omega \text{ are of class 1 with } \alpha' = H \circ \alpha,\ \beta' = H \circ \beta,$$

then $\{x: \alpha(x) = \beta(x)\}$ is open. To prove this we suppose

$$t \in U, \quad \alpha(t) = \beta(t), \quad r > 0, \quad r \cdot \operatorname{Lip}(H) < 1,$$

$$I = \{x: |x-t| \le r\} \subset U, \quad \mu = \sup\{|\alpha(x) - \beta(x)|: x \in I\},$$

and infer $|(\alpha-\beta)'(x)| = |H[\alpha(x)] - H[\beta(x)]| \le \operatorname{Lip}(H)\mu$ for $x \in I$,

$$|(\alpha-\beta)(x)| \le |x-t| \operatorname{Lip}(H)\mu \le r \operatorname{Lip}(H)\mu \text{ for } x \in I,$$

hence $\mu \le r \operatorname{Lip}(H)\mu$, $\mu = 0$ or $\mu < \mu$.

We deduce the proposition: *If U is an open interval, V is an open subset of $\mathbf{R}^{n-1}$, Φ is an elliptic integrand of degree 1 and class 2 on $U \times V$ and $f: U \to V$, $g: U \to V$ are maps of class 2 such that*

$$\int_U \langle [0, \theta(x), D\theta(x)], D\Phi^{\S}[x, f(x), Df(x)] \rangle \, d\mathscr{L}^1 x = 0,$$

$$\int_U \langle [0, \theta(x), D\theta(x)], D\Phi^{\S}[x, g(x), Dg(x)] \rangle \, d\mathscr{L}^1 x = 0$$

for $\theta \in \mathscr{D}(U, \mathbf{R}^{n-1})$, then $\{x: f(x) = g(x), f'(x) = g'(x)\}$ is open. Defining the nonparametric Legendre transformation

$$L: U \times V \times \operatorname{Hom}(\mathbf{R}, \mathbf{R}^{n-1}) \to U \times V \times \operatorname{Hom}[\operatorname{Hom}(\mathbf{R}, \mathbf{R}^{n-1}), \mathbf{R}],$$

$$L(x, y, \sigma) = (x, y, D\Phi^{\S}_{(x,y)}(\sigma)) \text{ for } (x, y, \sigma) \in U \times V \times \operatorname{Hom}(\mathbf{R}, \mathbf{R}^{n-1}),$$

we see that L is univalent and L^{-1} is of class 1, because 5.2.17, 5.2.3 and 5.1.10 imply that

$$|\tau - \sigma|^{-1} \|D\Phi^{\S}_z(\tau) - D\Phi^{\S}_z(\sigma)\| \ge |\tau-\sigma|^{-2} \langle (\tau - \sigma), D\Phi^{\S}_z(\tau) - D\Phi^{\S}_z(\sigma) \rangle$$

$$= \int_0^1 |\tau - \sigma|^{-2} \langle (\tau-\sigma)^2, D^2\Phi^{\S}_z[(1-t)\sigma + t\tau] \rangle \, d\mathscr{L}^1 t$$

has a positive lower bound when $(z, \sigma) \neq (z, \tau)$ are restricted to any compact subset of $(U \times V) \times \operatorname{Hom}(\mathbf{R}, \mathbf{R}^{n-1})$. Using the base vectors $\eta_1, \ldots, \eta_{n-1}$ of $\operatorname{Hom}[\operatorname{Hom}(\mathbf{R}, \mathbf{R}^{n-1}), \mathbf{R}]$ dual to the base vectors $X_1 v_1, \ldots, X_1 v_{n-1}$ of $\operatorname{Hom}(\mathbf{R}, \mathbf{R}^{n-1})$, we define

$$h(x, y, \sigma) = \left[1, v, \sum_{j=1}^{n-1} D_{1+j}\Phi^{\S}(x, y, \sigma)\, \eta_j\right]$$

whenever $x \in U$, $y \in V$, $v \in \mathbf{R}^{n-1}$, $\sigma = X_1 v \in \operatorname{Hom}(\mathbf{R}, \mathbf{R}^{n-1})$, and we take $H = h \circ L^{-1}$. Letting

$$\alpha(x) = L[x, f(x), Df(x)] \quad \text{and} \quad \beta(x) = L[x, g(x), Dg(x)]$$

for $x \in U$, one readily verifies that the Euler-Lagrange conditions in 5.1.11 imply $\alpha' = H \circ \alpha$ and $\beta' = H \circ \beta$.

5.3. Excess and smoothness

5.3.1. For use throughout Section 5.3 we readopt the notational conventions of 5.1.9 and 5.1.11. We will study those m dimensional rectifiable currents in $\mathbf{R}^m \times \mathbf{R}^{n-m}$ which are nearly parallel to $\mathbf{R}^m \times \{0\}$, measuring deviation from parallelism by means of the concepts of excess defined as follows:

Whenever $S \in \mathscr{R}_m(\mathbf{R}^m \times \mathbf{R}^{n-m})$, $b \in \mathbf{R}^m$, $0 < r < \infty$ we define the **excess** of S at (b, r) as

$$\operatorname{Exc}(S, b, r) = r^{-m}[\mathbf{M}(S_{b,r}) - \mathbf{M}(\mathbf{p}_\# S_{b,r})] \quad \text{where} \quad S_{b,r} = S \llcorner \mathbf{p}^{-1}\mathbf{U}(b, r).$$

We note that if $\mathbf{p}(\operatorname{spt} \partial S) \subset \mathbf{R}^m \sim \mathbf{U}(b, r)$, then

$$\mathbf{p}_\# S_{b,r} = \kappa\, \mathbf{E}^m \llcorner \mathbf{U}(b, r)$$

for some integer κ, by 4.1.7, hence

$$\mathbf{M}(\mathbf{p}_\# S_{b,r}) = \operatorname{sign}(\kappa)\, S_{b,r}(DZ_1 \wedge \cdots \wedge DZ_m),$$

$$r^m \operatorname{Exc}(S, b, r) = \int_{\mathbf{p}^{-1}\mathbf{U}(b,r)} [1 - \operatorname{sign}(\kappa) \langle \vec{S}, DZ_1 \wedge \cdots \wedge DZ_m \rangle]\, d\, \|S\|;$$

moreover, for $\|S\|$ almost all z in $\mathbf{R}^m \times \mathbf{R}^{n-m}$,

$$\begin{aligned} |\vec{S}(z) \llcorner Z_{m+j}|^2 &\le 1 - \langle \vec{S}(z), Z_1 \wedge \cdots \wedge Z_m \rangle^2 \\ &\le 2[1 - \operatorname{sign}(\kappa) \langle \vec{S}(z), Z_1 \wedge \cdots \wedge Z_m \rangle] \end{aligned}$$

whenever $j \in \{1, \ldots, n-m\}$, because 4.1.28 and 1.7.5 imply

$$1 = |\vec{S}(z)|^2 = \sum_{\lambda \in \Lambda(n,m)} \langle \vec{S}(z), Z_\lambda \rangle^2,$$

and because $1 - t^2 \le 2(1-t)$ for $t \in \mathbf{R}$.

We generalize the preceding notion by means of 4.2.26, defining for each nonnegative integer ν the **excess modulo** ν of S at (b, r) as

$$\operatorname{Exc}^\nu(S, b, r) = r^{-m}[\mathbf{M}^\nu(S_{b,r}) - \mathbf{M}^\nu(\mathbf{p}_\# S_{b,r})].$$

We observe that

$$\operatorname{Exc}^0(S, b, r) = \operatorname{Exc}(S, b, r) \quad \text{and} \quad \operatorname{Exc}^1(S, b, r) = 0.$$

If $\mathbf{p}_\# S_{b,r} = \kappa\, \mathbf{E}^m \llcorner \mathbf{U}(b, r)$ *and* $\kappa > 0$, *then*

$$\operatorname{Exc}^\kappa(S, b, r) = r^{-m} \mathbf{M}^\kappa(S_{b,r});$$

if also $\Theta^m(\|S\|, z) \ge \kappa$ *for* $\|S\|$ *almost all* z, *then*

$$\operatorname{Exc}^\kappa(S, b, r) \le \operatorname{Exc}(S, b, r).$$

To verify the last assertion we let

$$W = \mathbf{p}^{-1}\,\mathbf{U}(b,r) \cap \{z\colon\ \Theta^m(\|S\|, z) \geq \kappa\}$$

and use 4.1.28, 4.1.30 in computing

$$\begin{aligned}\mathbf{M}^\kappa(S_{b,r}) &\leq \mathbf{M}[S - \kappa(\mathscr{H}^m \llcorner W) \wedge \vec{S}] \\ &= \textstyle\int_W [\Theta^m(\|S\|, z) - \kappa]\, d\mathscr{H}^m z = \mathbf{M}(S_{b,r}) - \kappa\,\mathscr{H}^m(W) \\ &\leq \mathbf{M}(S_{b,r}) - \kappa\,\mathscr{L}^m[\mathbf{p}(W)] = \mathbf{M}(S_{b,r}) - \mathbf{M}(\mathbf{p}_\# S_{b,r}).\end{aligned}$$

Simple examples show that *without the supplementary hypothesis on* $\Theta^m(\|S\|, \cdot)$ *it can happen that*

$$\operatorname{Exc}^\nu(S, b, r) = \kappa\,\boldsymbol{\alpha}(m) > 0 = \operatorname{Exc}(S, b, r).$$

We note that $0 \leq r^m \operatorname{Exc}^\nu(S, b, r) \leq s^m \operatorname{Exc}^\nu(S, b, s)$ whenever ν is a nonnegative integer and $0 < r < s < \infty$.

The finite positive numbers $\Gamma_1, \Gamma_2, \ldots, \Gamma_{21}$ occurring in this section depend only on the dimensions m and n.

5.3.2. Lemma. *There exists a finite positive number* Γ_1 *such that*

$$[\|T\|\,\mathbf{U}(b,r)]^{1-1/m} \leq \Gamma_1\,\|\partial T\|\,\mathbf{U}(b,r)$$

whenever $b \in \mathbf{R}^m$, $0 < r < \infty$, $T \in \mathbf{I}_m^{\mathrm{loc}}(\mathbf{R}^m)$, $T \llcorner \mathbf{U}(b,r) \neq 0$ *and*

$$\mathscr{L}^m[\mathbf{U}(b,r) \cap \{x\colon\ \Theta^m(\|T\|, x) \neq 0\}] \leq 4\boldsymbol{\alpha}(m)\,r^m/5.$$

Proof. Using $\mu_{1/r} \circ \tau_{-b}$ we reduce the problem to the special case when $b = 0$ and $r = 1$.

In case $m > 1$ we then apply 4.5.2(1) with n, R replaced by $m, \mathbf{U}(0,1)$ to obtain an integer κ with

$$\|\kappa\,\mathbf{E}^m - T\|\,\mathbf{U}(0,1) \leq [\sigma\,\|\partial T\|\,\mathbf{U}(0,1)]^{m/(m-1)},$$

and observe that

$$\begin{aligned}&\|T\|\,\mathbf{U}(0,1) - \|T - \kappa\,\mathbf{E}^m\|\,\mathbf{U}(0,1) \leq \kappa\,\boldsymbol{\alpha}(m) \\ &\leq 5\kappa\,\mathscr{L}^m[\mathbf{U}(0,1) \cap \{x\colon\ \Theta^m(\|T\|, x) = 0\}] \leq 5\,\|\kappa\,\mathbf{E}^m - T\|\,\mathbf{U}(0,1);\end{aligned}$$

thus we can take $\Gamma_1 = 6^{(m-1)/m}\,\sigma$.

In case $m = 1$, then $\|\partial T\|\,\mathbf{U}(0,1) \geq 1$, and we let $\Gamma_1 = 1$.

5.3.3. Lemma. *In case $m>1$ there exist finite positive numbers Γ_2, Γ_3 with the following property:*

If $0<\rho<\infty$, $0<s<\infty$, $T\in \mathbf{I}_{m-1}(\mathbf{R}^m \times \mathbf{R}^{n-m})$,

$$\operatorname{spt} T \subset D = \{(x,y): |x|=\rho,\ |y|\le s\},\ \partial T = 0 \text{ and } \mathbf{M}(T) < \Gamma_2\, \rho^{m-1},$$

then there exists $R\in \mathbf{I}_m(\mathbf{R}^m \times \mathbf{R}^{n-m})$ *with*

$$\operatorname{spt} R \subset D, \quad \partial R = T \text{ and } \mathbf{M}(R) \le \Gamma_3\, \mathbf{M}(T)^{m/(m-1)}.$$

Proof. We will apply 4.2.9, 10 with m replaced by $m-1$. We let

$$\gamma = 2n^{2m}, \quad \Gamma_2 = (4n)^{1-m}\gamma^{-1}, \quad \Gamma_3 = 2^m \gamma^{m/(m-1)},$$

define the retraction f: $W=\{(x,y): |x|>\rho/2\} \to D$,

$$f(x,y) = (\rho\,|x|^{-1} x, y) \text{ for } (x,y)\in W \text{ with } |y|\le s,$$

$$f(x,y) = (\rho\,|x|^{-1} x, s\,|y|^{-1} y) \text{ for } (x,y)\in W \text{ with } |y|>s,$$

and note that $\operatorname{Lip}(f)\le 2$ (compare 4.1.16).

Given T we choose $\varepsilon>0$ so that $\varepsilon^{m-1} = \gamma\,\mathbf{M}(T)$ and obtain

$$S\in \mathbf{I}_m(\mathbf{R}^m \times \mathbf{R}^{n-m}) \text{ with } \partial S = T,\ \mathbf{M}(S)^{(m-1)/m} \le \gamma\,\mathbf{M}(T),$$

$$\operatorname{spt} S \subset \{(x,y): \operatorname{dist}[(x,y), \operatorname{spt} T] \le 2n\varepsilon\} \subset W$$

because $(2n\varepsilon)^{m-1} < (2n)^{m-1}\gamma\,\Gamma_2\,\rho^{m-1} = (\rho/2)^{m-1}$; then we take $R = f_{\#} S$.

5.3.4. Theorem. *There exist finite positive numbers $\Gamma_4, \ldots, \Gamma_{13}$ such that whenever κ is a positive integer,*

$$S\in \mathbf{I}_m(\mathbf{R}^m \times \mathbf{R}^{n-m}), \quad 0<r<\infty, \quad 0<s<\infty,$$

$$\operatorname{spt} S \subset A = \{(x,y): |x|\le r \text{ and } |y|\le s\},$$

$$\operatorname{spt} \partial S \subset \{(x,y): |x|=r\}, \quad \|S\|\,\{(x,y): |x|=r\} = 0,$$

$$\mathbf{p}_{\#} S = \kappa\,\mathbf{E}^m \llcorner \{x: |x|<r\},$$

$$E = \operatorname{Exc}(S,0,r) < \kappa\,\boldsymbol{\alpha}(m)/5, \quad \Omega = \operatorname{Exc}^{\kappa}(S,0,r) < \boldsymbol{\alpha}(m)/5,$$

$$W_j^+(t) = \{(x,y): y_j > t\} \textit{ for } j=1,\ldots,n-m \textit{ and } t\in\mathbf{R},$$

$$\alpha_j^+ = \inf\{t: r^{-m}\,\|S\|\, W_j^+(t) \le 3\kappa\,\boldsymbol{\alpha}(m)/5\},$$

$$\beta_j^+ = \inf\{t: r^{-m}\,\|S\|\, W_j^+(t) \le 2E\},$$

$$W_j^-(t) = \{(x,y): y_j < t\} \textit{ for } j=1,\ldots,n-m \textit{ and } t\in\mathbf{R},$$

$$\alpha_j^- = \sup\{t: r^{-m}\,\|S\|\, W_j^-(t) \le 3\kappa\,\boldsymbol{\alpha}(m)/5\},$$

$$\beta_j^- = \sup\{t: r^{-m}\,\|S\|\, W_j^-(t) \le 2E\}$$

the following ten statements hold:

(1) $-s \le \beta_j^- \le \alpha_j^+ \le \alpha_j^- \le \beta_j^+ \le s$.

(2) $(\beta_j^+ - \alpha_j^+)^2 \le \Gamma_4 (\kappa E)^{1/m} r^2$.

(3) $(\alpha_j^- - \beta_j^-)^2 \le \Gamma_4 (\kappa E)^{1/m} r^2$.

(4) $(\beta_j^+ - \beta_j^-)^2 \le 4\Gamma_4 (\kappa E)^{1/m} r^2$.

(5) *If* $\alpha_j^+ \le t \le \beta_j^+$, *then*

$$\int_{\{(x,y):\, t \le y_j \le \beta_j^+\}} (y_j - t)^2 \, d\,\|S\|\,(x,y) \le 2\Gamma_4 \kappa^{1/m} [r^{-m} \|S\| \, W_j^+(t) - E]^{1/m} E\, r^{m+2}.$$

(6) *If* $\alpha_j^- \ge t \ge \beta_j^-$, *then*

$$\int_{\{(x,y):\, t \ge y_j \ge \beta_j^-\}} (y_j - t)^2 \, d\,\|S\|\,(x,y) \le 2\Gamma_4 \kappa^{1/m} [r^{-m} \|S\| \, W_j^-(t) - E]^{1/m} E\, r^{m+2}.$$

(7) *If* $\alpha_j^+ \le t \le \alpha_j^-$, *then*

$$\int_{\{(x,y):\, \beta_j^- \le y_j \le \beta_j^+\}} (y_j - t)^2 \, d\,\|S\|\,(x,y) \le \Gamma_5 \kappa^{1/m} E\, r^{m+2}.$$

(8) *If* $t \in \mathbf{R}$, $\mu > 0$, $\omega > 0$ *and*

$$r^{-m} \|S\| \, \{(x,y):\, |y_j - t| \ge \omega r\} \le \mu^{4m} \le 3\kappa\, \boldsymbol{\alpha}(m)/5,$$

then

$$\int_{\{(x,y):\, \beta_j^- \le y_j \le \beta_j^+\}} (y_j - t)^2 \, d\,\|S\|\,(x,y) \le 4\kappa [\Gamma_4 \mu^4 E + \boldsymbol{\alpha}(m)\, \omega^2]\, r^{m+2}.$$

(9) *If* $0 < \mu \le \frac{1}{3}$, $E < \Gamma_6 \mu$ *and*

$$\eta \in C = \mathbf{R}^{n-m} \cap \{y:\, \beta_j^- \le y_j \le \beta_j^+ \text{ for } j = 1, \ldots, n-m\},$$

then there exist P, Q, R, ρ *for which*

$$P, Q, R \in \mathbf{I}_m(\mathbf{R}^m \times \mathbf{R}^{n-m}), \qquad (1-2\mu)\, r < \rho < (1-\mu)\, r,$$

$$\operatorname{spt} P \subset \{(x,y):\, |x| \le \rho \text{ and } y \in C\},$$

$$P \llcorner \{(x,y):\, |x| \le \rho - \mu^2 \rho\} = \kappa (\mathbf{E}^m \llcorner \{x:\, |x| \le \rho - \mu^2 \rho\}) \times \boldsymbol{\delta}_\eta,$$

$$\operatorname{spt} Q \subset \{(x,y):\, \rho \le |x| \le r \text{ and } y \in C\},$$

$$\operatorname{spt} R \subset \{(x,y):\, |x| = \rho \text{ and } |y| \le s\},$$

$$[\|S\| + \|P\| + \|Q\|]\, \{(x,y):\, |x| = \rho\} = 0,$$

$$\partial P + \partial R = \partial (S \llcorner \{(x,y):\, |x| < \rho\}),$$

$$\partial P + \partial Q = \kappa\, \partial [(\mathbf{E}^m \llcorner \{x:\, |x| < r\}) \times \boldsymbol{\delta}_\eta],$$

$$r^{-m} \mathbf{M}(R) \le \Gamma_7 (\mu^{-1} E)^{m/(m-1)} \text{ in case } m > 1, \qquad R = 0 \text{ in case } m = 1,$$

$$\operatorname{Exc}(P+Q, 0, r) \le \Gamma_8 [\mu + \mu^{-3} (\kappa E)^{1/m}]\, E + \Gamma_9 \mu^{-3} r^{-m-2} \int_{\mathbf{R}^m \times C} |y - \eta|^2 \, d\,\|S\|\,(x,y),$$

$$\operatorname{Exc}^\kappa (P+Q, 0, r) \le \Gamma_8 [\mu + \mu^{-3} (\kappa E)^{1/m}]\, \Omega,$$

and which satisfy the two conditions:

(I) *If* $\alpha_j^+ \leq \eta_j \leq \alpha_j^-$ *for* $j=1,\ldots,n-m$, *then*

$$\operatorname{Exc}(P+Q+R,0,r) \leq \Gamma_{10}\,\kappa^{1/m}\,\mu^{-3}\,E.$$

(II) *If* $\kappa\,E \leq \mu^{4m} \leq 3\kappa\,\boldsymbol{\alpha}(m)/5$, $\omega > 0$ *and*

$$r^{-m}\,\|S\|\,\{(x,y)\colon |y-\eta| \geq \omega\,r\} \leq \mu^{4m},$$

then

$$\operatorname{Exc}(P+Q+R,0,r) \leq \kappa(\Gamma_{11}\,\mu\,E + \Gamma_{12}\,\mu^{-3}\,\omega^2),$$
$$\operatorname{Exc}^\kappa(P+Q+R,0,r) \leq \Gamma_{10}\,\mu(E+\Omega).$$

(10) *Suppose* $0<\lambda<\infty$, Φ *is a parametric integrand of degree m such that*

$$\lambda^{-1}|\zeta| \leq \Phi(x,y,\zeta) \leq \lambda|\zeta| \ \text{ for } (x,y)\in A,\ \zeta\in\Lambda_m(\mathbf{R}^m\times\mathbf{R}^{n-m}),$$

S is absolutely Φ *minimizing with respect to A and*

$$\eta\in\mathbf{R}^{n-m} \text{ with } \alpha_j^+ \leq \eta_j \leq \alpha_j^- \text{ for } j\in\{1,\ldots,n-m\}.$$

(I) *If* $(x,y)\in\operatorname{spt} S$, *then either* $|x| \geq (1-\Gamma_{13}\,\lambda^{2-2/m}\,E^{1/m})\,r$ *or*

$$\beta_j^- - \Gamma_{13}\,\lambda^{2-2/m}\,E^{1/m}\,r \leq y_j \leq \beta_j^+ + \Gamma_{13}\,\lambda^{2-2/m}\,E^{1/m}\,r$$

for $j\in\{1,\ldots,n-m\}$; *the second alternative implies*

$$|y-\eta| \leq (n-m)^{\frac12}\,[\Gamma_4^{\frac12}\,\kappa^{1/(2m)} + \Gamma_{13}\,\lambda^{2-2/m}\,E^{1/(2m)}]\,E^{1/(2m)}\,r.$$

(II) $\int_{\mathbf{p}^{-1}\mathbf{U}[0,(1-\Gamma_{13}\lambda^{2-2/m}E^{1/m})r]} |y-\eta|^2\,d\|S\|(x,y)$

$$\leq (n-m)^2\,[(\Gamma_5 + 8\,\Gamma_4\,E^{1/m})\,\kappa^{1/m} + 8\,\Gamma_{13}^2\,\lambda^{4-4/m}\,E^{2/m}]\,E\,r^{m+2}.$$

Proof of (1) to (8). Noting that $r^{-m}\,\mathbf{M}(S) < 6\kappa\,\boldsymbol{\alpha}(m)/5$, one readily verifies (1).

To prove (2) and (5) we define

$$f(t) = \|S\|\,\{(x,y)\colon y_j \leq t\} = \mathbf{M}(S) - \|S\|\,W_j^+(t),$$
$$\phi(t) = \kappa\,\boldsymbol{\alpha}(m)\,r^m - f(t) = \|S\|\,W_j^+(t) - E\,r^m,$$
$$g(t) = \int_{\{(x,y)\colon y_j \leq t\}} [1 - \langle \vec{S}, D Z_1 \wedge \cdots \wedge D Z_m\rangle]\,d\|S\|$$

for $t\in\mathbf{R}$. The functions f and g are nondecreasing, ϕ is nonincreasing and

$$3\kappa\,\boldsymbol{\alpha}(m)\,r^m/5 \geq \|S\|\,W_j^+(t) \geq 2E\,r^m \ \text{ for } \alpha_j^+ < t < \beta_j^+,$$
$$[3\kappa\,\boldsymbol{\alpha}(m)/5 - E]\,r^m \geq \phi(t) \geq E\,r^m \ \text{ for } \alpha_j^+ < t < \beta_j^+.$$

We first observe that

$$\phi(t)^{1-1/m} \le \Gamma_1 \mathbf{M}\langle S, Z_{m+j}, t+\rangle \text{ for } \alpha_j^+ < t < \beta_j^+,$$

because

$$\begin{aligned}\phi(t) &= \mathbf{M}(\mathbf{p}_\# S) - \mathbf{M}[S - S \llcorner W_j^+(t)] \\ &\le \mathbf{M}(\mathbf{p}_\# S) - \mathbf{M}(\mathbf{p}_\# [S - S \llcorner W_j^+(t)]) \le \mathbf{M}(\mathbf{p}_\# [S \llcorner W_j^+(t)]),\end{aligned}$$

while 4.1.28, 4.1.30 and 4.2.26 imply

$$\begin{aligned}&\mathscr{L}^m\{x: \Theta^m(\|\mathbf{p}_\# [S \llcorner W_j^+(t)]\|, x) \ge 1\} \\ &\le \mathscr{H}^m(W_j^+(t) \cap \{(x,y): \Theta^m[\|S\|,(x,y)] \ge 1\}) \\ &\le \kappa^{-1} \|S\| \, W_j^+(t) + \|S\|^\kappa \, W_j^+(t) \le 3\,\alpha(m)\, r^m/5 + \Omega\, r^m < 4\alpha(m)\, r^m/5,\end{aligned}$$

hence 5.3.2 and 4.2.1 yield

$$\begin{aligned}\phi(t)^{1-1/m} &\le \Gamma_1 \|\partial \mathbf{p}_\# [S \llcorner W_j^+(t)]\| \, \mathbf{U}(0,r) \\ &\le \Gamma_1 \|\partial [S \llcorner W_j^+(t)]\| \, \mathbf{p}^{-1}\mathbf{U}(0,r) = \Gamma_1 \|\langle S, Z_{m+j}, t+\rangle\| \, \mathbf{p}^{-1}\mathbf{U}(0,r).\end{aligned}$$

Using 4.3.4, 4.3.2(2), Hölder's inequality and 5.3.1 we infer that if $\alpha_j^+ < u < v < \beta_j^+$ and $B = \{(x,y): u < y_j \le v\}$, then

$$\begin{aligned}(v-u)\,\phi(v)^{1-1/m} &\le \Gamma_1 \textstyle\int_u^v \mathbf{M}\langle S, Z_{m+j}, t+\rangle \, d\mathscr{L}^1 t \\ &= \Gamma_1 \|S \llcorner DZ_{m+j}\| \, B = \Gamma_1 \textstyle\int_B |\vec{S} \llcorner DZ_{m+j}| \, d\|S\| \\ &\le \Gamma_1 (\textstyle\int_B |\vec{S} \llcorner DZ_{m+j}|^2 \, d\|S\|)^{\frac12} (\|S\| \, B)^{\frac12} \\ &\le \Gamma_1 (2[g(v) - g(u)] \cdot [f(v) - f(u)])^{\frac12}.\end{aligned}$$

Consequently

$$\phi(v)^{1-1/m} \le \Gamma_1 [2g'(v)\, f'(v)]^{\frac12}$$

for $\mathscr{L}^1$ almost all numbers v between α_j^+ and β_j^+.

Next we assume $\alpha_j^+ \le t \le u < \beta_j^+$ and estimate

$$\begin{aligned}(u-t)^2 &\le (\textstyle\int_t^u \Gamma_1 [2g' f']^{\frac12} \phi^{1/m-1} \, d\mathscr{L}^1)^2 \\ &\le \Gamma_1^2 \, 2 \textstyle\int_t^u g' \phi^{1/m-1} \, d\mathscr{L}^1 \cdot \int_t^u f' \phi^{1/m-1} \, d\mathscr{L}^1;\end{aligned}$$

since $\int_t^u f' \phi^{1/m-1} \, d\mathscr{L}^1 = \int_t^u - m(\phi^{1/m})' \, d\mathscr{L}^1 \le m[\kappa\, \alpha(m)]^{1/m} r$, we conclude

$$(u-t)^2 \le \Gamma_4 \, \kappa^{1/m} \, r \textstyle\int_t^u g' \phi^{1/m-1} \, d\mathscr{L}^1$$

with $\Gamma_4 = \Gamma_1^2 \, 2m\, \alpha(m)^{1/m}$; noting that

$$\textstyle\int_t^u g' \phi^{1/m-1} \, d\mathscr{L}^1 \le (E\, r^m)^{1/m-1} \int_t^u g' \, d\mathscr{L}^1 \le (E\, r^m)^{1/m-1} E\, r^m = E^{1/m} r$$

we infer that $(u-t)^2 \le \Gamma_4 (\kappa\, E)^{1/m} r^2$, hence (2).

We deduce (5) with the help of 2.4.18 and 2.6.2 by computing

$$\begin{aligned}
&\int_{\{(x,y):\, t<y_j\le\beta_j^\dagger\}}(y_j-t)^2\, d\,\|S\|\,(x,y)=\int_t^{\beta_j^\dagger}(u-t)^2\, d(Y_{j\#}\,\|S\|)\,u\\
&\le\Gamma_4\,\kappa^{1/m}\, r\int_t^{\beta_j^\dagger}\int_t^u g'(v)\,\phi(v)^{1/m-1}\, d\mathscr{L}^1\, v\, d(Y_{j\#}\,\|S\|)\,u\\
&=\Gamma_4\,\kappa^{1/m}\, r\int_t^{\beta_j^\dagger}\int_v^{\beta_j^\dagger} g'(v)\,\phi(v)^{1/m-1}\, d(Y_{j\#}\,\|S\|)\,u\, d\mathscr{L}^1\, v\\
&=\Gamma_4\,\kappa^{1/m}\, r\int_t^{\beta_j^\dagger} g'(v)\,\phi(v)^{1/m-1}\,[f(\beta_j^+)-f(v)]\, d\mathscr{L}^1\, v\\
&\le\Gamma_4\,\kappa^{1/m}\, r\int_t^{\beta_j^\dagger} g'(v)\,\phi(v)^{1/m-1}\,[\phi(v)+E\, r^m]\, d\mathscr{L}^1\, v\\
&=\Gamma_4\,\kappa^{1/m}\, r\int_t^{\beta_j^\dagger} g'(v)\,\phi(v)^{1/m}\,[1+E\, r^m/\phi(v)]\, d\mathscr{L}^1\, v\\
&\le\Gamma_4\,\kappa^{1/m}\, r\int_t^{\beta_j^\dagger} g'(v)\,[\|S\|\, W_j^+(t)-E\, r^m]^{1/m}\, 2\, d\mathscr{L}^1\, v\\
&\le\Gamma_4\,\kappa^{1/m}\, r\,[\|S\|\, W_j^+(t)-E\, r^m]^{1/m}\, 2\, E\, r^m.
\end{aligned}$$

The assertions (3) and (6) may be verified similarly. Clearly (4) follows from (1), (2) and (3), while (7) follows from (5) and (6) with $\Gamma_5=4\,\Gamma_4\,\alpha(m)^{1/m}$.

The hypothesis of (8) implies

$$t^+=t+\omega\, r\ge\alpha_j^+ \quad\text{and}\quad t^-=t-\omega\, r\le\alpha_j^-;$$

we then define $M=\{(x,y):\ t^-<y_j<t^+\}$,

$$N^+=\{(x,y):\ t^+\le y_j\le\beta_j^+\} \quad\text{and}\quad N^-=\{(x,y):\ t^-\ge y_j\ge\beta_j^-\},$$

use (5) and (6) in estimating

$$\begin{aligned}
&\int_M (y_j-t)^2\, d\,\|S\|\,(x,y)\le\omega^2\, r^2\, 6\kappa\,\alpha(m)\, r^m/5,\\
&\int_{N^+}(y_j-t)^2\, d\,\|S\|\,(x,y)\le\int_{N^+} 2[(y_j-t^+)^2+\omega^2\, r^2]\, d\,\|S\|\,(x,y)\\
&\qquad\le 2\Gamma_4\,\kappa^{1/m}\,\mu^4\, E\, r^{m+2}+2\omega^2\,\mu^{4m}\, r^{m+2},\\
&\int_{N^-}(y_j-t)^2\, d\,\|S\|\,(x,y)\le 2\Gamma_4\,\kappa^{1/m}\,\mu^4\, E\, r^{m+2}+2\omega^2\,\mu^{4m}\, r^{m+2}.
\end{aligned}$$

Proof of (9). Defining $g(x,y)=|x|$ for $(x,y)\in\mathbf{R}^m\times\mathbf{R}^{n-m}$ we see from 4.2.1 that

$$\langle S, g, \rho-\rangle=\partial[S\llcorner\mathbf{p}^{-1}\,\mathbf{U}(0,\rho)] \text{ and}$$

$$\mathbf{p}_\#\,\langle S, g, \rho-\rangle=\partial[(\mathbf{p}_\#\, S)\llcorner\mathbf{U}(0,\rho)]=\kappa\,\partial[\mathbf{E}^m\llcorner\mathbf{U}(0,\rho)]$$

whenever $0<\rho<r$, hence

$$\begin{aligned}
&(\mu\, r)^{-1}\int_{r-2\mu r}^{r-\mu r}[\mathbf{M}\,\langle S, g, \rho-\rangle-\kappa\, m\,\alpha(m)\,\rho^{m-1}]\, d\mathscr{L}^1\,\rho\\
&\le(\mu\, r)^{-1}\int_0^r[\mathbf{M}\,\langle S, g, \rho-\rangle-\kappa\, m\,\alpha(m)\,\rho^{m-1}]\, d\mathscr{L}^1\,\rho\\
&\le(\mu\, r)^{-1}\,[\mathbf{M}(S)-\kappa\,\alpha(m)\, r^m]=\mu^{-1}\, E\, r^{m-1}.
\end{aligned}$$

Letting ϕ be the characteristic function of $\mathbf{R}^m \times (\mathbf{R}^{n-m} \sim C)$, we use 4.3.4, 4.3.2 (2) and the definition of β_j^+, β_j^- to estimate

$$(\mu r)^{-1} \int_{r-2\mu r}^{r-\mu r} \mathbf{M} \langle S \llcorner \phi, g, \rho \rangle \, d\mathscr{L}^1 \rho$$
$$\leq (\mu r)^{-1} \mathbf{M}(S \llcorner \phi) \leq \mu^{-1} 4(n-m) E r^{m-1}.$$

Defining $\psi(x, y) = |y - \eta|^2$ for $(x, y) \in \mathbf{R}^m \times C$,

$$\psi(x, y) = 0 \text{ for } (x, y) \in \mathbf{R}^m \times (\mathbf{R}^{n-m} \sim C),$$

we similarly obtain the inequality

$$(\mu r)^{-1} \int_{r-2\mu r}^{r-\mu r} \mathbf{M} \langle S \llcorner \psi, g, \rho \rangle \, d\mathscr{L}^1 \rho$$
$$\leq (\mu r)^{-1} \mathbf{M}(S \llcorner \psi) = (\mu r)^{-1} \int_{\mathbf{R}^m \times C} |y - \eta|^2 \, d\|S\|(x, y).$$

Moreover it follows from 4.2.26 that

$$(\mu r)^{-1} \int_{r-2\mu r}^{*r-\mu r} \mathbf{M}^\kappa \langle S, g, \rho - \rangle \, d\mathscr{L}^1 \rho$$
$$\leq (\mu r)^{-1} \mathbf{M}^\kappa(S) = \mu^{-1} \operatorname{Exc}^\kappa(S, 0, r) r^{m-1} = \mu^{-1} \Omega r^{m-1}.$$

We infer, reasoning as in the proof of 4.2.7 (2), that the set of all those numbers ρ between $r - 2\mu r$ and $r - \mu r$, which satisfy the five conditions

$$\mathbf{M} \langle S, g, \rho - \rangle - \kappa m \alpha(m) \rho^{m-1} \leq 4\mu^{-1} E r^{m-1},$$
$$\mathbf{M} \langle S \llcorner \phi, g, \rho \rangle \leq 16(n-m) \mu^{-1} E r^{m-1},$$
$$\mathbf{M} \langle S \llcorner \psi, g, \rho \rangle \leq 4\mu^{-1} r^{-1} \mathbf{M}(S \llcorner \psi),$$
$$\mathbf{M}^\kappa \langle S, g, \rho - \rangle \leq 4\mu^{-1} \Omega r^{m-1}, \qquad \|S\| \{(x, y) : |x| = \rho\} = 0,$$

has positive $\mathscr{L}^1$ measure. We henceforth fix such a number ρ, denote $L = \langle S, g, \rho - \rangle \in \mathbf{I}_{m-1}(\mathbf{R}^m \times \mathbf{R}^{n-m})$ and infer from 4.3.1 that

$$L \llcorner \phi = \langle S \llcorner \phi, g, \rho \rangle, \qquad L \llcorner \psi = \langle S \llcorner \psi, g, \rho \rangle.$$

Recalling 5.3.3 we take $\Gamma_6 = 2^{-5}(n-1)^{-1}$ in case $m = 1$,

$$\Gamma_6 = \inf\{2^{-5}(n-m)^{-1} 3^{1-m} \Gamma_2, 1\} \text{ in case } m > 1.$$

We let f be the nearest point retraction of $\mathbf{R}^{n-m}$ onto C (see 4.1.16), define

$$F(x, y) = (x, f(y)) \text{ for } (x, y) \in \mathbf{R}^m \times \mathbf{R}^{n-m},$$

and observe that $\operatorname{Lip}(f) = 1$, $L - F_\# L = (L \llcorner \phi) - F_\# (L \llcorner \phi)$,

$$\mathbf{M}(L - F_\# L) \leq 2\mathbf{M}(L \llcorner \phi) \leq 2^5 (n-m) \mu^{-1} E r^{m-1} < 2^5 (n-m) \Gamma_6 (3\rho)^{m-1}$$

with $\partial L=0$, $\partial(L-F_\# L)=0$. In case $m>1$ we apply 5.3.3 to obtain

$$R\in \mathbf{I}_m(\mathbf{R}^m\times\mathbf{R}^{n-m}) \text{ with } \operatorname{spt} R\subset D,$$
$$\partial R=L-F_\# L,\quad \mathbf{M}(R)\le\Gamma_3[2^5(n-m)\,\mu^{-1}E]^{m/(m-1)}\,r^m;$$

accordingly we take $\Gamma_7=\Gamma_3[2^5(n-m)]^{m/(m-1)}$. In case $m=1$ we find that $L-F_\# L=0$ and choose $R=0$.

Using the homotopy h and the embedding Δ defined by the formulae

$$h(t,x,y)=\big(t\,x,\,y+\mu^{-2}\,|t-1|\,(\eta-y)\big),$$
$$\Delta(x)=(x,\eta) \text{ for } t\in\mathbf{R},\ x\in\mathbf{R}^m,\ y\in\mathbf{R}^{n-m},$$

we construct the currents

$$P=h_\#([1-\mu^2,1]\times F_\# L)+\kappa\,\Delta_\#(\mathbf{E}^m\llcorner\{x\colon |x|<\rho-\mu^2\rho\}),$$
$$Q=h_\#([1,1+\mu^2]\times F_\# L)+\kappa\,\Delta_\#(\mathbf{E}^m\llcorner\{x\colon \rho+\mu^2\rho<|x|<r\}).$$

Noting the equations $h_1=\mathbf{1}$, $h_t=\Delta\circ\mu_t\circ\mathbf{p}$ for $t=1\pm\mu^2$, $\mathbf{p}\circ F=\mathbf{p}$ and $\mathbf{p}_\# L=\partial[\kappa\,\mathbf{E}^m\llcorner\mathbf{U}(0,\rho)]$ one sees from 4.1.8, 9 that

$$\partial P=F_\# L,\qquad \partial Q=-F_\# L+\kappa\,\partial\Delta_\#(\mathbf{E}^m\llcorner\{x\colon |x|<r\}).$$

Moreover, if $(t,x,y)\in\mathbf{R}\times\mathbf{R}^m\times\mathbf{R}^{n-m}$ with $0<|t-1|\le\mu^2$, then

$$|\dot h_t(x,y)|=(|x|^2+\mu^{-4}\,|\eta-y|^2)^{\frac12},$$
$$\|D\,h_t(x,y)\|\le\sup\{t,\,1-\mu^{-2}\,|t-1|\}=t.$$

Observing that $|x|=\rho$ for $(x,y)\in\operatorname{spt} F_\# L$,

$$(1+u)^{\frac12}\le 1+u \text{ for } u\ge 0,\qquad \|F_\# L\|\le\|L\|+F_\#\,\|L\llcorner\phi\|,$$

and applying (4) we obtain

$$\begin{aligned}
&\mathbf{M}\big[h_\#([1-\mu^2,1+\mu^2]\times F_\# L)\big]\\
&\le\textstyle\int_{1-\mu^2}^{1+\mu^2}\int\rho(1+\rho^{-2}\mu^{-4}\,|\eta-y|^2)^{\frac12}\,t^{m-1}\,d\,\|F_\# L\|\,(x,y)\,d\mathscr{L}^1 t\\
&\le\textstyle\int_{1-\mu^2}^{1+\mu^2} t^{m-1}\,d\mathscr{L}^1 t\cdot\int\rho(1+\rho^{-2}\mu^{-4}\,|\eta-y|^2)\,d\,\|F_\# L\|\,(x,y)\\
&=\frac{(1+\mu^2)^m-(1-\mu^2)^m}{m}\cdot\Big[\rho\,\mathbf{M}(F_\# L)+\int\frac{|\eta-y|^2}{\rho\,\mu^4}\,d\,\|F_\# L\|\,(x,y)\Big]\\
&\le\frac{(1+\mu^2)^m-(1-\mu^2)^m}{m}\cdot\big[\kappa\,m\,\boldsymbol{\alpha}(m)\,\rho^m+4\mu^{-1}E\,r^m\\
&\qquad+\rho^{-1}\mu^{-4}\,\|L\|\,(\psi)+\rho^{-1}\mu^{-4}\,4\cdot\Gamma_4(\kappa E)^{1/m}\,r^2\,\|L\|\,(\phi)\big]\\
&\le\mathbf{M}(\kappa\,\mathbf{E}^m\llcorner\{x\colon \rho-\mu^2\rho<|x|<\rho+\mu^2\rho\})+2^m\mu^2\,[4\mu^{-1}E\,r^m\\
&\qquad+12\mu^{-5}r^{-2}\,\|S\|\,(\psi)+96(n-m)\,\Gamma_4\,\mu^{-5}(\kappa E)^{1/m}E\,r^m]
\end{aligned}$$

and infer that the asserted estimate for $\operatorname{Exc}(P+Q, 0, r)$ holds with $\Gamma_8 = 2^m \sup\{4, 96(n-m)\Gamma_4\}$ and $\Gamma_9 = 2^{m+2}\, 3$.

Similarly we find that

$$\begin{aligned} r^m \operatorname{Exc}^\kappa(P+Q, 0, r) &= \mathbf{M}^\kappa\big[h_\#([1-\mu^2, 1+\mu^2] \times F_\# L)\big] \\ &\le \textstyle\int_{1-\mu^2}^{1+\mu^2} t^{m-1}\, d\mathscr{L}^1 t \cdot \int \rho(1+\rho^{-2}\mu^{-4}|\eta-y|^2)\, d\,\|F_\# L\|^\kappa(x,y) \\ &\le 2^m \mu^2 [\rho + \rho^{-1}\mu^{-4}\, 4\Gamma_4(\kappa E)^{1/m} r^2]\, \mathbf{M}^\kappa(L) \\ &\le 2^{m+2}[\mu + 12\Gamma_4 \mu^{-3}(\kappa E)^{1/m}]\, \Omega\, r^m. \end{aligned}$$

Finally we deduce (I) from (7), and (II) from (8), taking

$$\Gamma_{10} = 2\Gamma_8 + (n-m)\Gamma_5\Gamma_9 + \Gamma_7,$$

$$\Gamma_{11} = 2\Gamma_8 + (n-m)\, 4\Gamma_4\Gamma_9, \qquad \Gamma_{12} = 3\alpha(m)(n-m)\Gamma_9.$$

Proof of (10). If $(x,y) \in \operatorname{spt} S$, then

$$r - |x| \le \operatorname{dist}[(x,y), \operatorname{spt} \partial S];$$

in case $j \in \{1, \dots, n-m\}$ and $\rho = \inf\{r-|x|,\ y_j - \beta_j^+\} > 0$, the density property established in 5.1.6 implies

$$\lambda^{2-2m}\sigma^{-m} m^{-m} \rho^m \le \|S\|\ \mathbf{U}[(x,y), \rho] \le \|S\|\ W_j^+(\beta_j^+) \le 2E r^m,$$

hence $\rho \le \Gamma_{13}\lambda^{2-2/m} E^{1/m} r$ with $\Gamma_{13} = \sigma\, m\, 2^{1/m}$; similarly one estimates $\inf\{r-|x|,\ \beta_j^- - y_j\}$. The alternatives of (I) follow, and the postscript of (I) results by conjunction with (3) and (4).

We deduce (II) from (7) and (I).

5.3.5. Lemma. *There exists a finite positive number* Γ_{14} *such that whenever* κ *is a positive integer,*

$$S \in \mathscr{R}_m(\mathbf{R}^m \times \mathbf{R}^{n-m}), \qquad 0 < r < \infty, \qquad 0 < s < \infty,$$

$$\operatorname{spt} S \subset A = \{(x,y)\colon |x| \le r \text{ and } |y| \le s\},$$

$$\operatorname{spt} \partial S \subset \{(x,y)\colon |x| = r\}, \qquad E = \operatorname{Exc}(S, 0, r),$$

$$\mathbf{p}_\# S = \kappa\, \mathbf{E}^m \llcorner \{x\colon |x| < r\}, \qquad \Omega = \operatorname{Exc}^\kappa(S, 0, r),$$

$$f_j(x) = \kappa^{-1}\langle S, \mathbf{p}, x\rangle(Z_{m+j}) \text{ for } \mathscr{L}^m \text{ almost all } x,$$

$$U_j = \mathbf{p}_\#(S \llcorner Z_{m+j} \wedge DZ_1 \wedge \cdots \wedge DZ_m) \text{ for } j \in \{1, \dots, n-m\},$$

the following six statements hold:

(1) *If* $j \in \{1, \ldots, n-m\}$, *then*

$$\mathbf{p}_{\#}(S \llcorner Z_{m+j}) = \kappa\, \mathbf{E}^m \llcorner f_j, \qquad U_j = \kappa\, \mathscr{L}^m \llcorner f_j,$$
$$\|\partial(S \llcorner Z_{m+j})\|\, \mathbf{p}^{-1}\, \mathbf{U}(0,r) \le [2\kappa\, \alpha(m) + 2E]^{\frac{1}{2}}\, E^{\frac{1}{2}}\, r^m.$$

(2) *If* $\phi \in \mathscr{D}^0(\mathbf{R}^m)$, $\operatorname{spt}\phi \subset \mathbf{U}(0,r)$, $i \in \{1, \ldots, m\}$, $j \in \{1, \ldots, n-m\}$, *then*

$$D_i\, U_j(\phi) = S[(\phi \circ p) \wedge \Delta_{i,j}]$$

where $\Delta_{i,j} = DZ_1 \wedge \cdots \wedge DZ_{i-1} \wedge DZ_{m+j} \wedge DZ_{i+1} \wedge \cdots \wedge DZ_m$.

(3) *If* $j \in \{1, \ldots, n-m\}$ *and* $t \in \mathbf{R}$ *with*

$$\mathscr{L}^m[\mathbf{U}(0,r) \cap \{x\colon f_j(x) > t\}] \le \alpha(m)\, r^m/2,$$
$$\mathscr{L}^m[\mathbf{U}(0,r) \cap \{x\colon f_j(x) < t\}] \le \alpha(m)\, r^m/2,$$

then

$$\mathbf{M}[\mathbf{p}_{\#}(S \llcorner Z_{m+j}) - \kappa\, t\, \mathbf{E}^m \llcorner \mathbf{U}(0,r)]$$
$$= \kappa \int_{\mathbf{U}(0,r)} |f_j - t|\, d\mathscr{L}^m \le \Gamma_{14}[\kappa\, \alpha(m) + E]^{\frac{1}{2}}\, E^{\frac{1}{2}}\, r^{m+1}.$$

(4) *If* $0 < \omega < \infty$, *then there exist a map* $g\colon \mathbf{R}^m \to \mathbf{R}^{n-m}$ *of class* 1 *and a compact subset* C *of* $\mathbf{R}^m \cap \mathbf{U}(0,r)$ *such that*

$$\mathscr{L}^m[\mathbf{U}(0,r) \sim C] \le (5E + \Omega + \omega)\, r^m,$$
$$g_j | C = f_j | C \ \textit{with}\ g_j = Y_j \circ g \ \textit{for}\ j \in \{1, \ldots, n-m\},$$
$$S \llcorner \mathbf{p}^{-1}(C) = \kappa\, G_{\#}(\mathbf{E}^m \llcorner C) \ \textit{with}\ G(x) = (x, g(x)) \ \textit{for}\ x \in \mathbf{R}^m,$$
$$\mathbf{M}[S - \kappa\, G_{\#}(\mathbf{E}^m \llcorner C)] \le (6E + \kappa\, \Omega + \kappa\, \omega)\, r^m,$$
$$|D g(x)| \le 1 \ \textit{for}\ x \in C, \ \int_C |Dg|^2\, d\mathscr{L}^m \le 3\kappa^{-1} E\, r^m,$$
$$|D_i\, U_j(\phi) - \kappa \int_C \phi \cdot D_i\, g_j\, d\mathscr{L}^m| \le (6E + \kappa\, E + \kappa\, \omega)\, r^m\, \mathbf{M}(\phi)$$

for $i \in \{1, \ldots, m\}$, $j \in \{1, \ldots, n-m\}$, $\phi \in \mathscr{D}^0(\mathbf{R}^m)$ *with* $\operatorname{spt}\phi \subset \mathbf{U}(0,r)$.

(5) *If* $\eta \in \mathbf{R}^{n-m}$, $|\eta| \le s$, $0 < \lambda < \infty$ *and* Ψ *is a parametric integrand of degree* m *and class* 2 *on some neighborhood of* A *such that* $D\Psi^{\S}(0,0,0) = 0$ *and*

$$|\Psi(x,y,\gamma)| \le \lambda, \qquad \|D\Psi(x,y,\gamma)\| \le \lambda, \qquad \|D^2\, \Psi^{\S}(x,y,\tau)\| \le \lambda$$

for $(x,y) \in A$, $\gamma \in \wedge_m(\mathbf{R}^m \times \mathbf{R}^{n-m})$ *with* $|\gamma| = 1$, $\tau \in \operatorname{Hom}(\mathbf{R}^m, \mathbf{R}^{n-m})$ *with* $|\tau| \le 1$, *then*

$$r^{-m}\, |\langle \Psi, S\rangle - \langle \Psi, \kappa[\mathbf{E}^m \llcorner \mathbf{U}(0,r)] \times \boldsymbol{\delta}_\eta\rangle| \le \lambda\, \kappa[(r^2 + s^2)\, \alpha(m) + 13E + 2\Omega].$$

In case spt $\partial S \subset \mathbf{R}^m \times \{\eta\}$ *and* Ψ *is elliptic at* $(0,0)$ *with ellipticity bound* λ^{-1}, *then*

$$r^{-m}[\langle\Psi, S\rangle - \langle\Psi, \kappa[\mathbf{E}^m \llcorner \mathbf{U}(0,r)] \times \delta_\eta\rangle]$$
$$\geq \lambda^{-1} E - 2\kappa\lambda(r^2+s^2)^{\frac{1}{2}}[\alpha(m)^{\frac{1}{2}} E^{\frac{1}{2}} + 6E + \Omega] - 2\kappa\lambda(r^2+s^2)\alpha(m).$$

(6) *If* $0<s\leq r$, $0<\sigma<(4r)^{-1}$ *and* ψ *is a function of class* 2 *mapping some neighborhood of* A *into* $\mathbf{R}^m \times \mathbf{R}^{n-m}$ *with*

$$\psi(0,0)=(0,0), \quad D\psi(0,0)=\mathbf{1}_{\mathbf{R}^m \times \mathbf{R}^{n-m}}, \quad \|D^2\psi(x,y)\| \leq \sigma \text{ for } (x,y)\in A,$$

then $r-\sigma r^2>0$, $\psi\{(x,y): |x|=r\} \subset \{(x,y): |x|\geq r-\sigma r^2\}$,

$$\mathbf{p}_\#[(\psi_\# S)\llcorner \mathbf{p}^{-1}\mathbf{U}(0,r-\sigma r^2)] = \kappa\,\mathbf{E}^m \llcorner \mathbf{U}(0,r-\sigma r^2),$$
$$\operatorname{Exc}(\psi_\# S, 0, r-\sigma r^2) \leq E + 2^{m+2}\sigma r[\kappa\alpha(m)+E]$$

Proof. Noting that 4.3.2(1) implies

$$\int \phi(x)\kappa f_j(x)\,d\mathscr{L}^m x = S[Z_{m+j} \wedge \mathbf{p}^\#(\phi \wedge DX_1 \wedge \cdots \wedge DX_m)]$$

for $\phi\in\mathscr{D}^0(\mathbf{R}^m)$, we infer the first two conclusions of (1); in the case when spt $\phi \subset \mathbf{U}(0,r)$ we also obtain

$$\begin{aligned} D_i U_j(\phi) &= -U_j(D_i\phi) = -S[Z_{m+j} \wedge \mathbf{p}^\#(D_j\phi \wedge DX_1 \wedge \cdots \wedge DX_m)] \\ &= -S[Z_{m+j} \wedge \mathbf{p}^\#(DX_1 \wedge \cdots \wedge DX_{i-1} \wedge D\phi \wedge DX_{i+1} \wedge \cdots \wedge DX_m)] \\ &= -S[DZ_1 \wedge \cdots \wedge DZ_{i-1} \wedge Z_{m+j} \wedge D(\phi\circ\mathbf{p}) \wedge DZ_{i+1} \wedge \cdots \wedge DZ_m] \\ &= S[DZ_1 \wedge \cdots \wedge DZ_{i-1} \wedge DZ_{m+j} \wedge (\phi\circ\mathbf{p}) \wedge DZ_{i+1} \wedge \cdots \wedge DZ_m] \end{aligned}$$

because $(\partial S)\llcorner \mathbf{p}^{-1}\mathbf{U}(0,r)=0$, hence (2). To verify the third conclusion of (1) we abbreviate

$$\chi(z) = \langle \vec{S}(z), Z_1 \wedge \cdots \wedge Z_m\rangle \text{ for } z\in\operatorname{dmn}\vec{S},$$

then use Hölder's inequality and 5.3.1 in estimating

$$\begin{aligned} &\|\partial(S\llcorner Z_{m+j})\|\,\mathbf{p}^{-1}\mathbf{U}(0,r) = \|S\llcorner DZ_{m+j}\|\,\mathbf{p}^{-1}\mathbf{U}(0,r) \\ &\leq \int |\vec{S}\llcorner DZ_{m+j}|\,d\|S\| \leq \left(\int |\vec{S}\llcorner DZ_{m+j}|^2\,d\|S\| \cdot \mathbf{M}(S)\right)^{\frac{1}{2}} \\ &\leq \left(\int 2(1-\chi)\,d\|S\| \cdot \mathbf{M}(S)\right)^{\frac{1}{2}} = [2E(\kappa\alpha(m)+E)]^{\frac{1}{2}} r^m. \end{aligned}$$

Next suppose j and t satisfy the hypothesis of (3).

In case $m>1$ we replace n by m in 4.5.3 and define $\Gamma_{14}=2^{\frac{1}{m}}\sigma\,\alpha(m)^{1/m}$; applying Hölder's inequality and 4.5.9(19) with $T=\kappa^{-1}\,\mathbf{p}_\#(S\llcorner Z_{m+j})$ we obtain

$$\begin{aligned}
&r^{-m}\textstyle\int_{\mathbf{U}(0,r)}|f_j-t|\,d\mathscr{L}^m\\
&\leq(r^{-m}\textstyle\int_{\mathbf{U}(0,r)}|f_j-t|^{m/(m-1)}\,d\mathscr{L}^m)^{(m-1)/m}\,\alpha(m)^{1/m}\\
&\leq[\sigma\,r^{1-m}\,\|\partial(\mathbf{E}^m\llcorner f_j)\|\,\mathbf{U}(0,r)]\,\alpha(m)^{1/m},
\end{aligned}$$

hence deduce the conclusion of (3) from (1).

In case $m=1$ we take $\Gamma_{14}=2^{\frac{1}{2}}$ and infer the conclusion of (3) from (1) and 4.5.9(23).

To prove (4) we represent $S=(\mathscr{H}^m\llcorner W)\wedge\xi$, where W consists of all points z such that $\Theta^m(\|S\|,z)\geq 1$, $\vec{S}(z)$ is a simple m-vector, $|\vec{S}(z)|=1$, $\operatorname{Tan}^m(\|S\|,z)$ is the m dimensional vectorsubspace of $\mathbf{R}^m\times\mathbf{R}^{n-m}$ associated with $\vec{S}(z)$, and where

$$\xi(z)=\Theta^m(\|S\|,z)\,\vec{S}(z)\ \text{ for }\ z\in W.$$

We note that $\operatorname{ap}J_m(\mathbf{p}|W)(z)=|\chi(z)|$ for $z\in W$. Applying 4.3.8 with $F=\mathbf{p}$ we compute

$$\zeta(z)=\Theta^m(\|S\|,z)\operatorname{sign}\chi(z)\ \text{ for }\ z\in W,$$

and we find that, for $\mathscr{L}^m$ almost all x in $\mathbf{U}(0,r)$,

$$\langle S,\mathbf{p},x\rangle=(\mathscr{H}^0\llcorner W\cap\mathbf{p}^{-1}\{x\})\wedge\zeta=\sum_{z\in W\cap\mathbf{p}^{-1}\{x\}}\zeta(z)\,\delta_z,$$

$$f_j(x)=\kappa^{-1}\sum_{z\in W\cap\mathbf{p}^{-1}\{x\}}\zeta(z)\,z_{m+j}\ \text{ for }\ j\in\{1,\ldots,n-m\},$$

$$\kappa=\langle S,\mathbf{p},x\rangle(1)=\sum_{z\in W\cap\mathbf{p}^{-1}\{x\}}\zeta(z)$$

because $\kappa\,\mathbf{E}^m\llcorner\mathbf{U}(0,r)=\mathbf{p}_\#\,S=\mathbf{E}^m\llcorner\langle S,\mathbf{p},\cdot\rangle$ by 4.3.2(1). We note that $\mathscr{L}^m[\mathbf{U}(0,r)\sim\mathbf{p}(W)]=0$.

For $j\in\{1,\ldots,n-m\}$ we define B_j as the subset of $\operatorname{dmn}f_j$ consisting of those points where f_j is approximately differentiable, and note that $\mathscr{L}^m(\mathbf{R}^m\sim B_j)=0$ by 4.5.9(26). Letting

$$B=\bigcap_{j=1}^{n-m}B_j\cap\{x:\ \operatorname{card}(W\cap\mathbf{p}^{-1}\{x\})=1\}$$

we use 2.10.11 and 4.2.26 to infer

$$\begin{aligned}
&\mathscr{L}^m\,\mathbf{B}(0,r)+\mathscr{L}^m[\mathbf{B}(0,r)\sim B]\leq\textstyle\int\operatorname{card}(W\cap\mathbf{p}^{-1}\{x\})\,d\mathscr{L}^m\,x\\
&\leq\mathscr{H}^m(W)\leq\kappa^{-1}\,\|S\|\,(W)+\|S\|^\kappa(W)\leq[\alpha(m)+\kappa^{-1}\,E+\Omega]\,r^m,
\end{aligned}$$

hence $\mathscr{L}^m[\mathbf{B}(0,r)\sim B]\leq(\kappa^{-1}\,E+\Omega)\,r^m$.

If $x \in B$, then the unique $z \in W \cap \mathbf{p}^{-1}\{x\}$ satisfies the conditions $\zeta(z)=\kappa$ and $f_j(x)=z_{m+j}$ for $j \in \{1, \ldots, n-m\}$, hence

$$F(x)=(x_1, \ldots, x_m, f_1(x), \ldots, f_{n-m}(x)) \in W.$$

Defining $V=W \cap \{z: \chi(z) \geq 2^{-\frac{1}{2}}\}$ we see from 5.3.1 that

$$\|S\|(W \sim V) \leq (1-2^{-\frac{1}{2}})^{-1} \int (1-\chi)\, d\|S\| \leq 4 E r^m.$$

Next we apply 3.1.16 to obtain real valued functions $g_1, \ldots, g_{n-m}$ of class 1 on $\mathbf{R}^m$ such that

$$\mathscr{L}^m[\mathbf{U}(0,r) \sim M] < \omega r^m \quad \text{with} \quad M=\bigcap_{j=1}^{n} \{x: f_j(x)=g_j(x)\},$$

then use 2.9.11 and 2.2.2 to secure a compact subset C of $B \cap \mathbf{p}(V) \cap M$ such that

$$\mathscr{L}^m[B \cap \mathbf{p}(V) \sim C] < \omega r^m, \; \Theta^m(\mathscr{L}^m \llcorner \mathbf{R}^m \sim M, x)=0 \text{ for } x \in C.$$

We also define

$$g(x)=(g_1(x), \ldots, g_{n-m}(x)), \; G(x)=(x, g(x)) \text{ for } x \in \mathbf{R}^m.$$

If $x \in C$, then $G(x)=F(x) \in V$; employing 3.2.17 with $\psi=G$ we infer

$$\operatorname{im} D G(x) \subset \operatorname{Tan}^m[\mathscr{H}^m \llcorner W, G(x)]=\operatorname{Tan}^m[\|S\|, G(x)]$$

with $\dim \operatorname{im} D G(x)=m$, hence the vectorsubspace associated with $\vec{S}[G(x)]$ equals

$$\begin{aligned} \operatorname{im} D G(x) &= (\mathbf{R}^m \times \mathbf{R}^{n-m}) \cap \{(u,v): v=\langle u, D g(x)\rangle\} \\ &= \bigcap_{j=1}^{n-m} \ker \Big[Z_{m+j} - \sum_{i=1}^{m} D_i g_j(x)\, Z_i \Big], \end{aligned}$$

$$\langle \vec{S}[G(x)], Z_1 \wedge \cdots \wedge Z_{i-1} \wedge Z_{m+j} \wedge Z_{i+1} \wedge \cdots \wedge Z_m \rangle = D_i g_j(x)\, \chi[G(x)]$$

$$\text{for } i \in \{1, \ldots, m\} \text{ and } j \in \{1, \ldots, n-m\},$$

$$1=|\vec{S}[G(x)]|^2 \geq (\chi[G(x)])^2 [1+|D g(x)|^2],$$

$$|D g(x)| \leq 1 \quad \text{because} \quad \chi[G(x)] \geq 2^{-\frac{1}{2}}.$$

Since $W \cap \mathbf{p}^{-1}(C)=G(C) \subset V$ and $G[\mathbf{p}(z)]=z$ for $z \in G(C)$, we see from 4.1.15 that

$$S \llcorner \mathbf{p}^{-1}(C)=G_\# \, \mathbf{p}_\# [S \llcorner \mathbf{p}^{-1}(C)]=G_\#(\kappa \, \mathbf{E}^m \llcorner C),$$

and from 3.2.20 that

$$\begin{aligned}&\textstyle\int_C |D\,g(x)|^2\,d\mathscr{L}^m x = \int_{G(C)} |D\,g[\mathbf{p}(z)]|^2\,|\chi(z)|\,d\mathscr{H}^m z\\&\le 2^{\frac{1}{2}} \textstyle\int_{G(C)} |D\,g[\mathbf{p}(z)]|^2\,|\chi(z)|^2\,d\mathscr{H}^m z\\&\le 2^{\frac{1}{2}} \textstyle\int_{G(C)} [1-|\chi(z)|^2]\,d\mathscr{H}^m z \le 2^{\frac{3}{2}} \int_{G(C)} (1-\chi)\,d\mathscr{H}^m\\&= 2^{\frac{3}{2}}\,\kappa^{-1} \textstyle\int_{G(C)} (1-\chi)\,d\,\|S\| \le 3\kappa^{-1}\,E\,r^m.\end{aligned}$$

Finally we observe that

$$W\sim\mathbf{p}^{-1}(C)\subset(W\sim V)\cup\mathbf{p}^{-1}(N) \quad\text{with}\quad N=\mathbf{p}(V)\sim C,$$

$$N\subset[\mathbf{U}(0,r)\sim B]\cup[B\cap\mathbf{p}(V)\sim C], \quad \mathscr{L}^m(N)\le(\kappa^{-1}E+\Omega+\omega)\,r^m,$$

$$\mathscr{L}^m[\mathbf{U}(0,r)\sim\mathbf{p}(V)]\le\mathscr{L}^m[\mathbf{p}(W\sim V)]\le\mathscr{H}^m(W\sim V)\le\|S\|(W\sim V)\le 4E\,r^m,$$

$$\begin{aligned}\|S\|\,\mathbf{p}^{-1}(N)-\kappa\,\mathscr{L}^m(N)&=\|S\|\,\mathbf{p}^{-1}(N)-(\mathbf{p}_\#\,S)[N\wedge D\,X_1\wedge\cdots\wedge D\,X_m]\\&=\textstyle\int_{\mathbf{p}^{-1}(N)}(1-\chi)\,d\,\|S\|\le\int(1-\chi)\,d\,\|S\|=E\,r^m,\end{aligned}$$

$$\begin{aligned}&[G_\#(\mathbf{E}^m\llcorner C)]\,[(\phi\circ\mathbf{p})\wedge\Delta_{i,j}]=(\mathbf{E}^m\llcorner C)\,G^\#[(\phi\circ\mathbf{p})\wedge\Delta_{i,j}]\\&=(\mathbf{E}^m\llcorner C)\,[\phi\,D_i\,g_j\wedge D\,X_1\wedge\cdots\wedge D\,X_m]=\textstyle\int_C\phi\,D_i\,g_j\,d\mathscr{L}^m,\end{aligned}$$

hence (2) implies

$$|D_i\,U_j(\phi)-\kappa\textstyle\int_C\phi\,D_i\,g_j\,d\mathscr{L}^m|\le\|S\|\,[W\sim\mathbf{p}^{-1}(C)]\cdot\mathbf{M}(\phi).$$

To verify (5) we first estimate, by Taylor's theorem,

$$\|D\,\Psi^\S(x,y,\tau)\|\le\lambda(|x|^2+|y|^2+|\tau|^2)^{\frac{1}{2}},$$

$$|\Psi^\S(x,y,\tau)-\Psi^\S(0,0,0)|\le\lambda(|x|^2+|y|^2+|\tau|^2)/2$$

for $(x,y)\in A$ and $\tau\in\operatorname{Hom}(\mathbf{R}^m,\mathbf{R}^{n-m})$ with $|\tau|\le 1$, then infer from (4) that

$$\begin{aligned}&|\langle\Psi,S\rangle-\langle\Psi,\kappa[\mathbf{E}^m\llcorner\mathbf{U}(0,r)]\times\delta_\eta\rangle|\\&\le\kappa|\langle\Psi,G_\#(\mathbf{E}^m\llcorner C)\rangle-\langle\Psi,(\mathbf{E}^m\llcorner C)\times\delta_\eta\rangle|\\&\quad+\lambda\,\mathbf{M}[S-\kappa\,G_\#(\mathbf{E}^m\llcorner C)]+\lambda\,\kappa\,\mathscr{L}^m[\mathbf{U}(0,r)\sim C]\\&\le\kappa\,|\textstyle\int_C(\Psi^\S[x,g(x),D\,g(x)]-\Psi^\S[x,\eta,0])\,d\mathscr{L}^m x|\\&\quad+\lambda(6E+\kappa\,\Omega+\kappa\,\omega)\,r^m+\lambda\,\kappa(5E+\Omega+\omega)\,r^m\\&\le\kappa\textstyle\int_C\lambda[|x|^2+|g(x)|^2+|D\,g(x)|^2+|x|^2+|\eta|^2]/2\,d\mathscr{L}^m x\\&\quad+\kappa\,\lambda(11\,E+2\Omega+2\omega)\,r^m\\&\le\kappa\,\lambda[(r^2+s^2)\,\alpha(m)+2E+11\,E+2\Omega+2\omega]\,r^m.\end{aligned}$$

The first conclusion of (5) follows because ω may be chosen arbitrarily small. Next we assume the supplementary hypothesis of the second part, define

$$\Phi(x,y,\gamma)=\Psi(x,y,\gamma)-\Psi(0,0,\gamma)\quad\text{for}\quad(x,y,\gamma)\in\operatorname{dmn}\Psi,$$

observe that

$$\Phi|(x,y,\gamma)| \le \lambda(|x|^2+|y|^2)^{\frac{1}{2}} \text{ when } (x,y)\in A \text{ and } |\gamma|=1,$$

$$|\Phi^{\S}(x,y,\tau)| = |\Psi^{\S}(x,y,\tau)-\Psi^{\S}(0,0,\tau)|$$
$$\le (|x|^2+|y|^2)^{\frac{1}{2}}\cdot\lambda(|x|^2+|y|^2+|\tau|^2)^{\frac{1}{2}}$$

when $(x,y)\in A$ and $\tau\in\operatorname{Hom}(\mathbf{R}^m,\mathbf{R}^{n-m})$ with $|\tau|\le 1$, and combine (4) with Hölder's inequality to obtain

$$\begin{aligned}
&|\langle\Phi,S\rangle-\langle\Phi,\kappa[\mathbf{E}^m \llcorner \mathbf{U}(0,r)]\times\delta_\eta\rangle|\\
&\le \kappa\left|\int_C(\Phi^{\S}[x,g(x),D\,g(x)]-\Phi^{\S}[x,\eta,0])\,d\mathscr{L}^m x\right|\\
&\quad+\lambda(r^2+s^2)^{\frac{1}{2}}[6E+\kappa\Omega+\kappa\omega+\kappa(5E+\Omega+\omega)]\,r^m\\
&\le \kappa\int_C \lambda((r^2+s^2)^{\frac{1}{2}}[r^2+s^2+|D\,g(x)|^2]^{\frac{1}{2}}+(r^2+s^2))\,d\mathscr{L}^m x\\
&\quad+2\kappa\lambda(r^2+s^2)^{\frac{1}{2}}(6E+\Omega+\omega)\,r^m\\
&\le \kappa\lambda(r^2+s^2)^{\frac{1}{2}}\boldsymbol{\alpha}(m)^{\frac{1}{2}}[(r^2+s^2)\boldsymbol{\alpha}(m)+3E]^{\frac{1}{2}}\,r^m\\
&\quad+\kappa\lambda(r^2+s^2)\boldsymbol{\alpha}(m)\,r^m+2\kappa\lambda(r^2+s^2)^{\frac{1}{2}}(6E+\Omega+\omega)\,r^m.
\end{aligned}$$

Moreover the ellipticity of Ψ at $(0,0)$ implies

$$\langle\Psi-\Phi,S\rangle-\langle\Psi-\Phi,\kappa[\mathbf{E}^m \llcorner \mathbf{U}(0,r)]\times\delta_\eta\rangle \ge \lambda^{-1}E\,r^m,$$

and the second conclusion of (5) results by addition.

With regard to (6) we note that Taylor's theorem implies

$$|\psi(x,y)-(x,y)| \le \sigma(|x|^2+|y|^2)/2 \le \sigma r^2,$$
$$\|D\psi(x,y)-\mathbf{1}\| \le \sigma(|x|^2+|y|^2)^{\frac{1}{2}} \le 2\sigma r$$

for $(x,y)\in A$, hence $|\mathbf{p}[\psi(x,y)]| \ge r-\sigma r^2$ in case $|x|=r$. Letting h be the affine homotopy from $\mathbf{p}$ to $\mathbf{p}\circ\psi$ we infer that the support of

$$\mathbf{p}_\#(\psi_\# S)-\mathbf{p}_\# S = h_\#([0,1]\times\partial S)$$

is contained in $\mathbf{R}^m \sim \mathbf{U}(0,r-\sigma r^2)$, and deduce the third conclusion of (6). Then we verify the fourth conclusion by estimating

$$\begin{aligned}
\operatorname{Exc}(\psi_\# S,0,r-\sigma r^2) &\le (r-\sigma r^2)^{-m}\mathbf{M}(\psi_\# S)-\kappa\boldsymbol{\alpha}(m)\\
&\le r^{-m}(1-\sigma r)^{-m}(1+2\sigma r)^m\mathbf{M}(S)-\kappa\boldsymbol{\alpha}(m)\\
&\le \left(1+\frac{3\sigma r}{1-\sigma r}\right)^m[\kappa\boldsymbol{\alpha}(m)+E]-\kappa\boldsymbol{\alpha}(m)\\
&\le E+2^m\,3\sigma r(1-\sigma r)^{-1}[\kappa\boldsymbol{\alpha}(m)+E]
\end{aligned}$$

with $\sigma r<\frac{1}{4}$ and $1/(1-\sigma r)<\frac{4}{3}$.

5.3.6. Recalling 4.2.8, 4.1.1, 4.1.7, 2.7.16 one readily verifies the following statements for $0<\rho<\infty$:

If $T\in\mathscr{D}_m(\mathbf{R}^m)$, *then*

$$(\mu_{\rho\#} T)\llcorner(DX_1\wedge\cdots\wedge DX_m)=\rho^m\,\mu_{\rho\#}(T\llcorner DX_1\wedge\cdots\wedge DX_m).$$

If $U\in\mathscr{D}_0(\mathbf{R}^m)$ *and* $i\in\{1,\ldots,m\}$, *then* $\mu_{\rho\#}(D_i U)=\rho\, D_i(\mu_{\rho\#} U)$.

If f *is a real valued* $\mathscr{L}^m$ *summable function, then*

$$\mathbf{E}^m\llcorner(f\circ\mu_{1/\rho})=\mu_{\rho\#}(\mathbf{E}^m\llcorner f),\qquad \mathscr{L}^m\llcorner(f\circ\mu_{1/\rho})=\rho^m\,\mu_{\rho\#}(\mathscr{L}^m\llcorner f).$$

5.3.7. Lemma. *Suppose* κ *is a positive integer,* $1\le\lambda<\infty$ *and to each positive integer* ν *correspond:*

(1) *a parametric integrand* Ψ_ν *of degree* m *and class* 3 *on some neighborhood of*

$$Z=(\mathbf{R}^m\times\mathbf{R}^{n-m})\cap\{(x,y)\colon |x|\le\lambda^{-1},\ |y|\le\lambda^{-1}\}$$

such that $D\Psi_\nu^{\S}(0,0,0)=0$ *and*

$$|\Psi_\nu(x,y,\zeta)|\le\lambda,\qquad \|D^q\Psi_\nu^{\S}(x,y,\tau)\|\le\lambda$$

for $q\in\{0,1,2,3\}$, $(x,y)\in Z$, $\zeta\in\wedge_m(\mathbf{R}^m\times\mathbf{R}^{n-m})$ *with* $|\zeta|=1$, $\tau\in\operatorname{Hom}(\mathbf{R}^m,\mathbf{R}^{n-m})$ *with* $\|\tau\|\le 2$;

(2) $0<r_\nu<\lambda^{-1}$, $0<s_\nu<\lambda^{-1}$, $0<\varepsilon_\nu<1$;

(3) $S_\nu\in\mathscr{R}_m(\mathbf{R}^m\times\mathbf{R}^{n-m})$ *with* $\operatorname{spt} S_\nu\subset\{(x,y)\colon |x|\le r_\nu,\ |y|\le s_\nu\}$,

$$\operatorname{spt}\partial S_\nu\subset\{(x,y)\colon |x|=r_\nu\},\qquad \mathbf{p}_\# S_\nu=\kappa\,\mathbf{E}^m\llcorner\{x\colon |x|\le r_\nu\};$$

(4) $U_{\nu,j}=\mathbf{p}_\#(S_\nu\llcorner Z_{m+j}\wedge DZ_1\wedge\cdots\wedge DZ_m)\in\mathscr{D}_0(\mathbf{R}^m)$ *and*

$$H_{\nu,j}=\varepsilon_\nu^{-1}\,r_\nu^{-1-m}\,\mu_{1/r_\nu\#}\,U_{\nu,j}\in\mathscr{D}_0(\mathbf{R}^m)\ \text{ for } j\in\{1,\ldots,n-m\}.$$

Furthermore suppose:

(5) $\lim_{\nu\to\infty}(\Psi_\nu^{\S})_{(i,j;k,l)}(0,0,0)=\Upsilon_{i,j;k,l}\in\mathbf{R}$ *for* $\{i,k\}\subset\{1,\ldots,m\}$ *and* $\{j,l\}\subset\{1,\ldots,n-m\}$;

(6) $\lim_{\nu\to\infty}(\varepsilon_\nu+\varepsilon_\nu^{-1}r_\nu+\varepsilon_\nu^{-1}s_\nu)=0$;

(7) $\limsup_{\nu\to\infty}\varepsilon_\nu^{-2}[\operatorname{Exc}(S_\nu,0,r_\nu)+\operatorname{Exc}^\kappa(S_\nu,0,r_\nu)]=\alpha<\infty$;

(8) $H_j\in\mathscr{D}_0(\mathbf{R}^m\cap\{x\colon |x|<1\})$ *for* $j\in\{1,\ldots,n-m\}$ *with*

$$H_j(\phi)=\lim_{\nu\to\infty}H_{\nu,j}(\phi)\ \text{ for } \phi\in\mathscr{D}^0(\mathbf{R}^m\cap\{x\colon |x|<1\});$$

(9) $\theta\in\mathscr{D}(\mathbf{R}^m,\mathbf{R}^{n-m})$ *with* $\operatorname{spt}\theta\subset\mathbf{R}^m\cap\{x\colon |x|<1\}$,

$$\theta_j=Y_j\circ\theta\ \text{ for } j\in\{1,\ldots,n-m\},$$

$h_\nu(t,x,y)=h_{\nu,t}(x,y)=(x,y+t\,\varepsilon_\nu\,r_\nu\,\theta(r_\nu^{-1}x))$ *for* $(t,x,y)\in\mathbf{R}\times\mathbf{R}^m\times\mathbf{R}^{n-m}$;

(10) $J_\nu(t)=\varepsilon_\nu^{-2}\,r_\nu^{-m}\langle\Psi_\nu,h_{\nu,t\#}S_\nu\rangle$ *for* $t\in\mathbf{R}$.

Then the following four equations hold for every $t \in \mathbf{R}$, *with uniform convergence on each compact subset of* $\mathbf{R}$:

$$\lim_{\nu\to\infty} J_\nu'(t) = \sum_{i=1}^{m}\sum_{j=1}^{n-m}\sum_{k=1}^{m}\sum_{l=1}^{n-m} \Upsilon_{i,j;k,l}\left[D_k H_l(D_i\theta_j) + t\,\kappa \int D_i\theta_j \cdot D_k\theta_l \, d\mathscr{L}^m\right],$$

$$\lim_{\nu\to\infty} [J_\nu(t) - J_\nu(0)]$$

$$= \sum_{i=1}^{m}\sum_{j=1}^{n-m}\sum_{k=1}^{m}\sum_{l=1}^{n-m} \Upsilon_{i,j;k,l}\left[t\, D_k H_l(D_i\theta_j) + 2^{-1} t^2 \kappa \int D_i\theta_j \cdot D_k\theta_l \, d\mathscr{L}^m\right],$$

$$\lim_{\nu\to\infty} \varepsilon_\nu^{-2} r_\nu^{-m} [\mathbf{M}(h_{\nu,t\#} S_\nu) - \mathbf{M}(S_\nu)]$$

$$= \sum_{i=1}^{m}\sum_{j=1}^{n-m} \left[t\, D_i H_j(D_i\theta_j) + 2^{-1} t^2 \kappa \int (D_i\theta_j)^2 \, d\mathscr{L}^m\right],$$

$$\lim_{\nu\to\infty} \varepsilon_\nu^{-2} r_\nu^{-m} [\mathbf{M}^\kappa(h_{\nu,t\#} S_\nu) - \mathbf{M}^\kappa(S_\nu)] = 0.$$

Also, if $w \in \mathscr{D}[\mathbf{R}^m, \mathrm{Hom}(\mathbf{R}^m, \mathbf{R}^{n-m})]$ *with* $\operatorname{spt} w \subset \mathbf{U}(0,1)$, *and*

$$w_{i,j}(x) = \langle\langle e_i, w(x)\rangle, Y_j\rangle \text{ for } x \in \mathbf{R}^m,$$

then

$$\sum_{i=1}^{m}\sum_{j=1}^{n-m} D_i H_j(w_{i,j}) \le (3\alpha\kappa \int |w|^2 \, d\mathscr{L}^m)^{\frac{1}{2}}.$$

Proof. We suppose $\alpha < \gamma < \infty$ and choose $\beta < \infty$ so that

$$|\theta(x)| \le \beta \text{ and } \|D\theta(x)\| \le \beta \text{ for } x \in \mathbf{R}^m.$$

Noting that $\mathrm{Exc}(S_\nu, 0, r_\nu) + \mathrm{Exc}^\kappa(S_\nu, 0, r_\nu) < \gamma\,\varepsilon_\nu^2$ for large ν, we apply 5.3.5 (4) to obtain compact sets

$$C_\nu \subset \mathbf{R}^m \cap \mathbf{U}(0, r_\nu)$$

and maps $g_\nu \colon \mathbf{R}^m \to \mathbf{R}^{n-m}$ of class 1 such that

$$S_\nu \llcorner \mathbf{p}^{-1}(C_\nu) = \kappa\, G_{\nu\#}(\mathbf{E}^m \llcorner C_\nu),\ \mathbf{M}(\Xi_\nu) < 5\kappa\gamma\,\varepsilon_\nu^2 r_\nu^m$$

with

$$G_\nu(x) = (x, g_\nu(x)) \text{ for } x \in \mathbf{R}^m,$$

$$\Xi_\nu = S_\nu \llcorner \mathbf{p}^{-1}[\mathbf{U}(0, r_\nu) \sim C_\nu],$$

and which satisfy the conditions

$$\mathscr{L}^m[\mathbf{U}(0, r_\nu) \sim C_\nu] < 5\gamma\,\varepsilon_\nu^2 r_\nu^m,$$

$$|g_\nu(x)| \le s_\nu \text{ and } |D g_\nu(x)| \le 1 \text{ for } x \in C_\nu,$$

$$\int_{C_\nu} |D g_\nu|^2 \, d\mathscr{L}^m < 3\kappa^{-1}\gamma\,\varepsilon_\nu^2 r_\nu^m,$$

$$|D_i U_{\nu,j}(\phi) - \kappa \int_{C_\nu} \phi \cdot D_i g_{\nu,j} \, d\mathscr{L}^m| \le 5\kappa\gamma\,\varepsilon_\nu^2 r_\nu^m \mathbf{M}(\phi)$$

with $g_{\nu,j} = Y_j \circ g_\nu$ whenever $\phi \in \mathscr{D}^0(\mathbf{R}^m)$, $\operatorname{spt}\phi \subset \mathbf{U}(0, r_\nu)$, $i \in \{1, \ldots, m\}$ and $j \in \{1, \ldots, n-m\}$. For $t \in \mathbf{R}$ we express

$$J_\nu(t) = K_\nu(t) + \kappa\, L_\nu(t)$$

with

$$K_\nu(t) = \varepsilon_\nu^{-2} r_\nu^{-m} \langle \Psi_\nu, h_{\nu,t\,\#}\, \Xi_\nu \rangle, \quad L_\nu(t) = \varepsilon_\nu^{-2} r_\nu^{-m} \langle \Psi_\nu, (h_{\nu,t} \circ G_\nu)_\# (\mathbf{E}^m \llcorner C_\nu) \rangle,$$

assuming ν so large that

$$r_\nu < \varepsilon_\nu, \quad s_\nu < \varepsilon_\nu, \quad (1 + |t|\,\beta)\,\varepsilon_\nu < \lambda^{-1},$$

hence $h_{\nu,t}[(\operatorname{spt}\Xi_\nu) \cup G_\nu(C_\nu)] \subset Z$. Inasmuch as

$$\langle (u,v), D\,h_{\nu,t}(x,y) \rangle = (u, v + \langle u, t\,\varepsilon_\nu\, D\,\theta(r_\nu^{-1} x) \rangle),$$

$$\dot{h}_{\nu,t}(x,y) = (0, \varepsilon_\nu\, r_\nu\, \theta(r_\nu^{-1} x)), \qquad \langle (u,v), D\,\dot{h}_{\nu,t}(x,y) \rangle = (0, \langle u, \varepsilon_\nu\, D\,\theta(r_\nu^{-1} x) \rangle)$$

for $(x,y) \in \mathbf{R}^m \times \mathbf{R}^{n-m}$ and $(u,v) \in \mathbf{R}^m \times \mathbf{R}^{n-m}$, which implies

$$\|D\,h_{\nu,t}(x,y)\| \le 1 + |t|\,\varepsilon_\nu\,\beta \le 2, \quad |\dot{h}_{\nu,t}(x,y)| \le \varepsilon_\nu\, r_\nu\, \beta, \quad \|D\,\dot{h}_{\nu,t}(x,y)\| \le \varepsilon_\nu\, \beta,$$

we see from 5.1.7 that

$$\begin{aligned} |K_\nu'(t)| &= |\varepsilon_\nu^{-2} r_\nu^{-m}\, \delta^{(1)}(h_{\nu,t\,\#}\, \Xi_\nu, \Psi_\nu, h_\nu)| \\ &\le \varepsilon_\nu^{-2} r_\nu^{-m}\, \lambda(\varepsilon_\nu\, r_\nu\, \beta + m\,\varepsilon_\nu\,\beta)\, \mathbf{M}(h_{\nu,t\,\#}\, \Xi_\nu) \le \varepsilon_\nu^{-1} r_\nu^{-m}\, \lambda(1+m)\,\beta\, 2^m\, \mathbf{M}(\Xi_\nu) \\ &\le 5\,\kappa\,\lambda(1+m)\,\beta\, 2^m\, \gamma\, \varepsilon_\nu \to 0 \text{ as } \nu \to \infty. \end{aligned}$$

Next we observe that

$$(h_{\nu,t} \circ G_\nu)(x) = (x, V_{\nu,t}(x))$$

with

$$V_{\nu,t}(x) = g_\nu(x) + t\,\varepsilon_\nu\, r_\nu\, \theta(r_\nu^{-1} x), \qquad |V_{\nu,t}(x)| < \lambda^{-1},$$

$$D\,V_{\nu,t}(x) = D\,g_\nu(x) + t\,\varepsilon_\nu\, D\,\theta(r_\nu^{-1} x), \qquad \|D\,V_{\nu,t}(x)\| \le 2$$

for $x \in C_\nu$, and use 5.1.9 to compute

$$L_\nu(t) = \varepsilon_\nu^{-2} r_\nu^{-m} \int_{C_\nu} \Psi_\nu^\S[x, V_{\nu,t}(x), D\,V_{\nu,t}(x)]\, d\mathscr{L}^m x,$$

$$L_\nu'(t) = \varepsilon_\nu^{-1} r_\nu^{-m} \int_{C_\nu} \langle [0, r_\nu\, \theta(r_\nu^{-1} x), D\,\theta(r_\nu^{-1} x)], \; D\,\Psi_\nu^\S[x, V_{\nu,t}(x), D\,V_{\nu,t}(x)] \rangle\, d\mathscr{L}^m x.$$

Applying Taylor's theorem to $D\,\Psi_\nu^\S$ at $(0,0,0)$ we find that

$$D\,\Psi_\nu^\S[x, V_{\nu,t}(x), D\,V_{\nu,t}(x)] = \langle [x, V_{\nu,t}(x), D\,V_{\nu,t}(x)], D\,D\,\Psi_\nu^\S(0,0,0) \rangle + R_{\nu,t}(x)$$

for $x \in C_\nu$, where $R_{\nu,t}(x) \in \operatorname{Hom}[\mathbf{R}^m \times \mathbf{R}^{n-m} \times \operatorname{Hom}(\mathbf{R}^m, \mathbf{R}^{n-m}), \mathbf{R}]$ with

$$\|R_{\nu,t}(x)\| \le [|x|^2 + |V_{\nu,t}(x)|^2 + |D\,V_{\nu,t}(x)|^2]\, \lambda/2.$$

Then we use the linearity and bilinearity of the first and second differentials of $\Psi_\nu^\S$ to represent

$$L'_\nu(t) = M_{\nu,1}(t) + M_{\nu,2}(t) + M_{\nu,3}(t) + M_{\nu,4}(t)$$

where

$$M_{\nu,1}(t) = \varepsilon_\nu^{-1} r_\nu^{-m} \int_{C_\nu} \langle [0, r_\nu \theta(r_\nu^{-1} x), 0], D\Psi_\nu^\S [x, V_{\nu,t}(x), D V_{\nu,t}(x)] \rangle \, d\mathscr{L}^m x,$$

$$|M_{\nu,1}(t)| \le \alpha(m)\, \varepsilon_\nu^{-1} r_\nu \beta \lambda \to 0 \quad \text{as} \quad \nu \to \infty,$$

because $\mathscr{L}^m(C_\nu) \le \alpha(m)\, r_\nu^m$,

$$M_{\nu,2}(t) = \varepsilon_\nu^{-1} r_\nu^{-m} \int_{C_\nu} \langle [0, 0, D\theta(r_\nu^{-1} x)], R_{\nu,t}(x) \rangle \, d\mathscr{L}^m x,$$

$$\begin{aligned} |M_{\nu,2}(t)| &\le \varepsilon_\nu^{-1} r_\nu^{-m} \int_{C_\nu} \beta [r_\nu^2 + s_\nu^2 + (t\, \varepsilon_\nu r_\nu \beta)^2 \\ &\qquad + |D g_\nu(x)|^2 + (t\, \varepsilon_\nu)^2 m \beta^2] \lambda \, d\mathscr{L}^m x \\ &\le \varepsilon_\nu \beta \lambda \alpha(m) [2 + t^2(1+m) \beta^2 + 3\gamma] \to 0 \quad \text{as} \quad \nu \to \infty, \end{aligned}$$

$$M_{\nu,3}(t) = \varepsilon_\nu^{-1} r_\nu^{-m} \int_{C_\nu} \langle [0,0,D\theta(r_\nu^{-1} x)] \odot [x, V_{\nu,t}(x), 0], D^2 \Psi_\nu^\S(0,0,0) \rangle \, d\mathscr{L}^m x,$$

$$\begin{aligned} |M_{\nu,3}(t)| &\le \varepsilon_\nu^{-1} \alpha(m) \beta (r_\nu + s_\nu + |t|\, \varepsilon_\nu r_\nu \beta) \lambda \\ &= \alpha(m) \beta \lambda (\varepsilon_\nu^{-1} r_\nu + \varepsilon_\nu^{-1} s_\nu + |t|\, r_\nu \beta) \to 0 \quad \text{as} \quad \nu \to \infty, \end{aligned}$$

$$M_{\nu,4}(t) = \varepsilon_\nu^{-1} r_\nu^{-m} \int_{C_\nu} \langle [0,0,D\theta(r_\nu^{-1} x)] \odot [0,0,D V_{\nu,t}(x)], D^2 \Psi_\nu^\S(0,0,0) \rangle \, d\mathscr{L}^m x.$$

Recalling 5.1.11 we expand

$$D\theta(r_\nu^{-1} x) = \sum_{i=1}^{m} \sum_{j=1}^{n-m} D_i \theta_j(r_\nu^{-1} x) \cdot X_i v_j,$$

$$D V_{\nu,t}(x) = \sum_{k=1}^{m} \sum_{l=1}^{n-m} [D_k g_{\nu,l}(x) + t\, \varepsilon_\nu D_k \theta_l(r_\nu^{-1} x)] X_k v_l$$

for $x \in C_\nu$, and obtain

$$\begin{aligned} M_{\nu,4}(t) = \sum_{i=1}^{m} \sum_{j=1}^{n-m} \sum_{k=1}^{m} \sum_{l=1}^{n-m} &(\Psi_\nu^\S)_{(i,j;k,l)}(0,0,0)\, r_\nu^{-m} \int_{C_\nu} \\ &D_i \theta_j(r_\nu^{-1} x) [\varepsilon_\nu^{-1} D_k g_{\nu,l}(x) + t\, D_k \theta_l(r_\nu^{-1} x)] \, d\mathscr{L}^m x. \end{aligned}$$

Since $\mathscr{L}^m[\mathbf{U}(0,r_\nu) \sim C_\nu] \beta^2 r_\nu^{-m} \le 5\gamma \varepsilon_\nu^2 \beta^2 \to 0$ as $\nu \to \infty$, we find that

$$\begin{aligned} &\lim_{\nu\to\infty} \int_{C_\nu} D_i \theta_j(r_\nu^{-1} x)\, D_k \theta_l(r_\nu^{-1} x)\, r_\nu^{-m} \, d\mathscr{L}^m x \\ &= \lim_{\nu\to\infty} \int_{\mathbf{U}(0,r_\nu)} D_i \theta_j(r_\nu^{-1} x)\, D_k \theta_l(r_\nu^{-1} x)\, r_\nu^{-m} \, d\mathscr{L}^m x \\ &= \int_{\mathbf{U}(0,1)} D_i \theta_j(\xi)\, D_k \theta_l(\xi) \, d\mathscr{L}^m \xi. \end{aligned}$$

Next we observe that, for $\phi \in \mathscr{D}^0[\mathbf{R}^m \cap \mathbf{U}(0,1)]$,

$$r_\nu^{-m}\,\varepsilon_\nu^{-1}\,|D_k\,U_{\nu,t}(\phi\circ\mu_{1/r_\nu}) - \kappa \textstyle\int_{C_\nu}(\phi\circ\mu_{1/r_\nu})\,D_k\,g_{\nu,t}\,d\mathscr{L}^m| \\ \le 5\kappa\gamma\,\varepsilon_\nu\,\mathbf{M}(\phi)\to 0 \quad \text{as} \quad \nu\to\infty,$$

and infer with the help of 5.3.6 that

$$\lim_{\nu\to\infty} r_\nu^{-m}\,\kappa \textstyle\int_{C_\nu}\phi(r_\nu^{-1}x)\,\varepsilon_\nu^{-1}\,D_k\,g_{\nu,t}(x)\,d\mathscr{L}^m\,x \\ = \lim_{\nu\to\infty} r_\nu^{-m}\,\varepsilon_\nu^{-1}\,D_k\,U_{\nu,t}(\phi\circ\mu_{1/r_\nu}) = \lim_{\nu\to\infty} D_k\,H_{\nu,t}(\phi) = D_k\,H_t(\phi).$$

We combine these results, taking $\phi = D_i\,\theta_j$, to obtain the first conclusion of the lemma. The second conclusion follows by integration with respect to t.

Applying the second conclusion to the special case when all Ψ_ν equal the area integrand, as we may according to the next to last paragraph of 5.1.9, we obtain the third conclusion.

To verify the fourth conclusion we define

$$N_\nu(t) = \varepsilon_\nu^{-2}\,r_\nu^{-m}\,\mathbf{M}^\kappa(h_{\nu,t\,\#}\,S_\nu) \text{ for } t\in\mathbf{R},$$

and infer by the method of 5.1.7 that

$$|N_\nu'(t)| \le \varepsilon_\nu^{-2}\,r_\nu^{-m} \textstyle\int m\,\|D\,\dot h_{\nu,t}\|\,d\,\|S_\nu\|^\kappa \\ \le \varepsilon_\nu^{-2}\,r_\nu^{-m}\,m\,\beta\,\varepsilon_\nu\,\gamma\,\varepsilon_\nu^2\,r_\nu^m = m\,\beta\,\gamma\,\varepsilon_\nu \to 0 \quad \text{as} \quad \nu\to\infty,$$

hence $N_\nu(t) - N_\nu(0)\to 0$ as $\nu\to\infty$.

Moreover

$$\sum_{i=1}^{m}\sum_{j=1}^{n-m} r_\nu^{-m}\,\kappa \textstyle\int_{C_\nu} w_{i,j}(r_\nu^{-1}x)\,\varepsilon_\nu^{-1}\,D_i\,g_{\nu,j}(x)\,d\mathscr{L}^m\,x \\ = r_\nu^{-m}\,\kappa\,\varepsilon_\nu^{-1} \textstyle\int_{C_\nu} w(r_\nu^{-1}x)\bullet D\,g_\nu(x)\,d\mathscr{L}^m\,x \\ \le \left(r_\nu^{-m}\textstyle\int |w(r_\nu^{-1}x)|^2\,d\mathscr{L}^m\,x\right)^{\frac12}\cdot\left(r_\nu^{-m}\,\kappa^2\,\varepsilon_\nu^{-2}\textstyle\int_{C_\nu}|D\,g_\nu|^2\,d\mathscr{L}^m\right)^{\frac12} \\ \le \left(\textstyle\int |w|^2\,d\mathscr{L}^m\right)^{\frac12}\cdot(3\kappa\gamma)^{\frac12}.$$

5.3.8. Lemma. *If $1\le\lambda<\infty$, $0<\gamma<\infty$, $0<\delta<\frac15$ and κ is a positive integer, then there exists a positive integer ν with the following property:*

If Ψ is a parametric integrand of degree m and class 3 on some neighborhood of

$$Z = (\mathbf{R}^m\times\mathbf{R}^{n-m})\cap\{(x,y)\colon |x|\le\lambda^{-1},\,|y|\le\lambda^{-1}\}$$

such that $D\,\Psi^\S(0,0,0)=0$, Ψ is elliptic at $(0,0)$ with ellipticity bound λ^{-1},

$$\lambda^{-1}\le\Psi(x,y,\zeta)\le\lambda, \qquad \|D^q\,\Psi(x,y,\zeta)\|\le\lambda, \qquad \|D^q\,\Psi^\S(x,y,\tau)\|\le\lambda$$

for $q \in \{0,1,2,3\}$, $(x,y) \in Z$, $\zeta \in \wedge_m(\mathbf{R}^m \times \mathbf{R}^{n-m})$ *with* $|\zeta| = 1$, $\tau \in \operatorname{Hom}(\mathbf{R}^m, \mathbf{R}^{n-m})$ *with* $\|\tau\| \le 2$, *and if*

$$S \in \mathscr{R}_m(\mathbf{R}^m \times \mathbf{R}^{n-m}), \quad 0 < s \le r < \lambda^{-1}, \quad \operatorname{spt} S \subset \{(x,y)\colon |x| \le r, |y| \le s\},$$
$$\operatorname{spt} \partial S \subset \{(x,y)\colon |x| = r\}, \quad p_\# S = \kappa\, \mathbf{E}^m \llcorner \{x\colon |x| < r\},$$
$$\varepsilon \ge 0, \quad \varepsilon^2 = \operatorname{Exc}(S,0,r) + \operatorname{Exc}^\kappa(S,0,r) < v^{-1},$$
$$\int |y|^2\, d\|S\|(x,y) \le \gamma^2 \varepsilon^2 r^{m+2}, \quad b \in \mathbf{R}^m, \quad |b| \le r - 5\delta r,$$

S *is absolutely* Ψ *minimizing with respect to* Z,

then either $\varepsilon \le v\, r$ *or there exist linear maps*

$$g\colon \mathbf{R}^m \to \mathbf{R}^{n-m} \quad \text{and} \quad G\colon \mathbf{R}^m \times \mathbf{R}^{n-m} \to \mathbf{R}^m \times \mathbf{R}^{n-m}$$

with

$$G(x,y) = (x, y - g(x)) \text{ for } (x,y) \in \mathbf{R}^m \times \mathbf{R}^{n-m},$$
$$|g| < \Gamma_{15}\, \lambda^{1+m/2} (1 - r^{-1}|b| - \delta)^{-m/2} \kappa^{-\frac{1}{2}} \varepsilon,$$
$$\operatorname{Exc}(G_\# S, b, \delta r) \le \Gamma_{16}\, \lambda^{m+4} (1 - r^{-1}|b| - 4\delta)^{-m-2} \delta^2 \varepsilon^2,$$

where $\Gamma_{15} = e(4+m)^{2+m/2} \boldsymbol{\alpha}(m)^{-\frac{1}{2}}$, $\Gamma_{16} = 3^m (30)^2 \Gamma_{15}^2 \boldsymbol{\alpha}(m)$.

Proof. Assuming the lemma false, we choose $\Psi_v, S_v, s_v, r_v, \varepsilon_v, b_v$ for each positive integer v so that the hypothesis of the stated property holds, but the conclusion fails, hence

$$\varepsilon_v^{-1} s_v \le \varepsilon_v^{-1} r_v < v^{-1}.$$

For $j \in \{1, \ldots, n-m\}$ we recall 5.3.5 and 5.3.6 to express

$$\mathbf{p}_\# (S_v \llcorner Z_{m+j}) = \kappa\, \mathbf{E}^m \llcorner f_{v,j}$$

with

$$f_{v,j}(x) = \kappa^{-1} \langle S_v, \mathbf{p}, x \rangle (Z_{m+j}) \text{ for } \mathscr{L}^m \text{ almost all } x,$$

and consider the currents

$$T_{v,j} = \varepsilon_v^{-1} r_v^{-1} \mu_{1/r_v \#}\, p_\# (S_v \llcorner Z_{m+j}) = \kappa\, \varepsilon_v^{-1} r_v^{-1} \mathbf{E}^m \llcorner (f_{v,j} \circ \mu_{r_v}),$$
$$U_{v,j} = \mathbf{p}_\# (S \llcorner Z_{m+j} \wedge DZ_1 \wedge \cdots \wedge DZ_m) = \kappa\, \mathscr{L}^m \llcorner f_{v,j},$$
$$H_{v,j} = \varepsilon_v^{-1} r_v^{-1-m} \mu_{1/r_v \#}\, U_{v,j} = \kappa\, \varepsilon_v^{-1} r_v^{-1} \mathscr{L}^m \llcorner (f_{v,j} \circ \mu_{r_v})$$
$$= T_{v,j} \llcorner (DX_1 \wedge \cdots \wedge DX_m).$$

We use Hölder's inequality, 5.3.5 (1) and 4.2.1 to estimate

$$\begin{aligned}
\mathbf{M}(T_{\nu,j}) &\le \varepsilon_\nu^{-1} r_\nu^{-1-m}\,\mathbf{M}(S_\nu \llcorner Z_{m+j}) = \varepsilon_\nu^{-1} r_\nu^{-1-m} \int |Z_{m+j}|\, d\|S\| \\
&\le \varepsilon_\nu^{-1} r_\nu^{-1-m} \big(\textstyle\int |Z_{m+j}|^2\, d\|S\| \cdot \mathbf{M}(S)\big)^{\frac12} \\
&\le \varepsilon_\nu^{-1} r_\nu^{-1-m} \big(\gamma^2 \varepsilon_\nu^2 r_\nu^{m+2} [\kappa\,\boldsymbol{\alpha}(m) + \varepsilon_\nu^2]\, r_\nu^m\big)^{\frac12} \\
&= \gamma\,[\kappa\,\boldsymbol{\alpha}(m) + \varepsilon_\nu^2]^{\frac12} \le \gamma\,[\kappa\,\boldsymbol{\alpha}(m) + 1]^{\frac12}, \\
\|\partial T_{\nu,j}\|\, \mathbf{U}(0,1) &\le \varepsilon_\nu^{-1} r_\nu^{-m}\, \|\partial(S_\nu \llcorner Z_{m+j})\|\, \mathbf{p}^{-1}\, \mathbf{U}(0, r_\nu) \\
&\le \varepsilon_\nu^{-1} r_\nu^{-m} [2\kappa\,\boldsymbol{\alpha}(m) + 2\varepsilon_\nu^2]^{\frac12}\, \varepsilon_\nu\, r_\nu^m \\
&= [2\kappa\,\boldsymbol{\alpha}(m) + 2\varepsilon_\nu^2]^{\frac12} \le [2\kappa\,\boldsymbol{\alpha}(m) + 2]^{\frac12}, \\
\int_0^1 \mathbf{M}(\partial [T_{\nu,j} \llcorner \mathbf{U}(0,\alpha)])\, d\mathscr{L}^1 \alpha &\le [\|T_{\nu,j}\| + \|\partial T_{\nu,j}\|]\, \mathbf{U}(0,1),
\end{aligned}$$

and infer from 2.4.6 that $\{\alpha : 0 < \alpha < 1\}$ has a countable dense subset D with

$$\liminf_{\nu\to\infty} \mathbf{N}[T_{\nu,j} \llcorner \mathbf{U}(0,\alpha)] < \infty \quad \text{for } \alpha \in D.$$

Applying 4.2.17 (1) and Tychonoff's theorem we pass to a subsequence (without changing notation) for which there exist currents

$$T_j \in \mathbf{N}_m^{\mathrm{loc}}[\mathbf{R}^m \cap \mathbf{U}(0,1)] \quad \text{with} \quad \mathbf{M}(T_j) \le \gamma\,[\kappa\boldsymbol{\alpha}(m)]^{\frac12}, \mathbf{M}(\partial T_j) \le [2\kappa\boldsymbol{\alpha}(m)]^{\frac12},$$

$$\lim_{\nu\to\infty} \mathbf{F}_{\mathbf{B}(0,\alpha)}[(T_{\nu,j} - T_j) \llcorner \mathbf{U}(0,\alpha)] = 0 \quad \text{for } \alpha \in D.$$

It follows that 5.3.7 (8) holds with

$$H_j = T_j \llcorner (D X_1 \wedge \cdots \wedge D X_m).$$

Moreover 4.1.18 yields $\mathscr{L}^m \llcorner \mathbf{U}(0,1)$ measurable real valued functions v_j such that

$$H_j(\phi) = T_j(\phi \wedge D X_1 \wedge \cdots \wedge D X_m) = \kappa \int_{\mathbf{U}(0,1)} \phi\, v_j\, d\mathscr{L}^m$$

for $\phi \in \mathscr{D}^0[\mathbf{R}^m \cap \mathbf{U}(0,1)]$ and

$$\lim_{\nu\to\infty} \int_{\mathbf{U}(0,\alpha)} |\varepsilon_\nu^{-1} r_\nu^{-1} (f_{\nu,j} \circ \mu_{r_\nu}) - v_j|\, d\mathscr{L}^m = 0$$

for $\alpha \in D$. We define

$$f_\nu(x) = (f_{\nu,1}(x), \ldots, f_{\nu,n-m}(x)) \in \mathbf{R}^{n-m} \quad \text{for } x \in \bigcap_{j=1}^{n-m} \operatorname{dmn} f_{\nu,j},$$

$$v(x) = (v_1(x), \ldots, v_{n-m}(x)) \in \mathbf{R}^{n-m} \quad \text{for } x \in \bigcap_{j=1}^{n-m} \operatorname{dmn} v_j.$$

Since $|(\Psi_\nu^\S)_{(i,j;k,l)}(0,0,0)| \le \lambda$ whenever $\{i,k\} \subset \{1,\ldots,m\}$ and $\{j,l\} \subset \{1,\ldots,n-m\}$ we may also assume, after passage to another subsequence, that there exist numbers $\Upsilon_{i,j;k,l}$ satisfying 5.3.7 (5). We let

$$\Upsilon\colon \operatorname{Hom}(\mathbf{R}^m, \mathbf{R}^{n-m}) \times \operatorname{Hom}(\mathbf{R}^m, \mathbf{R}^{n-m}) \to \mathbf{R}$$

be the bilinear symmetric function characterized by the equations

$$\Upsilon(X_i\,v_j, X_k\,v_l) = \Upsilon_{i,j;k,l} = \lim_{\nu\to\infty} \langle X_i\,v_j \odot X_k\,v_l, D^2(\Psi_\nu^\S)_{(0,0)}(0)\rangle,$$

and infer from 5.1.10 that

$$\int \langle D\theta(x)\odot D\theta(x), \Upsilon\rangle\, d\mathscr{L}^m x = \lim_{\nu\to\infty} \int \langle D\theta(x)\odot D\theta(x), (D^2\Psi_\nu^\S)_{(0,0)}(0)\rangle\, d\mathscr{L}^m x$$
$$\ge \lambda^{-1} \int |D\theta(x)|^2\, d\mathscr{L}^m x$$

for every function $\theta\in\mathscr{D}(\mathbf{R}^m, \mathbf{R}^{n-m})$.

Whenever θ, θ_j, h_ν satisfy 5.3.7 (9) we see from 5.1.7 that

$$\delta^{(1)}(S_\nu, \Psi_\nu, h_\nu) = 0 \text{ for all positive integers } \nu,$$

and apply the first conclusion of 5.3.7 with $t=0$ to obtain

$$0 = \sum_{i=1}^{m} \sum_{j=1}^{n-m} \sum_{k=1}^{m} \sum_{l=1}^{n-m} \Upsilon_{i,j;k,l}\, D_k\, H_l(D_i\,\theta_j).$$

Recalling 5.2.5, 5.2.6 and the last conclusion of 5.3.7 we infer that the functions $\kappa\, v_j$ representing the distributions H_j may be assumed analytic on $\mathbf{R}^m\cap\mathbf{U}(0,1)$ with

$$\left[\int_{\mathbf{U}(0,1)} |D(\kappa\, v)|^2\, d\mathscr{L}^m\right]^{\frac{1}{2}} \le (3\kappa)^{\frac{1}{2}},$$

hence $\int |D\,v|^2\, d[\mathscr{L}^m \llcorner \mathbf{U}(0,1)] \le 3\kappa^{-1}$, and

$$|D\,v(x)| \le \Gamma_{15}\,\kappa^{-\frac{1}{2}}\,\lambda^{1+m/2}(1-|x|)^{-m/2},$$
$$|D^2\,v(x)| \le \Gamma_{15}\,\kappa^{-\frac{1}{2}}\,\lambda^{2+m/2}(1-|x|)^{-m/2-1}$$

for $x\in\mathbf{R}^m\cap\mathbf{U}(0,1)$.

Since $|r_\nu^{-1}\,b_\nu| \le 1-5\delta$ for all ν, we may again pass to a subsequence in order to assure the existence of

$$\lim_{\nu\to\infty} r_\nu^{-1}\,b_\nu = \beta\in\mathbf{R}^m\cap\mathbf{B}(0, 1-5\delta).$$

We will show that for some integer ν the second alternative of the conclusion of the stated property holds with

$$g_\nu = \varepsilon_\nu\, D\,v(\beta)\in\operatorname{Hom}(\mathbf{R}^m, \mathbf{R}^{n-m})$$

and

$$G_\nu(x,y) = \big(x, y - g_\nu(x)\big) \text{ for } (x,y)\in\mathbf{R}^m\times\mathbf{R}^{n-m}.$$

We now choose a particular map

$$\theta\in\mathscr{D}(\mathbf{R}^m, \mathbf{R}^{n-m}) \quad\text{with}\quad \operatorname{spt}\theta\subset\mathbf{R}^m\cap\mathbf{U}(0,1),$$
$$\theta(x) = v(x) \text{ for } x\in\mathbf{R}^m\cap\mathbf{U}(0, 1-\delta),$$

define θ_j and $h_{\nu,t}$ as in 5.3.7 (9) and study the rectifiable currents

$$S_\nu^1 = h_{\nu,-1\#}\, S_\nu.$$

Since spt $\partial S_\nu^1 \subset \{(x,y): |x| = r_\nu\}$ we can apply 4.2.1 and 4.2.16 to choose σ so that

$$1-2\delta<\sigma<1-\delta, \quad \text{hence} \quad \sigma^{-1}<\tfrac{5}{3}$$

and

$$S_\nu^2 = S_\nu^1 \llcorner \{(x,y): |x|<\sigma r_\nu\} \in \mathbf{I}_m(\mathbf{R}^m \times \mathbf{R}^{n-m})$$

for all positive integers ν. We let

$$M_1 = 2+\kappa \int_{\mathbf{U}(0,1)} (|Dv \bullet D\theta| + 2^{-1}|D\theta|^2)\, d\mathscr{L}^m,$$

$$M_2 = \kappa[\Gamma_{11} M_1 2^m + \Gamma_{12}(n-m)^4 2^4], \qquad M_3 = \Gamma_{10} M_1 2^m,$$

$$M_4 = \Gamma_{14}[\kappa\,\boldsymbol{\alpha}(m) + 2M_2]^{\frac{1}{2}} (2M_2)^{\frac{1}{2}} + (n-m)\,\kappa\,\boldsymbol{\alpha}(m),$$

$$M_5 = 2m^2(n-m)^2 \lambda M_2 \mathbf{M}(\|D^2\theta\|),$$

$$M_6 = 2^m M_1 + M_2 + M_3, \qquad M_7 = \kappa \lambda^2 (M_5 + 13M_2 + 2M_3 + 2M_2)$$

and suppose μ is any number satisfying the conditions

$$0<\mu<\tfrac{1}{3}, \qquad \mu^{4m} < \kappa\,\boldsymbol{\alpha}(m)/5,$$

$$1-2\delta < (1-2\mu)\,\sigma, \qquad (1-\mu^2)^m > (1-2\mu)^m > \tfrac{1}{2}.$$

We note that $\mathbf{p}_\# S_\nu^1 = \mathbf{p}_\# S_\nu = \kappa\, \mathbf{E}^m \llcorner \mathbf{U}(0, r_\nu)$ *and*

$$\begin{aligned} &\varepsilon_\nu^{-2}[\operatorname{Exc}(S_\nu^1, 0, r_\nu) + \operatorname{Exc}^\kappa(S_\nu^1, 0, r_\nu)] \\ &= 1 + \varepsilon_\nu^{-2} r_\nu^{-m} [\mathbf{M}(S_\nu^1) - \mathbf{M}(S_\nu) + \mathbf{M}^\kappa(S_\nu^1) - \mathbf{M}^\kappa(S_\nu)] \\ &\to 1 + \kappa \int_{\mathbf{U}(0,1)} (-Dv \bullet D\theta + 2^{-1}|D\theta|^2)\, d\mathscr{L}^m < M_1 \end{aligned}$$

as $\nu \to \infty$ by virtue of the last two conclusions of 5.3.7; accordingly

$$\operatorname{Exc}(S_\nu^1, 0, r_\nu) + \operatorname{Exc}^\kappa(S_\nu^1, 0, r_\nu) < M_1 \varepsilon_\nu^2$$

for large ν. Recalling 4.3.2(7) and 5.3.5 we compute

$$\begin{aligned} \langle S_\nu^1, \mathbf{p}, x\rangle (Z_{m+j}) &= \langle S_\nu, \mathbf{p}, x\rangle (Z_{m+j} \circ h_{\nu,-1}) \\ &= \langle S_\nu, \mathbf{p}, x\rangle [Z_{m+j} - \varepsilon_\nu r_\nu \theta_j(r_\nu^{-1} x)] \\ &= \kappa[f_{\nu,j}(x) - \varepsilon_\nu r_\nu \theta_j(r_\nu^{-1}x)] \end{aligned}$$

for $\mathscr{L}^m$ almost all x. Inasmuch as

$$\langle S^2, \mathbf{p}, x\rangle = \langle S^1, \mathbf{p}, x\rangle \text{ when } |x|<\sigma r_\nu, \quad \langle S^2, \mathbf{p}, x\rangle = 0 \text{ when } |x|>\sigma r_\nu,$$

we infer that

$$\begin{aligned} &r_\nu^{-m-1} \varepsilon_\nu^{-1} \kappa^{-1} \mathbf{M}[\mathbf{p}_\# (S_\nu^2 \llcorner Z_{m+j})] \\ &= r_\nu^{-m} \int |\varepsilon_\nu^{-1} r_\nu^{-1} \kappa^{-1} \langle S_\nu^2, \mathbf{p}, x\rangle (Z_{m+j})|\, d\mathscr{L}^m x \\ &\le \int_{\mathbf{U}(0,1-\delta)} |\varepsilon_\nu^{-1} r_\nu^{-1} f_{\nu,j}(r_\nu x) - v_j(x)|\, d\mathscr{L}^m x \to 0 \end{aligned}$$

as $\nu\to\infty$. Assuming ν so large that the last integral does not exceed $\mu^{4m+2}\,\sigma^m/[(n-m)\,2\kappa]$ for $j\in\{1,\ldots,n-m\}$, we define

$$B_{\nu,j}=\mathbf{U}(0,\sigma\,r_\nu)\cap\{x\colon |\kappa^{-1}\langle S_\nu^2,\mathbf{p},x\rangle\,Z_{m+j}|<\varepsilon_\nu\,r_\nu\,\mu^2\}$$

and obtain the inequality

$$\mathscr{L}^m\Big[\mathbf{U}(0,\sigma\,r_\nu)\sim\bigcap_{j=1}^{n-m}B_{\nu,j}\Big]\le\mu^{4m}\,\sigma^m\,r_\nu^m/(2\kappa).$$

Then we let

$$A_\nu=\bigcap_{j=1}^{n-m}\{(x,y)\colon x\in B_{\nu,j},\,y_j=\kappa^{-1}\langle S_\nu^2,\mathbf{p},x\rangle\,(Z_{m+j})\}$$

and apply 5.3.5(4) to S_ν^2 in estimating

$$\begin{aligned}\|S_\nu^2\|\,A_\nu&\ge\kappa\,\mathscr{L}^m[\mathbf{p}(A_\nu)]-\kappa[5\,\mathrm{Exc}(S_\nu^2,0,\sigma\,r_\nu)+\mathrm{Exc}^\kappa(S_\nu^2,0,\sigma\,r_\nu)]\,(\sigma\,r_\nu)^m\\&\ge(\sigma\,r_\nu)^m\,[\kappa\,\boldsymbol{\alpha}(m)-\mu^{4m}/2]-\kappa\,5\,M_1\,\varepsilon_\nu^2\,r_\nu^m,\end{aligned}$$

hence

$$\begin{aligned}\|S_\nu^2\|\,\{(x,y)\colon |y|\ge(n-m)\,\varepsilon_\nu\,r_\nu\,\mu^2\}&\le\mathbf{M}(S_\nu^2)-\|S_\nu^2\|\,A_\nu\\&\le(\sigma\,r_\nu)^m\,[\mathrm{Exc}(S_\nu^2,0,\sigma_\nu\,r_\nu)+\mu^{4m}/2]+\kappa\,5\,M_1\,\varepsilon_\nu^2\,r_\nu^m\\&\le(\sigma\,r_\nu)^m\,\mu^{4m}/2+\kappa\,6\,M_1\,\varepsilon_\nu^2\,r_\nu^m\le(\sigma\,r_\nu)^m\,\mu^{4m}\end{aligned}$$

for large ν. Recalling 5.3.4 we define

$$\alpha_{\nu,j}^+=\inf\{t\colon(\sigma\,r_\nu)^{-m}\,\|S_\nu^2\|\,W_j^+(t)\le3\,\kappa\,\boldsymbol{\alpha}(m)/5\},$$
$$\alpha_{\nu,j}^-=\sup\{t\colon(\sigma\,r_\nu)^{-m}\,\|S_\nu^2\|\,W_j^-(t)\le3\,\kappa\,\boldsymbol{\alpha}(m)/5\},$$

observe that $-(n-m)\,\varepsilon_\nu\,r_\nu\,\mu^2\le\alpha_{\nu,j}^+\le\alpha_{\nu,j}^-\le(n-m)\,\varepsilon_\nu\,r_\nu\,\mu^2$ for large ν, choose

$$\eta_\nu=(\eta_{\nu,1},\ldots,\eta_{\nu,n-m})\in\mathbf{R}^{n-m}$$

with

$$\alpha_{\nu,j}^+\le\eta_{\nu,j}\le\alpha_{\nu,j}^-\ \text{ for } j\in\{1,\ldots,n-m\},$$

let $\omega_\nu=2(n-m)^2\,\varepsilon_\nu\,\mu^2\,\sigma^{-1}$, infer

$$(\sigma\,r_\nu)^{-m}\,\|S_\nu^2\|\,\{(x,y)\colon |y-\eta|\ge\omega_\nu\,\sigma\,r_\nu\}\le\mu^{4m},$$

and apply 5.3.4(9, II) with S, r replaced by $S_\nu^2, \sigma\,r_\nu$. The resulting numbers ρ_ν and currents P_ν, Q_ν, R_ν satisfy the conditions

$$r_\nu-2\delta\,r_\nu<(1-2\mu)\,\sigma\,r_\nu<\rho_\nu<(1-\mu)\,\sigma\,r_\nu<r_\nu-\delta\,r_\nu,$$

$$\begin{aligned}\mathrm{Exc}(P_\nu+Q_\nu+R_\nu,0,\sigma\,r_\nu)&\le\kappa[\Gamma_{11}\,\mu\,\mathrm{Exc}(S_\nu^2,0,r_\nu)+\Gamma_{12}\,\mu^{-3}\,\omega_\nu^2]\\&\le\kappa[\Gamma_{11}\,M_1\,\sigma^{-m}+\Gamma_{12}\,4(n-m)^4\,\sigma^{-2}]\,\mu\,\varepsilon_\nu^2\le M_2\,\mu\,\varepsilon_\nu^2,\end{aligned}$$

$$\mathrm{Exc}^\kappa(P_\nu+Q_\nu+R_\nu,0,\sigma\,r_\nu)\le\Gamma_{10}\,\mu\,M_1\,\varepsilon_\nu^2\,\sigma^{-m}\le M_3\,\mu\,\varepsilon_\nu^2.$$

We further define

$$S_\nu^3 = P_\nu + R_\nu, \quad S_\nu^4 = P_\nu + R_\nu + S_\nu^1 \llcorner \{(x,y): |x| > \rho_\nu\},$$
$$S_\nu^5 = h_{\nu,1\#} S_\nu^4, \quad S_\nu^6 = S_\nu^1 \llcorner \{(x,y): |x| < \rho_\nu\} - R_\nu + Q_\nu,$$
$$S_\nu^7 = \kappa(\mathbf{E}^m \llcorner \{x: |x| < \sigma r_\nu\}) \times \boldsymbol{\delta}_{\eta_\nu},$$

and for $q \in \{1, \ldots, 7\}$, $j \in \{1, \ldots, n-m\}$ we let

$$T_{\nu,j}^q = \varepsilon_\nu^{-1} r_\nu^{-1} \mu_{1/r_\nu\#} \mathbf{p}_\# (S_\nu^q \llcorner Z_{m+j}),$$
$$U_{\nu,j}^q = \mathbf{p}_\# (S \llcorner Z_{m+j} \wedge D Z_1 \wedge \cdots \wedge D Z_m),$$
$$H_{\nu,j}^q = \varepsilon_\nu^{-1} r_\nu^{-1-m} \mu_{1/r_\nu\#} U_{\nu,j}^q = T_{\nu,j}^q \llcorner (D X_1 \wedge \cdots \wedge D X_m).$$

Our previous computation of $\langle S_\nu^1, \mathbf{p}, x \rangle (Z_{m+j})$ shows that

$$H_{\nu,j}^1 = \kappa \mathscr{L}^m \llcorner [\varepsilon_\nu^{-1} r_\nu^{-1} (f_{\nu,j} \circ \mu_{r_\nu}) - \theta_j],$$

hence for each j the distributions $H_{\nu,j}^1 | \mathscr{D}^0[\mathbf{R}^m \cap \mathbf{U}(0,1)]$ converge to $\mathscr{L}^m \llcorner (v_j - \theta_j) = H_j^1$ in $\mathscr{D}_0[\mathbf{R}^m \cap \mathbf{U}(0,1)]$ as $\nu \to \infty$, and that

$$\mathbf{M}(H_{\nu,j}^2) = \mathbf{M}(T_{\nu,j}^2) \to 0 \text{ as } \nu \to \infty.$$

Observing that

$$\langle S_\nu^3, \mathbf{p}, x \rangle (Z_{m+j}) = \kappa \eta_{\nu,j} \text{ for } |x| < \rho_\nu - \mu^2 \rho_\nu,$$

with $(\rho_\nu - \mu^2 \rho_\nu)^m = \rho_\nu^m (1-\mu^2)^m > \rho_\nu^m / 2$, hence

$$\mathscr{L}^m \{x: \kappa^{-1} \langle S_\nu^3, \mathbf{p}, x \rangle (Z_{m+j}) = \eta_{\nu,j}\} \ge \boldsymbol{\alpha}(m) \rho_\nu^m / 2,$$

and that $\operatorname{Exc}(S_\nu^3, 0, \rho_\nu) \le (1-2\mu)^{-m} M_2 \mu \varepsilon_\nu^2 < 2 M_2 \mu \varepsilon_\nu^2$ we see from 5.3.5 (3) that

$$\mathbf{M}[\mathbf{p}_\# (S_\nu^3 \llcorner Z_{m+j}) - \kappa \eta_{\nu,j} \mathbf{E}^m \llcorner \mathbf{U}(0, \rho_\nu)]$$
$$\le \Gamma_{14} [\kappa \boldsymbol{\alpha}(m) + 2 M_2 \mu \varepsilon_\nu^2]^{\frac{1}{2}} [2 M_2 \mu \varepsilon_\nu^2]^{\frac{1}{2}} \rho_\nu^{m+1},$$

and since $|\eta_{\nu,j}| \le (n-m) \varepsilon_\nu r_\nu \mu^2$ we obtain

$$\mathbf{M}(T_{\nu,j}^3) = \varepsilon_\nu^{-1} r_\nu^{-1-m} \mathbf{M}[p_\# (S_\nu^3 \llcorner Z_{m+j})] \le M_4 \mu^{\frac{1}{2}}.$$

Then we see that

$$T_{\nu,j}^4 = T_{\nu,j}^3 + T_{\nu,j}^1 \llcorner \{x: |x| > \rho_\nu / r_\nu\},$$

$$\mathbf{M}[T_{\nu,j}^4 \llcorner \mathbf{U}(0,\alpha)] \le M_4 + \| T_{\nu,j}^1 \| \mathbf{U}(0,\alpha) \to M_4 + \int_{\mathbf{U}(0,\alpha)} |v_j - \theta_j| \, d\mathscr{L}^m < \infty$$

as $\nu \to \infty$ whenever $0<\alpha<1$, and that $\partial S_\nu^4 = \partial S_\nu^1$,

$$r_\nu^m \operatorname{Exc}(S_\nu^4, 0, r_\nu) \le \rho_\nu^m \operatorname{Exc}(P_\nu + R_\nu, 0, \rho_\nu) + r_\nu^m \operatorname{Exc}(S_\nu^1, 0, r_\nu)$$
$$\le r_\nu^m (M_2\, \mu\, \varepsilon_\nu^2 + M_1\, \varepsilon_\nu^2),$$
$$\|\partial T_{\nu,j}^4\|\, \mathbf{U}(0,1) \le \varepsilon_\nu^{-1}\, r_\nu^{-m}\, \|\partial(S_\nu^4 \llcorner Z_{m+j})\|\, p^{-1}\, \mathbf{U}(0, r_\nu)$$
$$\le [2\kappa\, \alpha(m) + M_2 + M_1]^{\frac{1}{2}} (M_2 + M_1)^{\frac{1}{2}}$$

by 5.3.5(1); furthermore

$$\mathbf{M}^\kappa(S_\nu^4) \le \mathbf{M}^\kappa(P_\nu + Q_\nu) + \mathbf{M}^\kappa(S_\nu^1) \le (M_3\, \mu + M_1)\, \varepsilon_\nu^2\, r_\nu^m.$$

Once more we replace our sequences by subsequences, applying 4.2.1, 2.4.6, 4.2.17(1) and Tychonoff's theorem, so as to assure the existence of currents

$$T_j^4 \in \mathbf{N}_m^{\mathrm{loc}}[\mathbf{R}^m \cap \mathbf{U}(0,1)] \quad \text{with} \quad \lim_{\nu\to\infty} \mathbf{F}_{\mathbf{B}(0,\alpha)}[(T_{\nu,j}^4 - T_j^4) \llcorner \mathbf{U}(0,\alpha)] = 0$$

for all α in a dense subset of $\{\alpha\colon 0<\alpha<1\}$. It follows that the distributions $H_{\nu,j}^4 | \mathscr{D}^0[\mathbf{R}^m \cap \mathbf{U}(0,1)]$ converge to

$$T_j^4 \llcorner (D X_1 \wedge \cdots \wedge D X_m) = H_j^4 \in \mathscr{D}_0[\mathbf{R}^m \cap \mathbf{U}(0,1)]$$

as $\nu \to \infty$. Moreover

$$S_\nu^4 - S_\nu^1 = S_\nu^3 - S_\nu^2 \llcorner \{(x,y)\colon |x| < \rho_\nu\},$$
$$H_{\nu,j}^4 - H_{\nu,j}^1 = H_{\nu,j}^3 - H_{\nu,j}^2 \llcorner \mathbf{U}(0, \rho_\nu/r_\nu)\},$$
$$\operatorname{spt}(H_{\nu,j}^4 - H_{\nu,j}^1) \subset \mathbf{B}(0, \rho_\nu/r_\nu) \subset \mathbf{U}(0,\sigma),$$
$$\mathbf{M}(H_j^4 - H_j^1) \le \liminf_{\nu\to\infty} [\mathbf{M}(H_{\nu,j}^3) + \mathbf{M}(H_{\nu,j}^2)]$$
$$= \liminf_{\nu\to\infty} [\mathbf{M}(T_{\nu,j}^3) + \mathbf{M}(T_{\nu,j}^2)] \le M_4\, \mu^{\frac{1}{2}}.$$

Noting that $h_{1,t\,\#}\, S_\nu^1 = S_\nu$, $\partial S_\nu^5 = \partial S_\nu$, $\operatorname{spt} S_\nu^5 \subset Z$ for large ν, and S_ν is Ψ_ν minimizing with respect to Z, hence

$$\langle \Psi_\nu, S_\nu^5\rangle - \langle \Psi_\nu, S_\nu\rangle \ge 0,$$

we apply the second conclusion of 5.3.7 twice, replacing S_ν by S_ν^4 and by S_ν^1, and taking $t=1$, to estimate

$$\varepsilon_\nu^{-2}\, r_\nu^{-m} [\langle \Psi_\nu, S_\nu^1\rangle - \langle \Psi_\nu, S_\nu^4\rangle]$$
$$\le \varepsilon_\nu^{-2}\, r_\nu^{-m} ([\langle \Psi_\nu, S_\nu^5\rangle - \langle \Psi_\nu, S_\nu^4\rangle] - [\langle \Psi_\nu, S_\nu\rangle - \langle \Psi_\nu, S_\nu^1\rangle])$$
$$\to \sum_{i=1}^{m} \sum_{j=1}^{n-m} \sum_{k=1}^{m} \sum_{l=1}^{n-m} \Upsilon_{i,j;k,l} [D_k H_l^4 - D_k H_l^1](D_i \theta_j)$$
$$= \sum_{i=1}^{m} \sum_{j=1}^{n-m} \sum_{k=1}^{m} \sum_{l=1}^{n-m} \Upsilon_{i,j;k,l} [H_l^4 - H_l^1](-D_{k,i}\, \theta_j)$$
$$\le m^2 (n-m)^2\, \lambda\, M_2\, \mu^{\frac{1}{2}}\, \mathbf{M}(\|D^2 \theta\|) = M_5\, \mu^{\frac{1}{2}}/2$$

as $\nu \to \infty$, and infer with the help of 5.3.5(5) that, when ν is sufficiently large,

$$\begin{aligned}
\varepsilon_\nu^2 r_\nu^m M_5 \mu^{\frac{1}{2}} &\geq \langle \Psi_\nu, S_\nu^1 \rangle - \langle \Psi_\nu, S_\nu^4 \rangle \\
&= \langle \Psi_\nu, S_\nu^1 \llcorner \{(x,y): |x| < \rho_\nu\} \rangle - \langle \Psi_\nu, P_\nu \rangle - \langle \Psi_\nu, R_\nu \rangle \\
&= \langle \Psi_\nu, S_\nu^6 \rangle - \langle \Psi_\nu, P_\nu + Q_\nu \rangle - \langle \Psi_\nu, -R_\nu \rangle - \langle \Psi_\nu, R_\nu \rangle \\
&\geq [\langle \Psi_\nu, S_\nu^6 \rangle - \langle \Psi_\nu, S_\nu^7 \rangle] - [\langle \Psi_\nu, P_\nu + Q_\nu \rangle - \langle \Psi_\nu, S_\nu^7 \rangle] - 2\lambda \mathbf{M}(R_\nu) \\
&\geq \lambda^{-1} [\mathbf{M}(S_\nu^6) - \mathbf{M}(S_\nu^7)] - (4\kappa \lambda r_\nu \varepsilon_\nu [\alpha(m)^{\frac{1}{2}} M_6^{\frac{1}{2}} + 6 M_6] \\
&\quad + 8\kappa \lambda r_\nu^2 \alpha(m) + \lambda \kappa [4 r_\nu^2 \alpha(m) + (13 M_2 + 2 M_3) \mu \varepsilon_\nu^2] + 2\lambda M_2 \mu \varepsilon_\nu^2) r_\nu^m
\end{aligned}$$

because

$$\begin{aligned}
&\operatorname{Exc}(S_\nu^6, 0, \sigma r_\nu) + \operatorname{Exc}^\kappa(S_\nu^6, 0, \sigma r_\nu) \\
&\leq \sigma^{-m} [\operatorname{Exc}(S_\nu^1, 0, r_\nu) + \operatorname{Exc}^\kappa(S_\nu^1, 0, r_\nu)] \\
&\quad + \operatorname{Exc}(P_\nu + Q_\nu + R_\nu, 0, \sigma r_\nu) + \operatorname{Exc}^\kappa(P_\nu + Q_\nu + R_\nu, 0, \sigma r_\nu) \\
&\leq 2^m M_1 \varepsilon_\nu^2 + M_2 \mu \varepsilon_\nu^2 + M_3 \mu \varepsilon_\nu^2 \leq M_6 \varepsilon_\nu^2.
\end{aligned}$$

Since $\varepsilon_\nu^{-1} r_\nu < \nu^{-1}$ it follows that

$$\limsup_{\nu \to \infty} \varepsilon_\nu^{-2} r_\nu^{-m} [\mathbf{M}(S_\nu^6) - \mathbf{M}(S_\nu^7)] \leq M_7 \mu^{\frac{1}{2}}.$$

Observing that

$$(1 - 2\delta)^m \operatorname{Exc}(S_\nu^1, 0, r_\nu - 2\delta r_\nu) \leq \sigma^m \operatorname{Exc}(S_\nu^6, 0, \sigma r_\nu),$$

that δ, S_ν^1, σ, M_7 are independent of μ, and that μ may be taken arbitrarily small, we conclude

$$\lim_{\nu \to \infty} \varepsilon_\nu^{-2} \operatorname{Exc}(S_\nu^1, 0, r_\nu - 2\delta r_\nu) = 0.$$

Finally we choose $\psi \in \mathscr{D}^0(\mathbf{R}^m)$ with $\operatorname{spt} \psi \subset \mathbf{U}(\beta, 3\delta)$,

$$\psi(x) = 1 \text{ for } x \in \mathbf{B}(\beta, 2\delta), \quad \mathbf{M}(\psi) = 1, \quad \mathbf{M}(D\psi) \leq 2,$$

define $w \in \mathscr{D}(\mathbf{R}^m, \mathbf{R}^{n-m})$ with $\operatorname{spt} w \subset \operatorname{spt} \psi$ so that

$$w(x) = \psi(x) [v(x) - v(\beta) - \langle x - \beta, D v(\beta) \rangle] \text{ for } x \in \mathbf{U}(0,1),$$

and let $w_j = Y_j \circ w$ for $j \in \{1, \ldots, n-m\}$,

$$W_\nu(x,y) = (x, y + \varepsilon_\nu r_\nu w(r_\nu^{-1} x))$$

for $(x,y) \in \mathbf{R}^m \times \mathbf{R}^{n-m}$ and each positive integer ν. Since w_j and W_ν are constructed from w just as θ_j and h_1 were constructed from θ in 5.3.7(9), we can apply the third conclusion of 5.3.7 with S_ν replaced by S_ν^1 to compute

$$\lim_{\nu\to\infty} \varepsilon_\nu^{-2} r_\nu^{-m}[\mathbf{M}(W_{\nu\,\#} S_\nu^1)-\mathbf{M}(S_\nu^1)]$$

$$=\sum_{i=1}^{m}\sum_{j=1}^{n-m}[D_i H_j^1(D_i w_j)+2^{-1}\kappa\int (D_i w_j)^2\, d\mathscr{L}^m]=2^{-1}\kappa\int_{\mathbf{U}(\beta,3\delta)}|D w|^2\, d\mathscr{L}^m$$

$$\le 2^{-1}\Gamma_{15}^2\,\lambda^{4+m}(1-|\beta|-3\delta)^{-m-2}(30\delta)^2\,\boldsymbol{\alpha}(m)(3\delta)^m$$

because $\operatorname{spt} w\subset\mathbf{U}(\beta,3\delta)\subset\mathbf{U}(0,|\beta|+3\delta)\subset\mathbf{U}(0,1-2\delta)$,

$$\|H_j^1\|\,\mathbf{U}(0,1-2\delta)=\int_{\mathbf{U}(0,1-2\delta)}|v_j-\theta_j|\,d\mathscr{L}^m=0,$$

$$|D^2 v(x)|\le\Gamma_{15}\,\kappa^{-\frac12}\lambda^{2+m/2}(1-|\beta|-3\delta)^{-m/2-1}$$

for $x\in\mathbf{U}(\beta,3\delta)$, hence

$$|D w(x)|\le|D\psi(x)|\cdot|v(x)-v(\beta)-\langle x-b, D v(\beta)\rangle|+|\psi(x)|\cdot|D v(x)-D v(\beta)|$$

$$\le\Gamma_{15}\,\kappa^{-\frac12}\lambda^{2+m/2}(1-|\beta|-3\delta)^{-m/2-1}(18\delta^2+12\delta)$$

for $x\in\mathbf{U}(\beta,3\delta)$ by Taylor's theorem. Noting that

$$W_\nu(x,y)=(x,y) \text{ when } x\notin\mathbf{U}(r_\nu\beta,3\delta r_\nu),$$

$$(3\delta r_\nu)^m \operatorname{Exc}(W_{\nu\,\#} S_\nu^1, r_\nu\beta, 3\delta r_\nu)$$

$$\le\mathbf{M}(W_{\nu\,\#} S_\nu^1)-\mathbf{M}(S_\nu^1)+(r_\nu-2\delta r_\nu)^m \operatorname{Exc}(S_\nu^1,0,r_\nu-2\delta r_\nu),$$

we combine the two preceding limit formulae to estimate

$$\limsup_{\nu\to\infty}\varepsilon_\nu^{-2}\operatorname{Exc}(W_{\nu\,\#} S_\nu^1, r_\nu\beta, 3\delta r_\nu)$$

$$\le 2^{-1}(30)^2\,\Gamma_{15}^2\,\boldsymbol{\alpha}(m)\,\lambda^{m+4}(1-|\beta|-3\delta)^{-m-2}\delta^2.$$

If ν is sufficiently large, then $|\beta-r_\nu^{-1}b_\nu|<\delta$,

$$|g_\nu|=|\varepsilon_\nu D v(\beta)|<\varepsilon_\nu\Gamma_{15}\,\kappa^{-\frac12}\lambda^{1+m/2}(1-|r_\nu^{-1}b_\nu|-\delta)^{-m/2},$$

$$\mathbf{U}(b_\nu,\delta r_\nu)\subset\mathbf{U}(r_\nu\beta,2\delta r_\nu)\subset\mathbf{U}(0,r_\nu-\delta r_\nu),$$

hence it is true for all $(x,y)\in\mathbf{U}(b_\nu,\delta r_\nu)\times\mathbf{R}^{n-m}$ that

$$h_{\nu,-1}(x,y)=(x,y-\varepsilon_\nu r_\nu v(r_\nu^{-1}x)),$$

$$W_\nu(x,y)=(x,y+\varepsilon_\nu r_\nu v(r_\nu^{-1}x)-\varepsilon_\nu r_\nu v(\beta)-g_\nu(x-r_\nu\beta)),$$

$$(W_\nu\circ h_{\nu,-1})(x,y)=G_\nu(x,y)+(0,g_\nu(r_\nu\beta)-\varepsilon_\nu r_\nu v(\beta)),$$

and we infer

$$\operatorname{Exc}(G_{\#} S_\nu, b_\nu,\delta r_\nu)=\operatorname{Exc}(W_{\nu\,\#} S_\nu^1, b_\nu,\delta r_\nu)\le 3^m\operatorname{Exc}(W_{\nu\,\#} S_\nu^1, r_\nu\beta, 3\delta r_\nu)$$

$$<3^m(30)^2\,\Gamma_{15}^2\,\boldsymbol{\alpha}(m)\,\lambda^{m+4}(1-|\beta|-3\delta)^{-m-2}\delta^2\varepsilon_\nu^2.$$

5.3.9. Lemma. *There exists a positive integer Γ_{17} with the following property:*

If r, s, t, σ are finite positive numbers, ϕ is a function of class 2 mapping some neighborhood of

$$A = (\mathbf{R}^m \times \mathbf{R}^{n-m}) \cap \{(x, y): |x| \le r, |y| \le s\}$$

into $\mathbf{R}^m \times \mathbf{R}^{n-m}$ such that $\phi(0,0) = (0,0)$,

$$D\phi(0,0) \text{ is univalent}, \quad \mathbf{q} \circ D\phi(0,0) \circ \mathbf{p}^* = 0,$$

$$\|D\phi(0,0)\| \le t, \quad \|D\phi(0,0)^{-1}\| \le t, \quad \|D^2\phi(x,y)\| \le \sigma \text{ for } (x,y) \in A,$$

and if $S \in \mathbf{I}_m(\mathbf{R}^m \times \mathbf{R}^{n-m})$, $\operatorname{spt} S \subset A$, κ is a positive integer,

$$r^{-m} \|S\| \{(x,y): y_j > 0\} \le 3\kappa\,\boldsymbol{\alpha}(m)/5 \text{ for } j \in \{1, \ldots, n-m\},$$
$$r^{-m} \|S\| \{(x,y): y_j < 0\} \le 3\kappa\,\boldsymbol{\alpha}(m)/5 \text{ for } j \in \{1, \ldots, n-m\},$$
$$\operatorname{spt} \partial S \subset \{(x,y): |x| = r\}, \quad \|S\| \{(x,y): |x| = r\} = 0,$$
$$\mathbf{p}_\# S = \kappa\,\mathbf{E}^m \llcorner \mathbf{U}(0,r), \quad \operatorname{Exc}^\kappa(S,0,r) < \boldsymbol{\alpha}(m)/5,$$
$$E = \operatorname{Exc}(S,0,r) < \inf\{\kappa\,\boldsymbol{\alpha}(m)/5, \Gamma_6/3\}, \quad t\,\sigma\, r \le 5^{-2}, \quad t^2(n-m)\,s \le 5^{-2} r,$$

then $\operatorname{spt}(\partial \phi_\# S) \subset \{(x,y): |x| \ge r/(4t)\}$,

$$\mathbf{p}_\# [(\phi_\# S) \llcorner \{(x,y): |x| < r/(4t)\}]$$
$$= \operatorname{sign}(\det[\mathbf{p} \circ D\phi(0,0) \circ \mathbf{p}^*])\,\kappa\,\mathbf{E}^m \llcorner \{x: |x| < r/(4t)\},$$
$$\operatorname{Exc}[\phi_\# S, 0, r/(4t)] \le \Gamma_{17}\, t^{8m} \kappa (E + \kappa\,\sigma\, t\, r).$$

Proof. We assume $\det[\mathbf{p} \circ D\phi(0,0) \circ \mathbf{p}^*] \ge 0$. First we apply 5.3.4 (9, I) with $\eta = 0$ and $\mu = \frac{1}{3}$, let

$$T = [S \llcorner \mathbf{p}^{-1} \mathbf{U}(0,\rho)] - R + Q, \quad D = \kappa\,\mathbf{p}^*_\# [\mathbf{E}^m \llcorner \mathbf{U}(0,r)]$$

and infer that $\partial T = \partial D$,

$$\operatorname{spt} T \cup \operatorname{spt} D \subset Z = \{(x,y): |x| \le r, |y| \le (n-m)s\},$$
$$r^{-m}[\mathbf{M}(T) - \mathbf{M}(D)] = \operatorname{Exc}(T, 0, r) \le \alpha\,\kappa\, E$$

with $\alpha = (1 + 27\Gamma_{10})$. Abbreviating $\theta = D\phi(0,0)$ we next use the inequality in the last paragraph of 5.1.3 with f, R, S replaced by $\theta^{-1}, \theta_\# T, \theta_\# D$ to obtain

$$t^{7m}[\mathbf{M}(T) - \mathbf{M}(D)] \ge \mathbf{M}(\theta_\# T) - \mathbf{M}(\theta_\# D).$$

Since $\mathbf{q} \circ \theta \circ \mathbf{p}^* = 0$ we find that

$$|(\mathbf{p} \circ \theta \circ \mathbf{p}^*)\, x| = |(\theta \circ \mathbf{p}^*)\, x| \geq |x|/t \quad \text{for} \quad x \in \mathbf{R}^m,$$

hence $\mathbf{R}^m \cap \mathbf{U}(0, r/t) \subset (\mathbf{p} \circ \theta \circ \mathbf{p}^*)\, \mathbf{U}(0, r)$,

$$(\operatorname{spt} \partial \theta_{\#}\, T) \cap \mathbf{p}^{-1}\, \mathbf{U}(0, r/t) = \varnothing$$

because $\operatorname{spt} \partial \theta_{\#}\, T = (\theta \circ \mathbf{p}^*)\{x\colon |x| = r\}$. Inasmuch as $\partial \mathbf{p}_{\#}\, \theta_{\#}\, T = \partial \mathbf{p}_{\#}\, \theta_{\#} D$ and $\det(\mathbf{p} \circ \theta \circ \mathbf{p}^*) \geq 0$ we infer

$$\mathbf{p}_{\#}\, \theta_{\#}\, T = \mathbf{p}_{\#}\, \theta_{\#}\, D = \kappa (\mathbf{p} \circ \theta \circ \mathbf{p}^*)_{\#}\, \mathbf{E}^m \llcorner \mathbf{U}(0, r) = \kappa\, \mathbf{E}^m \llcorner [(\mathbf{p} \circ \theta \circ \mathbf{p}^*)\mathbf{U}(0, r)].$$

Defining $U = \theta_{\#}\, T \llcorner \mathbf{p}^{-1}\, \mathbf{U}(0, r/t)$ we note that

$$|\mathbf{q}[\theta(x, y)]| = |\mathbf{q}[\theta(0, y)]| \leq t\, |y| \quad \text{for} \quad (x, y) \in \mathbf{R}^m \times \mathbf{R}^{n-m},$$
$$\operatorname{spt} U \subset \{(x, y)\colon |x| \leq r/t,\ |y| \leq (n-m)\, s\, t\},$$
$$\operatorname{spt} \partial U \subset \{(x, y)\colon |x| = r/t\}, \qquad \mathbf{p}_{\#}\, U = \kappa\, \mathbf{E}^m \llcorner \mathbf{U}(0, r/t),$$
$$\operatorname{Exc}(U, 0, r/t) \leq (t/r)^m\, [\mathbf{M}(\theta_{\#}\, T) - \mathbf{M}(\theta_{\#}\, D)]$$
$$\leq t^{8m}\, r^{-m}\, [\mathbf{M}(T) - \mathbf{M}(D)] \leq t^{8m}\, \alpha\, \kappa\, E.$$

Now we apply 5.3.5(6) to the function $\psi = \phi \circ \theta^{-1}$ with S, r, s, σ replaced by $U, r/t, (n-m)\, s\, t, t^2\, \sigma$ and estimate

$$\operatorname{Exc}(\psi_{\#}\, U, 0, r/t - \sigma\, r^2) \leq t^{8m}\, \alpha\, \kappa\, E + 4m\, \sigma\, t\, r\, [\kappa\, \boldsymbol{\alpha}(m) + t^{8m}\, \alpha\, \kappa\, E].$$

Observing that, for $(x, y) \in Z$,

$$|\phi(x, y) - \theta(x, y)| \leq \sigma(|x|^2 + |y|^2)/2 \leq \sigma\, [r^2 + (n-m)^2\, s^2]/2 \leq \sigma\, r^2,$$

by Taylor's theorem, and also

$$|x| \leq t\, |\theta(x, y)| \leq t\, [|\mathbf{p}[\theta(x, y)]| + t(n-m)\, s],$$

we obtain the inclusions

$$Z \cap \{(x, y)\colon |\mathbf{p}[\phi(x, y)]| < r/(4t)\}$$
$$\subset Z \cap \{(x, y)\colon |\mathbf{p}[\theta(x, y)]| < r/(4t) + \sigma\, r^2\}$$
$$\subset Z \cap \{(x, y)\colon |x| \leq r/4 + t\, \sigma\, r^2 + t^2 (n-m)\, s\}.$$

Abbreviating $W = \mathbf{p}^{-1}\, \mathbf{U}[0, r/(4t)]$ and noting that

$$r/4 + t\, \sigma\, r^2 + t^2 (n-m)\, s < r/4 + 2 \cdot 5^{-2}\, r < r/3 < \rho,$$

which implies $r/(4t)+\sigma r^2<r/t$, we infer

$$Z\cap\phi^{-1}(W)\subset\mathbf{p}^{-1}\mathbf{U}(0,\rho),$$
$$\theta(Z)\cap\psi^{-1}(W)=\theta[Z\cap\phi^{-1}(W)]\subset\mathbf{p}^{-1}\mathbf{U}(0,r/t),$$

and compute

$$\begin{aligned}(\phi_\# S)\llcorner W&=\phi_\#[S\llcorner\phi^{-1}(W)]=\phi_\#[T\llcorner\phi^{-1}(W)]\\&=\psi_\#\theta_\#(T\llcorner\theta^{-1}[\psi^{-1}(W)])=\psi_\#[(\theta_\# T)\llcorner\psi^{-1}(W)]\\&=\psi_\#[U\llcorner\psi^{-1}(W)]=(\psi_\# U)\llcorner W.\end{aligned}$$

Since $r/(4t)\le r/t-5^{-2}r/t\le r/t-\sigma r^2\le 4r/(4t)$ we conclude

$$\begin{aligned}\operatorname{Exc}[\phi_\# S,0,r/(4t)]&\le 4^m\operatorname{Exc}(\psi_\# U,0,r/t-\sigma r^2)\\&\le 4^m t^{8m}\alpha\kappa\big(E+2^{m+2}\sigma t r[\alpha(m)+\kappa\alpha(m)/5]\big)\end{aligned}$$

and define Γ_{17} as the least integer greater than or equal to

$$4^m(1+27\Gamma_{10})\,2^{m+3}\alpha(m).$$

5.3.10. Lemma. *If $1\le\mu<\infty$ and Φ is a parametric integrand of degree m and class 1 on some neighborhood of the point a in $\mathbf{R}^m\times\mathbf{R}^{n-m}$ with*

$$|\tilde\Phi(a,\mathbf{p}^*)|\ge\mu^{-1}\quad\textit{and}\quad\|D\tilde\Phi(a,\mathbf{p}^*)\|\le\mu,$$

then there exists a polynomial function $F:\mathbf{R}^m\times\mathbf{R}^{n-m}\to\mathbf{R}^m\times\mathbf{R}^{n-m}$ of degree 2 such that

$$F(0,0)=a,\qquad DF(0,0)\ \textit{is univalent},\qquad DF(0,0)\circ\mathbf{p}^*=\mathbf{p}^*,$$
$$\|DF(0,0)\|\le 2\mu^2,\qquad\|DF(0,0)^{-1}\|\le 2\mu^2,\qquad\|D^2F(0,0)\|\le 4\mu^6,$$
$$D(F^\#\Phi)^\S(0,0,0)=0,\qquad\mathbf{q}\circ F=\mathbf{q}\circ\tau_a,$$

and the restriction of F to $W=\mathbf{U}[(0,0),2^{-2}\mu^{-4}]$ is a diffeomorphism with

$$\operatorname{Lip}(F|W)\le 3\mu^2,\qquad\mathbf{U}(a,2^{-4}\mu^{-6})\subset F(W),$$
$$\operatorname{Lip}[(F|W)^{-1}]\le 4\mu^2,\qquad\operatorname{Lip}[D(F|W)^{-1}]\le 4^4\mu^8.$$

Proof. We will construct F as a composition,

$$F=\tau_a\circ L\circ(\mathbf{1}+Q),$$

where L is a linear automorphism of $\mathbf{R}^m\times\mathbf{R}^{n-m}$ and Q is a homogeneous polynomial function of degree 2 (see 1.10.4).

Since $\tilde{\Phi}_a(\mathbf{p}^* \circ g) = \tilde{\Phi}_a(\mathbf{p}^*) \cdot \det(g)$ whenever $g \in \operatorname{Hom}(\mathbf{R}^m, \mathbf{R}^m)$ with $\det(g) \geq 0$, differentiation at $g = \mathbf{1}$ with use of 1.4.5 yields the equations

$$\begin{aligned}\langle \mathbf{p}^* \circ h, D\tilde{\Phi}_a(\mathbf{p}^*)\rangle &= \lim_{t \to 0} \tilde{\Phi}_a(\mathbf{p}^*)[\det(\mathbf{1} + t\,h) - 1]/t \\ &= \tilde{\Phi}(a, \mathbf{p}^*)\operatorname{trace}(h) \text{ for } h \in \operatorname{Hom}(\mathbf{R}^m, \mathbf{R}^m).\end{aligned}$$

Every real valued linear function on $\operatorname{Hom}(\mathbf{R}^m, \mathbf{R}^{n-m})$ is in the image of the polarity corresponding to the inner product defined in 1.7.9. Accordingly there exists $l \in \operatorname{Hom}(\mathbf{R}^m, \mathbf{R}^{n-m})$ such that

$$\langle \mathbf{q}^* \circ \tau, D\tilde{\Phi}_a(\mathbf{p}^*)\rangle = \operatorname{trace}(l^* \circ \tau) \text{ for } \tau \in \operatorname{Hom}(\mathbf{R}^m, \mathbf{R}^{n-m}).$$

We define

$$L = \mathbf{1} - \tilde{\Phi}(a, \mathbf{p}^*)^{-1}\,\mathbf{p}^* \circ l^* \circ \mathbf{q}, \qquad \Xi = (\tau_a \circ L)^{\#}\,\Phi$$

and use the last formula in 5.1.9 in computing

$$\begin{aligned}\Xi^{\S}(0, 0, \tau) &= \tilde{\Phi}(a, L \circ \mathbf{p}^* + L \circ \mathbf{q}^* \circ \tau) \\ &= \tilde{\Phi}[a, \mathbf{p}^* + \mathbf{q}^* \circ \tau - \tilde{\Phi}(a, \mathbf{p}^*)^{-1}\,\mathbf{p}^* \circ l^* \circ \tau],\end{aligned}$$

$$\langle (0, 0, \tau), D\Xi^{\S}(0, 0, 0)\rangle = \langle \mathbf{q}^* \circ \tau - \tilde{\Phi}(a, \mathbf{p})^{-1}\,\mathbf{p}^* \circ l^* \circ \tau, D\Phi_a(\mathbf{p}^*)\rangle = 0,$$

$$\tilde{\Xi}(z, \mathbf{p}^*) = \tilde{\Phi}[a + L(z), \mathbf{p}^*], \quad \langle (z, 0), D\tilde{\Xi}(0, \mathbf{p}^*)\rangle = \langle [L(z), 0], D\tilde{\Phi}(a, \mathbf{p}^*)\rangle$$

for $\tau \in \operatorname{Hom}(\mathbf{R}^m, \mathbf{R}^{n-m})$ and $z \in \mathbf{R}^m \times \mathbf{R}^{n-m}$. Moreover

$$\mathbf{q} \circ L = \mathbf{q}, \qquad L^{-1} = \mathbf{1} + \tilde{\Phi}(a, \mathbf{p}^*)^{-1}\,\mathbf{p}^* \circ l^* \circ \mathbf{q},$$

$$\sup\{\|L\|, \|L^{-1}\|\} \leq 1 + \|l\|/|\tilde{\Phi}(a, \mathbf{p}^*)| \leq 1 + \|D\tilde{\Phi}(a, \mathbf{p}^*)\|/|\tilde{\Phi}(a, \mathbf{p}^*)| \leq 2\mu^2.$$

Next we let

$$\begin{aligned}b_i &= \langle (e_i, 0, 0), D\tilde{\Xi}(0, 0, \mathbf{p}^*)\rangle \text{ for } i \in \{1, \ldots, m\}, \\ c_j &= \langle (0, v_j, 0), D\tilde{\Xi}(0, 0, \mathbf{p}^*)\rangle \text{ for } j \in \{1, \ldots, n-m\}\end{aligned}$$

and define $\alpha \in \odot^2(\mathbf{R}^m \times \mathbf{R}^{n-m}, \mathbf{R}^m)$ by the formula

$$\alpha = \sum_{i=1}^{m} b_i\, 2^{-1}(Z_i \odot Z_i)\, e_i + \sum_{j=1}^{n-m} c_j (Z_{m+j} \odot Z_m)\, e_m.$$

If $z = (x, y) \in \mathbf{R}^m \times \mathbf{R}^{n-m}$, then $z \,\lrcorner\, \alpha \in \operatorname{Hom}(\mathbf{R}^m \times \mathbf{R}^{n-m}, \mathbf{R}^m)$,

$$z \,\lrcorner\, \alpha = \sum_{i=1}^{m} b_i\, x_i\, Z_i\, e_i + \sum_{j=1}^{n-m} c_j (y_j\, Z_m + x_m\, Z_{m+j})\, e_m,$$

$$\operatorname{trace}[(z \,\lrcorner\, \alpha) \circ \mathbf{p}^*] = \sum_{i=1}^{m} b_i\, x_i + \sum_{j=1}^{n-m} c_j\, y_j = \langle (x, y, 0), D\tilde{\Xi}(0, 0, \mathbf{p}^*)\rangle.$$

Finally we define $Q: \mathbf{R}^m \times \mathbf{R}^{n-m} \to \mathbf{R}^m \times \mathbf{R}^{n-m}$,

$$Q(z) = -\tilde{\Phi}(a, \mathbf{p}^*)^{-1} \langle z \odot z/2, \mathbf{p}^* \circ \alpha \rangle \text{ for } z \in \mathbf{R}^m \times \mathbf{R}^{n-m},$$

$$\Psi = (\mathbf{1} + Q)^{\#}\, \Xi = [\tau_a \circ L \circ (\mathbf{1} + Q)]^{\#}\, \Phi,$$

and verify that, for $z = (x, y) \in \mathbf{R}^m \times \mathbf{R}^{n-m}$ and $\tau \in \operatorname{Hom}(\mathbf{R}^m, \mathbf{R}^{n-m})$,

$$D(\mathbf{1} + Q)(z) = \mathbf{1} - \tilde{\Phi}(a, \mathbf{p}^*)^{-1}\, \mathbf{p}^* \circ (z \,\lrcorner\, \alpha),$$

$$\begin{aligned} \Psi^{\S}(z, 0) &= \tilde{\Xi}[z + Q(z), \mathbf{p}^* - \tilde{\Phi}(a, \mathbf{p}^*)^{-1}\, \mathbf{p}^* \circ (z \,\lrcorner\, \alpha) \circ \mathbf{p}^*] \\ &= \tilde{\Xi}[z + Q(z), \mathbf{p}^*] \cdot \det[\mathbf{1}_{\mathbf{R}^m} - \tilde{\Phi}(a, \mathbf{p}^*)^{-1} (z \,\lrcorner\, \alpha) \circ \mathbf{p}^*] \end{aligned}$$

in case $|z|$ is small, and

$$\begin{aligned} \langle (x, y, 0), D\Psi^{\S}(0, 0, 0) \rangle &= \langle (x, y, 0), D\tilde{\Xi}(0, 0, \mathbf{p}^*) \rangle \cdot 1 \\ &\quad - \tilde{\Xi}(0, 0, \mathbf{p}^*) \cdot \operatorname{trace}[\tilde{\Phi}(a, \mathbf{p}^*)^{-1} (z \,\lrcorner\, \alpha) \circ \mathbf{p}^*] = 0, \end{aligned}$$

$$\Psi^{\S}(0, 0, \tau) = \tilde{\Xi}(0, 0, \mathbf{p}^* + \mathbf{q}^* \circ \tau) = \Xi^{\S}(0, 0, \tau),$$

$$\langle (0, 0, \tau), D\Psi^{\S}(0, 0, 0) \rangle = 0,$$

hence $D\Psi^{\S}(0, 0, 0) = 0$. Furthermore

$$\mathbf{q} \circ (\mathbf{1} + Q) = \mathbf{q}, \qquad \|\alpha\| \le \|L\| \cdot \|D\tilde{\Phi}(a, \mathbf{p}^*)\| \le 2\mu^3,$$

$$D^2(\mathbf{1} + Q)(z) = \tilde{\Phi}(a, \mathbf{p}^*)^{-1}\, \mathbf{p}^* \circ \alpha, \qquad \|D^2(\mathbf{1} + Q)(z)\| \le 2\mu^4$$

whenever $z \in \mathbf{R}^m \times \mathbf{R}^{n-m}$.

If z and w belong to W, then

$$\begin{aligned} |z + Q(z) - w - Q(w)| &\ge |z - w| - \mu |\langle z^2 - w^2, \alpha \rangle|/2 \\ &\ge |z - w| \cdot (1 - \mu^4 |z + w|) \ge |z - w|/2. \end{aligned}$$

Thus $\psi = (\mathbf{1} + Q) | W$ is univalent, $\operatorname{Lip}(\psi^{-1}) \le 2$, $\psi(W)$ is open by 3.1.1, and $\psi(0, 0) = (0, 0)$, hence $\psi(W) \supset \mathbf{U}[(0, 0), 2^{-3}\mu^{-4}]$. Using 3.1.1 and 2.7.16 we also estimate

$$\operatorname{Lip}(D\psi^{-1}) \le \operatorname{Lip}(\psi^{-1})^3 \operatorname{Lip}(D\psi) \le 32\mu^6.$$

5.3.11. *If* $g \in \operatorname{Hom}(\mathbf{R}^m, \mathbf{R}^{n-m})$, *then*

$$\mathbf{g} = \{(x, y) : g(x) = y\} = \ker(g \circ \mathbf{p} - \mathbf{q}) = \operatorname{im}(\mathbf{p}^* + \mathbf{q}^* \circ g)$$

is an m dimensional vectorsubspace of $\mathbf{R}^m \times \mathbf{R}^{n-m}$. *Moreover, if L is a linear automorphism of* $\mathbf{R}^m \times \mathbf{R}^{n-m}$ *such that*

$$\mathbf{q} \circ L = \mathbf{q} \quad \textit{and} \quad \|L\| \cdot \|g\| \cdot \|L^{-1}\| < 1,$$

then $L(g)=g\circ(\mathbf{p}\circ L\circ\mathbf{p}^*+\mathbf{p}\circ L\circ\mathbf{q}^*\circ g)^{-1}\in\operatorname{Hom}(\mathbf{R}^m,\mathbf{R}^{n-m})$,

$$\|L(g)\|\le\|g\|\cdot\|L^{-1}\|/(1-\|L\|\cdot\|g\|\cdot\|L^{-1}\|)$$

and

$$\mathbf{q}\circ(1-\mathbf{q}^*\circ L(g)\circ\mathbf{p})\circ L\circ(1+\mathbf{q}^*\circ g\circ\mathbf{p})=\mathbf{q}-L(g)\circ\mathbf{p}\circ L\circ\mathbf{q}^*\circ\mathbf{q}.$$

In fact $L=\mathbf{p}^*\circ\mathbf{p}\circ L+\mathbf{q}^*\circ\mathbf{q}$,

$$L\circ\mathbf{p}^*=\mathbf{p}^*\circ u \quad\text{with}\quad u=\mathbf{p}\circ L\circ\mathbf{p}^*,\quad u^{-1}=\mathbf{p}\circ L^{-1}\circ\mathbf{p}^*,$$

$$L\circ\mathbf{q}^*=\mathbf{p}^*\circ v+\mathbf{q}^* \quad\text{with}\quad v=\mathbf{p}\circ L\circ\mathbf{q}^*,$$

$$L\circ(\mathbf{p}^*+\mathbf{q}^*\circ g)=\mathbf{p}^*\circ(u+v\circ g)+\mathbf{q}^*\circ g=(\mathbf{p}^*+\mathbf{q}^*\circ h)\circ(u+v\circ g)$$

$$\text{with } h=g\circ(u+v\circ g)^{-1}=g\circ u^{-1}\circ\sum_{i=0}^{\infty}(-v\circ g\circ u^{-1})^i,$$

$$L(g)=L[\operatorname{im}(\mathbf{p}^*+\mathbf{q}^*\circ g)]=\operatorname{im}(\mathbf{p}^*+\mathbf{q}^*\circ h)=h,$$

$$(\mathbf{q}-h\circ\mathbf{p})\circ L\circ(\mathbf{p}^*\circ\mathbf{p}+\mathbf{q}^*\circ\mathbf{q}+\mathbf{q}^*\circ g\circ\mathbf{p})$$

$$=\mathbf{q}+g\circ\mathbf{p}-h\circ u\circ\mathbf{p}-h\circ v\circ\mathbf{q}-h\circ v\circ g\circ\mathbf{p}=\mathbf{q}-h\circ v\circ\mathbf{q}.$$

5.3.12. There exist positive integers Γ_{18}, Γ_{19} with the following property:

If $M\ge 1$ and Φ is a parametric integrand of degree m and class 3 on some neighborhood of the point a in $\mathbf{R}^n$ such that

$$\|D^q\Phi(a,\zeta)\|\le M$$

for $q\in\{0,1,2,3\}$ and $\zeta\in\wedge_m\mathbf{R}^n$ with $|\zeta|=1$, then

$$\|D^q\tilde{\Phi}(a,\sigma)\|\le\Gamma_{18}M$$

for $q\in\{0,1,2,3\}$ and $\sigma\in\operatorname{Hom}(\mathbf{R}^m,\mathbf{R}^n)$ with $1\le\|\wedge_m\sigma\|\le 3^m$, hence

$$\|D^q\Phi^{\S}(a,\tau)\|\le\Gamma_{18}M$$

for $q\in\{0,1,2,3\}$ and $\tau\in\operatorname{Hom}(\mathbf{R}^m,\mathbf{R}^{n-m})$ with $\|\tau\|\le 2$. If also $P\ge 1$ and F is a function of class 4 mapping some neighborhood of the point α in $\mathbf{R}^n$ into $\mathbf{R}^n$ so that

$$F(\alpha)=a,\quad DF(\alpha)\text{ is univalent},\quad \|DF(\alpha)^{-1}\|\le P,$$

$$\|D^qF(\alpha)\|\le P\ \text{ for } q\in\{1,2,3,4\},$$

then

$$\|D^q(F^{\#}\Phi)(\alpha,\zeta)\|\le\Gamma_{19}MP^{5m}$$

for $q\in\{0,1,2,3\}$ and $\zeta\in\wedge_m\mathbf{R}^n$ with $|\zeta|=1$.

One readily infers such estimates from 5.1.1, 5.1.9, and 3.1.11, observing that

$$D^q\Phi_a(\zeta)=|\zeta|^{1-q}D^q\Phi_a(|\zeta|^{-1}\zeta)\ \text{ for } 0\ne\zeta\in\wedge_m\mathbf{R}^n.$$

5.3.13. Theorem. *If* $1 \le M < \infty$, $0 < \theta < 2^{-16}(\Gamma_{18} M)^{-4}$ *and* κ *is a positive integer, then there exists a positive integer* N *with the following property:*

If Φ *is a parametric integrand of degree* m *and class* 3 *on some neighborhood of*

$$K = (\mathbf{R}^m \times \mathbf{R}^{n-m}) \cap \{(x, y): |x| \le M^{-1}, |y| \le n M^{-1}\}$$

such that Φ *is elliptic at* $(0,0)$ *with ellipticity bound* M^{-1} *and*

$$M^{-1} \le \Phi(x, y, \zeta) \le M, \qquad \|D^q \Phi(x, y, \zeta)\| \le M$$

for $q \in \{1, 2, 3\}$, $(x, y) \in K$, $\zeta \in \wedge_m(\mathbf{R}^m \times \mathbf{R}^{n-m})$ *with* $|\zeta| = 1$, *and if*

$$S \in \mathscr{R}_m(\mathbf{R}^m \times \mathbf{R}^{n-m}),\ 0 < r < M^{-1},\ \operatorname{spt} S \subset \{(x, y): |x| \le r,\ |y| \le M^{-1}\},$$

$$\operatorname{spt} \partial S \subset \{(x, y): |x| = r\}, \qquad \mathbf{p}_{\#} S = \kappa\, \mathbf{E}^m \llcorner \{x: |x| < r\},$$

$$d = \operatorname{Exc}(S, 0, r) + \operatorname{Exc}^\kappa(S, 0, r) < N^{-1}, \qquad \xi \in \mathbf{R}^m,\ |\xi| \le 2^{-9}(\Gamma_{18} M)^{-4} r,$$

S is absolutely Φ minimizing with respect to K,

then there exist linear maps

$$h: \mathbf{R}^m \to \mathbf{R}^{n-m}, \qquad H: \mathbf{R}^m \times \mathbf{R}^{n-m} \to \mathbf{R}^m \times \mathbf{R}^{n-m}$$

with

$$H(x, y) = (x, y - h(x)) \text{ for } (x, y) \in \mathbf{R}^m \times \mathbf{R}^{n-m},$$

$$\|h\| \le \Gamma_{20} M^{88 m^2} \kappa^2 (d + r)^{\frac{1}{4}},$$

$$\operatorname{Exc}(H_{\#} S, \xi, \theta r) \le \Gamma_{21} M^{200 m^2} \kappa^3 (\theta^2 d + \theta r),$$

where $\Gamma_{20} = \Gamma_{15} \Gamma_{17} \Gamma_{18}^{90 m^2} \Gamma_{19}^{1+m} 2^{41 m^2}$ *and* $\Gamma_{21} = \Gamma_{16} \Gamma_{17}^2 \Gamma_{18}^{205 m^2} \Gamma_{19}^{m+4} 2^{132 m^2}$.

Proof. We define

$$\Delta_1 = (n-m)(\Gamma_4^{\frac{1}{2}} \kappa + \Gamma_{13} M^2), \qquad \Delta_2 = (n-m)^2 [(\Gamma_5 + 8\Gamma_4)\kappa + 8\Gamma_{13}^2 M^4],$$

$$\mu = \Gamma_{18} M, \qquad W = (\mathbf{R}^m \times \mathbf{R}^{n-m}) \cap \mathbf{U}[(0,0), 2^{-2}\mu^{-4}],$$

$$\Delta_3 = \Gamma_{17} 2^{17m} \mu^{26m} \kappa^2, \qquad \delta = 2^{12} \mu^4 \theta \le 2^{-4},$$

$$\lambda = \Gamma_{18} \Gamma_{19} (4\mu^6)^{5m} = \Gamma_{18} \Gamma_{19} 2^{10m} \mu^{30m},$$

$$Z = (\mathbf{R}^m \times \mathbf{R}^{n-m}) \cap \{(x, y): |x| \le \lambda^{-1}, |y| \le \lambda^{-1}\},$$

$$\Delta_4 = \Gamma_{15} \lambda^{1+m} 2^m \Delta_3 \le \Gamma_{15} \Gamma_{17} (\Gamma_{18} \Gamma_{19})^{1+m} 2^{38m^2} \mu^{86m^2} \kappa^2$$

$$\Delta_5 = \Gamma_{16} \lambda^{m+4} 2^{m+2} \Delta_3 \le \Gamma_{16} \Gamma_{17} (\Gamma_{18} \Gamma_{19})^{m+4} 2^{70m^2} \mu^{176m^2} \kappa^2$$

$$\gamma = [(4\mu^2)^m \Delta_2 (2^4 \mu^2)^{m+2} \Delta_5^{-1} \delta^{-m-2} 2^m]^{\frac{1}{2}},$$

$$\Delta_6 = n(\Delta_1 2^5 \mu^2 + \Delta_4 2\theta^{-m-1}) \delta^{-1},$$

$$\Delta_7 = \Gamma_{17} 2^{32m} \mu^{16m} \kappa \Delta_5 2^{24} \mu^8 \le \Gamma_{16} \Gamma_{17}^2 (\Gamma_{18} \Gamma_{19})^{m+4} 2^{126m^2} \mu^{200m^2} \kappa^3,$$

$$\Delta_8 = \Gamma_{17} 2^{38m} \mu^{22m} \kappa^2 2^{12} \mu^4 \le \Gamma_{17} 2^{50m} \mu^{26m} \kappa^2,$$

choose ν according to 5.3.8, and let N be the least integer such that

$$2^{9m} n \mu^{10} \delta^{-m-2} \theta^{-m-2} \lambda \sum_{i=1}^{6} \Delta_i N^{-1/(2m)}$$
$$< \inf\{\nu^{-1}, \alpha(m)/5, \Gamma_6/3, \Delta_5 \nu^{-2}, 5^{-2}\}.$$

We now suppose Φ, S, r, d, ξ satisfy the hypothesis of the asserted property.

If $r \geq \theta^{-m-1} N^{-1}$, **then** $\operatorname{Exc}(S, \xi, \theta r) \leq \theta^{-m} d \leq \theta^{-m} N^{-1} \leq \theta r$ **and we take** $h=0$, $H=1$. **Hereafter we assume** $r < \theta^{-m-1} N^{-1}$.

Using 4.2.1, 4.2.16 we choose r_1 so that

$$r/2 < r_1 < r, \qquad S_1 = S \llcorner \mathbf{p}^{-1} \mathbf{U}(0, r_1) \in \mathbf{I}_m(\mathbf{R}^m \times \mathbf{R}^{n-m})$$

and estimate

$$d_1 = \operatorname{Exc}(S_1, 0, r_1) + \operatorname{Exc}^\kappa(S_1, 0, r_1) \leq 2^m d < 2^m N^{-1} < \alpha(m)/5.$$

Next we select $\eta \in \mathbf{R}^{n-m}$ with

$$r_1^{-m} \|S_1\| \{(x, y): y_j > \eta_j\} \leq 3 \kappa \alpha(m)/5,$$
$$r_1^{-m} \|S_1\| \{(x, y): y_j < \eta_j\} \leq 3 \kappa \alpha(m)/5$$

for $j \in \{1, \ldots, n-m\}$, hence $|\eta| \leq (n-m) M^{-1}$, choose r_2 so that

$$r_1/2 < r_2 < 3 r_1/4, \qquad S_2 = S \llcorner \mathbf{p}^{-1} \mathbf{U}(0, r_2) \in \mathbf{I}_m(\mathbf{R}^m \times \mathbf{R}^{n-m}),$$

note that $\Gamma_{13} M^2 d_1^{1/m} \leq \Delta_1 2 N^{-1/m} \leq \frac{1}{4}$, let

$$s_2 = \Delta_1 d_1^{1/(2m)} r_1,$$

and apply 5.3.4(10) with S, r, s, λ replaced by S_1, r_1, M^{-1}, M to conclude

$$\operatorname{spt} S_2 \subset C = \{(x, y): |x| \leq r_2, |y - \eta| \leq s_2\},$$
$$\int |y - \eta|^2 \, d\|S_2\|(x, y) \leq \Delta_2 d_1 r_1^{m+2}.$$

Recalling 5.3.12 we use 5.3.10 to construct F with $a = (0, \eta) \in K$ and $\mu = \Gamma_{18} M$. The we apply 5.3.9 with r, s, t, σ, S, ϕ replaced by

$$r_2, \quad s_2, \quad t_2 = 2\mu^2, \quad \sigma_2 = 4^4 \mu^8, \quad \tau_{(0,-\eta)\#} S_2, \quad (F|W)^{-1} \circ \tau_{(0,\eta)},$$

as we may because

$$s_2 \leq \Delta_1 (2^m d)^{1/(2m)} 2 r_2 \leq \Delta_1 4 N^{-1/(2m)} r_2 \leq r_2,$$

$$2 r_2 \leq 2 r \leq 2 \theta^{-m-1} N^{-1} \leq 2^{-4} \mu^{-6} < M^{-1}, \quad C \subset \mathbf{U}[(0, \eta), 2 r_2] \subset F(W) \cap K,$$

$$d_2 = \operatorname{Exc}(S_2, 0, r_2) + \operatorname{Exc}^\kappa(S_2, 0, r_2) \leq 2^{2m} d \leq 2^{2m} N^{-1} \leq \inf\{\alpha(m)/5, \Gamma_6/3\},$$

$$t_2 \sigma_2 r_2 \leq 2^9 \mu^{10} \theta^{-m-1} N^{-1} \leq 5^{-2},$$

$$t_2^2 (n-m) s_2 \leq 4 \mu^4 (n-m) \Delta_1 4 N^{-1/(2m)} r_2 \leq 5^{-2} r_2,$$

and we find that

$$r_3 = r_2/(8\mu^2), \quad S_3 = [(F|W)_\#^{-1} S_2] \llcorner \mathbf{p}^{-1} \mathbf{U}(0, r_3)$$

satisfy the conditions $\mathbf{p}_\# S_3 = \kappa \mathbf{E}^m \llcorner \mathbf{U}(0, r_3)$ and

$$\begin{aligned} d_3 &= \operatorname{Exc}(S_3, 0, r_3) + \operatorname{Exc}^\kappa(S_3, 0, r_3) \\ &\leq \Gamma_{17}(2\mu^2)^{8m} \kappa(d_2 + \kappa 4^4 \mu^8 2\mu^2 r_2) + (4\mu^2)^m d_2 \\ &\leq \Delta_3(d+r) \leq \Delta_3 2\theta^{-m-1} N^{-1} < \nu^{-1}. \end{aligned}$$

We observe that $|\xi| \leq 2^{-9} \mu^{-4} r < 2^{-7} \mu^{-4} r_2$, $(\xi, \eta) \in C$ and $\mathbf{q} \circ F = \tau_\eta \circ \mathbf{q}$, hence there exists $b \in \mathbf{R}^m$ with

$$(b, 0) \in W \quad \text{and} \quad F(b, 0) = (\xi, \eta);$$

moreover $5\delta \leq \frac{1}{2}$, $|b| \leq 4\mu^2 |\xi| < r_3/4 < (1 - 5\delta) r_3$,

$$1 - r_3^{-1} |b| - \delta \geq 1 - \tfrac{1}{4} - 1/(10) > \tfrac{1}{2}.$$

We also estimate $\|S_3\| \leq (4\mu^2)^m (F|W)_\#^{-1} \|S_2\|$,

$$\begin{aligned} \int |y|^2 \, d\|S_3\|(x, y) &\leq (4\mu^2)^m \int |y - \eta|^2 \, d\|S_2\|(x, y) \\ &\leq (4\mu^2)^m \Delta_2 d_1 (2^4 \mu^2 r_3)^{m+2} \end{aligned}$$

and consider two alternatives:

In case $d_1 > \Delta_5^{-1} \delta^{-m-2} 2^m d_3$ **we take** $g = 0$, $G = 1$ **and obtain the inequality**

$$\operatorname{Exc}(G_\# S_3, b, \delta r_3) \leq \delta^{-m} d_3 < \Delta_5 \delta^2 2^{-m} d_1 \leq \Delta_5 \delta^2 d.$$

In case $d_1 \leq \Delta_5^{-1} \delta^{-m-2} 2^m d_3$ **we find that**

$$\int |y|^2 \, d\|S_3\|(x, y) \leq \gamma^2 d_3 r_3^{m+2}$$

and apply 5.3.8 **with** $S, r, s, \varepsilon, \Psi$ **replaced by** $S_3, r_3, s_2, d_3^{\frac{1}{2}}, F^\# \Phi$, as we may because $F(Z) \subset K$,

$$\begin{aligned} r_3 \leq 2^{-3} \mu^{-2} r \leq 2^{-3} \mu^{-2} \theta^{-m-1} N^{-1} \leq \lambda^{-1} \leq 2^{-3} \mu^{-4}, \\ s_2 \leq \Delta_1 4 N^{-1/(2m)} 8\mu^2 r_3 \leq r_3, \\ \operatorname{spt} S_3 \subset \{(x, y) : |x| \leq r_3, |y| \leq s_2\} \subset Z \subset W, \end{aligned}$$

5.1.4 implies that Ψ is elliptic at $(0, 0)$ with ellipticity bound $M^{-1}(2\mu^2)^{-7m} \geq \lambda^{-1}$, while 5.3.12 with $P = 4\mu^6$ shows λ to be an appropriate bound for $D^q \Psi$ and $D^q \Psi^\S$. The conclusion of 5.3.8 leads to two subcases:

Either $d_3^{\frac{1}{2}} \leq \nu r_3$, **in which event we take** $g = 0$, $G = 1$ **and estimate**

$$\begin{aligned} \operatorname{Exc}(G_\# S_3, b, \delta r_3) &\leq \delta^{-m} d_3 \leq \delta^{-m} \nu^2 r_3^2 \leq \delta^{-m} \nu^2 r^2 \\ &\leq \delta^{-m} \nu^2 \theta^{-m-1} N^{-1} r \leq \Delta_5 \delta^2 r. \end{aligned}$$

Or there exists a map $g \in \operatorname{Hom}(\mathbf{R}^m, \mathbf{R}^{n-m})$ and a linear automorphism $G = \mathbf{1} - \mathbf{q}^* \circ g \circ \mathbf{p}$ of $\mathbf{R}^m \times \mathbf{R}^{n-m}$ with

$$|g| \le \Gamma_{15}\, \lambda^{1+m/2}\, 2^{m/2}\, d_3^{\frac{1}{2}} \le \Delta_4 (d+r)^{\frac{1}{2}},$$
$$\operatorname{Exc}(G_{\#} S_3, b, \delta r_3) \le \Gamma_{16}\, \lambda^{m+4}\, 2^{m+2}\, \delta^2\, d_3 \le \Delta_5\, \delta^2 (d+r).$$

In each of the three above cases and subcases it is true that

$$|g| \le \Delta_4 (d+r)^{\frac{1}{2}}, \qquad G = \mathbf{1} - \mathbf{q}^* \circ g \circ \mathbf{p},$$
$$\operatorname{Exc}(S_4, b, r_4) \le \Delta_5\, \delta^2 (d+r) \quad \textbf{with} \quad r_4 = \delta r_3, \quad S_4 = (G_{\#} S_3) \llcorner \mathbf{p}^{-1}\, \mathbf{U}(b, r_4).$$

Since $d + r \le 2\theta^{-m-1} N^{-1}$ we infer

$$\|g\| \le |g| \le \Delta_4\, 2\theta^{-m-1} N^{-\frac{1}{2}} \le 1, \qquad \|G\| \le 1 + \|g\| \le 2,$$
$$\operatorname{Exc}(S_4, b, r_4) \le \Delta_5\, 2\theta^{-m-1} N^{-1} < \inf\{\alpha(m)/5,\ \Gamma_6/3\},$$
$$\operatorname{Exc}^{\kappa}(S_4, b, r_4) \le \delta^{-m} \|G\|^m d_3 \le \delta^{-m} 2^m \Delta_3\, 2\theta^{-m-1} N^{-1} < \alpha(m)/5,$$
$$\operatorname{spt} S_4 \subset \{(x,y)\colon |x-b| \le r_4,\ |y| \le s_2 + \|g\|\, r_3\}.$$

Next we select $c \in \mathbf{R}^{n-m}$ with

$$r_4^{-m}\, \|S_4\|\, \{(x,y)\colon y_j > c_j\} \le 3\kappa\, \alpha(m)/5,$$
$$r_4^{-m}\, \|S_4\|\, \{(x,y)\colon y_j < c_j\} \le 3\kappa\, \alpha(m)/5$$

for $j \in \{1, \ldots, n-m\}$ and note that $|c| \le (n-m)(s_2 + \|g\|\, r_3)$,

$$\operatorname{spt} S_4 \subset D = \{(x,y)\colon |x-b| \le r_4,\ |y-c| \le s_4\} \quad \text{where}$$
$$s_4 = n(s_2 + \|g\|\, r_3) = n(s_2/r_3 + \|g\|)\, r_3$$
$$\le n(\Delta_1\, 2^5 \mu^2 N^{-1/(2m)} + \Delta_4\, 2\theta^{-m-1} N^{-\frac{1}{2}})\, \delta^{-1} r_4 \le \Delta_6\, N^{-1/(2m)} r_4 \le r_4 < r_3/2,$$
$$G^{-1}(D) \subset \{(x,y)\colon |x| \le r_3,\ |y| \le r_3\} \subset Z \subset W.$$

Letting $L = DF[b, c + g(b)]$ we estimate

$$\|L\| \cdot \|g\| \cdot \|L^{-1}\| \le 3\mu^2\, \Delta_4\, 2\theta^{-m-1} N^{-\frac{1}{2}}\, 4\mu^2 \le \tfrac{1}{2},$$

use 5.3.11 to obtain $h = L(g) \in \operatorname{Hom}(\mathbf{R}^m, \mathbf{R}^{n-m})$ with

$$\|h\| \le 2\, \|L^{-1}\| \cdot \|g\| \le 8\mu^2\, \Delta_4 (d+r)^{\frac{1}{2}} \le \Gamma_{20}\, M^{88m^2} \kappa^2 (d+r)^{\frac{1}{2}},$$

define $H = \mathbf{1} - \mathbf{q}^* \circ h \circ \mathbf{p}$, $u = (\mathbf{p} \circ F)[b, c + g(b)]$ and note that

$$D[H \circ F \circ G^{-1}](b,c) = H \circ L \circ G^{-1}, \qquad \mathbf{q} \circ H \circ L \circ G^{-1} = \mathbf{q} - h \circ \mathbf{p} \circ L \circ \mathbf{q}^* \circ \mathbf{q},$$
$$(H \circ F \circ G^{-1})(b,c) = (H \circ F)(b, c + g(b)) = H(u, \eta + c + g(b))$$
$$= (u, \eta + c + g(b) - h(u)).$$

Then we apply 5.3.9 with r, s, t, σ, S, ϕ replaced by

$$r_4, \quad s_4, \quad t_4 = (1+\|h\|)\cdot 4\mu^2\cdot(1+\|g\|), \qquad \sigma_4 = (1+\|h\|)\,4\mu^6(1+\|g\|)^2,$$
$$\tau_{(-b,\,-c)\#}\, S_4, \qquad \tau_{(-u,\,-\eta-c-g(b)+h(u))}\circ H\circ F\circ G^{-1}\circ\tau_{(b,\,c)},$$

as we may because $\|h\| \le 8\mu^2 \Delta_4\, 2\theta^{-m-1} N^{-\frac{1}{2}} \le 1$,

$$t_4\,\sigma_4\, r_4 \le 2^4\mu^2\, 2^5\mu^6\,\delta\, 2^{-3}\mu^{-2}\theta^{-m-1}N^{-1} \le 5^{-2},$$
$$t_4^2(n-m)\, s_4 \le 2^8\mu^4(n-m)\,\Delta_6\, N^{-1/(2m)}\, r_4 \le 5^{-2}\, r_4,$$

and we find that

$$r_5 = r_4/(4t_4), \qquad S_5 = [(H\circ F\circ G^{-1})_\#\, S_4] \llcorner \mathbf{p}^{-1}\,\mathbf{U}(u, r_5)$$

satisfy the conditions $\mathbf{p}_\#\, S_5 = \kappa\,\mathbf{E}^m \llcorner \mathbf{U}(0, r_5)$ and

$$\begin{aligned}\operatorname{Exc}(S_5, u, r_5) &\le \Gamma_{17}\, t_4^{8m}\,\kappa[\operatorname{Exc}(S_4, b, r_4) + \kappa\,\sigma_4\, t_4\, r_4]\\ &\le \Gamma_{17}(2^4\mu^2)^{8m}\kappa[\Delta_5\,\delta^2(d+r) + \kappa\, 2^5\mu^6\, 2^4\mu^2\,\delta\, 2^{-3}\mu^{-2} r]\\ &\le \Delta_7\,\theta^2(d+r) + \Delta_8\,\theta r \le (\Delta_7+\Delta_8)(\theta^2 d + \theta r).\end{aligned}$$

Inasmuch as

$$r_5 = \delta r_3\, 2^{-2} t_4^{-1} = \delta r_2\, 2^{-5}\mu^{-2} t_4^{-1} = \theta r_2\, 2^7/\mu^2/t_4$$

with $2^{-2} r < r_2 < r$ and $2^4\mu^2 \ge t_4 \ge 2^2\mu^2$, hence

$$2\theta r \le r_5 \le 2^5\theta r, \qquad r_4 = 4t_4\, r_5 \le 2^6\mu^2\, r_5,$$

and also $|c+g(b)| \le |c| + \|g\|\, r_3 \le s_4 \le \Delta_6 N^{-1/(2m)} r_4$,

$$\begin{aligned}|u-\xi| &\le |F[b, c+g(b)] - F(b,0)| \le 3\mu^2\,|c+g(b)|\\ &\le 3\mu^2\Delta_6 N^{-1/(2m)}\, 2^6\mu^2\, r_5 < r_5 < r_5/2,\end{aligned}$$

hence $\theta r + |u-\xi| < r_5$, $\mathbf{U}(\xi, \theta r) \subset \mathbf{U}(u, r_5)$, we infer

$$\operatorname{Exc}(S_5, \xi, \theta r) \le 2^{5m}\operatorname{Exc}(S_5, u, r_5) \le \Gamma_{21}\, M^{200m^2}\kappa^3(\theta^2 d + \theta r).$$

Observing that $|\xi| + r_5 \le (2^{-7}\mu^{-4} + 2^{-11}\mu^{-4})\, r_2 \le r_2$,

$$4\mu^2(r_5+s_2) \le [4^{-1} + \Delta_1\, 2^7\mu^4 N^{-1/(2m)}\delta^{-1}]\, r_4 \le r_4,$$
$$\begin{aligned}(\operatorname{spt} S_2)\cap\mathbf{p}^{-1}\,\mathbf{U}(\xi,\theta r) &\subset C\cap\mathbf{p}^{-1}\,\mathbf{U}(\xi, r_5) \subset F(W)\cap\mathbf{U}[(\xi,\eta), r_5+s_2]\\ &\subset F(W\cap\mathbf{U}[(b,0), 4\mu^2(r_5+s_2)]) \subset F[W\cap p^{-1}\,\mathbf{U}(b, r_4)]\end{aligned}$$

and $\mathbf{p}\circ G = \mathbf{p} = \mathbf{p}\circ H$, we finally compute

$$G_\#^{-1}\, S_4 = S_3 \llcorner \mathbf{p}^{-1}\,\mathbf{U}(b, r_4) = [(F|W)_\#^{-1}\, S_2] \llcorner \mathbf{p}^{-1}\,\mathbf{U}(b, r_4),$$
$$(F\circ G^{-1})_\#\, S_4 = S_2 \llcorner F[W\cap\mathbf{p}^{-1}\,\mathbf{U}(b, r_4)],$$
$$[(F\circ G^{-1})_\#\, S_4] \llcorner \mathbf{p}^{-1}\,\mathbf{U}(\xi,\theta r) = S_2 \llcorner \mathbf{p}^{-1}\,\mathbf{U}(\xi,\theta r) = S \llcorner \mathbf{p}^{-1}\,\mathbf{U}(\xi,\theta r),$$
$$S_5 \llcorner \mathbf{p}^{-1}\,\mathbf{U}(\xi,\theta r) = (H_\#\, S) \llcorner \mathbf{p}^{-1}\,\mathbf{U}(\xi,\theta r).$$

5.3.14. Theorem. *If* $1 \leq \lambda < \infty$ *and* κ *is a positive integer, then there exist numbers* $0 < \varepsilon < 1$, $0 < \gamma < \infty$ *with the following property:*

If Ψ *is a parametric integrand of degree* m *and class* 3 *on some neighborhood of*

$$Z = (\mathbf{R}^m \times \mathbf{R}^{n-m}) \cap \{(x, y) \colon |x| \leq \lambda^{-1},\ |y| \leq \lambda^{-1}\}$$

such that Ψ *is elliptic at* (x, y) *with ellipticity bound* λ^{-1} *and*

$$\lambda^{-1} \leq \Psi(x, y, \zeta) \leq \lambda, \qquad \|D^i \Psi(x, y, \zeta)\| \leq \lambda$$

for $i \in \{1, 2, 3\}$, $(x, y) \in Z$, $\zeta \in \wedge_m(\mathbf{R}^m \times \mathbf{R}^{n-m})$ *with* $|\zeta| = 1$, *and if*

$$T \in \mathscr{R}_m(\mathbf{R}^m \times \mathbf{R}^{n-m}), \qquad 0 < \rho < \varepsilon, \qquad \operatorname{Exc}(T, 0, \rho) < \varepsilon,$$

$$(0, 0) \in \operatorname{spt} T \subset \{(x, y) \colon |x| \leq \rho,\ |y| \leq \lambda^{-1}\},$$

$$\operatorname{spt} \partial T \subset \{(x, y) \colon |x| = \rho\}, \qquad \mathbf{p}_\# T = \kappa\, \mathbf{E}^m \llcorner \{x \colon |x| < \rho\},$$

$$\Theta^m[T, (x, y)] \geq \kappa \text{ for } \|T\| \text{ almost all } (x, y),$$

$$T \text{ is absolutely } \Psi \text{ minimizing with respect to } Z,$$

$$f = (\operatorname{spt} T) \cap \{(x, y) \colon |x| < \varepsilon \rho\},$$

then f *is a function of class* 1 *satisfying the conditions*

$$\|D f(x)\| \leq \gamma [\sup\{\operatorname{Exc}(T, 0, \rho), \rho\}]^{\frac{1}{4}} \text{ for } |x| < \varepsilon \rho,$$

$$\mathbf{h}_{\frac{1}{4}}(D f) \leq \gamma [\rho^{-1} \sup\{\operatorname{Exc}(T, 0, \rho), \rho\}]^{\frac{1}{4}},$$

$$T \llcorner \{(x, y) \colon |x| < \varepsilon \rho\} = (\mathbf{p}^* + \mathbf{q}^* \circ f)_\# (\kappa\, \mathbf{E}^m \llcorner \{x \colon |x| < \varepsilon \rho\}).$$

Moreover, if Ψ *is an integrand of class* $q+1$, *then* f *is a function of class* q *and, in case* $q < \infty$, $\mathbf{h}_{\frac{1}{4}}(D^q f | K) < \infty$ *for every compact subset* K *of* $\mathbf{R}^m \cap \mathbf{U}(0, \varepsilon \rho)$.

Proof. We let $\alpha = \Gamma_{19}\, 2^{7m}\, n$, $M = \alpha \lambda$,

$$\Delta_1 = \Gamma_{13} M^2, \qquad \Delta_2 = 4(n-m)[\Gamma_4^{\frac{1}{2}} \kappa + \Delta_1],$$

$$\Delta_3 = 2^{-16} (\Gamma_{18} M)^{-4}, \qquad \Delta_4 = \Gamma_{20} M^{88 m^2} \kappa^2, \qquad \Delta_5 = 2 \Gamma_{21} M^{200 m^2} \kappa^3,$$

choose a number θ for which

$$0 < \theta < \Delta_3, \qquad \Delta_5\, \theta^{\frac{1}{4}} < 1, \qquad (1 - \theta^{\frac{1}{4}})^{-1} + \theta^{\frac{1}{4}} < 2,$$

choose a positive integer N satisfying the conditions of 5.3.13 and also

$$N^{-1} < \alpha(m)/5, \qquad \Delta_1 N^{-1/m} < \tfrac{1}{2}, \qquad 2 \Delta_4 N^{-\frac{1}{4}} < 1,$$

choose ε so that

$$0 < 4 \theta^{-\frac{1}{4}} \alpha^m \varepsilon < N^{-1}, \qquad \Delta_2\, \alpha\, \varepsilon^{1/(2m)} < 1, \qquad 2 \alpha \varepsilon < \Delta_3,$$

and take $\gamma = 4 \Delta_4 (2 \theta^{-\frac{1}{4}} \alpha^{m+1} \Delta_3^{-1})^{\frac{1}{4}}$.

Given Ψ, T, ρ as in the hypothesis of the asserted property we define

$$\sigma=\alpha^{-1}\rho, \quad \delta=\sup\{2\operatorname{Exc}(T,0,\sigma),\theta^{-\frac{1}{2}}\sigma\},$$
$$t_\nu=\delta\theta^\nu\sum_{i=0}^{\nu}\theta^{i/2} \text{ for all nonnegative integers } \nu,$$

and note that $\sigma<\theta^{-\frac{1}{2}}\sigma<\theta^{-\frac{1}{2}}\alpha^{-1}\varepsilon<2^{-1}N^{-1}$,

$$2\operatorname{Exc}(T,0,\sigma)\le 2\alpha^m\operatorname{Exc}(T,0,\rho)<2\alpha^m\varepsilon<2^{-1}N^{-1},$$
$$\delta<2^{-1}N^{-1}, \quad t_\nu\le\delta\theta^\nu(1-\theta^{\frac{1}{2}})^{-1}\le\delta 2<N^{-1}.$$

From 5.3.1 we see that $\operatorname{Exc}^\kappa(T,0,\rho)\le\operatorname{Exc}(T,0,\rho)$, and applying 5.3.4 (4), (10) with S, r, s replaced by $T\llcorner\mathbf{p}^{-1}\mathbf{U}(0,\rho), \rho, \lambda^{-1}$ we infer

$$(\operatorname{spt} T)\cap\{(x,y): |x|\le\sigma\}\subset W=\{(x,y): |x|\le\sigma, |y|\le\sigma\}$$

because $\sigma<2^{-1}\rho<(1-\Delta_1\varepsilon^{1/m})\rho$ and $\Delta_2\varepsilon^{1/(2m)}\rho<\sigma$.

We will prove the following statement:

For every sequence of points $a_\nu=(b_\nu,c_\nu)\in\operatorname{spt} T$ such that $a_0=(0,0)$,

$$|b_\nu-b_{\nu-1}|<\Delta_3\theta^{\nu-1}\sigma \text{ if } \nu\ge 1,$$

there exist linear maps g_ν: $\mathbf{R}^m\to\mathbf{R}^{n-m}$ such that $g_0=0$,

$$\|g_\nu-g_{\nu-1}\|<\Delta_4(2\delta\theta^{\nu-1})^{\frac{1}{2}} \text{ if } \nu\ge 1,$$

and the endomorphisms $G_\nu=1-\mathbf{q}^*\circ g_\nu\circ\mathbf{p}$ of $\mathbf{R}^m\times\mathbf{R}^{n-m}$ satisfy the conditions

$$2\operatorname{Exc}(G_{\nu\#}T,b_\nu,\theta^\nu\sigma)<t_\nu,$$
$$(\operatorname{spt} T)\cap\{(x,y): |x-b_\nu|<2^{-1}\theta^\nu\sigma\}$$
$$\subset\{(x,y): |y-c_\nu-g_\nu(x-b_\nu)|\le\Delta_2 t_\nu^{1/(2m)}\theta^\nu\sigma\}.$$

We inductively construct g_ν for $\nu\ge 1$ by applying 5.3.13 with

$$\Phi=F^*\Psi, \quad r=\theta^{\nu-1}\sigma, \quad \xi=b_\nu-b_{\nu-1}$$
$$S=F_\#^{-1}[T\llcorner\{(x,y): |x-b_{\nu-1}|<\theta^{\nu-1}\sigma\}],$$
$$F=\tau_{a_{\nu-1}}\circ G_{\nu-1}^{-1}, \quad F^{-1}=\tau_{G_{\nu-1}(-a_{\nu-1})}\circ G_{\nu-1};$$

to verify that Φ, S have the required properties we recall 5.1.4, 5.3.12, 4.1.28, 4.1.30, 5.3.1 and observe that

$$\begin{gathered}\|g_{v-1}\| \le \sum_{i=1}^{v-1} \|g_i - g_{i-1}\| \le \sum_{i=1}^{v-1} \Delta_4 (2\delta\theta^{i-1})^{\frac{1}{2}} \\ \le \Delta_4 (2\delta)^{\frac{1}{2}} (1-\theta^{\frac{1}{2}})^{-1} \le \Delta_4 N^{-\frac{1}{2}} 2 < 1, \\ \sup\{\operatorname{Lip}(F), \operatorname{Lip}(F^{-1})\} \le 1 + \|g_{v-1}\| < 2, \\ F^{-1}(W) \subset \{(x,y): |x| \le 4\sigma, |y| \le 4\sigma\} \text{ with } 4\sigma < M^{-1}, \\ F\{(x,y): |x| \le M^{-1}, |y| \le n M^{-1}\} \subset Z\end{gathered}$$

because $F(0,0) = a_{v-1} \in W$ and $4(\sigma + n M^{-1}) < \lambda^{-1}$,

$$\begin{gathered}\Theta^m[\|S\|, (x,y)] \ge \kappa \text{ for } \|S\| \text{ almost all } (x,y), \\ \operatorname{Exc}(S,0,r) + \operatorname{Exc}^\kappa(S,0,r) \le 2\operatorname{Exc}(S,0,r) \\ = 2\operatorname{Exc}(G_{v-1\#} T, b_{v-1}, \theta^{v-1}\sigma) \le t_{v-1} < N^{-1};\end{gathered}$$

we use the resulting maps h, H to define $g_v = g_{v-1} + h$ and conclude

$$\begin{gathered}G_v = H \circ G_{v-1} = \tau_{H[G_{v-1}(a_{v-1})]} \circ H \circ F^{-1}, \\ 2\operatorname{Exc}(G_{v\#} T, b_v, \theta^v \sigma) = 2\operatorname{Exc}(H_\# S, \xi, \theta r) \\ \le \Delta_5(\theta^2 t_{v-1} + \theta^v \sigma) \le \theta^{\frac{3}{2}} t_{v-1} + \theta^{v-\frac{1}{2}} \sigma \\ \le \delta\theta^{v+\frac{1}{2}} \sum_{i=0}^{v-1} \theta^{i/2} + \delta\theta^v = t_v, \\ \|g_v - g_{v-1}\| \le \Delta_4 (t_{v-1} + \theta^{v-1}\sigma)^{\frac{1}{2}} \\ \le \Delta_4 [\delta\theta^{v-1}(1-\theta^{\frac{1}{2}})^{-1} + \delta\theta^{v-1}\theta^{\frac{1}{2}}]^{\frac{1}{2}} \le \Delta_4 (2\delta\theta^{v-1})^{\frac{1}{2}}.\end{gathered}$$

For every nonnegative integer v we infer from 5.3.4(10), with λ and S replaced by M and

$$\tau_{-a_v\#}[(G_{v\#} T) \llcorner \{(u,v): |u - b_v| < \theta^v \sigma\}],$$

that the diameter of the set

$$\begin{gathered}\{y - g_v(x): (x,y) \in \operatorname{spt} T, |x - b_v| < 2^{-1}\theta^v\sigma\} \\ = \{v: (u,v) \in \operatorname{spt} G_{v\#} T, |u - b_v| < 2^{-1}\theta^v \sigma\}\end{gathered}$$

does not exceed $\Delta_2 t_v^{1/2m} \theta^v \sigma$; since $c_v - g_v(b)$ belongs to this set, the final portion of our statement follows.

We note that $\mathbf{R}^m \cap \mathbf{B}(0,\rho) = \operatorname{spt}(\mathbf{p}_\# T) \subset \mathbf{p}(\operatorname{spt} T)$, hence

$$\mathbf{R}^m \cap \mathbf{U}(0,\varepsilon\rho) = \mathbf{p}(f),$$

and that $\varepsilon\rho < \Delta_3\sigma$. For each $u \in \mathbf{R}^m \cap \mathbf{U}(0,\varepsilon\rho)$ and each positive integer μ we apply the above statement with

$$b_\nu = u \text{ for all integers } \nu \geq \mu$$

to infer that $\operatorname{diam}(f \cap \mathbf{p}^{-1}\{u\}) = 0$; consequently the relation f is a function. Using the same sequence again we also obtain the existence of

$$l = \lim_{\nu\to\infty} g_\nu \in \operatorname{Hom}(\mathbf{R}^m, \mathbf{R}^{n-m}),$$

and we find that

$$|f(x) - f(u) - g_\nu(x-u)| \leq \Delta_2\, t_\nu^{1/(2m)}\, \theta^\nu \sigma$$

whenever $x \in \mathbf{U}(0,\varepsilon\rho)$, $\nu \geq \mu$ and $|x-u| < 2^{-1}\theta^\nu\sigma$; we infer that for each $x \in \mathbf{U}(0,\varepsilon\rho) \cap \mathbf{U}(u, 2^{-1}\theta^\mu\sigma) \sim \{u\}$ there exists an integer $\nu \geq \mu$ such that

$$2^{-1}\theta^{\nu+1}\sigma \leq |x-u| < 2^{-1}\theta^\nu\sigma,$$

$$|f(x) - f(u) - l(x-u)| \leq [\Delta_2\, t_\nu^{1/(2m)}\, \theta^{-1}\, 2 + \|g_\nu - l\|] \cdot |x-u|;$$

consequently f has the differential l at u, and

$$\|D f(u) - g_{\mu-1}\| \leq \sum_{i=\mu}^{\infty} \|g_i - g_{i-1}\|$$

$$\leq \sum_{i=\mu}^{\infty} \Delta_4 (2\delta\theta^{i-1})^{\frac{1}{2}} = \Delta_4 (2\delta\theta^{\mu-1})^{\frac{1}{2}} (1-\theta^{\frac{1}{2}})^{-1}.$$

Taking $\mu = 1$ we infer $\|D f(u)\| \leq \Delta_4 (2\delta\theta^{-1})^{\frac{1}{2}}\, 2$.

Whenever u_1 and u_2 are distinct points in $\mathbf{R}^m \cap \mathbf{U}(0,\varepsilon\rho)$ there exists a positive integer μ such that

$$\Delta_3\,\theta^\mu\sigma \leq |u_1 - u_2| < \Delta_3\,\theta^{\mu-1}\sigma,$$

and we can construct sequences for u_1 and u_2 as above with equal initial segments corresponding to all $\nu < \mu$; it follows that

$$\|D f(u_1) - D f(u_2)\| \leq 2\Delta_4 (2\delta\theta^{\mu-1})^{\frac{1}{2}} (1-\theta^{\frac{1}{2}})^{-1}$$

$$\leq 4\Delta_4 (2\delta\theta^{-1}\Delta_3^{-1}\sigma^{-1}|u_1-u_2|)^{\frac{1}{2}}.$$

Noting that $\delta \leq \theta^{-\frac{1}{2}}\alpha^m \sup\{\operatorname{Exc}(T,0,\rho),\rho\}$ we obtain the asserted bounds for $\|D f\|$ and $\mathbf{h}_{\frac{1}{2}}(D f)$.

Since $(\mathbf{p}^* + \mathbf{q}^* \circ f) \circ \mathbf{p} \,|\, f = \mathbf{1} \,|\, f$ we see from 4.1.15 that

$$T \llcorner f = [(\mathbf{p}^* + \mathbf{q}^* \circ f) \circ \mathbf{p}]_\# [T \llcorner p^{-1}\mathbf{U}(0,\varepsilon\rho)]$$

$$= (\mathbf{p}^* + \mathbf{q}^* \circ f)_\# [(\mathbf{p}_\# T) \llcorner \mathbf{U}(0,\varepsilon\rho)].$$

The final conclusion of the theorem follows from 5.2.18.

5.3.15. Corollary. *If Φ is a parametric integrand of degree m and class $q+1\geq 3$ on some neighborhood of*

$$A=(\mathbf{R}^m\times\mathbf{R}^{n-m})\cap\{(x,y)\colon |x|<2\lambda^{-1}, |y|\leq 2\lambda^{-1}\}$$

such that Φ is elliptic at (x,y) with ellipticity bound λ^{-1} and

$$\lambda^{-1}\leq\Phi(x,y,\zeta)\leq\lambda,\qquad \|D^i\Phi(x,y,\zeta)\|\leq\lambda$$

for $i\in\{1,2,3\}$, $(x,y)\in A$, $\zeta\in\wedge_m(\mathbf{R}^m\times\mathbf{R}^{n-m})$ with $|\zeta|=1$, and if

$$S\in\mathscr{R}_m(\mathbf{R}^m\times\mathbf{R}^{n-m}),\qquad 0<r<\lambda^{-1},$$

$$(0,0)\in\operatorname{spt}S\subset\{(x,y)\colon |x|\leq r, |y|\leq(2\lambda)^{-1}\},$$

$$\operatorname{spt}\partial S\subset\{(x,y)\colon |x|=r\},\qquad \mathbf{p}_\# S=\kappa\,\mathbf{E}^m\llcorner\{x\colon |x|<r\},$$

$$\Theta^m[S,(x,y)]\geq\kappa \text{ for } \|S\| \text{ almost all } (x,y),$$

$$S \text{ is absolutely } \Phi \text{ minimizing with respect to } A,$$

$$\sup\{0,r-\varepsilon\}<s<r,\qquad \operatorname{Exc}(S,0,r)\leq(1-s/r)^m\,\varepsilon,$$

$$g=(\operatorname{spt}S)\cap\{(x,y)\colon |x|<s\},$$

then g is a function of class q satisfying the conditions

$$\|D\,g\|\leq\gamma[\sup\{(1-s/r)^{-m}\operatorname{Exc}(S,0,r), r-s\}]^{\frac{1}{4}},$$

$$\mathbf{h}_{\frac{1}{4}}(D\,g)\leq 2\gamma[\sup\{(1-s/r)^{-m}\operatorname{Exc}(S,0,r), r-s\}]^{\frac{1}{4}}/(r-s),$$

$$\mathbf{h}_{\frac{1}{4}}(D^q\,g)<\infty \text{ in case } q<\infty,$$

$$S\llcorner\{(x,y)\colon |x|<s\}=(\mathbf{p}^*+\mathbf{q}^*\circ g)_\#(\kappa\,\mathbf{E}^m\llcorner\{x\colon |x|<s\}).$$

Proof. For each $(b,c)\in g$ we apply 5.3.14 with

$$\Psi=\tau_{(b,c)\#}\Phi,\qquad \rho=r-s,\qquad T=\tau_{(-b,-c)\#}[S\llcorner\{(x,y)\colon |x-b|<\rho\}],$$

$$\operatorname{Exc}(T,0,\rho)\leq(r/\rho)^m\operatorname{Exc}(S,0,r)\leq\varepsilon.$$

5.3.16. Theorem. *If Ψ is a positive elliptic integrand of degree m and class $q+1\geq 3$ on the open subset A of $\mathbf{R}^n$,*

$$S\in\mathscr{R}_m(A),\qquad a\in(\operatorname{spt}S)\sim\operatorname{spt}\partial S,$$

$\operatorname{Tan}^m(\|S\|,a)$ *is contained in some m dimensional vectorsubspace of* $\mathbf{R}^n$, *and a has a neighborhood V in A such that $S\llcorner V$ is absolutely Ψ minimizing with respect to V,*

$$\Theta^m(\|S\|,z)\geq\Theta^{*m}(\|S\|,a) \text{ for } \|S\| \text{ almost all } z \text{ in } V,$$

then a has a neighborhood W in V such that $W\cap\operatorname{spt}S$ is a connected m dimensional submanifold of class q of $\mathbf{R}^n$.

Proof. Identifying $\mathbf{R}^n$ with $\mathbf{R}^m \times \mathbf{R}^{n-m}$ and recalling the corollary of the density property in 5.1.6 we assume

$$a=(0,0), \quad \operatorname{Tan}(\operatorname{spt} S, a)=\operatorname{Tan}^m(\|S\|, a)\subset \mathbf{R}^m \times \{0\},$$

hence $\eta(r)=\inf\{s: |y|\le s|x| \text{ for all } (x,y)\in \mathbf{B}(a,r)\cap \operatorname{spt} S\}\to 0$ as $r\to 0+$, and we choose λ so that Ψ satisfies the conditions of 5.3.14 with $Z\subset V$.

Next we choose $t>0$ so that $\eta(2t)<1$ and $\mathbf{B}(a,2t)\subset Z\sim \operatorname{spt}\partial S$, which implies

$$\begin{aligned}\operatorname{spt}\partial[S\llcorner \mathbf{U}(a,2t)] &\subset (\operatorname{spt} S)\cap[\mathbf{B}(a,2t)\sim \mathbf{U}(a,2t)]\\ &\subset \{(x,y): |y|\le \eta(2t)|x|, |x|^2+|y|^2=4t^2\}\subset\{(x,y): |x|>t\},\end{aligned}$$

define $T_\rho=[S\llcorner \mathbf{U}(a,2t)]\llcorner\{(x,y): |x|<\rho\}$ for $0<\rho<t$, and observe that $\operatorname{spt}\partial T_\rho\subset\{(x,y): |x|=\rho\}$,

$$\operatorname{spt} T_\rho\subset(\operatorname{spt} S)\cap\{(x,y): |y|\le|x|\le\rho\}\subset\{(x,y): |x|\le\rho, |y|\le\eta(2\rho)|x|\}.$$

Consequently there exists an integer κ such that

$$\mathbf{p}_\# T_\rho=\kappa\,\mathbf{E}^m\llcorner\{x: |x|<\rho\} \text{ for } 0<\rho<t.$$

We may assume $\kappa\ge 0$, because otherwise we could replace S by $-S$, and $\Psi(z,\zeta)$ by $\Psi(z,-\zeta)$. Inasmuch as

$$\|S\|\,\mathbf{U}(a,\rho)\le \mathbf{M}(T_\rho)\le \|S\|\,\mathbf{U}[a,\rho+\eta(2\rho)\rho]$$

and $\mathbf{M}(T_\rho)\ge\mathbf{M}(\mathbf{p}_\# T_\rho)=\kappa\,\boldsymbol{\alpha}(m)\rho^m$ for $0<\rho<t$, we infer

$$\Theta_*^m(\|S\|,a)\ge\liminf_{\rho\to 0+}\mathbf{M}(T_\rho)/[\boldsymbol{\alpha}(m)\rho^m]\ge\kappa.$$

Moreover the hypothesis of the theorem implies $\Theta^{*m}(\|S\|,a)<\infty$, hence

$$\mu=\sup\{\rho^{-m}\mathbf{M}(T_\rho): 0<\rho<t\}<\infty.$$

For any number σ between 0 and $t/2$ we see from 4.2.1 that

$$\sigma^{-1}\int_\sigma^{*2\sigma}\mathbf{M}(\partial T_\rho)\,d\mathscr{L}^1\rho\le\sigma^{-1}\mathbf{M}(T_{2\sigma})\le\mu\,2^m\sigma^{m-1}$$

and choose ρ between σ and 2σ with

$$\mathbf{M}(\partial T_\rho)\le\mu\,2^m\sigma^{m-1}\le\mu\,2^m\rho^{m-1}, \text{ hence } T_\rho\in\mathbf{I}_m(A),$$

by 4.2.16. We let h be the affine homotopy from $\mathbf{p}^*\circ\mathbf{p}|Z$ to $\mathbf{1}_Z$, infer from 4.1.9 that

$$\begin{aligned}&\partial[h_\#([0,1]\times\partial T_\rho)+\mathbf{p}^*_\#(\mathbf{p}_\# T_\rho)-T_\rho]=0,\\ &\mathbf{M}[h_\#([0,1]\times\partial T_\rho)]\le\eta(2\rho)\,\rho\,\mathbf{M}(\partial T_\rho),\end{aligned}$$

and use the fact that T_ρ is absolutely Ψ minimizing with respect to Z in estimating

$$\begin{aligned}&\int \Psi[a, \vec{T}_\rho(z)]\, d\|T_\rho\|\, z - \lambda\, 2\rho\, \mathbf{M}(T_\rho)\\ &\le \langle \Psi, T_\rho\rangle \le \langle \Psi, \mathbf{p}_\#^*(\mathbf{p}_\# T_\rho) + h_\#([0,1]\times \partial T_\rho)\rangle\\ &\le \kappa \int_{\{x:\, |x|<\rho\}} \Psi[\mathbf{p}^*(x), \zeta]\, d\mathscr{L}^m x + \lambda\, \eta(2\rho)\, \rho\, \mathbf{M}(\partial T_\rho)\\ &\le \kappa[\Psi(a,\zeta) + \lambda\, \rho]\, \boldsymbol{\alpha}(m)\, \rho^m + \lambda\, \eta(2\rho)\, \rho\, \mathbf{M}(\partial T_\rho)\end{aligned}$$

where $\zeta = \mathbf{p}^*(e_1)\wedge\cdots\wedge\mathbf{p}^*(e_m)$. We also define

$$R_\rho = T_\rho - h_\#([0,1]\times\partial T_\rho)$$

and use the fact that λ^{-1} is an ellipticity bound for Ψ at a in estimating

$$\begin{aligned}&\lambda^{-1}[\rho^m\, \mathrm{Exc}(T_\rho, 0, \rho) - \eta(2\rho)\, \rho\, \mathbf{M}(\partial T_\rho)] \le \lambda^{-1}(\mathbf{M}(R_\rho) - \mathbf{M}[\mathbf{p}_\#^*(\mathbf{p}_\# T_\rho)])\\ &\le \int \Psi[a, \vec{R}_\rho(z)]\, d\|R_\rho\|\, z - \kappa\, \Psi(a,\zeta)\, \boldsymbol{\alpha}(m)\, \rho^m\\ &\le \int \Psi[a, \vec{T}_\rho(z)]\, d\|T_\rho\|\, z + \lambda\, \eta(2\rho)\, \rho\, \mathbf{M}(\partial T_\rho) - \kappa\, \Psi(a,\zeta)\, \boldsymbol{\alpha}(m)\, \rho^m.\end{aligned}$$

Combining these estimates we obtain

$$\begin{aligned}&\lambda^{-1} 2^{-m}\, \mathrm{Exc}(T_\sigma, 0, \sigma) \le \lambda^{-1}\, \mathrm{Exc}(T_\rho, 0, \rho)\\ &\le (\lambda^{-1} + 2\lambda)\, \eta(2\rho)\, \rho^{1-m}\, \mathbf{M}(\partial T_\rho) + 2\lambda\, \rho^{1-m}\, \mathbf{M}(T_\rho) + \kappa\, \lambda\, \rho\, \boldsymbol{\alpha}(m)\\ &\le (\lambda^{-1} + 2\lambda)\, \eta(2\rho)\, \mu\, 2^m + 2\lambda\, \rho\, \mu + \kappa\, \lambda\, \rho\, \boldsymbol{\alpha}(m).\end{aligned}$$

We infer that

$$\lim_{\sigma\to 0+} \mathrm{Exc}(T_\sigma, 0, \sigma) = 0, \quad \text{hence} \quad \Theta^m(\|S\|, a) = \kappa.$$

Finally we choose ρ so that both ρ and $\mathrm{Exc}(T_\rho, 0, \rho)$ are smaller than the number ε corresponding to λ and κ according to 5.3.14, and we let

$$W = \mathbf{U}(a, 2t) \cap \{x\colon |x| < \varepsilon\, \rho\}, \quad \text{hence} \quad W \cap \operatorname{spt} S = f.$$

5.3.17. Suppose $U \subset A$ are open subsets of $\mathbf{R}^n$, Ψ is a positive elliptic integrand of degree m and class 3 on A, $S \in \mathscr{R}_m(A)$ and $S \llcorner U$ is absolutely Ψ minimizing with respect to U.

A point $a \in (U \cap \operatorname{spt} S) \sim \operatorname{spt} \partial S$ is termed **regular** or **singular** with respect to S according to whether or not it has a neighborhood W in U such that $W \cap \operatorname{spt} S$ is a connected m dimensional submanifold of class 2 of $\mathbf{R}^n$. According to 4.1.31 the regularity condition implies that the manifold $B = W \cap \operatorname{spt} S$ is oriented by $\vec{S}|B$, $\Theta^m(\|S\|, a)$ is a positive integer and

$$S \llcorner W = \Theta^m(\|S\|, a)\cdot(\mathscr{H}^m \llcorner B) \wedge \vec{S}.$$

From 4.1.28 and 5.3.16 it follows that *the set of regular points is dense in* $(U \cap \operatorname{spt} S) \sim \operatorname{spt} \partial S$. However *it is not known whether the set of singular points must have m dimensional Hausdorff measure* 0. One might even conjecture that the intersection of the singular set with every compact subset of $U \sim \operatorname{spt} \partial S$ has finite k dimensional measure, where $k=m-1$ or $k=m-2$, depending on one's degree of optimism. In 5.4.19 we will show by examples that the singular set can have positive $m-2$ dimensional measure.

A major difficulty impeding immediate further progress is the hypothesis that $\Theta^m(\|S\|, z) \geq \Theta^{*m}(\|S\|, a)$ for $\|S\|$ almost all z in some neighborhood of a, which we were forced to require in 5.3.16, but which may be redundant.

In case Ψ is an analytic integrand one may further ask whether $U \cap \operatorname{spt} S$ must be analytic in U (see 3.4.5, 4.2.28, 5.2.16).

While the general problem regarding the nature of $U \cap \operatorname{spt} S$ is still quite baffling, the following results about important special cases are known:

If $m=1$, *then the singular set is empty* (see 5.3.20).

If $m=n-1$, *then the singular set has m dimensional measure* 0 (see 5.3.19).

If $m=n-1 \leq 6$ *and* Ψ *is the area integrand of degree m, then the singular set is empty* (see 5.4.15).

5.3.18. Theorem. *If* Ψ *is a positive elliptic integrand of degree m and class* $q+1 \geq 3$ *on* $\mathbf{R}^{m+1}$ *such that*

$$\Psi(z, \zeta) = \Psi(0, \zeta) \text{ for } (z, \zeta) \in \mathbf{R}^{m+1} \times \wedge_m \mathbf{R}^{m+1},$$

and if $S \in \mathscr{R}_m(\mathbf{R}^{m+1})$, U *is an open subset of* $\mathbf{R}^{m+1}$, $S \llcorner U$ *is absolutely* Ψ *minimizing with respect to U,*

$$a \in U \cap (\operatorname{spt} S) \sim \operatorname{spt} \partial S,$$

$\operatorname{Tan}^m(\|S\|, a)$ *is contained in some m dimensional vectorsubspace of* $\mathbf{R}^{m+1}$, *then a has a neighborhood W in U such that* $W \cap \operatorname{spt} S$ *is a connected m dimensional submanifold of class q of* $\mathbf{R}^{m+1}$.

Proof. First we choose finite positive numbers λ and ρ such that λ^{-1} is an ellipticity bound for Ψ at 0,

$$\lambda^{-1}|\zeta| \leq \Psi(0, \zeta) \leq \lambda|\zeta| \text{ for } \zeta \in \wedge_m \mathbf{R}^{m+1},$$

$$\mathbf{B}(a, \rho) \subset U \sim \operatorname{spt} \partial S \quad \text{and} \quad S \llcorner \mathbf{U}(a, \rho) \in \mathbf{I}_m(\mathbf{R}^{m+1}).$$

Since $\operatorname{spt} \partial[S \llcorner \mathbf{U}(a,\rho)] \subset \Omega = \mathbf{R}^{m+1} \cap \{z\colon |z-a| = \rho\}$ and

$$S \llcorner \mathbf{U}(a,\rho) \in \mathscr{Z}_m[\mathbf{B}(a,\rho),\Omega] = \mathscr{B}_m[\mathbf{B}(a,\rho),\Omega]$$

there exists $\Xi \in \mathbf{I}_m(\mathbf{R}^{m+1})$ with

$$\operatorname{spt} \Xi \subset \Omega \quad \text{and} \quad \partial \Xi = \partial[S \llcorner \mathbf{U}(a,\rho)].$$

Taking $R = [S \llcorner \mathbf{U}(a,\rho)] - \Xi$, $n = m+1$ we apply 4.5.17 and use the resulting sets M_i to construct the rectifiable currents

$$S_i = [\partial(\mathbf{E}^n \llcorner M_i)] \llcorner \mathbf{U}(a,\rho) \quad \text{with} \quad \operatorname{spt} \partial S_i \subset \Omega,$$

because $\partial S_i = -\partial([\partial(\mathbf{E}^n \llcorner M_i)] \llcorner \Omega)$, and

$$S \llcorner \mathbf{U}(a,\rho) = \sum_{i \in \mathbf{Z}} S_i, \qquad \|S \llcorner \mathbf{U}(a,\rho)\| = \sum_{i \in \mathbf{Z}} \|S_i\|.$$

It follows that each S_i is absolutely Ψ minimizing with respect to $A = \mathbf{B}(a,\rho)$, and the density property established in 5.1.6 implies

$$\|S_i\| \, \mathbf{U}(a,\rho) \geq \lambda^{2-2m} \sigma^{-m} m^{-m} [\rho - \operatorname{dist}(a, \Omega \cup \operatorname{spt} S_i)]^m;$$

consequently the set $\Delta = \{i\colon a \in \operatorname{spt} S_i\}$ is finite, and

$$\alpha = \inf\{\operatorname{dist}(a, \Omega \cup \operatorname{spt} S_i)\colon i \in \mathbf{Z} \sim \Delta\} > 0.$$

Identifying $\mathbf{R}^{m+1}$ with $\mathbf{R}^m \times \mathbf{R}$ we assume

$$a = (0,0), \qquad \operatorname{Tan}(\operatorname{spt} S, a) = \operatorname{Tan}^m(\|S\|, a) \subset \mathbf{R}^m \times \{0\}$$

and will prove next that $\Theta^{*m}(\|S_i\|, a) \leq 1$ for each integer i. Given any number ω between 0 and 1 we define

$$C_r^+ = (\mathbf{R}^m \times \mathbf{R}) \cap \{(x,y)\colon |x| < r, \omega r/2 < y < r\},$$

$$C_r^- = (\mathbf{R}^m \times \mathbf{R}) \cap \{(x,y)\colon |x| < r, -r < y < -\omega r/2\},$$

$$D_r = (\mathbf{R}^m \times \mathbf{R}) \cap \{(x,y)\colon |x| < r, -\omega r < y < \omega r\},$$

$$E_r = (\mathbf{R}^m \times \mathbf{R}) \cap \{(x,y)\colon |x| = r, -\omega r \leq y \leq \omega r\}$$

for $0 < r < \rho/2$, and observe that

$$(C_r^+ \cup C_r^-) \cap \operatorname{spt} S_i = \varnothing$$

when r is sufficiently small; then it follows from the constancy theorem in 4.1.7 that

$$\begin{aligned}
&\text{either} && (1)\ \mathscr{L}^n(C_r^+ \sim M_i) = 0 = \mathscr{L}^n(C_r^- \cap M_i),\\
&\text{or} && (2)\ \mathscr{L}^n(C_r^+ \cap M_i) = 0 = \mathscr{L}^n(C_r^- \sim M_i),\\
&\text{or} && (3)\ \mathscr{L}^n(C_r^+ \cap M_i) = 0 = \mathscr{L}^n(C_r^- \cap M_i),\\
&\text{or} && (4)\ \mathscr{L}^n(C_r^+ \sim M_i) = 0 = \mathscr{L}^n(C_r^- \sim M_i).
\end{aligned}$$

In case (1) holds we let $T_r = \partial[\mathbf{E}^n \llcorner (D_r \cap M_i)]$, use 4.5.6, 4.5.7 to represent

$$T_r = S_i \llcorner D_r + P_r + Q_r \text{ with}$$

$$P_r = [\mathbf{E}^m \llcorner \mathbf{U}(0,r)] \times \delta_{\omega r} \quad \text{and} \quad \|Q_r\| \le \mathscr{H}^m \llcorner E_r,$$

apply the Ψ minimizing property of $S_i \llcorner D_r$ and the ellipticity of Ψ to estimate

$$\langle \Psi, S_i \llcorner D_r \rangle \le \langle \Psi, S_i \llcorner D_r - T_r \rangle \le \langle \Psi, -P_r \rangle + \langle \Psi, -Q_r \rangle,$$

$$\lambda^{-1}[\mathbf{M}(S_i \llcorner D_r + Q_r) - \mathbf{M}(-P_r)] \le \langle \Psi, S_i \llcorner D_r \rangle + \langle \Psi, Q_r \rangle - \langle \Psi, -P_r \rangle,$$

$$\mathbf{M}(S_i \llcorner D_r) \le \mathbf{M}(Q_r) + \mathbf{M}(P_r) + \lambda[\langle \Psi, -Q_r \rangle + \langle \Psi, Q_r \rangle]$$

$$\le \mathbf{M}(P_r) + (1 + 2\lambda^2)\mathscr{H}^m(E_r)$$

and infer that, since $\mathbf{U}(a,r) \subset C_r^+ \cup C_r^- \cup D_r$ and $\lambda \ge 1$,

$$\|S_i\| \, \mathbf{U}(a,r) \le \|S_i\| \, D_r \le \alpha(m)\, r^m (1 + 6\lambda^2 m\, \omega).$$

In case (2) the same conclusion is obtained by a similar argument with $P_r = -[\mathbf{E}^m \llcorner \mathbf{U}(0,r)] \times \delta_{-\omega r}$. In case (3) one can take $P_r = 0$ and verify that

$$\|S_i\| \, \mathbf{U}(a,r) \le \|S_i\| \, D_r \le \alpha(m)\, r^m\, 6\lambda^2 m\, \omega.$$

This result is true also in case (4), as seen by analogous reasoning with $T_r = -\partial[\mathbf{E}^n \llcorner (D_r \sim M_i)]$ and $P_r = 0$. Having thus shown that

$$\Theta^{*m}(\|S_i\|, a) \le 1 + 6\lambda^2 m\, \omega \quad \text{whenever} \quad 0 < \omega < 1,$$

and recalling 4.1.28, we find that

$$\Theta^{*m}(\|S_i\|, a) \le 1 \le \Theta^m(\|S_i\|, z) \text{ for } \|S_i\| \text{ almost all } z.$$

Now we apply 5.3.16 with S replaced by S_i for each $i \in \Delta$, to obtain positive numbers β, γ and maps

$$f_i \colon \mathbf{R}^m \cap \mathbf{U}(0,\beta) \to \mathbf{R} \text{ of class } q$$

such that $W = (\mathbf{R}^m \times \mathbf{R}) \cap \{(x,y)\colon |x| < \beta, |y| < \gamma\} \subset \mathbf{U}(a,\rho)$,

$$W \cap \operatorname{spt} S_i = f_i \text{ for all } i \in \Delta.$$

Defining $W_i^+ = W \cap \{(x,y)\colon y > f_i(x)\}$, $W_i^- = W \cap \{(x,y)\colon y < f_i(x)\}$ we observe that, for each $i \in \Delta$,

$$\text{either} \quad (i^+) \ \mathscr{L}^n(W_i^+ \sim M_i) = 0 = \mathscr{L}^n(W_i^- \cap M_i)$$

$$\text{or} \quad (i^-) \ \mathscr{L}^n(W_i^+ \cap M_i) = 0 = \mathscr{L}^n(W_i^- \sim M_i).$$

If $i \in \Delta$, $j \in \Delta$ and $i > j$, then $M_i \subset M_j$, hence

$$(i^+) \text{ implies } (j^+) \text{ and } f_i \geq f_j, \quad (i^-) \text{ implies } (j^-) \text{ and } f_i \leq f_j;$$

in both cases we can apply the corollary of the maximum principle in 5.2.19 with

$$H(x,\tau) = \Psi^{\S}(0,0,\tau) \text{ for } (x,\tau) \in \mathbf{R}^m \times \operatorname{Hom}(\mathbf{R}^m, \mathbf{R}),$$

as we may by 5.1.9 and 5.2.17, to infer that $\{x: f_i(x) = f_j(x)\}$ is open; since this set is relatively closed in $\mathbf{R}^m \cap \mathbf{U}(0,\beta)$ and

$$(0,0) = a \in (\operatorname{spt} S_i) \cap (\operatorname{spt} S_j), \quad f_i(0) = 0 = f_j(0),$$

we conclude $f_i = f_j$. Thus $W \cap \operatorname{spt} S = f_i$ whenever $i \in \Delta$.

5.3.19. Theorem. *If Ψ is a positive elliptic integrand of degree m and class $q+1 \geq 3$ on the open subset A of $\mathbf{R}^{m+1}$, $S \in \mathscr{R}_m(A)$ and S is absolutely Ψ minimizing with respect to A, then A contains an open set Z such that*

$$\mathscr{H}^m[(\operatorname{spt} S \sim \operatorname{spt} \partial S) \sim Z] = 0$$

and $Z \cap \operatorname{spt} S$ is an m dimensional submanifold of class q of A.

Proof. Given $a \in \operatorname{spt} S \sim \operatorname{spt} \partial S$, we construct $\rho, \Omega, \Xi, R, M_i, S_i, \Delta, \alpha$ as in the first paragraph of the proof of 5.3.18. From 4.5.6 and 4.1.28 we infer for $\mathscr{H}^m$ almost all b in $\mathbf{U}(a,\alpha)$ that

$$\Theta^m(\|S_i\|, b) = \Theta^m(\|\partial(\mathbf{E}^{m+1} \llcorner M_i)\|, b) \leq 1 \text{ whenever } i \in \mathbf{Z},$$

$\operatorname{Tan}^m(\|S\|, b)$ is obtained in some m dimensional vectorsubspace of $\mathbf{R}^{m+1}$, and $\Delta_b = \{i: b \in \operatorname{spt} S_i\} \subset \Delta$.

To study the behavior of S near a particular point b which satisfies the above conditions, we identify $\mathbf{R}^{m+1}$ with $\mathbf{R}^m \times \mathbf{R}$ and assume

$$b = (0,0), \quad \operatorname{Tan}^m(\|S\|, b) \subset \mathbf{R}^m \times \{0\}.$$

Applying 5.3.16 to S_i at b, for each $i \in \Delta_b$, we construct f_i, β, γ, W as in the last paragraph of the proof of 5.3.18 (with a, Δ replaced by b, Δ_b). Using the corollary of the second proposition in 5.2.19 with $G = \Psi^{\S}$ (as a substitute for the corollary of the maximum principle) we find that if $i \in \Delta_b$ and $j \in \Delta_b$, then either $f_i = f_j$ or

$$\mathscr{H}^{m-1}\{x: f_i(x) = f_j(x)\} = 0, \text{ hence } \mathscr{H}^{m-1}(f_i \cap f_j) = 0.$$

Defining $D = \bigcup \{f_i \cap f_j : i \in \Delta_b, j \in \Delta_b, f_i \neq f_j\}$ we conclude that $\mathscr{H}^{m-1}(D) = 0$, $W \sim D$ is open in $\mathbf{R}^{m+1}$ and $(W \sim D) \cap \operatorname{spt} S$ is an m dimensional submanifold of class q of $\mathbf{R}^{m+1}$.

5.3.20. Theorem. *If Ψ is a positive elliptic integrand of degree* 1 *and class* $q+1\geq 3$ *on the open subset A of* $\mathbf{R}^n$, $S\in\mathbf{I}_1(A)$ *and S is absolutely Ψ minimizing with respect to A, then* (spt S)$\sim$spt ∂S *is a* 1 *dimensional submanifold of class q of* $\mathbf{R}^n$.

Proof. Since the problem is local we may assume that there exists a compact convex subset K of A and a finite positive number λ such that spt $S\subset\operatorname{Int}K$, Ψ has the ellipticity bound λ^{-1} at each point of K and

$$\lambda^{-1}\leq\Psi(z,\zeta)\leq\lambda,\ \ \|D^i\Psi(z,\zeta)\|\leq\lambda\ \text{ for } i\in\{1,2,3\}$$

whenever $(z,\zeta)\in K\times\mathbf{S}^{n-1}$. Taking $\kappa=1$ we choose ε according to 5.3.14.

First we consider the special case when

$$S=g_\#[0,\mathbf{M}(S)]\ \text{ with } g\colon\mathbf{R}\to\operatorname{Int}K \text{ and } \operatorname{Lip}(g)\leq 1,$$

hence $\mathbf{M}(g_\#[\alpha,\beta])=\beta-\alpha$ for $0\leq\alpha\leq\beta\leq\mathbf{M}(S)$; it follows from the Ψ minimality of $g_\#[\alpha,\beta]$ that

$$\lambda^{-1}(\beta-\alpha)\leq\langle\Psi,g_\#[\alpha,\beta]\rangle\leq\langle\Psi,[g(\alpha),g(\beta)]\rangle\leq\lambda|g(\beta)-g(\alpha)|,$$

and from the ellipticity of Ψ at $g(\alpha)$ that

$$\begin{aligned}0&\geq\langle\Psi,g_\#[\alpha,\beta]\rangle-\langle\Psi,[g(\alpha),g(\beta)]\rangle\\&\geq\langle\Phi,g_\#[\alpha,\beta]\rangle-\langle\Phi,[g(\alpha),g(\beta)]\rangle-2\lambda(\beta-\alpha)^2\\&\geq\lambda^{-1}[\beta-\alpha-|g(\beta)-g(\alpha)|]-2\lambda(\beta-\alpha)^2,\end{aligned}$$

where $\Phi(z,\zeta)=\Psi[g(\alpha),\zeta]$ for $z,\zeta\in\mathbf{R}^n$; consequently

$$|g(\beta)-g(\alpha)|\geq(\beta-\alpha)[1-2\lambda^4|g(\beta)-g(\alpha)|].$$

Given $a\in$(spt S)$\sim$spt ∂S and $r>0$ with

$$r<\operatorname{dist}(a,\operatorname{spt}\partial S),\ 8\lambda^4r<1,\ (1-8\lambda^4r)^{-2}<1+\varepsilon,$$

we let $C=\{t\colon 0\leq t\leq\mathbf{M}(S)\}\cap g^{-1}\mathbf{B}(a,r)$, $\alpha=\inf C$, $\beta=\sup C$ and choose $s\in C$ so that $g(s)=a$; then

$$|g(\alpha)-a|=r=|g(\beta)-a|,\ \alpha+r\leq s\leq\beta-r,\ \beta-\alpha\geq 2r,$$

hence $|g(\beta)-g(\alpha)|\geq 2r(1-4\lambda^4r)>0$; defining

$$u=|g(\beta)-g(\alpha)|^{-1}[g(\beta)-g(\alpha)],\ p(z)=z\bullet u\ \text{ for } z\in\mathbf{R}^n,$$

we note that $|p[g(\beta)]-p(a)|\leq r$, $|p(a)-p[g(\alpha)]|\leq r$ and

$$\begin{gathered}p[g(\beta)]-p(a)+p(a)-p[g(\alpha)]=|g(\beta)-g(\alpha)|\geq 2r-8\lambda^4r^2,\\p[g(\beta)]-p(a)\geq\rho,\ p(a)-p[g(\alpha)]\geq\rho\ \text{ with } \rho=r-8\lambda^4r^2,\end{gathered}$$

hence infer that the current $T=(g_{\#}[\alpha,\beta])\llcorner p^{-1}\mathbf{B}[p(a),\rho]$ satisfies the conditions $p_{\#}T=[p(a)-\rho,p(a)+\rho]$,

$$\begin{aligned}&(2\rho)^{-1}[\mathbf{M}(T)-\mathbf{M}(p_{\#}T)]\le(2\rho)^{-1}[\beta-\alpha-2\rho]\\&\le(2\rho)^{-1}|g(\beta)-g(\alpha)|\,[1-2\lambda^4|g(\beta)-g(\alpha)|]^{-1}-1\\&\le(2\rho)^{-1}2r(1-4\lambda^4r)^{-1}-1\le(1-8\lambda^4r)^{-2}-1<\varepsilon\end{aligned}$$

and $a\notin\operatorname{spt}(S-T)$. Applying 5.3.14 to an isometric image of T we obtain a neighborhood W of a such that $W\cap\operatorname{spt}S$ is a 1 dimensional submanifold of class q of $\mathbf{R}^n$.

To treat the general case we recall 4.2.25 and select indecomposable currents $S_i\in\mathbf{I}_1(A)$ such that

$$S=\sum_{i=1}^{\nu}S_i\quad\text{and}\quad\mathbf{N}(S)=\sum_{i=1}^{\nu}\mathbf{N}(S_i);$$

only finitely many summands occur here because the numbers $\mathbf{M}(\partial S_i)$ are positive integers. We also represent

$$S_i=g_{i\#}[0,\mathbf{M}(S_i)]\ \text{ with }\ g_i\colon\mathbf{R}\to\operatorname{Int}K,\ \operatorname{Lip}(g_i)=1$$

and infer from the preceding discussion that $(\operatorname{spt}S_i)\sim\operatorname{spt}\partial S_i$ is a 1 dimensional submanifold of class q of $\mathbf{R}^n$.

If $a\in[(\operatorname{spt}S_i)\sim\operatorname{spt}\partial S_i]\cap[(\operatorname{spt}S_j)\sim\operatorname{spt}\partial S_j]$, then there exist numbers s_i,s_j with

$$0<s_i<\mathbf{M}(S_i),\qquad 0<s_j<\mathbf{M}(S_j),\qquad g_i(s_i)=a=g_j(s_j).$$

Observing that $S_{i,j}=g_{i\#}[0,s_i]+g_{j\#}[s_j,M(S_j)]$ is absolutely Ψ minimizing with respect to A, we define

$$g_{i,j}(t)=g_i(t)\ \text{ for }\ t\le s_i,\qquad g_{i,j}(t)=g_j(t-s_i+s_j)\ \text{ for }\ t>s_i,$$

note that $\operatorname{Lip}(g_{i,j})\le 1$, $S_{i,j}=g_{i,j\#}[0,\mathbf{M}(S_{i,j})]$, and infer that $(\operatorname{spt}S_{i,j})\sim\operatorname{spt}\partial S_{i,j}$ is a 1 dimensional submanifold of class q of $\mathbf{R}^n$. Accordingly

$$V_i=\operatorname{Tan}(S_i,a),\qquad V_j=\operatorname{Tan}(S_j,a),\qquad V_{i,j}=\operatorname{Tan}(S_{i,j},a)$$

are 1 dimensional vectorsubspaces of $\mathbf{R}^n$ with

$$V_i\cap V_{i,j}\neq\{0\}\neq V_j\cap V_{i,j},\ \text{ hence }\ V_i=V_{i,j}=V_j.$$

Representing S_i and S_j nonparametrically near a we apply 5.2.20 to conclude $S_i\llcorner W=S_j\llcorner W$ for some neighborhood W of a.

5.3.21. The methods of this section remain applicable when rectifiable currents are replaced by flat chains modulo 2 (see 4.2.26) of finite mass, and yield even better results in this case.

Assuming Ψ is a positive parametric integrand of degree m and class $q+1\geq 3$ on the open subset A of $\mathbf{R}^n$ such that

$$\Psi(z,-\zeta)=\Psi(z,\zeta) \text{ for } z\in\mathbf{R}^n \text{ and } \zeta\in\wedge_m\mathbf{R}^n,$$

we define

$$\langle\Psi,T\rangle^{(2)}=\int\Psi[z,\vec{T}(z)]\,d\,\|T\|^2\,z$$

for $T\in\mathscr{R}_m(A)$, and note that this integral depends only on the equivalence class of T modulo 2. We suppose also that Ψ is **elliptic modulo 2**, *which means that for each compact $K\subset A$ there exists a $c>0$ such that*

$$\|R\|^2(\Psi_a\circ\vec{R})-\|S\|^2(\Psi_a\circ\vec{S})\geq c[\mathbf{M}^2(R)-\mathbf{M}^2(S)]$$

whenever $a\in K$, $R\in\mathscr{R}_m(\mathbf{R}^n)$, $S\in\mathscr{R}_m(\mathbf{R}^n)$, $\partial R\equiv\partial S \bmod 2$ and $\operatorname{spt} S$ is contained in some m dimensional vectorsubspace of $\mathbf{R}^n$.

If $S\in\mathscr{R}_m(A)$ and $\langle\Psi,S+R\rangle^{(2)}\geq\langle\Psi,S\rangle^{(2)}$ for all $R\in\mathscr{R}_m(A)$ with $\partial R\equiv 0$ modulo 2, then A contains an open set W such that

$$\|S\|\,[(\operatorname{spt} S\sim\operatorname{spt}\partial S)\sim W]=0$$

and $W\cap\operatorname{spt} S$ is an m dimensional submanifold of class q of $\mathbf{R}^n$.

The preceding proposition is one of the principal results of [AF 6], the paper from which the basic ideas of this section originate. In the case of flat chains modulo 2 one faces an additional difficulty when constructing $U_{\nu,j}$ from S_ν for the analogue of 5.3.7; this is overcome by [AF 6, 6.1]. On the other hand one now needs only a single excess, the excess modulo 2.

For flat chains modulo 3 the situation is more complicated. Here the analogue of 5.3.19 becomes false. For example, if $1\neq\omega\in\mathbf{C}$ with $\omega^3=1$, then

$$T=[0,\omega]+[0,\omega^2]+[0,1]$$

minimizes length modulo 3, but 0 is a singular point of $\operatorname{spt} T\sim\operatorname{spt}\partial T$.

5.4. Further results on area minimizing currents

5.4.1. Throughout this section we let Ψ be the parametric area integrand of degree m on $\mathbf{R}^n$, abbreviate

$$\delta^{(i)}(T,\Psi,h)=\delta^{(i)}(T,h)$$

and refer to Ψ minimizing currents as **area minimizing currents.**

We will discuss various special methods applicable to the area integrand, by which the general smoothness theory developed in Section 5.3

can be sharpened in this case, and which yield a complete interior regularity theorem (5.4.15) for $m=n-1\le 6$. We also construct examples (5.4.19) showing that singularities of area minimizing currents occur naturally when $1<m<n-1$.

5.4.2. *If U is an open subset of $\mathbf{R}^n$, $Q_i\in\mathscr{R}_m^{\mathrm{loc}}(U)$ and Q_i is absolutely area minimizing with respect to U for $i=1,2,3,\ldots$, and if*

$$Q_i\to Q \text{ in } \mathscr{F}_m^{\mathrm{loc}}(U) \quad \text{as} \quad i\to\infty \qquad \text{(see 4.3.16)},$$

then $Q\in\mathscr{R}_m^{\mathrm{loc}}(U)$, Q is absolutely area minimizing with respect to U and

$$\|Q_i\|\to\|Q\| \quad \text{weakly as} \quad i\to\infty \qquad \text{(see 2.5.19)};$$

moreover, if $L=\mathrm{Clos}(\mathrm{spt}\,Q\cup\bigcup\{\mathrm{spt}\,\partial Q_i\colon 0<i\in\mathbf{Z}\})$, then

$$\{i\colon H\cap\mathrm{spt}\,Q_i\}\neq\varnothing \text{ is finite for every compact } H\subset U\sim L.$$

In fact, whenever K is a compact subset of U and W is a neighborhood of K in U such that $U\cap\mathrm{Clos}\,W$ is compact, we can choose

$$\phi\in\mathscr{D}^0(U) \quad \text{with} \quad \mathrm{spt}\,\phi\subset W, \quad \mathrm{spt}(1-\phi)\subset U\sim K,$$
$$R_i\in\mathscr{R}_m(U) \quad \text{and} \quad S_i\in\mathscr{R}_{m+1}(U)$$

with

$$\mathrm{spt}(Q-Q_i-R_i-\partial S_i)\subset U\sim W, \qquad \mathbf{M}(R_i)+\mathbf{M}(S_i)\to 0 \quad \text{as} \quad i\to\infty,$$

recall 4.2.1 to estimate (in case $\mathbf{M}(S_i)>0$ and $\mathbf{M}(S_1)>0$)

$$\int[\mathbf{M}\langle S_i,\phi,r-\rangle/\mathbf{M}(S_i)+\mathbf{M}\langle S_1,\phi,r-\rangle/\mathbf{M}(S_1)]\,d\mathscr{L}^1 r\le 2\,\mathrm{Lip}(\phi),$$

choose numbers r_i for which $0<r_i<1$,

$$\mathbf{M}\langle S_i,\phi,r_i-\rangle\le 2\,\mathrm{Lip}(\phi)\,\mathbf{M}(S_i) \text{ and } \mathbf{M}\langle S_1,\phi,r_i-\rangle\le 2\,\mathrm{Lip}(\phi)\,\mathbf{M}(S_1),$$

let $K_i=\{z\colon \phi(z)\ge r_i\}$ and obtain the two equations

$$Q_i\llcorner K_i+\partial(S_i\llcorner K_i)=Q\llcorner K_i-R_i\llcorner K_i-\langle S_i,\phi,r_i-\rangle,$$
$$Q_i\llcorner K_i+\partial(S_i\llcorner K_i-S_1\llcorner K_i)$$
$$=Q_1\llcorner K_i+R_1\llcorner K_i-R_i\llcorner K_i+\langle S_1-S_i,\phi,r_i-\rangle.$$

Since $K\subset\mathrm{Int}\{z\colon\phi(z)=1\}\subset K_i$ and $Q_i\llcorner K_i$ is area minimizing, we see from 4.1.5 and the second equation that

$$\|Q\|(K)\le\liminf_{i\to\infty}\|Q_i\|(K_i)$$
$$\le\liminf_{i\to\infty}\mathbf{M}[Q_1\llcorner K_i+R_1\llcorner K_i-R_i\llcorner K_i+\langle S_1-S_i,\phi,r_i-\rangle]$$
$$\le\|Q_1\|(W)+\mathbf{M}(R_1)+2\,\mathrm{Lip}(\Phi)\,\mathbf{M}(S_1)<\infty.$$

It follows that $\|Q\|$ is a Radon measure and Q is a locally rectifiable current, by 4.2.16(3). Now we observe that, given $\varepsilon>0$ and $P\in\mathscr{R}_m(U)$ with $\partial P=0$, we can choose W so that $\|Q\|(W\sim K)<\varepsilon$, hence infer from the first equation that

$$\begin{aligned}\mathbf{M}(Q_i\llcorner K_i)&\le\mathbf{M}[Q_i\llcorner K_i+\partial(S_i\llcorner K_i)+P]\\&=\mathbf{M}[Q\llcorner K_i-R_i\llcorner K_i-\langle S_i,\phi,r_i-\rangle+P]\\&\le\mathbf{M}(Q\llcorner K+P)+\varepsilon+\mathbf{M}(R_i)+2\operatorname{Lip}(\phi)\,\mathbf{M}(S_i)\end{aligned}$$

for all i, and conclude

$$\|Q\|(K)\le\liminf_{i\to\infty}\|Q_i\|(K_i)\le\mathbf{M}(Q\llcorner K+P)+\varepsilon.$$

Thus $Q\llcorner K$ is absolutely area minimizing with respect to U. On the other hand we can take $P=0$ and obtain the inequality

$$\limsup_{i\to\infty}\|Q_i\|(K)\le\|Q\|(K)+\varepsilon.$$

Now the weak convergence of $\|Q_i\|$ to $\|Q\|$ is readily verified through approximation of integrals by sums. In order to prove the final assertion we let

$$\rho=2^{-1}\operatorname{dist}[H,L\cup(\mathbf{R}^n\sim U)],\qquad K=\{z\colon \operatorname{dist}(z,H)\le\rho\}$$

and observe that the density property established in 5.1.6 implies

$$\|Q_i\|(K)\ge\rho^m\sigma^{-m}m^{-m}\quad\text{whenever}\quad H\cap\operatorname{spt}Q_i\ne\varnothing.$$

5.4.3. Theorem. *If* $T\in\mathscr{R}_m^{\mathrm{loc}}(\mathbf{R}^n)$, *$T$ is absolutely area minimizing with respect to* $\mathbf{R}^n$,

$$a\in\operatorname{spt}T,\qquad \delta>0,\qquad \operatorname{spt}\partial T\subset\mathbf{R}^n\sim\mathbf{U}(a,\delta),$$

and $u(z)=|z-a|$ *for* $z\in\mathbf{R}^n$, *then the following eight statements hold:*

(1) $\|T\|\,\mathbf{U}(a,r)=m^{-1}r\int|\vec{T}\llcorner Du|\,d\|\langle T,u,r\rangle\|$ *for* $\mathscr{L}^1$ *almost all numbers r between 0 and δ.*

(2) $\int_{\mathbf{U}(a,s)\sim\mathbf{U}(a,r)}|z-a|^{-m-2}\,|\vec{T}(z)\wedge(z-a)|^2\,d\|T\|\,z$
$=s^{-m}\|T\|\,\mathbf{U}(a,s)-r^{-m}\|T\|\,\mathbf{U}(a,r)$ *for* $0<r<s<\delta$.

(3) $0<\alpha(m)\,\Theta^m(\|T\|,a)\le s^{-m}\|T\|\,\mathbf{U}(a,s)$ *for* $0<s<\delta$.

(4) $s^{-1}\int_0^s|y^{1-m}\mathbf{M}\langle T,u,y\rangle-m\,\alpha(m)\,\Theta^m(\|T\|,a)|\,d\mathscr{L}^1y$
$\le m[s^{-m}\|T\|\,\mathbf{U}(b,s)-\alpha(m)\,\Theta^m(\|T\|,a)]$ *for* $0<s<\delta$.

(5) *Every oriented tangent cone C of T at a is absolutely area minimizing with respect to $\mathbf{R}^n$ and satisfies the conditions*

$$\Theta^m(\|C\|,0)=\Theta^m(\|T\|,a),\qquad \partial C=0.$$

(6) *There exists an oriented tangent cone of T at a.*

(7) *If $d>0$ and a is a clusterpoint of $\{b: \Theta^m(\|T\|, b)\geq d\}$, then there exists an oriented tangent cone C of T at a such that*

$$\Theta^m(\|C\|, w)\geq d \text{ for some } w\in \mathbf{S}^{n-1}.$$

(8) *If $m=n-1$, $T\llcorner \mathbf{U}(a,\delta)=[\partial(\mathbf{E}^n\llcorner M)]\llcorner \mathbf{U}(a,\delta)$ for some $\mathscr{L}^n$ measurable set M and C is an oriented tangent cone of T at a, then there exists an $\mathscr{L}^n$ measurable set N such that $\mathbf{E}^n\llcorner N$ is an oriented tangent cone of $\mathbf{E}^n\llcorner M$ at a and $\partial(\mathbf{E}^n\llcorner N)=C$.*

Proof. We assume $a=0$, let $\Gamma=\{r: 0<r<\delta\}$,

$$f(r)=\|T\|\{z: |z|<r\} \text{ for } r\in\mathbf{R},$$

and define $\Xi\in\mathscr{D}_0(\mathbf{R})$ by the formula

$$\Xi(\phi)=\int(\phi\circ u)\, u\, |\vec{T}\llcorner D u|^2\, d\|T\|-\int m f \phi\, d\mathscr{L}^1$$

for $\phi\in\mathscr{D}^0(\mathbf{R})$. First we will prove that

$$\Xi(\psi')=0 \text{ whenever } \psi\in\mathscr{D}^0(\mathbf{R}) \text{ with } \operatorname{spt}\psi\subset\Gamma.$$

To this end we consider the isotopic deformation h of class ∞ such that

$$h(t,z)=[1+t\,\psi(|z|)]\,z \text{ for } (t,z)\in\mathbf{R}\times\mathbf{R}^n \text{ with } |t|\,\mathbf{M}(\psi)<1,$$

and note that $h(t,z)=z$ if $z\in\operatorname{spt}\partial T$, hence 5.1.7 implies

$$\delta^{(1)}(T,h)=0.$$

Recalling 5.1.8 we compute

$$h_0(z)=\psi[u(z)]\,z, \qquad \operatorname{grad} u(z)=|z|^{-1}z,$$
$$\langle v, D h_0(z)\rangle\bullet v=\psi'(|z|)\,|z|\,\langle v, D u(z)\rangle^2+\psi(|z|)\,|v|^2$$

for $z\in\mathbf{R}^n\sim\{0\}$, $v\in\mathbf{R}^n$ and we find that

$$\operatorname{trace}(M_z)=\psi'(|z|)\,|z|\cdot|\vec{T}(z)\llcorner D u(z)|^2+m\,\psi(|z|)$$

for $\|\vec{T}\|$ almost all z, because $\vec{T}(z)=v_1\wedge\cdots\wedge v_m$ where $v_1,\ldots,v_m$ are orthonormal and $v_i\bullet z=0$ for $i>1$, hence

$$\sum_{i=1}^m\langle v_i, D u(z)\rangle^2=\langle v_1, D u(z)\rangle^2=|\vec{T}(z)\llcorner D u(z)|^2;$$

we observe also that

$$|\vec{T}(z)\llcorner D u(z)|^2+|\vec{T}(z)\wedge\operatorname{grad} u(z)|^2$$
$$=|v_1\bullet\operatorname{grad} u(z)|^2+|v_1\wedge\operatorname{grad} u(z)|^2=1.$$

Accordingly we obtain the equation

$$0=\delta^{(1)}(T,h)=\int[(\psi'\circ u)\,u\,|\vec{T}\llcorner Du|^2+m(\psi\circ u)]\,d\|T\|.$$

Furthermore it follows from 2.5.18 (3) and 2.9.24 that

$$\int(\psi\circ u)\,d\|T\|=\int_0^\delta\psi\,df=-\int f\,\psi'\,d\mathscr{L}^1.$$

We conclude $0=\Xi(\psi')=-D_1\,\Xi(\psi)$. Thus $\operatorname{spt} D_1\,\Xi\subset\mathbf{R}\sim\Gamma$ and 4.1.4 yields a real number c such that

$$\Xi(\phi)=\int c\,\phi\,d\mathscr{L}^1 \text{ whenever } \phi\in\mathscr{D}^0(\mathbf{R}) \text{ with } \operatorname{spt}\phi\subset\Gamma.$$

Moreover $c=0$, because for $0<s<\delta$ we can choose $\phi\in\mathscr{D}^0(\mathbf{R})$ so that

$$\int\phi\,d\mathscr{L}^1=1,\quad \operatorname{spt}\phi\subset\{r\colon 0<r<s\},\quad \mathbf{M}(\phi)\le 2/s,$$
$$|c|=|\Xi(\phi)|\le\mathbf{M}(\phi)[s\,f(s)+m\,f(s)\,s]\le 2(1+m)\,f(s),$$

and because $f(s)\to 0$ as $s\to 0$ by 4.1.21. Thus $\operatorname{spt}\Xi\subset\mathbf{R}\sim\Gamma$. Next we apply 4.3.2 (2) to compute

$$\int(\phi\circ u)\,u\,|\vec{T}\llcorner Du|^2\,d\|T\|=\|T\llcorner Du\|\,[(\phi\circ u)\,u\,|\vec{T}\llcorner Du|]$$
$$=\int\|\langle T,u,y\rangle\|\,[(\phi\circ u)u|\vec{T}\llcorner Du|]\,d\mathscr{L}^1y=\int\phi(y)\,y\,\|\langle T,u,y\rangle\|(|\vec{T}\llcorner Du|)\,d\mathscr{L}^1y$$

for $\phi\in\mathscr{D}^0(\mathbf{R})$, because $\operatorname{spt}\langle T,u,y\rangle\subset\{z\colon u(z)=y\}$ for $\mathscr{L}^1$ almost all y, and obtain

$$\Xi(\phi)=\int\phi(y)[y\int|\vec{T}\llcorner Du|\,d\|\langle T,u,y\rangle\|-m\,f(y)]\,d\mathscr{L}^1y.$$

From this formula and the fact that $\operatorname{spt}\Xi\subset\mathbf{R}\sim\Gamma$ we deduce (1). We similarly verify, using (1), that

$$\int_{\{z\colon r\le|z|<s\}}u^{-m}\,|\vec{T}\wedge\operatorname{grad}u|^2\,d\|T\|$$
$$=\int_{\{z\colon r\le|z|<s\}}u^{-m}(1-|\vec{T}\llcorner Du|^2)\,d\|T\|$$
$$=\int_r^s y^{-m}\,d_y f(y)-\int_{\mathbf{U}(a,s)\sim\mathbf{U}(a,r)}u^{-m}\,|\vec{T}\llcorner Du|\,d\|T\llcorner Du\|$$
$$=s^{-m}f(s)-r^{-m}f(r)+\int_r^s f(y)\,m\,y^{-m-1}\,d\mathscr{L}^1y$$
$$-\int_r^s y^{-m}\,\|\langle T,u,y\rangle\|(|\vec{T}\llcorner Du|)\,d\mathscr{L}^1y=s^{-m}f(s)-r^{-m}f(r)$$

whenever $0<r<s<\delta$, as asserted in (2). Accordingly the function mapping r onto $r^{-m}f(r)$ is nondecreasing on Γ, and (3) becomes evident (see 5.1.6). Next we observe that

$$y^{1-m}\,\mathbf{M}\langle T,u,y\rangle\ge m\,y^{-m}f(y)\ge m\,\alpha(m)\,\Theta^m(\|T\|,0)$$

for $\mathscr{L}^1$ almost all y in Γ, by (1) and (3), and that

$$\begin{aligned}
&\int_r^s y^{1-m}\,\mathbf{M}\langle T,u,y\rangle\, d\mathscr{L}^1 y = \int_{\{z:\, r\le|z|<s\}} u^{1-m}\, d\,\|T \llcorner D u\| \\
&\le \int_{\{z:\, r\le|z|<s\}} u^{1-m}\, d\,\|T\| = \int_r^s y^{1-m}\, d_y f(y) \\
&= s^{1-m} f(s) - r^{1-m} f(r) + (m-1)\int_r^s f(y)\, y^{-m}\, d\mathscr{L}^1 y \\
&\le s^{1-m} f(s) + (m-1)\, s\, f(s)\, s^{-m} = m\, s^{1-m} f(s)
\end{aligned}$$

whenever $0<r<s<\infty$; letting r approach 0, subtracting and dividing by s we obtain (4).

For any sequence of positive numbers β_i approaching ∞, such that the corresponding currents

$$Q_i = \mu_{\beta_i\,\#}\, T$$

have a limit Q in $\mathscr{F}_m^{\mathrm{loc}}(\mathbf{R}^n)$, we see from (3) that

$$\begin{aligned}
r^{-m}\,\|Q_i\|\,\mathbf{U}(0,r) &= (r/\beta_i)^{-m}\,\|T\|\,\mathbf{U}(0,r/\beta_i) \\
&\to \alpha(m)\,\Theta^m(\|T\|,0) \quad \text{as} \quad i\to\infty \text{ for } 0<r<\infty,
\end{aligned}$$

hence from 5.4.2 that Q is absolutely area minimizing with respect to $\mathbf{R}^n$ and

$$r^{-m}\,\|Q\|\,\mathbf{U}(0,r) = \alpha(m)\,\Theta^m(\|T\|,0) \text{ for } 0<r<\infty.$$

It follows that $\Theta^m(\|Q\|,0) = \Theta^m(\|T\|,0)$. Moreover

$$\operatorname{spt} \partial Q_i \subset \mu_{\beta_i}(\operatorname{spt} \partial T) \subset \mathbf{R}^n \sim \mathbf{U}(0,\beta_i\,\delta),$$

hence $\partial Q = 0$. Applying (2) with T, δ replaced by Q, ∞ we find that $\vec{Q}(z)\wedge z = 0$ for $\|Q\|$ almost all z, and infer from 4.3.14 that Q is an oriented cone.

The preceding result implies (5) and reduces the proof of (6) to the construction of a sequence of numbers β_i approaching ∞ such that the corresponding sequence of currents Q_i is convergent in $\mathscr{F}_m^{\mathrm{loc}}(\mathbf{R}^n)$.

We abbreviate $\gamma = \inf\{1,\delta\}$ and $\lambda = m\,\alpha(m)\,\Theta^m(\|T\|,0)$. For each positive integer i we define

$$F_i = \{y\colon 0<y<\gamma \text{ and } y^{1-m}\,\mathbf{M}\langle T,u,y-\rangle < \lambda + i^{-1}\},$$

$$G_i = \bigcap_{j=1}^{i} \{r\colon j\,r \in F_i\},$$

and apply (4) to choose a positive number σ_i so that

$$\mathscr{L}^1\{y\colon 0<y<s,\ y\notin F_i\} < 2^{-i} s \text{ whenever } 0<s<\sigma_i.$$

It follows that if $0<\varepsilon<i^{-1}\sigma_i$, then

$$\{r: 0<r<\varepsilon, r\notin G_i\}=\bigcup_{j=1}^{i}\{j^{-1}y: 0<y<j\varepsilon, y\notin F_i\},$$

$$\mathscr{L}^1\{r: 0<r<\varepsilon, r\notin G_i\}<\sum_{j=1}^{i}j^{-1}2^{-i}j\varepsilon=i\,2^{-i}\varepsilon,$$

$$G_i\cap\{r: (1-i\,2^{-i})\varepsilon<r<\varepsilon\}\neq\emptyset.$$

We consider sequences of numbers β_i with the property that

$$1/\beta_i\in G_i \text{ for all positive integers } i.$$

This property is inherited by subsequences because

$$F_k\subset F_i \text{ and } G_k\subset G_i \text{ for } k>i.$$

The currents Q_i corresponding to such numbers β_i satisfy the conditions

$$Q_i\llcorner\mathbf{U}(0,j)=\mu_{\beta_i\#}[T\llcorner\mathbf{U}(j/\beta_i)],\quad \mathbf{M}[Q_i\llcorner\mathbf{U}(0,j)]=j^m(j/\beta_i)^{-m}f(j/\beta_i),$$
$$\mathbf{M}\,\partial[Q_i\llcorner\mathbf{U}(0,j)]=j^{m-1}(j/\beta_i)^{1-m}\mathbf{M}\langle T,u,j/\beta_i-\rangle\le j^{m-1}(\lambda+i^{-1})$$

for all integers $i\ge j\ge 1$, hence

$$\limsup_{i\to\infty}\mathbf{N}[Q_i\llcorner\mathbf{U}(0,j)]\le(j^m+j^{m-1}m)\,\alpha(m)\,\Theta^m(\|T\|,0)$$

for all positive integers j. Applying 4.2.17 (2) with $K=\mathbf{B}(0,j)$ for each j and using Cantor's diagonal process, we pass to a subsequence (without changing notation) such that for every positive integer j there exists a current

$$\Omega_j\in\mathbf{I}_{m,\,\mathbf{B}(0,j)}(\mathbf{R}^n)\quad\text{with}\quad\lim_{i\to\infty}\mathscr{F}_{\mathbf{B}(0,j)}[Q_i\llcorner\mathbf{B}(0,j)-\Omega_j]=0.$$

It follows that $\Omega_j\llcorner\mathbf{U}(0,j)=\Omega_l\llcorner\mathbf{U}(0,l)$ whenever $j<l$, hence the currents Q_i converge in $\mathscr{F}_m^{\text{loc}}(\mathbf{R}^n)$ to a current Q such that $Q\llcorner\mathbf{U}(0,j)=\Omega_j\llcorner\mathbf{U}(0,j)$ for all j. The proof of (6) is now complete.

In case the hypothesis of (7) holds we can first choose a sequence of points

$$b_i\in\mathbf{U}(0,i^{-1}\sigma_i)\sim\{0\}\quad\text{with}\quad\Theta^m(\|T\|,b_i)\ge d,$$

then choose β_i so that

$$1/\beta_i\in G_i\quad\text{and}\quad(1-i\,2^{-i})|b_i|<1/\beta_i<|b_i|.$$

The additional property is also inherited by subsequences because $k\,2^{-k}<i\,2^{-i}$ for $k>i$. Since

$$1<|\beta_i b_i|<(1-i\,2^{-i})^{-1}\to 1 \text{ as } i\to\infty,$$

we may pass to a subsequence such that

$$\beta_i b_i \to w \text{ as } i \to \infty \text{ for some } w \in \mathbf{S}^{n-1}.$$

Applying (3) with T, a, δ replaced by $Q_i, \beta_i b_i, \beta_i \delta - |\beta_i b_i|$ we obtain

$$\alpha(m)\, d \le s^{-m} \|Q_i\| \mathbf{U}(\beta_i b_i, s) \text{ for } 0 < s < \beta_i \delta - |\beta_i b_i|,$$

and use 5.4.2 to conclude

$$\alpha(m)\, d \le s^{-m} \|Q\| \mathbf{U}(w, s) \text{ for } 0 < s < \infty,$$

hence $d \le \Theta^m(\|Q\|, w)$.

Finally we assume the hypothesis of (8) and consider an arbitrary sequence of positive numbers β_i approaching ∞ such that the corresponding currents Q_i converge to C in $\mathscr{F}_{n-1}^{\mathrm{loc}}(\mathbf{R}^n)$. We note that

$$\begin{aligned} Q_i \llcorner \mathbf{U}(0, s) &= (\partial [\mathbf{E}^n \llcorner \mu_{\beta_i}(M)]) \llcorner \mathbf{U}(0, s) \\ &= -(\partial [\mathbf{E}^n \llcorner \mathbf{R}^n \sim \mu_{\beta_i}(M)]) \llcorner \mathbf{U}(0, s) \end{aligned}$$

for $0 < s < \beta_i \delta$, define

$$\tau_i(s) = \inf\{\mathscr{L}^n[\mathbf{U}(0, s) \cap \mu_{\beta_i}(M)], \mathscr{L}^n[\mathbf{U}(0, s) \sim \mu_{\beta_i}(M)]\},$$

infer that

$$\liminf_{i \to \infty} \tau_i(s) > 0$$

because $C \llcorner \mathbf{U}(0, s) \neq 0$ and

$$|C(\phi)| = \lim_{i \to \infty} |Q_i(\phi)| \le \liminf_{i \to \infty} \mathbf{M}(d\phi)\, \tau_i(s)$$

whenever $\phi \in \mathscr{D}^{n-1}(\mathbf{R}^n)$ with $\operatorname{spt} \phi \subset \mathbf{U}(0, s)$, hence use 4.5.2 (3) to conclude that

$$\|\mathbf{E}^n \llcorner \mu_{\beta_i}(M) - \mathbf{E}^n \llcorner \mu_{\beta_k}(M)\|\, \mathbf{U}(0, s) \to 0$$

as i and k approach ∞. Accordingly there exists an $\mathscr{L}^n$ measurable set N such that

$$\mathbf{E}^n \llcorner \mu_{\beta_i}(M) \to \mathbf{E}^n \llcorner N \text{ in } \mathscr{F}_n^{\mathrm{loc}}(\mathbf{R}^n) \text{ as } i \to \infty.$$

Since ∂ is continuous, $\partial(\mathbf{E}^n \llcorner N) = C$. If $0 < r < \infty$, then

$$\partial [\mathbf{E}^n \llcorner \mu_r(N)] = \partial \mu_{r\,\#}(\mathbf{E}^n \llcorner N) = \mu_{r\,\#}\, C = C$$

and the constancy theorem implies $\mathbf{E}^n \llcorner \mu_r(N) = \mathbf{E}^n \llcorner N$. Thus $\mathbf{E}^n \llcorner N$ is an oriented cone.

5.4.4. In the proof of the preceding theorem we deduced the monotonicity of $r^{-m}\,\|T\|\,\mathbf{U}(a,r)$, for $0<r<\delta$, from the explicit formula (2). Here we will describe an alternate method, which applies also to currents minimizing area with respect to certain subsets of $\mathbf{R}^n$ (for example sets with positive reach [F 15, p. 435], in particular submanifolds of class 2). Specifically we assume that

$$a\in A\subset\mathbf{R}^n,\ U \text{ is a neighborhood of } A \text{ in } \mathbf{R}^n,$$
$$\rho \text{ retracts } U \text{ onto } A,\ 0<\alpha<\infty,\ 0\le\tau<\infty,\ 0<\delta,$$
$$\operatorname{Lip}[\rho\,|\,\mathbf{B}(a,r)]\le 1+\tau r^\alpha \text{ for } 0<r<\delta,$$
$$T\in\mathscr{R}_m^{\text{loc}}(\mathbf{R}^n),\quad a\in\operatorname{spt} T,\quad \operatorname{spt}\partial T\subset A\sim\mathbf{U}(a,\delta),$$
$$T \text{ is absolutely area minimizing with respect to } A.$$

Defining u, f as in 5.4.3 and recalling 4.2.1, 4.1.11 we find that

$$\partial[T\llcorner\mathbf{U}(a,r)]=\langle T,u,r-\rangle=\partial\rho_\#(\delta_a\times\langle T,u,r-\rangle),$$
$$f(r)=\mathbf{M}[T\llcorner\mathbf{U}(a,r)]\le\mathbf{M}\,\rho_\#(\delta_a\times\langle T,u,r-\rangle)$$
$$\le(1+\tau r^\alpha)^m m^{-1} r\,\mathbf{M}\langle T,u,r-\rangle\le(1+2^m\tau r^\alpha)\,m^{-1} r f'(r)$$

for $\mathscr{L}^1$ almost all r in $\Delta=\{r: 0<r<\inf\{\delta,\tau^{-1/\alpha}\}\}$. Since f is nondecreasing and positive on Δ, $\log\circ f$ is nondecreasing on Δ. The above inequality implies

$$(\log\circ f)'(r)=\frac{f'(r)}{f(r)}\ge\frac{m}{r}-\frac{m\,2^m\tau r^{\alpha-1}}{1+2^m\tau r^\alpha}$$

for $\mathscr{L}^1$ almost all r in Δ, and integration using 2.9.19 shows that the function χ, defined by the formula

$$\chi(r)=f(r)\,r^{-m}(1+2^m\tau r^\alpha)^{m/\alpha}\ \text{ for } r\in\Delta,$$

is nondecreasing on Δ. It follows that

$$\boldsymbol{\alpha}(m)\,\Theta^m(\|T\|,a)=\lim_{r\to 0+} f(r)\,r^{-m}=\lim_{r\to 0+}\chi(r)\in\mathbf{R}.$$

Having thus proved a modified version of 5.4.3 (3), one easily generalizes 5.4.3 (4), and one can verify that *the conclusions* 5.4.3 (5) *and* (6) *remain true* (as stated) *under the present hypotheses.* For this purpose the reasoning of 5.4.2 may be adapted by consideration of the currents

$$\rho_\#[Q_i\llcorner K_i+\partial(S_i\llcorner K_i)+P]\quad\text{with}\quad Q_i=\mu_{\beta_i\,\#}T,\qquad W\cup\operatorname{spt}P\subset\mu_{\beta_i}(U).$$

It is also relevant that in the proof of 5.4.3 (6) the formula 5.4.3 (2) is applied only with T replaced by the limiting current Q, after Q is known to be area minimizing with respect to $\mathbf{R}^n$.

We are not including all details of these arguments for $A \neq \mathbf{R}^n$, because up to now the principal theorem 5.4.15 of this section has not been extended to $A \neq \mathbf{R}^n$. The scope of the present method is limited by use of the maximum principle 5.2.19 in the proof of 5.3.18. On the other hand the theorems 5.4.6 and 5.4.7 can be extended to currents minimizing area with respect to a submanifold of class ∞ of $\mathbf{R}^n$.

5.4.5. Assuming that $T \in \mathscr{R}_m^{\mathrm{loc}}(\mathbf{R}^n)$ *and* T *is absolutely area minimizing with respect to* $\mathbf{R}^n$ we derive the following corollaries of Theorem 5.4.3:

(1) $\limsup_{a \to b} \Theta^m(\|T\|, a) \leq \Theta^m(\|T\|, b)$ *for* $b \in \mathbf{R}^n \sim \operatorname{spt} \partial T$.

(2) $1 \leq \Theta^m(\|T\|, b)$ *for* $b \in \operatorname{spt} T \sim \operatorname{spt} \partial T$.

(3) *If* $\partial T = 0$ *and there exists a point* $b \in \operatorname{spt} T$ *with*

$$r^{-m} \|T\| \mathbf{U}(b, r) = \alpha(m) \text{ for } 0 < r < \infty,$$

then T *is an oriented* m *dimensional affine subspace of* $\mathbf{R}^m$.

In fact we see from 5.4.3 (3) that

$$\begin{aligned}\alpha(m)\,\Theta^m(\|T\|, a) &\leq s^{-m} \|T\| \mathbf{U}(a, s) \leq s^{-m} \|T\| \mathbf{U}(b, s + |a-b|)\\ &= (1 + |a-b|/s)^m (s + |a-b|)^{-m} \|T\| \mathbf{U}(b, s + |a-b|)\end{aligned}$$

whenever $a \in \operatorname{spt} T$, $0 < s < \operatorname{dist}(a, \operatorname{spt} \partial T)$ and $b \in \mathbf{R}^n$. The assertion (1) follows immediately from this estimate. Then (2) follows from (1) and 4.1.28 (5). Under the additional hypothesis of (3) the estimate implies

$$\limsup_{s \to \infty} s^{-m} \|T\| \mathbf{U}(a, s) \leq \alpha(m),$$

hence $s^{-m} \|T\| \mathbf{U}(a, s) = \alpha(m)$ for $a \in \operatorname{spt} T$ and $0 < s < \infty$, and we infer from 5.4.3 (2) that

$$\vec{T}(z) \wedge (z - a) = 0 \text{ for } \|T\| \text{ almost all } z.$$

Applying 2.6.2 we conclude that $\|T\|$ almost all points z in $\mathbf{R}^n$ satisfy the condition

$$\vec{T}(z) \wedge (z - a) = 0 \text{ for } \|T\| \text{ almost all } a,$$

hence $\operatorname{spt} T \subset \{a : \vec{T}(z) \wedge (a - z) = 0\} = \tau_z \{w : \vec{T}(z) \wedge w = 0\}$.

5.4.6. Theorem. *If* $T \in \mathscr{R}_m^{\mathrm{loc}}(\mathbf{R}^n)$, $a \in \operatorname{spt} T \sim \operatorname{spt} \partial T$, V *is a neighborhood of* a *in* $\mathbf{R}^n$, $T \llcorner V$ *is absolutely area minimizing with respect to* $\mathbf{R}^n$,

$$\Theta^m(\|T\|, z) \geq \Theta^m(\|T\|, a) \text{ for } z \in V \cap \operatorname{spt} T,$$

and there exists an oriented tangent cone C *of* T *at* a *such that* spt C *is an* m *dimensional vectorsubspace of* $\mathbf{R}^n$, *then* a *has a neighborhood* W *in* V *such that* $W \cap \operatorname{spt} T$ *is an* m *dimensional submanifold of class* ∞ *of* $\mathbf{R}^n$.

Proof. Recalling 5.1.11 and 5.4.3(5) we assume

$$a=0, \quad C=\kappa\, \mathbf{p}_\#^* \, \mathbf{E}^m \quad \text{with} \quad \kappa=\Theta^m(\|T\|,0),$$

choose a sequence of positive numbers β_i for which

$$\beta_i\to\infty \text{ and } Q_i=\boldsymbol{\mu}_{\beta_i\,\#}\, T\to C \text{ in } \mathscr{F}_m^{\mathrm{loc}}(\mathbf{R}^n) \text{ as } i\to\infty,$$

choose λ and ε so that the conclusion of 5.3.14 holds for the area integrand Ψ, choose τ so that

$$0<\tau<\lambda^{-1}, \quad \kappa\,\boldsymbol{\alpha}(m)\,[(1+\tau)^m-1]<\varepsilon,$$

and let $u(x,y)=\sup\{|x|,|y|\}$ for $(x,y)\in\mathbf{R}^m\times\mathbf{R}^{n-m}\simeq\mathbf{R}^n$. If i is sufficiently large, then

$$\{(x,y)\colon u(x,y)\le 2\varepsilon\}\subset\boldsymbol{\mu}_{\beta_i}(V\sim\operatorname{spt}\partial T),$$

$$(\operatorname{spt} Q_i)\cap\{(x,y)\colon |x|\le 2\varepsilon,\ \tau\,\varepsilon/2\le|y|\le 2\varepsilon\}=\emptyset$$

by virtue of the last conclusion in 5.4.2, and there exist rectifiable currents $R_i\in\mathscr{R}_m(\mathbf{R}^n)$, $S_i\in\mathscr{R}_{m+1}(\mathbf{R}^n)$ with

$$\operatorname{spt}(Q_i-C-R_i-\partial S_i)\subset\{(x,y)\colon u(x,y)\ge 2\varepsilon\},$$

$$\mathbf{M}(R_i)+\mathbf{M}(S_i)<\boldsymbol{\alpha}(m)\,(\varepsilon/2)^{m+2},$$

hence $\|\partial S_i\|\,\{(x,y)\colon u(x,y)<2\varepsilon\}<\infty$ and 4.2.1 yields numbers ρ_i for which

$$\varepsilon/2<\rho_i<\varepsilon, \quad \mathbf{M}\langle S_i, u, \rho_i-\rangle\le\boldsymbol{\alpha}(m)\,(\varepsilon/2)^{m+1}.$$

Defining $U_i=\{(x,y)\colon u(x,y)<\rho_i\}$, $T_i=Q_i\llcorner U_i$ we infer that T_i is absolutely area minimizing with respect to $\mathbf{R}^n$ and

$$\operatorname{spt}\partial T_i\subset(\operatorname{spt} Q_i)\cap\operatorname{Bdry} U_i\subset\{(x,y)\colon |x|=\rho_i\},$$

$$T_i-C\llcorner U_i-\partial(S_i\llcorner U_i)=R_i\llcorner U_i-\langle S_i,u,\rho_i-\rangle,$$

$$\mathbf{M}\,[\mathbf{p}_\#\, T_i-\kappa\,\mathbf{E}^m\llcorner\mathbf{U}(0,\rho_i)]<2\boldsymbol{\alpha}(m)\,(\varepsilon/2)^{m+1}<\boldsymbol{\alpha}(m)\,\rho_i^m,$$

$$\operatorname{spt}\partial\mathbf{p}_\#\, T_i\subset\operatorname{Bdry}\mathbf{U}(0,\rho_i), \quad \mathbf{p}_\#\, T_i=\kappa\,\mathbf{E}^m\llcorner\mathbf{U}(0,\rho_i),$$

$$\operatorname{spt} T_i\subset\{(x,y)\colon |x|\le\rho_i,\ |y|<\tau\,\varepsilon/2\}\subset\mathbf{U}(0,\rho_i+\tau\,\rho_i),$$

$$\mathbf{M}(T_i)\le\|Q_i\|\,\mathbf{U}(0,\rho_i+\tau\,\rho_i)=\beta_i^m\,\|T\|\,\mathbf{U}[0,(1+\tau)\,\rho_i/\beta_i],$$

$$\begin{aligned}\operatorname{Exc}(T_i,0,\rho_i)&=\rho_i^{-m}\,\mathbf{M}(T_i)-\kappa\,\boldsymbol{\alpha}(m)\\&\le(1+\tau)^m\,[(1+\tau)\,\rho_i/\beta_i]^{-m}\,\|T\|\,\mathbf{U}[0,(1+\tau)\,\rho_i/\beta_i]-\kappa\,\boldsymbol{\alpha}(m)\\&\to(1+\tau)^m\,\kappa\,\boldsymbol{\alpha}(m)-\kappa\,\boldsymbol{\alpha}(m)<\varepsilon \text{ as } i\to\infty.\end{aligned}$$

It follows from 5.3.14 that, for large i,

$$(\operatorname{spt} T_i)\cap\{(x,y)\colon |x|<\varepsilon\,\rho_i\}$$

is an m dimensional submanifold of class ∞ of $\mathbf{R}^n$, and we can take

$$W=\boldsymbol{\mu}_{1/\beta_i}\{(x,y)\colon |x|<\varepsilon\,\rho_i,\ |y|<\rho_i\}.$$

5.4.7. Theorem. *For any integers $n \geq m \geq 1$ there exists a number $\Upsilon > 1$ with the following property:*

If $S \in \mathscr{R}_m^{\mathrm{loc}}(\mathbf{R}^n)$, S is absolutely area minimizing with respect to $\mathbf{R}^n$ and

$$a \in \operatorname{spt} S \sim \operatorname{spt} \partial S \quad \text{with} \quad \Theta^m(\|S\|, a) < \Upsilon,$$

then a has a neighborhood W in $\mathbf{R}^n$ such that $W \cap \operatorname{spt} S$ is an m dimensional submanifold of class ∞ of $\mathbf{R}^n$.

Proof. Recalling 5.4.5 (2), we consider sequences of currents $S_i \in \mathscr{R}_m^{\mathrm{loc}}(\mathbf{R}^n)$ and points $a_i \in \operatorname{spt} S_i \sim \operatorname{spt} \partial S_i$ such that S_i is absolutely area minimizing with respect to $\mathbf{R}^n$ and

$$\Theta^m(\|S_i\|, a_i) \to 1 \text{ as } i \to \infty.$$

For each i we apply 5.4.3 (6), (5), (4) to choose oriented tangent cones C_i of S_i at a_i with $\partial C_i = 0$,

$$r^{-m} \mathbf{M}[C_i \llcorner \mathbf{U}(a, r)] = \alpha(m)\, \Theta^m(\|S_i\|, a_i)$$

and

$$r^{1-m} \mathbf{M}\, \partial[C_i \llcorner \mathbf{U}(a, r)] = m\, \alpha(m)\, \Theta^m(\|S_i\|, a_i)$$

for $0 < r < \infty$. Applying 4.2.17 we pass to a subsequence (without changing notation) such that the sequence of currents $C_i \llcorner \mathbf{U}(0,1)$ is $\mathscr{F}_{\mathbf{B}(0,1)}$ convergent in $\mathbf{I}_{m,\mathbf{B}(0,1)}(\mathbf{R}^n)$, then use the invariance of C_i under the transformations μ_r to obtain a current $C \in \mathscr{R}_m^{\mathrm{loc}}(\mathbf{R}^n)$ such that

$$\mathscr{F}_{\mathbf{B}(0,r)}[C_i \llcorner \mathbf{U}(0,r) - C \llcorner \mathbf{U}(0,r)] \to 0 \text{ as } i \to \infty$$

for $0 < r < \infty$. Clearly $\partial C = 0$. We see from 5.4.2 that C_i and C are absolutely area minimizing with respect to $\mathbf{R}^n$, and that

$$\|C\|\, \mathbf{U}(0,r) = \lim_{r \to \infty} \|C_i\|\, \mathbf{U}(0,r) = \alpha(m)\, r^m$$

for $0 < r < \infty$, hence from 5.4.5 (3) that C is an oriented m dimensional vectorsubspace of $\mathbf{R}^m$.

Recalling 5.1.11 we assume $C = \mathbf{p}_{\#}^{*}\, \mathbf{E}^m$. We choose λ and ε so that the conclusion of 5.3.14 holds for the area integrand Ψ with $\kappa = 1$, then choose τ so that

$$0 < \tau < 1, \qquad 0 < \varepsilon\, \tau < \lambda^{-1}, \qquad \alpha(m)\,[(1+\tau)^m - 1] < \varepsilon$$

and let $\rho = \varepsilon/2$, $T_i = C_i \llcorner \mathbf{p}^{-1}\, \mathbf{U}(0, \rho)$ for all positive integers i. It follows from 5.4.2 that

$$(\operatorname{spt} C_i) \cap \mathbf{B}(0, \varepsilon) \cap \{(x, y) : |y| \geq \rho\, \tau\} = \varnothing$$

when i is sufficiently large, hence

$$\operatorname{spt} T_i \subset \{(x, y) : |x| < \rho,\ |y| < \rho\, \tau\}$$

because the assumption that $(x,y)\in\operatorname{spt} C_i$ with $|x|<\rho$, $|y|\geq\rho\tau$ would imply $|x|^2+|y|^2>\varepsilon^2$, $|y|>\rho$,

$$(u,v)=\rho\,\tau\,|y|^{-1}(x,y)\in\operatorname{spt} C_i$$

with $|u|<\rho\,\tau$, $|v|=\rho\,\tau$, $(u,v)\in\mathbf{B}(0,\varepsilon)$, which has been excluded. Noting that $\operatorname{spt}\partial T_i\subset\{(x,y)\colon |x|=\rho\}$ and

$$\begin{aligned}\mathbf{M}[\mathbf{p}_{\#}T_i-\mathbf{E}^m\llcorner\mathbf{U}(0,\rho)]&=\|\mathbf{p}_{\#}[(C_i-C)\llcorner\mathbf{U}(0,\varepsilon)]\|\,\mathbf{U}(0,\rho)\\&\leq\mathscr{F}_{\mathbf{B}(0,\varepsilon)}[(C_i-C)\llcorner\mathbf{U}(0,\varepsilon)]\to 0\text{ as } i\to\infty,\end{aligned}$$

hence $\mathbf{p}_{\#}T_i=\mathbf{E}^m\llcorner\mathbf{U}(0,\rho)$ for large i, we estimate

$$\begin{aligned}\operatorname{Exc}(T_i,0,\rho)&\leq\rho^{-m}\,\|C_i\|\,\mathbf{U}(0,\rho+\rho\,\tau)-\boldsymbol{\alpha}(m)\\&\to(1+\tau)^m\boldsymbol{\alpha}(m)-\boldsymbol{\alpha}(m)<\varepsilon\text{ as } i\to\infty.\end{aligned}$$

When i is sufficiently large we see from the propositions 5.3.14 and 5.4.5(2) that $(\operatorname{spt} T_i)\cap\mathbf{p}^{-1}\mathbf{U}(0,\varepsilon\,\rho)$ is an m dimensional submanifold of class ∞ of $\mathbf{R}^n$, $\operatorname{spt} C_i=\operatorname{Tan}(\operatorname{spt} T_i,0)$ is an m dimensional vectorsubspace of $\mathbf{R}^n$ and $\Theta^m(\|S_i\|,a_i)=\Theta^m(\|T_i\|,0)=1$; then we apply 5.4.6 and 5.4.5(2) with T replaced by S_i to obtain a neighborhood W_i of a_i in $\mathbf{R}^n$ such that $W_i\cap\operatorname{spt} S_i$ is an m dimensional submanifold of class ∞ of $\mathbf{R}^n$.

5.4.8. Theorem. *If $m\geq 2$, $Q\in\mathscr{R}^{\text{loc}}_{m-1}(\mathbf{R}^{n-1})$ and $\mathbf{E}^1\times Q$ is absolutely m area minimizing with respect to $\mathbf{R}\times\mathbf{R}^{n-1}$, then Q is absolutely $m-1$ area minimizing with respect to $\mathbf{R}^{n-1}$.*

Proof. Otherwise there would exist a compact subset K of $\mathbf{R}^{n-1}$ and a current

$$R\in\mathbf{I}_m(\mathbf{R}^{n-1})\quad\text{with}\quad\mathbf{M}(Q\llcorner K+\partial R)<\mathbf{M}(Q\llcorner K).$$

Choosing $u,v\in\mathbf{R}$ so that $u<v$ and

$$2\mathbf{M}(R)<(v-u)\,[\mathbf{M}(Q\llcorner K)-\mathbf{M}(Q\llcorner K+\partial R)],$$

we would infer $[u,v]\times R\in\mathbf{I}_{m+1}(\mathbf{R}\times\mathbf{R}^{n-1})$ and

$$\begin{aligned}&(\mathbf{E}^1\times Q)\llcorner(\{t\colon u\leq t\leq v\}\times K)=[u,v]\times(Q\llcorner K),\\&\mathbf{M}[[u,v]\times(Q\llcorner K)-\partial([u,v]\times R)]\\&=\mathbf{M}[[u,v]\times(Q\llcorner K+\partial R)-\boldsymbol{\delta}_v\times R+\boldsymbol{\delta}_u\times R]\\&\leq(v-u)\,\mathbf{M}(Q\llcorner K+\partial R)+2\mathbf{M}(R)\\&<(v-u)\,\mathbf{M}(Q\llcorner K)=\mathbf{M}[[u,v]\times(Q\llcorner K)].\end{aligned}$$

5.4.9. Theorem. *If $Q \in \mathscr{R}_{m-1}(\mathbf{R}^{n-1})$, Q is absolutely $m-1$ area minimizing with respect to $\mathbf{R}^{n-1}$ and $u, v \in \mathbf{R}$, then $[u,v] \times Q$ is absolutely m area minimizing with respect to $\mathbf{R}^n$.*

Proof. Suppose $u<v$ and $T \in \mathbf{I}_m(\mathbf{R} \times \mathbf{R}^{n-1})$ with $\partial T = 0$. Defining $f(t,w)=t$ and $g(t,w)=w$ for $(t,w) \in \mathbf{R} \times \mathbf{R}^{n-1}$, we see from 4.1.8, 4.3.1, 4.3.6 that

$$\langle [u,v] \times Q, f, t \rangle = \delta_t \times Q, \quad g_\#(\delta_t \times Q) = Q,$$
$$\langle T, f, t \rangle \in \mathbf{I}_{m-1}(\mathbf{R} \times \mathbf{R}^{n-1}), \quad \partial \langle T, f, t \rangle = 0$$

for $\mathscr{L}^1$ almost all t between u and v; since g maps $f^{-1}\{t\}$ isometrically onto $\mathbf{R}^{n-1}$ we infer

$$\mathbf{M}[\langle ([u,v] \times Q) + T, f, t \rangle] = \mathbf{M}(Q + g_\# \langle T, f, t \rangle) \geq \mathbf{M}(Q).$$

Applying 4.3.2(2) we conclude

$$\mathbf{M}[([u,v] \times Q) + T] \geq \int_u^v \mathbf{M}(Q)\, d\mathscr{L}^1 t = \mathbf{M}([u,v] \times Q).$$

5.4.10. *If $\alpha \in \mathbf{R}$ and $\beta > \alpha^2/4$, then there exists a function $\psi \in \mathscr{D}^0(\mathbf{R})$ with $\operatorname{spt} \psi \subset \{r : 0 < r < 1\}$ and*

$$\int [\psi'(r)]^2 r^{\alpha+1} d\mathscr{L}^1 r < \beta \int [\psi(r)]^2 r^{\alpha-1} d\mathscr{L}^1 r.$$

Choosing σ so that $0<\sigma<1$ and $\alpha^2/4 + \pi^2 [\log(\sigma)]^{-2} < \beta$, we will show that the asserted inequality holds for some Lipschitzian function ψ with $\operatorname{spt} \psi \subset \{r : \sigma \leq r \leq 1\}$. We observe that if $g \in \mathscr{E}^0(\mathbf{R})$ with $g[\log(\sigma)] = 0 = g(0)$, then

$$\int_\sigma^1 [(g \circ \log)'(r)]^2 r^{\alpha+1} d\mathscr{L}^1 r = \int_{\log(\sigma)}^0 [g'(t)]^2 \exp(t\,\alpha)\, d\mathscr{L}^1 t$$
$$= \int_{\log(\sigma)}^0 g(t)[-g''(t) - \alpha\, g'(t)] \exp(t\,\alpha)\, d\mathscr{L}^1 t,$$
$$\int_\sigma^1 [(g \circ \log)(r)]^2 r^{\alpha-1} d\mathscr{L}^1 r = \int_{\log(\sigma)}^0 [g(t)]^2 \exp(t\,\alpha)\, d\mathscr{L}^1 t.$$

Taking $g(t) = \exp(-t\,\alpha/2) \sin[t\,\pi/\log(\sigma)]$ for $t \in \mathbf{R}$, so that

$$-g'' - \alpha\, g' = [\alpha^2/4 + \pi^2 \log(\sigma)^{-2}]\, g,$$

we find that the ratio of the above integrals is less than β, and let

$$\psi(r) = g[\log(r)] \text{ for } \sigma \leq r \leq 1, \quad \psi(r) = 0 \text{ for other } r \in \mathbf{R}.$$

5.4.11. We define $\exp\colon \operatorname{Hom}(\mathbf{R}^n, \mathbf{R}^n) \to \operatorname{Hom}(\mathbf{R}^n, \mathbf{R}^n)$,

$$\exp(T) = \sum_{i=0}^{\infty} i!^{-1} T^i \quad \text{for } T \in \operatorname{Hom}(\mathbf{R}^n, \mathbf{R}^n);$$

clearly $\|\exp(T)\| \le \exp(\|T\|)$, $\exp(T^*) = \exp(T)^*$ and

$$\exp(r\,T + s\,T) = \exp(r\,T) \circ \exp(s\,T) \quad \text{for } r, s \in \mathbf{R};$$

if T is skewsymmetric, then $\exp(T)$ is orthogonal.

Suppose U is a neighborhood of 0 in $\mathbf{R}^n$ and $g\colon U \to \mathbf{O}(n)$ is a map of class ∞. Since $g(x)^* \circ g(x) = 1$ for $x \in U$,

$$g(x)^* \circ \langle v, D\,g(x)\rangle \text{ is skewsymmetric for } x \in U,\ v \in \mathbf{R}^n.$$

Corresponding to any linear function

$$L\colon \mathbf{R}^n \to \operatorname{Hom}(\mathbf{R}^n, \mathbf{R}^n) \cap \{T\colon T^* = -T\}$$

we construct a map of class ∞,

$$f\colon U \to \mathbf{O}(n), \quad f(x) = g(x) \circ \exp[L(x)] \text{ for } x \in U.$$

We compute $f(0) = g(0)$ and, for all $v \in \mathbf{R}^n$,

$$\langle v, D\,f(0)\rangle = \langle v, D\,g(0)\rangle + g(0) \circ \langle v, L\rangle,$$
$$f(0)^* \circ \langle v, D\,f(0)\rangle = g(0)^* \circ \langle v, D\,g(0)\rangle + \langle v, L\rangle.$$

Given $m \in \{1, \dots, n\}$ we can choose L so that, for all $v \in \mathbf{R}^n$,

$$\langle e_i, g(0)^* \circ \langle v, D\,g(0)\rangle\rangle \bullet e_j + \langle e_i, \langle v, L\rangle\rangle \bullet e_j = 0$$
$$\text{in case } i \le m \text{ and } j \le m, \text{ or } i > m \text{ and } j > m,$$
$$\langle e_i, \langle v, L\rangle\rangle \bullet e_j = 0 \text{ in case } i \le m < j, \text{ or } j \le m < i,$$

where $e_1, \dots, e_n$ are the standard base vectors of $\mathbf{R}^n$; we note that $\langle v, L\rangle$ and $\exp(\langle v, L\rangle)$ map the vectorsubspace of $\mathbf{R}^n$ spanned by $\{e_1, \dots, e_m\}$ into itself. Defining

$$f_i\colon U \to \mathbf{R}^n, \quad f_i(x) = \langle e_i\, f(x)\rangle \text{ for } x \in U,$$

we conclude that

$$\langle v, D\,f_i(0)\rangle \bullet f_j(0) = \langle e_i, \langle v, D\,f(0)\rangle\rangle \bullet \langle e_j, f(0)\rangle$$
$$= \langle\langle e_i, \langle v, D\,f(0)\rangle\rangle, f(0)^*\rangle \bullet e_j = \langle e_i, f(0)^* \circ \langle v, D\,f(0)\rangle\rangle \bullet e_j = 0$$

whenever $v \in \mathbf{R}^n$ and $\{i, j\} \subset \{1, \dots, m\}$ or $\{i, j\} \subset \{m+1, \dots, n\}$, and that for each $x \in U$ the vectorspace spanned by $\{e_1, \dots, e_m\}$ has equal images under $g(x)$ and $f(x)$.

5.4.12. Here we discuss some basic facts concerning *the differential geometry of an m dimensional submanifold B of class ∞ of* $\mathbf{R}^n$. Recalling 3.1.18 – 3.1.22 we see that the inclusion map $P: B \to \mathbf{R}^n$ is of class ∞ and that, for each $b \in B$, $DP(b)$ is the inclusion map of $\operatorname{Tan}(B, b)$ into $\mathbf{R}^n$, hence $DP(b)^*$ is the orthogonal projection retracting $\mathbf{R}^n$ onto $\operatorname{Tan}(B, b)$ with kernel $\operatorname{Nor}(B, b)$. Moreover there exist a neighborhood Z of b in $\mathbf{R}^n$ and vectorfields $f_1, \ldots, f_n \in \mathscr{E}(Z, \mathbf{R}^n)$ such that

$$f_1(z), \ldots, f_n(z) \text{ are orthonormal for } z \in Z,$$
$$f_i(z) \in \operatorname{Tan}(B, z) \text{ for } z \in B \cap Z \text{ and } i \le m,$$
$$f_i(z) \in \operatorname{Nor}(B, z) \text{ for } z \in B \cap Z \text{ and } i > m;$$

such a sequence of vectorfields will be called a **Cartan frame for B in Z.** In view of 5.4.11 we may require also that

$$\langle v, Df_i(b) \rangle \bullet f_j(b) = 0 \text{ whenever } v \in \mathbf{R}^n \text{ and}$$
$$\{i, j\} \subset \{1, \ldots, m\} \text{ or } \{i, j\} \subset \{m+1, \ldots, n\};$$

such a Cartan frame is termed **osculating to B at b.**

For every Banach space W we define

$$\mathscr{E}(B, W) = \{\alpha | B : \alpha \in \mathscr{E}(U, W) \text{ for some neighborhood } U \text{ of } B \text{ in } \mathbf{R}^n\},$$

and note that $\mathscr{E}(B, W)$ is a module over the ring $\mathscr{E}(B, \mathbf{R})$. Letting

$$\mathfrak{T}(B) = \mathscr{E}(B, \mathbf{R}^n) \cap \{\xi : \xi(b) \in \operatorname{Tan}(B, b) \text{ for } b \in B\},$$
$$\mathfrak{N}(B) = \mathscr{E}(B, \mathbf{R}^n) \cap \{\nu : \nu(b) \in \operatorname{Nor}(B, b) \text{ for } b \in B\}$$

be the modules of infinitely differentiable **tangent and normal vectorfields of B in R^n**, we see that

$$\mathscr{E}(B, \mathbf{R}^n) = \mathfrak{T}(B) \oplus \mathfrak{N}(B).$$

In fact, if $\lambda \in \mathscr{E}(B, \mathbf{R}^n)$ and $\xi(b) = \langle \lambda(b), DP(b)^* \rangle$ for all $b \in B$, then $\xi = \langle \lambda, (DP)^* \rangle \in \mathfrak{T}(B)$ and $\lambda - \xi \in \mathfrak{N}(B)$, because

$$\xi | Z = \sum_{i=1}^{m} (\lambda \bullet f_i) f_i \quad \text{and} \quad (\lambda - \xi) | Z = \sum_{i=m+1}^{n} (\lambda \bullet f_i) f_i$$

whenever $f_1, \ldots, f_n$ form a Cartan frame for B in Z.

Whenever $b \in B$ there exist $\tau_1, \ldots, \tau_m \in \mathfrak{T}(B)$ and $\nu_1, \ldots, \nu_{n-m} \in \mathfrak{N}(B)$ such that $\tau_1(z), \ldots, \tau_m(z), \nu_1(z), \ldots, \nu_{n-m}(z)$ are orthonormal for all z in some neighborhood of b relative to B, and

$$\langle v, D\tau_i(b) \rangle \in \operatorname{Nor}(B, b), \quad \langle v, D\nu_j(b) \rangle \in \operatorname{Tan}(B, b)$$

for $v \in \operatorname{Tan}(B, b)$, $i \in \{1, \ldots, m\}$, $j \in \{1, \ldots, n-m\}$. Vectorfields with these properties can be constructed from a Cartan frame for B in Z, which is osculating to B at b, through multiplication by a function

$$\phi \in \mathscr{E}(B, \mathbf{R}) \quad \text{with} \quad \operatorname{spt} \phi \subset Z \quad \text{and} \quad b \notin \operatorname{spt}(1-\phi).$$

We will say that $\tau_1, \ldots, \tau_m$, $v_1, \ldots, v_{n-m}$ are **orthonormal near** b **and osculating at** b.

We define $\mathfrak{P}(B)$ as the module of all $\mathscr{E}(B, \mathbf{R})$ linear endomorphisms of $\mathfrak{T}(B)$. *A map* $F: \mathfrak{T}(B) \to \mathfrak{T}(B)$ *belongs to* $\mathfrak{P}(B)$ *if and only if*

$$\langle \xi + \eta, F \rangle = \langle \xi, F \rangle + \langle \eta, F \rangle \quad \textit{and} \quad \langle \psi\, \xi, F \rangle = \psi \langle \xi, F \rangle$$

whenever $\xi, \eta \in \mathfrak{T}(B)$ *and* $\psi \in \mathscr{E}(B, \mathbf{R})$; *this condition is equivalent to the existence of* $\mathbf{R}$ *linear endomorphisms* $F(b)$ *of* $\operatorname{Tan}(B, b)$ *corresponding to all* $b \in B$ *such that*

$$\langle \xi, F \rangle (b) = \langle \xi(b), F(b) \rangle \quad \textit{for } \xi \in \mathfrak{T}(B).$$

To construct $F(b)$ we choose $\tau_1, \ldots, \tau_m \in \mathfrak{T}(B)$ and a neighborhood E of b in B so that $\tau_1(z), \ldots, \tau_m(z)$ are orthonormal whenever $z \in E$, and we define

$$\langle v, F(b) \rangle = \sum_{i=1}^{m} (v \bullet \tau_i(b)) \langle \tau_i, F \rangle (b)$$

for $v \in \operatorname{Tan}(B, b)$; if $\xi \in \mathfrak{T}(B)$, then there exists

$$\psi \in \mathscr{E}(B, \mathbf{R}) \quad \text{with} \quad \psi(b) = 1 \quad \text{and} \quad \operatorname{spt} \psi \subset E,$$

hence $\langle \xi - \psi\, \xi, F \rangle = (1 - \psi) \langle \xi, F \rangle$, $\langle \xi - \psi\, \xi, F \rangle (b) = 0$,

$$\psi\, \xi = \sum_{i=1}^{m} (\psi\, \xi \bullet \tau_i)\, \tau_i, \qquad \langle \psi\, \xi, F \rangle = \sum_{i=1}^{m} (\psi\, \xi \bullet \tau_i) \langle \tau_i, F \rangle$$

and $\langle \xi, F \rangle (b) = \langle \psi\, \xi, F \rangle (b) = \langle \xi(b), F(b) \rangle$.

With each $F \in \mathfrak{P}(B)$ we associate $F^* \in \mathfrak{P}(B)$ so that

$$\langle \xi, F^* \rangle \bullet \eta = \langle \eta, F \rangle \bullet \xi \quad \text{for all } \xi, \eta \in \mathfrak{T}(B),$$

which is equivalent to $F^*(b) = F(b)^*$ for $b \in B$. If $F, G \in \mathfrak{P}(B)$, then $F \circ G \in \mathfrak{P}(B)$ and $(F \circ G)(b) = F(b) \circ G(b)$ for $b \in B$. We also define trace $F \in \mathscr{E}(B, \mathbf{R})$ so that

$$(\operatorname{trace} F)(b) = \operatorname{trace}[F(b)] \quad \text{for } b \in B,$$

and let $F \bullet G = \operatorname{trace}(F^* \circ G) \in \mathscr{E}(B, \mathbf{R})$, hence

$$(F \bullet G)(b) = F(b) \bullet G(b) \quad \text{for } b \in B.$$

Now we describe the process of **covariant differentiation.** *To each $\xi \in \mathfrak{X}(B)$ corresponds the first order differential operator ∇_ξ acting on $\mathscr{E}(B, \mathbf{R})$, on $\mathfrak{X}(B)$ and on $\mathfrak{P}(B)$ as follows:*

If $\phi \in \mathscr{E}(B, \mathbf{R})$, then $\nabla_\xi \phi = \langle \xi, D\phi \rangle \in \mathscr{E}(B, \mathbf{R})$.

If $\eta \in \mathfrak{X}(B)$, then $\nabla_\xi \eta = \langle \langle \xi, D\eta \rangle, (DP)^* \rangle \in \mathfrak{X}(B)$.

If $F \in \mathfrak{P}(B)$, then $\nabla_\xi F = \nabla_\xi \circ F - F \circ \nabla_\xi \in \mathfrak{P}(B)$; this definition is equivalent to the requirement

$$\nabla_\xi \langle \eta, F \rangle = \langle \nabla_\xi \eta, F \rangle + \langle \eta, \nabla_\xi F \rangle \text{ for } \eta \in \mathfrak{X}(B).$$

One readily verifies in each of the above three cases that ∇_ξ acts as a derivation; this means that ∇_ξ induces $\mathbf{R}$ linear endomorphisms of the vectorspaces $\mathscr{E}(B, \mathbf{R})$, $\mathfrak{X}(B)$, $\mathfrak{P}(B)$ and

$$\nabla_\xi(\psi \mu) = (\nabla_\xi \psi)\mu + \psi \nabla_\xi \mu$$

whenever $\psi \in \mathscr{E}(B, \mathbf{R})$ and $\mu \in \mathscr{E}(B, \mathbf{R})$ or $\mu \in \mathfrak{X}(B)$ or $\mu \in \mathfrak{P}(B)$. We note also that

$$\nabla_{\psi\xi} \mu = \psi \nabla_\xi \mu,$$

and that $(\nabla_\xi \mu)(b)$ is determined by $\xi(b)$ and the behavior of μ near b in B.

If $\xi, \eta, \zeta \in \mathfrak{X}(B)$ and $F, G \in \mathfrak{P}(B)$, then

$$\nabla_\xi(\eta \bullet \zeta) = (\nabla_\xi \eta) \bullet \zeta + \eta \bullet (\nabla_\xi \zeta), \qquad \nabla_\xi(F \circ G) = (\nabla_\xi F) \circ G + F \circ (\nabla_\xi G),$$

$$\nabla_\xi(F^*) = (\nabla_\xi F)^*, \qquad \nabla_\xi(\operatorname{trace} F) = \operatorname{trace}(\nabla_\xi F),$$

$$\nabla_\xi(F \bullet G) = (\nabla_\xi F) \bullet G + F \bullet (\nabla_\xi G).$$

To verify the formula for $\nabla_\xi(\operatorname{trace} F)$ we suppose $b \in B$, choose vectorfields $\tau_1, \ldots, \tau_m \in \mathfrak{X}(B)$ which are orthonormal near b and osculating at b, infer that

$$(\operatorname{trace} F)(z) = \sum_{i=1}^{m} (\langle \tau_i, F \rangle \bullet \tau_i)(z)$$

for all z sufficiently near b in B, and conclude

$$(\nabla_\xi \operatorname{trace} F)(b) = \sum_{i=1}^{m} (\langle \tau_i, \nabla_\xi F \rangle \bullet \tau_i)(b) = (\operatorname{trace} \nabla_\xi F)(b)$$

because $(\nabla_\xi \tau_i)(b) = 0$ for $i = 1, \ldots, m$.

Defining $[\xi, \eta] = \langle \xi, D\eta \rangle - \langle \eta, D\xi \rangle$ for $\xi, \eta \in \mathfrak{X}(B)$ we prove that

$$[\xi, \eta] \in \mathfrak{X}(B), \qquad \nabla_\xi \eta - \nabla_\eta \xi = [\xi, \eta],$$

$$\langle \xi, D\langle \eta, D\psi \rangle \rangle - \langle \eta, D\langle \xi, D\psi \rangle \rangle = \langle [\xi, \eta], D\psi \rangle$$

whenever $\psi \in \mathscr{E}(B, W)$ for some Banach space W. For this purpose we choose a neighborhood U of B in $\mathbf{R}^n$ and functions

$$\alpha, \beta \in \mathscr{E}(U, \mathbf{R}^n), \quad g \in \mathscr{E}(U, W) \quad \text{with} \quad \alpha | B = \xi, \quad \beta | B = \eta, \quad g | B = \psi$$

and infer from 4.1.34 that $[\alpha, \beta] | B = [\xi, \eta]$,

$$\langle \xi, D\langle \eta, Dg\rangle\rangle - \langle \eta, D\langle \xi, Dg\rangle\rangle = \langle [\xi, \eta], Dg\rangle.$$

In view of 3.1.20 we can in particular let g be a retraction of U onto B and conclude

$$\langle \xi, Dg\rangle = \xi, \quad \langle \eta, Dg\rangle = \eta, \quad [\xi, \eta] = \langle [\xi, \eta], Dg\rangle,$$

hence $[\xi, \eta] \in \mathfrak{T}(B)$, $\nabla_\xi \eta - \nabla_\eta \xi = \langle [\xi, \eta], (DP)^*\rangle = [\xi, \eta]$.

For $\xi, \eta \in \mathfrak{T}(B)$ we define the second order differential operator

$$\nabla_{\xi,\eta} = \nabla_\xi \nabla_\eta - \nabla_{\nabla_\xi \eta}$$

which acts $\mathbf{R}$ linearly and satisfies the conditions

$$\nabla_{\xi,\eta}(\psi\,\mu) = (\nabla_{\xi,\eta}\psi)\,\mu + \psi\,\nabla_{\xi,\eta}\mu + (\nabla_\xi \psi)(\nabla_\eta \mu) + (\nabla_\eta \psi)(\nabla_\xi \mu),$$

$$\nabla_{\psi\xi,\eta}\,\mu = \psi\,\nabla_{\xi,\eta}\,\mu = \nabla_{\xi,\psi\eta}\,\mu$$

whenever $\psi \in \mathscr{E}(B, \mathbf{R})$ and $\mu \in \mathscr{E}(B, \mathbf{R})$ or $\mu \in \mathfrak{T}(B)$ or $\mu \in \mathfrak{P}(B)$,

$$\nabla_{\xi,\eta} F = \nabla_\xi \circ \nabla_\eta \circ F - \nabla_\xi \circ F \circ \nabla_\eta - \nabla_\eta \circ F \circ \nabla_\xi + F \circ \nabla_\eta \circ \nabla_\xi - \nabla_{\nabla_\xi \eta} \circ F + F \circ \nabla_{\nabla_\xi \eta}$$

whenever $F \in \mathfrak{P}(B)$. Then we see that the **Riemannian curvature operator**

$$R_{\xi,\eta} = \nabla_{\xi,\eta} - \nabla_{\eta,\xi} = \nabla_\xi \nabla_\eta - \nabla_\eta \nabla_\xi - \nabla_{[\xi,\eta]}$$

annuls $\mathscr{E}(B, \mathbf{R})$, acts $\mathscr{E}(B, \mathbf{R})$ linearly on $\mathfrak{T}(B)$ and $\mathfrak{P}(B)$, and satisfies

$$R_{\xi,\eta} F = R_{\xi,\eta} \circ F - F \circ R_{\xi,\eta} \text{ for } F \in \mathfrak{P}(B).$$

We will use the **Laplace operator** *characterized by the property that*

$$(\operatorname{Lap} \mu)(b) = \sum_{i=1}^{m} (\nabla_{\tau_i,\tau_i} \mu)(b)$$

whenever $b \in B$ and $\tau_1, \ldots, \tau_m \in \mathfrak{T}(B)$ and $\tau_1(b), \ldots, \tau_m(b)$ are orthonormal; this definition is valid because if also $\sigma_1, \ldots, \sigma_m \in \mathfrak{T}(B)$ and $\sigma_1(b), \ldots, \sigma_m(b)$ are orthonormal, then there exist functions ψ and $\phi_{i,j}$ in $\mathscr{E}(B, \mathbf{R})$ such that

$$\psi\,\tau_i = \sum_{j=1}^{m} \phi_{i,j}\,\sigma_j \text{ for } i \in \{1, \ldots, m\},$$

$\psi(b) = 1$ and the numbers $\phi_{i,j}(b)$ form an orthogonal matrix, hence

$$\sum_{i=1}^{m} \psi^2\,\nabla_{\tau_i,\tau_i} = \sum_{i=1}^{m} \sum_{j=1}^{m} \sum_{k=1}^{m} \phi_{i,j}\,\phi_{i,k}\,\nabla_{\sigma_j,\sigma_k},$$

$$\sum_{i=1}^{m} (\nabla_{\tau_i,\tau_i} \mu)(b) = \sum_{j=1}^{m} (\nabla_{\sigma_j,\sigma_j} \mu)(b).$$

With each normal vectorfield $\nu \in \mathfrak{N}(B)$ *we associate an* $\mathscr{E}(B, \mathbf{R})$ *linear endomorphism* $Q \in \mathfrak{P}(B)$ *by letting*

$$\langle \xi, Q \rangle = -\langle \langle \xi, D\nu \rangle, (DP)^* \rangle \quad \textit{for} \quad \xi \in \mathfrak{T}(B).$$

If $\xi, \eta \in \mathfrak{T}(B)$, then $\nu \bullet \eta = 0$ and

$$\langle \xi, Q \rangle \bullet \eta = -\langle \xi, D\nu \rangle \bullet \eta = \nu \bullet \langle \xi, D\eta \rangle;$$

interchanging ξ and η and recalling that $[\xi, \eta] \in \mathfrak{T}(B)$, hence $\nu \bullet [\xi, \eta] = 0$, we infer

$$\langle \xi, Q \rangle \bullet \eta = \langle \eta, Q \rangle \bullet \xi.$$

Thus $Q^* = Q$ and for each $b \in B$ the $\mathbf{R}$ linear endomorphism $Q(b)$ of $\operatorname{Tan}(B, b)$ is symmetric; the corresponding bilinear symmetric form, which has the value

$$\langle v, Q(b) \rangle \bullet w = -\langle v, D\nu(b) \rangle \bullet w$$

at $(v, w) \in \operatorname{Tan}(B, b) \times \operatorname{Tan}(B, b)$, is called the **second fundamental form of** ν **at** b.

If $\nu_1, \ldots, \nu_{n-m} \in \mathfrak{N}(B)$ *with the associated endomorphisms* $Q_1, \ldots, Q_{n-m} \in \mathfrak{P}(B)$, *and if* $\nu_1(z), \ldots, \nu_{n-m}(z)$ *are orthonormal whenever* $z \in B$, *then the Riemannian curvature operator is given by the* **Gauss equation**

$$R_{\xi,\eta}\zeta = \sum_{j=1}^{n-m} \left((\langle \eta, Q_j \rangle \bullet \zeta) \langle \xi, Q_j \rangle - (\langle \xi, Q_j \rangle \bullet \zeta) \langle \eta, Q_j \rangle \right)$$

for $\xi, \eta, \zeta \in \mathfrak{T}(B)$, *and the covariant derivatives of* $Q_1, \ldots, Q_{n-m}$ *satisfy the* **Codazzi equations**

$$\langle \xi, \nabla_\eta Q_l \rangle - \langle \eta, \nabla_\xi Q_l \rangle = \sum_{j=1}^{n-m} \left((\nu_l \bullet \langle \xi, D\nu_j \rangle) \langle \eta, Q_j \rangle - (\nu_l \bullet \langle \eta, D\nu_j \rangle) \langle \xi, Q_j \rangle \right)$$

for $\xi, \eta \in \mathfrak{T}(B)$ *and* $l = 1, \ldots, n-m$. To prove these assertions we observe first that

$$\langle \eta, D\zeta \rangle = \nabla_\eta \zeta + \sum_{j=1}^{n-m} (\nu_j \bullet \langle \eta, D\zeta \rangle)\, \nu_j = \nabla_\eta \zeta + \sum_{j=1}^{n-m} (\langle \eta, Q_j \rangle \bullet \zeta)\, \nu_j.$$

Replacing η by ξ and ζ by $\nabla_\eta \zeta$, we compute

$$\begin{aligned}
\langle \xi, D\langle \eta, D\zeta \rangle \rangle = \nabla_\xi \nabla_\eta \zeta &+ \sum_{j=1}^{n-m} (\langle \xi, Q_j \rangle \bullet \nabla_\eta \zeta)\, \nu_j \\
&+ \sum_{j=1}^{n-m} (\langle \nabla_\xi \eta, Q_j \rangle \bullet \zeta + \langle \eta, \nabla_\xi Q_j \rangle \bullet \zeta + \langle \eta, Q_j \rangle \bullet \nabla_\xi \zeta)\, \nu_j \\
&+ \sum_{j=1}^{n-m} (\langle \eta, Q_j \rangle \bullet \zeta) \langle \xi, D\nu_j \rangle.
\end{aligned}$$

We subtract from the preceding equation the similar equation obtained by interchanging ξ and η, and also subtract the equation

$$\langle[\xi,\eta], D\zeta\rangle = \nabla_{[\xi,\eta]}\zeta + \sum_{j=1}^{n-m} (\langle[\xi,\eta], Q_j\rangle \bullet \zeta)\, v_j.$$

Since $\langle\xi, D\langle\eta,\zeta\rangle\rangle - \langle\eta, D\langle\xi, D\zeta\rangle\rangle = \langle[\xi,\eta], D\zeta\rangle$, the left member of the equation resulting from these subtractions equals 0. Considering the components in $\mathfrak{T}(B)$ of the terms on the right we find that

$$0 = R_{\xi,\eta}\zeta - \sum_{j=1}^{n-m} ((\langle\eta, Q_j\rangle \bullet \zeta)\langle\xi, Q_j\rangle - (\langle\xi, Q_j\rangle \bullet \zeta)\langle\eta, Q_j\rangle),$$

and taking components in $\mathfrak{N}(B)$ we obtain

$$0 = \sum_{j=1}^{n-m} (\langle\eta, \nabla_\xi Q_j\rangle \bullet \zeta - \langle\xi, \nabla_\eta Q_j\rangle \bullet \zeta)\, v_j$$

$$+ \sum_{j=1}^{n-m}\sum_{l=1}^{n-m} ((\langle\eta, Q_j\rangle \bullet \zeta)(v_l \bullet \langle\xi, D v_j\rangle) - (\langle\xi, Q_j\rangle \bullet \zeta)(v_l \bullet \langle\eta, D v_j\rangle))\, v_l$$

because $\nabla_\xi\eta - \nabla_\eta\xi = [\xi,\eta]$.

Whenever $\xi \in \mathfrak{T}(B)$ we define $\nabla\xi \in \mathfrak{P}(B)$ by the formula

$$\langle\eta, \nabla\xi\rangle = \nabla_\eta\xi \quad \textit{for } \eta \in \mathfrak{T}(B).$$

With each $\phi \in \mathscr{E}(B, \mathbf{R})$ we associate grad $\phi \in \mathfrak{T}(B)$ *so that*

$$\eta \bullet \operatorname{grad}\phi = \langle\eta, D\phi\rangle \quad \textit{for } \eta \in \mathfrak{T}(B),$$

and we verify that

$$\operatorname{Lap}\phi = \operatorname{trace}(\nabla \operatorname{grad}\phi);$$

choosing $\tau_1, \ldots, \tau_m \in \mathfrak{T}(B)$ orthonormal near b and osculating at b we compute

$$\operatorname{grad}\phi(z) = \sum_{i=1}^{m} \nabla_{\tau_i}\phi(z)\,\tau_i(z) \text{ for } z \text{ near } b,$$

$$\langle\tau_j, \nabla \operatorname{grad}\phi\rangle(b) = \sum_{i=1}^{m} \nabla_{\tau_j}\nabla_{\tau_i}\phi(b)\,\tau_i(b)$$

for $j \in \{1, \ldots, m\}$, hence

$$\operatorname{trace}(\nabla \operatorname{grad}\phi)(b) = \sum_{j=1}^{m} \langle\tau_j, \nabla \operatorname{grad}\phi\rangle(b) \bullet \tau_j(b)$$

$$= \sum_{j=1}^{m} \nabla_{\tau_j,\tau_j}\phi(b) = \operatorname{Lap}\phi(b).$$

Also, if $\xi \in \mathfrak{T}(B)$ and $\phi, \psi \in \mathbf{E}(B, \mathbf{R})$, then

$$\operatorname{trace}\nabla(\phi\,\xi) = (\operatorname{grad}\phi) \bullet \xi + \phi \operatorname{trace}(\nabla\xi),$$

$$\operatorname{trace}\nabla(\phi \operatorname{grad}\psi) = (D\phi) \bullet (D\psi) + \phi \operatorname{Lap}\psi.$$

Next we prove that

$$\int_B \operatorname{trace}(\nabla \xi)\, d\mathscr{H}^m = 0$$

whenever $\xi \in \mathfrak{X}(B)$ *and* $\operatorname{spt} \xi = B \cap \operatorname{Clos}\{z: \xi(z) \neq 0\}$ *is compact.* In view of additivity with respect to ξ, it suffices to consider the special case when $\operatorname{spt} \xi$ is contained in an open subset Z of $\mathbf{R}^n$ such that there exist a diffeomorphism $\sigma: Z \to \mathbf{R}^n$ of class ∞ and an m dimensional vectorsubspace V of $\mathbf{R}^n$ with $\sigma(B \cap Z) = V \cap \operatorname{im} \sigma$. Using σ one readily constructs an m vectorfield χ orienting $B \cap Z$ and an isotopic deformation h of class ∞ of Z in Z such that $\dot{h}_0 | B \cap Z = \xi$ and, for all sufficiently small numbers t,

$$h_t(B \cap Z) = B \cap Z, \quad h_t(z) = z \text{ whenever } z \in Z \cap B \sim \operatorname{spt} \xi.$$

Then we choose an open subset W of $\mathbf{R}^n$ with

$$\operatorname{spt} \xi \subset W, \quad \operatorname{Clos} W \subset Z, \quad \mathscr{H}^m(B \cap W) < \infty$$

and let $T = (\mathscr{H}^m \llcorner B \cap W) \wedge \chi$. Noting that

$$\operatorname{spt} h_\#([0,t] \times T) \subset B, \quad \operatorname{spt} \partial T \subset Z \cap B \sim \operatorname{spt} \xi$$

by 4.1.31 (1), we see from 4.1.9 and 4.1.20 that $h_{t\#} T = T$ for small t, hence $\delta^{(1)}(T, h) = 0$. Finally we recall 5.1.8, observe that $M_z = (\nabla \xi)(z)$ for $z \in B \cap W$, and conclude

$$0 = \int \operatorname{trace}(M_z)\, d\|T\|\, z = \int_{B \cap W} \operatorname{trace}(\nabla \xi)\, d\mathscr{H}^m.$$

Taking $\xi = \phi \operatorname{grad} \psi$ *we obtain the corollary*

$$\int_B (D\phi \bullet D\psi + \phi \operatorname{Lap} \psi)\, d\mathscr{H}^m = 0$$

whenever $\phi, \psi \in \mathscr{E}(B, \mathbf{R})$ *and* $\operatorname{spt}(\phi \operatorname{grad} \psi)$ *is a compact subset of* B.

5.4.13. Assuming now that B *is an* $n-2$ *dimensional submanifold of class* ∞ *of* $\mathbf{S}^{n-1}$, *and* χ *is an* $n-2$ *vectorfield orienting* B, we choose $v_1, v_2 \in \mathfrak{N}(B)$ so that

$$v_1(z) = z, \quad v_1(z) \bullet v_2(z) = 0 \quad \text{and} \quad \chi(z) \wedge v_1(z) \wedge v_2(z) = e_1 \wedge \cdots \wedge e_n$$

for all $z \in B$, where $e_1, \ldots, e_n$ are the standard base vectors of $\mathbf{R}^n$, and we study the endomorphisms $Q_1, Q_2 \in \mathfrak{P}(B)$ associated with v_1, v_2. If $\xi, \eta, \zeta \in \mathfrak{X}(B)$, then

$$\langle \xi, D v_1 \rangle = \xi \in \mathfrak{X}(B), \quad \langle \xi, Q_1 \rangle = -\xi,$$

$$v_1 \bullet \langle \xi, D v_2 \rangle = 0 \quad \text{because} \quad v_1 \bullet v_2 = 0 \quad \text{and} \quad \xi \bullet v_2 = 0,$$

$$v_2 \bullet \langle \xi, D v_2 \rangle = 0 \quad \text{because} \quad v_2 \bullet v_2 = 1,$$

$$\langle \xi, D v_2 \rangle \in \mathfrak{X}(B), \quad \langle \xi, Q_2 \rangle = -\langle \xi, D v_2 \rangle,$$

$$\langle \xi, \nabla_\eta Q_1 \rangle = \nabla_\eta \langle \xi, Q_1 \rangle - \langle \nabla_\eta \xi, Q_1 \rangle = 0,$$

hence the equations of Gauss and Codazzi take the special form

$$R_{\xi,\eta}\zeta=(\eta\bullet\zeta)\xi-(\xi\bullet\zeta)\eta+(\langle\eta,Q_2\rangle\bullet\zeta)\langle\xi,Q_2\rangle-(\langle\xi,Q_2\rangle\bullet\zeta)\langle\eta,Q_2\rangle,$$
$$\langle\xi,\nabla_\eta Q_l\rangle-\langle\eta,\nabla_\xi Q_l\rangle=0 \text{ for } l\in\{1,2\}.$$

Abbreviating $Q_2=F$ we will prove the proposition:

If trace $F=0$, $0<\varepsilon\in\mathbf{R}$ *and* $\gamma=(F\bullet F+\varepsilon)^{\frac{1}{2}}$, *then*

$$\operatorname{Lap} F=(n-2)F-(F\bullet F)F \quad \textit{and} \quad \gamma\operatorname{Lap}\gamma\geq(n-2)(F\bullet F)-(F\bullet F)^2.$$

Fixing $b\in B$ and replacing B by a sufficiently small relative neighborhood of b, we choose vectorfields $\tau_1,\ldots,\tau_{n-2}\in\mathfrak{X}(B)$ which are orthonormal everywhere in B and osculating at b, and we abbreviate

$$\nabla_{\tau_i}=\nabla_i,\quad \nabla_{\tau_i,\tau_j}=\nabla_{i,j},\quad R_{\tau_i,\tau_j}=R_{i,j}.$$

From the fact that $F^*=F$, $(\nabla_i F)^*=\nabla_i F$ and the equation of Codazzi we see that

$$\sum_{i=1}^{n-2}\langle\tau_i,\nabla_i F\rangle\bullet\tau_j=\sum_{i=1}^{n-2}\langle\tau_j,\nabla_i F\rangle\bullet\tau_i$$
$$=\sum_{i=1}^{n-2}\langle\tau_i,\nabla_j F\rangle\bullet\tau_i=\operatorname{trace}\nabla_j F=\nabla_j\operatorname{trace} F=0$$

for all $j\in\{1,\ldots,n-2\}$, hence

$$\sum_{i=1}^{n-2}\langle\tau_i,\nabla_i F\rangle=0.$$

Inasmuch as $\nabla_j\tau_i(b)=0$ it follows that

$$\sum_{i=1}^{n-2}\langle\tau_i,\nabla_j\nabla_i F\rangle(b)=0$$

for $j\in\{1,\ldots,n-2\}$. Moreover

$$\langle\tau_j,\nabla_i\nabla_i F\rangle(b)=\nabla_i\langle\tau_j,\nabla_i F\rangle(b)=\nabla_i\langle\tau_i,\nabla_j F\rangle(b)=\langle\tau_i,\nabla_i\nabla_j F\rangle(b),$$
$$(\nabla_i\nabla_j F-\nabla_j\nabla_i F)(b)=(\nabla_{i,j}F-\nabla_{j,i}F)(b)=R_{i,j}F(b),$$

and $\nabla_{i,i}F(b)=\nabla_i\nabla_i F(b)$ for $\{i,j\}\subset\{1,\ldots,n-2\}$, hence

$$\langle\tau_j,\operatorname{Lap} F\rangle(b)=\sum_{i=1}^{n-2}\langle\tau_j,\nabla_{i,i}F\rangle(b)=\sum_{i=1}^{n-2}\langle\tau_i,R_{i,j}F\rangle(b).$$

Abbreviating $\langle \tau_k, F\rangle = F_k$ we use the Gauss equation to compute

$$\begin{aligned}\langle \tau_i, R_{i,j}\circ F\rangle = R_{i,j}F_i &= (\tau_j\bullet F_i)\,\tau_i-(\tau_i\bullet F_i)\,\tau_j+(F_j\bullet F_i)\,F_i-(F_i\bullet F_i)\,F_j\\ &=(\tau_i\bullet F_j)\,\tau_i-(\tau_i\bullet F_i)\,\tau_j+(\langle\tau_j, F^*\circ F\rangle\bullet\tau_i)\,F_i-(F_i\bullet F_i)\,F_j,\end{aligned}$$

$$\langle \tau_i, F\circ R_{i,j}\rangle = \langle R_{i,j}\tau_i, F\rangle = \langle(\tau_j\bullet\tau_i)\,\tau_i-\tau_j+(F_j\bullet\tau_i)\,F_i-(F_i\bullet\tau_i)\,F_j, F\rangle,$$

$$\begin{aligned}\sum_{i=1}^{n-2}\langle\tau_i, R_{i,j}F\rangle &= \sum_{i=1}^{n-2}\langle\tau_i, R_{i,j}\circ F-F\circ R_{i,j}\rangle\\ &=\langle\tau_j, F+F\circ F^*\circ F-(F\bullet F)\,F-F+(n-2)\,F-F\circ F\circ F\rangle,\end{aligned}$$

and conclude $\langle\tau_j, \operatorname{Lap} F\rangle(b)=\langle\tau_j,(n-2)\,F-(F\bullet F)\,F\rangle(b)$. Furthermore $\nabla_i\gamma=\gamma^{-1}F\bullet\nabla_i F$ and we infer

$$\begin{aligned}\nabla_i\nabla_i\gamma &= -\gamma^{-3}(F\bullet\nabla_i F)^2+\gamma^{-1}(\nabla_i F\bullet\nabla_i F+F\bullet\nabla_i\nabla_i F)\\ &\geq\gamma^{-3}\big(-(F\bullet\nabla_i F)^2+|F|^2\,|\nabla_i F|^2\big)+\gamma^{-1}F\bullet\nabla_i\nabla_i F,\end{aligned}$$

$$\begin{aligned}\operatorname{Lap}\gamma(b) &= \sum_{i=1}^{n-2}\nabla_i\nabla_i\gamma(b)\geq\sum_{i=1}^{n-2}(\gamma^{-1}F\bullet\nabla_i\nabla_i F)(b)\\ &=(\gamma^{-1}F\bullet\operatorname{Lap}F)(b)=\big(\gamma^{-1}F\bullet((n-2)\,F-(F\bullet F)\,F)\big)(b).\end{aligned}$$

5.4.14. Lemma. *If $3\leq n\leq 7$, B is a compact connected $n-2$ dimensional submanifold of class ∞ of $\mathbf{S}^{n-1}$, χ is an $n-2$ vectorfield orienting B, and if the $n-1$ dimensional integral current*

$$T=\delta_0 \times [(\mathscr{H}^{n-2}\llcorner B)\wedge\chi]$$

is absolutely area minimizing with respect to $\mathbf{R}^n$, then

$$B=\mathbf{S}^{n-1}\cap\{z\colon u\bullet z=0\}\ \textit{for some}\ u\in\mathbf{S}^{n-1}.$$

Proof. We take $v_1, v_2, Q_1, Q_2=F, \varepsilon, \gamma$ as in 5.4.13 and let

$$C=\{r\,b\colon 0\leq r\in\mathbf{R}, b\in B\}.$$

For any two functions

$$\phi\in\mathscr{E}(B,\mathbf{R})\quad\text{and}\quad\psi\in\mathscr{D}^0(\mathbf{R})\quad\text{with}\quad\operatorname{spt}\psi\subset\{r\colon 0<r<1\}$$

there exists an isotopic deformation $h\colon I\times W\to\mathbf{R}^n$ such that W is a neighborhood of C in $\mathbf{R}^n$ and

$$h(t, r\,b)=r\,b+t\,\phi(b)\,\psi(r)\,v_2(b)$$

whenever $t\in I$, $0<r\in\mathbf{R}$, $b\in B$; it follows that

$$\begin{aligned}&C\cap\{z\colon h(t,z)\neq z\text{ for some }t\in I\}\\ &\subset C\cap\{z\colon 0<|z|<1\}\subset\mathbf{R}^n\sim B=\mathbf{R}^n\sim\operatorname{spt}\partial T\end{aligned}$$

by 4.1.11 and 4.1.31 (1), hence

$$\delta^{(1)}(T,h)=0 \quad\text{and}\quad \delta^{(2)}(T,h)\geq 0$$

by 5.1.7. For $0<r<1$ and $b\in B$ we find that

$$\dot{h}_0(r\,b)=\phi(b)\,\psi(r)\,v_2(b), \qquad v_1(b)=b,$$

$$\operatorname{Tan}(\|T\|,r\,b)=\operatorname{Tan}(C,r\,b)=\operatorname{Tan}(B,b)\oplus\mathbf{R}\,b,$$

$$\operatorname{Nor}(\|T\|,r\,b)=\operatorname{Nor}(C,r\,b)=\mathbf{R}\,v_2(b),$$

$$\langle b,D\,\dot{h}_0(r\,b)\rangle=\phi(b)\,\psi'(r)\,v_2(b),$$

$$\langle r\,v,D\,\dot{h}_0(r\,b)\rangle=\langle v,D\,\phi(b)\rangle\,\psi(r)\,v_2(b)+\phi(b)\,\psi(r)\,\langle v,D\,v_2(b)\rangle$$

for $v\in\operatorname{Tan}(B,b)$, and recalling 5.1.8 we compute

$$\langle b,M_{rb}\rangle=0,\ \langle b,N_{rb}\rangle=\phi(b)\,\psi'(r)\,v_2(b),$$

$$\langle v,M_{rb}\rangle=-r^{-1}\,\phi(b)\,\psi(r)\,\langle v,F(b)\rangle \text{ and}$$

$$\langle v,N_{rb}\rangle=r^{-1}\langle v,D\,\phi(b)\rangle\,\psi(r)\,v_2(b) \text{ for } v\in\operatorname{Tan}(B,b),$$

$$\operatorname{trace}M_{rb}=-r^{-1}\,\phi(b)\,\psi(r)\operatorname{trace}F(b),$$

$$\operatorname{trace}(M_{rb}\circ M_{rb})=r^{-2}\,|\phi(b)\,\psi(r)\,F(b)|^2 \text{ because } F^*=F,$$

$$\operatorname{trace}(N_{rb}^*\circ N_{rb})=|\phi(b)\,\psi'(r)|^2+r^{-2}\,|\psi(r)\,D\,\phi(b)|^2,$$

and $\Xi_{rb}=0$ because $\ddot{h}_0=0$. Since

$$\|T\|=\mathscr{H}^{n-1}\llcorner\{r\,b\colon 0<r<1, b\in B\},$$

we see with the help of 3.2.20 and 3.2.23 that

$$\begin{aligned}0=\delta^{(1)}(T,h)&=\textstyle\int_0^1\int_B(\operatorname{trace}M_{rb})\,r^{n-2}\,d\mathscr{H}^{n-2}\,b\,d\mathscr{L}^1\,r\\&=-\textstyle\int_0^1\psi(r)\,r^{n-3}\,d\mathscr{L}^1\,r\int_B\phi\operatorname{trace}F\,d\mathscr{H}^{n-2},\end{aligned}$$

use the arbitrariness with which ϕ and ψ can be chosen to infer that $\operatorname{trace}F=0$, and obtain

$$\begin{aligned}0\leq\delta^{(2)}(T,h)&=\textstyle\int_0^1\int_B\operatorname{trace}(N_{rb}^*\circ N_{rb}-M_{rb}\circ M_{rb})\,r^{n-2}\,d\mathscr{H}^{n-2}\,b\,d\mathscr{L}^1\,r\\&=\textstyle\int_0^1[\psi'(r)]^2\,r^{n-2}\,d\mathscr{L}^1\,r\int_B\phi^2\,d\mathscr{H}^{n-2}\\&\quad+\textstyle\int_0^1[\psi(r)]^2\,r^{n-4}\,d\mathscr{L}^1\,r\int_B(|D\,\phi|^2-|\phi\,F|^2)\,d\mathscr{H}^{n-2}.\end{aligned}$$

Now we take $\phi=\gamma$ and select ψ according to 5.4.10 with $\alpha=n-3\leq 4$ and $\beta=\alpha+\varepsilon>\alpha\geq\alpha^2/4$; inasmuch as

$$\textstyle\int_B|D\,\gamma|^2\,d\mathscr{H}^{n-2}=-\int_B\gamma\operatorname{Lap}\gamma\,d\mathscr{H}^{n-2}\leq\int_B[|F|^4-(n-2)\,|F|^2]\,d\mathscr{H}^{n-2}$$

by 5.4.12 and 5.4.13, we conclude

$$\begin{aligned}0&\le \delta^{(2)}(T,h)/\textstyle\int_0^1 [\psi(r)]^2\, r^{n-4}\, d\mathscr{L}^1 r\\ &\le \textstyle\int_B [\beta\gamma^2+|F|^4-(n-2)\,|F|^2-\gamma^2\,|F|^2]\, d\mathscr{H}^{n-2}\\ &=\textstyle\int_B (\beta\,\varepsilon-|F|^2)\, d\mathscr{H}^{n-2}.\end{aligned}$$

Since ε can be chosen arbitrarily small it follows that $F=0$, $D\,v_2=0$, hence v_2 maps B onto a single point $u\in\mathbf{S}^{n-1}$. Defining

$$f(z)=u\bullet z \ \text{ for } \ z\in C\cap\{z\colon 0<|z|<1\},$$

we infer $\langle w, D\,f(z)\rangle=u\bullet w=0$ for $w\in\operatorname{Tan}(C,z)$, hence $\operatorname{im} f$ consists of a single number, which must be 0 because f has the limit 0 at 0. We infer that

$$B\subset E=\mathbf{S}^{n-1}\cap\{z\colon u\bullet z=0\}.$$

Noting that B and E are $n-2$ dimensional submanifolds of $\mathbf{R}^n$, B is open and closed in E, and E is connected, we conclude $B=E$.

5.4.15. Theorem. *If $2\le n\le 7$, $T\in\mathscr{R}_{n-1}^{\mathrm{loc}}(\mathbf{R}^n)$ and T is absolutely area minimizing with respect to $\mathbf{R}^n$, then* $\operatorname{spt} T\sim\operatorname{spt}\partial T$ *is an $n-1$ dimensional submanifold of class ∞ of $\mathbf{R}^n$.*

Proof. We use induction with respect to n. In case $n=2$ the assertion follows from 5.3.20. Assuming henceforth that $3\le n\le 7$ we will first prove the following statement:

If M is an $\mathscr{L}^n$ measurable set, V is an open subset of $\mathbf{R}^n$,

$$S=[\partial(\mathbf{E}^n \llcorner M)]\llcorner V\in\mathscr{R}_{n-1}(\mathbf{R}^n)$$

and S is absolutely area minimizing with respect to $\mathbf{R}^n$, then $V\cap\operatorname{spt} S$ is an $n-1$ dimensional submanifold of class ∞ of R^n.

Suppose $a\in V\cap\operatorname{spt} S$. According to 5.4.3 (6), (8), (5) there exists an $\mathscr{L}^n$ measurable set N such that $C=\partial(\mathbf{E}^n\llcorner N)$ is an oriented tangent cone of S at a, C is absolutely area minimizing with respect to $\mathbf{R}^n$ and $\Theta^{n-1}(\|C\|,0)=\Theta^{n-1}(\|S\|,a)$.

For each $b\in(\operatorname{spt} C)\sim\{0\}$ we similarly find an $\mathscr{L}^n$ measurable set P such that $D=\partial(\mathbf{E}^n\llcorner P)$ is an oriented tangent cone of C at b, D is absolutely area minimizing with respect to $\mathbf{R}^n$ and $\Theta^{n-1}(\|D\|,0)=\Theta^{n-1}(\|C\|,b)$. We infer from 4.3.16 that D is a cylinder with direction $b/|b|$, from 4.3.15 that there exist an isometry H mapping $\mathbf{R}\times\mathbf{R}^{n-1}$ onto $\mathbf{R}^n$ and a current

$$Q\in\mathscr{R}_{n-2}^{\mathrm{loc}}(\mathbf{R}^{n-1}) \quad\text{with}\quad D=H_\#(\mathbf{E}^1\times Q),$$

from 5.4.8 that Q is absolutely $n-2$ area minimizing with respect to $\mathbf{R}^{n-1}$, by induction that spt Q is an $n-2$ dimensional submanifold of class ∞ of $\mathbf{R}^{n-1}$, hence spt D is an $n-1$ dimensional submanifold of class ∞ of $\mathbf{R}^n$. Since D is an oriented cone it follows that $\operatorname{spt} D = \operatorname{Tan}(\operatorname{spt} D, 0)$ is an $n-1$ dimensional vectorsubspace of $\mathbf{R}^n$. Consequently P is $\mathscr{L}^n$ almost equal to a halfspace bounded by spt D, and $\Theta^{n-1}(\|D\|, 0) = 1$. Applying 5.4.6 and 5.4.5 (2) with T, a, V and b replaced by C, b, $\mathbf{R}^n$ and any $z \in \operatorname{spt} C$ we find that b has a neighborhood W in $\mathbf{R}^n$ such that $W \cap \operatorname{spt} C$ is an $n-1$ dimensional submanifold of class ∞ of $\mathbf{R}^n$.

Since C is an oriented cone and $(\operatorname{spt} C) \sim \{0\}$ is an $n-1$ dimensional submanifold of class ∞ of $\mathbf{R}^n$, $\mathbf{S}^{n-1} \cap \operatorname{spt} C$ is a compact $n-2$ dimensional submanifold of class ∞ of $\mathbf{S}^{n-1}$, with finitely many components $B_1, \ldots, B_\nu$. We see from 4.3.14 that

$$C \llcorner \mathbf{U}(0,1) = \delta_0 \times \langle C, f, 1 \rangle \quad \text{with} \quad f(z) = |z| \text{ for } z \in \mathbf{R}^n,$$
$$0 \neq C \llcorner [\mathbf{U}(0,1) \cap E_i] = \delta_0 \times (\langle C, f, 1 \rangle \llcorner B_i)$$
$$\text{with } E_i = \{t\,b : 0 \le t \in \mathbf{R},\, b \in B_i\} \text{ for } i \in \{1, \ldots, \nu\},$$

and $\partial \langle C, f, 1 \rangle = \langle \partial C, f, 1 \rangle = 0$, from 4.1.31 (2) that

$$\langle C, f, 1 \rangle \llcorner B_i = r_i (\mathscr{H}^{n-2} \llcorner B_i) \wedge \chi_i$$

where r_i is a positive integer and χ_i is an $n-2$ vectorfield orienting B_i, hence that $\delta_0 \times (\mathscr{H}^{n-2} \llcorner B_i) \wedge \chi_i$ is absolutely area minimizing with respect to $\mathbf{R}^n$, and use 5.4.14 to infer that E_i is an $n-1$ dimensional vectorsubspace of $\mathbf{R}^n$. If $\{i, j\} \subset \{1, \ldots, \nu\}$, then

$$\dim(E_i \cap E_j) \ge 2(n-1) - n = n - 2 \ge 1,$$

hence $B_i \cap B_j \neq \varnothing$ and $i = j$. Thus $\nu = 1$, $\operatorname{spt} C = E_1$, N is $\mathscr{L}^n$ almost equal to a halfspace bounded by E_1, and $\Theta^{n-1}(\|C\|, 0) = 1$. Applying 5.4.6 and 5.4.5 (2) with T replaced by S we find that a has a neighborhood W in $\mathbf{R}^n$ such that $W \cap \operatorname{spt} S$ is an $n-1$ dimensional submanifold of class ∞ of $\mathbf{R}^n$.

Next we suppose $T \in \mathscr{R}_{n-1}^{\text{loc}}(\mathbf{R}^n)$, T is absolutely area minimizing with respect to $\mathbf{R}^n$ and $a \in \operatorname{spt} T \sim \operatorname{spt} \partial T$. Proceeding as in the proof of 5.3.18 we choose ρ, Ξ so that

$$0 < \rho < \operatorname{dist}(a, \operatorname{spt} \partial T), \qquad T \llcorner \mathbf{U}(a, \rho) \in \mathbf{I}_{n-1}(\mathbf{R}^n),$$
$$\Xi \in \mathbf{I}_{n-1}(\mathbf{R}^n), \qquad \operatorname{spt} \Xi \subset \mathbf{R}^n \sim \mathbf{U}(a, \rho), \qquad \partial \Xi = \partial [T \llcorner \mathbf{U}(a, \rho)],$$

apply 4.5.17 with $R = [T \llcorner \mathbf{U}(a, \rho)] - \Xi$ and use the resulting sets M_i to construct the currents

$$S_i = [\partial(\mathbf{E}^n \llcorner M_i)] \llcorner \mathbf{U}(a, \rho) \in \mathscr{R}_{n-1}(\mathbf{R}^n),$$

which are absolutely area minimizing with respect to $\mathbf{R}^n$ because

$$T \llcorner \mathbf{U}(a,\rho) = \sum_{i \in \mathbf{Z}} S_i, \qquad \|T\| \llcorner \mathbf{U}(a,\rho) = \sum_{i \in \mathbf{Z}} \|S_i\|.$$

From the statement proved earlier it follows that, for each integer i, $\mathbf{U}(a,\rho) \cap \operatorname{spt} S_i$ is an $n-1$ dimensional submanifold of class ∞ of $\mathbf{R}^n$. Moreover 5.4.5 (2) and 5.4.3 (3) imply that, for each i, either $S_i = 0$ or

$$\mathbf{M}(S_i) \geq \boldsymbol{\alpha}(n-1)\,[\rho - \operatorname{dist}(a, \operatorname{spt} S_i)]^{n-1};$$

consequently the set $\Delta = \{i\colon\ a \in \operatorname{spt} S_i\}$ is finite, and

$$\inf\{\operatorname{dist}(a, \operatorname{spt} S_i)\colon\ i \in \mathbf{Z} \sim \Delta,\ S_i \neq 0\} > 0.$$

Observing that $\operatorname{Tan}(M_i, a)$ is a halfspace bounded by $\operatorname{Tan}(\operatorname{spt} S_i, a)$ whenever $i \in \Delta$, and that

$$M_j \subset M_i, \quad \operatorname{Tan}(M_j, a) \subset \operatorname{Tan}(M_i, a) \text{ for } j > i,$$

we see that $\operatorname{Tan}(\operatorname{spt} S_i, a) = \operatorname{Tan}(\operatorname{spt} S_j, a)$ whenever $i \in \Delta$ and $j \in \Delta$. We infer that $\operatorname{Tan}(\operatorname{spt} T, a)$ is an $n-1$ dimensional vectorsubspace of $\mathbf{R}^n$. Applying 5.3.18 we conclude that a has a neighborhood W in $\mathbf{U}(a,\rho)$ such that $W \cap \operatorname{spt} T$ is an $n-1$ dimensional submanifold of class ∞ of $\mathbf{R}^{n-1}$.

5.4.16. We do not know whether in 5.4.15 the condition $n \leq 7$ is essential. To remove this restriction one would have to extend 5.4.14 by using criteria more precise than first and second variations. In fact consider the following example:

If λ is a positive integer and

$$G = (\mathbf{R}^\lambda \times \mathbf{R}^\lambda) \cap \{(x,y)\colon\ |x| = |y|\},$$

then $G \sim \{(0,0)\}$ is a $2\lambda - 1$ dimensional analytic submanifold of $\mathbf{R}^\lambda \times \mathbf{R}^\lambda$, oriented by a $2\lambda - 1$ vectorfield χ, the current

$$C = (\mathscr{H}^{2\lambda-1} \llcorner G) \wedge \chi \in \mathscr{R}^{\text{loc}}_{2\lambda-1}(\mathbf{R}^\lambda \times \mathbf{R}^\lambda)$$

is an oriented cone, $\partial C = 0$, and for every compact subset K of $\mathbf{R}^\lambda \times \mathbf{R}^\lambda$ and every isotopic deformation h of class 3 of a neighborhood W of K in $\mathbf{R}^\lambda \times \mathbf{R}^\lambda$ with

$$K \not\subset \{(x,y)\colon\ h_0(x,y) = 0\} \supset W \sim K$$

it is true that

$$\delta^{(1)}(C \llcorner K, h) = 0, \qquad \delta^{(2)}(C \llcorner K, h) > 0 \text{ in case } \lambda \geq 4.$$

One can verify this assertion by computations similar to those occurring in 5.4.10 and 5.4.14, with

$$B = G \cap \{(x, y): |x|^2 + |y|^2 = 1\} = \{(x, y): |x| = 2^{-\frac{1}{2}} = |y|\},$$

hence $|F(x, y)|^2 = 2\lambda$ for $(x, y) \in B$; one also uses expansions in characteristic functions of the differential operator mapping g onto $g'' + (2\lambda - 3)\, g'$ on $\{t: \log(\sigma) \le t \le 0\}$, and of the Laplace operator on B (see [SJ, Section 6]).

We observe that $(0, 0)$ is a singular point of $G = \operatorname{spt} C$. From 5.4.15 it follows that C is not absolutely area minimizing with respect to $\mathbf{R}^{2\lambda}$ in case $\lambda \le 3$. It is not known whether C is area minimizing for any $\lambda \ge 4$. Accordingly currents which minimize area are more regular than currents whose first variation vanishes for all admissible isotopic deformations, but the relative importance of positivity of the second variation is not yet fully understood.

Next we prove the following partial extension of 5.4.15 to the case $n = 8$:

If $T \in \mathscr{R}_7^{\mathrm{loc}}(\mathbf{R}^8)$ is absolutely area minimizing with respect to $\mathbf{R}^8$, then the singular subset of $\operatorname{spt} T \sim \operatorname{spt} \partial T$ *has no clusterpoint in* $\operatorname{spt} T \sim \operatorname{spt} \partial T$.

Taking $a \in \operatorname{spt} T \sim \operatorname{spt} \partial T$ we construct currents S_i as in the proof of 5.4.15 and assume there exists an integer i such that a is a clusterpoint of the singular set of $\operatorname{spt} S_i$. We choose Υ according to 5.4.7 with $n = 8$ and $m = 7$, then apply 5.4.3 (7) with T, d replaced by S_i, Υ to obtain an oriented tangent cone C of S_i at a such that

$$\Theta^{n-1}(\|C\|, b) \ge \Upsilon > 1 \text{ for some } b \in \mathbf{S}^{n-1}.$$

However the first part of the inductive reasoning in the proof of 5.4.15 remains applicable when $n = 8$, and shows that $\Theta^{n-1}(\|C\|, b) = 1$.

5.4.17. The following example shows that an area minimizing current need not be uniquely determined by its boundary, and need not inherit all symmetries of its boundary.

$$\text{If } \lambda \in \{1, 2, 3\},\ S = \partial[\mathbf{E}^\lambda \llcorner \mathbf{U}(0, 1)],\ T \in \mathbf{I}_{2\lambda-1}(\mathbf{R}^\lambda \times \mathbf{R}^\lambda),\ \partial T = S \times S,$$

T is absolutely area minimizing with respect to $\mathbf{R}^\lambda \times \mathbf{R}^\lambda$, and Γ is the group of those orthogonal automorphisms γ of $\mathbf{R}^\lambda \times \mathbf{R}^\lambda$ for which

$$\gamma_\#(S \times S) = S \times S,$$

then there exists a map $\gamma \in \Gamma$ such that $\gamma_\# T \ne T$.

In sketching a proof of this statement we exclude the trivial case $\lambda=1$, and suppose $\gamma_\# T=T$ for all $\gamma\in\Gamma$. We let

$$E=\{r:\ 0<r\in\mathbf{R}\},\quad h:\ E\times E\times\mathbf{R}^\lambda\times\mathbf{R}^\lambda\to\mathbf{R}^\lambda\times\mathbf{R}^\lambda,$$
$$h(s,t,x,y)=(s\,x,t\,y)\ \text{ for }\ (s,t,x,y)\in E\times E\times\mathbf{R}^\lambda\times\mathbf{R}^\lambda$$

and note that h induces a diffeomorphism

$$E\times E\times\mathbf{S}^{\lambda-1}\times\mathbf{S}^{\lambda-1}\simeq(\mathbf{R}^\lambda\sim\{0\})\times(\mathbf{R}^\lambda\sim\{0\}).$$

For $\mathscr{L}^1$ almost all numbers ρ between 0 and 1 we have

$$T_\rho=T\,\llcorner\,\{(x,y):\ \inf\{|x|,\ |y|\}>\rho\}\in\mathbf{I}_{2\lambda-1}(\mathbf{R}^\lambda\times\mathbf{R}^\lambda),$$
$$\operatorname{spt}(\partial T_\rho-S\times S)\subset\{(x,y):\ \inf\{|x|,\ |y|\}=\rho\}.$$

Since $\Delta=\{f\times g:\ f\in\mathbf{SO}(\lambda),\ g\in\mathbf{SO}(\lambda)\}\subset\Gamma$, $\gamma_\# T_\rho=T_\rho$ for all $\gamma\in\Delta$, and

$$\{\gamma(x,y):\ \gamma\in\Delta\}\supset h(\{|x|\}\times\{|y|\}\times\mathbf{S}^{\lambda-1}\times\mathbf{S}^{\lambda-1})$$

whenever $(x,y)\in(\mathbf{R}^\lambda\sim\{0\})\times(\mathbf{R}^\lambda\sim\{0\})$, it can be shown (compare 4.3.14) that

$$T_\rho=h_\#(Q_\rho\times S\times S)\ \text{ for some }\ Q_\rho\in\mathscr{R}_1(E\times E),$$

hence $\partial T_\rho=h_\#[(\partial Q_\rho)\times S\times S]$, $Q_\rho\in\mathbf{I}_1(E\times E)$, and

$$\operatorname{spt}[\partial Q_\rho-\boldsymbol{\delta}_{(1,1)}]\subset\{(s,t):\ \inf\{s,t\}=\rho\}$$

because $S\times S=h_\#[\boldsymbol{\delta}_{(1,1)}\times S\times S]$. Defining

$$\sigma(x)=(x_1,-x_2,\ldots,-x_\lambda)\in\mathbf{R}^\lambda\ \text{ for }\ x\in\mathbf{R}^\lambda,$$
$$\alpha(x,y)=(y,\sigma(x))\in\mathbf{R}^\lambda\times\mathbf{R}^\lambda\ \text{ for }\ (x,y)\in\mathbf{R}^\lambda\times\mathbf{R}^\lambda,$$
$$\beta(s,t)=(t,s)\in E\times E\ \text{ for }\ (s,t)\in E\times E,$$

we find that $\alpha_\#(S\times S)=(-1)^{\lambda-1}S\times\sigma_\# S=S\times S$, $\alpha\in\Gamma$, and furthermore that $h\circ(\beta\times\alpha)=\alpha\circ h$,

$$h_\#[(\beta_\# Q_\rho)\times S\times S]=h_\#[(\beta_\# Q_\rho)\times\alpha_\#(S\times S)]=\alpha_\# T_\rho=T_\rho,$$
$$\beta_\# Q_\rho=Q_\rho,\ \ \beta_\#[\partial Q_\rho-\boldsymbol{\delta}_{(1,1)}]=\partial Q_\rho-\boldsymbol{\delta}_{(1,1)}.$$

Inasmuch as $\mathbf{M}[\partial Q_\rho-\boldsymbol{\delta}_{(1,1)}]$ is an odd integer we infer that $(\rho,\rho)\in\operatorname{spt}\partial Q_\rho$ and conclude

$$h(\{\rho\}\times\{\rho\}\times\mathbf{S}^{\lambda-1}\times\mathbf{S}^{\lambda-1})\subset\operatorname{spt}\partial T_\rho\subset\operatorname{spt}T,$$
$$\rho^{2\lambda-2}\mathbf{M}(S\times S)\le\mathscr{H}^{2\lambda-2}[(\operatorname{spt}T)\cap\{(x,y):\ |x|^2+|y|^2=2\rho^2\}].$$

Accordingly

$$\begin{aligned}\mathbf{M}(T) &\geq \mathscr{H}^{2\lambda-1}(\operatorname{spt} T)\\ &\geq \int_0^\infty \mathscr{H}^{2\lambda-2}[(\operatorname{spt} T)\cap\{(x,y)\colon |x|^2+|y|^2=r^2\}]\, d\mathscr{L}^1 r\\ &\geq \int_0^{2^{\frac{1}{2}}}(2^{-\frac{1}{2}}r)^{2\lambda-2}\,\mathbf{M}(S\times S)\, d\mathscr{L}^1 r\\ &= 2^{\frac{1}{2}}(2\lambda-1)^{-1}\mathbf{M}(S\times S)=\mathbf{M}[\boldsymbol{\delta}_{(0,0)} \times (S\times S)]\end{aligned}$$

with $\partial[\boldsymbol{\delta}_{(0,0)} \times (S\times S)]=S\times S$, hence $\boldsymbol{\delta}_{(0,0)} \times (S\times S)$ is absolutely area minimizing with respect to $\mathbf{R}^\lambda\times\mathbf{R}^\lambda$. However this contradicts 5.4.15, because $2\lambda<7$.

Of course one encounters lack of uniqueness and invariance quite readily when one considers currents which are area minimizing with respect to a proper subset of $\mathbf{R}^n$. For instance an oriented semicircle minimizes length with respect to the whole circle.

5.4.18. Here we suppose *U is an open subset of $\mathbf{R}^{n-1}$ and $f\colon U\to\mathbf{R}$ is of class* 2. Defining

$$F\colon U\to\mathbf{R}^{n-1}\times\mathbf{R},\quad F(x)=(x,f(x)) \text{ for } x\in U,$$

and noting that $J_{n-1}F=|DF|=(1+|Df|^2)^{\frac{1}{2}}$, we will prove that the **minimal surface equation**

$$\sum_{i=1}^{n-1} D_i(|DF|^{-1}D_i f)=0$$

holds if and only if the current $S=F_\#(\mathbf{E}^{n-1}\llcorner U)$ is absolutely area minimizing with respect to $U\times\mathbf{R}$.

From 5.1.7, 5.1.9, 5.1.11 we see that if S is absolutely area minimizing with respect to $U\times\mathbf{R}$, then

$$\delta^{(1)}[F_\#(\mathbf{E}^{n-1}\llcorner K),h]=0$$

for all compact $K\subset U$ and all nonparametric isotopic deformations h corresponding to $\theta\in\mathscr{D}(U,\mathbf{R})$ with $\operatorname{spt}\theta\subset\operatorname{Int}K$, hence the Euler-Lagrange formula implies the minimal surface equation.

Recalling 4.1.7 we define the vectorfield v on $U\times\mathbf{R}$,

$$v(x,y)=|DF(x)|^{-1}(-D_1 f(x),\ldots,-D_{n-1}f(x),1)$$

for $(x,y)\in U\times\mathbf{R}$, and compute $|v(x,y)|=1$,

$$\operatorname{div} v(x,y)=-\sum_{i=1}^{n-1} D_i(|DF|^{-1}D_i f)(x),$$

$$D_1F(x)\wedge\cdots\wedge D_{n-1}F(x)\wedge v(x)=|DF(x)|\, e_1\wedge\cdots\wedge e_n,$$

where $e_1, \dots, e_n$ are the standard base vectors of $\mathbf{R}^{n-1} \times \mathbf{R}$,

$$\vec{S}(x,y) \wedge v(x,y) = e_1 \wedge \dots \wedge e_n, \qquad \langle \vec{S}(x,y), \mathbf{D}_1 v(x,y) \rangle = 1.$$

Thus $\mathbf{D}_1 v$ is a differential form of degree $n-1$ and class 1 on $U \times \mathbf{R}$ with $|\mathbf{D}_1 v| = |v| = 1$ and

$$(S \llcorner C)\, \mathbf{D}_1 v = \int_C \langle \vec{S}, \mathbf{D}_1 v \rangle \, d\|S\| = \mathbf{M}(S | C)$$

for every compact $C \subset U \times \mathbf{R}$. If the minimal surface equation holds, then $\operatorname{div} v = 0$, $d(\mathbf{D}_1 v) = 0$, hence

$$\mathbf{M}(S \llcorner C + \partial T) \geq (S \llcorner C + \partial T)\, \mathbf{D}_1 v = \mathbf{M}(S \llcorner C)$$

whenever $T \in \mathbf{I}_n(U \times \mathbf{R})$. Using the homotopy H defined by the formula

$$H(t,x,y) = (x, t\,y) \text{ for } (t,x,y) \in \mathbf{R} \times U \times \mathbf{R},$$

we also see that if $Q \in \mathbf{I}_{n-1}(U \times \mathbf{R})$ with $\partial Q = 0$, then

$$Q = \partial H_\# ([0,1] \times Q)$$

because $H_1 = \mathbf{1}$ and $H_{0\#} Q$ is an $n-1$ dimensional cycle with compact support in $U \times \{0\} \simeq U \subset \mathbf{R}^{n-1}$. It follows that S is absolutely area minimizing with respect to $U \times \mathbf{R}$.

The following proposition extends a theorem of S. Bernstein (who discovered it in case $n=3$):

If $2 \leq n \leq 7$ and f is a real valued function of class 2 which satisfies the minimal surface equation on $\mathbf{R}^{n-1}$, then f is an $n-1$ dimensional affine subspace of $\mathbf{R}^{n-1} \times \mathbf{R}$.

To prove this we assume $f(0) = 0$ and let

$$G = (\mathbf{R}^{n-1} \times \mathbf{R}) \cap \{(x,y) \colon f(x) \geq y\},$$

$$u(x,y) = (|x|^2 + |y|^2)^{\frac{1}{2}} \text{ for } (x,y) \in \mathbf{R}^{n-1} \times \mathbf{R},$$

$$B_r = \{(x,y) \colon u(x,y) \leq r\} \text{ for } 0 < r < \infty.$$

We see from 4.5.9(1) that $S = (-1)^{n-1} \partial(\mathbf{E}^n \llcorner G)$, from 4.2.1 that

$$\partial(\mathbf{E}^n \llcorner G \cap B_r) = (-1)^{n-1} S \llcorner B_r + \langle \mathbf{E}^n \llcorner G, u, r+ \rangle,$$

$$\mathbf{M}(S \llcorner B_r) \leq \mathbf{M} \langle \mathbf{E}^n \llcorner G, u, r+ \rangle \leq \liminf_{h \to 0+} \mathscr{L}^n(G \cap B_{r+h} \sim B_r)/h \leq \alpha(n)\, n\, r^{n-1}$$

whenever $0 < r < \infty$, from 5.4.3(3) and 5.4.5(2) that there exists a number β for which

$$\alpha(n-1) \leq r^{1-n} \|S\| B_r \uparrow \beta \leq \alpha(n)\, n \text{ as } r \uparrow \infty,$$

hence the currents $R_\varepsilon = \mu_{\varepsilon\#}(\mathbf{E}^n \llcorner G)$ satisfy the conditions

$$R_\varepsilon \llcorner B_r = \mu_{\varepsilon\#}(\mathbf{E}^n \llcorner G \cap B_{r/\varepsilon}), \quad \mathbf{M}(R_\varepsilon \llcorner B_r) \le \alpha(n)\, r^n,$$
$$M[\partial(R_\varepsilon \llcorner B_r)] \le \varepsilon^{n-1}\, 2\alpha(n)\, n (r/\varepsilon)^{n-1} = 2\alpha(n)\, n\, r^{n-1}$$

whenever $0<\varepsilon<\infty$ and $0<r<\infty$. Applying 4.2.17(2) with $K=B_r$ for each positive r, and using Cantor's diagonal process, we construct a sequence of positive numbers ε_i approaching 0 and a current R such that

$$R_{\varepsilon_i} \to R \text{ in } \mathscr{F}_n^{\mathrm{loc}}(\mathbf{R}^{n-1} \times \mathbf{R}) \text{ as } i \to \infty, \text{ hence}$$
$$\mu_{\varepsilon_i\#} S \to (-1)^{n-1}\, \partial R \text{ in } \mathscr{F}_{n-1}^{\mathrm{loc}}(\mathbf{R}^{n-1} \times \mathbf{R}) \text{ as } i \to \infty.$$

We infer from 5.4.2 that ∂R is absolutely area minimizing with respect to $\mathbf{R}^{n-1} \times \mathbf{R}$ and

$$r^{1-n}\, \|\partial R\|\, B_r = \lim_{i\to\infty} (r/\varepsilon_i)^{1-n}\, \|S\|\, B_{r/\varepsilon_i} = \beta$$

whenever $0<r<\infty$, and from 5.4.15 that spt ∂R is an $n-1$ dimensional submanifold of class ∞ of $\mathbf{R}^{n-1} \times \mathbf{R}$, hence $\beta/\alpha(n-1) = \Theta^{n-1}[\partial R, (0,0)]$ is a positive integer. Observing that $R = \mathbf{E}^n \llcorner A$ for some $\mathscr{L}^n$ measurable set A, we see from 4.1.31 that $\beta = \alpha(n-1)$, and apply 5.4.5(3) to conclude that S is an oriented $n-1$ dimensional affine subspace of $\mathbf{R}^{n-1} \times \mathbf{R}$.

It is known that *Bernstein's theorem can be extended also to* $n=8$, by the method of [DG 7]. The modified argument replaces 5.4.15 by its partial extension in 5.4.16, and involves judicious applications of the maximum principle to the partial derivatives of functions used in nonparametric representations of pieces of spt ∂R.

We have presented here only very few facts on solutions of the minimal surface equation – those facts most closely related to the main topic of the section. Much more information on $n-1$ dimensional nonparametric minimal surfaces in $\mathbf{R}^n$, including many important recent results, may be found in [SG], [GD], [MCB 4, §4.2], [MM1], [DGS], [BDGM] and [JS].

5.4.19. Here we take m and n even, say $m=2\mu$ and $n=2\nu$. We identify $\mathbf{C}^\nu$ with $\mathbf{R}^{2\nu}$ so that the standard complex coordinate functions $Z_1, \ldots, Z_\nu$ on $\mathbf{C}^\nu$ are related to the real coordinate functions $X_1, Y_1, \ldots, X_\nu, Y_\nu$ on $\mathbf{R}^{2\nu}$ by the equations

$$Z_j = X_j + \mathbf{i}\, Y_j \text{ for } j=1, \ldots, \nu.$$

We endow $\mathbf{C}^\nu$ with the hermitian product H and the corresponding alternating 2 form A as in 1.8.2, and we define the **fundamental differential form**

$$\Omega \in \mathscr{E}^2(\mathbf{C}^\nu), \quad \Omega(z) = A \text{ for } z \in \mathbf{C}^\nu.$$

From 1.8.2 and 1.6.6 we see that

$$\Omega=(\mathbf{i}/2)\sum_{j=1}^{\nu} DZ_j \wedge D\bar{Z}_j = \sum_{j=1}^{\nu} DX_j \wedge DY_j\,,$$

$$\Omega = dY \text{ with } Y=\sum_{j=1}^{\nu} X_j \wedge DY_j \in \mathscr{E}^1(\mathbf{C}^\nu),\ \Omega^\mu \in \mathscr{E}^{2\mu}(\mathbf{C}^\nu),\ d(Y \wedge \Omega^{\mu-1}) = \Omega^\mu,$$

$$S(\Omega^\mu) = (\partial S)(Y \wedge \Omega^{\mu-1}) = 0 \text{ for } S \in \mathscr{Z}_{2\mu}(\mathbf{C}^\nu).$$

Combining the above observation with Wirtinger's inequality we deduce the following proposition:

If $T \in \mathscr{R}_{2\mu}^{\text{loc}}(\mathbf{C}^\nu)$ and for $\|T\|$ almost all z the simple 2μ vector $\vec{T}(z)$ is complex and positive (see 1.6.6), *then*

$$\|T\| = T \llcorner \Omega^\mu/\mu!$$

and T is absolutely area minimizing with respect to $\mathbf{C}^\nu$.

In fact, for every compact set $K \subset \operatorname{spt} T$ and every cycle $S \in \mathscr{Z}_{2\mu}(\mathbf{C}^\nu)$,

$$(T \llcorner \Omega^\mu)\,K = \textstyle\int_K \langle \vec{T}(z), A\rangle\, d\|T\|\,z = \int_K \mu!\, d\|T\|\,z = \mu!\,\|T\|\,K,$$

$$\mu!\,\mathbf{M}[(T \llcorner K)+S] = \mathbf{M}(\Omega^\mu)\,\mathbf{M}[(T \llcorner K)+S]$$

$$\geq [(T \llcorner K)+S]\,\Omega^\mu = (T \llcorner K)\,\Omega^\mu = \mu!\,\mathbf{M}(T \llcorner K).$$

We also observe that

$$\Omega^\mu/\mu! = \sum_{\lambda \in \Lambda(\nu,\mu)} DX_{\lambda(1)} \wedge DY_{\lambda(1)} \wedge \cdots \wedge DX_{\lambda(\mu)} \wedge DY_{\lambda(\mu)},$$

$$\mathbf{M}(T \llcorner K) = \sum_{\lambda \in \Lambda(\nu,\mu)} \mathbf{M}[p_{\lambda\#}(T \llcorner K)]$$

where $p_\lambda\colon \mathbf{C}^\nu \to \mathbf{C}^\mu$, $p_\lambda(z) = (z_{\lambda(1)}, \ldots, z_{\lambda(\mu)})$ for $z \in \mathbf{C}^\nu$; in the case when $\|T\| = \mathscr{H}^{2\mu} \llcorner \operatorname{spt} T$ we use 4.1.30 to infer

$$\mathscr{H}^{2\mu}(K) = \sum_{\lambda \in \Lambda(\nu,\mu)} \int N(p_\lambda | K, y)\, d\mathscr{L}^{2\mu} y.$$

The preceding considerations are applicable to currents constructed from holomorphic varieties (see 3.4.12, 4.2.29). *If $U \subset W$ are open subsets of $\mathbf{C}^\nu$ and T is a positive complex μ dimensional holomorphic chain of W in U, then $T \llcorner U$ is absolutely area minimizing with respect to W.*

For instance, if W is an open subset of $\mathbf{C}^\nu$, V is a holomorphic subvariety of W and $\mu = \dim_{\mathbf{C}} V$, then there exists a unique current

$$T \in \mathscr{R}_{2\mu}^{\text{loc}}(W) \text{ with } \partial T = 0,\ \|T\| = \mathscr{H}^{2\mu} \llcorner V,$$

such that the simple 2μ vector $\vec{T}(z)$ is complex and positive for $\mathscr{H}^{2m}$ almost all z in V, hence T is absolutely area minimizing with respect to W. We observe that $V = \operatorname{spt} T \sim \operatorname{spt} \partial T$ *need not be a submanifold of* $\mathbf{C}^\nu$, *but can have a singular subset of complex dimension* $\mu - 1$.

To illustrate we consider the elementary example where

$$W = \mathbf{C}^2 \text{ and } V = \mathbf{C}^2 \cap \{z : (z_1)^3 = (z_2)^2\}.$$

Corresponding to $0 < r < \infty$ we define

$$f_r : \mathbf{C} \to \mathbf{C}^2,\ f_r(w) = (w^2, r^{-\frac{1}{2}} w^3) \text{ for } w \in \mathbf{C},$$

and verify that f_r is a univalent holomorphic function with

$$\mu_r(V) = \operatorname{im} f_r,\ \mu_{r\#} T = f_{r\#} \mathbf{E}^2.$$

We conclude that

$$\mu_r(T) \to 2\mathbf{E}^2 \times \delta_0 \text{ in } \mathscr{F}_2^{\mathrm{loc}}(\mathbf{C}^2) \text{ as } r \to \infty,$$

hence the unique oriented tangent cone of T at the singular point $(0,0)$ equals twice an oriented plane, and $\Theta^2[\|T\|, (0,0)] = 2$. It is no accident that these results agree with the notions of algebraic geometry, according to which the complex 1 dimensional variety V has a double tangent line at the singular point $(0,0)$, and $(0,0)$ has multiplicity 2 on V.

From 4.3.19 we know that *every complex μ dimensional holomorphic chain T of W in U has at each point $a \in U$ a unique holomorphic oriented tangent cone C. If T is positive, then C is positive,* because

$$\|C\| \psi \le \lim_{r \to \infty} \|(\mu_r \circ \tau_{-a})_\# T\| \psi = \lim_{r \to \infty} [(\mu_r \circ \tau_{-a})_\# T](\psi\, \Omega^\mu/\mu!) = C(\psi\, \Omega^\mu/\mu!)$$

for all $\psi \in \mathscr{D}^0(\mathbf{C}^\nu)$, hence $\|C\| = C \llcorner \Omega^\mu/\mu!$. Using the complex version of 4.3.11 and the fact that the regular part of any irreducible complex algebraic variety is connected, one can express C as a sum of integral multiples of currents corresponding to homogeneous irreducible complex algebraic subvarieties of $\mathbf{C}^\nu$. It was shown in [F 19, §4] that if T corresponds to an algebraic subvariety V of $\mathbf{C}^\nu$, then C corresponds to the virtual variety (algebraic cycle) which algebraic geometers call the tangent cone of V at a.

The preceding theory generalizes readily from $\mathbf{C}^\nu$ to arbitrary Kähler manifolds (defined in [W 2]), for example complex projective spaces. Then it is still true that $d\Omega = 0$ (though Ω need not be the exterior derivative of any differential form of degree 1) and all positive holomorphic currents are homologically (though not always absolutely) area minimizing.

All presently known types of singularities of area minimizing currents T at points of $\operatorname{spt} T \sim \operatorname{spt} \partial T$ arise from holomorphic varieties and cartesian products of such varieties with intervals (see 5.4.9). The search for other types of singularities presents a challenging problem.

5.4.20. Up to now most research on the regularity of minimizing currents T has dealt with the problem of interior regularity (smoothness of $\operatorname{spt} T \sim \operatorname{spt} \partial T$). However there has also been recent progress on the problem of boundary regularity (smoothness of $\operatorname{spt} T$ near and on $\operatorname{spt} \partial T$); we will summarize the results of [AW 1, 2].

Suppose $T \in \mathbf{I}_m(\mathbf{R}^n)$, T is absolutely area minimizing with respect to $\mathbf{R}^n$, $b \in \operatorname{spt} \partial T$ and there exists an $m-1$ dimensional submanifold B of class $q \geq 2$ of $\mathbf{R}^n$ such that

$$\|\partial T\| \llcorner V = \mathscr{H}^{m-1} \llcorner B$$

for some neighborhood V of b in $\mathbf{R}^n$. Then:

(1) $\Theta^m(\|T\|, b) \geq \frac{1}{2}$.

(2) *If $\Theta^m(\|T\|, b) = \frac{1}{2}$, then there exists an m dimensional submanifold A of class q of $\mathbf{R}^n$ such that*

$$W \cap \operatorname{spt} T \subset A \quad \text{and} \quad \|T\| \llcorner W \leq \mathscr{H}^m \llcorner A$$

for some neighborhood W of b in V, hence $W \cap B = W \cap \operatorname{spt} \partial T \subset A$.

(3) *If there exist an oriented tangent cone C of T at b and a real valued linear function α on $\mathbf{R}^n$ such that*

$$\operatorname{spt} C \subset \operatorname{Tan}(B, b) \cup \{z\colon \alpha(z) > 0\},$$

then $\Theta^m(\|C\|, b) = \frac{1}{2}$.

(4) *If there exist linearly independent real valued linear functions $\beta_1, \ldots, \beta_{n-m+1}$ such that*

$$\operatorname{spt} \partial T \subset \{z\colon \beta_i(z-b) \geq 0 \text{ for } i = 1, \ldots, n-m+1\},$$

then the hypothesis of (3) *holds with $\alpha = \beta_1 + \cdots + \beta_{n-m+1}$.*

The conclusion of (2) implies that $W \cap \operatorname{spt} T$ is a manifold with boundary $W \cap \operatorname{spt} \partial T$ in the sense of differential topology (see [MJR]). The hypothesis of (4) is a natural generalization of the bounded slope condition occurring in classical work on nonparametric minimal surfaces.

Bibliography

S. S. Abhyankar

[AB] *Local Analytic Geometry.* Academic Press, New York, 1964.

L. V. Ahlfors

[AL] *Complex Analysis.* McGraw-Hill, New York, 1966.

L. Alaoglu

[A] *Weak topologies of normed linear spaces.* Ann. of Math. (2) vol. 41 (1940) pp. 252–267.

W. K. Allard

[AW 1] *On boundary regularity for the Plateau problem.* Brown University dissertation, 1968.

[AW 2] *On boundary regularity for Plateau's problem.* Bull. Am. Math. Soc. vol. 75 (1969).

F. J. Almgren, Jr.

[AF 1] *The homotopy groups of the integral cycle groups.* Topology vol. 1 (1962) pp. 257–299.

[AF 2] *An isoperimetric inequality.* Proc. Am. Math. Soc. vol. 15 (1964) pp. 284–285.

[AF 3] *The theory of varifolds.* Princeton, 1965.

[AF 4] *Plateau's Problem.* W. A. Benjamin, New York, 1966.

[AF 5] *Some interior regularity theorems for minimal surfaces and an extension of Bernstein's theorem.* Ann. of Math. vol. 84 (1966) pp. 277–292.

[AF 6] *Existence and regularity almost everywhere of solutions to elliptic variational problems among surfaces of varying topological type and singularity structure.* Ann. of Math. vol. 87 (1968) pp. 321–391.

[AF 7] *A maximum principle for elliptic variational problems.* J. Funct. Analysis.

S. Banach

[BA] *Théorie des opérations linéaires.* Warsaw, 1932.

E. M. Beesley and A. P. Morse

[BEM] *ϕ Cantorian functions and their convex moduli.* Duke Math. J. vol. 12 (1945) pp. 585–619.

A. S. Besicovitch

[B 1] *On the fundamental geometric properties of linearly measurable plane sets of points (I), (II), (III).* Math. Ann. vol. 98 (1927) pp. 422–464, vol. 115 (1938) pp. 296–329, vol. 116 (1939) pp. 349–357.

[B 2] *Concentrated and rarified sets of points.* Acta. Math. vol. 62 (1933) pp. 289–300.

[B 3] *On the Kolmogoroff maximum and minimum measures.* Math. Ann. vol. 113 (1936) pp. 416–423.

[B 4] *A general form of the covering principle and relative differentiation of additive functions (I), (II).* Proc. Cambridge Phil. Soc. vol. 41 (1945) pp. 103–110, vol. 42 (1946) pp. 1–10.

A. S. Besicovitch

[B 5] *On the definition and value of the area of a surface.* Quart. J. Math. vol. 16 (1945) pp. 86–102.

[B 6] *On surfaces of minimum area.* Proc. Cambridge Phil. Soc. vol. 44 (1948) pp. 313–334.

[B 7] *Parametric surfaces.* (I), (II) Proc. Cambridge Phil. Soc. vol. 45 (1949) pp. 5–13, pp. 14–23. (III) J. London Math. Soc. vol. 23 (1948) pp. 241–246. (III, 1) Indag. Math. vol. 16 (1954) pp. 169–174. (IV) Quart. J. Math. vol. 20 (1949) pp. 1–7.

[B 8] *On existence of subsets of finite measure of sets of infinite measure.* Indag. Math. vol. 14 (1952) pp. 339–344.

A. S. Besicovitch and P. A. P. Moran

[B M] *The measure of product and cylinder sets.* J. London Math. Soc. vol. 20 (1945) pp. 110–120.

A. S. Besicovitch and H. D. Ursell

[B U] *Sets of fractional dimensions* (V): *On dimensional numbers of some continuous curves.* J. London Math. Soc. vol. 12 (1937) pp. 18–25.

A. S. Besicovitch and G. Walker

[B W] *On the density of irregular linearly measurable sets of points.* Proc. London Math. Soc. vol. 32 (1931) pp. 142–153.

W. Blaschke

[B L] *Vorlesungen über Differentialgeometrie, II, Affine Differentialgeometrie.* Julius Springer, Berlin, 1923.

M. Bôcher

[B O C] *On the regions of convergence of power series which represent two dimensional harmonic functions.* Trans. Am. Math. Soc. vol. 10 (1909) pp. 271–278.

S. Bochner

[B S] *Summation of multiple Fourier series by spherical means.* Trans. Am. Math. Soc. vol. 40 (1936) pp. 175–207.

E. Bombieri, E. de Giorgi and M. Miranda

[B D G M] *Una maggiorazione a priori relativa alle ipersuperfici minimali non parametriche.* Arch. Rat. Mech. Analysis vol. 32 (1969).

T. Bonnesen and W. Fenchel

[B F] *Theorie der konvexen Körper.* Julius Springer, Berlin, 1934.

A. Borel and A. Haefliger

[B H] *La classe d'homologie fondamentale d'un espace analytique.* Bull. Soc. Math. France vol. 89 (1961) pp. 461–513.

N. Bourbaki

[B O] *Eléments de mathématiques.* Act. Sci. et Ind., Hermann, Paris.

J. E. Brothers

[B J 1] *Integral geometry in homogeneous spaces.* Trans Am. Math. Soc. vol. 124 (1966) pp. 480–517.

[B J 2] *The (ϕ, k) rectifiable subsets of a homogeneous space.* Acta Math.

F. Bruhat and H. Cartan

[B C] *Sur la structure des sousensembles analytiques réels.* C. R. Acad. Sci. Paris vol. 244 (1957) pp. 988–990.

L. Bungart

[B N] *Stokes' theorem on real analytic varieties.* Proc. Nat. Acad. Sci. U.S.A. vol. 54 (1965) pp. 343–344.

A. P. Calderón
[CA] *On the differentiability of absolutely continuous functions.* Rev. Mat. Univ. Parma, vol. 2 (1951) pp. 202–213.

A. P. Calderón and A. Zygmund
[CZ] *On the differentiability of functions which are of bounded variation in Tonelli's sense.* Rev. Union Mat. Argentina, vol. 20 (1960) pp. 102–121.

C. Carathéodory
[C 1] *Über das lineare Maß von Punktmengen, eine Verallgemeinerung des Längenbegriffs.* Nachr. Ges. Wiss. Göttingen 1914, pp. 404–426.
[C 2] *Vorlesungen über reelle Funktionen.* Teubner, Leipzig, 1927.
[C 3] *Variationsrechnung und partielle Differentialgleichungen erster Ordnung.* Teubner, Leipzig, 1935.

L. Carleson
[CLE] *Selected problems on exceptional sets.* Van Nostrand, Princeton, 1968.

É. Cartan
[CE] *Leçons sur la géométrie des espaces de Riemann.* Gauthier-Villars, Paris, 1946.

H. Cartan
[CH] *Variétés analytiques réelles et variétés analytiques complexes.* Bull. Soc. Math. France vol. 85 (1957) pp. 77–99.

L. Cesari
[CL 1] *Sulle funzione a variazione limitata.* Ann. Scuola Norm. Sup. Pisa Ser. 2, vol. 5 (1936) pp. 299–313.
[CL 2] *Caratterizzazione analitica delle superficie continue di area finita secondo Lebesgue.* Ann. Scuola Norm. Sup. Pisa Ser. 2, vol. 10–11 (1941–1942).

C. Chevalley
[CC] *Fundamental concepts of algebra.* Academic Press. New York, 1956.

A. Chiffi
[CHA] *Correnti quasi-normali.* Ann. Scuola Norm. Sup. Pisa (3) 19 (1965) pp. 185–205.

G. Choquet
[CG] *L'isométrie des ensembles dans ses rapports avec la théorie du contact et la théorie de la mesure.* Mathematica vol. 20 (1944) pp. 29–64 (Bucharest).

R. Courant
[CR] *Dirichlet's principle, conformal mapping, and minimal surfaces.* Interscience, New York, 1950.

R. O. Davies
[D 1] *Subsets of finite measure in analytic sets.* Indag. Math. vol. 14 (1952) pp. 488–489.
[D 2] *A property of Hausdorff measure.* Proc. Cambridge Phil. Soc. vol. 52 (1956) pp. 30–34.
[D 3] *Non σ-finite closed subsets of analytic sets.* Proc. Cambridge Phil. Soc. vol. 52 (1956) pp. 174–177.

E. de Giorgi
[DG 1] *Su una teoria generale della misura (r – 1)-dimensionale in uno spacio ad r dimensioni.* Annali di mat. pura ed appl. Ser. 4, vol. 36 (1954) pp. 191–213.
[DG 2] *Nuovi teoremi relativi alle misure (r – 1)-dimensionale in un spazio ad r dimensioni.* Ricerche di matematica, vol. 4 (1955) pp. 95–113.
[DG 3] *Sulla differenziabilita e l'analiticita delle estremali degli integrali multipli regolari.* Mem. Acad. Ser. Torino vol. 143 (1957) pp. 25–43.

E. DE GIORGI

[DG 4] *Sulla proprietà isoperimetrica dell'ipersfera, nelle classe degli insiemi aventi frontiera orientata di misura finita.* Memorie Acc. Naz. Lincei, Ser. 8 vol. 5 (1958) pp. 33–44.

[DG 5] *Complementi alla teoria della misure (n – 1)-dimensionale in uno spazio n-dimensionale.* Seminario Mat. Scuola Norm. Sup. Pisa, 1961.

[DG 6] *Frontiere orientate di misura minima.* Seminario Mat. Scuola Norm. Sup. Pisa, 1961.

[DG 7] *Una estensione del teorema di Bernstein.* Ann. Scuola Norm. Sup. Pisa, Ser. 3, vol. 19 (1965) pp. 79–85.

[DG 8] *Un esempio di estremali discontinue per un problema varizionale di tipo ellittico.* Boll. Un. Mat. Ital. vol. 1 (1968) pp. 135–137.

E. DE GIORGI and G. STAMPACCHIA

[DGS] *Sulle singularità eliminabili delle ipersuperficie minimale.* Atti Accad. Naz. Lincei Rend. Cl. Sci. Fis. Mat. Natur. (8) 38 (1965) pp. 352–357.

M. R. DEMERS and H. FEDERER

[DF] *On Lebesgue area (II).* Trans. Am. Math. Soc. vol. 90 (1959) pp. 499–522.

G. DE RHAM

[DR 1] *Variétés differentiables, formes, courants, formes harmoniques.* Act. Sci. et Ind. vol. 1222, Hermann, Paris, 1955.

[DR 2] *On the area of complex manifolds.* Seminar on several complex variables, Institute for Advanced Study, Princeton, 1957.

D. R. DICKINSON

[DD] *Study of extreme cases with respect to the densities of irregular linearly measurable plane sets of points.* Math. Ann. vol. 116 (1939) pp. 359–373.

A. DINGHAS

[DA] *Einfacher Beweis der isoperimetrischen Eigenschaft der Kugel in Riemannschen Räumen konstanter Krümmung.* Math. Nachr. vol. 2 (1949) pp. 148–162.

J. DOUGLAS

[DJ] *Solution of the problem of Plateau.* Trans. Am. Math. Soc. vol. 33 (1931) pp. 263–321.

N. DUNFORD and J. T. SCHWARTZ

[DS] *Linear operators I, II.* Interscience, New York, 1957, 1963.

H. G. EGGLESTON

[EG 1] *A measureless one-dimensional set.* Proc. Cambridge Philos. Soc. vol. 50 (1954) pp. 391–393.

[EG 2] *Convexity.* Cambridge Univ. Press, 1963.

S. EILENBERG

[E] *On ϕ measures.* Ann. Soc. Pol. de Math. vol. 17 (1938) pp. 251–252.

S. EILENBERG and O. G. HARROLD, JR.

[EH] *Continua of finite linear measure I.* Am. J. Math. vol. 65 (1943) pp. 137–146.

S. EILENBERG and N. STEENROD

[ES] *Foundations of algebraic topology.* Princeton University Press, 1952.

T. ESTERMANN

[ET] *Über Carathéodory's und Minkowski's Verallgemeinerung des Längenbegriffs.* Abh. Math. Sem. Hamburg vol. 4 (1925) pp. 73–116.

G. C. EVANS

[EGC] *Fundamental points of potential theory.* Rice Institute Pamphlet vol. 7 (1920) pp. 252–329.

J. FAVARD

[FA] *Une définition de la longueur et de l'aire.* C. R. Acad. Sci. Paris vol. 194 (1932) pp. 344–346.

H. FEDERER

[F 1] *Surface Area (I), (II).* Trans Am. Math. Soc. vol. 55 (1944) pp. 420–456.

[F 2] *The Gauss-Green theorem.* Trans. Am. Math. Soc. vol. 58 (1945) pp. 44–76.

[F 3] *Coincidence functions and their integrals.* Trans. Am. Math. Soc. vol. 59 (1946) pp. 441–466.

[F 4] *The (ϕ, k) rectifiable subsets of n space.* Trans. Am. Math. Soc. vol. 62 (1947) pp. 114–192.

[F 5] *Dimension and measure.* Trans. Am. Math. Soc. vol. 62 (1947) pp. 536–547.

[F 6] *An introduction to differential geometry.* Brown University, 1948.

[F 7] *Essential multiplicity and Lebesgue area.* Proc. Nat. Ac. Sci. U.S.A. vol. 34 (1948) pp. 611–616.

[F 8] *Hausdorff measure and Lebesgue area.* Proc. Nat. Ac. Sci. U.S.A. vol. 37 (1951) pp. 90–94.

[F 9] *Measure and area.* Bull. Am. Math. Soc. vol. 58 (1952) pp. 306–378.

[F 10] *Some integralgeometric theorems.* Trans. Am. Math. Soc. vol. 77 (1954) pp. 238–261.

[F 11] *An analytic characterization of distributions whose partial derivatives are representable by measures.* Bull. Am. Math. Soc. vol. 60 (1954) p. 339.

[F 12] *On Lebesgue area.* Ann. Math. vol. 61 (1955) pp. 289–353.

[F 13] *An addition theorem for Lebesgue area.* Proc. Am. Math. Soc. vol. 6 (1955) pp. 911–914

[F 14] *A note on the Gauss-Green theorem.* Proc. Am. Math. Soc. vol. 9 (1958) pp. 447–451.

[F 15] *Curvature measures.* Trans. Am. Math. Soc. vol. 93 (1959) pp. 418–491.

[F 16] *The area of a nonparametric surface.* Proc. Am. Math. Soc. vol. 11 (1960) pp. 436–439.

[F 17] *Currents and area.* Trans. Am. Math. Soc. vol. 98 (1961) pp. 204–233.

[F 18] *Approximation of integral currents by cycles.* Proc. Am. Math. Soc. vol. 12 (1961) pp. 882–884.

[F 19] *Some theorems on integral currents.* Trans. Am. Math. Soc. vol. 117 (1965) pp. 43–67.

[F 20] *Two theorems in geometric measure theory.* Bull. Am. Math. Soc. vol. 72 (1966) p. 719.

[F 21] *Some properties of distributions whose partial derivatives are representable by integration.* Bull. Am. Math. Soc. vol. 74 (1968) p. 183–186.

H. FEDERER and W. H. FLEMING

[FF] *Normal and integral currents.* Ann. of Math. vol. 72 (1960) pp. 458–520.

H. FEDERER and A. P. MORSE

[FM] *Some properties of measurable functions.* Bull. Am. Math. Soc. vol. 49 (1943) pp. 270–277.

W. H. FLEMING

[FL 1] *An example in the problem of least area.* Proc. Am. Math. Soc. vol. 7 (1956) pp. 1063–1074.

[FL 2] *Functions whose partial derivatives are measures.* Illinois J. Math. vol. 4 (1960) pp. 452–478.

[FL 3] *On the oriented Plateau problem.* Rend. Circ. Mat. Palermo Ser. 2, vol. 11 (1962) pp. 1–22.

[FL 4] *Flat chains over a coefficient group.* Trans. Am. Math. Soc. vol. 121 (1966) pp. 160–186.

W. H. FLEMING and R. RISHEL

[FLR] *An integral formula for total gradient variation.* Arch. Math. vol. 11 (1960) pp. 218–222.

W. H. FLEMING and L. C. YOUNG

[FY 1] *A generalized notion of boundary.* Trans. Am. Math. Soc. vol. 76 (1954) pp. 457–484.

[FY 2] *Representation of generalized surfaces as mixtures.* Rend. Cir. Mat. Palermo, Ser. 2, vol. 5 (1956) pp. 117–144.

G. FREILICH

[FR 1] *On the measure of cartesian product sets.* Trans. Am. Math. Soc. vol. 69 (1950) pp. 232–275.

[FR 2] *Carathéodory measure of cylinders.* Trans. Am. Math. Soc. vol. 114 (1965) pp. 384–400.

A. FRIEDMAN

[FRA] *On the regularity of the solutions of nonlinear elliptic and parabolic systems of differential equations.* Jour. Math. Mech. vol. 7 (1958) pp. 43–59

O. FROSTMAN

[FO] *Potentiel d'équilibre et capacité des ensembles avec quelques applications à la théorie des fonctions.* Lund, 1935.

I. M. GELFAND

[G] *Abstrakte Funktionen und lineare Operatoren.* Mat. Sbornik N. S. vol. 4 (46) (1938) pp. 235–286.

I. M. GELFAND, M. I. GRAEV and N. YA. VILENKIN

[GGS] *Generalized functions V.* Moscow, 1962.

I. M. GELFAND and G. E. SILOV

[GS] *Generalized functions I, II, III.* Moscow, 1958.

I. M. GELFAND and N. YA. VILENKIN

[GV] *Generalized functions IV.* Moscow, 1961.

D. GILBARG

[GD] *Boundary value problems for non-linear elliptic equations in n variables.* Proc. Symp. Non-linear Problems. Univ. of Wisconsin Press, Madison, 1963.

J. GILLIS

[GJ 1] *On the projection of irregular linearly measurable plane sets of points.* Proc. Cambridge Philos. Soc. vol. 30 (1934) pp. 47–54.

[GJ 2] *A theorem on irregular linearly measurable sets of points.* J. London Math. Soc. vol. 10 (1935) pp. 234–240.

E. GIUSTI and M. MIRANDA

[GM 1] *Un esempio di soluzioni discontinue per un problema di minimo relativo ad un integrale regolare del calcolo delle variazione.* Boll. Un. Mat. Ital. vol. 2 (1968) pp. 1–8.

[GM 2] *Sulla regolarità delle soluzioni deboli di una classe di sistemi ellittici quasi-lineari.* Arch. Rat. Mech. Analysis vol. 31 (1968).

G. GLAESER

[GG] *Études de quelques algèbres tayloriennes.* J. d'analyse math. vol. 6 (1958) pp. 1–124.

C. GOFFMAN

[GC 1] *Nonparametric surfaces given by linearly continuous functions.* Acta Mat. vol. 103 (1960) pp. 269–291.

[GC 2] *A characterization of linearly continuous functions whose partial derivatives are measures.* Acta Mat. vol. 117 (1967) pp. 165–190.

W. GROSS

[GR 1] *Über das Flächenmaß von Punktmengen.* Monatsheft für Math. und Physik, vol. 29 (1918) pp. 145–176.

[GR 2] *Über das lineare Maß von Punktmengen.* Monatshefte für Math. und Physik vol. 29 (1918) pp. 177–193.

R. C. GUNNING and H. ROSSI

[GUR] *Analytic functions of several complex variables.* Prentice Hall, Englewood Cliffs, N.J., 1965.

W. GUSTIN

[GW] *Boxing Inequalities.* J. Math. Mech. vol. 9 (1960) pp. 229–239.

H. HADWIGER

[HH 1] *Volumschätzung für die einen Eikörper überdeckenden und unterdeckenden Parallelotope.* Elem. Math. vol. 10 (1955) pp. 122–124.

[HH 2] *Vorlesungen über Inhalt, Oberfläche und Isoperimetrie.* Springer-Verlag, Heidelberg, 1957.

P. HALMOS

[HA] *Measure theory.* Van Nostrand, New York, 1950.

F. HAUSDORFF

[HF] *Dimension und äußeres Maß.* Math. Ann. vol. 79 (1918) pp. 157–179.

C. A. HAYES, JR. and A. P. MORSE

[HM] *Some properties of annular blankets.* Proc. Am. Math. Soc. vol. 1 (1950) pp. 107–126.

C. A. HAYES, JR. and C. Y. PAUC

[HP] *Full individual and class differentiation theorems in their relation to halo and Vitali properties.* Can. J. Math. vol. 7 (1955) pp. 221–274.

S. HELGASON

[HE] *Differential geometry and symmetric spaces.* Academic Press, New York, 1962.

M. HERRERA

[HR] *Intégration sur un ensemble semi-analytique.* C. R. Acad. Sc. Paris, vol. 260 (1965) pp. 763–765.

M. HERVÉ

[HEM] *Several complex variables.* Oxford University Press, 1963.

M. H. HESTENES

[HS] *Extension of the range of a differentiable function.* Duke Math. J. vol. 8 (1941) pp. 183–192.

J. C. HOLY

[HO] *Sur l'ensemble des valeurs stationaires d'une application différentiable.* Comm. Math. Helv. vol. 41 (1966–67) pp. 157–169.

E. HOPF

[HOE 1] *Elementare Bemerkungen über die Lösungen partieller Differentialgleichungen zweiter Ordnung vom elliptischen Typus.* Sitzber. Preuss. Akad. Wiss. Berlin vol. 19 (1927) pp. 147–152.

[HOE 2] *Über den funktionalen, insbesondere den analytischen Charakter der Lösungen elliptischer Differentialgleichungen zweiter Ordnung.* Math. Zeit. vol. 34 (1932) pp. 194–233.

L. HÖRMANDER

[H 1] *On a theorem of Grace.* Math. Scand. 2 (1954) pp. 55–64.

[H 2] *Linear partial differential operators.* Springer-Verlag, Heidelberg, 1963.

[H 3] *An introduction to complex analysis in several variables.* Van Nostrand, Princeton, 1966.

W. HUREWICZ and H. WALLMAN

[H W] *Dimension theory.* Princeton Univ. Press, 1941.

H. JENKINS and J. SERRIN

[J S] *The Dirichlet problem for the minimal surface equation in higher dimensions.* J. Reine Angew. Math. vol. 223 (1968) pp. 170 – 187.

F. JOHN

[J] *Plane waves and spherical means.* Interscience, New York, 1955.

J. L. KELLEY

[K] *General Topology.* Van Nostrand, New York, 1955.

O. D. KELLOGG

[K O] *On bounded polynomials in several variables.* Math. Zeit. vol. 27 (1928) pp. 55 – 64.

M. D. KIRSZBRAUN

[K I] *Über die zusammenziehenden und Lipschitzschen Transformationen.* Fund. Math. vol. 22 (1934) pp. 77 – 108.

S. A. KLINE

[K S] *On curves of fractional dimension.* J. London Math. Soc. vol. 20 (1945) pp. 79 – 86.

M. KNESER

[K M 1] *Einige Bemerkungen über das Minkowskische Flächenmaß.* Archiv der Math. vol. 6 (1955) pp. 382 – 390.

[K M] *Summenmengen in lokalkompakten abelschen Gruppen.* Math. Zeitschr. vol. 66 (1956) pp. 88 – 110.

S. KOBAYASHI and K. NOMIZU

[K N] *Foundations of differential geometry I, II.* Interscience, New York, 1963, 1969.

A. KOLMOGOROFF

[K A] *Beiträge zur Maßtheorie.* Math. Ann. vol. 107 (1932) pp. 351 – 366.

J. KRAL

[K R] *The Fredholm method in potential theory.* Trans. Am. Math. Soc. vol. 125 (1966) pp. 511 – 547.

K. KRICKEBERG

[K K] *Distributionen, Funktionen beschränkter Variation und Lebesguescher Inhalt nichtparametrischer Flächen.* Annali di mat. pura ed appl. (IV) vol. 44 (1957) pp. 105 – 134.

I. KUPKA

[K P] *Counterexample to the Morse-Sard theorem in the case of infinite dimensional manifolds.* Proc. Am. Math. Soc. vol. 16 (1965) pp. 954 – 957.

C. KURATOWSKI

[K U] *Topologie I, II.* Warsaw, 1952, 1950.

P. LELONG

[L P] *Intégration sur un ensemble analytique complexe.* Bull. Soc. Math. France vol. 85 (1957) pp. 239 – 262.

S. LOJASIEWICZ

[L S 1] *Sur le problème de la division.* Rozprawy Matematyczne, No. 22, Warsaw, 1961.

[L S 2] *Une propriété topologique des sous-ensembles analytiques réels.* Colloques internationaux du Centre National de la Recherche Scientifique, No. 117, Les Équations aux Dérivées Partielles. Paris, 1962, pp. 87 – 89.

L. H. Loomis

[L 1] *The intrinsic measure theory of Riemannian and Euclidean metric spaces.* Ann. Math. vol. 45 (1944) pp. 367 – 374.

[L 2] *Haar measure in uniform structures.* Duke Math. J. vol. 16 (1949) pp. 193 – 208.

[L 3] *An introduction to abstract harmonic analysis.* Van Nostrand, New York, 1953.

S. MacLane

[MCL] *Homology.* Springer-Verlag, Heidelberg, 1963.

B. Malgrange

[MB 1] *Division des distributions.* Séminaire Schwartz 1959/60, Exp. 21 – 25. Faculté des Sciences de Paris.

[MB 2] *Ideals of differentiable functions.* Oxford University Press, 1966.

E. Marczewski and R. Sikorski

[MS] *Measures in nonseparable metric spaces.* Coll. Math. vol. 1 (1948), pp. 133 – 139.

J. M. Marstrand

[MJ 1] *Some fundamental geometrical properties of plane sets of fractional dimensions.* Proc. London Math. Soc. (3) 4 (1954) pp. 257 – 302.

[MJ 2] *The dimension of Cartesian product sets.* Proc. Cambridge Phil. Soc. vol. 50 (1954) pp. 198 – 202.

[MJ 3] *Circular density of plane sets.* J. London Math. Soc. vol. 30 (1955) pp. 238 – 246.

[MJ 4] *Hausdorff 2 dimensional measure in 3 space.* Proc. London Math. Soc. (3) 11 (1961) pp. 91 – 108.

[MJ 5] *The (ϕ, s) regular subsets of n space.* Trans. Am. Math. Soc. vol. 113 (1964) pp. 369 – 392.

S. Mazurkiewicz and S. Saks

[MAS] *Sur les projections d'un ensemble fermé.* Fund. Math. vol. 8 (1926) pp. 109 – 113.

E. J. McShane

[MCS] *Parametrizations of saddle surfaces with applications to the problem of Plateau.* Trans. Am. Math. Soc. vol. 35 (1934) pp. 718 – 733.

J. H. Michael

[MJH] *Lipschitz approximations to summable functions.* Acta Math. vol. 111 (1964) pp. 73 – 95.

E. J. Mickle

[MI 1] *On the extension of a transformation.* Bull. Am. Math. Soc. vol. 55 (1949) pp. 160 – 164.

[MI 2] *On a decomposition theorem of Federer.* Trans. Am. Math. Soc. vol. 92 (1959) pp. 322 – 335.

E. J. Mickle and T. Radó

[MIR] *Density theorems for outer measures in n-space.* Proc. Am. Math. Soc. vol. 9 (1958) pp. 433 – 439.

M. Miranda

[MM 1] *Un teorema di esistenza e unicità per il problema dell'area minima in n variabili.* Ann. Scuola Nor. Sup. Pisa, Ser. III vol. 19 (1965) pp. 233 – 249.

[MM 2] *Una maggiorazione integrale per le curvature delle ipersuperfici minimali.* Rend. Sem. Mat. Univ. Padova vol. 38 (1967) pp. 91 – 107.

D. Montgomery

[MD] *Nonseparable metric spaces.* Fund. Math. vol. 25 (1935) pp. 527 – 533.

E. F. Moore

[ME 1] *Density ratios and $(\phi, 1)$ rectifiability in n-space.* Trans. Am. Math. Soc. vol. 69 (1950) pp. 324 – 334.

E. F. Moore

[ME 2] *Convexly generated k dimensional measures.* Proc. Am. Math. Soc. vol. 2 (1951) pp. 597 – 606.

G. W. Morgan

[MO] *The density directions of irregular measurable sets.* Proc. London Math. Soc. vol. 38 (1935) pp. 481 – 494.

C. B. Morrey, Jr.

[MCB 1] *Multiple integral problems in the calculus of variations and related topics.* Univ. of California Publ. in Math., new ser. 1 (1943) pp. 1 – 130.

[MCB 2] *Second order elliptic systems of differential equations.* Annals of Math. Studies vol. 33 (1954) pp. 101 – 159.

[MCB 3] *On the analyticity of the solutions of nonlinear elliptic systems of partial differential equations.* Am. Jour. Math. vol. 80 (1958) pp. 198 – 234.

[MCB 4] *Multiple integrals in the calculus of variations.* Springer-Verlag, New York, 1966.

A. P. Morse

[M 1] *The behavior of a function on its critical set.* Ann. of Math. vol. 40 (1939) pp. 62 – 70.

[M 2] *A theory of covering and differentiation.* Trans. Am. Math. Soc. vol. 55 (1944) pp. 205 – 235.

[M 3] *Perfect blankets.* Trans. Am. Math. Soc. vol. 61 (1947) pp. 418 – 442.

[M 4] *On intervals of prescribed length.* Proc. Am. Math. Soc. vol. 5 (1954) pp. 407 – 414.

A. P. Morse and J. F. Randolph

[MR 1] *Gillespie measure.* Duke Math. J. vol. 6 (1940) pp. 408 – 419.

[MR 2] *The ϕ rectifiable subsets of the plane.* Trans. Am. Math. Soc. vol. 55 (1944) pp. 236 – 305.

B. J. Mueller

[MU] *Three results for locally compact groups connected with the Haar measure density theorem.* Proc. Am. Math. Soc. vol. 16 (1965) pp. 1414 – 1416.

J. R. Munkres

[MJR] *Elementary differential topology.* Ann. of Math. Studies No. 54 (1963).

L. Nachbin

[NA] *Lectures on the theory of distributions.* Universidade do Recife, Rio de Janeiro, 1964.

R. Narashimhan

[NR] *Introduction to the theory of analytic spaces.* Springer-Verlag, 1966.

W. C. Nemitz

[NE] *On a decomposition theorem for measures in Euclidean n-space.* Trans. Am. Math. Soc. vol. 98 (1961) pp. 306 – 333.

L. Nirenberg

[NI 1] *Remarks on strongly elliptic differential equations.* Comm. Pure Appl. Math. vol. 6 (1953) pp. 648 – 674.

[NI 2] *On elliptic partial differential equations.* Ann. Scuola Norm. Sup. Pisa (III) vol. 13 (1959) pp. 115 – 162.

J. C. C. Nitsche

[NJ] *On new results in the theory of minimal surfaces.* Bull. Am. Math. Soc. vol. 71 (1965) pp. 195 – 270.

G. NÖBELING

[N 1] *Über den Flächeninhalt dehnungsbeschränkter Flächen.* Math. Zeit. vol. 48 (1943) pp. 747–771.

[N 2] *Über die Flächenmaße im Euklidischen Raum.* Math. Ann. vol. 118 (1943) pp. 687–701.

O. PERRON

[PO] *Irrationalzahlen.* Chelsea, New York, 1948.

I. PETROWSKI

[PE] *Sur l'analyticité des solutions des systèmes d'équations differentielles.* Rec. Math. N. S. Mat. Sbornik vol. 5 (47) (1939) pp. 3–70.

B. J. PETTIS

[P 1] *On integration in vector spaces.* Trans. Am. Math. Soc. vol. 44 (1938) pp. 277–304.

[P 2] *Differentiation in Banach spaces.* Duke Math. J. vol. 5 (1939) pp. 254–269.

H. RADEMACHER

[RH] *Über partielle und totale Differenzierbarkeit I.* Math. Ann. vol. 79 (1919) pp. 340–359.

T. RADÓ

[RA 1] *Über des Flächenmaß rektifizierbaren Flächen.* Math. Ann. vol. 100 (1928) pp. 445–479.

[RA 2] *On the problem of Plateau.* Julius Springer, Berlin, 1933.

J. F. RANDOLPH

[RJ 1] *On generalizations of length and area.* Bull. Am. Math. Soc. vol. 42 (1936) pp. 268–274.

[RJ 2] *Some properties of sets of the Cantor type.* J. London Math. Soc. vol. 16 (1941) pp. 38–42.

R. E. REIFENBERG

[R 1] *Parametric surfaces.* (I), (II) Proc. Cambridge Phil. Soc. vol. 47 (1951) pp. 687–698, vol. 48 (1952) pp. 46–69. (III) Quart. J. Math. Oxford (2), 3 (1952) pp. 227–234. (IV) J. London Math. Soc. vol. 27 (1952) pp. 448–456. (V) Proc. London Math. Soc. (3) 5 (1955) pp. 341–357.

[R 2] *Solution of the Plateau Problem for m-dimensional surfaces of varying topological type.* Acta Mathematica vol. 104 (1960) pp. 1–92.

[R 3] *An epiperimetric inequality related to the analyticity of minimal surfaces. On the analyticity of minimal surfaces.* Ann. of Math. vol. 80 (1964) pp. 1–21.

W. RUDIN

[RU] *Real and Complex Analysis.* McGraw-Hill, New York, 1966.

S. SAKS

[S 1] *Theory of the integral.* Warsaw, 1937.

[S 2] *Remarque sur la mesure linéaire des ensembles plans.* Fund. Math. vol. 9 (1927) pp. 16–24.

P. SAMUEL

[SAM] *Méthodes d'algèbre abstraite en géométrie algébrique.* Springer-Verlag, Heidelberg, 1955.

A. SARD

[SA 1] *The measure of the critical values of differentiable maps.* Bull. Am. Math. Soc. vol. 48 (1942) pp. 883–890.

[SA 2] *Images of critical sets.* Ann. of Math. vol. 68 (1958) pp. 247–259.

[SA 3] *Hausdorff measure of critical images on Banach manifolds.* Am. J. Math. vol. 87 (1965) pp. 158–174.

I. J. SCHOENBERG
[SC] *On certain metric spaces arising from Euclidean spaces by a change of metric and their embedding in Hilbert space.* Ann. Math. vol. 38 (1937) pp. 787 – 793.

L. SCHWARTZ
[SCH] *Théorie des distributions I, II.* Act. Sci. et Ind. vol. 1245, 1122. Hermann, Paris, 1957, 1951.

S. SHERMAN
[SS] *A comparison of linear measures in the plane.* Duke Math. J. vol. 9 (1942) pp. 1 – 9.

B. SHIFFMAN
[SB] *On the removal of singularities of analytic sets.* Michigan Math. J. vol. 15 (1968) pp. 111 – 120.

W. SIERPINKSKI
[SP 1] *Sur la densité linéaire des ensembles plans.* Fund. Math. vol. 9 (1927) pp. 172 – 185.
[SP 2] *Sur le produit combinatoire de deux ensembles jouissant de la propriété C.* Fund. Math. vol. 24 (1935) pp. 48 – 50.

J. SIMONS
[SJ] *Minimal varieties in Riemannian manifolds.* Ann. of Math. vol. 88 (1968) pp. 62 – 105.

M. SION
[SI] *On the existence of functions having given partial derivatives on a curve.* Trans. Am. Math. Soc. vol. 77 (1954) pp. 179 – 201.

S. L. SOBOLEV
[SOB] *On a theorem of functional analysis.* Mat. Sbornik N. S. 4 (46) pp. 471 – 497 (1938).

E. H. SPANIER
[SE] *Algebraic topology.* McGraw-Hill, New York, 1966.

G. STAMPACCHIA
[SG] *On some regular multiple integral problems in the calculus of variations.* Comm. Pure Appl. Math. vol. 16 (1963) pp. 383 – 421.

W. STEPANOFF
[SW 1] *Über totale Differenzierbarkeit.* Math. Ann. vol. 90 (1923) pp. 318 – 320.
[SW 2] *Sur les conditions de l'existence de la differentielle totale.* Rec. Math. Soc. Math. Moscou vol. 32 (1925) pp. 511 – 526.

S. STERNBERG
[ST] *Lectures on differential geometry.* Prentice Hall, Englewood Cliffs, N.J., 1964.

A. TARSKI
[TA] *Über unerreichbare Kardinalzahlen.* Fund. Math. vol. 30 (1938) pp. 68 – 89.

S. J. TAYLOR
[TSJ] *On Cartesian product sets.* J. London Math. Soc. vol. 27 (1952) pp. 295 – 304.

R. THOM
[TH] *Quelques propriétés globales des variétés differentiables.* Comm. Math. Helv. vol. 28 (1954) pp. 17 – 86.

R. N. TOMPSON
[T] *Areas of k dimensional nonparametric surfaces in k+1 space.* Trans. Am. Math. Soc. vol. 77 (1954) pp. 374 – 407.

L. Tonelli

[T L] *Sulla quudratura delle superficie.* Rend. Accad. Naz. Lincei (6) vol. 3 (1926) pp. 357–363, 445–450, 633–638.

D. Triscari

[T D 1] *Sulle singularità delle frontiere orientate di misura minima.* Ann. Scula Nor. Sup. Pisa Ser. 3 vol. 17 (1963) pp. 349–371.

[T D 2] *Sull'estistenza di cilindri con frontiere di misura minima.* Ann. Scula Nor. Sup. Pisa Ser. 3 vol. 17 (1963) pp. 387–399.

[T D 3] *Sulle singularità delle frontiere orientate di misura minima nello spazio euclideo a 4 dimensione.* Le Matematiche, vol. 18 (1963) pp. 139–163.

S. Ulam

[U] *Zur Maßtheorie in der allgemeinen Mengenlehre.* Fund. Math. vol. 16 (1930) pp. 140–150.

J. Väisälä

[V J] *Two new characterizations for quasiconformality.* Ann. Acad. Sci. Fenn. AI 362 (1965).

F. A. Valentine

[V] *A Lipschitz condition preserving extension for a vector function.* Am. J. Math. vol. 67 (1945) pp. 83–93.

L. Van Hove

[V H 1] *Sur l'extension de la condition de Legendre du Calcul des Variations aux intégrales multiples à plusieurs fonctions inconnues.* Indagationes mathematicae, vol. 9 (1947) pp. 3–8.

[V H 2] *Sur le signe de la variation seconde des intégrales multiples à plusieurs fonctions inconnues.* Acad. royale de Belgique, Mémoires vol. 24 (1949) Fasc. 5.

W. Velte

[V W] *Zur Variationsrechnung mehrfacher Integrale.* Math. Zeitschr. vol. 60 (1954) pp. 367–383.

G. Walker

[W G] *On the density of irregular linearly measurable plane sets.* Proc. London Math. Soc. (2) vol. 30 (1929) pp. 481–499.

D. J. Ward

[W A] *A counterexample in area theory.* Proc. Cambridge Philos. Soc. vol. 60 (1964) pp. 821–845.

A. Weil

[W 1] *L'intégration dans les groupes topologiques et ses applications.* Act. Sci. et Ind. vol. 869, Hermann, Paris, 1938.

[W 2] *Introduction à l'étude des variétés Kähleriennes.* Act. Sci. et Ind. vol. 1267, Hermann, Paris, 1958.

H. Whitney

[W H 1] *Analytic extensions of differentiable functions defined in closed sets.* Trans. Am. Math. Soc. vol. 36 (1934) pp. 63–89.

[W H 2] *A function not constant on a connected set of critical points.* Duke Math. J. vol. 1 (1935) pp. 514–517.

[W H 3] *On totally differentiable and smooth functions.* Pac. J. Math. vol. 1 (1951) pp. 143–159.

[W H 4] *Geometric integration theory.* Princeton Univ. Press, 1957.

[W H 5] *Elementary structure of real algebraic varieties.* Ann. of Math. vol. 66 (1957) pp. 545–556.

H. Whitney

[W H 6] *Local properties of analytic varieties.* Differential and combinatorial topology, Princeton Univ. Press, 1965.

[W H 7] *Tangents to an analytic variety.* Ann. of Math. vol. 81 (1965) pp. 496–549.

H. Whitney and F. Bruhat

[W B] *Quelques propriétés fondamentales des ensembles analytiques-réels.* Comm. Math. Helv. vol. 33 (1959) pp. 132–160.

W. Wirtinger

[W W] *Eine Determinantenidentität und ihre Anwendung auf analytische Gebilde und Hermitesche Maßbestimmung.* Monatsh. f. Math. u. Physik vol. 44 (1936) pp. 343–365.

L. C. Young

[Y 1] *Generalized curves and the existence of an attained absolute minimum in the calculus of variations.* C. R. Soc. Sci. Lett. Varsovie, Classe III vol. 30 (1937) pp. 212–234.

[Y 2] *Surfaces paramétriques généralisées.* Bull. Soc. Math. France vol. 79 (1951) pp. 59–84.

[Y 3] *Generalized surfaces in the calculus of variations I, II.* Ann. of Math. vol. 43 (1942) pp. 84–103, 530–544.

O. Zariski and P. Samuel

[Z S] *Commutative algebra I.* Van Nostrand, Princeton, 1958.

W. P. Ziemer

[Z 1] *Integral currents mod 2.* Trans. Am. Math. Soc. vol. 105 (1962) pp. 496–524.

[Z 2] *Extremal length and conformal capacity.* Trans. Am. Math. Soc. vol. 126 (1967) pp. 460–473.

[Z 3] *The area and variation of linearly continuous functions.* Proc. Am. Math. Soc. vol. 20 (1969) pp. 81–87.

[Z 4] *Extremal length and p-capacity.* Michigan Math. J.

Glossary of some standard notations

(which are used but not defined in the text)

Membership: $\in$ Inclusion: $\subset$ Empty set: $\varnothing$

Classification: $\{x: ---\}$ = the class of all sets x for which $---$

Classes with listed elements: $\{a\}, \{a, b\}, \{a, b, c\}, \ldots$

Union, intersection, difference of A and B: $A \cup B, A \cap B, A \sim B$

Union of F: $\bigcup F = \bigcup_{A \in F} A = \{x: x \in A \text{ for some } A \in F\}$

Intersection of F: $\bigcap F = \bigcap_{A \in F} A = \{x: x \in A \text{ for all } A \in F\}$

Ordered pair: (x, y) Sequence: $s = (s_1, s_2, s_3, \ldots)$

Domain, image of f: $\operatorname{dmn} f$, $\operatorname{im} f$

Inverse of f: $f^{-1} = \operatorname{inv} f = \{(y, x): (x, y) \in f\}$

Composition: $g \circ f = \{(x, z): (x, y) \in f \text{ and } (y, z) \in g \text{ for some } y\}$

Mapping: f maps X into Y ($f: X \to Y$) if and only if
f is a function, $\operatorname{dmn} f = X$ and $\operatorname{im} f \subset Y$

$\langle x, f \rangle = f(x)$ = the unique y such that $(x, y) \in f$ $\quad \langle \cdot, f \rangle = f(\cdot) = f$

Identity map of A: $\mathbf{1}_A$ Restriction: $f|A = f \cap \{(x, y): x \in A\}$

Cartesian product: $X \times Y = \{(x, y): x \in X \text{ and } y \in Y\}$

If $f: A \to B$ and $g: C \to D$, then $f \times g: A \times C \to B \times D$,
$(f \times g)(a, c) = (f(a), g(c))$ for $(a, c) \in A \times C$.

n-fold Cartesian product of X: X^n

Class of all functions mapping X into Y: Y^X

Cardinal number of S: card S m-th infinite cardinal number: $\aleph_{m-1}$

Interior, closure, boundary of T: Int T, Clos T, Bdry T

Isomorphism: $\simeq$ Class of cosets: G/H

Kernel of homomorphism f: $\ker f$ Direct sums: $V \oplus W$, $\bigoplus_{j \in J} V_j$

Rings of integers, integers modulo ν: $\mathbf{Z}$, $\mathbf{Z}_\nu = \mathbf{Z}/(\nu\,\mathbf{Z})$

Fields of rational, real, complex numbers: $\mathbf{Q}, \mathbf{R}, \mathbf{C}$

Square root of -1: $\mathbf{i} = (0, 1) \in \mathbf{C} = \mathbf{R} \times \mathbf{R}$

Dimension of vectorspace V over $\mathbf{R}$, over $\mathbf{C}$: $\dim V$, $\dim_{\mathbf{C}} V$

Class of all $\mathbf{R}$ linear functions mapping V into W: $\operatorname{Hom}(V, W)$

List of basic notations defined in the text

(in the order of their appearance)

1.1.1	$\otimes$
1.2.3	$\otimes_*$ $\otimes_m$
1.3.1	$\mathfrak{A}$ $\wedge_*$ $\wedge_m$ $\wedge$
1.3.2	e_λ $\Lambda(n, m)$
1.3.3	$\operatorname{Sh}(p, q)$ $\operatorname{index}(\sigma)$
1.3.4	$\det(f)$
1.4.1	$\wedge^m$ Hom^m $\wedge^*$ $\langle \xi, f \rangle$
1.4.2	$\wedge$
1.4.3	ω_λ
1.4.5	$\Gamma(f)$ trace
1.5.1	$\lrcorner$ $\llcorner$
1.5.2	$\mathbf{D}_p$ $\mathbf{D}^p$
1.6.2	$\mathbf{G}(n, m)$ $\mathbf{G}_0(n, m)$
1.6.6	$\wedge^m_{\mathbf{C}}$ $\wedge^*_{\mathbf{C}}$
1.7.1	$x \bullet y$ $\lvert x \rvert$
1.7.2	$\mathbf{O}(n, m)$
1.7.4	f^* $\mathbf{O}^*(n, m)$ $\mathbf{p}_\lambda$
1.7.5	$\xi \bullet \eta$ $\lvert \xi \rvert$ $\phi \bullet \psi$ $\lvert \phi \rvert$
1.7.6	$\lVert f \rVert$
1.7.8	$*$
1.7.10	discr trace
1.8.1	$\lVert \phi \rVert$ $\lVert \xi \rVert$
1.9.1	$\mathfrak{B}$ $\odot_*$ $\odot_m$ $\odot$
1.9.2	e^α $\Xi(n, m)$
1.9.3	t^α v^α $\alpha!$
1.10.1	$\odot^m$ $\langle \xi, f \rangle$ Hom^m $\odot^*$
1.10.2	$\odot$ ω^α
1.10.3	$\lrcorner$ $\llcorner$
1.10.5	$\lvert \phi \rvert$ $\lvert \xi \rvert$ $\lVert \phi \rVert$ $\lVert \xi \rVert$ $\lVert P \rVert$
1.10.6	$\mathscr{S}(n, m)$
2.1.1	$\overline{\mathbf{R}}$ inf sup $\uparrow$ $\downarrow$
	sign f^+ f^- Σ
2.1.2	$\mathbf{2}^X$ $\phi \llcorner Y$ $f_\#$
2.2.1	spt ϕ
2.2.6	$\mathscr{P}$ $\mathscr{N}$
2.2.7	$\operatorname{Lip}(f)$
2.3.8	$\mathbf{A}(\phi, Y)$ $\lvert f \rvert_\phi$
2.4.2	$\int^* f\, d\phi$ $\int_* f\, d\phi$ $\int f\, d\phi$
	$\phi_x[f(x)]$ $\langle f, \phi \rangle$
2.4.10	$\int^*_A$ $\int_{*A}$ $\int_A$ $\int_a^b$
2.4.12	$\phi_{(p)}(f)$ $\mathbf{L}_p(\phi, Y)$ $\int f\, d\phi$ Y^*
2.5.1	L^+
2.5.5	μ^+ μ^-
2.5.6	$\mu_x[f(x)]$
2.5.13	spt f
2.5.14	$\mathscr{K}(X)$ spt μ
2.5.16	$\mathbf{V}_a^b$
2.5.17	$\int_a^b f\, dg$ $\int_a^b f(x)\, d_x g(x)$ $\mathscr{L}^1$
2.5.19	L^* $\boldsymbol{\delta}_x$ $u_\#$
2.6.1	$\alpha \times \beta$
2.6.5	$\mathscr{L}^n$
2.7.1	A^x A_g G^x u_g
2.7.8	Δ_G
2.7.16	$\mathscr{A}(V, W)$ $\boldsymbol{\tau}_z$ $\mathbf{GL}(n, \mathbf{R})$
	$\mathbf{SL}(n, \mathbf{R})$ $\boldsymbol{\alpha}(n)$ $\boldsymbol{\theta}_{n,m}$
	$\boldsymbol{\theta}^*_{n,m}$ $\boldsymbol{\gamma}_{n,m}$ $\boldsymbol{\beta}_t(n, m)$
2.8.1	$\mathbf{B}(a, r)$ $\mathbf{U}(a, r)$
2.8.4	$\hat{S}$
2.8.8	diam
2.8.16	$(C) \limsup_{x} f$ $(C) \limsup_{S \to x} f(S)$
2.9.1	ψ_ϕ $\mathbf{D}(\psi, \phi, V, x)$
2.9.12	$(\phi, V) \operatorname{ap} \lim_{x} f$ $(\phi, V) \operatorname{ap} \lim_{z \to x} f(z)$
2.9.19	$g'(x)$
2.10.2	$\mathscr{H}^m$ $\mathscr{S}^m$
2.10.3	$\mathscr{T}^m$
2.10.4	$\mathscr{G}^m$ $\mathscr{C}^m$
2.10.5	$\mathscr{I}_t^m$ $\mathscr{Q}_t^m$

2.10.9	$N(f,y)$
2.10.19	$\Theta^{*m}(\mu,a)$ $\Theta^m_*(\mu,a)$ $\Theta^m(\mu,a)$
3.1.1	Df $\operatorname{grad} f$ $D_i f$
3.1.2	$\operatorname{ap} Df$ $\operatorname{ap} D_i f$
3.1.11	$D^k f$ $f^{(k)}$ $D_v f$ $D^\alpha f$
3.1.21	$\operatorname{Tan}(S,a)$ $\operatorname{Tan}(S)$
	$\operatorname{Nor}(S,a)$ $\operatorname{Nor}(S)$
3.1.22	Df
3.2.1	$J_k f$ $\operatorname{ap} J_k f$
3.2.13	$\mathbf{S}^{m-1}$ $\Gamma(s)$
3.2.16	$\operatorname{Tan}^m(\phi,a)$ $\mathbf{E}(a,v,\varepsilon)$
	$\operatorname{Nor}^m(\phi,a)$ $(\phi,m)\operatorname{ap} Df$
3.2.20	$\operatorname{ap} J_m f$
3.2.22	$\operatorname{ap} J_\mu f$
3.2.28	$\mathbf{SO}(m)$
3.2.37	$\mathscr{M}^{*m}$ $\mathscr{M}^m_*$ $\mathscr{M}^m$
3.3.1	$\mathbf{X}(a,r,V,s)$
3.4.5	$\mathscr{O}_a$ γ_a ideal $\mathscr{V}_a$ var
3.4.8	depth dim
3.4.12	$\mathscr{O}^{\mathbf{C}}_a$ $\text{ideal}^{\mathbf{C}}$ $\mathscr{V}^{\mathbf{C}}_a$ $\mathbf{U}(n)$ $\dim_{\mathbf{C}}$
4.1.1	$\mathscr{E}$ v^i_K $\mathscr{E}'$ spt $\mathscr{E}'_K$
	$\mathscr{D}_K$ $\mathscr{D}$ $\mathscr{D}'$ $T \llcorner v$
	TD_j TD^α $D_j T$ $D^\alpha T$ $T_x \phi(x)$
4.1.2	Φ_ε T_ε $*$
4.1.5	$\\|S\\|$
4.1.6	$\wedge$ $\lrcorner$ $\llcorner$ $\langle \xi,\phi\rangle$ $\\|\phi\\|$ $\\|\xi\\|$
	X_λ e_λ $(DX)_\lambda$ $\mathbf{e}_\lambda$
	$\mathbf{D}_p\xi$ $\mathbf{D}^p\phi$ $*\xi$ $*\phi$
	$d\phi$ $\operatorname{div}\xi$ $f^\#$
4.1.7	$\mathscr{E}^m$ $\mathscr{E}_m$ $\mathscr{D}^m$ $\mathscr{D}_m$
	$T\llcorner\phi$ $T\wedge\xi$ ∂T $\mathbf{E}^n$
	$\vec{S}$ $\mathbf{M}$ $\mathbf{N}$ $\mathbf{N}_m$ $\mathbf{N}^{\text{loc}}_m$ $\mathbf{N}_{m,K}$ $f_\#$
4.1.8	$S\times T$ $[u,v]$
4.1.9	h_t $\dot{h}_t$
4.1.11	$S \times T$ $[u_0,\dots,u_m]$
4.1.12	$\mathbf{F}_K$ $\mathbf{F}_{m,K}$ $\mathbf{F}_m$ $\mathbf{F}^{\text{loc}}_m$
4.1.14	$f_\#$
4.1.19	$\mathbf{F}$

4.1.22	$\mathscr{P}_{m,K}$ $\mathbf{P}_{m,K}$ $\mathscr{P}_m$ $\mathbf{P}_m$
4.1.24	$\mathscr{R}_{m,K}$ $\mathscr{R}_m$ $\mathbf{I}_{m,K}$ $\mathbf{I}_m$
	$\mathscr{F}_{m,K}$ $\mathscr{F}_m$ $\mathscr{R}^{\text{loc}}_m$ $\mathbf{I}^{\text{loc}}_m$ $\mathscr{F}^{\text{loc}}_m$ $\mathscr{F}_K$
4.1.31	$\int_B\phi$ $\int_{\partial A}\psi$
4.1.34	$[\alpha,\beta]$
4.2.1	$\langle T,u,r+\rangle$ $\langle T,u,r-\rangle$
4.2.2	$f_{\#u}$ $H_u(f,g)$
4.2.5	$\mathbf{Z}^n_j$ $\mathbf{W}'(z)$ $\mathbf{W}''(z)$ $\mathbf{W}'_k$ $\mathbf{W}''_k$
4.2.8	μ_ε
4.2.26	$\mathscr{F}^\nu_K$ $\mathscr{F}^\nu_{m,K}$ $T_1\equiv T_2 \bmod \nu$ $\mathscr{F}^\nu_m$
	spt^ν $(T)^\nu_K$ $(T)^\nu$ $\mathscr{R}^\nu_m$ $\mathbf{I}^\nu_m$
	$\mathscr{R}^\nu_{m,K}$ $\mathbf{I}^\nu_{m,K}$ $\mathscr{P}^\nu_m$ $\mathbf{M}^\nu$ $\\|T\\|^\nu$
	$(T)^\nu\llcorner A$ $\mathbf{N}^\nu$ $\langle (T)^\nu,u,r+\rangle$
4.3.1	$T\llcorner f^\#\phi$ $\langle T,f,y\rangle$
4.3.20	$S\cap T$
4.4.1	$\mathscr{Z}_m$ $\mathscr{B}_m$ $\mathbf{H}_m$ ∂
4.4.4	$\mathscr{F}_{K,B}$
4.4.5	$\mathbf{H}_m(A,B;\mathbf{Z})$
4.4.6	$\mathscr{Z}^\nu_m$ $\mathscr{B}^\nu_m$ $\mathbf{H}_m(A,B;\mathbf{Z}_\nu)$
	$\mathbf{Z}_m$ $\mathbf{B}_m$ $\mathbf{H}_m(A,B;\mathbf{R})$
4.4.8	$\tilde{\mathscr{Z}}_k$
4.5.5	$\mathbf{n}(A,b)$
4.5.10	$\text{ess}\,\mathbf{V}^\beta_\alpha$
5.1.1	$\int_T\Phi$ $\langle\Phi,T\rangle$ $G^\#\Psi$
5.1.2	Φ_a
5.1.7	$\delta^{(i)}(T,\Phi,h)$
5.1.9	$\tilde\Phi$ $\tilde\Phi_a$ $\mathbf{p}$ $\mathbf{q}$ Φ^δ Φ^δ_a
5.1.11	$D_i\phi$ $D_{m+j}\phi$ $\phi_{(i,j)}$ $\phi_{(i,j;k,l)}$
5.2.1	$(f,g)_U$ $\lvert f\rvert_U$ (f,g) $\lvert f\rvert$
	$(f,g)_{b,r}$ $\lvert f\rvert_{b,r}$ $\mathbf{h}_\delta(f)$
5.2.7	$*$
5.2.10	Lap $\mathbf{r}$ $\mathbf{c}(m)$
5.3.1	$\operatorname{Exc}(S,b,r)$ $\operatorname{Exc}^\nu(S,b,r)$
5.4.1	$\delta^{(i)}(T,h)$
5.4.11	$\exp(T)$
5.4.12	$\mathscr{E}(B,W)$ $\mathfrak{T}$ $\mathfrak{N}$ $\mathfrak{P}$
	∇_ξ $[\xi,\eta]$ $\nabla_{\xi,\eta}$ $R_{\xi,\eta}$
	Lap $\nabla\xi$ $\operatorname{grad}\phi$

Index

Subsection numbers are followed by page numbers in parentheses

Universitätsdruckerei H. Stürtz AG Würzburg

Springer-Verlag and the Environment

We at Springer-Verlag firmly believe that an international science publisher has a special obligation to the environment, and our corporate policies consistently reflect this conviction.

We also expect our business partners – paper mills, printers, packaging manufacturers, etc. – to commit themselves to using environmentally friendly materials and production processes.

The paper in this book is made from low- or no-chlorine pulp and is acid free, in conformance with international standards for paper permanency.

DRUCK: STRAUSS OFFSETDRUCK, MÖRLENBACH
BINDEN: TRILTSCH, WÜRZBURG

CIM

M. Aigner Combinatorial Theory ISBN 978-3-540-61787-7
A. L. Besse Einstein Manifolds ISBN 978-3-540-74120-6
N. P. Bhatia, G. P. Szegő Stability Theory of Dynamical Systems ISBN 978-3-540-42748-3
J. W. S. Cassels An Introduction to the Geometry of Numbers ISBN 978-3-540-61788-4
R. Courant, F. John Introduction to Calculus and Analysis I ISBN 978-3-540-65058-4
R. Courant, F. John Introduction to Calculus and Analysis II/1 ISBN 978-3-540-66569-4
R. Courant, F. John Introduction to Calculus and Analysis II/2 ISBN 978-3-540-66570-0
P. Dembowski Finite Geometries ISBN 978-3-540-61786-0
A. Dold Lectures on Algebraic Topology ISBN 978-3-540-58660-9
J. L. Doob Classical Potential Theory and Its Probabilistic Counterpart ISBN 978-3-540-41206-9
R. S. Ellis Entropy, Large Deviations, and Statistical Mechanics ISBN 978-3-540-29059-9
H. Federer Geometric Measure Theory ISBN 978-3-540-60656-7
S. Flügge Practical Quantum Mechanics ISBN 978-3-540-65035-5
L. D. Faddeev, L. A. Takhtajan Hamiltonian Methods in the Theory of Solitons ISBN 978-3-540-69843-2
I. I. Gikhman, A. V. Skorokhod The Theory of Stochastic Processes I ISBN 978-3-540-20284-4
I. I. Gikhman, A. V. Skorokhod The Theory of Stochastic Processes II ISBN 978-3-540-20285-1
I. I. Gikhman, A. V. Skorokhod The Theory of Stochastic Processes III ISBN 978-3-540-49940-4
D. Gilbarg, N. S. Trudinger Elliptic Partial Differential Equations of Second Order ISBN 978-3-540-41160-4
H. Grauert, R. Remmert Theory of Stein Spaces ISBN 978-3-540-00373-1
H. Hasse Number Theory ISBN 978-3-540-42749-0
F. Hirzebruch Topological Methods in Algebraic Geometry ISBN 978-3-540-58663-0
L. Hörmander The Analysis of Linear Partial Differential Operators I – Distribution Theory and Fourier Analysis ISBN 978-3-540-00662-6
L. Hörmander The Analysis of Linear Partial Differential Operators II – Differential Operators with Constant Coefficients ISBN 978-3-540-22516-4
L. Hörmander The Analysis of Linear Partial Differential Operators III – Pseudo-Differential Operators ISBN 978-3-540-49937-4
L. Hörmander The Analysis of Linear Partial Differential Operators IV – Fourier Integral Operators ISBN 978-3-642-00117-8
K. Itô, H. P. McKean, Jr. Diffusion Processes and Their Sample Paths ISBN 978-3-540-60629-1
T. Kato Perturbation Theory for Linear Operators ISBN 978-3-540-58661-6
S. Kobayashi Transformation Groups in Differential Geometry ISBN 978-3-540-58659-3
K. Kodaira Complex Manifolds and Deformation of Complex Structures ISBN 978-3-540-22614-7
Th. M. Liggett Interacting Particle Systems ISBN 978-3-540-22617-8
J. Lindenstrauss, L. Tzafriri Classical Banach Spaces I and II ISBN 978-3-540-60628-4
R. C. Lyndon, P. E Schupp Combinatorial Group Theory ISBN 978-3-540-41158-1
S. Mac Lane Homology ISBN 978-3-540-58662-3
C. B. Morrey Jr. Multiple Integrals in the Calculus of Variations ISBN 978-3-540-69915-6
D. Mumford Algebraic Geometry I – Complex Projective Varieties ISBN 978-3-540-58657-9
O. T. O'Meara Introduction to Quadratic Forms ISBN 978-3-540-66564-9
G. Pólya, G. Szegő Problems and Theorems in Analysis I – Series. Integral Calculus. Theory of Functions ISBN 978-3-540-63640-3
G. Pólya, G. Szegő Problems and Theorems in Analysis II – Theory of Functions. Zeros. Polynomials. Determinants. Number Theory. Geometry ISBN 978-3-540-63686-1
W. Rudin Function Theory in the Unit Ball of $\mathbb{C}^n$ ISBN 978-3-540-68272-1
S. Sakai C*-Algebras and W*-Algebras ISBN 978-3-540-63633-5
C. L. Siegel, J. K. Moser Lectures on Celestial Mechanics ISBN 978-3-540-58656-2
T. A. Springer Jordan Algebras and Algebraic Groups ISBN 978-3-540-63632-8
D. W. Stroock, S. R. S. Varadhan Multidimensional Diffusion Processes ISBN 978-3-540-28998-2
R. R. Switzer Algebraic Topology: Homology and Homotopy ISBN 978-3-540-42750-6
A. Weil Basic Number Theory ISBN 978-3-540-58655-5
A. Weil Elliptic Functions According to Eisenstein and Kronecker ISBN 978-3-540-65036-2
K. Yosida Functional Analysis ISBN 978-3-540-58654-8
O. Zariski Algebraic Surfaces ISBN 978-3-540-58658-6